Nanostructures for Novel Therapy

Nanostructures in Therapeutic Medicine Series

Nanostructures for Novel Therapy

Synthesis, Characterization and Applications

Edited by

Denisa Ficai
University Politehnica of Bucharest, Bucharest, Romania

Alexandru Mihai Grumezescu
University Politehnica of Bucharest, Bucharest, Romania

ELSEVIER

Elsevier
Radarweg 29, PO Box 211, 1000 AE Amsterdam, Netherlands
The Boulevard, Langford Lane, Kidlington, Oxford OX5 1GB, United Kingdom
50 Hampshire Street, 5th Floor, Cambridge, MA 02139, United States

Library of Congress Cataloging-in-Publication Data
A catalog record for this book is available from the Library of Congress

British Library Cataloguing-in-Publication Data
A catalogue record for this book is available from the British Library

ISBN: 978-0-323-46142-9

For information on all Elsevier publications visit our website at
https://www.elsevier.com/books-and-journals

Publisher: Matthew Deans
Acquisition Editor: Simon Holt
Editorial Project Manager: Sabrina Webber
Production Project Manager: Julie-Ann Stansfield
Designer: Greg Harris

Typeset by Thomson Digital

Contents

List of Contributors

Giorgia Adamo
University of Palermo, Palermo, Italy

Andreea Aiacoboae
University Politehnica of Bucharest, Bucharest, Romania

Parvez Alam
University of Edinburgh, Edinburgh, United Kingdom;
Abo Akademi University, Turku, Finland

Ecaterina Andronescu
University Politehnica of Bucharest, Bucharest, Romania

Hiromichi Aono
Ehime University, Matsuyama, Japan

Ilaria Armentano
Tuscia University, Viterbo, Italy

Danielle C. Arruda
Minas Gerais Federal University, Belo Horizonte, Brasil

Petar A. Atanasov
Bulgarian Academy of Sciences, Sofia, Bulgaria

Fatemeh Atyabi
Tehran University of Medical Sciences, Tehran, Iran

Muhammad A. Azmi
Isra University Karachi Campus, Karachi, Pakistan

Patrick Babczyk
Bonn-Rhine-Sieg University of Applied Sciences, Rheinbach, Germany

Jalal Barzin
Iran Polymer & Petrochemical Institute, Tehran, Iran

Gamze Bektas
Private Clinic, Istanbul, Turkey

Luciano A. Benedini
Southern National University, Bahía Blanca, Argentina

Pascal Bigey
CNRS UMR 8258; INSERM UMR-S 1022; Paris Descartes University; Chimie ParisTech;
Paris, France

Fernanda B. Borghi-Pangoni
State University of Maringá, Maringá, Brazil

Lydia M. Bouchet
National University of Cordoba, Córdoba, Argentina

Verónica Brunetti
National University of Cordoba, Córdoba, Argentina

Marcos L. Bruschi
State University of Maringá, Maringá, Brazil

Simona Campora
University of Palermo, Palermo, Italy

Christine Charrueau
CNRS UMR 8258; INSERM UMR-S 1022; Paris Descartes University; Chimie ParisTech; Paris, France

Murthy Y. Chavali
Technology and Research University, Guntur, India

Mariana C. Chifiriuc
Research Institute of the University of Bucharest (ICUB), Bucharest, Romania

Anı Cinpolat
Private Clinic, Istanbul, Turkey

Carmen Curutiu
Research Institute of the University of Bucharest (ICUB), Bucharest, Romania

Noelia L. D'Elía
Southern National University, Bahía Blanca, Argentina

Nily Dan
Drexel University, Philadelphia, PA, United States

Sabrina B. de Souza Ferreira
State University of Maringá, Maringá, Brazil

Corinne Dejous
University of Bordeaux, Bordeaux, France

Tanmoy K. Dey
University of Calcutta, Kolkata, India

Pubali Dhar
University of Calcutta, Kolkata, India

Vijaya R. Dirisala
Technology and Research University, Guntur, India

Slavomira Doktorovova
University of Coimbra, Coimbra, Portugal

Surbhi Dubey
Dr. Hari Singh Gour Central University, Sagar, India

Fatma Elsayed
Bonn-Rhine-Sieg University of Applied Sciences, Rheinbach, Germany

Fuyuaki Endo
Keio University, Yokohama, Japan

Virginie Escriou
CNRS UMR 8258; INSERM UMR-S 1022; Paris Descartes University; Chimie ParisTech; Paris, France

Abdol-Rahim Faramarzi
Iran Polymer & Petrochemical Institute, Tehran, Iran

Eliana D. Farias
National University of Cordoba, Córdoba, Argentina

Eelna Fortunati
University of Perugia, Terni, Italy

Irina Gheorghe
Research Institute of the University of Bucharest (ICUB), Bucharest, Romania

Tania Gheorghe
University Politehnica of Bucharest, Bucharest, Romania

Giulio Ghersi
University of Palermo, Palermo, Italy

Mohsen Gorji
Amirkabir University of Technology, Tehran, Iran

Alexandru Mihai Grumezescu
Research Institute of the University of Bucharest (ICUB), Bucharest, Romania;
University Politehnica of Bucharest, Bucharest, Romania

Hamida Hallil
University of Bordeaux, Bordeaux, France

Céline Hoffmann
CNRS UMR 8258; INSERM UMR-S 1022; Paris Descartes University; Chimie ParisTech; Paris, France

Alina M. Holban
University Politehnica of Bucharest, Bucharest, Romania;
Research Institute of the University of Bucharest (ICUB), Bucharest, Romania

Atsushi Hotta
Keio University, Yokohama, Japan

Daniel Iglesias
University of Trieste, Trieste, Italy

Cristina Ş. Iosub
University Politehnica of Bucharest, Bucharest, Romania

Mariana V. Junqueira
State University of Maringá, Maringá, Brazil

Josè M. Kenny
University of Perugia, Terni, Italy
Institute of Polymer Science and Technology, CSIC, Madrid, Spain

Somayeh Khezrian
University of Tehran, Tehran, Iran

Sepideh Khoee
University of Tehran, Tehran, Iran

Stephanie E. Klein
Bonn-Rhine-Sieg University of Applied Sciences, Rheinbach, Germany

Sengo Kobayashi
Ehime University, Matsuyama, Japan

Naruki Kurokawa
Keio University, Yokohama, Japan

Iulia I. Lungu
University Politehnica of Bucharest, Bucharest, Romania

Tomoki Maeda
Keio University, Yokohama, Japan

Silvia Marchesan
University of Trieste, Trieste, Italy

Sabata Martino
University of Perugia, Perugia, Italy

Samantha Mattioli
University of Perugia, Terni, Italy

Paula V. Messina
Southern National University, Bahía Blanca, Argentina

Hamid Mobedi
Iran Polymer & Petrochemical Institute, Tehran, Iran

Nishi Mody
Dr. Hari Singh Gour Central University, Sagar, India

Francesco Morena
University of Perugia, Perugia, Italy

Fatemeh Mottaghitalab
Tehran University of Medical Sciences, Tehran, Iran

Deboshree Mukherjee
CSIR-Indian Institute of Chemical Technology, Hyderabad, India

Alieh Naraghi
Shahid Rajaee Teacher Training University, Tehran, Iran

Sampath K. Nune
Technology and Research University, Guntur, India

Elena Olăreț
University Politehnica of Bucharest, Bucharest, Romania

Saeed Olyaee
Shahid Rajaee Teacher Training University, Tehran, Iran

Ulvan Ozad
Near East University Hospital, Nicosia, North Cyprus

Debjyoti Paul
University of Calcutta, Kolkata, India

Vincent Raimbault
University of Bordeaux, Bordeaux, France
LAAS-CNRS, Toulouse, France

Krupanidhi S. Rama
Technology and Research University, Guntur, India

Bolla G. Rao
CSIR-Indian Institute of Chemical Technology, Hyderabad, India

Ali Rastegari
Tehran University of Medical Sciences, Tehran, Iran

Benjaram M. Reddy
CSIR-Indian Institute of Chemical Technology, Hyderabad, India

Nicoletta Rescignano
Institute of Polymer Science and Technology, CSIC, Madrid, Spain

Zumreta Rizvanovic
Private Clinic, Istanbul, Turkey

Rajagopalan Rukkumani
Pondicherry University, Kalapet, India

Avneet Saini
Panjab University, Chandigarh, India

Dorothee Schipper
Bonn-Rhine-Sieg University of Applied Sciences, Rheinbach, Germany

Margit Schulze
Bonn-Rhine-Sieg University of Applied Sciences, Rheinbach, Germany

Kaneez F. Shad
University of Technology Sydney (UTS), Sydney, NSW, Australia

Rajeev Sharma
Dr. Hari Singh Gour Central University, Sagar, India

Ranjita Shegokar
Freie Universität Berlin, Berlin, Germany

Jalpa Soni
Indian Institute of Science Education and Research, Kolkata, India

Eliana B. Souto
University of Coimbra, Coimbra, Portugal

Miriam C. Strumia
National University of Cordoba, Córdoba, Argentina

Hiromichi Takebe
Ehime University, Matsuyama, Japan

Edda Tobiasch
Bonn-Rhine-Sieg University of Applied Sciences, Rheinbach, Germany

Luigi Torre
University of Perugia, Terni, Italy

Gaurav Verma
Panjab University, Chandigarh, India

George Vlasceanu
University Politehnica of Bucharest, Bucharest, Romania

Suresh P. Vyas
Dr. Hari Singh Gour Central University, Sagar, India

Jatinder Vir Yakhmi
Homi Bhabha National Institute, Anushaktinagar, India

Saeki Yamamuro
Ehime University, Matsuyama, Japan

Foreword of the Series

Material science and engineering at the nanoscale have brought revolutionary advances to the biomedical sciences, overturning many of the known traditional approaches. Nanotechnology has driven many of the most successful innovative technologies, and their impressive record of accomplishment has made nanostructures promising candidates for future therapy. The advantages that nanomaterials have already provided to therapeutics, such as targeted and controlled delivery, wide accessibility, high specificity, low side effects, improved efficiency, and impressive versatility are currently considered key elements in designing personalized medicine approaches for prophylaxis, diagnosis, and therapy.

Therapeutic nanostructures can be highly diverse, and their unique properties have led to the development of highly specialized biosensors, more efficient drug delivery vehicles, and controlled release targeting systems to fight severe or incurable diseases, such as cancer, infections, and cardiovascular disease.

In view of the astounding progress made in the field of therapeutic nanotechnology, and its rapidly progressing expansion, this book aims to collect together in one place all the most recent and innovative aspects regarding the impact of nanomaterials in both current and future therapy. The book is organized in five volumes, covering the main areas that are relevant for the design and implementation of nanostructures in medical therapies.

The first volume, *Nanostructures for Novel Therapy: Synthesis, Characterization, and Applications* describes methods to obtain and characterize nanosystems, emphasizing their biomedical applications. Special attention in this volume was paid to modern synthesis methods to reduce side effects and limit the toxicity of nanomaterials in biomedical applications. Numerous examples of nanostructures designed for therapy, as well as the most efficient synthesis and characterization routes for these materials are clearly described and critically analyzed.

The second volume, entitled *Nanostructures for Drug Delivery* covers one of the most widely utilized and investigated applications of nanomaterials in the biomedical field, namely drug delivery. The design of nanoscale agents in order to specifically and safely carry therapeutic agents to their final destination is an intriguing approach to future targeted therapeutics. This approach could provide a treatment for previously incurable diseases, as well as reducing the side effects of current drugs. Many highly active drugs are severely limited by side effects related to their unspecific sites of action. This book introduces the readers to the amazing field of nanomedicine by discussing the versatility and variety of nanovehicles for drug delivery and targeting. Moreover, readers will find numerous examples, and will learn about the currently used or investigational drug delivery agents for therapy, prophylaxis, and diagnosis.

The third volume, *Nanostructures for Antimicrobial Therapy* highlights the impressive progress made by nanotechnology in the design of novel antimicrobial approaches. Since microbial resistance to antibiotics is a very real and worrying issue growing throughout the world, the development of more efficient antimicrobial agents has a high priority to allow control of infections in the future. Antimicrobial nanosystems have proved to be highly efficient against drug-resistant microorganisms, are able to fight biofilm-associated infections and control the social behavior of microbial communities. Nanostructures can also reduce microbial virulence factors and reduce pathogenesis mechanisms thus offering a promising alternative for future therapy.

The fourth volume, entitled *Nanostructures for Cancer Therapy* covers the applications of nanomedicine in cancer diagnosis and treatment. The use of nanoparticles for cancer therapy is not in itself a new approach, but numerous advances have been recently made in this area, and the aim of this book is to cover the most interesting new approaches in the management of this deadly disease. Nanosized drugs are currently believed to represent the most efficient approach in cancer chemotherapy, and this volume provides coverage of the latest novel findings, while also discussing possible improvements in more established types of nanosystems to increase the efficiency of cancer therapy.

Last but not least, the fifth volume of this series entitled *Nanostructures for Oral Medicine* covers the progress made in applications of nanotechnology in treating various diseases of the oral cavity and in dentistry. Readers will have the chance to learn about the most efficient modern materials used to treat or to prevent widely encountered oral diseases, such as gingivitis, periodontitis, caries, and dental plaque. Moreover, restorative dentistry also now makes wide use of nanomaterials.

Overall, this book series will provide a state-of-the-art compendium of knowledge, and a crystal ball to see into the future of biomedical nanotechology and nanomedicine. It will appeal to researchers, clinicians, engineers, pharmacologists, pharmacists, oncologists, infectious disease experts, and dentists. More many interested general readers may discover the impact, current progress and future applications of nanotechnology in therapeutics and diagnosis. Taken together, nanoscale approaches will improve the efficiency of personalized medicine for better management of diseases in the 21st century.

Michael R. Hamblin
Wellman Center for Photomedicine, Massachusetts General Hospital, Boston, MA, United States
Department of Dermatology, Harvard Medical School, Boston, MA, United States
Harvard-MIT Division of Health Sciences and Technology, Cambridge, MA, United States

Preface

ABOUT THE SERIES (VOLUMES I–V)

In our permanently changing world, novel therapeutics are constantly required to manage health and well-being of population. Although numerous diseases are currently considered incurable, impressive progress made in biomedicine and associated fields, such as chemistry, physics, engineering, pharmacology, and materials science, offers a new light to the therapeutics domain. In this context, most physicists and researchers believe that a personalized and adequate treatment may significantly improve the outcome of severe diseases and ensure a faster healing. Nanotechnology offers great perspectives for personalized medicine, since nanostructured therapeutics proved their efficiency and amazing impact in improving therapy, prophylaxis, and diagnosis. The emerging field of nanosized materials has numerous applications in the biomedical field, especially in therapy. This series of five volumes came out by the need of learning about recent progress of the science of nanostructured materials to improve current therapy and led to the next level. The books offer an interesting and updated perspective regarding applications of nanomaterials in therapy of most-investigated and difficult-to-treat diseases, such as cancer and severe infections. The presentation approach of each chapter contained in these five volumes is clear and easy to understand by most readers and interesting for biomedical specialists, researchers, and engineers. The series is organized in an attractive manner for students and academics on the field, starting with a volume dealing with the synthesis, characterization, and main applications of nanostructures, emphasizing on their impact in therapy. Next volume reveals the most recent progress made on a very investigated field, which is considered a key element in personalized medicine and future therapy, namely nanostructured drug-delivery systems. Their impact in antimicrobial therapy is also widely discussed and suggestive examples are given and explained. Moreover, a whole volume is dedicated to the management of the disease of the century—cancer—revealing the huge value added by the utilization of nanosystems in the therapy of this deadly disease. Important aspects related to improved diagnosis and prophylaxis are highlighted. In the last volume, the progress and novel applications of nanotechnology in oral medicine are dissected. The field of oral diseases represents a wide-interest and priority field since both physicists and researchers believe that they can be prevented and treated more easily with targeting systems and nanofunctional prosthetics. All chapters are clearly illustrated to highlight important or more difficult-to-understand aspects and suggestive examples are often enumerated in organized tables, which are explained and discussed. Overall, the series contains very recent but accessible information regarding the progress of nanostructures in therapeutics and gives a novel perspective about future therapy of severe diseases.

Alexandru M. Grumezescu
University Politehnica of Bucharest, Bucharest, Romania

ABOUT VOLUME I

First volume of the series *Therapeutic nanostructures* introduces the readers in the amazing field of nanostructured size materials, describing types, synthesis and characterization approaches, and emphasizing on their applications. The utility of nanosized materials in the biomedical field is highly influenced by the physicochemical properties of nanostructured materials utilized for therapy. Therefore, knowing their qualities ensure a more successful approach in designing effective agents for the management of a particular disease. This book offers an updated perspective regarding the most interesting advances of nanoarchitectured therapeutics, discusses the wide variety of their synthesis routes, preferential functionalities, and most supported applications for the design of future therapeutics.

Volume I contains 30 chapters, prepared by outstanding international researchers from USA, Brazil, Argentina, Australia, Portugal, Spain, Finland, France, Italy, Germany, Bulgaria, Romania, Turkey, Cyprus, Iran, Pakistan, India, Taiwan, China, and Japan.

In Chapter 1, entitled *Novel Approaches for Preparation of Nanoparticles*, Bolla Govinda Rao et al., give an overview about new developments in various preparation methodologies for the fabrication of nanoparticles. Also, this first chapter offers a great introductory basis in the design of functional nanostructures aimed for particular purposes.

Andreea Aiacoboae et al., in Chapter 2, entitled *Applications of Nanoscale Drugs Carriers in the Treatment of Chronic Diseases*, discuss about smart nanosized drug-delivery systems which made or have the potential to make a great impact in the management of chronic illness. The fact that numerous nanosystems are able to respond to various stimuli, such as pH, temperature, light, ultrasounds, electrical, and magnetical fields, make them special candidates for targeted and efficient therapy. The chapter discusses most investigated examples of nanoparticles which are under consideration for nanotherapy, such as gold nanoparticles (NPs), silver NPs, magnetic NPs, quantum dots, and mesoporous silica NPs, useful in targeting and delivering applications.

Chapter 3, prepared by Giorgia Adamo et al., entitled *Functionalization of Nanoparticles in Specific Targeting and Mechanism Release*, reviews the potential of smart multifunctional nanostructures as customizable, targeted drug-delivery vehicles capable of carrying large doses of therapeutic agents into target cells. Covalent and noncovalent chemical linking using different molecules have been reported for nanoparticles surface functionalization, since this approach confers them specific properties, such as more specific targeting ability.

In Chapter 4, entitled *Fabrication and Characterization of Natural/Synthesized, Micro-, and Nanostructured Materials for Biomedical Applications*, prepared by Hiromichi Takebe et al., is presented an up-to-date overview about the preparation of modified micro/nanoparticles such as, magnetic oxide nanoparticles for thermal coagulation therapy, iron oxide nanocubes and their shape-induced self-assembly and micro/submicrosized, imprinted inorganic glasses for nanointegration. This chapter is a great opportunity to review the properties of most investigated nanosystems for biomedical applications.

In chapter 5, entitled *Multifunctional Nanostructured Biopolymeric Materials for Therapeutic Applications*, prepared by Armentano et al., is revealed the current state and future prospects of the new generation of multifunctional bionanomaterials, based on different natural or synthetic biopolymers, together with their role in development of personalized therapy.

Chapter 6, entitled *Polymeric Pharmaceutical Nanoparticles Developed by Electrospray*, prepared by Faramarzi et al., gives an overview about advantages and disadvantages, basic principles and production methods of electrospray based approaches.

Debjyoti Paul et al., in Chapter 7, entitled *Nanoformulation and Administration of PUFA-Rich Systems for Applications in Modern Healthcare*, explore the realms of nanotechnology that deals with the fabrication methods of polyunsaturated fatty acids (PUFA) rich nanoformulations, the holistic changes in stability, bioavailability, and bioactivity as a consequence of different administration routes (oral, topical, and parenteral) of such nanosystems.

Naruki Kurokawa et al., in Chapter 8, entitled *Electrospinning and Surface Modification Methods for Functionalized Cell Scaffolds*, present an extensive overview of effective techniques for making optimal scaffolds, including electrospinning and surface modification methods.

Daniel Iglesias, et al., in Chapter 9, entitled *Short Peptide Self-Assembled Nanostructures for Therapeutics Innovative Delivery*, described the high innovation potential offered by self-assembling short peptides in the field of drug delivery. Several options for such use are possible, ranging from physical drug entrapment in the hydrogel matrix, through drug noncovalent binding to the supramolecular structure by the covalent binding of the drug to the self-assembling building block.

Avneet Saini et al., in Chapter 10, entitled *Peptoids: Tomorrow's Therapeutics*, summarize the recent research efforts in the field of peptoid biopolymers from synthesis to folded structural motifs. Also, the authors present their major biological applications discovered in the last two decades.

Chapter 11, prepared by N.S. Sampath Kumar, *Electrospinning of Collagen Nanofiber Scaffolds for Tissue Repair and Regeneration*, provide information about tissue repair and regeneration using biomimetic scaffolds and a brief overview of state-of-the-art methods for fabricating nanofibrous scaffolds, including phase-separation, freeze-drying, self-assembly, and electrospinning.

Chapter 12, entitled *Nanostructured Therapeutic Systems With Bioadhesive and Thermoresponsive Properties*, prepared by Marcos Luciano Bruschi et al., discussed the importance, applications, and perspectives of Mucoadhesive Thermoresponsive Systems as potentially useful therapeutic approach.

Chapter 13, entitled *Design Considerations in the Development of Wound Healing Bionanomaterials*, prepared by Parvez Alam et al., provide a comprehensive overview about fundamental considerations in the design of bionanomaterials with a focus on skin properties, material types, surface structures and methods for nanofabrication.

In the Chapter 14, *Role of Nanostructure Molecules in Enhancing the Bioavailability of Oral Drugs*, prepared by Muhammad Ahmed Azmi et al., the main focus is on the role of nanostructured molecules in the enhancement of the bioavailability of drugs. The authors show recent results for different compounds, which are promising candidates for anti-inflammatory, immunosuppressive, antifertility, anticytogenesis, and anticancer responses.

Chapter 15, prepared by Ulvan Ozad, entitled *Improvement Steps of Plastic Surgery to Tissue Engineering by Nanotechnology*, give an overview about the application of nanotechnology in plastic and reconstructive surgery field. The effects of nanotechnology are observed in numerous such procedures; ranging from the nanomaterials used in implantable materials to wound closure, wound healing, and wound dressing, emphasizing on tissue engineering and regeneration.

Eliana D. Farías et al., in Chapter 16, entitled *Dendrimers and Dendronized Materials as Nanocarriers*, describe general characteristics of dendrimers and dendronized materials, synthetic methodologies and their properties. Successful applications of dendrimers and novel dendritic structures to obtain efficient nanocarriers are also discussed.

Petar A. Atanasov, in Chapter 17, entitled *Gold Nanostructures: Preparation, Properties, Application in SERS, and Biophotonics*, present an up to date overview about the development of several advanced methods for fabrication of nanoparticles and nanosized arrays and examples of their application in SERS and biophotonics. Also, advanced laser systems which are used for controlling and manipulating the properties of the structures are presented and dissected.

Chapter 18, entitled *Quantum Dots for Bioimaging and Therapeutic Applications*, prepared by George Vlasceanu et al., presents an up to date review about Quantum Dots synthesis methods and bioimaging techniques, with the focus on the latter. Bioimaging emerged as a relatively recent field, as the integration of chemistry, physics, and biology in an intricate overlapping. This intriguing field proved a great impact in modern therapeutics and the potential for future therapy seems endless.

Nily Dan et al., in Chapter 19, entitled *Lipid-Based Synthetic Gene Carriers*, review the equilibrium properties of lipoplexes and the kinetics pathways for their formation. Better understanding of lipoplex formation can not only lead to the design of effective nucleic acid lipoplex-based gene delivery or silencing agents, but is also of interest as a fundamental multicomponent, multilength scale, and multitime scale process.

Chapter 20 prepared by Corinne Dejous et al., *Using Microsensors to Promote Development of Innovative Therapeutic Nanostructures*, gives an overview of the existing micro- and nanosensors, and their principles and describe microsensors that can be developed for fast evaluation of efficiency and toxicity of innovative therapeutic nanostructures. The authors also describe their key applications and analytical techniques employed.

D. Schipper et al., in Chapter 21, entitled *The effect of Nanostructured Surfaces on Stem Cell Fate*, present details about the chemical structure of various nanomaterials used as scaffolds for stem cell differentiation including bulk and surface properties and corresponding analytical methods for surface characterization. Furthermore, the authors present recently developed methods for the design of tailored nanomaterials used in stem cell differentiation.

Chapter 22, prepared by Sepideh Khoee, entitled *Applications of Aptamers for the Diagnosis and Therapy of Different Diseases*, gives an up to date overview about aptamers and their application as biotechnological tools in biosensor development and therapeutic agents for drug delivery and diagnosis of diseases.

Chapter 23, entitled *Recent Progress in Therapeutic Diagnosis Using Photonic Crystal Nanostructures*, prepared by Saeed Olyaee, discusses about photonic crystals and their properties, as well as the fabrication procedure. The chapter presents several demonstrations using photonic crystals and their laser biosensors module, as lab on chip devices.

Chapter 24, prepared by Surbhi Dubey, entitled *Novel Carriers and Approaches: Insight for Psoriasis Management*, presents recent progress in the field of psoriasis in terms of pathogenesis, role of cytokines, currently available treatment options pertaining to mode of action, pharmacokinetics, major hindrances in psoriasis treatment, side effects of individual antipsoriatic drugs, and recent developments in the delivery of various antipsoriatic drugs through lipid-based novel colloidal drug carriers. Recent nano-based approaches bring a new light in the management of this disease.

Jalpa Soni et al., in Chapter 25, entitled *Use of Nanostructures Based on Noble Metals in Nanobiomedicine*, present new developments in the nanobiotechnology with a general overview on the impact of various types of nanostructures based on noble metals for various biomedical applications.

Chapter 26, entitled *Innovative Nonviral Vectors for Small-Interfering RNA Delivery and Therapy*, prepared by Danielle Campiol Arruda, introduces the biological barriers to siRNA delivery *in vivo* and discusses recent advances in material sciences, nanotechnology, and nucleic acid chemistry that have yielded promising nonviral delivery systems, some of which are currently undergoing testing in clinical trials. The diversity of these systems highlights the recent progress of siRNA-based therapy using nonviral approaches.

Ali Rastegari et al., in Chapter 27, entitled *siRNA-Based Nucleoceuticals for Tissue Regeneration*, provide an in-depth discussion of the development of siRNA-based nucleoceuticals for tissue engineering applications.

Paula V. Messina et al., in Chapter 28, entitled *Bone Tissue Regenerative Medicine Via Bioactive Nanomaterials*, summarize the most recent developments related to the application of nanomaterials to the bone regenerative medicine and discusses their commercialization projection in consideration to their toxicological risks.

Chapter 29, prepared by Stefana Iosub et al., entitled *Toxicity of Nanostructures—A General Approach*, provides a general picture of the toxicological responses that nanomaterials can generate. Beside the size and the concentration of nanoparticles, there are several factors that influence their cytotoxicity, such as: aggregation state, shape, chemistry, biodistribution, etc. However, the toxic effects could be diminished by capping or coating the nanoparticles with various molecules to enhance their biocompatibility, which also may increase their specificity.

Chapter 30, entitled *Role of Excipients in Formulation Development and Biocompatibility of Lipid Nanoparticles*, prepared by Slavomira Doktorovova, discusses the role of excipients in the production of lipid nanoparticles. An overview of commonly used and upcoming promising excipients, which can facilitate the development of lipid nanoparticles and lipid-based formulations, is presented. A review of new research outcomes and market status of lipid nanoparticles is also addressed in this chapter.

Denisa Ficai
University Politehnica of Bucharest, Bucharest, Romania

Alexandru M. Grumezescu
University Politehnica of Bucharest, Bucharest, Romania

CHAPTER

NOVEL APPROACHES FOR PREPARATION OF NANOPARTICLES

1

Bolla G. Rao, Deboshree Mukherjee, Benjaram M. Reddy
CSIR-Indian Institute of Chemical Technology, Hyderabad, India

CHAPTER OUTLINE

1 INTRODUCTION

Since the beginning of the 21st century, with the rising concern of multidrug resistance and scarcity of new antibiotics, the use of metal nanoparticles (NPs) in medicine is undergoing renaissance. In contrast to bulk, nanomaterials exhibit huge surface area per unit volume and tunable optical, electronic, magnetic, and biological properties. Metal NPs can be engineered to have different sizes, shapes, and surface characteristics. The size and shape tunable properties of metal nanoparticles and their wide scope of applicability in pharmacy and biotechnology have drawn global attention toward their size and shape-controlled synthesis. However, while dealing with metal nanoparticle synthesis, few things should be considered seriously. First, the chosen method must be simple, less expensive, ecofriendly,

Nanostructures for Novel Therapy

and commercially viable. Second, the simultaneous control of particle size and shape along with their uniformity is another key objective (Kwon and Hyeon, 2008). Moreover, the NPs are kinetically unstable, they should be stabilized against aggregation into larger particles (Richter et al., 2010). Micelles, polymers, and coordinative ligands are frequently used as stabilizers to control the growth of NPs (Kim et al., 2004). Solution-based nanofabrication methods usually offer more control and reproducibility over the metal NPs. A wide range of nanofabrication methods, including precipitation, deposition-precipitation, sol–gel, liquid–liquid interface technique, hydrothermal and solvothermal syntheses, microwave-assisted processes, polyol method, template-directed synthesis, ionic-liquid assisted methods, and so on are reported in the literature. In this chapter, the frequently used fabrication techniques are addressed in detail with appropriate literature references. A careful study of their usefulness and drawbacks with special importance on the laboratory scale synthesis has also been discussed with appropriate illustrations.

1.1 EVOLUTION OF METAL NANOPARTICLES IN PHARMACY AND BIOTECHNOLOGY

Medicinal application of metals is already known from the ancient time. The Ebers Papyrus from 1500 BC is the earliest written account of the use of metals for treatment and describes more than 800 recipes. It has mentioned the use of copper to reduce inflammation and the use of iron to treat anemia. Sushruta Samhita, the ancient book on Indian Ayurvedic treatment methodology written around 7th century BC, mentioned the use of some primary metals like gold, silver, copper, lead, tin, zinc, and iron in medicines (Prakash, 1997). Sodium vanadate has also been used since the early 20th century to treat rheumatoid arthritis. Antimicrobial properties of metals are also known, and have been in use for thousands of years. For example, copper and silver vessels have been used for water disinfection and food preservation since the time of the Persian kings. This practice was later adopted by the Phoenicians, Greeks, Romans, and Egyptians (Alexander, 2009). Settlers of North America used to drop Ag coins into transport containers to preserve water, wine, milk, and vinegar, and a similar strategy was used by Japanese soldiers during the Second World War to prevent the spread of dysentery (Borkow and Gabbay, 2009). The medicinal use of metals was prevalent until the discovery of antibiotics by Nobel laureate Sir Alexander Fleming in the 1920s.

Reducing the metal particle size to nanorange (<100 nm), however, has been demonstrated as an efficient and reliable tool for improving biocompatibility (Taton et al., 2000). Earlier studies have shown that antimicrobial formulations in the form of NPs could be used as effective bactericidal materials (Fresta et al., 1995). Later on, it was revealed that highly reactive metal oxide NPs exhibit excellent biocidal action against Gram-positive and Gram-negative bacteria (Stoimenov et al., 2002). Zero-valent metal nanoparticles like, Au, Ag, and FeNPs, as well as, metal oxide NPs like, zinc oxide (ZnO), titanium dioxide (TiO_2), magnesium oxide (MgO), etc., exhibit bactericidal activity and have been found to be stable and safe for human beings (Lemire et al., 2013). The intrinsic properties and related possible applications of different metal NPs are mainly dependent on their size, shape, composition, crystallinity, and morphology (Dickson and Lyon, 2000). NP toxicity may arise from several attributes, which include the traits that are specific to the properties of the NPs, such as size, shape, or surface charge, and traits that control the release of metal ions (El Badawy et al., 2011; Morones et al., 2005; Pal et al., 2007). The toxic mode of action of NPs has also been connected with the generation of reactive oxygen species (ROS) and membrane disruption (Fig. 1.1) (Bandyopadhyay et al., 2012; Gunawan et al., 2011). A striking capability that has been reported for many NPs is their

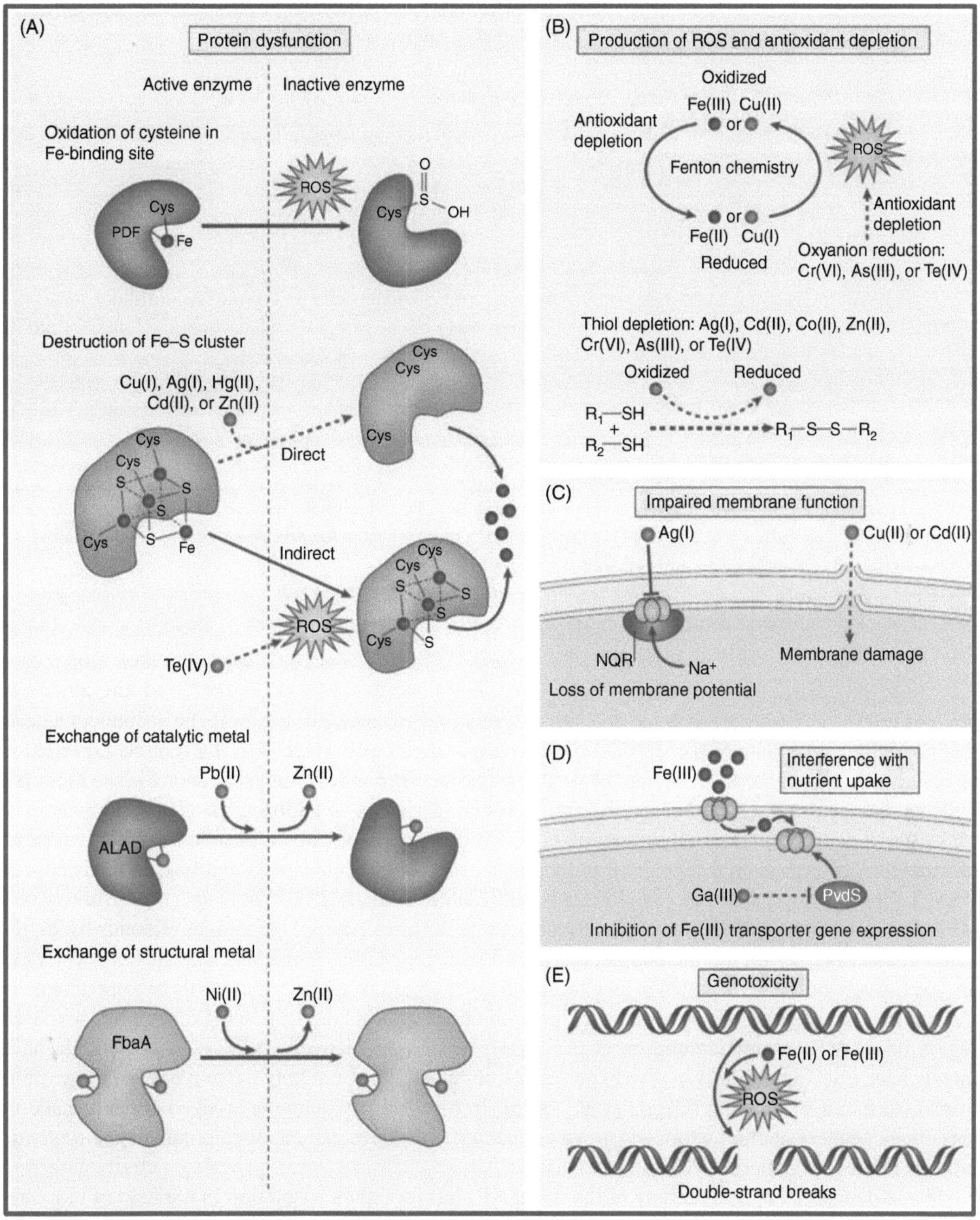

FIGURE 1.1 Antibacterial Mechanisms of Metal Toxicity

(A) Metals can lead to protein dysfunction. (B) Production of reactive oxygen species (ROS) and depletion of antioxidants. (C) Certain metals have been shown to impair membrane function. (D) Interference with nutrient assimilation. (E) They can also be genotoxic. *Solid arrows* represent elucidated biochemical mechanism, whereas, *dashed arrows* represent a route of toxicity for which the underlying biochemical mechanism is unclear. *ALAD*, δ-Aminolevulinic acid dehydratase; *FbaA*, fructose-1,6-bisphosphate aldolase; *NQR, NADH*; quinone oxidoreductase; *PDF*, peptide deformylase; *PvdS*, a σ-factor (σ24) from *Pseudomonas aeruginosa*.

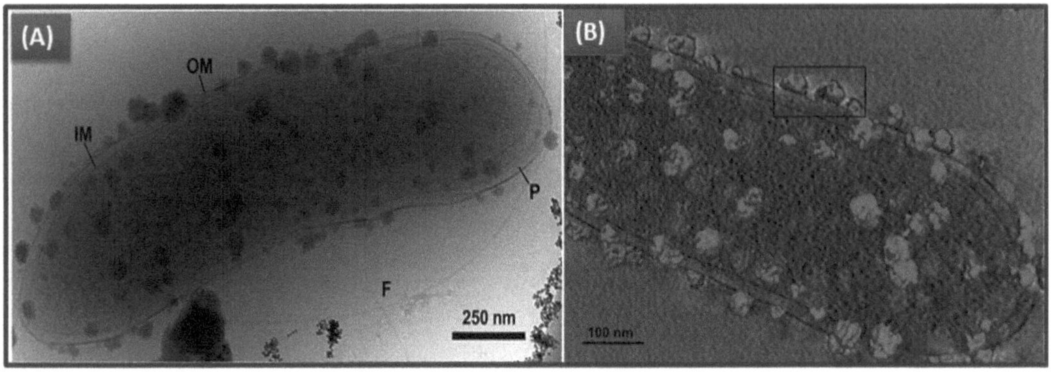

FIGURE 1.2

(A) The figure shows cryo-TEM image of a single cell planktonic Fe-reducing bacterium isolated from groundwater at a site in Rifle, Colorado, USA. (B) A 3-D construction of nanoaggregates *[orange (mid-gray in print version)]* attached to the groundwater bacterium cell wall (Luef et al., 2013).

Reprinted with permission from Bao, Y., An, W., Turner, C.W., Krishnan, K.M., 2010. The critical role of surfactants in the growth of cobalt nanoparticles. Langmuir 26, 478–483. Copyright 2013, International Society for Microbial Ecology.

ability to physically interact with the cell surfaces of some bacteria. For example, Fe-reducing bacteria that naturally accumulate NPs of ferric oxyhydroxide at their cell surface (Fig. 1.2) can be exploited in targeting a specific bacterial cell (Luef et al., 2013). However, a large number of reports in the literature indicate that the driving force behind the antimicrobial properties of antibacterial NPs is the release of ions (Xiu et al., 2012). Tactically designed NPs, which release ions in a controlled manner or can be targeted to specific bacterial cells, have numerous antimicrobial applications and bagged a business of US$17.5 billion in the year 2011 in global market (Lemire et al., 2013).

Metal and metal oxide NPs have also found applications in advanced techniques of therapeutics, diagnosis, and drug delivery. For example, there are a wide range of paramagnetic metal NPs, mostly but not exclusively based on iron; when an external magnetic field is applied, these NPs become magnetic themselves, and align themselves with the direction of the external field, showing up as a hypoexposed region on an MRI scan (Edmundson et al., 2003). Iron-based magnetic NPs, such as Feridex, have mostly been used in in vitro or in vivo experiments, for example, in tracking the movement of stem cells implanted into a wound site (Okada et al., 2005). In recent times, nanometer-sized crystals made of metallic or semiconductor materials, known as quantum dots have gained much popularity in synthetic biology (Nath and Banerjee, 2013).

Shape-dependent optical activity of the metal NPs has brought a revolution in the field of bioimaging and diagnosis. Huang and coworkers demonstrated that due to longitudinal absorption band in the near infrared (NIR), Au-nanorods are effective in acoustic imaging and are suitable as photothermal agents for hyperthermia therapy of cancer cells. Small diameter Au-nanorods are being used as photothermal converters of near infrared radiation (NIR) for in vivo applications due to their high absorption cross-sections beyond the tissue absorption spectra (Huang et al., 2006a,b). Gold nanocages and nanoframes developed by Xia and coworkers also exhibit potential biomedical applications due to their desirable optical properties and cargo-holding hollow structures (Skrabalak et al., 2008). Silica-gold core–shell NPs, or gold nanoshells, have attracted much attention due to their interesting optical properties and numerous biomedical applications. Aden and Kerker (1951) predicted that concentric spherical particles could exhibit tunable plasmon resonance which vary as a function of the ratio of shell thickness to core radius. Other gold nanostructures, such as Au-nanoshells, Au-nanocages, and spherical AuNPs have also demonstrated effective photothermal destruction of cancer cells and tissue (Conde et al., 2011). Novel optical properties of other related structures, such as asymmetric "nanoeggs" and quadruply concentric "nanomatryushkas" have also been explored (Wang et al., 2007).

2 SYNTHETIC METHODS FOR PREPARATION OF METAL NANOPARTICLES

2.1 THERMAL DECOMPOSITION METHOD

Thermal decomposition method is an excellent synthetic route to fabricate metal NPs. This method is facile and involves single step process. It is inexpensive, environmentally benign, and provides higher quality of metal NPs in terms of morphology, size, and particle-size distribution (Luo et al., 2009). It is a well-known fact from literature that, nucleation step and particle growth are the crucial factors to achieve monodisperse NPs (Kashchiev and van Rosmalen, 2003). The size and shape of the NP can be tuned in thermal decomposition method by controlling the previously mentioned factors by the use of appropriate surfactants. This method involves thermal decomposition of organometallic precursor, like metal carbonyls and metal surfactant complex in solution resulting in metallic NPs (Kwon and Hyeon, 2008). For example, Bao et al. (2010) have synthesized monodisperse cobalt NPs using $CO_2(CO)_8$ as the precursor and oleic acid (OA), tri-*n*-octylphosphine oxide (TOPO), and di-*n*-octylamine (DOA) as the surfactants. Fig. 1.3 shows the TEM images of CoNPs at different time periods in the presence of OA and TOPO surfactants. The particle size was found to rapidly increase with respect to time period. Growth pathway was also investigated in terms of the influence of the surfactant. Combination of OA and TOPO provides diffusional growth pathway, whereas OA and DOA combination and TOPO alone allowed aggregation and ripening growth pathways, respectively. In a similar way, metallic Cu and PdNPs were prepared using precursors, copper cupferronate $Cu(cupf)_2$ complex, and Pd–trioctylphosphine complex, respectively in the presence of trioctylphosphine (TOP) and DOA surfactant. The concentration of stabilizing or capping agent (TOP) played a vital role in controlling the particle size (Diab et al., 2011; Kim et al., 2003).

Chen et al. (2007) have reported the preparation of monodisperse spherical NiNPs via thermal decomposition of nickel (II) acetylacetonate (Ni $(acac)_2$) complex in the presence of various alkyl amines. Reaction temperature and solvent type exhibit profound influence on the crystalline phase of NiNPs, whereas, surfactants played a significant role in controlling the particle size as well as morphology. Kura et al. (2010) successfully synthesized monodisperse FeNPs with high saturation magnetization from Fe(CO)x-OAm precursor, in which CO ligands are partially replaced with OAm. During the course of the reaction the precursor decomposed and yielded monodispersed FeNPs. However, OAm played a significant role as a ligand and surfactant to produce small Fe particles by covering the surface of the metal particle.

Kim et al. (2007) have reported the synthesis of hollow iron nanoframes by thermal decomposition of Fe (II)–oleate complex, yielding uniform size Fe nanocubes in the presence of oleic acid. Fig. 1.4 shows various morphologies of iron and inset shows the HRTEM image of FeNPs. However, the addition of a little amount of sodium oleate to the reaction solution, resulted in a remarkable change in the

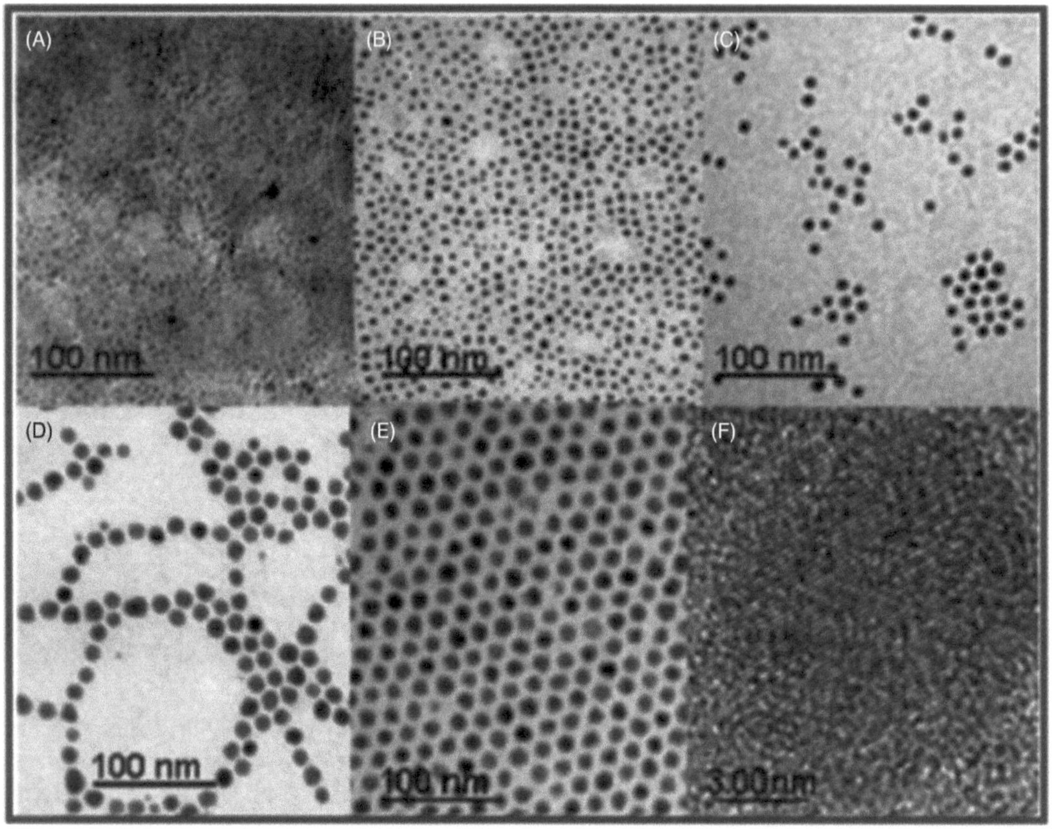

FIGURE 1.3 TEM Images of Co Nanoparticles (NPs) in the Presence of Surfactants OA and TOPO at Different Stages

(A) 1, (B) 5, (C) 10, (D) 15, and (E) 30 min. (F) High resolution of TEM image of Co NPs.

Reprinted with permission from Bao, Y., An, W., Turner, C.W., Krishnan, K.M., 2010. The critical role of surfactants in the growth of cobalt nanoparticles. Langmuir 26, 478–483. Copyright 2010, American Chemical Society.

morphology of FeNPs, that is, nanocubes to nanoframes. Therefore, it is believed that sodium oleate played a significant role in controlling the morphology of the Fe NPs and the obtained Fe nanoframes showed exceptional biomedical applications, especially in drug delivery. A wide variety of bimetallic NPs have also been prepared by thermal decomposition method (Samia et al., 2006). For instance, Cu–Pt bimetallic NPs with 1.2-nm size were synthesized by thermolysis of corresponding metal precursors—$Pt(acac)_2$ and $Cu(acac)_2$ at 498 K under oleylamine (Zheng et al., 2003). EXAFS studies revealed that, Pt atoms occupied the Cu sites. Charge transfer from Cu to Pt was observed during the formation of Cu–Pt bimetallic system. Kang and Murray (2010) have reported the synthesis of Mn–Pt bimetallic nanocubes using $[Mn(acac)_2]$ or $Mn_2(CO)_{10}$ with platinum acetylacetonate $[Pt(acac)_2]$ in the presence of oleylamine and oleic acid stabilizers.

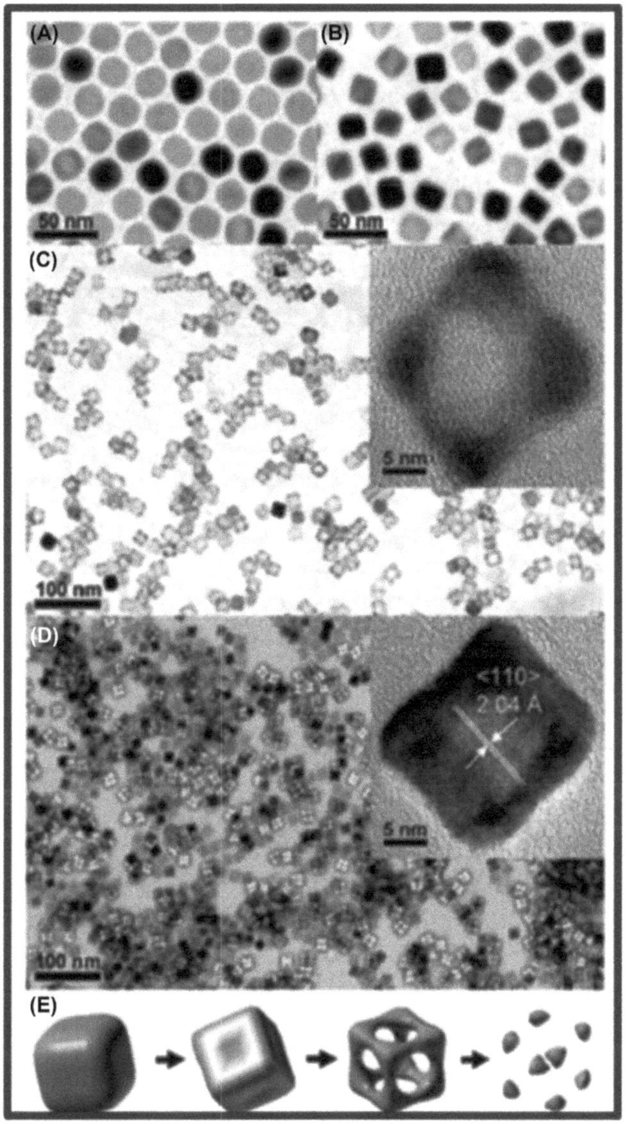

FIGURE 1.4 TEM Images of FeNPs

(A) 23-nm sized spheres with slightly faceted shape, (B) 21-nm sized nanocubes, (C) 21-nm sized nanoframes (inset HRTEM), (D) 17-nm sized particles with intermediate shape between *solid* and *hollow* nanocubes (inset HRTEM), (E) the overall shape evolution of the Fe nanoparticles.

Reprinted with permission from Kim, D., Park, J., An, K.,Yang, N.K., Park, J.G., Hyeon, T., 2007. Synthesis of hollow iron nanoframes. J. Am. Chem. Soc. 129, 5812–5813. Copyright 2007, American Chemical Society.

2.2 SOL–GEL METHOD

Sol–gel method is one of the well-established synthetic approaches to prepare novel metal oxide NPs as well as mixed oxide composites. This method has potential control over the textural and surface properties of the materials. Sol–gel method mainly undergoes in few steps to deliver the final metal oxide protocols and those are hydrolysis, condensation, and drying process. The formation of metal oxide involves different consecutive steps, initially the corresponding metal precursor undergoes rapid hydrolysis to produce the metal hydroxide solution, followed by immediate condensation which leads to the formation of three-dimensional gels. Afterward, obtained gel is subjected to drying process, and the resulting product is readily converted to Xerogel or Aerogel based on the mode of drying. Sol–gel method can be classified into two routes, such as aqueous sol–gel and nonaqueous sol–gel method depending on the nature of the solvent utilized.

If water is used as reaction medium it is known as aqueous sol–gel method; and use of organic solvent as reaction medium for sol–gel process is termed as nonaqueous sol–gel route. The reaction pathway for the production of metal oxide nanostructures in the sol–gel method is shown in Fig. 1.5. In the sol–gel approach, nature of metal precursor and solvent plays a remarkable role in the synthesis of metal oxides NPs.

2.2.1 Aqueous sol–gel method

In aqueous sol–gel method, oxygen is necessary for the formation of metal oxide, which is supplied by the water solvent. Generally, metal acetates, nitrates, sulfates, chlorides, and metal alkoxides are employed as the metal precursors for this method. However, metal alkoxides are widely used as the

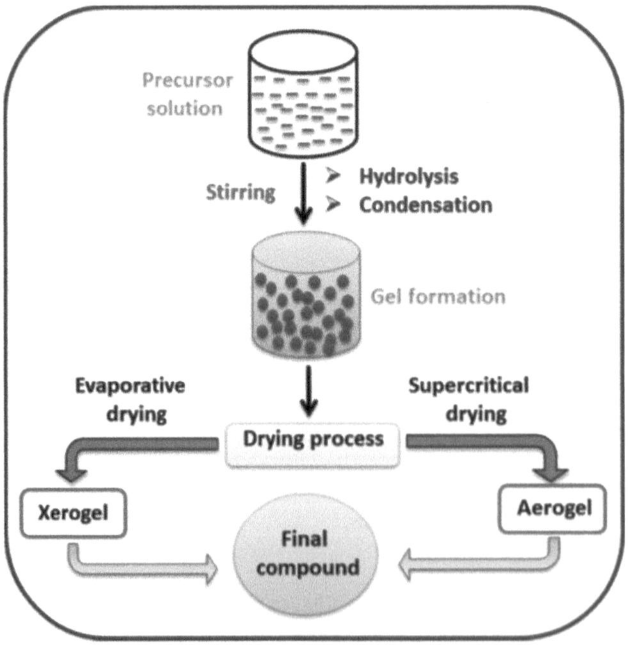

FIGURE 1.5 The Reaction Pathway for the Production of Metal Oxide Nanostructures in the Sol–Gel Method

precursors for the production of metal oxide NPs, due to high reaction affinity of alkoxides toward water (Bradley et al., 2001; Turova and Turevskaya, 2002). However, some difficulties are associated with the aqueous sol–gel method. The key steps, such as hydrolysis, condensation, and drying take place simultaneously in a number of cases resulting in difficulty in controlling particle morphology, and reproducibility of the final protocol during the sol–gel process (Corriu and Leclercq, 1996). The aforementioned difficulties, however, do not affect much of the synthesis of metal oxides in bulk, but strongly affect the preparation of nanooxides. Therefore, it is believed that the aqueous sol–gel route is highly recommended for the synthesis of bulk metal oxides rather than their nanoscale counterparts (Niederberger, 2007).

2.2.2 Nonaqueous sol–gel method

Nonaqueous or nonhydrolytic sol–gel method is devoid of some of the major drawbacks found in aqueous sol–gel method. In nonaqueous sol–gel process, oxygen required for the formation of metal oxide is supplied from the solvents, such as alcohols, ketones, aldehydes, or by the metal precursors. Furthermore, these organic solvents not only serve as oxygen providers but also offer a versatile tool for tuning several key components like morphology, surface properties, particle size, and composition of the final oxide material. Although, nonaqueous sol–gel approach is not as popular as aqueous sol–gel method; nonaqueous sol–gel routes have shown excellent impact on the production of nanooxides compared to that of aqueous sol–gel route. The nonaqueous sol–gel route can be divided into two important methodologies, namely, surfactant-controlled and solvent-controlled approaches for the production of metal oxide NPs. Surfactant-controlled strategy involves direct transformation of metal precursor into the respective metal oxide at higher temperature range in hot injection method. This method permits outstanding control over the shape, growth of the NP, and avoids the agglomeration of particles. Few examples of surfactant-controlled synthesized NPs are mentioned here for understanding. Song and Zhang (2004) have demonstrated the simple nonhydrolytic route to synthesize high-quality spherical-shaped $CoFe_2O_4$ NPs with 8-nm size. However, the spherical morphology can be changed to cubic shape with 10-nm edge length during the seed-mediated growth. Heating rate and growth temperature played a pivotal role in controlling the shape of $CoFe_2O_4$ nanomaterial (Fig. 1.6A–B).

The resulting materials were subjected to shape-dependent magnetic properties. Zeng et al. (2004) have extensively studied the shape-controlled synthesis of $MnFe_2O_4$ nanomaterial. The relative ratio between surfactant and $Fe(acac)_3$ showed a remarkable role in controlling the final morphology of $MnFe_2O_4$. TEM analysis revealed the formation of cube-like or polyhedron-type morphology for $MnFe_2O_4$ (Fig. 1.6C–D). In addition, size of $MnFe_2O_4$ particle is dependent on the concentration of metal precursors. Novel cone-shaped ZnO was obtained by decomposition of TOPO–$Zn(OAc)_2$ complex resulting in the formation of hierarchically ordered spheres of cone-shaped ZnO nanocrystals (Joo et al., 2005). Li et al. (2006) fabricated titanium oxide nanorods with 3.3-nm diameter and a length of 25 nm using appropriate amounts of reaction ingredients, such as titanium butoxide, triethylamine, linoleic acid, and cyclohexane. Reaction temperature, time, and concentration of the reactant were found to show huge effect on the shape and size of the TiONPs. Preparation of high-quality single crystalline MnO multipods with homogeneous size and shape, involved decomposition of $Mn(oleate)_2$ in the presence of oleic acid and *n*-trioctylamine (Fig. 1.6E) (Zitoun et al., 2005). Tungsten oxide nanorods were generated by treatment of WCl_4 with oleylamine and oleic acid (Fig. 1.6F) (Seo et al., 2005). Solvent-controlled sol–gel route, involves the reaction between metal halide and alcohols to produce metal oxide nanostructures. For example, porous SnO_2 NPs were prepared by the addition of tin chloride to benzylalcohol under stirring condition, which was immediately dispersed in THF solution, producing

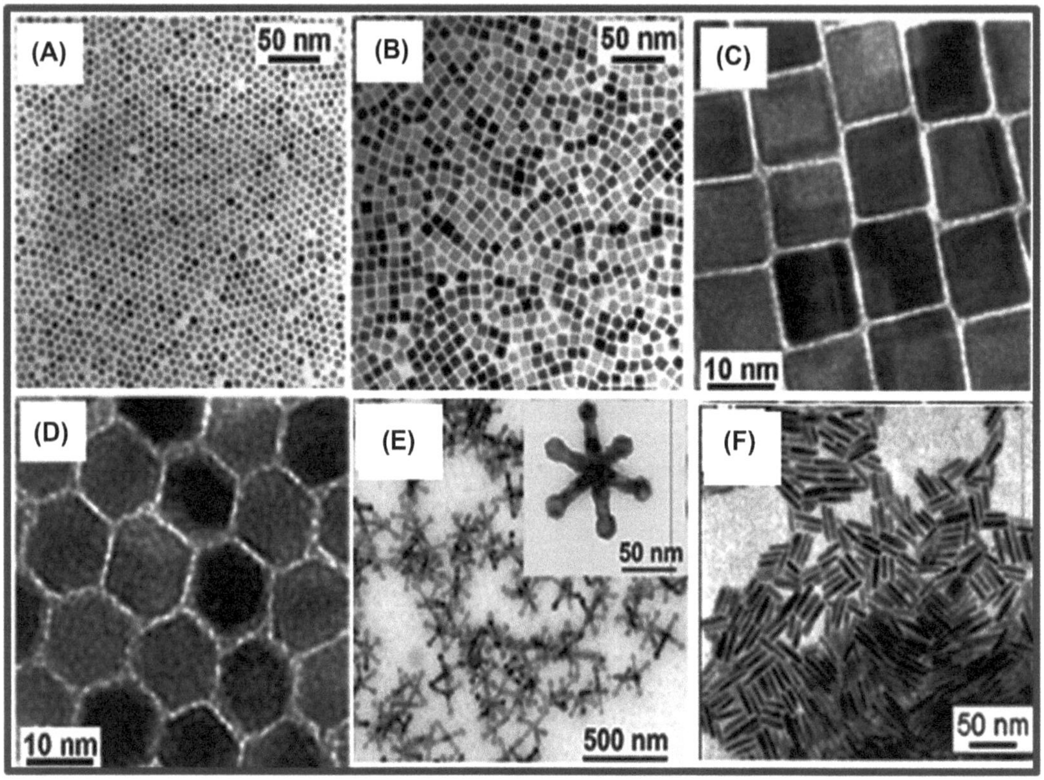

FIGURE 1.6

TEM images of (A) 8-nm sized spherical $CoFe_2O_4$ NPs and (B) cube-like $CoFe_2O_4$ NPs. TEM images of (C) cube-like and (D) polyhedron-shaped $MnFe_2O_4$NPs. (E) TEM image of MnO multipods (inset, hexapod). (F) TEM image of Tungsten oxide nanorods.

Part (A–B): Reproduced from Song, Q., Zhang, Z.J., 2004. Shape control and associated magnetic properties of spinel cobalt ferrite nanocrystals. J. Am. Chem. Soc. 126, 6164–6168. Copyright 2004, American Chemical Society. Part (C–D): Reprinted from Zeng, H., Rice, P.M., Wang, S.X., Sun, S.H., 2004. Shape-controlled synthesis and shape-induced texture of $MnFe_2O4$ nanoparticles. J. Am. Chem. Soc. 126, 11458–11459. Copyright 2004, American Chemical Society. Part (E): Reproduced from Zitoun, D., Pinna, N., Frolet, N., Belin, C., 2005. Single crystal manganese oxide multipods by oriented attachment. J. Am. Chem. Soc. 127, 15034–15035. Copyright 2005, American Chemical Society. Part (F): Reproduced from Seo, J.-W., Jun, Y.-W., Ko, S.J., Cheon, J., 2005. In situ one-pot synthesis of 1-dimensional transition metal oxide nanocrystals. J. Phys. Chem. B. 109, 5389–5391. Copyright 2005, American Chemical Society.

sol. The subsequent addition of block polymer to sol allowed mesoporous nanostructure for SnO_2 by the elimination of solvent molecule (Ba et al., 2005).

2.3 HYDROTHERMAL AND SOLVOTHERMAL METHOD

Hydrothermal or solvothermal method is one of the most common and effective synthetic routes to fabricate the nanomaterial with a variety of morphologies. In this method, the reactants are placed into an autoclave filled with water or organic compound to carry out the reaction under high temperature and pressure conditions. If the nonaqueous solvents are utilized as reaction medium, it is termed as solvothermal method; whereas, in case the preparation is carried out in the presence of water, it is known as hydrothermal process (Cushing et al., 2004; Wu et al., 2002). Different kinds of autoclaves and their functions are deeply discussed in the literature (Hakuta et al., 2005; Rabenau, 1985). Generally, Teflon-lined autoclaves are capable of working at high temperature and pressure. In addition, it sustains alkaline media and exhibits a strong resistance to hydrofluoric acid when compared to glass and quartz autoclaves. Therefore, Teflon-lined autoclave is chosen as an ideal container to perform the reaction under desired conditions. Precise control in hydrothermal process is the key factor that enables the synthesis of various nanostructured inorganic materials (Shi et al., 2013). This method can facilitate and accelerate the reaction among the reactants, promote hydrolysis, followed by crystal growth resulting in self-assembly of nanomaterials in the solution. Moreover, the properties, morphology, size, and structure of nanomaterials can be tailored easily by varying the different reaction parameters, such as reaction time, temperature, reaction medium, pressure, pH, and concentration of the reactants and filled volume of autoclave. This method can be suitable for the preparation of nanomaterials with a variety of shapes as compared to other methodologies.

Yan et al. (2008) have fabricated ceria nanooctahedrons and nanorods without assistance of surfactant or template via simple and facile hydrothermal approach. High quality of ceria nanorods with 20-nm width and few hundred nanometer length were produced by treating $Ce(NO_3)_2.6H_2O$ with $Na_3PO_4.6H_2O$ in water medium (Fig. 1.7A). The change in the morphology of the ceria NPs was achieved by tuning the hydrothermal reaction time. It was demonstrated that, Na_3PO_4 played a crucial role in the formation of

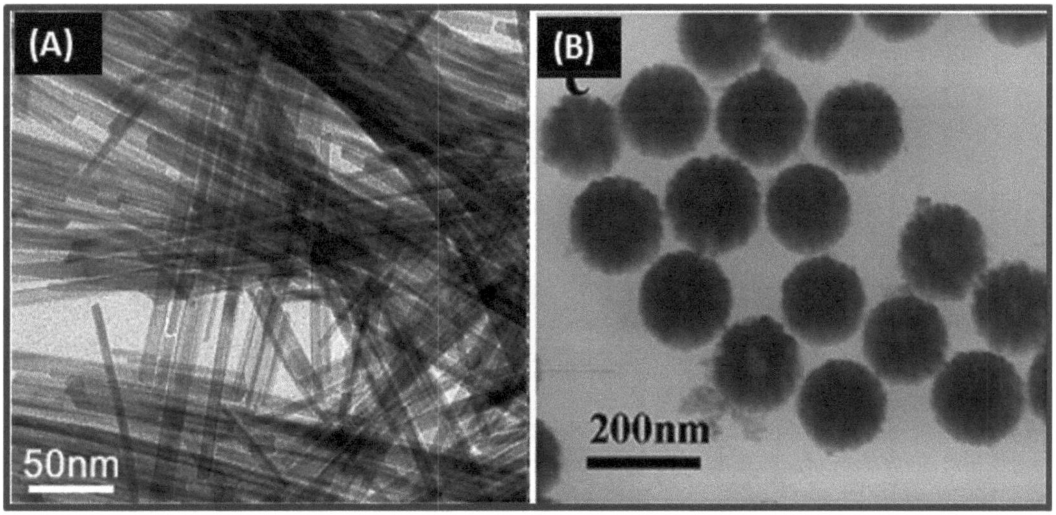

FIGURE 1.7 TEM image of (A) CeO_2 nanorod. (B) TEM image of CeO_2 *hollow sphere.*

Part (A): Reprinted with permission from Yan, L., Yu, R., Chen, J., Xing, X., 2008. Template-free hydrothermal synthesis of CeO_2 nanooctahedrons and nanorods: investigation of the morphology evolution. Cryst. Growth Des. 8, 1474–1477. Copyright 2008, American Chemical Society. Part (B): Reprinted with permission from Yang, Z., Han, D., Ma, D., Liang, H., Liu, L., Yang, Y., 2010. Fabrication of monodisperse CeO_2 hollow spheres assembled by nanooctahedra, Cryst. Growth Des., 10, 291–295. Copyright 2010, American Chemical Society.

ceria nanostructures. A simple hydrothermal method has been developed to prepare ceria nanocubes and octahedra by employing $Ce(NO_3)_2.6H_2O$, $Na_3PO_4.6H_2O$, and NaOH as precursors. The obtained nanostructured ceria was effectively studied for CO oxidation reaction. Interestingly, after the exposure of these nanoshapes to CO oxidation at 673 K, they have retained their original shape without losing their individual nanoshape, indicating that these nanostructures are thermally stable up to 673 K (Zili et al., 2012).

Yang et al. (2010) have demonstrated that nanooctahedra are the building blocks for the production of monodisperse ceria hollow spheres that are obtained by mild hydrothermal route (Fig. 1.7B). H_2O_2 played a significant role in the production of novel hollow sphere nanostructures, whereas polyvinylpyrrolidone (PVP) played an important role in promoting the nucleation of nanocrystals during the course of reaction. The alignment of hollow sphere nanostructure of ceria was explained based on the Ostwald ripening mechanism. Van et al. (2012) have reported the synthesis of Fe_2O_3 nanostructure with a variety of morphologies by using capping agents like sodium carboxymethyl cellulose (CMC) and hydrazine (N_2H_4) assisted hydrothermal method (Fig. 1.8). Capping agents can produce different morphologies, such as spheres and truncated hexagonal pyramid-shaped Fe_2O_3 nanocrystals. However, without the aid of capping agent dendrite shape of Fe_2O_3 nanocrystals were obtained. Therefore, it is believed that capping agents play a vital role in controlling the morphology of final nanomaterial. Magnetic and optical properties of nanostructured hematite phase Fe_2O_3 nanocrystals were extensively studied.

FIGURE 1.8 Morphology of α-Fe_2O_3 With Various Reactants Under Hydrothermal Conditions at 433 K for 8 h

(A) $K_3[Fe(CN)_6]$, 0.02 M, in the absence of CMC and N_2H_4; (B) $K_3[Fe(CN)_6]$, in the absence of CMC, 3.5 gL^{-1}; (C) $K_3[Fe(CN)_6]$, 0.02 M, with the addition of CMC, 3.5 gL^{-1} and N_2H_4, 1.5 wt.%; (D) $K_3[Fe(CN)_6]$, 0.02 M, with the addition of CMC, 3.5 gL^{-1}, and N_2H_4, 4.5 wt.%; (E) $K_3[Fe(CN)_6]$, 0.02 M, with the addition of CMC, 3.5 gL^{-1}, and N_2H_4, 6.0 wt.%; (F) $K_3[Fe(CN)_6]$, 0.02 M, with the addition of CMC, 3.5 gL^{-1}, and N_2H_4, 8.0 wt.%.

Reprinted with permission from Van, T.-K., Cha, H.G., Nguyen, C.K., Kim, S.-W., Jung, M.-H., Kang, Y.S., 2012. Nanocystals of hematite with unconventional shape-truncated hexagonal bipyramid and its optical and magnetic properties. Cryst. Growth Des. 12, 862–868. Copyright 2011, American Chemical Society.

2.4 MICROWAVE-ASSISTED METHOD

Generally, solution-based approaches utilize conventional heating, which is the driving force for chemical reactions. In this process, heat energy is transferred from the source to the solvent and then transferred to reactants during the course of the reaction. However, conventional heating process inevitably suffers from various drawbacks, including high thermal gradient effects, slow reaction kinetics, nonconsistent and undesirable reaction conditions throughout the bulk (Gerbec et al., 2005). Particularly, for large-scale production of NPs, the aforesaid problematic aspects, such as inhomogeneity, poor crystallization, and thermal gradient effects may be magnified tremendously resulting in poor nucleation and wide-size distribution (Hu et al., 2008; Hu and Yu, 2008). From this point of view, microwave method is one of the excellent alternative routes to address the previously mentioned issues raised in conventional heating process.

Microwaves are nothing but electromagnetic energy with frequency in the range of 300 MHz to 300 GHz. Generally, interaction of microwaves with materials during the reaction is based on two important mechanisms: dipole interactions and ionic conduction. However, these two mechanisms can effectively work, when coupling takes place between components of target compound and oscillating electric field of microwave. Fig. 1.9 depicts the production of heat energy during the interaction of microwaves with polar water molecule (Tsuji et al., 2005). In microwave frequency range, water molecules try to orient with the electric field and the two polar ends try to reorient with respect to the oscillating electric field, as a result they lose energy in the form of heat by molecular collision and friction. Other polar molecules, such as alcohols, DMF, ethylene glycol are used as ideal solvents for microwave-assisted synthesis of metal NPs, due to their high dielectric loss and high reduction ability.

Microwave method received considerable attention as a new, promising, and environmental friendly method to synthesize the metallic nanostructures as well as metal oxides with a variety of morphologies. In addition, microwave strategy offers several benefits, such as high efficiency in the utilization of heat. Moreover, it is clean, cheap, and produces higher yields of desired materials within shorter reaction times. Fast and uniform heating of reaction mixture throughout the solution are the beneficial factors. These factors always allow the rapid decomposition of respective metal precursors followed by the nucleation and subsequent growth to obtain the required nanostructures (Gerbec et al., 2005). A wide variety of metal NPs and metal oxide NP with control over size and morphology of the NPs to tailor their properties have been synthesized by adopting the microwave-assisted method.

Mohamed et al. (2010) have explored various morphologies, such as spherical, truncated prisms, and hexagons for AuNP by using $HAuCl_4$ in HCl as gold precursor via microwave-assisted strategy

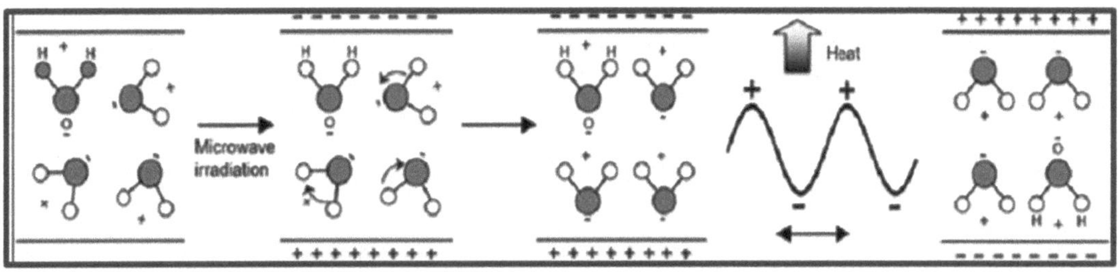

FIGURE 1.9 Production of Heat Energy by Microwave Irradiation of Water Molecule

Reprinted with permission from Tsuji, M., Hashimoto, M., Nishizawa, Y., Kubokawa, M., Tsuji, T., 2005. Microwave-assisted synthesis of metallic nanostructures in solution. Chem. Eur. J. 11, 440–452. Copyright 2005, Wiley-VCH.

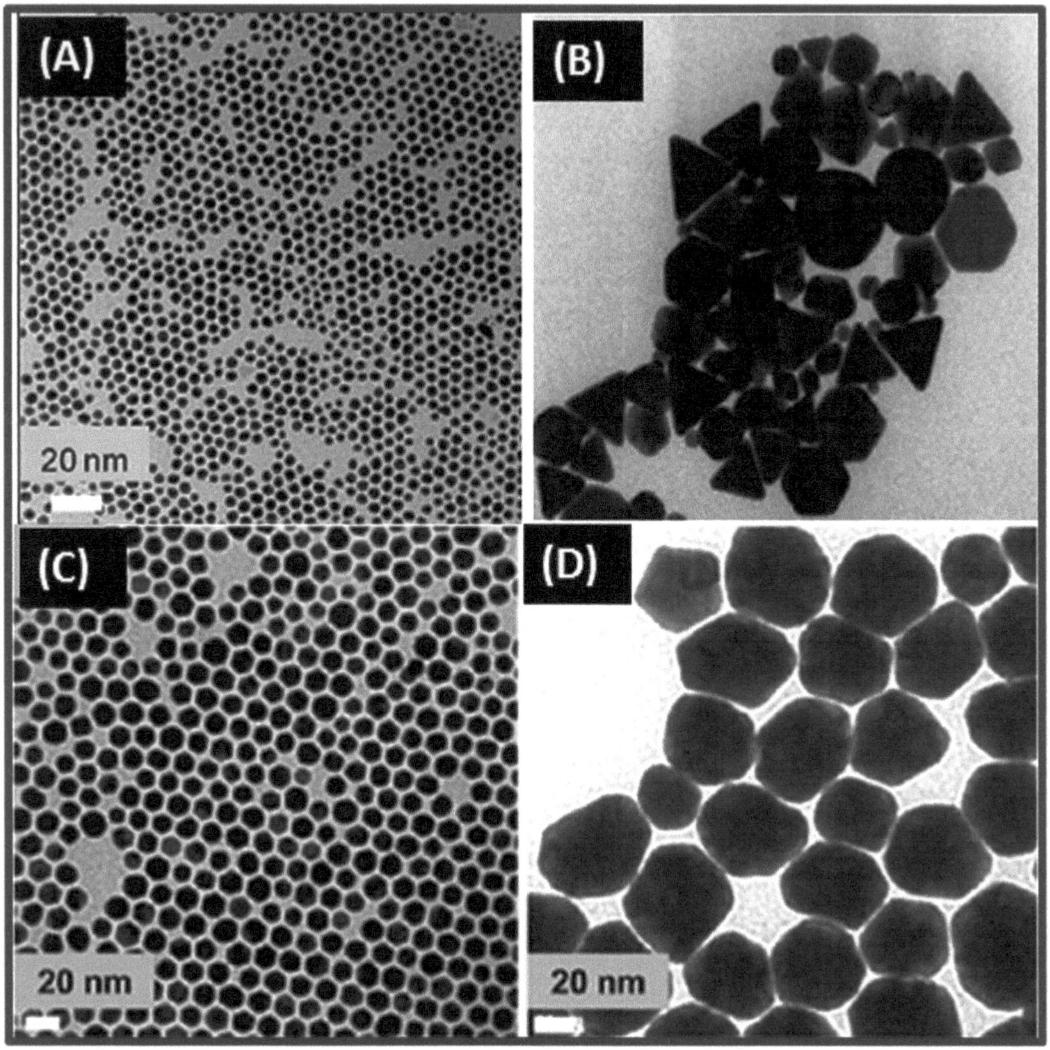

FIGURE 1.10

TEM image of (A) AuNPs prepared using pure amine, (B) AuNPs prepared in 70% oleic acid, (C–D) AuNPs formed using 9 and 36 mM of $HAuCl_4$, respectively.

Reprinted with permission from Mohamed, M.B., AbouZeid, K.M., Abdelsayed, V., Aljarash, A.A., Samy El-Shall, M., 2010. Growth mechanism of anisotropic gold nanocrystals via microwave synthesis: formation of dioleamide by gold nanocatalysis. ACS Nano 4, 2766–2772. Copyright 2010, American Chemical Society.

(Fig. 1.10). Oleylamine and oleic acid mixture served as capping agents to control the particle size. The oleic acid and oleylamine ratio, concentration of gold precursor and microwave exposure time profoundly influence the size and morphology of Au nanocrystals. Also, these factors show a favorable role in tuning the surface resonance properties of the resulting nanocrystals in visible and near-infrared regions. Mallikarjuna and Varma (2007) have successfully generated various nanostructures of noble metal NPs, such as gold, silver, platinum, and palladium in bulk quantities, with particle size in the range of 2–15 nm, by reduction of respective metal salts using D-glucose, maltose, and sucrose solutions via microwave-assisted method. The type of reducing sugar was found to be a vital parameter in determining the morphology of the resultant nanomaterial. As prepared noble metal NPs showed versatile applications in various fields including catalysis, biosensors, energy storage systems, and nanodevices.

Microwave-assisted strategy offers precise control over the various key parameters, such as morphology, crystallinity, nucleation, and growth of the monodisperse Pd, Pt, and RhNPs compared to conventional heating method. Recently, our group prepared different ceria-based mixed oxides, such as CeO_2–ZrO_2, CeO_2–SiO_2, and CeO_2–Al_2O_3 and others by microwave-induced combustion process. In this method, stoichiometric ratios of corresponding metal nitrate precursors were added to the desired quantity of urea. The resulting mixtures were heated in a microwave oven for 40 min; as a result, mixed oxides were generated. The metal mixed oxides prepared by MW showed high oxygen vacancies and lattice defects compared to the coprecipitation method (Reddy et al., 2009a,b,c, 2010).

2.5 POLYOL METHOD

A number of promising synthetic routes have been employed to synthesize different kinds of NPs with a variety of morphologies. Among them, polyol method is a versatile liquid-phase method utilizing high boiling and multivalent alcohols to produce the NPs. Polyols played a dual role as a reducing agent and solvent, also they are capable of controlling the particle growth. The polyol synthetic route was first introduced in 1989 by Fievets group where they used the term polyol for metal NPs synthesis. Several kinds of polyols that include, ethylene glycol (EG), propylene glycol (PG), butylene glycol (BG), diethylene glycol (DG), triethylene glycol (TrEG), tetraethylene glycol (TEG), and so on up to polyethylene glycol (PEG) have been utilized in this process (Carroll et al., 2011; Grisaru et al., 2003). However, the polyols provide excellent advantages in terms of various aspects. The high boiling point of polyols can allow the synthesis in the temperature range of 473–593 K without high pressure and autoclave. The good capabilities of polyols to solubilize the starting materials allow the use of simple and cheap metal precursors as starting compounds. The chelating ability of polyol is a beneficial factor to control key features, such as nucleation, growth, and agglomeration of the particles. Another important advantage of polyols is that the reductive ability at elevated temperatures can reduce the metal solution readily to form metal NPs. The selection of polyol for the preparation of NPs is highly dependent on two aspects, those are boiling point and reduction potential of the polyol.

Wang et al. (2013) developed a robust polyol method for preparation of Ag nanocubes with controlled edge length in the range between 18 and 32 nm by taking silver trifluoroacetate as a metal precursor in DEG medium. Fig. 1.11 shows the TEM images of silver nanocubes obtained at different reaction times in DEG. The edge length of silver nanocubes can be tailored by quenching the reaction at different reaction times with the aid of UV-Visible spectrometer. DEG not only serves as a solvent medium but is also responsible for achievement of nucleation in early stage to generate a large number

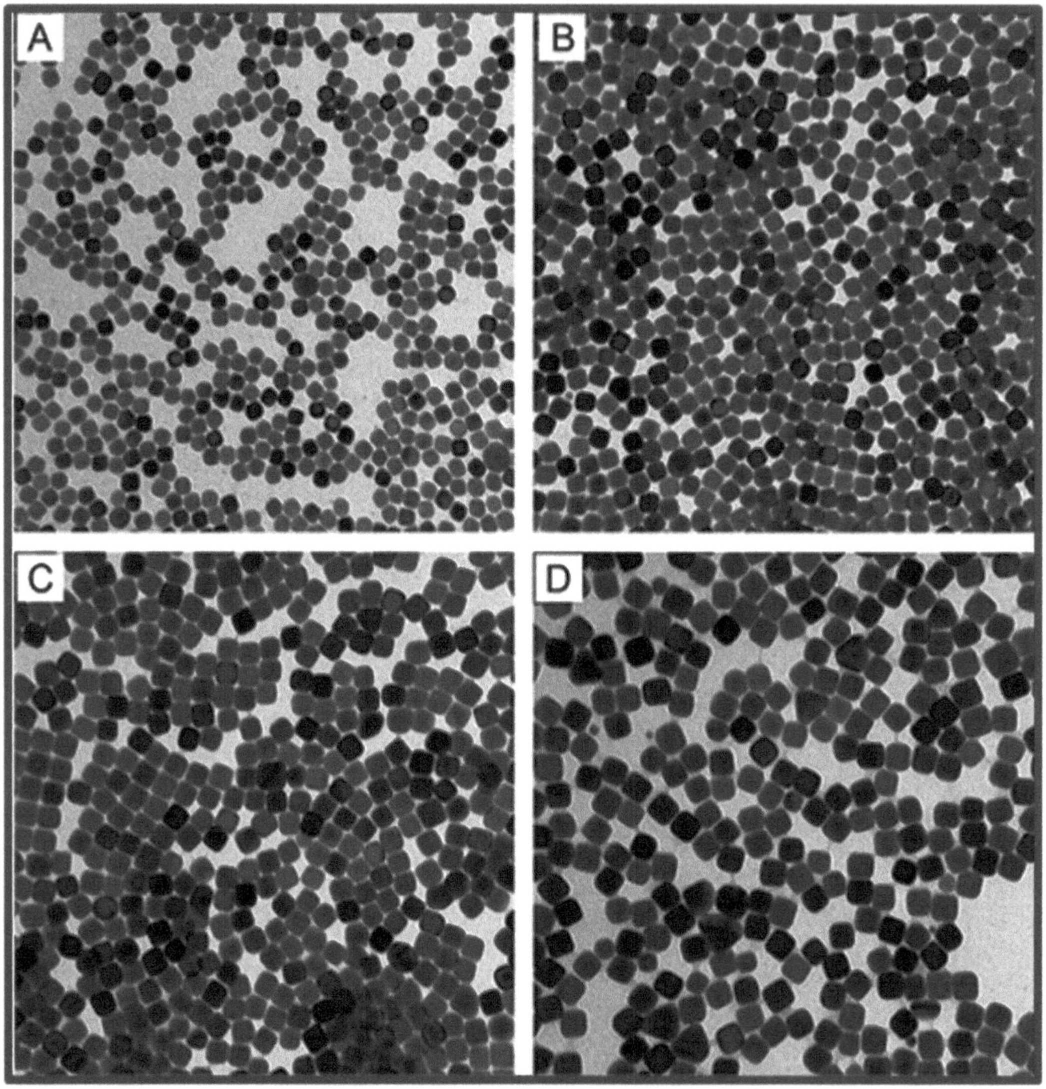

FIGURE 1.11 TEM Images of Ag Nanocubes Obtained at Different Time Intervals in DEG

(A) 30, (B) 60, (C) 120, and (D) 180 min.

Reprinted with permission from Wang, Y., Zheng, Y., Huang, C.Z., Xia, Y., 2013. Synthesis of Ag Nanocubes 18–32 nm in edge length: the effects of polyol on reduction kinetics, size control, and reproducibility. J. Am. Chem. Soc. 135, 1941–1951. Copyright 2013, American Chemical Society.

of Ag seeds with slow down the growth rate. The nucleation and growth of the Ag cubes could be monitored by TEM analysis.

Carroll et al. (2011) have extensively studied the effect of polyol for the final morphology of Cu by simple and robust polyol method. Various polyols, such as EG, PG, BG, DEG, and TEG have been utilized with NaOH and methanol. TEM results revealed that the morphology of Cu is strongly influenced by the nature of the employed polyols that alter the important parameters like nucleation and growth steps during the preparation. Li et al. (2013) developed a straightforward polyol method for the fabrication of Au@Ag heterostructure with nanorod morphology. High quality of Au@Ag heterogeneous nanorods of various aspect ratios were produced by varying the initial concentration of poly(diallyldimethylammonium) chloride (PDDA).

TEM results shown in Fig. 1.12 revealed that the aspect ratios of Au@Ag heterogeneous nanorods are manipulated by the initial concentration of PDDA. The prepared heterogeneous Au@Ag nanorods are stable and found to show size-dependent optical properties, as a result maintained optical properties up to long time.

2.6 SONOCHEMICAL METHOD

The sonochemical method is one of the recently emerged techniques that provide a versatile route to obtain nanostructures for biomedical applications. Fundamentally, it signifies the utilization of high-intensity ultrasounds, which induce transient conditions that are distinct from other conventional synthetic routes like hydrothermal route, spray pyrolysis method, and wet chemical method (Bang and Suslick, 2010; Suslick, 1990; Suslick and Price, 1999). Generally, these transient conditions are generated during the acoustic cavitation, which produce high temperature and high pressure at the center of the bubbles. The bubbles are usually obtained when solutions are exposed to high-intensity ultrasound. The employed conditions in sonochemical method are as follows: (1) speed of sound in liquid should be in the range 1000–1500 ms^{-1}, (2) ultrasonic wavelength is required between 10 cm and 100 μm, and (3) the frequency range of 20 kHz to 15 MHz is necessary. In sonochemical method, two kinds of materials, such as volatile and nonvolatile precursors have been used as starting material for the production of nanomaterials through different reaction mechanisms (Xu et al., 2013). If the volatile compounds are utilized as precursors in sonochemical method, it is called as primary sonochemistry and in case nonvolatile compounds are used, it is termed as secondary sonochemistry. In primary sonochemistry, free metal atoms are generated from volatile organometallic compounds via bond dissociation facilitated by high temperature produced during bubble collapse. The obtained atoms are injected into the liquid phase and subsequent nucleation yielded NPs of various materials in the presence of appropriate template. In the latter case, sonochemical reactions occur even outside of bubble, which undergoes reaction with radicals. The radicals diffuse into liquid, as a result reaction occurs and nanomaterials are generated. By employing this method various metal and metal oxide NPs have been prepared with wide variety of morphologies.

Okitsu et al. (2009) developed an efficient and novel one-pot sonochemical method for the production of Au nanorods via reduction of $HAuCl_4$ in the presence of $AgNO_3$, CTAB, and ascorbic acid. Aspect ratio of AuNPs decreased with increasing pH of the solution. Fig. 1.13 illustrates the TEM images of AuNPs formed at different pH values. It was concluded that morphology and aspect ratio of AuNPs are strongly dependent on pH. At higher pH values, spherical-shaped AuNPs with small particle size were formed, whereas, at low pH Au nanorods are exclusively obtained.

2.7 LIQUID–LIQUID INTERFACE METHOD

Liquid–liquid interface strategy is an excellent, simple, and straightforward method to produce nanocrystalline films of metals, metal chalcogenides, and metal oxides by the process of self-assembly at liquid–liquid interface. Self-assembly is a powerful strategy to create novel structures of nanomaterials with great academic interest. The nanomaterial in the form of film is considered important, due to its potential

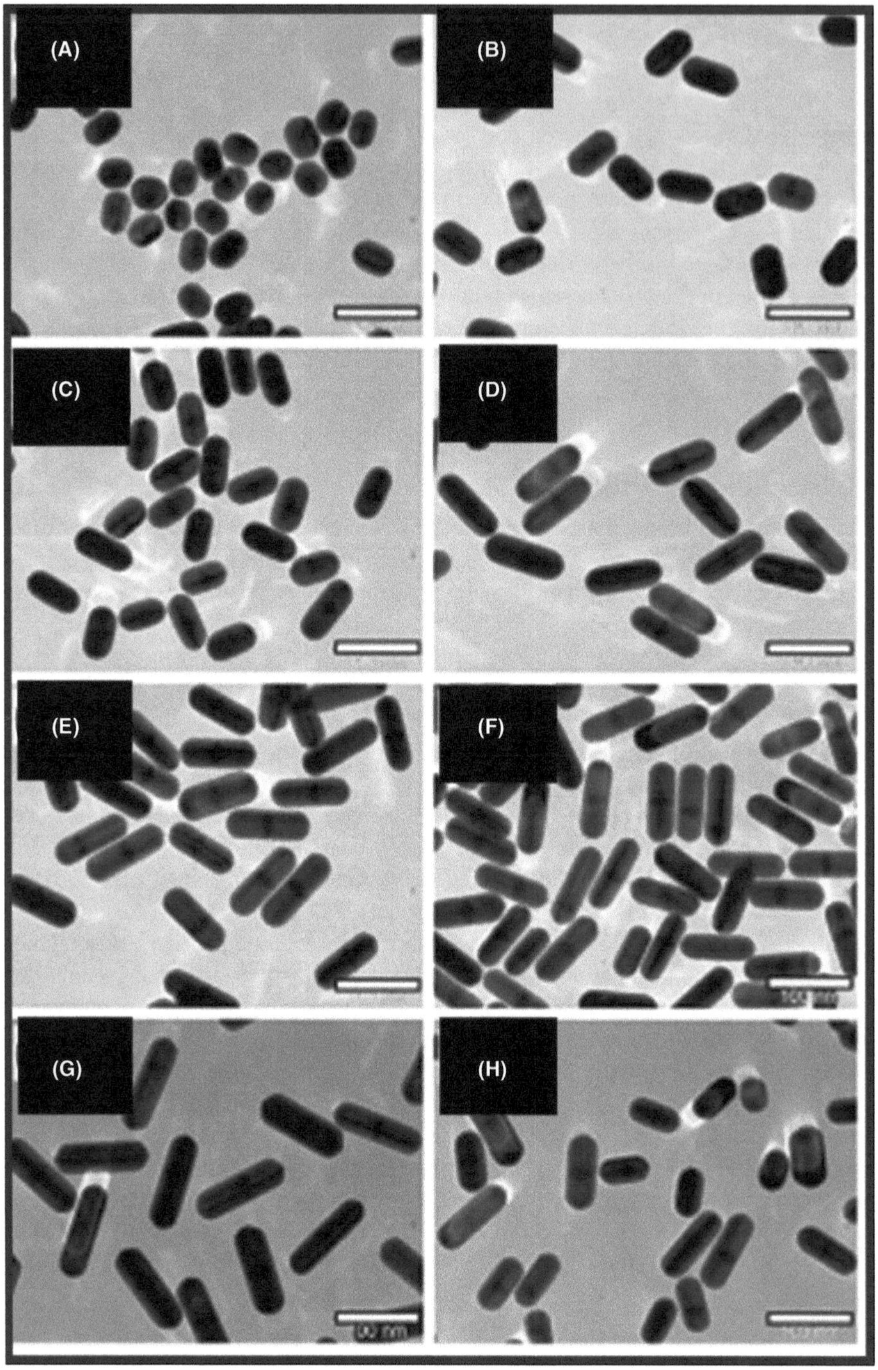

FIGURE 1.12 TEM Image of Au@Ag Heterogeneous Nanorods Obtained at 473 K for 60 h in an EG Solution Containing 0.25 mM $HAuCl_4$, 2.5 mM $AgNO_3$, and Different PDDA Concentrations

(A) 5, (B) 10, (C) 15, (D) 20, (E) 25, (F) 30, (G) 35, and (H) 40 mM. All scale bars are 100 nm.

Reprinted with permission from Li, C., Sun, L., Sun, Y., Teranishi, T., 2013. One-pot controllable synthesis of Au@Ag heterogeneous nanorods with highly tunable plasmonic absorption, Chem. Mater. 25, 2580–2590. Copyright 2013, American Chemical Society.

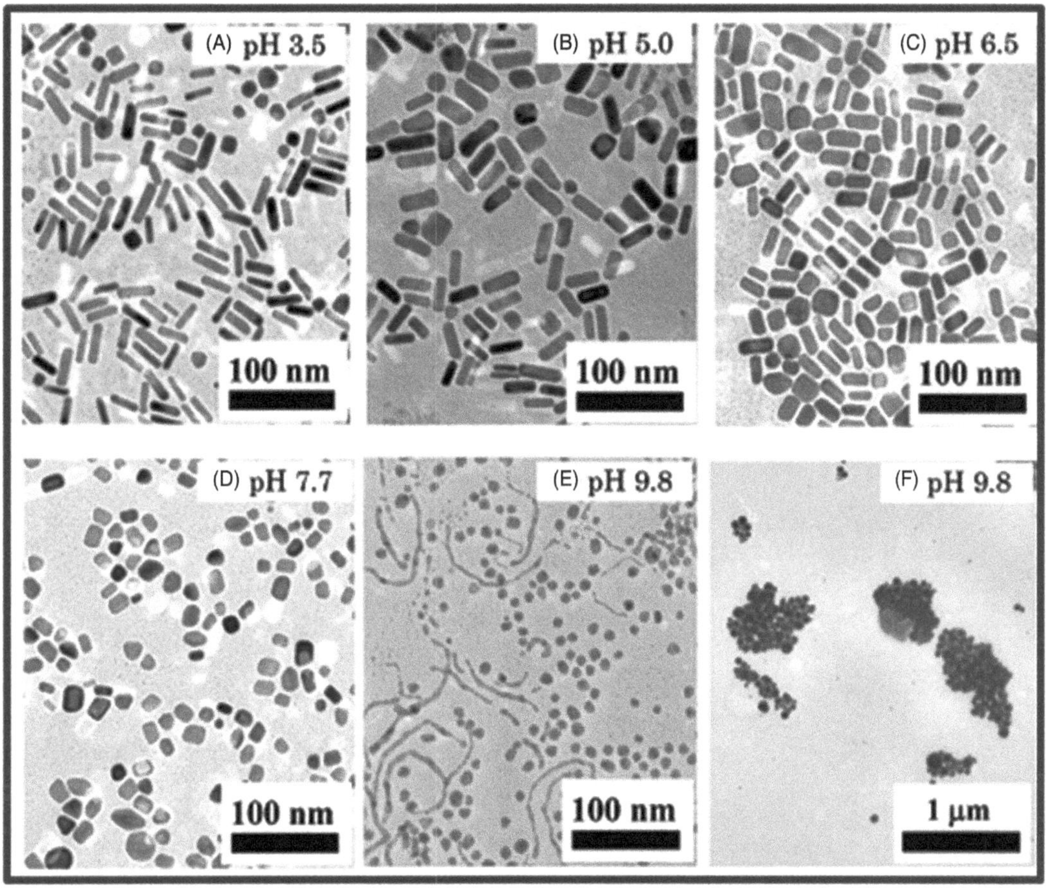

FIGURE 1.13

TEM images of Au nanorods and particles formed in different pH solutions of (A) pH 3.5, (B) pH 5.0, (C) pH 6.5, (D) pH 7.7, (E) pH 9.8 after 180 min irradiation under argon. (F) TEM images of AuNP formed at a pH 9.8 solution at 180 min standing without ultrasonic irradiation.

Reprinted with permission from Okitsu, K., Sharyo, K., Nishimura, R., 2009. One-pot synthesis of gold nanorods by ultrasonic irradiation: the effect of pH on the shape of the gold nanorods and nanoparticles. Langmuir 25, 7786–7790. Copyright 2009, American Chemical Society.

application in nanodevices (Rao et al., 2003). This method generally proceeds through a single step to form nanocrystalline films at the interface of two immiscible liquids, where one liquid is water and the other one is organic liquid. The interface is nonhomogeneous, usually has the thickness of a few nanometers, and offers an excellent path for the self-assembly and chemical manipulation of nanocrystals. In this method, metal precursors are taken in organic layer, which on reaction with appropriate reagent containing aqueous layer formed an ultrathin nanocrystalline film at the interface by self-assembly process. The obtained film was readily transferred on solid substrate by using an appropriate solvent. The film was broken into hydrosol or organosol for further characterization studies. The interfacial energy of the

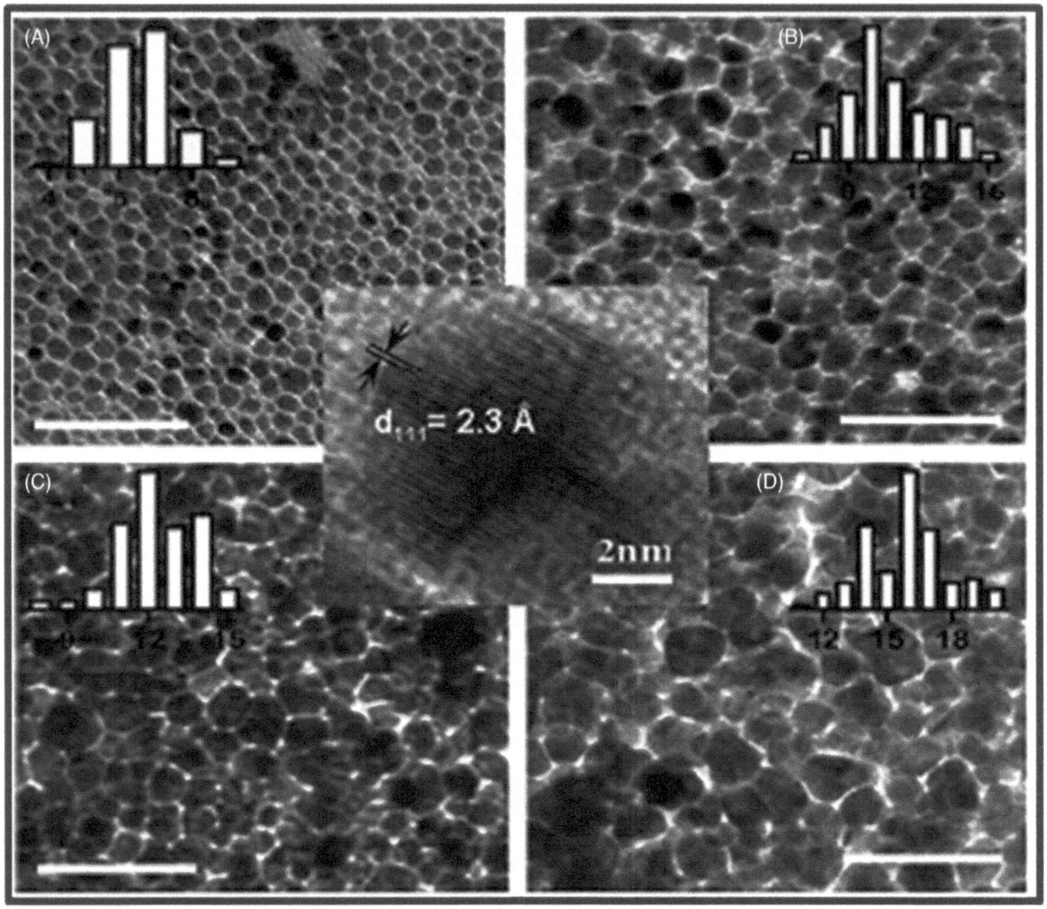

FIGURE 1.14

TEM image of ultrathin nanocrystalline Au films obtained at the liquid–liquid interface at (A) 303, (B) 318, (C) 333, and (D) 348 K. Histograms of particle size distribution are shown in insets. All scale bars are 50 nm. A high-resolution image of an individual particle is shown at the center.

Reprinted with permission from Agrawal, V.V., Kulkarni, G.U., Rao, C.N.R., 2005. Nature and properties of ultrathin nanocrystalline gold films formed at the organic-aqueous interface. J. Phys. Chem. B. 109, 7300–7305. Copyright 2005, American Chemical Society.

assembly process at liquid–liquid interface is influenced by the following factors: (1) nature of the interface, (2) alteration of surface of the NPs at interface, and (3) the effective radius of the NPs, smaller NPs are weakly attached to the interface than larger particles (Rao and Kalyanikutty, 2008). Rao et al. have proposed and developed a facile one-pot liquid–liquid interface method for the fabrication of a variety of noble metal NPs (Fig. 1.14) (Agrawal et al., 2005). The preparation of nanocrystalline ultrathin Au films was carried out by taking appropriate concentration of corresponding metal precursor Au $(PPh_3)Cl$ in toluene, later allowed to contact with aqueous solution containing NaOH at 300 K. After the formation of two layers, tetrakishydroxy-methylphosphonium chloride (THPC) which is acting as a reducing agent was injected; as a result yellowish colored transparent Au nanocrystalline film was formed at the liquid–liquid interface, which was thoroughly characterized by employing XRD, transmission electron microscopy, and so on. According to their study, various parameters significantly influenced the formation of nanocrystalline Au films. In this respect, temperature profoundly affected the particle size of Au film and the mean diameter of particles increased with respect to employed temperature. The effect of temperature on the size distribution of film is shown in Fig. 1.8. In addition, concentration of metal precursor played a crucial role in terms of quality, stability, and thickness of the film. Furthermore, the electronic properties of the resulted Au film could be affected by the reaction temperature. In addition to Au, it is possible to produce nanocrystalline films of other metals, such as Ag, Cu, and Pd at the interface by adopting liquid–liquid interface strategy (Rao et al., 2003).

The nanocrystalline films are obtained by taking respective precursors, such as $Ag_2(PPh_3)_4$, $Cu(PPh_3)Cl$, and palladium acetate in the organic layer and appropriate reducing agent in aqueous layer. Nanocrystalline films of binary and ternary alloys, such as Au–Ag, Au–Cu and Au–Ag–Cu have been generated at toluene–water interface by taking mixtures of corresponding metal precursors and reducing agents (Agrawal et al., 2006). The nature of the reducing agent showed noticeable effect on the morphology of the final nanocrystalline film.

2.8 PHASE-TRANSFER METHOD

Phase-transfer method is a widely used synthetic technique to produce metal NPs. Phase-transfer method represents a good approach to prepare size and shape-tunable metal NPs. This method involves the transfer of reactant from polar medium to nonpolar medium and vice versa (Rao et al., 2012). However, the transfer of metal NPs to organic medium is an excellent approach; because, yield of NPs in aqueous phase suffers from several drawbacks. Whereas, preparation of metal NPs directly in nonpolar organic medium was limited due to poor solubility of the corresponding metal precursors. Therefore, phase transfer of metal NPs in aqueous phase to nonpolar organic phase has few advantages, such as avoiding the use of organometallic precursor, lowering of interfacial energy, and use of relatively high concentrations thereby producing narrow size distribution of particles with controllable morphologies (Yang et al., 2011). Brust et al. (1994) were the first to report the alkanethiol-capped AuNPs by adopting phase-transfer method using tetraalkylammonium bromide as the phase-transfer agent. This method involves the phase transfer of chloroaurate ions into toluene in presence of a phase-transfer agent, followed by the subsequent reduction of gold ions by $NaBH_4$ yielding 1–3 nm sized capped AuNPs with homogeneous size distribution. A simple and efficient phase-transfer method was developed for the preparation of AOT-capped AgNPs, which are highly dispersible in both aqueous and organic solvents. Initially hydrophilic AOT-capped AgNPs are prepared by the reduction of corresponding silver precursor. The resulting aqueous AgNPs are readily transferred to organic phase (cyclohexane) by the aid of phosphoric acid. The concentration of surfactant played vast effect on the particle size of AOT-capped aqueous AgNPs (Prasad et al., 2005). Phase transfer of Pt, Au, and AgNPs to organic phase was achieved by simple phase-transfer technique. The size and morphology of metal NPs before and after phase transfer was thoroughly examined by TEM analysis. Some interesting findings were revealed from TEM analysis; those are, the particle size and shape of Pt didn't change after phase transfer, whereas Ag showed slight variations in the particle size after phase transfer. However, gold showed a variety of morphologies with different particle size (Feng et al., 2006). Yang et al. (2009) have demonstrated a protocol for phase

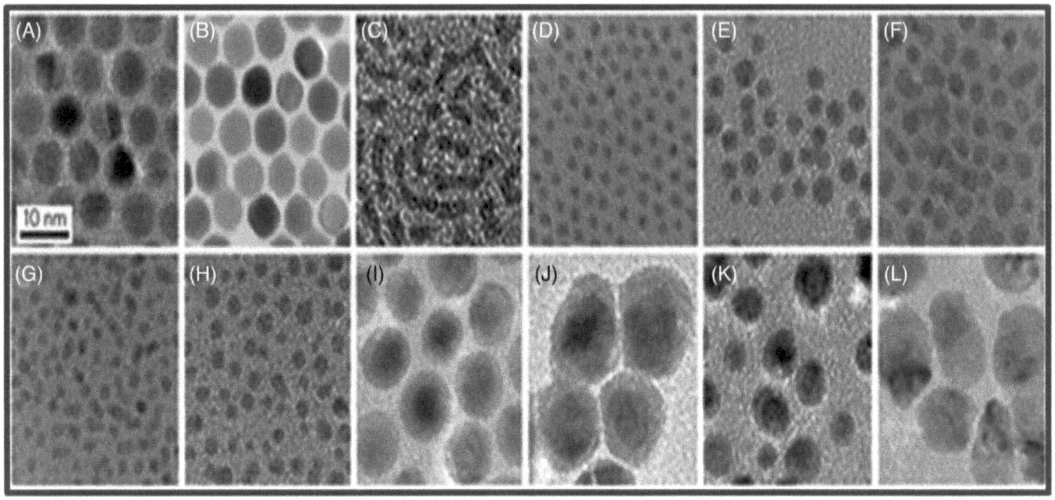

FIGURE 1.15

(A) AgNPs derived with HDD, (B) Au, (C) warm like Pd and, (D) Pt from Pt (IV), obtained with TBAB, (E) Ag–Au, (F) Pd–Pt, (G) Pt–Rh, (H) Pt–Ru alloy NPs synthesized by coreduction of the metal precursors with TBAB. (I) 7.4 nm Au@Ag, (J) 12.7 nm Au@Ag, (K) 3.9 nm Pt@Ag, and (L) 9.2 nm Pt@Ag core–shell NPs prepared by seed-mediated growth.

Reprinted with permission from Yang, J., Sargent, E., Kelley S., Ying, J.Y., 2009. A general phase-transfer protocol for metal ions and its application in nanocrystal synthesis. Nat. Mater. 2009, 8, 683. Copyright 2009, Nature Publishing Group.

transfer of various metal ions to organic medium for the production of NPs (Fig. 1.15). The presence of ethanolic dodecylamine (DDA) solution can facilitate the transfer of metal ions from aqueous phase to organic medium.

2.9 BIOSYNTHESIS METHOD

Till date numerous methods including physical and chemical methods have been developed for metal NP synthesis in bulk quantities. However, these methods, usually, often use toxic chemicals, drastic conditions, and additives. Therefore, development of reliable, biocompatible, and green method for the production of metal NPs has gained more attention from the scientific community (Jain et al., 2011).

In this context, biosynthesis of metal NPs using plant extract or plant tissue or living plant has opened new opportunity in the production of NPs in large-scale quantities. Biosynthesis approach is an environmental friendly, cost effective, and fast method, which avoids the use of harsh conditions, toxic reagents, and expensive chemicals. Few challenges have been encountered in the biosynthesis method for production of NPs that are of the size and shape of the NPs. Njagi et al. (2011) have demonstrated a simple, one-step biosynthetic method for the preparation of Ag and FeNPs using aqueous sorghum bran extract as the reducing and capping agent at room temperature. The corresponding metal precursors are reduced to respective metal ions by the assistance of aqueous sorghum bran extract producing 10-nm AgNP with face-centered cubic structure and 50-nm diameter FeNPs. The phenolic compounds presented in sorghum bran extract can be used as reducing as well as capping agents during the formation

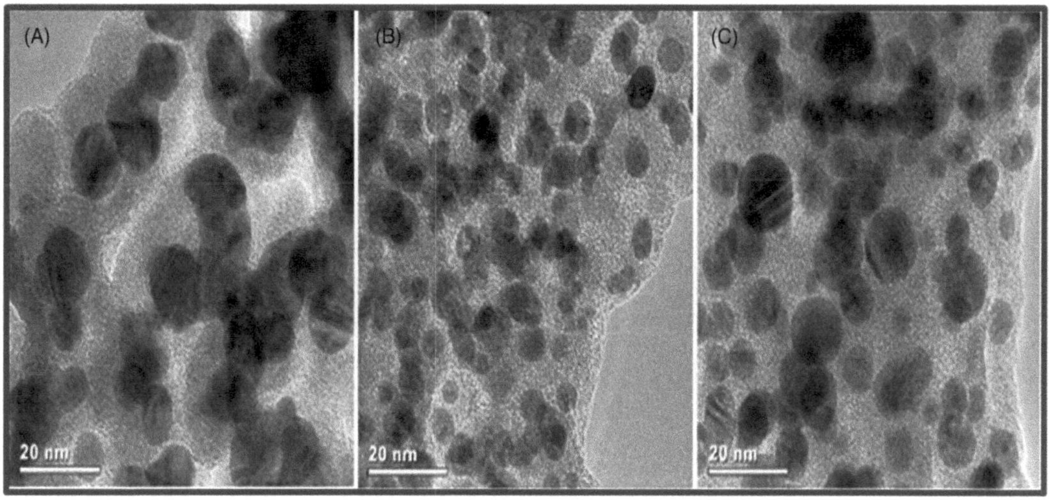

FIGURE 1.16

TEM images of AgNPs prepared by sorghum bran extract at (A) 25, (B) 50, and (C) 80°C.

Reprinted with permission from Njagi, E.C., Huang, H., Stafford, L., Genuino, H., Galindo, H.M., Collins, J.B., Hoag, G.E., Steven, L.S., 2011. Biosynthesis of iron and silver nanoparticles at room temperature using aqueous sorghum bran extracts. Langmuir 27, 264–271. Copyright 2011, American Chemical Society.

of NPs. The phenolic compounds exhibit several advantages, such as water solubility, nontoxicity, and biodegradability and offer green synthetic route for the preparation of metal NPs. Fig. 1.16 shows the TEM images of AgNPs synthesized at different temperatures. TEM analysis revealed the formation of uniform size spherical AgNPs. As prepared FeNPs are examined for bromothymol degradation. Shiv Shankar et al. (2003) have reported different shapes of AuNPs by exposing the geranium leaves and its endophytic fungus colletotrichum sp. to chloroaurate. The rapid reduction of respective metal ions had taken place by terpenoids present in the geranium leaves; whereas, polypeptides or enzymes in colletotrichum sp. were responsible for the reduction resulting in the formation of AuNPs. The TEM results of AuNPs obtained by both the sources provided excellent features. The AuNPs synthesized by using fungus were spherical in shape whereas in case of geranium leaves different morphologies including rods, flat sheets, and triangles were obtained. The crucial reaction parameters, such as temperature, pH, and reaction time profoundly influenced the size and shape of the final nanomaterial. It is obvious from the literature that, pH may play a crucial role in controlling the size and morphology of NPs. Castro et al. (2011) reported that the morphology of AuNPs obtained by sugar beet pulp was highly dependent on the pH. The size and shape of the NPs with respect to pH was examined by UV-Vis spectroscopy and TEM study. At, pH 9 smaller NPs with 10-nm diameter were obtained and with gradual increase of pH to 10, rod-like shape was achieved. Finally, nanowire shape was attained by adjusting pH to 11. Sathishkumar et al. (2009) reported the production of PdNPs by reduction of palladium chloride using *Cinnamomum zeylanicum* bark extract as biomaterial. They found that, temperature shows a notable effect on the particle size but not on the morphology. Li et al. (2007) studied the effect of reaction time on the preparation of AgNPs by treating silver ions with *Capsicum annuum* L. extract. A spherical-shaped

NP with small size was obtained at 5 h, with gradual increase of reaction time to 9 and 11 h, particle size increased and showed crystalline and polycrystalline morphology, respectively for AgNPs. Therefore, it is believed that, reaction time played a significant effect on the shape and size of the particles.

2.10 TEMPLATE-DIRECTED SYNTHETIC METHOD

The template-directed method is practically a straightforward method for the production of high-quality novel metal and metal oxide NPs through the mediation of appropriate template. The template is generally used to manipulate the formation and growth of the NPs with desired morphology. Importantly pore size, grain morphology, phase composition, and size of the nanomaterial can be improved by employing suitable template material. In this method, the solid product is formed over the sacrificial template via the liquid–solid interfacial reaction (Bao et al., 2014). The main function of the template is to provide a desired morphology, which is similar to that of template morphology to the final nanomaterial during the synthetic procedure. Additionally, template-directed method offers several benefits, such as lower reaction temperature, shorter reaction time, less requirement of toxic precursors and solvents resulting in nonformation of toxic-side products, and the use of external energy source can be avoided (Patete et al., 2011). Production of preferred nanostructures within the template and the complete removal of template material by suitable routes, such as chemical etching and calcination are the crucial steps in the aforementioned method. Generally, template methodology can be categorized as hard and soft template approaches based on their structure. Hard templates usually are porous silica, porous carbon, porous anodic alumina, polystyrene beads, block polymers, carbon NPs, metals, metal oxides, and mica. On the other hand, emulsion droplets, surfactant micelles, polymer vesicles, bubbles, and amphiphilic surfactants are often adopted as soft templates. However, in most of the cases, template method is associated with a major drawback that the template employed in the process must be completely removed at the end of the synthesis, which rigorously affects the purity of the resulted nanomaterial and makes the process very complicated.

ZnO hollow spheres were fabricated by a novel and simple method with the use of sulfonated polystyrene core–shell spheres as template (PS) (Deng et al., 2008). In a typical approach, requisite quantity of $Zn(Ac)_2.2H_2O$ was mixed with ethanol solution containing sulfonated polystyrene core–shell (PS) template; as a result, Zn ions adhered on the surface of template spheres through the electrostatic interaction. Once NaOH was added to the above mixture, the Zn ions readily reacted with NaOH to form ZnO crystal nuclei, which was followed by the formation of ZnO nanoshells. Afterward, the template spheres were dissolved in the same solution to produce ZnO hollow spheres. A systematic analysis for the formation of ZnO hollow spheres was undertaken by varying the reaction parameters, such as concentration of NaOH, concentration of $Zn(Ac)_2.2H_2O$, and size of templates employed. TEM images of ZnO hollow spheres prepared by different NaOH concentrations showed that lower NaOH concentration either deformed or broke the ZnO hollow spheres (Fig. 1.17). However, NaOH here played a dual role: assisted the formation of ZnO hollow spheres as a reactant and helped to dissolve the template spheres. On the other hand, the concentration of $Zn(Ac)_2.2H_2O$ showed a significant effect on the wall thickness of the ZnO hollow spheres. At lower concentrations, very thin hollow spheres were obtained, whereas, increase of the concentration enhanced the wall thickness tremendously; but some agglomerated ZnONPs were also observed along with the spheres. Therefore, it was proved that the variation in the concentration of $Zn(Ac)_2.2H_2O$ could facilitate the changes in the wall thickness of ZnO hollow spheres obtained in the course of reaction. The synthesized ZnO hollow spheres were examined for

FIGURE 1.17 TEM Images of ZnO Hollow Spheres Prepared by Various NaOH Concentrations

(A) 0.025 mol/L, (B) 0.1 mol/L, (C) 0.3 mol/L, (D) 0.4 mol/L, and $Zn(Ac)_2.2H_2O$: 0.036 mol/L.

Reprinted with permission from Deng, Z., Chen, M., Gu, G., Wu, L., 2008. A facile method to fabricate ZnO hollow spheres and their photocatalytic property. J. Phys. Chem. B 112,16–22. Copyright 2008, American Chemical Society.

photocatalytic activity and found to show excellent catalytic performance. Interestingly, the size of the template employed had significant control over the size of the ZnO hollow sphere.

Gundiah et al. (2003) have prepared a variety of oxide nanotubes, such as TiO_2, ZrO_2, WO_3, and ZnO by employing tripodal cholamide-based hydrogel as template. For example, nanotubes are synthesized by taking tetrabutyl orthotitanate (TBOT) as titanium source and the synthetic method involves hydrolysis of the titanium precursor in the presence of template followed by deposition of TiO_2 on the surface of the template and removal of the template material. The TEM analysis revealed the formation of TiO_2 nanotubes with 4–7 nm and 10–20 nm inner diameter and outer diameter, respectively. Similarly, ZrO_2, WO_3, and ZnO nanotubes with desired diameter can be easily synthesized by adopting the aforementioned procedure in the presence of tripodal cholamide-based hydrogel template. Irrespective of metal oxide nanotubes, all are produced with excellent yields in the course of the reaction. Apart from the metal oxide nanotubes, a variety of metal oxide nanorods, such as V_2O_5,WO_3, MoO_3, RuO_2, IrO_2, and Sb_2O_5 were prepared by the same group employing carbon nanotube as template agent. Interestingly, the obtained metal oxide nanotubes are longer than the carbon nanotube template (Satishkumar et al., 2000).

Bao et al. (2014) have developed an efficient template-assisted method for synthesis of CeO_2@Cu_2O nanocomposites, employing Cu_2O cubes and octahedra as the sacrificial template materials. The morphologies of the CeO_2@Cu_2O nanocomposites were examined by transmission electron microscopy (TEM) and scanning electron microscopy (SEM). Fig. 1.18 shows the TEM and SEM images of CeO_2@Cu_2O nanocomposites.

FIGURE 1.18

SEM and TEM images of (A–B) CeO_2@c-Cu_2O-0.045, (C–D) CeO_2@c-Cu_2O-0.09, (E–F) CeO_2@c-Cu_2O-0.18, and (G–H) CeO_2@c-Cu_2O-0.36.

Reprinted with permission from Bao, H., Zhang, Z., Hua, Q., Huang, W., 2014. Compositions, structures, and catalytic activities of CeO_2@Cu_2O nanocomposites prepared by the template-assisted method. Langmuir 30, 6427–6436. Copyright 2014, American Chemical Society.

SEM images demonstrated that $CeO_2@Cu_2O$ nanocomposites with various compositions exhibited cubic nanostructure, whereas, TEM analysis demonstrated the cubic core/shell nanostructure for $CeO_2@Cu_2O$ nanocomposites. The cubic core had shrunk with respect to the increase in Ce content, thereby enhancing the space between the shell and the core, which is clearly shown in the Fig. 1.18. The resulting $CeO_2@Cu_2O$ nanocomposites were examined for CO oxidation and the activity is highly dependent on their composition and structure.

Formation of triangular Ag nanoplates was achieved by the reduction of Ag ions with ascorbic acid in the presence of a soft template (CTAB) (Chen and Carroll, 2002). The resulting particles had an average edge size of 8 nm, thickness of 24 nm, and the degree of truncation of 0.35. Particles with different morphologies like rods, discs, spheres, cubes, tetrahedral, and squares were detected. Liu et al. (2005) have prepared the hierarchical iron oxide with controllable porous structure with the assistance of wood template. Hierarchical iron oxide nanostructures with a pore size of 20 nm to 50 μm were prepared by taking inorganic precursor solution in wood template. The structural property of hierarchical iron oxide was investigated with the help of various characterization techniques, such as SEM, FESEM, EDAX, XRD, and TEM. Moreover, the pore-size distribution of wood-templated hierarchical iron oxide was controlled by varying the calcination temperature as well as nature of template. Metal oxide nanostructures were fabricated with the aid of porous carbonaceous coating and multiwalled carbon nanotube core as dual template. Variety of metal oxide nanostructures were fabricated with the utility of porous carbonaceous coating and multiwalled carbon nanotube core as dual templates (Zhang et al., 2010a). For example, the preparation of SnO_2, ZrO_2, Fe_2O_3, and CeO_2 hollow nanostructures involves multiple key steps. The initial step comprises the coating of carbonaceous layer over the surface of the nanotube, to form a dual template during the thermal treatment. Then the metal cations were allowed to be adsorbed on the dual template followed by removal of the carbonaceous layer by calcination, at relatively lower temperature, resulting in the formation of MWCNTs@MO composite. The prepared composite was subsequently subjected to calcination at 550°C to remove the MWCNTs, resulting in the production of hollow structured metal oxides.

3 APPLICATION OF METAL NANOPARTICLES IN THERANOSTICS

Theranostics is a new and innovative process in the field of cancer therapy. It combines multiple steps of the therapeutic system, such as therapy, diagnosis, and monitoring of therapy by imaging (Fig. 1.19). Theranostics provides a means to follow up the success rate of the therapy in individual patients by imaging the tumors throughout the entire period of treatment. The metal NPs exhibiting physical properties, such as, plasmonic resonance and fluorescent enhancement and chemical property of catalytic activity enhancement provide the suitable means of targeting and drug delivery simultaneously.

3.1 DIAGNOSIS AND DRUG DELIVERY

Nanoscale structures can exhibit widely different properties to bulk materials or small molecules, which renders them applicable in the fields of medical imaging. AuNPs, such as rods or shells exhibit strikingly strong fluorescence with a quantum yield of up to 10^{-3} (Payne et al., 2006). Such fluorescence can be readily applied in the biomedical field, especially as it can be tuned to the NIR window (650–900 nm) by changing nanostructure size and morphology. Tissue absorbs light weakly in the

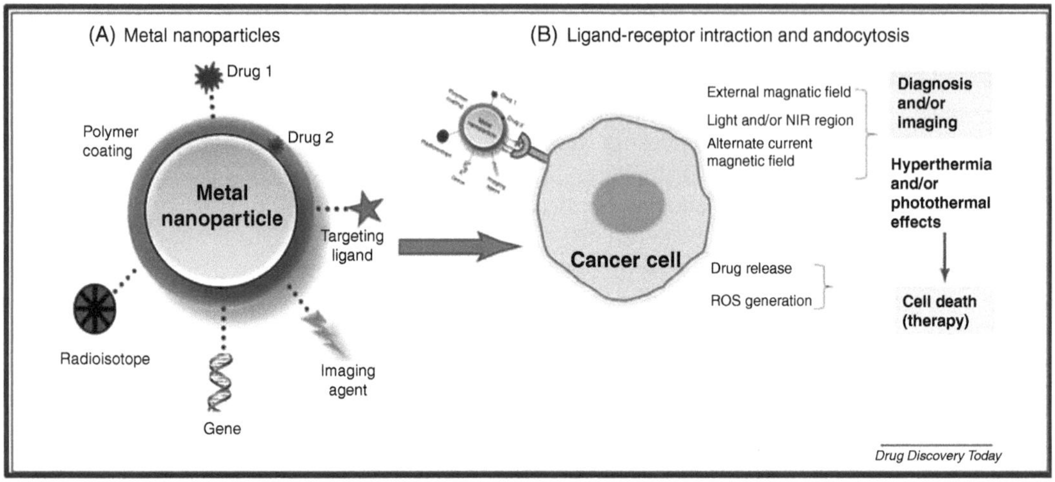

FIGURE 1.19 Designing Multifunctional Metal NPs for Cancer Therapy

(A) Conjugation of drugs molecules over the metal NP surface of particles using chemical functional groups present on the surface of particle or on the coated polymeric matrix. (B) Interaction of Particles with cancer cells via ligand–receptor interactions, and endocytosis.

Reprinted with permission from Sharma, H., Mishra, P.K., Talegaonkar, S., Vaidya, B., 2015. Metal nanoparticles: a theranostic nanotool against cancer. Drug Discov. Today 20, 1143–1151. Copyright 2015, Science Direct.

NIR window, making this window ideal for optically based applications (Zhang et al., 2010b). AuNPs absorb X-rays strongly and can thus be used as contrast agents for X-ray based imaging techniques and as adjuvants for radiotherapy (Galper et al., 2012).

Qian et al. (2015) have demonstrated the use of TiS_2–PEG (titanium sulfide–polyethylene glycol) nanosheet as a photothermal agent for in vivo cancer treatment. Mice bearing 4T1 tumors were injected with a TiS_2–PEG solution (2 mg/mL, 200 μL) or saline as the control was exposed to an 808-nm laser at a power density of 0.8 W/cm^2 for 5 min at 24 h postinjection. Owing to the strong NIR absorbance and efficient tumor accumulation of TiS_2–PEG, the tumor temperatures of mice injected with TiS_2–PEG quickly increased to ~65°C within 5 min under NIR laser irradiation (Fig. 1.20). In contrast, the surface temperature of tumors in mice injected with saline showed little change under the same irradiation conditions (Qian et al., 2015).

Another highly useful feature is the electromagnetic field enhancement by sharp and spiky edges of the nanostructures, such as stars or nanorods, which can be used in surface-enhanced Raman spectroscopy imaging (von Maltzahn et al., 2009). Fleischmann et al. (1974) showed that Raman signals from different molecular sources can be enhanced by their adsorption on silver metal surface. Later, the same was found with other metals as well. Since then, metal NPs mostly gold and silver were employed in surface-enhanced Raman scattering (SERS), leading to the development of a wide variety of new biosensors for detection of biomolecules (Rycenga et al., 2009). Afterward, conjugated noble metal NPs were also introduced in SERS-based biosensors.

Metal and metal oxide NPs are being actively investigated as carriers for targeted drug delivery in cancer therapeutics (Dobson, 2006). They are designed to contain tumor-targeting ligands that bind to

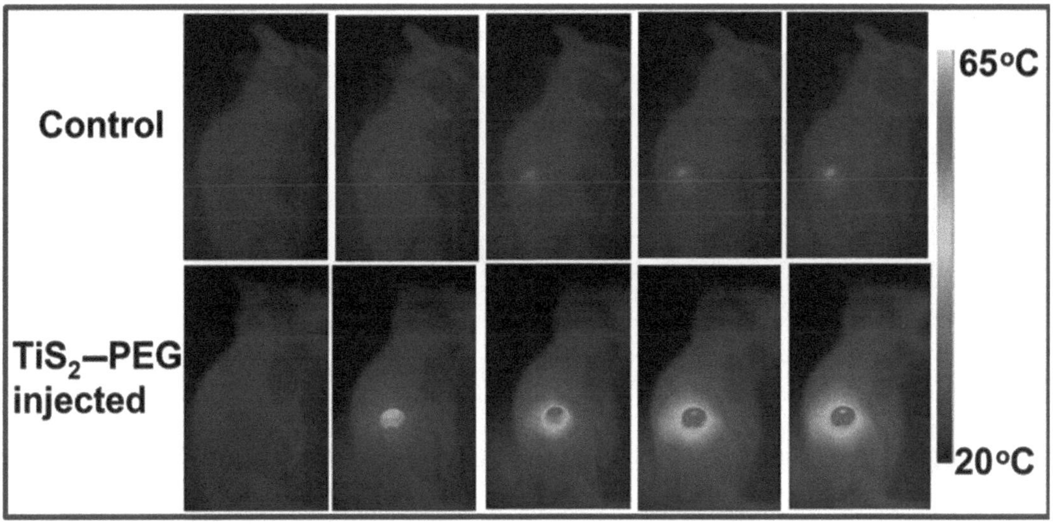

FIGURE 1.20 Infrared Thermal Images of 4T1 Tumor-Bearing Mice With i.v. Injection of a TiS_2–PEG Solution (2 mg/mL, 200 μL) Under Laser Irradiation at a Power Density of 0.8 W/cm^2 for 5 min (24 h Postinjection)

Reprinted with permission from Qian, X., Shen, S., Liu, T., Cheng, L., Liu, Z., 2015. Two-dimensional TiS_2 nanosheets for in vivo photoacoustic imaging and photothermal cancer therapy. Nanoscale 7, 6380–6387. Copyright 2015, Royal Society of Chemistry.

particular cells within the tumor to fasten the NP within solid tumor. In this way, metallic NP drug-delivery systems are capable of sequestering anticancer drugs exclusively within the tumor and thereby reduce the accumulation of drugs in healthy organs. Their large surface area-to-volume ratio provides a surface for chemical modification, which can improve cell entry, protect the therapeutic agent in the biological milieu, and improve bioavailability of the anticancer agent (Moghimi and Patel, 1998). For example, Zhang et al. have reported multifunctional mesoporous silica NPs for cancer-targeted and controlled drug delivery, which have three components—the mesoporous silica NP core, the amino-b-cyclodextrin, and the PEG polymers functionalized with an adamantane (Ad) unit at one end and a folate (FA) unit at the other end (Ad–PEG–FA). The surface of mesoporous silica NPs is first functionalized with amino-b-cyclodextrin rings bridged by cleavable disulfide bonds, blocking drugs inside the mesopores of the NPs.

The Ad–PEG–FA polymers are immobilized onto the NP surface through strong b-cyclodextrin/adamantane complexation. The multifunctional NPs can be efficiently trapped by folate-receptor-rich cancer cells through receptor-mediated endocytosis, where they then rapidly release the loaded anticancer drug (Zhang et al., 2012). AuNP targeting can be achieved in a passive fashion, where long circulating AuNPs can accumulate in cancerous tissues by penetrating the leaky tumor vasculature. This is known as the enhanced permeability and retention (EPR) effect (Maeda et al., 2000). Alternatively, AuNPs can be targeted to specific cell types, receptors, or proteins via attachment of targeting ligands, such as antibodies, proteins, peptides, aptamers, and small molecules (Fig. 1.21) (Cormode et al., 2008). Thus far, Doxil and Abraxane are the most successful nanotechnology-based treatments in the world market. Both nanocarriers showed reduction of the toxicity allowing for elevated doses (Nichols and Bae, 2012).

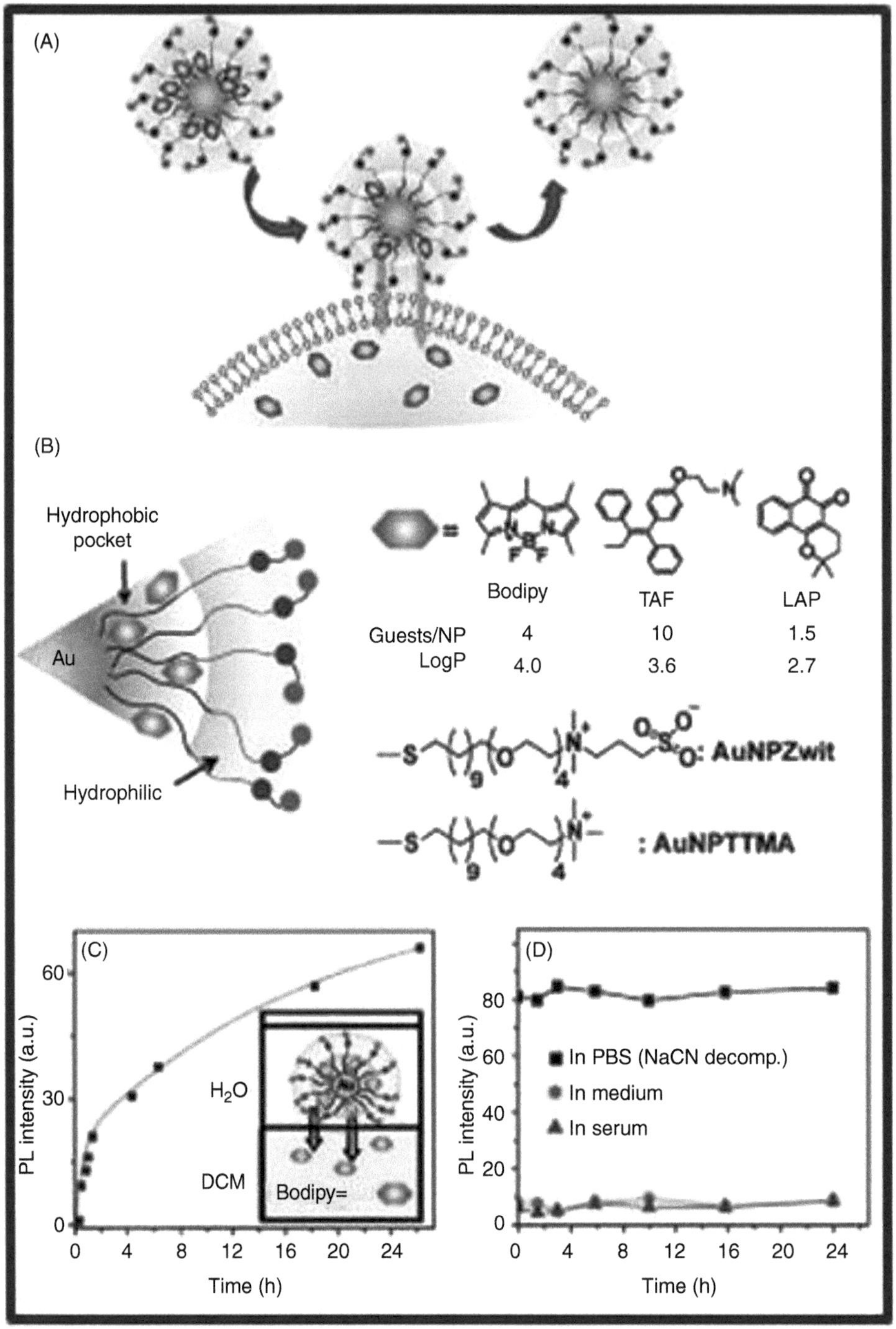

FIGURE 1.21

(A) Delivery of payload to cell through monolayer–membrane interactions. (B) Structure of particles and guest compounds: Bodipy, TAF, and LAP, the number of encapsulated guests per particle, and log P of the guests. (C) Release of Bodipy from AuNPZwit-Bodipy in DCM-aqueous solution two-phase systems ($\lambda_{ex} = 499$ nm; $\lambda_{em} = 517$ nm). (D) PL intensity of AuNPZwit-Bodipy in cell culture medium and 100% serum, indicating little or no release relative to AuNPZwit-Bodipy in PBS after NaCN-induced release of guest molecules ($\lambda_{ex} = 499$ nm; $\lambda_{em} = 510$ nm).

Reprinted with permission from Kim, C.K., Ghosh, P., Pagliuca, C., Zhu, Z.-J., Menichetti, S., Rotello, V.M., 2009. Entrapment of hydrophobic drugs in nanoparticle monolayers with efficient release into cancer cells. J. Am. Chem. Soc. 131, 1360–1361. Copyright 2009, American Chemical Society.

4 CONCLUSIONS

The past couple of decades have witnessed an exponential growth of activities in the field of nanotechnology worldwide, which came up with the hope of diverse possible applications in pharmacy and biotechnology. A wide range of synthesis methods have been studied so far to fabricate various metal and metal oxide NPs. Mainly, wet chemical approaches offer potential reaction pathways to obtain better materials in terms of chemical homogeneity and size-tunable properties. Nevertheless, it is difficult to separately control the crystallization, surface structure, and agglomeration of particles. Biosynthetic techniques are also being explored. This method is superior in terms of toxicity and deduction of use of hazardous chemicals in synthesis; although challenges regarding size and shape controllability are yet to be achieved.

However, understanding of metal NP toxicity is also necessary with the rapid growth of this field. The normal mode of action of metal NPs is the production of ROS. Excessive oxidative stress is harmful and can lead to atherosclerosis, pulmonary diseases, and further cancers. The toxicity of metal NPs can be subdued by targeting the specific target site. Or else, alternative approaches like application of biodegradable polymers as reducing and functionalizing agents, can also suppress the toxicity of metal NPs. Hence, before their clinical trial several factors have to be settled, which include preparation method, reproducibility and stability of the metal NPs, etc. More importantly, the safe level of application, extent of accumulation of metal NPs in the patient body, and their toxicological effects in the long run are to be considered very seriously. However, once these problems are sorted out, metal NPs can become a potential clinical weapon to fight against serious illness.

ACKNOWLEDGMENTS

We wish to specially acknowledge all the researchers whose work is described in this chapter for their valuable contributions. B.G and D.M thank the Council of Scientific and Industrial Research (CSIR), New Delhi for the research fellowships. Financial support for this project was received from the Department of Science and Technology, New Delhi, under SERB Scheme (SB/S1/PC-106/2012).

REFERENCES

Aden, A.L., Kerker, M., 1951. Scattering of electromagnetic waves from two concentric spheres. J. Appl. Phys. 22, 1242–1246.

Agrawal, V.V., Kulkarni, G.U., Rao, C.N.R., 2005. Nature and properties of ultrathin nanocrystalline gold films formed at the organic-aqueous interface. J. Phys. Chem. B 109, 7300–7305.

Agrawal, V.V., Mahalakshmi, P., Kulkarni, G.U., Rao, C.N.R., 2006. Nanocrystalline films of Au-Ag, Au-Cu, and Au-Ag-Cu alloys formed at the organic-aqueous interface. Langmuir 22, 1846–1851.

Alexander, J.W., 2009. History of the medical use of silver. Surg. Infect. 10, 289–292.

Ba, J., Polleux, J., Antonietti, M., Niederberger, M., 2005. Nonaqueous synthesis of tin oxide nanocrystals and their assembly into ordered porous mesostructures. Adv. Mater. 17, 2509–2512.

Bandyopadhyay, S., Peralta-Videa, J.R., PlascenciaVilla, G., Jose-Yacaman, M., Gardea-Torresdey, J.L., 2012. Comparative toxicity assessment of CeO_2 and ZnO nanoparticles towards *Sinorhizobium meliloti*, a symbiotic alfalfa associated bacterium: use of advanced microscopic and spectroscopic techniques. J. Hazard. Mater. 241–242, 379–386.

Bang, J.H., Suslick, K.S., 2010. Applications of ultrasound to the synthesis of nanostructured materials. Adv. Mater. 22, 1039.
Bao, Y., An, W., Turner, C.W., Krishnan, K.M., 2010. The critical role of surfactants in the growth of cobalt nanoparticles. Langmuir 26, 478–483.
Bao, H., Zhang, Z., Hua, Q., Huang, W., 2014. Compositions, structures, and catalytic activities of CeO_2@Cu_2O nanocomposites prepared by the template-assisted method. Langmuir 30, 6427–6436.
Borkow, G., Gabbay, J., 2009. Copper an ancient remedy returning to fight microbial, fungal and viral infections. Curr. Chem. Biol. 3, 272–278.
Bradley, D.C., Mehrotra, R.C., Rothwell, I.P., Singh, A., 2001. Alkoxo and Aryloxo Derivatives of Metals. Academic Press, London.
Brust, M., Walker, M., Bethell, D., Schiffrin, D.J., Whyman, R., 1994. Synthesis of thiol-derivatised gold nanoparticles in a two-phase liquid-liquid system. J. Chem. Soc. Chem. Commun. 7, 801.
Carroll, K.J., Reveles, J.U., Shultz, M.D., Khanna, S.N., Carpenter, E.E., 2011. Preparation of elemental Cu and Ni nanoparticles by the polyol method: an experimental and theoretical approach. J. Phys. Chem. C 115, 2656–2664.
Castro, L., Blazquez, Munoz, M.L., Gonzalez, J.A.F., Garcia-Balboa, C., Ballester, A., 2011. Biosynthesis of gold nanowires using sugar beet pulp. Process Biochem. 46, 1076–1082.
Chen, S., Carroll, D.L., 2002. Synthesis and characterization of truncated triangular silver nanoplates. Nano Lett. 2, 1003–1007.
Chen, Y., Peng, D.L., Lin, D., Luo, X., 2007. Preparation and magnetic properties of nickel nanoparticles via the thermal decomposition of nickel organometallic precursor in alkylamines. Nanotechnology 18, 505703.
Conde, J., Doria, G., Baptista, P., 2011. Noble metal nanoparticles applications in cancer. J. Drug Del. 2012, 1–12.
Cormode, D.P., Skajaa, T., van Schooneveld, M.M., Koole, R., Jarzyna, P., Lobatto, M.E., Calcagno, C., Barazza, A., Gordon, R.E., Zanzonico, P., Fisher, E.A., Fayad, Z.A., Mulder, W.J., 2008. Nanocrystal core high-density lipoproteins: a multimodality contrast agent platform. Nano Lett. 8, 3715–3723.
Corriu, R.J.P., Leclercq, D., 1996. Recent developments of molecular chemistry for sol-gel processes. Angew. Chem. Int. Ed. 35, 1420–1436.
Cushing, B.L., Kolesnichenko, V.L., O'Connor, C.J., 2004. Recent advances in the liquid-phase syntheses of inorganic nanoparticles. Chem. Rev. 104, 3893–3946.
Deng, Z., Chen, M., Gu, G., Wu, L., 2008. A facile method to fabricate ZnO hollow spheres and their photocatalytic property. J. Phys. Chem. B 112, 16–22.
Diab, M., Moshofsky, B., Plante, I.J.L., Mokari, T., 2011. A facile one-step approach for the synthesis and assembly of copper and copper-oxide nanocrystals. J. Mater. Chem. 21, 11626–11630.
Dickson, R.M., Lyon, L.A., 2000. Unidirectional plasmon propagation in metallic nanowires. J. Phys. Chem. B 104, 6095–6098.
Dobson, J., 2006. Magnetic nanoparticles for drug delivery. Drug Dev. Res. 67, 55–60.
Edmundson, M., Thanh, N.T., Song, B., 2003. Nanoparticles based stem cell tracking in regenerative medicine. Theranostics 3, 573–582.
El Badawy, A.M., Silva, R.G., Morris, B., Scheckel, K.G., Suidan, M.T., Tolaymat, T.M., 2011. Surface charge-dependent toxicity of silver nanoparticles. Environ. Sci. Technol. 1, 283–287.
Feng, X., Ma, H., Huang, S., Pan, W., Zhang, X., Tian, F., Gao, C., Cheng, Y., Luo, J., 2006. Aqueous-organic phase-transfer of highly stable gold, silver, and platinum nanoparticles and new route for fabrication of gold nanofilms at the oil/water interface and on solid supports. J. Phys. Chem. B 110, 12311–12317.
Fleischmann, M., Hendra, P.J., McQuillan, A.J., 1974. Raman spectra of pyridine adsorbed at a silver electrode. Chem. Phys. Lett. 26, 163–166.
Fresta, M., Wehrli, E., Puglisi, G., 1995. Enhanced therapeutic effect of cytidine-5¢-diphosphate choline when associated with G (M1) containing small liposomes as demonstrated in a rat ischemia model. Pharm. Res. 12, 1769–1774.

Galper, M.W., Saung, M.T., Fuster, V., Roessl, E., Thran, A., Proksa, R., Fayad, Z.A., Cormode, D.P., 2012. Effect of computed tomography scanning parameters on gold nanoparticle and iodine contrast. Invest. Radiol. 47, 475–481.

Gerbec, J.A., Magana, D., Washington, A., Strouse, G.F., 2005. Microwave enhanced reaction rates for nanoparticle synthesis. J. Am. Chem. Soc. 127, 15791–15800.

Grisaru, H., Palchik, O., Gedanken, A., Palchik, V., Slifkin, M.A., Weiss, A.M., 2003. Microwave-assisted polyol synthesis of $CuInTe_2$ and $CuInSe_2$ nanoparticles. Inorg. Chem. 42, 7148–7155.

Gunawan, C., Teoh, W.Y., Marquis, C.P., Amal, R., 2011. Cytotoxic origin of copper (II) oxide nanoparticles: comparative studies with micron-sized particles, leachate, and metal salts. ACS Nano 5, 7214–7225.

Gundiah, G., Mukhopadhyay, S., Tumkurkar, U.G., Govindaraj, A., Maitra, U., Rao, C.N.R., 2003. Hydrogel route to nanotubes of metal oxides and sulfates. J. Mater. Chem. 13, 2118–2122.

Hakuta, Y., Ura, H., Hayashi, H., Arai, K., 2005. Continuous production of $BaTiO_3$ nanoparticles by hydrothermal synthesis. Ind. Eng. Chem. Res. 44, 840–846.

Hu, X., Gong, J., Zhang, L.Z., Yu, J.C., 2008. Continuous size tuning of monodisperse ZnO colloidal nanocrystal clusters by a microwave polyol process and their application for humidity sensing. Adv. Mater. 20, 4845–4850.

Hu, X.L., Yu, J.C., 2008. Continuous aspect-ratio tuning and fine shape control of monodisperse alpha-Fe_2O_3 nanocrystals by a programmed microwave-hydrothermal method. Adv. Funct. Mater. 18, 880–887.

Huang, X., El-Sayed, I.H., Qian, W., El-Sayed, M.A., 2006b. Cancer cell imaging and photothermal therapy in the nearinfrared region by using gold nanorods. J. Am. Chem. Soc. 128, 2115–2120.

Huang, X., Jain, P.K., El-Sayed, I.H., El-Sayed, M.A., 2006a. Determination of the minimum temperature required for selective photothermal destruction of cancer cells with the use of immunotargeted gold nanoparticles. Photobioch. Photobiop. 82, 412–417.

Jain, N., Bhargava, A., Majumdar, S., Panwar, J., 2011. Extracellular biosynthesis and characterization of silver nanoparticles using Aspergillus flavusNJP08: a mechanism prospective. Nanoscale 3, 635–641.

Joo, J., Kwon, S.G., Yu, J.H., Hyeon, T., 2005. Synthesis of ZnO nanocrystals with cone, hexagonal cone, and rod shapes via nonhydrolytic ester elimination sol-gel reactions. Adv. Mater. 17, 1873–1877.

Kang, Y., Murray, C.B., 2010. Synthesis and electrocatalytic properties of cubic Mn-Pt nanocrystals. J. Am. Chem. Soc. 132, 7568–7569.

Kashchiev, D., van Rosmalen, G.M., 2003. Review-Nucleation in solutions revisited. Cryst. Res. Technol. 38, 555.

Kim, D., Park, J., An, K., Yang, N.K., Park, J.G., Hyeon, T., 2007. Synthesis of hollow iron nanoframes. J. Am. Chem. Soc. 129, 5812–5813.

Kim, K.S., Demberelnyamba, D., Lee, H., 2004. Size-selective synthesis of gold and platinum nanoparticles using novel thiol functionalized ionic liquids. Langmuir 20, 556–560.

Kim, S.W., Park, J., Jang, Y., Chung, Y., Hwang, S., Hyeon, T., Kim, Y.W., 2003. Synthesis of Monodisperse Palladium Nnaoparticles. Nano Lett. 3, 1289–1291.

Kura, H., Takahashi, M., Ogawa, T., 2010. Synthesis of monodisperse iron nanoparticles with a high saturation magnetization using an Fe(CO)x-oleylamine reacted precursor. J. Phys. Chem. C 114, 5835–5838.

Kwon, S.G., Hyeon, T., 2008. Colloidal chemical synthesis and formation kinetics of uniformly sized nanocrystals of metals, oxides, and chalcogenides. Acc. Chem. Res. 41, 1696–1709.

Lemire, J.A., Harrison, J.J., Raymond, J., 2013. Antimicrobial activity of metals: mechanisms, molecular targets and applications. Nat. Rev. Microbiol. 11, 371–384.

Li, C., Sun, L., Sun, Y., Teranishi, T., 2013. One-pot controllable synthesis of Au@Ag heterogeneous nanorods with highly tunable plasmonic absorption. Chem. Mater. 25, 2580–2590.

Li, S., Shen, Y., Xie, A., Yu, X., Qiu, L., Zhang, L., Zhang, Q., 2007. Green synthesis of silver nanoparticles using *Capsicum annuum* L. extract. Green Chem. 9, 852–858.

Li, X.L., Peng, Q., Yi, J.X., Wang, X., Li, Y., 2006. Near monodisperse TiO_2 nanoparticles and nanorods. Chem. Eur. J. 12, 2383–2391.

Liu, Z., Fan, T., Zhang, W., Zhang, D., 2005. The synthesis of hierarchical porous iron oxide with wood templates. Micropor. Mesopor. Mater. 85, 82–88.

Luef, B., Fakra, S.C., Csencsits, R., Wrighton, K.C., Williams, K.H., Wilkins, M.J., Downing, K.H., Long, P.E., Comolli, L.R., Banfield, J.F., 2013. Iron-reducing bacteria accumulate ferric oxyhydroxide nanoparticle aggregates that may support planktonic growth. ISME J. 7, 338–350.

Luo, X., Chen, Y., Yue, G.-H., Peng, D.-L., Luo, X., 2009. Preparation of hexagonal close-packed nickel nanoparticles via a thermal decomposition approach using nickel acetate tetrahydrate as a precursor. J. Alloy Compd. 476, 864–868.

Maeda, H., Wu, J., Sawa, T., Matsumura, Y., Hori, K., 2000. Tumor vascular permeability and the EPR effect in macromolecular therapeutics: a review. J. Control. Release 65, 271–284.

Mallikarjuna, N.N., Varma, R.S., 2007. Microwave-assisted shape-controlled bulk synthesis of noble nanocrystals and their catalytic properties. Cryst. Growth Des. 7, 686–690.

Moghimi, S.M., Patel, H.M., 1998. Serum-mediated recognition of liposomes by phagocytic cells of the reticuloendothelial system-the concept of tissue specificity. Adv. Drug Deliv. Rev. 32, 45–60.

Mohamed, M.B., AbouZeid, K.M., Abdelsayed, V., Aljarash, A.A., Samy El-Shall, M., 2010. Growth mechanism of anisotropic gold nanocrystals via microwave synthesis: formation of dioleamide by gold nanocatalysis. ACS Nano 4, 2766–2772.

Morones, J.R., Elechiguerra, J.L., Camacho, A., Holt, K., Kouri, J.B., Ramirez, J.T., Yacaman, M.J., 2005. The bactericidal effect of silver nanoparticles. Nanotechnology 16, 2346–2353.

Nath, D., Banerjee, P., 2013. Green nanotechnology—a new hope for medical biology. Environ. Toxicol. Pharmacol. 36, 997–1014.

Nichols, J.W., Bae, Y.H., 2012. Odyssey of a cancer nanoparticle from injection site to site of action. Nano Today 7, 606–618.

Niederberger, M., 2007. Nonaqueous sol-gel routes to metal oxide nanoparticles. Acc. Chem. Res. 40, 793–800.

Njagi, E.C., Huang, H., Stafford, L., Genuino, H., Galindo, H.M., Collins, J.B., Hoag, G.E., Steven, L.S., 2011. Biosynthesis of iron and silver nanoparticles at room temperature using aqueous sorghum bran extracts. Langmuir 27, 264–271.

Okada, T., Sasaki, F., Kamiyama, T., Nakagawa, T., Nakanishi, K., Kobayashi, R., 2005. Focal nodular hyperplasia of the liver: usefulness of super paramagnetic iron oxide-enhanced magnetic resonance imaging. J. Pediatr. Surg. 40, E21–E25.

Okitsu, K., Sharyo, K., Nishimura, R., 2009. One-pot synthesis of gold nanorods by ultrasonic irradiation: the effect of ph on the shape of the gold nanorods and nanoparticles. Langmuir 25, 7786–7790.

Pal, S., Tak, Y.K., Song, J.M., 2007. Does the antibacterial activity of silver nanoparticles depend on the shape of the nanoparticle? A study of the Gram-negative bacterium *Escherichia coli*. Appl. Environ. Microbiol. 73, 1712–1720.

Patete, J.M., Peng, X., Koenigsmann, C., Xu, Y., Karn, B., Wong, S.S., 2011. Viable methodologies for the synthesis of high-quality nanostructures. Green Chem. 13, 482–519.

Payne, E.K., Shuford, K.L., Park, S., Schatz, G.C., Mirkin, C.A., 2006. Multipole plasmon resonances in gold nanorods. J. Phys. Chem. B 110, 2150–2154.

Prakash, B., 1997. Use of metals in Ayurvedic medicine. Indian J. History Sci. 32, 1–28.

Prasad, B.L.V., Arumugam, S.K., Bala, T., Sastry, M., 2005. Solvent-adaptable silver nanoparticles. Langmuir 21, 822–826.

Qian, X., Shen, S., Liu, T., Cheng, L., Liu, Z., 2015. Two-dimensional TiS_2 nanosheets for in vivo photoacoustic imaging and photothermal cancer therapy. Nanoscale 7, 6380–6387.

Rabenau, A., 1985. The role of hydrothermal synthesis in preparative chemistry. Angew. Chem. Int. Ed. Engl. 24, 1026–1040.

Rao, C.N.R., Kalyanikutty, K.P., 2008. The liquid–liquid interface as a medium to generate nanocrystalline films of inorganic materials. Acc. Chem. Res. 41, 489–499.

Rao, C.N.R., Kulkarni, G.U., John Thomas, P., Agrawal, V.V., Saravanan, P., 2003. Films of metal nanocrystals formed at aqueous-organic interfaces. J. Phys. Chem. B 107, 7391–7395.

Rao, C.N.R., Ramakrishna Matte, H.S.S., Voggu, R., Govindaraj, A., 2012. Recent progress in the synthesis of inorganic nanoparticles. Dalton Trans. 41, 5089.
Reddy, B.M., Reddy, G.K., Ganesh, I., Ferreira, J.M.F., 2009a. Microwave-assisted synthesis and structural characterization of nanosized $Ce_{0.5}Zr_{0.5}O_2$ for CO oxidation. Catal. Lett. 130, 227–234.
Reddy, B.M., Reddy, G.K., Ganesh, I., Ferreira, J.M.F., 2009b. Single step synthesis of nanosized CeO_2–M_xO_y mixed oxides (M_xO_y=SiO_2, TiO_2, ZrO_2, and Al_2O_3) by microwave induced solution combustion synthesis: characterization and CO oxidation. J. Mater. Sci. 44, 2743–2751.
Reddy, B.M., Reddy, G.K., Rao, K.N., Ganesh, I., Ferreira, J.M.F., 2009c. Characterization and photocatalytic activity of TiO_2–M_xO_y (M_xO_y=SiO_2, Al_2O_3, and ZrO_2) mixed oxides synthesized by microwave-induced solution combustion technique. J. Mater. Sci. 44, 4874–4882.
Reddy, B.M., Reddy, G.K., Katta, L., 2010. Structural characterization and dehydration activity of CeO_2–SiO_2 and CeO_2–ZrO_2 mixed oxides prepared by a rapid microwave-assisted combustion synthesis method. J. Mol. Catalysis A Chem. 319, 52–57.
Richter, K., Birkner, A., Mudring, A.V., 2010. Stabilizer-free metal nanoparticles and metal-metal oxide nanocomposites with long-term stability prepared by physical vapor deposition into ionic liquids. Angew. Chem. Int. Ed. 49, 2431–2435.
Rycenga, M., Wang, Z., Gordon, E., Cobley, C.M., Schwartz, A.G., Lo, C.S., Xia, Y., 2009. Probing the photothermal effect of gold-based nanocages with surface-enhanced Raman scattering (SERS). Angew. Chem. Int. Ed 48, 9924–9927.
Samia, A.C.S., Schlueter, J.A., Jiang, J.S., Bader, S.D., Qin, C.J., Lin, X.M., 2006. Effect of ligand-metal interactions on the growth of transition-metal and alloy nanoparticles. Chem. Mater. 18, 5203–5212.
Sathishkumar, M., Sneha, K., Kwak, I.S., Mao, J., Tripathy, S.J., Yun, Y.-S., 2009. Phyto-crystallization of palladium through reduction process using *Cinnamom zeylanicum* bark extract. J. Hazard. Mater. 171, 400–404.
Satishkumar, B.C., Govindaraj, A., Nath, M., Rao, C.N.R., 2000. Synthesis of metal oxide nanorods using carbon nanotubes as templates. J. Mater. Chem. 10, 2115–2119.
Seo, J.-W., Jun, Y.-W., Ko, S.J., Cheon, J., 2005. In situ one-pot synthesis of 1-dimensional transition metal oxide nanocrystals. J. Phys. Chem. B 109, 5389–5391.
Shi, W., Song, S., Zhang, H., 2013. Hydrothermal synthetic strategies of inorganic semiconducting nanostructures. Chem. Soc. Rev. 42, 5714–5743.
Shiv Shankar, S., Ahmad, A., Pasricha, R., Sastry, M., 2003. Bioreduction of chloroaurate ions by geranium leaves and its endophytic fungus yields gold nanoparticles of different shapes. J. Mater. Chem. 13, 1822–1826.
Skrabalak, S.E., Chen, J., Sun, Y., Lu, X., Au, L., Cobley, C.M., Xia, Y., 2008. Gold nanocages: synthesis, properties, and applications. Acc. Chem. Res. 41, 1587–1595.
Song, Q., Zhang, Z.J., 2004. Shape control and associated magnetic properties of spinel cobalt ferrite nanocrystals. J. Am. Chem. Soc. 126, 6164–6168.
Stoimenov, P.K., Klinger, R.L., Marchin, G.L., Klabunde, K.J., 2002. Metal oxide nanoparticles as bactericidal agents. Langmuir 18, 6679–6686.
Suslick, K.S., 1990. Sonochemistry. Science 247, 1439.
Suslick, K.S., Price, G.J., 1999. Applications of ultrasound to materials chemistry. Annu. Rev. Mater. Sci. 29, 295.
Taton, T.A., Mirkin, C.A., Letsinger, R.L., 2000. Scanometric DNA array detection with nanoparticle probes. Science 289, 1757–1760.
Tsuji, M., Hashimoto, M., Nishizawa, Y., Kubokawa, M., Tsuji, T., 2005. Microwave-assisted synthesis of metallic nanostructures in solution. Chem. Eur. J. 11, 440–452.
Turova, N.Y., Turevskaya, E.P., 2002. The Chemistry of Metal Alkoxides. Kluwer Academic Publishers, Boston.
Van, T.-K., Cha, H.G., Nguyen, C.K., Kim, S.-W., Jung, M.-H., Kang, Y.S., 2012. Nanocystals of hematite with unconventional shape-truncated hexagonal bipyramid and its optical and magnetic properties. Cryst. Growth Des. 12, 862–868.

von Maltzahn, G., Park, J.-H., Agrawal, A., 2009. Computationally guided photothermal tumor therapy using long-circulating gold nanorod antennas. Cancer Res. 69, 3892–3900.
Wang, H., Brandl, D.W., Nordlander, P., Halas, N.J., 2007. Plasmonic nanostructures: artificial molecules. Acc. Chem. Res. 40, 53–62.
Wang, Y., Zheng, Y., Huang, C.Z., Xia, Y., 2013. Synthesis of Ag nanocubes 18–32 nm in edge length: the effects of polyol on reduction kinetics, size control, and reproducibility. J. Am. Chem. Soc. 135, 1941–1951.
Wu, M., Lin, G., Chen, D., Wang, G., He, D., Feng, S., Xu, R., 2002. Sol-hydrothermal synthesis and hydrothermally structural evolution of nanocrystal titanium dioxide. Chem. Mater. 14, 1974–1980.
Xiu, Z.M., Zhang, Q.B., Puppala, H.L., Colvin, V.L., Alvarez, P.J., 2012. Negligible particle-specific antibacterial activity of silver nanoparticles. Nano Lett. 12, 4271–4275.
Xu, H., Zeiger, B.W., Suslick, K.S., 2013. Sonochemical synthesis of nanomaterials. Chem. Soc. Rev. 42, 2555–2567.
Yan, L., Yu, R., Chen, J., Xing, X., 2008. Template-free hydrothermal synthesis of CeO_2 nano-octahedrons and nanorods: investigation of the morphology evolution. Cryst. Growth Des. 8, 1474–1477.
Yang, J., Lee, J.Y., Ying, J.Y., 2011. Phase transfer and its applications in nanotechnology. Chem. Soc. Rev. 40, 1672–1696.
Yang, J., Sargent, E., Kelley, S., Ying, J.Y., 2009. A general phase-transfer protocol for metal ions and its application in nanocrystal synthesis. Nat. Mater. 8, 683.
Yang, Z., Han, D., Ma, D., Liang, H., Liu, L., Yang, Y., 2010. Fabrication of monodisperse CeO_2 hollow spheres assembled by nano-octahedra. Cryst. Growth Des. 10, 291–295.
Zeng, H., Rice, P.M., Wang, S.X., Sun, S.H., 2004. Shape-controlled synthesis and shape-induced texture of $MnFe_2O4$ nanoparticles. J. Am. Chem. Soc. 126, 11458–11459.
Zhang, L.-S., Jiang, L.-Y., Chen, C.-Q., Li, W., Song, W.-G., Guo, Y.-G., 2010a. Programmed fabrication of metal oxides nanostructures using dual templates to spatially disperse metal oxide nanocrystals. Chem. Mater. 22, 414–441.
Zhang, Q., Liu, F., Nguyen, K.T., Ma, X., Wang, X., Xing, B., Zhao, Y., 2012. Multifunctional mesoporous silica nanoparticles for cancer-targeted and controlled drug delivery. Adv. Funct. Mater. 22, 5144–5156.
Zhang, Y., Yu, J., Birch, D.J., Chen, Y., 2010b. Gold nanorods for fluorescence lifetime imaging in biology. J. Biomed. Opt. 15, 020504.
Zheng, U., Liu, S., Chen, X., Cheng, J., Si, C., Pan, Z., Marcelli, A., Chu, W., Wu, Z., 2003. Synthesis and characterization of cu-pt bimetallic nanoparticles. J. Phys. 430, 012037.
Zili, W., Li, M., Overbury, S.H., 2012. On the structure dependence of CO oxidation over CeO_2 nanocrystals with well-defined surface planes. J. Catalysis 285, 61–73.
Zitoun, D., Pinna, N., Frolet, N., Belin, C., 2005. Single crystal manganese oxide multipods by oriented attachment. J. Am. Chem. Soc. 127, 15034–15035.

CHAPTER

2 APPLICATIONS OF NANOSCALE DRUGS CARRIERS IN THE TREATMENT OF CHRONIC DISEASES

Andreea Aiacoboae*, Tania Gheorghe*, Iulia I. Lungu*, Carmen Curutiu, Mariana C. Chifiriuc**, Alexandru Mihai Grumezescu**, Alina M. Holban*,****

**University Politehnica of Bucharest, Bucharest, Romania*

***Research Institute of the University of Bucharest (ICUB), Bucharest, Romania*

CHAPTER OUTLINE

1 INTRODUCTION IN DRUG DELIVERY AND TARGETING

Paul Ehrlich (1854–1915) proposed that if a compound could selectively reach an organism responsible for certain diseases in the body, then along with this compound some toxins could be delivered to destroy the respective disease-causing organism. This was called the "magic bullet," a complex made from two components: one that identifies and attaches to the targeted site and the other one that delivers the pharmaceutical drug into the site (Torchilin, 2010).

Targeted drug delivery can be associated with an increased concentration of active pharmaceutical drugs in a specific body site, comparing with the rest of the body. For most pharmaceutical agents, their ability to specifically target damaged sites depends on the active pharmaceutical drug aggregation to interact with biological pathways. Commonly, the distribution of the pharmaceutical drug released by the targeted drug-delivery system (DDS) is equally proportioned in the blood flow. One main concern is that in order to reach the desired site, the drug carrier must overcome some biological barriers that can induce severe side effects. It is desirable that the concentration of the drug is higher in the targeted site, and to significantly decrease in nontargeted sites in order to avoid unwanted side effects (Torchilin, 2010).

Since Ehrlich's statement, many approaches were proposed for drugs delivery, multiple systems that can specifically and precisely deliver the drugs have been developed in the last decade, with great impact in medicine, in fields, such as oncology, cardiology, and immunology. The architecture and development of effective new DDSs has been well investigated in the last few years. The main focus was on accomplishing new methods for controlled drug release. In traditional drug delivery, the concentration of the drug in the blood increases rapidly and then suddenly drops. Every drug has a well-established plasma level. If the drug has a higher amount than the plasma level it becomes toxic, however, if the amount of drug is below the plasma level it considerably reduces its efficiency. The main goal is to develop an optimal DDS and maintain it in the chosen therapeutic range (Liu et al., 2014a).

Targeted DDSs come with such advantages, maintaining drug levels in a desirable range that is not harmful. Moreover, side effects are diminished because through DDSs specific site is targeted, leading to patient compliance due to the less amount of drug needed (Langer, 1998).

In order to achieve drug safety and greater efficiency, pharmaceutical drugs can be either encapsulated, or attached to a polymer or a lipid. For a prolonged period of time to a specifically injured body site, targeted drug delivery is also necessary. To achieve targeted DDSs, there are few criteria that must be taken into consideration: the properties of the drug, the side effects that can be induced by the drug, the manner in which the drug is introduced in the body, minimal leakage during delivery etc. (Rani and Paliwal, 2014).

Ideal targeted DDSs should be nontoxic, nonimmunogenic, physically, and chemically stable both in vivo and in vitro. The vehicles should be biodegradable or at least easily removable from the body. Moreover, the preparation technique should be facile, easily reproductive, and cost effective. Pharmaceutic, pharmacokinetic, and pharmacodynamic reasons justify choosing targeted drug delivery over traditional drug delivery. The pharmaceutical reasons are the instability of the drugs and also their low solubility. The pharmacokinetic reasons are based on the distribution panel and the short half-life. The last reason, as aforementioned, is the low specificity of traditional drugs (Rani and Paliwal, 2014).

A targeted DDS is formed from a drug vehicle and a target. The target can be either a particular group of cells or an organ that are damaged and need therapy. It is very important that the drug does not "leak" from the drug carrier in nonspecific sites in order to minimize the severe side effects. An ideal drug carrier must be capable to overpass the blood–brain barrier (BBB). These carriers must be identified by specific cells and also should control surface specificity of the ligands. The ligand–drug mix must be stable in biofluids, such as plasma. After the drug carrier is recognized, it should release its charge into the targeted site (Rani and Paliwal, 2014).

In the last years, researchers were looking for new methods of drug delivery, and also for new drugs (Brannon-Peppas and Blanchette, 2012). A desirable situation in drug delivery is to have smart DDS that can be integrated in the human body (Jain, 2011).

A smart system is a complex that can reshape its properties when perceiving changes in the environment in which it operates. As in nature, the chameleon changes its colors in the concern of integrating into the surrounding area and not to be noticed by the predators, smartness can be observed in all the cells in the biological system. For example, in the immune system, cells immediately drift to a specific site when they seize a change in the concentration of cytokines (Lammers, 2013).

A smart pharmaceutical drug could mimic the systems inside the body in order to effectively attain progresses in healthcare. Nanosized systems have proved to be smart due to the fact that they can easily infiltrate into cells. These systems, besides the fact that, by functionalizing them, they can overcome the biological barriers of the body, can also be manipulated in order to respond to stimuli. This is an important property that makes them suitable for therapeutic applications (Lammers, 2013).

Smart DDSs need to meet numerous requirements, such as promoting the pharmaceutical drugs solubility and stability, decrease the quantity of the drug needed, diminishing side effects. The drug vehicle must be biocompatible and should not induce responses of the immune system. The major property of smart DDSs is the ability to deliver the pharmaceutical drug to the desired site during a prolonged period of time. For this, the drug carrier must have its surface modified so that it can attach to the interest cells and deliver its charge by means of provided stimuli. The stimuli can be either external, such as electric and magnetic field, ultrasound and temperature, or internal, by environmental systems, such as temperature, pH, and so on. Encapsulating gold, silica, or magnetic nanoparticles into smart polymers increases the advantages of the complexes (Joglekar and Trewyn, 2013).

1.1 PASSIVE AND ACTIVE TARGETING

Passive targeting refers to drug accumulation in a specific body part due to physicochemical or pharmacological factors of the disease. For example, in cancer therapy, surface properties and size ranges of the nanoparticles must be carefully controlled in order to refrain from the uptake of reticulum-endothelial system (RES) and also to magnify targeting ability (Rani and Paliwal, 2014).

This kind of targeting has the advantage of particular pathophysiological properties of tumor vessels; they permit carriers to accumulate in the tumor tissue. Generally, tumor vessels have a very chaotic structure and a great pore distribution, which results in expanded gap junctions. Due to this type of vascularization, based on the enhanced permeability and retention effect, macromolecules (>400 nm) can easily migrate in the enclosing tumor site. Another fact that backs passive targeting is the different microenvironments of the adjacent tumor tissue, from healthy tissues. Even though passive targeting approaches have been used in clinical therapy, they do have various limitations (Bamrungsap et al., 2012).

Active targeting refers to specific ligand–receptor interaction with intracellular localization that takes place only after extravasations. There are three categories of active targeting. The first one is designed for limited distribution of the drug vehicles to the capillary network of an already known location. The second one is based on the particularity of the drug to attack specific cell types (e.g., cancerous cells), without harming the normal cells. The latter refers to the delivery of drug into the intracellular site of cancerous cells (Maity and Stepensky, 2015; Rani and Paliwal, 2014).

In order to overcome the disadvantages of the passive targeting approach, specific ligands can be attached to the exterior of the drug delivery carriers by various conjugation chemistries. The carriers will recognize and bind the targeted cells over ligand–receptor interactions. For a high specificity, these receptors must be highly expressed on cancerous cells, but not on healthy cells (Bamrungsap et al., 2012).

In conclusion, as it was mentioned earlier, there are two drug targeting mechanisms: active targeting and passive targeting. Active targeting of drug-loaded nanoparticles enhances the effects of passive targeting and makes the nanoparticles more specific to a target site (Galvin et al., 2012). Active targeting strategies uses antibodies against the tumor targets selected for amplification, specifically these therapeutic nanoparticles compared to the mechanisms of passive targeting (Cho et al., 2008). It is necessary to find new ways to convey these passive delivery systems in active delivery systems that can recognize cancerous cells and deliver drugs directly to the targeted site (Lobenberg, 2003).

1.2 SMART DRUG DELIVERY SYSTEMS—DRUG CARRIERS

The most used targeted drug delivery carriers are microspheres and nanoparticles. Nanoparticles have dimensions that range between 0.2 and 0.5 μm, apart from microspheres that have a size range between 30 and 200 μm. Relying on the physicochemical properties of the drug, it can be either encapsulated in the nanoparticles nucleus, or attached to its surface providing protection of the drug against the unfavorable properties of the environment (Bamrungsap et al., 2012; Rani and Paliwal, 2014).

Nanoparticles are being used in the pharmaceutical field for over 10 years in order to reduce side effects and toxic effects of drugs. Researchers try to obtain an appropriate biological response to nanoparticles because the development of nanostructurated safe DDSs is absolutely necessary in the future (Webster, 2006). Some research shows that nanoparticles used in targeted drug delivery could improve organ damages (Lobenberg, 2003).

In order to assemble functionalized nanosized carriers proteins, peptides, or polymers must be attached to the drug carrier surface. Some functional elements can be also encapsulated into nanosized carriers, such as, for example, magnetic nanoparticles encapsulated into polymeric shells. Surface attachment can be accomplished either by noncovalent procedures, or by chemical mechanism. Chemically attachments are the ones that are mostly used (Torchilin, 2010).

Controlled DDSs can be made of hydrogels, nanoparticles, or multifunctional nanostructures, carbon nanotubes and radioactive microspheres. In addition, multifunctional nanoparticles are used for targeting tumors, attacking and destroying cancerous cells (Kleinstreuer et al., 2013).

Regardless of passive or active targeting, an essential property of nanosized particles is longevity in the blood flow. A modality of achieving this is to preserve the pharmaceutical drug in the blood flow for a lengthened period of time until it reaches the desired site. The most used solution for this is the chemical adjustment of drug carriers with specific synthetic polymers. For example, promising results against different solutes have been attained with the use of hydrophilic polymers (Torchilin, 2010).

1.3 MICROPARTICLE CARRIERS

Microparticles consist of a matrix that encapsulates pharmaceutical drugs. According to the latter description, microparticles can be represented by microspheres, that have the drug, dissipated in the matrix, or microcapsules, which have the drug encapsulated in a core. So, their physicochemical properties depend on the encapsulation parameters, which can increase blood barrier permeability (Liu et al., 2014b).

1.4 SMART STIMULI RESPONSIVE DRUG DELIVERY SYSTEMS

Nanocarriers that have pharmaceutical drugs encapsulated in them, mostly anticancer agents, have been widely investigated in the concern of improving drug bioavailability. Furthermore, functionalizing these nanoparticles with stimuli responsive polymers makes them "smarter" DDSs and help to the improvement of effectiveness of the drug (Cukierman and Khan, 2010). Most of such smart DDSs are developed to maintain the specific drug inside, until one stimulus provokes changes that lead to the releasing of the drug (Alvarez-Lorenzo and Concheiro, 2014).

1.4.1 pH

pH-sensitive DDSs have been used in the medical area due to their unique properties. In the field of cancer therapy, anticancer agents can be attached to the surface or encapsulated into pH-responsive polymers (Balamuralidhara et al., 2011). Materials that are pH sensitive present changes in structure and conformation as disequilibrium of ionization and deionization occur. Blood and unaffected tissues present a higher pH than that found in tumor tissues, mainly because the production of lactic acid is extremely high, as a result of hypoxic environment and the fast development and spread of tumor cells (Alvarez-Lorenzo and Concheiro, 2014). For example, the pH of a pathological tissue has a value of approximately 7.4, but when a specific site is affected by inflammation, its pH drops from 7.4 down to 6.5 (Ganta et al., 2008). Thus, smart DDSs must be manufactured in such a manner, that the drug can be released when it reaches the acidity of the tumor microenvironment (He et al., 2013; Liu et al., 2014a).

Usually a pH smart drug delivery complex is made from a magnetic nanocore and a pH responsive polymer shell. An example of core–shell pH responsive smart drug systems are superparamagnetic iron oxide nanoparticles coated with poly(acrylic acid) conjugated with folic acid. Poly(acrylic acid) is used to achieve a good pH responsive delivery profile and the folic acid provides efficient encapsulation (Shen and Jiang, 2014; Yang et al., 2014).

1.4.2 Temperature

Under healthy conditions, the temperature of the body does not fluctuate much. However, under pathological condition, the body temperature rises locally. In order to increase the heat in a specific area, external sources can be used. Irradiation of metals can be induced for converting the energy into heat, or the source can be applied directly on the surface of the skin (Alvarez-Lorenzo and Concheiro, 2014).

In order to successfully treat cancerous tumors, a large amount of drugs need to be delivered directly into the carcinoma. Most of the drugs used for treating cancer are hydrophobic and tend to interact with other hydrophobic specific locations in the body (e.g., blood proteins) (Lee et al., 2015). A proper method for such application consists in the use of liposomes or polymer micelles. One of the most encouraging systems of smart drug delivery is the one using thermosensitive liposomes. Liposomes are well-known and highly spread among systems for targeted drug delivery, because of their biocompatibility and improved effect in terms of permeability and retention (Li et al., 2015). Temperature can be increased in the tumor location to 41°C for 1 h or ultrasounds can be applied for 20 min in order to cause hyper permeable tumor vessels, resulting in a fast deliverance of the specific drug. Keeping the drug inside the delivery system has a major importance, as it needs to be released near the tumor location, simultaneous with the appliance of stimulus (Lee et al., 2015).

An example for smart DDSs that are responsive to temperature stimuli is the use of superparamagnetic iron oxide nanoparticles through hyperthermia treatments (Mahmoudi et al., 2011). Hyperthermia is a noninvasive method that is strictly related to cancer. This method is based on the fact that the target tissue could be subjected to a high temperature for killing cancerous cells. The temperature at which cancerous cells die is around 42–46°C. However, this technique comes with some disadvantages, such as: the heat sources could not be controlled, nor the quantity or the site where the heat is applied (Bañobre-López et al., 2013; Giustini et al., 2010).

1.4.3 Light

DDSs, which are responsive to light have gained much attention. Visible light and UV (ultraviolet) can activate drug release. Near infrared radiation can go deeper in the body and possess the ability to either discharge the drug or not in a focalized place of a tissue (Alvarez-Lorenzo and Concheiro, 2014).

Systems which contain gold nanoparticles (AuNPs) are a perfect option in terms of the combination between chemotherapy and thermal ablation for tumor treatment. AuNPs, due to their surface plasmon resonance may rise the body temperature up to several degrees, after near-infrared radiation is applied. Those nanoparticles can also be conjugated with peptides or oligonucleotides as they have a suitable surface. A good example of drug release using infrared radiation consists in using PLGA—poly(lactic-*co*-glycolic acid) matrix particles that contain doxorubicin (anticancer drug) and that is covered with gold. Irradiation will cause the delivery of doxorubicin that will lead to cancerous cells death (Alvarez-Lorenzo and Concheiro, 2014).

1.4.4 Electrical field

Networks that are electrically sensitive can be either injected or inserted under the skin. Furthermore, a patch can be attached over the place of implantation. If the battery is on, protons will travel toward the cathode and will cause a pH change. After that, the drug will be released. Release rate or duration of the drug can be adjusted by modifying the intensity of the electrical field (Alvarez-Lorenzo and Concheiro, 2014).

1.4.5 Magnetic field

A great amount of different magnetic nanoparticles is already used in magnetic resonance imaging (MRI) as contrast agents. Most of those particles are based on magnetite (Fe_3O_4). Studies were conducted in order to obtain nanoparticles which present chemical and physical stability and long circulation time for cancer MRI. Nanoparticles around 30 nm proved to be suitable as contrast agents, the only negative aspect being represented by their tendency to aggregate, producing clusters. This problem was resolved using a proper dispersing agent, which offered high contrast effects regarding magnetic resonance. Ferro fluids represent the term used for colloidal suspension that results. Such nanoparticles possess important characteristics and they can integrate fluorophores, which can be optically detected, in order to obtain information about the tumor and to deliver therapeutic agents to the cancerous site. If a high- gradient magnetic field is applied, the magnetic nanoparticle carrier will be directed to a specific region (Tietze et al., 2015). Even though the injected dose is not high, the local concentration will be. Drug release occurs when magnetic field is alternating. While the alternating field is on, magnetic nanoparticles rotate and the heat is dissipating in tissues located near that area. When the temperature is in the range of 46–56°C, thermoablation occurs, leading to cell necrosis. If the temperature is increased up to 41–46°C the apoptosis mechanism will be activated. This last situation is approved to be used as an additional technique to other cancer treatments modalities (chemotherapy, radiotherapy, surgery). Magnetic hyperthermia represents the only source of producing of heat that does not affect healthy tissues, but only cancerous cells, the target of high heat energy. There are some products available in the market and others that are still being tested. One of the existing products is represented by the ferrofluid nanotherm that has water dispersible nanoparticles with a diameter of 15 nm. Nanothermis consists of an amino saline coating and a core–shell structure. A syringe is used for distributing nanoparticles, which will stay in the tumor tissue, due to the shell–core complex. A device, which applies alternative magnetic field and software for therapy planning are monitoring the distribution of the temperature inside the tumor and also help to manage the strength of the applied magnetic field and to repeat the treatment until the tumor is eradicated. Drug release rate depends on the applied temperature, if it is strongly increased, the carrier will disintegrate, and the drug will be rapidly eliminated. If it is the other way around, and the temperature rises only in a small amount, the drug will be released gradually (Alvarez-Lorenzo and Concheiro, 2014).

1.4.6 Ultrasounds

Ultrasounds can help with the nanostructures penetration into well-determined areas and with the drug release activation. They are applied to the body by physiotherapy equipment and can monitor the therapy in an efficient way. Low-frequency ultrasounds can get through only some centimeters into the body and because of this, the scattering is not high. This method contains several unknown aspects, such as its influence on cell apoptosis and also the effect over the genetic information (Alvarez-Lorenzo and Concheiro, 2014).

Recently, metal nanoparticles received much attention due to their exclusive physical and chemical properties, which promote them for applications as smart DDSs. In cancer, due to their specific surface area and great area volume ratio, such nanoparticles are great for diagnosis and imaging, not only for disease treatment. They possess also magnetic and photoluminescence properties and their capacity of producing reactive oxygen species (ROS) help, as well for killing cancerous cells. Even though it has not been reported any typical mechanism of destroying unhealthy cells, there have been more suggestions, including hyperthermia, production of ROS, or electromagnetic radiation, depending on the nanoparticle shape, size, surface properties, and the metal used (Sharma et al., 2015).

Metal nanoparticles are biocompatible and they can be easily evacuated from the human body. Imaging agents as radioisotopes, or agents for optical imaging can be attached on the surface of metal nanoparticles, for obtaining multimodal images. Even though those types of nanoparticles reduce patient's disturbance and adverse effects, they cannot be used in clinical application until issues about toxicology are being resolved (Sharma et al., 2015).

Metal nanoparticles respond to stimuli from the exterior, such as pH, light, or temperature and this represents one of the main reasons for why are highly investigated, not only for targeted drug therapy, but also for developing nanodevices or biosensors. As they are functionalized, properties of the aggregates and the aggregation process are better controlled (Alvarez-Lorenzo and Concheiro, 2014).

2 APPLICATIONS OF NANOTECHNOLOGY IN DRUG DELIVERY AND TARGETING

Nanotechnology is widely spread in almost every industry field (Farokhzad and Langer, 2006). Even though there are various ways to define the term "nanotechnology", it mainly refers to the dimensions, which need to be in the range of 0.1–100 nm. In the year 2000, in the USA, a more specific explanation of the term described nanotechnology as a field where the attention is focused on materials and systems that can improve different types of properties due to their size (Riehemann et al., 2009). At first, nanotechnology focused only on improving the existing therapeutic approaches, but recently it came with new ones, due to the nanoscale engineering, which had a major contribution. Although it appears to be new, nanotechnology has been applied to medical field since 1965. As it gained more and more interest, it received funding in order to develop new nanoscale systems (Farokhzad and Langer, 2006).

Recently, there are researches trying to develop scaffolds to induce the growth of skin, bones, or body organs (Liu et al., 2007).

Nanomedicine attempts to provide, instruments and devices for research and practice in the near future, which could revolutionize the current way of thinking (preventive and diagnostic) and taking action (applied treatments) in the medical field, especially in the sphere of influence of chronic degenerative diseases. Nanomedicine offers new and innovative solutions to solve health problems, with a positive impact on society through contributions made for improving the quality of life and lifespan and reducing costs in the healthcare system (De Jong and Borm, 2008).

Nanotechnology provides four main interest areas: targeted pharmaceutics, which deliver drugs to the desired site; tissue engineering, which is capable of forming new tissues; molecular imaging, which can detect specific diseases by means of intelligent imaging; and biosensors and diagnostics, which can detect dysfunctions in the body, and also develop diagnostic tools for lab research (Dvir et al., 2011).

Nanomedicine represents an area, which requires the knowledge from many different disciplines, is a field of continuous development and changes and it is expected to contribute in improving drug delivery, technologies for scientific discovery acceleration, imaging, or diagnosis methods (Farokhzad and Langer, 2006). Nanomedicine uses other approaches than the conventional ones, based on killing only the diseased cells instead of both diseased and healthy, as chemotherapy does (Riehemann et al., 2009).

2.1 TARGETED DRUG DELIVERY FOR CANCER THERAPY

Targeted delivery of drugs can be used to cure many diseases; however, it is mostly used to cure cancerous tumors. It is desirable to provide controlled release systems that provide drugs that target only tumor sites. In this aspect, the use of nanomaterials is a promising and exciting field. Development of truly targeted DDSs will require an improved understanding of multiple factors, regulation of distribution in the blood, dynamic aspects of tumor spatial and temporal heterogeneity (Bae and Park, 2011). The major progress in targeted drug delivery was attributed to nanotechnology, which resulted in the development of nanosized drug delivery platform with many advances in cancer therapy (Chakravarty et al., 2014).

Cancer is a major cause of death in the world, competing cardiovascular diseases (Liu et al., 2007). Cancer represents an abnormal and uncontrolled growth and proliferation of cells, which leads to tumor formation. Tumors are divided into two major categories, depending on their mode of growth: benign and malign. The first ones develop local, exclusively at the appearance site, while the others can spread among other tissues or body parts. Cancer occurs when DNA suffers a change due to lifestyle or environmental factors. Humans exposed constantly to UV radiations or different chemical substances, alcohol, polluted air, or tobaccos are predisposed to DNA changes. When the tumor is evolving, it slowly isolates itself from blood vessels, which lead to oxygen and nutritive substances loss. These substances are necessary for tumor cells to survive and develop. The tumor sends signals to activate angiogenesis, which will lead to a faster growth rate of cancerous cells. The pressure applied on the surrounding tissue will help tumor development.

It is extremely difficult to cure cancer and even if patients are properly treated with modern medical techniques, usually the result is not a positive one. Surgery can be used to remove parts of the tumor, but it does not help when the tumor is already spread (Zhang et al., 2011). Beside surgery, there are other traditional techniques used for cancer treatment, such as chemotherapy or radiation therapy

(Aguilar et al., 2011). Presently, the aim of chemotherapy is to destroy primary tumors, despite the fact that cancer deaths are caused by metastatic disease (Lee et al., 2015).

Targeted DDSs represent a very attractive and new technique used more and more often in the medical field, due to its beneficial effects. It is described as an approach, which leads the specific drug delivered at the targeted area, without affecting the normal cells. In order to obtain positive results, the carriers used for such applications need to possess special characteristics: they should have proper size in order to enter into the targeted tissues and cells through the blood circulation and deliver the drug (Lobenberg, 2003).

Nanoparticles used in drug delivery lead to an immediate response in cancer treatment. They can be attached in different ways to specific drugs and travel to the tumor through blood vessels (Yih and Wei, 2015). Functionalized nanoparticles present more control in terms of aggregation process, properties of the aggregates and structure (Alvarez-Lorenzo and Concheiro, 2014). Although small molecule drugs are successfully used as therapeutic agents, there are some aspects, which still need to be resolved, in terms of toxicity and solubility (Li et al., 2013).

Smart nanoparticles used in targeted DDSs have the ability to release, under stimulation, drug molecules in the cancerous area. There are many types of stimuli including physical, chemical, or biological constituents, such as temperature and electricity for the first category, pH and ions for the second one and antibodies or small molecules for the last one. Smart DDSs are designed to react to factors from the exterior, such as light, temperature, or ultrasound, which significantly improve the results of the therapy (Lee et al., 2015).

In conclusion, targeted drug delivery has gained more and more interest and it is proved to be an amazing option for disease treatment, especially for cancer therapy.

Even though various cancer treatments were developed, such as chemotherapy or surgical intervention, none can actually treat cancer; therefore the mortality rates associated with this disease remain increased. Chemotherapy is a method based on radiation. Both types of cells, normal and affected ones, are suffering from this technique. In order to destroy only cancerous cells, targeted DDSs were created.

Smart DDSs can encapsulate special anticancer drugs, such as doxorubicin and release them under stimuli conditions. Light, temperature conditions, pH conditions, ultrasound, electric, and magnetic field participate directly to drug release, under well-established conditions. In this way, only unhealthy cells will be affected. For positive results, nanocarriers need properly size in order to penetrate target tissues. There are several types of nanoparticles that are successfully used for such applications. Gold, silver, magnetic particles, or quantum dots (QD), due to their unique properties, are suitable for drug delivering and targeting. Combining those types of particles with polymers, an improved response is given.

2.1.1 Drug nanocarriers

2.1.1.1 Gold nanoparticles

There are various types of nanoparticles used in drug targeting and one of the most efficient is represented by AuNPs (Kumar et al., 2013). They play an important role in cancer treatment. AuNPs possess optical properties, which are unique and promote them for being used in cancer photothermal therapy and diagnosis. They are given by the interaction of light and electrons existing on their surface. The electromagnetic wave, make contact with free electrons when the surface of the metal is being irradiated, generating a fluctuation of electron charge, depending on the visible light frequency (Sharma et al., 2015). AuNPs absorb near infrared light, optimal for thermal therapy as it assures excellent light penetration and minimal disturbance of normal tissues. They present low toxicity levels and molecular stability due to the gold surface, which allows them to be bioconjugated with markers for tumor targeting. Even if no targeting agents exist on their surface, they are still able to gather at the tumor location, as a result of their permeability (Park et al., 2015).

Their toxic effect depends on nanoparticles size, as it is suggested in different studies. Size also affects the penetration success of AuNPs into the tissues altered by tumors. It has been demonstrated that AuNPs need to possess a dimension of maximum 10 nm in order to reach the inner nucleus. Otherwise, they will form clusters in the cytoplasm (Kumar et al., 2013). It is important for antitumor agents to reach the nucleus from tumor cells, because this is the most utilized target of drug activity (Zhong et al., 2015). AuNPs represent the most suitable and remarkable option among metal nanoparticles and this represents the reason of their large use in different applications which depend not only on the size of nanoparticles, but also on their shape, crystallinity, or structure (Murugan et al., 2014).

AuNPs with ultrasmall dimensions are also helpful in targeted drug delivery. For example, they were functionalized with an anticancer drug named doxorubicin which was successfully delivered into the nucleus of affected cells and its efficiency in also killing cancer cells which are more resistant, was also proved (Kumar et al., 2013).

AuNPs can be designed in order to recognize a stimulus and to instantly respond, releasing the drug into the tissue or cells. These nanosized particles can reach cancerous compartments at various stages of the disease. This accumulation does not affect cell viability, but it can produce a modification in the metabolism of normal cells. AuNPs are known for a long time as being suitable for arthritis or epilepsy treatment (Kumar et al., 2013).

Unfortunately, they have two challenges to defeat: the first one is about avoiding reticuloendothelial system approval and the second represents endothelial penetration. When nanoparticles are injected into the blood, they may be perceived as being foreign and phagocytized by macrophage cells. In order to avoid such situations, AuNPs are combined with different molecules that prevent the nanoparticles surface from being covered with plasma proteins. Heparin, dextran, or polysaccharides are only a few examples of such molecules (Kumar et al., 2013).

2.1.1.2 Silver nanoparticles

Silver nanoparticles can also absorb and disperse some of the light, which can be used for thermal induction of apoptosis of cancerous cells. Silver, as well as gold, possess a number of positive ions, which is the same as the ones for conduction electrons. When irradiation with an electromagnetic wave is applied, free electrons are transported by the electric field. Plasma oscillations will interact with the visible light. Silver has the highest efficiency in terms of plasmon excitation and its resonance can be turned into any wavelength of the visible spectrum. When it comes to their therapeutic activity, silver nanoparticles affect those proteins, which are responsible for ROS neutralization, after the interaction with cells. Inflammatory response and perturbation are initiated by the accumulation of ROS, which also provoke destruction of mitochondria and cell apoptosis (Sharma et al., 2015).

2.1.1.3 Magnetic nanoparticles

Magnetic nanoparticles have gained extremely high attention, due to the fact that they present large surface area, paramagnetic properties and also increased biocompatibility (Dutta et al., 2016). Moreover, their usage in medical applications is also caused by their small dimensions, that is, 1–100 nm. These kinds of nanoparticles are made of various magnetic elements, such as Fe, Ni, Co, and their oxides. If they are not functionalized, they tend to form clusters because of magnetic dipole–dipole attraction and also as a result of superparamagnetic behavior. Among the advantages that come with its usage there can be counted: improved biological performance, reduced immunogenicity, great colloidal stability. The major advantage of magnetic nanoparticles is that they can transport a pharmaceutical drug to a specific site. Moreover, these nanoparticles can also deliver their content in a controlled way without affecting untargeted areas. Their use in smart DDSs, magnetic nanoparticles can be controlled by means of an external magnetic field applied. The drug is either attached to the surface of the nanoparticles, or encapsulated into a polymeric shell that is used to coat the core (Dutta et al., 2016; Mody et al., 2013; Rajeswari, 2014).

The most frequently used materials for magnetic drug delivery consist of metal or metal oxide nanoparticles, such as superparamagnetic iron oxide nanoparticles, made of a core of iron oxide which is frequently covered with organic materials: fatty acids, polymers, or polysaccharides. Coating of magnetic nanoparticles comes with many advantages, such as increase stabilization of pH environments. For example, silica coating promotes biomolecules binding and also, due to the high porosity of the silica, pharmaceutical drugs can be easily encapsulated in its pores. However, while passing through the blood flow phagocitation of the nanoparticles may occur. To avoid this inconvenience, organic or inorganic coating must be used. The organic materials are beneficial for colloidal stability improvement and for prevention of segregation into particles. Many magnetic delivery systems have been produced although drug delivery based on such particles represents a sophisticated concept (Arruebo et al., 2007; Tietze et al., 2015).

As an application, drug carriers composed from polymer responsive to heat and magnetic nanosized particle were obtained due to the structural changes that occur in the polymer when heat is applied, making it to release the encapsulated drug molecule. This combination is perfect for drug targeting application. The polymer maintains the therapeutic drug trapped until they cross through the blood flow and when the physiological temperature is reached, the drug is discharged, when a slightly amount of heat is locally applied. Magnetic nanoparticles can produce heat if a magnetic field is applied. This way, conformational changes are induced to the thermoresponsive polymer, which will deliver the drug to the target, helping in this way in cancer therapy (Dutta et al., 2016).

Another method of cancer treatment using DDSs consists of using two drugs, both hydrophobic: doxorubicin and verapamil. The first one is known as being an anticancer drug, and the second one is used as an antiangiocardiopathy agent. Both are combined with a chitosan shell covered with magnetic nanoparticles and then trapped into lactic-*co*-glycolic acid (PLGA). Nanoparticles obtained with polymeric materials have been designed as double DDSs. PLGA proved to be biocompatible, nontoxic, and bioabsorbable. Utilizing the properties of polymeric coating and also the superparamagnetic properties of magnetic nanoparticles will lead to a better medical imaging applications, hyperthermia therapy, and response to magnetic stimuli. The fabrication of magnetic nanoparticles and PLGA particles consists in an aqueous phase and an organic one, which are favorable when encapsulating hydrophobic and hydrophilic drugs is desired. Hydrophilic drugs will dissolve in aqueous phase, while the organic phase will be used for incorporating hydrophobic phase. The difficulty in this case consists in releasing each drug separately at the cancerous site. However, under proper pH and temperature condition, PLGA will be degraded. The acidic microenvironment around tumor tissues will participate to the acceleration of PLGA degradation, causing the drug release around the tumor site (Shen et al., 2013).

2.1.1.4 Quantum dots

QDs containing tumor-targeting ligands are used as imaging agents (Piao et al., 2008). QDs represent colloidal nanoparticles with electrical properties. Their size varies between 2 and 100 nm. They have various fluorescence spectra and lifetime dependent on the size and present some advantages, such as: narrower emission spectra, higher quantum yield, and photo stability. In addition, QDs are perfect candidates for applications regarding cancer imaging, as their broad excitation spectrum promotes them for being used in multicolor imaging applications. Due to such broad spectrum, QDs can be excited by applying the same light source for different emission peaks (Liu et al., 2013). They can be used for the detection of prostate or breast cancer (Liu et al., 2013).

2.1.1.5 Mesoporous silica nanoparticles

Mesoporous silica nanoparticles have gained a high interest in diagnosis and therapy. Several great properties promote them for being used in targeted drug delivery, for example, uncommon mesoporous structure, tunable size, chemical stability, and biocompatibility (Xie et al., 2016).

Even though 10 years have passed since silica based delivery systems were first mentioned in a report, none of the systems are used in clinical applications. The reason is mainly the fact that silica's behavior in vivo is unknown and its biocompatibility in discussed (Yang and Yu, 2016).

Despite this aspect, with no doubt, mesoporous silica nanoparticles proved their efficiency as drug nanocarriers for cancer therapy, for the reason that they possess highly ordered mesoporous surface that can be easily functionalized and also high surface areas (Chen et al., 2015).

Producing synthetic systems of mesoporous silica nanoparticles has been a challenge, as this requested huge effort. Their surface contain a great amount of reactive silanol groups which are able to conjugate with a wide variety of functional group types, in order to obtain properties that allow them being used in applications for tumor targeting, bioimaging, or stimuli responsive release (Yang and Yu, 2016).

An application example consists in targeting drug delivery based on mesoporous silica nanoparticles combined with anticancer drug, doxorubicin, for cancer treatment. A specific colon cancer therapy application used this type of system, which also incorporated aptamer to prevent epithelial cell adhesion molecule, when doxorubicin is delivered to the cancerous site. Nucleic acid aptamer can identify a particular target molecule and, unlike antibody, it presents some advantages, offered by aptamer's small size. Aptamer is also nontoxic or immunogenic in contact with human organism. Epithelial cell adhesion molecule is the main antigen on colon carcinoma. This transmembrane protein is correlated with proliferation, migration, and invasion of cells (Yang and Yu, 2016).

2.2 TARGETED DRUG DELIVERY FOR CARDIOVASCULAR DISEASES

Cardiovascular diseases are conditions affecting heart function. To achieve controlled release systems for cardiovascular diseases, physiological properties of the cardiovascular system should be considered. Drug therapy of the cardiovascular system is different from other systems because of the anatomy and phsysiology of the vascular system (Jain, 2011).

It has been noted that over 7.9 million Americans have myocardial infarctions (MI), and 15.8 suffer from ischemic heart diseases. MI is the main cause for congestive heart failure (CHF) and is one of the major reasons of death in the USA. MI is induced by the obstruction of a coronary vessel leading to cellular death. Ischemia, which is the restrain in coronary flow, leads to myocardial necrosis. All these factors result in compromising a certain region. If this region is considerable, the ultimate result is death (Scott et al., 2008).

The advanced delivery strategies are very important for completing regeneration therapies for the treatment of ischemic heart disease. Research made on the interactions between the toxin and targeted cells lead to the development of molecular pharmaceutics (Hastings et al., 2015). For a controlled delivery system is necessary to understand the fundamentals of various cardiac diseases, their etiology, pathological manifestation, clinical characteristics, and desired therapy and transport operators; heart appropriate target drug must be selected and the properties of active targeting of these carriers must be regulated in order to improve the stability of the drug loaded (Jain, 2011).

Usually, cardiovascular diseases occur due to a restriction in blood perfusion. If neoangiogenesis occurs in the proximity of the ischemic site, this induces the delivery of nutrients and oxygen, which leads to a decreased cell death. Targeted drug delivery to the endothelium has been widely investigated. Heart target carriers have an important role in conserving pharmaceutical drugs from deteriorating in the blood flow (Liu et al., 2014b).

Intracoronary or intravenous (IV) delivery of small-scale molecules is the most used delivery system for the treatment of MI because is a minimal invasive method. However, this kind of method come with the downside of repeatedly dosing for a long time, in order to increase the concentration in the infarct zone leading to sever side effects. For achieve delivery of pharmaceutics to targeted site, specially designed surgical tools can be used to inject therapeutic drug directly into the myocardium. This method also shows a significant disadvantage: the local injection permit takeover of the drug in the blood flow very rapid, thus the pharmaceutical drugs are being altered and are unable to achieve localized delivery (Sy and Davis, 2010).

In order to overcome the problems, bioengineering has developed, evolved, and created new nano-sized tools. Pharmaceutical loaded nanoparticles have gained much interest since they are DDSs able to detect and distinguish early stages of the diseases.

Polymeric microparticles with encapsulated antiinflammatory pharmaceutics have been widely investigated in regard with intracardiac injection, and had promising results in clinical therapies of MI (Liu et al., 2014b).

In the treatment of cardiovascular diseases, the use of nanoparticles is more advantageous than microparticles due to their size, making them suitable for IV delivery. Capillaries have a diameter in the range of 5–6 μm. As the nanoparticles are injected into the blood flow, they have to be significantly smaller than 5 μm; otherwise, they can form clusters that can result in embolism (Dvir et al., 2011).

Magnetic nanoparticles have a substantial role in cardiovascular molecular imaging. These nanoparticles have the property of being superparamagnetic and are composed of an iron oxide core, which is encapsulated into a polymeric shell. Magnetic nanoparticles are appropriate in the use of myocardial injury due to their dimensions, physical, and pharmacokinetic properties (Sosnovik et al., 2008).

The formation of atherosclerotic plaque is mainly due to inflammation, which is composed of many processes that serve as targeting points for detecting specific diseases by molecular imaging. There is a vast number of imaging modalities used in cardiovascular treatment. However, they do present limitations. For example, cardiac MRI has been intensively used in cardiac dysfunction processes, such as cardiomyopathy, coronary artery disorders, and so on. Even though many types of nanoparticles have not been used in clinical trials, studies reported that they could substantially improve the applications of MRI in cardiac dysfunctions. For instance, there have been conducted studies on different nanoparticles for the detection of atherosclerosis in its early stages. The best results were attained from superparamagnetic iron oxide nanoparticles for the reason that they gather into the macrophages in atherosclerotic plaques, and therefore they can be observed by MRI. Moreover, superparamagnetic iron oxide nanoparticles have proven their applicability in myocardial angiogenesis and ischemia imaging (Kumar and Kumar, 2014).

Table 2.1 Nanoparticles Used in Cardiovascular Applications

Cardiovascular Applications	Types of Nanoparticles
MRI	Gold nanoparticles, nanoK, superparamagnetic iron oxide nanoparticles, liposomes
Positron emission tomography	Fe_3O_4 nanoparticles
Atherosclerosis	Magnetic-fluorescent nanoparticles, liposomes
MI	Liposomes
Cardiac arrhythmias	Superparamagnetic iron oxide nanoparticles
Aortic stenosis	Magnetic-fluorescent nanoparticles

Kumar, N., Kumar, R., 2014. Nanomedicine for treatment of cardiovascular diseases and stroke. Nanotechnology and Nanomaterials in the Treatment of Life-threatening Diseases, William Andrew Publishing, Oxford, (Chapter 5). Reprinted with kind permission from Elsevier.

Liposomes have also been studied as therapeutically nanosized vehicles for drug delivery and diagnostics. It has been concluded that liposomes are the most used nanosized carriers due to the fact that they have a size range of 50–1000 nm and can encapsulate several pharmaceutical drugs. Moreover, they exhibit suitable properties as drug carriers, such as biocompatibility, biodegradability, and reduced toxicity levels (Sawant and Torchilin, 2010) (Table 2.1).

Many applications in nanotechnology are currently under investigation in the concern of treating atherosclerosis with a special interest to mechanical stents, which contain nanosized components. This type of stents provided new approaches for drug delivery. The main challenge is to ensure revascularization of restricted arteries. Many industries have developed drug-eluting stents that were able to deliver drugs, such as paclitaxel. The main advantage of drug eluting stents, over classical metal stents, is the small incidence rate of restenosis (Godin et al., 2010). However, an issue that occurs at the use of drug eluting stents is a late stent thrombosis that could appear in prolonged time after the stent is implanted into the body, and is associated with the slow rate of curing of the endothelium that is followed by the denudation of the endothelium as a result of stent and the balloon with which is inserted. There have been studies that showed that 5 years after the drug eluted stents were introduced it remained un-endothelialized. As a comparison, it has been showed that classical metal stents were covered by new endothelium in less than 6 months after the stent was introduced in the body. Paclitaxel and sirolimus are the most common used drugs for drug eluting stents and they have the advantage of promoting smooth muscle cells proliferation and therefore preventing vascular restenosis (Bozsak et al., 2015).

Studies, that have been conducted in order to fabricate a computational model of a drug eluting stent which assumed that the arterial wall is composed from two porous sheets. The drug was delivered from the stent by means of diffusion. The results showed no negative effects on the endothelium in area were the stent was implanted. Moreover, the results were strictly dependent on the parameters of the drug, the release rate in the arterial wall and the primary stent concentration (Bozsak et al., 2014).

2.3 TARGETED DRUG DELIVERY FOR DIABETES

Diabetes has become thefifth cause of death worldwide, in the last decade (Rai et al., 2015), affecting almost 230 million people. The number is expecting to reach approximately 366 million by 2030

(Rahiman and Ba, 2012). It represents a chronic condition, expressing a metabolic disorder, which is characterized by not suitable fat, protein, and carbohydrate metabolism (Bittencourt and Hajjar, 2015). Diabetes is divided into two major categories: type 1 and type 2 diabetes. Type 1diabetes occurs when the body cannot produce insulin anymore and it is correlated with irregular hyperglycemia and ketosis. Type 2 diabetes represents the most common type that appears when insulin is not being used properly by cells. It usually occurs in overweight adults, but obese children are predisposed, as well. It has been assumed since 1921 that both types can be treated, as insulin became available. In order to obtain a better control of type 2 diabetes, hypoglycemic substances was used for oral administration. There were also some other approaches, which registered success, such as pancreas transplant or gastric bypass (Rai et al., 2015).

Even though there are medications for diabetes, it still cannot be cured because of several side effects that appear. Some of them are represented by gastric irritation or injection phobia, but much more examples exist (Rai et al., 2015). High levels of molecular oxygen or ROS provoke an imbalance in normal oxidative metabolism and this will lead to an abnormally high level of glucose in blood. Thus, oxidative stress is produced and chronic complications appeared. The use of insulin for therapy presents some negative aspects and in chronic treatment, it can produce brain atrophy or fatty liver (Rahiman and Ba, 2012).

A problem in diabetes also represents the fact that wounds are healing slowly, but it can be resolved by using nanoparticles. For example, yttrium oxide nanoparticles can save cells suffering from oxidative stress. It is important the particle structure, but not its size. This specific metal oxide presents no toxicity and it suffers small changes under temperature and pressure normal conditions (Rahiman and Ba, 2012).

Due to limited medication in diabetes treatment, delivery systems had to be designed to improve therapeutic efficiency. Recently, they gained more and more interest because of their advantages in terms of bioavailability, which is increased, reduced side effects, avoidance of degradation in gastric environment Those systems are made so that they can be able to deliver the specific drug safely and precisely to the target, for a well-established period of time in order to enhance better control and therapeutic effectiveness over diabetes disease (Rai et al., 2015).

Carbon nanotubes are used to frequently measure the production of insulin. Small samples are collected and insulin levels are measured. Carbon nanotubes represent tubular structures that consist in a sheet of graphite, which is rolled in a cylinder shape. They are divided into three categories, depending on the number of sheets used: single walled carbon nanotube (formed of only one sheet of graphite), double walled carbon nanotubes (two sheets of graphite), and multi walled carbon nanotubes (more than two sheets of graphite). The nanotubes can conduct electricity. Therefore, an electrode can be used to conduct the concentration of insulin. If the level of insulin molecules is high, the current coming from the sensor increases (Rahiman and Ba, 2012).

Medicines for diabetes adapt to each patient. Researchers are trying to bring advances in insulin therapy. The number of people diagnosed with diabetes is very high; therefore we see that insulin treatment helps to delay long-term complications of diabetes. Currently there are several kinds of formulations, of insulin and noninvasive approaches, for applying insulin having potential to be used for delivering the therapeutic levels of insulin in order to improve the control of sugar levels. The long-term medical benefits and cost effectiveness of these approaches remain to be seen. As such strategies continue to evolve and researchers find new approaches, insulin therapy will also evolve and insulin therapy for patients with type 1 or 2 diabetes may look quite different in the future (Holden et al., 2015).

Insulin remains one of the best means of balancing the intermediary metabolism disrupted in diabetes, in over 30%–40% of all patients. Insulin delivery systems form de Pombal, which are disposable, have been widely investigated. These types of pumps may provide an attractive opportunity to improve sugar control in the future, particularly for individuals with type 2 diabetes. Delivery of insulin through a pump is an easy and flexible method (Heinemann, 2007).

Researchers have developed a technique for blood distribution of medicines useful in the treatment of diabetes; the technique takes a sponge-like material core containing insulin. This method could be used in the case of medicine for cancer treatment (Lobenberg, 2003).

In conclusion, drugs can be introduced in the vascular system targeting an organ from the regional blood supply. To mitigate these impairments, controlled release of drugs were designed that reach the target and seek to destroy damaged cells (Hastings et al., 2015; Jain, 2011).

Cardiovascular diseases are a major cause of death worldwide. In the concern of overcoming the problems in the medical field, drug loaded nanoparticles have been realized due to their ability, as DDSs, to detect diseases in their infancy. Magnetic nanoparticles and liposomes are the most used nanoparticles in cardiovascular applications due to their unique properties and small sized ranges.

3 CONCLUSIONS

In conclusion, nanotechnology aims to spread its use in all medical areas (De Jong and Borm, 2008). Drug delivery technologies represent a vast, vital area of research and development in pharmaceuticals (Ranade, 1991). To manage and treat diseases, medicine is based on the use of medication, but its benefits are related to the drug delivery and its release. To improve the delivery of drugs to treat cancer, there have been designed nanoparticles, to target cancer cells.

There has been an increasing interest in targeted DDSs in the medical field due to the fact that they can specifically release their charge into desired sites without affecting the healthy tissue. For the treatment of a disease, physicochemical properties of controlled drug release systems are very important (Webster, 2006). In order to achieve an ideal targeted drug-delivery system there are some goals that should be taken into consideration: they have to be nonimmunogenic, nontoxic, and also be physically and chemically stable in the body. Smart DDSs have a significant advantage over classical DDSs: the property of responding to stimuli, no matter if they are external or internal.

Over the years, drugs were designed to improve health and extend lifetime. Researchers are trying to develop a large number of new ways to adjust DDSs. It is important to understand the transport method of the drug in the body. It passes through tissues and cells, being necessary to identify and destroy cancer cells without affecting healthy ones (Cho et al., 2008).

Within health systems, nanomedicine is a segment whose development and exploitation may lead to new solutions cost effective and cost efficient. In many application fields of nanomedicine, researchers already achieved promising results, which show very interesting premises for further research on the usefulness and applicability of nanotechnology in medicine. Market demand and pharmaceutical area are growing and nanomedicine can play an important role, contributing to economic growth, along with their great impact in medicine (Webster, 2006).

REFERENCES

Aguilar, L.K., Guzik, B.W., Aguilar-Cordova, E., 2011. Cytotoxic immunotherapy strategies for cancer: mechanisms and clinical development. J. Cell. Biochem. 112, 1969–1977.

Alvarez-Lorenzo, C., Concheiro, A., 2014. Smart drug delivery systems: from fundamentals to the clinic. Chem. Commun. 50, 7743–7765.

Arruebo, M., Fernández-Pacheco, R., Ibarra, M.R., Santamaría, J., 2007. Magnetic nanoparticles for drug delivery. Nano Today 2, 22–32.

Bae, Y.H., Park, K., 2011. Targeted drug delivery to tumors: myths, reality and possibility. J. Control. Release 153, 198–205.

Balamuralidhara, V., Kumar, T.M.P., Srujana, N., Venkatesh, M.P., Gupta, N.V., Krishna, K.L., Gangadharappa, H.V., 2011. pH sensitive drug delivery systems: a review. Am. J. Drug Discov. Dev. 1, 24–48.

Bamrungsap, S., Zhao, Z., Chen, T., Wang, L., Li, C., Fu, T., Tan, W., 2012. Nanotechnology in therapeutics: a focus on nanoparticles as a drug delivery system. Nanomedicine 7, 1253–1271.

Bañobre-López, M., Teijeiro, A., Rivas, J., 2013. Magnetic nanoparticle-based hyperthermia for cancer treatment. Rep. Pract. Oncol. Radiother. 18, 397–400.

Bittencourt, M.S., Hajjar, L.A., 2015. Insulin therapy in insulin resistance: could it be part of a lethal pathway? Atherosclerosis 240, 400–401.

Bozsak, F., Chomaz, J.M., Barakat, A.I., 2014. Modeling the transport of drugs eluted from stents: physical phenomena driving drug distribution in the arterial wall. Biomech. Model Mechanobiol. 13, 327–347.

Bozsak, F., Gonzalez-Rodriguez, D., Sternberger, Z., Belitz, P., Bewley, T., Chomaz, J.M., Barakat, A.I., 2015. Optimization of drug delivery by drug-eluting stents. PLoS One 10, e0130182.

Brannon-Peppas, L., Blanchette, J.O., 2012. Nanoparticle and targeted systems for cancer therapy. Adv. Drug Deliv. Rev. 64 (Supplement), 206–212.

Chakravarty, R., Hong, H., Cai, W., 2014. Positron emission tomography image-guided drug delivery: current status and future perspectives. Mol. Pharm. 11, 3777–3797.

Chen, X., Yao, X., Wang, C., Chen, L., Chen, X., 2015. Mesoporous silica nanoparticles capped with fluorescence-conjugated cyclodextrin for pH-activated controlled drug delivery and imaging. Micropor. Mesopor. Mater. 217, 46–53.

Cho, K., Wang, X., Nie, S., Chen, Z.G., Shin, D.M., 2008. Therapeutic nanoparticles for drug delivery in cancer. Clin. Cancer Res. 14, 1310–1316.

Cukierman, E., Khan, D.R., 2010. The benefits and challenges associated with the use of drug delivery systems in cancer therapy. Biochem. Pharmacol. 80, 762–770.

De Jong, W.H., Borm, P.J.A., 2008. Drug delivery and nanoparticles: applications and hazards. Int. J. Nanomed. 3, 133–149.

Dutta, S., Parida, S., Maiti, C., Banerjee, R., Mandal, M., Dhara, D., 2016. Polymer grafted magnetic nanoparticles for delivery of anticancer drug at lower pH and elevated temperature. J. Colloid Interface Sci. 467, 70–80.

Dvir, T., Bauer, M., Schroeder, A., Tsui, J.H., Anderson, D.G., Langer, R., Liao, R., Kohane, D.S., 2011. Nanoparticles Targeting the Infarcted Heart. Nano Lett. 11, 4411–4414.

Farokhzad, O.C., Langer, R., 2006. Nanomedicine: developing smarter therapeutic and diagnostic modalities. Adv. Drug Deliv. Rev. 58, 1456–1459.

Galvin, P., Thompson, D., Ryan, K.B., Mccarthy, A., Moore, A.C., Burke, C.S., Dyson, M., MacCraith, B.D., Gun'ko, Y.K., Byrne, M.T., Volkov, Y., Keely, C., Keehan, E., Howe, M., Duffy, C., MacLoughlin, R., 2012. Nanoparticle-based drug delivery: case studies for cancer and cardiovascular applications. Cell. Mol. Life Sci. 69, 389–404.

Ganta, S., Devalapally, H., Shahiwala, A., Amiji, M., 2008. A review of stimuli-responsive nanocarriers for drug and gene delivery. J. Control. Rel. 126, 187–204.

Giustini, A.J., Petryk, A.A., Cassim, S.M., Tate, J.A., Baker, I., Hoopes, P.J., 2010. Magnetic nanoparticle hyperthermia in cancer treatment. Nano Life 1.
Godin, B., Sakamoto, J.H., Serda, R.E., Grattoni, A., Bouamrani, A., Ferrari, M., 2010. Emerging applications of nanomedicine for the diagnosis and treatment of cardiovascular diseases. Trends in pharmacological sciences 31, 199–205.
Hastings, C.L., Roche, E.T., Ruiz-Hernandez, E., Schenke-Layland, K., Walsh, C.J., Duffy, G.P., 2015. Drug and cell delivery for cardiac regeneration. Adv. Drug Deliv. Rev. 84, 85–106.
He, X., Li, J., An, S., Jiang, C., 2013. pH-sensitive drug-delivery systems for tumor targeting. Ther. Deliv. 4, 1499–1510.
Heinemann, L., 2007. Future directions for insulin therapy and diabetes treatment. Endocrinol. Metabol. Clinics North America 36 (Suppl. 2), 69–79.
Holden, S.E., Jenkins-Jones, S., Morgan, C.L., Schernthaner, G., Currie, C.J., 2015. Glucose-lowering with exogenous insulin monotherapy in type 2 diabetes: dose association with all-cause mortality, cardiovascular events and cancer. Diabetes, Obes. Metab. 17, 350–362.
Jain, K., 2011. Applications of Biotechnology in Cardiovascular Therapeutics. Humana Press, New York, NY, USA.
Joglekar, M., Trewyn, B.G., 2013. Polymer-based stimuli-responsive nanosystems for biomedical applications. Biotechnol. J. 8, 931–945.
Kleinstreuer, C., Childress, E., Kennedy, A., 2013. Targeted drug delivery: multifunctional nanoparticles and direct micro-drug delivery to tumors A2. In: Kuznetsov, A.V., Becker, S.M. (Eds.), Transport in Biological Media. Elsevier, Boston, (Chapter 10).
Kumar, A., Zhang, X., Liang, X.-J., 2013. Gold nanoparticles: emerging paradigm for targeted drug delivery system. Biotechnol. Adv. 31, 593–606.
Kumar, N., Kumar, R., 2014. Nanomedicine for treatment of cardiovascular diseases and stroke. William Andrew Publishing, Oxford, (Chapter 5).
Lammers, T., 2013. Smart drug delivery systems: back to the future vs. clinical reality. Int. J. Pharm. 454, 527–529.
Langer, R., 1998. Drug delivery and targeting. Nature 392, 5–10.
Lee, B.K., Yun, Y.H., Park, K., 2015. Smart nanoparticles for drug delivery: boundaries and opportunities. Chem. Eng. Sci. 125, 158–164.
Li, J., Fan, C., Pei, H., Shi, J., Huang, Q., 2013. Smart drug delivery nanocarriers with self-assembled DNA nanostructures. Adv. Mater. 25, 4386–4396.
Li, L., An, X., Yan, X., 2015. Folate-polydiacetylene-liposome for tumor targeted drug delivery and fluorescent tracing. Colloids Surf. B 134, 235–239.
Liu, Y., Miyoshi, H., Nakamura, M., 2007. Nanomedicine for drug delivery and imaging: a promising avenue for cancer therapy and diagnosis using targeted functional nanoparticles. Int. J. Cancer 120, 2527–2537.
Liu, Y., Solomon, M., Achilefu, S., 2013. Perspectives and potential applications of nanomedicine in breast and prostate cancer. Med. Res. Rev. 33, 3–32.
Liu, J., Huang, Y., Kumar, A., Tan, A., Jin, S., Mozhi, A., Liang, X.J., 2014a. pH-sensitive nano-systems for drug delivery in cancer therapy. Biotechnol. Adv. 32, 693–710.
Liu, M., Li, M., Wang, G., Liu, X., Liu, D., Peng, H., Wang, Q., 2014b. Heart-targeted nanoscale drug delivery systems. J. Biomed. Nanotechnol. 10, 2038–2062.
Lobenberg, R., 2003. Smart materials: applications of nanotechnology in drug delivery and drug targeting. MEMS, NANO and Smart Systems. Proceedings. International Conference on, July 20–23, 2003. pp. 82–83.
Mahmoudi, M., Sant, S., Wang, B., Laurent, S., Sen, T., 2011. Superparamagnetic iron oxide nanoparticles (SPIONs): development, surface modification and applications in chemotherapy. Adv. Drug Deliv. Rev. 63, 24–46.
Maity, A.R., Stepensky, D., 2015. Delivery of drugs to intracellular organelles using drug delivery systems: analysis of research trends and targeting efficiencies. Int. J. Pharm. 496, 268–274.
Mody, V.V., Cox, A., Shah, S., Singh, A., Bevins, W., Parihar, H., 2013. Magnetic nanoparticle drug delivery systems for targeting tumor. Appl. Nanosci. 4, 385–392.

Murugan, M., Anthony, K.J.P., Jeyaraj, M., Rathinam, N.K., Gurunathan, S., 2014. Biofabrication of gold nanoparticles and its biocompatibility in human breast adenocarcinoma cells (MCF-7). J. Ind. Eng. Chem. 20, 1713–1719.

Park, J., Park, J., Ju, E.J., Park, S.S., Choi, J., Lee, J.H., Lee, K.J., Shin, S.H., Ko, E.J., Park, I., Kim, C., Hwang, J.J., Lee, J.S., Song, S.Y., Jeong, S.-Y., Choi, E.K., 2015. Multifunctional hollow gold nanoparticles designed for triple combination therapy and CT imaging. J. Control. Rel. 207, 77–85.

Piao, Y., Burns, A., Kim, J., Wiesner, U., Hyeon, T., 2008. Designed fabrication of silica-based nanostructured particle systems for nanomedicine applications. Adv. Funct. Mater. 18, 3745–3758.

Rahiman, S., Ba, T., 2012. Nanomedicine current trends in diabetes management. Nanomed. Nanotechol. 3, 137.

Rai, V.K., Mishra, N., Agrawal, A.K., Jain, S., Yadav, N.P., 2015. Novel drug delivery system: an immense hope for diabetics. Drug Deliv., 1–20.

Ranade, V.V., 1991. Drug delivery systems 5A. Oral drug delivery. J. Clinic. Pharmacol. 31, 2–16.

Rani, K., Paliwal, S., 2014. A review on targeted drug delivery: its entire focus on advanced therapeutics and diagnostics. Sch. J. App. Med. Sci. 2 (1C), 328–331.

Riehemann, K., Schneider, S.W., Luger, T.A., Godin, B., Ferrari, M., Fuchs, H., 2009. Nanomedicine—challenge and perspectives. Angew. Chem. Int. Ed. 48, 872–897.

R. Rajeswari, R.J., 2014. Magnetic nanoparticles as drug carriers: review. Mater. Sci. Forum 807, 1–12.

Sawant, R.R., Torchilin, V.P., 2010. Liposomes as "smart" pharmaceutical nanocarriers. Soft Matter 6, 4026–4044.

Scott, R.C., Crabbe, D., Krynska, B., Ansari, R., Kiani, M.F., 2008. Aiming for the heart: targeted delivery of drugs to diseased cardiac tissue. Expert Opin. Drug Deliv. 5, 459–470.

Sharma, H., Mishra, P.K., Talegaonkar, S., Vaidya, B., 2015. Metal nanoparticles: a theranostic nanotool against cancer. Drug Discov. Today 20, 1143–1151.

Shen, J.-M., Gao, F.-Y., Yin, T., Zhang, H.-X., Ma, M., Yang, Y.-J., Yue, F., 2013. cRGD-functionalized polymeric magnetic nanoparticles as a dual-drug delivery system for safe targeted cancer therapy. Pharmacol. Res. 70, 102–115.

Shen, Y.Q., Pu., M.L., Jiang, W., 2014. Polymer-tethered pH-sensitive magnetic nanoparticles for targeted delivery of anti-cancer drug. Adv. Mater. Res., 936.

Sosnovik, D.E., Nahrendorf, M., Weissleder, R., 2008. Magnetic nanoparticles for MR imaging: agents, techniques and cardiovascular applications. Basic Res. Cardiol. 103, 122–130.

Sy, J.C., Davis, M.E., 2010. Delivering regenerative cues to the heart: cardiac drug delivery by microspheres and peptide nanofibers. J. Cardiovasc. Transl. Res. 3, 461–468.

Tietze, R., Zaloga, J., Unterweger, H., Lyer, S., Friedrich, R.P., Janko, C., Pöttler, M., Dürr, S., Alexiou, C., 2015. Magnetic nanoparticle-based drug delivery for cancer therapy. Biochem. Biophys. Res. Commun. 468, 463–470.

Torchilin, V., 2010. Passive and active drug targeting: drug delivery to tumors as an example. In: Schäfer-Korting, M. (Ed.), Drug Delivery. Springer, Berlin Heidelberg.

Webster, T.J., 2006. Nanomedicine: what's in a definition? Int. J. Nanomed. 1, 115–116.

Xie, X., Li, F., Zhang, H., Lu, Y., Lian, S., Lin, H., Gao, Y., Jia, L., 2016. EpCAM aptamer-functionalized mesoporous silica nanoparticles for efficient colon cancer cell-targeted drug delivery. Eur. J. Pharma. Sci. 83, 28–35.

Yang, C., Guo, W., Cui, L., An, N., Zhang, T., Lin, H., Qu, F., 2014. pH-responsive magnetic core–shell nanocomposites for drug delivery. Langmuir 30, 9819–9827.

Yang, Y., Yu, C., 2016. Advances in silica based nanoparticles for targeted cancer therapy. Nanomedicine 12 (2), 317–332.

Yih, T.C., Wei, C., 2015. Nanomedicine in cancer treatment. Nanomedicine 1, 191–192.

Zhang, W., Zhang, Z., Zhang, Y., 2011. The application of carbon nanotubes in target drug delivery systems for cancer therapies. Nanoscale Res. Lett. 6, 1–22.

Zhong, J., Li, L., Zhu, X., Guan, S., Yang, Q., Zhou, Z., Zhang, Z., Huang, Y., 2015. A smart polymeric platform for multistage nucleus-targeted anticancer drug delivery. Biomaterials 65, 43–55.

CHAPTER

FUNCTIONALIZATION OF NANOPARTICLES IN SPECIFIC TARGETING AND MECHANISM RELEASE

3

Giorgia Adamo, Simona Campora, Giulio Ghersi
University of Palermo, Palermo, Italy

CHAPTER OUTLINE

1 INTRODUCTION

The current focus of nanotech research is to design a nanocarrier that can specifically deliver drugs to the site of disease. In fact, the conventional therapies are characterized by limited effectiveness, poor drug biodistribution, and lack of selectivity. All these disadvantages and limitations can be overcome by regulating drug delivery. This concept is applied in modern therapies, based on drug-delivery systems (DDS), in which the drug is transported in the body by controlling the rate, time, and place of release, improve its efficacy and safety, and minimize undesirable side effects. In addition, DDS protects the drug from fast degradation or clearance and enhances drug efficiency in target tissues. The NPs, structures with size in the range of 1–100 nm in at least one dimension, have a great potential as carriers for drug delivery, even if they must have basic features: they must be biocompatible (able to integrate them with a biological system, eluding immune response) and nontoxic (Wilczewska et al., 2012). Indeed, their physicochemical properties can be finely engineered by customizing the chemical compositions, sizes, shapes, structures, morphologies, surface properties, and stability. In general, the release of drug by NPs can be regulated through different routes: by diffusion of the therapeutic molecules through the matrix of the nanocarriers or by their differential surface and bulk erosion. Alternatively, drug release can be triggered by specific microenvironment changes in the body (pH, temperature, and enzymatic activities) or manipulated by external events (electric field, magnetic field, and ultrasound). These nanoconstructs should serve as targeted drug-delivery vehicles able to carry high doses of chemotherapeutic agents into malignant cells while sparing healthy cells, by using active or passive mechanisms. For tumor cells, the passive targeting occurs by the NP's accumulation in the leaky tissues of tumors, through the enhanced permeability and retention (EPR) effect and it is largely determined by the physicochemical properties of NPs and their long circulation half-life (Farokhzad and Langer, 2009). While, the active strategy relies on the nanocarrier's attraction to the affected site by using recognition ligands, attached to the NPs-surface, such as monoclonal antibodies, carbohydrates (CHs), glycoprotein, aptamers, and small molecular ligands (folic acids, peptides) (Jaracz et al., 2005). More recently, targeted NPs have been employed in solving the complex problems of multidrug resistance.

In this chapter, we focus on the principal classes of NPs (i.e., polymeric NPs, nanogels, liposomes, dendrimers, lipid–polymer hybrid NPs, and magnetic NPs), their rule in targeted and controlled delivery system.

2 CONTROLLED-RELEASE MECHANISMS

Controlled drug release offers multiple benefits compared to the conventional administration based on free drugs. Over the past few decades, different NP platforms have been developed to enhance the pharmacological properties and therapeutic index of many types of drugs. Controlled release, combined with nanosystems, have been devised to transport drugs with high loading efficiency, to assist them in crossing physiological barriers, to shield them from premature elimination, coming intact to the target site, while minimizing drug exposure elsewhere in the body.

To design NPs as vehicles for delivery, the in-depth knowledge of features of the drug play a key role, and permit the achievement of the best-loading efficiency and optimum release profile.

A drug can be adsorbed or covalently attached to the NPs surface or else it may be encapsulated inside (Liu et al., 2006). Various methods of drug-loading have been developed to improve the efficiency, which depends upon the method of NP preparation, as well as the physiochemical properties of the drug and the nanocarrier both. Maximum loading can be obtained by incorporating the drug during the synthesis of NPs, but procedural parameters, such as method of preparation, presence of additives, and step of purification may affect it. The DDS based on NPs offers other attractive features, including a better protection of drug from harsh environments. Different studies, indeed, have demonstrated a higher drug half-life in the presence of low pH values (stomach or intracellular compartments, such as lysosomes) or of proteolytic enzymes in the plasma (Couvreur, 2013). Commonly, it is possible to catalog the drugs into different classes in relation to their aqueous solubility (hydrophobic and hydrophilic drugs) or to their electrostatic properties (highly charged or neutral drugs) (Sun et al., 2014). Many anticancer drugs, widely used for treating various types of cancers, are hydrophobic, such as paclitaxel, methotrexate, cisplatin, doxorubicin, and others. These molecules may not be soluble enough to cross the aqueous extracellular environment and then pass through the plasma membrane of target cells. Moreover, their strong tendency to aggregate in the bloodstream can lead to complications, such as embolisms and local toxicity (Owen et al., 2012). Several nanocarriers were designed for delivering

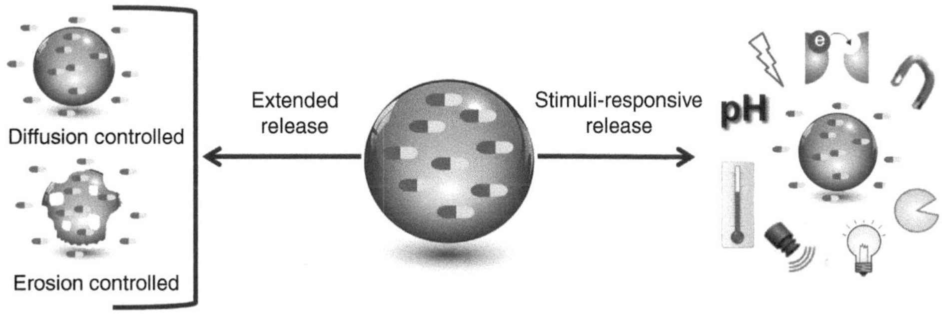

FIGURE 3.1 General Scheme for Extended and Stimuli-Responsive Release of Active Agents from Nanocarriers

hydrophobic drugs, especially those based on micelles and polymeric NPs composed of hydrophobic or amphiphilic material. On the contrary, hydrophilic drugs, including biomacromolecules, such as proteins, peptides, and nucleic acids and many small molecules, also play an important role in treating various types of diseases. For example, monoclonal antibodies (such as trastuzumab, cetuximab, bevacizumab) are now commonly used in cancer treatment; or nucleoside analogs, for instance gemcitabine or 5-fluorouracil, had been used to treat bladder, pancreas, ovarian, and breast cancers (Torchilin et al., 2003). A special class of hydrophilic drugs are represented by siRNA, micro-RNA, and DNA, with high densities of charges, highly used in gene therapy (Dispenza et al., 2014). It must be highlighted that there has been no clinical success in gene therapy, since the systemic administration of these hydrophilic drug is impeded due to rapid clearance by the mononuclear phagocyte system (MPS) and kidney filtration (Ku et al., 2014). In this case, the drug loading occurs through electrostatic interactions between the nucleic acids (negatively charged) and nanocarriers (positively charged), such as liposomes or polymeric NPs, consisting of cationic building blocks (Woodrow et al., 2009). Alternatively, the direct conjugation of gene-drug to the NPs surface has been proved as another effective approach. Mirkin and coworkers have coated AuNPs with a layer of short DNA chains containing specific sequences, able to recognize and bind a target siRNA involved in the development of specific diseases (Seferos et al., 2007). However, the use of hydrophilic drugs can induce a lower cellular uptake level due to the hydrophobic nature of plasma membrane. Moreover, they show reduced bioavailability and low stability in the presence of proteolytic enzymes and hydrolytic phenomena, as well as a short half-life in the bloodstream (Torchilin et al., 2003). Also in this case, the drug loading increases, but considering the hydrophobic nature of many nanomaterials used for fabricating the nanocarriers, it is not always straightforward due to the poor miscibility between these two phases. Several research groups have solved these problems making structural drug changes that make them amphiphilic, or adjusting the pH value of the external aqueous phase to the isoelectric point of protein-based drug, that could increase the drug loading (Leo et al., 1998). However, the NP-based systems for controlled release can be enclosed into two major categories: extended and stimuli-responsive release (Fig. 3.1).

2.1 EXTENDED RELEASE

The aim of the extended release is to deliver drugs over a sustained period, at determinate rates. It permits to keep stable the amount of drug in the bloodstream or in the specific target sites by regulating the balance between the drug release rate and its clearance. This approach is advantageous for those therapeutic agents that, after administration, show a faster catabolism. The extended release can maintain constant, the drug concentration in the plasma or target tissue by matching the rate of release and clearance of drug (Robitzki and Kurz, 2010).

One possible approach to obtain the extended release is to prevent the contact of drug in the aqueous body fluid, for determined times, by controlling the degradation rate of the nanocarrier or the drug diffusion through an insoluble polymer shell or matrix.

In the diffusion-controlled release, the drugs inside the NPs are protected from the aqueous environment by an insoluble barrier, usually made of an organic polymer. The diffusion-controlled release can be realized using a core that contains the drug with a membrane surrounding it: the drug molecules can only diffuse out through the membrane at a determined rate, depending on its physicochemical properties, such as porosity or thickness. For instance, it was demonstrated that Fe_3O_4 hollow NPs (pores size of 3 nm) are able to gradually delivery cisplatin through diffusion-controlled release, by maintaining unchanged the porous shell, over time (Cheng et al., 2009). Currently, the nanocarriers most used in this approach are made of a water-insoluble polymer, such as poly(methyl methacrylate) (PMMA) and polyurethane (PUA). These nanosystems show an initial substantial release (namely "burst release"), due to the detachment of drug molecules adsorbed on the NPs surface. Next, the drug release will be delayed due to diffusion across the polymer matrix, up to the surface. In one study, it was shown that the docetaxel-loaded NPs, based on poly-ε-caprolactone (PCL)-Tween 80 copolymer, released only 35% of the encapsulated drugs, over a period of 28 days (Ma et al., 2011).

Alternatively, another approach is the erosion-controlled release, in which nanocarriers are made of erodible or degradable polymers that do not require changes or clearance mechanisms, after the drug is completely released (Chen et al., 2006).

The drug-release profile can be controlled by matching it with the erosion kinetics of the polymeric matrix. Many biodegradable polymers, both natural and synthetic, have been used to formulate erodible polymeric NPs. For example, poly(glycolic acid) (PGA), poly(lactic acid) (PLA), and poly(lactic-*co*-glycolic acid) (PLGA) are some synthetic polymers that release the preloaded drugs for several weeks. Furthermore, in the body, these polyesters can be hydrolyzed breaking the ester linkage between glycolic and lactic acid and expelled as carbon dioxide and water (Göpferich, 1996). The degree of polymer hydrophilicity defines the erosion level in the hydrolytic process, depending on the closer presence of proton donors and acceptors, and on the affinity of the polymeric matrix for water molecules. The polymer degradation, that occurs by surface or bulk erosion, can be determined by adding of acidic or basic excipients (Välimaa and Laaksovirta, 2004).

The surface way happen if the water molecules go inside the polymer slower than the erosion speed. This permits the achievement of a reproducible system, especially by maintaining constant the surface area of nanocarriers or by using a very thin polymer to obtain the steady-state release of drug. Bulk erosion takes place when water molecules can quickly penetrate the polymer and the chain scission occurs throughout the matrix. For instance, polyesters that work by bulk erosion, release the drug through three steps: in the first, the drug adsorbed was released from the surface or from pores connected to the surface; in the second step, the polymer gradually degrades and releases the remaining drug, at a slow rate; at the end the polymer matrix is completely destructed so that the leftover drug is released rapidly. One example is the bioerodible polyanhydride, that was used as a pH-sensitive nanocarrier of *p*-nitroaniline, in which the high hydrophobicity retards water permeation (Sun et al., 2014).

2.2 STIMULI-RESPONSIVE RELEASE

Stimuli-responsive NPs are specialized active delivery carriers that react with an external signal, with "load-and-release" features within their constituting units. These nanovectors are developed to respond to specific stimulus of chemical, biochemical, or physical origin, modifying their structural composition or conformation, thereby triggering release of the drug to the target site. These changes include the isomerization, polymerization, decomposition, the activation of supramolecular aggregation, and others (Fleige et al., 2012).

The general concept of the stimuli-responsive release, according to the nature of the interaction between the drug and the NPs, can be achieved by a complexation approach, where the therapeutic agent is entrapped within the nanocarrier and the release can be triggered by structural change within the carrier network. While, in the nanocarrier-conjugate approach, the mechanism of release involves the cleavage of a linker between the carrier and the drug.

The classification of biological stimuli that can trigger the delivery of drugs, genes, or diagnostic agents from the NPs are various and manifold in nature. Though overlapping in many instances, the stimuli can be broadly classified with respect to the biological systems as either internal/biochemical (exploiting physiological, pathological, and pathochemical conditions) or external (which are also classified as physical stimuli).

Biochemical stimuli include pH, redox, and ionic variations in the microenvironment of specific tissues and in the enzyme overexpression in certain pathological states. Although, physical stimuli that can be applied to trigger drug release may involve temperature, light, and strength of magnetic or electrical fields.

2.2.1 pH-responsive nanocarriers

A well-established fact is that in a pathological situation, the normal pH value existing between extra and intracellular microenvironment undergoes changes. Inflamed, infected, or cancerous tissues usually show lower pH values. For example, in solid tumors, the extracellular pH can be more acidic (about 6–7) than physiological pH (7.4), due to poor vasculature and consequent anaerobic conditions (Vaupel et al., 1989). Moreover, some cellular organelles also exhibit sharp pH differences, for instance, between the cytosolic, endosomal, and lysosomal compartments. In particular, after cellular internalization via endocytosis, the nanocarriers may route through various defined compartments with different pH, such as the early endosome that has a pH of about 5–6, while the late lysosome, which has the most acidic environment, has a pH of around 4–5 (Mura et al., 2013). These extra/intracellular pH gradients can be exploited to design a controlled release system that triggers in the target site. Several synthetic polymers were developed as pH-sensitive nanocarriers, such as poly(acrylamide) (PAAm), poly(acrylic acid) (PAA), poly(methyl acrylate) (PMA), and poly(diethylaminoethylmethacrylate) (PDEAEMA), that can perceive such pH variation and modify their structure and hydrophobicity as a result of protonation or deprotonation of internal functional groups or due to the cleavage of the shell (Bawa et al., 2009). Polymersomes (Ps), self-assembly polymeric nanocarriers, have attracted attention because, under acidic conditions, they can alter their conformation or solubility upon protonation/deprotonation. Moreover, it was described that the solubility change, depending on pH, can be used to form the polymersomes: an aqueous solution of poly(butadiene)-*b*-poly(L-glutamic acid) (PBD-*b*-PGA) formed well-defined polymersomes, upon the addition of NaOH. Furthermore, a so-called "schizophrenic" polymersome, was constituted by poly(L-glutamic acid)-*b*-poly(L-lysine) (PGA-*b*-PLL), which, depending on the pH changes, had one or the other polymer block as corona.

In another work, the protonation of a PEG-block-poly(2-vinylpyridine) copolymer, which shows a sharp transition at pH of 4.9, results in a burst-drug release in earlier stages of endocytosis (Meng et al., 2009). In vivo studies show that, upon the combinatorial encapsulation of the drugs doxorubicin and taxol, following the pH-sensitive degradation of the polyester block, it induced the shrinkage of a fast growing tumor in nude mice, after a single injection (Ahmed et al., 2006). The pH-sensitive release of guest molecules from a poly(propylene imine) (PPI) dendrimer was reported by Pistolis et al. (1999) who demonstrated that the protonation of tertiary amines of a PPI dendrimer facilitated the release of encapsulated pyrene molecules. Another example sees the attachment of small fluorinated alkyl chains on the outside of poly(amidoamine) (PAMAM) dendrimers led to self-assembling pH-sensitive NPs, which furthermore could be used in vivo via noninvasive ^{19}F magnetic resonance imaging (Criscione et al., 2009).

Another approach is to conjugate the drug through pH-sensitive linkages that can be incorporated into nanocarrier structures and then broken to release the therapeutic molecules, under specific stimuli. Some of these pH-responsive cleavable linkers, include imine, ketal, oxime, citraconylamide, hydrazone, hydrazide, and acetal, have been adopted to bind drug molecules to a chemical group of nanocarriers. Ríhová and Kubácková (2003) developed a polymeric nanosystem named HPMA (*N*-(2-hydroxypropyl) methacrylamide), functionalized with a hydrazone bridge to link doxorubicin that is released at pH of 5 (Ulbrich et al., 2004). It was seen that the same linker permits a drug release also if conjugated with dendrimers: the drug is rapidly released at low pH reaching 100% within 48 h (Lee et al., 2006). A new interesting class of nanocomposites is the so-called dendrimersomes, which are amphiphilic molecules bearing a dendron. Recently, a whole library of dendrimersomes was developed, containing different variants that were connected via a diacetal linkage, to develop pH-sensitive nanodevices for drug delivery (Percec et al., 2010). Another example was represented by an inulin-based random copolymer bearing high dose doxorubicin, that was prepared by coupling doxorubicin with inulin, though citraconylamide bridge used as pH-sensitive spacer (Mauro et al., 2015). Furthermore, it was seen that the pH responsiveness of nanogels is based on the change of their ionization state that leads to a swelling/deswelling, which can be used to release or entrap various molecules. While, a nanogel useful for endosomal release was synthesized by copolymerization of 2-(*N*,*N*-diethylamino)ethyl methacrylate (EAMA) with heterobifunctional poly(ethylene glycol) bearing a 4-vinylbenzyl and a carboxylic acid group, in the presence of a ethylene glycol dimethacrylate cross-linker. The doxorubicin loaded to this nanogel showed a 3 times enhanced release at pH 5.3 compared to the release at physiological pH (Oishi et al., 2007).

2.2.2 Redox-responsive nanocarriers

Redox-responsive nanocarriers are a promising approach for intracellular delivery, considering the difference of redox potential (~100–1000-fold) existing between the extracellular and intracellular space. It has been well established that extracellular environment is oxidative while the intracellular one is reductive, in accordance with the glutathione (GSH) concentration. In fact, the cytosolic concentration of GSH is higher (1–10 mM) than the micromolar range present in the extracellular space or in the plasma. Moreover, the level of cytosolic GSH in cells with an enhanced oxidative stress (such as tumor cells) were found to be 7–10-fold higher than in normal cells (West and Otto, 2005). For instance, doxorubicin-loaded poly(*N*-vinyl pyrrolidone) (PVP) nanogels for glutathione-mediated delivery were generated. These nanogels, synthetized by pulsed electron-beam irradiation of semidilute poly(*N*-vinyl pyrrolidone)/acrylic acid aqueous solutions, have been conjugated to doxorubicin, through a linker containing a cleavable disulphide bridge, aminoethyldithiopropionic acid (AEDP). The controlled release mechanism proposed relies on the cleavage of the bridge presented in the spacer interposed between the nanogels and the drug (Adamo et al., 2014). The disulfide bond can be broken down into two sulfhydryl moieties and the GSH is oxidized to GSSG, which can be achieved either by cell-surface protein disulfide isomerase (PDI) or reductive cleavage in the cytoplasm. In vitro release studies showed that GSH is able to trigger an enhanced release of doxorubicin at a concentration comparable to its levels in cytosol (10 mM).

This controlled release system is especially effective for gene delivery. In fact, for DNA or siRNA delivery, it is crucial that they are protected outside the cell and released after the cell-internalization. Carlisle and coworkers made a direct complex of a plasmidic DNA with linear PEI (cross-linked with a disulfide or with a thiol group), which resulted in degradable disulfide or stable thioether bonds. The disulfide containing complexes showed a 40–100-fold higher transfection efficiency than the stable complexes (Neu et al., 2007). In another approach, reversible disulfide cross-linked micelles were synthesized to trigger the drug release in cancer cells: thiolated linear dendritic polymers (telodendrimers) were generated by introducing cysteines to the dendritic oligo-lysine bone, composed of linear PEG and a dendritic cluster of cholic acids. Reversible cross-linked micelles were then produced through a disulfide bond with the self-assembly of thiolated telodendrimers. The resulting micelles are loaded with paclitaxel (PTX) in the core. The release of PTX from disulfide cross-linked micelles was drastically slower than that from noncross-linked ones. Moreover, the micelles showed a higher in vivo blood circulation time, low hemolytic activities, and superior toxicity profiles in nude mice, when compared to noncross-linked species (Li et al., 2011). The disulfide bridge is the most used linkage to generate redox-sensitive nanocarriers. However, there are some disadvantages about the in vivo stability, because GSH and cysteine are also present outside the cells and they can cause an early breakdown of the nanocarriers. This could be overcome by using multiple disulfide bonds, by controlling the amount of disulfide cross-links due to high degrees of cross-linkage (Miyata et al., 2004).

2.2.3 Enzyme-responsive nanocarriers

Enzymes perform a wide range of important biochemical functions. They regulate our metabolism, can break specific bonds causing disassembly or destruction of different protein structures, or they can link polymer chains together with covalent bonds, to generate self-assembled structures. Additionally, the overexpression of a specific enzyme associated with a pathological condition can be used to achieve enzyme-mediated drug release. The controlled-release by enzyme-responsive nanocarriers is based on the cleavage of short peptide sequences by esterases or proteases and can be achieved by introducing specific enzyme substrate sequence either into the nanocarrier scaffold or in the linker segment by which the drug is conjugate to the nanocarrier (Law and Tung, 2009). One of the most used linkers is the tetrapeptide Gly–Phe–Leu–Gly, which can be cleaved by the cathepsin B enzyme that degrades proteins in lysosomes and is overexpressed in tumor cells. Many polymeric NPs used copolymers as backbone and the Gly–Phe–Leu–Gly linker for the attachment of various active agents, such as doxorubicin, diaminoplatin, and diaminocyclohexylplatin (Duncan, 2009). Many NPs can be used to develop enzyme-responsive nanosystems, such as nanogels, micelles and liposomes, porous silica NPs with an enzyme-sensible polymer coating as the gatekeeper. In recent studies, a micelle capable of changing its morphology, in response to certain enzymes, was reported. These micelles present specific peptide sequences, which are known to be substrates for particular enzymes in the hydrophilic block. This hydrophilic block could be phosphorylated by protein kinase A and dephosphorylated with phosphatase-1, leading micelles to a reversible dimensional change, or could be selectively cleaved by matrix metalloproteinases (MMPs), which modify them irreversibly (Ku et al., 2011). Tumor-associated enzymes have been an interesting class for designing these responsive nanocarrier systems. In fact, the extracellular proteases, such as the MMPs, are specific biomarkers of malignant tissues and are responsible for the proteolysis of the extracellular matrix and basement membranes, required during tissue remodeling, and angiogenesis. For instance, it was reported that short peptides with sequences cleavable by MMPs have been used as linkers between PEG chains and liposomes, polymeric NPs, or iron oxide NPs (Zhu et al., 2012). In particular, a bioresponsive cisplatin (CDDP) delivery system with a self-assembling peptide amphiphile (PA), with a MMP-2-sensitive sequence, was produced. In vitro results show that MMP-2 is able to trigger the cisplatin release from such biomimetic gels and this release was dependent on the enzyme concentration in the medium (Kim et al., 2009). Shabat and coworkers developed a so-called "self-immolative dendrimer": after the enzymatic cleavage of substrate, there is a complete dendrimer degradation, in a cascade of reactions into small building blocks, which simultaneously released upto three different drug molecules (Haba et al., 2005). Moreover, there have been numerous reports on enzyme-responsive hydrogels using enzyme substrate as cross-linker.

2.2.4 Thermoresponsive nanocarriers

Some pathological conditions, such as tumor, inflammation, or infection show a local temperature increase. Nanocarriers containing thermoresponsive elements are advantageous for constructing smart DDS, which reversibly respond to changes in temperature, within a very narrow range. Thermoresponsive nanocarriers are constituted by temperature-sensitive materials able to change their physical properties and trigger drug release, following a local variation of temperature close to the vector. To develop these kinds of systems, it is opportune to use a range of temperatures between 37 and 42°C, because higher or lower temperatures could cause protein denaturation or loss of function (Aluri et al., 2009). With the onset of advanced research in the field of hyperthermia and their purported advantage in increasing vascular perfusion and permeability, thermoresponsive nanocarriers have become an attractive candidate for designing therapeutic vehicles, usually made of liposomes or NPs composed of thermosensitive polymers. For liposomes, thermo responsiveness usually depends on the properties of the constituent lipids, which undergo conformational structural changes associated with variations in the permeability of the lipid bilayers. Doxorubicin loaded to thermosensitive liposomes (ThermoDox, Celsion Corporation) are in phase II trial for the breast and colorectal cancers therapy, and have overcome phase III trial for the treatment of hepatocellular carcinoma (Al-Ahmady et al., 2012). More recently, a thermoresponsive bubble-generating liposomal system has been developed, which exploits the generation of CO_2 bubbles as a disruptive force to filtrate the lipid bilayer membranes, through fast decomposition of ammonium bicarbonate, upon heating to 40°C. While, thermosensitive polymers undergo to a phase of transition if they are in a lower critical solution temperature (LCST) or an upper critical solution temperature (UCST), such as critical temperatures which the polymer is totally miscible with the solvent. (Chen et al., 2014). Therefore, the polymer, at LCST, is hydrophobic and undergoes a conformational change from the extended (soluble) to the globular (insoluble) state. Some of the commonly used thermosensitive polymers include: poly(*N*-vinylcaprolactam) (PVCL), poly(*N*-isopropylacrylamide) (PNIPAAm), poly(methyl vinylether) (PMVE), poly(*N*,*N* diethylacrylamide) (PDEAAm), and poly(ethyleneoxide)-poly(propylene oxide) poly(ethylene oxide) (PEO–PPO–PEO, also named Pluronics) (Martin et al., 2009). Block-copolymers of two thermoresponsive monomers allow not only to set the LCST to a selected point, but also to form micelles in water. For instance, PEG-*b*-PNIPAAm copolymers, synthesized by atom transfer radical polymerization (ATRP), formed micelles when heated above their LCST. Other copolymers, for example, a PNIPAM-*b*-poly(acrylic acid) (PNIPAM-*b*-PAA), was synthesized through radical addition fragmentation transfer polymerization. Another copolymer was obtained by condensation of both polymer parts, by activating the terminal PMMA end group and linking it with an amine from the PNIPAM polymer. The micelles resulting (PNIPAM-*b*-PMMA) showed accelerated drug release when heated above the LCST of 38°C (Wei et al., 2006). Self-assembled thermoresponsive NPs prepared from poly(*N*,*N*-diethylacrylamide-*co*-acrylamide)-block-poly(γ-benzyl L-glutamate) were designed to release their cargo at the onset of hyperthermia (c. 40–45°C). The LCST of NPs was set at the physiological body temperature (37°C) and the local hyperthermia (about 43°C) was used as stimulus. The in vivo application of thermoresponsive nanocarriers may be difficult because they might start to flocculate or even precipitate once heated above the LCST. To prevent this flocculation, Zheng and coworkers attached poly(dimethylaminoethylmethacrylate) PDMAEMA and poly(ethyleneoxide) (PEO)-block-poly(tetrahydrofuran) arms to a Boltorn H_20 core. The PEO avoids the nanocarrier aggregation and precipitation, after heating it over its LCST (Zheng et al., 2010). Other examples are thermoresponsive nanogels that can shrink or expand upon heating, depending on their composition. This system realizes an on–off release kinetic, due to the close packing of the fully collapsed gel, especially at the surface. An example is the interpenetrating polymer network (IPN) of PNIPAAM and poly(teramethylene ether glycol) (PTMEG), that release the incorporated indomethacin only at lower temperatures. The temperature of response of these systems can be obtained by copolymerization with other monomers, such as acrylic acids, making them pH sensitive (Yin et al., 2006). Recently, a new method to target solid tumors using a temperature-responsive macromolecular drug carrier, an elastin-like polypeptide (ELP), was reported. The authors showed that the ELP forms microaggregates that adhere to the tumor vasculature only when tumors are heated to 41.5°C. The results indicated that thermal cycling enhanced the drug exposure of tumor cells (Dreher et al., 2007).

2.2.5 Photoresponsive nanocarriers

The use of light as an external stimulus offers a range of advantages, including ease of application and the spatial and temporal control. The photoresponsiveness is usually archived by functional groups or linkers, which can change their conformations or other properties upon irradiation by light with a proper wavelength or by destabilizing the NPs by molecular switch. Its high resolution on spatiotemporal control is governed by the fact that the encapsulated drugs are released after they have been irradiated with a light source from outside of the body. To prevent an early release, the patient would have to stay in the dark for a certain period of time, also known as "dark toxicity" which is one of the major obstacles of photodynamic therapy (Jiang et al., 2005). Some groups, and their derivatives, include nitrobenzene, azobenzene, cinnamoyl, pyrene, and spirobenzopyran. For example, the reversible molecular switching of the azobenzene group from *trans* to *cis* can be achieved by UV irradiation (300–380 nm) and from *cis* to *trans* by visible light, allowing for photoresponsive control. Therefore, the hydrophobic–hydrophilic transition, resulting by a *trans–cis* switching, has been used for the synthesis of phototriggered nanodrugs. In a study, an amphiphilic block-copolymer of PEO and a hydrophobic poly(methacrylate), decorated with a pyrene derivative, was described. Upon UV light irradiation, the pyrene groups were cleaved turning the hydrophobic core of the formed micelles into hydrophilic PMA, resulting in distraction of the micelle and release of encapsulated Nile red (Jiang et al., 2005). Besides these systems involving structural changes, UV light has also been used to trigger the drug release by introducing a photocleavable linker in the polymer backbone, among which the most used is the 2-nitrobenzyl ester linker. The conjugation of a PAA block through a nitrobenzyl linker to a poly(methyl caprolactone) block allows the formation of photoresponsive polymersomes, which successfully encapsulated fluorescein and released it when irradiated (Cabane et al., 2011). A combination of sugar-based targeting and two-photon sensitivity within the same nanocarrier was described. In such constructs, the light-responsive groups can be modularly conjugated and/or altered. Kabanov and coworkers described a very attractive concept for designing photosensitive polymeric nanocarriers through the electrostatic association between an azobenzene containing surfactant and the double-hydrophilic block ionomer poly(ethylene glycol)-*b*-poly(acrylic acid) (Wang et al., 2010). Such complex self-assembled in aqueous solution and formed vesicle-like aggregates composed of a poly(ethylene glycol) corona and a poly(acrylic acid) shell attached to azobenzene. The photoisomerization of azobenzene, controlled by light irradiation, reversibly altered the amphiphilicity of the surfactants that induce the disassembly of the vesicles. The oligoaminesbis-(3-aminopropyl)methylamine (AMPA) and pentaethylenehexamine (PEHA) were conjugated by photocleavable 2-nitrobenzyl ester (a photoswitchable azobenzene unit), to the polyglycerol core. The cleavage was complete after 2 min upon irradiation at 350 nm, which shows the high potential of light-sensitive, multivalent amine functionalized polymers for drug and gene-delivery applications (Fischer et al., 2011).

Despite many reports on these light-sensitive nanosystems, the superficial penetration depth (<200 mm) caused by the strong scattering of soft tissues has restricted these UV-based systems for in vivo applications. Moreover, even for in vitro applications, the UV light causes genetic damages to the cells. An alternative might be the near-infrared (NIR, 700–900 nm) irradiation, that can penetrate more deeply into soft tissues, without causing damage in the area of application.

Currently, the capability of plasmonic NPs to transduce the absorbed NIR light into heat has been adopted to trigger the release of drug molecules. Moreover, it was reported that a photothermal effect of Au nanorods upon NIR irradiation can cause a fast increase of local temperature, which can be exploited to induce denaturation of DNA helices attached to the nanocarriers and thereby release drug molecules bound to consecutive cytosine–guanine base pairs (You et al., 2012).

2.2.6 Sonic and ultrasonic-responsive nanocarriers

Applying ultrasound to trigger drug release is an interesting possibility, since it allows spatial and temporal control and it is already used as a biomedical imaging method. Usually, the ultrasonic-responsive nanocarriers, most used, are constituted mainly by PEG-lipid based liposomes. Novel ultrasound-responsive doxorubicin (DOX)-loaded NPs were developed, using poly(D,L-lactide-*co*-glycolide)-methoxypoly(ethylene glycol) (PLGA-mPEG), encapsulating doxorubicin with a high efficiency (Du et al., 2011). Moreover, ultrasound-mediated delivery of genetic material from microbubbles was used as a method for gene delivery to vascular smooth muscle cells in vitro, using ultrasound that triggers the delivery of a plasmidic DNA, which was linked by electrostatic interactions. These microbubbles were acoustically ruptured near smooth muscle cells, under a range of acoustic pressures (0–950 kPa) and pulse durations (0–100 cycles) (Phillips et al., 2010).

Wu et al. has synthesized a novel fluoride anion-modified gelatin nanogels and investigated its characteristics for ultrasound-sensitive drug release applications. Their results show that the ultrasound-triggered drug release was most efficient at a frequency of 20 kHz and 51.5% of drug was released under this condition within 1–2 min, with respect to 8.2% of drug that was released from nanogels within 2–3 min without sonication (Wu and Wan, 2008).

2.2.7 Other stimuli-responsive nanocarriers

Over the main classes of stimulus-responsive nanocarriers mentioned earlier, there still exists a wide range able to respond to other stimulus and trigger a controlled release. For instance, electrical field mediated drug release is another attractive approach for designing stimuli-responsive drug-delivery nanocarriers. The generation of nanoscale thin films that can release precise quantities of drug in response to application of a small, anodic electric potential of at least +0.5 V was described. These films bring Prussian Blue (PB) NPs (negatively charged) and gentamicin (positively charged). When the PB NPs were oxidized, they shifted from negatively charged to neutral, inducing dissolution of the film. While, the control of the drug released was achieved by tuning the film thickness, as well as the magnitude of the applied voltage. Kinetics analysis of drug release ranging from triggered burst release to on/off, or pulsatile release, were obtained by applying different electric profiles (Schmidt et al., 2010).

Magnetic field-responsive nanocarrier, generally made of paramagnetic or super-paramagnetic materials either embedded into a polymeric scaffold forming liposome, micelle, or supramolecular aggregates, represents another category. Besides, alternating magnetic fields can be used to control timing and dose of repeatedly released drug from such NPs. Recently, it was described that a siRNA–polyethylene glycol conjugates with embedded magnetite nanocrystals, that formed stable nanosized polyelectrolyte complexes (siRNA–S–S–PEG) bearing a cleavable disulfide linkage. These complexes were efficiently internalized by cancer cells upon exposure to a magnet, by enhancing intracellular uptake and silencing effect of siRNA (Lee et al., 2010).

When the interaction between the nanocarrier and the target is based on coordinative bonds, the release can be triggered by competitive ligands or ions. For example, PEG–PAsp and PEG–PLglu micelles have been used to carry the cisplatin molecule, which is coordinated to the carboxyl groups, presented in the core of the micelle. No release was observed using a noncompetitive ligand, while the cisplatin release was about 50% both from PEG–PAsp (after 24 h) and from PEG–PGlu (after 72 h) in phosphate buffered saline containing chloride ions which are better ligands for cisplatin with respect to carboxyl groups. Thus, these micelles can be considered as ion-responsive nanocarriers.

Furthermore, the trigger release through metabolic molecules has always been extremely interesting for designing responsive nanocarriers. For example, a glucose-responsive nanocarrier was generated by incorporating the enzymes, glucose oxidase (GOD) and glucose catalase (CAT) with imine bonds in a nanocapsule, containing an insulin solution. It was seen that, upon treatment with glucose solution, the capsule disintegrates due to the conversion of glucose to gluconic acid, thereby releasing the insulin (Qi et al., 2009). At last, dual-responsive nanocarriers are a new way for designing stimuli-sensitive DDS. These systems can respond to two different stimuli, such as pH and ionic strength or thermoresponsiveness. Most often, these dual-responsive systems are produced by the combination of two monomers having responsiveness to different stimuli. Ultrasonic/pH or magnetic/thermocoupled responsive nanocarriers have attracted much attraction because one stimulus can be used to load the NPs with the drug while a second one can be used to trigger the release.

Generally, it is possible to obtain dual-responsive systems for all types of stimuli. This can be achieved by choosing the specific-related monomers and combining them with the appropriate linkers. However, the complexity of dual-responsive systems increases and certainly complicates the phase of the synthesis and the clinical approval of these systems.

3 PROJECT OF NANOPARTICLES

NPs need to be synthetized, engineered, and optimized to increase the circulating half-life and to reach site-specific release of drugs at therapeutically optimal rates and dose regimes. The size, shape, surface properties, composition, biocompatibility, and degradation profile should be carefully considered for the optimal design of the NPs for therapeutic therapy.

3.1 SIZE AND SHAPE

The NP size plays a key role in their long circulation, biodistribution, and clearance. Typically, they have a size between 1 and 100 nm that permits a longer circulation half-life in vivo and experience reduced hepatic filtration. Furthermore, their small size allows a higher intracellular uptake compared with microparticles as reported by Desai et al. (1997) who carried out uptake studies in a Caco-2 cell line by contrasting 100-nm NPs with 1 and 10-μm microparticles: NPs had a 2.5- and 6-fold greater uptake than microparticles.

NPs with size smaller than 10 nm can easily pass through the blood vessels and be cleared by kidneys, while larger NPs can be captured by cells of the mononuclear phagocyte system (MPS). NPs can also be used for brain diseases, such as Parkinson, Alzheimer, or glioma, thanks to their ability to cross the blood–brain barrier (BBB) by passing the opening of tight junctions treated with a hyperosmotic mannitol solution (Kroll et al., 1998). Magnitude can also affect the drug release because the smaller particles present a larger surface area so that most of the drugs associated would be at or near the particle surface, leading to fast drug release.

The shape of NPs (sphere, cube, rod, plate) plays, also, a fundamental role in internalization and drug release processes. Different studies show that spherical particles are good candidates for drug delivery because they can be internalized with high likelihood and faster than rod-shaped NPs (Chithrani et al., 2006). Champion and Mitragotri focused their studies on the importance of the rod NPs orientation to the cellular membrane: in fact, they proposed that these kind of particles have a higher probability of internalization when their major axis is perpendicular to the cell membrane (Champion and Mitragotri, 2006). In some cases, the larger ratios of surface area to volume of anisotropic structures determine higher efficiencies in drug delivery. Shape can also influence their accumulation site in different tissues of the organism: for example, cylindrical and regular spherical NPs gather in the liver; discoidal-shaped NPs in the heart; and irregular spherical NPs in the spleen (Park et al., 2009; Devarajan et al., 2010).

3.2 SURFACE PROPERTIES

The hydrophobic character of the NPs can reduce the circulation time in vivo because of their opsonization: small molecules, called opsonins, adsorb to the surface of NPs increasing phagocytosis by marking an antigen for an immune response. Even the surface charge could concern their uptake by the MPS cells, in fact, positively charged NPs induce higher complement activation compared to negatively and neutrally charged NPs. In particular, nanocarriers with a zeta potential above (+/−) 30 mV have been shown to be stable in suspension and prevent the aggregation process. Therefore, it is necessary to prolong the circulation time of the particles by directly synthetizing NPs from block polymers presenting both hydrophobic and hydrophilic fragments (Harris et al., 2001; Adams et al., 2003) or by coating them with hydrophilic protective layer like polyethylene glycol (PEG) (Mohanraj and Chen, 2006; Kamaly et al., 2012; Xia et al., 2012). Different researches demonstrated the effect of PEG configuration on the NPs surface: a brush-like conformation limits phagocytosis and complements activation; while a, mushroom-like arrangement favors both of the processes (Yang and Ma, 2010). Even though the presence of PEG shell on the NPs increases their half-life in bloodstream, it can interfere with the interaction with the target cells and hinder the endosomal escape, as demonstrated by many studies comparing the effect of the same NPs PEGylated or not (Du et al., 1997). For this reason, the researchers are prompting new strategies for increasing the circulation life of NPs including the use of synthetic or natural polymers, or coating of biomimetic molecules, or red blood cell (RBC) membranes to evade the immune response cascade (Sambruy et al., 2001).

It is also possible to increase the NPs uptake by coating them, for example, with chitosan then help to disrupt the integrity of intracellular tight junctions (TJs) involved in paracellular transport. In particular, it modifies the cytoskeletal arrangement in the apical membrane and this induces the opening of TJs that facilitate the delivery of higher amount of nanocarriers to the targeted site (Luk et al., 2014).

4 NANOPARTICLES TARGETING

One of the major problems of the classical pharmacological treatments is the induction of diffused side effects due to their nonspecific action because they can act both on normal and specific (like tumor) cells. Therefore, the use of NPs for DDS can resolve this problem because they can be synthetized and/or functionalized for a specific targeting.

The physicochemical features of NPs, such as size, shape, surface charge, surface properties, and conjugation with specific biological molecules, can all be used to improve the delivery of drug to the target tissue through two different ways: passive and active targeting.

4.1 PASSIVE TARGETING

In passive targeting, the location of the particle is entrusted only to morphological, chemical, and physical properties of the carrier and the anatomical and physiological characteristics of the target site. Small molecules, including NPs, cannot cross the tight junctions between endothelial cells of normal vascular linings, so they are retained in the bloodstream. Tumor tissue needs extensive vascularization to deliver sufficient supplies of nutrients and oxygen; therefore, the rapid growth lends itself to a leaky and defective architecture. In fact, tumor microvasculature is characterized by abnormal branching and enlarged interendothelial gaps due to damage of TJ between endothelial cells and a disrupted basement membrane. Consequently, NPs with right size can efficiently cross through the large gaps and extravasate from the surrounding vessels to the malignant tissue via a passive targeting mechanism named EPR (Enhanced Permeation and Retention) effect. This phenomenon is very different between several tumors but also between various parts of the same tumor in relation to the dissimilar size of the gaps that is typically between 100 and 800 nm. Furthermore, the lymphatic drainage inside the malignant microenvironment is dysfunctional or totally absent and induces the accumulation of the NPs conjugated with drugs. In this way, concentration of drugs released from vectors in the malignant site can reach levels 10–100 times higher than that obtained from the administration of free drug with the classical pharmacological treatments. EPR effect was described, for the first time, in 1986 by Matsumara and

Maeda who conducted localization experiments by comparing the effect of a polymer conjugated with an anticancer molecule named poly(styrene-*co*-maleic acid)-Neo Carrzino Stain (SMANCS) (16 kDa) and the complex formed by its noncovalent bind with Albumin (SMANCS-Albumin) (80 kDa). Using labeled Albumin, they demonstrated that proteins larger than 30 kDa could preferentially accumulate into tumor interstitium and remain there for prolonged period of time (Matsumura and Maeda, 1986). For this reason, NP size plays a fundamental role in EPR-based tumor targeting because only the NPs smaller than the gap between adjacent endothelial cells can extravasate from the bloodstream. In particular, different studies show that particles with a diameter smaller than 200 nm work most effectively because the larger NPs were cleared faster from the bloodstream by the spleen (Aruna et al., 2013). Size is also important for the opposite process because they can travel back to the blood vessels: NPs with a size range of 30–200 nm present most excellent retention by the tissue resistance that shifts the equilibrium toward the extravasation and so induce their accumulation in the tumor site. In this context, the shape also plays a key role associated with tumor interstitial pressure in the malignant interstitium that takes position against the extravasation process. Many studies demonstrated that spherical shape could help follow a laminar flow pattern so that in close contact with the vessel wall, it can cross the gaps and go into the tumor microenvironment; while rod or bar-shaped NPs are hydrodynamically more unstable and present high difficulty to follow a laminar flow pattern and to linger into the bloodstream.

Since EPR effect is due to the high permeability of tumor vasculature, different vascular mediators, such as NO, CO, VEGF, and others, can increase this phenomenon. Nitric oxide (NO) is a signaling messenger involved in various pathways and produced by NO synthase that is more expressed in tumor cells than in healthy ones, indicating that tumors produce more NO involved probably in tumor growth as demonstrated by NOS knockout mice experiments (Kisley et al., 2002). In particular, NO scavenger inhibits intensely the extravasation process, suggesting its involvement in EPR effect and same result can be obtained by inhibition of oxidized products of NO. In addition, carbon monoxide (CO) plays an important role in this mechanism: in cancer sites, CO induces angiogenesis and acts as a vasodilator increasing vascular permeability and blood flow as shown in experiments in which the production of CO is induced by using a CO donor and this determines an increase of EPR effect. Although often EPR effect promotes NPs targeting tumor site and it is widely exploited for cancer therapy, passive targeting efficiency will be rather limited because of the nature of the tumor. For example, small tumor or metastatic lesion does not present strong and diffused angiogenesis, so that vessels are not evenly distributed and size of gaps can be different in several parts of the same malignant site. Hence, it is necessary to improve specific targeting by conjugating specific molecules for an active targeting.

4.2 ACTIVE TARGETING

Tumor cells overexpress many molecules in their membrane that can link growth factors or other kinds of antigens and induce different tumorigenic pathways, such as angiogenesis or growth metabolic pathways. This characteristic is fundamental for an active targeting using NPs conjugated to a targeting moiety that can purposely bind to the surface of specific types of cells, such as tumor cells. Beyond the function in the identification of the malignant cells, active targeting, also known as ligand-mediated targeting, facilitates cellular uptake as a result of receptor-mediated endocytosis, but only after that, NPs are in tumor microenvironment.

Therefore, active and passive targeting can be used in an independent way or, also, combined to improve the NPs efficiency (Fig. 3.2). The conjugation process is very complicated because it is

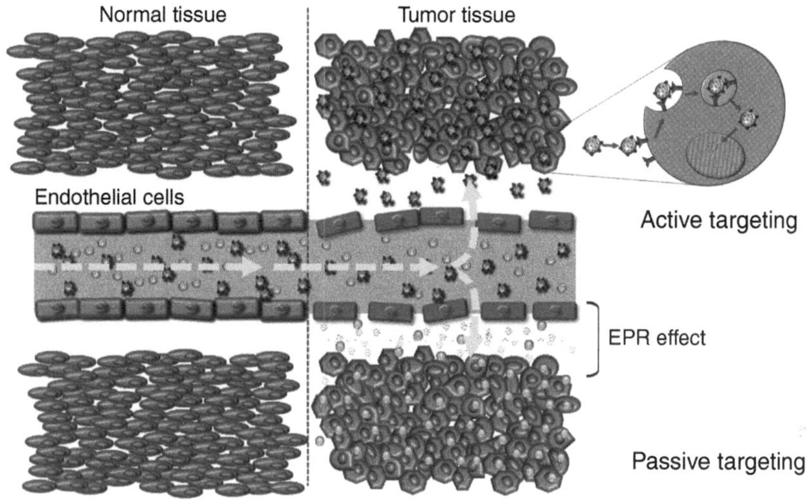

FIGURE 3.2 Schematic Presentation of the Two Strategies (Passive and Active Targeting) Adopted for Targeting Nanocarriers to Tumor Tissues

In active targeting, the ligand-conjugated nanoparticles *(NPs)*, interact with the receptors present on the tumoral target cell or tissue, resulting in cellular uptake through receptor-mediated endocytosis. In passive targeting, NPs extravasate through the leaky vasculature and preferentially accumulate through the enhanced permeability and retention effect (*EPR* effect). In this case the drugs may be released in the extracellular matrix and diffuse throughout the tissue.

very important to keep unchanged efficiency of the system that can be modified by physicochemical properties, such as size and shape of nanomaterials, but also density, orientation, or charge of ligands (Bertrand et al., 2014).

The attachment strategies depend on the NPs nature: usually it is preferable a covalent bond of the ligand, but, sometimes, it is better a physical adsorption as in case of liposomes in which it is necessary to use a bulkier hydrophobic anchor to be insert into the lipid bilayer ligands to increase their stability especially in vivo. Furthermore, some inorganic polymers require the interposition of functional groups like NH_2 or OH to bind not only targeting ligands, but also the dye molecules for localization studies and/or drugs.

In this context, density and orientation of ligands and their affinity for the substrate play a key role in particular for cellular uptake, because the multiple interactions of NPs to target membrane induce the clustering of receptors that facilitate endocytosis. In other situations, high density of ligand conjugated can improve the macrophage internalization and sequester them to tumor cells, therefore it is very important to identify the optimal concentration.

Independently of conjugation approaches used, NPs can be functionalized using targeting ligands of different nature, such as aptamers, small molecules, peptides, antibodies, CHs, or glycoproteins independent by the characteristics of the target cells (Fig. 3.3).

On the other hand, some data also suggest an alternative targeting mechanism named active vascular targeting (Neri and Bicknell, 2005). In this case, NPs are bound with specific molecules to bind to angiogenic endothelial cell-surface receptors including fibronectin (extradomain B), integrins, annexin

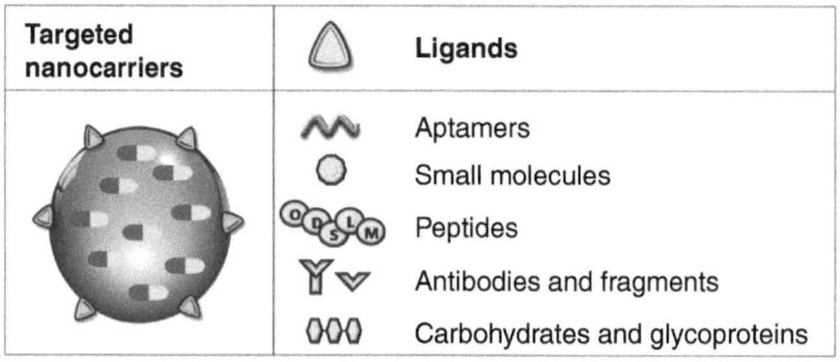

FIGURE 3.3 Representation of Targeted NPs and Their Ligands, Such as Aptamers, Antibodies, Peptides, Carbohydrates, and Various Other Small Molecules

A1, and growth factors or their receptors like VEGFs (vascular endothelial growth factors) and VEG-FR. Nanosystems are also linked to specific drugs or biological molecules that induce apoptotic events on target cells so as to kill endothelial cells that cannot supply oxygen and nutrients to tumor: in this way cancer mass is eradicated.

4.2.1 Aptamers

Aptamers (Apts) are small single-stranded DNA or RNA sequences of oligonucleotides (20–80 nucleotides) that can fold into unique tertiary conformations capable of binding proteins, phospholipids, acid nucleic, or CHs expressed by target cells. Furthermore, they are nonimmunogenic and are stable so that they can tolerate small variations of temperature or pH and these characteristics make them optimal candidates for cancer targeting if conjugated with NPs. Thanks to the development of a specific chemical process known as "Systemic evolution of ligands by exponential enrichment" (SELEX), it is possible to select and amplify the aptamers with higher binding affinity for the target molecules starting from a random library of single-stranded DNA or RNA (Daniels et al., 2003). Different studies reported this technology to identify specific aptamers to conjugate NPs for drug delivery, such as fluoropyrimidine RNA that bind PSMA (prostate-specific membrane antigen) on prostate cancer cells (Swami et al., 2012). In some cases, it is also possible to improve the half-life of aptamer–NPs in bloodstream and delay their clearance by the kidneys by functionalizing with PEG or cholesterol (Nimjee et al., 2005).

It was identified a role of Apts conjugated with nanosystems as pharmaceutical agents because some of them can link and inhibit proteins involved in pathological pathways. For example, Pegaptanib sodium aptamer recognize VEGF or AS1411 aptamer bind a protein preventing its link with Bcl-2 mRNA and, in this way, the AS1411–NPs complex can induce apoptotic events (Bates et al., 2009). However, because of their high cost of production, it is preferable to employ Apts only as targeting agents. Another possible purpose can be their use as antidotes for anticancer drugs to regulate antitumoral effects because they can inhibit cytotoxic activity of some molecules.

4.2.2 Small molecules

Small molecules are identified as organic molecules with a molecular weight less than 500 Da. Their small size, together with other properties, such as high stability, chemical management, and low cost production, confer them the fundamental characteristics to be an optimum candidate as molecules for specific targeting if conjugated with NPs. Indeed, they present a high binding affinity for different substrates of interest. The most famous example is folic acid (FA), also named vitamin B9, that is involved in different metabolic pathways, such as DNA or RNA synthesis and cellular proliferation. It is able to bind, with very high specificity and affinity, its receptor (folate receptor, FR) that is overexpressed in majority of human tumors including brain, colon, ovarian, breast, renal, and lung cancers; although FR expression seems to be different in relation with the patient and with the kind or status of cancer. When FA binds its receptor, the complex is internalized into endocytic vesicles and fused to lysosomes, the presence of acid pH induces the release of conjugate FA from FR. In this context, FA conjugates escape the endosome and this permits the deposit of the drug into the cytoplasm where it can act. Some experimental results suggest a higher uptake of FA–NPs compared with NPs alone (Zhang et al., 2015) and so a higher cytotoxic effect when the nanosystem is conjugated, also, with an anticancer drug capable of killing cells.

TPP (triphenyl-phosphonium) represents another example of small molecules involved in targeting process tanks to its properties. In particular, it is a hydrophobic relatively large and positively charged molecule that can cross the mitochondrial membrane and accumulate some hundred folds inside this cellular compartment (Smith et al., 2003). Different studies demonstrated that TPP can in part confer positive charge to NPs that, in this mode, can escape the endosomal compartment and accumulate in mitochondria more than NPs alone when their size was around 100 nm and their zeta-potential was more than +22 mV (Marrache and Dhar, 2012).

4.2.3 Peptides

Unlike proteins, peptides that have linear or cyclic sequence of amino acid (not more than 50 residues), can be used as targeting moieties because they can fold into three-dimensional structures conferring high stability and resistance to the immune system.

The most used is peptide family RGD (arginine-glycine-aspartic acid) like cRGD (cyclic RGD). It can recognize and bind to $\alpha_V\beta_3$ integrin receptors expressed on both tumor and angiogenic endothelial cells and, also, on normal or inflamed tissues even if at low levels: its nonspecific activity is the only limitation observed. In particular, some data demonstrated that cyclic form presents 10-fold higher anticancer activity (Colombo et al., 2002). Furthermore, $\alpha_V\beta_3$ is typically associated with VEGFR-2, so that the binding of protein with RGD sequence to its receptor induces upregulation of VEGF signaling. By using inhibitor of $\alpha_V\beta_3$ binding site there is a reduction of VEGF signaling so that it can be employed for an antiangionesisis mechanism (Ruoslahti, 2002). In this context, the use of cRGD conjugated with NPs induced two advantages: it can improve specific targeting but it can also, indirectly, downregulate the VEGF pathway and induce an antiangiogenetic effect (Blanco et al., 2012). Several types of nanosystems were synthetized using RGD as targeting molecule and a variety of moieties as therapeutic agents among which drugs, such as doxorubicin or paclitaxel, but also siRNA to block mRNA involved in tumor growth pathways.

Han et al. (2010) performed studies using a system composed by RGD peptide-labeled chitosan NP (RGD–CH–NP) as a tumor-targeted delivery system for siRNA (siRNA/RGD–CH–NP). They show that the presence of cyclic peptide significantly increased selective intratumoral delivery in orthotopic animal models of ovarian cancer that overexpress integrin $\alpha_V\beta_3$ (Han et al., 2010).

4.2.4 Antibodies and their fragments

The most common class of targeting moieties used is certainly composed by monoclonal antibodies (mAb) because of their very high specificity of action and capacity of binding strongly the complementary antigen expressed by targeted cells.

Antibodies are very large glycoproteins (150 kDa) and their principle role is to recognize antigens from the circulation, facilitating the macrophages activity, therefore, they confer a high immunogenicity to the NPs after conjugation. For this reason, is diffused the use of engineered antibodies, to obtain Ab-fragments that can escape the immune system. In particular, it is necessary to delete Fc portion that confers higher immunogenicity to get only Fab region able to recognize specific antigen. Single-chain variable fragments (scFv) give one example. They are fusion proteins composed by the variable regions of both heavy and light chains of the Ab linked with a short peptide of around 15 amino acid. In this way, thanks to low molecular weight and absence of Fc fragment, it is resistant to MPS activity.

MCC-465 and STG-53 are two examples of Ab-fragment-targeted NPs: the first one shows excellent results in preclinical studies in doxorubicin delivery for stomach tumor cells (Jain, 2010); while the second one is able to bind transferrin receptors (Tf-R) on cancer cell membrane and release into the cytoplasm a p53 inhibitor (Nemunaitis and Nemunaitis, 2008).

Despite these challenges, since the 1980s, more than 30 different types of mAb and their derivatives have been accepted for clinical use, such as cetuximab, rituximab, bevacizumab, or trastuzumab. The last one, for example, is an antihuman epidermal receptor 2 (HER-2) mAb that was used to produce PTX conjugated NPs for the active targeting of ovarian cancer cells that overexpress HER2 receptors. Specific targeting induced by mAbs or scFVs are demonstrated by Chen and coworkers in their studies about mammalian carcinoma. Using magnetic IONPs (iron oxide NPs) conjugated with HER2 antibody (Herceptin) or with a single-chain fragment (ScFv) of antibody against epidermal growth factor receptor (ScFvEGFR), they demonstrated specific efficiently targeting breast cancer cells that overexpress HER2 or EGFR, respectively and IONPs alone cannot go into the same cells (Chen et al., 2013).

4.2.5 Carbohydrates and glycoproteins

CHs are very important mediators of cell–cell interactions; in fact, they are involved in processes, such as cellular differentiation, cell signaling regulation, and immune response. They present high affinity of link for lectins, cellular membrane proteins expressed ubiquitously. Different studies reported that tumor cells contain lectins similar in size and sugar-binding specificity to the lectins of the normal cells so that CHs can be used as targeting ligands (Lotan and Raz, 1988). In 2002, Seymour and coworkers evaluated one of the first actively targeted antitumor nanotherapy in human using doxorubicin conjugated NPs carrying galactosamine as targeting molecule. The hexosamine can bind specific membrane receptors into primary and metastatic liver cancer cells and permit NPs uptake via receptor-mediated endocytosis that was not observed in normal cells (Seymour et al., 2002). Beyond galactose, it is documented as an active targeting of nanosystems also by exploiting other CH moieties, such as mannose or glucose and their derivatives. Nevertheless, CHs are not good targeting ligands because they are difficult to synthetize and, because of their labile glycosidic linkages, can be easily degraded by enzymes. On the other hand, glycoproteins, such as Lactoferrin (Lf) and Transferrin (Tf) are widely employed for this purpose, even if Lf–NPs present high targeting efficiency for tumor cells compared with Tf–NPs as reported by Choe et al. (1998) in their uptake and transfection studies using DNA encapsulated liposomes targeted to cervical cancer Hela cells. Other experimental results investigated the role of Lf-nanocarriers that were efficiently taken up by colon cancer (Caco-2) and did not alter the mucosal integrity during transcytosis. In particular, if these nanosystems were given orally to mice before Caco-2 cell injections, none of the mice fed with nanodiet would have developed tumors, in contrast to the mice fed with the control diet that showed normal tumor growth (Kanwar et al., 2011).

Another important limit of the use of classical chemotherapeutic agents is the transport across the blood–brain barrier (BBB) to act against neural cells. Since many transporters are expressed at the BBB, it is possible to use an active targeting to permit the NPs to reach the target site in the brain. For example, BBB express transferrin receptors so that Tr–NPs can cross the barrier by receptor- or transporter-mediated endocytosis. Once there NPs functionalized with lactoferrin can target, for example, to glioma cells that overexpress Lt receptor as reported by two very similar studies about glioma treatment with Lt-liposomes (Pang et al., 2010; Chen et al., 2011). Liposomes were also conjugated with two different drugs that are Doxorubicin (DOX) and Tetrandrine (TEX) and the data revealed that the complex NPs–Lt–Drug showed an increased rate of 2.3 times more than the unconjugated TEX- and DOX-loaded polymersomes.

5 CONCLUSIONS

The use of nanocarriers as DDS improve the pharmacological and therapeutic properties of traditional drugs. In this context, it is of fundamental importance that the drug is transported and released close to the diseased tissue in a specific manner, both temporally and spatially. Therefore, the need of developing smart and innovative nanosystems has become quite critical, because they are more effective, safer, selective, and they can also reduce the toxicity and side effects in normal tissues, by accumulating drugs in target sites. Stimuli-responsive abilities that can be incorporated into the nanocarrier scaffolds depend on the architecture and chemical composition of NPs. Each selected modality or a combination of them may have their own advantages and/or disadvantages to consider for nanocarriers design. As reported in this chapter, different studies are focusing on the use of several types of nanosystems for cancer treatments thanks to continuous advances in identifying new cancer biomarkers as targeting agents. In particular, NP-based therapeutic agents are currently being developed and are under preclinical evaluation, for example, liposomal conjugates: Doxil (doxorubicin) or DaunoXome (daunorubicin).

In conclusion, the potential to generate a wide range of nanomaterials with tunable properties that can interact in a unique fashion with biological systems is providing exciting new opportunities for designing advanced NPs for therapeutic applications.

REFERENCES

Adamo, G., Grimaldi, N., Campora, S., Sabatino, M., Dispenza, C., Ghersi, G., 2014. Glutathione-sensitive nanogels for drug release. Chem. Eng. Transact. 38, 457–462.

Adams, M.L., Lavasanifar, A., Kwon, G.S., 2003. Amphiphilic block copolymers for drug delivery. J. Pharm. Sci. 92, 1343–1355.

Ahmed, F., Pakunlu, R.I., et al., 2006. Shrinkage of a rapidly growing tumor by drug-loaded polymersomes: pH-triggered release through copolymer degradation. Mol. Pharm. 3 (3), 340–350.

Al-Ahmady, Z.S., Al-Jamal, W.T., et al., 2012. Lipid-peptide vesicle nanoscale hybrids for triggered drug release by mild hyperthermia in vitro and in vivo. ACS Nano 6 (10), 9335–9346.

Aluri, S., Janib, S.M., et al., 2009. Environmentally responsive peptides as anticancer drug carriers. Adv. Drug Deliv. Rev. 61 (11), 940–952.

Aruna, U., Rajalakshmi, R., Muzib, Y.I., Vinesha, V., Sushma, M., Vandana, K.R., Kumar, N.V., 2013. Formulation and evaluation of Felodipine mouth dissolving tables by using solubility enhancing technique. Int. J. Innov. Pharm. Res. 4, 318–324.

Bates, P.J., Laber, D., Miller, D.M., Thomas, S.D., Trent, J.O., 2009. Discovery and development of the G-rich oligonucleotide AS1411 as a novel treatment for cancer. Exp. Mol. Pathol. 86, 151–164.

Bawa, P., Pillay, V., et al., 2009. Stimuli-responsive polymers and their applications in drug delivery. Biomed. Mater. 4 (2), 022001.

Bertrand, N., Wu, J., Xu, X., Kamaly, N., Farokhzad, O.C., 2014. Cancer nanotechnology: the impact of passive and active targeting in theera of modern cancer biology. Adv. Drug Deliv. Rev. 66, 2–25.

Blanco, M.D., Teijón, C., Olmo, R.M., Teijón, J.M. 2012. Targeted Nanoparticles for Cancer Therapy, Recent Advances in Novel Drug Carrier Systems. In: Sezer, A.D., (Ed.). InTech. DOI: 10.5772/51382. Available from: http://www.intechopen.com/books/recent-advances-in-novel-drug-carrier-systems/targeted-nanoparticles-for-cancer-therapy.

Cabane, E., Malinova, V., Menon, S., Palivan, C.G., Meier, W., 2011. Photoresponsive polymersomes as smart, triggerable nanocarriers. Soft Matter 7, 9167–9176.

Champion, J.A., Mitragotri, S., 2006. Role of target geometry in phagocytosis. Proc. Natl. Acad. Sci. USA 103 (13), 4930–4934.

Chen, H., Qin, Y., Zhang, Q., Jiang, W., Tang, L., et al., 2011. Lactoferrin modified doxorubicin-loaded procationic liposomes for the treatment of gliomas. Eur. J. Pharm. Sci. 44, 164–173.

Chen, H., Wang, L., Yu, Q., Qian, W., Tiwari, D., Yi, H., Wang, A.Y., Huang, J., Yang, L., Mao, H., 2013. Anti-HER2 antibody and ScFvEGFR-conjugated antifouling magnetic iron oxide nanoparticles for targeting and magnetic resonance imaging of breast cancer. Int. J. Nanomed. 8, 3781–3794.

Cheng, K., Peng, S., et al., 2009. Porous hollow Fe(3)O(4) nanoparticles for targeted delivery and controlled release of cisplatin. J. Am. Chem. Soc. 131 (30), 10637–10644.

Chen, K.J., Chaung, E.Y., et al., 2014. Hyperthermia-mediated local drug delivery by a bubble-generating liposomal system for tumor-specific chemotherapy. ACS Nano 8 (5), 5105–5115.

Chen, X., Ooi, C.P., et al., 2006. Effect of ganciclovir on the hydrolytic degradation of poly(lactide-*co*-glycolide) microspheres. J. Biomater. Appl. 20 (3), 287–302.

Chithrani, B.D., Ghazani, A.A., Chan, W.C., 2006. Determining the size and shape dependence of gold nanoparticle uptake into mammalian cells. Nano Lett. 6 (4), 662–668.

Choe, T.B., Park, I.C., Hong, S.I., 1998. Enhancement of cationic liposome-mediatedtransfection by lactoferrin. Biotechnol. Tech. 12, 577–581.

Colombo, G., Curnis, F., De Mori, G.M.S., Gasparri, A., Longoni, C., Sacchi, A., Longhi, R., Corti, A., 2002. Structure–activity relationships of linear and cyclic peptides containing the NGR tumor-homing motif. J. Biol. Chem. 277, 47891–47897.

Couvreur, P., 2013. Nanoparticles in drug delivery: past, present and future. Adv. Drug Deliv. Rev. 65 (1), 21–23.

Criscione, J.M., Le, B.L., et al., 2009. Self-assembly of pH-responsive fluorinated dendrimer-based particulates for drug delivery and noninvasive imaging. Biomaterials 30 (23–24), 3946–3955.

Daniels, D.A., Chen, H., Hicke, B.J., Swiderek, K.M., Gold, L., 2003. A tenascin-C aptamer identified by tumor cell SELEX: systematic evolution of ligands by exponential enrichment. Proc. Natl. Acad. Sci. USA 100 (26), 15416–15421.

Desai, M.P., Labhasetwar, V., Walter, E., Levy, R.J., Amidon, G.L., 1997. The mechanism of uptake of biodegradable microparticles in Caco-2 cells is size dependent. Pharm. Res. 14, 1568–1573.

Devarajan, P.V., Jindal, A.B., Patil, R.R., Mulla, F., Gaikwad, R.V., Samad, A., 2010. Particle shape: a new design parameter for passive targeting in splenotropic drug delivery. J. Pharm. Sci. 99 (6), 2576–2581.

Dispenza, C., Adamo, G., Sabatino, M., Grimaldi, N., Bulone, D., Bondì, M., Rigogliuso, S., Ghersi, G., 2014. Oligonucleotides-decorated-poly(N-vinyl pyrrolidone) nanogels for gene delivery. J. Appl. Polym. Sci. 131, 1–8.

Dreher, M.R., Liu, W., et al., 2007. Thermal cycling enhances the accumulation of a temperature-sensitive biopolymer in solid tumors. Cancer Res. 67 (9), 4418–4424.

Du, H., Chandaroy, P., Hui, S.W., 1997. Biochim. Biophys. Acta Biomembr. 1326, 236–248.

Du, L., Jin, Y., et al., 2011. Ultrasound-triggered drug release and enhanced anticancer effect of doxorubicin-loaded poly(D,L-lactide-*co*-glycolide)-methoxy-poly(ethylene glycol) nanodroplets. Ultrasound Med. Biol. 37 (8), 1252–1258.

Duncan, R., 2009. Development of HPMA copolymer-anticancer conjugates: clinical experience and lessons learnt. Adv. Drug Deliv. Rev. 61 (13), 1131–1148.

Farokhzad, O.C., Langer, R., 2009. Impact of nanotechnology on drug delivery. ACS Nano 3 (1), 16–20.

Fischer, W., Quadir, M.A., et al., 2011. Controlled release of DNA from photoresponsive hyperbranched polyglycerols with oligoamine shells. Macromol. Biosci. 11 (12), 1736–1746.

Fleige, E., Quadir, M.A., et al., 2012. Stimuli-responsive polymeric nanocarriers for the controlled transport of active compounds: concepts and applications. Adv. Drug Deliv. Rev. 64 (9), 866–884.

Göpferich, A., 1996. Mechanisms of polymer degradation and erosion. Biomaterials 17 (2), 103–114.

Haba, K., Popkov, M., et al., 2005. Single-triggered trimeric prodrugs. Angew. Chem. Int. Ed. Engl. 44 (5), 716–720.

Han, D., Mangala, L.S., Lee, J.W., Shahzad, M.M., Kim, H.S., Shen, D., Nam, E.J., Mora, E.M., Stone, R.L., Lu, C., Lee, S.J., Rho, J.W., Nick, A.M., Lopez-Berestein, G., Sood, A.K., 2010. Targeted gene silencing using RGD-labeled chitosan nanoparticles. Clin. Cancer Res. 16 (15), 3910–3922.

Harris, J.M., Martin, N.E., Modi, M., 2001. Pegylation: a novel process for modifying pharmacokinetics. Clin. Pharmacokinet. 40, 539–551.

Jain, K.K., 2010. Advances in the field of nanooncology. BMC Med. 8, 83.

Jaracz, S., Chen, J., et al., 2005. Recent advances in tumor-targeting anticancer drug conjugates. Bioorg. Med. Chem. 13 (17), 5043–5054.

Jiang, J., Tong, X., et al., 2005. A new design for light-breakable polymer micelles. J. Am. Chem. Soc. 127 (23), 8290–8291.

Kamaly, N., Xiao, Z., Valencia, P.M., Radovic-Moreno, A.F., Farokhzad, O.C., 2012. Targeted polymeric therapeutic nanoparticles: design, development and clinical translation. Chem. Soc. Rev. 41 (7), 2971–3010.

Kanwar, J.R., Mahidhara, G., Kanwar, R.K., 2011. Novel alginate enclosedchitosan-calcium phosphate-loaded iron saturated bovine lactoferrinnanocarriers for oral delivery in colon cancer. Nanomedicine (Lond.) 7 (10), 1521–1550.

Kim, J.K., Anderson, J., et al., 2009. Self-assembling peptide amphiphile-based nanofiber gel for bioresponsive cisplatin delivery. Mol. Pharm. 6 (3), 978–985.

Kisley, L.R., Barrett, B.S., Bauer, A.K., Dwyer-Nield, L.D., Barthel, B., Meyer, A.M., Thompson, D.C., Malkinson, A.M., 2002. Genetic ablation of inducible nitric oxide synthase decreases mouse lung tumorigenesis. Cancer Res. 62 (23), 6850–6856.

Kroll, R.A., Pagel, M.A., Muldoon, L.L., Roman-Goldstein, S., Fiamengo, S.A., Neuwelt, E.A., 1998. Improving drug delivery to intracerebral tumor and surrounding brain in a rodent model: a comparison of osmotic versus bradykinin modification of the blood-brain and/or blood-tumor barriers. Neurosurgery 43 (4), 879–886.

Ku, S.H., Kim, K., et al., 2014. Tumor-targeting multifunctional nanoparticles for siRNA delivery: recent advances in cancer therapy. Adv. Healthcare Mater. 3 (8), 1182–1193.

Ku, T.H., Chien, M.P., et al., 2011. Controlling and switching the morphology of micellar nanoparticles with enzymes. J. Am. Chem. Soc. 133 (22), 8392–8395.

Law, B., Tung, C.H., 2009. Proteolysis: a biological process adapted in drug delivery, therapy, and imaging. Bioconjug. Chem. 20 (9), 1683–1695.

Lee, C.C., Gillies, E.R., et al., 2006. A single dose of doxorubicin-functionalized bow-tie dendrimer cures mice bearing C-26 colon carcinomas. Proc. Natl. Acad. Sci. USA 103 (45), 16649–16654.

Lee, K., Bae, K.H., et al., 2010. Pluronic/polyethylenimine shell crosslinked nanocapsules with embedded magnetite nanocrystals for magnetically triggered delivery of siRNA. Macromol. Biosci. 10 (3), 239–245.

Leo, E., Pecquet, S., et al., 1998. Changing the pH of the external aqueous phase may modulate protein entrapment and delivery from poly(lactide-*co*-glycolide) microspheres prepared by a w/o/w solvent evaporation method. J. Microencapsul. 15 (4), 421–430.

Li, Y., Xiao, K., et al., 2011. Well-defined, reversible disulfide cross-linked micelles for on-demand paclitaxel delivery. Biomaterials 32 (27), 6633–6645.

Liu, J., Lee, H., et al., 2006. Liposome formulation of a novel hydrophobic aryl-imidazole compound for anti-cancer therapy. Cancer Chemother. Pharmacol. 58 (3), 306–318.

Lotan, R., Raz, A., 1988. Lectins in cancer cells. Ann. N Y Acad. Sci. 551, 385–396.

Luk, B.T., Hu, C.M., Fang, R.H., Dehaini, D., Carpenter, C., Gao, W., Zhang, L., 2014. Interfacial interactions between natural RBC membranes and synthetic polymeric nanoparticles. Nanoscale 6 (5), 2730–2737.

Ma, Y., Zheng, Y., et al., 2011. Novel docetaxel-loaded nanoparticles based on PCL-Tween 80 copolymer for cancer treatment. Int. J. Nanomed. 6, 2679–2688.

Marrache, S., Dhar, S., 2012. Engineering of blended nanoparticle platform for delivery of mitochondria-acting therapeutics. Proc. Natl. Acad. Sci. USA 109, 16288–16293.

Martin, L., Alonso, M., et al., 2009. Synthesis and characterization of macroporous thermosensitive hydrogels from recombinant elastin-like polymers. Biomacromolecules 10 (11), 3015–3022.

Matsumura, Y., Maeda, H., 1986. A new concept for macromolecular therapeutics in cancer chemotherapy: mechanism of tumoritropic accumulation of proteins and the antitumor agent smancs. Cancer Res. 46, 6387–6392.

Mauro, N., Campora, S., Scialabba, C., Adamo, G., Licciardi, M., Ghersi, G., et al., 2015. Self-organized environment-sensitive inulin–doxorubicin conjugate with a selective cytotoxic effect towards cancer cells. RSC Adv. 5 (41), 32421–32430.

Meng, F., Zhong, Z., et al., 2009. Stimuli-responsive polymersomes for programmed drug delivery. Biomacromolecules 10 (2), 197–209.

Miyata, K., Kakizawa, Y., et al., 2004. Block catiomer polyplexes with regulated densities of charge and disulfide cross-linking directed to enhance gene expression. J. Am. Chem. Soc. 126 (8), 2355–2361.

Mohanraj, V.J., Chen, Y., 2006. Nanoparticles—a review. Trop. J. Pharm. Res. 5 (1), 561–573.

Mura, S., Nicolas, J., et al., 2013. Stimuli-responsive nanocarriers for drug delivery. Nat. Mater. 12 (11), 991–1003.

Nemunaitis, J.M., Nemunaitis, J., 2008. Potential of Advexin®: a p53 gene-replacementtherapy in Li–Fraumeni syndrome. Future Oncol. 4, 759–768.

Neri, D., Bicknell, R., 2005. Tumour vascular targeting. Nat. Rev. Cancer 5, 436–446.

Neu, M., Germershaus, O., et al., 2007. Crosslinked nanocarriers based upon poly(ethylene imine) for systemic plasmid delivery: in vitro characterization and in vivo studies in mice. J. Control Release 118 (3), 370–380.

Nimjee, S.M., Rusconi, C.P., Sullenger, B.A., 2005. Aptamers: an emerging class of therapeutics. Ann. Rev. Med. 56, 555–583.

Oishi, M., Hayashi, H., Iijima, Y., Nagasaki, M., 2007. Endosomal release and intracellular delivery of anticancer drugs using pH-sensitive PEGylated nanogels. J. Mater. Chem. 17, 3720–3725.

Owen, S.C., Doak, A.K., et al., 2012. Colloidal aggregation affects the efficacy of anticancer drugs in cell culture. ACS Chem. Biol. 7 (8), 1429–1435.

Pang, Z., Feng, L., Hua, R., Chen, J., Gao, H., et al., 2010. Lactoferrin-conjugated biodegradable polymersome holding doxorubicin and tetrandrine for chemotherapy of glioma rats. Mol. Pharm. 7, 1995–2005.

Park, J.H., Von Maltzahn, G., Zhang, L., Derfus, A.M., Simberg, D., Harris, T.J., Ruoslahti, E., Bhatia, S.N., Sailor, M.J., 2009. Systematic surface engineering of magnetic nanoworms for in vivo tumor targeting. Small 5 (6), 694–700.

Percec, V., Wilson, D.A., et al., 2010. Self-assembly of Janus dendrimers into uniform dendrimersomes and other complex architectures. Science 328 (5981), 1009–1014.

Phillips, L.C., Klibanov, A.L., et al., 2010. Targeted gene transfection from microbubbles into vascular smooth muscle cells using focused, ultrasound-mediated delivery. Ultrasound Med. Biol. 36 (9), 1470–1480.

Pistolis, G., Malliaris, A., Tsiourvas, D., Paleos, C.M., 1999. Polypropylenimine dendrimers as pH-sensitive controlled-release systems. Chem. Eur. J. 5, 1440–1444.

Qi, W., Yan, X., et al., 2009. Triggered release of insulin from glucose-sensitive enzyme multilayer shells. Biomaterials 30 (14), 2799–2806.

Ríhová, B., Kubácková, K., 2003. Clinical implications of *N*-(2-hydroxypropyl)methacrylamide copolymers. Curr. Pharm. Biotechnol. 4 (5), 311–322.

Robitzki, A.A., Kurz, R., 2010. Biosensing and drug delivery at the microscale: novel devices for controlled and responsive drug delivery. Handb. Exp. Pharmacol. 197, 87–112.

Ruoslahti, E., 2002. Specialization of tumour vasculature. Nat. Rev. Cancer 2 (2), 83–90.

Sambruy, Y., Ferruzza, S., Ranaldi, G., De Angelis, I., 2001. Intestinal cell culture models: applications in toxicology and pharmacology. Cell Biol. Toxicol. 17 (4–5), 301–317.

Schmidt, D.J., Moskowitz, J.S., et al., 2010. Electrically triggered release of a small molecule drug from a polyelectrolyte multilayer coating. Chem. Mater. 22 (23), 6416–6425.

Seferos, D.S., Giljohann, D.A., et al., 2007. Nano-flares: probes for transfection and mRNA detection in living cells. J. Am. Chem. Soc. 129 (50), 15477–15479.

Seymour, L.W., Ferry, D.R., Anderson, D., Hesslewood, S., Julyan, P.J., Poyner, R., Doran, J., Young, A.M., Buetles, S., Kerr, D.J., Cancer Research Campaign Phase I/II Clinical Trials Committee, 2002. Hepatic drug targeting: phase I evaluation of polymer-bound doxorubicin. J. Clin. Oncol. 20 (6), 1668–1676.

Smith, R.A.J., Porteous, C.M., Gane, A.M., Murphy, M.P., 2003. Delivery of bioactive molecules to mitochondria in vivo. Proc. Natl. Acad. Sci. USA 100, 5407–5412.

Sun, T., Zhang, Y.S., et al., 2014. Engineered nanoparticles for drug delivery in cancer therapy. Angew. Chem. Int. Ed. Engl 53 (46), 12320–12364.

Swami, A., Shi, J., Gadde, S., Votruba, A.R., Kolishetti, N., Farokhzad, O.C., 2012. Nanoparticles for targeted and temporally controlled drug delivery. In: Svenson, S., Prud'homme, R.K., (Eds.), Multifunctional Nanoparticles for Drug Delivery Applications: Imaging, Targeting, and Delivery, Nanostructure Science and Technology. Part of the Series Nanostructure Science and Technology. Springer Science+Business Media, LLC, pp. 9-29. DOI 10.1007/978-1-4614-2305-8_2.

Torchilin, V.P., Lukyanov, A.N., et al., 2003. Immunomicelles: targeted pharmaceutical carriers for poorly soluble drugs. Proc. Natl. Acad. Sci. USA 100 (10), 6039–6044.

Ulbrich, K., Etrych, T., et al., 2004. Antibody-targeted polymer-doxorubicin conjugates with pH-controlled activation. J. Drug Target 12 (8), 477–489.

Välimaa, T., Laaksovirta, S., 2004. Degradation behaviour of self-reinforced 80L/20G PLGA devices in vitro. Biomaterials 25 (7–8), 1225–1232.

Vaupel, P., Kallinowski, F., et al., 1989. Blood flow, oxygen and nutrient supply, and metabolic microenvironment of human tumors: a review. Cancer Res. 49 (23), 6449–6465.

Wang, Y., Han, P., et al., 2010. Photocontrolled self-assembly and disassembly of block ionomer complex vesicles: a facile approach toward supramolecular polymer nanocontainers. Langmuir 26 (2), 709–715.

Wei, H., Zhang, X.Z., et al., 2006. Self-assembled thermoresponsive micelles of poly(N-isopropylacrylamide-*b*-methyl methacrylate). Biomaterials 27 (9), 2028–2034.

West, K.R., Otto, S., 2005. Reversible covalent chemistry in drug delivery. Curr. Drug Discov. Technol. 2 (3), 123–160.

Wilczewska, A.Z., Niemirowicz, K., Markiewicz, K.H., Car, H., 2012. Nanoparticles as drug delivery systems. Pharmacol. Rep. 64 (5), 1020–1037.

Woodrow, K.A., Cu, Y., et al., 2009. Intravaginal gene silencing using biodegradable polymer nanoparticles densely loaded with small-interfering RNA. Nat. Mater. 8 (6), 526–533.

Wu, D., Wan, M., 2008. A novel fluoride anion modified gelatin nanogel system for ultrasound-triggered drug release. J. Pharm. Pharm. Sci. 11 (4), 32–45.

Xia, X.H., Yang, M.X., Wang, Y.C., Zheng, Y.Q., Li, Q.G., Chen, J.Y., Xia, Y., 2012. Quantifying the coverage density of poly(ethylene glycol) chains on the surface of gold nanostructures. ACS Nano 6, 512–522.

Yang, K., Ma, Y.Q., 2010. Computer simulation of the translocation of nanoparticles with different shapes across a lipid bilayer. Nat. Nanotechnol. 5, 579–583.

Yin, X., Hoffman, A.S., et al., 2006. Poly(*N*-isopropylacrylamide-*co*-propylacrylic acid) copolymers that respond sharply to temperature and pH. Biomacromolecules 7 (5), 1381–1385.

You, J., Zhang, R., et al., 2012. Effective photothermal chemotherapy using doxorubicin-loaded gold nanospheres that target EphB4 receptors in tumors. Cancer Res. 72 (18), 4777–4786.

Zhang, L., Zhu, D., Dong, X., Sun, H., Song, C., Wang, C., Kong, D., 2015. Folate-modified lipid–polymer hybrid nanoparticles for targeted paclitaxel delivery. Int. J. Nanomed. 10, 2101–2114.

Zheng, Y., Zhong, L., Huang, W., Zhou, Y., Yan, D., 2010. Flocculation-resistant multimolecular micelles with thermoresponsive corona from dendritic heteroarm star copolymers. J. Polym. Sci. Part A 48, 4428–4438.

Zhu, L., Kate, P., et al., 2012. Matrix metalloprotease 2-responsive multifunctional liposomal nanocarrier for enhanced tumor targeting. ACS Nano 6 (4), 3491–3498.

CHAPTER

4

FABRICATION AND CHARACTERIZATION OF NATURAL/SYNTHESIZED, MICRO-, AND NANOSTRUCTURED MATERIALS FOR BIOMEDICAL APPLICATIONS

Hiromichi Takebe, Sengo Kobayashi, Hiromichi Aono, Saeki Yamamuro
Ehime University, Matsuyama, Japan

CHAPTER OUTLINE

1 INTRODUCTION

Micro- and nanostructured materials are recently of growing interest for therapeutic structured ones, for example, magnetic nanoparticles for thermal coagulation therapy (Aono, 2014; Grumezescu et al., 2013; Popescu et al., 2014), periodic structure for cell formation (Barb et al., 2014; Tao and Desai, 2007), and enhancement of epithelial permeability (Wu et al., 2015).

These materials are fabricated by both natural and synthesizing rotes. The former representative examples are biomimetics (McPhedran and Parker, 2015) and pearl oyster shell (Sato et al., 2013). The biomimetics has been interested for optics, for example, photonic crystals, iridescent felt, layered structure of butterfly wings, opal structure, such as regular array of solid submicron particles. Their optical properties result from the interaction between their micro-/nanostructure and light wave with similar sizes. If we consider an analogy between optical and biomedical applications, micro- and nanostructured materials are expected for interactions between material structure and surroundings, for example, water molecule, ions, human cells, blood vessel, pore, and so on.

The latter pearl shell has been also interested as biomedical materials due to its periodic structure consisting of inorganic calcium carbonate and organic substance. The Cleopatra legendary has taken pearl powders for her beauty. In the Mayan Civilization, a piece of shells had been used for the implant of human teeth. Recycled microparticles of pearl shell have been already commercialized for polishing additive of toothpaste.

The pearl shell is yielded as a waste of jewel pearl culture with 500–1000 ton/year even only in Ehime prefecture, Japan. The pearl is one of famous and important products in Ehime prefecture. The pearl shell has micro- and nanostructure by combination of organic/inorganic substances. The structure is prepared by the mollusk in natural sea environments.

We have also fabricated various micro- and nanostructured inorganic materials, such as magnetic iron oxide, its relating compound particles and HAp. The structural features and their fabricating methods are as follows:

- Micro-/nanoparticles including modified pearl ones;
- Self-assembling of shape-controlled nanocubes;
- Apatite formation on heat-treated nacre (pearl layer) powder;
- Periodic structure formed by thermal imprinting on an inorganic glass.

Fig. 4.1 summarizes our research on mico/nanostrcutured materials for biomedical applications. In this chapter, recent research results relating to biomedical applications are introduced.

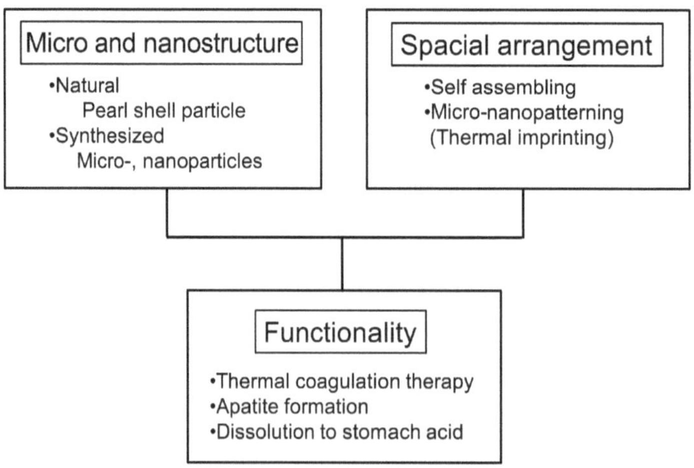

FIGURE 4.1 Research Keywords on Micro-/Nanostructured Materials at Ehime University for Biomedical Applications

2 PEARL SHELL

2.1 MACRO- AND MICROSTRUCTURES

Pearl shell is one of micro- and nanostructured materials fabricated in natural sea environments by pearl oyster. One of pearl oysters, named *Pinctada fucata*, for jewel pearl has been actively cultured at Uwajima sea area in Ehime prefecture (Fig. 4.2A). The pearl oysters are hanged in baskets for pearl culture at the sea, as shown in Fig. 4.2B–C. Two pieces of pearl shells with ~30 g in weight are formed as a by-product per pearl with ~1 g, namely ~97 mass% of pearl oyster is yielded as a waste. The utilization of the wasted pearl shell is one of important subjects for sustainable pearl culture.

In the meantime, the pearl shell consisting of outer prismatic and inner pearl layers has one of potential candidates for biomedical applications with micro- and nanostructures. Fig. 4.3A shows the macroscopic appearance from the inside: prismatic and pearl layers. Both layers are composites consisting of inorganic $CaCO_3$ with different polymorphs and organic substances.

The prismatic layer consists of calcite and organic membrane with columnar morphology (Fig. 4.3B). Each column has growth line in the perpendicular direction of the organic membrane between calcite columns (Arma et al., 2014).

The pearl layer (nacre) consists of aragonite and organic layer with plate morphology (Fig. 4.3C). Each aragonite plate has the thickness of 500–600 nm approximately. There are organic thin layers between the aragonite plates. This submicron-layered structure results in iridescent color of pearl. A small (~a few millimeter square) piece of the outer mantle epithelial cell to produce pearl layers due to biomineralization process is attached on a spherically grinded and polished shell as a core for jewel pearl culture. Therefore, both the pearl layers of the shell and of jewel pearl compose apparently same constituents of aragonite and organic substances (Wada, 1999).

Fig. 4.4 shows (A) TG and (B) DTA curves of powdered prismatic layer sample and $CaCO_3$ reagent (Arma et al., 2014). Five stages of weight losses and three exothermic peaks exist in powder sample of prismatic layer due to thermal decompositions. The initial weight loss (~1%) at 25–190°C was due to absorbed water. The second, third, and fourth weight losses relating exothermic reactions I, II, and III between 190 and 670°C are attributed to the decompositions of organic substances. The first to fourth weight losses were not observed in the TG curve of $CaCO_3$ reagent. The fifth weight loss at 670–850°C is due to the phase transition from $CaCO_3$ to CaO. Both the prismatic layer powder and the $CaCO_3$

FIGURE 4.2

(A) Pearl cultivation at Uwajima sea, Ehime-prefecture, Japan, (B) hanging basket in the sea, (C) pearl oyster.

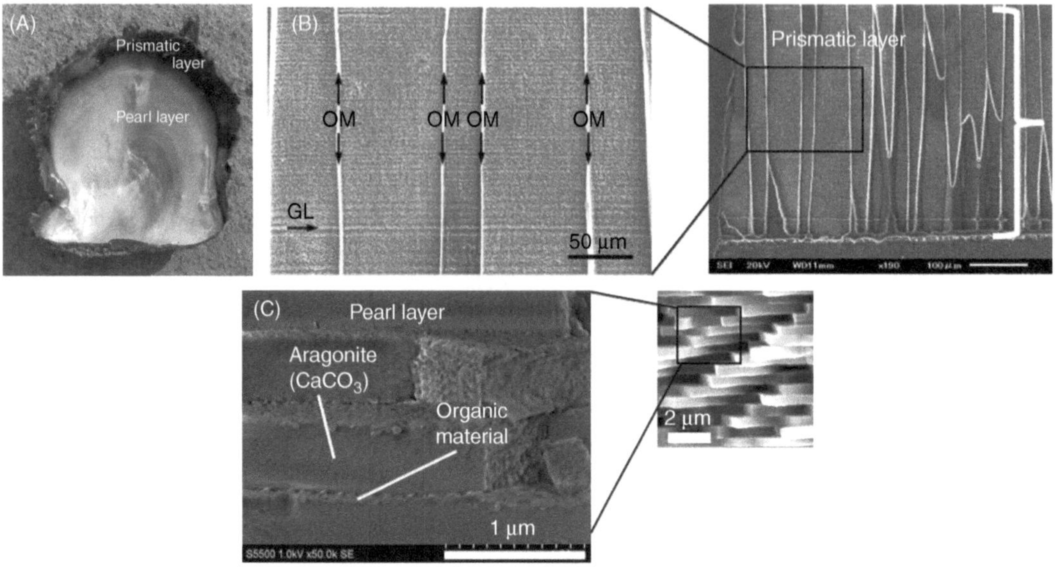

FIGURE 4.3

(A) Macroscopic appearance of a pearl shell and SEM micrographs of (B) prismatic and (C) pearl layers.

Reprinted with the permission from Optical Materials Express 2014, 4, 1816, Fig. 2. Copyright 2014, Optical Society of America.

reagent exhibit the fifth weight loss. The TG–DTA data characterizes well the decomposition processes of organic substances and $CaCO_3$ for the prismatic layer of pearl shell.

3 MICROSTRUCTURE ANALYSIS OF NACRE BY TRANSMISSION ELECTRON MICROSCOPY

Nacre (pearl layer) is the inner iridescent layer of molluscan shells, which is composed of alternating layers of aragonite platelet and organic materials film. The aragonite platelets occupying ~95 mass% of nacre are polymorph of calcium carbonate with a thickness of about 500 nm (Fritz et al., 1994; Jackson et al., 1990). The organic materials taking up 5 mass% of nacre with about 40 nm in thickness are composed of proteins and β-chitin (Levi-Kalisman et al., 2001; Nakahara et al., 1982). Microstructure of nacre in nanoscale and atomic scale is explained in this section.

The shell of Akoya pearl oyster (*P. fucata*) was cleaned by brushing and the outer prismatic layer of the shell was removed with a grinding machine to obtain nacre layer. Thin foil of nacre for transmission electron microscopy (TEM) was obtained by mechanical thinning with diamond rapping disks and ion milling (Duo-mill model 600, Gatan). The foils were examined in an analytical JEM 2100 and a high-resolution JEM 3010 microscopes operated at 200 and 300 kV, respectively.

Fig. 4.5A shows a plan view of the nacre observed by TEM. Thin foil for TEM was carefully fabricated so as to become the surface of the TEM foil was parallel to nacre surface. Selected area diffraction pattern taken from dotted circle area is shown as an insert. The diffraction pattern can be indexed as aragonite phase whose zone axis is nearly parallel to the [001] direction. Enlarged image of Fig. 4.5A is

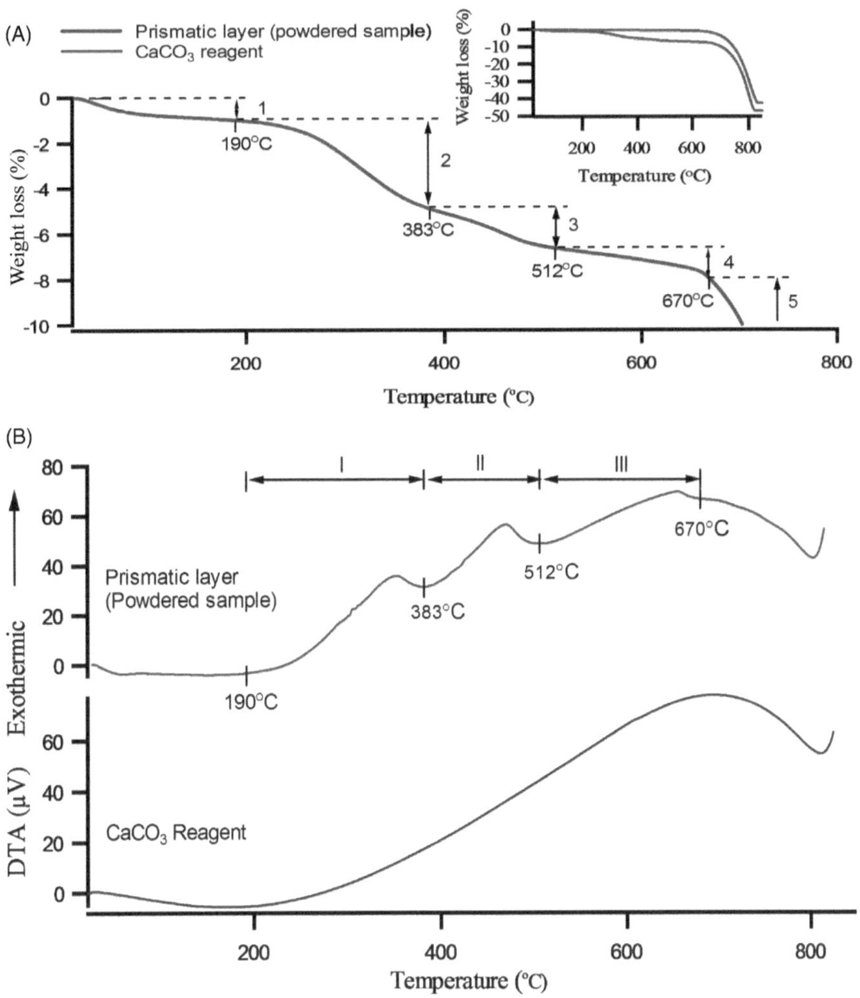

FIGURE 4.4

(A) TG and (B) DTA curves of prismatic layer powdered sample and $CaCO_3$ reagent.

Reprinted with the permission from Optical Materials Express 2014, 4, 1817, Figs. 4 and 5. Copyright 2014, Optical Society of America.

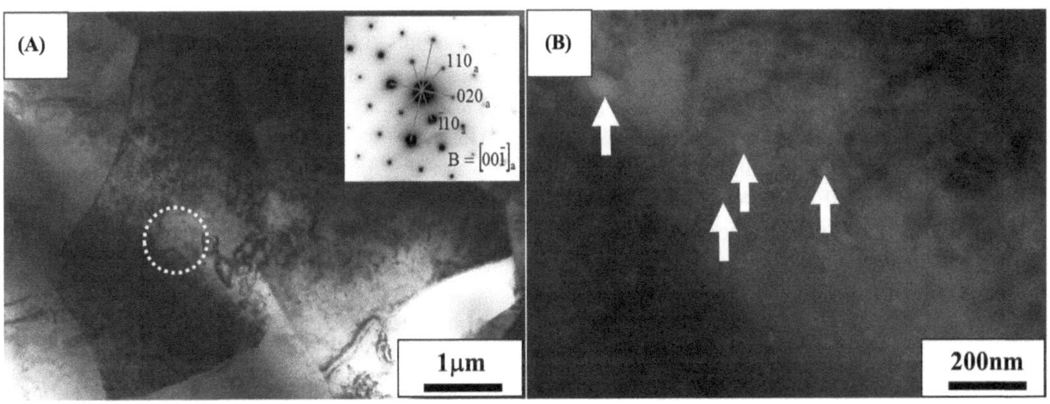

FIGURE 4.5

(A) Bright field image of a plan view of the nacre. Selected area diffraction pattern taken from dotted circle area is shown as an insert. Enlarged image of A is shown in B. Arrows in (B) indicate pores.

Reprinted with the permission from Materials Science Forum 2012, 706–709, 526–531. Copyright 2012, Trans Tech Publications.

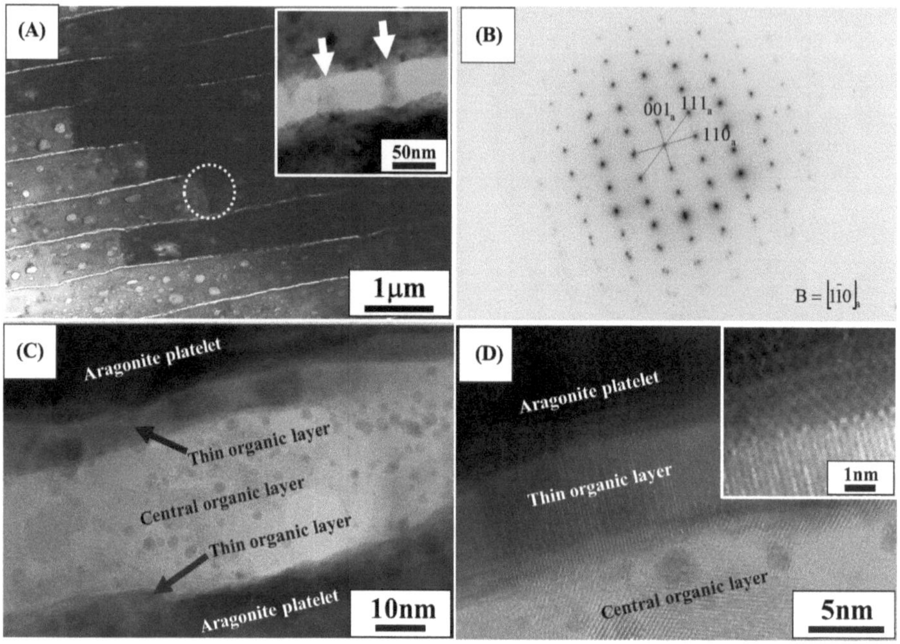

FIGURE 4.6

(A) Cross sectional view of the nacre taken by using TEM. Layered structure of aragonite platelets about 500 nm thick can be observed. Many pores exist within aragonite platelets. (B) Selected area diffraction pattern taken from dotted circle area in Fig. 4.6A. (C) High resolution TEM image of organic material between aragonite platelets. (D) An enlarged image of (C), and a further enlarged image around aragonite and organic interface is illustrated as an insert.

Reprinted with the permission from Materials Science Forum 2012, 706–709, 526–531. Copyright 2012, Trans Tech Publications.

shown in Fig. 4.5B. Pores indicated by arrows in Fig. 4.5B densely existed in the aragonite grains. Size of the pores varied from about several dozen to 100 nm in diameter. Fig. 4.6A is a cross sectional view of the nacre observed by TEM. Selected area diffraction pattern taken from dotted circle area in Fig. 4.6A is shown in Fig. 4.6B. Layered structure of aragonite platelets can be observed in Fig. 4.6A. Average thickness of aragonite platelets is about 500 nm. It should be noted that diffraction patterns taken from another aragonite platelet in Fig. 4.6A exhibits almost the same as that in Fig. 4.6B. By comparing Figs. 4.6A and B, the *c*-axis of the aragonite phase, [001] direction, is almost perpendicular to broad surface of the platelets and thus to the surface of the nacre. The pores are also detected in aragonite platelets as shown in Fig. 4.6A. Thin films about 50 nm thick between aragonite platelets are found in Fig. 4.6A. The inserted image in Fig. 4.6A, which is a magnified image of the thin film, shows the so-called *mineral bridge* between vertically stacked aragonite platelets as indicated by arrows. The mineral bridge has almost the same composition as that of aragonite platelets. The composition analysis of the film except for the mineral bridge revealed that the film contains an increased amount of carbon, which suggests the existence of organic material in the film. High resolution TEM image of the organic film between aragonite platelets is shown in Fig. 4.6C. The organic film seems to consist of very thin organic layers of about 10 nm thick attached to aragonite platelets and central organic layer of about 30 nm thick. Atomic arrangement can be analyzed from an enlarged image of Fig. 4.6C, as shown in Fig. 4.6D. Further enlarged image near the aragonite/the thin organic layer boundary is shown as an insert in Fig. 4.6D. Lattice continuity among aragonite platelet, the thin organic layer and the central organic layer is found to be good. This implies that well-organized crystallographical orientation of aragonite platelets in nacre is attributed to not only mineral bridge but also the organic layers. The organic layers would control crystallographical orientation of aragonite platelets during forming process of nacre.

4 MICRO-/NANOPARTICLES

Two types of methods of built-up and breakdown processes have been utilized for the preparation of micro- and nanoparticles. For the built-up method, there are many types of chemical methods, such as the coprecipitation method, sol–gel method, heteronuclear complex method, alkoxide production method, polymerization method, hydrothermal crystallization process, and so on. The coprecipitation method is a common chemical method for fine polymetalic materials, such as MFe_2O_4, $R_3Fe_5O_{12}$ ferrites, and perovskite-type $RFeO_3$ oxides containing a divalent metal (M) and rare earth (R) (Aono et al., 2001, 2011a; Hirazawa et al., 2008). For example, Fe_3O_4 was prepared by the addition of a NaOH solution to the mixed iron solution of $FeCl_2{\cdot}4H_2O$ and $FeCl_3{\cdot}6H_2O$ with 1:2 molar ratio. The nanosized Fe_3O_4 particles can be directly prepared in a solution at low temperature by a chemical method involving a conventional coprecipitation method as the reaction ($Fe^{2+} + 2Fe^{3+} + 8OH^- \rightarrow Fe_3O_4 + 4H_2O$). However, a heat treatment at low temperature for the solid precursor is necessary for the other materials, because metal hydroxides are stable in the solution except for the case of Fe_3O_4.

On the other hand, the breakdown methods of the milling process are also very effective to obtain fine particles. The ball and beads milling methods involve a powder mixture placed in the milling-pod being subjected to high-energy collisions from the balls and beads. The bead mill is a wet grinding and dispersing machine designed to grind and/or disperse particles in a liquid to nano and submicron sizes.

The grinding chamber is filled with the beads, which are circulated by rotating the shaft at high speed. The slurry, an aqueous fluid containing particles, is then released into the grinding chamber through a pump. The particles are ground and/or dispersed by the impact and friction of the beads. Fig. 4.7 shows SEM photograph of the pearl powder obtained by the bead milling using 0.3 mmϕ beads for 5 h. Although the particle size is 30–100 nm for this powder, it can be easily reduced by additional milling using small beads, such as 0.1 mmϕ.

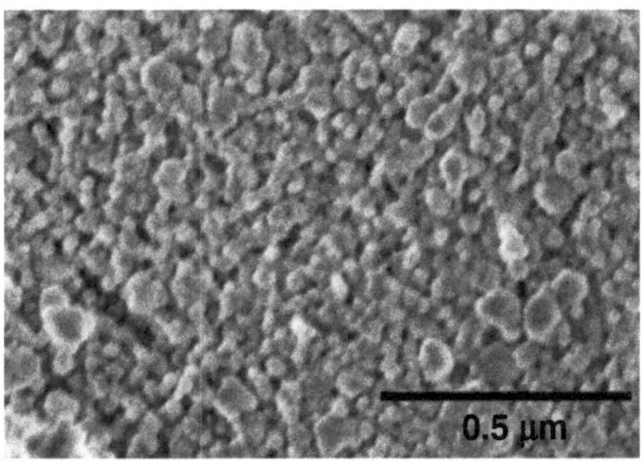

FIGURE 4.7 SEM Photograph of the Pearl Powder Obtained by the Bead Milling Using 0.3 mmϕ Beads for 5 h

5 SURFACE MODIFICATION OF NANOPARTICLES

5.1 SHAPE CONTROL: IRON OXIDE NANOCUBES

Chemical synthesis of nanoparticles has experienced great advances in their size and morphological controls over the past 2 decades. This is largely owing to a combined utilization of nonreactive organic solvents and stabilizing agents (surfactants) (Jun et al., 2006; Lee et al., 2007; Park et al., 2007). The usage of organic solvent with high boiling point provides a wide-range controllability of reaction temperature, enabling to decompose a metal precursor into nanoparticles at high temperature. The surfactant molecules adsorb to the surfaces of formed nanoparticles and suppress their coagulation and coalescence by steric effects. The surfactant also binds to the precursor to form metal complex and assists the particle growth via delivering the constituent solute to the surfaces of growing particles. The formation of highly monodisperse nanoparticles is strongly demanded and generally understandable from the viewpoint of LaMer's mechanism, where a burst nucleation is followed by a diffusional growth of the nuclei without agglomeration (LaMer and Dinege, 1950; Sugimoto, 1987). Though this mechanism can account for many cases, there also exist some cases that do not follow this mechanism, such as spontaneous self-narrowing of polydisperse particle size distribution during the growth stage (Hines and Scholes, 2003; Kuzuya et al., 2005; Prasad et al., 2002; Puntes et al., 2002).

Recent synthetic progress enables to generate nonspherical nanoparticles with peculiar crystal habits, such as nanorod, nanowire, nanodisc, and other complicated shapes (Jun et al., 2006). A typical example can be seen for cubic nanoparticles (nanocubes) of iron oxides (Yamamuro and Sumiyama, 2006; Yamamuro et al., 2008). The nanocubes are prepared via the partial alcohol reduction of Fe(III) acetylacetonate in a high-temperature organic solution in the presence of stabilizing agent. In a typical synthesis, 0.5 mmol of Fe(III) acetylacetonate was dissolved in 10 mL of dioctyl ether with 3.75 mmol of 1,2-hexadecanediol and 1.5 mmol of lauric acid in a flask. The air in the flask was exchanged by Ar gas, and the solution mixture was then rapidly heated to the boiling point of the solvent (286°C) under severe stirring. The temperature was kept for 150 min to reflux. After the reaction was finished, the black colloidal solution was air-cooled to room temperature. The nanoparticles formed were collected via centrifugation with the addition of ethanol, then, redispersed in hexane. Such purification process was repeated 2–3 times. The resulting nanoparticles were finally redispersed in hexane. The TEM image of typical iron oxide nanocubes is shown in Fig. 4.8A. These nanocubes have the average edge length of 7.5 nm and the standard deviation of ~3%, indicating that the particles are monodisperse in size. The X-ray diffraction measurement showed that the crystal structure of these nanocubes is matched with the spinel-type Fe_3O_4 phase. The high-resolution TEM observation showed that each nanocube has a {100} facet. Such a crystal habit reflects the symmetry of cubic crystal structure of Fe_3O_4. This feature is also confirmed by the fact that similar cubic nanoparticles have been prepared in various systems with cubic crystal structures (Quan and Fang, 2010). The formation of cubic shape strongly depends on the particle size; the distinct cubic facet usually appears for the particles beyond 6–7 nm in size. Since other shapes also appear depending on the particle size, the cubic facet seems to be a nonequilibrium shape that appears during the particle growth.

5.2 SHAPE-INDUCED SELF-ASSEMBLY OF NANOCUBES

When the size of colloidal nanoparticles becomes monodisperse, they have a tendency to self-assemble into a variety of two- (2D) and three-dimensional (3D) ordered structures. A good monodispersity is

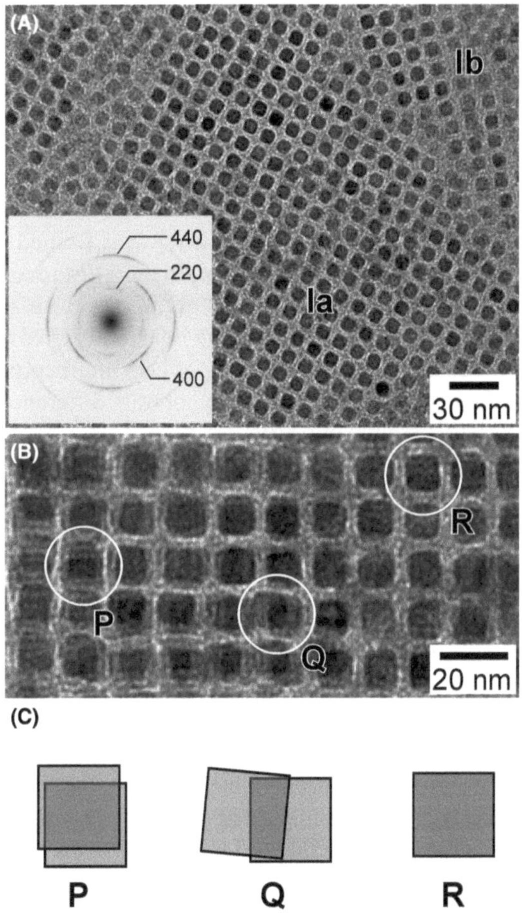

FIGURE 4.8

(A) Low- and (B) high-magnification TEM images of a thin square array of iron oxide nanocubes, and (C) the stacking manner of underlayer and overlayer particles. The inset of Fig. 4.8A shows a typical selected area electron diffraction pattern from the same specimen. The selected area is ~600 nm in diameter.

Reprinted with the permission from Yamamuro, S., Sumiyama, K., Kamiyama, T., 2008. Shape-induced simple cubic arrangement in three-dimensional nanocube self-assemblies. Appl. Phys. Lett. 92, 113108. Copyright 2008, American Institute of Physics.

highly desirable for the formation of long-range ordering (>100 nm) in their self-assembly. In the assemblies, each nanoparticle is coated with a surfactant monolayer, interacting with surrounding particles via van der Waals (vdW) attraction and steric repulsion (Korgel and Fitzmaurice, 1999). These isotropic interactions typically give rise to preferences for close-packed arrangements to minimize the potential energy of the system (Pusey et al., 1989). Spherical nanoparticles generally favor a hexagonal arrangement in 2D monolayer arrays (Motte et al., 1997) and either face-centered cubic (fcc) or hexagonal close-packed (hcp) structure in 3D arrays (Korgel and Fitzmaurice, 1999; Pusey et al., 1989).

Any anisotropy in the interactions (e.g., directional bonding) usually leads to more open structures like those commonly seen in the crystals of covalently bonded atoms. This is achieved by using nonspherical shapes of nanoparticles, inducing to generate peculiar packing in their arrays that reflect the particle shape symmetry. A typical example is represented by a square lattice arrangement of nanocubes rather than a hexagonal one. We studied the packing arrangement of these self-assemblies by a combined analysis of real-space TEM observation and reciprocal-space small-angle X-ray scattering (SAXS) measurement (Pusey et al., 1989). Fig. 4.8A shows the TEM image of a thin array (monolayer and bilayer) of iron oxide nanocubes. The array was formed by slowly evaporating the organic solvent from the nanocube suspension in hexane. In the TEM image, each nanocube attaches to the neighboring particles in a face-to-face manner, giving rise to a square lattice arrangement with a fourfold symmetry. Within the projected charge-density approximation model (Yamamuro et al., 2002), where multiple scattering and defocusing effects are incorporated in a limited way, a thicker specimen results in a greater contrast. Thus, the contrast difference in the image arises from the difference in the layer number of stacking particles: for example, an enhanced dark contrast like region Ia and a less dark contrast like region Ib. The stacking manner of the particles was carefully analyzed in the enhanced dark contrast regions. Fig. 4.8B shows the enlarged image of the region Ia, where clearly outlined cubic nanoparticles (e.g., particle R) can be seen together with vaguely outlined, fringed ones (particles in the left side of the image). The significant cases are highlighted with white circles in the image (e.g., particles P and Q). These heterogeneous contrasts are caused by a slight misarrangement of lateral positions between underlayer and overlayer particles, as schematically illustrated in Fig. 4.8C. In this stacking, the overlayer particles sit on almost the same positions as the particles in the underlayer. This stacking corresponds to the simple cubic arrangement. The SAXS measurement has also revealed the formation of macroscopic simple cubic arrangement for a micrometer-thick self-assembly of the iron oxide nanocubes (Pusey et al., 1989). The formation of square arrays has been also reported for other nanocube assemblies regardless of kinds of constituting substances and surfactants (Quan and Fang, 2010). This suggests that the phenomenon has a universal origin. These results indicate that the simple cubic stacking is exclusively favored in assemblies of nanocubes.

It is scientifically fascinating to clarify the underlying mechanism because the simple cubic arrangement rarely appears in nature due to its low packing density. For cubic nanoparticles, the face-to-face attachment among adjacent nanocubes is the most effective way to increase the packing density. If the interparticle spacing caused by a surfactant layer is ignored, the face-to-face attached square array can posses the ultimate packing fraction of 100%. Thus, the square array is one of the close-packed arrangements of all the nanocube-assembled monolayers. Simple calculations of vdW potential energy were carried out to explain the preference for 2D and 3D square arrays in cubic nanoparticle self-assemblies, where the vdW attractions among the particle cores were took into account. It turned out that the nearest neighbor particles decisively determine the lattice symmetry of the arrays due to the short-ranged nature of the interactions. Hence, the highly coordinated, square lattice arrangement is energetically the most favorable. Similar calculation was made for a bilayer configuration that is the simplest model of 3D arrays. As shown in the inset of Fig. 4.9, an isolated particle (the filled squares) is placed on the underlayer array (the open squares) having a square lattice symmetry. The overlayer particle is assumed to take three different lateral positions: the fourfold top site (site A), the twofold bridge site (site B), and the fourfold bridge site (site C). The calculation result shows that the site A stacking (i.e., the simple cubic stacking) leads to the

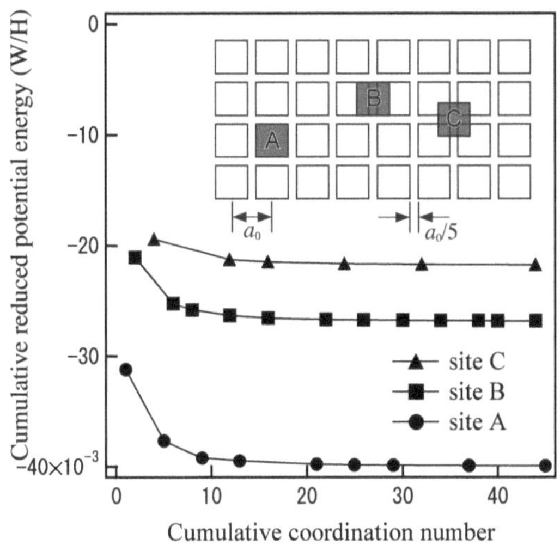

FIGURE 4.9 The Cumulative Potential Energy Normalized by the Hamaker Constant as a Function of the Cumulative Number of Neighboring Particles Within a Given Coordination Number

The inset shows the schematics of a bilayer model for the calculation of van der Waals potential energy between different stacking positions. The gap between the neighboring particles is assumed to be 1/5 of the interparticle distance.

Reprinted with the permission from Yamamuro, S., Sumiyama, K., Kamiyama, T., 2008. Shape-induced simple cubic arrangement in three-dimensional nanocube self-assemblies. Appl. Phys. Lett. 92, 113108. Copyright 2008, American Institute of Physics.

lowest potential energy (Fig. 4.9). Since such stacking manner leads to the highest coordination, it can be said that the simple cubic packing is energetically the most stable from the viewpoint of vdW interaction.

Another intriguing result induced by nanocubes is the appearance of preferred crystallographic orientation of the particles, though it is not achievable for spherical nanoparticles even when the particles form ordered structures. The preferred orientation originates from the fact that the crystal facet is closely related to the crystal structure of the particles. For instance, when nanocubes are self-assembled on a substrate, every nanocube attaches to the substrate surface via a face of the cube. Such assembling manner assists the nanocubes to stick to a substrate with <100> orientation along the surface normal direction, because the <100> planes of all the nanocubes are parallel to the substrate surface. Along with the out-of-plane preferred orientation, the in-plane crystallographic alignment like a single crystal is also confirmed in the selected area electron diffraction pattern. Though the long-range order is still limited at most within a micrometer order, the discontinuous rings with a fourfold symmetry is observed in the electron diffraction pattern (see the inset of Fig. 4.8A). The preferred orientation along the surface normal direction is deteriorated to some extent for thicker assembly, however, it still remains for a micrometer-thick specimen. Fig. 4.10 shows the results of powder X-ray diffraction measurements for the specimens of a micrometer-thick assembly of the iron oxide nanocubes and commercially available micrometer-sized Fe_3O_4 powder (Yamamuro and Sumiyama, 2008). Since the <100> direction is preferentially oriented to the surface normal direction of the substrate, the peak intensity of the {400} reflection for the nanocube assembly is significantly enhanced, comparing to that for Fe_3O_4 powder.

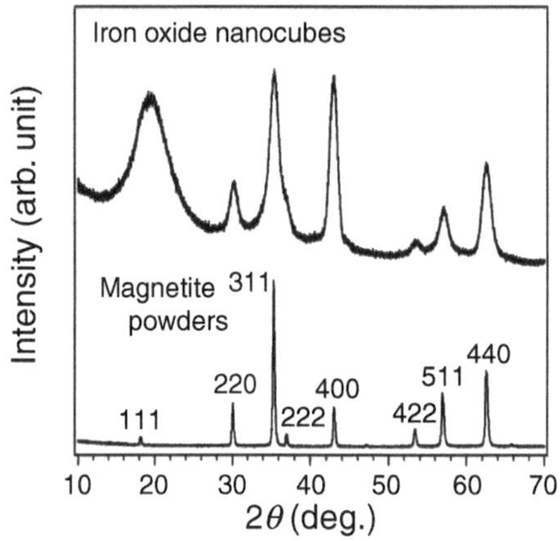

FIGURE 4.10 Powder X-Ray Diffraction Patterns for a Thick Assembly of Iron Oxide Nanocubes and Commercially Available Micrometer-Sized Fe_3O_4 Powder

Reprinted with the permission from Yamamuro, S., Sumiyama, K., 2008. Solution-phase synthesis of size- and shape-controlled magnetic nanoparticles of transition metal systems and their recent development. Mater. Sci. Tech. 45, 246. Copyright 2008, The Materials Science Society of Japan.

6 2D PERIODIC STRUCTURE BY THERMAL IMPRINTING PROCESS

Micro-/nanoimprinting is widely used for patterning organic thin films, such as polypropylene. Thermal imprinting is a process to fabricate micro-/nanopatterns on inorganic glasses at high temperatures around their glass transition and softening temperatures. Micro-/nanopatterned substrates and thin films may have applications for integrated biochemical devices and drug release and/or cell adhesion controlling ones. In organic glasses based on phosphate system are attractive for thermal imprinting due to their low softening temperature (Takebe et al., 2007a) and degradable properties to control various ion releases for bioactive glasses (Abou Neel et al., 2005).

We have developed a new type of lead-free, low-softening tin phosphate ($SnO–P_2O_5$) glasses with low glass transition temperatures of less than 300°C. Various periodic patterns with micro-/nanosizes are successfully fabricated on a $SnO–P_2O_5$ glass (Fig. 4.11) (Takebe et al., 2007b). The imprinted glass sample was illuminated by white light fiber sources in reflection modes. The patterns exhibit various visible colors due to the light interaction with different periodic micro-/submicroscale patterns. Fig. 4.12 shows (A) SEM and (B) AFM photographs of thermal-imprinted square grid patterns (Takebe et al., 2007b).

Micro-/nanopatterned glasses are attractive for integrated devices for biomedical applications. The assembled magnetic nanocubes (Fig. 4.8), pearl particles (Fig. 4.7), and other functional particles are expecting for the integration of patterned glass substrates.

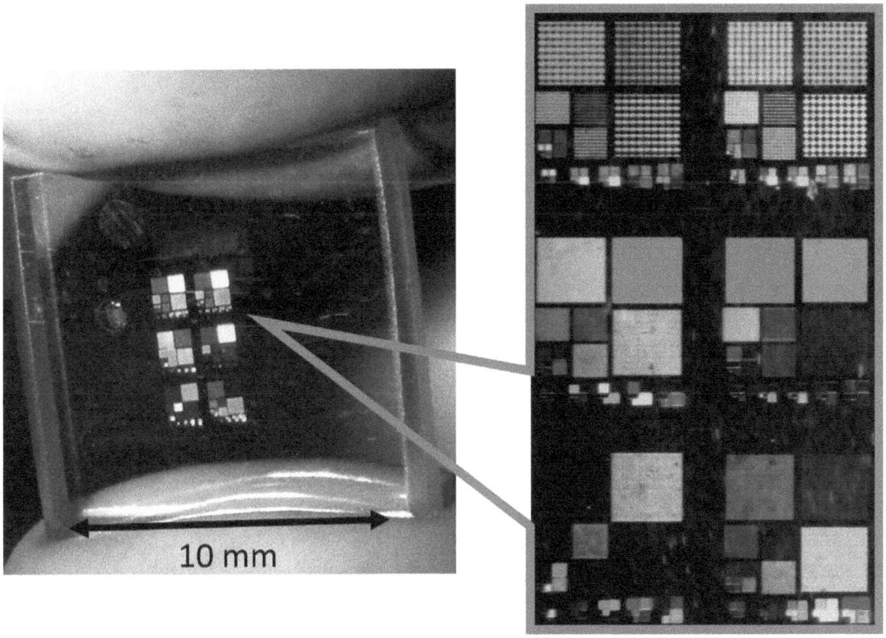

FIGURE 4.11 Thermal-Imprinted Tin Phosphate Glass

Reprinted with the permission from Optics Lett. 32, 2751, Fig. 1. Copyright 2007, Optical Society of America.

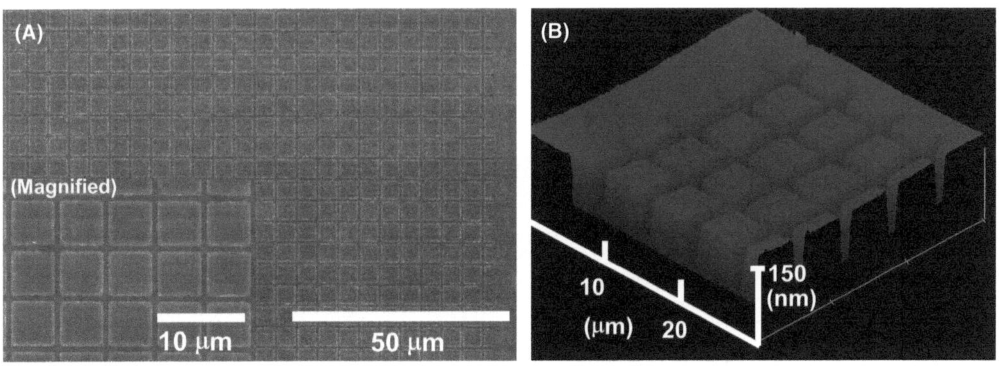

FIGURE 4.12

(A) SEM and (B) AFM photographs of a square grid pattern thermally imprinted on the tin phosphate glass.

Reprinted with the permission from Optics Lett. 2007, 32, 2751, Figs. 2 and 3. Copyright 2007, Optical Society of America.

7 FUNCTIONALITIES FOR BIOMEDICAL APPLICATIONS

7.1 THERMAL COAGULATION THERAPY

7.1.1 Mechanism for the heat generation

Hyperthermia is a type of thermal therapy for cancer treatment in which body tissue is exposed to high temperatures (up to 42–43°C) using a high frequency voltage because cancer cells can undergo damage at this temperature while normal cells remain unaffected. The device for cancer therapy using a radio frequency (8 MHz) has been commercialized as the "Thermotron RF-8" from Yamamoto Vinyter Co., Ltd (Abe et al., 1986).

Thermal coagulation therapy (at 60–70°C) using powdered magnetic materials in an alternating current (AC) magnetic field has been expected as a new method for the treatment of cancerous tissues (Jordan et al., 1999). A drug-delivery system (DDS) to target tumor cells using nanosized magnetic particles encapsulated in a liposome with an antibody (<100 nm) is possible as a cancer therapy in the future. For this method, Fe_3O_4 (magnetite) nanoparticles have been mainly investigated as the candidate material for this type of therapy, because the nanosized Fe_3O_4 particles can be easily prepared in a solution at low temperature by a conventional coprecipitation method (Aono et al., 2005). Fig. 4.13 presents the approximate heat generation mechanism for the various magnetic particles diameters. For large particle size of ferro- and ferrimagnetic materials, the mechanism for the heat generation is attributed to the hysteresis loss in the magnetization curve (*B–H* curve). This hysteresis loss is ascribed to the rotation of the magnetic moment within the magnetic domains in the AC magnetic field. In the case of a superparamagnetic material having a nanosized particle size, the reason for the heat generation ability is due to the Néel relaxation and the Brownian relaxation (Jeyadevan, 2010; Rosensweig, 2002).

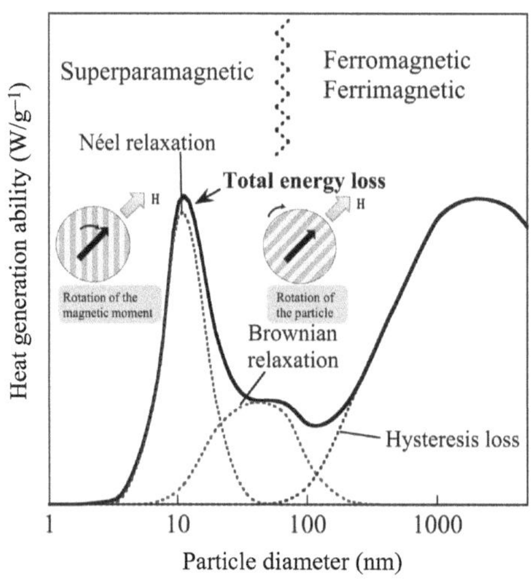

FIGURE 4.13 Approximate Heat Generation Mechanism for the Magnetic Materials With Various Particle Diameters

Reprinted with the permission from Aono, H., Ebara, H., Senba, R., Naohara, T., Maehara, T., Hirazawa, H., Watanabe, Y., 2012. High heat generation ability in AC magnetic field for nano-sized magnetic $Y_3Fe_5O_{12}$ powder prepared by bead milling. J. Magn. Magn. Mater. 324, 1985–1991. Copyright 2012, Elsevier.

The energy losses are attributed to the magnetic moment rotation within the nanosized crystal and the particle rotation in the dispersed solvent for the Néel relaxation and the Brownian relaxation, respectively. The energy losses (heat generation ability) of the superparamagnetic nanoparticles due to the Néel relaxation and Brownian relaxation are proportional to the square of the magnetic field (H^2) and its frequency (f). The equation is simply expressed as:

$$\text{Heat generation ability (W/g)} = k \cdot f \cdot H^2 \tag{4.1}$$

where, k, f, and H are a constant value, frequency (kHz), and magnetic field (kA/m), respectively. In general, the nanosized materials have been synthesized by a chemical method (built-up process), such as coprecipitation. Bead milling as a physical method (break down process) was also effective for the preparation of the nanosized particles.

7.1.2 Ferrite materials for thermal coagulation therapy

Ferrites are usually iron-containing ferrimagnetic oxides, such as MFe_2O_4, $MFe_{12}O_{19}$ (M, divalent cation) and $R_3Fe_5O_{12}$ (R, rare earth). Many ferrites have been reported as heat generation materials in an AC magnetic field. The chemical method (built-up process), such as the coprecipitation methods was utilized for the preparation of the superparamagnetic nanosized materials. A magnetite (Fe_3O_4) nanoparticle has been mainly investigated as the candidate material for this application, because the nanosized Fe_3O_4 particles can be easily prepared in a solution at low temperature by the conventional coprecipitation method. An additional reason for the utilization of the magnetite is its good biocompatibility, because an agent containing the magnetite has been commercialized for contrast-enhanced MRI. The maximum increasing temperature in an AC magnetic field is for a particle with a 13-nm crystal diameter (Aono et al., 2005). For other ferrite materials, nanosized materials have been reported as having a particle-size dependence regarding the heat generation properties. However, a heat treatment involving the decomposition of the precursors for the chemical method is necessary in order to form the ferrite materials, except for Fe_3O_4 (Magnetite). This calcination process decreases the dispersibility of the ferrite particles. These ferrites are possible for the utilization of the thermal coagulation therapy, because the Curie temperatures (T_c) are considerably high compared with 70–80°C.

The beads milling as a physical method (break down process) is very effective for all the ferrite materials. The heat generation ability in an AC magnetic field was studied using fine Fe_3O_4, $MgFe_2O_4$, and $Y_3Fe_5O_{12}$ powders that were physically prepared using the bead-milling of a commercial powder having a large particle size (Aono et al., 2011b, 2012; Hirazawa et al., 2011).

7.1.3 Heat generation ability for $Y_3Fe_5O_{12}$

To study the mechanism of the heat generation for the best sample, the heat generation ability (W/g) was quantitatively measured for three different methods (Fig. 4.14), that is, (A) a magnetic material powder (2.0 g) with 10 mL of water, (B) a hardened magnetic material powder sample (2.0 g) with 10 mL of water using agar, and (C) solidified samples containing the magnetic material powder (2.0 g) using an epoxy resin adhesive with 10 mL of water. Air was slowly bubbled into the glass container for stirring the water for (A) and (C) in order to have the same temperature distribution and the homogeneous magnetic material particle dispersion for (A). In this measurement, the heat generation ability was calculated using the temperature enhance ratio (dT/dt = K/s) for the initial 5 min using the following equation:

$$\text{Heat generation ability (W/g)} = C \cdot (dT/dt)/M \tag{4.2}$$

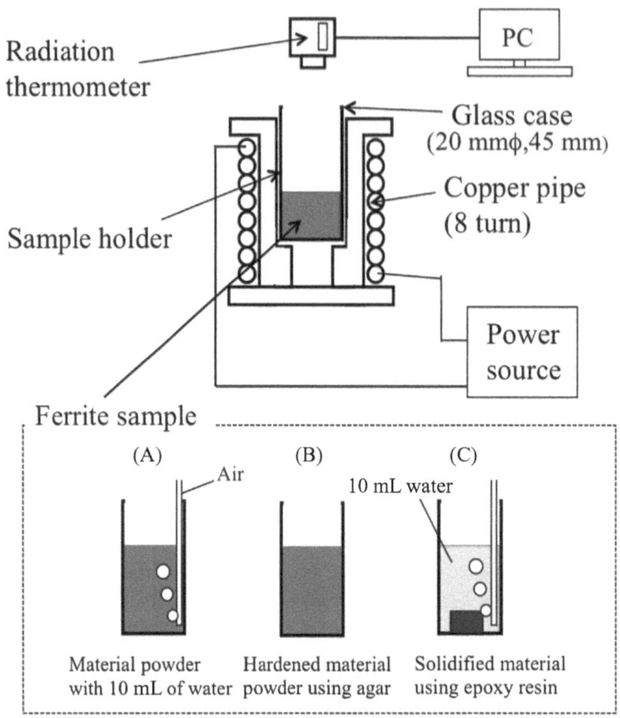

FIGURE 4.14 Apparatus for Measuring the Heat Generation Ability

The samples for the measurement of the heat generation ability, (A) ferrite powder (2.0 g) with 10 mL of water, (B) hardened ferrite powder (2.0 g) with 10 mL of water using agar, (C) solidified ferrite powder (2.0 g) using epoxy resin.

Reprinted with the permission from Aono, H., 2014. Development of nano-sized superparamagnetic ferrites having heat generation ability in an AC magnetic field for thermal coagulation therapy. J. Ceram. Soc. Jpn. 122(1), 237–243. Copyright 2014, The Ceramic Society of Japan.

where, M and C are the sample weight (g), and the estimated total heat capacity (J/K) of 10 mL of water with the glass container, respectively.

The bead-milled $Y_3Fe_5O_{12}$ ferrite had the highest heat generation ability of the nanosized superparamagnetic materials. The $Y_3Fe_5O_{12}$ showed poor heat generation ability in an AC magnetic field for the starting powder material using a commercial powder sample with c. 2.9 μm particle size. The heat generation ability in the AC magnetic field improved with the milling time of a decrease in the average particle size. The maximum heat generation ability was obtained for the samples milled for 4 h using the 0.1 mmϕ beads (after milling using the 0.3 mmϕ beads for 2 h) having c. 15 nm crystallite size, and c. 26 nm particle size. The particle size was 1.5–2 times greater than the crystallite size for all the samples due to the cohesion of the small crystallites, which resulted in larger particles. The heat generation ability decreased with a decrease in the crystallite diameter below c. 15 nm. Fig. 4.15 shows the relationship between the square of the magnetic field (H^2) and the heat generation ability for the three types of methods as shown in Fig. 4.14 for the bead-milled $Y_3Fe_5O_{12}$ ferrite having the maximum heat generation

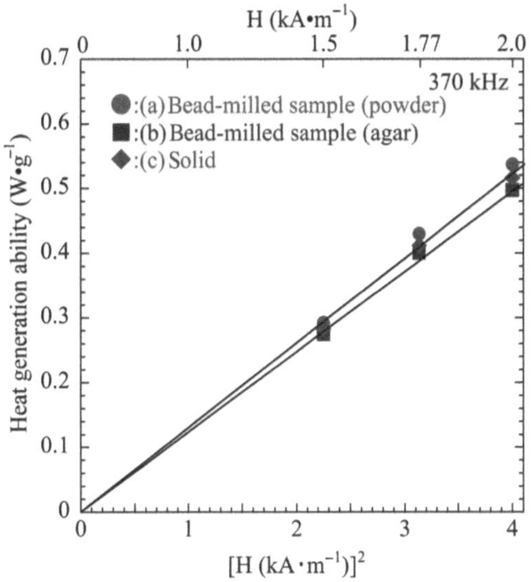

FIGURE 4.15 Relationship Between the Square of the Magnetic Field (370 kHz) and Heat Generation Ability for Three Types (Fig. 4.14) of $Y_3Fe_5O_{12}$ Powder Samples Prepared by Bead Milling (Using 0.1 mmϕ Beads for 4 h)

Reprinted with the permission from Aono, H., Ebara, H., Senba, R., Naohara, T., Maehara, T., Hirazawa, H., Watanabe, Y., 2012. High heat generation ability in AC magnetic field for nano-sized magnetic $Y_3Fe_5O_{12}$ powder prepared by bead milling. J. Magn. Magn. Mater. 324, 1985–1991. Copyright 2012, Elsevier.

ability. The energy losses for all the samples were proportional to the square of the magnetic field. For these samples, the energy losses for all the samples were also proportional to the frequency of the AC magnetic field. The result of sample (A) in water is the total energy loss value. The estimated k value in Eq. (4.1) using these results was 3.58×10^{-4} for the total loss of this sample having the best heat generation ability. When the heat generation of sample (A) was compared to that of sample (B), this difference due to the Brownian relaxation was 5%–6%. The main reason for the high heat generation ability of the bead-milled samples was attributed to an increase in the Néel relaxation. For this nano-sized sample, the concentration of the ferrite in the sample does not affect to the heat generation ability in an AC magnetic field because of the similar heat abilities of the samples (B) and (C) (Aono, 2014).

7.1.4 Preparation of microspheres for embolization method

As another type of magnetic material, a microsphere for the embolization of blood vessels can be considered as one of the methods for cancer treatment. The microspheres also act to block the bloodstream (nourishment) of a cancer tumor. Radioactive ^{90}Y containing microspheres using TheraSphere and SIR-Sphere have already been used for the radioembolization of liver tumors (Murthy et al., 2005; Salem et al., 2005). For the thermal coagulation therapy, the magnetic materials containing glass ceramics have been studied for the embolization method. Fe_3O_4 microspheres with a 20–30 μm diameter range were synthesized by melting the powder in a high frequency induction thermal plasma at high temperature and by precipitation from an aqueous solution (Kawashita et al., 2005). As a conventional method, the preparation of microspheres was possible by a spray drying technique at low temperature using nanoparticles.

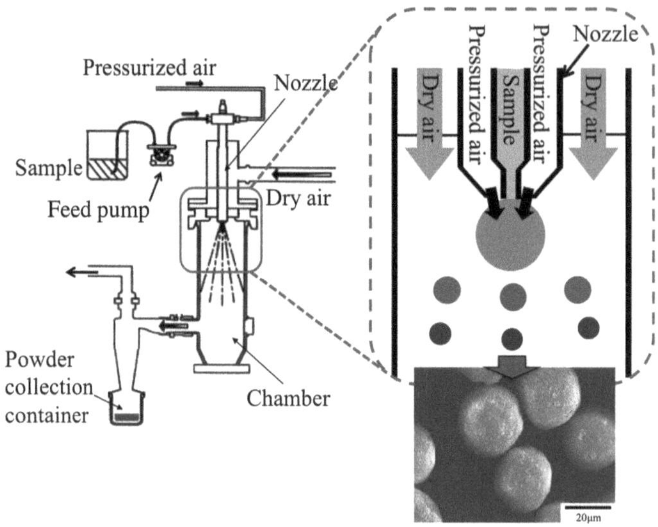

FIGURE 4.16 The Spray Dryer Apparatus Used for the Preparation of Ferrite Microspheres and SEM Photo of the Spray Dried Particles

Reprinted with the permission from Aono, H., Senba, F R., Nishimori, T., Naohara, T., 2013. Preparation of $Y_3Fe_5O_{12}$ microsphere using bead-milled nanosize powder for embolization therapy application. J. Am. Ceram. Soc. 96(11), 3483–3488. Copyright 2013, The American Ceramic Society.

For the embolization method used for cancer therapy, the preparation of $Y_3Fe_5O_{12}$ microspheres with a 20–32 μm diameter range was investigated by the spray-drying method using the bead-milled nanosized particles. The preparation of microspheres using a spray-dry method has been specifically reported for organic materials, such as chitosan for colonic drug delivery. Bead-milled $Y_3Fe_5O_{12}$ was tested for the preparation of microspheres using the spray-drying method (Aono et al., 2013). Fig. 4.16 shows a spray-drier used for the preparation of the microspheres and SEM photos of the spray dried particles after sifting. The slurry for the spray-drier was a mixture of the bead-milled $Y_3Fe_5O_{12}$ and purified water, and then sprayed by heated air at 200°C as small droplets by the spray-drier equipped with a nozzle atomizer. The droplets were isotropically dried in the chamber to form ferrite microspheres. The spray-dried powder was washed using purified water on a 32-μm sieve to screen out the larger particles. The sifted slurry containing the particles under 32-μm and purified water was then thoroughly washed on a 20-μm sieve to remove any of the smaller particles. The powder on the 20-μm sieve, which contained the 20–32 μm particles, was then dried at 100°C. The homogeneous microspheres of 20–32 μm in size were obtained using the spray-dry method and sifting. These microspheres are formed by the cohesive power of the bead-milled nanosized powder. The heat generation ability of the microsphere sample was close to that for the bead-milled sample. A low temperature treatment at 200°C using the spray-drier for the preparation of the microspheres did not influence the heat generation ability. The energy losses for both samples were also proportional to the frequency of the AC magnetic field. The result of the bead-milled sample in water is the total energy loss value of the Néel relaxation and Brownian relaxation. The estimated k value in Eq. (4.1) using these results was 3.50×10^{-4} for the total loss of this sample having the best heat generation ability. The heat generation for the Brownian relaxation of the particle rotation could be ignored for the microsphere sample, because the particle rotation hardly occurred for the large microspheres. This result also suggested that the main reason for the high heat generation property of the milled samples was an increase in the Néel relaxation.

7.2 APATITE FORMATION

7.2.1 TEM observation of HAp formation on nacre

Westbroek and Marin have shown that nacre possesses bioactive properties to contact well with bone (Westbroek and Marin, 1998). Atlan et al. (1999) demonstrated that nacre pieces were accepted in vivo after implantation in the sheep femur and they stimulated bone remodeling. The implanted nacre kept in osteointegration without interposition of fibrous tissue was confirmed (Berland et al., 2005). Nacre also contains molecules, which are capable of reducing osteoclast bone resorption activity by inhibiting cathepsin-K (Duplat et al., 2007). Nacre increasingly attracts a great deal of attention as a bioactive material. In Section 6.2.1, HAp formation on nacre will be explained. The HAp is a main ceramic component of living bone. The bioactive property of material for bone formation would be evaluated by confirming HAp formation on the material in simulated body fluid (SBF).

The shells of Akoya pearl oyster (*Pinctada imbricata*), were cleaned by brushing and the outer prismatic layer of the shell was removed with a grinding machine to obtain nacre plate. Nacre plates were heat-treated at 100, 200, and 300°C for 3.6 ks in air. Nacre plates with or without heat treatment were soaked in SBF for 259.2 ks. A pure Ti plate 10×10 mm^2 was also soaked in SBF for 259.2 ks. They were gently washed under a flow of water. HAp formation on nacre plate was observed by scanning electron and transmission electron microscopies. Thin foils for TEM to observe the cross section between nacre plate and HAp formed in SBF were obtained by mechanical thinning with diamond rapping disks and ion milling (Duo-mill model 600, Gatan). The foils were examined in an analytical JEM 2100 microscope operated at 200 kV.

Fig. 4.17A–B are scanning electron micrographs for the Ti and the nacre after soaked in SBF for 259.2 ks. Surface of the Ti seems to be unchanged, while particles can be observed on the nacre surface. X-ray diffractometry revealed that the particles were HAp. It should be noted that HAp particles were

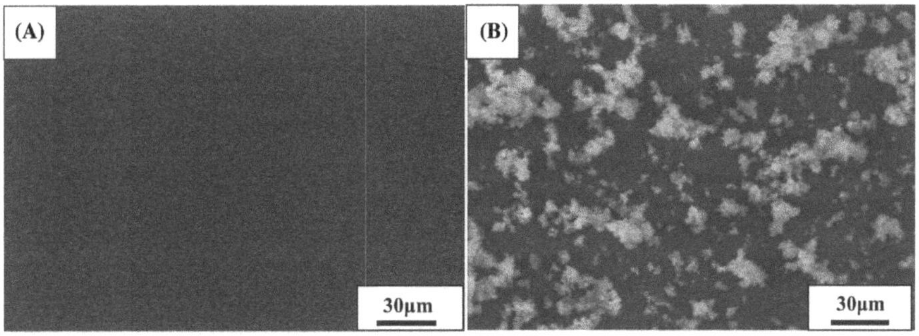

FIGURE 4.17

Scanning electron micrographs of (A) Ti and (B) nacre surfaces after soaked in simulated body fluid (SBF) for 259.2 ks. HAp particles only formed on the nacre surface.

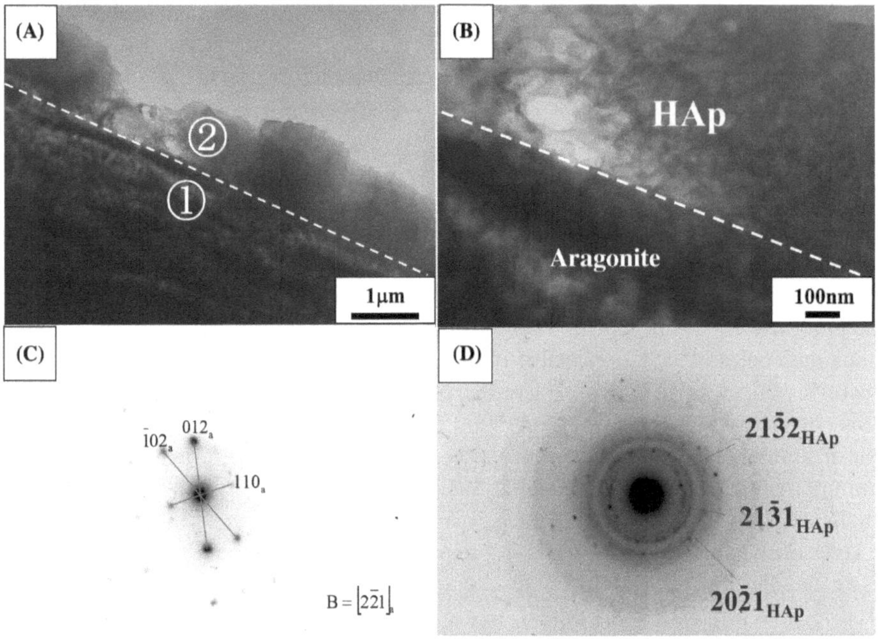

FIGURE 4.18

(A) Transmission electron micrograph for the cross-sectional view of the nacre near surface soaked in simulated body fluid (SBF) for 259.2 ks. Enlarged image of Fig. 4.18(A) is shown in Fig. 4.18(B). Selected area diffraction pattern taken from the circle areas of 1 and 2 are shown in Fig. 4.18(C) and (D). Dotted line in (A) and (B) indicates the interface between the nacre and HAp particles formed in SBF.

Reprinted with the permission from Materials Science Forum 2012, 706–709, 526–531. Copyright 2012, Trans Tech Publications.

formed only on the nacre surface, indicating a high bioactivity of the nacre. Fig. 4.18A is a bright field image of a cross-sectional view of the nacre after soaked in SBF for 259.2 ks. Enlarged image near the dotted line in Fig. 4.18A is shown in Fig. 4.18B. Selected area diffraction patterns taken from the circle areas of 1 and 2 are shown in Figs. 4.18C–D, respectively. Single-phase diffraction pattern of aragonite can be seen in Fig. 4.18C, while the Debye ring pattern of HAp phase is found in Fig. 4.18D. This indicates that polycrystalline HAp particles formed on the single crystal aragonite, implying no epitaxial growth of HAp on the nacre. HAp formation is well induced in SBF on the nacre surface comparing with that on Ti surface, indicating good bioactivity of the nacre.

7.2.2 Effect of heat treatment of nacre on the HAp formation in SBF

The color of nacre without heat treatment was iridescence. The nacre heated at 100°C still exhibited iridescent color. The color of nacre changed from iridescence to brown by heat treatments at 200 and 300°C. It should be noted that the nacre heated at 300°C yielded to a slight force, implying disappearance of organic materials that connect aragonite platelets. Fig. 4.19A is a low magnification image of Fig. 4.17B, that is, showing HAp formation on nacre without heat treatment. Fig. 4.19B–D illustrate HAp formation on the nacres, which were heated at 100, 200, and 300°C, after soaked in SBF for

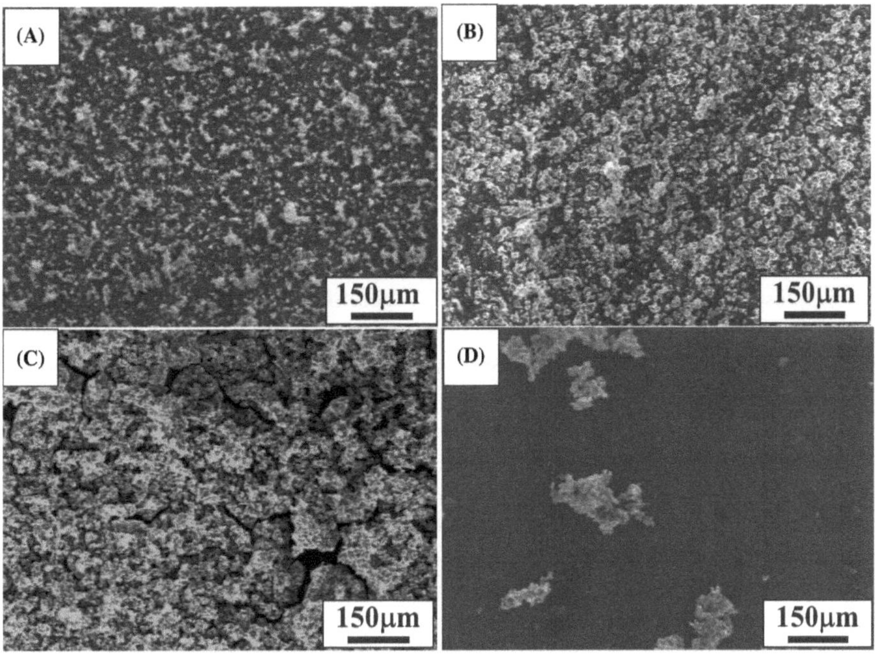

FIGURE 4.19 Scanning Electron Micrographs of the Nacres After Soaked in SBF for 259.2 ks

The nacres are (A) without heat treatment and (B–D) with heat treatment at (B) 100°C, (C) 200°C, and (D) 300°C, respectively, for 3.6 ks before soaking in SBF.

Reprinted with the permission from Materials Science Forum 2012, 706–709, 526–531. Copyright 2012, Trans Tech Publications.

259.2 ks. HAp formation on the nacres heated at 100 and 200°C was accelerated. On the other hand, HAp formation was suppressed on the nacre heated at 300°C. Nacre is composed of alternating layers of aragonite platelet and organic materials as shown in Fig. 4.6A. Stacking cycle of the layers is about 550 nm. The stacking direction of the alternating layers is almost perpendicular to the surface of nacre. When we grind and polish a nacre to obtain a flat plate of nacre without any special attention, we will get surface of the plate inclined to the alternating layers. Therefore, not only aragonite platelets but also organic materials do exist on the polished flat nacre surface. Furthermore, there are some studies indicating existence of biopolymer inside aragonite platelets. Li et al. (2000, 2006) and Li (2007) revealed that individual aragonite platelets are composed of millions of nanosized grains with an average grain size of 32 nm. And thin biopolymer layer exist between these nanosized grains. Bourrat et al. (2005) also confirmed that aragonite platelets in nacre are made up of an intracrystalline continuous organic matrix, which breaks the mineral (calcium carbonate) up into coherent nanosized grains, about 45 nm. Furthermore, Gries et al. (2009) found faceted pores several nanometers wide containing organic materials in the aragonite platelets. They analyzed pores by electron tomography, energy-dispersive X-ray analysis and electron energy-loss spectroscopy, and revealed that pores contain much carbon implying existence of organic materials in the pores. Such pores are also confirmed in the TEM micrographs shown in Figs. 4.5B and 4.6A. It can be said that the polished surface of nacre contains not only aragonite mineral but also plenty of organic materials.

The polished surface of nacre has an ability to induce HAp formation in SBF as shown in Fig. 4.17B, suggesting aragonite or organic materials or both may accelerate the nucleation of HAp. The nacre heated at 300°C became brittle, suggesting organic materials, which connect aragonite platelets started to disappear. Yoshioka and Kitano (1985) examined the effect of heat treatment on various type of aragonite, such as coral, shell, and inorganic aragonites. They showed that transformation temperatures of various shell's aragonite to calcite are 330–350°C. They also detected that emission of water at 140°C and both water and carbon dioxide at and above 250°C. The emission of both water and carbon dioxide implies decomposition or combustion of organic materials. These studies support that heating nacre at 300°C induces disappearance of organic materials without changing aragonite platelets in nacre. The HAp formation in SBF was suppressed on the nacre heated at 300°C. Therefore, we can conclude that organic materials play a critical role for HAp formation in SBF. It is said that organic materials of nacre are composed of insoluble and soluble proteins. After the first report on shell organic material about insoluble protein named *conchiolin* by Frémy many studies have been done on the organic materials in the shells. There are several models explaining biomineralization in the shells associated with organic materials (Furuhashi et al., 2009; Levi-Kalisman et al., 2001; Nudelman et al., 2007). Recently, critical review about shell calcification hypothesis has been done by Furuhashi et al. (2009). They suggested that chitin, silk fibroin protein, and acidic macromolecules are important components for shell formation. And the acidic macromolecules with a negative charge are important for the calcium carbonate crystallization process. The main acidic macromolecules are thought to be aspartic acid-rich proteins (Frédéric and Gilles, 2007) and sulfated GAG (Dauphin, 2003). Carboxyl residue of aspartic acid in the organic materials of nacre would provide negative charge bonds for attracting calcium ion in SBF, which could act as a trigger of HAp nucleation. Acceleration of HAp formation in SBF was found on the nacre heated at 100 and 200°C as shown in Fig. 4.19. Although combustion of organic materials would not occur at 100 and 200°C, as shown in Fig. 4.4 for prismatic layer sample, denaturation of protein or partial debonding of polypeptide chains could occur at these temperatures. Denaturation of proteins involves the disruption of both the secondary and tertiary structures, for example, disruption of the beta sheets in a protein. Since the denaturation occurs because the bonding interactions responsible for the secondary structure and tertiary structure are disrupted, the denaturation provides free molecules that could attract calcium ion in SBF resulting in promotion of HAp formation. The acceleration of HAp formation on the nacre heated at 100 and 200°C might be attributed to the denaturation of organic materials in nacre. HAp formation in SBF was suppressed on the nacre heated at 300°C due to loss of organic materials in nacre.

7.3 DISSOLUTION OF PEARL PARTICLES

Pearl layer consists of ~95 mass% $CaCO_3$ (aragonite) and ~5 mass% organic materials, for example, proteins and β-chitin. Therefore, the pearl layer has a possibility for Ca nutrients to human beings. The dissolution behavior of micrometer-sized pearl powder into 15–20 mmol/L HCl aqueous solution with pH = 1.6–1.8 as an artificial stomach acid was studied at 37°C. Table 4.1 summarizes the variations of solution pH and of dissolution percent of pearl powder with immersion time. The ~35% of pearl powder was dissolved into the HCl aqueous solution within 30 min and gradually up to ~42% with 180 min, corresponding to pH change. The pearl nanoparticles (Fig. 4.7) have approximately two orders, smaller size, compared to the pearl microparticles used in dissolution experiments. The pearl nanoparticles may have much faster dissolution rate into stomach acid. Further study is necessary to distinguish pearl nanoparticles from pearl microparticles.

Table 4.1 Dissolution Experiment Results of Pearl Microparticles

Time (min)	0	30	60	180
pH	1.8	6.5	6.8	7.4
Dissolved (mass%)	0	35.2	40.7	41.8

Pearl powder sample with 0.3 g was added to 15–20 mmol/L HCl aqueous solution with 100 mL at 37°C. The solution was stirred at 200 rpm.

8 SUMMARY AND FUTURE DIRECTIONS

We have introduced our research on micro- and nanostructured materials for biomedical applications. Various types of micro- and nanostructured materials are fabricated at our laboratories. The materials features are modified micro-/nanoparticles for HAp formation and Ca nutrients, magnetic oxide nanoparticles for thermal coagulation therapy, iron oxide nanocubes, and their shape-induced self-assembly and micro-/submicrosized, imprinted inorganic glasses for nanointegration. Further studies have been continuously performed to develop micro-/nanostructured materials for new biomedical applications.

REFERENCES

Abe, M., Hiraoka, M., Takahashi, M., Egawa, S., Matsuda, C., Onoyama, Y., Morita, K., Kakehi, M., 1986. Multiinstitutional studies on hyperthermia using an 8-mhz radiofrequency capacitive heating device (thermotron rf-8) in combination with radiation for cancer-therapy. Cancer 58, 1589–1595.

Abou Neel, E.A., Ahmed, I., Pratten, J., Nazhat, S.N., Knowles, J.C., 2005. Characterization of anitibacterial copper releasing degradable phosphate glass fibres. Biomaterials 26, 2247–2254.

Aono, H., 2014. Development of nano-sized superparamagnetic ferrites having heat generation ability in an AC magnetic field for thermal coagulation therapy. J. Ceram. Soc. Jpn. 122, 237–243.

Aono, H., Sato, M., Traversa, E., Sakamoto, M., Sadaoka, Y., 2001. Design of ceramic materials for chemical sensors: effect of $SmFeO_3$ processing on its surface and electrical properties. J. Am. Ceram. Soc. 84, 341–347.

Aono, H., Hirazawa, H., Naohara, T., Maehara, T., Kikkawa, H., Watanabe, Y., 2005. Synthesis of fine magnetite powder using reverse coprecipitation method and its heating properties by applying AC magnetic field. Mat. Res. Bull. 40, 1126–1135.

Aono, H., Izumi, J., Tomida, M., Sadaoka, Y., 2011a. Synthesis for fine perovskite-type materials using precursor prepared by metal nitrates solution mixed with organic solvent. Maters. Chem. Phys. 130, 973–979.

Aono, H., Watanabe, Y., Naohara, T., Maehara, T., Hirazawa, H., Watanabe, Y., 2011b. Effect of bead milling on heat generation ability in AC magnetic field of $FeFe_2O_4$ powder. Mater. Chem. Phys. 129, 1081–1088.

Aono, H., Ebara, H., Senba, R., Naohara, T., Maehara, T., Hirazawa, H., Watanabe, Y., 2012. High heat generation ability in AC magnetic field for nano-sized magnetic $Y_3Fe_5O_{12}$ powder prepared by bead milling. J. Magn. Magn. Mater. 324, 1985–1991.

Aono, H., Senba, R., Nishimori, T., Naohara, T., 2013. Preparation of $Y_3Fe_5O_{12}$ microsphere using bead-milled nanosize powder for embolization therapy application. J. Am. Ceram. Soc. 96, 3483–3488.

Arma, L.H., Saitoh, A., Ishibashi, Y., Asahi, T., Sueoka, Y., Sakakibara, M., Takebe, H., 2014. Red fluorescence lamellae in calcitic prismatic layer of *Pinctada vulgaris* shell (Mollusc, bivalvia). Opt. Mater. Ex. 4, 1813–1823.

Atlan, G., Delattre, O., Berland, S., LeFaou, A., Nabias, G., Cot, D., Lopez, E., 1999. Interface between bone and nacre implants in sheep. Biomaterials 20, 1017–1022.

Barb, R.-A., Hrelescu, C., Dong, L., Heitz, J., Siegel, J., Slepicka, P., Vosmanska, V., Svorcik, V., Magnis, B., Marksteiner, R., Schernthaner, M., Groschner, K., 2014. Laser-induced periodic surface structures on polymers for formation of gold nanowires and activation of human cells. Appl. Phys. A 117, 295–300.

Berland, S., Delattre, O., Borzeix, S., Catonne, Y., Lopez, E., 2005. Nacre/bone interface changes in durable nacre endosseous implants in sheep. Biomaterials 26, 2767–2773.

Bourrat, X., Rousseau, M., Lopez, E., Stempflé, P., 2005. Nano-composite structure of nacre biocrystal. BioMin 2005, 9th International Symposium on Biomineralization, pp. 6–9.

Dauphin, Y., 2003. Soluble organic matrices of the calcitic prismatic shell layers of two Pteriomorphid bivalves. J. Biol. Chem. 278, 15168–15177.

Duplat, D., Gallet, M., Berland, S., Marie, A., Dubost, L., Rousseau, M., Kamel, S., Milet, C., Brazier, M., Lopez, E., Bédouet, L., 2007. The effect of molecules in mother-of-pearl on the decrease in bone resorption through the inhibition of osteoclast cathepsin K. Biomaterials 28, 4769–4778.

Frédéric, M., Gilles, L., 2007. Unusually acidic proteins in biomineralization. In: Edmund, B. (Ed.), Handbook of Biomineralization. Wiley-VCH, Weinheim, Germany, pp. 273–290.

Fritz, M., Belcher, A.M., Radmacher, M., Walters, D.A., Hansma, P.K., Stucky, G.D., Morse, D.E., Mann, S., 1994. Flat pearls from biofabrication of organized composites on inorganic substrates. Nature 371, 49–51.

Furuhashi, T., Schwarzinger, C., Miksik, I., Smrz, M., Beran, A., 2009. Molluscan shell evolution with review of shell calcification hypothesis. Comp. Biochem. Phys. B 154, 351–371.

Gries, K., Kröger, R., Kübel, C., Fritz, M., Rosenauer, A., 2009. Investigations of voids in the aragonite platelets of nacre. Acta Biomater. 5, 3038–3044.

Grumezescu, A.M., Andronescu, E., Holban, A.M., Ficai, A., Ficai, D., Voicu, G., Grumezescu, V., Balaure, P.C., Chifiriuc, C.M., 2013. Water dispersible cross-linked magnetic chitosan beads for increasing the antimicrobial efficiency of aminoglycoside antibiotics. Int. J. Pharm. 454, 233–240.

Hines, M.A., Scholes, G.D., 2003. Colloidal PbS nanocrystals with size-tunable near-infrared emission: observation of post-synthesis self-narrowing of the particle size distribution. Adv. Mater. 15, 1844–1849.

Hirazawa, H., Kusamoto, S., Aono, H., Naohara, T., Mori, K., Hattori, Y., Maehara, T., Watanabe, Y., 2008. Preparation of fine $Mg_{1-x}Ca_xFe_2O_4$ powder using reverse coprecipitation method for thermal coagulation therapy in an AC magnetic field. J. Alloys Compd. 461, 467–473.

Hirazawa, H., Aono, H., Naohara, T., Maehara, T., Watanabe, Y., 2011. Heat generation ability in AC magnetic field of nano $MgFe_2O_4$-based ferrite powder prepared by bead milling. J. Magn. Magn. Mater. 323, 675–680.

Jackson, A.P., Vincent, J.F.V., Turner, R.M., 1990. Comparison of nacre with other ceramic composites. J. Mater. Sci. 25, 3173–3178.

Jeyadevan, B., 2010. Present status and prospects of magnetite nanoparticles-based hyperthermia. J. Ceram. Soc. Jpn. 118, 391–401.

Jordan, A., Scholz, R., Wust, P., Fähling, H., Felix, R., 1999. Magnetic fluid hyperthermia (MFH): cancer treatment with AC magnetic field induced excitation of biocompatible superparamagnetic nanoparticles. J. Magn. Magn. Mater. 201, 413–419.

Jun, Y.-W., Choi, J.-S., Cheon, J., 2006. Shape control of semiconductor and metal oxide nanocrystals through nonhydrolytic colloidal routes. Angew. Chem. Int. Ed. 45, 3414–3439.

Kawashita, M., Tanaka, M., Kokubo, T., Inoue, Y., Takeshi, Y., Hamada, S., Shinjo, T., 2005. Preparation of ferrimagnetic magnetite microspheres for in situ hyperthermic treatment of cancer. Biomaterials 26, 2231–2238.

Korgel, B.A., Fitzmaurice, D., 1999. Small-angle x-ray-scattering study of silver-nanocrystal disorder-order phase transitions. Phys. Rev. B 59, 14191–14201.

Kuzuya, T., Tai, Y., Yamamuro, S., Sumiyama, K., 2005. Size-focusing of ZnS nanocrystals observed by MALDI-TOF mass spectroscopy. Chem. Phys. Lett. 407, 460–463.

LaMer, V.K., Dinege, R.H., 1950. Theory, production and mechanism of formation of monodispersed hydrosols. J. Am. Chem. Soc. 72, 4847–4854.

Lee, D.C., Smith, D.K., Heitscha, A.T., Korgel, B.A., 2007. Colloidal magnetic nanocrystals: synthesis, properties and applications. Annu. Rep. Prog. Chem., Sect. C 103, 351–402.

Levi-Kalisman, Y., Falini, G., Addadi, L., Weiner, S., 2001. Structure of the nacreous organic matrix of a bivalve mollusk shell examined in the hydrated state using cryo-TEM. J. Struct. Biol. 135, 8–17.

Li, X., 2007. Nanoscale structural and mechanical characterization of natural nanocomposites: seashells. JOM 59, 71–74.

Li, X., Chang, W.-C., Chao, Y.J., Wang, R., Chang, M., 2000. Nanoscale structural and mechanical characterization of a natural nanocomposite material: the shell of red abalone. Nano Lett. 4, 613–617.

Li, X., Xu, Z.-H., Wang, R., 2006. In situ observation of nanograin rotation and deformation in nacre. Nano Lett. 6, 2301–2304.

McPhedran, R.C., Parker, A.R., 2015. Biomimetics: lessons on optics from nature's school. Phys. Today 2015, 32–37.

Motte, L., Billoudet, F., Lacaze, E., Douin, J., Pileni, M.P., 1997. Self-organization into 2D and 3D superlattices of nanosized particles differing by their size. J. Phys. Chem. B 101, 138–144.

Murthy, R., Nunez, R., Szklaruk, F., Erwin, W., Madoff, D.C., Gupta, S., Ahrar, K., Wallace, M.F., Cohen, A., Coldwell, D.M., Kennedy, A.S., Hicks, M.E., 2005. Yttrium-90 microsphere therapy for hepatic malignancy: devices, indications, technical considerations, and potential complications. Radiographics 25 (SPEC. ISS.), S41–S55.

Nakahara, H., Bevelander, G., Kakei, M., 1982. Electron microscopic and amino acid studies on the outer and inner shell layers of *Haliotis rufescens*. Venus Jpn. J. Malacol. 41, 33–46.

Nudelman, F., Chen, H.H., Goldberg, H.A., Weiner, S., Addadi, L., 2007. Spiers memorial lecture. Lessons from biomineralization: comparing the growth strategies of mollusc shell prismatic and nacreous layers in *Atrina rigida*. Faraday Discuss 136, 9–25.

Park, J., Joo, J., Kwon, S.G., Jang, Y., Hyeon, T., 2007. Synthesis of monodisperse spherical nanocrystals. Angew. Chem. Int. Ed. 46, 4630–4660.

Popescu, R.C., Andronescu, E., Grumezescu, A.M., 2014. In vivo evaluation of Fe_3O_4 nanoparticles. Rom. J. Morphol. Embryol. 55 (3 Suppl), 1013–1018.

Prasad, B.L.V., Stoeva, S.I., Sorensen, C.M., Klabunde, K.J., 2002. Digestive ripening of thiolated gold nanoparticles: the effect of alkyl chain length. Langmuir 18, 7515–7520.

Puntes, V.F., Zanchet, D., Erdonmez, C.K., Alivisatos, P., 2002. Synthesis of hcp-Co Nanodisks. J. Am. Chem. Soc. 124, 12874–12880.

Pusey, P.N., van Megen, W., Bartlett, P., Ackerson, B.J., Rarity, J.G., Underwood, S.M., 1989. Structure of crystals of hard colloidal spheres. Phys. Rev. Lett. 63, 2753–2756.

Quan, Z., Fang, J., 2010. Superlattices with non-spherical building blocks. Nano Today 5, 390–411.

Rosensweig, R.E., 2002. Heating magnetic fluid with alternating magnetic field. J. Magn. Magn. Mater. 252, 370–374.

Salem, R., Lewandowski, R.J., Atassi, B., Gordon, S.C., Gates, V.L., Barakat, O., Sergie, Z., Wong, C.-Y., Thurston, K.G., 2005. Treatment of unresectable hepatocellular carcinoma with use of Y-90 microspheres (TheraSphere): safety, tumor response, and survival. J. Vasc. Interv. Radiol. 16, 1627–1639.

Sato, Y., Inoue, N., Ishikawa, T., Ishibashi, R., Obata, M., Aoki, H., Atsumi, T., Komatsu, A., 2013. Pearl microstructure and expression of shell matrix protein genes MSI31 and MSI60 in the pearl sac epithelium of *Pinctada fucata* by in situ hybridization. PLoS One 8, e52372.

Sugimoto, T., 1987. Preparation of monodispersed colloidal particles. Adv. Colloid Interface Sci. 28, 65–108.

Takebe, H., Nonaka, W., Kubo, T., Cha, J., Kuwabara, M., 2007a. Preparation and properties of transparent SnO–P_2O_5 glasses. J. Phys. Chem. Solids 68, 983–986.

Takebe, H., Kuwabara, M., Komori, M., Fukugami, N., Soma, M., Kusuura, T., 2007b. Imprinted optical pattern of low-softening phosphate glass. Opt. Lett. 32, 2750–2752.

Tao, S.L., Desai, T.A., 2007. Aligned arrays of biodegradable poly (ε-caprolactone) nanowires and nanofibers by template synthesis. Nano Lett. 7, 1463–1468.

Wada, K., 1999. Science of Pearl. Shinju Shimbun, Tokyo. (in Japanese).

Westbroek, P., Marin, F., 1998. A marriage of bone and nacre. Nature 392, 861–862.

Wu, C., Wang, C., Zheng, J., Luo, C., Li, Y., Guo, S., Zhang, J., 2015. Vacuolization in Ccytoplasm and cell membrane permeability enhancement triggered by micrometer-sized graphene oxide. ACS Nano 9, 7913–7924.

Yamamuro, S., Sumiyama, K., 2006. Why do cubic nanoparticles favor a square array? Mechanism of shape-dependent arrangement in nanocube self-assemblies. Chem. Phys. Lett. 418, 166–169.

Yamamuro, S., Sumiyama, K., 2008. Solution-phase synthesis of size- and shape-controlled magnetic nanoparticles of transition metal systems and their recent development. Mater. Sci. Tech. 45, 246–251.

Yamamuro, S., Farrell, D.F., Majetich, S.A., 2002. Direct imaging of self-assembled magnetic nanoparticle arrays: phase stability and magnetic effects on morphology. Phys. Rev. B 65, 224431.

Yamamuro, S., Sumiyama, K., Kamiyama, T., 2008. Shape-induced simple cubic arrangement in three-dimensional nanocube self-assemblies. Appl. Phys. Lett. 92, 113108.

Yoshioka, S., Kitano, Y., 1985. Transformation of aragonite to calcite through heating. Geochem. J. 19, 245–249.

CHAPTER

5 MULTIFUNCTIONAL NANOSTRUCTURED BIOPOLYMERIC MATERIALS FOR THERAPEUTIC APPLICATIONS

Ilaria Armentano[‡], Nicoletta Rescignano, Eelna Fortunati*, Samantha Mattioli*, Francesco Morena[†], Sabata Martino[†], Luigi Torre*, Josè M. Kenny*,****

**University of Perugia, Terni, Italy*

***Institute of Polymer Science and Technology, CSIC, Madrid, Spain*

[†]University of Perugia, Perugia, Italy

[‡]Tuscia University, Viterbo, Italy

CHAPTER OUTLINE

1 INTRODUCTION

The application of polymeric nanostructures (Elsabahy et al., 2015) for different therapies has received important attention in recent years (Goldberg et al., 2007; Zhang et al., 2015). They show great promise in the treatment of a wide range of diseases, due to the flexibility with which their structures can be modified, with intricate definition over their shape, chemical compositions, micro- and nanostructures and properties. Advances in polymerization chemistry, engineering materials, and nanotechnology have enabled the engineering of new multifunctional polymeric systems with precise control over the architectures of individual polymer components, so as to direct their assembly and subsequent transformations into nanosystems of selective overall shapes, sizes, internal morphologies, external surface charges, and functionalizations. In addition, incorporation of certain functionalities can modulate the responsiveness of these nanostructures to specific stimuli through the use of remote activation (Elsabahy and Wooley, 2012).

In order to have therapeutic benefits, the drugs, or the drug load, have to reach their specific destinations, one of the sites of action that are at cellular or molecular levels. Although nanocarriers hold promise for therapeutic applications, as tissue engineering (TE) and cancer therapies, their intracellular drug-delivery pathways are not fully understood. In particular, the influence of nanocarrier stability on cellular uptake is still uncertain (Grant et al., 2008; Ta et al., 2008; Wischke and Borchert, 2008). The use of polymeric nanostructures has demonstrated efficiency in improving delivery of diagnostic and therapeutic agents to the target sites, and the feasibility of incorporating several therapeutic/diagnostic/targeting moieties within specific compartments of the nanoparticles (NPs), with control of their navigation in the body, and to the target sites (Subramanian et al., 2015). Further understanding of the nature and microenvironments of biological systems (e.g., different pH, temperature, permeability, drainage, or overexpressing proteins, enzymes, or receptors), and the barriers toward the delivery of various moieties to destinations, which could be either intra- or extracellular has aided the design of nanomaterials that could evade various physiological barriers. Selective delivery to the site of the disease can increase the therapeutic efficacy, imaging contrast, and accuracy, while reducing adverse reactions, and the dose and cost of medications (Elsabahy et al., 2015). The advantages of polymeric nanostructures over other types of nanomaterials are based on the flexibility with which their structures can be modified to yield materials, with the possibility of hierarchical assembly of several nanomaterials of various components into one construct that can be accommodated with a variety of therapeutic, diagnostic, and/or targeting moieties, within selective compartments of the nanodevices. High efficiency in diagnosis and treatment of diseases and improving patient quality of life and compliance can be achieved through understanding the molecular events associated with various diseases, and combining the advances in the design of therapeutic and diagnostic agents and nanomaterials, together with the innovative instruments utilized for monitoring these agents. The use of combinations of different nanostructured materials will allow the development of novel multifunctional nanomedical platforms for multimodal imaging, and simultaneous diagnosis and therapy (Fortunati et al., 2011a; Lai et al., 2003). Nanotechnology offers high potential applications for future biomedical technology. Due to their unique characteristics, including small size comparable to biomolecules, nanostructured materials have emerged as novel bioimaging, diagnostic, and therapeutic agents for future medical fields. The development of nanotechnology provides opportunities to characterize, manipulate, and organize matter systematically at the nanometer scale. Nanotechnology starts with Richard Feynman, in 1959, in his famous historical talk "There's Plenty of Room at the Bottom," where the possibility to manipulate materials at the atomic and molecular levels was introduced for the first time (Toumey, 2009). The word "nanotechnology" was for the first time introduced by Norio Taniguchi, at Tokyo University, referring to the design of materials on the *nanoscale* (Taniguchi, 1974). Up to the present day, the use of different natures of nanomaterials (organic and inorganic) and for various applications has been greatly expanded (Brambilla et al., 2014; Devadasu et al., 2013). Moreover, "*nanomedicine*" has emerged in the medical field to introduce noninvasive systems for the probing of disease, and is also capable of carrying cargo for localized high concentration delivery, with reduction of off-target effects.

Biomaterials with nanoscale organization have been used as controlled release reservoirs for drug delivery, and artificial matrices for TE (Williams, 2009). Drug-delivery systems can be synthesized with controlled composition, shape, size, and morphology. Their surface properties can be manipulated to increase solubility, immunocompatibility, and cellular uptake. The limitations of current drug-delivery systems include suboptimal bioavailability, limited effective targeting, and potential cytotoxicity. Promising and versatile nanoscale drug-delivery systems include NPs, nanocapsules, nanotubes, nanogels, and dendrimers. They can be used to deliver both small-molecule drugs and various classes of biomacromolecules. Whereas traditional tissue-engineering scaffolds were based on hydrolytically degradable macroporous materials, current approaches emphasize the control over cell behaviors and tissue formation by nanoscale topography that closely mimics the natural extracellular matrix (ECM) (Armentano et al., 2010; Pooyan et al., 2012; Rescignano et al., 2014). This chapter will focus on several recent advances in the design of multifunctional polymeric nanostructures that have been utilized for the delivery of diagnostic and/or therapeutic agents, as well as the various barriers toward the clinical development of these materials. After an analysis of the materials and processing technology used to develop the new multifunctional materials, the chapter summarizes the most recent development in utilizing nanostructured materials for applications in drug delivery, cancer therapy, and TE.

2 BIOPOLYMERS

A biomaterial is defined by Williams (2009) as any natural or synthetic material that is able to interact with biological systems. Biomaterials are usually biocompatible (Williams, 2009) and, in order to answer to the needs of the biomedical sector, formulations composed of metals, ceramics, glasses, and/or polymers have been researched for this specific application. In this context, polymer-based materials possess important potential in terms of physical and mechanical properties. Moreover, degradable polymers are of interest because they are able to be degraded, excreted, or resorbed without removal or surgical revision (Ulery et al., 2011). In the design of biodegradable materials for specific therapeutic applications, many important properties must be considered. These systems must (1) not evoke a sustained inflammatory response; (2) possess an appropriate degradation time; (3) have specific and tunable mechanical properties; (4) produce nontoxic products during degradation; and (5) include appropriate permeability and processability (Lloyd, 2002; Ulery et al., 2011). These characteristics are greatly affected by a number of features of degradable polymers, including material chemistry, molecular weight, water adsorption, surface charge, hydrophobicity, erosion, and degradation mechanisms.

Concerning the degradation and erosion mechanism that are of interest for the biomedical application field, hydrolytically degradable polymers possess hydrolytically labile chemical bonds in their backbone, and can be broken down without secondary influence. A number of degradable polymers possess bonds that are susceptible to hydrolysis, and one of the major features that convey significant impact on the capacity of these polymeric families to function as biomaterials is their degradation rates and erosion mechanisms. Table 5.1 shows an extensive investigation of different degradable polymeric families (Burkersroda et al., 2002).

2.1 NATURAL BIOPOLYMERS: POLYSACCHARIDES AND PROTEINS

2.1.1 Chitosan

Chitin, the precursor of chitosan, is a linear polymer highly acetylated and insoluble in water. Chitin is mainly isolated from crustacean wastes (Alishahi and Aider, 2012), it is bound in complexes with

Table 5.1 Summary of Different Polymeric Families' Applications, Advantages, Disadvantages, Degradation Rate, and Structure

Polymer	Applications	Advantages	Disadvantages	λ Degradation Rate Constant (s^{-1})	Structures
Polyphosphazenes	Tissue engineering; vaccine adjuvant	Synthetic flexibility; controllable mechanical properties	Complex synthesis	4.5×10^{-2} to 1.4×10^{-7}	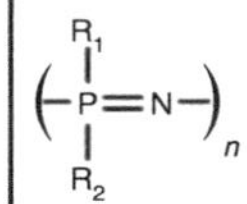
Polyanhydrides	Drug delivery; tissue engineering	Significant monomer flexibility; controllable degradation rates	Low molecular weights; weak mechanical properties	1.9×10^{-3} to 9.4×10^{-9}	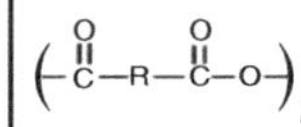
Polyacetals	Drug delivery	Mild pH-degradation products; pH-sensitive degradation	Low molecular weights; complex synthesis	6.4×10^{-5}	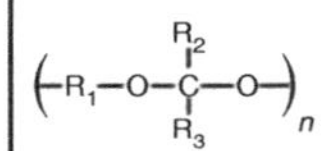
Poly(ortho esters)	Drug delivery	Controllable degradation rates; pH-sensitive degradation	Weak mechanical properties; complex synthesis	4.8×10^{-5}	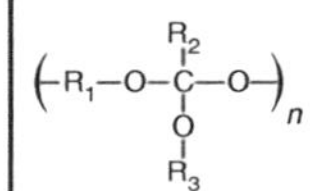
Polyphosphoesters	Drug delivery; tissue engineering	Biomolecule compatibility; highly biocompatible degradation products	Complex synthesis	1.4×10^{-6}	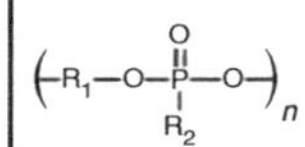
Polycaprolactone	Tissue engineering	Highly processable; many commercial vendors available	Limited degradation	3.5×10^{-8}	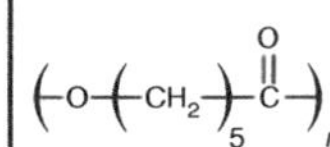
Polyurethanes	Prostheses; tissue engineering	Mechanically strong; handle physical stresses well	Limited degradation; require copolymerization with other polymers	8.3×10^{-9}	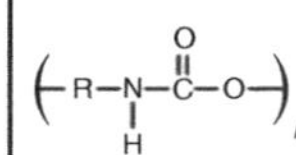
Polylactide	Tissue engineering; drug delivery	Highly processable; many commercial vendors available	Limited degradation; highly acidic degradation products	6.6×10^{-9}	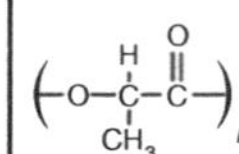
Polycarbonates	Drug delivery; tissue engineering; fixators	Chemistry-dependent mechanical properties; surface eroding	Limited degradation; require copolymerization with other polymers	4.1×10^{-10}	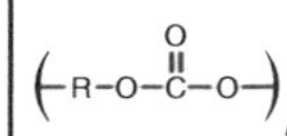
Polyamides	Drug delivery	Conjugatable side group; highly biocompatible degradation products	Very limited degradation; charge induced toxicity	2.6×10^{-13}	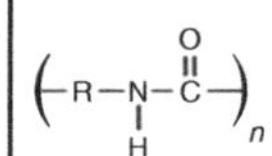

Reprinted with permission from Ulery, B.D., Nair, L.S., Laurencin, C.T., 2011. Biomedical applications of biodegradable polymers. J. Polym. Sci. Part B Polym. Phys. 49, 832–864.

proteins and minerals and, for its production at industrial scale, an acid treatment (decalcification) followed by an alkaline treatment (deproteination), and a decolorization step are used. In order to synthesize chitosan, an additional alkaline treatment is performed to deacetylate chitin. At this moment, the research activities are directed to more environment friendly methods that involved the use of enzymes and/or microorganisms for the different isolation steps, while different factors, like alkali concentration, the chitin: alkali ratio, incubation time, atmosphere, temperature, type of chitin source, particle size, heterogeneous/homogeneous *N*-deacetylation, and single or multiple process play a role in the alkaline deacetylation of chitosan. Most chitosans are insoluble in water, and also in most common organic solvents; however, they can be easily dissolved in acidic aqueous solutions. The good film-forming properties of chitosan allow the production of films, coating materials, or membranes semipermeable to gases, while chitosan has excellent properties, such as biodegradability, biocompatibility, and nontoxicity (Barikani et al., 2014). For biomedical applications, chitosan hydrogels and networks (for instance, interpenetrating polymer networks) formed by aggregation or complexation have been deeply investigated (Berger et al., 2004; Wang and Stegemann, 2010). In farming applications, chitosan is widely used in veterinary drug delivery, while chitosan-based products have been used for the delivery of chemotherapeutics, such as antibiotics, antiparasitics, anesthetics, painkillers, and growth promotants to mucosal epithelium for absorption for local or systemic activity (Sevda and McClure, 2004). However, it is not clear whether these applications are really commercialized. Chitosan-based products are widely applied in cosmetic formulations, and they are highly commercialized there. Interesting properties, such as minimal foreign body reaction, an intrinsic antibacterial nature, and the ability to be molded into various geometries and forms, such as porous structures suitable for cell ingrowth and osteoconduction, make chitosan suitable for biomedical applications. Covalent cross-linking of chitosan guarantees the formation of hydrogels with a permanent network structure, since irreversible chemical links are formed. This type of linkage allows absorption of water and/or bioactive compounds without dissolution, and permits drug release by diffusion under pH-controlled conditions. Chitosan derivatives can also be applied in various tissue-engineering applications, namely skin, bone, cartilage, liver, nerve, and blood vessels (Kim et al., 2008). Injectable chitosan hydrogels can be used in cancer treatment. In literature, formulation protocols for in situ hydrogel systems, their cytotoxic properties, loading, and in vitro release of drugs, their effect on cell growth in vitro, and inhibition of tumor growth in vivo, using mouse models are often reported (Grant et al., 2008; Ta et al., 2008; Wischke and Borchert, 2008).

2.1.2 Cellulose

Cellulose is the most abundant natural polymer on Earth, and—like starch—is also comprised of glucose monomer units. The glucose units in cellulose are joined together via β-1,4 glycosidic linkages that enable cellulose chains to pack tightly together and form strong interchain hydrogen bonds. Because of its unique structural and mechanical properties, combined with a highly hydrophilic nature, biocompatibility, transparency, and nontoxicity, cellulose-based structures make attractive candidates for a wide range of applications in various fields, especially those related to biomedical and biotechnology applications (Dahman, 2009; Rescignano et al., 2014, 2015). Recent studies demonstrated that different types of cellulose-based nanostructured materials present desired properties and functions to be applied as nanocomposite films, membranes, and hydrogels, with potential application in TE and therapeutic applications. A significant effort is under way to match the properties of the lignocellulosic materials with the ones required for specific applications. The use of lignocellulosic components in drug-delivery systems has been studied for several years, while the use of cellulose derivatives as excipients in drug formulations is already a standard in the pharmaceutical industry. Recently, Rescignano et al. reported on the possibility to use cellulose nanocrystals (CNC), obtained from microcrystalline cellulose by an acid hydrolysis procedure, as stabilizers during the synthesis of biopolymeric poly(D,L-lactide-*co*-glycolide) PLGA NPs. The results obtained suggest that PLGA NPs with CNC as emulsion stabilizer might represent promising formulations, opening new perspectives in the field of self-assembly of biodegradable nanomaterials for medical and pharmaceutical applications (Rescignano et al., 2015).

Still in this context, several water-soluble cellulose derivatives can be used as a monocomponent of multicomponent systems, and they can be used also as reinforcements phases in some other biodegradable polymers to form hydrogel networks with specific properties, in terms of swelling capability and sensitivity to external stimuli (Sannino et al., 2009). Recently Sanna et al. (2013) reported on the synthesis and characterization of thermoresponsive poly(*N*-vinylcaprolactam)—PNVCL, nanocomposite hydrogels containing nanocrystalline cellulose produced by a frontal polymerization technique; this is a convenient, easy, and low energy-consuming method of macromolecular synthesis. The authors proved that the presence of CNC resulted in a significant increase of the mechanical properties, even at very low CNC concentrations, as confirmed by rheological tests indicating that the nanocellulose has great potential to reinforce PNVCL polymer hydrogels. Finally, considering the real and practical application of cellulose-based devices for the biomedical sector, wound dressing systems based on pure oxidized regenerated cellulose (ORC) have been developed, such as Tabotamp, as well as a mixture from collagen and ORC named Promogran. Tabotamp is a thin gauze layer used in acute wounds, such as trauma, surgical injuries, and burns, while Promogran is a commercial spongy collagen matrix and cellulose introduced in the US and European markets for the treatment of exuding diabetic and ulcer wounds (Fernandes et al., 2013). Moreover, bacterial cellulose-based porous membranes have been applied for guided bone regeneration in bone defects, periodontal lesions, and as a resorbable barrier membrane for preventing the invasion of fibroblast cells and fibrous connective tissue into bone defects (Fernandes et al., 2013).

2.1.3 Collagen

Collagen is a protein present in the human body, and it represents the major component of skin, ligaments, tendon, bone, and cartilage. Collagen is composed of polypeptide strands bearing triamino acid blocks of glycine-X-Y, where X and Y are most commonly proline and hydroxyproline (Ramshaw et al., 1998). In addition, collagen is processable, and shows high solubility in acidic aqueous solutions, allowing for the fabrication of collagen sponges; moreover, it shows advantages, such as good mechanical strength, and enzymatic degradability by collagenases and metalloproteinases (Przybyla and Chmielewski, 2010). Collagen is used in surgery biomedical applications as a suture material (Yilmaz et al., 2010), as synthetic sutures, to reduce inflammation and infection rates. Collagen is used to realize drug-delivery devices for the release of antibiotics (Bushnell et al., 2008). The Food and Drug Administration (FDA) approved the use of a number of collagen-containing solutions as a treatment for bleeding during surgery. In this scenario, collagen was used also as a TE cartilage (Mueller-Rath et al., 2010; Pulkkinen et al., 2010), tendon (Gurkan et al., 2010), and skin engineering (Reckenrich et al., 2011), in particular in load-bearing applications. Moreover, composites of hydroxyapatite and collagen are used to produce composites for realizing natural bone (Gleeson et al., 2010). These advances promise the possibility to regenerate more complex structures, such as organs' three-dimensional (3D) patterning of multiple tissue types. The potential properties of collagen could be improved or modulated combining different biodegradable polymers, or associating bioactive molecules (Casper et al., 2007; Sahoo et al., 2010). Collagen is also combined with others polymers, such as polylactide-*co*-glycolide

(PLGA) (Sahoo et al., 2007), polylactic acid (PLA) (Liu et al., 2010b), polycaprolactone (PCL) and/or chitosan, in order to improve functional properties (Wang et al., 2010a).

2.1.4 Alginate

Alginates are generally referred to as a family of polyanionic copolymers derived from marine kelp, mainly the brown sea algae (Goh et al., 2012). As a commercial entity, Stanford patented an impure alginate salt in the early 1880s, and this natural anionic polysaccharide subsequently received great attention. Its interesting chemical and physical properties resulted in many commercial applications (Cheetham, 1979). In the more recent period, alginates are of growing importance in the healthcare and pharmaceutical industries. Since the first successful encapsulation of islet cells by Lim and Sun (1980), alginate matrices were extensively employed for cell culture and transplantation (Stevens Molly et al., 2004). Other workers further explored the use of hybridoma cells in monoclonal antibodies production, and in immobilizing drugs for sustained release systems (Dubrot et al., 2010; Wang et al., 2010b). Oral administration of low molecular weight sodium alginate (10–100 kDa) could potentially combat hypercholesterolemia and diabetes mellitus, by encapsulation of cholesterol and blood glucose following physiological gelation of alginic acid in the stomach (Kimura et al., 1996). Moreover, the hemostatic and absorbent properties make alginates suitable for wound dressing management (Lloyd et al., 1998). A recent study suggested that the uronate content of alginates might activate prothrombotic coagulation in the formation of blood clots, while zinc-containing alginate dressings were shown to possess greater potentiating effect on prothrombotic coagulation and platelet activation than calcium (Segal et al., 1998).

2.2 BIOPOLYMERS FROM BIOTECHNOLOGY

2.2.1 Polyglycolide

Polyglycolide (PGA) was one of the first degradable polymers investigated for biomedical and therapeutic applications, and it was used as degradable suture, such as Dexon, due to its interesting thermal and mechanical properties [a melting point (T_m) greater than 200°C, a glass transition temperature (T_g) of 35–40°C, and high tensile strength (12.5 GPa)] (Katz and Turner, 1970; Maurus and Kaeding, 2004). From 1984 to 1996, PGA was marketed as an internal bone pin under the name Biofix, but since 1996 Biofix has been converted to a poly(L-lactide) base for better long-term stability (Burns, 1995). However, the fast degradation characteristics and the insolubility in common solvents limited the research activities that involved PGA as a long-term delivery system. On the contrary, recent researches are focused on short-term TE scaffolds. PGA is often fabricated into a mesh network, and has been used as a scaffold for bone (Wang et al., 2010b), cartilage (Mahmoudifar and Doran, 2010), tendon (Cao et al., 2010), tooth (Ohara et al., 2010), vaginal (Sayasneh and Johnson, 2010), intestinal (Aysan et al., 2010), lymphatic, and spinal regeneration (Aysan et al., 2010; Dai et al., 2010). Although a lot of different research is conducted with PGA, and many commercial products are made from it, the rapid degradation leads to the loss of mechanical strength and significant local production of glycolic acid that have been linked to a strong, undesired inflammatory response, limiting, sometimes, its application in the biomedical field (Ulery et al., 2011).

2.2.2 Polylactic acid

PLA is a biodegradable aliphatic polyester, obtained from agricultural products, such as corn, sugarcane, and others sources. In the process of fermentation of sugars, various monomers are produced that are converted to polymers. The PLA pellets are obtained through direct polycondensation of lactic acid monomers, or through ring-opening polymerization of lactide. Since lactic acid exists as two optical isomers, L- and D-lactic acid, there exist four forms: poly(L-lactic acid) (PLLA), poly(D-lactic acid) (PDLA), poly(D,L-lactic acid) (PDLLA), and meso-poly(lactic acid), but due to their intrinsic properties, only PLLA and PDLLA were commonly studied and used in the biomedical field. High molecular weight PLLA has been shown to take longer than 5 years to be completely resorbed in vivo. Because of the slow degradation time, limited research has been recently conducted into drug delivery by PLLA systems alone (Zielhuis et al., 2007). Recently, scientists have either developed modification techniques (shortening of the polymer and decrease in crystallinity facilitate a more rapid degradation phenomenon), or have blended or copolymerized PLLA with other polymers in order to reduce its degradation time. Under the product name Fixsorb, PLLA has been used as a bone fixator (Ueda and Tabata, 2003). PLLA is also extensively used in TE applications as scaffolds for bone, cartilage, tendon, neural, and vascular regeneration (Ulery et al., 2011). Poly(L-lactic acid)/keratin tridimensional scaffolds were recently developed in our lab by using two main solvent based techniques: solvent casting particulate leaching in the presence of paraffin microspheres as porogen, and thermally induced phase-separation process. Keratins from different sources were used proving that they can be considered as suitable fillers able to modify the pore architecture and interconnection of 3D PLLA/keratin porous composites useful as biocompatible and nontoxic support for adult bone-marrow mesenchymal stem cells (Puglia et al., 2015). Composite materials include PLLA combined with PDLLA, poly(lactide-*co*-glycolide) (PLGA), PCL, poly(ethylene glycol) (PEG), collagen, and chitosan (Ulery et al., 2011).

2.2.3 Polylactide-co-glycolide

PLGA is produced by random copolymerization of PLA (L- and D,L-lactide forms) and poly(glycolic acid) (PGA). It represents one of the most used biodegradable polymers for biomedical applications (as suture, drug-delivery devices, and tissue-engineering scaffolds). The PLGA characteristics can be changed and modulated modifying the PLA and PGA copolymer composition; moreover, it is an amorphous polymer and it degrades rapidly, so PLGA has been utilized extensively in drug-delivery applications. PLGA has been used to deliver chemotherapeutics (Liu et al., 2010a), antibiotics (Farazuddin et al., 2010), vaccines (Thomas et al., 2010), analgesics (Vega et al., 2008), and antiinflammatory drugs (Zolnik and Burgess, 2008). Moreover, to facilitate and to control drug delivery into the human body, PLGA is used to produce microspheres (Zolnik and Burgess, 2008), microcapsules, nanofibers (Ionescu et al., 2010), or nanospheres (Gavenis et al., 2010). The release rate of drugs depends on the composition of PLGA, on bulk erosion, and on water diffusion of the coating.

PLGA demonstrates great cell adhesion and proliferation properties; this is why it is used in TE in order to realize scaffolds, using different techniques: porogen leaching (Narayan and Venkatraman, 2008), gas foaming (Zhu et al., 2008), electrospinning (Aviss et al., 2010), microsphere sintering (Jabbarzadeh et al., 2008), polymer printing (Lee et al., 2008), or a combination of these techniques (Perron et al., 2009). Finally, in more recent years, novelty and innovation in TE was introduced utilizing 3D printing in order to design and generate structured tissue-like villi for smooth muscle (Lee et al., 2008).

2.3 BIOPOLYMERS FROM MICROORGANISMS

2.3.1 Polyhydroxyalkanoates

The family of bacterially synthesized biopolyesters, that is, the family of polyhydroxyalkanoates (PHAs), represents biodegradable, biocompatible, thermoprocessable polymers with flexible strengths and important mechanical response that have found applications in the form of packaging materials, including films, boxes, coatings, fibers and foam materials, biofuels, medical implants, and drug-delivery carriers. However, the cost of these polymers limits their involvement in large-scale applications (Halami, 2008). The PHAs can be produced using various renewable raw materials. Currently, the majority of PHA is produced using plant sugar, which can be easily obtained from sugarcane, corn sugar, and sugar beet, making it an ideal raw material for PHA. A considerable amount of PHA is also produced using plant or vegetable oils, such as soybean oil, palm oil, and corn oil.

The most common used polymer of PHA family is poly(3-hydroxybutyrate) (PHB), a semicrystalline isotactic polymer that undergoes surface erosion due to its hydrophobicity, degradation properties, and crystallinity (Abe and Doi, 1999). The biocompatibility, processability, and degradability of PHB make it an excellent candidate for use in long-term TE applications (Marcal et al., 2010). Unfortunately, the stability of PHB makes it a poor candidate for controlled delivery applications. To widen the applicability of PHB as a biomaterial, PHB is most commonly copolymerized with 3-hydroxyvalerate to create poly(3-hydroxybutyrate-*co*-3-hydroxyvalerate) (PHBV). Although the addition of HV content improves the biomaterial potential of PHB, degradation is still too slow for other biomedical applications.

2.4 BIOPOLYMERS FROM PETROCHEMICAL SOURCES

2.4.1 Polycaprolactone

PCL is a semicrystalline polyester with great organic solvent solubility, and it has found favor as a long-term implant delivery device due to the very low in vivo degradation rate, and its high drug permeability. Moreover, it possesses a melting temperature of 55–60°C, and a glass transition temperature at around 54°C (Patlolla et al., 2010). In this context, Capronor is a commercial PCL-based product able to deliver levonorgestrel in vivo for over a year, and has been on the market for over 25 years. Moreover, research activities are focused into the development of microsized and nanosized drug-delivery vehicles based on PCL, although the slow degradation rate (2–3 years) represents an important limitation to obtaining the FDA approval for pure PCL-based products. For this reason, PCL is often blended or copolymerized with other polymers, such as PLLA, PDLLA, PLGA, and polyether to expedite overall polymer erosion, while its processability allows the easy design and development of its scaffolds, electrospun fibers (Luciani et al., 2008), and it can be used for the production of porous structures by porogen leaching (Guarino and Ambrosio, 2008). Moreover, PCL and PCL composites have been used as tissue-engineering scaffolds for the regeneration of bone, ligament, cartilage, skin, nerve, and vascular tissues (Ulery et al., 2011).

2.4.2 Polyurethanes

Polyurethanes (PU) are biocompatible, moldable, biostable polymers typically synthesized by polycondensation of diisocyanates with alcohols and amines, and are composed of hard and soft segments (Guelcher, 2008) (Table 5.1). PU were used in prostheses, such as cardiac assist devices, small vascular shunts, and tracheal tubes. A commercial polyurethane product, NovoSorb (PolyNovo), is a two-component system that cures in situ. Pure PU are degradation resistant and, for this reason, they are poor candidates for drug delivery and many TE applications, while they are usually utilized in multidegradable group polymers to expand their biomedical potential. Finally, biobased PU have been recently considered and deeply studied.

3 MULTIFUNCTIONAL NANOSTRUCTURED MATERIALS

3.1 SHAPE-CONTROLLED NANOSTRUCTURES

The development of new nanomaterials for the treatment of different diseases has been receiving greater attention in recent years, and has now become an important field of medical research. One major advantage of nanomaterials is their potential to be used as noninvasive therapeutic tools. NPs have been suggested to be ideal for several therapies (Brannon-Peppas and Blanchette, 2012) because of their favorable biodistribution and clearance/accumulation behavior (Longmire et al., 2008).

A wide range of NPs has been designed to reach tumors, including liposomes, quantum dots (QDs), polymers, dendrimers, and metal (Arvizo et al., 2012; Azzazy et al., 2007; Baker, 2009; Brown and Khan, 2012). Although all of these types of NPs have been studied for cancer diagnosis and tumor drug delivery and release, each type of NPs can exhibit different and sometimes unique properties, which make them useful for different applications.

Chitosan, albumin, heparin, dextran, gelatin, alginate, and collagen, as well as synthetic polymers, such as polyglutamic acid, PLA, PCL, and poly-D,L-lactide-*co*-glycolide have been widely used to prepare NPs and encapsulate drugs for several therapies (Liu et al., 2008; Rao and Geckeler, 2011). Based on their structural features, polymeric NPs can be further classified into nanocapsules and nanospheres (Letchford and Burt, 2007). In nanocapsules, the shells are usually filled with an aqueous or oil solution, which can contain a solubilized drug. Nanospheres consist of a solid mass, which may be impregnated with a bioactive agent (Letchford and Burt, 2007). Other classes of polymeric NPs include polymeric micelles, consisting in amphiphilic block copolymers that self-assemble into micelles in aqueous solutions (Thipparaboina et al., 2015), and dendrimers, which consist of extensive branching of oligomeric units to form a single entity with a myriad of surface groups (Kesharwani et al., 2014). Polymeric NPs can be synthesized either by polymerization of monomer units, or from a preformed polymer (Pinto Reis et al., 2006). Drugs can be loaded into polymeric NPs by physical entrapment or chemical conjugation (Wang et al., 2009). In nanocapsules, the drug can be physically loaded in the interior, if soluble in the liquid phase contained in the nanocapsule, or via conjugation to the polymer chains. In nanospheres, the drug can also be dispersed or covalently bound to the polymer matrix, while in micelles, hydrophobic drugs are generally encapsulated in their hydrophobic interior (Letchford and Burt, 2007). The choice of material impacts various properties, including drug encapsulation, immunogenicity, and targeting. At the same time, the design of NPs, such as size, shape, flexibility, and compartmentalization will also impact NP performance. These two attributes (choice of material and NP design) collectively determine the therapeutic outcome.

The utilization of these materials in the field of biomedicine requires a controlled interaction with biomacromolecules, as well as a high degree of biocompatibility. Several issues should be taken in consideration with regard to nanomedicines' progress into clinical practice (Walkey and Chan, 2012). In particular, hydrophobic/hydrophilic balance, hydrodynamic diameter, and surface charge are important physicochemical parameters that can affect the NP release and in vivo biodistribution (Alexis et al., 2008). For example, the use of PEG to obtain a chemical surface functionalization of NPs allows a reduction of opsonization phenomena, and uptake processes in the reticuloendothelial system, as well as improving NPs and/or their metabolites release from the body (Spek et al., 2015).

3.2 NANOCOMPOSITES

Nanocomposites represent an important class of nanomaterials, and there have been some fascinating developments in this area in the past decade (Jorfi et al., 2013). Polymeric nanocomposites consisting of inorganic NP and organic polymers represent a new class of materials that exhibit improved performance, compared to their microparticle counterparts. Incorporation of inorganic NPs into a polymer matrix can significantly affect the properties of the matrix (Armentano et al., 2010). The resulting nanocomposite might exhibit improved thermal, mechanical, rheological, electrical, catalytic, fire retardancy, and optical properties. The properties of polymer composites depend on the type of NPs that are incorporated, their size and shape, their concentration, their dispersion, and their interactions with the polymer matrix. The main problem with polymer nanocomposites is the prevention of particle aggregation. It is difficult to produce monodispersed NPs in a polymer matrix because NPs agglomerate due to their specific surface area and volume effects. This problem can be overcome by modification of the surface of the inorganic particles. The modification improves the interfacial interactions between the inorganic particles and the polymer matrix (Armentano et al., 2010; Jorfi et al., 2013). One of the major challenges in the synthesis of nanocomposites is how to design and prepare the desired structure with various functions, such as catalytic, magnetic, electronic, antibacterial, and optical properties, as well as surface functionality coupled with biocompatibility, low toxicity, and high stability (Armentano et al., 2010; Fernandes et al., 2013). Different processing methods can be used to generate polymer nanocomposite biomaterials, and these can be divided in solution-based or melt-blending systems.

In the context of a nanocomposite approach, considering their safety and efficacy, cellulose nanostructures and, in particular, cellulose nanocrystals have recently attracted increasing attention in biomedical applications, as reinforcements in composite systems. Dense two-dimensional (2D) films and 3D porous scaffolds with different properties, architectures, and pore sizes can be produced from synthetic or naturally occurring materials and cellulose nanofibers, cellulose nanocrystals, and bacterial nanocellulose; these are widely used as biobased nanoreinforcements in several polymeric matrices owing to their superior mechanical properties (Fernandes et al., 2013). A lot of research work has been performed all over the world on the use of cellulose fibers as reinforcing material for the preparation of various types of composites. However, certain drawbacks, such as incompatibility with the hydrophobic polymer matrix, the tendency to form aggregates during processing and the water-swellable nature of cellulose, especially in its amorphous regions, greatly reduce the potential of natural fibers as reinforcements in polymers.

Recently, Pooyan et al. (2012) have developed a cellulose-based nanohybrid system in which CNC were dispersed within a cellulose acetate propionate matrix to form a 3D rigid percolating network, aimed at fabricating a potential scaffold of small diameter vascular grafts (Pooyan et al., 2012). The obtained bionanocomposites exhibited excellent mechanical performance at body temperature, claiming that CNC impart significant strength and directional rigidity even at very low concentration. In a multifunctional nanocomposite approach, Rescignano et al. (2014) recently described the production of a new generation of hybrid bionanocomposites dense films based on a poly(vinyl alcohol) (PVA) matrix reinforced with a combination of cellulose nanocrystals and poly(D,L-lactide-*co*-glycolide) NPs loaded with bovine serum albumin fluorescein isothiocynate conjugate (FITC-BSA). The authors demonstrated that the studied nanocomposites are suitable to vehiculate biopolymeric NPs to adult bone marrow mesenchymal stem cells successfully, representing a new tool for drug-delivery strategies. In different ternary nanocomposite systems based on PVA matrix, George et al. (2012) combined bacterial CNC

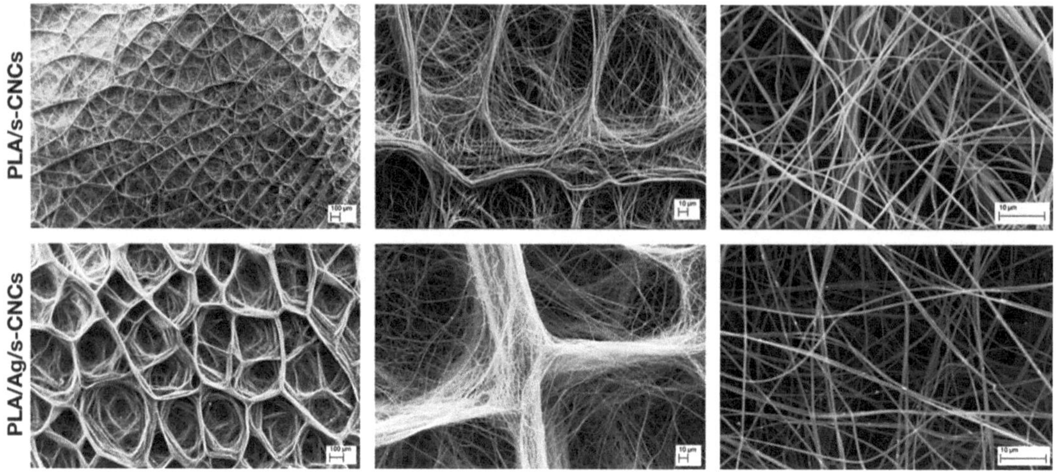

FIGURE 5.1 SEM Micrographs of Electrospun PLA/s-CNC and PLA/Ag/s-CNCs Porous Mats

Reprinted with permission from Cacciotti, I., Fortunati, E., Puglia, D., et al., 2014. Effect of silver nanoparticles and cellulose nanocrystals on electrospun poly(lactic) acid mats: morphology, thermal properties and mechanical behavior. Carbohydr. Polym. 103, 22–31.

and silver (Ag) NPs (1 wt.%) as nanofillers (George et al., 2012). The fabrication of ternary fibrous mats based on poly(lactic) acid, cellulose nanocrystals, both pristine (p-CNC) and modified with a commercial surfactant (s-CNC) and AgNPs by electrospinning was instead reported by Cacciotti et al. (2014). The addition of Ag and/or pristine p-CNC fillers did not remarkably affect the fiber morphology and the average size dimension. The study demonstrated that it is possible to modulate the thermal, mechanical properties, and especially morphological properties of PLA fibers, properly selecting the cellulose-based reinforcements able to induce specific topography useful for the final applications. On the other hand, the authors proved that the presence of surfactant modified s-CNC promoted the self-assembling of submicrometric fibers in bundles randomly organized in polygonal structures, such as pentagons and hexagons, resulting in a honeycomb-like network (Fig. 5.1).

Nanocomposite films containing antimicrobial agents as, for example, AgNPs, have been deeply investigated and considered in biomedical and therapeutic sectors to prevent the attack of a broad spectrum of microorganisms (Rai et al., 2009). Despite, in fact, considerable research and development efforts (Schneider et al., 2008; Xu et al., 2008), the problem of infections related to biomedical devices and implants, persists. Bacteria evidently can readily colonize surfaces of synthetic materials, such as those used for the fabrication of catheters, hip and knee implants, and many other devices.

There have been substantial research efforts toward engineering nanocomposite films loaded with Ag and/or AgNPs in order to achieve a more prolonged release. Ho et al. (2004) presented an interesting approach, first forming AgNPs within a thin film of polyethylene imine. The polymer was copolymerized with 2-hydroxyethyl acrylate, which allowed further surface grafting of PEG. In this manner, coatings with two functions were generated, with the PEG layer preventing bacterial adhesion and the silver ions (Ag^+) released from the coating toxic against *Staphylococcus aureus* (*S. aureus*). Kong and Jang (2008) synthesized poly(methyl methacrylate) nanofibers containing AgNPs by radical-mediated dispersion polymerization, and tested antibacterial properties against Gram-negative (*Escherichia coli*) and Gram-positive *S. aureus* bacteria, showing that the Ag-loaded nanofibers had superior antimicrobial efficacy compared with that of Ag sulfadiazine and Ag nitrate at the same Ag concentration. Fortunati et al. (2011a) demonstrated that the dispersion of a nanofiller phase as inorganic NPs, and particularly AgNPs, in the polymer can modify the matrix surface and the thermal properties, inducing changes in the polymer degradation process directly related to the efficacy as antibacterial formulation. The in vitro degradation studies of PLGA/Ag nanocomposites were conducted by the authors under physiological conditions, over a 5-week period, and compared to the behavior of the neat polymer. Furthermore, the Ag^+ released upon degradation were monitored to obtain information on the properties of the nanocomposites during the incubation (Fortunati et al., 2011a). The results imply that PLGA properties can be modified introducing a small percentage of AgNPs, and that the AgNPs do not affect the overall degradation mechanism of PLGA. Furthermore, the presence of AgNPs induces a mineralization process when nanocomposite samples are immersed in the buffer, as a consequence of the combined effect of AgNPs and the induced nanocomposite surface topography. The Ag^+ release in solution can be controlled by the polymer degradation processes, evidencing a possible prolonged antibacterial effect, during PLGA immersion. The results of the research suggest that the combination of biodegradable polymers and AgNPs opens a new perspective for the use of nanomaterials with tunable properties, obtaining antimicrobial surfaces for biomedical applications.

In more multifunctional nanocomposite approach, we recently investigated the possibility to use a poly(ε-caprolactone) (PCL) biodegradable matrix reinforced with carbon nanotubes (SWCNTs) and AgNPs as a potential support for primary human bone marrow–mesenchymal stem cells (hBM–MSCs) (Fortunati et al., 2011b). As previously discussed, the chance to combine two different nanostructures with specific properties in a new advanced functional system represents an innovative strategy to transfer two definite properties in a polymer, developing multifunctional nanocomposites. The new designed SWCNTs and AgNPs-based materials possessed the unique properties offered by the synergistic interaction of the two different reinforcement phases. The suitability of these conductive composite films as support for hBM–MSCs cells was demonstrated, showing comparable viability and cell/material interaction in the culture period. AgNPs and SWCNTs do not have a cytotoxic effect on hBM–MSCs in the produced binary and ternary composites, guarantying, at the same time, an antimicrobial response of the produced formulations that encourage the application of the AgNPs as conductive antimicrobial nanostructures, inducing stem cell activation.

In another work, we generated nanostructured PLLA-based biomaterials for the study of the effect of physical, chemical, and mechanical properties, on critical mechanosensitive molecules involved on the stem cell functions. Interestingly, hBM–MSCs and human umbilical cord mesenchymal stem cells (hUCMS) interacted differently when seeded on nanostructured PLLA-based biomaterials. While hBM–MSCs showed a canonical mesenchymal morphology with F-actin-containing fibers arranged on the major cellular axis, hUCMS changed their fibroblast-like morphology and acquired a sphere-like structure (Armentano et al., 2014). 3D biomaterials are often used as scaffolds for cell–material interaction in order to guide cell adhesion, migration, and proliferation processes. In particular, 3D hydrogels mimic the native tissue microenvironment due to their porous and hydrated molecular structure. An emerging approach is to reinforce polymeric hydrogels and to include multiple functionalities focused on incorporating NPs within the hydrogel network. A wide range of NPs, such as carbon-based, polymeric, ceramic, and metallic nanomaterials, can be integrated within the hydrogel networks to obtain nanocomposites with superior properties and tailored functionality. Nanocomposite hydrogels can be engineered to possess superior physical, chemical, electrical, and biological properties (Gaharwar et al., 2014).

3.3 NANOSTRUCTURED SURFACE

The surface topography and the physicochemical surface characteristics of biopolymeric characteristics are important parameters that influence the basic reactions of cells on material surfaces. Substratum topography has direct effects on the abilities of cells to orient themselves, migrate, and produce organized cytoskeletal arrangements (Flemming et al., 1999), in particular, to influence cell proliferation (Dalby et al., 2002; Trinkaus-Randall et al., 1988), cell morphology (D'Angelo et al., 2010; Martino et al., 2009a; Mattioli et al., 2014), differentiation (Dalby et al., 2007; Popat et al., 2007; Reid, 1994), gene expression (Andersson et al., 2003; Chiquet et al., 1996; Chou et al., 1995), and function (Flemming et al., 1999; Teixeira et al., 2003). Numerous recent studies have highlighted the importance of nano- and microscale structure, substrate mechanics, and 3D culture environments in regulating cell adhesion, morphology, migration, signaling, and differentiation (Romano et al., 2011). The cell reactions to a material surface are complex, and cover different phenomena, such as protein adsorption at the material surface, and physical–chemical interactions between cells and materials.

There are different techniques for producing surface topography, and they can be divided in two categories: top-down and bottom-up (McHugh et al., 2013).

Top-down methods rely on removing molecules from the starting substrate to produce topography, and are typically flexible, but expensive. Photolithography, X-ray lithography, and electron-beam lithography are three techniques used to transfer patterns into photoresist, a photosensitive polymer. These processes use ultraviolet light, X-rays, or electrons, respectively, to alter photoresist solubility in a pattern. The minimum resolution of photolithography is approximately 220 nm, and limited by the wavelength of light used, while X-ray and electron-beam lithography resolution can reach 1 nm, albeit at a substantially higher cost. These processes typically involve coating a substrate (traditionally a silicon wafer) with photoresist, and exposing the photoresist to light, X-ray, or electrons through a patterned photomask (McHugh et al., 2013).

Fig. 5.2 shows a scheme of microfabrication processes capable of producing topography by using a top-down approach. Photo- or electron-beam lithography (arrows) is used to transfer a pattern into photoresist. This pattern can then be used with subsequent dry (reactive ion) etching, or directly through soft lithography. These processes can be used to produce varying biomaterial topographies, even when using the same initial photomask pattern.

Alternately, bottom-up methods involve the addition of material onto the substrate, and are less expensive, but offer fewer patterning options (Deok-Ho et al., 2008). Among the different techniques that can be used to modify material surfaces, plasma-based processes represent a very versatile tool, since it is possible to develop nanotopographical surfaces both in top-down and bottom-up approaches. Plasma enhanced-chemical vapor deposition (PE-CVD) treatment, deposition, and etching processes play a very important role in the biomaterial field because of their ability to tune the surface chemistry and properties of materials, at room temperature and in a controlled way, without altering their bulk properties, for a wide range of applications. (Armentano et al., 2010; d'Agostino et al., 2005). Plasma treatment of polymers is now a widely used industrial technique to modify the surfaces of many different materials (metals, semiconductors, polymers, ceramics), and it permits to modify uniformly the surface without affecting the bulk properties. Fig. 5.3 shows a schematic representation of the main processes that can be obtained by using plasma on the biopolymeric surface. Furthermore, in order to improve the interaction between cell and biomaterial, the nanostructured surface can be coated with distinct biomacromolecules, for example, particular ECM-proteins, such as laminin, fibronectin, or collagens (Romano et al., 2011).

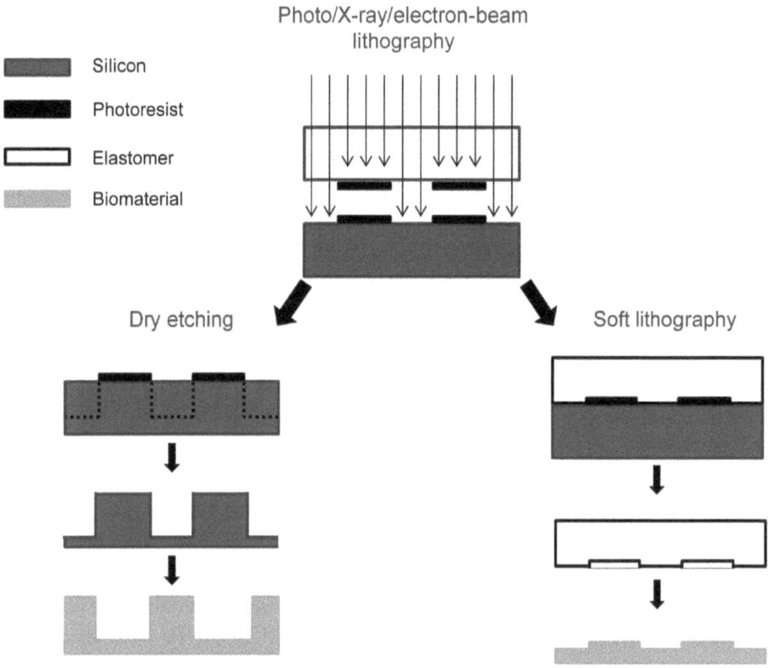

FIGURE 5.2 Description of Microfabrication Processes Capable of Producing Topography

Photo- or electron-beam lithography *(arrows)* is used to transfer a pattern into photoresist. This pattern can then be used with subsequent dry (reactive ion) etching, or directly through soft lithography. These processes can be used to produce varying biomaterial topographies even when using the same initial photomask pattern.

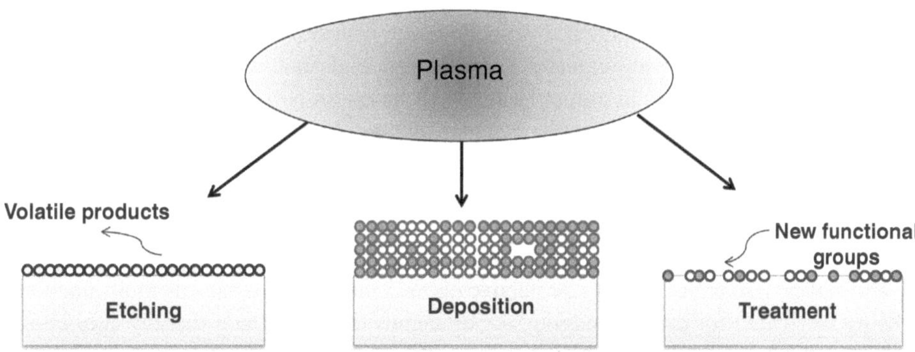

FIGURE 5.3 Scheme of the Plasma Processes: Etching, Deposition, and Treatment

4 THERAPEUTIC APPLICATIONS

4.1 TARGETED DELIVERY

The ability of NPs to target diseased tissues is perhaps their most appealing feature. Targeting has the potential to reduce off-target effects, and enhance accumulation at the diseased site. The interaction with a biological target can be achieved using several approaches, such as passive and active drug targeting, active drug targeting to endothelial cells, and triggered drug delivery (Ganta et al., 2008).

"Passive drug targeting" happens when the NPs circulate in the blood, and passively extravasate in the tissue thanks to the better permeability of tumor blood vessels, compared to normal tissue. This hyperpermeability is due to the formation of tumor blood vessels that proceeds in a chaotic manner, leading to a defective architecture with large gaps in the endothelium (EPR effect).

"Active drug delivery" consists in the following: when the NPs passively extravasate and accumulate in the target tissue via the EPR effect, the surface functionality of NPs allow the interaction between NPs and receptors that are overexpressed on tumor cells or tissue, causing an enhanced uptake and internalization via receptor-mediated endocytosis. The mechanism is based on the fact that fictionalized NPs link with the cell surface receptor, and the endosomes trigger the NP internalization; as such, when the NPs come into contact with the internal acidic pH of cells, they release the drug in the cytoplasm.

"Triggered drug delivery by stimuli-sensitive nanomedicines" consists in the following: thanks to the EPR effect, the NPs passively accumulate in the tumor. The NPs can be activated by external stimuli, such as hyperthermia, magnetic or electric field, and so on, when they arrive at the target site or are circulating in the tumor vasculature, inducing the release of drugs (Danhier et al., 2010; Torchilin, 2010).

4.2 CANCER THERAPY

Cancer is the second leading cause of mortality around the globe, despite the various advancements in science. The success of cancer treatment lies in early diagnosis and effective therapy, making them inseparable. In recent years, there has been an extraordinary development in the field of nanomedicine with the development of new and multifunctional nanostructured biomaterials for the diagnosis and treatment of cancer. It was revealed that nanobiomaterial-based diagnosis engrosses nanosensors for detection and conjugation of biomarkers, quantum dots, radiolabeling, and delivery of contrast agent for better imaging of cancer development. In cancer therapy, nanocomposites hold enormous potential in maximizing the benefits of both targeted chemotherapy and photodynamic therapy (Subramanian et al., 2015). Nanobiotechnology, in particular methods based on NPs, has made significant contributions to the improvement of drug delivery in cancer therapy over the past decades. Several types of NP-based drug-delivery systems have been developed for this purpose (Davis et al., 2008; Kamaly et al., 2012; Peer et al., 2007). Among these, the concept of multifunctional NPs has attracted increasing attention (Kim et al., 2006). Smart combinations of different types of functional nanomaterials—for example, by combining functional inorganic nanomaterials, including quantum dots, AuNPs, magnetic particles, and carbon nanotubes—with a protective organic matrix led to various multifunctional nanocarriers being fabricated for cancer imaging, simultaneous diagnosis, and therapy (Lai et al., 2003). Recently, mesoporous silica NPs (MSNs; size typically <500 nm) started to play an important role in cancer therapy due to their advantageous structural properties, such as high internal surface area and pore volume, tunable pore sizes, colloidal stability, and the possibility to specifically functionalize the inner pore system and/or the external particle surface (Ashley et al., 2011; Slowing et al., 2007; Zhang et al., 2015). For example, superparamagnetic iron oxide NPs with dual imaging and therapeutic capabilities hold great promise for the noninvasive detection and treatment of tumors. When conjugated with tumor-specific targeting ligands, these multifunctional NPs can be used to specifically deliver anticancer drugs to tumors, thereby minimizing severe side effects. Santra et al. show a new method to synthesize multimodal and theranostic poly(acrylic acid)-iron oxide NPs (PAA-IONPs) for the potential in vivo target-specific detection and treatment of tumors. The IONPs are biocompatible and biodegradable, as they are synthesized from biodegradable and biocompatible components. These functional IONPs are stable in aqueous buffered solutions, possess good cellular targeting ability, and their simple synthesis process is amenable to scale-up. In addition, this method can easily be used to generate libraries of targeted theranostic NPs with different targeting ligands or encapsulated agents, and even include different metallic cores. Furthermore, the drug-encapsulating IONPs, when conjugated with folic acid (using click chemistry), provide targeted drug delivery to cancer cells that overexpress the folate receptor, while avoiding normal cells that do not overexpress this receptor. These multimodal (magnetic and fluorescent) and multifunctional (imaging and therapeutic) IONPs will open many exciting opportunities for the targeted delivery of therapeutic agents to tumors. In addition, the dual optical and magnetic properties of the synthesized NPs will allow for the dual fluorescence- and MR-based imaging and monitoring of drug efficacy. All these positive attributes make the functional IONPs a promising drug-delivery vehicle for further in vivo evaluation (Santra et al., 2009). New systems on nanotechnology-based materials have the possibility to increase the solubility of poorly soluble drugs. For example, curcumin, a natural pigment, has a lot of beneficial properties, such as anticancer effects, as validated by various in vitro and in vivo cancer research studies. The low bioavailability and water solubility of curcumin hinder practical applications. Encapsulation in specific NPs, liposomes, micelles, and systems could increase and encourage curcumin delivery. Research studies confirmed that nanoformulations of this natural anticancer agent can improve the effect, compared to the normal system. In order to apply nanocurcumin anticancer systems in clinical and preclinical models, further studies are needed (Shehzad et al., 2014). Hence, nanotechnology can improve the bioavailability of specific systems, such as curcumin but, more importantly, it permits the increase of the concept of personalized medicine, by working at the molecular level (Amit et al., 2014).

Cancer as a disease is an ever-evolving medical condition that severely limits the successful outcome of therapeutic approaches, primarily due to anatomical and physiological complexities, as well as tumor heterogeneity. Surgical debulking followed by radiation and chemotherapy is a classical approach for the containment of the disease, but the emergence of acquired multidrug resistance (MDR) in the majority of cancer types renders many of the frontline drugs ineffective, leading to poor prognosis. Therefore, the central paradigm of cancer therapy has been modified to accommodate novel approaches that counter the predicaments of conventional therapy, and exploit the benefits of biological macromolecules (biologics), such as antibodies, plasmids, and small oligonucleotides, as standalone therapy or in combination with anticancer drugs. However, similar to chemotherapeutics, these biologics suffer from poor in vivo stability, short half-life, poor bioavailability, and off-target effects.

Although nanobiomaterials hold promise for cancer chemotherapy, their intracellular drug-delivery pathways are not fully understood. In particular, the influence of nanocarrier stability on cellular uptake is still uncertain. Nanocarrier-based chemotherapy has been studied not only for maximizing the therapeutic effect of chemodrugs, but also for minimizing nonselective toxicity to normal tissue. In order to achieve such goals, the nanocarrier must effectively deliver a chemoagent to the intracellular pharmacological target organelle in a cancer cell (Petros and DeSimone, 2010; Rajendran et al., 2010). In spite of the great potential for nanocarrier-based intracellular drug delivery to achieve sufficient effects for cancer therapy, the mechanism and key parameters remain elusive (Bae and Park, 2011; Gullotti and Yeo, 2012; Ruenraroengsak et al., 2010). Thus, a comprehensive understanding of the cellular uptake and intracellular trafficking of a drug-loaded nanocarrier is essential for successful chemotherapy (Bonner et al., 2011; Conner and Schmid, 2003; Duncan and Richardson, 2012; Sahay et al., 2010). Therefore, the results show that a synergistic effect in improving therapeutic applications was found by combined photothermal therapy and photodynamic cancer therapy. This design might find potential applications in new drug development, existing drug improvement, and drug delivery for cancer therapy.

4.3 TISSUE ENGINEERING

The term tissue engineering (TE) highlights a specific biomedical intervention area where stem cell type are combined with a biomaterial specifically designed in order to generate a specific tissue/organ (Katari et al., 2015 ; Martino et al., 2012). The TE paradigm includes the following.

4.3.1 Selection of a stem cell type

Generally, adult stem cells (ASCs) or induced pluripotent stem cells (iPSCs) are preferred to embryonic stem cells because of the avoidance of ethical implications. Moreover, ASCs and iPSCs are favorable to somatic cells based on the high capability of those cells to be differentiated toward several types of tissues, either in vitro or in vivo systems. The latter are a consequence of stem cells' biological properties of self-renewal and pluri/multipotency (Lindvall and Kokaia, 2010; Orlacchio et al., 2010a). Stem cell self-renewal is due to a peculiar cell division that could generate either a daughter cell with stem cell characteristic and a progenitor daughter cell (asymmetric division), or two stem cell daughters (symmetric division), depending on the signaling present in the microenvironment where the stem reside (Cheng et al., 2008; Rusan and Peifer, 2007; Yamashita, 2009; Yamashita and Fuller, 2008). It is noteworthy, thanks to asymmetric division that stem cells give rise the balanced replacement of progenitor cells that, upon specific molecular signals, generate a differentiated progeny (Cheng et al., 2008; Orlacchio et al., 2010b; Rusan and Peifer, 2007; Yamashita, 2009; Yamashita and Fuller, 2008). Stem cell pluri/multipotency property is referred to as the capability of stem to generate all types of differentiated cells (pluripotency) present in adult tissues/organs, as well as a more restricted cell lineage (multipotency) (Cheng et al., 2008; Orlacchio et al., 2010a,b; Yamashita, 2009; Yamashita and Fuller, 2008).

Thus, in respect to embryonic stem cells, ASCs exhibit similar self-renewal capacity, but have a more restricted differentiation potential that results in a limited specialized tissue-specific cell lineage (Fig. 5.4) (Cossu and Bianco, 2003; Martino et al., 2009b; McKay, 2004). This could be explained since ASCs maintain the characteristics of their embryonic layer of origin, but persist within the adult animal tissues/organs (Gritti et al., 2008; Lee and Park, 2009; Uccelli et al., 2008), in order to provide cell replacement to tissues/organs under physiological and pathological conditions. Conversely, iPSCs are a new class of stem cells generated in vitro, starting from somatic differentiated cells (Fig. 5.4), according to the original work of Takahashi and Yamanaka (2006) that demonstrated how overexpression of Oct3/4, Sox2, c-Myc, and Klf4 transcription factors was capable to revert the epigenetic state of somatic differentiated cells to pluripotent cells with embryonic-like properties (Martino et al., 2012, 2014; Takahashi and Yamanaka, 2006).

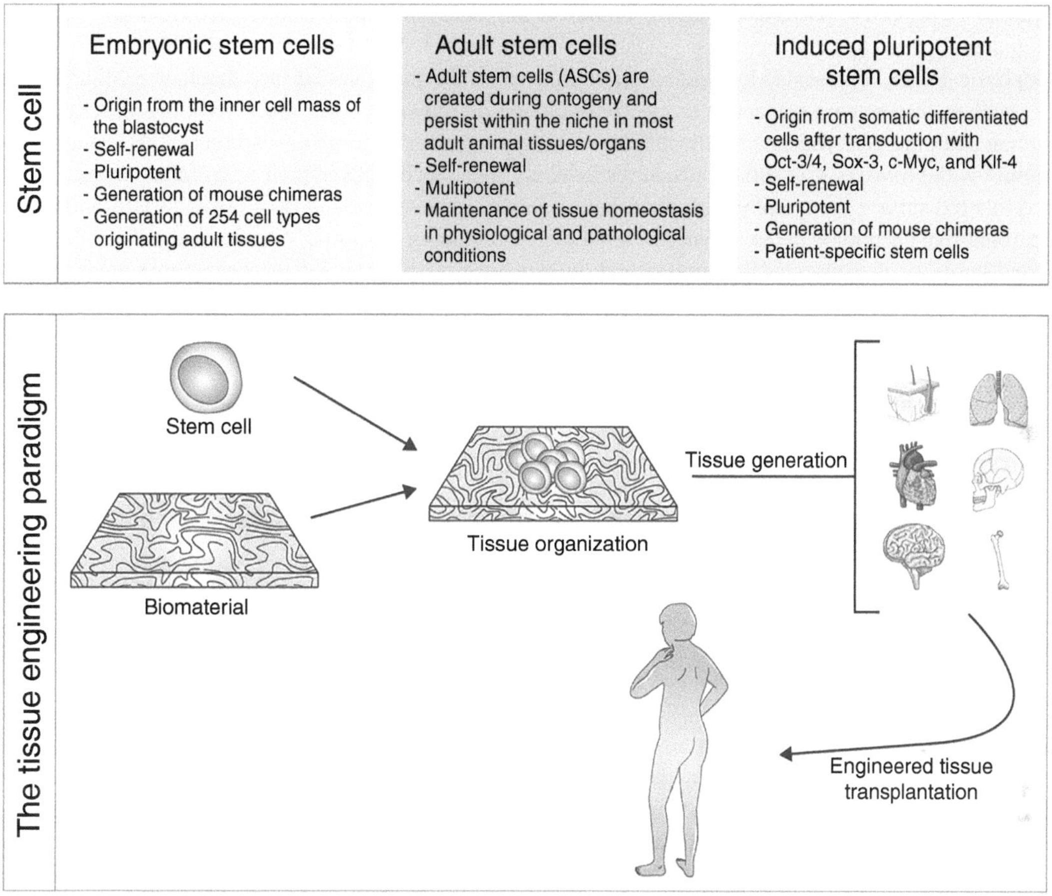

FIGURE 5.4 Schematic of Embryonic, Adult, and Induced Pluripotent Stem Cells Biological properties, and Their Application in Tissue Engineering

4.3.2 Selection of a biomaterial with specific properties

Selection of a biomaterial with specific properties (e.g., surface, mechanical, electrical properties, etc., as above) could serve as a 3D scaffold mimicking the tissue/organ architecture and signaling, allowing stem cells adhesion, proliferation, and differentiation (D'Angelo et al., 2011). Indeed, fabrication of biomaterials with selected properties could represent a suitable approach to modulate and control the stem cell's fate. Many reports have validated the assumption by which physical and chemical materials properties, such as size, shape, mechanical properties, surface texture, etc., have a deep impact on the molecular events leading to stem cells' biological activities. (D'Angelo et al., 2010, 2012; Martino et al., 2009a). In this regard, innovative biotechnologies are dedicated to identify so called "mechanotransduction" pathways (Engler et al., 2006). It is likely that interaction of stem cells with biomaterials by adhesion cell membrane proteins generated traction forces that are transmitted to cytoskeleton fibers, and by these to the nucleus where chromatin is remodeled and gene expression influenced (D'Angelo et al., 2011; Martino et al., 2012;

Murphy et al., 2014; Swift et al., 2013; Treiser et al., 2010; Wen et al., 2005). A direct link between biomaterial properties and stem fate guidance was demonstrated by Downing et al. (2013), uncovering biophysical cues, represented by parallel microgrooves on the substrates surface that have similar effects of small-molecules with epigenetic recognized function. In this regard, our group have demonstrated that human adult mesenchymal stem cells responded to hydrogenated amorphous carbon (a-C:H) nanotopographies with groove or grid surface structures inducing specific changes of their microtubule organization, and that the surface topography with micropatterned nanogrooves width/spacing of 40/30 μm induced a neuronal-like phenotype in the absence of soluble differentiating agents (D'Angelo et al., 2010). In a different study, we also observed that multipotent (human bone marrow-derived mesenchymal stem cells) and pluripotent stem cells (murine-induced pluripotent stem cells and murine embryonic stem cells) responded to nanocomposite fibrous mats of poly(L-lactic acid) (PLLA) loaded with 1 or 8 wt.% of calcium-deficient nanohydroxyapatite (d-HAp), with expression of osteogenic markers in the absence of soluble osteogenic factors. Interestingly, this phenomenon was confined only in stem cells grown on direct contact with the PLLA/d-HAp mats, and was absent in the same cells growth in neat PLLA, thus associating the phenomenon to mechanical properties acquired by the PLLA/d-HAp nanocomposites (D'Angelo et al., 2012).

4.3.3 Selection of an "in vivo" or "ex vivo" strategy for generating/regenerating the tissue/organ (Fig. 5.4)

This step is crucial in order to reestablish tissue/organ functionality. Therefore, depending on the tissue damage, a good TE strategy could consist in the generation of a biohybrid organ in an in vitro system, and then implanting it in vivo (ex vivo), or in a direct implantation of chosen stem cells and biomaterial in the surgical site (in vivo). Of course, all manufacturing challenges are correctly warranted stringent GMP regulations (Guilak et al., 2014). Generation of bone tissue represents one of the goal standards of TE. In fact, bone TE-based systems are beginning to be used in clinical interventions, and thus could be expected to be available to patients in the next few years (Liu et al., 2013). This includes the treatment of craniofacial tissues that could take advantage by engineering personalized bone grafts, customized to the patient with selected scaffold (Bhumiratana and Vunjak-Novakovic, 2012). Bone TE paradigm consists in a combination of adult mesenchymal stem cells isolated from the patients and their seeding onto porous biodegradable matrices (scaffolds). During this ex vivo strategy, stem cells, seeded within the scaffold, could be exposed to exogenous molecules (e.g., growth factors and other osteoinductive cues), in order to support stem cell osteogenic lineage differentiation. At maturation, the engineered bone is implanted into the intervention site in order to reestablish new bone as the scaffold degrades (Salgado et al., 2004). Good results have been obtained for the generation of cartilage. Generally, neocartilage is created ex vivo. Autologous cells are deposited throughout the scaffold, and it can be introduced into the origin patient only when mature cartilage construct is formed (Teh et al., 2013; Watson and Reuther, 2014). This include the treatment of pediatric osteoarticular diseases with severe congenital bone or local cartilage defects, not responding to conventional surgery treatment (Gentili et al., 2014). TE application is a good option for providing artificial organs as substitute for donor organs, alleviating the shortage, reducing healthcare costs, and improving quality of life. For instance, TE of heart valves represents a promising alternative for the replacement of damaged heart valves, escaping the restrictions of currently available bioprosthetic and mechanical heart valves (Jana et al., 2014). Moreover, engineering kidneys based on the implantation of autologous grafts on-demand (stem cells and biomaterials) is highly investigated. In this regard, an interesting approach is based on the use of decellularized kidney matrices that are then repopulated by implantation of iPSCs (Uzarski et al., 2014; Vaitkuviene et al., 2015). The TE of nervous system is more complex. Advances are made on the regeneration of peripheral nerve injuries. Carriel et al. (2015), starting from a combination of a nanostructured fibrin-agarose hydrogels and human adult adipose-derived mesenchymal stem cells, generated biologically active and mechanically stable nerve tissue-like substitutes (Carriel et al., 2015).

5 CONCLUSIONS AND PERSPECTIVES

This chapter deals with multifunctional nanostructured biomaterials and their potential use for various therapeutic applications. This review has covered only a few of the important nanostructured systems. Numerous advanced polymeric nanostructured systems, possessing uniquely controlled frameworks at the molecular scale and well-defined structures at the nanoscale, are expected to emerge and will manifest themselves with unprecedented properties. The increasing relevance of the potential of various nanocarriers in drug delivery emphasizes the need to explore the routes by which they can be administered.

Nanotechnology-based platforms had an important impact on the development of therapeutic systems in the past decades. A high number of organic and inorganic NPs, and micelles, have been engineered and developed, permitting to encapsulate or bond specific drugs.

Further work should focus on the synthesis of nanomaterials with high NP loading and large pore sizes. This review will encourage the need for in-depth molecular level examinations in relation to the cytotoxicity, and biodistribution of the developed nanocomposite should be evaluated in the clinical setting for better understanding. These supplementary studies on nanooncology would help in personalizing cancer theranostics, making nanooncology a future trend.

REFERENCES

Abe, H., Doi, Y., 1999. 1999 Structural effects on enzymatic degradabilities for poly [(R)-3-hydroxybutyric acid] and its copolymers. Int. J. Biol. Macromol. 25, 185–192.

Alexis, F., Pridgen, E., Molnar, L.K., Farokhzad, O.C., 2008. Factors affecting the clearance and biodistribution of polymeric nanoparticles. Mol. Pharm. 5, 505–515.

Alishahi, A., Aider, M., 2012. Applications of chitosan in the seafood industry and aquaculture: a review. Food Bioprocess Technol. 5, 817–830.

Amit, S., Meghn, T., Thanh-Huyen, T., Abishek, S., Ravi, S., Mansoor, A., 2014. Combinatorial approach in the design of multifunctional polymeric nano-delivery systems for cancer therapy. J. Mater. Chem. B 2, 8069.

Andersson, A.S., Bäckhed, F., Von Euler, A., Richter-Dahlfors, A., Sutherland, D., Kasemo, B., 2003. Nanoscale features influence epithelial cell morphology and cytokine production. Biomaterials 24, 3427–3436.

Armentano, I., Dottori, M., Fortunati, E., Mattioli, S., Kenny, J.M., 2010. Biodegradable polymer matrix nanocomposites for tissue engineering: a review. Polym. Degrad. Stab. 95, 2126–2146.

Armentano, I., Montanucci, P., Morena, F., Bicchi, I., Basta, G., Fortunati, E., Mattioli, S., Calafiore, R., Martino, S., Kenny, J.M., 2014. Nano-engineered PLLA based biomaterial drives stem cell responses. J. Tissue Eng. Regen. Med. 8, 501–502.

Arvizo, R.R., Bhattacharyya, S., Kudgus, R.A., Giri, K., Bhattacharya, R., Mukherjee, P., 2012. Intrinsic therapeutic applications of noble metal nanoparticles: past, present and future. Chem. Soc. Rev. 41, 2943–2970.

Ashley, C.E., Carnes, E.C., Phillips, G.K., Padilla, D., Durfee, P.N., Brown, P.A., Hanna, T.N., Liu, J.W., Phillips, B., Carter, M.B., 2011. The targeted delivery of multicomponent cargos to cancer cells by nanoporous particle-supported lipid bilayers. Nat. Mater. 10, 389–397.

Aviss, K.J., Gough, J.E., Downes, S., 2010. Aligned electrospun polymer fibres for skeletal muscle regeneration. Eur. Cells Mater. 19, 193–204.

Aysan, E., Bektas, H., Ersoz, F., Sari, S., Kaygusuz, A., 2010. A novel colonic anastomosis technique involving fixed polyglycolic acid mesh. Int. J. Clin. Exp. Med. 3, 341–346.

Azzazy, H.M.E., Mansour, M.M.H., Kazmierczak, S.C., 2007. From diagnostics to therapy: prospects of quantum dots. Clin. Biochem. 40, 917–927.

Bae, Y.H., Park, K., 2011. Targeted drug delivery to tumors: myths, reality and possibility. J. Control. Release 153, 198–205.

Baker, J.R., 2009. Dendrimer-based nanoparticles for cancer therapy. ASH Educ. Program 2009, 708–719.

Barikani, M., Oliaei, E., Seddiqi, H., Honarkar, H., 2014. Preparation and application of chitin and its derivatives: a review. Iran. Polym. J. 23, 307–326.

Berger, J., Reist, M., Mayer, J.M., Felt, O., Gurny, R., 2004. Structure and interactions in chitosan hydrogels formed by complexation or aggregation for biomedical applications. Eur. J. Pharm. Biopharm. 57, 35–52.

Bhumiratana, S., Vunjak-Novakovic, G., 2012. Concise review: personalized human bone grafts for reconstructing head and face. Stem Cells Transl. Med. 1, 64–69.

Bonner, D.K., Leung, C., Chen-Liang, J., Chingozha, L., Langer, R., Hammond, P.T., 2011. Intracellular trafficking of polyamidoamine-poly(ethylene glycol) block copolymers in DNA delivery. Bioconjug. Chem. 22, 1519–1525.

Brambilla, D., Luciani, P., Leroux, J.C., 2014. Breakthrough discoveries in drug delivery technologies: the next 30 years. J. Control. Release 190, 9–14.

Brannon-Peppas, L., Blanchette, J.O., 2012. Nanoparticle and targeted systems for cancer therapy. Adv. Drug Deliv. Rev. 64 (Suppl.), 206–212.

Brown, S., Khan, D.R., 2012. The treatment of breast cancer using liposome technology. J. Drug Deliv. 2012, 212965, 6 p.

Burkersroda, F.v., Schedl, L., Gopferich, A., 2002. Why degradable polymers undergo surface erosion or bulk erosion. Biomaterials 23, 4221–4231.

Burns, A.E., 1995. Biofix fixation techniques and results in foot surgery. J. Foot Ankle Surg. 34, 276–282.

Bushnell, B.D., McWilliams, A.D., Whitener, G.B., Messer, T.J., 2008. Early clinical experience with collagen nerve tubes in digital nerve repair. Hand Surg. 33, 1081–1087.

Cacciotti, I., Fortunati, E., Puglia, D., et al., 2014. Effect of silver nanoparticles and cellulose nanocrystals on electrospun poly(lactic) acid mats: morphology, thermal properties and mechanical behavior. Carbohydr. Polym. 103, 22–31.

Cao, D., Liu, W., Zhou, G., Zhang, W.J., Cao, Y., 2010. In vivo engineering of a functional tendon sheath in a hen model. Biomaterials 31, 3894–3902.

Carriel, V., Scionti, G., Campos, F., Roda, O., Castro, B., Cornelissen, M., Garzón, I., Alaminos, M., 2015. In vitro characterization of a nanostructured fibrin agarose bio-artificial nerve substitute. J. Tissue Eng. Regen. Med, DOI: 10.1002/term.2039.

Casper, C.L., Yang, W., Farach-Carson, M.C., Rabolt, J.F., 2007. Coating electrospun collagen and gelatin fibers with perlecan domain I for increased growth factor binding. Biomacromolecules 8, 1116–1123.

Cheetham, P.S.J., 1979. Physical studies on the mechanical stability of columns of calcium alginate gel pellets containing entrapped microbial cells. Enzyme Microb. Technol. 1, 183–188.

Cheng, J., Türkel, N., Hemati, N., Fuller, M.T., Hunt, A.J., Yamashita, Y.M., 2008. Centrosome misorientation reduces stem cell division during ageing. Nature 456, 599–604.

Chiquet, M., Matthisson, M., Koch, M., Tannheimer, M., Chiquet-Ehrismann, R., 1996. Regulation of extracellular matrix synthesis by mechanical stress. Biochem. Cell Biol. 74, 737–744.

Chou, L., Firth, J.D., Uitto, V.J., Brunette, D.M., 1995. Substratum surface topography alters cell shape and regulates fibronectin mRNA level, mRNA stability, secretion and assembly in human fibroblasts. J. Cell Sci. 108, 1563–1573.

Conner, S.D., Schmid, S.L., 2003. Regulated portals of entry into the cell. Nature 422, 37–44.

Cossu, G., Bianco, P., 2003. Mesoangioblasts vascular progenitors for extravascular mesodermal tissues. Curr. Opin. Genet. Dev. 13, 537–542.

d'Agostino, R., Favia, P., Oehr, C., Wertheimer, M.R., 2005. Low temperature plasma processing of materials: past, present and future. Plasma Process Polym. 2, 7–15.

D'Angelo, F., Armentano, I., Cacciotti, I., Tiribuzi, R., Quattrocelli, M., Del Gaudio, C., Fortunati, E., Saino, E., Caraffa, A., Cerulli, G.G., Visai, L., Kenny, J.M., Sampaolesi, M., Bianco, A., Martino, S., Orlacchio, A., 2012. Tuning multi/pluri-potent stem cell fate by electrospun poly(l-lactic acid)-calcium-deficient hydroxyapatite nanocomposite mats. Biomacromolecules 13, 1350–1360.

D'Angelo, F., Armentano, I., Mattioli, S., Crispoltoni, L., Tiribuzi, R., Cerulli, G.G., Palmerini, C.A., Kenny, J.M., Martino, S., Orlacchio, A., 2010. Micropatterned hydrogenated amorphous carbon guides mesenchymal stem cells towards neuronal differentiation. Eur. Cell Mater. 20, 231–244.

D'Angelo, F., Tiribuzi, R., Armentano, I., Kenny, J.M., Martino, S., Orlacchio, A., 2011. Mechanotransduction: tuning stem cells fate. J. Funct. Biomater. 2, 67–87.

Dahman, Y., 2009. Nanostructured biomaterials and biocomposites from bacterial cellulose nanofibres. J. Nanosci. Nanotechnol. 9, 5105–5122.

Dai, T.T., Jiang, Z.H., Li, S.L., Zhou, G.D., Kretlow, J.D., Cao, W.G., Liu, W., Cao, Y.L.J., 2010. Reconstruction of lymph vessel by lymphatic endothelial cells combined with polyglycolic acid scaffolds: a pilot study. J. Biotechnol. 150, 182–189.

Dalby, M.J., Gadegaard, N., Tare, R., Andar, A., Riehle, M.O., Herzyk, P., Wilkinson, C.D.W., Oreffo, R.O.C., 2007. The control of human mesenchymal cell differentiation using nanoscale symmetry and disorder. Nat. Mater. 6, 997–1003.

Dalby, M.J., Riehle, M.O., Johnstone, J.H., Affrossman, S., Curtis, A.S.G., 2002. Polymer-demixed nanotopography: control of fibroblast spreading and proliferation. Tissue Eng. 8, 1099–1107.

Danhier, F., Feron, O., Préat, V., 2010. To exploit the tumor microenvironment: passive and active tumor targeting of nanocarriers for anti-cancer drug delivery. J. Control. Release 148, 135–146.

Davis, M.E., Chen, Z., Shin, D.M., 2008. Nanoparticle therapeutics: an emerging treatment modality for cancer. Nat. Rev. Drug Discov. 7, 771–782.

Deok-Ho, K., Levchenko, A., Suh, K.Y., 2008. Engineered surface nanotopography for controlling cell–substrate interactions. In: Khademhosseini, A., Borenstein, J., Toner, M., Takayama, S. (Eds.), Micro and Nanoengineering of the Cell Microenvironment: Technologies and Applications. Artech House Publishers, Norwood, MA, pp. 185–208.

Devadasu, V.R., Bhardwaj, V., Kumar, M.N., 2013. Can controversial nanotechnology promise drug delivery? Chem. Rev. 113, 1686–1735.

Downing, T.L., Soto, J., Morez, C., Houssin, T., Fritz, A., Yuan, F., Chu, J., Patel, S., Schaffer, D.V., Li, S., 2013. Biophysical regulation of epigenetic state and cell reprogramming. Nat. Mater. 12, 1154–1162.

Dubrot, J., Portero, A., Orive, G., Hernandez, R.M., Palazon, A., Rouzaut, A., Perez-Gracia, J.L., Hervás-Stubbs, S., Pedraz, J.L., Melero, I., 2010. Delivery of immunostimulatory monoclonal antibodies by encapsulated hybridoma cells. Cancer Immunol. Immunother. 59, 1621–1631.

Duncan, R., Richardson, S.C.W., 2012. Endocytosis and intracellular trafficking as gateways for nanomedicine delivery: opportunities and challenges. Mol. Pharm. 9, 2380–2402.

Elsabahy, M., Heo, G.S., Lim, S.M., Sun, G., Wooley, K.L., 2015. Polymeric nanostructures for imaging and therapy. Chem. Rev 115 (19), 10967–11011.

Elsabahy, M., Wooley, K.L., 2012. Design of polymeric nanoparticles for biomedical delivery applications. Chem. Soc. Rev. 41, 2545–2561.

Engler, A.J., Sen, S., Sweeney, H.L., Discher, D.E., 2006. Matrix elasticity directs stem cell lineage specification. Cell 126, 677–689.

Farazuddin, M., Alam, M., Khan, A.A., Khan, N., Parvez, S., Dutt, G.U., Mohammad, O.J., 2010. Efficacy of amoxicillin bearing microsphere formulation in treatment of *Listeria monocytogenes* infection in Swiss albino mice. Drug Target 18, 45–52.

Fernandes, E.M., Pires, R.A., Mano, J.F., et al., 2013. Bionanocomposites from lignocellulosic resources: properties, applications and future trends for their use in the biomedical field. Progr. Polym. Sci. 38, 1415–1441.

Flemming, R.G., Murphy, C.J., Abrams, G.A., Goodma, S.L., Nealey, P.F., 1999. Effects of synthetic micro- and nano-structured surfaces on cell behavior. Biomaterials 20, 573–588.

Fortunati, E., D'Angelo, F., Martino, S., Orlacchio, A., Kenny, J.M., Armentano, I., 2011a. Carbon nanotubes and silver nanoparticles for multifunctional conductive biopolymer composites. Carbon 49, 2370–2379.

Fortunati, E., Latterini, L., Rinaldi, Kenny, J.M., Armentano, I., 2011b. PLGA/Ag nanocomposites: in-vitro degradation study and silver ion release. J. Mater. Sci. Mater. Med. 12, 2735–2744.

Gaharwar, A.K., Peppas, N.A., Khademhosseini, A., 2014. Nanocomposite hydrogels for biomedical applications. Biotechnol. Bioeng. 111, 441–453.

Ganta, S., Devalapally, H., Shahiwala, A., Amiji, M., 2008. A review of stimuli-responsive nanocarriers for drug and gene delivery. J. Control. Release 126, 187–204.

Gavenis, K., Schneider, U., Groll, J., Schmidt-Rohlfing, B., 2010. BMP-7-loaded PGLA microspheres as a new delivery system for the cultivation of human chondrocytes in a collagen type I gel: the common nude mouse model. Int. J. Artif. Organs 33, 45–53.

Gentili, C., Torre, M., Cancedda, R., 2014. Tissue engineering approaches in skeletal pediatric disorders. Eur. J Pediatr. Surg. 24, 263–269.

George, J., Sajeevkumar, V.A., Ramana, K.V., et al., 2012. Augmented properties of PVA hybrid nanocomposites containing cellulose nanocrystals and silver nanoparticles. J. Mater. Chem. 22, 22433–22439.

Gleeson, J.P., Plunkett, N.A., O'Brien, F., 2010. Addition of hydroxyapatite improves stiffness, interconnectivity and osteogenic potential of a highly porous collagen-based scaffold for bone tissue regeneration. J. Eur. Cells Mater. 20, 218–230.

Goh, C.H., Heng, P.W.S., Chan, L.W., 2012. Alginates as a useful natural polymer for microencapsulation and therapeutic applications. Carbohydr. Polym. 88, 1–12.

Goldberg, M., Langer, R., Jia, X., 2007. Nanostructured materials for applications in drug delivery and tissue engineering. J. Biomater. Sci. Polym. Ed. 18, 241–268.

Grant, J., Lee, H., Soo, P., Lim, C.J., Piquette-Miller, M., Allen, C., 2008. Influence of molecular organization and interactions on drug release for an injectable polymer–lipid blend. Int. J. Pharm. 360, 83–90.

Gritti, A., Galli, R., Vescovi, A.L., 2008. Clonal analyses and cryopreservation of neural stem cell cultures. Methods Mol. Biol. 438, 173–184.

Guarino, V., Ambrosio, L., 2008. The synergic effect of polylactide fiber and calcium phosphate particle reinforcement in poly ε-caprolactone-based composite scaffolds. Acta Biomater. 4, 1778–1787.

Guelcher, S.A., 2008. Biodegradable polyurethanes: synthesis and applications in regenerative medicine. Tissue Eng. Part B Rev. 14, 3–17.

Guilak, F., Butler, D.L., Goldstein, S.A., Baaijens, F.P., 2014. Biomechanics and mechanobiology in functional tissue engineering. J. Biomech. 7, 1933–1940.

Gullotti, E., Yeo, Y., 2012. Beyond the imaging: limitations of cellular uptake study in the evaluation of nanoparticles. J. Control. Release 164, 170–176.

Gurkan, U.A., Cheng, X., Kishore, V., Uquillas, J.A., Akkus, O., 2010. Comparison of morphology, orientation, and migration of tendon derived fibroblasts and bone marrow stromal cells on electrochemically aligned collagen constructs. J. Biomed. Mater. Res. A 94, 1070–1079.

Halami, P.M., 2008. Production of polyhydroxyalkanoate from starch by the native isolate Bacillus cereus CFR06. World J. Microb. Biotechnol. 24, 805–812.

Ho, C.H., Tobis, J., Sprich, C., Thomann, R., Tiller, J.C., 2004. Nanoseparated polymeric networks with multiple antimicrobial properties. Adv. Mater. 16, 957–964.

Ionescu, L.C., Lee, G.C., Sennett, B.J., Burdick, J.A., Mauck, R.L., 2010. An anisotropic nanofiber/microsphere composite with controlled release of biomolecules for fibrous tissue engineering. Biomaterials 31, 4113–4120.

Jabbarzadeh, E., Starnes, T., Khan, Y.M., Jiang, T., Wirtel, A.J., Deng, M., Lv, Q., Nair, L.S., Doty, S.B., Laurencin, C.T., 2008. Induction of angiogenesis in tissue-engineered scaffolds designed for bone repair: a combined gene therapy–cell transplantation approach. Proc. Natl. Acad. Sci. USA 105, 11099–11104.

Jana, S., Tefft, B.J., Spoon, D.B., Simari, R.D., 2014. Scaffolds for tissue engineering of cardiac valves. Acta Biomater. 10, 2877–2893.

Jorfi, M., Roberts, M.N., Foster, E.J., Weder, C., 2013. Physiologically responsive, mechanically adaptive bio-nanocomposites for biomedical applications. ACS Appl. Mater. Interfaces 5, 1517–1526.

Kamaly, N., Xiao, Z., Valencia, P.M., Radovic-Moreno, A.F., Farokhzad, O.C., 2012. Targeted polymeric therapeutic nanoparticles: design, development and clinical translation. Chem. Soc. Rev. 41, 2971–3010.

Katari, R., Peloso, A., Orlando, G., 2015. Tissue engineering and regenerative medicine: semantic considerations for an evolving paradigm. Front. Bioeng. Biotechnol. 12, 57.

Katz, A.R., Turner, R.J., 1970. Evaluation of tensile and absorption properties of polyglycolic acid sutures. Surg. Gynecol. Obstet. 131, 701–716.

Kesharwani, P., Jain, K., Jain, N.K., 2014. Dendrimer as nanocarrier for drug delivery. Prog. Polym. Sci. 39, 268–307.

Kim, I.Y., Seo, S.J., Moon, H.S., Yoo, M.K., Park, I.Y., Kim, B.C., Cho, C.S., 2008. Chitosan and its derivatives for tissue engineering applications. Biotechnol. Adv. 26, 1–21.

Kim, J., Park, S., Lee, J.E., Jin, S.M., Lee, J.H., Lee, I.S., Yang, I., Kim, J.S., Kim, S.K., Cho, M.H., 2006. Designed fabrication of multifunctional magnetic gold nanoshells and their application to magnetic resonance imaging and photothermal therapy. Angew. Chem. Int. Ed. Engl. 45, 7754–7758.

Kimura, Y., Watanabe, K., Okuda, H., 1996. Effects of soluble sodium alginate on cholesterol excretion and glucose tolerance in rats. J. Ethnopharmacol. 54, 47–54.

Kong, H., Jang, J., 2008. Antibacterial properties of novel poly(methyl methacrylate) nanofiber containing silver nanoparticles. Langmuir 24, 2051–2056.

Lai, C.Y., Trewyn, B.G., Jeftinija, D.M., Jeftinija, K., Xu, S., Jeftinija, S., Lin, V.S.Y., 2003. A mesoporous silica nanosphere-based carrier system with chemically removable CdS nanoparticle caps for stimuli-responsive controlled release of neurotransmitters and drug molecules. J. Am. Chem. Soc. 125, 4451–4459.

Lee, M., Wu, B.M., Dunn, J.C.Y.J., 2008. Effect of scaffold architecture and pore size on smooth muscle cell growth. Biomed. Mater. Res. A 87, 1010–1016.

Lee, P.H., Park, H.J., 2009. Bone marrow-derived mesenchymal stem cell therapy as a candidate disease-modifying strategy in Parkinson's disease and multiple system atrophy. J. Clin. Neurol. 5, 1–10.

Letchford, K., Burt, H., 2007. A review of the formation and classification of amphiphilic block copolymer nanoparticulate structures: micelles, nanospheres, nanocapsules and polymersomes. Eur. J. Pharm. Biopharm. 65, 259–269.

Lim, F., Sun, A.M., 1980. Microencapsulated islets as bioartificial endocrine pancreas. Science. 210, 908–910.

Lindvall, O., Kokaia, Z., 2010. Stem cells in human neurodegenerative disorders—time for clinical translation? J. Clin. Invest. 120, 29–40.

Liu, J., Qiu, Z., Wang, S., Zhou, L., Zhang, S., 2010a. A modified double-emulsion method for the preparation of daunorubicin-loaded polymeric nanoparticle with enhanced in vitro anti-tumor activity. Biomed. Mater. 5, 065002.

Liu, X., Huang, C., Feng, Y., Liang, J., Fan, Y., Gu, Z., Zhang, X., 2010b. Reinforcement of a porous collagen scaffold with surface-activated PLA fibers. J. Biomater. Sci. Polym. Ed. 21, 963–977.

Liu, Y., Lim, J., Teoh, S.H., 2013. Review: development of clinically relevant scaffolds for vascularised bone tissue engineering. Biotechnol. Adv. 31, 688–705.

Liu, Z., Jiao, Y., Wang, Y., Zhou, C., Zhang, Z., 2008. Polysaccharides-based nanoparticles as drug delivery systems. Adv. Drug Deliv. Rev. 60, 1650–1662.

Lloyd, A.W., 2002. Biomedical applications of biodegradable polymers. Med. Device Technol. 13, 18–21.

Lloyd, L.L., Kennedy, J.F., Methacanon, P., Paterson, M., Knill, C.J., 1998. Carbohydrate polymers as wound management aids. Carbohydr. Polym. 37, 315–322.

Longmire, M., Choyke, P.L., Kobayashi, H., 2008. Clearance properties of nano-sized particles and molecules as imaging agents: considerations and caveats. Nanomedicine (Lond.) 3, 703–717.

Luciani, A., Coccoli, V., Orsi, S., Ambrosio, L., Netti, P.A., 2008. PCL microspheres based functional scaffolds by bottom-up approach with predefined microstructural properties and release profiles. Biomaterials 29, 4800–4807.

Mahmoudifar, N., Doran, P.M., 2010. Chondrogenic differentiation of human adipose-derived stem cells in polyglycolic acid mesh scaffolds under dynamic culture conditions. Biomaterials 31, 3858–3867.

Marcal, H., Lawless, M., Wanandy, N.S., Chiu, A., Foster, L.J.R., 2010. Polyhydroxybutyrate and its copolymer with polyhydroxyvalerate as biomaterials: influence on progression of stem cell cycle. Biomacromolecules 11, 2707–2715.

Martino, S., D'Angelo, F., Armentano, I., Kenny, J.M., Orlacchio, A., 2012. Stem cell-biomaterial interactions for regenerative medicine. Biotechnol. Adv. 30, 338–351.

Martino, S., D'Angelo, F., Armentano, I., Tiribuzi, R., Pennacchi, M., Dottori, M., Mattioli, S., Caraffa, A., Cerulli, G.G., Kenny, J.M., Orlacchio, A., 2009a. Hydrogenated amorphous carbon nanopatterned film designs drive human bone marrow mesenchymal stem cell cytoskeleton architecture. Tissue Eng. Part A 5, 3139–3149.

Martino, S., di Girolamo, I., Orlacchio, A., Datti, A., Orlacchio, A., 2009b. MicroRNA implications across neurodevelopment and neuropathology. J. Biomed. Biotechnol. 2009, 654346.

Martino, S., Morena, F., Barola, C., Bicchi, I., Emiliani, C., 2014. Proteomics and epigenetic mechanisms in stem cells. Curr. Proteomics 1, 193–209.

Mattioli, S., Martino, S., D'Angelo, F., Emiliani, C., Kenny, J.M., Armentano, I., 2014. Nanostructured polystyrene films engineered by plasma processes: surface characterization and stem cell interaction. J. Appl. Polym. Sci. 131, 40427.

Maurus, P.B., Kaeding, C.C., 2004. Bioabsorbable implant material review. Oper. Tech. Sports Med. 12, 158–160.

McHugh, K.J., Saint-Geniez, M., Tao, S.L., 2013. Topographical control of ocular cell types for tissue engineering. J. Biomed. Mater. Res. Part B 101B, 1571–1584.

McKay, R.D., 2004. Stem cell biology and neurodegenerative disease. Philos. Trans. R. Soc. Lond. B Biol. Sci. 359, 851–856.

Mueller-Rath, R., Gavenis, K., Andereya, S., Mumme, T., Albrand, M., Stoffel, M., Weichert, D., Schneider, U., 2010. Condensed cellular seeded collagen gel as an improved biomaterial for tissue engineering of articular cartilage. Biomed. Mater. Eng. 20, 317–328.

Murphy, W.L., McDevitt, T.C., Engler, A.J., 2014. Materials as stem cell regulators. Nat. Mater. 13, 547–557.

Narayan, D., Venkatraman, S.S.J., 2008. Effect of pore size and interpore distance on endothelial cell growth on polymers. Biomed. Mater. Res. A 87, 710–718.

Ohara, T., Itaya, T., Usami, K., Ando, Y., Sakurai, H., Honda, M.J., Ueda, M., Kagami, H., 2010. Evaluation of scaffold materials for tooth tissue engineering. J. Biomed. Mater. Res. A 94, 800–805.

Orlacchio, A., Bernardi, G., Orlacchio, A., Martino, S., 2010a. Stem cells: an overview of the current status of therapies for central and peripheral nervous system diseases. Curr. Med. Chem. 17, 595–608.

Orlacchio, A., Bernardi, G., Orlacchio, A., Martino, S., 2010b. Stem cells and neurological diseases. Discov. Med. 9, 546–553.

Patlolla, A., Collins, G., Arinzeh, T.L., 2010. Solvent-dependent properties of electrospun fibrous composites for bone tissue regeneration. Acta Biomater. 6, 90–101.

Peer, D., Karp, J.M., Hong, S., Farokhzad, O.C., Margalit, R., Langer, R., 2007. Nanocarriers as an emerging platform for cancer therapy. Nat. Nanotechnol. 2, 751–760.

Perron, J.K., Naguib, H.E., Daka, J., Chawla, A., Wilkins, R., 2009. A study on the effect of degradation media on the physical and mechanical properties of porous PLGA 85/15 scaffolds. J. Biomed. Mater. Res. Part B Appl. Biomater. 2009 (91B), 876–886.

Petros, R.A., DeSimone, J.M., 2010. Strategies in the design of nanoparticles for therapeutic applications. Nat. Rev. Drug Discov. 9, 615–627.

Pinto Reis, C., Neufeld, R.J., Ribeiro, A.J., Veiga, F., 2006. Nanoencapsulation I. Methods for preparation of drug-loaded polymeric nanoparticles. Nanomed. Nanotechnol. Biol. Med. 2, 8–21.

Pooyan, P., Tannenbaum, R., Garmestani, H., 2012. Mechanical behavior of a cellulose-reinforced scaffold in vascular tissue engineering. J. Mech. Behav. Biomed. Mater. 7, 50–59.

Popat, K.C., Chatvanichkul, K.I., Barnes, G.L., Latempa, T.J., Grimes, C.A., Desai, T.A., 2007. Osteogenic differentiation of marrow stromal cells cultured on nanoporous alumina surfaces. J. Biomed. Mater. Res. A 80, 955–964.

Przybyla, D.E., Chmielewski, J., 2010. Higher-order assembly of collagen peptides into nano- and microscale materials. Biochemistry 49, 4411–4419.

Puglia, D., Ceccolini, R., Fortunati, E., Armentano, I., Morena, F., Martino, S., Aluigi, A., Torre, L., Kenny, J.M., 2015. Effect of processing techniques on the 3D microstructure of poly (L-lactic acid) scaffolds reinforced with wool keratin from different sources. J Appl. Polym. Sci. 132, 48.

Pulkkinen, H.J., Tiitu, V., Valonen, P., Jurvelin, J.S., Lammi, M.J., Kiviranta, I., 2010. Engineering of cartilage in recombinant human type II collagen gel in nude mouse model in vivo. Osteoarthr. Cartil. 18, 1077–1087.

Rai, M., Yadav, A., Gade, A., 2009. Silver nanoparticles as a new generation of antimicrobials. Biotechnol. Adv. 27, 76–83.

Rajendran, L., Knolker, H.-J., Simons, K., 2010. Subcellular targeting strategies for drug design and delivery. Nat. Rev. Drug Discov. 9, 29–42.

Ramshaw, J.A.M., Shah, N.K., Brodsky, B., 1998. Gly-X-Y tripeptide frequencies in collagen: a context for host–guest triple-helical peptides. J. Struct. Biol. 122, 86–91.

Rao, J.P., Geckeler, K.E., 2011. Polymer nanoparticles: preparation techniques and size-control parameters. Progr. Polym. Sci. 36, 887–913.

Reckenrich, A.K., Hopfner, U., Krotz, F., Zhang, Z., Koch, C., Kremer, M., Machens, H.-G., Plank, C., Egana, J.T., 2011. Bioactivation of dermal scaffolds with a non-viral copolymer-protected gene vector. Biomaterials 32, 1996–2003.

Reid, T.W., 1994. Growth control of cornea and lens epithelial cells. Prog. Retinal Eye Res. 13, 507–554.

Rescignano, N., Fortunati, E., Armentano, I., Hernandez, R., Mijangos, C., Pasquino, R., Kenny, J.M., 2015. Use of alginate, chitosan and cellulose nanocrystals as emulsion stabilizers in the synthesis of biodegradable polymeric nanoparticles. J. Colloid Interface Sci. 445, 31–39.

Rescignano, N., Fortunati, E., Montesano, S., Emiliani, C., Kenny, J.M., Martino, S., Armentano, I., 2014. PVA bio-nanocomposites: a new take-off using cellulose nanocrystals and PLGA nanoparticles. Carbohydr. Polym. 99, 47–58.

Romano, N.H., Sengupta, D., Chung, C., Heilshorn, S.C., 2011. Protein-engineered biomaterials: nanoscale mimics of the extracellular matrix. Biochim. Biophys. Acta 1810, 339–349.

Ruenraroengsak, P., Cook, J.M., Florence, A.T., 2010. Nanosystem drug targeting: facing up to complex realities. J. Control. Release 141, 265–276.

Rusan, N.M., Peifer, M., 2007. A role for a novel centrosome cycle in asymmetric cell division. J. Cell Biol. 177, 13–20.

Sahay, G., Alakhova, D.Y., Kabanov, A.V., 2010. Endocytosis of nanomedicines. J. Control. Release 145, 182–195.

Sahoo, S., Ang, L.-T., Goh, J.C.-H., Toh, S.-L., 2010. Bioactive nanofibers for fibroblastic differentiation of mesenchymal precursor cells for ligament/tendon tissue engineering applications. Differentiation 79, 102–110.

Sahoo, S., Cho-Hong, J.G., Siew-Lok, T., 2007. Development of hybrid polymer scaffolds for potential applications in ligament and tendon tissue engineering. Biomed. Mater. 2, 169–173.

Salgado, A.J., Coutinho, O.P., Reis, R.L., 2004. Bone tissue engineering: state of the art and future trends. Macromol. Biosci. 4, 743–765.

Sanna, R., Fortunati, E., Alzari, V., Nuvoli, D., Terenzi, A., Casula, M.F., Kenny, J.M., Mariani, A., 2013. Poly(N-vinylcaprolactam) nanocomposites containing nanocrystalline cellulose: a green approach to thermoresponsive hydrogels. Cellulose 20, 2393–2402.

Sannino, A., Demitri, C., Madaghiele, M., 2009. Biodegradable cellulose-based hydrogels: design and applications. Materials 2, 353–373.
Santra, S., Kaittanis, C., Grimm, J., Perez, J.M., 2009. Drug/dye-loaded, multifunctional iron oxide nanoparticles for combined targeted cancer therapy and dual optical/magnetic resonance imaging. Small 5, 1862–1868.
Sayasneh, A., Johnson, H.J., 2010. Risk factors for mesh erosion complicating vaginal reconstructive surgery. Obstet. Gynaecol. 30, 721–724.
Schneider, O.D., Loher, S., Brunner, T.J., Schmidlin, P., Stark, W.J., 2008. Flexible, silver containing nanocomposites for the repair of bone defects: antimicrobial effect against E. coli infection and comparison to tetracycline containing scaffolds. J. Mater. Chem. 18, 2679–2684.
Segal, H.C., Hunt, B.J., Gilding, K., 1998. The effects of alginate and non-alginate wound dressings on blood coagulation and platelet activation. J. Biomater. Appl. 12, 249–257.
Sevda, S., McClure, S.J., 2004. Potential applications of chitosan in veterinary medicine. Adv. Drug Deliv. Rev. 56, 1467–1480.
Shehzad, A., Ul-Islam, M., Wahid, F., Lee, Y.S., 2014. Multifunctional polymeric nanocurcumin for cancer therapy. J. Nanosci. Nanotechnol. 14, 803–814.
Slowing, I.I., Trewyn, B.G., Giri, S., Lin, V.S.Y., 2007. Mesoporous silica nanoparticles for drug delivery and biosensing applications. Adv. Funct. Mater. 17, 1225–1236.
Spek, S., Haeuser, M., Schaefer, M., Langer, K., 2015. Characterisation of PEGylated PLGA nanoparticles comparing the nanoparticle bulk to the particle surface using UV/vis spectroscopy, SEC, 1 H NMR spectroscopy, and X-ray photoelectron spectroscopy. Appl. Surf. Sci. 347, 378–385.
Stevens Molly, M., Qanadilo Hala, F., Langer, R., Prasad Shastri, V., 2004. A rapid-curing alginate gel system: utility in periosteum-derived cartilage tissue engineering. Biomaterials 25, 887–894.
Subramanian, A.P., Jaganathan, S.K., Supriyanto, E., 2015. Overview on in vitro and in vivo investigations of nanocomposite based cancer diagnosis and therapeutics. RSC Adv. 5, 72638–72652.
Swift, J., Ivanovska, I.L., Buxboim, A., Harada, T., Dingal, P.C.D.P., Pinter, J., Pajierowski, J.D., Spinler, K.R., Shin, J.W., Tewari, M., Rehfeldt, F., Speicher, D.W., Discher, D.E., 2013. Nuclear lamin-A scales with tissue stiffness and enhances matrix-directed differentiation. Science 341, 1240104.
Ta, H.T., Dass, C.R., Dunstan, D.E., 2008. Injectable chitosan hydrogels for localised cancer therapy. J. Control. Release 126, 205–216.
Takahashi, K., Yamanaka, S., 2006. Induction of pluripotent stem cells from mouse embryonic and adult fibroblast cultures by defined factors. Cell 126, 663–676.
Taniguchi, N., 1974. On the basic concept of "NanoTechnology". In: Proceedings of the International Conference on Production Engineering, Part II. Japan Society of Precision Engineering, Tokyo, Japan.
Teh, B.M., Marano, R.J., Shen, Y., Friedland, P.L., Dilley, R.J., Atlas, M.D., 2013. Tissue engineering of the tympanic membrane. Tissue Eng. Part B Rev. 19, 116–132.
Teixeira, A.I., Abrams, G.A., Bertics, P.J., Murphy, C.J., Nealy, P.F., 2003. Epithelial contact guidance on well-defined micro- and nanostructured substrates. J. Cell Sci. 116, 1881–1892.
Thipparaboina, R., Chavan, R.B., Kumar, D., Modugula, S., Shastri, N.R., 2015. Micellar carriers for the delivery of multiple therapeutic agents. Colloids Surf. B Biointerfaces 135, 291–308.
Thomas, C., Gupta, V., Ahsan, F., 2010. Particle size influences the immune response produced by hepatitis B vaccine formulated in inhalable particles. Pharm. Res. 27, 905–919.
Torchilin, V.P., 2010. Passive and active drug targeting: drug delivery to tumors as an example. Handb. Exp. Pharmacol. 197, 3–53.
Toumey, C., 2009. Plenty of room, plenty of history. Nat. Nanotechnol. 4, 783–784.
Treiser, M.D., Yanga, E.H., Gordonova, S., Cohend, D.M., Androulakis, I.P., Kohn, J., Chen, C.S., Moghe, P.V., 2010. Cytoskeleton-based forecasting of stem cell lineage fates. Proc. Natl Acad. Sci. USA 107, 610–615.
Trinkaus-Randall, V.C.J., Newton, A., Vadasz, A., Leibowitz, H.M., Franzblau, C., 1988. Development of a biopolymeric keratoprosthesis material: evaluation in vitro and in vivo. Invest. Ophthalmol. Vis. Sci. 29, 293–300.

Uccelli, A., Moretta, L., Pistoia, V., 2008. Mesenchymal stem cells in health and disease. Nat. Rev. Immunol. 8, 726–736.

Ueda, H., Tabata, Y., 2003. Polyhydroxyalkanonate derivatives in current clinical applications and trials. Adv. Drug Deliv. Rev. 55, 501–518.

Ulery, B.D., Nair, L.S., Laurencin, C.T., 2011. Biomedical applications of biodegradable polymers. J. Polym. Sci. Part B Polym. Phys. 49, 832–864.

Uzarski, J.S., Xia, Y., Belmonte, J.C., Wertheim, J.A., 2014. New strategies in kidney regeneration and tissue engineering. Curr. Opin. Nephrol. Hypertens. 23, 399–405.

Vaitkuviene, A., McDonald, M., Vahidpour, F., Noben, J.P., Sanen, K., Ameloot, M., Ratautaite, V., Kaseta, V., Biziuleviciene, G., Ramanaviciene, A., Nesladek, M., Ramanavicius, A., 2015. Impact of differently modified nanocrystalline diamond on the growth of neuroblastoma cells. N. Biotechnol. 25, 7–12.

Vega, E., Gamisans, F., Garcia, M.L., Chauvet, A., Lacou- lonche, F., Egea, M.A.J., 2008. PLGA nanospheres for the ocular delivery of flurbiprofen: drug release and interactions. Pharm. Sci. 97, 5306–5317.

Walkey, C.D., Chan, W.C., 2012. Understanding and controlling the interaction of nanomaterials with proteins in a physiological environment. Chem. Soc. Rev. 41, 2780–2799.

Wang, L., Dormer, N.H., Bonewald, L.F., Detamore, M.S., 2010a. Osteogenic differentiation of human umbilical cord mesenchymal stromal cells in polyglycolic acid scaffolds. Tissue Eng. Part A 16, 1937–1948.

Wang, L., Stegemann, J.P., 2010. Thermogelling chitosan and collagen composite hydrogels initiated with β-glycerophosphate for bone tissue engineering. Biomaterials 31, 3976–3985.

Wang, W., Zhou, S., Sun, L., Huang, C., 2010b. Controlled delivery of paracetamol and protein at different stages from core–shell biodegradable microspheres. Carbohydr. Polym. 79, 437–444.

Wang, X., Wang, Y., Chen, Z.G., Shin, D.M., 2009. Advances of cancer therapy by nanotechnology. Cancer Res. Treat. 41, 1–11.

Watson, D., Reuther, M.S., 2014. Tissue-engineered cartilage for facial plastic surgery. Curr. Opin. Otolaryngol. Head Neck Surg. 22, 300–306.

Wen, J.H., Vincent, L.G., Fuhrmann, A., Choi, Y.S., Hribar, K.C., Taylor-Weiner, H., Chen, S., Engler, A.J., 2005. Interplay of matrix stiffness and protein tethering in stem cell differentiation. Nat. Mater. 13, 979–987.

Williams, D.F., 2009. On the nature of biomaterials. Biomaterials 30, 5897–5909.

Wischke, C., Borchert, H.H., 2008. Increased sensitivity of chitosan determination by a dye binding method. Carbohydr. Res. 341, 2978–2979.

Xu, X., Yang, Q., Bai, J., Lu, T., Li, Y., Jing, X., 2008. Fabrication of biodegradable electrospun poly(L-lactide-*co*-glycolide) fibers with antimicrobial nanosilver particles. J. Nanosci. Nanotechnol. 8, 5066–5070.

Yamashita, Y.M., 2009. The centrosome and asymmetric cell division. Prion 3, 84–88.

Yamashita, Y.M., Fuller, M.T., 2008. Asymmetric centrosome behavior and the mechanisms of stem cell division. J. Cell Biol. 180, 261–266.

Yilmaz, N., Inal, S., Muglali, M., Guvenc, T., Bas, B., 2010. Effects of polyglecaprone 25, silk and catgut suture materials on oral mucosa wound healing in diabetic rats: an evaluation of nitric oxide dynamics. Med. Oral Patol. Oral Cir. Bucal 15, e526–e530.

Zhang, Z., Liu, C., Bai, J., Wu, C., Xiao, Y., Li, Y., Zheng, J., Yang, R., Tan, W., 2015. Silver nanoparticle gated, mesoporous silica coated gold nanorods (AuNR@MS@AgNPs): low premature release and multifunctional cancer theranostic platform. ACS Appl. Mater. Interfaces 7, 6211–6219.

Zhu, X.H., Lee, L.Y., Jackson, J.S.H., Tong, Y.W., Wang, C.-H., 2008. Characterization of porous poly(D,L-lactic-*co*-glycolic acid) sponges fabricated by supercritical CO2 gas-foaming method as a scaffold for three-dimensional growth of Hep3B cells. Biotechnol. Bioeng. 100, 998–1009.

Zielhuis, S.W., Nijsen, J.F.W., Seppenwoolde, J.H., Bakker, C.J.G., Krijger, G.C., Dullens, H.F.J., Zonnenbery, B.A., Rijk, P.P.V., Hennink, W.E., Schip, A.D.V.G., 2007. Long-term toxicity of holmium-loaded poly(L-lactic acid) microspheres in rats. Biomaterials 28, 4591–4599.

Zolnik, B.S., Burgess, D. J.J., 2008. Evaluation of in vivo–in vitro release of dexamethasone from PLGA microspheres. Control. Release 127, 137–145.

CHAPTER

6 POLYMERIC PHARMACEUTICAL NANOPARTICLES DEVELOPED BY ELECTROSPRAY

Abdol-Rahim Faramarzi*, Jalal Barzin*, Hamid Mobedi*, Mohsen Gorji**

**Iran Polymer & Petrochemical Institute, Tehran, Iran*

***Amirkabir University of Technology, Tehran, Iran*

CHAPTER OUTLINE

1 INTRODUCTION

During the past several decades, polymers have been extensively used as biomaterials due to their positive properties, such as good biocompatibility, easy planning and production, diversity of structures, and biomimetic nature. Particularly in the field of smart drug delivery, polymers played an important part because of their roles in delivering pharmaceutical agents into desired sites (Sezer, 2014). Moreover, due to the important dissimilarities in the physical properties, such as electrical, optical, magnetic, chemical, and mechanical characteristics realized at nanoscale, the focus on particle technology changed from macro to micro, and from micro- to nanoscales. Designing medication or drug delivery vehicles to provide enhanced efficacy, higher stability, solubility, or biocompatibility are some of advantages of this shift. By making well-defined particle sizes and size distributions, it is possible to produce a lot of industrial products. Nowadays, researchers understand that therapeutics of distinct and narrow particle size distribution has strong properties, with a predictable time release to the body (Yurteri et al., 2010). These polymeric particles can reduce the proliferation of drugs in the nontargeted sites of the body, lowering unwanted effects, and increasing the delivery efficiency to the site of actions (Freiberg and Zhu, 2004; Lassalle and Ferreira, 2007; Park et al., 2007; Ye et al., 2010). Aiming to use bioactive materials, targeting, and release profile, further study in the field of encapsulation of solid, liquid, or gas materials through efficient technique is essential (Luan et al., 2006).

To design a useful nanoparticle (NP) for application in drug delivery systems, it is necessary to control particle size, surface characteristics, increasing permeation, flexibility, solubility, and release of active agents in order to attain the target and definite activity at a preset rate and time (Sezer, 2014). About the importance of size, it was found that cell uptake of silver and gold NPs, coated by antibodies, is size-dependent (Peltonen et al., 2010). An important reason for producing polymeric NPs and microparticles (MPs) is the increase of blood circulation half-life of drug-loaded particles, and inhibiting phagocytosis by cells of the reticuloendothelial system (RES) of the body (Anderson and Shive, 2012). For example, by adding PEG and PEG-containing copolymers to the surface of NPs, the blood circulation half-life of the particles will increase by several orders of magnitude (Owens and Peppas, 2006). Among particles with the same material, because of their diverse properties, nonoparticles have showed potential in creating advantages in several fields of studies and applications (Wu et al., 2004). In addition to size, size distribution is a fundamental parameter in the fabrication of particles. By controlling size distribution, it is possible to control the release profile of drugs, and bioavailability of the loaded drug in the site of actions (Bock et al., 2011).

Using nanotechnology in medicine (nanomedicine) has had significant effects in improving healthcare (Wu et al., 2012). Biodegradable polymeric MPs and NPs in the form of carriers have found many applications in the field of therapeutic delivery and controlled release drugs (Bourges et al., 2003; Hall et al., 2007; Jalón et al., 2001; Mundargi et al., 2008). This kind of particles can be linked to peptides, antibodies, drugs, and other targeting moieties with the aim of diagnosis and therapy (Byrne et al., 2008). It is possible to enhance the contrast of images in a variety of imaging systems, such as ultrasonography, fluorescence imaging, photoacoustic imaging, and MRI, by the encapsulation of different imaging agents in micro- and NPs. In addition, image-guided therapy, image-guided drug delivery, and other theranostic applications are achievable through loading both imaging and therapeutic agents in multifunctional MPs and NPs (Zhang et al., 2012). For example, by using super paramagnetic NPs of magnetite (Fe_3O_4) it is possible to produce NPs that are applicable in magnetic resonance imaging (MRI) (Kim et al., 2013). Biodegradable polymers loaded with Fe_3O_4 enhance the contrast of images in MRI, in addition to use in the field of drug delivery. Dispersion of these particles in the liquid phase makes ferrofluids that have a variety of applications (Leea et al., 2004; Nasongkla et al., 2006; Nkansah et al., 2011; Ren et al., 2014; Sun et al., 2008). Moreover, NPs loaded by solid lipid NPs, lipoplexes, and polyplexes, are promising carriers for delivering drugs or nucleic acid-based therapeutics (Wu et al., 2012).

There are some conventional methods for the production of particles and capsules; these consist of microemulsification (Djordjevic et al., 2004), polymerization (Limmatvapirat et al., 2008), self-assembly (Guo et al., 2009), solvent evaporation (Yang et al., 2001), and spray drying (Gaete et al., 2008). However, biodegradable polymeric NPs are usually fabricated by methods, such as emulsification–solvent evaporation, salting out, solvent displacement, double emulsion solvent evaporation, and emulsification–solvent diffusion. The synthesizing techniques consisting in the salting out method are based on the separation of a water miscible solvent from aqueous solution through the salting out effect, solvent displacement method, phase separation method, evaporation precipitation, antisolvent precipitation, and electrospray (ES) techniques (Sezer, 2014). However, emulsification is the most commonly used method for microencapsulation/nanoencapsulation (Zhang et al., 2012).

However, even if much progress has been made in the field of drug delivery and preparing polymeric particulate systems, just a few of them are being commercialized. The most important reason for this lack of usage of produced particles is the deficit related to these techniques. Sometimes the degradation of materials and instability happens during the process. For example, in the emulsion method, aqueous/organic interface and shear stress are the major problems. Long-lasting contact with organic solvents and residual materials, as well as with processing agents, are the main anxiety in the other techniques. Resulting defects of these limitations can affect the final usage of produced particles in vitro and in vivo (Bock et al., 2012). Moreover, sometimes it is necessary to make multifunctional particles with different agents and releases in one particle, and producing such structure through common techniques is not easy. As an example, delivering a suitable dose of the active pharmaceutical ingredient (API) at the precise time to the exact part of body is the most important dilemma in the science of drug delivery and therapeutic applications (Peltonen et al., 2010).

To overcome these limitations, one suggestion proposed by researchers, made in the past 20 years, has been ES. However, this technique is not fully developed, and it is necessary to optimize its effective parameters (Ciach, 2007; Jaworek and Sobczyk, 2008). In brief, ES is a technique that produces micro- and nanoparticles (MPs and NPs) using applied voltage on the liquid phase passing through a capillary. The interaction of forces on the liquid leads to breaking the droplets, and the production of particles (Bock et al., 2012).

2 POLYMERIC PARTICLES AND ATOMIZERS

2.1 POLYMER NANOPARTICLES

In general, particles produced for different applications can be categorize in polymeric and non-polymeric particles. Polymeric particles consist of particles that have all or part of their structure made up of polymers. Among polymer particles, biodegradable polymers and their copolymers are popular polymers for the production of polymeric particles. Micelles, capsules, platelets, spheroids colloids, dendrimers, core–shells, NP incorporated polymer matrix and fibers are some example of polymeric NPs. The first pharmaceutical polymeric NP produced between 1960 and 1970 was a micelle. Moreover, depending on the loading style of agents and the size of fabricated nano- or microparticles, these structures can be classified into nano/microspheres and nano/microcapsules (Allemann et al., 1993; Fresta et al., 1996; Legrand et al., 1999). The ultrafine particles have sizes between 1 and 100 nm, and fine particles are 100–2,500 nm; both of these groups are identified as NPs (Sezer, 2014).

2.2 PREPARATION OF PARTICLES

To produce polymeric particles/powders, the first step is to make a solution composed of solvent, polymer, and other required substances. The next step is the disintegration of this liquid phase. As mentioned in the previous section, there are several techniques for the atomization of a liquid. The third step in forming polymeric particles is the drying of the atomized droplets, and there are some methods for this process, such as spray-freeze drying (sublimation), spray drying (evaporation), supercritical fluid drying (precipitation), and spray pyrolysis. After the drying of droplets, the size and morphology of particles are interpretable. The size of particles mainly depends on the concentration of the primary droplets. To produce particles smaller in size it is necessary to produce smaller primary droplets, and to prepare solution in lower concentrations. As an example, to produce a nanosized solid particle, the initial size of the droplet should be less than 1 μm, and a solution concentration of about 0.1 wt.% is required. In addition, the selection of atomizer chiefly affects the primary size of droplets, and the size distribution of produced particles. This factor is extremely vital for the quality and evenness of the dried powder (Yurteri et al., 2010).

2.3 ATOMIZATION

Based on the forces applied in order to break up primary liquid to the small droplets, these might classify into pneumatic, ultrasonic, and electrical. All these techniques can produce primary droplets with a size distribution between 1 and 100 μm. Among these methods, the pneumatic system produces polydisperse droplets, and droplet size depends on viscosity, surface tension density, atomization pressure, and orifice size. In ultrasonic atomizers, the vibration of the air-liquid border of liquid through an ultrasonic apparatus breaks up the droplets, and the diameter of the produced droplets is in relation to the frequency of vibration and the physical characteristics of the solution, such as density, viscosity, and surface tension. This method is capable to produce relatively monodisperse droplets, compared to pneumatic atomizers. A long-lasting operation in this system might lead to an increase in temperature (Biskos et al., 2008; Yurteri et al., 2010).

In ES, liquid solution can break up into evenly sized droplets through a fully controlled electrical force. Compared to the usual methods, ES creates lots of capabilities for the production of novel polymeric particles. Because of the narrow size distribution of produced particles, this method is an appropriate choice for laboratories and industries. The common range of frequency for the ES method is 104–109 Hz, with which an ES apparatus could produce particles ranging in size from micrometers to nanometers, respectively. This technique can produce monodisperse droplets. The property of coulombic repulsion leads to the self-dispersing nature of ES. In addition, in order to reduce the probability of clogging the needle, it is possible to control the rout of flying droplets in the ES (Yurteri et al., 2010).

2.4 ELECTROSPRAY ADVANTAGES AND DISADVANTAGES

To produce drug-loaded polymeric particles, the best method will be chosen depending on the end use and materials. However, it is necessary to mention the advantages and disadvantages of common techniques. Emulsion-based methods have been used broadly for the production of particulate drug-delivery system (DDS) (Soppimath et al., 2001). Solvent evaporation and single or double emulsification are the most popular methods to produce drug-loaded polymeric particles, but broad size distribution

(Kissel et al., 1995), use of some nondegradable emulsifiers (Freitas et al., 2005), low encapsulation efficiency (Zhang et al., 2012), inhomogeneous size, lack of reproducibility (Bock et al., 2011), changing the nature of drugs, and biomacromolecules are some of disadvantages of these methods. The high temperature of the process, and shear stress, are related to these faults (Kohane et al., 2006). The pointed faults lead to limitations in the clinical use of produced particles (Bock et al., 2011).

In contrast, ES has some potential advantages. First, it produces monodisperse droplets with sizes from nanometers to micrometers; second, the free charges that gather on the surface of droplet do not affect the biomolecules. Finally, it is possible to control the size and encapsulation efficiency of particles in ES (Wu et al., 2012). Moreover, there are some other advantages that consists in the ease of operation, in cost-effectiveness, the obviation of elevated shearing forces resulted from stirring or sonication, high loading level, high-throughput if there are multiple electrospinning/ES spinnerets, and uniform drug distribution in the polymer matrix; one can also get rid of residual surfactants, and the technique is easily scalable. In addition, it is possible to check the quality of particles by halting the process and checking the expenditure of drugs and biopolymers by terminating the process. Because of the aforementioned characteristics and the ability to spray/spin, virtually any polymer into NPs and fibers, without altering the fundamental setup of ES/spinning, it seems that ES particles and electrospun fibers are suitable drug delivery vehicles (Chakraborty et al., 2009).

Even though ES is a promising method for the fabrication of particles, there are some limitations to this method. As a major problem, the high number of processing parameters and apparatus parameters can make its optimization highly complex. In addition, the characteristics of ES particles are still not completely studied. To achieve the appropriate results and desired particles, it is better to examine the step-wise manner of process in order to determine the relationship between processing parameters and characteristics of ES particles, before delivering biomolecules into the structure of particles (Bock et al., 2011).

3 ELECTROSPRAY

3.1 HISTORY OF ELECTROHYDRODYNAMIC ATOMIZATION (ELECTROSPRAYING AND ELECTROSPINNING)

ES and electrospinning refer to electro-hydro-dynamic atomization (EHDA), meaning the atomization of a moving (dynamic) liquid (hydro) in an electric (electro) field (Geerse and Marijnissen, 2003). The history of EHDA goes back to the 16th century, when the effect of electricity on liquid menisci was recognized. Conical shaped droplets resulted from the interaction between a water droplet and a piece of amber close to it; this was first reported by William Gilbert in 1600 (Yurteri et al., 2010). Gilbert wrote a book entitled *"De Magnete,"* a revealing work of art about the research on electrostatics and magnetism. He described in this book the electrostatic spraying of water from the nozzle connected to the high voltage power supply (Ciach, 2007). Establishing the change of a hemispherical meniscus droplet to the conical shape under an electrical stress was done by Zeleny, at the beginning of the 20th century (Yurteri et al., 2010). Zeleny published the first photograph of electrostatic atomization, and showed the conical shape of droplets in the electric field, and the production of a mist of water in the cone apex. In his definition, an electrostatic force is the only energy source leading to the atomization of liquid (Ciach, 2007). In 1952, some theoretical and experimental studies were performed by Vonnegut and Neubauer, and some work was carried out by Taylor, in 1964. Researchers followed this trend of study, and have published many papers and books in the field of ES up to the present day. For the precise study, and in order to predict features of particles produced by an ES system, scaling laws for the size of particles in the cone jet mode of ES were set by some researchers. De la Mora and Loscertales (1994); Gañán-Calvo (1997); Hartman et al. (1997); and Chen and Pui (1997) have published their results in this field. Nowadays, there are some methods that use both electrostatic and other atomizers; but electrohydrodynamic atomization implies techniques that uses only electrostatics in order to atomize liquids (Ciach, 2007).

3.2 PRINCIPLES AND THEORETICAL ASPECTS

3.2.1 Basic principles

Generally, the coulombic interaction of charges located on the liquid surface and applied electric field leads to the atomization of the liquid, and this is the principle of EHDA. The interaction causes acceleration and subsequent disruption of the liquid, producing droplets (Geerse and Marijnissen, 2003). In the EHDA process, applying high voltage – several kV – to a liquid droplet that forms at the tip of a nozzle leads to the creation of electrostatic force inside it, called coulomb force. On the other hand, the cohesive force intrinsic to the droplet is in competition with the cohesive force, and if the coulomb force overcomes the cohesive force, the droplet will undergo a breakup into smaller droplets in the micro- to nanoscales. Surface tension of a liquid droplet could be a good representative for cohesive force. After applying potential difference, and before the formation of the broken up droplets, the hemispherical shape of primary droplets changes to the cone form that is known as a Taylor cone (Bock et al., 2011). If the cone is then disturbed by further charge (Peltonen et al., 2010) by Coulomb force, this overcomes the surface tension of the liquid and the smaller charged droplets will be ejected. Rayleigh predicted a limit to establish the disintegration of the droplet, known as Rayleigh limit, L_R, Eq. (6.1) (Bock et al., 2011):

$$L_R = q(64\pi^2 \varepsilon \gamma R^3) \tag{6.1}$$

L_R is a function of q, the surface charge of the droplet, ε is the permittivity of the surrounding medium, γ the surface tension of the liquid, and R the radius of the droplet. The maximum surface charge of the droplet is given by Eq. (6.2) (Bock et al., 2011):

$$q = 8\pi\sqrt{(\varepsilon_0 \gamma R^3)} \tag{6.2}$$

Choosing appropriate parameters to produce monodisperse ES particles is necessary. The schematic diagram of the experimental setup for ES is shown in Fig. 6.1. In real conditions, the range of applied voltage is up to +30 or –30 kV, and the distance between the tip of nozzle/needle and collector is between 7 and 30 cm. Needles of the gauges 16–26 are more popular in the EHDA process. After the separation of droplets from the Taylor cone, the evaporation of solvent will start, and solid MPs or NPs will be created (Bock et al., 2011).

The highly charged generated droplets travel along the electrical field toward the collector. As they are travelling in the gas phase, they get smaller as the solvent evaporates. This can lead to droplet fission if the Rayleigh limit is reached, thus particles will break down and free ions will be generated. This mode is applicable in mass spectrometry (Peltonen et al., 2010).

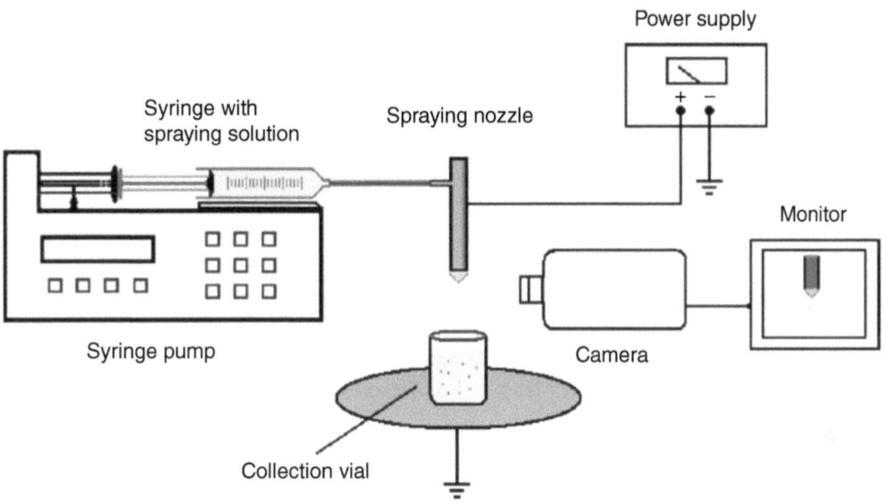

FIGURE 6.1 Schematic Diagram of the Experimental Setup for Electrospraying (ES) (Bohr et al., 2011)

3.2.2 Scaling laws for the size

Many researchers have tried to formulate the EHDA process, and scaling laws for the size of droplets and produced particles represent the major factors that have been investigated. Cloupeau and Prunet-Foch (1994); Rosell-Llompart and de la Mora (1994); Chen and Pui (1997) and Gañán-Calvo (1997) have proposed the first semiexperimental models for the average dimension of a droplet. In addition, Hartman et al. (1998) proposed scaling laws for the droplet size in the cone-jet mode. Based on studies, the mechanism of breaking-up of the jet depends on the ratio of the electric normal stress over the surface tension stress. Varicose instabilities lead to the jet breaking up at lower values. However, at higher values, the whipping of the jet will start. Then, both varicose and kink instabilities will affect the jet breakup (Yurteri et al., 2010). The scaling relations developed are useful in the drug delivery framework, and they allow the prediction of the effects that different parameters have on particle and fiber geometry (Chakraborty et al., 2009). Based on formulas suggested by Hartman et al. in the varicose jet breakup for the droplet size (Yurteri et al., 2010):

$$d_{\mathrm{d,varicose}} = \left(\frac{16\rho\varepsilon_0 Q^3}{\gamma K} \right)^{1/6} \tag{6.3}$$

where ρ is liquid density, vacuum permittivity, Q is the flow rate, γ is the conductivity of liquid, and K is the liquid conductivity. Moreover, $d_{\mathrm{d,\ varicose}}$ is close to the primary (main) droplet size (Yurteri et al., 2010).

And

$$d_{\mathrm{d,whipping}} = \left(0.8 \frac{288\varepsilon_0 Q^2}{I^2} \right)^{1/3} \tag{6.4}$$

where, $d_{\mathrm{d,whipping}}$ is the droplet size predicted for a jet that breaks up in the whipping jet breakup mode, I is electric current for liquids, where the radial velocity profile is not flat (Yurteri et al., 2010).

In addition, a theatrical model was introduced by Gañán-Calvo (1997), with a good agreement with experimental results:

$$d_D = \alpha \left(\frac{Q^3 \varepsilon \rho}{\pi^4 \sigma \gamma} \right)^{1/6} \quad (6.5)$$

where α is a constant, Q is the liquid flow rate, ε is dielectric constant in vacuum, ρ is the liquid density, γ is the conductivity of liquid, and σ is its surface tension (Gañán-Calvo, 1997). Like in the equation as shown, the size of the droplet is proportional with the root square of Q; it means that if the particles' diameter plot against $Q^{0.5}$ they would fit to a linear line with a good R^2 (Jafari-Nodoushan et al., 2015b).

By electing appropriate liquid properties and flow, it is possible to fabricate droplets with the desired size in a well-controlled way. The initial solution contains both solute and solvent, thus d_p—final particle size—is related to the initial droplet size d_D by the following equation (Wu et al., 2012):

$$d_p = d_D \left(\frac{c\rho}{\rho p} \right)^{1/3} \quad (6.6)$$

where c is the concentration of solute, and ρ_p is the density of the particle (Wu et al., 2012).

3.3 ELECTROSPRAY MODES

Generally, the EHDA mode will most importantly determine by a variety of parameters the physical properties of the liquid (e.g., surface tension, density, electrical conductivity and viscosity) and system parameters (e.g., flow rate, collecting distance and electric field strength). The mode of ES will be changed by varying the potential difference. At lower voltages, there is intermittent jet formation that creates the "dripping," "microdripping," "spindle," "multispindle," and "ramified-meniscus" modes. By increasing the voltage, a sustained and a continuous jet is formed, and the "stable cone-jet," "oscillating jet," "multijet," and "ramified-jet" modes can be observed (Fig. 6.2). Among ES modes, the stable cone-jet mode is the best choice for the production of fine particles, and by adjusting the flow rate (µL/h to mL/h) and voltage (several kV) the formation of a liquid cone is not difficult (Enayati et al., 2010; Yurteri et al., 2010).

3.4 EHDA CONFIGURATIONS

There are four major configurations for EHDA setup. Needle–plate (Fig. 6.3A) and needle–ring extractor–plate (Fig. 6.3B) configurations are suitable methods for the direct coating of medical substrates, such as individual microparticles, tablets, and medical inserts (Yurteri et al., 2010). In the second configuration, there is a charge neutralization mechanism, which forms ions with opposite charges through a radioactive source or corona discharge. For the fabrication of drugs that apply in inhalation, an air-flow transfers particles to the collector. In the third configuration (Fig. 6.3C), two oppositely charged EHDAs create oppositely charged droplets, and the electrical attraction between them leads to the coagulation of the droplets (bipolar coagulation). It is possible to produce new products, such as composite spheres by using this system (Camelot et al., 1999). This method is also applicable in the neutralization of the droplets and coating of one material with another. For encapsulation of particles in one step, the fourth configuration, coaxial EHDA (Fig. 6.3D), is more useful. Nozzles are fed

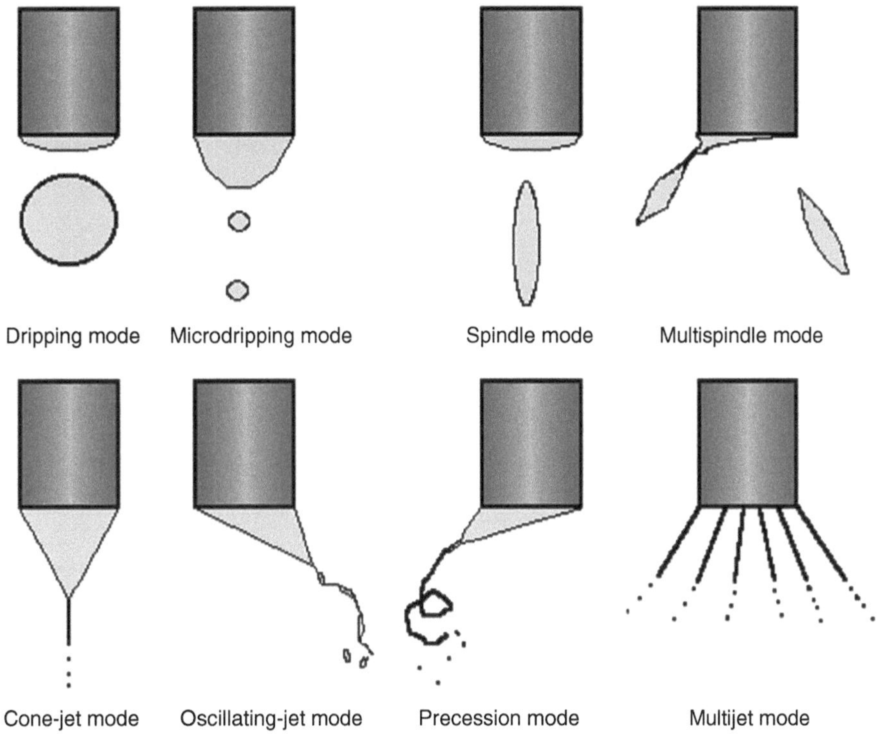

FIGURE 6.2 Flow of 5 wt.% PCL Solution Under an Electric Field Showing, with Flow Rate set at 10 μL min^{-1}

(A) No flow; (B) dripping mode; (C) microdripping mode; (D) rapid dripping mode; (E) unstable cone-jet mode; (F) stable cone-jet mode; (G) multijet mode; and (H) irregular instabilities mode. Scale bars, 600 mm (Enayati et al., 2010).

separately, and protrude slightly off the central one, and lead to making easy the coating by the outer fluid. In this approach, control of encapsulation is more convenient. The spraying process is mainly controlled by the outer solution properties. Moreover, multilayer capsules can be produced by aligning more nozzles concentrically (Yurteri et al., 2010).

3.5 PARTICLE PRODUCTION BY EHDA

3.5.1 Particle production by monoaxial EHDA

The classic setup of EHDA solution or liquid is injected to the nozzle/needle by a syringe pump. On the other hand, the nozzle is connected to the high voltage power supply, and there is potential difference between nozzle and collector (Jafari-Nodoushan et al., 2015a). Mostly DC voltage is used as a power supply; however, the possibility of production of particles through AC voltage has been examined (Yeo et al., 2005). Sometimes, for stabilizing the process, a ring is connected to an intermediate voltage and is located below the nozzle. However, a pipe ring surrounding the needle might be used for better control of the process (Jafari-Nodoushan et al., 2015a). The needle diameter is

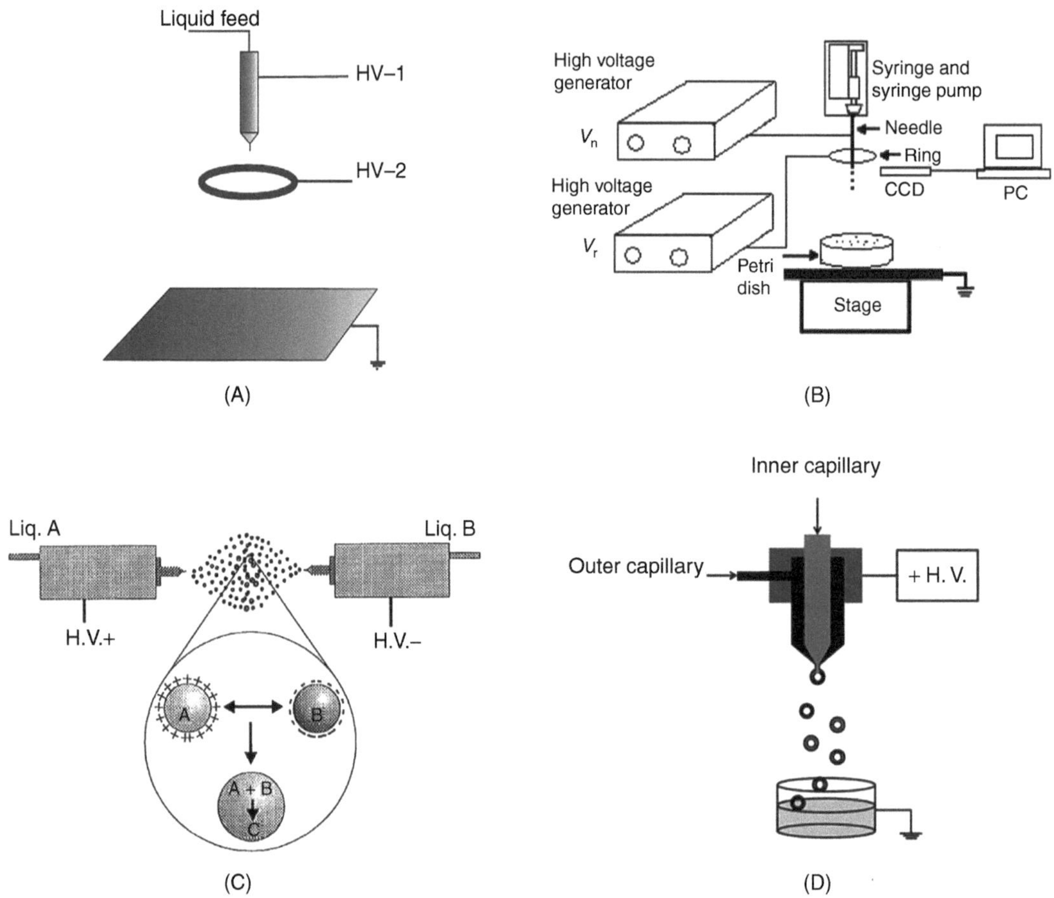

FIGURE 6.3 Configurations of Electrospaying (ES) (Yurteri et al., 2010)

between 1 and 10 mm, and in order to work in cone-jet mode, voltage is set to 10–15 kV. However, the voltage of the intermediate ring should be set to half to three-quarters of the nozzle voltage. Depending on the evaporation time, the liquid collector should be located from a few centimeters to 30 cm from the nozzle (Ciach, 2007). To produce particles by EHDA it is essential to prepare a suitable solution. It should have an appropriate level of conductivity. By adding appropriate additives to the solution, this is achievable. Moreover, before starting the process, it is necessary to measure the conductivity and surface tension of the solution (Yurteri et al., 2010). After the production of droplets from the polymer solution, solvent evaporation and polymer diffusion will start at the same time. This process leads to chain entanglement that is accountable for the ultimate particles' morphology (Bock et al., 2012). In general, the initial micron size droplets contain both solvent and a nonvolatile solute, and as soon as evaporation of solvent from the high surface area aerosol takes place, the particle is formed (Wu et al., 2012). It is also possible to introduce a map, a drawing of the area, by use of data measured from the EHDA process. As shown in Fig. 6.4, the cone-jet mode

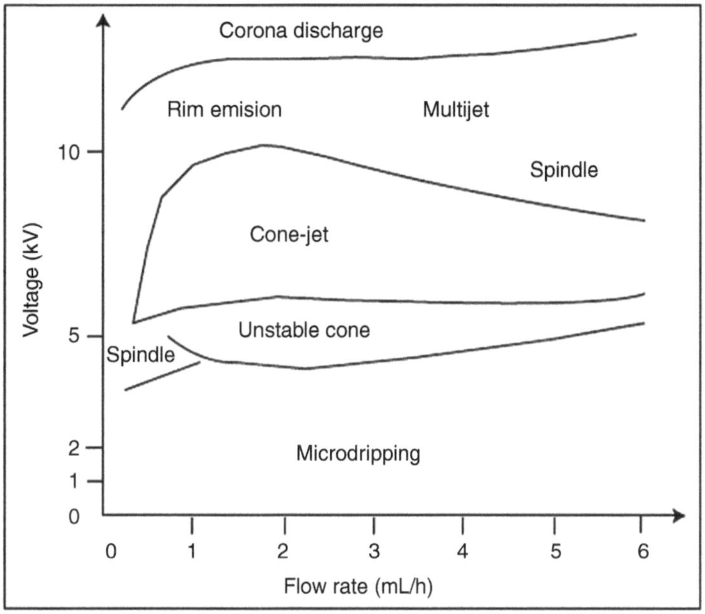

FIGURE 6.4 A Typical Map of EHDA Behavior (Nozzle Diameter 4 mm, 2% Ethanol Solution of Acetaminophen) (Ciach, 2007)

leaving operation zone leads to the creation of small hysteresis in the borders of the presented EHDA zones (Ciach, 2007).

Collecting solution is usually applied in the EHDA setup that operates in microdripping mode. Solid particles are formed in the collecting solution due to one of the following mechanisms: evaporation of solvent (in immiscible liquids), the process of solvent–antisolvent, and chemical coagulation (Ciach, 2007). In the process of solvent–antisolvent, the precursor of particles is dissolved in the solvent that is soluble in the collecting liquid, and vice versa. While immersing a droplet in the collecting bath, liquids diffuse into each other. This phenomenon leads to the precipitation of dissolved solids, and formation of highly porous particles (Fig. 6.5). Furthermore, the chemical coagulation process can be applied in the encapsulation of living cells.

3.5.2 Particle production by coaxial EHDA (CEHDA)

The principle of CEHDA is the same as monoaxial electrospray, and the major difference is using coaxial nozzles. Two immiscible liquids form a "cone in the cone" system at the tip of the nozzles, and an outer jet surrounds the inner jet. Particles produced by a coaxial system have a core and a shell (Ciach, 2007). However, liquid and gas can be fed to the nozzles of EHDA setup to form a bubble or hollow sphere (Zhang et al., 2012). CEHDA is a suitable method for the encapsulation of proteins, because it is difficult to encapsulate them by the other methods. Obviously, a water solution of protein is fed to the inner nozzle by a syringe pump, and the organic phase is fed to the outer nozzle (Ciach, 2007). In the CEHDA configuration, the properties of the jet and resulting droplets depend on the liquid that has the smaller(est) electrical relaxation time. Either inner or outer liquid could play this role (Enayati et al., 2011).

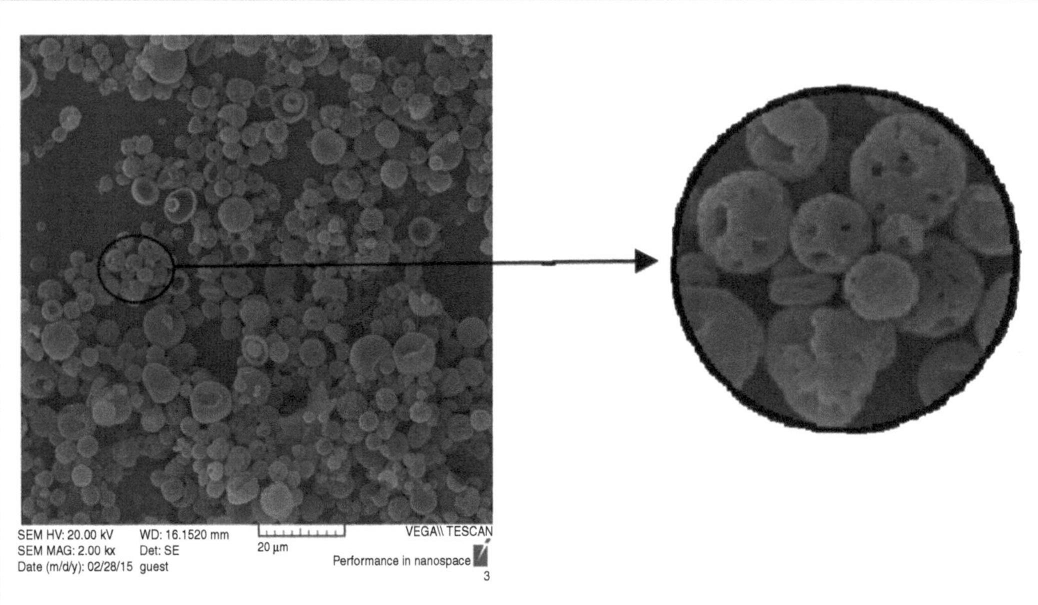

FIGURE 6.5 Electrosprayed (ES) PLGA Particles Produced by Solvent–Antisolvent Coagulation (Chloroform, Water, Saturate; Mean Diameter: 5.22 μm) (Faramarzi et al., unpublished work)

The same as monoaxial ES, this system can fabricate particles ranging in size from tens of nanometers to hundreds of micrometers (Farook et al., 2009). Jet breaking in CEHDA makes it possible to achieve an encapsulation rate of nearly 100%, exact control of the geometry of the core–shell, and to prevent the aggregation and denaturation of the brittle therapeutic cargos. In addition, it is possible to use this system for the mass production of drug-loaded MPs and NPs. Three positions of same height, below and above the outer needle, have been tested for the inner needle by the researchers (Zhang et al., 2012).

3.6 EFFECTIVE PARAMETERS IN ELECTROSPRAY

Although ES is a powerful method for the production of fine particles, the large number of process variables limits productivity of the technique. To control the morphology and size of particles, it is necessary to minimize variations in solution and system parameters. To produce ES pharmaceutical particles, parameters include process parameters such as polymer, drug, solvent, surfactant, protein:polymer ratio, and organic:aqueous ratio, as well as apparatus parameters, such as flow rate, voltage, needle gauge, and distance to collector. The mentioned parameters have an interdependent influence on electrical conductivity, viscosity, distribution, particle size, encapsulation efficiencies, loading capacities, and in vitro release profiles (Bock et al., 2011).

The effect of major parameters known on the size of produced particles/fibers in the EHDA process has been summarized in Fig. 6.6. (Chakraborty et al., 2009).

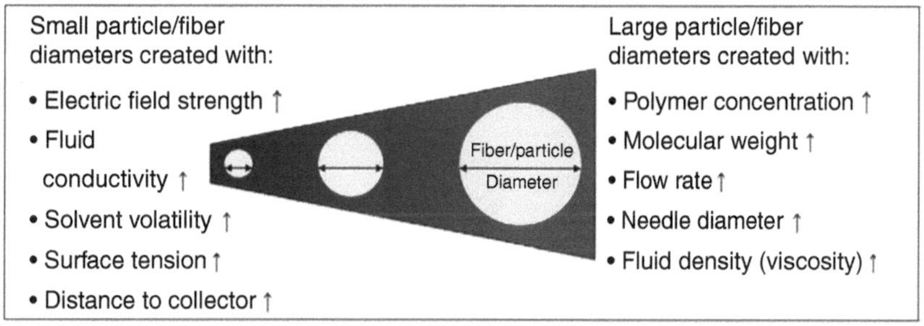

FIGURE 6.6 Schematic Diagram of Effect of EHDA Parameters on the Size of Fiber/Particle (↑, Increase) (Chakraborty et al., 2009)

Faramarzi et al. plotted particle size by the root square of the flow rate, for the particles produced by ES a PLGA solution. Based on Eq.(6.5), the results show good agreement with the suggested scaling law for the size of particles (Faramarzi et al., unpublished work) (Fig. 6.7).

4 THERAPEUTIC NANOPARTICLES LOADED BY ELECTROSPRAY

As aforementioned, ES and electrospinning are promising EHDA fabrication techniques that can fabricate MPs and NPs loaded with pharmaceuticals and diagnostic moieties through a one-step process. The nanosized particles and fibers fabricated by these methods recommend tunable release kinetics applicable to a wide range of biomedical applications. Moreover, CEHDA offer near-zero release kinetics

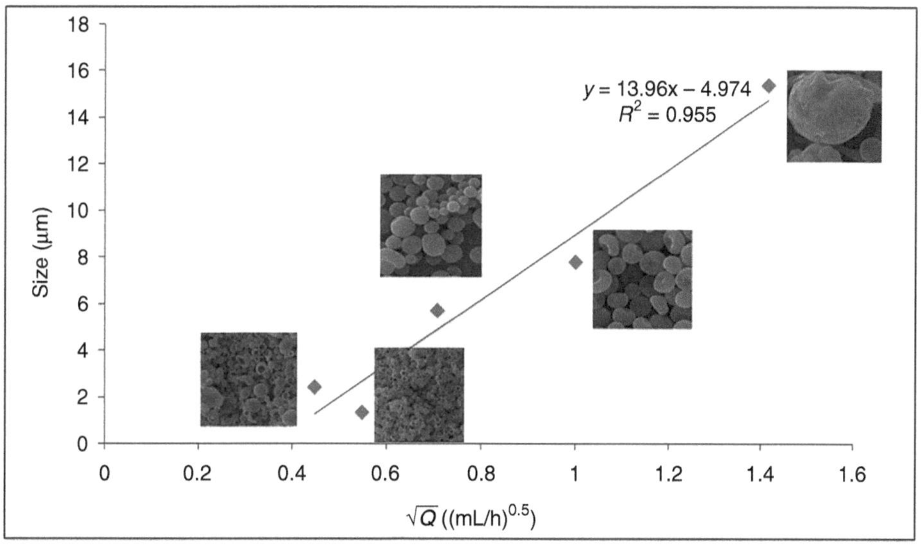

FIGURE 6.7 The Effect of Flow Rate on the Saunter Mean Diameter of PLGA Particles (Polymer Concentration: 2%, Voltage: 11 kV, Collecting Distance: 13 cm) (Faramarzi et al., unpublished work)

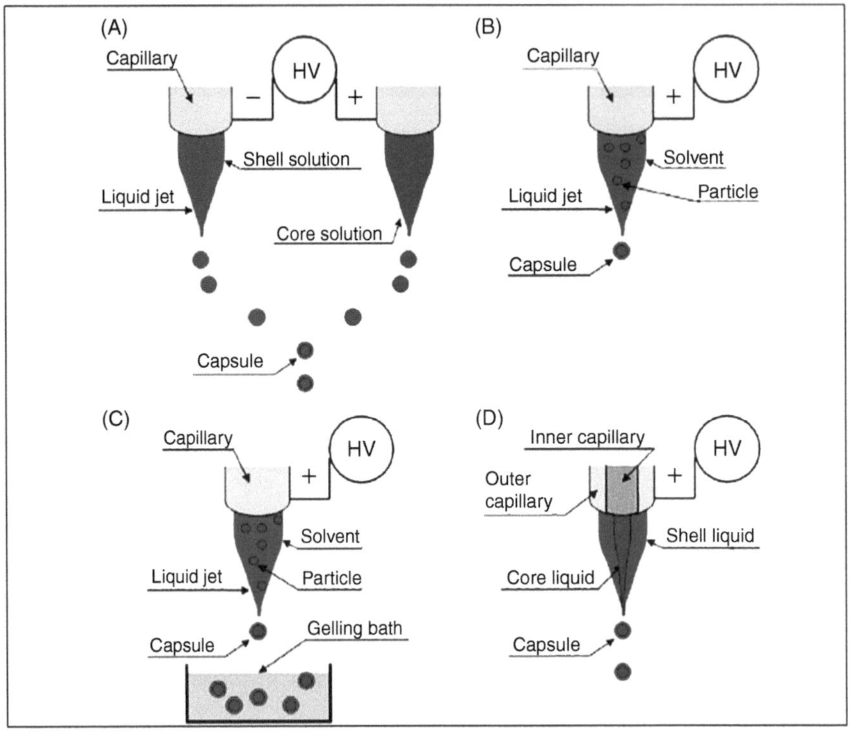

FIGURE 6.8 Schematics of Various Electroencapsulation Processes

(A) Collision of droplets of opposite polarity; (B) ES of a drug dissolved/suspended in polymer with solidification by evaporation; (C) ES of a drug dissolved/suspended in polymer with solidification by a chemical or ionic crosslinker; and (D) coaxial ES (Chakraborty et al., 2009).

in drug delivery systems. The variety of drugs, such as pharmaceutical agents, proteins, peptides and DNA molecules, can be delivered to the polymer matrix in one step, in the EHDA process. Most researches on pharmaceutical polymeric particles are PLGA-based. Even so, for a slower degrading desire, such as orthopedic applications, PCL is a better option (Ciach, 2006). It is possible to produce MPs and NPs using this polymer. By a decrease in polymer concentration, the porosity of the particles also decreases, and hollows structures are formed (Wu et al., 2009; Xie et al., 2006a).

Pharmaceutical polymeric NPs thus produced may be gathered in different collectors: water without excipients (Xu and Hanna, 2006), polar solvent with a stabilizer (Valo et al., 2009), or gelatinizing or polymerizing baths (Lewinska et al., 2004; Xie and Wang, 2007b) being some of the investigated collectors.

In general, there are two major techniques to produce drug-loaded MPs or NPs; these are adsorption and encapsulation. In the adsorption technique, the drug is adsorbed onto the sprayed carrier by exposing the carrier to a drug solution. The detaching of drug burst release is the disadvantage of this method. As shown in Fig. 6.8, there are several techniques for the encapsulation of pharmaceutical agents, consisting in the collision of droplets of opposite polarities (Fig. 6.8A), electrospraying (ES) of a drug dissolved/suspended in polymer solution with solidification by evaporation (Fig. 6.8B), ES of a drug dissolved/suspended in polymer with solidification by a chemical or ionic crosslinker (Fig. 6.8C), and coaxial ES (Fig. 6.8D) (Chakraborty et al., 2009).

Most of the drug-loaded particles have been produced through monoaxial and coaxial nozzle array, thus we propose in this section medical applications of drug-loaded particles by focusing on these configurations.

4.1 SINGLE NEEDLE EHDA PROCESSING

Different drug-loaded polymeric particles with sizes ranging from several tens of nanometres to ~100 μm can be produced by single needle ES. A therapeutic agent can be added to the polymer solution, and is dispersed homogeneously in the polymeric matrix without aggregation, owing to residual charges for the duration of drying (Enayati et al., 2011).

4.1.1 Particles

A wide range of research work has been carried out to investigate pharmaceutical polymeric NPs and MPs. Many scientists have tried to load different agents and drugs to the polymeric particles in order to develop new pharmaceutical and medical polymeric particles.

Bovine serum albumin (BSA) is a model protein drug that was applied by some researchers. Poly(lactide) (PLA)-loaded BSA fine particles have been produced, and particles with smooth morphology and size range from 800 to 4000 nm and encapsulation efficiency between 22 and 80% were obtained (Xu et al., 2006). In another study performed by Xu and Hanna (2007), BSA-loaded tripolyphosphate (TPP) crosslinked chitosan capsules were produced, and the effect of processing parameters of ES on the size of fine particles have been studied. In another study, biodegradable PCL NPs with the size range between 300 to 4500 nm have been prepared by EHDA, and the size distribution of NPs has been studied. In the solutions with higher viscosity, the upper and lower limits of voltage for the fabrication of particles from stable cone-jet mode have been between 7.5 and 11 kV. Especially in the higher flow rates, by increase of voltage, the size of NPs has been decreased. Moreover, by increasing the viscosity from 2.5 to 11 mPa, the average size of NPs has been increased from 340 to 4400 nm. By increasing the flow rate in the cone-jet region, the polydispersivity of NPs has been increased. Adding of 15 wt.% β-estradiol as a model drug to the polymer solution has not had significant effects on morphology; however, it has increased the size. Drug-loaded PCL particles have showed sustained release of the loaded drug (Enayati et al., 2010).

During EHDA processing, solvent evaporation may lead to coulomb fission. To avoid this phenomenon, neutralization has been accomplished by a corona discharge system that was created at the tip of a sharp, grounded needle positioned opposite the high voltage ring and needle. BSA-loaded PLGA microparticles were produced using this method. By adjusting the quantity of the additive Pluronic F127 and processing parameters, the encapsulation efficiency could reach 76%. Furthermore, after release from the PLGA matrix, BSA maintained integrity (Xie and Wang, 2007a).

With the aim of fabricating near-monodisperse particles in a stable amorphous form, celecoxib, a low solubility drug, was loaded in the PLGA MPs and NPs. It has been showed that different polymer:drug ratios give different release profiles. By increasing the loaded celecoxib, the release rate has also been increased. In addition, dissolution studies have shown that, after an initial burst, release has been controlled by degradation of PLGA (Bohr et al., 2011).

The possibility of delivery of chemotherapeutic agents to the site of action has been studied by many researchers. However, in order to achieve efficient delivery of chemotherapeutic agents and to decrease side effects, it is necessary to determine the exact location of cancer cells. MRI is one of the conventional imaging techniques for the detection of tumors in the body. On the other hand, super paramagnetic colloidal iron oxide NPs, such as super paramagnetic iron oxide (SPIO), are NPs that have had many applications in biomedicine. For example, they can be used as contrast enhancement agents in MRI (Faramarzi et al., 2014b) and drug delivery (Faramarzi et al., 2015a). Particle size is an important parameter for the use of fine particles as contrast agents in MRI, and sizes of 100–3000 nm, depending on the intended use of MRI, can be used. However, in the field of drug delivery, SPIO-loaded particles up to the size of 7 μm have been applied. Fe_3O_4-loaded PLGA MPs and NPs have been fabricated by using ES, and the effects of solution and process parameters have been investigated. The aim of the research has been producing dual functional polymeric particles. Loading both SPIOs and drugs to the particle, and delivering it to the site of action makes it possible to use the contrast enhancement and drug delivery capabilities of the produced particles. Results of the research have indicated that in both PLGA and magnetite-loaded PLGA fine particles, by decrease of flow rate, the size of particles has been reduced (Faramarzi et al., 2015a). A ferrofluid is produced by the dispersion of fine particles in a liquid phase; thus, MRI tests could be done (Faramarzi et al., 2014a, 2015b) (Fig. 6.9).

In order to increase the capability of polymeric particles, it is possible to add anticancer drugs, such as paclitaxel and doxorubicine to the polymer solution, and to fabricate SPIO and drug-loaded fine particles.

Some research work has been done to study the anticancer drug-loaded polymeric particles. Doxorubicin-loaded elastin-like polypeptides (ELPs), a biodegradable and bioresponsive polymer, has been fabricated by ES. The fabricated NPs' size ranges from 300 to 400 nm. The loaded drug has application in cancer therapy. However, it was indicated that Doxorubicin did not have a significant influence on the morphology of produced NPs. Concentration, voltage, and polymer molecular weight have had significant effects on the morphology, size, and polydispersivity (ratio of standard deviation to mean size) of NPs (Wu et al., 2009).

PLGA-loaded paclitaxel MPs and NPs have been produced using ES apparatus with the aim of aiding cancer therapy. Controllable morphologies, such as spheres, donut shapes, and corrugated shapes with particle sizes ranging from several tens of microns to hundred nanometers have been observed by scanning electron microscopy (SEM) and field emission electron microscope (FSEM). By decreasing the flow rate, NPs of around 300 and 600 nm have been produced when the polymer concentrations have been 2 and 8%, respectively. Despite encapsulation, paclitaxel has kept its biological function and has inhibited most of C6 glioma cells. As the sustained release properties, the cytotoxicity of microparticles to C6 glioma cells has been higher than taxols, after 4–5 days treatment (Xie et al., 2006b).

In the research work performed by Ranganath et al. (2009), a combination of EHDA and gelation has been used to produce paclitaxel-loaded PLGA microspheres entrapped in a gel matrix with sustained drug release properties and implantability advantages for local glioma chemotherapy. Cytotoxicity tests have been assessed by the quantification of IC50 and caspase-3 activity against C6 glioma cells in vitro. In addition, in vitro cellular apoptosis of C6 glioma cells indicated that the formulations could provide sustainable paclitaxel-induced apoptosis in comparison to Taxol. Furthermore, subcutaneously implanted gel beads in mice have been able to reduce the tumor size in nude mice, more effectively than that of Taxol).

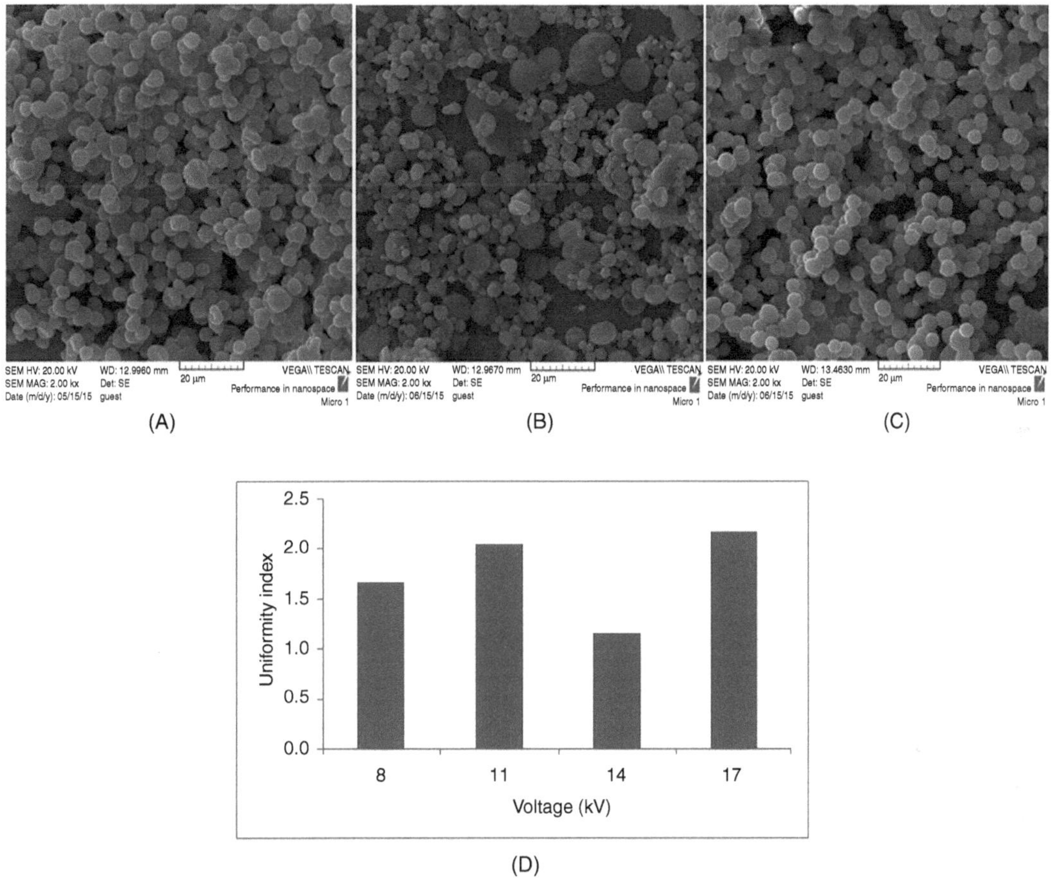

FIGURE 6.9 SEM Image of Magnetite Loaded PLGA Fine Particles Produced by ES

Polymer concentration (%) and voltage, respectively are: (A) 2, 8; (B) 2, 11; and (C) 2, 14. (D) The effect of voltage on uniformity of produced magnetite loaded PLGA fine particles.

In another study, submicron/nanoscale intracranial implants loaded with paclitaxel have been produced by EHDA (ES and electrospinning). The same as in many researches, PLGA was the biodegradable polymer matrix. Produced mats consisting of fibers and particles have been intracranially implanted in BALB/c mice. Moreover, significant tumor inhibition has been observed (Ranganath et al., 2010).

With the aim of decreasing the sense of pain, research work has been done by Jafari-Nodoushan et al. (2014a). They have fabricated morphine-loaded PLGA particles. Three types of biodegradable drug delivery systems for controlled release of morphine with the aim of application in treatment of moderate to severe pain were fabricated. The systems have been in situ-forming, ES microspheres, and a combination of these two drug delivery systems. Results have indicated that release profile of the combined system is more favorable (Fig. 6.10) (Jafari-Nodoushan et al., 2016).

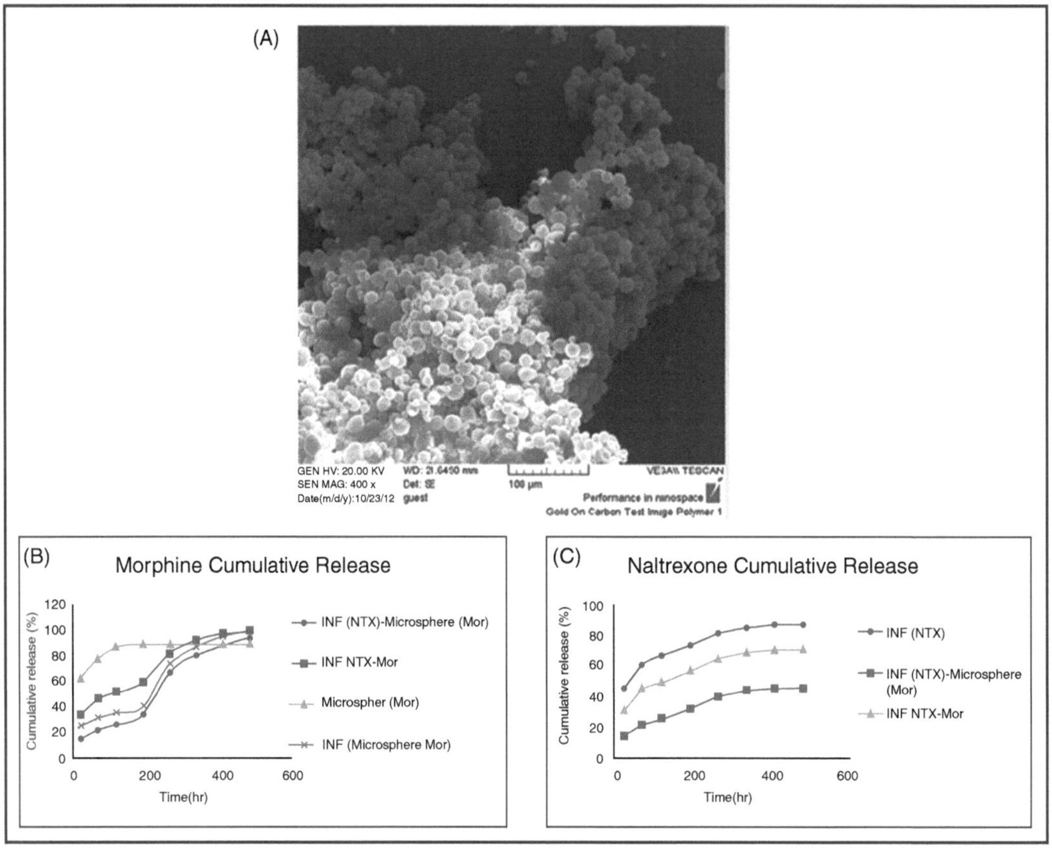

FIGURE 6.10

(A) Morphine particles produced by EHDA; (B) release profile of morphine from prepared formulations; (C) release profile of Naltrexone from prepared formulations (Jafari-Nodoushan et al., 2014b).

4.1.2 Aerosols

Aerosols are another category of pharmaceutical particles, producible by single needle ES. Application of aerosols to treat pulmonary disease started in 1990. Monodispersity is an important factor in the fabrication of aerosols, fabricating them as an inhaler drug. The final size of particles should be between 2000 and 5000 nm (Enayati et al., 2011). ES has been applied to the production of antiasthmatic drugs. A modified apparatus 2-meters in length has been designed to produce particles with narrow size distribution. To prevent the side effects, controlling the dose of the drug in inhalers is very important, and by producing such particles the dose will have been controlled (Ijsebaert et al., 1999, 2001).

Hong et al. (2008) have produced quasi-monodisperse and size-controllable PLGA aerosol particles with matrix-distributed Rifampicin (RIF) by using ES. Chloroform has been used as a solvent. In their fabrication approach, particles were passed over a size separator, and primary droplets were collected, and satellite droplets were removed from the process. In comparison to conventional emulsion technique, this method showed three advantages: (1) control of particle size and narrow size distribution;

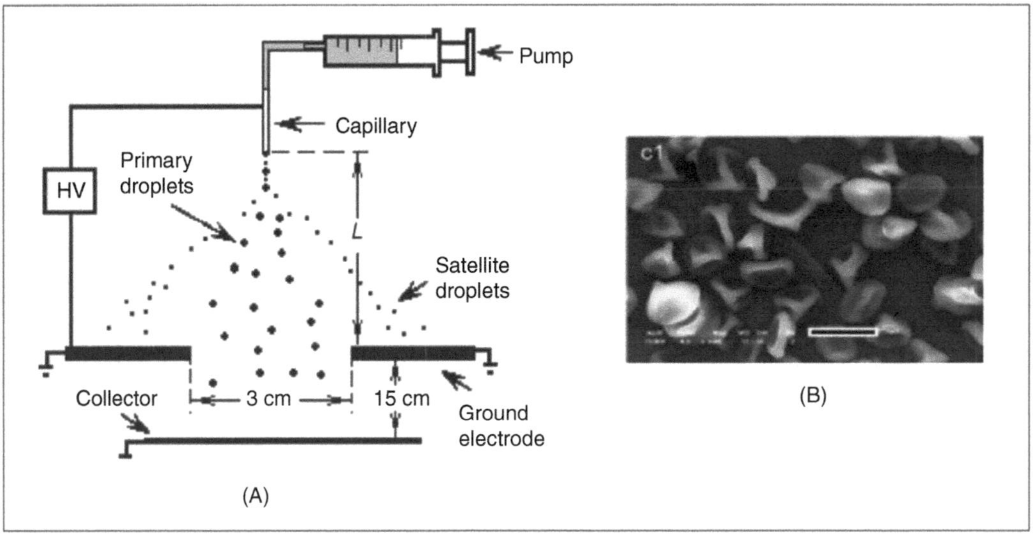

FIGURE 6.11

(A) Schematics of the experimental ES setup; (B) RIF-loaded PLGA aerosol particles (applied voltage 10 kV, and liquid flow rate of 1.0 mL/h) (Hong et al., 2008).

(2) reduction of toxicity because nondegradable additives have been needed; (3) no tedious separation process has been needed to remove the particles from the solvent. By the use of low and high drug loading, spherical, and wrinkled microparticles have been produced, respectively (Fig. 6.11).

4.1.3 Porous particles

Porous structures are another useful carrier for pharmaceutical applications and their morphology and high surface area to volume ratio is unique. Porous polymer particles have been of great interest to nanotechnology and bioengineering in the last decade (Wu and Clark, 2007). In addition to conventional methods, such as solvent evaporation, colloidal particle self-assembly, supercritical carbon dioxide emulsification, and aerosol-spray techniques, they can be fabricated by EHDA technique. Studies have showed that solvent evaporation and phase separation are the main factors that affect pore structure. A low boiling point solvent produces nonuniform morphology, along with large pores (Enayati et al., 2011).

Wu and Clark (2007) have produced porous PCL particles with different structures. They showed that it is possible to manage particle microstructure by controlling solvent evaporation during the ES process.

In another work, by using a combination of ES deposition techniques, biodegradable films along with controllable thickness for sustained release applications have been produced. Paclitaxel was loaded in PLGA films, and significant differences of release profiles were observed between 5 and 30% drug-loading samples in vitro. Moreover, cell viability has been determined by the ratio of absorbance of test cells and absorbance of control cells at the wavelength of 560 nm. By increasing the ratio of paclitaxel to PLGA, the cytotoxicity was increased. Cell viability has been determined by the ratio of absorbance of test cells and absorbance of control cells at the wavelength of 560 nm (Xie et al., 2008b).

4.2 TWO-NEEDLE COAXIAL ELECTROSPRAY

CEHDA consists of ES through two or more concentric or off-center needles, facilitating the encapsulation of a wide range of ingredients. For medical applications, a variety of pharmaceutical agents, such as drugs, proteins, cells, enzymes, etc., could be encapsulated in a protective shell. Many researchers have worked on encapsulation through coaxial nozzles in order to fabricate new particle structures and capabilities.

4.2.1 Capsules

Encapsulation of BSA and lysozyme as a model protein drugs has been done by CEHDA. By changing the solvent evaporation rate, irregular-shape to spherical microparticles were produced. Particles had a rough surface morphology that might be due to rapid solvent evaporation from the droplets. PLGA has been used as a polymer matrix and dichloromeran has been applied to make polymer solution. The in vitro release profile has been studied, and phosphate buffer saline (PBS) was used as a release medium. Sustained release of protein of more than 30 days has been observed, and more than 90% bioactivity of released lysozyme has been maintained (Xie et al., 2008a).

Budesonide (or EGCG) as a model drug has been loaded in PLGA by CEHDA. Acetonitrile was applied as a solvent. The release profile of the drug showed that the release rate of the drug in smaller NPs has been more rapid than that of larger ones (Fig. 6.12) (Lee et al., 2010).

As shown in Fig. 6.12C, in the release rate profile, Stage I strongly depends on the rate of water penetration and drug diffusion, and Stage II came first from the local rupture of the PLGA shell. There is a completely deteriorated release at this stage (Lee et al., 2010).

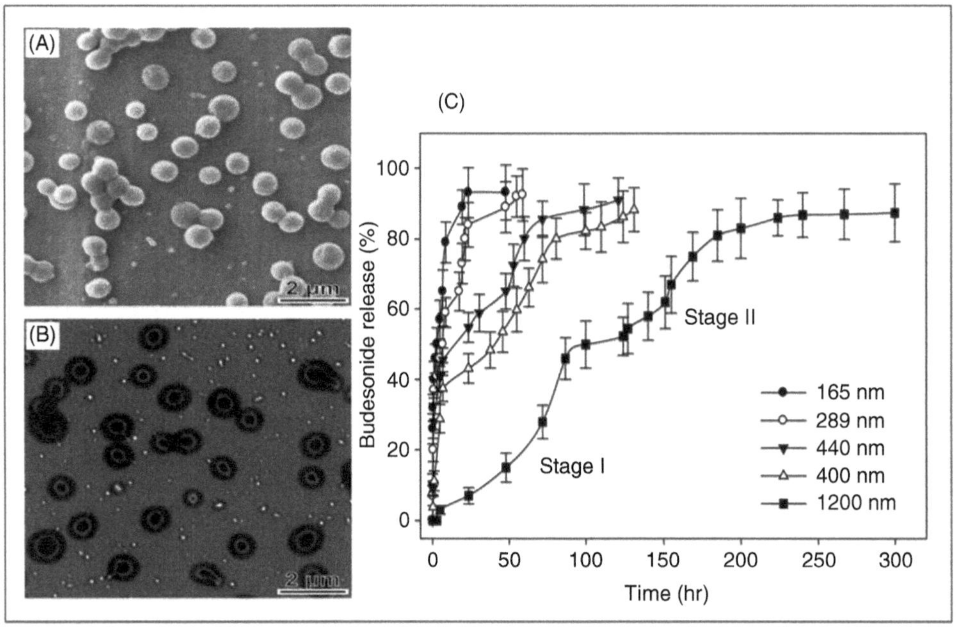

FIGURE 6.12

(A) SEM image, and (B) TEM image of Budesonide-loaded PLGA particles prepared by dual-capillary ES (with 3 wt.% PLGA at outer flow rates of 5 μL/min, and 0.1 wt.% Budesonide at inner flow rate of 0.8 μL/min); (c) the drug release profile of Budesonide-loaded PLGA particles with different particle size (Lee et al., 2010).

Starch/BAS loaded polydimethylsiloxane (PDMS) microspheres in the size range of 5–6 μm have been fabricated by CHDEA. The inner capillary has been kept slightly below the outer capillary. The prepared microspheres have been monodisperse, and about 75% of the protein has been released over a period of 7 days. As an advantage, in this approach, protein has not been exposed to harmful organic solvents (Pareta and Edirisinghe, 2006).

With the aim to study the field of enhancement of image contrast in MRI, nanocapsules of polysorbate 80 coated with water-stable magnetic NPs (MNPs) have been fabricated by coaxial ES in the cone-jet mode. Monodispersed MNPs (Fe_3O_4, 8.2 nm) were synthesized by thermal decomposition and seed-mediated growth. The produced PEGylated MNPs have been stable in water. The droplets had a geometric mean diameter of 96.1 nm, and a geometric standard deviation of 2.2. Moreover, the average size of the water-stable MNPs has been 10.7 ± 0.7 nm. The MRI potential of these MNPs is established by in vivo imaging using a gastric cancer xenograft mouse model. This CEHDA system has demonstrated notable potential for the continuous encapsulation of MNPs to render them water-stable while maintaining their properties as contrast enhancement in MRI (Fig. 6.13) (Kim et al., 2013).

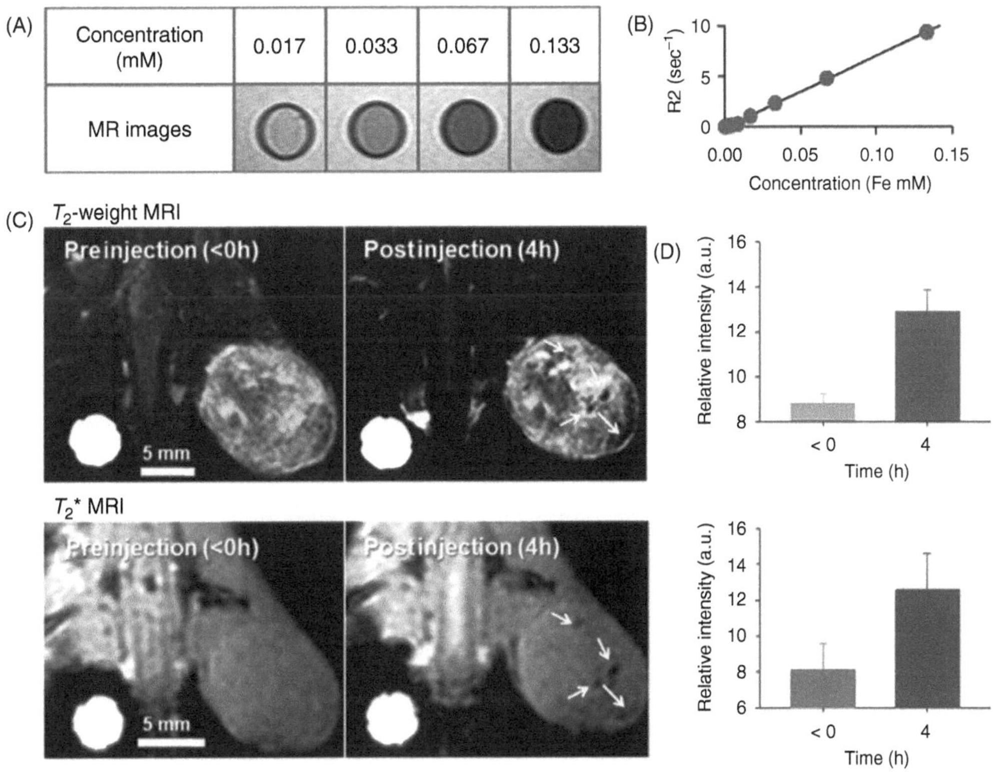

FIGURE 6.13

(A) T_2-weighted MR images of water-stable MNPs in an aqueous phase at various concentrations of Fe; (B) T_2 relaxivity (R_2) graph; (C) in vivo MR images (upper: T_2-weighted MR images; lower: T_2^* -weighted MR images) of the xenograft mouse model implanted with MKN_{45} cells at the proximal thigh region of the model before (left), and after (right) injection of water-stable MNPs into the tail vein; (D) graph of the relative intensity (Kim et al., 2013).

Composite structures have been fabricated by CEHDA. In this approach, a doxorubicin-loaded PLGA core has been surrounded by a drug-free poly(D,L-lactic acid) (PDLLA) shell layer. Capsules with high encapsulation efficiency have been produced, and after a short duration of lag phase, the capsules showed two stages of constant drug release rates (Xu et al., 2013).

Coaxial ES has been used to fabricate polyethylenimines pDNA/polyethylenimines (PEI) polyplexes for pulmonary gene delivery, using green fluorescent protein (pGFP) and pSEAP for qualitative and quantitative studies, respectively. Both pGFP/PEI and pSEAP/PEI polyplexes have been successfully fabricated, with sizes of less than 1000 nm, and the efficiency of delivery of pDNA/PEI polyplexes has been assessed in NIH 3T3 cells (Wu et al., 2012).

4.2.2 Bubble particles

Microbubbles have exhibited the capability for use in biomedical contexts. They can be used in a range of application, from use as contrast agents for ultrasound imaging to drug delivery. It was proved that, compared to the conventional methods, EHDA has notable advantages (Enayati et al., 2011). Microbubbles of water-insoluble polymethylsilsesquioxane (PMSQ) have been fabricated by Farook et al. (2008) using the CEHDA method. A solution of PMSQ was injected from the outer needle, and air was delivered to the inner needle. The liquid and air flow rates have been varied at μL/s for both materials, and at 5.7 kV stable microbubbles were fabricated. Distilled water has been used as a collector bath and, after sinking, they have been transformed quickly into polymer-shelled microspheres containing 60 wt.% liquid. After 300 s of production time, they had a spherical morphology and a mean size of 6 μm. After 48 h, the particles were collected at the bottom of the vial. Results have showed that with increasing weight percentage of polymer in ethanol, the electrical conductivity of solutions decreases, while viscosity increases.

4.2.3 Porous particles

The same as with single needle EHDA, producing porous structures by CEHDA and applying them to drug delivery is possible. In a study carried out by Enayati et al. (2009), estradiol-loaded PLGA MPs and NPs with porous internal structures have been fabricated using a CEHDA. The mean size of the particles ranged from less than 120 nm to a few micrometers, and had an encapsulation efficiency of about 70%. Moreover, the release of the drugs has been sustained for a period of over 20 days. By exposure of porous NPs to 22.5 kHz ultrasound for 30 s, the release rate and porosity has been increased.

4.2.4 Hollow particles

Hollow spheres are another category of particles that are produced by CEHDA. Generally, in order to fabricate hollow particles, a raw material particle is dissolved in a very volatile solvent, or a low soluble substrate is used. While solvent evaporation starts, the concentration of the solution increases, the solvent emerges from the inner part of the droplet to come out, and the polymer diffuses into the core of particle. Accordingly, the outer crust or shell is formed (Ciach, 2007).

Perfluorated hydrocarbons (PFH)-filled capsules have been fabricated via CEHDA, with a mean diameter of ~200 nm, and a polydispersity index of ~30%. It has been found that capsules with a mean diameter of 37 nm and a polydispersity index of 23% can be achieved by changing the processing parameters. It has been indicated that this is an appropriate technique to produce both micro- and nanoscale capsules. It has been also indicated that it is possible to control the size and shell thickness of the hollow capsules by changing the concentration of the polymer, voltage, and flow rate

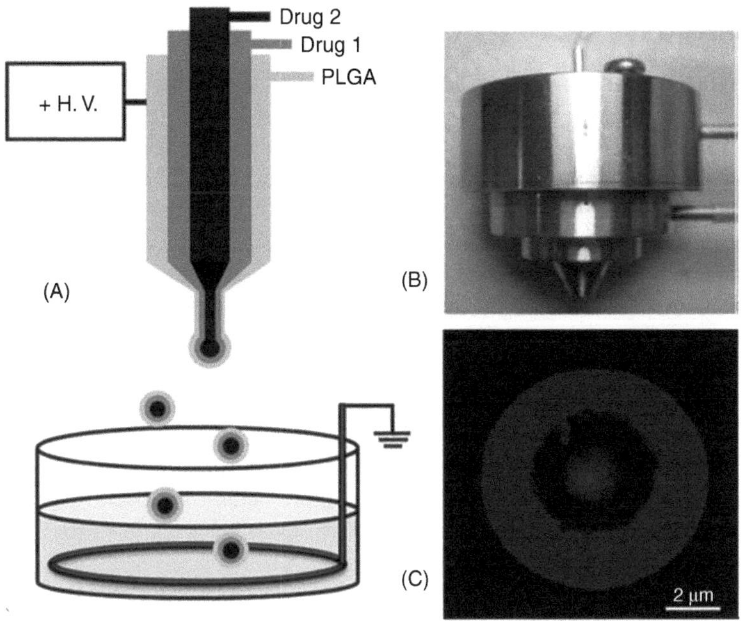

FIGURE 6.14

(A) A schematic diagram of the tricapillary ES system for the synthesis of trilayered particles; (B) digital photographs of the tricapillary spray nozzle; (C) LSCM image of PSS/PLGA/PSS microcapsule (inner/middle/outer flow rate: 10/10/20 μL/min) (Lee et al., 2011).

(Chang et al., 2009). Hollow spheres have application for inhalation drug delivery. They are efficiently deposited inside the lungs (Ciach, 2007).

4.3 MULTICAPILLARY ELECTROSPRAY

Different particle combinations have been produced by the use of three capillary nozzle EHDA. It has been indicated that controlling flow rate and voltage makes it possible to produce double-layered bubbles, porous encapsulated threads, and nanocapsules containing three layers (Ahmad et al., 2008). Lee et al. (2011) have investigated the possibility of producing PLGA particles containing various drugs (EGCG and Budesonide) in one step by applying a coaxial tricapillary ES setup. By applying this system, the complicated steps involved in emulsion processes will be removed. The loaded drugs were shown to have different release profiles (Fig. 6.14) (Lee et al., 2011).

The comparisons have shown that three-layered particles can release multiple drugs in different kinetic phases. Moreover, the release profile of the produced particles could be controlled by changing the thickness and chemical composition of their individual layers (Lee et al., 2011).

In another research, multishell capsules have been fabricated by the use of two conducting fluids, ethylene glycol (EG) and 4-hydroxybutyl acrylate (4-HBA), and one nonconducting fluid, olive oil. The capsules produced had monodisperse size distributions, and the shell thickness and dimension of particles has been controlled by the selection of appropriate values of flow rate and voltage (Kim and Kim, 2010).

5 CONCLUSIONS

There have been many considerable innovations in the production of polymeric MPs and NPs over the past decade. As well as different polymers, particles have been also been produced by a variety of applications. Using polymeric particles as a carrier for drug delivery is a new approach for achieving high efficiency and targeting in drug delivery. Using polymeric matrixes makes it possible to load a variety of diagnostic and therapeutic agents into the particles.

Among different methods of preparation of polymeric particles, ES seems to be an attractive technique for researchers; as such, a wide range of particles have been designed and fabricated using of this method. By using ES, drug-loaded polymeric particles with both solid matrix and core–shell structures can be fabricated from different materials in one step, and under normal ambient conditions. Solid particles with homogeneous distribution of loaded drugs can be fabricated by single needle ES. For the production of particles with core–shell configuration or hollow structure, two-needle ES can be applied. By arranging more coaxial nozzles, it is possible to produce particles with multilayered and multicomponent encapsulated structures. In addition, ES has the potential to produce particles in a wide range of sizes, from tens of nanometer to hundreds of micrometer. Moreover, by changing the process parameters it is possible to produce different structures, from particles to fibrous structures. One of the most important challenges in ES is the wide range of variables affecting the process. Thus, more research is necessary in order to overcome the limitations.

REFERENCES

Ahmad, Z., Zhang, H.B., Farook, U., Edirisinghe, M., Stride, E., Colombo, P., 2008. Generation of multilayered structures for biomedical applications using a novel tri-needle coaxial device and electrohydrodynamic flow. J. R. Soc. Interface 5, 1255–1261.

Allemann, E., Gurny, R., Doelker, E., 1993. Drug-loaded nanoparticles: preparation methods and drug targeting issues. Eur. J. Pharm. Biopharm. 39, 173–191.

Anderson, J.M., Shive, M.S., 2012. Biodegradation and biocompatibility of PLA and PLGA microspheres. Adv. Drug Deliv. Rev. 64, 72–82.

Biskos, G., Vons, V., Yurteri, C.U., Schmidt-Ott, A., 2008. Generation and sizing of particles for aerosol-based nanotechnology. KONA Powder Part. J. 26, 13–35.

Bock, N., Woodruff, M.A., Hutmacher, D.W., Dargaville, T.R., 2011. Electrospraying, a reproducible method for production of polymeric microspheres for biomedical applications. Polymers 3, 131–149.

Bock, N., Dargaville, T.R., Woodruff, M.A., 2012. Electrospraying of polymers with therapeutic molecules: state of the art. Prog. Polym. Sci. 37, 1510–1551.

Bohr, A., Kristensen, J., Stride, E., Dyas, M., Edirisinghe, M., 2011. Preparation of microspheres containing low solubility drug compound by electrohydrodynamic spraying. Int. J. Pharm. 412, 59–67.

Bourges, J.L., Gautier, S.E., Delie, F., Bejjani, R.A., Jeanny, J.C., Gurny, R., BenEzra, D., Behar-Cohen, F.F., 2003. Ocular drug delivery targeting the retina and retinal pigment epithelium using polylactide nanoparticles. Invest. Ophthalmol. Vis. Sci. 44 (8), 3562–3569.

Byrne, J., Betancourt, T., Brannon-Peppas, L., 2008. Active targeting schemes for nanoparticle systems in cancer therapeutics. Adv. Drug Deliv. Rev. 60 (15), 1615–1626.

Camelot, D., Marijnissen, J.C.M., Scarlett, B., 1999. Bipolar coagulation process for the production of powders. Ind. Eng. Chem. Res. 38, 631–638.

Chakraborty, S., Liao, I.-C., Adler, A., Leong, K.W., 2009. Electrohydrodynamics: a facile technique to fabricate drug delivery systems. Adv.Drug Deliv. Rev. 61, 1043–1054.

Chang, M.-W., Stride, E., Edirisinghe, M., 2009. A novel process for drug encapsulation using a liquid to vapour phase change material. Soft Matt. 5, 5029–5036.

Chen, D.-R., Pui, D.Y.H., 1997. Experimental Investigation of scaling laws for electrospraying: dielectric constant effect. Aerosol. Sci. Technol. 27, 367–380.

Ciach, T., 2006. Microencapsulation of drugs by electro-hydro-dynamic atomization. Int. J. Pharm. 324, 51–55.

Ciach, T., 2007. Application of electro-hydro-dynamic atomization in drug delivery. J. Drug Deliv. Sci. Tech. 17 (6), 367–375.

Cloupeau, M., Prunet-Foch, B., 1994. Electrohydrodynamic spraying functioning modes: a critical review. J. Aerosol. Sci. 25, 1021–1036.

Djordjevic, L., Primorac, M., Stupar, M., Krajisnik, D., 2004. Characterization of capry-locaproyl macrogolglycerides based microemulsion drug delivery vehicles for an amphiphilic drug. Int. J. Pharm. 271, 11–19.

Enayati, M., Ahmad, Z., Stride, E., Edirisinghe, M., 2009. One-step electrohydrodynamic production of drug-loaded micro- and nanoparticles. J. R. Soc. Interface 1, 1–9.

Enayati, M., Ahmad, Z., Stride, E., Edirisinghe, M., 2010. Size mapping of electric field-assisted production of polycaprolactone particles. J. R. Soc. Interface 7, S393–S402.

Enayati, M., Chang, M.-W., Bragman, F., Edirisinghe, M., Stride, E., 2011. Electrohydrodynamic preparation of particles, capsules and bubbles for biomedical engineering applications. Colloids Surf. A, Physicochem. Eng. Aspects 382, 154–164.

Faramarzi, A.R., Barzin, J., Mobedi, H. 2014a. Preparation of biocompatible PLGA fine particles containing magnetite in MRI. The 1th Iranian Nanomedicine Congress. 2014, Tehran, Iran.

Faramarzi, A.R., Barzin, J., Mobedi H. 2014b. The use of biocompatible polymer nanoparticles containing magnetic nanoparticles as contrast agents in MRI. The 1th Iranian Nanomedicine Congress, 2014, Tehran, Iran.

Faramarzi, A.R., Barzin, J., Mobedi H. 2015a. Preparation and characterization of in situ forming PLGA polymeic fine particles for drug delivery by electrospray. Iran Membrane, 2015, Tehran, Iran.

Faramarzi, A.R., Barzin J., Mobedi H. 2015b. Processing parameters of electrosprayed Fe3O4 loaded PLGA nanoparticles for utilizing in M.R.I. 14th Iranian Pharmaceutical Science Congress and 1th Symposium of Biopharmaceutics and Pharmacokinetics (IPSC 2015). Tehran, Iran.

Farook, U., Edirisinghe, M.J., Stride, E., Colombo, P., 2008. Novel co-axial electrohydrodynamic in-situ preparation of liquid-filled polymer-shell microspheres for biomedical applications. J. Microencapsul. 25 (4), 241–247.

Farook, U., Stride, E., Edirisinghe, M., 2009. Stability of microbubbles prepared by co-axial electrohydrodynamic atomisation. Eur. Biophys. J. 38, 713–718.

Freiberg, S., Zhu, X., 2004. Polymer microspheres for controlled drug release. Int. J. Pharm. 282, 1–18.

Freitas, S., Merkle, H.P., Gander, B., 2005. Microencapsulation by solvent extraction/evaporation: reviewing the state of the art of microsphere preparation process technology. J. Control. Release 102, 313–332.

Fresta, M., Cavallaro, G., Giammona, G., Wehrli, E., Puglisi, G., 1996. Preparation and characterization of polyethyl-2-cyanoacrylate nanocapsules containing antiepileptic drugs. Biomaterials 17, 751–758.

Gaete, C.G., Tsapis, N., Silva, L., Bourgaux, C., Fattal, E., 2008. Morphology, structure and supramolecular organization of hybrid 1,2-dipalmitoyl-*sn*-glycero-3-phosphatidylcholine-hyaluronic acid microparticles prepared by spray drying. Eur. J. Pharm. Sci. 34, 12–21.

Gañán-Calvo, A.M., 1997. Cone-Jet analytical extension of Taylor's electrostatic solution. The asymptotic universal scaling laws in electrospraying of liquids. APS Division of Fluid Dynamics Meeting Abstracts. American Physics Society, College Park, MD.

Geerse, K.B., Marijnissen, J.C.M., 2003. Electrospray as means to produce monodisperse drug particles. Optimization of Aerosol Drug Delivery. Kluwer Academic Publishers, Delft, The Netherlands, Particle Technology Group of Delft University of Technology.

Guo, X.D., Tan, J.P.K., Kim, S.H., Zhang, L.J., Zhang, Y., Hedrick, J.L., Yang, Y.Y., Qian, Y., 2009. Computational studies on self-assembled paclitaxel structures: templates for hierarchical block copolymer assemblies and sustained drug release. Biomaterials 30, 6556–6563.

Hall, J., Dobrovolskaia, M., Patri, A., Mcneil, S., 2007. Characterization of nanoparticles for therapeutics. Nanomedicine 2 (6), 789–803.

Hartman, R.P.A., Borra, J.P., Marijnissen, J.C.M., Scarlett, B. 1997. Electrohydrodynamic atomization in the cone-jet mode. ESF Workshop on Electrospray, Sevilla.

Hartman, R.P.A., Brunner, D.J., Marijnissen, J.C.M., Scarlett, B., 1998. Scaling laws for droplet size and current produced in the cone-jet mode. J. Aerosol. Sci. 29, S977–S978.

Hong, Y., Lib, Y., Yin, Y., Lia, D., Zou, G., 2008. Electrohydrodynamic atomization of quasi-monodisperse drug-loaded spherical/wrinkled microparticles. Aerosol Sci. 39, 525–536.

Ijsebaert, J.C., Geerse, K.B., Marijnissen, J.C.M., Scarlett, B., 1999. Electrohydrodynamic spraying of inhalation medicine. J. Aerosol Sci. 30, 825–826.

Ijsebaert, J.C., Geerse, K.B., Marijnissen, J.C.M., Lammers, J.-W.J., Zanen, P., 2001. Electro-hydrodynamic atomization of drug solutions for inhalation purposes. J. Appl. Physiol. 91, 2735–2741.

Jafari-Nodoushan, M., Barzin, J., Mobedi, H., 2014a. Design of a combined morphine delivery system. The 1st Middle East and the 6th Iranian Controlled Release Conference, Tehran, Iran

Jafari-Nodoushan, M., Barzin, J., Mobedi, H., 2014b, microsphere embedded matrix: a new method for drug co-therapy, In: 11th International Seminar on Polymer Science and Technology, Tehran, Iran.

Jafari-Nodoushan, M., Barzin, J., Mobedi, H., 2015a. Size and morphology controlling of PLGA microparticles produced by electro hydrodynamic atomization. Polym. Adv. Technol. 26, 502–513.

Jafari-Nodoushan, M., Barzin, J., Mobedi, H., 2016. A stability-indicating HPLC method for simultaneous determination of morphine and naltrexone. J. Chromatogr. B. 1011, 163–170.

Jafari-Nodoushan, M., Mobedi, H., Barzin, J., 2015b. Encapsulation via electrohydrodynamic atomization spray technology (Electrospray). Handbook of encapsulation and controlled release. CRC Press, Boca Raton, FL.

Jalón, E.D., Blanco-Príeto, M., Ygartua, P., Santoyo, S., 2001. PLGA microparticles: possible vehicles for topical drug delivery. Int. J. Pharm. 226 (1-2), 181–184.

Jaworek, A., Sobczyk, A., 2008. Electrospraying route to nanotechnology: an overview. J. Electrostat. 66, 197–219.

Kim, W., Kim, S.S., 2010. Multishell encapsulation using a triple coaxial electrospray system. Anal. Chem. 82, 4644–4647.

Kim, S.-Y., Yang, J., Kim, B., Park, J., Suh, J.-S., Huh, Y.-M., Haam, S., Hwang, J., 2013. Continuous coaxial electrohydrodynamic atomization system for water-stable wrapping of magnetic nanoparticles. Small 9 (13), 2325–2330.

Kissel, T., Li, Y.X., Volland, C., Görich, S., Koneberg, R., 1995. Parenteral protein delivery systems using biodegradable polyesters of ABA block structure, containing hydrophobic poly(lactide-*co*-glycolide) A blocks and hydrophilic poly(ethylene oxide) B blocks. J. Control. Release 39, 315–326.

Kohane, D.S., Tse, J.Y., Yeo, Y., Padera, R., Shubina, M., Langer, R., 2006. Biodegradable polymeric microspheres and nanospheres for drug delivery in the peritoneum. J. Biomed. Mater. Res. A77A, 351–361.

Lassalle, V., Ferreira, M.L., 2007. PLA nano and microparticles for drug delivery: an overview of the methods of preparation. Macromol. Biosci. 7, 767–783.

Lee, Y.-H., Mei, F., Bai, M.-Y., Zhao, S., Chen, D.-R., 2010. Release profile characteristics of biodegradable-polymer-coated drug particles fabricated by dual-capillary electrospray. J. Control. Release 145, 58–65.

Lee, Y.-H., Baib, M.-Y., Chena, D.-R., 2011. Multidrug encapsulation by coaxial tri-capillary electrospray. Colloids Surf. B, Biointerfaces 82, 104–110.

Leea, S.-J., Jeonga, J.-R., Shina, S.-C., Kimb, J.-C., Chang, Y.-H., Chang, Y.-M., Kima, J.-D., 2004. Nanoparticles of magnetic ferric oxides encapsulated with poly(D,L latide-*co*-glycolide) and their applications to magnetic resonance imaging contrast agent. J. Magnet. Magnetic Mater. 272–276, 2432–2433.

Legrand, P., Barratt, G.V.M., Fessi, H., Devissaguet, J., 1999. Polymeric nanocapsules as drug delivery systems. A review. STP Pharma Sci. 9, 411–418.

Lewinska, D., Rosinski, S., Werynski, A., 2004. Influence of process conditions during impulsed electrostatic formation on size distribution of hydrogel beads. Artif. Cells Blood Substit. Immobil. Biotechnol. 32, 41–53.

Limmatvapirat, S., Limmatvapirat, C., Puttipipatkhachorn, S., Nunthanid, J., Luangtana-Anan, M., Sriamornsak, P., 2008. Modulation of drug release kinetics of shellac-based matrix tablets by in situ polymerization through annealing process. Eur. J. Pharm. Biopharm. 69, 1004–1013.

Luan, X., Skupin, M., Siepmann, J., Bodmeier, R., 2006. Key parameters affecting the initial release (burst) and encapsulation efficiency of peptide-containing poly(lactide-*co*-glycolide) microparticles. Int. J. Pharm. 324, 168–175.

Mora, J.D.L., Loscertales, I., 1994. The current emitted by highly conducting Taylor cones. J. Fluid Mech. 260, 155–184.

Mundargi, R., Babu, V., Rangaswamy, V., Patel, P., Aminabhavi, T., 2008. Nano/micro technologies for delivering macromolecular therapeutics using poly(D,L-lactide-*co*-glycolide) and its derivatives. J. Control. Release 125, 193–209.

Nasongkla, N., Bey, E., Ren, J., Ai, H., Khemtong, C., Guthi, J.S., Chin, S.-F., Sherry, A.D., Boothman, D.A., Gao, J., 2006. Multifunctional polymeric micelles as cancer-targeted, MRI-ultrasensitive drug delivery systems. Nano Lett., 6.

Nkansah, M.K., Thakral, D., Shapiro, E.M., 2011. Magnetic poly(lactide-*co*-glycolide) and cellulose particles for MRI-based cell tTracking. Magn. Reson. Med. 65, 1776–1785.

Owens, III, D.F., Peppas, N.A., 2006. Opsonization, biodistribution, and pharmacokinetics of polymeric nanoparticles. Int. J. Pharm. 307, 93–102.

Pareta, R., Edirisinghe, M.J., 2006. A novel method for the preparation of biodegradable microspheres for protein drug delivery. J. R. Soc. Interface 3, 573–582.

Park, J.S., Lee, J.H., Shin, H.S., Lee, T.W., Kim, M.S., Khang, G., Rhee, J.M., Lee, H.K., Lee, H.B., 2007. Biodegradable polymer microspheres for controlled drug release. Tissue Eng. Regen. Med. 4, 347–359.

Peltonen, L., Valo, H., Kolakovic, R., Laaksonen, T., Hirvonen, J., 2010. Electrospraying, spray drying and related techniques for production and formulation of drug nanoparticles. Expert Opin. Drug Deliv. 7 (6).

Ranganath, S.H., Kee, I., Krantz, W.B., Chow, P.K.-H., Wang, C.-H., 2009. Hydrogel matrix entrapping plga-paclitaxel microspheres: drug delivery with near zero-order release and implantability advantages for malignant brain tumour chemotherapy. Pharm. Res., 26.

Ranganath, S.H., Fu, Y., Arifin, D.Y., Kee, I., Zheng, L., Lee, H.-S., Chowd, P.K.-H., Wang, C.-H., 2010. The use of submicron/nanoscale PLGA implants to deliver paclitaxel with enhanced pharmacokinetics and therapeutic efficacy in intracranial glioblastoma in mice. Biomaterials 31, 5199–5207.

Ren, X., Chen, H., Yang, V., Sun, D., 2014. Iron oxide nanoparticle-based theranostics for cancer imaging and therapy. Front. Chem. Sci. Eng. 8 (3), 253–264.

Rosell-Llompart, J., Mora, J.F.D.L., 1994. Generation of monodisperse droplets 0.3 to 4 m in diameter from electrified cone-jets of highly conducting and viscous liquids. J. Aerosol. Sci 25, 1093–1119.

Sezer, A.D., 2014. Application of nanotechnology in drug delivery. InTech, Rijeka, Croatia, CC by 3.0 license.

Soppimath, K.S., Aminabhavi, T.M., Kulkarni, A.R., Rudzinski, W.E., 2001. Biodegradable polymeric nanoparticles as drug delivery device. J. Control. Release 70, 1–20.

Sun, C., Veiseh, O., Gunn, J., Fang, C., Hansen, S., Lee, D., Sze, R., Ellenbogen, R.G., Olson, J., Zhang, M., 2008. In vivo MRI detection of gliomas by chlorotoxin-conjugated superparamagnetic nanoprobes. Small 4, 372–379.

Valo, H., Peltonen, L., Vehviläinen, S., Karjalainen, M., Kostiainen, R., Laaksonen, T., Hirvonen, J., 2009. Electrospray encapsulation of hydrophilic and hydrophobic drugs in poly(l-lactic acid) nanoparticles. Small 5, 1791–1798.

Wu, Y., Clark, R.L., 2007. Controllable porous polymer particles generated by electrospraying. J. Colloid Interface Sci. 310, 529–535.

Wu, Y., Duong, A., Lee, L.J., Wyslouzil, B.E., 2012. Electrospray production of nanoparticles for drug/nucleic acid delivery. In: Hashim, A.A. (Ed.), The Delivery of Nanoparticles. InTech, Rijeka, Croatia.

Wu, Y., Loperb, A., Landisb, E., 2004. The role of biopharmaceutics in the development of a clinical nanoparticle formulation of MK-0869: a beagle dog model predicts improved bioavailability and diminished food effect on absorption in human. Int. J. Pharm. 285, 135–146.

Wu, Y., Mackay, J.A., Mcdaniel, J.R., Chilkoti, A., Clark, R.L., 2009. Fabrication of elastin-like polypeptide nanoparticles for drug delivery by electrospraying. Biomacromolecules, 12.

Xie, J., Wang, C.-H., 2007a. Encapsulation of proteins in biodegradable polymeric microparticles using electrospray in the Taylor cone-jet mode. Biotechnol. Bioeng. 97, 178–1290.

Xie, J., Wang, C., 2007b. Electrospray in the dripping mode for cell microencapsulation. J. Colloid Interface Sci. 312, 247–255.

Xie, J., Lim, L.K., Phua, Y., Hua, J., Wang, C.H., 2006a. Electrohydrodynamic atomization for biodegradable polymeric particle production. J. Colloid Interface Sci. 302, 103–112.

Xie, J., Marijnissen, J.C.M., Wanga, C.-H., 2006b. Microparticles developed by electrohydrodynamic atomization for the local delivery of anticancer drug to treat C6 glioma in vitro. Biomaterials 27, 3321–3332.

Xie, J., Ng, W.J., Lee, L.Y., Wang, C.-H., 2008a. Encapsulation of protein drugs in biodegradable microparticles by co-axial electrospray. J. Colloid Interface Sci. 317, 469–476.

Xie, J., Tan, J.C., Wang, C.-H., 2008b. Biodegradable films developed by electrospray deposition for sustained drug delivery. J. Pharm. Sci. 97, 3109–3122.

Xu, Y., Hanna, M.A., 2006. Electrospray encapsulation of water-soluble protein with polylactide. Effects of formulations on morphology, encapsulation efficiency and release profile of particles. Int. J. Pharm. 320, 30–36.

Xu, Y., Hanna, M.A., 2007. Electrosprayed bovineserumalbumin-loaded tripolyphos-phate cross-linked chitosan capsules: synthesis and characterization. J. Microencapsul. 24, 143–151.

Xu, Y., Skotak, M., Hanna, M., 2006. Electrospray encapsulation of water-soluble protein with polylactide. I. Effects of formulations and process on morphology and particle size. J. Microencapsul. 23 (1), 69–78.

Xu, Q., Qin, H., Yin, Z., Hua, J., Pack, D-W., Wang, C-H., 2013. Coaxial electrohydrodynamic atomization process for production of polymeric composite microspheres. Chem. Eng. Sci. 104, 330–346.

Yang, Y.Y., Chung, T.S., Ng, N.P., 2001. Morphology, drug distribution, and in vitro release profiles of biodegradable polymeric microspheres containing protein fabricated by double-emulsion solvent extraction/evaporation method. Biomaterials 22, 231–241.

Ye, M., Kim, S., Park, K., 2010. Issues in long-term protein delivery using biodegradable microparticles. J. Control. Release 146, 241–260.

Yeo, L.Y., Gagnon, Z., Chang, H.-C., 2005. AC electrospray biomaterials synthesis. Biomaterials 26, 6122–6128.

Yurteri, C.U., Hartman, R.P.A., Marijnissen, J.C.M., 2010. Producing pharmaceutical particles via electrospraying with an emphasis on nano and nano structured particles—a review. KONA Powder Part. J. 28, 91–115.

Zhang, L., Huang, J., Si, T., Xu, R.X., 2012. Coaxial electrospray of microparticles and nanoparticles for biomedical applications. Expert Rev. Med. Devices 9 (6), 595–612.

CHAPTER

7

NANOFORMULATION AND ADMINISTRATION OF PUFA-RICH SYSTEMS FOR APPLICATIONS IN MODERN HEALTHCARE

Debjyoti Paul, Tanmoy K. Dey, Pubali Dhar
University of Calcutta, Kolkata, India

CHAPTER OUTLINE

1 INTRODUCTION

In the present context and recent paradigm in the designing of novel functional food and food formulations, a lot of emphasis is being laid on developing food structure and matrices for enhanced and targeted delivery of the bioactives responsible for the intended functionalities in the human system. Consequently, a host of solutions toward this objective have also come to the fore and are taken into consideration. The solution ranges from modifying the administrative routes to incorporating novel excipients, to varying the particle sizes of the core compound(s) for the better delivery and efficiency of the functional bioactive compounds in question. Efforts are also being made to develop formulations which can bypass certain specific physiological barriers and intermediate metabolic conversions which interfere with the optimum activity and therapeutic opportunity of the key bioactives integrated into the formulations. Nanotechnology is one of the key components in this field of research to obtain desired objectives toward this end. Recent findings to explore the molecular mechanisms to effect prophylactic and therapeutic potential of already established micronutrients, macronutrients, and natural derivatives in designed stable and nontoxic formulations have highlighted the fact that nanoscaling of such bioactives enhances bioavailability and bioactivity (Ganta and Amiji, 2009; Paul et al., 2014; Vyas et al., 2008). Consequently, interest in the food and pharma sectors for nano- and microencapsulation technology is rapidly increasing.

A group of functional bioactive compounds which are now routinely encapsulated as food structures and nutraceuticals involve: carotenoids, phytosterols, antioxidants (including vitamins), and fatty acids. As such, a major portion of such functional bioactives involves compounds that are lipophilic. Importantly, the targeted delivery of such lipophilic compounds in the form of regular and coarse food formulations is very low. The formulation fabrication is often more challenging in the light of the fact that only limited excipients with generally recommended as safe (GRAS) status are available, and it is only these excipients that can be incorporated for a sustainable formulation for the targeted delivery of lipophilic functional compounds under physiological conditions. Efforts are thus also underway to explore novel surfactants and stabilizers with practically no reported toxicity, and higher feasibility in terms of applicability and sustenance that can be incorporated in nanofood and nutraceutical formulations (Holmberg, 2003).

In the field of lipid technology, a lot of progress has been made to develop lipid nanoparticles in the form of nanolipid carriers (NLCs), solid–lipid nanoparticles (SLN), liquid-in-liquid dispersion systems (including oil-in-water, water-in-oil, and pickering nanoemulsions), nanoliposomes, and nanoethosomes (including compositionally similar vesicular structures, such as transfersomes and niosomes) (Attama et al., 2012). In light of recent research with such nanostructures, it has now become a truism that, due to the unique properties of lipid particles at nanoscale with higher surface to mass ratio (compared to other colloidal particles), such lipid nanoparticles are excellent carriers to deliver hydrophobic functional bioactives of interest to the target biological systems.

One of the most essential micronutrients which have been attributed to have nutraceutical potential refers to polyunsaturated fatty acids (PUFA), a key component in food fortification. Often, as a global trend, dietary intake of PUFA has been found to be lower than the recommended dietary allowances (RDA) in everyday meals. Consequently, as a long-term effect, the absence of such essential micronutrients in optimized levels in our daily diet led to increased susceptibility to inflammation (as well as proinflammatory mediators), subdued cognitive ability, osteodystrophy-like conditions, and myriad other poor-health conditions. To overcome such deficiencies, nowadays food supplements and food products fortified with formulations comprising micronutrients including PUFAs are attracting a lot interest among consumers, in light of increased awareness on healthy diet and lifestyle options. It is now being strongly felt that efficacious delivery and activity of such bioactives in our human system would require further interventions in the form of food and pharma technology heavily inspired by the progresses made in this relevant field of lipid nanotechnology. Efforts are thus now being directed toward this end. This chapter aims in its subsequent section to explore these recent findings in the field of lipid nanotechnology for the nanoencapsulation of PUFA and PUFA-based systems for the improvement of health conditions. Emphasis has also been laid to determine the empirical effects created by such formulations when administered either parentally or orally.

This chapter intends to explore the available literature on lipid nanoparticles rich in PUFA, and the vast expanse of potential of novel nanoformulations of PUFA that is till untapped. This chapter has also been restricted in content to identify the key areas that need to be addressed for the delivery of optimum functionality of such formulations at only the clinical level. The new methodologies that are coming up or are being propagated should also be developed by juxtaposing their relevance with the intricacies involved in the human metabolism of lipid nanoparticles. It also needs to be highlighted that, in the present context, not much data are as yet available to show the differences in metabolism of PUFA-rich nanolipid formulations compared to conventional formulations of lipids.

2 LIPID NANOPARTICLES: TYPES AND FABRICATION TECHNOLOGY

The dynamic physicochemical interactions, energetics, and kinetics between biological systems and nanomaterial surfaces, in a more potent manner than conventional systems, have led to the emergence of a novel and more efficacious targeted therapy. At the nanolevel, while the surface-to-volume potency of therapeutic structures in influencing the biological system generally increases, on the other hand the molecular traits of such structures also vary on the basis of particle size/shape, surface charge, rate of receptor diffusion, and elasticity of biological membranes. Lipid-based nanostructures also comply with this general rule, and thus, on the basis of intended target/functionality, suitable formulations and fabrication techniques of such lipid-based systems must be employed. Hence, while nanomaterials traditionally implied inorganic, metal, and ceramic structures, with the emergence of lipid nanostructures this field is being redefined for therapeutic applications, owing to its lower cytotoxicity consequences. Various diseases, including hypercholesterolemia, cardiovascular diseases, diabetes, inflammatory disorders, neurodegenerative diseases, cancer, and many other painstakingly curable diseases can now be treated with ease by employing such nature-based nanostructures.

Conceptually, lipid-based nanoparticle vehicles are developed to deliver poorly water-soluble drugs and bioactives in order to overcome the challenges of low bioavailability, chemical oxidation, and crystallization. It has been inferred that consumption of nutraceutical-added functional foods reduces the incidence of certain diseases, such as cardiovascular complications, cancer, diabetes, arthritis, immunomodulatory diseases, high blood pressure, and disorders of the central nervous system (CNS) (Babbar et al., 2015; Thompkinson et al., 2014). These nutraceuticals and lipophilic life-saving drugs get solubilized into the lipophilic core of the nanostructures, and are delivered in vivo. The drug lipid carriers are used for the targeted delivery of drugs to specific disease-affected tissues or cells, crossing different biological membranes, such as the gastrointestinal (GI) epithelium, the blood–brain barrier (BBB), and dermal epithelium. The lipid nanoparticles designed for drug delivery are designated as nanoliposomes, nanoethosomes, phytosomes, lipospheres, niosomes, cubosomes, virosomes, iscoms, nanoemulsions, cochleates, and transfersomes. The advantages of production and application of these nanoformulations are as follows (Fig. 7.1):

- The stability of the drug or nutraceutical is enhanced,
- The drug or bioactives are released in a controlled manner,
- Drug-loading capacity is raised,
- Both lipophilic and hydrophilic drugs and bioactives can be loaded,
- Lipid carriers are biodegradable, and are soon metabolized in the system,
- Lipid molecules are nonirritant, nonallergenic, nontoxic,

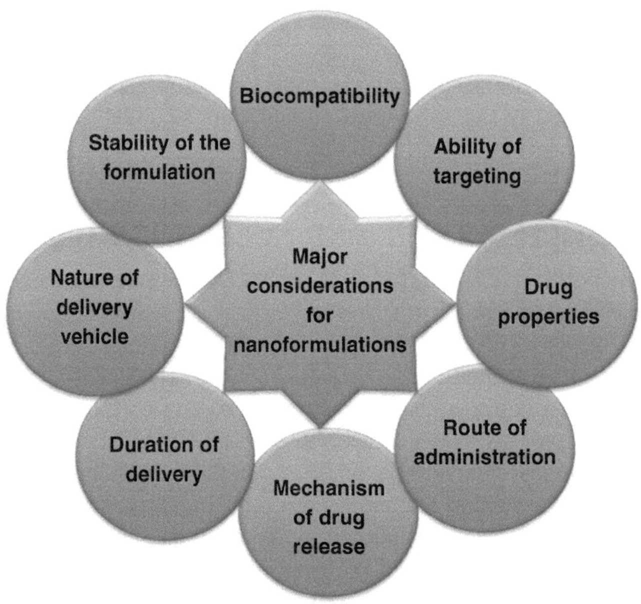

FIGURE 7.1 Major Considerations for Successful Preparation of Lipid Nanoformulation for Biological Delivery of Drugs, Pharmaceutical Products, and Nutrients

- Toxic organic solvents are avoided, and water-based technologies used,
- They can easily be scaled up,
- They are economically viable, and cheaper than polymeric surfactants.

2.1 NANOENCAPSULATION

Nanoencapsulation is defined as a technology to encapsulate bioactive substances within nanometric dimensions, and refers to bioactive packaging in order to form stable formulations (Lopez et al., 2006). Encapsulation technologies include the preparation of two explicitly distinctive structures, that is, nanocapsules and nanospheres (Konan et al., 2002). The term "nanocapsules" means nanosized vesicular systems with the bioactives encapsulated within a confined core surrounded by a polymeric lamellar structure. But "nanospheres" are known to be matrix-based systems within which the encapsulated products are dispersed (Couvreur et al., 1995). Nanoencapsulation techniques are employed to generate encapsulated nutraceuticals and drugs for protection against oxidation, chemical as well as microbial degradation, and the increase of bioavailability. Lipid-based nanoencapsulation of nutraceuticals improves the solubility, stability, and controlled release of the nutraceuticals, and reduces their interaction with other food components. Various applications of lipid-based encapsulation forms are nanoliposomes, nanocochleates, and archaeosomes.

2.2 LIPOSOMES

Initially termed "banghasomes," liposomes are enclosed phospholipid bilayer structures consisting of a single bilayer with hydrophilic or aqueous core (Deamer, 2010). Lipophilic nutraceuticals and drugs are entrapped in the lipid bilayer, while the hydrophilic drugs are filled in the aqueous matrix. The drugs are targeted to the damaged tissues, thus reducing systematic toxicity. A single bilayer liposome is called unilamellar liposome, whereas there are concentric bilayer shells in a multilamellar liposome (Reineccius, 2007; Taylor et al., 2005). The hydrophilic bioactives prone to oxidation may be delivered in liposomes encapsulated by single- or multilayers of surfactants. Researchers have reviewed the liposomal delivery of lipophilic bioactives (Mozafari et al., 2008). Multicompartment liposomes have been developed to deliver drugs for an extended period of time (Coldren et al., 2004).

Preparation of liposomes was originally proposed to be formed in a low pressure, low throughput manner (Olson et al., 1979). But later, with the advent of ultrasonication and high pressure homogenization systems, the liposome size was also successfully brought down to 50 nm (Hope et al., 1985; Huang, 1969; Mayer et al., 1986). Recently, microfluidization technologies have enabled the production of liposomes within a diameter range of 20–50 nm (Zhigaltsev et al., 2012). These nanoliposomes (<100 nm) offer enhanced bioavailability of the entrapped nutraceuticals, compared to their conventional counterparts (Moussaoui et al., 2002).

2.3 NANOCOCHLEATES

Nanocochleates are principally synthesized from negatively charged lipid and divalent cations to form stable lipid-based nanocarrier systems. These are basically spiral rod-like structures made up of a solid lipid bilamellar sheet spirally arranged around a central axis. Various negatively or positively charged hydrophobic molecules can be packed within the lamellar structures, whereas amphipathic molecules can be inserted within the interlamellar space. Hence, the versatility of these structures as a carrier for bioactive delivery has been increased significantly (Kumar and Lal, 2014). Moreover, the augmented

GI stability of these structures has made them a preferred choice for the oral delivery of various bioactives (Zarif, 2006). Cholesterol cochleates have been developed for the treatment of hemophilia-A, and in vivo tests in mice have shown promising results (Miclea et al., 2007). However, the utilization of these nanocochleates in food and pharmaceutical industries are somewhat restricted by the fact that the production of such structures involves the application of nonfood grade ingredients (Zarif, 2005), which contribute to the failure of the system in meeting biocompatibility criteria.

2.4 NANODISPERSIONS

Nanoencapsulation technologies are employed to produce nanodispersions in which lipid soluble vitamins, antimicrobials, antioxidants, food preservatives and food colors are dispersed in an aqueous medium. They are loaded in soy lecithin carriers with high loading capacity. Encapsulated nutraceuticals are easily dispersed in an aqueous medium.

2.5 MICELLES

The amphipathic surfactant molecules with large polar head groups and nonpolar tail, when dispersed in an aqueous medium, at a concentration above the critical micellar concentration (CMC) (Leser et al., 2006), form self-assemblies called micelles. The formation of the micelle is mainly dependent on the bipolarity of the surfactant monomers. The hydrophilic sections of the monomer structures get associated together in a polar media, and are directed to the external hydrophilic environment, whereas the hydrophobic blocks come closer in order to form a hydrophobic core. A micelle is termed a normal phase micelle when the hydrophilic blocks are directed to the exterior polar environment, whereas a micelle formulation with the hydrophobic matrix as a continuous phase is called a reversed phase micelle. Micelles are thermodynamically stable structures formed of surfactants, cosurfactants, or cosolvents, and serve as vectors to deliver hydrophobic bioactives. The micelle-forming surfactants have high hydrophilic-to-lipophilic (HLB) ratio. To deliver a lipophilic bioactive, micelles of the size of 5–20 nm are prepared. Phospholipids with long chain PUFAs are used to produce micelles with applications in the food industry.

2.6 NANOEMULSIONS

Emulsions are generally known to be mixtures of two immiscible liquids like oil (O) and water (W), of which one is dispersed into the other, in the presence of a surfactant(s). The droplet size of the dispersed phase is generally in the range of 10^{-6} to 10^{-8} m. Nanoemulsions can form various structurally unique systems, depending on their composition and preparation protocol. They may form one, two, three, or more separate phases that are in equilibrium with each other. The continuous phase may be either water or oil, and the dispersed phase vice versa. A bicontinuous phase can also be formed, depending on the concentrations, the nature of the constituents, and the arrangements of the molecules present in the system. Depending on stability and biodelivery, emulsion systems can be divided into three groups, that is, conventional, nano, and microemulsion.

"Microemulsion," by definition, refers to thermodynamically stable colloidal systems constituted by mixing oil, water, and surfactants together (Fanun, 2008; Jonsson et al., 1998). Various structures, such as spheroids (e.g., micelles or reverse micelles), cylinder-like (such as rod-micelles or reverse micelles), or planar (e.g., lamellar structures) are formed in the microemulsion systems (Jonsson et al., 1998). Nanoemulsion generally consists of very small droplets ($r < 100$ nm) (McClements, 2012c), dispersed in the continuous phase. Both conventional emulsions (with sizes > μm) and nanoemulsions are thermodynamically nonequilibrium systems. These systems are, however, endowed with kinetic stabilization, mainly due to their very small size and high Brownian motion that prevents droplet flocculation and coalescence. The Ostwald ripening alone governs the destabilizing process in these systems (Anton and Vandamme, 2011).

2.7 MULTILAYER NANOEMULSIONS

Oil-in-water (O/W) emulsions with charged shells are prepared. The charged droplets are coated with an oppositely charged biopolymer, forming multilayer emulsions having radii in the range of 100 nm to 100 μm. These biopolymer coated droplets are stable against aggregation. Multilayer oil-in-water (M-O/W) emulsions have been developed by adsorptive coating of oppositely charged biopolymers onto charged core droplets formed of ionic emulsifiers. This technique was termed as layer-by-layer (LBL) by Guzey and McClements (2006). The strong steric and electrostatic forces between the coated layers stabilize the formulation against various adverse chemical environments (Guzey et al., 2004; Moreau et al., 2003). The stability of the M-O/W emulsion has been dependent upon the correct ratio of the layer components. Reports suggest that encapsulation of nonpolar components (e.g., limonene) within M-O/W emulsions has improved their oxidative stability (Djordjevic et al., 2007). Another advantage of this complex interfacial system is its potential to deliver and release in a controlled manner bioactive components by the modification of the physicochemical parameters of the emulsion layers (e.g., composition, permeability, thickness, and charge), and the chemical environments (Gu et al., 2005).

2.8 FILLED HYDROGEL PARTICLES

Hydrogels are, by definition, three dimensional networks that have high water/biological fluid absorbing capacity, attributed by the hydrophilic groups present in constituent polymers (Peppas and Khare, 1993). These systems are usually formed from an emulsion, with a biopolymer solution. The fat droplets serving as carriers of lipophilic active ingredients are dispersed in the hydrogel matrix within which the hydrophilic active ingredients can be trapped. These structures offer advantages, such as nontoxic, biocompatible design, and versatile applicability in drug-delivery systems (Coviello et al., 2007; Lin and Metters, 2006). Chung et al. (2013) have reported the production of filled hydrogel particles using biopolymer phase separation methods for the development of reduced calorie food products. In these systems, the fat droplets partitioned into the dispersed aqueous phase are entrapped within the hydrogel matrix (O/W2/W1) (Chung et al., 2013).

2.9 SOLID-LIPID NANOPARTICLES

The thought behind using solid lipids instead of liquid oils is to achieve controlled drug release. Due to the solid state of the constituent lipid, the molecular drug mobility with the lipid matrix is significantly lower compared to liquid oil (Mehnert and Mader, 2001). Solid lipids have been considered for the preparation of SLN for the oral administration of drugs with improved payload, enhanced physical-chemical stability during storage, and retarded drug release. SLN are bioengineered as an alternative to liposomes for the administration of fourth generation drugs, even for transdermal application. In the process of designing of SLN, the liquid oil from a nanoemulsion is replaced by a solid lipid, without using organic solvents (Muller et al., 2002; Radtke and Muller, 2001). Triglycerides, glyceryl palmitostearate, glyceryl behenate, and cetyl palmitate from wax are used for SLN production of a size variation from 50 to 1000 nm. The lipid concentration of SLN ranges from 5% to 40%, and the surfactant ranges from 0.5% to 5%. It has high penetrability and it is hence used in the cosmetics industry. Another advantage of SLN is the ability of slow release of lipophilic bioactives.

2.10 NANOSTRUCTURED LIPID CARRIERS

Being of large scale industrial production, SLNs have certain limitations; that is, reduced solubility of drug in the solid lipid matrix leads to lower drug loading, lipid crystallization due to the stable β-form resulting in drug expulsion, low particle concentration in the aqueous dispersions. These limitations have been resolved only recently by introducing a new lipid carrier, the nanostructured lipid carrier (NLC) (Muller et al., 2002, 2012; Selvamuthukumar and Velmurugan, 2012). In NLC, solid–liquid blended lipids are used to form the matrix, enclosed by a surfactant. As the melting temperature of the matrix is reduced, due to the blending of liquid oils with a solid lipid, the porosity of the matrix is amplified. During the cooling of the lipid matrix, crystals are formed with multiple imperfections. Within the crystal assembly, the nutraceuticals are accommodated and form a "perfect solution within the imperfections." The NLC formulations can be of three different types, that is, (1) amorphous type, (2) imperfect crystal type, and (3) multiple O/F/W type. The amorphous and the imperfect crystal type, because of their elongated release profile and flexibility, are preferred most (Muller et al., 2002) (Fig. 7.2).

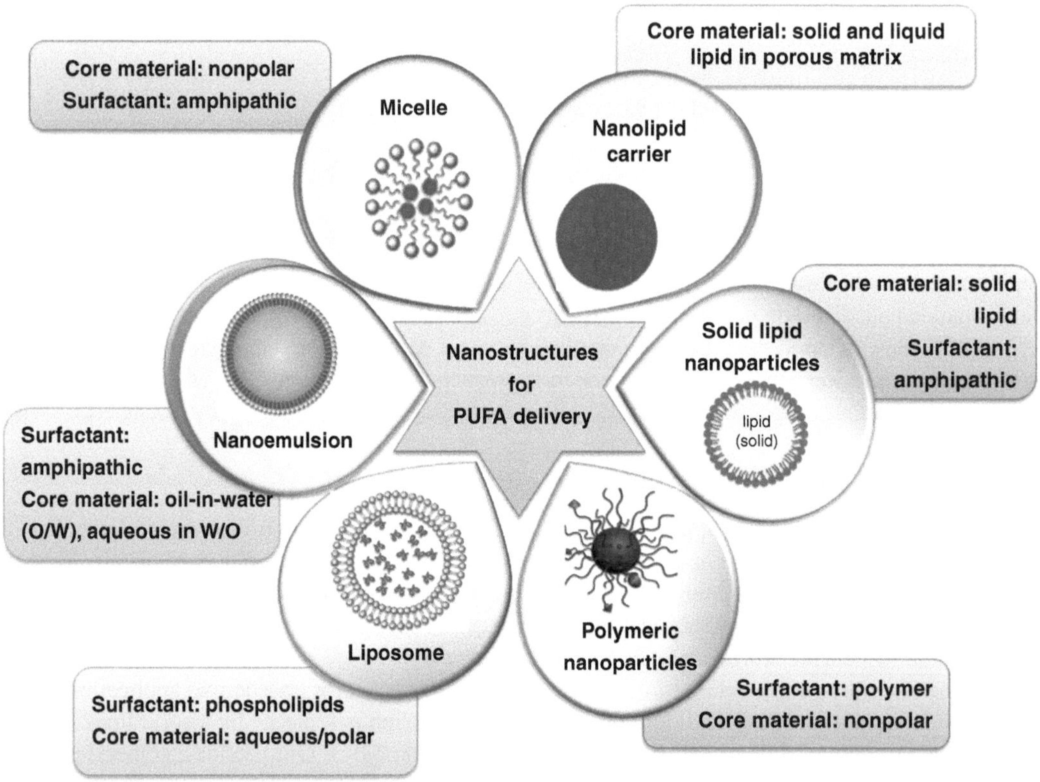

FIGURE 7.2 Major Nanostructures Applied for Delivery of PUFA, Through Oral, Parenteral, and Dermal Routes

3 UNDERSTANDING THE PHYSIOLOGICAL SYSTEM FOR THERAPY: THE SIGNIFICANCE OF LIPID NANOTECHNOLOGY AT TISSUE AND CELLULAR LEVEL

3.1 THE POTENTIAL OF NANOPARTICLES FOR TARGETED ACTION AT TISSUE LEVEL

Prior to the synthesis of any therapeutic formulation with the intention of enhanced potency and targeted action, it is of prime importance to understand the intermittent barriers which interfere with the desired action. Broadly, the challenges of delivery of therapeutic formulations, including nanosized systems, can be described as in the subsequent sections.

3.1.1 The gastrointestinal barrier

Oral administration of drug is the most popular and easy route of drug delivery. But lipophilic drugs are poorly soluble in the GI tract environment, unstable due to interactions with food components, and are prevented from crossing the gastrointestinal barrier. Paclitaxel, a lipophilic anticancer drug, is 6% absorbed when administered through the oral route. The mucosal layer over the apical polar GI epithelial lining acts as a barrier. The drugs are transported across the intestinal enterocyte lining through two transport mechanisms, namely, transcellular and paracellular (Salama et al., 2006). The drug or a nutraceutical traverse the apical cell membrane, as well as the basal cell membrane of the enterocyte. The possible mechanisms that are involved in the transcellular transport include (1) passive diffusion, (2) facilitated diffusion, (3) ATP dependent transport, (4) transcytosis (Fig. 7.3).

Macromolecules are transported across the GI epithelial lining by endocytosis and exocytosis processes, but the lipophilic pharmaceuticals administered orally through nanovehicles cross the enterocyte lining through the process of endocytosis (Florence, 2005). Both cationic and anionic nanoparticles are targeted mainly to the clathrin-coated endocytic pathway, but a portion of them are internalized through macropinocytosis. Anionic nanoparticles undergo lysosomal degradation, whereas cationic nanoparticles transcytose across polarized MDCK cells, as observed by Harush-Frenkel et al. (2008). The hydrophilic pharmaceuticals cross the GI epithelium through the process of passive diffusion, due to their concentration, electrochemical, or hydrostatic gradients. The small molecules cross the tight junctions by the process of paracellular transport. The leaky nature of the tight junctions decreases from

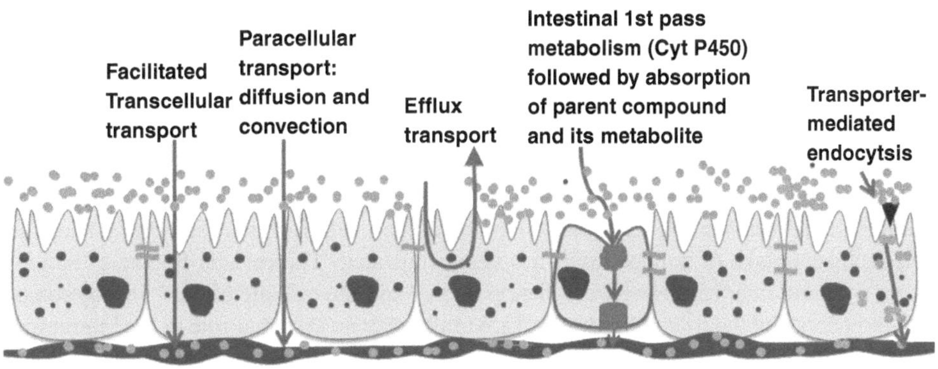

FIGURE 7.3 Different Absorption Mechanism Through GI Tract Epithelium

the duodenum to the colon. The tight junctions are composed of >40 membrane proteins and claudin, a family of transmembrane protein responsible for its high selectivity. The claudin-based pores that are 4 Å in radius are charge-selective. Another pathway involves larger discontinuities in the barrier that are not charge and size dependent. The ATP-binding cassette (ABC) efflux transporters are integral membrane proteins that are responsible for the efflux of molecules. The apical membrane of enterocytes is lined by p-glycoproteins (Pgp). The absorption of drugs is inhibited by these ABC proteins, and they serve as a barrier for the drugs administered through the oral route (Zhang and Bent, 2001). The active excipients are used to design food formulations with higher bioavailability. Tween 20 and PEG-300 Pluronic P85 are regarded as active excipients that affect the efflux transporters and, ultimately, oral bioavailability. Many autoimmune disease (rheumatoid arthritis, type I diabetes, Crohn's disease) pathogeneses are interlinked with the abrogation of the intestinal epithelium by zonulin pathway (Arrieta et al., 2006).

3.1.2 The blood–brain barrier

The CNS comprises of a highly selective physiological barrier which separates the circulating blood components from the brain extracellular fluid (BECF), commonly referred to as the BBB. The endothelial lining of the blood capillaries in the brain is semipermeable, protecting the neuronal microenvironment and maintaining tissue homeostasis, a balance between blood coagulation and fibrinolysis in the capillaries. An important aspect of neuropharmacology relates to the blood–brain permeability barriers. Two physiological barriers, BBB and the blood–cerebrospinal fluid barrier (BCSF), restrict the influx of therapeutics and noxious compounds. Physically, the brain capillary endothelium lacks fenestra. The tight junctions are endowed with a double seal joining the cerebral endothelial cells to each other in the capillary walls, and there exist foot processes of the protoplasmic astrocytes and pericytes that rest on the basement membrane (Egleton and Davis, 1997). There exists a high electrical resistance of 1500–2000 Ω cm^2 across the BBB compared to 3.33 Ω cm^2 in other tissue barriers (Crone and Olesen, 1982). The chemical barriers include charges borne on ion-transport membrane channels, and solvent partitions. The therapeutics developed for brain disorders encounter another BCSFB which is located on the choroid plexus. Alterations in the function of choroid plexus epithelium is evinced in aging and neurological disorders, such as Alzheimer's, Parkinson's, HIV, and disorders of CSF circulation (Szmydynger-Chodobska et al., 2009). In the circumventricular organs (CVO), BBB is absent, and is highly vascularized. The BBB expresses peptidases, the cytosolic enzymes and Pgp transporters that effluxes drugs from the endothelial cell into the blood. Fig. 7.4 describes the various transport mechanisms at the BBB that are involved in the transport of nutrients.

3.2 THE FACETS OF NANOPARTICLE PLAY AT CELLULAR LEVEL

Macromolecular uptake mechanisms identified to date include caveolae-mediated endocytosis, clathrin-mediated endocytosis, macropinocytosis, and clathrin- and caveolae-independent endocytosis (Conner and Schmid, 2003; Johannes and Lamaze, 2002; Khalil et al., 2006). Few articles have been identified where the cellular uptake of nanoparticles has been illustrated. The cellular uptake of nanoparticles depends on dynamic interactions between nanoparticles and the cell membrane. In this context, it is important to understand the role of lipid rafts. Lipid rafts play an essential role in the internalization of nanoparticles. Cholesterol and sphingolipid-enriched lipid rafts are microdomains present in the cellular membrane, involved in membrane trafficking. When the nanoparticles touch the cells, the membrane

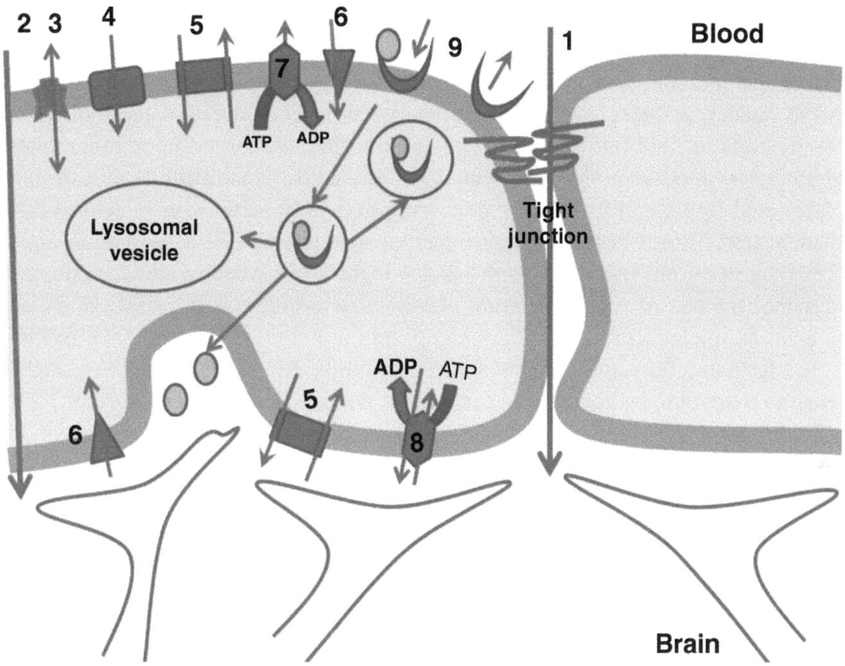

FIGURE 7.4 Different Absorption Mechanisms Through the BBB

1, Paracellular diffusion; 2, transcellular diffusion; 3, ion channel (K^+ gated); 4, ion-simport channel ($Na^+/K^+/Cl^-$ cotransporter); 5, ion-antiport channel (Na^+/H^+ exchange); 6, facilitated diffusion; 7, active efflux pump (P glycoprotein); 8, active-antiport transport (Na^+/K^+ ATPases); 9, receptor-mediated endocytosis.

curves and phagocytose the drug-loaded nanoparticle. The adhesion strength, membrane tension, and bending energy are the determining factors of membrane penetrability of the particle, and thus affect the final uptake (Panyam and Labhasetwar 2003; Whitehead et al., 2009). A fluorescence tracking study by Chiu et al. (2010) has found that depletion of cholesterol reduces the internalization of nanoparticles. Caveolae are flask-shaped invaginations present in the plasma membrane. It is a specific type of lipid raft enriched with cholesterol, sphingolipids, and a family of caveolin proteins. Nanoparticle uptake decreased significantly when cells are treated with genistein, an inhibitor of caveolae-mediated endocytosis. The endocytosed cargo (caveosome) was transported to the golgi/ER, or into early endosome. Fluorescent rhodamine conjugated palmitoyl chitosan nanoparticles are coupled to caveosomes, and cellular trafficking takes place through the endocytic pathway (Chiu et al., 2010). The outer surface of a cell is anionic, compared to the cell interior. Thus, cationic nanoparticles bind readily to the outer cell surface, and are readily endocytosed (Qiu et al., 2010). Tumor targeting with polymeric nanoparticles has also been studied extensively. Nam et al. (2009) fabricated hydrophobically modified glycol chitosan (HGC) polymeric nanoparticles (359 nm) for the targeted delivery of therapeutics. Fluorescent chromophore was conjugated with HGC and incubated with Hela H2B-GEP culture cells. Localization of the HGC nanoparticles has been studied with fluorescent confocal microscope. HGC nanoparticles get localized in the cellular membrane. Cellular uptake of HGC nanoparticles was time and dose dependent. Treatment with endocytic pathway inhibitors showed that several mechanisms are involved in cellular uptake. TEM image confirmed that HGC nanoparticles were taken up by the lysosomal compartment. The study inferred that cells internalize HGC nanoparticles by diverse mechanisms.

Disorders of the immune system are responsible for the complications of cardiovascular, rheumatoid arthritis, cancer, diabetes, and several other diseases. The course of these diseases can be altered by interventions of lipid nutrition. Rheumatoid arthritis, an autoimmune inflammatory disease, is improved by dietary modifications employing ω-3 fatty acids. Nanoformulations of ω-3 fatty acids have been fabricated by several groups for delivering ω-3 fatty acids with increased bioavailability. Eicosapentaenoic acid (EPA, C20:5), docosapentaenoic acid (DPA, C22:5), and docosahexaenoic acid (DHA, C22:6), sourced from fish lipid, in the diet safeguard the cardiovascular system, the immune system, and reduce the risk of type 2 diabetes, obesity, and autoimmune diseases. Data on the effects of ω-3 fatty acids supplementation on inflammatory and immune parameters have already been well established, but these ω-3 fatty acids are oxidatively unstable. Lipophilic bioactives, such as EPA and DHA are prone to oxidation, and need lipid carriers for delivery into the body system.

A study published in 2013 reported of an NLC comprising fish oil and tristearin, with low and high molecular weight lecithin as the surfactant molecule (Salminen et al., 2013). Variation in lecithin types altered crystallization temperature, melting temperature, and the melting enthalpy of tristearin. Different percentages of tristearin (0, 1.5, 4, and 5%) were melted, and ω-3 fatty acids were dispersed at 90°C using high speed homogenization (24,000 rpm) to produce the preemulsion. The hot high-pressure homogenization method using microfluidizer was employed for the preparation of the emulsion. Cooling of the emulsion to 20°C leads to the production of SLN containing 100, 40, and 20, or 0% ω-3 fish oil. The different surfactants are employed for the preparation of SLN. These nanoparticles endocytosed through the endosomal pathway may be used for the delivery of drugs in the treatment of lysosomal storage diseases (cancer, Alzheimer's disease) (Fig. 7.5)

4 THERAPEUTIC LIPID NANOSTRUCTURES

4.1 FOR THE DELIVERY OF PUFAs

The functional characteristics of nanoemulsions are dependent on the composition of edible oils used to prepare the nanoformulations, and the size of the droplets (Sarker, 2005). The nanoemulsions using ω-3 long chain polyunsaturated fatty acids (LCPUFA) are found to cross the physiological barriers more efficiently than other synthetic lipids. The imaging agents and the drugs are easily incorporated in the oil phase (Tiwari et al., 2006; Vyas et al., 2008).

A growing interest in preventive/prophylactic medicine of natural origins had led to the scientific discovery of the dietary effects of LCPUFAs, for example, linoleic acid (LA; 18:2; *n*-6), α-linolenic acid (ALA; 18:3; *n*-3), γ-linolenic acid (GLA; 18:3; *n*-6), eicosapentaenoic acid (EPA; 20:5; *n*-3), as well as docosahexaenoic acid (DHA; 22:6; *n*-3), and arachidonic acid (ARA, 20:4, *n*-6), in mitigating health problems. The multifaceted health benefits (Kris-Etherton et al., 2004) of these lipid bioactives have been reviewed. In addition, conjugated forms of these fatty acids, especially LA and linolenic acid (LLA), were proved to have significant beneficial effects on various lifestyle-related ailments (Bhattacharya et al., 2006; Carvalho et al., 2010). There are reports that attest the versatile beneficial role of these fatty acids in mammalian physiology. Hence, considerable scientific efforts have been put into bringing these nutraceuticals into formulated functional foods, as a prophylactic strategy against lifestyle related diseases. Despite its multitude benefits, the use of PUFAs is generally confronted with

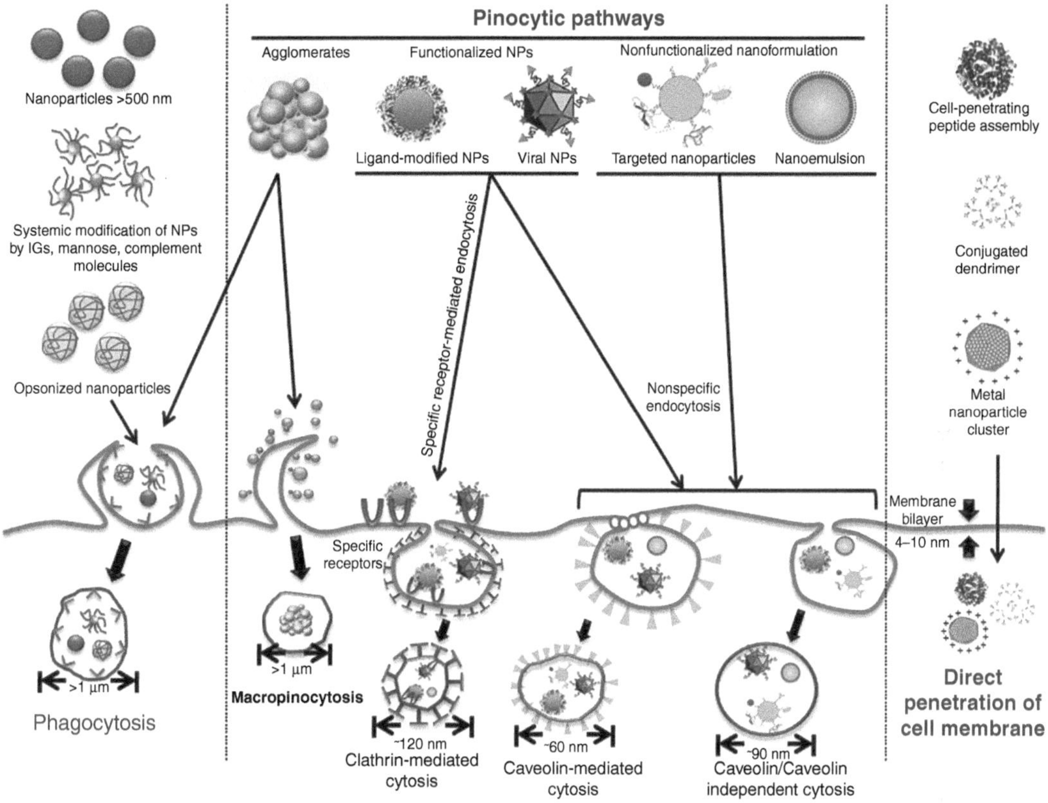

FIGURE 7.5 Cellular Uptake of Nanoparticles Designed With Heterogeneous Surface and Dispersion Characteristics: Understanding the Biophysicochemical Interactions at the Nanocell Interface

Natural size rules the cellular entry process for nanoparticle internalization within a mammalian cell, crossing the lipid-bilayer membrane (thickness 4–10 nm). Phagocytes absorb large nanoparticle agglomerates, opsonized nanoparticles, and nanoparticles with certain surface modification, by the means of simple phagocytosis. The other route of nanoparticle internalization in nonphagocytic cells mainly involves pinocytosis or direct penetration. Nanoparticles can be targeted to specific cell-surface receptors for receptor mediated endocytosis or nonspecific endocytosis by various surface modifications.

several hiccups, such as instability and vulnerability toward oxidative degradation, poor and inconsistent bioavailability via oral as well as topical administration. The bioavailability of lipophilic drugs and nutraceuticals depends on particle size, composition, and interfacial properties. The bioavailability of β-carotene is enhanced with nanoemulsion of LCT, rather than MCT, and other with digestible lipids (Qian et al., 2012; Rao et al., 2013). The potential use of nanoformulations in the delivery of PUFAs has gained wider attention owing to their promising bioavailability enhancement characteristics, improved stability, and better efficacy. Various formulations, such as nanoemulsions (NE), SLN, NLC, self-nanoemulsifying drug-delivery systems (SNEDDs), etc., have been employed toward the successful encapsulation and efficient delivery of these lipophilic nutraceuticals in the mammalian system, either through oral, parenteral, or topical route. CLA, GLA have been successfully nanoencapsulated by zinc oxide (ZnO) to increase their stability at elevated temperatures (Choy et al., 2010; Won et al., 2008).

Mammals obtain DHA either as DHA itself, or in its precursor forms of ALA, or from metabolic intermediates EPA. The major dietary sources of ALA are soybean and canola oils, flax seed oils, and some edible nuts. Radioisotopic studies in clinical subjects have established low conversion rates of DHA synthesis from ALA, and this suggests a general characteristic of the human metabolism, and n-3 fatty acid desaturation, that is, conversion of EPA to DHA, could be attributed as the rate-determining step (Uauy et al., 2000; Williams and Burdge, 2006). Dietary DHA, however, is well absorbed and readily incorporated into plasma and blood cell lipids in humans. However, instances of metabolic conversions during GIT absorption have also been reported. For instance, Tsuzuki et al. (2006) have reported that CLnA isomers are quickly converted to CLA isomers within the intestines of rats, prior to reaching systemic circulation. This could lead to physiological ramifications at in vivo level which interfere with the desired results of in vitro studies. Lipid nanoparticles are coated with impermeable barriers that prevent the chemical interaction of different biomolecules, and thereby lipid oxidation. Methodologies employed for preparation also influence lipid structure stability. PUFA liposomal stability, for instance, prepared with Mozafari method (direct hydration without using organic solvents) was found to be significantly higher than liposomes prepared with the Bagham thin film hydration method which uses organic solvents and conventional rotary evaporation (Rasti et al., 2012). This emphasizes the need to account for formulation techniques and conditions for the efficient fabrication of nanostructures.

In the past, we had successfully formulated a stable fish lipid oil nanoemulsion with EPA–DHA, using two-stage homogenization steps, of size distributions ranging from 76.02 to 94.20 nm, and polydispersity indices (PdI) in the range of 0.174–0.344 during the entire storage period. Intestinal absorption of lipid from nano- and conventional emulsions was demonstrated in single pass perfusion animal model (Dey et al., 2012). Lipid digestion by lipolytic enzymes at the lipid–water interface and emulsification in the gastrointestinal tract are considered to be the guiding factors for fat absorption. Particle size distribution of nanoformulations was found to affect lipid absorption. In our perfusion model, the small intestine was cannulized proximally and distally, and the perfusate containing the nanoformulation of fish oil was entrapped in the small intestine for different residence periods, as depicted in Fig. 7.6.

The influence of droplet size on lipid absorption through the enterocyte cell layer was also investigated. The intestinal absorption of lipid and the volume of nanoemulsion was compared and correlated between conventional and nanoemulsion (Figs. 7.6 and 7.7).

The larger lipid water interface, as in the case of nanoemulsion, enhanced the lipase activity, resulting in a higher rate of formation of free fatty acid and 2-monoacylglycerol that consequently served as emulsifiers. The higher rate of lipid absorption took place by passive diffusion and carrier-mediated processes in gastrointestinal epithelium with nanoemulsion, compared to conventional. It was concluded that digestion and absorption in the enteric nutrition is particle-size dependent. Thus, our nanoformulation of fish lipid oil may be administered to patients suffering from pancreatic diseases with insufficient digestive enzymes for optimized delivery (Fig. 7.8).

4.2 NANOSTRUCTURES FOR CARDIOVASCULAR DISEASE TREATMENT

Cardiovascular disease leading to arterial stenosis is still a predominant cause of death throughout the world. In the damaged vascular lumen, inflammatory responses cause the production of various

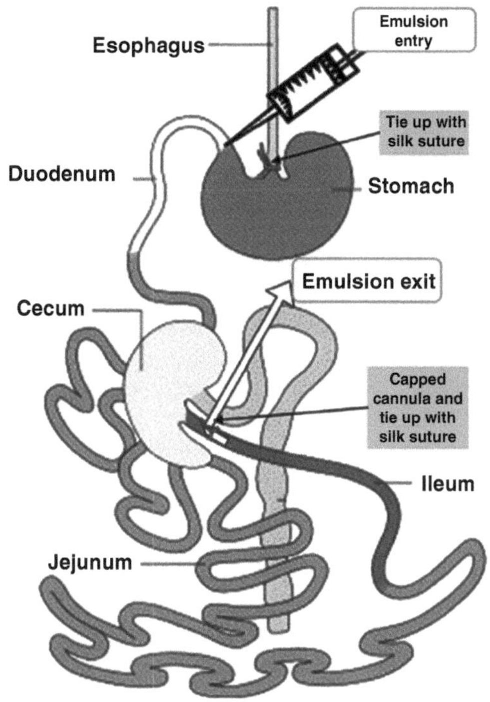

FIGURE 7.6 Schematic Model Showing the In Vivo Experimental Setup, With the Stretch of Small Intestine Flushed With the Emulsion

Reproduced from Dey, T.K., Ghosh, S., Ghosh, M., Koley, H., Dhar, P., 2012. Comparative study of gastrointestinal absorption of EPA & DHA rich fish oil from nano and microemulsion formulation in rats. Food Res. Int. 49, 72–79.

signaling molecules, such as PDGF secreted by activated macrophages, endothelium (EC) and vascular smooth muscle cell (VSMC). This leads to hyperplasia, and downstream mitogen-activated protein kinase (MAPK) pathway is activated. Estrogens and ceramides with cardioprotective properties were utilized to bioengineer a O/W nanoemulsion system. Flaxseed oil rich in ω-3 PUFA, ALA were used as organic phase carrying lipophilic bioactives, 17β-estradiol, and ceramides. Human aortic ECs and VSMCs were cultured, and cellular uptake and intracellular distribution of 17β-estradiol and ceramides nanoemulsion was assessed under fluorescence microscopy. The biological effect on occlusive vasculopathies of bioactives-ω-3 PUFA, ALA, 17β estradiol and ceramides in nanoemulsion form significantly inhibited PDGF stimulated VSMCs proliferation. The efficacy of inhibition of VSMC proliferation by 17β-estradiol and ceramides increased when delivered in ω-3 PUFA cargo. Ceramide nanoemulsion induced cellular apoptosis in VSMCs as well as ECs, induced by caspase 3/7 activity by the ω-3 PUFA rich cargo-free blank nanoemulsions. PDGF stimulated p38 MAPK phosphorylation which was inhibited by 17β-estradiol-ω-3 PUFA nanoemulsion, as well as ω-3 PUFA rich blank nanoemulsion. Amiji and coworkers have successfully formulated nanoemulsions of 17β-estradiol-ω-3 PUFA and ceramides that provides a novel therapeutic structure against occlusive vasculopathies and restenosis in animal model (Deshpande et al., 2013).

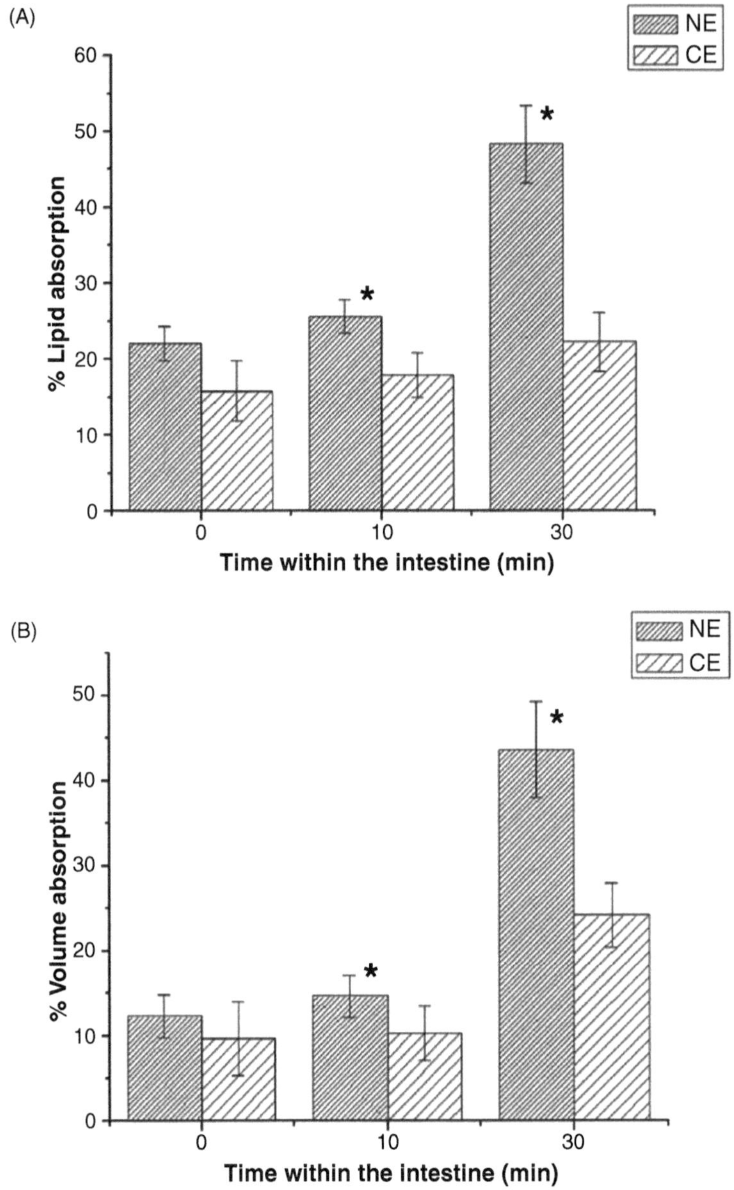

FIGURE 7.7

Comparison of intestinal absorption of D22 nanoemulsion (NE) and conventional emulsion (CE), with EPA–DHA in rats, with respect to absorbed lipid percentage (A) and absorbed volume percentage (B) of emulsion at three time intervals within the intestine (Mean ± SD, $n = 6$) The difference between the absorbed volume percentage from NE and CE is proved to be statistically significant at significance limit of $P > 0.05$.

Data reproduced from Dey, T.K., Ghosh, S., Ghosh, M., Koley, H., Dhar, P., 2012. Comparative study of gastrointestinal absorption of EPA & DHA rich fish oil from nano and microemulsion formulation in rats. Food Res. Int. 49, 72–79.

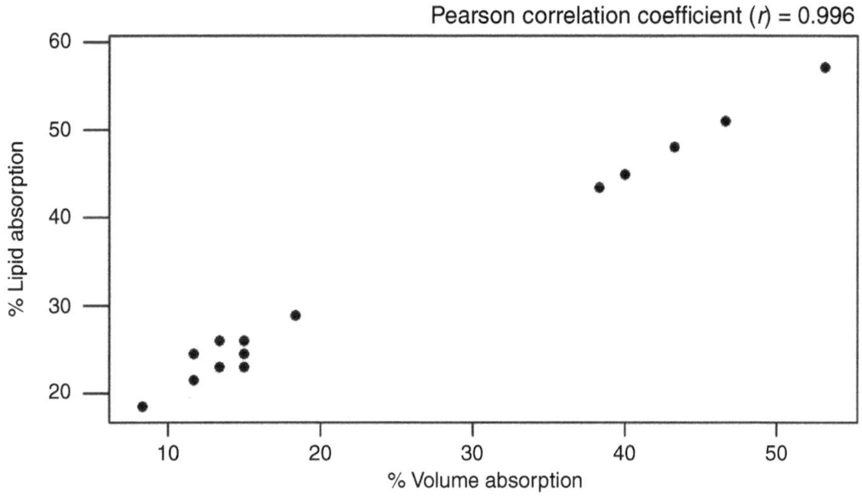

FIGURE 7.8 Correlation Between Percentage Lipid Absorption and Percentage Volume Absorption From Nanoemulsion System (MINITAB 13.31)

Data reproduced from Dey, T.K., Ghosh, S., Ghosh, M., Koley, H., Dhar, P., 2012. Comparative study of gastrointestinal absorption of EPA & DHA rich fish oil from nano and microemulsion formulation in rats. Food Res. Int. 49, 72–79.

Low calorie fat products are on the market for commercial use against obesity and liver diseases. Nanoemulsions are possible solutions to produce gel-like viscous materials with droplet concentration lower than conventional emulsion (Mason et al., 2006; Sonneville-Aubrun et al., 2004; Tadros et al., 2004). The transport efficiency of a drug for dermal application will likely be increased due to the nanosizing of droplets.

Nanoemulsions (O/W) of conjugated LAs are synthesized by the modification of LA. CLA possesses anticancer, antioxidative, and antibacterial activity. It also prevents the absorption and accumulation of fat in adipocytes, increasing lipolysis in adipocytes. CLA has been described as a hypolipidemic fatty acid. The stabilized functional nanoemulsion of CLA has been formulated with remarkable storage stability, pH stability, and transparency. A weight of CLA of 0.5%–5% has been incorporated in the nanoemulsions, with 0.01%–5% weight of lecithin as emulsifier. The nanoemulsions of CLA as described in the patent (Yu et al., 2010) has been found to be hypoadipogenic, lipolytic, and to affect leptin secretion from adipocyte. Tween 20, Tween 80 were used as coemulsifiers for the production of nanoemulsions which were proposed as functional food. Cell viability, lipolytic effect, adipogenesis inhibitory effect, secretion of leptin, the hormone responsible for appetite regulation, were examined in 3T3-L1 preadipocyte cells in an in vitro assay. In an in vivo experiment, CLA nanoemulsion diet prevented the increase of body weight even in high fat diets, compared to non-CLA nanoemulsion in Sprague Dawley rats.

Additionally, the delivery of PUFA nanoparticles has been extensively studied by Nacka et al. (2001) for better release kinetics in various therapies, including cardiovascular. Nacka and his group fabricated marine-lipid based liposomes, terming them marinosomes, and characterized the influence of gastrointestinal fluid within an in vitro simulated model. They studied the oxidative stability and the hydrolytic effect of phospholipase A2 on the structured liposomes in vitro. The effect of variation of pH and temperature on the agglomeration of liposomes were also studied. The marine-lipid based liposome developed by the group was found to be a vector for the delivery of α-tocopherol through oral route. The absorption of α-tocopherol delivered through marine-lipid based liposomes was examined in a lymph cannulated rat model, and the results showed a threefold increase in α-tocopherol recovery in lymph (Nacka et al., 2001). Marine-lipid liposomes comprising fish-lipid oil increased the bioavailability of ω-3 fatty acids when administered in rats for in vivo activity. Liposome ingestion increased fatty acid absorption (98 ± 1%), compared to fish oil (73 ± 6%) (Cansell et al., 2003). Fecal lipid analysis revealed absence of triacylglycerol, but presence of free fatty acids and monoglycerides, indicating partial absorption of digestion products.

4.3 NANOSTRUCTURES IN NEURODEGENERATIVE DISEASE TREATMENT

Investigations have shown that 100% of drug molecules and 98% of small molecules are prevented from influx into the brain by BBB. The lipophilic drugs of <400 kDa are suitable for the CNS. Various strategies, such as intravenous, intraarterial, transnasal, intracerebral (intraparenchymal), intraventricular (transcranial drug delivery) delivery are employed in order for the drugs to cross the BBB. Antibody-directed enzyme prodrug therapy, biotin–avidin conjugated system, cyclodextrin complexes, and cationic proteins are some of the strategies for CNS drug application. With the advance in nanotechnology, lipid-based surface modified liposomes are employed to encapsulate lipophilic drugs targeted to brain tissue. Liposomes conjugated with brain targeting vectors, such as monoclonal Ab against insulin or transferrin receptors, cationized protein cross the BBB through absorptive-mediated transcytosis, or by receptor-mediated transcytosis (Schnyder and Huwyler, 2005). Lipid nanoparticles, such as nanospheres, nanosuspensions, nanoemulsions, nanogels, nanomicelles, and nanoliposomes, SLN, NLC, and lipid drug conjugates (LDC) designed to deliver lipophilic CNS drugs are presented in Table 7.1. Nanoemulsions can be fabricated with oils of different fatty acid composition and surface modifiers. ω-3 PUFA-rich oil nanoemulsions overcome the BBB, and may be utilized as lipophilic drug carriers. Brain imaging agents soluble in PUFA-rich oil nanoemulsions have practical applications (Tiwari and Amiji, 2006; Vyas et al., 2008) in diagnostics, and may serve as theranostic agents. The transporters located in the capillary endothelium are responsible for the paracellular movement, transcellular movement, carrier-mediated, and receptor-mediated movements. But the tight junctions and limited number of vesicles inhibit the transport of drugs through the BBB. Nanoemulsions fabricated with ω-3 and ω-6 PUFA containing flaxseed, pine-nut, hemp fish oils, safflower, and wheat-germ oils were used for drug delivery to the CNS. Egg phosphatidylcholine as surfactants and stearylamine, deoxycholic acid, dioleoyl-trimethyl-ammonium-propane (DOTAP) as cosurfactants are used for increasing stability and imparting surface charges.

4.4 NANOSTRUCTURES IN DIABETES MANAGEMENT

Diabetes mellitus (DM) is a multifactorial heterogeneous chronic disorder associated with a number of dysfunctions, such as high blood pressure, retinopathy, nephropathy, neuropathy, along with cardiovascular problems, such as atherosclerosis, ischemic heart disease, stroke, etc. The systemic damages are caused by generation of cellular ROS (reactive oxygen species) and RNS (reactive nitrogen species) like O^{2-}, NO, $ONOO^-$, H_2O_2, and NO. ROS/RNS induced activation of polyol pathway increases hexosamine flux, advanced glycation end-products (AGE) formation, and protein kinase C (PKC). AGE and

Table 7.1 Lipid-Based Nanodelivery Systems for Efficient Delivery of Therapeutics Across Blood–Brain Barrier

Formulation	Drug Molecule	Lipid Carrier, Encapsulant	Mechanism/Inference	References
Nanoemulsion	Risperidone	Capmul MCM as the oil phase along with mucoadhesive polymers	Intranasal administration showed the presence of drug in cerebral, as well as CSF	Kumar et al. (2008)
	Paclitaxel	Pine-nut oil containing high concentrations of PUFA	Cytotoxic against human glioblastoma brain tumor cells	Desai et al. (2008)
	Saquinavir	Edible oils rich in PUFA; lipoid-80 and deoxycholic acid	Enhanced oral bioavailability of effective antiretroviral therapy	Vyas et al. (2008)
Solid lipid nanoparticle	Etoposide	Tripalmitin (solid fat)	Applied in Dalton's lymphoma, injected by subcutaneous, IV, or IP routes	Harivardhan et al. (2005)
	5-Fluoro-2'-deoxyuridine (FUdR)	Caprylic acid, pluronic F-68 and F-127	Blood–brain brain penetration	Wang et al. (2002)
	Camptothecin	Stearic acid, soybean lecithin, and poloxamer 188	Targeted delivery to brain, sustained release, reduced dose, and systemic toxicity	Yang et al. (1999)
	Temozolomide	Stearic acid, lecithin, and poloxamer 188 and Tween 80	Cerebral concentration increased, reduced cardiac and nephrocytotoxicity	Huang et al. (2008)
Nanoliposome	Tempamine	Lipid	Neurodegenerative diseases involving oxidative damage (multiple sclerosis)	Kizelsztein et al. (2009)

PCK are the cause of inflammatory responses associated with diabetes, triggering NF-κβ which in turn expresses the genes of proinflammatory cytokines like IL-6, TNF-α, and growth factors like VEGF. These molecules are responsible for the progressive complications in the late stages of diabetes (Evans et al., 2002; Yamamoto and Gaynor, 2001).

Prolonged use of oral antidiabetic drugs (OAD) leads to weight gain and requires frequent dosing of the drug. Development of insulin resistance leading to hyperglycemia is common in type 2 diabetes, and most of the patients develop microvascular and macrovascular difficulties. Dietary antioxidants prevent oxidative stress and tissue damages common in type 2 diabetes, but the bioavailability of these antioxidants is a concern to food scientists. Bioavailability of antioxidants that are present in complex food matrix needs a complicated process of liberation from the food matrix. But antioxidants are easily discharged from food nanoformulations with higher bioavailability. The traditional approaches of delivery of antioxidants have some disadvantages which lead to the development of novel approaches of delivery of OADs. Lipid-based microparticles of quercetin were fabricated by emulsion method using phosphatidylcholine and tristearin. Encapsulation of quercetin reduced the photodegradation rate and chemical instability, compared to uncapsulated one (Scalia and Mezzena, 2009). SNEDDS are formulated to deliver lipophilic drugs, to enhance their bioavailability. SNEDDS are more palatable and given in capsules. Coenzyme Q10 nanoemulsions with a size of 243.3 nm were bioengineered with oils, surfactants, and cosurfactants which enhanced bioavailability by twofold (Balakrishnan et al., 2009).

Phenolic compounds, one of the most widely occurring groups of phytochemicals, are of considerable interest as they are reported to exhibit antiallergenic, antiatherogenic, antiinflammatory, antimicrobial, antioxidant, antithrombotic, cardioprotective, and vasodilatory effects. Owing to their rather polar character, these molecules are not optimal in dispersed lipid systems, such as emulsions or liposome suspensions, since they tend to concentrate preferentially into the aqueous phase far away from the O/W interphase where lipid oxidation is assumed to occur. Based on this observation, lipophilization has been considered to improve the surface active properties of polar antioxidants, and thus their effectiveness in such dispersed lipid systems. Modification consists of grafting a lipid moiety by different reactions (etherification, esterification, transesterification, and amidation), and catalytic approaches (chemical, enzymatic, and chemoenzymatic), according to the structure and reactivity of the starting molecules. In recent years, there has been a growing interest to produce molecules by lyophilization of phenolic compounds with fatty acids, and the product molecules thus produced are reported to have improved function as antioxidants for food and pharma-applications. The molecules thus produced are designated as phenolipids, and can have the beneficial effects of both the parent compounds, and also help in the delivery and uptake of the compounds. Designing delivery vehicles for lipophilic nutraceuticals (Raatz et al., 2009) with therapeutic potentials has been developed.

Extensive reports are now available to highlight the ability of PUFAs to mitigate DM-health complications, even at nutrigenomic level. From mediating sensitization of factors causing insulin resistance, to upregulating glucose–homeostasis gene expression, empirical evidences corroborate the role of PUFAs in managing diabetes (Iwase et al., 2015). In our endeavor to study the ramifications of nanofabricating α-eleostearic acid (ESA) rich system against alloxan-induced diabetes, we found optimum activity at ~50% lower dosage (Paul et al., 2014) than the previously established dose (Saha and Ghosh, 2012) in attenuating DM-induced stress. Further, ESA, an isomer of conjugated linolenic acids, has been shown to possess immense health promoting effects, such as antitumorigenic, antiadipogenic, and antidiabetic mediators. Pertinent literature showed that ESA activates nuclear receptors, such as PPAR-γ, and inhibits transcription factors, such as ERK ½ in vitro. MAPK/ERK signaling pathway for progression of differentiation has been found to be modulated by ESA. Other cellular activities include inhibition of carcinogenesis, protection against arsenate, and methylmercury toxicity activation of signaling cascade impeding proinflammatory to antiinflammatory cytokine ratios during induced hypersensitive response. To explain our observations regarding the higher efficacy of ESA-NE, compared to conventional systems, we assume that the improved bioavailability of ESA in systemic circulation due to nanosizing of the system is the only logical explanation of the results of our work (Fig. 7.9).

To evaluate further the impact of nanosizing ESA-rich systems in mitigating oxidative stress during 100% bioavailability, we have also studied the transcellular uptake of the fabricated ESA-rich nanoemulsion in nonadherent human lymphocytes. To understand the real-time dynamics in attenuating stress, flow cytometric data analysis of the LPS-induced oxidatively challenged PBMCs, mean-fluorescence intensity (MFI) of a fluorochrome DCFDA for ROS quantification in a time dependent manner was carried out. The dose-dependent assay for therapeutic functions of ESA-NE revealed that ESA-NE at a lower concentration (70 μM) optimally modulates molecular parameters aggravated by

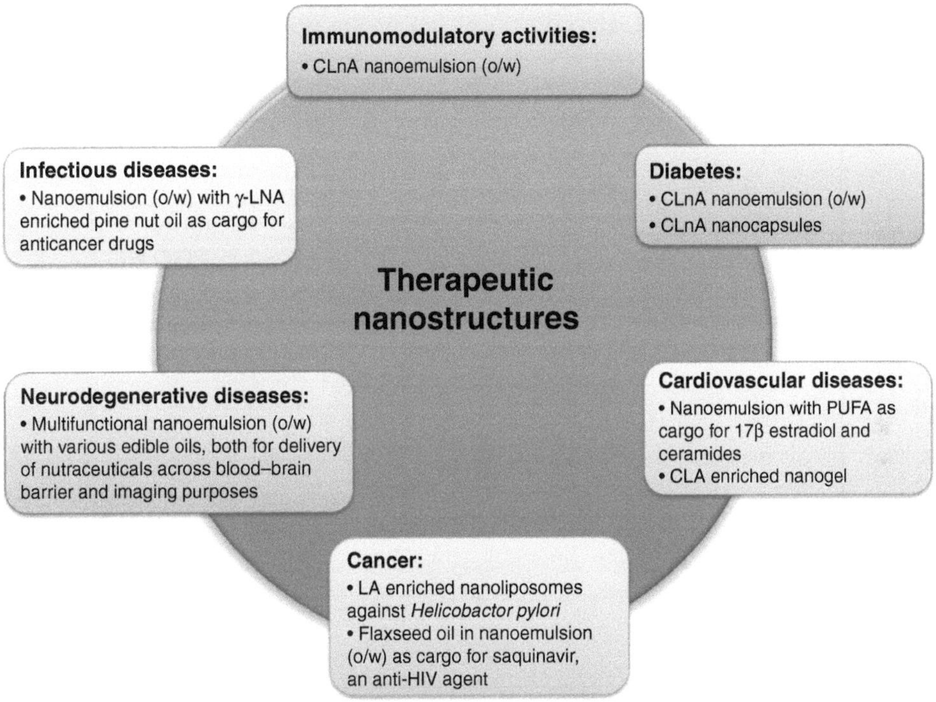

FIGURE 7.9 Major Nanostructures Applied for Delivery of PUFA, Through Oral, Parenteral, and Dermal Routes

disturbed redox equilibrium in nonphagocytic cells. This therapy for parenteral applications can well be administered for treatment of acute diabetic complications in future (Paul et al., 2015) (Fig. 7.10).

Similar to our work, Sengupta et al. (2013a,b) encapsulated CLnA with egg phospholipid and soy protein emulsifiers containing the encapsulant, in order to improve antiadipogenic effects of CLnA efficaciously. A high-speed disperser along with ultrasonication were used to formulate the nanoemulsion. The emulsion was frozen with liquid nitrogen (–196°C) and freeze-dried to remove crystallized water. The group has studied the comparative antioxidant activity of CLnA nanocapsules protected with lipoic acid and TBHQ in a rat model. The lipid peroxidations and antioxidative enzymes were examined in liver and heart tissues (Table 7.2).

The hypocholesterolemic activity of CLnA nanocapsules with lipoic acid was comparative found to be more effective than CLnA nanocapsules with TBHQ (Sengupta et al., 2013b).

4.5 NANOSTRUCTURES FOR CANCER TREATMENT

Cancer: the world population is suffering in different forms of it, and scientists are working hard to develop different chemotherapeutic agents to counter it. Tumor-specific chemotherapy requires development of novel drug-delivery systems which will deliver the chemotherapeutic agents targeted to the tumors, reducing the systemic toxicity. These highly active cancer cells overexpress a number of receptors that are tumor specific (Jaracz et al., 2005). For targeted delivery of drugs to the multidrug resistant

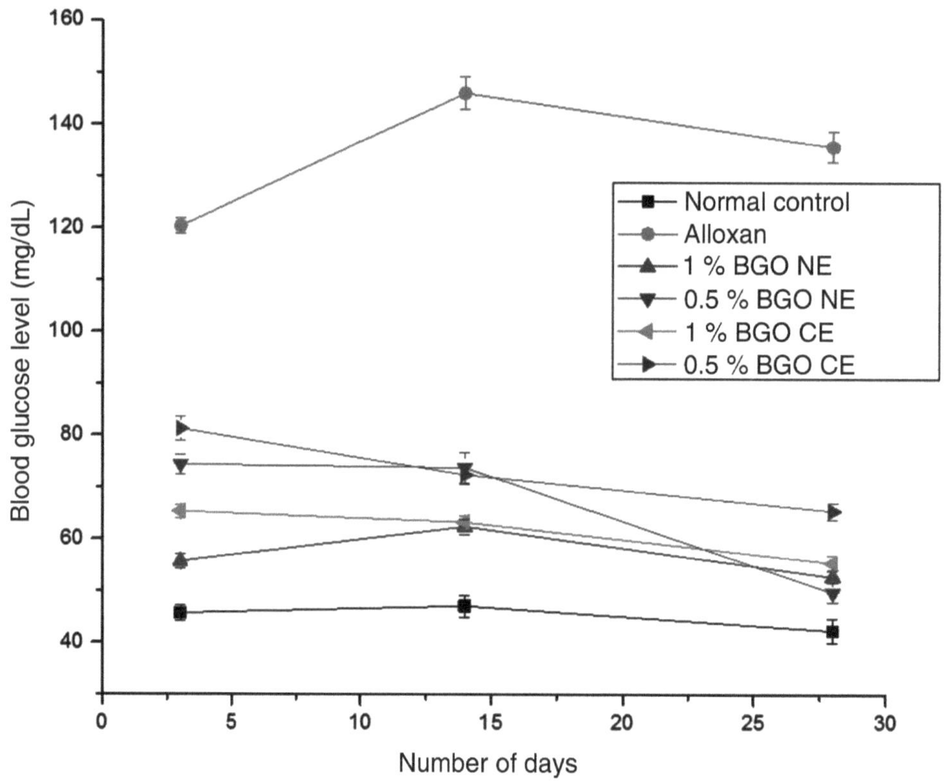

FIGURE 7.10 Effect of BGO Nanoemulsion Supplemented Along With Diet at Varying Levels

Normal indicates healthy normal rat diet fed group; ALLOXAN indicates alloxan (180 mg/kg body weight) induced untreated group; 1% BGO-NE refers to the diabetic group orally fed 1% of the fabricated BGO-NE containing 0.5% ESA; 0.5% BGO-NE refers to the diabetic group orally fed 0.5% of the fabricated BGO-NE containing 0.25% ESA; 0.5% BGO-CE refers to the diabetic group orally fed 0.5% BGO-CE containing 0.25% ESA, and 1% BGO-CE group refers to the diabetic group orally fed 1% BGO-CE containing 0.5% ESA. Values are presented as mean ± SEM ($n = 6$). At the end of 28 days, the blood glucose level in the experimental groups.

* Mean values were significantly different from those of alloxan.

Data reproduced from Paul, D., Dey, T.K., Mukherjee, S., Ghosh, M., Dhar, P., 2014. Comparative prophylactic effects of α-eleostearic acid rich nano and conventional emulsions in induced diabetic rats. J. Food Sci. Technol. 51, 1724–1736.

cancer cells, the delivery system must have tumor recognition moiety linked to the toxic chemical, conjugated by a linker molecule. Based on this concept, Ojima and coworkers developed a tumor targeted delivery system utilizing ω-3 PUFA. Monoclonal antibodies (Chari, 1998; Chen et al., 2005; Hamaan et al., 2002; Ojima et al., 2002; Wu and Ojima, 2004), hyaluronic acid (Luo et al., 2002), polyunsaturated fatty acids (Bradley et al., 2001; Kuznetsova et al., 2006), oligopeptides (Nagy et al., 1998), folic acid (Leamon and Reddy, 2004; Lu and Low, 2002), aptamers (Chu et al., 2006), are tumor-specific moieties that have been applied for targeting the tumor cells. PUFAs possess 18, 20, and 22 C atoms, naturally occurring, 2–6 unconjugated cis double bonds, and are excellent tumor-specific guiding

Table 7.2 Effect of Feeding of Nanoformulations of Conjugated Linolenic Acid Along With Natural and Synthetic Antioxidants on Tissue Lipid Profile

Parameters		Control	HCD	Lipoic Acid	TBHQ
Gain in weight		25.0 ± 0.28[a]	29.5 ± 0.76[b]	22.8 ± 0.12[a]	23.1 ± 0.55[a]
Food consumed		8.9 ± 0.87[a]	9.4 ± 0.07[a]	9.0 ± 0.12[a]	9.2 ± 0.49[a]
Food efficiency ratios		0.36 ± 0.02[a]	0.32 ± 0.01[b]	0.37 ± 0.03[a]	0.39 ± 0.03[a]
Liver weight (g)		3.6 ± 0.22[a]	5.6 ± 0.12[b]	39. ± 0.07[a]	4.9 ± 0.15[c]
Brain weight (g)		1.1 ± 0.01[a]	1.5 ± 0.01[b]	1.2 ± 0.01[c]	1.3 ± 0.03[d]
Plasma lipid profile (mg/dL)	TC	89.5 ± 1.22[a]	173.9 ± 2.30[b]	108.4 ± 2.31[c]	109.9 ± 1.56[c]
	HDL-C	28.2 ± 0.98[a]	25.3 ± 0.56[b]	28.1 ± 0.45[c]	26.4 ± 0.78[d]
	Non-HDL-C	49.0 ± 1.02[a]	126.4 ± 1.06[b]	61.1 ± 1.04[c]	64.9 ± 0.78[d]
Liver lipid profile (mg/g tissue)	Triglyceride	61.2 ± 1.55[a]	111.4 ± 1.40[b]	83.6 ± 1.39[c]	91.7 ± 0.97[d]
	TC	20.5 ± 0.95[a]	90.7 ± 0.67[b]	28.9 ± 0.95[c]	32.9 ± 0.45[d]
	Triglyceride	24.4 ± 1.55[a]	104.2 ± 2.34[b]	42.9 ± 0.87[c]	43.2 ± 1.01[c]
	Phospholipid	28.9 ± 0.65[a]	8.2 ± 0.50[b]	16.1 ± 0.43[c]	22.3 ± 0.22[d]
Brain lipid profile (mg/g tissue)	TC	62.6 ± 2.99[a]	65.9 ± 0.56[a]	62.6 ± 0.67[a]	62.9 ± 0.52[a]
	Phospholipid	24.1 ± 0.09[a]	16.7 ± 0.13[b]	21.1 ± 0.11[c]	18.4 ± 0.10[d]

Effect of feeding of dietary lipids on body weight gain, organ weights, plasma, liver, and brain lipid profiles of different dietary groups. C, Cholesterol; Control, fed with soybean oil; HCD, fed with 1% cholesterol (hypercholesterolemic); HDL, high density lipoprotein; lipoic acid, fed with conjugated linolenic acid rich oil nanocapsules with lipoic acid; TBHQ, fed with conjugated linolenic acid rich oil nanocapsules with tert-butylhydroquinone; TC, total cholesterol.
Values are mean ± S.E.M (n = 6 > *rats). Values not sharing a common superscript within a row are statistically significant (P < 0.05).*
Adapted with permission, from Sengupta et al. (2013b).

molecules, that some PUFAs, such as EPA DHA, LNA, LnA are readily taken tumor cells, compared to normal cells, as demonstrated by perfusion experiments (Sauer and Dauchy, 1992; Sauer et al., 2000). The membrane fluidity of tumor cells is disrupted due to incorporation of PUFA into the lipid bilayer (Grammatikos et al., 1994). Paclitaxel conjugates of DHA were synthesized by Bradley et al. (2001). PUFA-taxoid conjugates were assessed in severe combined immunodeficiency (SCID) mice for efficacy against a drug-resistant human colon tumor xenograft DLD-1, and a drug-sensitive human ovarian tumor xenograft A121. The experiments demonstrated that ω-3 FA, that is, DHA EPA conjugated paclitaxel, possess the efficacy to completely regress tumor growth, rather than ω-6 FA, LA conjugated drug. The application of nanoemulsion in cancer drug delivery has several advantages. Nanoemulsions with higher surface to volume ratio and energy than macroemulsions make it a better delivery system than macroemulsions. In addition, it is much more stable, and with less creaming, flocculation, coalescence, and sedimentation which makes a macroemulsion unsuitable for delivery. Nanoemulsions can be applied through different routes, such as enteric, and parenteral, and can also be applied on the skin, as they are nontoxic, and surfactants used for the production are GRAS. In the last decade, 90% of the cancer drugs that are discovered are lipophilic in nature (Kommuru et al., 2001), and thus pose in vivo consequences like decreased bioavailability, increased interaction with food, imperfect release of dosage of drug, and individual variability becomes a factor that influences drug action. With the advances of nanotechnology, formulation scientists have developed nanoemulsions, microemulsions, SLN, solid dispersions, and liposomes for drug delivery, with enhanced bioavailability of lipophilic drugs. Transdermal application of nanoemulsion of cancer drug has enhanced the bioavailability and permeability.

Amiji and coworkers (Tiwari et al., 2006) formulated nanoemulsions with pine-nut oil containing 20% γ-linolenic acid employing phosphatidylethanolamine conjugated to diethylene triaminepenta acetic acid as stabilizers, with high affinity for metal ions. The lipophilic oil phase was loaded with lipophilic anticancer drug paclitaxel and gadolinium (Gd3+) was tagged with the surface material with a diameter of 100 nm which enables it to pass through the vasculature. The fabricated drug-loaded nanoemulsion is capable of imparting and enhancing contrast in MRI to develop tissue images. Cellular uptake by MCF-7 estrogen-positive human breast adenocarcinoma cells was evaluated. GLA, ω-6 FA present in pine oil was utilized to produce the lipid core as GLA receptors located in the brain showed higher affinity for GLA associated surface coating molecules. Pgp efflux pumps located on tumor cells, BBB and vascular endothelial cells prevent the anticancerous activity of paclitaxel. When the drug was administered in GLA-filled nanoemulsion, Pgp receptors were prevented from effusing the drug. The effect of paclitaxel loaded nanostructures was evaluated in in vitro system, and needs in vivo authentication.

Hydrogel nanoparticles were prepared using LA modified carboxymethyl chitosan by the process of sonication. ADR, a cancer drug, was loaded in hydrogel nanoparticles and its in vitro anticancer activity was examined in Hela cancer cells (Tan and Liu, 2009).

Water extracts of certain mushrooms contain β-glycans. It has been previously found that marine phospholipids with *n*-3 PUFAs exert anticancerous effects. Okada et al. bioengineered liposomes with β-glycans in *n*-3 PUFAs containing marine phospholipid. In an in vivo experiment, *Agaricus blazei* Murrill water extract alone and squid phospholipid alone group showed moderate regression of tumor, but enhanced regression of tumor was demonstrated when liposome containing *A. blezei* Murrill extract and squid phospholipid was administered on sp2 myeloma cancer tumor cells implanted in BALB/c male mouse (Murakawa et al., 2007). The results showed the suppression of cancer by marine phospholipid liposomes in combination with mushroom immune activator.

Nanoemulsions are used as vehicles for drugs active against *Herpes labialis*, fungal infections, bacterial infections, vaginitis, and so on.

The present context of lipid nanotechnology involves encapsulation of such bioactives (such as sesame lignans) for its targeted delivery. Since the synergistic effect of PUFA, such as EPA–DHA and sesame lignans, against various types of stress and cancer models at in vitro level has been reported, perspicacious technological interventions will be required for more pragmatic applications to fortify human nutrition. Lipid-based nanoencapsulation systems can be used for protection and delivery.

4.6 NANOSTRUCTURES FOR INFECTIOUS DISEASES TREATMENT

Prevalence of *Helicobacter pylori* infection causes associated diseases, such as gastritis, peptic ulcers, and gastric cancer. About half of the world population suffers from *H. pylori* infection and develops antibiotic resistance. In the developed countries, there is a 40% increase in metronidazole-resistant *H. pylori* strains which raised to 90% in developing countries. Studies have shown that free fatty acids—lauric acid, myristoleic acid, LA, and LLA—possess antibacterial activity against *H. pylori* (Desbois and Smith, 2010; Jarboe et al., 2013). Due to esterification, oxidation, and lipid–protein complexation, the antibacterial activity of these FFAs diminish. To overcome this problem, Thamphiwatana et al. (2014) bioengineered a novel LA liposomal formulation of ~100 nm, a new regime of drug against *H. pylori*. Egg phosphatidylcholine, cholesterol and LLA were employed for the fabrication of nanoliposomes by standard vesicle extrusion method. The retention and distribution of liposomes in the gastric mucous wall was examined by fluorescence microscopy that revealed that, after 24 h of oral gavage, linolenic liposomes were retained in the stomach. C57BL/6 mice were infected with 3×10^8 CFU *H. pylori* SS1 strains. *H. pylori* count in mouse stomach were estimated in all the groups and triple-therapy antibiotics reduced the count significantly to 7.2×105 CFU/g, which is significantly lower than with negative controls. The LLA liposome treated group showed maximum reduction of bacterial load (5.5×104 CFU/g). IL-1β, IL-6, and TNFα are upregulated in *H. pylori* infection (Obonyo et al., 2007; Rad et al., 2007) while LLA reduced the mRNA expression of thesis proinflammatory cytokines. These LLA liposomes fuse with the *H. pylori* bacterial membrane, causing destruction of membrane integrity of the bacterium.

Vyas et al. (2008) has reported the formulation of nanoemulsion with PUFA-enriched flaxseed oil or safflower oil as vehicle for delivering saquinavir, an anti-HIV protease inhibitor, to assess the bioavailability across GI tract and BBB. The same strategy was applied by Tiwari and Amiji (2006) to encapsulate and biodeliver paclitaxel efficiently across the GI tract epithelium.

4.7 NANOSTRUCTURES IN OVERCOMING PROINFLAMMATORY CONDITIONS

Inflammatory responses are associated with a number of disorders, such as cardiovascular diseases, complications of diabetes, and rheumatoid arthritis. Several steroidal and nonsteroidal drugs are available to combat the progress of the disease. To retard the side-effects and control the dose increase of the drugs, the method of delivery of the drug was investigated systematically. Calle et al. (2015) have reported of a nanoliposome formulation comprising ω-3 PUFA ethyl ester in impeding proinflammatory cytokines and associated parameters. Liposomes of size 200 nm were successfully bioengineered, comprising ω-3 PUFA ethyl ester with paramagnetic nanoparticle, with or without rhodamine 100. Dye was used to track the particle in vivo. The nanoliposome was utilized for both diagnostic and therapeutic purposes. Colonic inflammation was induced in male C56BL6 mice. Treatment of nanoliposome reduced the inflammatory responses. MRI was carried to detect the accumulation of liposome in the affected tissue. The antiinflammatory activity of the same nanoparticle was also investigated in oncology inflammation model. Fluorescence imaging studies showed the accumulation of the drug in vivo conjugated with magneto-liposomes and ω-3 PUFA-EE within the C6 glioma in vitro models, and its subsequent regression. The previously mentioned study also concurs with previous reports of correlation between inflammation and cancer (Coussens and Werb, 2002), directly implying that therapeutics associated with nanoformulations of PUFA which have the ability to control cancer progression also mediates antiinflammatory activity. Thus, the potential of nanoformulations of PUFA in modulating inflammation is further reinforced, and merits further research on a holistic scale (Fig. 7.11).

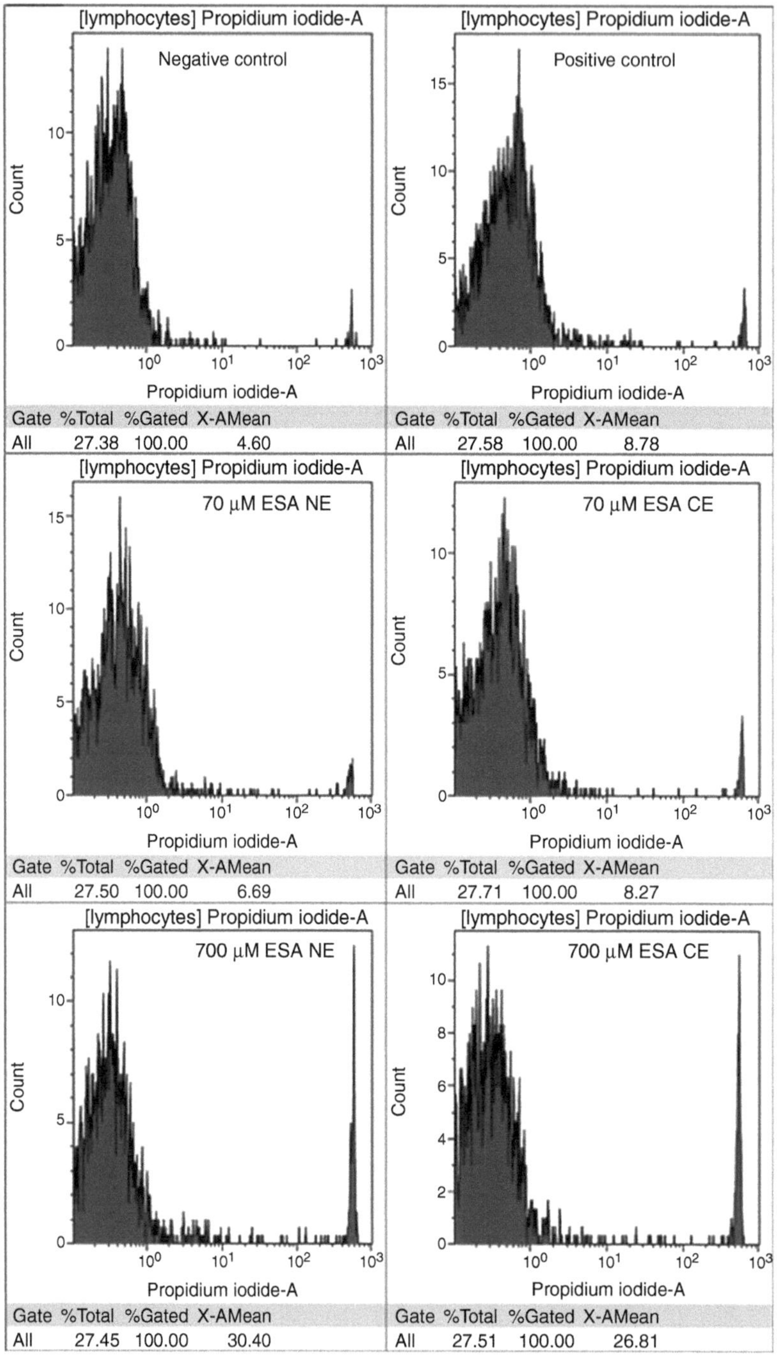

FIGURE 7.11 Therapeutic Implications of a Nanoformulation of ESA NE Compared to ESA Conventional Emulsion ▶

Electronically gated lymphocytes from a 100% population of 5×10^5 PBMCs (per well) by propidium iodide staining due to 120 min exposure to 5 μg/mL LPS exposure in variously treated marked groups (analysis for each group was carried out in triplicates), with reflected results indicating statistical significance of $P < 0.05$ for every group; negative control refers to the control group (of lymphocytes) that was not challenged with 5 μg/mL LPS to

5 RISK AND HAZARD ASSESSMENT

The chemical composition influences the toxicity ramifications of any formulation. It is thus of pivotal importance to assess the potential hazards of every excipient and byproduct involved at each step in the fabrication of the intended nano- and microstructures. Previous reports on the evaluation of nanostructures suggest a risk of greater damage to epithelial cells and soft tissue (Xia et al., 2004), as compared to conventional structures. The enhanced surface to volume ratio due to nanosizing not only has the potential to not only amplify the therapeutic index of the intended bioactive moiety, but also the toxicity of the excipients involved, if any. Further, the route of administration (i.e., either orally or parenterally) also impacts therapy differently, based on the bioavailability and metabolic interconversions of the bioactives in systemic circulation. Thus, it becomes all the more important to evaluate minutely the implications of the fabrication and excipients involved on human/humanized models, for clinical relevance. Broadly, the assessment should be more holistic if the study involves:

5.1 ORGANIC SOLVENTS FOR FABRICATION

Solvent evaporation technique contributes to one of the most common techniques involved in the fabrication of micro- and nanoparticles; more so, in the synthesis of hydrophobic lipid nanostructures. However, use of potentially hazardous organic solvents, such as ethyl acetate, dichloromethane, etc., particularly to dissolve biopolymers initially can lead to gradients of toxicity, including oxidative stress (Majuru and Oyewumi, 2009).

Certain nonvolatile solvents, albeit with GRAS status, commonly used in nanofabrication processes, such as polyethylene glycols (PEGs), in injectable medications have been reported to cause acute physiological effects, namely, CNS depression, hyperosmolality, lactic acidosis, intravenous acidosis, nephrotoxicity, seizures, and even coma in sensitive subjects, particularly among infants and the elderly (Varon and Marik, 1998; Wilson et al., 2000). Thus, adequate studies should be done as clinical studies prior to standardizing nanoformulations comprising such solvents of lipids for injectable applications.

Ekwall and Sandstrom (1978) have reported that use of organic solvents, such as dichloromethane, *t*-butanol, *n*-butanol, benzene, leads to cell detachment as a sigmoid function, and creates acute cytotoxicity in epithelial cells, fibroblasts, and lymphocytes. Using simple in vitro tests may be used to predict acute toxicity of certain organic solvents (Smith et al., 1963).

◀ determine natural death under culture conditions; positive control refers to the group challenged with 5 μg/mL LPS to determine death induced by the pathogenic mitogen under culture; 70 μM ESA NE refers to the positive control group preincubated with 70 μM ESA NE before being challenged with 5 μg/mL LPS to determine its impact on cell death induced by the pathogenic mitogen under culture; 70 μM ESA CE refers to the positive control group preincubated with 70 μM ESA CE before being challenged with 5 μg/mL LPS to determine its impact on cell death induced by the pathogenic mitogen under culture; 700 μM ESA NE refers to the positive control group preincubated with 700 μM ESA NE before being challenged with 5 μg/mL LPS to determine its impact on cell death induced by the pathogenic mitogen under culture; 700 μM ESA CE refers to the positive control group preincubated with 700 μM ESA CE before being challenged with 5 μg/mL LPS to determine its impact on cell death induced by the pathogenic mitogen under culture.

Data reproduced from Paul, D., Mukherjee, S., Chakraborty, R., Mallick, S.K., Dhar, P., 2015. Comparative real-time study of cellular uptake of a formulated conjugated linolenic acid rich nano and conventional macro emulsions and their bioactivity in ex vivo models for parenteral applications. Colloids Surf. B Biointerfaces 126, 426–436.

5.2 POTENTIALLY HAZARDOUS STABILIZERS AND SURFACTANTS

By definition, surfactants have surface active properties with degreasing properties. Being amphipathic in nature, they reduce interfacial tension between hydrophobic core and aqueous environment, or vice versa. In the fabrication of solid lipid nanoparticles, in vitro cytotoxicity is directly proportional to the surfactant used in the stabilization of nanoparticles (Olbrich et al., 2004).

Dose-dependent studies by Song et al. (2012) have shown that surfactants in formulations, such as polyoxyethylene, allow amine, polyoxyethylene lauryl amine ether, and polyoxypropylene glycol block copolymer are much more cytotoxic than formulations comprising polyethylene glycol, polypropylene glycol, and monoethylene glycol at the same concentrations.

As a corollary, the concentration and type of surfactant involved in a formulation for nanostructures intended for health applications should always be considered with utmost caution.

5.3 ENCAPSULATING POLYMERS

One of the primary encapsulation polymers routinely used for the fabrication of nano- and microparticles involves poly(D,L-lactide-*co*-glycolide) or PLGA. However, it has been seen that the increase in concentration of such polymers in formulating nanostructures often generates significant cytotoxicity (Katsikari et al., 2009). It has been reported that IC_{50} concentrations of PLGA nanoparticles based on particle size, charge, surface properties, and even experimental models/methodology significantly altered cellular viabilities (Davda and Labhasetwar, 2002). However, the exact mechanism of such manifestations of PLGA in inducing cellular toxicity are as yet not clear.

However, reports by Kim et al. (2003) have indicated that not all encapsulating polymers are potentially toxic. Nanoparticles (>200 nm) composed of methoxypoly(ethylene glycol)/poly(ε-caprolactone) had lower in vitro cytotoxicity than free drugs. In vivo acute cytotoxicity studies with the fabricated nanoparticles also did not evince any toxicity to the soft tissues and organs (Kim et al., 2003).

6 CHALLENGES AND SCOPE

PUFAs present themselves as excellent bioactive mediators for various therapeutic purposes. They incorporate inside lipid bilayers readily, and form important biochemical precursors and energy sources (Ojima, 2007). Molecular docking studies have also corroborated the role of PUFAs at molecular level which interact both directly and indirectly with factors responsible in improving health and overcome disease etiology (Vetrivel et al., 2012). At the same time, the role of nanotechnology in the field of targeted delivery of prophylactic and therapeutic agents is also reaching a pivotal point. It has already been proven that PUFAs in conjugation with specific ligands and as prodrugs can optimize activity inside cellular systems by modulating parameters, even at genomic level. In vitro, in silico, and in situ studies provide empirical and verifiable evidences of molecular mechanisms to prove the health promoting effects of PUFAs. However, one of the common impeding parameters to translate these health promoting qualities of all natural drugs at clinical level involves the inability of such compounds to cross the BBB. One of the established solutions thus involves the fabrication of nanoformulations of such natural compounds with or without involving specific ligands, to overcome the BBB and other physiological barriers, thereby facilitating cure against diseases. PUFAs are also part of this initiative to be formulated into nanocompounds duly involving multitude of parameters, technology, and toxicological considerations prior to their subsequent applications. It is thus only prudent to discuss the merits and challenges of this most upcoming initiative.

A lot of physicochemical, energetic, and toxicological considerations are always taken into account prior to development of an encapsulation methodology protocol for bioactives. The most important factor in the inventory of these considerations involves excipient analysis. Since excipients get shortlisted only after factoring in equilibrium partition coefficients, original location, mass transfer coefficients in different phases, mechanical agitation, particle size, particle degradation, surface charge and layer thickness, etc., all of which determine the outcome of release profile of the encapsulated product (McClements, 2012a,b; McClements and Xiao, 2012; Nel et al., 2009). PUFAs are considered as lipophilic nutraceuticals, which offer an interesting alternative as encapsulated cargo for targeted delivery to bring about desired health benefits. However, every PUFA, owing to its specific physicochemical properties, such as water solubility, chemical degradation, oil–water partition coefficient, melting point and crystal morphology, offers unique challenges for nano- and microencapsulation for targeted delivery (Arab-Tehrany et al., 2012). Additionally, PUFAs, being long chain hydrophobic molecules, have lower water solubility, and therefore need to be mixed with an oil phase prior to processing in the form of stable colloidal delivery systems. PUFAs are also highly susceptible to lipid oxidation, rendering them unfeasible for long-term storage. However, these impediments can be countered by the inclusion of antioxidants, reduced exposure to light, interfacial engineering, and storage at reduced temperatures.

7 CONCLUSIONS

Advancement in this field of colloidal nanodelivery system offers a very promising direction to successfully incorporate PUFA-rich systems as nanoformulations in the form of emulsions, multilayer emulsions, multiple emulsions, and filled hydrogel particles using nontoxic excipients (Berton-Carabin et al., 2013; Djordjevic et al., 2007). Refined vegetable and fish lipid oils that are naturally rich in important PUFA(s) can provide for the oil-phase, and as carriers for other important hydrophobic taxoids, taxanes, flavonoids, minerals, and drugs in such nanoformulations, thereby providing enhanced functionality due to synergistic effect(s) of both PUFA and the encapsulated molecule against ailments. This field, however, requires further research to understand the implications of PUFA-rich nanoformulations at a clinical level which is hitherto lacking in practice.

ACKNOWLEDGMENTS

The authors gratefully acknowledge the help rendered by Miss Amrita Chakraborty in the final drafting of the manuscript.

REFERENCES

Anton, N., Vandamme, T.F., 2011. Nano-emulsions and micro-emulsions: clarifications of the critical differences. Pharm. Res. 28, 978–985.

Arab-Tehrany, E., Jacquot, M., Gaiani, C., Imran, M., Desobry, S., Linder, M., 2012. Beneficial effects and oxidative stability of omega-3 long-chain polyunsaturated fatty acids. Trends Food Sci. Technol. 25, 24–33.

Arrieta, M.C., Bistritz, L., Meddings, J.B., 2006. Recent advances in clinical practice alterations in intestinal permeability. Gut 55, 1512–1520.

Attama, A.A., Momoh, M.A., Builders, P.F., 2012. Lipid nanoparticulate drug delivery systems: a revolution in dosage form design and development. In: Sezer, A.D. (Ed.), Recent Advances in Novel Drug Carrier systems. InTech Open Access Publisher, Rijeka, Croatia.

Babbar, N., Oberoi, H.S., Sandhu, S.K., 2015. Therapeutic and nutraceutical potential of bioactive compounds extracted from fruit residues. Crit. Rev. Food Sci. Nutr. 55, 319–337.

Balakrishnan, P., Lee, B.J., Oh, D.H., 2009. Enhanced oral bioavailability of Coenzyme Q10 by self-emulsifying drug delivery systems. Int. J. Pharm. 374, 66–72.

Berton-Carabin, C., Genot, C., Gaillard, C., Guibert, D., Ropers, M.H., 2013. Design of interfacial films to control lipid oxidation in oil-in-water emulsions. Food Hydrocoll. 33, 99–105.

Bhattacharya, A., Banu, J., Rahman, M., Causey, J., Fernandes, G., 2006. Biological effects of conjugated linoleic acids in health and disease. J. Nutr. Biochem. 17, 789–810.

Bradley, M.O., Webb, N.L., Anthony, F.H., Devanesan, P., Witman, P.A., Hemamalini, S., Chander, M.C., Baker, S.D., He, L., Horowitz, S.B., Swindell, C.S., 2001. Tumor targeting by covalent conjugation of a natural fatty acid to taxol. Clin. Cancer Res. 7, 3229–3238.

Calle, D., Negri, V., Ballesteros, P., Cerdán, S., 2015. Magnetoliposomes loaded with poly-unsaturated fatty acids as novel theranostic anti-inflammatory formulations. Theranostics 5, 489–503.

Cansell, M., Nacka, F., Combe, N., 2003. Marine lipid-based liposomes increase in vivo FA availability. Lipids 38, 551–559.

Carvalho, E.B.T., Melo, I.L.P., Mancini-Filho, J., 2010. Chemical and physiological aspects of isomers of conjugated fatty acids. Ciênc. Tecnol. Aliment. 30, 295–307.

Chari, R.V.J., 1998. Targeted delivery of chemotherapeutics: tumor activated prodrug (TAP) therapy. Adv. Drug Deliv. 31, 89–104.

Chen, J., Jaracz, S., Zhao, X., Chen, S., Ojima, I., 2005. Antibody-based toxin conjugates for cancer therapy. Exp. Opin. Drug Deliv. 2, 873–890.

Chiu, Y.L., Ho, Y.C., Chen, Y.M., Peng, S.F., Ke, C.J., Chen, F.L., Mi, K.J., Sung, H.W., 2010. The characteristics, cellular uptake and intracellular trafficking of nanoparticles made of hydrophobically-modified chitosan. J. Control. Release 146, 152–159.

Choy, J.H., Shin, J., Lim, S.Y., Oh, J.M., Oh, M.H., Oh, S., 2010. Characterization and stability analysis of zinc oxide nanoencapsulated conjugated linoleic acid. J. Food Sci. 75, 63–68.

Chu, T.C., Marks, III, J.W., Lavery, L.A., Faulkner, S., Rosenblum, M.G., Ellington, A.D., Levy, M., 2006. Aptamer toxin conjugates that specifically target prostate tumor cells. Cancer Res. 66, 5989–5992.

Chung, C., Degner, B., Decker, E.A., McClements, D.J., 2013. Oil-filled hydrogel particles for reduced-fat food applications: fabrication, characterization, and properties. Innov. Food Sci. Emerg. Technol. 20, 324–334.

Coldren, B., Boyer, C., Kisak, E.T., Evans, C.A., Zasadzinski, J.A., 2004. The vesosome—a multicompartment drug delivery vehicle. Curr. Med. Chem. 11, 199–219.

Conner, S.D., Schmid, S.L., 2003. Regulated portals of entry into the cell. Nature 422, 37–44.

Coussens, L.M., Werb, Z., 2002. Inflammation and cancer. Nature 420, 860–867.

Couvreur, P., Dubernet, C., Puisieux, F., 1995. Controlled drug delivery with nanoparticles: current possibilities and future trends. Eur. J. Pharm. Biopharm. 41, 2–13.

Coviello, T., Matricardi, P., Marianecci, C., Alhaique, F., 2007. Polysaccharide hydrogels for modified release formulations. J. Control. Release 119, 5–24.

Crone, C., Olesen, S.P., 1982. Electrical resistance of brain microvascular endothelium. Brain Res. 241, 49–55.

Davda, J., Labhasetwar, V., 2002. Characterization of nanoparticle uptake by endothelial cells. Int. J. Pharm. 233, 51–59.

Deamer, D.W., 2010. From "banghasomes" to liposomes: a memoir of Alec Bangham, 1921–2010. FASEB J. 24, 1308–1310.

Desai, A., Vyas, T., Amiji, M., 2008. Cytotoxicity and apoptosis enhancement in brain tumor cells upon coadministration of paclitaxel and ceramide in nanoemulsion formulations. J. Pharm. Sci. 97, 2745–2756.

Desbois, A.P., Smith, V.J., 2010. Antibacterial free fatty acids: activities, mechanisms of action and biotechnological potential. Appl. Microbiol. Biotechnol. 85, 1629–1642.

Deshpande, D., Janero, D.R., Amiji, M.M., 2013. Engineering of an ω-3 polyunsaturated fatty acid-containing nanoemulsion system for combination C6-ceramide and 17β-estradiol delivery and bioactivity in human vascular endothelial and smooth muscle cells. Nanomed. Nanotechnol. Biol. Med. 9, 885–894.

Dey, T.K., Ghosh, S., Ghosh, M., Koley, H., Dhar, P., 2012. Comparative study of gastrointestinal absorption of EPA & DHA rich fish oil from nano and microemulsion formulation in rats. Food Res. Int. 49, 72–79.

Djordjevic, D., Cercaci, L., Alamed, J., McClements, D.J., Decker, E.A., 2007. Chemical and physical stability of citral and limonene in sodium dodecyl sulfate-chitosan and gum arabic-stabilized oil-in-water emulsions. J. Agric. Food Chem. 55, 3585–3591.

Egleton, R.D., Davis, T.P., 1997. Bioavailability and transport of peptides and peptide drugs into the brain. Peptides 18, 1431–1439.

Ekwall, B., Sandstrom, B., 1978. Combined toxicity to HeLa cells of 30 drug pairs, studied by a two-dimensional microtitre method. Toxicol. Lett. 2, 285–292.

Evans, J.L., Goldfine, I.D., Maddux, B.A., Grodsky, G.M., 2002. Oxidative stress and stress-activated signaling pathways: a unifying hypothesis of type 2 diabetes. Endocr. Rev. 23, 599–622.

Fanun, M., 2008. Microemulsions: Properties and Applications. CRC Press, Boca Raton, FL.

Florence, A.T., 2005. Nanoparticle uptake by the oral route: fulfilling its potential? Drug. Discov. Today. Technol. 2, 75–81.

Ganta, S., Amiji, M., 2009. Co administration of paclitaxel and curcumin in nanoemulsion formulations to overcome multidrug resistance in tumor cells. Mol. Pharm. 6, 928–939.

Grammatikos, S.I., Subbaiah, P.V., Victor, T.A., Miller, W.M., 1994. *n*-3 and *n*-6 fatty acid processing and growth effects in neoplastic and non-cancerous human mammary epithelial cell lines. Br. J. Cancer 70, 219–227.

Gu, Y.S., Decker, A.E., McClements, D.J., 2005. Production and characterization of oil-in-water emulsions containing droplets stabilized by multilayer membranes consisting of beta-lactoglobulin, iota-carrageenan and gelatin. Langmuir 21, 5752–5760.

Guzey, D., Kim, H.J., McClements, D.J., 2004. Factors influencing the production of o/w emulsions stabilized by beta-lactoglobulin-pectin membranes. Food Hydrocoll. 18, 967–975.

Guzey, D., McClements, D.J., 2006. Formation, stability and properties of multilayer emulsions for application in the food industry. Adv. Colloid Interface Sci. 128, 227–248.

Hamaan, P.R., Hinman, L.M., Hollander, I., Beyer, C.F., Lindh, D., Holcomb, R., Hallett, W., Tsou, H.H., Upeslacis, J., Shochat, D., Mountain, A., Flowers, D.A., Bernstein, I., 2002. Gemtuzumab Ozogamicin, a potent and selective anti-CD33 antibody calicheamicin conjugate for treatment of acute myeloid leukemia. Bioconjug. Chem. 13, 47–58.

Harivardhan, R.L., Sharma, R.K., Chuttani, K., Mishra, A.K., Murthy, R.S., 2005. Influence of administration route on tumor uptake and biodistribution of etoposide loaded solid lipid nanoparticles in Dalton's lymphoma tumor bearing mice. J. Control. Release 105, 185–198.

Harush-Frenkel, O., Rozentur, E., Benita, S., Altschuler, Y., 2008. Surface charge of NPs determines their endocytic and transcytotic pathway in polarized MDCK cells. Biomacromolecules 9, 435–443.

Holmberg, K. (Ed.), 2003. Novel Surfactants: Preparation Applications and Biodegradability, Revised and Expanded. CRC Press, Boca Raton, FL, 114.

Hope, M.J., Bally, M.B., Webb, G., Cullis, P.R., 1985. Production of large unilamellar vesicles by a rapid extrusion procedure. Characterization of size distribution, trapped volume and ability to maintain a membrane potential. Biochim. Biophys. Acta 812, 55–65.

Huang, C.H., 1969. Studies on phosphatidylcholine vesicles: formation and physical characteristics. Biochemistry 8, 344–352.

Huang, G., Zhang, N., Bi, X., Dou, M., 2008. Solid lipid nanoparticles of temozolomide: potential reduction of cardial and nephric toxicity. Int. J. Pharm. 1, 314–320.
Iwase, Y., Kamei, N., Takeda Morishita, M., 2015. Antidiabetic effects of omega-3 polyunsaturated fatty acids: from mechanism to therapeutic possibilities. Pharmacol. Pharm. 6, 190–200.
Jaracz, S., Chen, J., Kuznetsova, L.V., Ojima, I., 2005. Recent advances in tumor-targeting anticancer drug conjugates. Bioorg. Med. Chem. 13, 5043–5054.
Jarboe, L.R., Royce, L.A., Liu, P., 2013. Understanding biocatalyst inhibition by carboxylic acids. Front. Microbiol. 4, 272.
Johannes, L., Lamaze, C., 2002. Clathrin-dependent or not: is it still the question? Traffic 3, 443–451.
Jonsson, B., Lindman, B., Holmberg, K., Kronberg, B., 1998. Surfactants & Polymers in Aqueous Solutions. John Wiley & Sons, Chichester, UK.
Katsikari, A., Patronidou, C., Kiparissides, C., Arsenakis, M., 2009. Uptake and cytotoxicity of poly (D,L-lactide-*co*-glycolide) nanoparticles in human colon adenocarcinoma cells. J. Nanosci. Nanotechnol. 165, 160–164.
Khalil, I.A., Kogure, K., Akita, H., Harashima, H., 2006. Uptake pathways and subsequent intracellular trafficking in novel gene delivery. Pharmacol. Rev. 58, 32–45.
Kim, S.Y., Lee, Y.M., Baik, D.J., Kang, J.S., 2003. Toxic characteristics of methoxypoly(ethyleneglycol)/poly(ε-caprolactone) nanospheres; in vitro and in vivo studies in the normal mice. Biomaterials 24, 55–63.
Kizelsztein, P., Ovadia, H., Garbuzenko, O., Sigal, A., Barenholz, Y., 2009. Pegylated nano liposomes remote-loaded with the antioxidant tempamine ameliorate experimental autoimmune encephalomyelitis. J. Neuroimmunol. 213, 20–25.
Kommuru, T.R., Gurley, B., Khan, M.A., Reddy, I.K., 2001. Self-emulsifying drug delivery systems (SEDDS) of coenzyme Q10: formulation development and bioavailability assessment. Int. J. Pharm. 212, 233–246.
Konan, Y.N., Gurny, R., Allémann, E., 2002. Preparation and characterization of sterile and freeze-dried sub-200 nm nanoparticles. Int. J. Pharm. 233, 239–252.
Kris-Etherton, P., Hecker, K., Binkoski, A., 2004. Polyunsaturated fatty acids and cardiovascular health. Nutr. Rev. 62, 414–426.
Kumar, M., Misra, A., Babbar, A.K., Mishra, A.K., Mishra, P., Pathak, K., 2008. Intranasal nanoemulsion based brain targeting drug delivery system of risperidone. Int. J. Pharm. 358, 285–291.
Kumar, R., Lal, S., 2014. Synthesis of organic nanoparticles and their applications in drug delivery and food nanotechnology: a review. J. Nanomater. Mol. Nanotechnol. 3, 4.
Kuznetsova, L.V., Chen, J., Sun, L., Wu, X., Pepe, A., Veith, J.M., Pera, P., Bernacki, R.J., Ojima, I., 2006. Synthesis and evaluation of novel fatty acid-2nd generation taxoid conjugates as promising anticancer agents. Bioorg. Med. Chem. Lett. 16, 974–977.
Leamon, C.P., Reddy, J.A., 2004. Folate-targeted chemotherapy. Adv. Drug Deliv. Rev. 56, 1127–1141.
Leser, M.E., Sagalowicz, L., Michel, M., Watzke, H.J., 2006. Self-assembly of polar food lipids. Adv. Colloid Interface Sci. 123, 125–136.
Lin, C.C., Metters, A.T., 2006. Hydrogels in controlled release formulations: network design and mathematical modeling. Adv. Drug Deliv. Rev. 58, 1379–1408.
Lopez, A., Gavara, R., Lagaron, J., 2006. Bioactive packaging: turning foods into healthier foods through biomaterials. Trends Food Sci. Technol. 17, 567–575.
Lu, Y., Low, P.S., 2002. Folate-mediated delivery of macromolecular anticancer therapeutic agents. Adv. Drug Deliv. Rev. 54, 675–693.
Luo, Y., Bernshaw, N.J., Lu, Z.R., Kopecek, J., Prestwich, G.D., 2002. Targeted delivery of doxorubicin by HPMA copolymer–hyaluronan bioconjugates. Pharm. Res. 19, 396–402.
Majuru, S., Oyewumi, M.O., 2009. Nanotechnology in drug development and life cycle management. In: de Villiers, M.M., Aramwit, P., Kwon, G.S. (Eds.), Nanotechnology in Drug Delivery. Springer, New York, NY.
Mason, T.G., Wilking, J.N., Meleson, K., Chang, C.B., Graves, S.M., 2006. Nanoemulsions: formation, structure and physical properties. J. Phys. Condens. Matter. 18, R635–R666.

Mayer, L.D., Hope, M.J., Cullis, P.R., 1986. Vesicles of variable sizes produced by a rapid extrusion procedure. Biochim. Biophys. Acta 858, 161–168.

McClements, D.J., 2012a. Advances in fabrication of emulsions with enhanced functionality using structural design principles. Curr. Opin. Colloid Interface Sci. 17, 235–245.

McClements, D.J., 2012b. Crystals and crystallization in oil-in-water emulsions: Implications for emulsion-based delivery systems. Adv. Colloid Interface Sci. 174, 1–30.

McClements, D.J., 2012c. Nanoemulsions versus microemulsions: terminology, differences, and similarities. Soft Matter 8, 1719–1729.

McClements, D.J., Xiao, H., 2012. Potential biological fate of ingested nanoemulsions: influence of particle characteristics. Food Funct. 3, 202–220.

Mehnert, W., Mader, K., 2001. Solid lipid nanoparticles. Production, characterization and applications. Adv. Drug Deliv. Rev. 47, 165–196.

Miclea, R.D., Varma, P.R., Peng, A., Balu-Iyer, S.V., 2007. Development and characterization of lipidic cochleate containing recombinant factor VIII. Biochim. Biophys. Acta 1768, 2890–2898.

Moreau, L., Kim, H.J., Decker, E.A., McClements, D.J., 2003. Production and characterization of oil-in-water emulsions containing droplets stabilized by beta-lactoglobulin-pectin membranes. J. Agric. Food Chem. 51, 6612–6617.

Moussaoui, N., Cansell, M., Denizot, A., 2002. Marinosomes, marine lipid-based liposomes: physical characterization and potential application in cosmetics. Int. J. Pharm. 242, 361–365.

Mozafari, M.R., Johnson, C., Hatziantoniou, S., Demetzos, C., 2008. Nano-liposomes and their applications in food technology. J. Liposome Res. 18, 309–327.

Muller, R.H., Radtke, M., Wissing, S.A., 2002. Solid lipid nano-particles (SLN) and nanostructured lipid carriers (NLC) in cosmetic and dermatological preparations. Adv. Drug Deliv. Rev. Suppl. 54, S15–S131.

Muller, R.H., Radtke, M., Wissing, S.A., 2012. Nanostructured lipid matrices for improved microencapsulation of drugs. Int. J. Pharm. 242, 121–128.

Murakawa, K., Fukunaga, K., Tanouchi, M., Hosokawa, M., Hossain, Z., Takahashi, K., 2007. Therapy of myeloma in vivo using marine phospholipid in combination with *Agaricus blazei* Murill as an immunerespond activator. J. Oleo Sci. 56, 179–188.

Nacka, F., Cansell, M., Meleard, P., Combe, N., 2001. Incorporation of alpha-tocopherol in marine–lipid based liposome in vitro and in vivo studies. Lipids 36, 1313–1320.

Nagy, A., Schally, A.V., Halmos, G., Armatis, P., Cai, R.Z., Csernus, V., Kovacs, M., Koppan, M., Szepeshazi, K., Kahan, Z., 1998. Synthesis and biological evaluation of cytotoxic analogs of somatostatin containing doxorubicin or its intensely potent derivative, 2-pyrrolinodoxorubicin. Proc. Natl. Acad. Sci. USA 95, 1794–1799.

Nam, H.Y., Kwon, S.M., Chung, H., Lee, S.Y., Kwon, S.H., Jeon, H., Kim, Y., Park, J.H., Kim, J., Her, S., Oh, Y.K., Kwon, I.C., Kim, K., Jeong, S.Y., 2009. Cellular uptake mechanism and intracellular fate of hydrophobically modified glycol chitosan nanoparticles. J. Control. Release 135, 259–267.

Nel, A.E., Mädler, L., Velego, D., Xia, T., Hoek, E.M., Somasundaran, P., Klaessig, F., Castranova, V., Thompson, M., 2009. Understanding biophysicochemical interactions at the nano-bio interface. Nat. Mater. 8, 543–557.

Obonyo, M., Sabet, M., Cole, S.P., Ebmeyer, J., Uematsu, S., Akira, S., Guiney, D.G., 2007. Deficiencies of myeloid differentiation factor 88, Toll-like receptor 2 (TLR2), or TLR4 produce specific defects in macrophage cytokine secretion induced by *Helicobacter pylori*. Infect. Immun. 75, 2408–2414.

Ojima, I., 2007. Guided molecular missiles for tumor-targeting chemotherapy—case studies using the second-generation taxoids as warheads. Acc. Chem. Res. 41, 108–119.

Ojima, I., Geng, X., Wu, X., Qu, C., Borella, C.P., Xie, H., Wilhelm, S.D., Leece, B.A., Bartle, L.M., Goldmacher, V.S., Chari, R.V.J., 2002. Tumor-specific novel taxoid monoclonal antibody conjugates. J. Med. Chem. 45, 5620–5623.

Olbrich, C., Scholer, N., Tabatt, K., Kayser, O., Muller, R.H., 2004. Cytotoxicity studies of Dynasan 114 solid lipid nanoparticles (SLN) on RAW264.7 macrophages impact of phagocytosis on viability and cytokine production. J. Pharm. Pharmacol. 56, 883–891.

Olson, F., Hunt, C.A., Szoka, F.C., Vail, W.J., Papahadjopoulos, D., 1979. Preparation of liposomes of defined size distribution by extrusion through polycarbonate membranes. Biochim. Biophys. Acta 557, 9–23.

Panyam, J., Labhasetwar, V., 2003. Biogradable nanoparticles for drug and gene delivery to cells and tissue. Adv. Drug Deliv. Rev. 55, 329–347.

Paul, D., Dey, T.K., Mukherjee, S., Ghosh, M., Dhar, P., 2014. Comparative prophylactic effects of α-eleostearic acid rich nano and conventional emulsions in induced diabetic rats. J. Food Sci. Technol. 51, 1724–1736.

Paul, D., Mukherjee, S., Chakraborty, R., Mallick, S.K., Dhar, P., 2015. Comparative real-time study of cellular uptake of a formulated conjugated linolenic acid rich nano and conventional macro emulsions and their bioactivity in ex-vivo models for parenteral applications. Colloids Surf. B Biointerfaces 126, 426–436.

Peppas, N.A., Khare, A.R., 1993. Preparation, structure and diffusional behavior of hydrogels in controlled release. Adv. Drug Deliv. Rev. 11, 1–35.

Qian, C., Decker, E.A., Xiao, H., McClements, D.J., 2012. Inhibition of beta-carotene degradation in oil-in-water nanoemulsions: influence of oil-soluble and water-soluble antioxidants. Food Chem. 135, 1036–1043.

Qiu, Y., Liu, Y., Wang, L., Xu, L., Bal, R., Ji, Y., Wu, X., Zhao, Y., Li, Y., Chen, C., 2010. Surface chemistry and aspect ratio mediated cellular uptake of Au nanorods. Biomaterials 31, 7606–7619.

Raatz, S.K., Redmon, J.B., Wimmergren, N., Donadio, J.V., Bibus, D.M., 2009. Enhanced absorption of *n*-3 fatty acids from emulsified compared with encapsulated fish oil. J. Am. Diet Assoc. 109, 1076–1081.

Rad, R., Brenner, L., Krug, A., Voland, P., Mages, J., Lang, R., Schwendy, S., Reindl, W., Dossumbekova, A., Ballhorn, W., Wagner, H., Schmid, R.M., Bauer, S., Prinz, C., 2007. Toll-like receptor-dependent activation of antigen-presenting cells affects adaptive immunity to *Helicobacter pylori*. Gastroenterology 133, 150–163.

Radtke, M., Muller, R.H., 2001. Stability study of creams containing cyclosporine SLN. Int. Symp. Control. Release Bioact. Mater. 28, 472–473.

Rao, J.J., Decker, E.A., Xiao, H., McClements, D.J., 2013. Nutraceutical nanoemulsions: influence of carrier oil composition (digestible versus indigestible oil) on β-carotene bioavailability. J. Sci. Food Agric. 93, 3175–3183.

Rasti, B., Jinap, S., Mozafari, M.R., Manap, Y.A., 2012. Comparative study of the oxidative and physical stability of liposomal and nanoliposomal polyunsaturated fatty acids prepared with conventional and Mozafari methods. Food Chem. 135, 2761–2770.

Reineccius, G., 2007. Flavor Chemistry and Technology. CRC Press, Boca Raton, FL.

Saha, S.S., Ghosh, M., 2012. Antioxidant and anti-inflammatory effect of conjugated linolenic acid isomers against streptozotocin-induced diabetes. Br. J. Nutr. 108, 974–983.

Salama, N.N., Eddington, N.D., Fasano, A., 2006. Tight junction modulation and its relationship to drug delivery. Adv. Drug Deliv. Rev. 58, 15–28.

Salminen, H., Helgason, T., Kristinsson, B., Kristbergsson, K., Weiss, J., 2013. Formation of solid shell nanoparticles with liquid ω-3 fatty acid core. Food Chem. 141, 2934–2943.

Sarker, D.K., 2005. Engineering of nanoemulsion for drug delivery. Curr. Drug Deliv. 2, 297–310.

Sauer, L.A., Dauchy, R.T., 1992. The effect of w-6 and w-3 fatty acids on 3H-thymidine incorporation in hepatoma 7288CTC perfused in situ. Br. J. Cancer 66, 297–303.

Sauer, L.A., Dauchy, R.T., Blask, D.E., 2000. Mechanism for the antitumor and anticachectic effects of *n*-3 fatty acids. Cancer Res. 60, 5289–5295.

Scalia, S., Mezzena, M., 2009. Incorporation of quercetin in lipid microparticles: effect on photo-and chemical-stability. J. Pharm. Biomed. Anal. 49, 90–94.

Schnyder, A., Huwyler, J., 2005. Drug transport to brain with targeted liposomes. NeuroRx 2, 99–107.

Selvamuthukumar, S., Velmurugan, R., 2012. Nanostructured lipid carriers: a potential drug carrier for cancer chemotherapy. Lipids Health Dis. 11, 159.

Sengupta, A., Sen Gupta, S., Ghosh, M., 2013a. Optimization of process parameters in preparation of nanoemulsions of CLnA rich oil by response surface methodology. J. Inst. Eng. India Ser. 94, 23–28.

Sengupta, A., Sen Gupta, S., Nandi, I., Ghosh, M., 2013b. Conjugated linolenic acid nanoparticles inhibit hypercholesterolemia induced by feeding a high-fat diet in male albino rats. J. Food Sci. Technol. 52, 458–464.

Smith, C.G., Grady, J.E., Northam, J.I., 1963. Relationship between cytotoxicity in vitro and whole animal toxicity. Cancer Chemother. Rep. 30, 9.

Song, H.Y., Kim, Y.H., Seok, S.J., Gil, Y.W., Yang, J.O., Lee, E.Y., Hong, S.Y., 2012. Cellular toxicity of surfactants used as herbicide additives. J. Korean Med. Sci. 27, 3–9.

Sonneville-Aubrun, O., Simonnet, J.T., L'Alloret, F., 2004. Nanoemulsions: a new vehicle for skincare products. Adv. Colloid Interface Sci. 108–109, 145–149.

Szmydynger-Chodobska, J., Strazielle, N., Zink, B.J., Ghersi-Egea, J.F., Chodobski, A., 2009. The role of the choroid plexus in neutrophil invasion after traumatic brain injury. J. Cereb. Blood Flow Metab. 29, 1503–1516.

Tadros, T., Izquierdo, P., Esquena, J., Solans, C., 2004. Formation and stability of nano-emulsions. Adv. Colloid Interface Sci. 108–109, 303–318.

Tan, Y., Liu, C., 2009. Self-aggregated nanoparticles from linoleic acid modified carboxymethyl chitosan: synthesis, characterization and application in vitro. Colloids Surf. B Biointerfaces 69, 178–182.

Taylor, T.M., Davidson, P.M., Bruce, B.D., Weiss, J., 2005. Liposomal nanocapsules in food science and agriculture. Crit. Rev. Food Sci. Nutr. 45, 587–605.

Thamphiwatana, S., Gaoa, W., Obonyob, M., Zhanga, L., 2014. In vivo treatment of *Helicobacter pylori* infection with liposomal linolenic acid reduces colonization and ameliorates inflammation. Proc. Natl. Acad. Sci. USA 111, 17600–17605.

Thompkinson, D.K., Bhavana, V., Kanika, P., 2014. Dietary approaches for management of cardio-vascular health—a review. J. Food Sci. Technol. 51, 2318–2330.

Tiwari, S.B., Amiji, M.M., 2006. Improved oral delivery of paclitaxel following administration in nanoemulsion formulations. J. Nanosci. Nanotechnol. 6, 15–21.

Tiwari, S., Tan, Y., Amiji, M., 2006. Preparation and in vitro characterization of multifunctional nanoemulsions for simultaneous MR imaging and targeted drug delivery. J. Biomed. Nanotechnol. 2, 1–8.

Tsuzuki, T., Kawakami, Y., Abe, R., Nakagawa, K., Koba, K., Imamura, J., Iwata, T., Ikeda, I., Miyazawa, T., 2006. Conjugated linolenic acid is slowly absorbed in rat intestine, but quickly converted to conjugated linoleic acid. J. Nutr. 136, 2153–2159.

Uauy, R., Mena, P., Wegher, B., Nieto, S., Salem, Jr., N., 2000. Long chain poly-unsaturated fatty acid formation in neonates: effect of gestational age and intrauterine growth. Pediatr. Res. 47, 127–135.

Varon, J., Marik, P., 1998. Etomidate and propylene glycol toxicity. J. Emerg. Med. 3, 485.

Vetrivel, U., Ravichandran, S.B., Kuppan, K., Mohanlal, J., Das, U.N., Narayanasamy, A., 2012. Agonistic effect of polyunsaturated fatty acids (PUFAs) and its metabolites on brain-derived neurotrophic factor (BDNF) through molecular docking simulation. Lipids Health Dis. 11, 109.

Vyas, T.K., Shahiwala, A., Amiji, M.M., 2008. Improved oral bioavailability and brain transport of saquinavir upon administration in novel nanoemulsion formulations. Int. J. Pharm. 347, 93–101.

Wang, J.X., Sun, X., Zhang, Z.R., 2002. Enhanced brain targeting by synthesis of 3′, 5′-dioctanoyl-5-fluoro-2′-deoxyuridine and incorporation into solid lipidnanoparticles. Eur. J. Pharm. Biopharm. 54, 285–290.

Whitehead, K.A., Langer, R., Anderson, D.G., 2009. Knocking down barrier: advances in siRNA drug delivery. Nat. Rev. 8, 129–138.

Williams, C.M., Burdge, G.C., 2006. Long–chain *n*-3 PUFA: plant v. marine sources. Proc. Nutr. Soc. 65, 42–50.

Wilson, K.C., Reardon, C., Farber, H.W., 2000. Propylene glycol toxicity in a patient receiving intravenous diazepam. N. Eng. J. Med. 343, 815–1815.

Won, J., Oh, M.H., Oh, J.M., Kang, M.S., Choy, J.H., Oh, S., 2008. Stability analysis of zinc oxide-nano encapsulated conjugated linoleic acid and gamma-linolenic acid. J. Food Sci. 73, 39–43.

Wu, X., Ojima, I., 2004. Tumor specific novel taxoid-monoclonal antibody conjugates. Curr. Med. Chem. 11, 429–438.

Xia, Y., Yu, H., Jansen, R., Seringhaus, M., Baxter, S., Greenbaum, D., Zhao, H., Gerstein, M., 2004. Analyzing cellular biochemistry in terms of molecular networks. Annu. Rev. Biochem. 73, 1051–1087.

Yamamoto, Y., Gaynor, R.B., 2001. Role of the NF-kB pathway in the pathogenesis of human disease states. Curr. Mol. Med. 1, 287–296.

Yang, S.C., Lu, L.F., Cai, Y., Zhu, J.B., Liang, B.W., Yang, C.Z., 1999. Body distribution n mice of intravenously injected camptothecin solid lipid nanoparticles and targeting effect on brain. J. Control. Release 59, 299–307.

Yu, H.G, Ji, H.G., Woo, H.K., Kim, S.D., 2010. Aqueous nanoemulsion composition containing conjugated linoleic acid. Patent no. EP 2398450A2.

Zarif, L., 2005. Drug delivery by lipid cochleates. Methods Enzymol. 391, 314–329.

Zarif, L., 2006. Nanocochleate cylinders for oral and parenteral delivery of drugs. J. Liposome Res. 13, 109–110.

Zhang, Y., Bent, L.Z., 2001. The gut as a barrier to drug absorption. Clin. Pharmacokinet. 40, 159–168.

Zhigaltsev, I.V., Belliveau, N., Hafez, I., Leung, A.K., Huft, J., Hansen, C., Cullis, P.R., 2012. Bottom-up design and synthesis of limit size lipid nanoparticle systems with aqueous and triglyceride cores using millisecond microfluidic mixing. Langmuir 28, 3633–3640.

CHAPTER

8 ELECTROSPINNING AND SURFACE MODIFICATION METHODS FOR FUNCTIONALIZED CELL SCAFFOLDS

Naruki Kurokawa, Fuyuaki Endo, Tomoki Maeda, Atsushi Hotta
Keio University, Yokohama, Japan

CHAPTER OUTLINE

1 ELECTROSPINNING FOR TISSUE ENGINEERING

Tissue engineering is an emerging discipline for the regeneration of physiological tissue inside and outside the human body. As defined by Langer and Vacanti (1993), tissue engineering is "an interdisciplinary field that applies the principles of engineering and life sciences toward the development of biological substitutes that restore, maintain, or improve tissue function." In tissue engineering, it is necessary to use artificial scaffolds that could function as the support for cells to regenerate the tissue that has possibly been destroyed by injury, disease, or congenital defects. The goal of tissue engineering is to furnish living structures that have the capability to integrate with surrounding tissue (Gomes and Reis, 2004). Also, in the field of tissue engineering, artificial scaffolds imitating the structure of extracellular matrix (ECM) are required (Agarwal et al., 2008). ECM is a supermolecular structure consisting of substances outside the cells in the human body. ECM plays an important role as a framework of tissue, such as a scaffold where cells could adhere and proliferate, as well as a biomaterial that holds and provides growth factors. Simply put, ECM is a "bed" for cells. ECM normally consists of 50–500 nm fibrous structures made of protein materials (Barnes et al., 2007). One of the important aspects of tissue engineering is the design of scaffolds with specific functionalities that would realize architectural and biological properties similar to those of a natural ECM.

The nanofibrous structures fabricated by electrospinning have been expected to be used for a tissue-engineering scaffold (Agarwal et al., 2008). Electrospinning has been recognized as a simple method creating continuous ultrafine polymer fibers from diverse polymer materials. The diameters of these fibers generally range from a few micrometers down to tens of nanometers—the size range that would otherwise be difficult to form using conventional techniques. In fact, electrospun fiber mats can imitate the structure of a natural ECM. Forming ultrafine and uniform fibers could be the key to make nanofibrous structures for tissue engineering scaffolds. In order to fabricate such fine fibers by electrospinning, it is essential to understand the fundamentals of the method to optimize the experimental conditions, and to achieve practical mastery of the electrospinning.

The essential properties when using an electrospun nanofibrous scaffold for tissue engineering are biocompatibility, biodegradability, cellular adhesiveness, and proliferativeness, and these properties can be achieved and improved by careful and appropriate selection of polymeric materials, with the proper control of fibrous structures, and surface modification. Among them, it should be most necessary to select the polymer materials with biocompatibility and biodegradability for electrospun scaffolds. These two properties will lead to greater integration of the manufactured scaffolds into the host tissue, without stimulating any immune response. In addition, in numerous instances, a tissue-engineering scaffold should be biodegradable, so that a second surgical intervention may not be requested to remove the implant. The scaffold should also mimic a natural ECM architecturally, in order to allow cell attachment and proliferation. The structures of the electrospun scaffolds can be characterized by fiber diameter, fiber orientation, and pore size. Tuning these structural parameters can provide the electrospun scaffolds with good cellular adhesiveness and proliferativeness.

Surface modification is an effective way to improve surface properties, especially when the electrospun fibers have still got inadequate surface properties as cell scaffolds. Synthetic polymers, in general, can offer easier and simpler processability during electrospinning, and more controllable nanofibrous morphology can be achieved, as compared with the morphology of natural polymers. However, the biocompatibility, cellular adhesiveness, and proliferativeness of synthetic polymers are usually inferior to those of natural polymers, for which the surface modification of the electrospun nanofibers with, for example, functional groups or biomolecules that can be one of the ideal solutions.

In this chapter, we will introduce and summarize the fundamentals of the electrospinning process, in order to fabricate uniform fibers, while presenting the typical polymeric materials used in tissue engineering. Control of the morphology of the electrospun scaffolds, and the surface modification techniques of the electrospun nanofibers were discussed for the preparation of the new highly-functionalized scaffolds.

2 FUNDAMENTALS OF THE ELECTROSPINNING PROCESS

Electrospinning is an effective method to fabricate scaffolds of polymeric fibers with very small diameters, a method known for more than a century. Especially during the last two decades, electrospinning has been applied to various engineering fields, such as filtrations, optical and chemical sensors, electrode

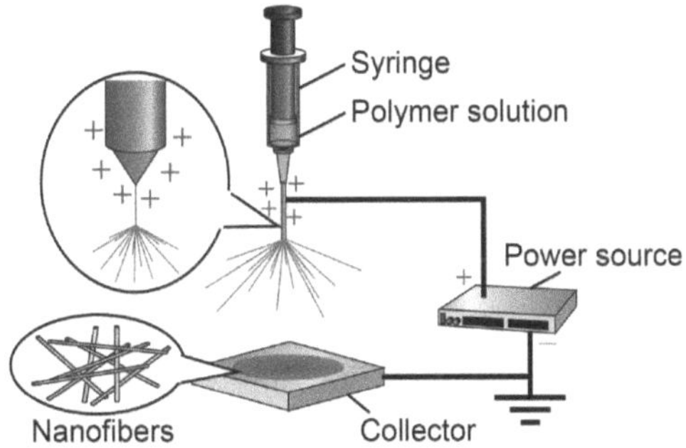

FIGURE 8.1 Electrospinning Method

materials, and tissue engineering scaffolds, since the methods were relatively straightforward, and a large variety of polymeric materials could be electrospun to form fibers (Teo and Ramakrishna, 2006). Another attractive feature of electrospinning may be the uncomplicated and inexpensive nature of the setup: the typical electrospinning setup consists of a syringe pump, a high voltage source, and a collector (Fig. 8.1). Once an electric voltage is applied between the syringe needle and the collector, the charged polymeric jets are sprinkled and injected at high velocities toward the collector. During the travel of the fibrous jets, the solvent in the jet gradually evaporates, generating a solidified nanofibrous structure on the collector.

In order to obtain uniform polymeric fibers, it is essential to optimize and control the electrospinning parameters. It has been actively reported that a number of electrospinning parameters can greatly influence the configurations of the obtained electrospun products. Different morphologies other than fibers, such as particles and even beads-like structures, can also be obtained by controlling the electrospinning parameters, but these are certainly unfavorable to be used as scaffolds. Electrospinning parameters can actually be classified into process parameters, and polymer-solution parameters. By adjusting these parameters adequately, many types of polymeric fibers with desired configurations can be obtained. In this section, the electrospinning parameters that could directly affect the resulting fiber configurations are discussed, while the typical polymers that have been electrospun to form fine fibers for the use in the tissue-engineering materials are introduced.

2.1 POLYMER SOLUTION PARAMETERS

The properties of polymer solution, including polymer concentration, solvent volatility, and solution conductivity largely affect the experimental outcome of the electrospinning process, and the resultant fiber configurations. We will separately discuss each solution parameter in terms of the electrospinning results.

2.1.1 Polymer concentration

Polymer concentration may have a most substantial influence on the configurations of the electrospun products. The formation of fibers through electrospinning is largely based on the uniaxial stretching of the ejected jet during the electrospinning. In order to fabricate continuous long fibers, polymer molecules need to become entangled to some extent in the jet. When the polymer concentration is too low, a misty spray would generally be formed, with the polymer turning into particles instead of forming fibers, due to the lack of entanglement of the polymer molecules in the jet. This phenomenon is also called electrospraying (Greiner and Wendorff, 2007; Venugopal et al., 2005). As the concentration increases, the polymer molecules become more entangled in the solution, and hence the obtained configurations would change from beads into fibers (sometimes with beads). Further increase in polymer concentration would often result in the formation of smooth fibers. However, excessive concentrations with exceedingly high viscosity could disrupt the flow of the polymer solution, resulting in counterproductive activity (Greiner and Wendorff, 2007; Venugopal et al., 2005).

Within an appropriate concentration range of the polymer solution, uniform fibers could be obtained, and their diameters can be increased with an increase in polymer concentration. Beachley and Wen (2009) electrospun poly(ε-caprolactone) (PCL) from a 7:3 mixture of dichloromethane (DCM) and methanol at concentrations ranging from 8 to 20 wt./vol., producing fine fibers with the diameter ranging from 350 to 1000 nm. Tan et al. (2005) fabricated poly(L-lactic acid) (PLLA) fibers from the mixture of DCM and pyridine at polymer concentrations ranging from 1.25 to 4.0 wt.%, producing extremely fine fibers with the diameter ranging from 50 to ~500 nm. Controlling the polymer concentration, therefore, could wisely change the diameter of the resulting synthesized fibers.

2.1.2 Solvent volatility

Solvent volatility is closely related to the boiling point of the solvent: if the boiling point of the solvent is too high, the solvent of the electrospun polymer solution in the jets cannot evaporate sufficiently during the short travel of the fibrous jets, causing wet jets, and hence wet distorted formless fibers, to be attached on the electrospinning collector. The higher boiling point would also cause the flattening and conglutination of fibers or fiber segments. However, the boiling point should not be too low, as the resulting polymer solution could dry out during the initial stage of electrospinning, causing ineffective uniaxial stretching, and even clogging at the capillary tip (Yamashita, 2007). If the choosing of a solvent with relatively low boiling point is inevitable, a solvent with relatively high boiling point can be added to adjust the evaporation rate of the resulting solvent mixture.

2.1.3 Solution conductivity

Solution conductivity actually regulates the tensile forces acting on ejected jets in an electric field. Here, the dielectric constant becomes another decisive parameter that needs to be deeply considered. More specifically, the higher dielectric constant of a solvent provides the higher conductivity of the resulting polymer solution. Adding an organic or inorganic salt, or a surfactant with a high dielectric constant, could improve the conductivity of the solution (Sill and Von Recum, 2008). Tan et al. (2005) investigated the effects of solution conductivity on the morphology of electrospun fibers. They used DCM as a main solvent to dissolve poly(L-lactide-*co*-caprolactone), and added N,N-dimethylformamide (DMF) or pyridine to increase the conductivity of the polymer solution. Eventually, there was a significant drop in the diameter of the electrospun fibers from 300 to 100 nm, when the conductivity of the solution increased from 0.0 to 13.1 μS/cm. The solution with high conductivity has greater charge-carrying capacity than the solution with lower conductivity. Therefore, the fiber jets produced from the solution with high conductivity were subjected to a greater tensile force in an electric field than the fiber jets from the solution with low conductivity (Sill and Von Recum, 2008). Also, in the presence of a strong electric field, a highly conductive solution became extremely unstable, causing dramatic bending instability, which produced fibers with broad diameter distribution (Hayati et al., 1987).

2.2 PROCESS PARAMETERS

Other important parameters that could significantly affect electrospinning and its results are the electrospinning process factors, such as the applied voltage, the polymer flow rate, and the tip–collector distance. These parameters have a crucial influence on the electrospinning jets and the fiber configurations, although they are relatively less significant in the diameter effectiveness than the aforementioned polymer solution parameters.

2.2.1 Applied voltage

The applied voltage can change the configuration of the obtained fibers due to its control over the stretching and acceleration of the electrospun jets. At a low applied voltage, a sufficient tensile force cannot be expected for the polymer solution, and thus the fiber jets cannot be efficiently ejected from the capillary tip. In fact, a voltage of approximately 5–20 kV can normally induce the effective ejection of the fiber jets from the capillary tip (Boland et al., 2001). Insufficient electric field strength can provoke bead defects in the electrospun fibers, or even serious failure in the jet formation. Based on various previous studies, it was evident that there was an optimal range of an electric field, when using a certain polymer solution. It was reported that a too weak or even too strong electric field would cause the formation of beaded fibers (Deitzel et al., 2001; Meechaisue et al., 2006).

2.2.2 Polymer flow rate

The polymer flow rate also had a nonnegligible impact on fiber configuration and its diameter (Megelski et al., 2002; Sill and Von Recum, 2008; Taylor, 1969). Megelski et al. (2002) changed the flow rate of polystyrene (PS) solution to determine its effects on fiber diameter and its morphology. Bead formation was observed when the flow rate was above 0.10 mL/min, largely due to the inadequate drying of the jet prior to its arrival at the collector. Incomplete drying also led to the formation of ribbon-like fibers, because the subsequent drying of the core layer resulted in the collapse of the already dried skin layer (Lannutti et al., 2007; Sill and Von Recum, 2008).

2.2.3 Tip–collector distance

The configuration of electrospun fibers was also affected by the tip–collector distance, because the distance simply changed the traveling time of the electrospun jets: the solvent in the jets gradually evaporated, and the wetness of the fibers decreased during the travel time of the fibrous jets. To assess the effects of the tip–collector distance on fiber morphology, Megelski et al. (2002) varied the gap between the capillary tip and the collector from 5 to 35 cm, keeping all other parameters constant. Although there was no marked change in fiber size, it was found that inhomogeneously distributed elongated beads started to form along the PS fibers after reducing the working distance down to only 5 cm. This phenomenon could be attributed to the inadequate drying of the polymer fibers prior to the arrival at the metal collector, due to the short time for solvent evaporation (Megelski et al., 2002).

The descriptions of the electrospinning process previously suggested that quite a few electrospinning parameters could indeed influence the configurations of the resulting electrospun fibers. By appropriately varying some of these experimental parameters, uniform fibers can be obtained.

2.3 ELECTROSPUN FIBER MATERIALS

The fiber materials for biomedical applications using electrospinning should be selected in terms of biocompatibility and biodegradability. For biocompatibility, the material should not stimulate any immune response to the surrounding tissue. Additionally, to regenerate tissue through the support of cells, the scaffolds should integrate the tissue. The scaffold should also degrade, possibly by itself, to prevent any surgical interventions for the removal of the implants. The biodegradation rate of the material should be carefully considered and controlled for its use as tissue engineering scaffolds: if the degradation rate is higher than the rate (speed) of tissue regeneration, the scaffolds could not support the tissue. In contrast, if it is too low, the scaffolds could remain in the tissue.

Previous studies have revealed that many biocompatible and biodegradable polymers could be electrospun into fibers for biomedical applications. These types of polymers can actually be classified into synthetic polymers and natural polymers. The representative examples of synthetic polymers include poly(glycolic acid) (PGA), poly(lactic acid) (PLA), poly(ε-caprolactone) (PCL), poly(2-methacryloyloxyethyl phosphorylcholine) (PMPC), while examples of natural polymers could be collagen and elastin.

2.3.1 Poly(glycolic acid) (PGA)

PGA is one of the interesting biodegradable aliphatic polyesters used for a variety of medical applications. For the material that would promptly degrade in the early phase, PGA can be a prospective candidate. PGA was initially developed as suture for medical use, in the 1970s, and was also found to possess good tissue compatibility. Its controllable and hence predictable bioabsorption gave PGA an advantage over other biomedical materials. Additionally, PGA monofilaments can degrade in a short period of time ranging from 2 to 4 weeks in vivo, owing to the hydrophilic nature of PGA (Boland et al., 2001). It should be noted that the glass transition temperature of PGA is very close to our body temperature (~37°C). At or above the glass transition temperature, water molecules diffuse quickly, while hydrolysis can easily occur, resulting in an instant degradation of PGA. Such a high degradation rate of PGA may drastically increase the value of localized pH that caused undesired tissue response (Wong and Mooney, 1997).

The largest problem in the electrospinning of PGA may be its low solubility in organic solvents. In fact, PGA can be dissolved only in highly fluorinated solvents, such as hexafluoroisopropanol (HFIP). The Bowlin group was the first to electrospin nanofibers of PGA at 1/7 wt./vol. from HFIP (Boland et al., 2001, 2004). Special care and attention are necessary for the use of such highly fluorinated solvent, as it is extremely deleterious to the human body.

2.3.2 Poly(lactic acid) (PLA)

PLA is another biodegradable aliphatic polyester that is a very popular polymer used in medical applications (Wong and Mooney, 1997). PLA is more harmless than PGA, since the L-lactic acid, a product of PLA hydrolysis, naturally emerges in the metabolic pathway of the human body. Moreover, the methyl groups of the PLA monomers make PLA more hydrophobic than PGA, also promoting the steric hindrance of PLA, introducing higher solubility of PLA with several organic solvents. Such characteristics of PLA significantly hinder the hydrolysis of PLA that would normally degrade within 30–50 weeks (Wong and Mooney, 1997).

PLA has been electrospun by many groups. Tan et al. (2005) fabricated PLA fibers from the mixture of DCM and pyridine at PLA concentrations from 1.25 to 4.0 wt.%, resulting in the production of fine fibers with the diameters ranging from 50 to ~500 nm. Neto et al. (2012) electrospun PLA at 0.07 g/mL using a 3/7 v/v mixture of DMF and DCM. PLA has also been successfully electrospun by Yang et al. (2005) at 5 wt.%, using a 7/3 wt./wt. mixture of chloroform and dimethyl sulfoxide for neural tissue engineering.

2.3.3 Poly(ε-caprolactone) (PCL)

PCL is a relatively inexpensive biodegradable polyester with nontoxic nature, possessing a relatively slow degradation rate (1–2 years) and high elasticity (Yoshimoto et al., 2003; Zeng et al., 2004). Electrospun PCL fibers were first reported by Reneker et al. (2002). The electrospun

PCL, using acetone as a solvent, by changing the concentrations from 14 to 18 wt.%. (Reneker et al., 2002). Yoshimoto et al. (2003) electrospun PCL using chloroform at a concentration of 10 wt.% as bone tissue engineering scaffold. Shin et al. (2004) used a 1:1 mixture of chloroform and methanol to obtain PCL fibers for a tissue-engineered cardiac graft. For the both applications, PCL was wisely selected due to its good mechanical properties (high modulus and high elasticity), and the slow degradation rate. Li et al. (2005) also successfully electrospun nanofibrous PCL scaffolds using a 1:1 mixture of tetrahydrofuran and DMF at the concentration of 0.14 g/mL. In addition, to demonstrate the potential of electrospun PCL as a scaffold for multiphasic tissue engineering, they seeded human mesenchymal stem cells (hMSCs) on the scaffolds, induced to differentiate along adipogenic, chondrogenic, and osteogenic lineages with specific differentiation media (Li et al., 2005). Fujihara et al. (2005) electrospun PCL scaffolds using a 75/25 wt./wt. mixture of chloroform and methanol by changing the concentrations from 3 to 7.5 wt.%, including calcium carbonate ($CaCO_3$) nanoparticles (PCL/$CaCO_3$ 75/25, and 25/75 wt./wt.) for the materials used in a guided bone regeneration scaffold. The scaffold could enhance the good attachment and proliferation of osteoblast cells (Fujihara et al., 2005).

2.3.4 Poly(2-methacryloyloxyethyl phosphorylcholine) (PMPC)

Specific biomaterials, such as vascular grafts, are normally used in contact with blood or body fluid, and thus need to be highly antithrombogenic. PMPC was designed for the use as an antithrombogenicity biomaterial, having phosphorylcholine group in its formula, imitating the surface properties of the endothelial cell membrane that would get into touch with blood in vessels. PMPC was a functional polymer developed by Ishihara et al. (1991, 1998, 1991). While PMPC does not possess biodegradability, it is expected to be used in vivo in the field of tissue engineering, owing to its valuable bio-related property.

Electrospun nanofibers of PMPC possessing excellent hemocompatibility were successfully fabricated by our group (Maeda et al., 2014). We varied the PMPC concentrations in ethanol from 1 to 10 wt.% to fabricate electrospun PMPC fibers. When the PMPC concentration was lower than 5 wt.%, beads were formed instead of fibers. When the PMPC concentration was higher than 5 wt.%, uniform fibers could be produced with the average fiber diameter of ~160 nm (5 wt.%), 640 nm (7.5 wt.%), and 1270 nm (10 wt.%). Such diameter-controlled PMPC fibers could be utilized as a new vascular graft material.

2.3.5 Collagen

Collagen is one of the glycoproteins that are the primary component of ECM. In fact, ECM consists of glycoproteins, such as collagen, proteoglycan, fibronectin, elastin, and laminin. In the native tissue, type I collagen is the principal structural element of the ECM (Matthews et al., 2002). The underlying alpha chains of these natural polymers are arranged into a repeating motif that forms a coiled configuration. The repeating motif possesses a 67 nm interval that provides a characteristic banding pattern to the collagen fibril. Collagen has been used as a material for tissue engineering due to its presence in natural ECM. Collagen can be isolated from natural products, and is relatively nonimmunogenic. However, the typical isolation procedures and the reprocess of collagen may impair its biological and structural properties (Matthews et al., 2002).

Research on electrospun type I collagen by Matthews et al. (2002) showed the capability of reproducing nanostructured collagen-based scaffolds that retained biological and structural properties of collagen. They electrospun collagen using HFIP as a solvent at concentrations between 0.03 and 0.10 g/mL, and obtained the scaffold using the fibers with diameters ranging from 100 nm to 5 μm. As a result, electrospun type I collagen fibers could construct the structure with 67 nm banding similar to the native collagen in natural ECM, meaning that the fibers could construct a truly biomimicking scaffold.

2.3.6 Elastin

Elastin is an extremely linearly-elastic biosolid, a highly insoluble and hydrophobic protein existing in the organs of the human body. Elastin composes a large part of artery and vein walls, providing the elastic recoil to prevent the transduction of transmitted pulsatile energy into heat. Elastin is also a component of ligaments, lung parenchyma, and skin, owing to its elastic properties (Huang et al., 2000; Fung, 1981). Such characteristics of diverse and essential proteins should be extremely effective to be applied to a tissue scaffold or a graft, particularly for vascular applications.

Huang and colleagues electrospun a synthetically-manufactured elastin-mimetic material composed of an 81 kDa recombinant protein that was based on the repeating elastomeric peptide sequence of elastin. They obtained nanofibers with the diameters ranging from 300 to 400 nm (Huang et al., 2000).

3 TUNING THE MORPHOLOGY OF ELECTROSPUN SCAFFOLDS AND ITS EFFECTS ON CELL BEHAVIOR

The architectural parameters of the scaffold, such as attachment, proliferation, configuration, and infiltration, are very important, and they would strongly affect cell behavior. In many previous studies, the structures of the synthesized electrospun scaffolds, such as fiber uniformity, fiber diameter, fiber orientation, and the pore size of the scaffolds, have been analyzed and tuned in order to investigate the effects of the electrospun scaffold microstructures on cell behavior.

In this section, we will focus on the effects of the scaffold structures on the attachment, proliferation, configuration, and infiltration properties of cells, and study how to control these nanofibrous structures.

3.1 FIBER UNIFORMITY

The cell cultivation on the electrospun scaffolds has been carried out on the assumption that the electrospun fibers were uniform without any beads in the fiber configurations. It is generally known that on the scaffolds with beaded structures, cells would not attach and proliferate well. Chen et al. (2007) cultivated NIH 3T3 (fibroblast cells) on the beaded PCL scaffold and the uniform PCL-fiber scaffold. It was found that the beaded scaffolds exhibited lower cell attachment and proliferation than the uniform fiber scaffolds (Chen et al., 2007). To fabricate uniform fibers by electrospinning, the parameters that were defined and mentioned in the Section 2 should be properly controlled.

3.2 FIBER DIAMETER

Previous studies revealed that the uniform and fine electrospun fibers were more suitable for cell adhesion and proliferation. Chen et al. (2007) published papers investigating the effects of fiber diameter (428–1647 nm) on the high porosity of the PCL scaffolds. The pore size of these scaffolds was almost uniformly determined as above 50 μm. NIH 3T3 were seeded on the electrospun PCL scaffolds. When the average fiber diameter was within the nanofiber range (428–1051 nm), cell adhesion and cell proliferation increased with the decrease in fiber diameter. When the average fiber diameter was in the micron range (1647 nm), although cell adhesion remained invariant, cell proliferation became slightly greater than that of the scaffolds made from the fibers of 900 nm in diameter. Chen et al. (2009) also correlated the improved cell proliferation rates and its attachment with the drastically enlarged specific surface area, brought about by the decrement of fiber diameter down to below 1 μm. In fact, the distribution of the specific surface area was changed from 2.24 to 18.79 μm^{-1} by changing the PCL fiber diameter. The concentration of serum, the bio-liquid including several types of proteins used to provide a simulated body environment, varied from 0% to 12% for the cell-cultivation experiments. The significant effects of the submicron PCL fibers on cell attachment were also confirmed at above 7.5% in serum concentrations. When the serum concentration was less than 5%, the cell attachment became low, confirming that the increase in the specific surface area failed to make a significant improvement in cell attachment, in this case. Thus, they concluded that cell attachment and proliferation were improved by enhanced protein adsorption, due to the higher specific surface area. The study by Christopherson et al. (2009) further corroborated the influence of protein adsorption. They reported the improved neural stem-cell proliferation observed in poly(ether sulfone) nanofibers, but not in microfibers, only in the limited condition with the presence of basic fibroblast growth factor (bFGF) (Christopherson et al., 2009).

In order to achieve a wider range of fiber diameter, varying the concentration of polymer solutions has been actively attempted as the simplest and easiest procedure by many research groups. Beachley and Wen (2009) electrospun PCL from a 7:3 mixture of DCM and methanol at concentrations from 8 to 20 wt./vol., resulting in the construction of fibers ranging from 350 to 1000 nm in diameter (Beachley and Wen, 2009). As discussed in Section 2, there were several parameters having considerable influence on electrospun fiber diameter. In this case, it was found that the concentration of polymer solution had the most prominent effect on fiber diameter among these parameters.

3.3 FIBER ORIENTATION

It was known that the angular direction of cell elongation corresponded to the direction of the underlying fibrous substrate, and thus cell populations cultured on the well-aligned nanofibrous substrates were well aligned. In fact, most cell types, including neural, muscle, fibroblast, endothelial, and mesenchymal stem cells (MSCs), adopted an aligned morphology when cultured on aligned nanofibrous substrates (Chew et al., 2008; Choi et al., 2008a; Chow et al., 2007; Li et al., 2007b; Rockwood et al., 2008; Wise et al., 2009; Yang et al., 2005). Since these cells expressed the aligned morphology in natural tissue, the cultivation on scaffolds, where cells were aligned, led to good cell growth.

Cell elongation and alignment are closely related to cell shape, which can also account for accelerated cell growth and maturation. Cell elongation is an especially important property in, for instance, skeletal muscles and neural tissue engineering, where native cells reach the lengths of several centimeters, and even more than a meter, respectively. Aligned nanofibrous substrates have consistently been reported to promote accelerated elongation in neural cells, as compared with randomly aligned substrates (He et al., 2010; Lam et al., 2010; Madduri et al., 2010; Patel et al., 2007). Myotube length on aligned nanofibers was also observed to be twice as high as that observed on randomly aligned scaffolds, while the thickness of the myotube remained similar (Choi et al., 2008a). According to several previous studies, the quantitative degree of cellular alignment generally corresponded to the degree of underlying fiber alignment (Bashur et al., 2006, 2009; Corey et al., 2007; Ma et al., 2011; Moffat et al., 2009). The

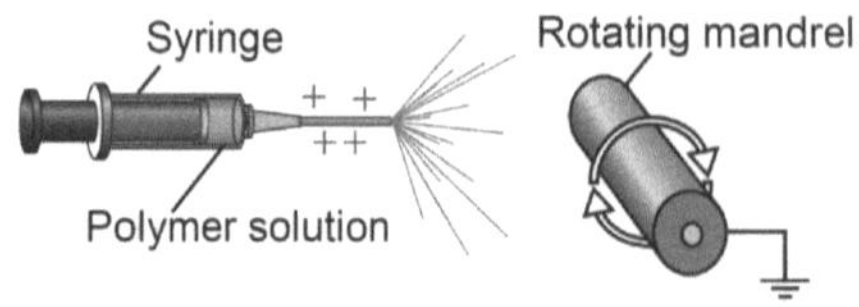

FIGURE 8.2 Rotating Mandrel Technique

elongated cell morphology could be quantified by its aspect ratio, the ratio of the length of the cell's long axis to its short axis (Bashur et al., 2006). The quantification of cell morphology varied greatly for specific cell types, substrates, and culture times. In one representative example, the cell aspect ratio on highly aligned nanofibers was actually quantified, and found twice as high as that observed on randomly oriented nanofibers, and three times as that seen on a glass control after 24 h of seeding (Aviss et al., 2010).

The aligned morphology of the scaffolds also affected the proliferation of cells that took aligned morphology in natural tissue. Kuppan et al. (2014) fabricated random and aligned PCL/gelatin (70/30) blend nanofibrous scaffolds. Nerve stem cells (C17.2 cells) were seeded on the scaffolds, and their proliferation rates were evaluated. In comparison to random nanofibers, the cell proliferation of C17.2 cells was significantly higher on the aligned nanofibrous scaffolds after 3, 7, and 10 days of culture (Kuppan et al., 2014). Ghasemi-Mobarakeh et al. (2008) have shown that PCL/gelatin (70/30) aligned nanofibrous scaffolds supported greater proliferation of nerve stem cells (C17.2 cells) than the random PCL/gelatin nanofibrous scaffolds, after 4 and 6 days of culture (Ghasemi-Mobarakeh et al., 2008). Likewise, Subramanian et al. (2010) also demonstrated that their aligned nanofibrous scaffolds also exhibited better attachment and proliferation using rat Schwann cells than the random nanofibrous scaffolds, after 1, 3, 5, and 7 days of culture.

The most common way of collecting aligned electrospun nanofibers, known as the rotating mandrel technique, utilized a high-speed grounded rotating mandrel as a collector (Fig. 8.2). The jet formed during electrospinning was ejected at a high-speed rate of several meters per second (Reneker and Chun, 1996). It was found that the deposition pattern of the electrospun fibers could be controlled by using a target moving at a high-speed rate. Mandrel diameters were generally a few centimeters, and were rotated at rotation speeds ranging from zero to several thousand revolutions per minute (rpm). The tangential velocity was the most informative value to use, when comparing different studies, because mandrel diameters could significantly vary between the studies.

The speed of the rotating collector had various effects on the microstructures of electrospun fibers, particularly the alignment. It was hypothesized that when the tangential velocity at the edge of the rotating collector matched the speed of the electrospun jet, the degree of fiber alignment would be maximized (Wu et al., 2007). Fiber orientation resulting from different mandrel velocities could be classified into three stages: (1) too low mandrel velocity to initiate fiber alignment, (2) medium to high mandrel velocity that initiated and increased fiber alignment, and (3) too high velocity that decreased fiber alignment due to fiber fracture (Edwards et al., 2010). The threshold speeds for fiber alignment were different from system to system. Several groups using various synthetic polymers observed the onset of fiber alignment at around 3 m/s (Aviss et al., 2010; Bashur et al., 2006; Wang et al., 2009), whereas the others saw the required speeds approaching even 10 m/s (Zhong et al., 2006) for the onset of the alignment. The degree of fiber alignment generally increased by increasing the tangential velocity after the onset of fiber alignment (Lee et al., 2010; Li et al., 2007b). As the mandrel speed continued to increase, fibers began to break due to the tensile force exerted by the mandrel on the fibers (Zhong et al., 2006). Such fiber breakage could result in a decrease in fiber alignment and quality (Wu et al., 2007). The mandrel speed associated with fiber breakage was dependent on the original material properties of the electrospun polymers, and it was shown that the nanofibers electrospun from brittle materials were more likely to break at lower mandrel speeds (Chan et al., 2010).

3.4 PORE SIZE OF ELECTROSPUN SCAFFOLD

The pore diameter, defined as the void area between the fibers, affected the cell infiltration and the cell proliferation. Guimaraes et al. (2010) investigated the effects of pore size of the PCL scaffolds on cell infiltration. Their research showed that, by doubling pore sizes from 2.6 to 5 μm, cell growth permeance made a significant improvement from permeance only at the surface to enhanced permeance of a complete infiltration into the scaffold (Guimaraes et al., 2010).

Lowery and colleagues independently investigated the effects of fiber diameter and pore size on cell proliferation and tissue formation, when using electrospun PCL scaffolds (Lowery et al., 2010). Their data showed a significantly increased cell count for the scaffolds, with larger pore size (using a similar fiber diameter), while the pore as within the size-range of affecting cell migration (6 vs. 14 μm). When pore size became too large to affect cell migration (20 vs. 36 μm), the changes in cell proliferation could no longer be observed.

However, for the scaffolds with very large pore diameters, reduced cell matrix deposition and an increase in nonviable cells were observed. Based on these findings, Lowery et al. (2010) concluded that cells in scaffolds with pores possessing very large diameters were no longer able to effectively bridge the gaps between fibers. In contrast, very small pore diameters of less than 1 μm were required to completely prevent cell infiltration. Therefore, an optimal range of pore size from 6 to 20 μm was proposed based on these findings. In fact, this range corresponded almost to ± 50% of the average suspended cell diameter (~15 μm).

To increase the pore size of nanofibrous scaffolds, and solve the general problem of low infiltration capacity of nanofiber scaffolds that was an intrinsic limitation of electrospun scaffolds, some researchers introduced sacrificial fibers into fiber mats (Fig. 8.3) (Frey and Li, 2007; Guimaraes et al., 2010). In this case, the sacrificial component was water-soluble, whereas the permanent component was insoluble in water. Pore size was adjusted by the subsequent removal of the sacrificial component by immersion in water. Baker and colleagues demonstrated that the selective removal of the sacrificial fibers produced an exponential increase in pore size, as the mass content of the sacrificial fibers increased (Baker et al., 2008). They found the optimal effects on pore size and cell infiltration at mass contents from 40% to 60%. Below the mentioned content, the effects on cell infiltration were not significant, while higher content ratios led to the mechanical instability of the fiber mat.

4 SURFACE MODIFICATION METHODS OF ELECTROSPUN NANOFIBERS

To add desirable functionalities, such as hydrophilicity and biocompatibility to the electrospun fibrous scaffolds, the surface modification to the original polymeric fibers can be highly beneficial, enhancing cell attachment and cell proliferation. In many cases, while synthetic polymers can offer easier

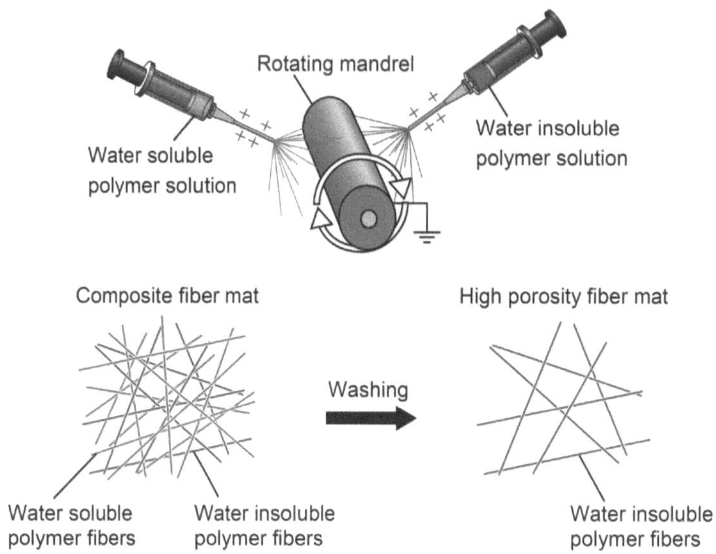

FIGURE 8.3 Sacrificial Fiber Technique for Making High Porosity Fiber Mats

processability for electrospinning, and more controllable nanofibrous morphology than natural polymers, biocompatibility, cellular adhesiveness, and proliferativeness of synthetic polymers are usually inferior to those of natural polymers. The surface modifications of such synthetic electrospun nanofibers by biomolecules, functional groups, and biocompatible polymers could improve these properties. For effective surface modifications, powerful methods such as physical adsorption of biomolecules, plasma treatment, and surface graft copolymerization have been introduced. We will discuss the several successful cases of well-controlled polymeric surfaces that may lead to the construction of new highly functionalized polymeric scaffolds.

4.1 LOADING BIOMOLECULES

Various biomolecules have been incorporated into tissue engineering scaffolds in order to enhance their functional properties, such as cell attachment and cell proliferation. The most frequently used biomolecules are proteins, such as gelatin, growth factors, and cytokines. We will introduce the powerful techniques to incorporate such biomolecules into the electrospun scaffolds, using the physical surface adsorption, blend electrospinning, core–shell electrospinning, and the covalent immobilization method with biomolecules.

4.1.1 Physical surface adsorption

The simplest way to load biomolecules into electrospun scaffolds may be to dip the scaffolds into an aqueous solution containing biomolecules, in order to provide the fiber surface with biomolecules directly. Such physical surface adsorption would be the simplest approach to the loading of biomolecules onto nanofibrous structures (Fig. 8.4). During such an approach, biomolecules can be in the form of pure solution or emulsion. Generally, the driving force for surface adsorption includes

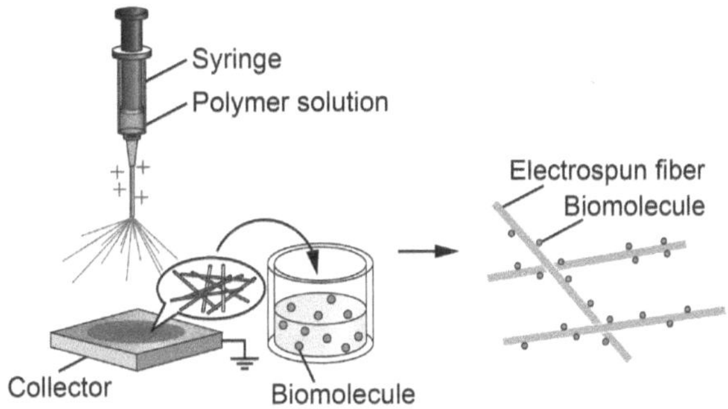

FIGURE 8.4 Physical Adsorption

electrostatic interaction, hydrogen bonding, hydrophobic interaction, and van der Waals interaction (Yoshida et al., 2006). Moreover, physical surface adsorption techniques would not interrupt the activity of loaded biomolecules. However, the techniques have not been frequently used so far, due to uncontrolled biomolecule release profile.

It was reported that the bone morphogenic protein-2 (BMP2) attached to poly(lactide-*co*-glycolide) (PLGA) scaffolds was released over 75% within just 5 days, and completed its release almost within 20 days. The release rate was much faster than that of the same amount of protein incorporated into PLGA scaffolds made by blend electrospinning (Nie et al., 2008).

A more developed example of physical adsorption may be the use of specific interaction between heparin and growth factor (Lode et al., 2008; Stendahl et al., 2008; Yoon et al., 2007). Heparin, a highly sulfated glycosaminoglycan, bound with various growth factors, such as fibroblast growth factor (FGF), vascular endothelial growth factor (VEGF), and transforming growth factor-β (TGF-β) that possessed high affinity with heparin. The approach also preserved essential biological activity by preventing the early degradation of growth factors. In fact, heparin immobilization on the biomaterial surface, and the subsequent attachment of the growth factor, should be the most efficient way for the successful local delivery of growth factors (Joung et al., 2008).

Heparin-functionalized electrospun fibers subsequently loaded with bFGF were also developed (Casper et al., 2005). Low molecular weight heparin (LMWH) was successfully incorporated into either poly(ethylene oxide) (PEO) or PLGA electrospun fibers, as the PEO–LMWH conjugate form with free LMWH.

The PEO nanofibers incorporated with LMWH were first prepared by electrospinning an aqueous solution to fibers, with the surface heparin content slightly increased, as compared to PLGA/LMWH fibers with significant degree of bFGF surface immobilization. This was mainly attributed to the slower dissolution of PEO–LMWH conjugates from the electrospun fibers. These surface-functionalized electrospun matrices could be a promising growth-factor delivery vehicle to induce local angiogenesis.

Aligned electrospun nanofibrous scaffolds with immobilized heparin on the surface loaded with growth factor were prepared (Patel et al., 2007). Initially, heparin was covalently immobilized onto the surface of PLLA nanofibers that became aligned either during the electrospinning process, or through mechanical stretch after electrospinning. Then, laminin, an ECM protein, and bioactive bFGF could be simultaneously adsorbed onto the nanofibers through specific heparin-mediated interaction. It was reported that neurite outgrowth and cell migration could be induced by the alignment of nanofibers that was even further enhanced by the immobilization of biochemical cues (laminin and bFGF). The results

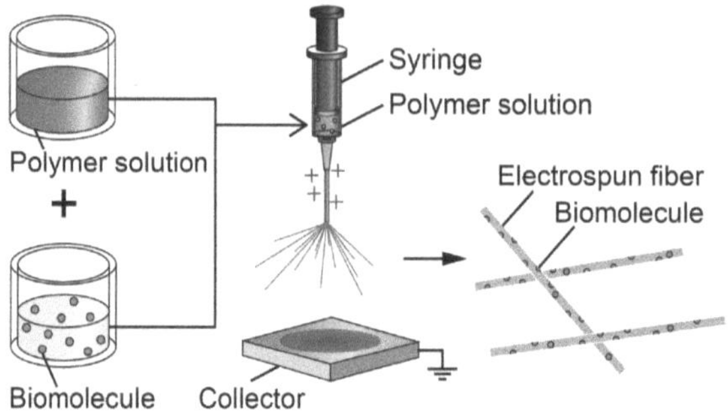

FIGURE 8.5 Blend Electrospinning

suggested that the combination of locally delivered bioactive agents and nanosized surface topography could synergistically promote therapeutic efficacy in the repair of damaged or malfunctioned tissue.

4.1.2 Blend electrospinning

The blend electrospinning technique was commonly employed to produce fibers with a drug-releasing functionality (Ignatious et al., 2010; Verreck et al., 2003; Xie and Buschle-Diller, 2010; Zeng et al., 2005b). In blend electrospinning, biomolecules were mixed with polymer solution, after which the mixed solution was electrospun to fabricate a hybrid scaffold (Fig. 8.5).

The drug release system by the blend electrospinning could be used in tissue engineering scaffolds to reduce the risk of infection (Kim et al., 2004), or to suppress adverse tissue reactions (Innocente et al., 2009). Many groups have reported that this strategy could be successfully utilized to functionalize fibers with biomolecules, such as dexamethasone, heparin, and plasmids (Luong-Van et al., 2006; Luu et al., 2003; Martins et al., 2010). In general, for proteins, water-soluble polymers, such as poly(vinyl alcohol) (PVA) and PEO, have been used for electrospun fibers (Assmann et al., 2010; Szentivanyi et al., 2009; Zeng et al., 2005a). In our laboratory, curcumin and PMPC blend fibers were electrospun to be used as new vascular graft materials, and their drug-release characteristics were also investigated (Maeda et al., 2014). Curcumin was a major chemical component of turmeric that inhibited platelet-derived growth factor and stimulated vascular smooth-muscle-cell migration, cell proliferation, and collagen synthesis that could be the key events for cardiovascular disease (Yang et al., 2006). In the experiments, PMPC was first dissolved in ethanol at concentrations ranging from 5 to 10 wt.%. Then, curcumin was added to the PMPC solution at 4 wt.% against the PMPC mass. Afterward, the curcumin/PMPC blend fibers were electrospun, resulting in a fiber formation with the fiber diameter ranging from 164 to 1270 nm, depending on the PMPC concentrations. The drug-release rates from the electrospun PMPC fibers were well controlled, especially when we decreased the diameter of the PMPC fibers. Our experimental results may be highly applicable to the construction of next-generation vascular grafts that may possess both biocompatibility and well-controlled drug-release characteristics to prevent the thrombosis.

Blend electrospinning, however, had its own limitations, when it was used with organic solvents. Most sensitive biomolecules would denature in the presence of organic solvents that would also eventually lose their bioactivity. Sahoo et al. (2010) demonstrated the increased cell proliferation for PLGA fibers functionalized by the bFGF spun from HFIP, but failed to quantify and maintain the total bioactivity of the growth factor.

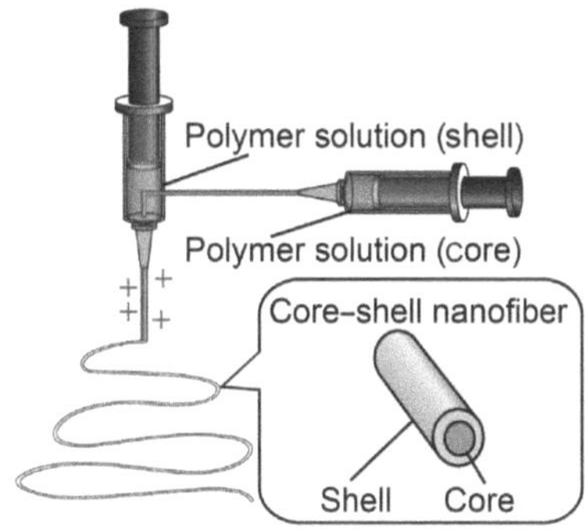

FIGURE 8.6 Core–Shell Electrospinning

4.1.3 Core–shell electrospinning

Core–shell electrospinning is a relatively new method to fabricate coaxial nanofibers with two polymeric compounds (Sun et al., 2003). The schematic illustration of the experimental setup of core–shell electrospinning is presented in Fig. 8.6 (Zhang et al., 2004). The introduction of an inner capillary tube was the main difference from the conventional electrospinning setup. The inner capillary tube was connected to an electrical potential ranging from several kV to tens of kV, relative to the collector. Two immiscible liquid solutions were injected at the appropriate flow rates through the two concentrically arranged capillary tubes. The flow rate was adjusted by a syringe pump, air pressure, and so on. At an appropriate flow rate and applied voltage, a coaxial jet could occur. After the evaporation of the solvent during the jet-flying, core–shell polymer nanofibers could be obtained.

The fabricated core–shell nanofibers had the average diameter of hundreds of nanometers, having the core part with the diameter of tens of nanometers that sometimes reached even a few micrometers (Sun et al., 2003; Wang et al., 2010). The ratio of the core diameter to the shell diameter (D_c/D_s) could change by some factors. Zhang and colleagues reported that increasing the concentration of the core compound specifically related to the ratio of D_c/D_s that increased from ~0.35 to above 0.7 (Zhang et al., 2004). Huang and colleagues also found that the core concentration affected the core and shell diameters, but there was no marked tendency in the obtained ratio of D_c/D_s (Huang et al., 2006).

The core–shell nanofibers fabricated by core–shell electrospinning were great candidates for drug delivery vehicles, as well as scaffolds for tissue culture (Wu et al., 2013; Yi and Lavan, 2008). Encapsulation of active agents such as growth factors in the core compound solution has been investigated. Wang et al. (2010) reported that the controlled drug release from their poly(D,L-lactic acid) (PDLLA) and poly(3-hydroxy butyrate) (PHB) core–shell nanofibers was observed (Wang et al., 2010), where dimethyloxalylglycine (DMOG) was blended into the PHB solution preliminarily as a model drug. They fabricated coaxial PDLLA/PHB (core/shell) nanofibers with the diameter of ~1.5 μm, and shell thickness of 100 or 200 nm. They found that the burst of dimethyloxalylglycine drug release, observed by using simple PDLLA or PHB nanofibers, was successfully suppressed by using coaxial PDLLA/PHB nanofibers.

4.1.4 Covalent immobilization of bioactive molecules

Covalent immobilization is a way to immobilize biomolecules onto the fiber surface via chemical bonding, for instance, peptide bonds through amino groups (Fig. 8.7) (Kim and Yoo, 2010). The chemical

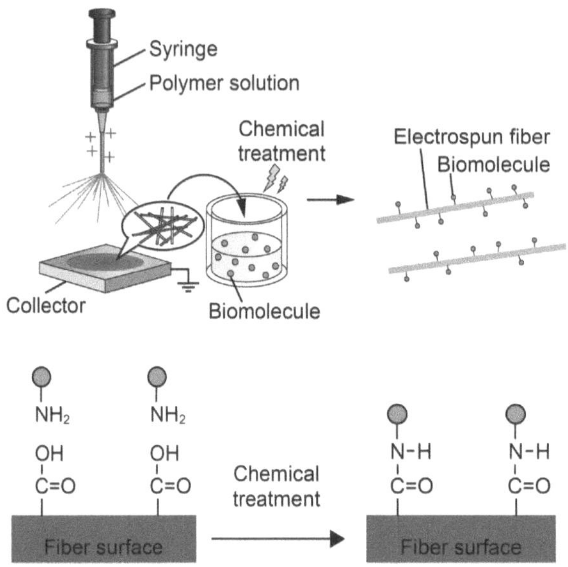

FIGURE 8.7 Chemical Immobilization of Bioactive Molecules

immobilization of bioactive molecules onto the surface of fibers was usually favored over the aforementioned physical adsorption, in tissue engineering applications. The preference was because the immobilized molecules were covalently and stably attached to the nanofibers, where they were not easily leached out from the surface-modified nanofibers, especially when incubated over an extended period of time. However, it should be noted that partial inactivation of the immobilized molecules could occur upon the covalent modification, when the active sites were chemically modified (Ye et al., 2006).

Primary amine and carboxylate groups were extensively used to immobilize bioactive molecules onto the surface of nanofibers. Upon the activation of the carboxylic acid groups by 1-ethyl-3(3-dimethylaminopropyl) carbodiimide (EDC) and *N*-hydroxysuccimide (NHS), nanofibers were subsequently conjugated to primary amine groups with bioactive molecules (Li et al., 2008; Ma et al., 2005a). The carboxylic groups on the surface of the polymeric nanofibers containing different amounts of poly(acrylic acid) could be utilized for conjugation with collagen (Park et al., 2007). In other studies, acrylic acid-immobilized nanofibers were actually conjugated to amine groups to prepare aminated nanofibers (Chua et al., 2006). These nanofibers were further tested for in vitro cultivation of umbilical cord blood cells (Chua et al., 2006, 2007). The mixture of poly(acrylonitrile-*co*-acrylic acid) and poly(acrylonitrile-*co*-maleic acid) was also electrospun before lipase was immobilized onto its surface (Li et al., 2007a; Ye et al., 2006). It was found that the carboxylic acid from the acrylic acid and the maleic acid units reacted with the amine groups of lipase, and the immobilized lipase showed a significant decrease in enzyme activities, as compared with native lipase.

Functional amine groups including primary and ε-amine groups (NH_3^+) were also great candidates for the covalent modifications of electrospun nanofibers, because of their high reactivity. However, the direct conjugation of bioactive molecules on the surface of nanofibers exhibited several limitations in cell adhesion and cell proliferation: the cells could not readily recognize the biological ligands because the immobilized ligands were not fully exposed to the surface, but instead they were often sterically buried within the mesh structure. Thus, several hydrophilic linkers between the biomolecules of interest and the nanofibrous mesh were introduced to promote cellular recognition. For this purpose, a hydrophilic polymer spacer, the poly(ethylene glycol) (PEG), was chemically conjugated to various hydrophobic polymeric segments for electrospinning (Choi et al., 2008b; Kim and Park, 2006). A diblock copolymer of PLGA–*b*–PEG–NH_2 was synthesized and electrospun to expose the terminal amine group of PEG segment in an aqueous medium. The electrospun nanofibrous meshes displayed the primary amine groups on the surface, and those amine groups were subsequently used to immobilize the cell adhesive peptides. For example, Gly-Arg-Gly-Asp-Tyr (GRGDY) peptide, containing a cell adhesive sequence (RGD), was covalently conjugated to the surface-amine groups of PLGA nanofibers to enhance the proliferation of NIH3T3 cells (Kim and Park, 2006). It was found that the amount of immobilized peptide linearly increased when the blend ratio of PLGA–*b*–PEG–NH_2/PLGA was increased. Also, the surface conjugation of RGD peptide significantly enhanced cell attachment, and the attachment rate of the RGD modified PLGA scaffold, reaching approximately 96%. The enhanced cell attachment could also be attributed to the positive surface charge with high wettability, and the topography of the modified nanofibers.

4.2 PLASMA TREATMENT

The adhesion and proliferation of cells were influenced by the wettability of the polymer surface. Many research groups have found that cell adhesion occurred preferentially on hydrophilic polymer surfaces (Yoo et al., 2009). Hence, surface hydrophilization of polymer materials has become an increasingly popular method to improve the biocharacteristics of the materials for good cell scaffolds (Venugopal et al., 2008).

Plasma treatment is an attractive method to modify polymer surface wettability (Fig. 8.8) (Egitto and Matienzo, 1994; Guruvenket et al., 2004; Hocker, 2002; Morent et al., 2008). Plasma treatment has been recognized as a versatile and effective method to modify the material surface, actively

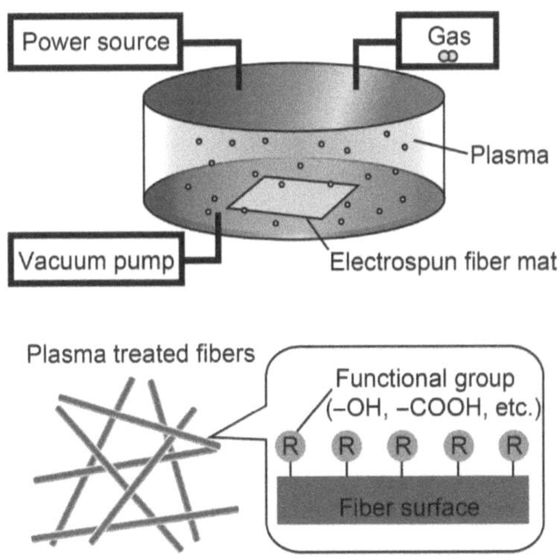

FIGURE 8.8 Plasma Treatment for Introducing Functional Group to Fiber Surface

investigated especially for bulk polymers, since 1990s (Egitto and Matienzo, 1994). Plasma treatment has been applied to the entire material substrate that was placed in the vacuum chamber. The chamber pressure and gas-flow rates often determined the residence time of the gases in the plasma (Egitto and Matienzo, 1994). The free electrons in the plasma interacted with atoms and molecules in the gas phase, leading to fragmentation (dissociation) and ionization that were highly chemically reactive. Plasma excitation sources usually came from either direct current or radio frequency. Plasma could exist at a greater electrical potential over the surface, where positive ions in the plasma were hugely accelerated toward the surface of the substrate.

Altering surface wettability by plasma treatment was caused by the chemical structural change on polymer substrates. Several types of gases such as oxygen (O_2), argon (Ar), ammonia (NH_3), water vapor (H_2O), and even air were commonly used (Venugopal et al., 2008; Yoo et al., 2009; Zheng et al., 2012; Zhu et al., 2005). These reactive dissociation products of the parent gas could break the chemical bonds in the target polymer, changing the surface chemical composition of the polymer. It is widely known that, by plasma treatment, various chemical configurations such as –C–C–, –C=C–, or –RX, where X is an element attached to an organic function group R, will be effectively reduced (Yang et al., 2002). In addition to the removal effects against organic materials, plasma treatment could result in the formation of hydrophilic functional groups such as hydroxyl group (–OH), carboxylic group (–COOH), and carbonyl group (C=O) (Lai et al., 2006) on polymer substrates. Such effects can be clearly confirmed by the results showing that the atomic ratios of O and N increased, while the ratio of C decreased (Cui and Brown, 2002; Lai et al., 2006). Thus, the introduction of the functional groups induced a wet surface on the plasma treated polymer.

Plasma treatment has also been applied for electrospun polymer nanofibers since 2000s, aiming to obtain a more desirable surface, especially on cell culture scaffolds (Jeong et al., 2009; Martins et al., 2009; Nandakumar et al., 2013; Prabhakaran et al., 2008; Yoo et al., 2009; Zanden et al., 2012). The modifications of various kinds of biomedical polymer nanofibers, including PCL, PLA, polyurethane (PU), PA, silk and so on, by plasma have been actively investigated and reported. Prabhakaran et al. (2008), concluded that surface modification by air plasma treatment increased the wettability of PLC fibers, leading to the enhancement of cell proliferation—one of the intriguing examples showing the effectiveness of plasma treatment on biodegradable nanofibers. In their research, the water contact angle of 122°, originally observed for untreated PLC-nanofiber mats, decreased to $\sim$110°, and the contact angle of O_2-treated specimens also drastically decreased even down to 0° after plasma treatment. This was mainly caused by the decrease in the C atomic ratio (from 75% to 62%), and the increase in the O atomic ratio (from 25% to 37%) on the surface of the nanofiber mats. In fact, the proliferation of Schwann cells, one of the neural cells, shown by the absorbance index, enhanced by 1.5 times using plasma treatment, as compared with untreated nanofiber samples.

4.3 SURFACE GRAFT COPOLYMERIZATION

Virtually all types of synthetic biodegradable polymers can retain their hydrophobic surface nature, so they often require a hydrophilic surface modification for the desired cellular responses. To achieve cell adhesion and cell proliferation, surface graft polymerization has been introduced to confer not only surface hydrophilicity, but also multifunctional groups on the surface for the covalent immobilization of bioactive molecules (Kou et al., 2003; Liu et al., 2004; Mori et al., 1994; Turmanova et al., 2008). Surface graft polymerization was often initiated with plasma and UV radiation treatment to generate

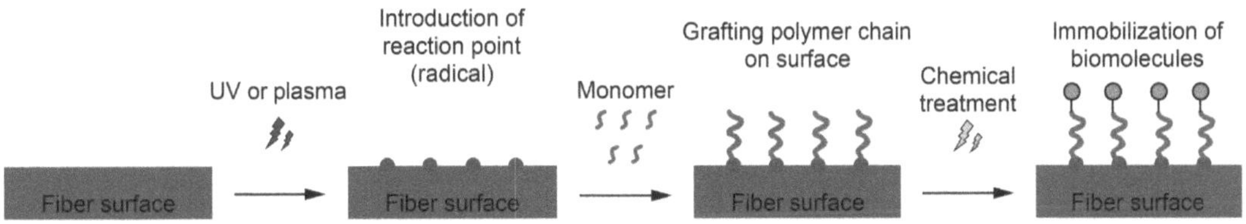

FIGURE 8.9 Surface Graft Copolymerization and Immobilization of Biomolecules

free radicals. The radicals produced on the polymer surface could immediately initiate the copolymerization of the monomer (Fig. 8.9).

Electrospun polyethylene terephthalate (PET) nanofibers were modified with poly(methacrylic acid), by graft polymerization in a mild condition, without any structural damage in the bulk phase (Ma et al., 2005b). The PET nanofibers were pretreated with formaldehyde to generate hydroxyls groups by Ce (IV) producing free radicals, thereby initiating the polymerization of methacrylic acid monomers from the surface. The carboxylic acid moieties on the surface were then conjugated to immobilize gelatin on the nanofibers. The surface density of carboxylic acid was shown to increase with the reaction time, and with monomer concentration.

For antibacterial applications, electrospun PU nanofibers were modified with poly(4-vinyl-*N*-hexyl pyridinium bromide) on the surface (Yao et al., 2008). In this study, the PU fibers were first treated with argon plasma that produced surface oxide and peroxide groups. When the plasma-treated PU fibers were immersed in a 4-vinylpyridine monomer solution, with exposure to UV irradiation, poly(4-vinilpyridine)-grafted PU fibers were successfully produced. Through the quaternization of the grafted pyridine groups with hexylbromide, the surface-modified PU fibers would start with antibacterial activities. The viability of Gram-positive *Staphylococcus aureus* and Gram-negative *Escherichia coli* after contact with the PU fibers was evaluated. The antibacterial efficacy of the modified PU fibers for *S. aureus* and *E. coli* were 99.999% and 99.9%, respectively, after 4 h of the contact, indicating a highly effective antibacterial activity.

5 CONCLUSIONS

In tissue engineering, it is necessary to use artificial scaffolds that imitate the structure of ECM. The scaffolds may function as the support for cells to regenerate the tissue that has been destroyed by injuries, diseases, or congenital defects. In this chapter, we introduced the electrospinning method and its techniques to control the morphologies of the electrospun scaffolds, and to modify the surface properties of electrospun fibers for the construction of new highly functionalized scaffolds.

Electrospinning is a valid method to fabricate polymeric fibrous scaffolds for biomaterials in tissue engineering. It is necessary to fabricate uniform fibers for use as fine fibers in tissue-engineering scaffolds. Many biocompatible and biodegradable polymeric fibers have been fabricated by electrospinning. The structures and morphologies of the electrospun scaffolds, such as fiber diameter, fiber orientation, and pore size, were found to be directly connected with cell attachment, cell proliferation, and cell infiltration. In addition, it was found that several surface modification methods, including loading biomolecules, plasma treatment, and surface graft copolymerization, could improve biocompatibility, cell attachment, and cell proliferation. The multiple morphology tuning methods and surface modification methods we proposed in this chapter may create excellent biocompatible scaffolds that may be used in future tissue engineering products.

REFERENCES

Agarwal, S., Wendorff, J.H., Greiner, A., 2008. Use of electrospinning technique for biomedical applications. Polymer 49, 5603–5621.

Assmann, U., Szentivanyi, A., Stark, Y., Scheper, T., Berski, S., Drager, G., Schuster, R.H., 2010. Fiber scaffolds of polysialic acid via electrospinning for peripheral nerve regeneration. J. Mater. Sci. Mater. Med. 21, 2115–2124.

Aviss, K.J., Gough, J.E., Downes, S., 2010. Aligned electrospun polymer fibres for skeletal muscle regeneration. Eur. Cell. Mater. 19, 193–204.

Baker, B.M., Gee, A.O., Metter, R.B., Nathan, A.S., Marklein, R.A., Burdick, J.A., Mauck, R.L., 2008. The potential to improve cell infiltration in composite fiber-aligned electrospun scaffolds by the selective removal of sacrificial fibers. Biomaterials 29, 2348–2358.

Barnes, C.P., Sell, S.A., Boland, E.D., Simpson, D.G., Bowlin, G.L., 2007. Nanofiber technology: designing the next generation of tissue engineering scaffolds. Adv. Drug Deliv. Rev. 59, 1413–1433.

Bashur, C.A., Dahlgren, L.A., Goldstein, A.S., 2006. Effect of fiber diameter and orientation on fibroblast morphology and proliferation on electrospun poly(D,L-lactic-*co*-glycolic acid) meshes. Biomaterials 27, 5681–5688.

Bashur, C.A., Shaffer, R.D., Dahlgren, L.A., Guelcher, S.A., Goldstein, A.S., 2009. Effect of fiber diameter and alignment of electrospun polyurethane meshes on mesenchymal progenitor cells. Tissue Eng. A 15, 2435–2445.

Beachley, V., Wen, X.J., 2009. Effect of electrospinning parameters on the nanofiber diameter and length. Mater. Sci. Eng. C 29, 663–668.

Boland, E.D., Wnek, G.E., Simpson, D.G., Pawlowski, K.J., Bowlin, G.L., 2001. Tailoring tissue engineering scaffolds using electrostatic processing techniques: a study of poly(glycolic acid) electrospinning. J. Macromol. Sci. Pure Appl. Chem. 38, 1231–1243.

Boland, E.D., Telemeco, T.A., Simpson, D.G., Wnek, G.E., Bowlin, G.L., 2004. Utilizing acid pretreatment and electrospinning to improve biocompatibility of poly(glycolic acid) for tissue engineering. J. Biomed. Mater. Res. B Appl. Biomater. 71B, 144–152.

Casper, C.L., Yamaguchi, N., Kiick, K.L., Rabolt, J.F., 2005. Functionalizing electrospun fibers with biologically relevant macromolecules. Biomacromolecules 6, 1998–2007.

Chan, K.H.K., Wong, S.Y., Tiju, W.C., Li, X., Kotaki, M., Bin He.F C., 2010. Morphologies and electrical properties of electrospun poly (R)-3-hydroxybutyrate-*co*-(R)-3-hydroxyvalerate/multiwalled carbon nanotubes fibers. J. Appl. Polym. Sci. 116, 1030–1035.

Chen, M., Patra, P.K., Warner, S.B., Bhowmick, S., 2007. Role of fiber diameter in adhesion and proliferation of NIH 3T3 fibroblast on electrospun polycaprolactone scaffolds. Tissue Eng. 13, 579–587.

Chen, M., Patra, P.K., Lovett, M.L., Kaplan, D.L., Bhowmick, S., 2009. Role of electrospun fibre diameter and corresponding specific surface area (SSA) on cell attachment. J. Tissue Eng. Regen. Med. 3, 269–279.

Chew, S.Y., Mi, R., Hoke, A., Leong, K.W., 2008. The effect of the alignment of electrospun fibrous scaffolds on Schwann cell maturation. Biomaterials 29, 653–661.

Choi, J.S., Lee, S.J., Christ, G.J., Atala, A., Yoo, J.J., 2008a. The influence of electrospun aligned poly(epsilon-caprolactone)/collagen nanofiber meshes on the formation of self-aligned skeletal muscle myotubes. Biomaterials 29, 2899–2906.

Choi, J.S., Leong, K.W., Yoo, H.S., 2008b. In vivo wound healing of diabetic ulcers using electrospun nanofibers immobilized with human epidermal growth factor (EGF). Biomaterials 29, 587–596.

Chow, W.N., Simpson, D.G., Bigbee, J.W., Colello, R.J., 2007. Evaluating neuronal and glial growth on electrospun polarized matrices: bridging the gap in percussive spinal cord injuries. Neuron Glia Biol. 3, 119–126.

Christopherson, G.T., Song, H., Mao, H.Q., 2009. The influence of fiber diameter of electrospun substrates on neural stem cell differentiation and proliferation. Biomaterials 30, 556–564.

Chua, K.N., Chai, C., Lee, P.C., Tang, Y.N., Ramakrishna, S., Leong, K.W., Mao, H.Q., 2006. Surface-aminated electrospun nanofibers enhance adhesion and expansion of human umbilical cord blood hematopoietic stem/progenitor cells. Biomaterials 27, 6043–6051.

Chua, K.N., Chai, C., Lee, P.C., Ramakrishna, S., Leong, K.W., Mao, H.Q., 2007. Functional nanofiber scaffolds with different spacers modulate adhesion and expansion of cryopreserved umbilical cord blood hematopoietic stem/progenitor cells. Exp. Hematol. 35, 771–781.

Corey, J.M., Lin, D.Y., Mycek, K.B., Chen, Q., Samuel, S., Feldman, E.L., Martin, D.C., 2007. Aligned electrospun nanofibers specify the direction of dorsal root ganglia neurite growth. J. Biomed. Mater. Res. A 83A, 636–645.

Cui, N.Y., Brown, N.M.D., 2002. Modification of the surface properties of a polypropylene (PP) film using an air dielectric barrier discharge plasma. Appl. Surf. Sci. 189, 31–38.

Deitzel, J.M., Kleinmeyer, J., Harris, D., Tan, N.C.B., 2001. The effect of processing variables on the morphology of electrospun nanofibers and textiles. Polymer 42, 261–272.

Edwards, M.D., Mitchell, G.R., Mohan, S.D., Olley, R.H., 2010. Development of orientation during electrospinning of fibres of poly(epsilon-caprolactone). Eur. Polym. J. 46, 1175–1183.

Egitto, F.D., Matienzo, L.J., 1994. Plasma modification of polymer surfaces for adhesion improvement. IBM J. Res. Dev. 38, 423–439.

Frey, M.W., Li, L., 2007. Electrospinning and porosity measurements of nylon-6/poly(ethylene oxide) blended nonwovens. J. Eng. Fibers Fabrics 2, 31–37.

Fujihara, K., Kotaki, M., Ramakrishna, S., 2005. Guided bone regeneration membrane made of polycaprolactone/calcium carbonate composite nano-fibers. Biomaterials 26, 4139–4147.

Fung, Y.C., 1981. Biomechanics: mechanical properties of living tissues. Springer-Verlag, New York, NY.

Ghasemi-Mobarakeh, L., Prabhakaran, M.P., Morshed, M., Nasr-Esfahani, M.H., Ramakrishna, S., 2008. Electrospun poly(epsilon-caprolactone)/gelatin nanofibrous scaffolds for nerve tissue engineering. Biomaterials 29, 4532–4539.

Gomes, M.E., Reis, R.L., 2004. Tissue engineering: key elements and some trends. Macromol. Biosci. 4, 737–742.

Greiner, A., Wendorff, J.H., 2007. Electrospinning: a fascinating method for the preparation of ultrathin fibres. Angew. Chem. Int. Ed. 46, 5670–5703.

Guimaraes, A., Martins, A., Pinho, E.D., Faria, S., Reis, R.L., Neves, N.M., 2010. Solving cell infiltration limitations of electrospun nanofiber meshes for tissue engineering applications. Nanomedicine 5, 539–554.

Guruvenket, S., Rao, G.M., Komath, M., Raichur, A.M., 2004. Plasma surface modification of polystyrene and polyethylene. Appl. Surf. Sci. 236, 278–284.

Hayati, I., Bailey, A.I., Tadros, T.F., 1987. Investigations into the mechanisms of electrohydrodynamic spraying of liquids.1. Effect of electric-field and the environment on pendant drops and factors affecting the formation of stable jets and atomization. J. Colloid Interface Sci. 117, 205–221.

He, L.M., Liao, S.S., Quan, D.P., Ma, K., Chan, C., Ramakrishna, S., Lu, J.A., 2010. Synergistic effects of electrospun PLLA fiber dimension and pattern on neonatal mouse cerebellum C17.2 stem cells. Acta Biomater. 6, 2960–2969.

Hocker, H., 2002. Plasma treatment of textile fibers. Pure Appl. Chem. 74, 423–427.

Huang, L., Mcmillan, R.A., Apkarian, R.P., Pourdeyhimi, B., Conticello, V.P., Chaikof, E.L., 2000. Generation of synthetic elastin-mimetic small diameter fibers and fiber networks. Macromolecules 33, 2989–2997.

Huang, Z.M., He, C.L., Yang, A.Z., Zhang, Y.Z., Hang, X.J., Yin, J.L., Wu, Q.S., 2006. Encapsulating drugs in biodegradable ultrafine fibers through co-axial electrospinning. J. Biomed. Mater. Res. A 77A, 169–179.

Ignatious, F., Sun, L.H., Lee, C.P., Baldoni, J., 2010. Electrospun nanofibers in oral drug delivery. Pharm. Res. 27, 576–588.

Innocente, F., Mandracchia, D., Pektok, E., Nottelet, B., Tille, J.C., De Valence, S., Faggian, G., Mazzucco, A., Kalangos, A., Gurny, R., Moeller, M., Walpoth, B.H., 2009. Paclitaxel-eluting biodegradable synthetic vascular prostheses: a step towards reduction of neointima formation? Circulation 120, S37–S45.

Ishihara, K., Ziats, N.P., Tierney, B.P., Nakabayashi, N., Anderson, J.M., 1991. Protein adsorption from human plasma is reduced on phospholipid polymers. J. Biomed. Mat. Res. 25, 1397–1407.

Ishihara, K., Nomura, H., Mihara, T., Kurita, K., Iwasaki, Y., Nakabayashi, N., 1998. Why do phospholipid polymers reduce protein adsorption? J. Biomed. Mater. Res. 39, 323–330.

Jeong, L., Yeo, I.-S., Kim, H.N., Yoon, Y.I., Jang, D.H., Jung, S.Y., Min, B.-M., Park, W.H., 2009. Plasma-treated silk fibroin nanofibers for skin regeneration. Int. J. Biol. Macromol. 44, 222–228.

Joung, Y.K., Bae, J.W., Park, K.D., 2008. Controlled release of heparin-binding growth factors using heparin-containing particulate systems for tissue regeneration. Exp. Opin. Drug Deliv. 5, 1173–1184.

Kim, T.G., Park, T.G., 2006. Biomimicking extracellular matrix: cell adhesive RGD peptide modified electrospun poly(D,L-lactic-*co*-glycolic acid) nanofiber mesh. Tissue Eng. 12, 221–233.

Kim, H.S., Yoo, H.S., 2010. MMPs-responsive release of DNA from electrospun nanofibrous matrix for local gene therapy: In vitro and in vivo evaluation. J. Controlled Release 145, 264–271.

Kim, K., Luu, Y.K., Chang, C., Fang, D.F., Hsiao, B.S., Chu, B., Hadjiargyrou, M., 2004. Incorporation and controlled release of a hydrophilic antibiotic using poly(lactide-*co*-glycolide)-based electrospun nanofibrous scaffolds. J. Control. Rel. 98, 47–56.

Kou, R.Q., Xu, Z.K., Deng, H.T., Liu, Z.M., Seta, P., Xu, Y.Y., 2003. Surface modification of microporous polypropylene membranes by plasma-induced graft polymerization of alpha-allyl glucoside. Langmuir 19, 6869–6875.

Kuppan, P., Sethuraman, S., Krishnan, U.M., 2014. Interaction of human smooth muscle cells with nanofibrous scaffolds: effect of fiber orientation on cell adhesion, proliferation, and functional gene expression. J. Biomed. Mater. Res. A 103, 2236–2250.

Lai, J.N., Sunderland, B., Xue, J.M., Yan, S., Zhao, W.J., Folkard, M., Michael, B.D., Wang, Y.G., 2006. Study on hydrophilicity of polymer surfaces improved by plasma treatment. Appl. Surf. Sci. 252, 3375–3379.

Lam, H.J., Patel, S., Wang, A.J., Chu, J., Li, S., 2010. In vitro regulation of neural differentiation and axon growth by growth factors and bioactive nanofibers. Tissue Eng. A 16, 2641–2648.

Langer, R., Vacanti, J.P., 1993. Tissue engineering. Science 260, 920–926.

Lannutti, J., Reneker, D., Ma, T., Tomasko, D., Farson, D.F., 2007. Electrospinning for tissue engineering scaffolds. Mater. Sci. Eng. C 27, 504–509.

Lee, J.Y., Bashur, C.A., Gomez, N., Goldstein, A.S., Schmidt, C.E., 2010. Enhanced polarization of embryonic hippocampal neurons on micron scale electrospun fibers. J. Biomed. Mater. Res. A 92A, 1398–1406.

Li, W.J., Tuli, R., Huang, X.X., Laquerriere, P., Tuan, R.S., 2005. Multilineage differentiation of human mesenchymal stem cells in a three-dimensional nanofibrous scaffold. Biomaterials 26, 5158–5166.

Li, S.F., Chen, J.P., Wu, W.T., 2007a. Electrospun polyacrylonitrile nanofibrous membranes for lipase immobilization. J. Mol. Catal. B Enzym. 47, 117–124.

Li, W.J., Mauck, R.L., Cooper, J.A., Yuan, X.N., Tuan, R.S., 2007b. Engineering controllable anisotropy in electrospun biodegradable nanofibrous scaffolds for musculoskeletal tissue engineering. J. Biomech. 40, 1686–1693.

Li, W.S., Guo, Y., Wang, H., Shi, D.J., Liang, C.F., Ye, Z.P., Qing, F., Gong, J., 2008. Electrospun nanofibers immobilized with collagen for neural stem cells culture. J. Mater. Sci. Mater. Med. 19, 847–854.

Liu, Z.M., Xu, Z.K., Wang, J.Q., Wu, J., Fu, J.J., 2004. Surface modification of polypropylene microfiltration membranes by graft polymerization of N-vinyl-2-pyrrolidone. Eur. Polym. J. 40, 2077–2087.

Lode, A., Reinstorf, A., Bernhardt, A., Wolf-Brandstetter, C., Konig, U., Gelinsky, M., 2008. Heparin modification of calcium phosphate bone cements for VEGF functionalization. J. Biomed. Mater. Res. A 86A, 749–759.

Lowery, J.L., Datta, N., Rutledge, G.C., 2010. Effect of fiber diameter, pore size and seeding method on growth of human dermal fibroblasts in electrospun poly(epsilon-caprolactone) fibrous mats. Biomaterials 31, 491–504.

Luong-Van, E., Grondahl, L., Chua, K.N., Leong, K.W., Nurcombe, V., Cool, S.M., 2006. Controlled release of heparin from poly(epsilon-caprolactone) electrospun fibers. Biomaterials 27, 2042–2050.

Luu, Y.K., Kim, K., Hsiao, B.S., Chu, B., Hadjiargyrou, M., 2003. Development of a nanostructured DNA delivery scaffold via electrospinning of PLGA and PLA-PEG block copolymers. J. Control. Rel. 89, 341–353.

Ma, Z.W., He, W., Yong, T., Ramakrishna, S., 2005a. Grafting of gelatin on electrospun poly(caprolactone) nanofibers to improve endothelial cell spreading and proliferation and to control cell orientation. Tissue Eng. 11, 1149–1158.

Ma, Z.W., Kotaki, M., Yong, T., He, W., Ramakrishna, S., 2005b. Surface engineering of electrospun polyethylene terephthalate (PET) nanofibers towards development of a new material for blood vessel engineering. Biomaterials 26, 2527–2536.

Ma, J.Y., He, X.Z., Jabbari, E., 2011. Osteogenic differentiation of marrow stromal cells on random and aligned electrospun poly(L-lactide) nanofibers. Ann. Biomed. Eng. 39, 14–25.

Madduri, S., Papaloizos, M., Gander, B., 2010. Trophically and topographically functionalized silk fibroin nerve conduits for guided peripheral nerve regeneration. Biomaterials 31, 2323–2334.

Maeda, T., Hagiwara, K., Yoshida, S., Hasebe, T., Hotta, A., 2014. Preparation and characterization of 2-methacryloyloxyethyl phosphorylcholine polymer nanofibers prepared via electrospinning for biomedical materials. J. Appl. Polym. Sci. 131, 6.

Martins, A., Pinho, E.D., Faria, S., Pashkuleva, I., Marques, A.P., Reis, R.L., Neves, N.M., 2009. Surface modification of electrospun polycaprolactone nanofiber meshes by plasma treatment to enhance biological performance. Small 5, 1195–1206.

Martins, A., Duarte, A.R.C., Faria, S., Marques, A.P., Reis, R.L., Neves, N.M., 2010. Osteogenic induction of hBMSCs by electrospun scaffolds with dexamethasone release functionality. Biomaterials 31, 5875–5885.

Matthews, J.A., Wnek, G.E., Simpson, D.G., Bowlin, G.L., 2002. Electrospinning of collagen nanofibers. Biomacromolecules 3, 232–238.

Meechaisue, C., Dubin, R., Supaphol, P., Hoven, V.P., Kohn, J., 2006. Electrospun mat of tyrosine-derived polycarbonate fibers for potential use as tissue scaffolding material. J. Biomater. Sci. Polym. Ed. 17, 1039–1056.

Megelski, S., Stephens, J.S., Chase, D.B., Rabolt, J.F., 2002. Micro- and nanostructured surface morphology on electrospun polymer fibers. Macromolecules 35, 8456–8466.

Moffat, K.L., Kwei, A.S.P., Spalazzi, J.P., Doty, S.B., Levine, W.N., Lu, H.H., 2009. Novel nanofiber-based scaffold for rotator cuff repair and augmentation. Tissue Eng. A 15, 115–126.

Morent, R., De Geyter, N., Verschuren, J., De Clerck, K., Kiekens, P., Leys, C., 2008. Non-thermal plasma treatment of textiles. Surface Coatings Technol. 202, 3427–3449.

Mori, M., Uyama, Y., Ikada, Y., 1994. Surface modification of polyethylene fiber by graft-polymerization. J. Polym. Sci.A Polym. Chem. 32, 1683–1690.

Nandakumar, A., Birgani, Z.T., Santos, D., Mentink, A., Auffermann, N., Van Der Werf, K., Bennink, M., Moroni, L., Van Blitterswijk, C., Habibovic, P., 2013. Surface modification of electrospun fibre meshes by oxygen plasma for bone regeneration. Biofabrication 5.

Neto, W.A.R., Pereira, I.H.L., Ayres, E., De Paula, A.C.C., Averous, L., Goes, A.M., Orefice, R.L., Bretas, R.E.S., 2012. Influence of the microstructure and mechanical strength of nanofibers of biodegradable polymers with hydroxyapatite in stem cells growth. Electrospinning, characterization and cell viability. Polym. Degrad. Stabil. 97, 2037–2051.

Nie, H., Soh, B.W., Fu, Y.C., Wang, C.H., 2008. Three-dimensional fibrous PLGA/HAp composite scaffold for BMP-2 delivery. Biotechnol. Bioeng. 99, 223–234.

Park, K., Ju, Y.M., Son, J.S., Ahn, K.D., Han, D.K., 2007. Surface modification of biodegradable electrospun nanofiber scaffolds and their interaction with fibroblasts. J. Biomater. Sci. Polym. Ed. 18, 369–382.

Patel, S., Kurpinski, K., Quigley, R., Gao, H.F., Hsiao, B.S., Poo, M.M., Li, S., 2007. Bioactive nanofibers: synergistic effects of nanotopography and chemical signaling on cell guidance. Nano Lett. 7, 2122–2128.

Prabhakaran, M.P., Venugopal, J., Chan, C.K., Ramakrishna, S., 2008. Surface modified electrospun nanofibrous scaffolds for nerve tissue engineering. Nanotechnology, 19.

Reneker, D.H., Chun, I., 1996. Nanometre diameter fibres of polymer, produced by electrospinning. Nanotechnology 7, 216–223.

Reneker, D.H., Kataphinan, W., Theron, A., Zussman, E., Yarin, A.L., 2002. Nanofiber garlands of polycaprolactone by electrospinning. Polymer 43, 6785–6794.

Rockwood, D.N., Akins, R.E., Parrag, I.C., Woodhouse, K.A., Rabolt, J.F., 2008. Culture on electrospun polyurethane scaffolds decreases atrial natriuretic peptide expression by cardiomyocytes in vitro. Biomaterials 29, 4783–4791.

Sahoo, S., Ang, L.T., Goh, J.C.H., Toh, S.L., 2010. Growth factor delivery through electrospun nanofibers in scaffolds for tissue engineering applications. J. Biomed. Mater. Res. A 93A, 1539–1550.

Shin, M., Ishii, O., Sueda, T., Vacanti, J.P., 2004. Contractile cardiac grafts using a novel nanofibrous mesh. Biomaterials 25, 3717–3723.

Sill, T.J., Von Recum, H.A., 2008. Electro spinning: applications in drug delivery and tissue engineering. Biomaterials 29, 1989–2006.

Stendahl, J.C., Wang, L.J., Chow, L.W., Kaufman, D.B., Stupp, S.I., 2008. Growth factor delivery from self-assembling nanofibers to facilitate islet transplantation. Transplantation 86, 478–481.

Subramanian, A., Krishnan, U.M., Sethuraman, S., 2010. Fabrication, characterization and in vitro evaluation of aligned PLGA-PCL nanofibers for neural regeneration. Annals of Biomedical Engineering 40, 2098–2110.

Sun, Z.C., Zussman, E., Yarin, A.L., Wendorff, J.H., Greiner, A., 2003. Compound core–shell polymer nanofibers by co-electrospinning. Adv. Mater. 15, 1929–1932.

Szentivanyi, A., Assmann, U., Schuster, R., Glasmacher, B., 2009. Production of biohybrid protein/PEO scaffolds by electrospinning. Materialwissenschaft und Werkstofftechnik 40, 65–72.

Tan, S.H., Inai, R., Kotaki, M., Ramakrishna, S., 2005. Systematic parameter study for ultra-fine fiber fabrication via electrospinning process. Polymer 46, 6128–6134.

Taylor, G., 1969. Electrically driven jets. Proc. R. Soc. Lond. A Math. Phys. Sci. 313, 453.

Teo, W.E., Ramakrishna, S., 2006. A review on electrospinning design and nanofibre assemblies. Nanotechnology 17, R89–R106.

Turmanova, S., Minchev, M., Vassilev, K., Danev, G., 2008. Surface grafting polymerization of vinyl monomers on poly(tetrafluoroethylene) films by plasma treatment. J. Polym. Res. 15, 309–318.

Venugopal, J., Zhang, Y.Z., Ramakrishna, S., 2005. Electrospun nanofibres: biomedical applications. J. Nanoeng. Nanosys. 218, 35–45.

Venugopal, J., Low, S., Choon, A.T., Kumar, A.B., Ramakrishna, S., 2008. Electrospun-modified nanofibrous scaffolds for the mineralization of osteoblast cells. J. Biomed. Mater. Res. A 85A, 408–417.

Verreck, G., Chun, I., Rosenblatt, J., Peeters, J., Van Dijck, A., Mensch, J., Noppe, M., Brewster, M.E., 2003. Incorporation of drugs in an amorphous state into electrospun nanofibers composed of a water-insoluble, nonbiodegradable polymer. J. Control. Rel. 92, 349–360.

Wang, H.B., Mullins, M.E., Cregg, J.M., Hurtado, A., Oudega, M., Trombley, M.T., Gilbert, R.J., 2009. Creation of highly aligned electrospun poly-L-lactic acid fibers for nerve regeneration applications. J. Neur. Eng. 6, 15.

Wang, C., Yan, K.-W., Lin, Y.-D., Hsieh, P.C.H., 2010. Biodegradable core/shell fibers by coaxial electrospinning: processing, fiber characterization, and its application in sustained drug release. Macromolecules 43, 6389–6397.

Wise, J.K., Yarin, A.L., Megaridis, C.M., Cho, M., 2009. Chondrogenic differentiation of human mesenchymal stem cells on oriented nanofibrous scaffolds: engineering the superficial zone of articular cartilage. Tissue Eng. A 15, 913–921.

Wong, W.H., Mooney, D.J., 1997. Synthesis and properties of biodegradable polymers used as synthetic matrices for tissue engineering. Birkhauser, Boston, MA.

Wu, Y.Q., Carnell, L.A., Clark, R.L., 2007. Control of electrospun mat width through the use of parallel auxiliary electrodes. Polymer 48, 5653–5661.

Wu, J., Wang, N., Zhao, Y., Jiang, L., 2013. Electrospinning of multilevel structured functional micro-/nanofibers and their applications. J. Mater. Chem. A 1, 7290–7305.

Xie, Z.W., Buschle-Diller, G., 2010. Electrospun poly(D,L-lactide) fibers for drug delivery: the influence of cosolvent and the mechanism of drug release. J. Appl. Polym. Sci. 115, 1–8.

Yamashita, Y., 2007. Electrospinning - the latest in nanotechnology. Senisha, Osaka, Japan.

Yang, J., Bei, J.Z., Wang, S.G., 2002. Improving cell affinity of poly(D,L-lactide) film modified by anhydrous ammonia plasma treatment. Polym. Adv. Technol. 13, 220–226.

Yang, F., Murugan, R., Wang, S., Ramakrishna, S., 2005. Electrospinning of nano/micro scale poly(L-lactic acid) aligned fibers and their potential in neural tissue engineering. Biomaterials 26, 2603–2610.

Yang, X.P., Thomas, D.P., Zhang, X.C., Culver, B.W., Alexander, B.M., Murdoch, W.J., Rao, M.N.A., Tulis, D.A., Ren, J., Sreejayan, N., 2006. Curcumin inhibits platelet-derived growth factor-stimulated vascular smooth muscle cell function and injury-induced neointima formation. Arterioscler. Thromb. Vascular Biol. 26, 85–90.

Yao, C., Li, X.S., Neoh, K.G., Shi, Z.L., Kang, E.T., 2008. Surface modification and antibacterial activity of electrospun polyurethane fibrous membranes with quaternary ammonium moieties. J. Membr. Sci. 320, 259–267.

Ye, P., Xu, Z.K., Wu, J., Innocent, C., Seta, P., 2006. Nanofibrous membranes containing reactive groups: electrospinning from poly(acrylonitrile-*co*-maleic acid) for lipase immobilization. Macromolecules 39, 1041–1045.

Yi, F., Lavan, D.A., 2008. Poly(glycerol sebacate) nanofiber scaffolds by core/shell electrospinning. Macromol. Biosci. 8, 803–806.

Yoo, H.S., Kim, T.G., Park, T.G., 2009. Surface-functionalized electrospun nanofibers for tissue engineering and drug delivery. Adv. Drug Deliv. Rev. 61, 1033–1042.

Yoon, J.J., Chung, H.J., Park, T.G., 2007. Photo-crosslinkable and biodegradable pluronic/heparin hydrogels for local and sustained delivery of angiogenic growth factor. J. Biomed. Mater. Res. A 83A, 597–605.

Yoshida, M., Langer, R., Lendlein, A., Lahann, J., 2006. From advanced biomedical coatings to multi-functionalized biomaterials. Polym. Rev. 46, 347–375.

Yoshimoto, H., Shin, Y.M., Terai, H., Vacanti, J.P., 2003. A biodegradable nanofiber scaffold by electrospinning and its potential for bone tissue engineering. Biomaterials 24, 2077–2082.

Zanden, C., Voinova, M., Gold, J., Moersdorf, D., Bernhardt, I., Liu, J., 2012. Surface characterisation of oxygen plasma treated electrospun polyurethane fibres and their interaction with red blood cells. Eur. Polym. J. 48, 472–482.

Zeng, J., Chen, X.S., Liang, Q.Z., Xu, X.L., Jing, X.B., 2004. Enzymatic degradation of poly(L-lactide) and poly (epsilon-caprolactone) electrospun fibers. Macromol. Biosci. 4, 1118–1125.

Zeng, J., Aigner, A., Czubayko, F., Kissel, T., Wendorff, J.H., Greiner, A., 2005a. Poly(vinyl alcohol) nanofibers by electrospinning as a protein delivery system and the retardation of enzyme release by additional polymer coatings. Biomacromolecules 6, 1484–1488.

Zeng, J., Yang, L.X., Liang, Q.Z., Zhang, X.F., Guan, H.L., Xu, X.L., Chen, X.S., Jing, X.B., 2005b. Influence of the drug compatibility with polymer solution on the release kinetics of electrospun fiber formulation. J. Control. Rel. 105, 43–51.

Zhang, Y.Z., Huang, Z.M., Xu, X.J., Lim, C.T., Ramakrishna, S., 2004. Preparation of core–shell structured PCL-r-gelatin bi-component nanofibers by coaxial electrospinning. Chem. Mater. 16, 3406–3409.

Zheng, Z., Ren, L., Feng, W., Zhai, Z., Wang, Y., 2012. Surface characterization of polyethylene terephthalate films treated by ammonia low-temperature plasma. Appl. Surf. Sci. 258, 7207–7212.

Zhong, S.P., Teo, W.E., Zhu, X., Beuerman, R.W., Ramakrishna, S., Yung, L.Y.L., 2006. An aligned nanofibrous collagen scaffold by electrospinning and its effects on in vitro fibroblast culture. J. Biomed. Mater. Res. A 79A, 456–463.

Zhu, X., Chian, K.S., Chan-Park, M.B.E., Lee, S.T., 2005. Effect of argon-plasma treatment on proliferation of human-skin-derived fibroblast on chitosan membrane in vitro. J. Biomed. Mater. Res. A 73A, 264–274.

CHAPTER

9

SHORT PEPTIDE SELF-ASSEMBLED NANOSTRUCTURES FOR THERAPEUTICS INNOVATIVE DELIVERY

Daniel Iglesias, Silvia Marchesan
University of Trieste, Trieste, Italy

CHAPTER OUTLINE

1 INTRODUCTION

1.1 THE ADVANTAGES OF NANOMATERIALS

The advent of nanomaterials set an important milestone for the progress of medicine, thanks to the unprecedented advantages and properties that arise from working at nanoscale (Marchesan and Prato, 2013). It is not just a matter of a quantitative enhancement, in terms of early disease diagnosis, and efficient targeted drug delivery that allows for reduced therapeutic doses. The unique physical and chemical properties of nanostructures allow for innovative performance, in terms of a qualitative leap in therapy that can be realized through unprecedented paradigms, as we discuss in this chapter.

In addition to specific properties of certain nanomaterials (e.g., optical, magnetic, electronic properties), interesting added benefits can be achieved by the means of supramolecular behavior, and especially through the self-organization of small molecules in discrete nanomorphologies. Innovative means of drug delivery can be realized through interaction with self-assembling building blocks, and/or drug incorporation into supramolecular structures, or nanomaterial matrices. Drugs can also be covalently linked to specific self-assembling motifs to achieve a multifunctional building block, capable of functional (e.g., via the drug therapeutic effect) and structural (e.g., via the self-assembling behavior) features.

Moreover, the supramolecular structures achieved by self-organization of small molecules can also exert a desired biological effect per se, without the need for the inclusion of a small molecule drug. For instance, there are emerging reports on the exploitation of a cytotoxic effect exerted by supramolecular nanofibers in cancer therapy, in order to achieve selective cell death at a disease site (Kuang et al., 2014). In line of principle, this interesting phenomenon can be combined with the enhanced retention permeability (EPR) effect, which translates into enhanced uptake of nanomaterials and nanostructures from cancer cells, relative to healthy cells, due to generally increased vasculature permeability, as well as lack of efficient draining from the lymphatic system (Maeda, 2015).

Overall, nanomaterials and supramolecular nanostructures are revealing their full potential in nanomedicine, with an increasingly higher number of scientific reports on the exploitation of their interesting properties for added benefits in nanomedicine, to realize a significant progress in the treatment of unmet medical needs that was simply unthinkable just a few years back. Clearly, there are still many unsolved questions that need to be addressed, and many challenges to be overcome, including appropriate regulation and assessment of such new nanoscale entities (Scott-Fordsmand et al., 2014). Nevertheless, the field is clearly making rapid progress, and holds promise to provide new, effective solutions in medicine.

1.2 SELF-ASSEMBLING SHORT PEPTIDES AS NANOMATERIAL BUILDING BLOCKS

Among the wide variety of nanostructured materials that have been developed in recent years, self-assembling peptides have played an elective role in medicine, thanks to their inherent ability to encode diverse biological messages for cells or their subcomponents, such as proteins. However, peptides tend to suffer from the well-known disadvantage of high-cost production and purification, sensitivity to environmental conditions leading to limited durability in biological settings, and presence of many different functional groups that may render their chemical derivatization quite a challenge. The scientific effort dedicated to the identification of simple and short peptide sequences devoid of these drawbacks, and capable of elaborate supramolecular behavior, is thus not surprising.

A leading role in the field has certainly been played by fragments of the amyloid beta peptide, identified as a component of amyloid plaques present in the pathological condition associated with Alzheimer's disease. Many research groups have attempted to identify the core sequences that are responsible for the marked self-assembling ability, and, among these, the pentapeptide Lys-Leu-Val-Phe-Phe is probably the most well-known. Interestingly, this was also the shortest L-peptide devoid of unnatural chemical derivatizations that was reported to form a hydrogel in phosphate-buffered-saline solution in 2008 (Krysmann et al., 2008). However, already in 2003, through a reductionist approach aimed at identifying the shortest self-assembling peptide motif from the amyloid beta peptide sequence, the diphenylalanine dipeptide was reported as a powerful self-assembling motif. Phe-Phe was shown to be capable of self-organization into persistent nanotubes (but not a hydrogel) in aqueous solutions, upon dilution of a highly concentrated sample dissolved in hexafluoroisopropanol (Reches and Gazit, 2003). This seminal work opened up an entirely new field based on the exploitation of the Phe-Phe motif to drive self-assembly and hydrogelation of small molecules and peptide derivatives (Yan et al., 2010).

For example, addition of a third, hydrophobic amino acid to the Phe-Phe sequence in a tripeptide composed of both D- and L-amino acids was reported as a useful approach to exploit appropriate choice of amino acid chirality, to obtain hydrogelling tripeptides at physiological conditions under which the L-homochiral peptide stereoisomers did not gel (Marchesan et al., 2012a). The concept can also be extended to enantiomers and other stereoisomers, thus widening the scope of the method (Marchesan et al., 2015). In line of principle, use of D-amino acids is also advantageous to achieve resistance toward protease-mediated peptide digestion, and therefore to increase the lifetime of such self-assembling building blocks in biological settings, as well as other interesting effects (Melchionna et al., 2016). An alternative strategy employs the use of amino acid or peptide synthetic analogs, such as dehydrophenylalanine (Panda et al., 2008) or beta-peptides (Bonetti et al., 2015).

Another successful approach to derive self-assembling short peptides, or even single amino acids, is the conjugation at the N-terminal to an aromatic, rigid template, such as fluorenylmethyloxycarbonyl (Fmoc), or naphthalene (Nap) (Fleming and Ulijn, 2014). Use of the Fmoc moiety is particularly attractive, since it is typically employed as amino-protecting group in solid-phase peptide synthesis (SPPS), therefore every commercially available amino acid for SPPS can be bought as a Fmoc-derivative. However, examples of cytotoxicity have been reported with this class of hydrogelators, raising concern over the use of aromatic moieties as capping groups (Martin et al., 2014; Orbach et al., 2009; Wang et al., 2011).

Not surprisingly, there is a very high interest in the identification of key principles to design ex novo uncapped, self-assembling di- and tripeptides as building blocks for hydrogel biomaterials; however, this objective is extremely challenging, with a survey of all $20^3 = 8{,}000$ combinations of natural L-amino acids in tripeptides leading only to a very few hits (Frederixpim et al., 2015). Such short sequences made of up to three amino acids have the obvious advantage of simple and low-cost preparation by means of liquid-phase chemistry or biocatalysis, in contrast with more elaborate and longer peptide sequences (Adams, 2011). Indeed, addition of each amino acid to produce longer sequences significantly increases the cost of production, and renders the cheap liquid-phase synthesis more and more challenging. Nevertheless, it is generally considered that short peptides composed of up to approximately eight amino acids are still attractive building blocks for nanomaterials, especially when compared to sequences made of 20 or 30 amino acids, or even more residues, to form polypeptides.

Although the area of short self-assembling peptides (especially di- and tripeptides) is dominated by the beta-sheet forming sequences bearing the Phe-Phe motif, and/or those peptides N-capped with aromatic moieties, such as Fmoc or Nap, other classes of short peptides with supramolecular behavior are recently being identified through a reductionist approach (Fig. 9.1). Interestingly, the amino acid phenylalanine is a recurrent building block also in these other classes of self-assembling peptides, and this fact is not too surprising, considering its high propensity toward self-organization; this is shown by the fact that, also on its own, Phe is able to form nanofibers in biological settings

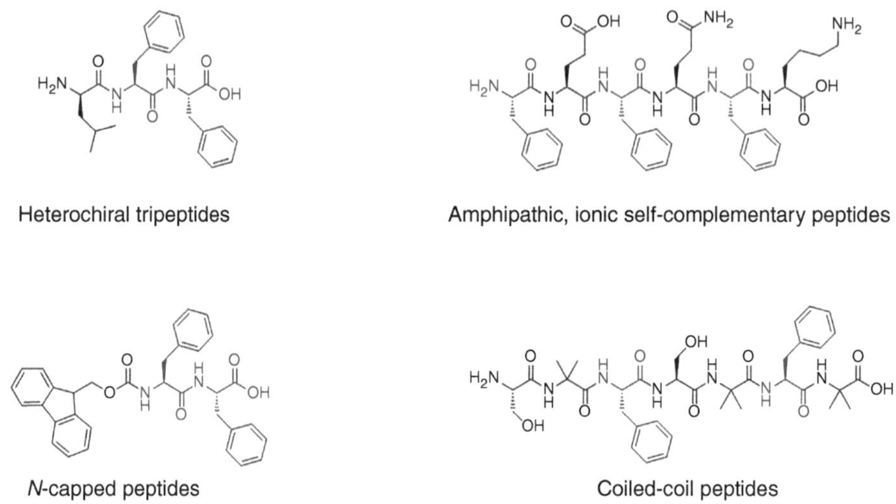

FIGURE 9.1 Examples of Hydrogelling Short Peptides With the Recurrent Phe Amino Acid

(Adler-Abramovich et al., 2012). For instance, the amphipathic, ionic, self-assembling EAK16 peptide reported by Zhang et al. (1993) was composed of the sequence Ac-Ala-Glu-Ala-Glu-Ala-Lys-Ala-Lys-Ala-Glu-Ala-Glu-Ala-Lys-Ala-Lys, and was reported to form beta-sheets, and self-assemble into a macroscopic membrane upon the addition of salt to an aqueous solution of the peptide. Based on this pioneering work, other groups reported shorter amino acid sequences capable of self-assembling and hydrogelling behavior, such as the octamer Phe-Glu-Phe-Lys-Phe-Glu-Phe-Lys (Mohammed et al., 2007), or Phe-Lys-Phe-Glu-Phe-Lys-Phe-Glu (Bowerman et al., 2009). Recently, a shorter hydrogelling sequence of this kind was reported to be the hexamer Phe-Glu-Phe-Gln-Phe-Lys (Bibian et al., 2015).

Another motif that is widely studied in peptide self-assembly is the α-helical coiled-coil motif. The typical motif is based on the repetition of heptads of amino acids, for which the $(abcdefg)_n$ nomenclature is used. Positions *a,d,e,* and *g* are involved in the noncovalent interactions that direct dimer interface, while positions *b,c*, and *f* are exposed on the surface of the coil. Usually, positions *a* and *d* are occupied by hydrophobic amino acids that are involved in the hydrophobic interactions between helices in a "knob into hole" fashion. Several geometries exist for the supramolecular arrangements of helices (Apostolovic et al., 2010). In any case, being composed of heptad-repeats, coiled-coils are generally formed by long peptides. However, one example was reported recently of a minimalist coiled-coil composed of a peptide heptamer that can, therefore, be included in the realm of short self-assembling motifs, discussed herein. To overcome the fact that a helical conformation is energetically disfavored for a 7-residue peptide, the sequence was designed to contain a residue that is known to stabilize short helices, that is, the noncoded amino acid α-aminoisobutyric acid (Aib), an extension of alanine with a dimethylated α-carbon atom. The presence of the additional α-methyl group, in fact, reduces the number of possible conformations. Taking inspiration from leucine zipper proteins, the heptad was designed to display leucine residues at the *a* and *d* positions; however, leucine substitution with phenylalanine afforded a more rigid structure which gave the characteristic α-helical signal in the circular dichroism spectrum. Single crystal X-ray analysis confirmed a supramolecular architecture held together by an "aromatic zipper" composed of packed phenylalanine residues at the interface between helices (Mondal et al., 2015). The overall supramolecular arrangement of phenylalanine residues was similar to what was previously reported for the gelled heterochiral tripeptide ^{D}Leu-Phe-Phe, where a phenylalanine zipper was proposed as a powerful motif for the self-assembly of short peptides (Marchesan et al., 2012b).

1.3 SELF-ASSEMBLING SHORT PEPTIDES FOR DRUG DELIVERY

Overall, capped and uncapped amino acids, or short peptides and their derivatives, constitute highly advantageous building blocks for the self-assembly of nanostructured biomaterials. Although a variety of biological applications can be envisaged, drug delivery is probably one of the key uses for such systems, and rapid progress is being made worldwide in this area of research, holding promise to translate into the clinic innovative nanotechnological solutions for the efficient and safe delivery of therapeutics. This can be achieved by several means, namely: (1) physical entrapment in the hydrogel; (2) noncovalent drug binding; (3) covalent linkage of the drug (Fig. 9.2). We shall now analyze more in detail the successful application of the different approaches for the delivery of therapeutics or imaging agents.

Besides hydrogels, peptide-based supramolecular nanostructures can also be used for the delivery of therapeutics, considering that they can be able to penetrate cells, and localize in specific subcellular regions and, despite their molecular simplicity, can display remarkable stability. As an example, fluorinated beta-dipeptides that are derivatives inspired to the Phe-Phe motif can form nanotubes that enter cells and locate in the perinuclear region (Bonetti et al., 2015). Interestingly, their supramolecular arrangement leads to nanotubes that are stable to protease degradation, and heating up to 120°C. Besides, preliminary cytotoxicity data showed very promising results for their use as systemic drug delivery nanovehicles. Indeed, assays were performed on cultured primary human smooth muscle cells that are derived from human vessel, and therefore offer an appropriate model of the most abundant cell type that is directly exposed to delivery systems present in the bloodstream. After 2 days of cell culture, no cytotoxic effect was noted for peptide concentrations ranging from 1 to 100 μM, while at 200 μM cell viability was reduced to about 80%, relative to the control (Bonetti et al., 2015).

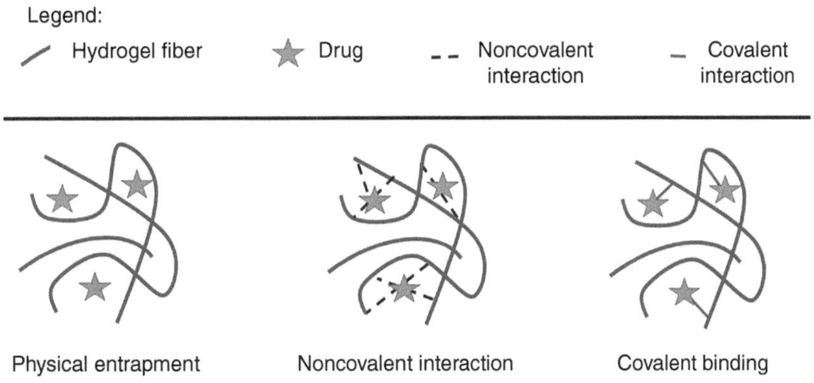

FIGURE 9.2 Different Modes for Drug Entrapment in a Hydrogel

2 PREPARATION AND CHARACTERIZATION OF SELF-ASSEMBLING SHORT PEPTIDES

2.1 PREPARATION OF SHORT PEPTIDES

The preparation of short peptides, especially those composed of only two to three amino acids, is particularly convenient, relative to that of longer analogs. Polypeptides are typically obtained by means of expensive solid-phase peptide synthesis (SPPS), or through recombinant DNA techniques; in either case, there are several technical challenges that need to be addressed for their successful preparation that typically requires specialized expertise, and dedicated equipment. By contrast, short peptides can be prepared by means of many different techniques, also at low cost, and, depending on the sequence, their ease of preparation makes them accessible also to scientists with a nonsynthetic background. For these reasons, they are very attractive (bio)nanomaterial building blocks not only to chemists, but also to materials scientists, and even to biochemists. In addition, there is a plethora of companies currently offering custom-made peptides at very reasonable prices, and it is not uncommon to see scientific works that make use of such services, and simply buy the peptides, to avoid the inconvenience of preparation, purification, and characterization of the compounds.

If we look at the different options to prepare short peptides, the most used technique is probably SPPS, even though this approach makes use of a large amount of reagents and solvents, and thus it would be more appropriate to use it for polypeptides that cannot be easily accessed by other means. The main reasons that make this approach so popular are probably the high yields, and simplicity of the experimental protocols, which can rapidly be applied for the preparation of different peptide sequences, and can easily be used to instruct young scientists. The convenient application of colorimetric tests as "check-points" to verify the success of each synthetic step is another advantageous feature that simplifies monitoring of the process, and trouble-shooting. The main disadvantage of SPPS is the elevated cost that arises partly from the use of specialized resins, and partly from the use of a large excess of reagents and high-quality solvents (i.e., dry dichloromethane and amine-free dimethylformamide). In addition, the large amount of toxic waste generated from SPPS, including halogenated solvents, requires appropriate management.

Liquid-phase synthesis is probably a more sensible approach to prepare short peptides, especially those that contain only two to three or four amino acids. This technique is by far more economical than SPPS, and generates notably less waste. The simplicity of scaling up this class of experimental protocols (relative to SPPS) makes it also particularly attractive for commercial purposes. However, preparation of peptides following this chemical route requires more care and technical skills, making it more suitable to the experienced chemist, or at least to the attentive and diligent research student. Scientists with a background different from organic chemistry will likely find the technical barriers posed by this approach too high to be easily overcome. Compound purification and close monitoring to avoid the occurrence of undesired side reactions are also critical steps.

A different route gaining more and more approval among scientists is the use of biocatalysis, that is, the use of enzymes to produce peptides. This route is obviously far more ecological relative to the previous ones, typically making use of aqueous media or, sometimes, aqueous mixtures containing nonhalogenated solvents. In the past, enzyme availability was a limiting factor for the widespread application of biocatalysis, but industrial enzymes at very different levels of purity (and thus cost) are currently offered by a high number of companies, according to the taste and needs of almost every customer. From a biochemical point of view, the most obvious choice to catalyze oligopeptide synthesis is an enzyme that belongs to the class of ligases (Arai and Kino, 2010). However, these proteins are ATP-dependent, and this is a disadvantage both in terms of costs, and in terms of technical expertise, since ATP has well-known issues of stability, being prone to hydrolysis, if no appropriate care is given to its storage and use. Therefore, other enzyme classes that do not require nucleoside triphosphates for their activity would be preferable options. Besides, there is an increasing number of reports on the promiscuity of enzymes that are tolerant to unnatural substrates, or to substrates that are different to the ones those enzymes are designed to process. Moreover, depending on the applied experimental conditions, enzymes can catalyze reactions also in reverse, relative to their physiological activity (Toledano et al., 2006). This translates into the opening of many possibilities for the use of enzymes to prepare unnatural peptides, or different peptides using the same enzyme. Importantly, it opens the way for using convenient hydrolases for peptide synthesis, instead of the ATP-dependent ligases. For instance, a series of Fmoc-Phe-Phe-Phe stereoisomers, including D-amino acids, such as Fmoc-^{D}Phe-Phe-Phe, Fmoc-Phe-^{D}Phe-^{D}Phe, and Fmoc-^{D}Phe-^{D}Phe-^{D}Phe, was reported from the corresponding Fmoc-phenylalanine and diphenylalanine isomers by employing a promiscuous lipase in reverse. The interesting aspect is that the synthesis of Fmoc-^{D}Phe-^{D}Phe-^{D}Phe occurred in higher yield, relative to the L-peptide, Fmoc-Phe-Phe-Phe. This fact can be rationalized considering that there exists an equilibrium, and the peptide hydrolysis is favored for the L-peptide derivative, relative to the D-analog (Chronopoulou et al., 2014).

Finally, another possible avenue for short peptide preparation makes use of recombinant techniques, which are usually a preferred choice for the production of higher molecular weight compounds that are not easily obtained by alternative means (e.g., long polypeptides or full-length proteins). Nevertheless, bacterial recombinant expression of peptides composed of less than 10 amino acids is possible, and in high yield (Zorko et al., 2009). Biotechnology also offers other possibilities for the production of peptides and amino acids, including those with D-stereoconfiguration, through the creation of new metabolic pathways (Taylor et al., 2001). All of these biotech routes are very attractive, especially for commercial applications, in light of the fact that they allow for large-scale production in a highly reproducible manner. The time-consuming step is the optimization of experimental conditions, especially due to the fact that it may be difficult to predict those that allow for a successful outcome in terms of recombinant peptide expression. A common issue is peptide degradation, difficulties in product isolation, or bacteria cytotoxicity with consequent low-yield in peptide production. Therefore, this approach is preferable for the systematic preparation of a specific sequence, rather than the small-scale expression of different peptides. Another aspect to consider is the training and adoption of biosafety measures that are required for the correct handling of genetically modified organisms. Biotech dedicated equipment is another factor that makes this route not accessible to many.

2.2 CHARACTERIZATION OF SELF-ASSEMBLING SHORT PEPTIDES

Supramolecular characterization of self-assembling short peptides is not always a trivial task, and scientific videos can be very helpful to assist the nonexpert in the field in the technical protocols to handle such systems for correct sample preparation and successful characterization (Yuran and Reches, 2013). Ideally, the systems should be analyzed in their self-organized state to gain truthful information about the noncovalent interactions, and the peptide conformation that occur in the supramolecular architecture. However, this can be a real challenge, and quite often the systems are analyzed under conditions that are different, compared to those required for self-assembly (e.g., in solution, as opposed to a gelled state, or in diluted systems at low concentrations, as opposed to more concentrated systems to achieve self-assembly). Therefore, many assumptions have to be made, and this kind of approach probably raises more questions than the answers it provides (e.g., is the peptide conformation in diluted samples the same as that in the self-assembled state?). For instance, circular dichroism is a well-established and very useful technique to gain insights of peptide conformation, and of chirality of supramolecular arrangements. Unfortunately, this technique performs at best for samples that are very diluted in aqueous solutions, and even the presence of salts and buffer can hinder data acquisition in highly informative regions of the spectra. A compromise must often be made, and through careful selection of experimental conditions for both self-assembly and data acquisition, supramolecular systems can be analyzed, although typically not below 200 nm, where the spectrometers are very sensitive, and noise becomes easily an issue.

From a materials science perspective, there are several techniques that are commonly applied to polymers and that can also be transferred to supramolecular systems composed of short peptides. An obvious characterization technique to employ for systems that form macroscopic hydrogels is rheometry that probes viscoelastic properties, and provides valuable data about material resistance to stress and strain, and thus about material failure, and recovery, as well as gel to sol transitions, and vice versa. Thermogravimetric analysis and differential scanning calorimetry are also very informative of the material behavior at different temperatures, with the added benefit of offering the possibility to measure the enthalpy of defined transitions. Despite the high relevancy of these characterization methods, they are not all widely adopted in the field of self-assembled peptide systems.

Far more common is the use of microscopy methods, possibly also because of the urge to visualize the nanomorphologies of the assemblies. In particular, transmission electron microscopy (TEM), both at cryogenic temperatures and at room temperature, with the aid of specific stains that increase contrast, is a popular option thanks to the high amount of data that can be generated in relatively short sessions. With TEM, one can observe a large area of the sample, and then switch to high magnification to capture fine details of a nanomorphological feature within a very reasonable time. Not surprisingly, the technique is well established for self-assembling peptides (Gras et al., 2011). Atomic force microscopy (AFM) is another very useful characterization method for nanoscale-systems of this kind, but suffers from the fact that it is very time-consuming, not only to set up the instrument and correctly observe a certain area of the sample, but also to switch from an area to another in distant regions of the sample. Correct data acquisition with AFM is more challenging than with TEM, and a proof of this fact is given by the large amount of image artefacts that can be found across scientific literature. However, AFM has the advantage to be able to probe even single nanostructures for different properties, not just physical dimensions, but also elasticity, conductivity, etc., depending on the instrument setup. AFM is indeed a technique of choice for the characterization of self-assembling peptides (Adamcik and Mezzenga, 2012). Similarly, scanning tunneling microscopy (STM) can provide compelling images with fine level of detail, as well as very accurate information about molecules' spatial organization in the supramolecular arrangements. STM is often combined with AFM for the characterization of self-assembling peptides (Wang et al., 2003). Scanning electron microscopy (SEM) is less common in the field, since it is usually preferred for the characterization of thin coatings and, generally, surfaces. SEM is also very useful for the study of self-assembled systems (Takai et al., 2014). Optical microscopy is not traditionally considered a first choice to visualize nanostructures due to the size limits of the magnification that can be achieved with this method. Things may start to be different as the latest developments in the field have made super-resolution instrumentation a viable option (Schermelleh et al., 2010). Besides, light microscopy can be used in combination with specific stains that are well-known structural markers, for instance, to confirm the amyloid nature of self-assembled systems (Biancalana and Koide, 2010; Howie and Brewer, 2009).

There are also other techniques that can be highly informative and/or are gaining recent consensus to characterize self-assembled peptides. X-ray techniques, including diffraction, and small-angle and wide-angle scattering, provide precious tools to investigate the fine structure of peptides, but are not easily accessible to many, be it for their high cost (also of maintenance), for the expertise required for correct data acquisition and processing, or for the technical challenges posed by sample preparation. Nuclear magnetic resonance is a far more accessible technique that is becoming more and more popular, as new methods are being developed to study supramolecular systems (Dudukovic and Zukoski, 2015; Pudakalakatti et al., 2014).

3 DRUG-DELIVERY APPLICATIONS

3.1 SUPRAMOLECULAR HYDROGELS AS SYSTEMS FOR DRUG DELIVERY

Already in 1960, it was evident that conventional plastic materials and human tissue are likely not compatible, both from a structural and biological point of view (Wichterle and Lim, 1960). Evidently, materials with different properties are required for biomedical use. Hydrogels stand out for their highly hydrated and porous structure capable of transporting nutrients and metabolites effectively, and of accommodating drugs, bioactive molecules, or imaging agents, in the reservoirs present within their matrix (Hoffman, 2002). They can be considered as a network of entangled fibers, composed of macromolecules or self-assembled architectures built from small molecules. The network is characterized by a mesh size that can be correlated to the ability of the hydrogel to entrap and release drug molecules. An additional advantage is the possibility to fine-tune the physical and chemical properties of the soft materials through changes in the chemical structure of their components, the gelation process, the hydrogel matrix architecture, and thus mesh size, or their microenvironment (i.e., pH, ionic strength, viscosity, and so on). Therefore, hydrogels are an appropriate choice for localized drug delivery, a fact particularly advisable not only to reduce side effects that would arise from systemic administration, but also to improve the bioavailability of drugs that are metabolized rapidly, or that need to be administered to a site that presents physiological barriers for systemic delivery. Besides, they offer ideal solutions to maintain a moist environment, as required, for instance, for wound dressings to be applied to chronic wounds or severe burns.

Although preformed hydrogels can thus find certain application in the biomedical field, current research is focused on injectable systems that gel in situ, for instance, driven by a change of pH, temperature, ionic strength, or exposure to light, electromagnetic field, etc. Such stimuli-responsive systems offer the advantageous possibility to be mixed with cells or other biological components in a low-viscosity state (i.e., in solution), which can then be injected at a target site to form a gel depot for sustained release, and/or to allow for cell infiltration and tissue regeneration (Singh and Lee, 2014). The majority of injectable systems are composed of biodegradable components, such as natural polymers (e.g., chitosan, alginate, gelatin, chondroitin sulfate, and so on) or synthetic polymers (e.g., polypeptides, polyesters, and so on) (Li et al., 2012). Self-assembled systems are rising as an attractive alternative (Chung and Park, 2009), especially in light of the fact that their gelation typically occurs through physiologically benign processes, as opposed to chemically-cross-linked polymers that may involve use of reactive radicals or toxic monomers. In addition, supramolecular systems employ non-covalent interactions, which require significantly less energy, relative to covalent bonds, to be formed or broken. This means that, in line of principle, supramolecular hydrogels can be reversible, and this is an attractive feature, especially for smart systems that can adapt to specific conditions (e.g., storage, injection, on-demand drug delivery). Hybrid systems that exploit the synergy between different components (e.g., self-assembled peptides and other macromolecules or nanomaterials) are also interesting options to achieve unprecedented functional materials with hierarchical organization (Iglesias et al., 2016; Kopecek and Yang, 2012).

3.2 PHYSICAL DRUG ENTRAPMENT IN THE SUPRAMOLECULAR HYDROGEL MATRIX

Hydrogels are well-established vehicles for drug delivery, and the simplest method to incorporate a drug is through physical entrapment in the hydrogel matrix. The diffusion of small molecules within and out of the hydrogel matrix is a crucial aspect in the context of drug delivery. This kind of vehicle is well-suited for hydrophobic, poorly-water soluble drugs, but can also be employed for hydrophilic components.

Fmoc-Arg-Gly-Asp is a simple, yet effective, hydrogelator that is capable of forming rigid and porous hydrogels, of remarkable stability at 10 wt.%. The materials can be prepared as monoliths that persist in water for nearly 40 days, although release of the building blocks can be observed in the form of Fmoc-peptide monomers first, and self-assembled nanofibers later. Encapsulation (and subsequent release) of bioactive molecules or imaging agents was successfully performed in the monoliths, using various model compounds, such as salicylic acid, riboflavin, methylene blue. Loading of small molecules into the hydrogel was achieved through their dissolution in an aqueous medium that was added to the Fmoc-peptide, either during or after the self-assembly process. Presence of the integrin binding motif Arg-Gly-Asp that is capable of promoting cell adhesion also offered interesting avenues for the biological applications of such hydrogels (Castelletto et al., 2012).

Poly(lactic-*co*-glycolic acid) (PLGA) nanoparticles with or without a chitosan coating were used as nanostructured depots for dexamethasone, as a model steroid drug. Mixing of such components with the self-assembling peptide Fmoc-^{D}Phe-^{D}Phe-^{D}Phe precursors (i.e., Fmoc-^{D}Phe and ^{D}Phe-^{D}Phe) during the gelation process, allowed to incorporate effectively the drug-loaded nanoparticles in the hydrogel that was formed. An interesting aspect of this approach is the use in reverse of a promiscuous lipase enzyme for the biosynthesis of the unnatural Fmoc-tripeptide, synthesized in a higher yield (i.e., 50%), relative to its L-enantiomer (i.e., 30%), likely due to the presence of a dynamic equilibrium, where the competing reaction of enzyme-catalyzed hydrolysis of the latter is favored. Drug release with such hybrid system was achieved in a controlled manner over 168 h, relative to just a couple of hours for the nanoparticles alone. Despite it being an unnatural peptide of D-stereoconfiguration, Fmoc-^{D}Phe-^{D}Phe-^{D}Phe performed well in cytotoxicity assays on microglial cells in vitro (Chronopoulou et al., 2014).

Nap-^{D}Phe-^{D}Phe is another interesting gelling candidate that was chosen over its L-enantiomer for a release study in vivo, due to its excellent stability to protease degradation, as expected for a D-peptide. A 0.8 wt.% peptide gel containing ^{125}I-NaI, a single-photon emission computed tomography (SPECT) tracer, encapsulated within the hydrogel matrix was subcutaneously injected in a rat model. Remarkably, after 42 days, no inflammation response was observed in the skin tissue surrounding the site of injection, and no sign of toxicity was observed either, be it clinical, hematological, or biochemical. Kinetics studies also confirmed the usefulness of the delivery vehicle, since the use of the gel allowed to triplicate the time required to reach the maximal drug concentration in blood (i.e., C_{max}). In particular, the control (i.e., ^{125}I-NaI solution) reached a C_{max} of 136.5 μCi/L within 1 h, while the ^{125}I-NaI-encapsulated gel reached a C_{max} of 113.0 μCi/L within 3 h. In addition, the hydrogel allowed to maintain the mean blood concentration of the SPECT tracer in a narrow range (i.e., 94.7– 61.3 μCi/L) over the first 12 h, whereas that of the control group diminished rapidly (i.e., from 136.5 to 38.1 μCi/L) during the same time period. Similar results were obtained by using dopamine D_2 receptor imaging agent for clinical diagnosis, that is, ^{125}I-epidepride, thus confirming the usefulness of the dipeptide hydrogel as a vehicle for slow release (Liang et al., 2009).

The self-assembling peptide Nap-Gly-^{D}Phe-^{D}Phe-^{D}Tyr-Gly-Arg-Gly-Asp was chosen in a study as a drug delivery vehicle. The D-peptide assembling motif offers stability against protease-mediated degradation, while the cancer-targeting L-peptide motif represents a fibronectin epitope for integrin protein binding. In this case, 10-hydroxycamptothecin (HCPT) was chosen as a drug model, HCPT being an indole alkaloid that has a broad spectrum of anticancer activity. The hydrophobic drug was encapsulated in the self-assembled nanofibers with 66% efficiency, allowing to incorporate the drug at a level that was nearly 50 times higher than its water solubility limit. Interestingly, the drug-peptide system in solution appeared more stable than its L-peptide analog, with no sedimentation observed up to 4 weeks (relative to 2 weeks for drug-L-peptide system). Self-assembled nanofiber morphology did not appear altered in the presence of the drug, as observed by transmission electron microscopy (TEM) analysis. In vitro release studies in a dialysis system confirmed a slow drug delivery from the nanostructures over a 1-week period, with less than 5% of the drug being released from the nanofibers during the first hour, in contrast with nearly 50% for the free drug control. Viability studies on several cell lines, including models of both physiological and pathological (i.e., cancer) states, confirmed a low-cytotoxicity profile for the self-assembling peptide, and enhanced cancer cell selectivity for the HCPT-peptide system, relative to the free drug. Use of the fluorescent stain Nile Blue allowed visualization of the nanofibers by optical microscopy, which revealed cell internalization, and localization in the cytoplasm. Finally, antitumor activity was evaluated in vivo using a xenograft model of breast cancer in mice. The overall animal health did not seem negatively affected, since no significant body weight change was observed during various treatments. Survival rate studies, as well as tumor growth monitoring, confirmed a superior therapeutic effect for the peptide-drug system, relative to the drug alone, possibly thanks to the combination of enhanced permeability and retention (EPR) effect at the tumor tissue, and of cancer tissue targeting mediated by the Arg-Gly-Asp motif. In addition, histopathological analysis did not reveal any abnormality in liver and spleen tissue arising from use of the nanofibers (Liu et al., 2014).

The hydrogelator Fmoc-Phe-Phe was also used as a self-assembling building block to form hydrogel nanoparticles for drug delivery. The nanocarriers were obtained with an inverse emulsion method (Fig. 9.3), and were stabilized with the addition of a biocompatible and biodegradable surfactant, that is, D-α-tocopheryl (vitamin E), polyethylene glycol 1000 succinate. The peptide was first dissolved as a monomer in dimethylsulfoxide prior to dilution in water. The aqueous solution was then added dropwise to an organic phase composed of mineral oil and containing the surfactant. The resulting heterogeneous mixture was homogenized at 30°C. During this step, the increase in the energy input allowed the peptide mixture to disperse and self-assemble into nanoparticle clusters. Self-assembly was next promoted by gentle stirring at 4°C to allow the formation of a monolayer of surfactant around the nanoparticles. Purification was performed by centrifugation, and a series of washings with hexane. The nanoparticles were collected and resuspended in phosphate-buffered saline or water prior to use. Optimization of the parameters involved in the emulsion process, such as stirring speed and volume ratios of the two liquid components, allowed for fine-control over the nanoparticle size distribution. In particular, additional shear forces were needed to achieve smaller nanoparticles. Dynamic light scattering revealed the presence of two discrete populations with an average diameter of 21.5 ± 1.3 and

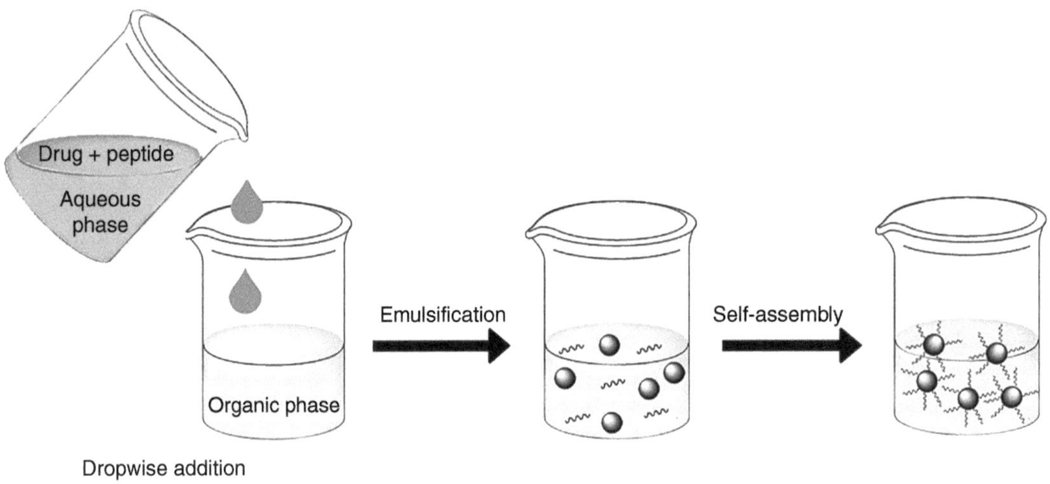

FIGURE 9.3 Inverse Emulsion Method to Obtain Hydrogel Nanoparticles

225.9 ± 0.8 nm, respectively. Such bimodal distribution was likely the result of equilibrium of two opposite phenomena, such as droplet fracture and coalescence. The hydrogel nanoparticles were used to encapsulate hydrophilic gold nanoparticles or chemotherapeutics. In the latter case, doxorubicin and 5-fluorouracil were chosen as two model compounds of different molecular weight, chemical structure, and hydrophobicity. The release studies showed that 5-fluorouracil was delivered more rapidly, possibly due to faster diffusion owing to its smaller molecular weight. Finally, the authors probed the nanoparticles for freeze-drying, showing that this process could be a viable option for the long-term storage of the nanoparticles that preserved their size-distribution characteristics (Ischakov et al., 2013).

The hydrogelator Fmoc-Phe-Phe was mixed to a glucomannan polysaccharide to obtain a hybrid system where a hydrogel was obtained from the entanglement of self-assembled fibers of Fmoc-Phe-Phe, and the polysaccharide chains. As a result, the hybrid system displayed enhanced viscoelastic properties and stability, relative to the peptide alone. The self-recognition ability of the dipeptide was evident, considering that spectroscopic analyses did not reveal significant differences in the supramolecular arrangements in the presence of the polysaccharide (other than a thinning effect on the supramolecular fibers), although the kinetics of peptide self-assembly were notably slower, relative to the system containing the peptide alone. Docetaxel was chosen as a model hydrophobic drug for entrapment and release studies. The rate of the release could be finely controlled by adjusting the concentration of the polymer, as well as its molecular weight, which could be tailored upon adjustment of experimental conditions for enzymatic hydrolysis mediated by β-mannanase. Interestingly, tests on the peptide stability toward protease digestion performed by treating the hydrogels with an equal volume of simulated gastric fluid (containing pepsin) or intestinal fluid (containing trypsin) confirmed peptide's remarkable stability over 24 h. Drug encapsulation did not alter the morphology of the supramolecular fibers. Drug release from the gel composed of the peptide alone was much faster, relative to the hybrid gels, due to disassembly of the supramolecular structure in the release medium. In the latter case, release kinetics were slower for systems that employed smaller molecular weight polysaccharide, likely due to the higher density of the inner structure, relative to those with higher molecular weight. In addition, release was promoted by beta-glucosidase activity, and considering that such enzymes are naturally present in the colon, this may offer a viable option for targeted colon drug release (Huang et al., 2011).

The self-assembling peptide phenylacetyl-Phe-Phe-Ala-Gly-Leu-Asp-Asp was designed to include (1) a hydrophobic, self-assembling region at the N-terminal, useful also to promote retainment of hydrophobic drugs, (2) a matrix metalloprotease (MMP-9) cleavable site, and (3) a hydrophilic unit at the C-terminal. In this manner, the system formed micelles that were prone to protease-mediated hydrolysis, and after loss of the hydrophilic component, the self-assembling molecules rearranged into supramolecular fibers. MMP-9 is a matrix metalloprotease that is overexpressed in cervical, colon, breast, and other types of cancer, therefore it constitutes a convenient biological trigger to achieve targeted drug release. The hydrophobic anticancer drug doxorubicin was chosen as a model to study its loading in the micelles, and its release driven by passive diffusion. When tested with MMP-9, the release of drug was followed by uptake into the supramolecular fibers, thus confirming the ability of the system to further temporarily withhold the drug. Optimization of such system could provide attractive solutions for the targeted drug delivery of anticancer drugs, by exploiting different release kinetic profiles based on switches on supramolecular behavior (Kalafatovic et al., 2015).

The ionic self-complementary peptide Phe-Glu-Phe-Gln-Phe-Lys was also employed to achieve sustained drug delivery. Relative to longer analogs, this peptide was designed to have a glutamine residue replacing the Lys-Phe-Glu sequence. The rationale was that the amide functionality of the glutamine side chain could act both as a donor and as an acceptor of hydrogen bonding, thus offering an interesting alternative to the three residues bearing lysine as a hydrogen bond donor, and glutamate as an acceptor. This kind of amphipathic peptides effectively display one hydrophobic face with aromatic residues that can be involved in π–π interactions, and one hydrophilic face with the ionic side chains for electrostatic interactions between complementary charges. Aqueous solutions of the peptide yield gelled upon mixing with salt or phosphate buffer, thanks to a salting out effect. It was proposed that this approach could be exploited to achieve hydrogelation in situ by injecting an aqueous peptide solution in vivo, so that peptide gelation would occur upon contact with a biological environment of increased ionic strength. This system was also tested for drug release studies, by dissolving peptide and a hydrophobic drug (i.e., the antibiotic ciprofloxacin) in water, followed by addition of phosphate-buffered saline to trigger self-assembly, and the system was left unperturbed overnight to allow gelation to occur. Static release experiments showed an initial burst effect, with approximately 40% of the drug delivered within the first 30–50 h (depending on the drug loading), followed by a more controlled and sustained release of up to 50% of the encapsulated drug within 8 days (Bibian et al., 2015).

3.3 NONCOVALENT INTERACTIONS FOR DRUG DELIVERY

Drug entrapment into the supramolecular structures composed of self-assembled peptides can be achieved not only through physical interactions, but also through chemical noncovalent interactions. In particular, if these are of the same nature as those that hold together the peptide molecules within the supramolecular architectures, it is possible that the drug intercalates with the peptide to participate in the building of such superstructures. In this case, not only the nanomorphology could be affected, but also the spectroscopic signature of the system could be probed to reveal whether such noncovalent interactions are effectively taking place. The concept is interesting because active drug participation into the supramolecular structure would allow for release kinetics that are different from those achieved through more conventional systems. An attractive opportunity would be the possibility to actively control such interactions, for instance through a switch-on/off mechanism, through the application of a certain stimulus, or change in the experimental conditions.

Although there are different reports in the literature that call for drug-peptide interactions, typically no experimental evidence is shown to unequivocally prove its taking place. To the best of our knowledge, the first example of a drug actively participating in the formation of a supramolecular architecture of a short peptide that builds a nanostructured hydrogel was a study reported in 2013. In this case, the heterochiral self-assembling peptide ^{D}Leu-Phe-Phe was first dissolved with the drug under alkaline conditions, in a phosphate buffered solution. Subsequent dilution with an equal volume of a more acidic buffer to reach neutral pH led to the formation of a hydrogel within just a few seconds, which is remarkable, especially in light of the fact that typically short self-assembling peptides require hours, or even the whole night, to form a three-dimensional matrix that builds a macroscopic hydrogel. The circular dichroism signature of the peptide assemblies with and without the drug were significantly different, as well as the nanomorphologies observed by TEM analysis. The rheological properties and fluorescence signal of the system were also clearly affected by an interaction between drug and peptide, and it was proposed that π–π interactions between the two hydrophobic, aromatic molecules were playing a key role in the self-assembly process, as shown by the altered circular dichroism signal. However, it cannot be excluded that other interactions, for instance, of ionic nature, were taking place. Interestingly, sustained drug release was achieved with this system, with approximately just over 40% of the drug being delivered within a 6-day timecourse (Marchesan et al., 2013a).

A similar effect was also reported for the interaction between the self-assembling tripeptide and a fluorescent dye as a model of a small molecule. In this case, the effect of the protocol used for self-assembly was probed. The approach employed was similar to the one described earlier, making use of a pH trigger so that the peptide was firstly dissolved under alkaline conditions, and then diluted with a more acidic buffer to reach physiological pH. Rhodamine B was dissolved either with the peptide in the first buffer, and thus before self-assembly, or in the second buffer, so that the dye would come in contact with the peptide at the time of self-assembly (Fig. 9.4). Interestingly, such difference led to separate outcomes in terms of nanomorphology, circular dichroism spectroscopic signature, and rheological properties of the system. Importantly, it was observed that the gel breaking point was shifted toward higher values of applied strain upon addition of the drug to the system, and such an effect was increased by augmenting the amount of dye, relative to the peptide. This suggests that the dye participates in the noncovalent interactions that hold together the supramolecular architecture that composes the hydrogel. In conclusion, it appears that the conditions employed to achieve self-assembly of multicomponent systems could also have interesting effects on the supramolecular architecture, and thus offer interesting options to fine-tune the properties of the resulting nanostructured (bio)materials (Marchesan et al., 2013b).

N,N'-dibenzoyl-L-cystine (DBC) is a well-known low-molecular weight gelator that was tested for the entrapment and subsequent release of 8-aminoquinoline (AQ) and 2-hydroxyquinoline (HQ). The choice for such model bioactive compounds was motivated by several factors, namely: (1) the toxicity typically associated with their use, which could be improved by use of a hydrogel system for local delivery; (2) the ability of both AQ and HQ to interact with the hydrogelator through Van der Waals and hydrogen bonding interactions; (3) the ability of AQ to interact with the carboxylic group of the hydrogelator via the amine moiety. Drug entrapment was achieved by heating with DBC and water until all the components dissolved, followed by slow cooling to room temperature to achieve gelation. Drug delivery kinetics showed that the initial release of AQ appeared to follow the kinetics of gel degradation,

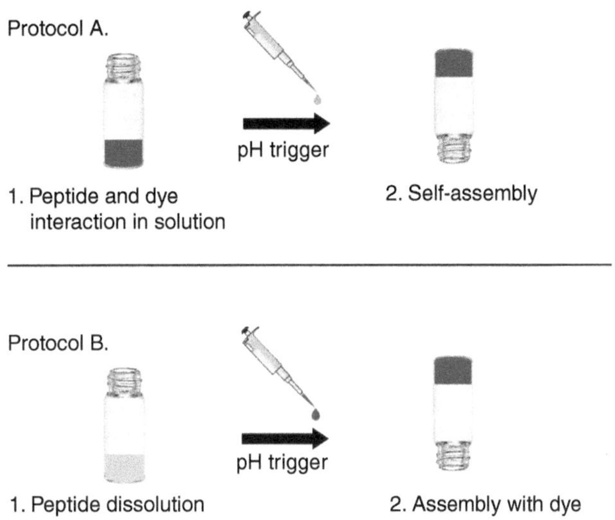

FIGURE 9.4 Different Protocols That Employ a pH Trigger to Achieve Dye-Peptide Noncovalent Interactions Within a Supramolecular Hydrogel

whereas the release of HQ did not, presumably due to the ionic interaction between AQ and the gelator. Indeed, AQ was released slower than HQ, and the process could be accelerated through promotion of gel dissolution via change of pH or temperature (Friggeri et al., 2004). A separate study used UV-promoted thiol-ene click chemistry to cross-link supramolecular hydrogels composed of DBC and analogs bearing a terminal alkene unit in the presence of polyvalent thiol-containing cross-linkers (in equimolar ratio relative to the alkene group), and a water-soluble photoinitiator (i.e., lithium phenyl-2,4,6-trimethylbenzoyl phosphinate). While such multicomponent hydrogels did not display any improvement in the viscoelastic properties, upon photoirradiation at 365 nm for 10 min, a significant increase in the storage and, to a minor extent, the loss moduli was observed. Changes in the gel microstructure morphology were also observed to some extent, depending on the experimental conditions applied. HQ physical entrapment and release revealed different rates of drug delivery that could be correlated to the expected cross-link density of the hydrogels during the first 14 h, after which time appreciable gel degradation was observed. This work thus illustrated how photoinduced click chemistry can be applied to supramolecular hydrogels to fine-tune drug release rate, according to the intended application (Diaz Diaz et al., 2011).

3.4 COVALENT DRUG BINDING TO A SELF-ASSEMBLING PEPTIDE WITHIN A SUPRAMOLECULAR HYDROGEL

Covalent binding of a hydrogelator with a drug is certainly a more challenging approach, since it usually involves changing the physicochemical properties of the self-assembling molecule, which may not self-organize as efficiently as a result of the process. Besides, important parameters, such as solubility have to be appropriately fine-tuned to ensure applicability of the resulting system in biological settings. Covalent binding of the drug to the self-assembling building block can be achieved by using peptide side chain functional groups as anchors, or by using the peptide termini (i.e., C- or N-terminal). Drug release is achieved upon hydrolysis of the covalent linkage between drug and hydrogelator, and this may be mediated by biological events (e.g., enzymatic activity), or physicochemical triggers (e.g., pH-induced hydrolysis).

Curcumin is a good model of hydrophobic drugs. It is a bioactive compound of natural origin that can be extracted from turmeric root. This hydrophobic polyphenol has well-known antioxidant, antiinflammatory, and antitumor properties. However, its application is severely limited by its lack of water solubility, and relatively low bioavailability. A hydrogelator was designed to display curcumin bound to the hydrophobic tripeptide Phe-Phe-Glu, liked via a disulfide bridge to the targeting peptide Glu-Arg-Gly-Asp. Under chemical reduction conditions (e.g., in the presence of naturally occurring glutathione), the disulfide bridge was cleaved to form a hydrogelling building block, from which sustained and long-term curcumin release was observed in vitro. Cytotoxicity assays performed on different cell lines did not highlight any toxic effect form the peptide alone, while the cell proliferation inhibitory activity displayed by self-assembling compound in its reduced form was unexpectedly significantly lower than that displayed by the precursor in the oxidized form. In vivo studies confirmed the trend, suggesting that the precursor would be a better candidate for anticancer treatment, and the increased activity was correlated to a reduced cellular uptake, relative to the nanofibers (Yang et al., 2014).

As a proof of principle, the self-assembling unit Fmoc-lysine was conjugated via a *p*-aminobenzyloxy-carbonyl (PABC) linker to a small molecule drug model (e.g., phenethylamine), as shown in Fig. 9.5 (top). The peptidic bond formed between the amino acid C-terminal and the linker primary amine can conveniently be hydrolyzed through action of a protease (e.g., trypsin). Once deprotected, the linker rapidly underwent 1,6-elimination to form unstable products that decomposed to *p*-amino benzyl alcohol, carbon dioxide, and the free drug (Fig 9.5, bottom). The process was followed by NMR to reveal drug delivery kinetics that were clearly a reflection of the rate of the enzymatic hydrolysis, thus giving scope for using the system for sustained release devoid of burst effects (Sáez et al., 2010).

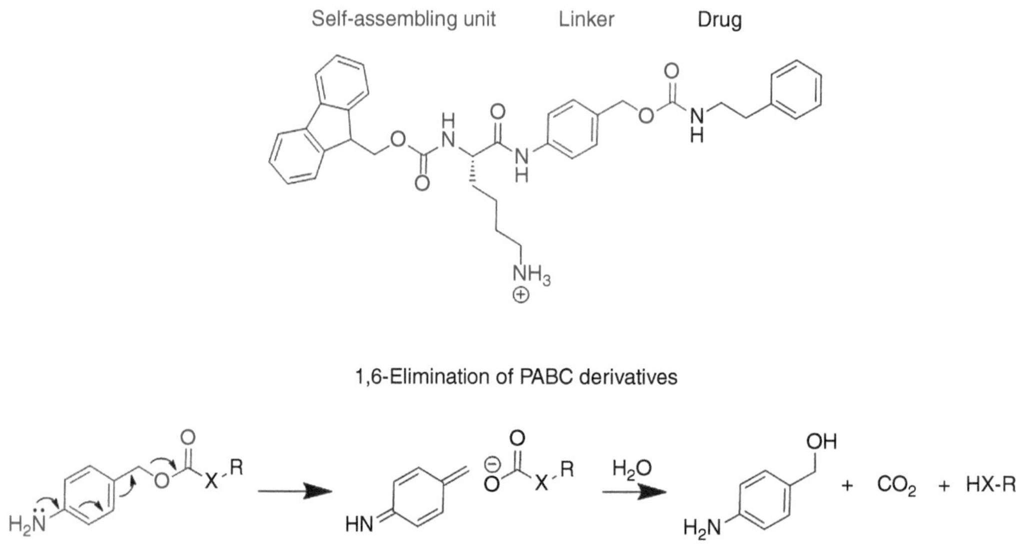

FIGURE 9.5 Drug Covalent Binding to a Self-Assembling Unit via PABC Linker Leads to a Compound (Top) That Undergoes Hydrolysis to Release the Drug (Bottom)

The phosphorylated D-peptide derivative Nap-^{D}Phe-^{D}Phe-^{D}Lys-^{D}Tyr(p) is a useful precursor of the hydrogelator Nap-^{D}Phe-^{D}Phe-^{D}Lys-^{D}Tyr, since it was shown that, despite it being of unnatural D-stereoconfiguration, it can be enzymatically processed by a phosphatase, while being resistant toward protease-mediated hydrolysis. The compound displays a useful primary amine functional group on the side chain of lysine that can be exploited to link drugs or imaging agents covalently. Coupling of the fluorophore 4-nitro-2,1,3-benzoxadiazole (NBD) allowed for intracellular imaging of self-assembled structures, since only the presence of NBD within the nanofibers is associated with fluorescence, while that of NBD as a free molecule is not. Treatment of HeLa cells with such compound revealed fluorescence in the perinuclear region that was significantly reduced upon addition of a tyrosine phosphatase-1B inhibitor, suggesting that the enzyme localized on the outer membrane of the endoplasmic reticulum played a key role in the conversion of the peptide-fluorophore precursor into the fluorescent hydrogelator. Alternatively, the amine group on the lysine side chain was exploited to bind small molecule drugs covalently, such as anti-HIV prodrugs (Li et al., 2013c) or anticancer taxol (Li et al., 2013a), through the use of a linker. In the case of the antitumoral drug taxol, its activity was preserved in vitro after drug conjugation to both the phosphorylated and the dephosphorylated D-peptide, as observed through viability assays on HeLa cells. The activity of the taxol conjugates was confirmed also in vivo through intratumoral injection in a mouse model, and the encouraging results give scope for the local use of taxol-bearing hydrogels to achieve long term drug delivery with enhanced efficacy and biocompatibility, relative to the intravenous injection of taxol alone (Li et al., 2013a).

Taxol was used also as a hydrophobic self-assembling motif. The drug was covalently linked through an ester functional group to one end of succinic acid, while the other end was linked through an amide to the reduced form of glutathione. The compound was designed to include a hydrophilic and biocompatible component to improve the solubility of taxol, and to allow its use in vivo. Interestingly, the amphiphilic compound did not self-assemble per se, but, in phosphate-buffered saline solutions at neutral pH, it underwent hydrolysis to release the hydrogelling taxol. Local administration was used to inhibit tumor growth and formation of metastasis in two tumor models in mice (i.e., lymphoma and breast tumor). The hydrogels inhibited tumor growth in a dose-dependent manner during the first 6 days, and delayed tumor growth during the following 6 days. Results were more promising for the breast tumor model, relative to the lymphoma model, and in both cases the hydrogels performed better than taxol on its own, injected at the same dosage (Wang et al., 2012).

The tetrapeptide Phe-Phe-Lys-Glu constitutes the core of a hydrogelator that can be covalently bound to two drugs, for dual action. The peptide was linked to the antiinflammatory drug dexamethasone through the N-terminal, to either taxol or HCPT through the amino group of the lysine side chain, and to a hydrophilic unit composed of the dipeptide Glu-Glu, and a linker with a disulfide bridge through the C-terminal. This approach allowed for hydrogel formation upon addition of a reducing agent, such as dithiothreitol that led to the separation of the hydrophilic unit containing the dipeptide Glu-Glu from the hydrophobic hydrogelator. The obtained hydrogels displayed different stabilities at 37°C, with the one containing taxol being stable for only 2 days, as opposed to over 2 weeks for the HCPT-bearing hydrogel. Mixing of the two components or addition of bovine serum albumin was a useful method to improve the stability of the hydrogels. Drug release studies in vitro did not reveal the presence of any hydrogelator as a free molecule in solution, but only of the free drugs, confirming the strong self-assembling ability of the designed compounds. Sustained release was obtained with each hydrogelator, and also with a mixture of both components (i.e., the one bearing taxol with the one bearing HCPT), however, a burst effect was noted. The dual delivery system was also successfully tested on a cancer-model cell line, and no cytotoxic effect was observed to arise from the self-assembling peptide alone for concentrations up to 100 μM, over 48 h. Encouraged by such promising results, the authors envisage a scope for such systems as injectable hydrogels for the sustained release of anticancer agents after surgical removal of a tumor mass (Mao et al., 2012).

The similarity of the chemical structures of Nap and the drug naproxen (Npx) was exploited to obtain drug-loaded hydrogels from Npx-peptides, instead of Nap-peptides. A series of D-peptides was analyzed in order to obtain protease-resistant systems that could offer prolonged stability in a biological setting, including ^{D}Phe-^{D}Phe, ^{D}Phe-^{D}Phe-^{D}Lys, and ^{D}Phe-^{D}Phe-^{D}Lys-^{D}Tyr. The short peptides were conjugated to Npx through either the N-terminal or the amino group of the lysine side chain. Interestingly, all these approaches led to the formation of hydrogels, albeit with different viscoelastic properties (e.g., stiffness, rigidity, and critical strains). An interesting effect due to the presence of the D-peptide covalently bound to the drug was the improved selectivity for COX-2, relative to COX-1, an important aspect to reduce drug side effects. Indeed, when the L-peptide analogs were tested, such an effect on selectivity was not observed. The presence of the peptide, however, increased the IC_{50} relative to free Npx in all the tested cases. Nevertheless, use of the peptide ^{D}Phe-^{D}Phe-^{D}Lys-^{D}Tyr bound to Npx either via the N-terminal or via the lysine side chain, led to the best results, with IC_{50} comparable to free Npx and significantly higher COX-2 selectivities. Besides, presence of the tyrosine was a useful advantage, as it allowed to employ a nongelling phosphorylated precursor that could be conveniently converted into the hydrogelator through enzymatic processing in situ. Cytotoxicity assays on HeLa cells over 72 h did not highlight any toxic effect for peptide concentrations as high as 500 μM, in almost all cases (Li et al., 2013b). This approach can also be combined with the covalent binding of other drugs, in order to achieve a dual effect; indeed, Npx-^{D}Phe-^{D}Phe-^{D}Lys-^{D}Tyr bound to anti-HIV prodrugs via a linker anchored to the side chain of lysine, led to a hydrogel for the sustained delivery of both bioactive components (Li et al., 2013c).

4 SELF-ASSEMBLY TO ACHIEVE A THERAPEUTIC EFFECT

We have discussed in previous sections how self-assembling short peptides can be used as nanostructured innovative vehicles for drug delivery. An even more revolutionizing application in therapy is to exploit their supramolecular effect to achieve a therapeutic outcome, for instance, targeted cell death in anticancer therapy. In other words, instead of engineering a system where the self-assembling peptide is used as a vehicle to deliver a chemotherapeutic agent, scientists have designed the appropriate use of the peptide to achieve targeted cytotoxic effect through its supramolecular behavior, without the need for an added chemotherapeutic agent. The hydrophobic dipeptide derivative Nap-Phe-Phe-$NHCH_2CH_2OH$ is a convenient gelator for the purpose. Covalent attachment of a hydrophilic butyric diacid to the terminal OH group of the peptide, thus forming an ester, gave a nongelling precursor. HeLa cells treated with the precursor uptook 32% of the compound within 3 days. Intracellular esterases hydrolyzed it to release the self-assembling unit, which self-organized into supramolecular fibers leading to cell death. In fact, this approach can provide a convenient option for anticancer therapy, considering that fibroblasts cells, used as a model for the physiological state, did not suffer from cytotoxicity under analogous conditions. This fact has been rationalized in terms of overexpression of esterases in HeLa cells, relative to fibroblasts, thus leading to self-assembly (and cytotoxicity) only in the former, when treated with as little as 0.04 wt.% of the amphiphilic precursor (Yang et al., 2007).

5 CONCLUSIONS AND FUTURE OUTLOOK

In conclusion, we have described the high innovation potential offered by self-assembling short peptides in the field of drug delivery. Several options for such use are possible, ranging from physical drug entrapment in the hydrogel matrix, through drug noncovalent binding to the supramolecular structure (be it in the hydrogel, nanoparticle, or nanofiber forms), to covalent binding of the drug to the self-assembling building block. We have illustrated a series of proof-of-concept examples reported in the literature that exemplify the applicability of such approaches. The progress achieved in this area is remarkable, and shows promise for further development of these systems toward clinical applications. However, there are still challenges to overcome. While the hydrogel application as wound dressing seems the most viable option, their use in vivo, for instance, after surgical removal of diseased tissue, would be highly appealing. To ensure optimal filling of a cavity generated from surgery, such systems should be injected as solutions, and self-assemble in situ. Technological optimization is still required to achieve fine-control over such process.

An attractive option is to control self-assembly and disassembly of supramolecular hydrogels by applying an external stimulus. The most established methods employ changes of pH, ionic strength, temperature, solvent, or enzymatic catalysis to form the hydrogelator from suitable precursors (Chen et al., 2011; Raeburn et al., 2013). Recently, examples have been reported of short peptides that respond to ultrasounds or light, a fact that would be particularly relevant if controlled by using clinical tools currently available to doctors. A key advantage by using these methods is the possibility to achieve gelation only in a specific area, rather than in bulk, upon focusing of the ultrasound or light source. For instance, light-driven gelation of Fmoc-dipeptides or Nap-dipeptides was achieved upon addition of a photoacid generator (i.e., diphenyliodonium nitrate) and UV-irradiation, leading to a lowering of pH below the pK_a of the hydrogelator (Raeburn et al., 2012). UV-light irradiation was also used to fine-tune the rheological properties of cumarin-dipeptide derivatives, thus inducing dimerization of the cumarin units within a self-assembled fiber (Draper et al., 2015; Kim et al., 2015). Alternatively, supramolecular behavior can be controlled by introducing azobenzene units that undergo *cis–trans* isomerization upon UV irradiation (Sahoo et al., 2014). Ultrasound energy offers another attractive stimulus to control self-assembly, for instance by using an ultrasonicator bath, with different outcomes depending on the peptide sequence. For instance, gelation of heterochiral tripeptides ^{D}Phe-Phe-Val and ^{D}Val-Phe-Phe under physiological conditions can be accelerated or slowed down, respectively, by treatment in an ultrasonicator water bath for a few minutes (Marchesan et al., 2012a). Ultrasonication waves were shown to promote the formation of an anisotropic hydrogel in phosphate buffer from the aligned supramolecular arrangement of self-assembling tripeptide ^{D}Phe-Phe-Ile (Pappas et al., 2015a). An ultrasound-driven transient effect was reported on Fmoc-Phe-Leu and Fmoc-Tyr-Leu, resulting in the induction of supramolecular reconfiguration from tapes to coiled fibers, or from straight fibers to spherical aggregates, respectively (Pappas et al., 2015b).

In light of the high interest raised for the commercialization of long self-assembling peptides, witnessed by a series of patents and spin-off companies based on such technology, we believe that short peptide analogs will benefit from even higher popularity due to their inherent advantages in terms of ease and low cost of preparation.

ABBREVIATIONS

Ac Acetyl
AFM Atomic force microscopy
Aib α-Aminoisobutyric acid

Ala	Alanine
AQ	8-Aminoquinoline
Arg	Arginine
Asp	Aspartic acid
COX	Cyclooxygenase
DBC	*N,N'*-dibenzoyl-L-cystine
^{D}Leu	D-Leucine
^{D}Lys	D-Lysine
^{D}Phe	D-Phenylalanine
^{D}Tyr	D-Tyrosine
^{D}Tyr(p)	Phosphorylated D-tyrosine
EPR	Enhanced permeability effect
Fmoc	Fluorenylmethyloxycarbonyl
Gln	Glutamine
Glu	Glutamic acid
Gly	Glycine
HCPT	10-Hydroxycamptothecin
HQ	2-Hydroxyquinoline
I	Iodine
IC_{50}	Half maximal inhibitory concentration
Ile	Isoleucine
Leu	Leucine
Lys	Lysine
MMP-9	Matrix metalloprotease 9
NaI	Sodium iodide
Nap	Naphthalene
NBD	4-Nitro-2,1,3-benzoxadiazole
Npx	Naproxen
PABC	*p*-Aminobenzyloxy-carbonyl
Phe	Phenylalanine
PLGA	Poly(lactic-*co*-glycolic acid)
SEM	Scanning electron microscopy
SPECT	Single-photon emission computed tomography
SPPS	Solid-phase peptide synthesis
STM	Scanning tunneling microscopy
TEM	Transmission electron microscopy
UV	Ultraviolet
Val	Valine
wt.	Weight

REFERENCES

Adamcik, J., Mezzenga, R., 2012. Study of amyloid fibrils via atomic force microscopy. Curr. Opin. Colloid Interface Sci. 17, 369–376.

Adams, D.J., 2011. Dipeptide and tripeptide conjugates as low-molecular-weight hydrogelators. Macromol. Biosci. 11, 160–173.

Adler-Abramovich, L., Vaks, L., Carny, O., Trudler, D., Magno, A., Caflisch, A., Frenkel, D., Gazit, E., 2012. Phenylalanine assembly into toxic fibrils suggests amyloid etiology in phenylketonuria. Nat. Chem. Biol. 8, 701–706.

Apostolovic, B., Danial, M., Klok, H.A., 2010. Coiled coils: attractive protein folding motifs for the fabrication of self-assembled, responsive and bioactive materials. Chem. Soc. Rev. 39, 3541–3575.

Arai, T., Kino, K., 2010. New L-amino acid ligases catalyzing oligopeptide synthesis from various microorganisms. Biosci. Biotechnol. Biochem. 74, 1572–1577.

Biancalana, M., Koide, S., 2010. Molecular mechanism of thioflavin-T binding to amyloid fibrils. Biochim. Biophys. Acta 1804, 1405–1412.

Bibian, M., Mangelschots, J., Gardiner, J., Waddington, L., Diaz Acevedo, M.M., De Geest, B.G., Van Mele, B., Madder, A., Hoogenboom, R., Ballet, S., 2015. Rational design of a hexapeptide hydrogelator for controlled-release drug delivery. J. Mater. Chem. B 3, 759–765.

Bonetti, A., Pellegrino, S., Das, P., Yuran, S., Bucci, R., Ferri, N., Meneghetti, F., Castellano, C., Reches, M., Gelmi, M.L., 2015. Dipeptide nanotubes containing unnatural fluorine-substituted beta-diarylamino acid and L-alanine as candidates for biomedical applications. Org. Lett. 17, 4468–4471.

Bowerman, C.J., Ryan, D.M., Nissan, D.A., Nilsson, B.L., 2009. The effect of increasing hydrophobicity on the self-assembly of amphipathic [small beta]-sheet peptides. Mol. BioSyst. 5, 1058–1069.

Castelletto, V., Hamley, I.W., Stain, C., Connon, C., 2012. Slow-release RGD-peptide hydrogel monoliths. Langmuir 28, 12575–12580.

Chen, L., Pont, G., Morris, K., Lotze, G., Squires, A., Serpell, L.C., Adams, D.J., 2011. Salt-induced hydrogelation of functionalised-dipeptides at high pH. Chem. Commun. 47, 12071–12073.

Chronopoulou, L., Sennato, S., Bordi, F., Giannella, D., Di Nitto, A., Barbetta, A., Dentini, M., Togna, A.R., Togna, G.I., Moschini, S., Palocci, C., 2014. Designing unconventional Fmoc-peptide-based biomaterials: structure and related properties. Soft Matter 10, 1944–1952.

Chung, H.J., Park, T.G., 2009. Self-assembled and nanostructured hydrogels for drug delivery and tissue engineering. Nano Today 4, 429–437.

Diaz Diaz, D., Morin, E., Schoen, E.M., Budin, G., Wagner, A., Remy, J.S., 2011. Tailoring drug release profile of low-molecular-weight hydrogels by supramolecular co-assembly and thiol–ene orthogonal coupling. J. Mater. Chem. 21, 641–644.

Draper, E.R., Mcdonald, T.O., Adams, D.J., 2015. Photodimerisation of a coumarin-dipeptide gelator. Chem. Commun. 51, 12827–12830.

Dudukovic, N.A., Zukoski, C.F., 2015. Gelation of Fmoc-diphenylalanine is a first order phase transition. Soft Matter 11, 7663–7673.

Fleming, S., Ulijn, R.V., 2014. Design of nanostructures based on aromatic peptide amphiphiles. Chem. Soc. Rev. 43, 8150–8177.

Frederixpim, W.J.M., Scott, G.G., Abul-Haija, Y.M., Kalafatovic, D., Pappas, C.G., Javid, N., Hunt, N.T., Ulijn, R.V., Tuttle, T., 2015. Exploring the sequence space for (tri-)peptide self-assembly to design and discover new hydrogels. Nat. Chem. 7, 30–37.

Friggeri, A., Feringa, B.L., van Esch, J., 2004. Entrapment and release of quinoline derivatives using a hydrogel of a low molecular weight gelator. J. Control. Release 97, 241–248.

Gras, S.L., Waddington, L.J., Goldie, K.N., 2011. Transmission electron microscopy of amyloid fibrils. Methods Mol. Biol. 752, 197–214.

Hoffman, A.S., 2002. Hydrogels for biomedical applications. Adv. Drug Deliv. Rev. 54, 3–12.

Howie, A.J., Brewer, D.B., 2009. Optical properties of amyloid stained by Congo red: history and mechanisms. Micron 40, 285–301.

Huang, R., Qi, W., Feng, L., Su, R., He, Z., 2011. Self-assembling peptide-polysaccharide hybrid hydrogel as a potential carrier for drug delivery. Soft Matter 7, 6222–6230.

Iglesias, M., Bosi, S., Melchionna, M., Da Ros, T., Marchesan, S., 2016. The glitter of carbon nanostructures in hybrid/composite hydrogels for medicinal use. Curr. Top. Med. Chem. 16, 1976-L1989.

Ischakov, R., Adler-Abramovich, L., Buzhansky, L., Shekhter, T., Gazit, E., 2013. Peptide-based hydrogel nanoparticles as effective drug delivery agents. Bioorg. Med. Chem. 21, 3517–3522.

Kalafatovic, D., Nobis, M., Javid, N., Frederix, P.W.J.M., Anderson, K.I., Saunders, B.R., Ulijn, R.V., 2015. MMP-9 triggered micelle-to-fibre transitions for slow release of doxorubicin. Biomater. Sci. 3, 246–249.

Kim, S.H., Sun, Y., Kaplan, J.A., Grinstaff, M.W., Parquette, J.R., 2015. Photo-crosslinking of a self-assembled coumarin-dipeptide hydrogel. New J. Chem. 39, 3225–3228.

Kopecek, J., Yang, J., 2012. Smart self-assembled hybrid hydrogel biomaterials. Angew. Chem. Int. Ed. 51, 7396–7417.

Krysmann, M.J., Castelletto, V., Kelarakis, A., Hamley, I.W., Hule, R.A., Pochan, D.J., 2008. Self-assembly and hydrogelation of an amyloid peptide fragment. Biochemistry 47, 4597–4605.

Kuang, Y., Shi, J., Li, J., Yuan, D., Alberti, K.A., Xu, Q., Xu, B., 2014. Pericellular hydrogel/nanonets inhibit cancer cells. Angew. Chem. Int. Ed. 53, 8104–8107.

Li, J., Gao, Y., Kuang, Y., Shi, J., Du, X., Zhou, J., Wang, H., Yang, Z., Xu, B., 2013a. Dephosphorylation of D-peptide derivatives to form biofunctional, supramolecular nanofibers/hydrogels and their potential applications for intracellular imaging and intratumoral chemotherapy. J. Am. Chem. Soc. 135, 9907–9914.

Li, J., Kuang, Y., Gao, Y., Du, X., Shi, J., Xu, B., 2013b. D-amino acids boost the selectivity and confer supramolecular hydrogels of a nonsteroidal anti-inflammatory drug (NSAID). J. Am. Chem. Soc. 135, 542–545.

Li, J., Li, X., Kuang, Y., Gao, Y., Du, X., Shi, J., Xu, B., 2013c. Self-delivery multifunctional anti-HIV hydrogels for sustained release. Adv. Healthc. Mater. 2, 1586–1590.

Li, Y., Rodrigues, J., Tomas, H., 2012. Injectable and biodegradable hydrogels: gelation, biodegradation and biomedical applications. Chem. Soc. Rev. 41, 2193–2221.

Liang, G., Yang, Z., Zhang, R., Li, L., Fan, Y., Kuang, Y., Gao, Y., Wang, T., Lu, W.W., Xu, B., 2009. Supramolecular hydrogel of a D-amino acid dipeptide for controlled drug release in vivo. Langmuir 25, 8419–8422.

Liu, J., Liu, J., Chu, L., Zhang, Y., Xu, H., Kong, D., Yang, Z., Yang, C., Ding, D., 2014. Self-assembling peptide of D-amino acids boosts selectivity and antitumor efficacy of 10-hydroxycamptothecin. ACS Appl. Mater. Interfaces 6, 5558–5565.

Maeda, H., 2015. Toward a full understanding of the EPR effect in primary and metastatic tumors as well as issues related to its heterogeneity. Adv. Drug Deliv. Rev. 91, 3–6.

Mao, L.N., Wang, H.M., Tan, M., Ou, L.L., Kong, D.L., Yang, Z.M., 2012. Conjugation of two complementary anti-cancer drugs confers molecular hydrogels as a co-delivery system. Chem. Commun. 48, 395–397.

Marchesan, S., Easton, C.D., Kushkaki, F., Waddington, L., Hartley, P.G., 2012a. Tripeptide self-assembled hydrogels: unexpected twists of chirality. Chem. Commun. 48, 2195–2197.

Marchesan, S., Prato, M., 2013. Nanomaterials for (nano)medicine. ACS Med. Chem. Lett. 4, 147–149.

Marchesan, S., Qu, Y., Waddington, L.J., Easton, C.D., Glattauer, V., Lithgow, T.J., Mclean, K.M., Forsythe, J.S., Hartley, P.G., 2013a. Self-assembly of ciprofloxacin and a tripeptide into an antimicrobial nanostructured hydrogel. Biomaterials 34, 3678–3687.

Marchesan, S., Styan, K., Easton, C.D., Waddington, L., Vargiu, A.V., 2015. Higher and lower supramolecular order for the design of self-assembled heterochiral tripeptide hydrogel biomaterials. J. Mater. Chem. B 3, 8123–8132.

Marchesan, S., Waddington, L., Easton, C.D., Kushkaki, F., Mclean, K.M., Forsythe, J.S., Hartley, P.G., 2013b. Tripeptide self-assembled hydrogels: soft nanomaterials for biological applications. BioNanoScience 3, 21–29.

Marchesan, S., Waddington, L., Easton, C.D., Winkler, D.A., Goodall, L., Forsythe, J., Hartley, P.G., 2012b. Unzipping the role of chirality in nanoscale self-assembly of tripeptide hydrogels. Nanoscale 4, 6752–6760.

Martin, A.D., Robinson, A.B., Mason, A.F., Wojciechowski, J.P., Thordarson, P., 2014. Exceptionally strong hydrogels through self-assembly of an indole-capped dipeptide. Chem. Commun. 50, 15541–15544.

Melchionna, M., Styan, K.E., Marchesan, S., 2016. The unexpected advantages of using D-amino acids in self-assembled peptide hydrogels for medicine. Curr. Top. Med. Chem. 16, 2009–2018.

Mohammed, A., Miller, A.F., Saiani, A., 2007. 3D networks from self-assembling ionic-complementary octapeptides. Macromol. Symp. 251, 88–95.

Mondal, S., Adler-Abramovich, L., Lampel, A., Bram, Y., Lipstman, S., Gazit, E., 2015. Formation of functional super-helical assemblies by constrained single heptad repeat. Nat. Commun. 6, 8615.

Orbach, R., Adler-Abramovich, L., Zigerson, S., Mironi-Harpaz, I., Seliktar, D., Gazit, E., 2009. Self-assembled Fmoc-peptides as a platform for the formation of nanostructures and hydrogels. Biomacromolecules 10, 2646–2651.

Panda, J.J., Mishra, A., Basu, A., Chauhan, V.S., 2008. Stimuli responsive self-assembled hydrogel of a low molecular weight free dipeptide with potential for tunable drug delivery. Biomacromolecules 9, 2244–2250.

Pappas, C.G., Frederix, P.W.J.M., Mutasa, T., Fleming, S., Abul-Haija, Y.M., Kelly, S.M., Gachagan, A., Kalafatovic, D., Trevino, J., Ulijn, R.V., Bai, S., 2015a. Alignment of nanostructured tripeptide gels by directional ultrasonication. Chem. Commun. 51, 8465–8468.

Pappas, C.G., Mutasa, T., Frederix, P.W.J.M., Fleming, S., Bai, S., Debnath, S., Kelly, S.M., Gachagan, A., Ulijn, R.V., 2015b. Transient supramolecular reconfiguration of peptide nanostructures using ultrasound. Mater. Horiz. 2, 198–202.

Pudakalakatti, S.M., Chandra, K., Thirupathi, R., Atreya, H.S., 2014. Rapid characterization of molecular diffusion by NMR spectroscopy. Chem. Eur. J. 20, 15719–15722.

Raeburn, J., Mcdonald, T.O., Adams, D.J., 2012. Dipeptide hydrogelation triggered via ultraviolet light. Chem. Commun. 48, 9355–9357.

Raeburn, J., Zamith Cardoso, A., Adams, D.J., 2013. The importance of the self-assembly process to control mechanical properties of low molecular weight hydrogels. Chem. Soc. Rev. 42, 5143–5156.

Reches, M., Gazit, E., 2003. Casting metal nanowires within discrete self-assembled peptide nanotubes. Science 300, 625–627.

Sáez, J.A., Escuder, B., Miravet, J.F., 2010. Supramolecular hydrogels for enzymatically triggered self-immolative drug delivery. Tetrahedron 66, 2614–2618.

Sahoo, J.K., Nalluri, S.K.M., Javid, N., Webb, H., Ulijn, R.V., 2014. Biocatalytic amide condensation and gelation controlled by light. Chem. Commun. 50, 5462–5464.

Schermelleh, L., Heintzmann, R., Leonhardt, H., 2010. A guide to super-resolution fluorescence microscopy. J. Cell Biol. 190, 165–175.

Scott-Fordsmand, J.J., Pozzi-Mucelli, S., Tran, L., Aschberger, K., Sabella, S., Vogel, U., Poland, C., Balharry, D., Fernandes, T., Gottardo, S., Hankin, S., Hartl, M.G.J., Hartmann, N.B., Hristozov, D., Hund-Rinke, K., Johnston, H., Marcomini, A., Panzer, O., Roncato, D., Saber, A.T., Wallin, H., Stone, V., 2014. A unified framework for nanosafety is needed. Nano Today 9, 546–549.

Singh, N.K., Lee, D.S., 2014. In situ gelling pH- and temperature-sensitive biodegradable block copolymer hydrogels for drug delivery. J. Control. Release 193, 214–227.

Takai, E., Ohashi, G., Ueki, R., Yamada, Y., Fujita, J.-I., Shiraki, K., 2014. Scanning electron microscope imaging of amyloid fibrils. Am. J. Biochem. Biotechnol. 10, 31.

Taylor, P.P., Grinter, N.J., McCarthy, S.L., Pantaleone, D.P., Ton, J.L., Yoshida, R.K., Fotheringham, I.G., 2001. D-phenylalanine biosynthesis using *Escherichia Coli*: creation of a new metabolic pathway. Applied Biocatalysis in Specialty Chemicals and Pharmaceuticals. American Chemical Society. .

Toledano, S., Williams, R.J., Jayawarna, V., Ulijn, R.V., 2006. Enzyme-triggered self-assembly of peptide hydrogels via reversed hydrolysis. J. Am. Chem. Soc. 128, 1070–1071.

Wang, H., Wei, J., Yang, C., Zhao, H., Li, D., Yin, Z., Yang, Z., 2012. The inhibition of tumor growth and metastasis by self-assembled nanofibers of taxol. Biomaterials 33, 5848–5853.

Wang, H.M., Yang, C.H., Tan, M., Wang, L., Kong, D.L., Yang, Z.M., 2011. A structure-gelation ability study in a short peptide-based 'Super Hydrogelator' system. Soft Matter 7, 3897–3905.

Wang, Z., Zhou, C., Wang, C., Wan, L., Fang, X., Bai, C., 2003. AFM and STM study of beta-amyloid aggregation on graphite. Ultramicroscopy 97, 73–79.

Wichterle, O., Lim, D., 1960. Hydrophilic gels for biological use. Nature 185, 117–118.
Yan, X., Zhu, P., Li, J., 2010. Self-assembly and application of diphenylalanine-based nanostructures. Chem. Soc. Rev. 39, 1877–1890.
Yang, C., Wang, Z., Ou, C., Chen, M., Wang, L., Yang, Z., 2014. A supramolecular hydrogelator of curcumin. Chem. Commun. 50, 9413–9415.
Yang, Z.M., Xu, K.M., Guo, Z.F., Guo, Z.H., Xu, B., 2007. Intracellular enzymatic formation of nanofibers results in hydrogelation and regulated cell death. Adv. Mater. 19, 3152–3156.
Yuran, S., Reches, M., 2013. Formation of ordered biomolecular structures by the self-assembly of short peptides. J. Vis. Exp. 81, e50946.
Zhang, S., Holmes, T., Lockshin, C., Rich, A., 1993. Spontaneous assembly of a self-complementary oligopeptide to form a stable macroscopic membrane. Proc. Nat. Acad. Sci. USA 90, 3334–3338.
Zorko, M., Japelj, B., Hafner-Bratkovic, I., Jerala, R., 2009. Expression, purification and structural studies of a short antimicrobial peptide. Biochim. Biophys. Acta Biomembr. 1788, 314–323.

CHAPTER

PEPTOIDS: TOMORROW'S THERAPEUTICS

10

Avneet Saini, Gaurav Verma
Panjab University, Chandigarh, India

CHAPTER OUTLINE

1 INTRODUCTION

The increasing knowledge of the manifold modes of action of bioactive peptides has led to an increased interest of researchers in this class of compounds. There has been a rapid expansion in the use of peptides as drugs over the last two decades, and it still continues. They are already being used therapeutically in areas as diverse as neurology, endocrinology, and hematology. Human diseases for which peptide-based drug therapies already exist include osteoporosis (calcitonin), diabetes (insulin), prostate cancer and endometriosis (gonadotropin-releasing hormone), acromegaly and ulcers (somatostatin), and hypothyroidism (thyrotropin-releasing hormone TRH) (Adessi and Soto, 2002).

However, most peptides cannot be administered orally as they are rapidly inactivated by gastrointestinal enzymes, so that subcutaneous or intravenous administration is required. In tissue or cellular compartments, the inactivation involves specific endopeptidases, and in the systemic circulation and peripheral organs it involves less specific exopeptidases (amino- and carboxy-peptidases). Moreover, as drugs are mostly administered far from the actual site of action, they must be capable to reach the target site that usually involves penetration of several biological barriers. Unmodified peptides do not circulate in the blood for more than a few minutes, because of their vulnerability to in vivo enzymatic degradation that reduces their bioavailability in tissues and organs, thereby limiting usefulness as therapeutic agents (Adessi and Soto, 2002). Thus, peptides offer inferior drug candidates mainly due to their low oral bioavailability, short half-life, potential immunogenicity, fluctuations in their pharmacological profile, and poor metabolic stability in vivo (Latham, 1999; Patch and Barron, 2002). So, despite their wide range of bioactivities, research is focused on alternate routes: either of peptide delivery, or to develop peptide mimics for biomedical applications.

Identification of bioinspired yet nonnatural, low-cost, and stable chemical scaffolds that recapitulate the desirable attributes of polypeptides has been a long-term goal for a plethora of applications, ranging from medicinal chemistry to material science. In peptides, chemical modifications maximize enzymatic stability and bioavailability, while simultaneously preserving or even enhancing their potency and selectivity (Latham, 1999; Patch and Barron, 2002). These may lead to bioinspired mimics with many unique properties that bridge the gap between proteins and bulk polymers (Barron and Zuckermann, 1999). Like proteins, they are sequence-specific heteropolymers, capable of folding into specific shapes, and exhibiting potent biological activity, and, just like bulk polymers, they are chemically and biologically stable, and relatively cheap to make.

Peptidomimetic compounds have been found to be efficient routes due to their enhanced pharmacological properties, and ability to retain the activities of their biological counterparts, while counteracting their pharmacological limitations (Patch et al., 2004). They are defined as molecules bearing identifiable resemblances to a peptide that, as a ligand of a biological receptor, it can imitate/inhibit the effect of a natural peptide. They are classified into two broad categories: (1) modified peptide mimetics—incorporating nonstandard peptide groups in order to mimic the peptide backbone and properties (including N-methylated amino acids, α-alkyl amino acids, N^{α}–C^{α} cyclized amino acids, β-amino acids, peptoids, peptide bond isosteres and D-peptides), and (2) structural mimetics—groups and structures intended to mimic or include elements of peptide secondary structure, that is, reverse turns, helices, and β-sheets. Peptoids (or oligomers of N-substituted glycine-NSGs) represent one such class of peptidomimetics with versatile properties to meet the goal of biomimicry. This has been one of the most symbolic approaches of the past two decades, due to their ability to adopt folded nanostructures and thus be designed to adopt specific conformations (Table 10.1).

Table 10.1 Pharmacological Properties of Peptoids

1	Easy and low cost synthesis process
2	Resistance against protease
3	Enhances bioavailability
4	High metabolic stability
5	High thermal stability
6	Stable to salt and pH variations

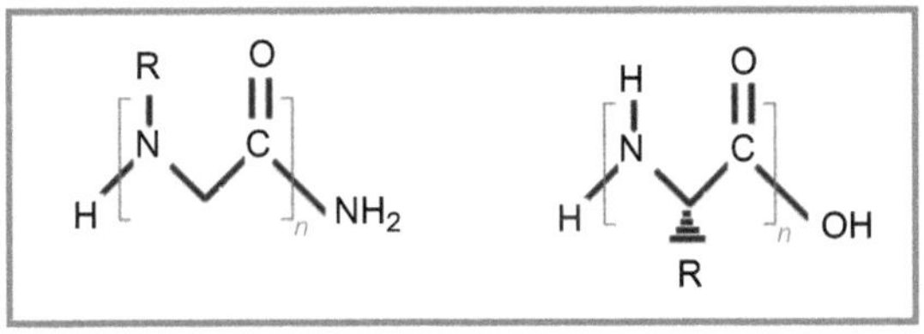

FIGURE 10.1 Generic Structure of a Peptoid and α-Peptide

Peptoids introduced in 1990s (Simon et al., 1992) differ from peptides in the manner of the side chain appendage, that is, the side chain is appended to the amide nitrogen rather than α-carbon, resulting in an N-substituted oligoglycine. This modification renders the backbone achiral and flexible, due to the methylene group at the α-carbon similar to glycine amino acid. At the same time, analogous to polyprolines and N-methylated polypeptides, the peptoid backbone is composed entirely of tertiary amides (Fig. 10.1). Thus, the presentation of peptide and peptoid side chains is roughly isosteric, potentially allowing for suitable mimicry of the spacing between the critical chemical functionalities of bioactive peptides (Patch et al., 2004). Further, this modification bestows protease resistance, as natural proteases do not recognize the N-substituted amide bond, thus conferring higher metabolic stability to the molecule (Miller et al., 1995). Further, as peptoids lack the amide hydrogen, formation of hydrogen bond networks that stabilize peptide secondary structures are absent. Instead, they are generally stabilized by backbone and steric constraints (Armand et al., 1997). The secondary structures of peptoids are guided by electronic and streic interactions that are stable over a wide range of solvents and thermal conditions (Nandel and Saini, 2011; Sanborn et al., 2002). Their thermal stability ranges to > 75°C, and are generally stable to salt and pH variations. In addition, NSGs are more hydrophobic than peptides, and hence possess superior cellular permeability (Figliozzi et al., 1996; Lim et al., 2009; Miller et al., 1994, 1995; Patch et al., 2004; Richter et al., 1996; Zuckermann and Kodadek, 2009).

Further, the ease and economy of synthesis, and the possibility to include varied side chain chemistries, make obvious their importance in terms of pharmacologically useful peptide mimics that are under extensive research for biomedical applications (Patch et al., 2004). The use of peptoids has been, therefore, one of the most symbolic approaches to meet the goal of biomimicry, due to their ability to adapt folded structures and thus be designed to adapt specific backbone conformations.

Consequently, peptoids have found application as (1) attractive targets for the drug discovery process (Konca et al., 2014), (2) a motif for combinatorial strategies (Garcia-Martinez et al., 2002; Gibbons et al., 1996; Heizmann et al., 1999; Miller et al., 1994; Thompson et al., 2003), and (3) to develop peptide mimics for biomedical applications (Brown et al., 2008; Nandel and Saini, 2007, 2011; Nandel et al., 2014; Rouhi, 2001; Seurynck-Servoss et al., 2006, 2007; Statz et al., 2009; Wu et al., 2003). New approaches to direct the peptoid backbone toward formation of specific secondary structures for the discovery of bioactive peptoid modules (Chongsiriwatana et al., 2008; Daelemans et al., 2000; Huang et al., 2014; Nandel et al., 2014; Ryge and Hansen, 2006; Patch and Barron, 2003; Yoo et al., 2010) that find applications as nanoparticles (Burkoth et al., 2002; Lee et al., 2005a, 2008; Nam et al., 2010), catalysts (Maayan and Ward, 2009), and sensors (Fisher and Naughton, 2003; Lee et al., 2010b; Shah and Kirshenbaum, 2008) reflect the considerable potential of peptoids in medicinal chemistry and biological tools.

In this chapter, we summarize the recent research efforts in the field of peptoid biopolymers, from synthesis to folded structural motifs. This will be followed by a section that lays emphasis on their major biological applications discovered in the past two decades.

2 PEPTOID SYNTHESIS

In the 1990s, the preferred length of oligopeptoids was a trimer and, as research progressed, the length extended to 48-mers, 50-mers for homo-oligomers with short linear side-chains, 60-mers via chemical ligation of 15-mers, and even 150-mers by bioligation using the cysteine protease, clostripain (Culf and Ouellette, 2010; Lee et al., 2005a,b; Murphy et al., 1998; Yoo and Kirshenbaum, 2005)

During the past few decades, advances in polymer synthesis have significantly increased the degree of control over their structure. There are presently many methods for the synthesis of peptoids/polypeptoids. The first to appear has a direct analogy with fluorenylmethyloxycarbonyl solid-phase peptide synthesis (Fmoc SPPS), and hence called the monomer approach. Here, previously prepared N-Fmoc, N-substituted glycine monomers are sequentially coupled, creating oligomers (De Cola et al., 2009; Diaz-Mochon et al., 2008; Schroeder et al., 2008; Simon et al., 1992; Unciti-Broceta et al., 2009). The other three methods are actually based on the submonomer method itself, where the acyl and amine functions of an amino acid monomeric unit are derived from two sequential chemical reactions of acylation, and displacement or amination.

The submonomer method introduced in 1992 greatly increased the synthetic efficiency, synthesis yields, and available side chain diversity, while also dramatically reducing time and costs (Simon et al., 1992). Any chemical functionality (not limited to usual amino acid side chains) viable as a primary amine can be incorporated via this synthetic method. This method allows a researcher molecular diversity, as there are over 1000 primary amines commercially available that can be incorporated as peptoid side chains by this strategy of peptoid synthesis, also at low costs. Despite this, researchers are currently using a small subset of this total for any given area of study (Zuckermann and Kodadek, 2009). Culf and Ouellette (2010) have compiled 230 amine submonomers that are currently being used in NSG synthesis (Culf and Ouellette, 2010).

In this strategy, peptoid monomers are synthesized by a two-step process, from a haloacetic acid and a primary amine. The first step is the acylation step, followed by the nucleophilic displacement step. Acylation reactions were performed by addition of bromoacetic acid in dimethyl formamide (DMF). Reaction mixtures were agitated at room temperature for 30 min. Each acylation was repeated once, and displacement reactions were performed by addition of primary amine as 2.5 M solution in DMSO or DMF, followed by agitation for 2 h at room temperature. Optimization of displacement reactions was performed by varying amine concentrations, from 0.25 to 2.5 M. Propagation of peptoid chains to substantial length is permitted by the high synthetic yields attained at each synthetic step. Moreover, postsynthetically peptoids can be analyzed by HPLC and mass spectroscopy, much like peptides. Also, they can be sequenced by Edman degradation (Boeijen and Liskamp, 1998; Thakkar et al., 2009), and by tandem mass spectrometry (Burkoth et al., 2002), as they conveniently fragment along the main chain amides (Heerma et al., 1996, 1997). Microwave irradiation has also been used to reduce the reaction time during coupling. The purity and yields of peptoids synthesized in this way were as good as, or better than, those achieved using the standard submonomer method (Gorske et al., 2005; Nnanabu and Burgess, 2006; Olivos et al., 2002).

There are a couple of methods for making peptoids completely in solution, as well. The lengths of the sequences made by this method are limited to less than 10 residues. One of the solution-phase peptoid syntheses is a submonomeric method using bromoacetylbromide (Hjelmgaard et al., 2009). The first step is the acylation reaction of the N-terminus of the existing peptoid chain with bromoacetylbromide. After filtration and evaporation, the crude bromoacetamide intermediate is then reacted with a primary amine. The newly formed peptoid is then isolated by flash chromatography, before the next cycle of submonomer addition. Thus, this method requires labor-intensive filtration, evaporation, and chromatography, although it generates reasonably large quantities.

Another method for solution-phase synthesis is the Ugi four component reaction (4CR). 4CR is an efficient process in which a primary amine, an oxo compound, a carboxylic acid, and an isocyanide react in one pot to form a dipeptoid backbone. The Ugi-4CR has been combined with macrocyclization, generating many useful compounds, such as functional macrocycles, cryptands, cages, etc. (Rivera and Wessjohann, 2006; Vercillo et al., 2008).

A more recent technology as a submonomer strategy for high throughout synthesis is the light-directed route of Kodadek (2006) (Li et al., 2004; Marthandan et al., 2008). The use of digital photolithography is a highly attractive path to the development of diagnostics (Culf and Ouellette, 2010). Peptoid polymers have also been synthesized on a small scale using biosynthetic machinery. An *Escherichia coli* cell-free translation system was used with genetically coded mRNA and artificial tRNAs charged with peptoid monomers (Kawakami et al., 2008). In this translation system, certain proteinogenic amino acids and/or cognate aminoacyl-tRNA synthetases are withdrawn, diminishing the competing background incorporation of the proteinogenic amino acids in the translation elongation event. They prepared a wide variety of tRNAs charged with nonproteinogenic peptoid monomers, using artificial tRNA acylation ribozymes. Using this method, six peptoid monomers were translated from the genetic code and purified (Seo et al., 2011).

In terms of instrumentation, a wide variety of platforms have been put to use for the synthesis of polypeptoids, including customized automated peptide synthesizers, robotic workstations, microwave synthesizers; Houghten teabags to contain pools of resin (Masip et al., 2005; Messeguer et al., 2008), as well as sonicator and manual equipment, such as fritted syringes or chromatography columns.

3 PEPTOID STRUCTURES

In contrast to the large amount of information available from peptide or protein structures, there is still a lack of experimental knowledge of peptoid structures. Investigating the conformational preferences of peptoids is a formidable challenge due to lack of conformational rigidity at backbone C_α, compared to α-peptides. However, Ramachandran-like plots of peptoids have already been calculated by Armand et al. (1997) and Simon et al. (1992). Mohle and Hofmann (1996) did another comparison between peptoids and peptides by calculating energy hypersurfaces with quantum mechanical methods for N-acetylglycine- N-methylamide, N-acetylalanine-N-methylamide, and N.N acetylmethylglycine-N', N'-dimethylamide (Mohle and Hofmann, 1996). Butterfoss et al. (2012) and Nandel et al. (2011, 2014) performed high-level quantum mechanical and molecular mechanical calculations on short peptoid oligomers, including Ramachandran-like plots, in comparison to experimental data.

3.1 PEPTOID CRYSTAL STRUCTURES

Access to a large database of known protein structures is an advantage for studying the folding characteristics of proteins. Unfortunately, much less structural data is available on peptoids than for peptide or protein systems. Structural analysis of peptoids is also hindered by their highly challenging crystallization (Fowler and Blackwell, 2009; Shah and Kirshenbaum, 2008; Wu et al., 2003), as they are more hydrophobic, compared to their corresponding amino acids. Still, several high resolution peptoid structures have been solved in recent years (Butterfoss et al., 2012; Crapster et al., 2011; Huang et al., 2006; Maulucci et al., 2008; Shah and Kirshenbaum, 2008; Shin et al., 2007; Wu et al., 2003).

The X-ray crystal structure of N-(1-cyclohexylethyl) glycine pentamer was among the first to be solved. It adopted a structure with repeating Φ, Ψ values similar to those of polyproline type I peptide helix with cisamide bonds, ~3 residues per turn, and a pitch of ~ 6.7 Å (Wu et al., 2003). Further, the crystal structure of cyclic peptoid hexamer and octamer of the repeating unit N-(1-phenylmethyl) glycine (Npm) and N-(methoxyethyl) glycine (Nme) revealed the capacity for the peptoid backbone to accommodate cis- and transamide bonds with polyproline and inverse-polyproline type torsion angles (Shin et al., 2007). All transamide bond geometries were reported in the crystal of N-aryl peptoid dimer (Shah and Kirshenbaum, 2008), N-hydroxy amides, and also N-alkoxy amides (Crapster et al., 2011). These studies clearly demonstrated that the nature of amide bond geometry may be trans or cis in peptoids. Further, crystal structure of o-nitro acetamide monomer showed that molecules were arranged in the P43 space group, a tetrad repeat in which each molecule was rotated 90° relative to the neighboring molecule. Intermolecular hydrogen bonds were present between adjacent amide protons and amide oxygens (Fowler and Blackwell, 2009).

3.2 HELICES IN PEPTOIDS

Peptide helices are characterized by a regular set of torsion angles, hydrogen bond interactions and a characteristic distance the helix rises along its axis per turn. Though, peptoids are nonnatural oligomers but being derived from a polypeptide backbone presence of regularly repeating secondary structures is evident.

In one of the earliest reports, formation of helices with cisamide bonds was reported by incorporation of side chains with branched carbons immediately adjacent to the main chain nitrogen ($C_{N\alpha}$) (Armand et al., 1998). A model proposed for peptoid octamer of (s)-N-(1-phenylethyl) glycine (Nspe) also showed that substitution with bulky, chiral groups significantly reduced the energetically accessible conformational space (Armand et al., 1997). These conformationally restricted peptoids were predicted to be stabilized by a combination of side chain–backbone steric interactions and dipole–dipole interactions between main chain amide carbonyl electrons, and for specific peptoid sequences aromatic side chains (Armand et al., 1997).

CD results have shown stable helical structures in peptoids with peaks at 192, 202, and 218 nm, in the presence of α-chiral aromatic side chains (Armand et al., 1998; Wu et al., 2001a,b). This pattern of peaks was particularly used in the diagnostics for helical structurin peptoids (Fig. 10.2). Many other chiral polypeptoids showed similar CD spectra, suggesting that the whole family of peptoids containing chiral side chains was capable of adopting the same secondary structure motif.

CD spectra (in acetonitrile) of peptoids with α-chiral aliphatic side chains based on (s)-N-(1-cyclohexylethyl) glycine, that is, Nsch, and (s)-N-(2-butyl) glycine, that is, Nssb or Nile monomers, displayed polyproline type I-like spectra (Fig. 10.2). Meanwhile peptoids with Nspe monomers showed α-helix like CD spectrum (Sanborn et al., 2002; Wu et al., 2001a,b). However, it was also shown that they form helices of essentially the same type (Wu et al., 2003).

The discovery that peptoids with aliphatic side chains could also form biomimetic, polyproline-like helices (Wu et al., 2001b; Nandel et al., 2011; Nandel et al., 2014), in a manner similar to the aromatic

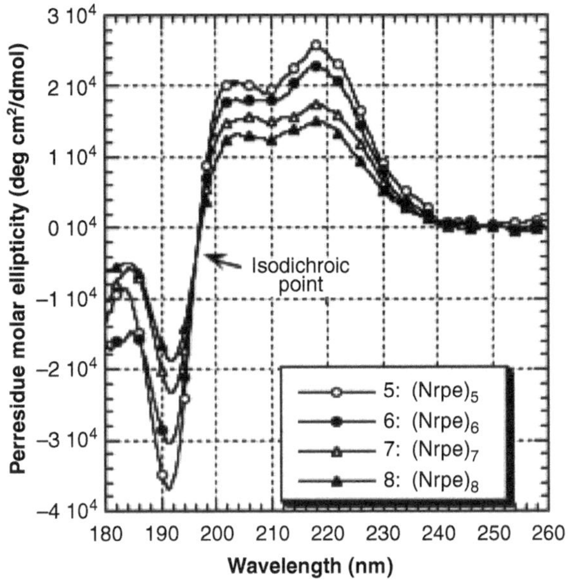

FIGURE 10.2 CD Spectra of Peptoid Helices Containing Solely Aromatic or Solely Aliphatic Residues (Wu et al., 2001b)

side chain containing peptoids, implied that peptoids had the capacity to even mimic predominantly aliphatic peptide α-helices that is essential for designing accurate nonnatural analogues of transmembrane proteins and other small, active protein domains.

The CD signals were also found to be independent of concentration in the range 50 μm to 1 mM in acetonitrile, and the ^{1}H one-dimensional NMR spectra for this molecule revealed sharp line widths. This indicated the absence of intermolecular associations, and the monomeric nature of the structure (Armand et al., 1998). Helix-like CD signals were preserved in sequences comprised of 33%–100% aromatic residues, and 0%–66% charged residues, without substantial change in the signal intensity. Shin et al. (2014) have recently provided an extensive study on peptoid models, and reported that helicity can be effectively modulated by position-specific incorporation of α-chiral aromatic monomers (Shin et al., 2014).

These studies laid the groundwork for the rational design of more complex polypeptoid molecules, with applications ranging through therapeutics and biomaterial designs. In the early 2000s, Barron and coworkers also developed a set of predictive rules for helix formation in peptoids (Wu et al., 2001b). They studied more than 30 different peptoid oligomers that incorporated achiral, as well as chiral, side chains, and found that short, hetero-oligomeric peptoids that included achiral side chains were typically stable if (1) at least 50% of the constituent monomers were α-chiral and aromatic; (2) an α-chiral, aromatic monomer is placed on the carboxy terminus, as it was less stable structurally, than the amino terminus (i.e., the peptoid helix was more easily frayed by the presence of a mobile, achiral side chain at carboxy-terminal position); and (3) sequences should be designed periodically so as to maximize the number of aromatic faces on the peptoid helix. These design criteria became less critical for longer peptoid sequences, particularly when chain length was increased beyond 12–15 residues, as a stable peptoid helix could be attained even for a sequence that was quite unstable at the hexamer level. This was presumed mainly due to the peptoid helix stabilization provided by cooperative interactions that propagate down the ordered chain (Wu et al., 2001b). Polyproline type I helices have also been reported for cyclic and linear poly-α peptoids of N-substituted N-carboxyanhydrides (Guo et al., 2011).

NMR study in 100 % D_2O, on di- and tripeptoids of various types with both N and C-terminals blocked has shown greater population of conformers with transamide bonds, relative to the cis conformers (Sui et al., 2007). Also, the rate of exchange between cis and trans configurations was strongly dependent on the location of the amide bond in the peptoid, and not on the monomer residues linked by the bond.

Barron and coworkers went on to discover a second well-defined peptoid structure, the threaded loop (Fowler and Blackwell, 2009). This structure was unique to peptoid nonamers with α-chiral side chains, unblocked N- and C-terminals, and was first identified in a homononamer of Nspe (Huang et al., 2006). The threaded loop structure of $(Nspe)_9$ was determined by solution-phase NMR spectroscopy in acetonitrile-d_3, and was found to be stabilized by three intramolecular hydrogen bonds from backbone carbonyls (residues 5, 7, and 9) to the N-terminal secondary ammonium, and one intramolecular hydrogen bond from a backbone carbonyl (residue 2) to the C-terminal primary amide. Interestingly, the threaded loop could be converted into a peptoid helix by the addition of a solvent capable of disrupting its set of intramolecular hydrogen bonds (e.g., 50% methanol in acetonitrile). The strength of these hydrogen bonds is therefore a key factor in stabilizing the loop conformation in acetonitrile. Strategies to stabilize the helical conformation over the threaded loop have also been reported. For example, both Kirshenbaum and coworkers, and Vaz and Brunsveld covalently linked peptoid side chains at the ith, ith + 3 positions to enforce helicity (Holub et al., 2007; Vaz and Brunsveld, 2008).

In the case of β-peptoids, helical secondary structures were also predicted by theoretical methods (Baldauf et al., 2006), and high-resolution X-ray crystal structures of homomeric β-peptoid hexamers have recently revealed right-handed helical conformations with three residues per turn and a helical pitch of 9.6–9.8 Å. The results were also supported by circular dichroism spectroscopy showing length and solvent dependency, and molecular dynamics simulations (Laursen et al., 2015).

3.2.1 Turns in peptoids

In peptides, turns are characterized by their typical backbone dihedral angles (Φ, Ψ) and hydrogen bonds, while in peptoids the nature of the amide bond (not Φ, Ψ) has been exploited to create cyclic structures that mimic the turns. Kirshenbaum and coworkers performed head-to-tail macrocyclization of achiral peptoid oligomers, and discovered that macrocyclic hexamer and octamer peptoids resembled peptide β-turns (Shin et al., 2007). These peptoids had an alternating pattern of benzyl and methoxyethyl side chains, and their crystal structures revealed that the aromatic and alkyl side chains were segregated to opposite faces of the macrocycle. In each macrocycle, the two backbone amides in the turn region were cis, and the remaining amides were trans, such that the pattern of *cis*- and *trans* amides was $(cis–cis–trans)_2$ for the hexamer, and $(cis–cis–trans–trans)_2$ for the octamer. Furthermore, the peptoids were overlaid with native peptide β-turns, and found to closely resemble both type I and type III β-turns. Elaboration of these macrocyclic peptoids via the introduction of functionalized aromatic and alkyl side chains could provide useful β-turn mimics, with the choice of a smaller hexamer or larger octamer scaffold. In the threaded loop structure of $(Nspe)_9$ the placement of four *cis*- and four *trans* amide bonds also created a turn-like structure in the loop regions (Huang et al., 2006).

To introduce the concept of turns analogous to peptides and proteins, a triazole monomer was designed to function as a turn mimic, and incorporated into peptoid oligomers. The triazole moiety introduced a constraint in the peptoid backbone, similar to that of a cis double bond, resulting in a tight turn in the peptoid structure. This triazole monomer was flanked by bulky α-chiral, aromatic monomers to further rigidify the turn motif, and it was observed that structural stability increased when two hydrophobic residues were incorporated to increase hydrophobic collapse. This work represents the first hairpin-like structure of a linear peptoid in aqueous solution. Such a strategy for turn motif stabilization may prove fruitful in the future design of biomimetic peptoids.

3.2.2 Cyclic peptoids

One potentially promising way to overcome peptoid flexibility, and restrict the conformational freedom of peptoids is cyclization of linear peptoids. It is generally considered that cyclic peptides may have enhanced binding affinity, selectivity, and metabolic stability, compared to their linear counterparts. Indeed, a number of cyclic peptides have been shown to exhibit highly potent biological activities, and some of them are clinically used. Similarly, cyclic peptoids are expected to have increased conformational rigidity and preorganized structures, potentially enabling them to bind more tightly to target proteins, without a major entropy loss. Thus, cyclic peptoids are of great interest as a promising class of peptidomimetics. Not surprisingly, there have been a growing number of reports on the synthesis of cyclic peptoids and cyclic peptoid/peptide hybrids (Butterfoss et al., 2012; Fowler et al., 2008; Guo and Zhang, 2009; Khan et al., 2011; Kwon and Kodadek, 2008; Lee et al., 2010a, 2011; Maulucci et al., 2008; Nnanabu and Burgess, 2006; Olsen, 2010; Roy et al., 2008; Shin et al., 2007; Vaz and Brunsveld, 2008; Yoo et al., 2010).

3.2.3 Biomimetic nanostructures

Compared to peptides, protein functional domains are typically 50–200 (average) residues in length, and utilize a sequence of amino acids with specific side chains to form folded structures with a compact hydrophobic core. The dominant driving force involved in the folding of polypeptide chains into protein structures in aqueous solution is the burial of hydrophobic amino-acid residues in an interior core, and the exposure of the hydrophilic side chains to solvent. These design principles can be applied to the folding of nonnatural polymers like peptoids to create nanostructures. Moreover, the study of folded proteins and self-assembling peptides has shown that the sequence pattern and periodicity of polar and nonpolar residues are the primary determinants of secondary structure (Nam et al., 2010).

Peptoid nanosheets are an emerging class of two-dimensional (2D) nanomaterials with potential applications ranging from molecular sensors to artificial enzymes. These biocompatible materials can be designed to assemble into 2D peptoid nanosheets (Nam et al., 2010). In one of the earliest studies on pepetoid nanosheets, peptoid polymers were self-assembled into 2D crystalline sheets that are 3-nm thick in aqueous environment, directed by periodic amphiphilicity, electrostatic recognition, and aromatic interactions (Nam et al., 2010). The bilayer structure consisted of a hydrophobic interior and two polar faces guided by electrostatic interactions. Such nanosheet structures may serve as chemically well-defined hydrophilic scaffolds on which biologically active ligands can be displayed for biomolecular recognition (Kudirka et al., 2011). These nanosheets are particularly interesting, as they serve as a versatile platform for constructing multifunctional, precisely ordered 2D nanostructures. (Peng et al., 2011; Sakamoto et al., 2009; Sanii et al., 2011; Tang et al., 2006), free-floating, ultrathin 2D crystals from sequence-specific peptoid polymers (Kudirka et al., 2011; Nam et al., 2010; Sanii et al., 2011; Sun and Zuckermann, 2013), peptoids that are rugged and highly designable. Further, the outer surfaces of these nanosheets can be readily functionalized with a controlled spatial density of binding sites (Olivier et al., 2013).

Peptoid nanosheets have been made through a unique assembly mechanism involving adsorption, compression, and organization of molecular units at the air–water interface (Payamyar et al., 2014; Sanii et al., 2011), and also at an oil–water interface (Robertson et al., 2014). At the air-water interface, peptoids containing an alternating sequence of polar and nonpolar monomers organize into a monolayer via adsorption from solution. Under compression, the monolayer collapses into a bilayer, and desorbs from the air-water interface. During collapse, the two hydrophobic faces contact each other, and physically bind to irreversibly form free-floating bilayer nanosheets. Under decompression, polymer chains from solution repopulate the interface, forming a new monolayer in equilibrium, with the reservoir of polymers dissolved in the water phase (Sanii et al., 2011). The packing of the polymers to form a nanosheet occurs at the monolayer (Sanii et al., 2014).

Zuckermann and coworkers have also developed model systems that allow controlled engineering of self-assembled structures in aqueous solution (Murnen et al., 2010; Sun et al., 2014a,b). In this regard Block copolymers that self-assemble into a number of nanostructured morphologies, including body-centered-cubic packed spheres, hexagonally packed cylinders, and lamellae are imperative structures. Zuckerman et al. synthesized and studied two ethyleneoxy-containing block copolypeptoids pNeh-*b*-pNte (poly-*N*-(2-ethyl)hexylglycine-block-poly-*N*-2-(2(2-methoxyethoxy)ethoxy)ethylglycine) and pNdc-*b*-pNte (poly-*N*-decylglycine-block-poly-*N*-2-(2-(2-methoxyethoxy)-ethoxy)ethylglycine) (Sun et al., 2013; Sun et al., 2014a,b). This diblock copolypeptoid system first self-assembles into bilayer sheets and further twists into a superhelix structure at pH 6.5. (Murnen et al., 2010) Ion-containing block copolymers are of interest for many applications, such as batteries, fuel cells, and photovoltaic devices. The structure-conductivity relationships of two types of monodisperse diblock copolypeptoids doped with LiTFSI have shown that such blocks dissolves salt and conduct ions (Sun et al., 2014a,b).

4 APPLICATIONS OF PEPTOIDS

The processability and robustness of peptoids and their capacity to form periodic structures, even over macroscopic length scales are advantages over polypeptide-based approaches. As a result peptoids have found utility in a vast array of applications as tools to study bio-molecular interactions (Hara et al., 2006; Udugamasooriya et al., 2008; Xiao et al., 2007a,b) and also hold significant promise for therapeutic applications relative to α-peptides. They are already used in many applications such as antibiotics, anticancer compounds, or as specific inhibitors for various proteins. (Au et al., 2006; Astle et al., 2008; Chongsiriwatana et al., 2008; Eggenberger et al., 2009; Gobbo et al., 2006; Green et al., 2006; Hara et al., 2006; Hoffmann et al., 2006; Huang et al., 2014; Jia et al., 2002; Kodadek, 2006; Kumar et al., 2006; Lim et al., 2006, 2007, 2008; Liu et al., 2006; MacKerell et al., 1998; Orzaez et al., 2007; Prasad et al., 2006; Ryge and Hansen, 2006; Schroder et al., 2007; Seurynck et al., 2005; Seurynck-Servoss et al., 2006; Shuey et al., 2006; Udugamasooriya et al., 2008; Vicent and Perez-Paya, 2006; Zhu et al., 2006). Along with the development of peptoid mimics, properties of certain naturally occurring peptides have also been improved by incorporating one or more peptoid residues to generate peptide-peptoid hybrids (or peptomers) (Fowler and Blackwell, 2009) or chimeras (Brandt et al., 2011). Peptoids have found numerous application as attractive targets for the drug discovery process (Miller et al., 1994), to develop peptide mimics for biomedical applications (Brown et al., 2008; Chongsiriwatana et al., 2008; Nandel and Saini, 2007; Patch and Barron, 2003; Seurynck-Servoss et al., 2006, 2007; Statz et al., 2009), high-density ligand arrays (Li et al., 2004), activation domain mimics (Liu et al., 2006), inhibitors of protein-protein interactions (Hara et al., 2006), vehicles for gene and drug delivery (Wender et al., 2000), asymmetric catalysts (Maayan et al., 2009), zinc binding peptoids (Lee et al., 2008), antifreeze protein mimics (Huang et al., 2012)., artificial light-harvesting complexes and (Kang et al., 2013), and antifouling polymers (Statz et al., 2009). In the field of diagnostics too peptoids have found utility as Biomarkers, for example, in Alzheimer's disease (Reddy et al., 2011).

4.1 PEPTOID COMBINATORIAL LIBRARIES

Combinatorial libraries are collections of large number of synthetic, chemical or biological compounds that are synthesized and stored in mixtures taking into account all possible combinations of the building blocks. This feature permits the synthesis of a large number of compounds in a relatively short time (Desai et al., 1994). The utility of having such libraries is that they can be screened to identify useful components, such as ligands and drug candidates. Due to the ease of synthesis and structural similarity to natural peptides, peptoids have been used as the basis for the combinatorial drug discovery process, gene delivery and for identification of new compounds. From a synthetic perspective, the modular nature of peptoids makes them amenable to combinatorial strategies through submonomer peptoid synthesis protocol, split-and-mix format, "one bead two compound" strategy and even by positional scanning format. In addition, the ability to screen and analyze peptoids rapidly and inexpensively from the libraries has been a primary tool in modern peptoid research.

Early on, Miller et al. (1994) designed, synthesized and screened a diverse peptoid library of 5000 compounds and discovered potent ligands that bind opiate and α_1-adrenergic receptors. The library was synthesized using the submonomer peptoid synthesis protocol (Simon et al., 1992). Peptoid libraries constructed by the split-and-mix format have been used to identify ligands for membrane receptors (Gibbons et al., 1996; Heizmann et al., 1999; Miller et al., 1994) that exhibited specific protein-binding activity (Alluri et al., 2006). A large library of 328,000 tripeptoids (Heizmann et al., 1999) was synthesized using 69 different primary amines to discover ligands for melanocortin type 1 and gastrin-releasing peptide/bombesin receptors. Another group screened a library of ~ 350,000 peptoid trimer compounds against a variety of agrochemical targets to provide a submicromolar antagonist of the nicotinic acetylcholine receptor (Kleschick et al., 2001). Furthermore, the inherent conformational flexibility of these oligomers has led to their use in the disruption of protein–protein, protein–nucleic acid and/or protein–membrane interactions (Daelemans et al., 2000). To evaluate binding affinities for the coactivator CREB-binding protein (CBP), a library of ~100,000 peptoid hexamers was created (Alluri et al., 2006). Further, peptoids that antagonize the vascular endothelial growth factor receptor 2 were identified from a library of 300,000 peptoid nonamers (Udugamasooriya et al., 2008). Such peptoid libraries have also been used for the identification of new antibacterial compounds (Goodson et al., 1999).

The modular composition of peptoids raised the question of whether libraries of these oligomers could be synthesized and screened in a positional scanning format (Blondelle et al., 1996). The latter is an important tool for sequence optimization as it identifies residues of interest at a given position or positions and substitutes them at that position with other residues one at a time. This generates high value data by locating potentially more favorable residue(s) at specified position(s) for enhanced activity. It was envisaged that a single peptoid library in this format could be used to identify hit compounds for multiple biological targets. In this context, N-alkylglycine trimers were constructed under the positional scanning format and used to discover antagonists of the vanilloid receptor subunit 1 (Garcia-Martinez et al., 2002). Also, compounds exhibiting antimicrobial activity against a panel of Gram-positive and Gram-negative bacteria were reported. A library of 10,000 compounds provided TRPV1 channel (Garcia-Martinez et al., 2002) and NMDA receptor open-channel blockers when screened on a subset of membrane receptors. Library screening in a whole-cell based assay, also led to the discovery of peptoids with in vivo neuro-protectant activity (Montoliu et al., 2002). An optimized solid-phase synthesis of a library of 5120 trimeric N-alkylglycines (peptoids) used the positional scanning format and the submonomer strategy. Formation of undesirable side-products observed in previous library synthesis was averted by restricting the use of primary amines functionalized with tertiary amino groups to the third amination step. Screening of this new library led to the identification of chemosensitizers, with potent in vitro activity as multidrug resistance reversal agents (Masip et al., 2005).

A "one bead two compound" strategy for the creation of encoded combinatorial libraries for cyclic peptoid microarrays has also been given (Kwon and Kodadek, 2008). Microarrays of surface-bound small molecules are increasingly important tools for high-throughput screening and discovery of new ligands (Koehler et al., 2006). Microarray format offers potential benefits as a means to discover new ligands because thousands of compounds can be screened simultaneously on microarrays with low requirements for costly samples, and many replicate arrays can be printed from a single chemical library, allowing the same library to be screened against multiple targets or in multiple screening trials with the same target. Peptoid microarray libraries was synthesized for a system of RNA ligand discovery specifically for the RNA hairpin precursor of miR-21, a microRNA involved in cancer and heart disease (Chirayil et al., 2009).

Kodadek and coworkers used a two-color, cell-based screening method to isolate peptoid hits to exhibit specificity for a membrane protein, VEGFR-2 (vascular endothelial growth factor receptor-2) (Astle et al., 2008; Lim et al., 2007). Harbury and colleagues utilized the concept of evolutionary mechanism and DNA-programmed combinatorial chemistry to prepare a large library of peptoids individually associated with specific DNA sequences, enabling amplification of compound hits. The authors adopted the bioinspired but completely abiotic approach to prepare a collection of 100 million distinct peptoids. The library was then subjected to a selection process for binding to the N-terminal SH3 domain of the proto-oncogene Crk, and novel ligands were discovered after six amplification steps. The hits bind to the protein tightly with affinities similar to those of peptide SH3 ligands discovered from phage display libraries (Wrenn et al., 2007).

4.2 BIOMOLECULAR RECOGNITION AND INTERACTIONS

The objective behind using peptoids as tools to recognize bio-molecular interactions is that protein-protein interactions control many cellular processes, and selective inhibition of these interactions could lead to new therapies for several diseases. Peptoid-protein interactions were first demonstrated by Zuckermann and coworkers, who examined high-affinity binding of peptoid dimers and trimers to G-protein coupled receptors (Desai et al., 1994). Tripeptoid ligands for both the α_1-adrenergic receptor and the μ-opiate receptor were identified. Shortly thereafter, peptomers that could bind selectively to a Src homology 3 (SH3) domain and could be utilized to inhibit protein-protein interactions involving SH3 domains were identified (Michnick and Shakhnovich, 1998). In this context, proteins containing the SH3 domain that recognize peptides with a PXXP (P is proline, X is any amino acid) motif were investigated with a variety of substituants for proline. A hybrid oligomer bearing dimethoxybenzyl N-substitution showed a 100-fold increases in binding affinity, over the wild type peptide, accompanied by an increase in specificity for binding to the SH3 domain of the N-Grb2, relative to those of Src and Crk. Therefore, even a single site substitution by a peptoid residue could have a strong effect on the binding affinity. In another study, a peptoid 9-mer with a number of cationic groups (similar to peptides rich in lysine, or arginine amino acids) was shown to block the interaction of HIV-1 Tat protein with TAR RNA, suppressing HIV-1 replication in vivo (Daelemans et al., 2000). Further, a tripeptoid inhibitor of the apoptotic protease-activating factor Apaf-1 was also identified (Malet et al., 2006). Apaf-1 is a central protein component of the apoptosome, and the formation of this multiprotein complex is a key event in apoptosis activation.

In vitro studies by Kodadek and coworkers show three peptoid hexamers with binding affinities for the coactivator CREB-binding protein (CBP) (Alluri et al., 2006; Liu et al., 2006). CBP is involved in the transcription of a large number of mammalian genes, and hence served as a target for the isolation of peptoid activation domain mimics. This study also revealed that a more hydrophilic peptoid (with appropriate length of side chains) displayed better cell permeability, and presence of numerous aromatic, hydrophobic moieties increased nonselective protein binding.

Another study uncovered a peptoid that antagonized the vascular endothelial growth factor receptor 2 (VEGFR2) (Udugamasooriya et al., 2008). The binding of VEGF to VEGFR2 is a critical event in angiogenesis, and blocking this hormone receptor interaction represents a possible strategy for the treatment of cancers (especially those involving solid tumors) and macular degeneration. To date, monoclonal antibodies, protein-based molecules, and peptides have been shown to inhibit the VEGF pathway, but they have limited tumor penetration, and are often immunogenic. Small molecule-based treatments, on the other hand, often have side effects due to interaction with other receptors similar to VEGFR2 in the body. Five peptoid nonamers were shown to bind selectively to VEGFR2 in vitro, with low micromolar affinity. Each of these peptoids was rich in charged Nlys residues, and also contained hydrophobic aliphatic and aromatic residues. A relatively potent antiangiogenic peptoid agent named "GU40C4" disrupted VEGF-induced receptor autophosphorylation and, therefore, downstream events leading to angiogenesis (Udugamasooriya et al., 2008). Based on the knowledge that VEGFR2 functions as a homodimer, dimeric peptoid ligands were also designed, and found to be active in vivo, as they inhibited tumor growth in a mouse model.

In cancer, overexpression of the protein human double minute 2 (HDM2) that binds to and inactivates the p53 proteinhas been linked to tumor aggressiveness and drug resistance. Structural information of the HDM2-p53 complex has also been used to design an all-peptoid decamer that competes with the wild type, N-terminal p53 peptide, for the binding site of the N-terminal domain fragment of HDM2, thus effectively inhibiting HDM2-p53 binding (Hara et al., 2006). Inhibitors of the HDM2-p53 interaction therefore hold promise for cancer therapy. Proapoptotic activity has also been reported for a peptoid trimer family (Tarrason et al., 2004). Subsequently, a group of peptoid dimers was also associated with cytotoxic and antimitotic activity in a panel of representative human neoplastic cell lines (Moruno-Mas et al., 2007).

Inhibitors of proteasome function that can intercept proteins targeted for degradation would be valuable as both research tools and therapeutic agents. The 26S proteasome is responsible for most nonlysosomal protein degradation in eukaryotic cells. The 19S regulatory particle, a portion of the 26S proteasome, binds polyubiquitinated proteins, unfolds them, and guides them into the proteasome interior for proteolytic degradation. In 2007, the first chemical modulator of the 19S regulatory particle, a purine-capped peptoid heptamer, was identified (Lim et al., 2007, 2008) and put forward as a minimal pharmacophore of a peptoid inhibitor of the 19S regulatory particle.

The ability of these peptidomimetics to access unique conformations, and the variety of available monomers, make peptoids particularly valuable as ligands, and their metal binding capacity has also been brought to light (Izzo et al., 2013; Knight et al., 2013; Lee et al., 2008; Maayan et al., 2009; Nalband et al., 2014). Recently, peptoid-based ligands for the removal of cadmium from human serum were identified by Knight et al. (2015). The identification of peptoid metal ligands is a great advancement in the treatment of heavy metal poisoning, as it is very difficult to handle such poisoning (Knight et al., 2015).

4.3 CELLULAR UPTAKE AND DELIVERY

4.3.1 Cell penetrating peptoids

Novel peptidomimetic backbone designs with stability toward proteases are of interest for several pharmaceutical applications, including intracellular delivery. A drug must be either highly lipophilic or very small to cross the lipid bilayer. These restrictions limit the repertoire of possible drug molecules. Similarly, novel therapeutic approaches, such as gene and protein therapy, also have limited potential due to the cell-impermeable nature of peptides and oligonucleotides. Because the membrane acts as a protective barrier, cell-penetrating peptides (CPPs) are potential tools for drug delivery, as they are able to traverse the cell membrane and bring with them attached cargo in a seemingly energy-independent manner (Jarver and Langel, 2006; Lundberg and Langel, 2003; Viües et al., 2003). They are short amino acid sequences (<30 amino acids) that are highly cationic and hydrophilic (Drin et al., 2003), characterized by their ability to enter cells, independent of a membrane receptor (Derossi et al., 1996); they also show no cell-type specificity (Kerkis et al., 2006).

Subsequent work by Wender et al. (2000) on various truncated and alanine-substituted fluorescently labeled peptoid analogues of Tat_{49-57} demonstrated rapid cellular uptake of guanidinylated peptoids, relative to L- and D-arginine polypeptides in Jurkat cells. Presence of at least six arginine residues was found critical for rapid cell entry. Correspondingly, they synthesized a series of oligoguanidine peptoids, and compared their capacity of cellular uptake with peptide D-arginine oligomers between five and nine residues in length. The peptoid compound named "N-hxg9" was found to be superior to a D-arginine nonamer in cellular uptake, as it was 100-fold more rapidly translocated across the cell membrane than Tat_{49-57}.

A high-throughput assay for assessing mammalian cell permeability by synthetic molecules, based on assessing the relative cell permeability of synthetic molecule-dexamethasone derivative conjugates, was used to study the cell permeability of peptoids, relative to their corresponding α-peptide analogs (Kwon and Kodadek, 2008). Six peptoid sequences displayed 3–26-fold increase in cell permeability over the corresponding α-peptides. The relative permeability was inversely proportional to oligomer length, and average cell permeability of the peptoids was twice that of α-peptides.

A study where $(Nlys)_6$, $(Narg)_5$, and $(Narg)_6$ homopeptoids were conjugated to fluorophores demonstrated more rapid cellular uptake in guanidinium-functionalized peptoids (Narg), compared to amino-functionalized peptoids (Nlys) (Schroeder et al., 2008). Furthermore, the location of transport was found to differ, depending on amide side chain functionality, that is, amino peptoids were transported to the cytosol, while guanidinium peptoids had greater, but not exclusive, accumulation in the nucleus.

In 2008, Bradley and coworkers described peptoid dendrimers that functioned as efficient gene transfection agents (Diaz-Mochon et al., 2008). Dendrimers are synthetic, three-dimensional (3D) molecules with branching. Each step results in a new "generation" that has twice the complexity of the previous generation. They are formed using a nanoscale, multistep fabrication process. First-, second-, and third-generation peptoid dendrimers were synthesized containing lysine-like amide side chains that displayed 4, 8, or 16 amines, respectively. Efficient uptake of DNA by the third-generation peptoid dendrimer (~ threefold higher transfection relative to a polyamidoamine dendrimer) with no cytotoxicity in human cells was reported.

A study on intranasal administration (in rat models) that directly targets the central nervous system (CNS) and lymph nodes, has shown significant delivery throughout the CNS and deep cervical lymph nodes for the peptoid oligomer "CHIR5585," an antagonist of the urokinase plasminogen activator receptor (Rossa et al., 2008).

Membrane permeability of peptoids has led to the design of peptoid-based transcription factor mimics, allowing for upregulation of targeted genes. Kodadek and coworkers screened a peptoid library to identify oligomers capable of binding a transcriptional coactivator. A peptoid hit sequence was conjugated to a polyamide to provide specific binding to a target region of DNA. The biological activity of the conjugate was facilitated by the membrane permeability of cationic peptoid (Xiao et al., 2007a,b).

4.3.2 Lipitoids

Lipitoids, or lipid-peptoid conjugates, were inspired by peptide-lipid hybrids that have frequently been employed as cationic gene delivery agents (Pack et al., 2005) and drug delivery vehicles. To refine the activity of peptoids for applications in cellular delivery, properties of lipid molecules were utilized, and they were conjugated to the N-termini of peptoids. As expected, these peptoid-lipid conjugates, that is, lipitoids, were even more active in delivering plasmid DNA to cells, and showed reduced cellular toxicity (Huang et al., 1998). Thus, high in vitro activity and low toxicity make lipitoids ideal transfection reagents. In these studies, it was observed that peptoids with a repeating trimeric side chain motif (cationic, neutral) displayed the highest transfection activities. The transfection efficiency could not be directly correlated with size, surface charge, structure (as characterized by CD), or affinity of the synthetic polymer for DNA (as measured by differential scanning calorimetry or ITC) (Lobo et al., 2003).

Lipitoids have been used as a siRNA transfection agent, efficiently transfecting a variety of difficult cell lines, including primary human cells (Utku et al., 2006). The lack of toxicity to immortalized cell lines was confirmed, and efficient siRNA delivery was demonstrated by monitoring the target genes with Western blots and real-time PCR. Additional SEM and DLS characterization was performed in follow-up work to characterize the particles formed from lipitoids and siRNA (Konca et al., 2014), and the size of the particles formed was found to vary depending on the ionic strength of the environment, and the ratio of lipitoid to siRNA.

4.4 PEPTIDES MIMICS: PEPTOIDS, PEPTIDE-PEPTOID HYBRIDS, OR PEPTOID CHIMERAS

Peptoids have been rationally designed as mimics of HIV-Tat protein (Hamy et al., 1997), lung surfactant proteins (Brown et al., 2008; Seurynck-Servoss et al., 2006, 2007), antimicrobial peptides, anticancer peptides (Chongsiriwatana et al., 2008; Patch and Barron, 2003), and many more. Such biomimetic peptoid structures could potentially yield valuable therapeutic and bioactive compounds.

4.4.1 HIV-Tat mimics

The nuclear transcription activator protein (Tat) from HIV type I (HIV-1) is a 101 amino acid protein that interacts with a 59-base RNA stem loop structure, called the transactivation region (Tar) at the 5′ end of all nascent HIV-1 mRNA molecules, so that the virus can replicate. HIV-Tat is actively transported across the cell membrane, and the arginine rich region of Tar protein (residues 49–57) has been found responsible for this tranlocation activity. Several peptidomimetic forms of this Tat_{49-57} have been synthesized because of its membrane tranlocation activity, and also to prevent HIV-Tat/Tar interaction (Daelemans et al., 2000; Hamy et al., 1997; Kesavan et al., 2002).

4.4.2 Peptoids as lung surfactants

Peptides displaying a structure similar to that observed in surfactant molecules, that is, with a defined hydrophilic head group constituted of charged amino acids and a lipophilic tail made out of hydrophobic amino acids, such as alanine, valine, or leucine etc., are increasingly being used as natural surfactants or their mimics. Lung surfactant (LS) is a naturally-occurring material that is essential for proper respiration in humans, and is composed of lipids and proteins that reduce and regulate surface tension at the air-liquid interface in the lungs (Notter, 2000). A deficiency in functional LS leads to respiratory distress syndrome, a leading cause of mortality in premature infants, and of respiratory impairment or failure in children and adults. Treatment with current synthetic surfactant formulations is less effective than natural LS, and animal extracts of natural LS have several disadvantages, such as causing an immune response. Nonnatural functional mimics of the LS protein B and C (SP-B and SP-C, that is, surfactant protein-B and -C, respectively), based on peptoids that shall be used in a biomimetic LS replacement to treat respiratory distress syndrome in premature infants, is under extensive research, with the eventual goal of creating better synthetic LS replacements (Brown et al., 2008; Rouhi, 2001; Seurynck-Servoss et al., 2006, 2007; Wu et al., 2003).

Peptoid-based mimics of the lung surfactant protein SP-C for the 11 or 14 residue long C-terminal helix, with either all α-chiral aromatic (Nspe) or all α-chiral aliphatic (Nssb/Nile) residues, were also designed. Interestingly, in these mimics, aromatic peptoids were found to have better surface activity than the aliphatic peptoids. They reasoned that α-chiral, aromatic peptoids were the best mimics in the series studied because the more rigid helix modulated the surface film better, and the increased hydrophobicity mimicked SP-C protein better (Brown et al., 2008).

4.4.3 Antimicrobial peptoids

The class of linear, cationic, facially amphipathic and helical antimicrobial peptides is also under extensive research to develop peptoid mimics capable of the same, or enhanced, activity and specificity (Chongsiriwatana et al., 2008; Patch and Barron, 2003). Patch and Barron (2003) reported peptoid mimics of the helical antimicrobial peptide magainin-2 that had low micromolar activity against *E. coli* and *Bacillus subtilis* (Patch and Barron, 2003). Peptoids with strong helical CD signatures exhibited higher antibacterial activity, and greater hemolytic activity was associated with peptoids that were more hydrophobic. Thus, overall cationic charge (at least +3) and moderate hydrophobicity were demonstrated as key structural features that conferred good activity. Subsequently, they evaluated such antimicrobial peptoids for antibacterial activity and selectivity for bacterial cells over mammalian cells (Chongsiriwatana et al., 2008). Peptoids that displayed strong, as well as a weak, CD signal effectively killed bacteria; but peptoids with strong helical CD signatures had greater hemolytic activity. Also, hemolysis increased with increase in hydrophobicity. Therefore, it was suggested that peptoid helical structures were not important for antimicrobial activity and selectivity, but cationic charge and moderate hydrophobicity were crucial.

In another study, Zhu et al. (2006) described the incorporation of peptoid residues into antimicrobial peptide melittin, and identified cell-selective analogs. Melittin is a 26 amino acid protein (20 predominantly apolar residues, 4 charged, and 2 polar residues) known to adopt an amphipathic helical structure in a lipid environment, and contains a leucine zipper motif. This zipper motif is believed to promote self-association of melittin, leading to the formation of a transmembrane pore causing cell lysis. Melittin analogs with peptoid residue replacements (Nala, Nleu, Nphe, and Nlys) of key residues in the leucine zipper were designed. The analogs containing Nleu, Nphe, and Nlys showed strong antimicrobial activity against three Gram-negative bacterial strains, three Gram-positive bacterial strains, and four antibiotic-resistant bacterial strains. Also, they did not cause hemolysis of human erythrocytes at concentrations as high as 100 mM, while melittin was hemolytic even at 1 mM. This study showed that destabilizing helical structures in peptomers, at least in the absence of cell membranes, appeared to reduce toxicity in mammalian cells.

A peptomer backbone was also used to mimic protegrin-I, a cationic β-hairpin antibiotic (Shankaramma et al., 2003). D-Pro-L-Pro hairpin-stabilizing template was incorporated into a macrocyclic structure to mimic the disulphide bond induced β-hairpin motif of protegrin-I. The authors synthesized both the α-peptide and peptomer versions of the mimic, and found that incorporation of just one peptoid residue (substitution of Arg-6 with Nlys) into a protegrin-I mimic led to enhanced structural stability, antibacterial activity, and selectivity. In a very recent study, peptoid analogues of tryptophan-rich GN-2, GN-4, and GN-6 peptides were designed. The report highlights peptoids with an overall net charge of 4, and length of 8 or 9 residues that retain sufficient antimicrobial activity and toxicity profiles (Mojsoska et al., 2015).

4.4.4 Anticancer peptoids

New molecular platforms that employ novel mechanisms of action to overcome drug resistance, and provide more effective treatments, are the need of the hour. The well-known molecular classes in the modern drug development arena, such as small organic molecules, antibodies, peptides, and nanoparticles, carry both useful as well as undesired properties (Frangioni, 2008). Years ago, small organic molecules exclusively represented therapeutics as easy to handle and orally available chemical compounds. However, the synthesis of the majority of biologically active complex organic compounds remains a challenge to date. In addition, these molecules clear through the kidneys, and do not have adequate affinity and contact time for effective imaging or therapy of the tumor. During the past decade or so, antibodies have been developed as high affinity and specificity drugs, but this is a costly approach, and antibodies also have intermediate to poor biodistribution and tumor penetration. Other traditional macromolecules, such as nanoparticles, virtually have no clearance from the body, a fact that raises fundamental questions on their use as drugs. In the meantime, peptides as intermediate size molecules emerged as great therapeutics, but current development of peptide-based pharmaceuticals is hindered by their rapid in vivo degradation. Researchers have recently investigated alternative peptide-like constructs that may be able to circumvent such complications. This is the point where peptidomimetics or peptide-like molecular developments were initiated, and peptoids have emerged as one such promising molecular class, in particular in oncology applications (Udugamasooriya, 2013).

Over the past two decades, a growing number of studies have shown that some cationic, amphipathic peptides, including both natural host defense peptides and synthetic antimicrobial peptides, exhibit a broad spectrum of cytotoxic activity against cancer cells, and are effective in reducing tumor burdens in several cancer animal models (Afinsen and Scheraga, 1975; Ahmad et al., 1995; Alberts et al., 1994; Allen et al., 1991, 1998). The selectivity of these peptides toward cancer cells is not well understood, and is hypothesized to result from some altered membrane properties of cancer cells, as compared to normal cells. The peptidomimetic analogues of these antimicrobial peptides have also been developed and tested for their potential. Anticancer peptoids have been developed successfully with potent cytotoxicity toward a broad range of human cancer cell lines, MCF-7 (human breast cancer), LNCaP (human prostate cancer), and OVCAR-3 (human ovarian cancer), and their killing was not affected by multidrug resistance. The most potent peptoid—H–(NLys–Nspe–Nspe)$_4$–NH_2—greatly reduced tumor growth via intratumor injections in a human breast cancer xenograft mouse model, without causing any obvious acute side effects in mice, suggesting in vivo anticancer efficacy, and good local tolerance of peptoids. (Huang et al., 2014)

4.4.5 Peptoids as antifouling biomaterials

Controlling biointerfacial phenomena is crucial to the success of many biomedical technologies. For applications in biosensing, diagnostics, and medical devices, precise control of interactions between material surfaces and the biological milieu is an important but elusive goal (Ratner, 2000). Limiting nonspecific interaction of cells, proteins, and microorganisms with material surfaces is critical, since these interactions can prove highly problematic for device efficacy and safety. Thus, research continues to focus on the development of versatile, convenient, and cost effective strategies for rendering surfaces resistant to fouling by proteins, cells, and bacteria (Dalsin and Messersmith, 2005).

A common method to reduce cellular and proteinaceous fouling is the immobilization of antifouling (AF) polymers on biomaterial surfaces. AF polymers should be robustly conjugated to the anchoring peptides/peptoids, and should be water resistant (Deacon et al., 1998). As such, the composition of the anchored peptides/peptoids in terms of amino acid sequence is always the main consideration. Coating of polyethylene glycol (PEG) as AF polymer is well established (Zhang et al., 2001), but its long-term in vitro and in vivo success is not good (Deacon et al., 1998), due to oxidative degradation and enzymatic cleavage of PEG chains. To overcome these limitations, chimeric peptidomimetic polymer (PMP1) (Dalsin and Messersmith, 2005) consisting a peptoid oligomer coupled to a short functional peptide domain for surface adhesion (Statz et al., 2009) was synthesized. The hydrophilic methoxyethyl chain of PMP1 possesses hydrogen bond acceptor (Ostuni et al., 2001), but not hydrogen bond donor, and is water insoluble, bearing resemblance to the repeat unit of PEG.

The authors then conjugated the polypeptoids to biomimetic adhesive polypeptides, and immobilized them on titanium oxide (TiO_2) surfaces. All polypeptoid modified surfaces exhibited significant reduction in adsorption of lysozyme, fibrinogen, and serum proteins, and were resistant to 3T3 (cell line) fibroblast cell attachment, for up to 7 days (Statz et al., 2008). Moreover, attachment of Gram-positive and Gram-negative bacteria for up to 4 days under continuous flow conditions was significantly reduced on the polypeptoid-modified surfaces, compared to unmodified TiO_2 surfaces. The passive resistance of *N*-methoxyethyl glycine peptoids was then combined with an active antimicrobial sequence, and a reduction in protein adsorption and fibroblast cell attachment was demonstrated, while the immobilized antimicrobial peptoid remained active (Statz et al., 2008, 2009).

Zwitterionic antifouling polymer brushes have captured significant attention, and a targeted library of zwitterionic peptoids was quantitatively evaluated for antifouling properties through a range of protein adsorption and cell attachment assays. Specific zwitterionic brush designs are found to give rise to distinct but subtle differences in properties, and also point to the dominant roles of the grafted chain density and chain length in determining the performance of antifouling polymer brushes (Lau et al., 2015).

4.5 PEPTOID BIOMARKERS, BIOSENSORS, AND BIOCATALYSTS

Discovery of disease specific protein biomarkers is of great interest to researchers. A particularly interesting subproteome in this regard is the IgG antibody population (Blennow et al., 2010), described as a technology in which the entire complement of serum IgG antibodies was screened against a peptoid library in order to identify complexes of disease-specific antibodies and individual peptoids. The screening was particularly carried out using serum samples from six patients with Alzheimer's disease (AD) (three autopsy confirmed), six matched control individuals, and six patients with Parkinson's disease (PD). Three peptoids were identified that captured at least threefold higher levels of IgG antibodies from all six of the patients with AD, than any of the controls or patients with PD. In the process, at least two candidate auto-antibody biomarkers for AD have been discovered. Subsequent blinded studies were conducted, using samples from more patients with AD, controls, and patients with a different disease (lupus) to test the utility of the peptoids identified in the original screen. These studies constitute a fair and critical test of the utility of the peptoid-antibody complexes as biomarkers (Reddy et al., 2011).

In the context of biocatalysts, the peptoid polymer C792-5 was identified to have the capacity of mineralizing calcium ion and phosphate ion into an apatite equal to or more than tenfold, when compared to the formation of apatite in the absence of the peptoid polymer (Zuckermann et al., 2015).

4.6 PEPTOID NANOSTRUCTURES

In recent times, peptoid nanosheets have emerged as ideal scaffolds for peptoid and peptide loops to be displayed on the nanosheet surface (Olivier et al., 2013). These materials have been designed to mimic antibodies and/or enzymes, and participate in the multivalent binding of target molecules. It was envisioned that insertion of a hydrophilic sequence within a sheet-forming strand would lead to its exclusion from the bilayer scaffold during compression, and that positioning the loop insert in the middle would be less likely to disrupt sheet formation. In addition, a loop insert must not be too long or too hydrophobic in nature, as to not hinder the adsorption at the air-water interface. Thus, precise sequence control over the position, nature, and length of the insert was an important element in designing artificial antibody mimetics. In addition, phosphorylation of a peptide loop substrate followed by incubation with FITC-labeled antiphosphoserine antibody allowed fluorescent imaging of the nanosheets, thereby providing additional evidence of loop display. These materials hold great promise as molecular recognition elements for chemical and biological detection, and as templates for the growth of 2D materials. This work also highlights the importance of sequence control to predictably form a 3D, antibody-like architecture from a carefully designed information-rich, linear peptoid.

Nanosheets can serve as 2D scaffolds with high surface area that can be readily functionalized to display a high density of molecular binding sites on both faces of the sheet (Olivier et al., 2013). Moreover, peptoid nanosheets are stable in air and in water at a variety of pHs, and do not aggregate in solution. The peptoid nanosheet surface is hydrophilic and zwitterionic, enabling it to form stable free-floating suspensions in solution. Recently, Olivier et al. (2013) demonstrated that peptoid nanosheets, in which the surface was functionalized with a gold binding peptide, effectively nucleated the growth of gold films on the sheet surface, suggesting that nanosheets can serve as an effective platform for templating inorganic materials.

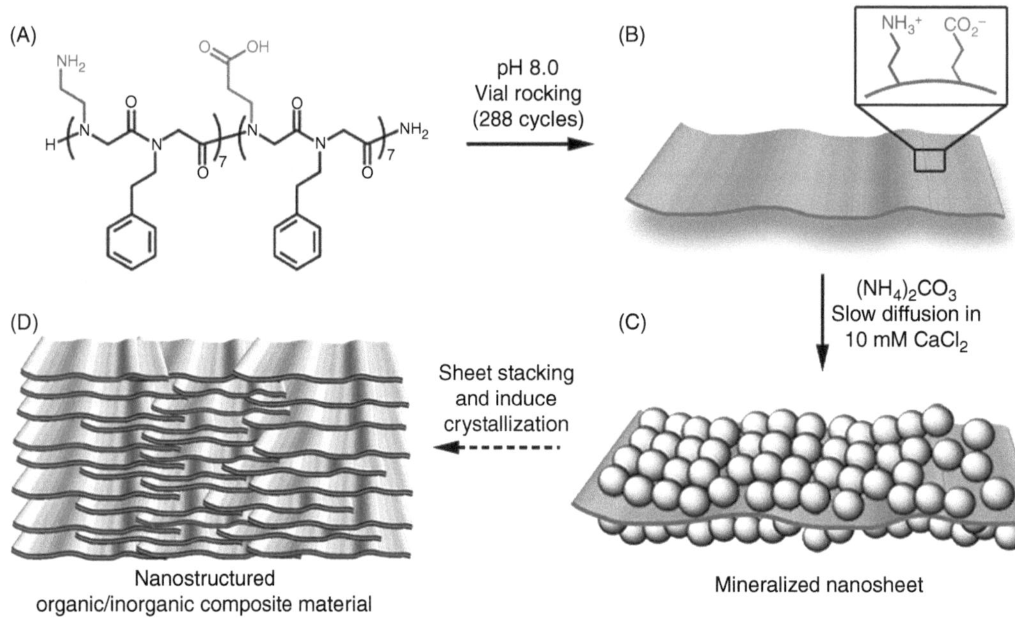

FIGURE 10.3 Representation of Pathways (A–D) for Formation of Peptoid Nanosheets Inspired by Nacre (Myung et al., 2015)

Natural $CaCO_3$ mineralization proteins are typically acidic, with a relatively large number of glutamic acid and aspartic acid residues (Jones and Ogden, 2010; Wang et al., 2008). Peptidomimetic oligomers bearing multiple acidic side chains have been shown to promote the nucleation of $CaCO_3$ crystal growth (Chen et al., 2014; Estroff et al., 2004; Donners et al., 2002). In fact, Chen et al. (2011) reported that short acidic peptoid oligomers can control the morphology of growing calcite crystals, and can even accelerate the crystal growth rate by a factor of more than 20-fold. Thus, peptoid nanosheets are promising template, as they display a high surface density of carboxylic acid groups, and can be synthesized in high yield. Furthermore, the nanosheets have a well-defined structure that can be readily engineered at the atomic level, allowing precise tunability of the material properties. Using a simple CO_2 diffusion method, free-floating zwitterionic peptoid nanosheets were mineralized with thin films of amorphous calcium carbonate (of 2–20 nm thickness) on their surface to produce planar nacre synthons. These can serve as tunable building blocks to produce layered brick and mortar nanoarchitectures, as shown in Fig. 10.3. (Myung et al., 2015)

5 CONCLUSIONS

Using peptoids in, and as, drugs will show promising results, when it comes to therapeutics. The advantages of using peptoids is their ease of mimicking natural structures, their ability to transform easily into new structures, and their compatibility with bimolecular surfaces. The possibility of using a wide variety of peptoid structures by tuning their chemistry can open up the gates for futuristic drug discoveries. This chapter has spelled out the various potential applications for peptoids in therapeutics. Antifouling, antimicrobial, and other useful functions of peptoids are ascribed to the specificity, biomolecular recognition, and interaction capability, and their ability to be designed or engineered according to application and need. Newer platforms for overcoming drug resistance and providing effective treatments can be realized by the use of peptoid nanostructures. Peptoid-based polymeric structures that can self-assemble into nature-inspired supramolecular nanostructures will truly be tomorrow's therapeutics.

REFERENCES

Adessi, C., Soto, C., 2002. Converting a peptide into a drug: strategies to improve stability and bioavailability. Curr. Med. Chem. 9, 963–978.

Afinsen, C.B., Scheraga, H.A., 1975. Experimental and theoretical aspects of protein folding. Adv. Protein. Chem. 29, 205–301.

Ahmad, I., Perkins, W.R., Lupan, D.M., Selsted, M.E., Janoff, A.S., 1995. Liposomal entrapment of the neutrophil-derived peptide indolicidin endows it with in vivo antifungal activity. Biochem. Biophys. Acta 1237, 109–114.

Alberts, B., Bray, D., Lewis, J., Raff, M., Roberts, K., Watson, J.D., 1994. Molecular biology of the cell, third ed. Garland Publishing, New York, NY.

Allen, F.H., Dacies, J.E., Galloy, J.J., Johnson, O., Kennard, O., Macrae, C.F., Mitchell, E.M., Mitchell, G.F., Smith, J.M., Watson, D.G., 1991. The development of version 3 and 4 of the Cambridge Structural Database System. J. Chem. Inf. Comput. Sci. 31, 187–204.

Allen, F.H., Baalham, C.A., Lommerse, J.P.M., Raithby, P.R., 1998. Carbonyl–carbonyl interactions can be competitive with hydrogen bonds. Acta Crystallogr. B54, 320–329.

Alluri, P.G., Liu, B., Yu, P., Xiao, X., Kodadek, T., 2006. Isolation and characterization of coactivator-binding peptoids from a combinatorial library. Mol. BioSyst. 2, 568–579.

Armand, P., Kirshenbaum, K., Falicov, A., Dunbrack, R.L.J., Dill, K.A., Zuckermann, R.N., Cohen, F.E., 1997. Chiral N-substituted glycines can form stable helical conformations. Fold. Des. 2, 369–375.

Armand, P., Kirshenbaum, K., Goldsmith, R.A., Barron, A.E., Truong, K.T.V., Dill, K.A., Cohen, F.E., Zuckermann, R.N., Bradley, E.K., 1998. NMR determination of the major solution conformation of a peptoid pentamer with chiral side chains. Proc. Natl. Acad. Sci. 95, 4309–4314.

Astle, J.M., Udugamasooriya, D.G., Smallshaw, J.E., Kodadek, T., 2008. A VEGFR2 antagonist and other peptoids evade immune recognition. Int. J. Pept. Res. Therap. 14, 223–227.

Au, V.S., Bremner, J.B., Coates, J., Keller, P.A., Pyne, S.G., 2006. Synthesis of some cyclic indolic peptoids as potential antibacterials. Tetrahedron 62, 9373–9382.

Baldauf, C., Guenther, R., Hofmann, H.-J., 2006. Helices in peptoids of alpha- and beta-peptides. Phys. Biol. 3, S1–S9.

Barron, A.E., Zuckermann, R.N., 1999. Bioinspired polymeric materials: in-between proteins and plastics. Curr. Opn. Chem. Biol. 3, 681–687.

Blennow, et al., 2010. Cerebrospinal fluid and plasma biomarkers in Alzheimer disease. Nat. Rev. Neurol. 6, 131–144.

Blondelle, S.E., Houghten, R.A., Perez-Paya, E., 1996. Identification of inhibitors of Melittin using non-support bound combinatorial libraries. J. Biol. Chem. 271, 4093–4099.

Boeijen, A., Liskamp, R.M.J., 1998. Sequencing of peptoid peptidomimetics by edman degradation. Tetrahedron Lett. 39, 3589–3592.

Brandt, W., Herberg, T., Wessjohann, L., 2011. Systematic conformational investigations of peptoids and peptoid–peptide chimeras. Pept. Sci. 96 (5), 651–668.

Brown, N.J., Wu, C.W., Seurynck-Servoss, S.L., Barron, A.E., 2008. Effects of hydrophobic helix length and side chain chemistry on biomimicry in peptoid analogues of SP-C. Biochemistry 47, 1808–1818.

Burkoth, T.S., Beausoleil, E., Kaur, S., Tang, D., Cohen, F.E., Zuckermann, R.N., 2002. Towards the synthesis of artificial proteins: the discovery of an amphipathic helical peptoid assembly. Chem. Biol. 9, 647–654.

Butterfoss, G.L., Yoo, B., Jaworski, J.N., Chorny, I., Dill, K.A., Zuckermann, R.N., Bonneau, R., Kirshenbaum, K., Voelz, V.A., 2012. De novo structure prediction and experimental characterization of folded peptoid oligomers. PNAS 109 (36), 14320–14325.

Chen, C.-L., Qi, J., Zuckermann, R.N., DeYoreo, J.J., 2011. Engineered biomimetic polymers as tunable agents for controlling $CaCO_3$ mineralization. J. Am. Chem. Soc. 133, 5214–5217.

Chen, C.-L., Qi, J., Tao, J., Zuckermann, R.N., DeYoreo, J.J., 2014. Tuning calcite morphology and growth acceleration by a rational design of highly stable protein-mimetics. Sci. Rep. 4, 6266.

Chirayil, S., Chirayil, R., Luebke, K.J., 2009. Discovering ligands for a microRNA precursor with peptoid microarrays. Nucl. Acids Res. 37, 5486–5497.

Chongsiriwatana, N.P., Patch, J.A., Czyzewski, A.M., Dohm, M.T., Ivankin, A., Gidalevitz, D., Zuckermann, R.N., Barron, A.E., 2008. Peptoids that mimic the structure, function, and mechanism of helical antimicrobial peptides. Proc. Natl. Acad. Sci. 105, 2794–2799.

Crapster, J.A., Stringer, J.R., Guzei, I.A., Blackwell, H.E., 2011. Design and conformational analysis of peptoids containing N-hydroxy amides reveals a unique sheet-like secondary structure. Pept. Sci. 96, 604–616.

Culf, A.S., Ouellette, R.J., 2010. Solid-phase synthesis of N-substituted glycine oligomers (a-peptoids) and derivatives. Molecules 15, 5282–5335.

Daelemans, D., Schois, D., Witvrouw, M., Rannecouque, C., Hatse, S., Dooren, S.V., Hamy, F., Klimkait, T., Clercq, E.D., Vandamme, A.-M., 2000. A second target for the peptoid Tat/transactivation response element inhibitor CGP64222: Inhibition of human immunodeficiency virus replication by blocking CXC-chemokine receptor 4-mediated virus entry. Mol. Pharmacol. 57, 116–124.

Dalsin, J., Messersmith, P.B., 2005. Antifouling polymers inspired by biology. Mater. Today 8, 38-36.

De Cola, C., Licen, S., Comegna, D., Cafaro, E., Bifulco, G., Izzo, I., Tecilla, P., De Riccardis, F., 2009. Size-dependent cation transport by cyclic a-peptoid ion carriers. Org. Biomol. Chem. 7, 2851–2854.

Deacon, M.P., Davis, S.S., Waite, J.H., Harding, S.E., 1998. Structure and mucoadhesion of mussel glue protein in dilute solution. Biochemistry 37, 14108–14112.

Derossi, D., Calvet, S., Trembleau, A., Brunissen, A., Chassaing, G., Prochiantz, A., 1996. Cell internalization of the third helix of the Antennapedia homeodomain is receptor-independent. J. Biol. Chem. 271, 18188–18193.

Desai, M.C., Zuckermann, R.N., Moos, W.H., 1994. Recent advances in the generation of chemical diversity libraries. Drug Devel. Res. 33, 174–188.

Diaz-Mochon, J.J., Fara, M.A., Sanchez-Martin, R.M., Bradley, M., 2008. Peptoid dendrimers-microwave-assisted solid-phase synthesis and transfection agent evaluation. Tetrahedron Lett. 49, 923–926.

Donners, J.J.J.M., Nolte, R.J.M., Sommerdijk, N.A.J.M., 2002. A shape-persistent polymeric crystallization template for $CaCO_3$. J. Am.Chem. Soc. 124, 9700–9701.

Drin, G., Cottin, S., Blanc, E., Rees, A.R., Temsamani, J., 2003. Studies on the internalization mechanism of cationic cell-penetrating peptides. J. Biol. Chem. 278, 31192–31201.

Eggenberger, K., Birtalan, E., Schroder, T., Brase, S., Nick, P., 2009. Passage of Trojan peptoids into plant cells. Chembiochem 10, 2504–2512.

Estroff, L.A., Incarvito, C.D., Hamilton, A.D., 2004. Design of a synthetic foldamer that modifies the growth of calcite crystals. J. Am. Chem. Soc. 126, 2–3.

Figliozzi, G.M., Goldsmith, R., Ng, S.C., Banville, S.C., Zuckermann, R.N., 1996. Synthesis of N-substituted glycine peptoid libraries. Methods Enzymol. 267, 437–447.

Fisher, A.E.O., Naughton, D.P., 2003. Novel peptoids for the detection and suppression of reactive oxygen and nitrogen species. Biochem. Soc. Trans. 31, 1302–1304.

Fowler, S.A., Blackwell, H.E., 2009. Structure-function relationships in peptoids: recent advances toward deciphering the structural requirements for biological function. Org. Biomol. Chem. 7, 1508–1524.

Fowler, S.A., Stacy, D.M., Blackwell, H.E., 2008. Design and synthesis of macrocyclic peptomers as mimics of a quorum sensing signal from *Staphylococcus aureus*. Org. Lett. 10, 2329–2332.

Frangioni, J.V., 2008. New technologies for human cancer imaging. J. Clin. Oncol. 26, 4012–4021.

Garcia-Martinez, C., Humet, M., Planells-Cases, R., Gomis, A., Caprini, M., Viana, F., Pena, E.D., Sanchez-Baeza, F., Carbonell, T., Felipe, C.D., Perez-Paya, E., Belmonte, C., Messeguer, A., Ferre-Montiel, A., 2002. Attenuation of thermal nociception and hyperalgesia by VR1 blockers. Proc. Natl. Acad. Sci. 99, 2374–2379.

Gibbons, J.A., Hancock, A.A., Vitt, C.R., Knepper, S., Buckner, S.A., Brune, M.E., Milicic, I., Kerwin, J.F., Richter, L.S., Taylor, E.W., Spear, K.L., Zuckermann, R.N., Spellmeyer, D.C., Braeckman, R.A., Moos, W.H., 1996. Pharmacologic characterization of CHIR 2279, an N-substituted glycine peptoid with high-affinity binding for alpha 1-adrenoceptors. J. Pharmacol. Exp. Ther. 277, 885–899.

Gobbo, M., Biondi, L., Rocchi, R., De Cian, V., Reddi, E., Bertoloni, G., 2006. Peptoid-scan in antimicrobial peptides: synthesis and antibacterial activity of [Narg] peptoid-peptide hybrids of apidaecin I beta. J. Pept. Sci. 12, 109.

Goodson, B., Ehrhardt, A., Ng, S., Nuss, J., Johnson, K., Giedlin, M., Yamamoto, R., Moos, W.H., Krebber, A., Ladner, M., Giacona, M.B., Vitt, C., Winter, J., 1999. Characterization of novel antimicrobial peptoids. Antimicrob. Agents Chemother. 43, 1429–1434.

Gorske, B.C., Jewell, S.A., Guerard, E.J., Blackwell, H.E., 2005. Expedient synthesis and design strategies for new peptoid construction. Org. Lett. 7, 1521–1524.

Green, B.R., Bayudan, W., Ellison, M.E., Zhang, M.M., Yoshikami, D., Legowska, A., Rolka, K., Olivera, B.M., Bulaj, G., 2006. Examples of conotoxin engineering: introduction of non-peptidic backbone spacers into conotoxins and peptide-peptoid chimeras (Conopeptoids). J. Pept. Sci. 12, 209.

Guo, L., Zhang, D.H., 2009. Cyclic poly(α-peptoid)s and their block copolymers from N-heterocyclic carbene-mediated ring-opening polymerizations of N-substituted N-carboxylanhydrides. J. Am. Chem. Soc. 131, 18072–18074.

Guo, L., Li, j., Brown, Z., Ghale, K., Zhang, D., 2011. Synthesis and characterization of cyclic and linear helical poly(α-peptoid)s by *N*-heterocyclic carbene-mediated ring-opening polymerizations of *N*-substituted *N*-carboxyanhydrides. Pept. Sci. 96, 596–603.

Hamy, F., Felder, E.R., Heizmann, G., Lazdins, J., Aboul-ela, F., Varani, G., Karn, J., Klimkait, T., 1997. An inhibitor of the Tat/TAR RNA interaction that effectively suppresses HIV-1 replication. Proc. Natl. Acad. Sci., USA 94, 3548–3553.

Hara, T., Durell, S.R., Myers, M.C., Appella, D.H., 2006. Probing the structural requirements of peptoids that inhibit HDM2-p53 interactions. J. Am. Chem. Soc. 128, 1995–2004.

Heerma, W., Boon, J.-P.J.L., Versluis, C., Kruijtzer, J.A.W., Hofmeyer, L.J.F., Liskamp, R.M.J., 1997. Comparing mass spectrometric characteristics of peptides and peptoids-2. J. Mass Spec. 32, 697–704.

Heerma, W., Versluis, C., Kruijtzer, C.G.D., Zigrovic, I., Liskamp, R.M.J., 1996. Comparing mass spectrometric characteristics of peptides and peptoids. Comm. Mass Spec. 10, 459–464.

Heizmann, G., Hildebrand, P., Tanner, H., Ketterer, S., Panksy, A., Froidevaux, S., Beglinger, C., Eberle, A.N., 1999. A combinatorial peptoid library for the identification of novel MSH and GRP/Bombesin receptor ligands. J. Recept. Signal Tranduct. Res. 19, 449–466.

Hjelmgaard, T., Faure, S., Caumes, C., De Santis, E., Edwards, A.A., Taillefumier, C., 2009. Convenient solution-phase synthesis and conformational studies of novel linear and cyclic α, β-alternating peptoids. Org. Lett. 11, 4100–4103.

Hoffmann, B., Ast, T., Polakowski, T., Reineke, U., Volkmer, R., 2006. Transformation of a biologically active peptide into peptoid analogs while retaining biological activity. Prot. Pept. Lett. 13, 829–833.

Holub, J.M., Jang, H., Kirshenbaum, K., 2007. Fit to be tied: conformation-directed macrocyclization of peptoid foldamers. Org. Lett. 9, 3275–3278.

Huang, C.Y., Uno, T., Murphy, J., Lee, S., Hamer, J., Escobedo, J., Cohen, F., Radhakrishnan, R., Dwarki, V., Zuckermann, R.N., 1998. Lipitoids-novel cationic lipids for cellular delivery of plasmid DNA in vitro. Chem. Biol. 5, 345–354.

Huang, K., Wu, C.W., Sanborn, T.J., Patch, J.A., Kirshenbaum, K., Zuckermann, R.N., Barron, A.E., Radhakrishnan, I., 2006. A threaded loop conformation adopted by a family of peptoid nonamers. J. Am. Chem. Soc. 128, 1733–1738.

Huang, M.L., Ehre, D., Jiang, Q., Hu, C., Kirshenbaum, K., Ward, M.D., 2012. Proc. Natl. Acad. Sci. 109, 19922–19927.

Huang, W., Seo, J., Willingham, S.B., Czyzewski, A.M., Gonzalgo, M.L., Weissman, I.L., Barron, A.E., 2014. Learning from host-defense peptides: cationic, amphipathic peptoids with potent anticancer activity. PLoS One 9 (2), e90397.

Izzo, I., Ianniello, G., De Cola, C., Nardone, B., Loredana, E., Vaughan, G., Tedesco, C., De Riccadis, F., 2013. Structural effects of proline substitution and metal binding on hexameric cyclic peptoids. Org. Lett. 15, 598–601.

Jarver, P., Langel, U., 2006. Cell-penetrating peptides–a brief introduction. Biochim. Biophys. Acta 1758, 260–263.

Jia, L., Sun, H.L., Shay, J.T., Allgeier, A.M., Hanton, S.D., 2002. Living alternating copolymerization of N-alkylaziridines and carbon monoxide as a route for synthesis of poly-β-peptoids. J. Am. Chem. Soc. 124, 7282–7283.

Jones, F., Ogden, M.I., 2010. Controlling crystal growth with modifiers. Cryst. Eng. Comm. 12, 1016–1023.

Kang, B., Chung, S., Ahn, Y.D., Lee, J., Seo, J., 2013. Porphyrin–peptoid conjugates: face-to-face display of porphyrins on peptoid helices. Org. Lett. 15, 1670–1673.

Kawakami, T., Murakami, H., Suga, H., 2008. Ribosomal synthesis of polypeptoids and peptoid-peptide hybrids. J. Am. Chem. Soc. 130, 16861–16863.

Kerkis, A., Hayashi, M.A., Yamane, T., Kerkis, I., 2006. Properties of cell penetrating peptides (CPPs). IUBMB Life 58, 7–13.

Kesavan, V., Tamilarasu, N., Cao, H., Rana, T.M., 2002. A new class of RNA-binding oligomers: peptoid amide and ester analogues. Bioconj. Chem. 13, 1171–1175.

Khan, S.N., Kim, A., Grubbs, R.H., Kwon, Y.U., 2011. Ring-closing metathesis approaches for the solid-phase synthesis of cyclic peptoids. Org. Lett. 13, 1582–1585.

Kleschick, W.A., Davis, L.N., Dick, M.R., Garlich, J.R., Martin, E.J., Orr, N., Ng, S.C., Pernich, D.J., Unger, S.H., Watson, G.B., Zuckermann, R.N., 2001. The application of combinatorial chemistry in agrochemical discovery. ACS Symposium Series, 774, American Chemical Society, Washington, DC.

Knight, A.S., Zhou, E.Y., Pelton, J.G., Francis, M.B., 2013. Selective chromium (VI) ligands identified using combinatorial peptoid libraries. J. Am. Chem. Soc. 135, 17488–17493.

Knight, A.S., Zhou, E.Y., Francis, M.B., 2015. Development of peptoid-based ligands for the removal of cadmium from biological media. Chem. Sci. 6, 4042–4048.

Kodadek, T., 2006. Peptoid microarrays as tools for monitoring and manipulating proteome. FASEB J. 20, A34.

Koehler, A.N., Shamji, A.F., Schreiber, S.L., 2006. Discovery of an inhibitor of a transcription factor using small molecule microarrays and diversity-oriented synthesis. J. Am. Chem. Soc. 125, 8420–8421.

Konca, Y.U., Kirshenbaum, K., Zuckermann, R.N., 2014. Nanometer-scale sirNa carriers incorporating peptidomimetic oligomers: physical characterization and biological activity. Int. J. Nanomed. 9, 2271.

Kudirka, R., Tran, H., Sanii, B., Nam, K.T., Choi, P.H., Venkateswaran, N., Chen, R., Whitelam, S., Zuckermann, R.N., 2011. Folding of a single-chain, information-rich polypeptoid sequence into a highly ordered nanosheet. Pept. Sci. 96, 586–595.

Kumar, P., Fara, M.A., Bradley, M., Friedmann, P.S., Healy, E., 2006. Cell-penetrable peptide-based delivery of oligonucleotide antisense/peptide nucleic acid directed against tyrosinase. Br. J. Dermat. 155, 237.

Kwon, Y.U., Kodadek, T., 2008. Encoded combinatorial libraries for the construction of cyclic peptoid microarrays. Chem. Commun., 5704–5706.

Latham, P.W., 1999. Therapeutic peptides revisited. Nat. Biotechnol. 17, 755–757.

Lau, K.H.A., Sileika, T.S., Park, S.H., Sousa, A.M.L., Burch, P., Szleifer, I., Messersmith, P.B., 2015. Molecular design of antifouling polymer brushes using sequence-specific peptoids. Adv. Mater. Inter. 2 (1).

Laursen, J.S., Harris, P., Fristrup, P., Olsen, C.A., 2015. Triangular prism-shaped β-peptoid helices as unique biomimetic scaffolds. Nat. Commun. 6, 7013.

Lee, B.-C., Dill, K.A., Zuckermann, R.N., 2005a. Synthesis of long non-natural sequence-specific heteropolymers. Polym. Preprints 46, 174–175.

Lee, B.-C., Zuckermann, R.N., Dill, K.A., 2005b. Folding a nonbiological polymer into a compact multihelical structure. J. Am.Chem. Soc. 127, 10999–11009.

Lee, B.-C., Chu, T.K., Dill, K.A., Zuckermann, R.N., 2008. Biomimetic nanostructures: creating a high-affinity zinc-binding site in a folded nonbiological polymer. J. Am. Chem. Soc. 130, 8847–8855.

Lee, J.H., Meyer, A.M., Lim, H.S., 2010a. A simple strategy for the construction of combinatorial cyclic peptoid libraries. Chem. Commun. 46, 8615–8617.

Lee, J., Udugamasooriya, D.G., Lim, H.-S., Kodadek, T., 2010b. Potent and selective photo-inactivation of proteins with peptoid-ruthenium conjugates. Nat. Chem. Biol. 6, 258–260.

Lee, J.H., Kim, H.-S., Lim, H.-S., 2011. Design and facile solid-phase Synthesis of conformationally constrained bicyclic peptoids. Org. Lett. 13, 5012–5015.

Li, S., Bowerman, D., Marthandan, N., Klyza, S., Luebke, K.J., Garner, H.R., Kodadek, T., 2004. Photolithographic synthesis of peptoids. J. Am. Chem. Soc 126, 4088–4089.

Lim, S.S., Yoon, S.P., Park, Y., Zhu, W.L., Park, I.S., Hahm, K.S., Shin, S.Y., 2006. Mechanism of antibacterial action of a synthetic peptide with an Ala-peptoid residue based on the scorpion-derived antimicrobial peptide IsCT. Biotech. Lett. 28, 1431–1437.

Lim, H.-S., Archer, C.T., Kodadek, T., 2007. Identification of a peptoid inhibitor of the proteasome 19S regulatory particle. J. Am. Chem. Soc. 129, 7750–7751.

Lim, H.-S., Archer, C.T., Kim, Y.-C., Hutchens, T., Kodadek, T., 2008. Rapid identification of the pharmacophore in a peptoid inhibitor of the proteasome regulatory particle. Chem. Commun. 9, 1064–1066.

Lim, H.-S., Muralidhar Reddy, M., Xiao, X., Wilson, J., Wilson, R., Connell, S., Kodadek, T., 2009. Rapid identification of improved protein ligands using peptoid microarrays. Bioorg. Med. Chem. Lett. 19, 3866–3869.

Liu, F., Stephen, A.G., Adamson, C.S., Gousset, K., Aman, M.J., Freed, E.O., Fisher, R.J., Burke, T.R., 2006. Hydrazone-and hydrazide-containing N-substituted glycines as peptoid surrogates for expedited library synthesis: application to the preparation of Tsg 101-directed HIV-1 budding antagonists. Org. Lett. 8, 5165–5168.

Lobo, B.A., Vetro, J.A., Suich, D.M., Zuckermann, R.N., Middaugh, C.R., 2003. Structure/function analysis of peptoid/lipitoid: DNA complexes. J. Pharm. Sci. 92, 1905–1918.

Lundberg, P., Langel, U.A., 2003. Brief introduction to cell-penetrating peptides. J. Mol. Recognit. 16, 227–233, 2003.

Maayan, G., Ward, M.D., Kirshenbaum, K., 2009. Folded biomimetic oligomers for nantioselective catalysis. Proc. Nat. Acad. Sci. 106, 13679–13684.

Maayan, G., Ward, M.D., Kirshenbaum, K., 2009. Metallopeptoids. Chem. Commun., 56–58.

MacKerell, A.D., Bashford, D., Bellott, M., Dunbrack, R.L., Evanseck, J.D., Field, M.J., Fischer, S., Gao, J., Guo, H., Ha, S., Joseph-McCarthy, D., Kuchnir, L., Kuczera, K., Lau, F.T.K., Mattos, C., Michnick, S., Ngo, T., Nguyen, D.T., Prodhom, B., Reiher, W.E., Roux, B., Schlenkrich, M., Smith, J.C., Stote, R., Straub, J., Watanabe, M., Wiorkiewicz-Kuczera, J., Yin, D., Karplus, M., 1998. All-atom empirical potential for molecular modeling and dynamics studies of proteins. J.Phys. Chem. B 102, 3586–3616.

Malet, G., Martin, A.G., Orzaez, M., Vicent, M.J., Masip, I., Sanclimens, G., Ferrer-Montiel, A., Mingarro, I., Messeguer, A., Fearnhead, H.O., Perez-Paya, E., 2006. Small molecule inhibitors of Apaf-1-related caspase-3/-9 activation that control mitochondrial-dependent apoptosis. Cell Death Differ. 13, 1523–1532.

Marthandan, N., Klyza, S., Li, S., Kwon, Y.U., Kodadek, T., Garner, H.R., 2008. Construction and evaluation of an automated light directed protein-detecting microarray synthesizer. IEEE Trans. Nanobiosci. 7, 20–27.

Masip, I., Cortes, N., Abad, M.-J., Guardiola, M., Perez-Paya, E., Ferragut, J., Ferrer-Montiel, A., Messeguer, A., 2005. Design and synthesis of an optimized positional scanning library of peptoids: identification of novel multidrug resistance reversal agents. Bioorg. Med. Chem. 13, 1923–1929.

Maulucci, N., Izzo, I., Bifulco, G., Aliberti, A., De Cola, C., Comegna, D., Gaeta, C., Napolitano, A., Pizza, C., Tedesco, C., Flot, D., De Riccardis, F., 2008. Synthesis, structures, and properties of nine-, twelve-, and eighteen-membered *N*-benzyloxyethyl cyclic α-peptoids. Chem. Commun. 33, 3927–3929.

Messeguer, J., Cortes, N., Garcia-Sanz, N., Navarro-Vendrell, G., Ferrer-Montiel, A., Messeguer, A., 2008. Synthesis of a positional scanning library of pentamers of *N*-alkylglycines assisted by microwave activation and validation via the identification of trypsin inhibitors. J. Comb. Chem. 10, 974–980.

Michnick, S.W., Shakhnovich, E., 1998. A strategy for detecting the conservation of folding-nucleus residues in protein superfamilies. Fold Des. 3 (4), 239–251.

Miller, S.M., Simon, R.J., Ng, S., Zuckermann, R.N., Kerr, J.M., Moos, W.H., 1994. Proteolytic studies of homologous peptide and *N*-substituted glycine peptoid oligomers. Bioorg. Med. Chem. Lett. 4, 2657–2662.

Miller, S.M., Simon, R.J., Ng, S., Zuckermann, R.N., Kerr, J.M., Moos, W.H., 1995. Comparison of the proteolytic susceptibilities of homologous L-amino acid, D-amino acid, and N-substituted glycine peptide and peptoid oligomers. Drug Develop. Res. 35, 20–32.

Mohle, K., Hofmann, H.J., 1996. Description of peptide and protein secondary structures employing semiempirical methods. Biopolymers 38, 781–790.

Mojsoska, B., Zuckermann, R.N., Jenssen, H., 2015. Structure-Activity relationship study of novel peptoids that mimic the structure of antimicrobial peptides. Antimicr. Agents Chemother. 59, 4112–4120.

Montoliu, C., Humet, M., Canales, J.-J., Burda, J., Planells-Cases, R., Anchez-Baeza, F., Carbonell, T., Perez-Paya, E., Messeguer, A., Ferrer-Montiel, A., Felipo, V., 2002. Prevention of in vivo excitotoxicity by a family of trialkylglycines, a novel class of neuroprotectants. J. Pharmacol. Exp. Ther. 301, 29–36.

Moruno-Mas, C., Cruz, L.J., Mora, P., Francesch, A., Messeguer, A., Perez-Paya, E., Albericio, F., 2007. Small peptoids with anti proliferative activity on human neoplastic cells. J. Med. Chem. 50, 2443–2449.

Murnen, H.K., Rosales, A.M., Jaworski, J.N., Segalman, R.A., Zuckermann, R.N., 2010. Hierarchical self-assembly of a biomimetic diblock copolypeptoid into homochiral superhelices. J. Am. Chem. Soc. 132, 16112–16119.

Murphy, J.E., Uno, T., Hamer, J.D., Cohen, F.E., Dwarki, V., Zuckermann, R.N., 1998. A combinatorial approach to the discovery of efficient cationic peptoid reagents for gene delivery. Proc. Nat. Acad. Sci. 95, 1517–1522.

Myung, J., Jun, V., Virginia, M., Altoe, P., Aloni, S., Zuckermann, R.N., 2015. Peptoid nanosheets as soluble, two-dimensional templates for calcium carbonate mineralization. Chem. Commun. 51, 10218–10221.

Nalband, D.M., Warner, B.P., Zahler, N.H., Kirshenbaum, K., 2014. Rapid identification of metal-binding peptoid oligomers by on-resin X-ray fluorescence screening. Pept. Sci. 102, 407–415.

Nam, K.T., Sarah, A.S., Philip, H.C., Amanda, B.M., Ritchie, C., Li, T., Tammy, K., Chu, et al., 2010. Free-floating ultrathin two-dimensional crystals from sequence-specific peptoid polymers. Nat. Mater. 9, 454–460.

Nandel, F.S., Saini, A., 2007. Conformational study of short peptoid models for future applications as potent antimicrobial compounds. Macromol. Theor. Simul. 16, 295–303.

Nandel, F.S., Saini, A., 2011. Peptoids with aliphatic sidechains as helical structures without hydrogen bonds and collagen/ inverse-collagen type structures. J. Biophys. Chem. 2, 37–48.

Nandel, F.S., Jaswal, R.R., Saini, A., Nandel, V., Shafique, M., 2014. Construction and conformational behavior of peptoids with *cis*-amide bond geometry: design of a peptoid with alternate ϕ, ψ values of inverse PP-II/PP-II and PP-I structures. J. Mole. Mod. 20, 1–9.

Nnanabu, E., Burgess, K., 2006. Cyclic semipeptoids: peptoid-organic hybrid macrocycles. Org. Lett. 8, 1259–1262.

Notter, R.H., 2000. Lung surfactants: basic science and clinical applications. Marcel Dekker, New York.

Olivier, G.K., Cho, A., Sanii, B., Connolly, M.D., Tran, H., Zuckermann, R.N., 2013. Antibody-mimetic peptoid nanosheets for molecular recognition. ACS Nano 7, 9276–9286.

Olivos, H.J., Alluri, P.G., Reddy, M.M., Salony, D., Kodadek, T., 2002. Microwave-assisted solid-phase synthesis of peptoids. Org. Lett. 4, 4057–4059.

Olsen, C.A., 2010. Peptoid–peptide hybrid backbone architectures. Chem. Bio. Chem. 11, 152–160.

Orzaez, M., Mondragon, L., Marzo, I., Sanclimens, G., Messeguer, A., Perez-Paya, E., Vicent, M.J., 2007. Conjugation of a novel Apaf-1 inhibitor to peptide-based cell-membrane transporters: effective methods to improve inhibition of mitochondria-mediated apoptosis. Peptides 28, 958–968.

Ostuni, E., Robert, G., Chapman, R., Holmlin, E., Takayama, S., Whitesides, G.M., 2001. A survey of structure-property relationships of surfaces that resist the adsorption of protein. Langmuir 17, 5605–5620.

Pack, D.W., Hoffman, A.S., Pun, S., Stayton, P.S., 2005. Design and development of polymers for gene delivery. Nat. Rev. Drug Discov. 4, 581–593.

Patch, J.A., Barron, A.E., 2002. Mimicry of bioactive peptides via non-natural, sequence-specific peptidomimetic oligomers. Curr. Opin. Chem. Biol. 6, 872–877.

Patch, J.A., Barron, A.E., 2003. Helical peptoid mimics of magainin-2 amide. J. Am. Chem. Soc. 125, 12092–12093.

Patch, J.A., Kirshenbaum, K., Seurynck, S.L., Zuckermann, R.N., Barron, A.E., 2004. Versatile Oligo(N-Subsitituted) Glycines: The Many Roles of Peptoids in Drug Discovery. In: Nielsen, P.E. (Ed.), Pseudopeptides in drug discovery. Wiley-VCH Verlag, GmbH & Co. KgaA, Weinheim, Germany, pp. 1–31.

Payamyar, P., et al., 2014. Synthesis of a covalent monolayer sheet by photochemical anthracene dimerization at the air/water interface and its mechanical characterization by AFM indentation. Adv. Mater. 26, 2052–2058.

Peng, C., Jiang, B., Liu, Q., Guo, Z., Xu, Z., Huang, Q., Xu, H., Tai, R., Fan, C., 2011. Graphene-templated formation of two-dimensional lepidocrocite nanostructures for high-efficiency catalytic degradation of phenols. Energy Environ. Sci. 4, 2035–2040.

Prasad, S., Mathur, A., Sharma, R., Gupta, N., Ahuja, R., Jaggi, M., Singh, A.T., Mukherjee, R., 2006. Octapeptide analogs of somatostatin containing α, α-dialkylated amino acids with potent anticancer activity. Int. J. Pept. Res. Ther. 12, 179–185.

Ratner, B.D., 2000. Titanium in medicine: material science, surface science, engineering, biological responses and medical applications. Heidelberg and Berlin, Springer.

Reddy, M.M., German, D., Wilson, R., Kodadek, T., Wilson, J., Connell, S., Gocke, A., Hynan, L., 2011. Identification of candidate IgG biomarkers for Alzheimer's disease via combinatorial library screening. Cell 144, 132–142.

Richter, L.S., Spellmeyer, D.C., Martin, E.J., Figliozzi, G.M., Zuckermann, R.N., 1996. Automated Synthesis of Nonnatural Oligomer Libraries: The Peptoid Concept. In: Jung, G. (Ed.), Combinatorial peptide and nonpeptide libraries. VCH, Weinheim, Germany, pp. 387–404, 1996.

Rivera, D.G., Wessjohann, L.A., 2006. Supramolecular compounds from multiple Ugi multicomponent macrocyclizations: peptoid-based cryptands, cages, and cryptophanes. J. Am. Chem. Soc. 128, 7122–7123.

Robertson, E.J., Olivier, G.K., Qian, M., Proulx, C., Zuckermann, R.N., Richmond, G.L., 2014. Assembly and molecular order of two-dimensional peptoid nanosheets through the oil–water interface. Proc. Natl. Acad. Sci. 111, 13284–13289.

Rossa, T.M., Zuckermann, R.N., William, C.R., Frey, H., 2008. Intranasal administration delivers peptoids to the rat central nervous system. Neur. Lett. 439, 30–33.

Rouhi, A.M., 2001. A peptoid promise: synthetic lung surfactant mimics may widen access to treatment of respiratory disease. Chem. Eng. News 79, 50–51.

Roy, O., Faure, S., Thery, V., Didierjean, C., Taillefumier, C., 2008. Cyclic beta-peptoids. Org. Lett. 10, 921–924.

Ryge, T.S., Hansen, P.R., 2006. Potent antibacterial lysine–peptoid hybrids identified from a positional scanning combinatorial library. Bioorg. Med. Chem. 14, 4444–4451.

Sakamoto, J., van Heijst, J., Lukin, O., Schlüter, A.D., 2009. Two-dimensional polymers: just a dream of synthetic chemists. Angew. Chem. Int. Ed. 48, 1030–1069.

Sanborn, T.J., Wu, C.W., Zuckermann, R.N., Barron, A.E., 2002. Extreme stability of helices formed by water-soluble poly-*N*-substituted glycines (polypeptoids) with a-chiral side chains. Biopolymers 63, 12–20.

Sanii, B., et al., 2011. Shaken, not stirred: collapsing a peptoid monolayer to produce free-floating, stable nanosheets. J. Am. Chem. Soc. 133, 20808–20815.

Sanii, B., Haxton, T.K., Olivier, G.K., Cho, A., Barton, B., Proulx, C., Whitelam, S., Zuckermann, R.N., 2014. ACS Nano 8, 11674–11684.

Schroder, T., Schmitz, K., Niemeier, N., Balaban, T.S., Krug, H.F., Schepers, U., Brase, S., 2007. Solid-phase synthesis, bioconjugation, and toxicology of novel cationic oligopeptoids for cellular drug delivery. Biocon. Chem. 18, 342–354.

Schroeder, T., Niemeier, N., Afonin, S., Ulrich, A.S., Krug, H.F., Bräse, S., 2008. Peptoidic amino- and guanidinium-carrier systems: targeted drug delivery into the cell cytosol or the nucleus. J. Med. Chem. 51, 376–379.

Seo, J., Lee, B.-C., Zuckermann, R.N., 2011. Peptoids: synthesis, characterization, and nanostructures. Ducheyne, P., Healy, K.E., Hutmacher, D.W., Grainger, D.W., Kirkpatrick, C.J. (Eds.), Comprehensive Biomaterials, 2, Elsevier, Amsterdam, The Netherlands, pp. 53–76.

Seurynck, S.L., Patch, J.A., Barron, A.E., 2005. Simple, helical peptoid analogs of lung surfactant protein B. Chem. Biol. 12 (1), 77–88.

Seurynck-Servoss, S.L., Dohm, M.T., Barron, A.E., 2006. Effects of including an N-terminal insertion region and arginine-mimetic side chains in helical peptoid analogues of lung surfactant protein B. Biochemistry 45, 11809–11811.

Seurynck-Servoss, S.L., Brown, N.J., Dohm, M.T., Wu, C.W., Barron, A.E., 2007. Lipid composition greatly affects the in vitro surface activity of lung surfactant protein mimics. Colloids Surf. B Biointer. 57, 37–55.

Shah, N.H., Kirshenbaum, K., 2008. Photoresponsive peptoid oligomers bearing azobenzene side chains. Org. Biomol. Chem. 6, 2516–2521.

Shankaramma, S.C., Moehle, K., James, S., Vrijbloed, J.W., Obrecht, D., Robinson, J.A., 2003. A family of macrocyclic antibiotics with a mixed peptide-peptoid ß-hairpin backbone conformation. Chem. Commun., 1842–1843.

Shin, S.B., Yoo, B., Todaro, L.J., Kirshenbaum, K., 2007. X-ray diffraction cyclic peptoids. J. Am. Chem. Soc 129, 3218–3225.

Shin, H.-M., Kang, C.-M., Yoon, M.-H., Seo, J., 2014. Peptoid helicity modulation: precise control of peptoid secondary structures via position-specific placement of chiral monomers. Chem. Commun. 50, 4465–4468, 2014.

Shuey, S.W., Delaney, W.J., Shah, M.C., Scialdone, M.A., 2006. Antimicrobial beta-peptoids by a block synthesis approach. Bioorg. Med. Chem. Lett. 16, 1245–1248.

Simon, R.J., Kania, R.S., Zuckermann, R.N., Huebner, V.D., Jewell, D.A., Banville, S., Ng, S., Wang, L., Rosenberg, S., Marlowe, C.K., Spellmeyer, D.C., Tan, R., Frankel, A.D., Santi, D.V., Cohen, F.E., Barlett, P.A., 1992. Peptoids: a modular approach to drug discovery. Proc. Natl. Acad. Sci. 89, 9367–9371.

Statz, A.R., Barron, A.E., Messersmith, P.B., 2008. Protein, cell and bacterial fouling resistance of polypeptoid-modified surfaces: effect of side-chain chemistry. Soft Matter 4, 131–139.

Statz, A.R., Kuang, J.H., Ren, C.L., Barron, A.E., Szleifer, I., Messersmith, P.B., 2009. Experimental and theoretical investigation of chain length and surface coverage on fouling of surface grafted polypeptoids. Biointerphases 4, 22–32.

Sui, Q., Borchardt, D., Rabenstein, D.L., 2007. Kinetics and equilibria of *cis*/*trans* isomerization of backbone amide bonds in peptoids. J. Am. Chem. Soc. 129, 12042–12048.

Sun, J., Zuckermann, R.N., 2013. Peptoid polymers: a highly designable bioinspired material. ACS Nano 7, 4715–4732.

Sun, J., Teran, A.A., Liao, X., Balsara, N.P., Zuckermann, R.N., 2013. Nanoscale phase separation in sequence-defined peptoid diblock copolymers. J. Am. Chem. Soc. 135, 14119–14124.

Sun, J., Liao, X., Minor, A.M., Balsara, N.P., Zuckermann, R.N., 2014a. Morphology-conductivity relationship in crystalline and amorphous sequence-defined peptoid block copolymer electrolytes. J. Am. Chem. Soc. 136, 14990–14997.

Sun, J., Teran, A.A., Liao, X., Balsara, N.P., Zuckermann, R.N., 2014b. Crystallization in sequence-defined peptoid diblock copolymers induced by microphase separation. J. Am. Chem. Soc. 136, 2070–2077.

Tang, Z., Zhang, Z., Wang, Y., Glotzer, S.C., Kotov, N.A., 2006. Self-assembly of CdTe nanocrystals into free-floating sheets. Science 314, 274–278.

Tarrason, G., Cases, C., Piulats, J., Messeguer, A., Cortes, N., Masip, I., 2004. Identification of *n*-alkylglycine trimers for induction of apoptosis in cancer cells. PCT Int. Appl., 39, WO 2004092204 A1 20041028 CAN 141:374707, AN, 902408.

Thakkar, A., Cohen, A.S., Connolly, M.D., Zuckermann, R.N., Pei, D., 2009. High-throughput sequencing of peptoids and peptide-peptoid hybrids by partial edman degradation and mass spectrometry. J. Comb. Chem. 11, 294–302.

Thompson, D.A., Chai, B.X., Rood, H.L.E., Siani, M.A., Douglas, N.R., Grantiz, I., Millhauser, G.L., 2003. Peptoid mimetics of agouti related protein. Bioorg. Med. Chem. Lett. 13, 1409–1413.

Udugamasooriya, G., 2013. Peptoids: an emerging class of peptidomimetics for cancer therapy and diagnostics. J. Biomol. Res. Ther. 3, e121.

Udugamasooriya, D.G., Ritchie, C., Brekken, R.A., Kodadek, T., 2008. A peptoid antagonist of VEGF receptor 2 recognizes a "Hotspot" in the extracellular domain distinct from the hormone-binding site. Bioorg. Med. Chem. 16, 6338–6343.

Unciti-Broceta, A., Diezmann, F., Ou-Yang, C.Y., Fara, M.A., Bradley, M., 2009. Synthesis, penetrability and intracellular targeting of fluorescein-tagged peptoids and peptide-peptoid hybrids. Bioorg. Med. Chem. 17, 959–966.

Utku, Y., Dehan, E., Ouerfelli, O., Piano, F., Zuckermann, R.N., Pagano, M., Kirshenbaum, K., 2006. A peptidomimetic siRNA transfection reagent for highly effective gene silencing. Mol. BioSyst. 2, 312–317.

Vaz, B., Brunsveld, L., 2008. Stable helical peptoids via covalent side chain to side chain cyclization. Org. Biomol. Chem. 6, 2988–2994.

Vercillo, O.E., Andrade, C.K.Z., Wessjohann, L.A., 2008. Design and synthesis of cyclic RGD pentapeptoids by consecutive Ugi reactions. Org. Lett. 10, 205–208.

Vicent, M.J., Perez-Paya, E., 2006. Poly-L-glutamic acid (PGA) aided inhibitors of poptotic protease activating factor 1 (Apaf-1): an antiapoptotic polymeric nanomedicine. J. Med. Chem. 49, 3763–3765.

Viòes, E., Richard, J.P., Rispal, C., Lebleu, B., 2003. TAT peptide internalization: seeking the mechanism of entry. Curr. Protein Pept. Sci. 4, 125–132.

Wang, C.-A., Long, B., Lin, W., Huang, Y., Sun, J., 2008. Poly (amic acid)–clay nacrelike composites prepared by electrophoretic deposition. J. Mater. Res. 23, 1706–1712.

Wender, P.A., Mitchell, D.J., Pattabiraman, K., Pelkey, E.T., Steinman, L., Rothbard, J.B., 2000. The design, synthesis, and evaluation of molecules that enable or enhance cellular uptake: peptoid molecular transporters. Proc. Natl. Acad. Sci. 97, 13003–13008.

Wrenn, S.J., Weisinger, R.M., Halpin, D.R., Harbury, P.B., 2007. Synthetic ligands discovered by in vitro selection. J. Am. Chem. Soc. 129, 13137–13143.

Wu, C.W., Sanborn, T.J., Zuckermann, R.N., Barron, A.E., 2011a. Peptoid oligomers with α-chiral, aromatic side chains: effect of chain length on secondary structure. J. Am. Chem. Soc. 123, 2958–2963.

Wu, C.W., Sanborn, T.J., Huang, K., Zuckermann, R.N., Barron, A.E., 2001b. Peptoid oligomers with α-chiral, aromatic side chains: sequence requirements for the formation of stable peptoid helices. J. Am. Chem. Soc. 123, 6778–6784.

Wu, C.W., Seurynck, S.L., Lee, K.Y.C., Barron, A.E., 2003. Helical peptoid mimics of lung surfactant protein C. Chem. Biol. 10, 1057–1063.

Xiao, X.S., Yu, P., Lim, H.S., Sikder, D., Kodadek, T., 2007a. Design and synthesis of a cell permeable synthetic transcription factor mimic. J. Comb. Chem. 9, 592–600.

Xiao, X., Yu, P., Lim, H.S., Sikder, D., Kodadek, T., 2007b. A cell-permeable synthetic transcription factor mimic. Angew. Chem. Int. Ed. Engl. 46, 2865–2868.

Yoo, B., Kirshenbaum, K., 2005. Protease-mediated ligation of abiotic oligomers. J. Am. Chem. Soc 127, 17132–17133.

Yoo, B., Shin, S.B.Y., Huang, M.L., Kirshenbaum, K., 2010. Peptoid macrocycles: making the rounds with peptidomimetic oligomers. Chem. Eur. J. 16, 5528–5537.

Zhang, F., Kang, E.T., Neoh, K.G., Wang, P., Tan, K.L., 2001. Modification of Si(100) surface by the grafting of poly(ethylene glycol) for reduction in protein adsorption and platelet adhesion. J. Biomed. Mater. Res. 56, 324–332.

Zhu, W.L., Park, Y., Park, I.S., Park, Y.S., Kim, Y., Hahm, K.S., Shin, S.Y., 2006. Improvement of bacterial cell selectivity of melittin by a single Trp mutation with a peptoid residue. Prot. Pept. Lett. 13 (7), 719–725.

Zuckermann, R.N., Kodadek, T., 2009. Peptoids as potential therapeutics. Curr. Opin. Mol. Therap. 11, 299–307.

Zuckermann, R.N., De Yoreo, J.J., Chen, C.-L., Chien, Y.-C., Marshall, G.W., Marshall, S.J., Saeki, K., 2015. Peptoids Useful for the Mineralization of Apatite. United States Patent Application 20,150,174,197.

CHAPTER

11

ELECTROSPINNING OF COLLAGEN NANOFIBER SCAFFOLDS FOR TISSUE REPAIR AND REGENERATION

Sampath K. Nune, Krupanidhi S. Rama, Vijaya R. Dirisala, Murthy Y. Chavali

Technology and Research University, Guntur, India

CHAPTER OUTLINE

1 INTRODUCTION

The world's population is rapidly aging, and unless researchers find effective strategies/solutions to address the health issues, the growing burden of chronic diseases will greatly affect the quality of life of older people (Abegunde et al., 2007; Avendano et al., 2009). As people across the world live longer, soaring levels of chronic illness and age-related diseases, such as obesity, diabetes, arthritis, osteoporosis, cancer, cardiovascular diseases, Alzheimer's, and Parkinson's disease can lead to tissue or organ dysfunction. Regenerative medicine holds the promise for the restoration of tissues and organs damaged by disease, trauma, congenital deformity, as well as fire or road accidents, using the principles of tissue engineering. Langer and Vacanti (1993) defined "tissue engineering is an interdisciplinary field that applies the principles of engineering and life sciences toward the development of biological substitutes that restore, maintain, or improve tissue function." Peter et al. (2007) state that "tissue engineering is the employment of biologic therapeutic strategies aimed at the replacement, repair, maintenance, and/or enhancement of tissue function."

In the past 20 years, the definition of tissue engineering has changed as driven by scientific progress but, in practice, this technology stands to represent applications that repair or replace structural and functional tissues with any deformities (Bagheri et al., 2011; Kim et al., 2014). Biomaterials (nanofiber scaffolds) is interdisciplinary science that aspires to design artefacts with specific characteristics, namely, to mimic the natural environment; temporarily fill the gap until tissue is regenerated; and serve as a guide for growing cells/tissue. With this aim, biotechnologists are applying the principles of nanotechnology to design and manufacture scaffolds that can replace the natural extracellular matrix until host cells or tissues regenerate. In most cases, the researchers confine their knowledge in using the instrumental services provided from nanotechnologists, without understanding the basic principles, the ground realities and constraints involved in the process.

This chapter tries to bridge the gap between biotechnology and nanotechnology by providing the preliminary information about tissue repair and regeneration using biomimetic scaffolds. It is followed by brief overview of state-of-the-art methods for fabricating nanofibrous scaffolds, including phase separation, freeze drying, self assembly, and electrospinning. Finally, there is a special emphasis on application of collagen scaffolds in the field of biomedical and tissue engineering.

2 TISSUE ENGINEERING

Tissue healing (or tissue repair) refers to the replacement of destroyed tissue by living tissue in a series of events (Walter and Israel, 1987). This process can be broadly separated into regeneration and repair. The differentiation between the two is based on the resultant tissue formation. Regeneration results in the complete replacement of lost or damaged tissue by the proliferation of surrounding undamaged specialized cells; repair may restore some original structures but involves collagen deposition and scar formation.

2.1 HISTORICAL PERSPECTIVE OF TISSUE ENGINEERING

The concept of tissue engineering is to create more complex organisms from simpler living individuals or at least parts of individuals is deeply embedded in the people's imaginary world. The biblical tale of Eve created from Adam's rib is perhaps the most well-known example of this concept (Fox, 1983). Human beings have dreamt and achieved today's advancement of producing successful and functional prosthetics to whole organisms through systematic research.

John Hunter, in his pioneering work, reported the transplantation of teeth for animals, and has experimentally set the basis for a scientific approach on transplantation medicine (Hunter, 1771). Almost 75 years later, Virchow's hypothesis of "cells arise from preexisting cells" (Virchow, 1858) has been the foundation for tissue culture, working on the concept that tissue regeneration is dependent on cell proliferation, and this increased the pace of research in this field, and attracted several researchers. Thiersch (1874) discovered the significance of granulation tissue on wound healing, during his attempt to grow skin cells. His observations have diversified the parameters required for tissue culture, which led to the new concept of developing extracellular matrix (ECM) for successful development of three-dimensional (3D) tissue grafts. Harrison (1907) was the pioneer of in vitro neuronal tissue culture, using frog ectodermal cells. The findings of Rous and Jones (1916) experimentally proved that trypsin is capable of degrading matrix proteins, thus separating cells. This research has changed the trend toward growing cells instead of complete tissues, and started generating and production of primary cell culture to maintain various cell lines. In the mid-1960s, synthetic fibers were used as artificial skin grafts for burn treatment (Hall, 1966) and, in the early 1970s, there were concerted efforts in developing blood coagulation agents like heparin, assessing biocompatibility of various organic polymers, and development of novel gels (Spira et al., 1969; Leininger et al., 1972). In the same year, collagen-based artificial skin was developed for treating oral mucosa injuries. It was followed in 1981 by a skin equivalent prosthesis of a silicone cover over a sponge of porous collagen, cross-linked with chondroitin that was used successfully to treat severe burns (Burke et al., 1981). In this line of research, several advancements paved the way to develop stem cell technology in the year 1998, forming the basis of modern tissue engineering (Amit et al., 2000). Since then, researchers have focused on identifying the differential abilities of embryonic and adult stem cells, and how to manipulate their differentiating pathway. Furthermore, fundamental research is in continuing progress to understand the cellular interactions that are of great importance in tissue engineering (Kanani and Bahrami, 2010). In this decade, several products reached the market with government approval, saving millions of patients around the world who suffer from end-stage organ failure or tissue loss.

2.2 TISSUE REPAIR AND REGENERATION

Probably the simplest way to illustrate the repair process is to divide it into four broad stages which are not mutually exclusive, and overlap considerably (Broughton et al., 2006). They are bleeding, inflammation, proliferation, and remodeling. Bleeding is the first and relatively short phase that occurs during injury, trauma, or other similar insult. The body has an intricately complex and balanced mechanism through which the repair and regeneration of damaged tissue takes place. Bleeding happens because of the damaged blood vessels, followed by thrombosis (Evans, 1980). Platelets form the initial thrombus release growth factors that induce the chemotaxis and proliferation of neutrophils and macrophages, which cooperate to remove necrotic tissue, debris, and bacteria from the wound. Macrophages then become the prominent cell of this phase, and release various growth factors and cytokines that change the relatively acellular wound into a highly cellular environment. Next, fibroblasts proliferate to become the dominant cell of the proliferative phase. Acute inflammation results from vasodilatation and vasopermeability of the blood vessels, initiated and controlled by a wide array of chemical mediators released by the damaged tissues (Watson, 2003).

Clinically, acute inflammation manifests mechanical irritation, minor trauma, excessive heating, infections, and a wide range of autoimmune disorders (Kannus et al., 2003). Typically, the inflammatory phase (Lag phase) has a rapid onset (within few hours), and swiftly increases in magnitude to its maximal reaction (1–3 days) before gradually resolving (couple of weeks) (Hunter, 1998). It is followed by the proliferative phase that begins with the formation of collagen (fibroplasia) and new local blood vessels (angiogenesis) (Watson, 2006). This phase has a rapid onset (24–48 h), but takes considerably (2–3 weeks) to reach its peak reactivity, and then initiates the remodeling phase; with maturity, the collagen remodels, becoming more obviously arranged in line with local stresses (Clark, 1996; Fine et al., 2005). A portion of the type III collagen is reabsorbed and is replaced by type I collagen, with greater tensile strength. This phase is the often overlooked stage of repair, especially in the context of rehabilitation, because it is neither rapid, nor highly reactive. But it results in an organized and functional scar which is capable of behaving in a similar way to the parent tissue (Mustoe, 2006; Mustoe et al., 2002). In most of the cases, tissue repair and regeneration will occur without the need for drugs, therapy, or other intervention. But, in certain organs or patients, depending on the magnitude of damage, some help may be required in order to facilitate the process. Support may be required to initiate cell proliferation, or provide required signal transducer molecules, or support with artificial ECM through tissue engineering.

2.3 TISSUE ENGINEERING STRATEGIES

A common strategy in tissue engineering is to seed cells on biodegradable scaffolds along with bioactive factors, in order to replicate natural processes of tissue regeneration. The interactions among these components are essential to achieve biologically functional engineered tissue. Conventionally, tissue engineering applies a "top-down" approach, in which cells are allowed to thrive within a biogrid called scaffold, and are expected to multiply and ultimately create the tissue. Tissue engineering techniques mimics the natural process of tissue repair, and ultimately provides structurally sound tissue. On the other hand, the bottom-up approach is an emerging technology that focuses on fabricating tissue building blocks with specific microarchitecture, such as the shape and composition of individual blocks, and combines them to create complex tissue constructs (Nichol and Khademhosseini, 2009). Although several tissues, such as bone, cartilage, skin, and bladder have been engineered successfully through top-down strategy, there are still problems in in vitro culturing of tissues, namely, liver and kidney, due to limited diffusion properties of biomimetic scaffolds. This problem can be overcome through a bottom-up approach, but a lot of research is still required in this area in order to fuse both the top-down and bottom-up strategies, and to develop scaffolds with a high vasculature system to supply oxygen and nutrients for the developing tissue. In brief, in vitro regeneration of tissue involves three steps: fabrication of 3D scaffolds; surface modification of the prepared scaffolds to enhance their biocompatibility; and coculturing of cells on scaffolds (Lu et al., 2013). Thus, the success of in vitro tissue engineering is dependent on choosing and fabricating a suitable and appropriate scaffold.

3 SCAFFOLD

Scaffolds for tissue engineering act as a surrogate for the ECM, and provide a substrate for cellular adhesion, proliferation, differentiation, and organization. Thus, tissue engineering scaffolds should have a strong resemblance to the natural ECM, which is comprised of nanometer-diameter protein fibers. The prerequisite physicochemical properties of scaffolds include: to support and deliver cells; stimulate, differentiate, and channel tissue growth; target cell-adhesion substrates; encourage cellular response; provide a wound-healing barrier; be biocompatible and biodegradable; processability and malleability into desired shapes; be highly porous with a large surface/volume ratio; possess mechanical strength and dimensional stability; and sterilizability (Kim et al., 2005a,b).

3.1 SIGNIFICANT FEATURES FOR SCAFFOLD

Many studies have tried to define the properties required for an optimal engineered scaffold, in particular for tissue regeneration. The requirements depend mainly on the tissue to be restored, the location and size of the defect to be treated. In this juncture, the scaffold should be biocompatible in the host tissue, and prevent any adverse response to the surrounding tissue (Morais et al., 2010). The scaffold should be a highly porous structure with interconnected geometry to provide access for cell penetration, tissue in growth, and vascularization. More than 60%–70% of open pores with a minimum pore size of 100 μm are required for TE (Mitra et al., 2013). The scaffold should have suitable chemical and topographical properties to promote cellular adhesion, spreading, and proliferation. Besides, a large surface area is required, because most of the primary cells are anchorage-dependent (Wei and Ma 2004). The scaffold should have sufficient mechanical strength to resist the biological forces and maintain cell integrity, which is particularly important for the regeneration of load-bearing tissues (Moutos et al., 2007, 2010). If the scaffold degrades slowly, tissue formation is hindered, and if the scaffold degrades rapidly, fibrous tissue invades the void space, and scar tissue formation occurs. As such, it is necessary to avoid this scaffold degrading at a rate that is compatible with the growth rate of the neotissue, and be completely degraded upon tissue regeneration (Place et al., 2009). The scaffold should be designed and manufactured with a technique that promotes high precision and reproducibility. The scaffold should also be sterilizable for clinical use, and processed economically (Almeida et al., 2013).

3.2 POLYMERS FOR SCAFFOLDS

Scaffolds for tissue engineering are commonly fabricated from biodegradable polymeric materials, which can be categorized as naturally derived polymers or synthetic polymers. Naturally-derived polymers, such as collagen (Mostafa et al., 2015), chitosan, hyaluronic acid, fibrin, alginate and silk fibroin (Almeida et al., 2013), have the advantage of biological recognition and potential bioactive behavior. However, there are several disadvantages associated with the use of naturally-derived polymers, namely, the immunogenicity issue, pathogenic contaminations, limited mechanical properties, rapid biodegradability, and lack of batch-to-batch consistency (Liu and Ma, 2004; Stevens, 2008). On the other hand, popular synthetic polymers, such as poly(lactic acid), PLA; poly(glycolic acid), PGA; poly(ε-caprolactone), PCL; poly(vinyl alcohol); poly(butylene succinate); polyanhydride;

polyethylene oxide; poly(L-lactide-*co*-3-caprolactone), PLCL; poly(ethylene oxide) etc., are considered to be efficient materials due to their high mechanical properties, well oriented microstructure, and controlled degradation rate (Naveena et al., 2012). In spite of the good biodegradability and biocompatibility, synthetic polymers lack the ability of biological adhesion sites (Li et al., 2012; Sui et al., 2007; Woodruff and Hutmacher 2010), and control over pH during cell metabolisms (Böstman and Pihlajamäki 2000; Sung et al., 2004). Elaborating on individual natural and synthetic polymers is beyond the scope of this chapter; hence, the following sections throw light on collagen, a popular and commercially important biomaterial.

3.3 COLLAGEN

Collagen is the main component of connective tissue, and is the most abundant protein in mammals, making up approximately 25%–35% of the whole body protein. It provides strength and support to skin, bone, cartilage, tendon, blood vessels, and plays a major role in ECM. There are 29 types of collagen in tissue, including collagen I, II, III, V, X, and XI (Erdmann et al., 2008; Mendis et al., 2005; Ngo et al., 2011), out of which type I collagen is the most frequent fiber-forming protein in the tissues (Ahmad and Benjakul, 2010). This indicates that presence of the right type of collagen is necessary, rather than having the right amount of collagen. Of all the types of collagen, type-I collagen is abundantly found in tendons, bones, skin and other tissues in the body. Other types of collagen, such as II, IX, X, XI are widely found in the cartilage, whereas type III collagen is commonly found in the fast growing tissue, particularly at the early stages of wound repair, and at the later stages is then replaced by the stronger and tougher type I collagen. Type IV collagen is confined to the basal lamina (filtration membrane of capillaries), whereas type V and VI are in general found along with type I. Type VII collagen is widely found in the epithelia (lining of GI tract, urinary tract, etc.). Type VIII collagen is richly found in the lining of blood vessels. Collagen type XII is commonly found along with types I and III, and is known to interact with them. Type I and III of collagen are most abundantly present in the skin, and their fibrils form the mesh which contributes largely to the mechanical properties of skin. Collagen types V, VI, and XII are found in smaller amounts, and appear to have a supportive role, but their details remain unclear (Birk and Bruckner, 2005; Gelse et al., 2003).

The name collagen means glue producer in Greek. It is the oldest known protein to humans; the Egyptians extracted collagen 4000 years back from skin and sinews of horses and cattle to obtain glue. It was used by ancestors for preparing bows, utensils, rope baskets, embroidered fabrics, and crisscross decorations on human skulls (Walker, 1998). Even today, collagen has a wide range of applications in the fields of leather, pharmaceutical, and biomedical industries. This hierarchically structured protein, built from basic amino acids, specifically glycine, proline, and hydroxyproline, was used in the inhibition of free radicals (Chow and Yang, 2011; Je et al., 2007; Khantaphant and Benjakul, 2008; Kim et al., 2001; Yang et al., 2008a,b), the treatment of hypertension (Gómez-Guillén et al., 2011; Kim and Mendis, 2006), urinary incontinence (Lee et al., 2001), microbial diseases (Brogden et al., 2003; Gómez-Guillén et al., 2010), and osteoarthritis (Nakatani et al., 2009; Raabe et al., 2010), as well as to enhance opioid-like activity (Ashmarin et al., 1998) and calciotropic activity (Martínez-Alvarez et al., 2007), and for implants in humans. It is also a very attractive substrate for fragrance, antiaging creams, and other cosmetic applications (Tzaphlidou, 2004). Its denatured form (gelatin) has been widely used in the food industry (Kittiphattanabawon et al., 2005).

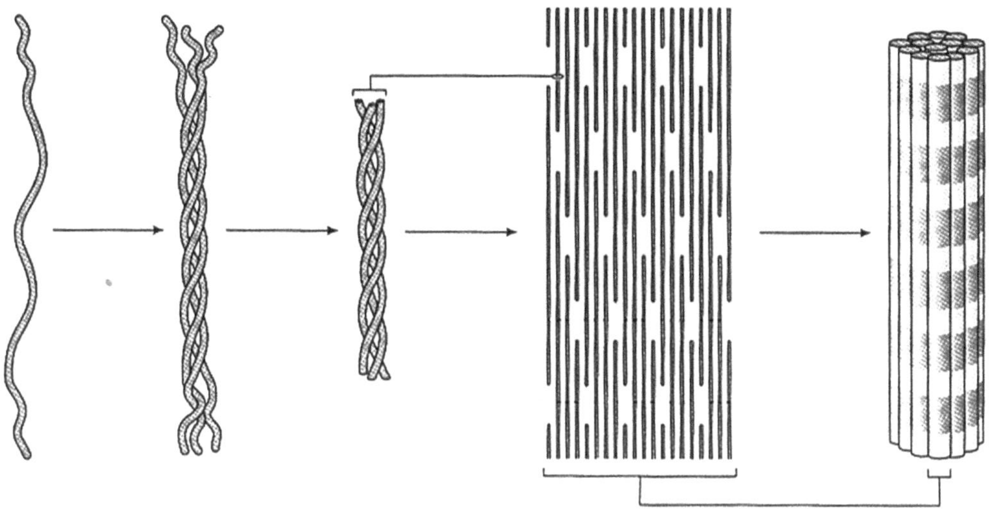

FIGURE 11.1 Structure of Collagen

3.3.1 Molecular structure

The tropocollagen or "collagen molecule" is a subunit of collagen fibrils which is approximately 300 nm long and 1.5 nm in diameter, made up of three polypeptide strands called alpha (α) chains. Each strand assigned with a left-handed helical conformation (its name is not to be confused with the commonly occurring alpha helix, a right-handed structure). These three left-handed helices are twisted together into a right-handed manner, forming a triple helix or "super helix." This cooperative quaternary structure is stabilized by numerous hydrogen bonds which are referred as the collagen microfibril, as shown in Fig. 11.1 (Fraser et al., 1979). These collagen triple helices aggregate to form collagen fibrils with an average diameter between 50 and 200 nm and quarter stagger molecular arrangement resulting in a 67 nm banding pattern, or D-periodicity. The banding pattern can be seen under high resolution transmission electron microscopy (Shoulders and Raines, 2009).

A salient feature of collagen is the regular arrangement of amino acids in each of the three chains of these collagen subunits. The sequence often follows the pattern Gly-Pro-Y or Gly-X-Hyp, where X and Y may be other amino acid residues. Proline (Pro) or hydroxyproline (Hyp) constitute about 1/6 of the total sequence, and glycine (Gly) is accounting for the 1/3 of the sequence, meaning that approximately half of the collagen sequence is made of glycine, proline, and hydroxyproline, a fact often missed due to the distraction of the unusual GXY character of collagen α-peptides (Heino et al., 2009). Glycine being the smallest amino acid with no side-chain, it plays a unique role in fibrous structural proteins. In collagen, glycine is required at every third position because the assembly of the triple helix puts this residue at the interior (axis) of the helix, where there is no space for a larger side group than glycine's single hydrogen atom. These two amino acids help to stabilize the triple helix, where hydroxyproline content is required more than proline in animals, such as fish, whose body temperatures are lower than warm-blooded animals (Harkness, 1966).

3.3.2 Collagen source

Collagen has a broad range of applications in the fields of food, pharmaceutical, cosmetic, biomedical materials, photographic film, and leather industries. Furthermore, it serves in tissue engineering because of its excellent biocompatibility and biodegradability (Zhang et al., 2009). Commercially, collagen was extracted from skin and bones of bovine or porcine origin. However, the outbreaks of bovine spongiform encephalopathy (BSE) and transmissible spongiform encephalopathy (TSE) have resulted in anxieties among the users of mammalian collagen-derived products (Jongjareonrak et al., 2005). So, collagen from aquatic animals, namely, octopus (Mizuta et al., 2003; Takema and Kimura, 1982), squid (Mingyan et al., 2009; Shanmugam et al., 2012; Thanonkaew et al., 2006), marine fishes (Kittiphattanabawon et al., 2005; Kumar and Nazeer, 2012, 2013; Zhang et al., 2009), fresh water fishes (Tang et al., 2015; Zhang et al., 2011) were considered as potential alternative sources. Moreover, fish processing waste, such as skin, viscera, bones, and scales are rich sources of collagen (Gómez-Guillén et al., 2002). Although the physical and chemical properties of fish collagen are different from those of mammalian collagen, it is unlikely to be related to transmittable diseases, and will not be forbidden for religious reasons (Zhang et al., 2009).

3.3.3 Extraction and characterization of collagen

The structures formed by many of the cross-linked collagen fibrils contribute to the strong, rigid nature of skins, tendons, and bones. Primarily, extraction of insoluble native collagen starts with pretreatment, before it gets converted into a suitable form for extraction, done by heating at 45°C or higher in water. In chemical pretreatment, noncovalent bonds will be broken, so as to disorganize the protein structure which produces adequate swelling, and will in turn facilitate collagen solubilization (Stainsby, 1987). Heat treatment in the subsequent step results in conversion into soluble gelatin (Djabourov et al., 1993; Gómez-Guillén et al., 2002). The conversion of collagen into gelatin is dependent on factors, such as pretreatment and the warm-water extraction process (Johnston Banks, 1990). By precise control of these parameters, extraction of mammalian collagen without converting into gelatin was successful.

Collagen from fish was extracted by soaking the source (skin/muscle/viscera/bone) in 0.1 M NaOH with a sample/solution ratio of 1:20 (w/v) to remove noncollagenous proteins, followed by addition of 0.5% nonionic detergent for 24 h to get rid of partial pigments and fat, followed by washing the samples with distilled water. Residual fat was removed in 15% (v/v) butyl alcohol with a sample/solution ratio of 1:20, for 24 h. Defatted skins were thoroughly washed with distilled water, and then stirred in 15 volumes of 0.5 M acetic acid for 24 h to solubilize collagen. The suspensions were centrifuged at 9000*g* for 15 min at 4°C, and the supernatants obtained were salted-out by the addition of NaCl to bring it to a final concentration of 0.7 M, which was then centrifuged at 9000*g* for 15 min at 4°C, and is then dissolved in 0.5 M acetic acid. Further, the solution obtained was dialyzed against 0.1 M acetic acid for 2 days, with change of solution once every 6 h, followed by 1 day with double distilled water, which is then lyophilized in a freeze dryer. Finally, acid-soluble collagen (ASC) was obtained.

The simple way to characterize collagen is to measure absorbance ranging from 200 to 400 nm (Piez, 1965) using UV-Vis spectrophotometer. As it is a known fact that the maximum absorption of any protein is at 280 nm, and since tryptophan does not exist in collagen, maximum absorbance at 230 nm is observed (Bama et al., 2010). Other analytical methods include running of sample on 8% resolving SDS–PAGE gel. Standard collagen shows at least two different α-chains, namely α1 (~ 135 kDa) and α2 (~ 115 kDa), and their cross-linked chains. The existence of α3 chain may or may not be seen under electrophoretic conditions. In general, structural characterization of collagen is performed using Fourier transformation infrared (FTIR) spectroscopy, in which spectra of type I collagen conform a free N-H stretching vibration in the range of 3400–3440 cm^{-1} (Li and Xia, 2004). The position of amide I and amide II bands will be typically found in the range of 1625–1690 cm^{-1} and a position of collagens were typically found in the range of 1625–1690 and 1550–1600 cm^{-1} (Duan et al., 2009). The IR spectra of collagen facilitates in conforming its helical structure by the presence of predominant peak at 1240 cm^{-1} (amide III). A protein possessing features, such as presence of ~30% glycine, along with 14%–18% of amino acids (proline + hydroxyproline) content, and lack of tryptophan can be conformed as collagen.

4 NANOFIBERS

Nanofibers are highly engineered fibers with a diameter of less than one micron, although the National Science Foundation (NSF) defines nanofibers as having at least one dimension of 100 nm or less. Nanofiber materials made from biopolymers are possible substrates for growing cells. The mechanical and structural properties of the nanofiber facilitate the fabrication of scaffolds which are suitable for implanting different types of cells (Ladd et al., 2011). Moreover, it is possible to incorporate different bioactive materials, namely, growth factor (Sahoo et al., 2010; Xie et al., 2013) and immunosuppressant during the preparation of nanofiber scaffolds. Generally, 3D porous scaffolds can be fabricated by various methods, such as drawing, template synthesis, freeze drying, phase separation, self-assembly, and electrospinning to imitate the structure and functional biology of native ECM (Table 11.1).

Table 11.1 Advantages and Disadvantages of Various Methods in Preparing Nanofibers

S. No.	Technique	Advantages	Disadvantages
1	Drawing	Produce very long single nanofibers	Limited to viscoelastic material; dependent on the orifice size of the extrusion mold; difficult to obtain fibers diameters less than 100 nm
2	Template synthesis	Uses nanoporous membrane as template	Cannot craft long continuous nanofibers
3	Freeze drying	Simple and cost effective	Uniform porosity cannot be maintained
4	Phase separation	Controlled pore size and structure; range of shapes and sizes	Long continuous fibers cannot be produced; limited to very few polymers
5	Self-assembly		The high cost of synthesis of biomaterials limits their applications; engineered peptide nanofibers can be fragmented and are susceptible to endocytosis
6	Electrospinning	Fibers with diameters size of nanometer to few microns; relatively inexpensive technique; high aspect ratio; enhanced mechanical properties	Difficult to make a large volume scaffold

4.1 DRAWING

Drawing is a process similar to dry spinning in the fiber industry which can produce very long single nanofibers (Amrinder et al., 2005). However, it requires a material with a distinct viscoelastic behavior to undergo strong deformations, while being cohesive enough to sustain the stresses developed during the pulling. It is a discontinuous process, which is always accompanied by a solidification of solvent that transforms the drawn fiber into a solid fiber by a micropipette dipped on a droplet. These steps are repeated many times on each droplet (Pascual et al., 1995). Comparatively, the drawing process is very disadvantageous, since the fiber size is dependent on the orifice size of the extrusion mold, and it is difficult to obtain fiber diameters less than 100 nm (Nain and Sitti, 2003; Ondarcuhu and Joachim, 1998).

4.2 TEMPLATE SYNTHESIS

Template synthesis is a method using a nanoporous membrane as a template to make nanofibers from various materials, namely, conductive polymers, metals, semiconductors, and carbon (Huczko, 2000; Hulteen and Martin, 1997; Toro and Buriak, 2014). This method cannot craft long continuous nanofibers. Extrusion of the polymer solution through the porous membrane is achieved by water pressure. The resultant fiber diameters vary from a few nanometers to hundreds of nanometers (Wadea and Wegrowe, 2005).

4.3 FREEZE DRYING

In the past two decades, the freeze-drying method has been widely investigated for the fabrication of 3D porous scaffolds which work on the basic principle of sublimation. Freeze drying involves three major steps: initially, the solution is frozen at a low temperature (-70 to $-8°C$) to convert the water content in the material to ice. This is followed by placing the frozen sample in a chamber connected to a vacuum pump with less pressure, known as the primary drying process. Ice in the material is removed by direct sublimation; unfrozen water in the material is removed by desorption in a secondary drying process. For instance, scaffolds engineered with collagen extracted from marine fish and gluteraldehyde were fabricated using the freeze-drying method (Kumar and Nazeer, 2012). Although there are several advantages to the freeze-drying method, it is still is a big challenge to engineer scaffolds for vascularized systems using this approach.

4.4 PHASE SEPARATION

Phase separation is one of the porous polymer membranes forming techniques that have been used for years (Moriya et al., 2009). This process can be induced either thermally or by a nonsolvent method, whereas a nonsolvent commonly results in scaffolds with a heterogeneous pore structure, and thermally-induced phase separation processes give a homogeneous porous scaffold suitable for the fabrication of tissue engineering (Ji et al., 2008; Mehta et al., 1995). The whole process occurs due to physical incompatibility, which indeed yields nanofibers. However, it takes a lot of time to transfer a solid polymer into nanoporous foam with fiber dimensions from 50 to 500 nm, and with a length of around 10 μm (McGuire et al., 1996). This process can be extended only for limited materials.

4.5 ELECTROSPINNING

Electrospinning is a simple and cost-effective method to produce scaffolds with an interconnected pore structure and fiber diameters in the submicron to nano range, and comparatively better technique (Chandrasekar et al., 2009). Extensive information regarding this technique is covered in following sections.

4.6 SELF-ASSEMBLY METHOD

Self-assembly presents a very attractive and practical strategy to construct a variety of nanostructures due to its simplicity in application (Palmer and Stupp, 2008; Stupp et al., 2000). In this method, molecules organize voluntarily into ordered structures at multiple length scales. The greatest merit of this method is that structural features of the final assemblies can be regulated by molecular chemistry and assembly kinetics (Cui et al., 2010). The universal method to make nanofiber scaffold is synthesizing the peptide amphiphile (PA). In particular, β-sheet forming peptides demonstrate the extraordinary ability to assemble into one-dimensional (1D) nanostructures through intermolecular hydrogen bonding (Hartgerink et al., 2001; Peck et al., 2011). These interactions help in the formation of 3D networks. Moreover, by changing the amino acid sequence, desired characteristics can be achieved in the hydrogel scaffolds to mimic the structure and function of native extracellular matrix. Liu et al. (2010a,b) successfully cultured chondrocytes in the self-assembled hydrogel which produced a high amount of GAG and type-II collagen (components of cartilage-specific ECM). Analysis of the gene expression of the ECM molecules also confirmed the chondrocytes in the peptide hydrogel maintained their phenotype during in vitro culture. However, a major problem with this technique is that mass production is not easy because of complicated manufacturing process and low productivity (Nagai et al., 2006; Tu and Tirrell, 2004).

5 ELECTROSPINNING

Electrospinning is an electrostatically-driven interaction of several instable processes (Bhardwaj and Kundu, 2010; Dan and Younan, 2014) that generate fibers of nanometer to micrometer diameters. The electrospinning unit was first fabricated by Cooley (1902) and Morton (1902). After almost 30 years, Formals designed the modern electrospinner (Formals, 1934) by placing the positive electrode in the polymer solution filled in a syringe with needle, and the negative electrode in the collector, as shown in Fig. 11.2. When the positively charged polymer left the needle, it was drawn by the negatively charged collector due to electromechanical stress. The jet continues to accelerate, and the polymer becomes thin due to the molecular cohesion, or chain entanglement. Finally, the solution evaporates from the charged jets to become nanofibers that are then collected.

In principle, the electrospinning unit needs three basic components, such as a syringe with metal spinneret, a voltage supplier, and a grounded collector. On supplying high voltage in the range of 10–50 kV to the metal spinneret, the charge is induced into the polymer, and a Taylor cone is formed by the balance of electrical force and surface tension of the solution (Deitzel et al., 2001; Reneker and Yarin, 2008). The Taylor cone on the spinneret is slowly stretched, as the voltage increases, and if the increase of the voltage is continued, a jet is formed from the deformed droplet, which moves toward the grounded collector, and becomes nanofibers, as shown in Fig. 11.2 (Theron et al., 2005).

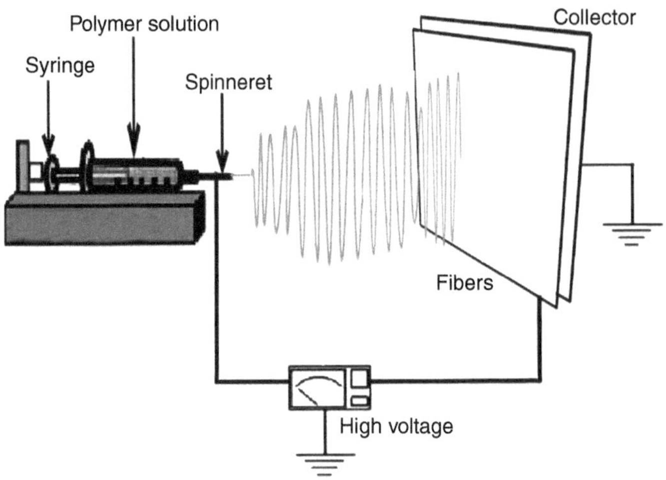

FIGURE 11.2 Schematic Diagram of Electrospinning Unit

5.1 PARAMETERS INFLUENCING ELECTROSPINNING

Formation of nanofibers is influenced by various parameters, which are mainly grouped in the following three categories.

5.1.1 Ambient parameters

Ambient parameters include temperature and humidity of the surrounding environment of the spinning area which affect the diameter and morphology of fibers (Su et al., 2011; Vrieze et al., 2009). Mit-Uppatham et al. (2004) has proven that increasing temperature favors the decrease in fiber diameter, using the polymer solution polyamide-6, and also observed the direct relationship between the solution viscosity and temperature. As the viscosity increased, a decrease in temperature is required to fabricate strong nanofibers (Dao and Jirsak, 2010). As for humidity, low humidity may dry the solvent totally, and increase the velocity of solvent evaporation. On the contrary, increasing humidity may lead to the pores coalescing on the surface of the fibers, and even change the morphology of fibers (Casper, 2004; Li and Xia, 2004). So, during the spinning process, it is necessary to have suitable ambient conditions for the polymer solution to reach expected fiber diameter and high throughput. This could be achieved by placing the whole electrospinning setup in a plexiglass box, as that helps in isolating the electrospinning process from unpredictable parameters (i.e., atmospheric temperature and humidity) that can alter the fiber production process (Pelipenko et al., 2013).

5.1.2 Process parameters

Process parameters include applied voltage, flow rate, distance between tip to collector, and conductivity of the collecting material.

1. Applied voltage is responsible for the formation and stability of the Taylor cone. At higher voltage, a greater amount of charge causes the jet to accelerate faster, leading to a smaller and unstable Taylor cone. Moreover, at high voltage, bead formation takes place, and these beads join to form a thick diameter fiber. At lower voltage, the flight time of the fiber to collector plate

increases, and that leads to the formation of fine fibers. As a point to be noted, for the successful spinning of nanofibers, AC voltage is preferred over DC voltage, in order to produce strong fibers.

2. Flow rate is considered as one of the key parameters in controlling fiber diameter during electrospinning (Fridrikh et al., 2003; Supaphol et al., 2005; Zhou et al., 2009). At high flow rates, large droplets are formed which directly correspond to the increase in the fiber diameters and bead size; this is because a larger volume of solution is drawn from the needle tip, and it needs a longer time to dry (Yuan et al., 2004). In this case, the residual solvent might induce the fibers to merge together, and make webs instead of fibers. So, lower flow rates are more desirable, as the solvent will have sufficient time for evaporation. Zargham et al. (2012) experimentally proved that 0.5 mL/h was the optimal flow rate for spinning Nylon 6 in formic acid. At this parameter, they have observed that Taylor cone formation is stable, resulting in the smallest average droplet size, and narrowest fiber diameter distribution (Chowdhury and Stylios, 2010).
3. Distance between tip to collector has a direct influence on fiber diameter and morphologies (Ki et al., 2005). In principle, if the distance is too short, the fiber will not have enough time to solidify before reaching the collector, whereas if the distance is too long, bead fiber formation takes place (Bosworth and Downes, 2012; Cha et al., 2006; Ding et al., 2010; Mazoochi et al., 2012). Moreover, the effect of decreasing the gap between tip and collector is almost the same as increasing of voltage (Frenot and Chronakis, 2003; Ramakrishna et al., 2005). In a typical electrospinning setup, this distance ranges from 10 to 15 cm, generally allowing sufficient flight time for the solvent to vaporize, such that a dry fiber strand is deposited (Chowdhury and Stylios, 2010).
4. Collector material should be always conductive, because there is a stable potential difference between needle to collector, and nonconducting material fails to produce fiber deposition (Pham et al., 2006a,b). However, in case of conducting materials, accumulation of closely packed fibers with higher packing density can be achieved. Several researchers have found that increasing the solution conductivity helps to produce more uniform fibers (Jiang et al., 2004; Kim et al., 2005a,b; Mit-Uppatham et al., 2004; Zong et al., 2002). Solution conductivity can be improved through the addition of salt, namely, NaCl, KH_2PO_4, NaH_2PO_4 (Fong et al., 1999; Huang et al., 2001), alcohol (Zuo et al., 2005), and cationic surfactant, namely, dodecyltrimethyl ammonium bromide, and tetrabutylammonium chloride (Lin et al., 2004).

5.1.3 Solution parameters

Solution parameters include viscosity of solution, surface tension, and polymer molecular weight.

1. Solution viscosity is the key factor in determining the fiber morphology, and the viscosity range varies between different polymers. Continuous and smooth fibers cannot be obtained either with low or high viscous solutions (Ding et al., 2002; Ki et al., 2005; Kim et al., 2005a,b; Larrondo and Manley, 1981). High viscosity results in the hard ejection of jet, leading to bead formation, and low viscous solutions lack the strength to form continues fibers (Lee et al., 2004; Zhang et al., 2005a,b). Generally, solution viscosity can be adjusted by altering the polymer concentration. Because solution concentration and viscosity are two closely correlated factors, increase in solution concentration always results in increase in solution viscosity, and decrease in solution concentration always results in decrease in solution viscosity (Sukigara et al., 2003).
2. Impact of surface tension on the morphology of electrospun fibers has also been well established. Basically, surface tension determines the upper and lower boundaries of the electrospinning

window (Haghi and Akbari, 2007; Pham et al., 2006a,b; Zhang et al., 2005a,b). When low molecular weight polymers are used, viscoelastic forces radically weaken, and surface tension plays a strong role in the morphology of the fibers. So, beaded fibers tend to form for higher surface tension solvents like water, low viscosity, and low conductivity/charge density systems. Yang et al. (2004) reported the influence of surface tensions on the morphologies of fibers with polyvinylpyrrolidone (PVP) as model, with ethanol and dimethylformamide as solvents. They observed that different solvents may contribute to different surface tensions, and it is always inversely proportional to the concentration of polymer. On the other hand, selection of suitable solvent is also very important. As such, addition of ethanol to PEO and polyvinyl alcohol (PVA) solutions lowered the surface tension, but significant variation in beading was reported (Fong et al., 1999; Zhang et al., 2005a,b). In the case of PEO, the solution containing ethanol exhibited less beading; and beading was increased when ethanol was added to PVA solutions. The difference in the effect of adding ethanol to these systems was attributed to the fact that it is a nonsolvent for PVA, and a solvent for PEO (Pham et al., 2006a,b).

3. Molecular weight of the polymer is one of the parameters considered to fabricate electrospun fiber, and is strongly influenced by the rheological properties of the solution (McKee et al., 2006; Zhao et al., 2005). In principle, molecular weight reflects the entanglement of polymer chains in solutions, and increase in its viscosity. Polymers with low molecular weight tend to form beads, and increasing the molecular weight of certain polymers leads to microribbon formation (Koski et al., 2004); this cannot be varied by the change in concentration of polymer. However, the breakup of polymer jets into droplets and fibers is one of the practical problems, and this can be avoided by adding high molecular weight polymers to solutions (Jones et al., 2002).

5.2 VARIOUS TYPES OF ELECTROSPINNING UNITS

Electrospinning process makes use of electrostatic and mechanical force to spin fibers from the tip of a fine spinneret (Liang et al., 2007). The spinneret is maintained at positive or negative charge by a DC power supply; when the electrostatic repelling force overcomes the surface tension of polymer solution, the liquid spills out of the spinneret and forms extremely fine continuous nanofibers on a grounded collection target with opposite charge. With the diversified needs for various applications, the design of the electrospinning unit was changed, and major modifications happened for two important components, that is, (1) the spinneret, and (2) the geometry of collectors.

5.2.1 Coaxial electrospinning

Nanofibers fabricated for tissue engineering purposes should create native/natural environments for growing cells that require the presence of functionalizing agents, such as enzymes, proteins, drugs, viruses, and bacteria. On the other hand, mixing of polymer solution with biomolecules is not possible before electrospinning. In order to overcome this problem, core–shell nanofibers are manufactured by a process called coaxial electrospinning. As per the requirements, the following modifications were done for the regular electrospinning unit, as shown in the Fig. 11.3. Two syringes were arranged interseparated, and coaxial "inner fluid" and "outer fluid" are drawn out from the spinneret, and form a composite Taylor cone with a core–shell structure (Loscertales et al., 2002). Fiber diameter is maintained by outer liquid flow rate, and the bead diameter can be controlled by the inner liquid flow rate. This technique has wide applications, ranging from complex catalyst system to drug delivery.

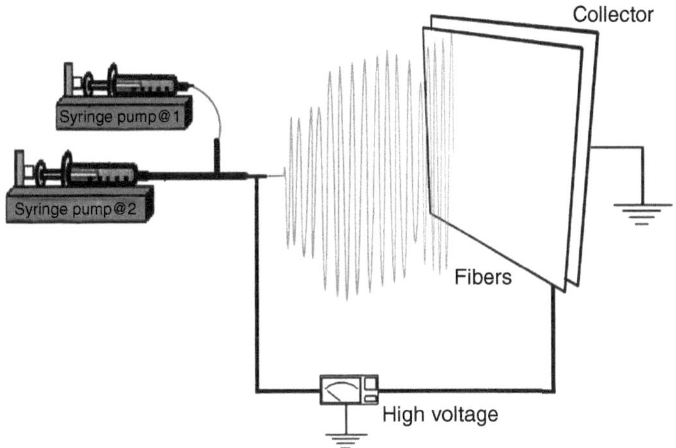

FIGURE 11.3 Schematic Diagram of Coaxial Electrospinning Unit

5.2.2 Multichannel coaxial electrospinning

Multichannel coaxial electrospinning unit is an extension to coaxial electrospinning in which the fibers can be spinned with more than two fluidic compounds at the same time. As shown in Fig. 11.4, capillaries with varying diameters are arranged at different vertices of a triangle. These capillaries are inserted into a syringe, maintaining the spatial distance between individual inner capillaries, as well as with the syringe. The immiscible inner and outer fluid are carried out separately from individual capillaries, and

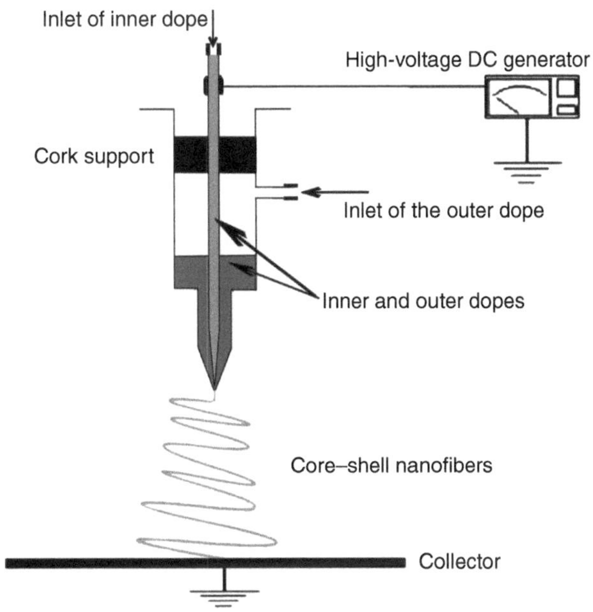

FIGURE 11.4 Schematic Diagram of Multiaxial Electrospinning Unit

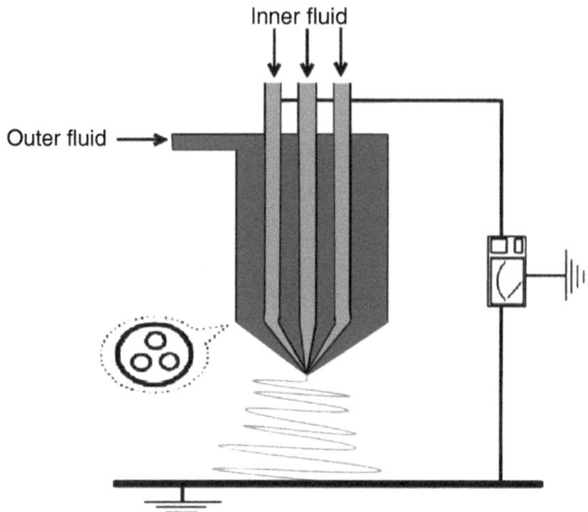

FIGURE 11.5 Schematic Diagram of Modified Electrospinning Unit

with the application of suitable voltage, a whipping compound is formed, and then a fibrous membrane is formed on the counter electrode (Li et al., 2010).

5.2.3 Modified electrospinning processes

Electrospinning procedure is further modified to address the needs of materials for biomedical applications. Dual syringe reactive electrospinning as shown in Fig. 11.5 is one of such modifications (Ji et al., 2006). The cross-linking reaction occurs simultaneously during the electrospinning process, using a dual syringe mixing technique.

5.2.4 Collector composition and geometry

Collector geometry influences the fiber alignment and pattern. As such, various types of collectors with unique geometrics were designed, such as plate collectors, or disk, drum, or mandrel collectors, as shown in Fig. 11.5 (Huang et al., 2003). Fibers have also been collected using a stationary or a rotating collector; in general, randomly oriented fibers are collected on stationary targets, whereas aligned fibers are collected on spinning substrates. Electrospun fibers were collected on number of materials, namely, metal collectors, copper mesh, aluminum foil, paper, methanol collectors, and water reservoir (Sahay et al., 2012). Collection on the surface of methanol caused the polymer fibers to swell, while water caused shrinking of the fibers. Instead of spinning on a single collector, researchers have used a series of parallel electrodes to generate aligned fibers, rather than rotating collectors (Beachley and Wen, 2009; Liu and Dzenis, 2008). For instance, Rutledge and coworkers used two parallel plates when spinning their fibers in order to generate uniform electric fields as shown in Fig. 11.6. Li and Xia used a pattern of four electrodes to generate a cross-bar array of aligned nanofibers (Li and Xia, 2004). PEO was also spun using a multiple field method in which the polymer jet passed through three parallel rings, each connected to an independent power supply, so that it was possible to produce fine fibers. Always selecting a suitable

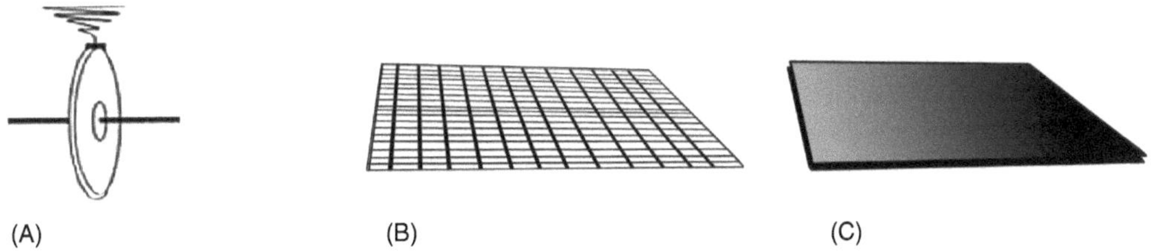

FIGURE 11.6 Schematic Diagrams of Collectors

(A) Rotating Disc (B) grid and (C) static plate.

composition and geometry of collector depends on the application aimed at, because a polymer, such as poly(ethylene-*co*-vinyl alcohol) was even spun directly onto a human hand (Matthews et al., 2002).

6 APPLICATIONS OF ELECTROSPUN COLLAGEN

Scaffolds fabricated by electrospinning technology have special properties and advantages, compared to the conventional methods (Zhang et al., 2005a,b). Electrospun scaffolds with small pores and high effective surface area are believed to promote hemostasis without the usage of other hemostatic agents (Wnek et al., 2003). The high surface area to volume ratio of electrospun fibers facilitates higher uptake capacity of wound fluids and nutrients (Augustine et al., 2014), and provides good respiration for the growing cells (Huang et al., 2003; Hunter, 1998). A most desirable property of the electrospun fibers is that they are easier to fit to a complicated wound with irregular architecture, and facilitate incorporation of bioactive agents, such as drugs or antimicrobial agents (Liang et al., 2007). Because of these desirable and supportive characters, electrospun nanofibers especially made of collagen were found suitable for repairing and regeneration of various tissue culture grafts.

6.1 CARTILAGE GRAFTS

Articular cartilage is a specialized connective tissue that covers the ends of the bones in articulating joints, providing low friction and a highly elastic surface to withstand the loads of body weight (Izadifar et al., 2012; Moutos and Guilak, 2008; Mow et al., 1992). Usually, the ECM of cartilage is secreted by the chondrocyte cells that reside within, but the healing ability of this tissue is extremely limited, posing significant clinical challenges for the treatment of joint cartilage defects (Haleem and Chu, 2010). Tissue engineering aims to replace damaged articular cartilage with a healthy tissue, a biocompatible scaffold, live chondrocytes, or multipotent mesenchymal cells capable of developing into chondrocytes, and a combination of appropriate bioactive molecules [e.g., growth factor proteins (GFs)] are three main components required in achieving successful treatment of joint repair (Mikos et al., 2006; Lavik and Langer, 2004). Although cartilage ECM contains a plethora of molecular components (Athanasiou et al., 2009, 1996), it is primarily composed of type II collagen fibril with a cartilage-specific proteoglycan and aggrecan. So, scaffold engineered by collagen makes the whole tissue engineering process easy by providing natural ambience. Shields et al. (2004) reported electrospun type II collagen to form fiber diameter between 70 and 496 nm. These scaffolds cultured with chondrocytes demonstrated the ability of the cells to infiltrate the scaffold surface, and supported the cell growth.

Successful seeding of chondrocytes was done on electrospun type II collagen scaffolds extracted from chicken sterna, with a fiber diameter ranging from 0.11 to 1.8 μm (Matthews et al., 2003). However, naturally extracted collagen has weak mechanical properties, and this problem can be overcome by cross-linking the collagen scaffolds with glutaraldehyde vapor. These scaffolds seeded with human articular chondrocytes revealed the potential of articular cartilage repair (Shields et al., 2004). On the other hand, scaffolds fabricated from other biodegradable materials, such as poly(L-lactic acid-*co*-ε-caprolactone) also require collagen coating in order to improve cartilage cell attachment and growth (Chen et al., 2010; He et al., 2013). Fertala et al. (2001) also developed 3D nanofibrous matrices from poly-L-lactic acid with free $-NH_2$ groups for the covalent binding of recombinant collagen II variants for culturing of human chondrocytes.

6.2 SKIN GRAFTS

Skin is the largest organ of the body in vertebrates, with many essential functions that would help to avoid loss of water and protein, and bacterial invasion to the underlying tissue. Since it is in direct contact with the external environment, this renders skin highly prone to damage or injury (Augustine et al., 2014). Skin grafts provide a new strategy to treat a wide variety of skin defects. Since the skin and is composed of collagen with some elastin and glycosaminoglycans (GAGs), electrospun collagen nanofibers possess the advantage of resembling the native topographical features of the natural skin ECM, and thus promote cell growth in a biomimetic fashion. Polycaprolactone (PCL) and collagen represent a suitable matrix for preparing a dermal substitute for engineering skin (Dai et al., 2004). PCL nanofiber membrane coated with collagen demonstrated good cell attachment, growth, proliferation, and migration inside the matrix of human dermal fibroblasts and keratinocytes (Venugopal and Ramakrishna 2005; Venugopal et al., 2006; Zhao et al., 2007). On the other hand, collagen (type I) nanofibers cross-linked by glutaraldehyde vapor showed relatively low adhesion of keratinocytes (Rho et al., 2006). Cell adhesion and spreading of cells can be improved by coating ECM protein to this matrix. Zhang et al. (2005a,b) developed human fibroblast skin graft using a coaxial electrospinning unit, where the shell material, collagen type I, was wrapped in PLC core component, and finally postcoated again with collagen solution. This kind of collagen-coated nanofibers resemble the natural ECM, and encourage fibroblast migration into the scaffold, and facilitate cell proliferation. Lin et al. (2013) fabricated electrospun collagen blended with PVA and chitosan for skin grafts. Skin substitute made of collagen blends showed better structural stability, tensile strength, and improved biocompatibility than the grafts made of pure collagen.

6.3 VASCULAR GRAFTS

The vascular system or circulatory system is made up of the vessels that carry blood and lymph through the body. The arteries and veins carry blood, delivering oxygen and nutrients to the body tissues, and taking away tissue waste matter (Zhang et al., 2007). The lymph vessels carry lymphatic fluid which helps to protect and maintain the fluid environment of the body. Cardiovascular malfunction and atherosclerosis are the two main problems that arise in the domain of vascular diseases, and that require a suitable scaffold material to rectify the damaged tissue (He et al., 2009). Similar to other tissues, the main part of a vascular tissue is its ECM which consists of collagen (types I and III), elastin, some proteoglycans, and glycoproteins. But collagen constructs have limited applicability in the development of vascular grafts due to the complex structural architecture of the vessels, and other immunological issues (Daniel et al., 2005). Lack of structural integrity of collagen was overcome by the production of multilayered scaffolds with controlled morphology, and mechanical properties (Buttafoco et al., 2006). On the other hand, vascular scaffolds have been cross-linked using electrospun collagen with poly(D,L-lactide-*co*-glycolide), poly(L-lactide-*co*-ε-caprolactone), elastin, poly(lactic-*co*-glycolic acid), *N*-(3 dimethylaminopropyl)-*N'*-ethylcarbodiimide hydrochloride, and *N*-hydroxy succinimide to improve the mechanical properties, and enhance cell attachment and proliferation (Buttafoco et al., 2006; Kwon et al., 2005; Stitzel et al., 2006). In this way, Matthews et al. (2002) fabricated collagen-elecrospun nanofibers, and reported the relationship between electrospinning parameters and their influence on fiber diameter. Besides, they experimentally proved that electrospun collagen promotes cell growth, and investigated the feasibility of cell penetration into the engineered matrix. In another study, a biological vascular substitute was fabricated from marine source collagen and PLGA, using electrospinning (Jeong et al., 2007). These hybrid scaffolds formed a fibrous mat with improved mechanical strength, and an average pore size of 150 ± 50 μm. In a recent study, Fu et al. (2014) fabricated collagen and PLCL nanofibers by the electrospinning method for tissue engineering applications, such as blood vessel grafts. The results of the experiments indicated that the hematoxylin and eosin staining showed that the engineered blood vessels constructed by collagen/PLCL electrospun membranes formed relatively homogenous vessel-like tissues. Kanani and Bahrami (2010) suggested that electrospun scaffolds combined with vascular cells may become an alternative to prosthetic vascular grafts for vascular reconstruction.

6.4 BONE GRAFTS

Bone reconstruction is an important topic of modern medicine, either for functional or esthetic surgery. Bone grafts have been used to fill bone defects caused by disease or trauma, such as bone fractures, infections, and tumors (Weiner and Wagner, 1998; Zheng et al., 2010). Bone is a highly complex and well organized organ, with an architecture composed mainly of inorganic substance, hydroxyapatite, and an organic fraction type I collagen (Fratzl et al., 2004). Selecting an ideal bone graft substance should mimic both physical and chemical characteristics of bone. Besides, osteoconductivity and osteoinductivity are the two main characteristics for an ideal scaffold material, along with the vital capability to support the in-growth of cells and newly formed bone tissue (Calori et al., 2011; Janicki and Schmidmaier, 2011). Collagen-based biomaterial implantation is essential when autograft has to be avoided for practical or pathological reasons. Matthews et al. (2002) initially tried to develop nanofibrous pure collagen fibers through electrospinning to develop biodegradable and biomimetic scaffolds. However, pure collagen protein scaffolds have certain limitations, namely, they are easily denatured, have poor resistance to collagenase, and lack thermal stability (Yang et al., 2008a,b; Zeugolis et al., 2008). Thus, as an alternative, bone tissue engineering relies on hardening of a collagen biomaterial by mineralization with calcium phosphate (Schofer et al., 2009), or cross-linking with substances like hydroxyapatite (HA) (Boudriot et al., 2006; Katta et al., 2004; Theron et al., 2001). On the other hand, glutaraldehyde (GTA) is used as a cross-linking agent in clinical use for fixing collagenous tissues (Buttafoco et al., 2006; Gonçalves et al., 2015; Zhong et al., 2005). Similar to GTA, 1,6-diisocyanatohexane (Li et al., 2005), carbodiimide (Olde Damink et al., 1996), and poly(ε-capro lactone) are reported as cross-linking agents for protein fibers. Most of the times, these synthetic cross-linkers may cause potential cytotoxic effects, and even lead to breakdown of collagen helical structure. Thus, natural cross-linking agents, such as transglutaminases (Damodaran et al., 2009; Zeugois et al., 2010), D,L-glyceraldehyde (Wollensak et al., 2004), and genipin derived from geniposide were considered advantageous due to their low cytotoxicity. Besides, incorporation of minerals into polymer nanofibers may create more biomimetic constructions, and improve the mechanical properties of the grafts. Thomas et al. (2007) fabricated nanostructured type I collagen and HA, using electrostatic cospinning. The strong adhesion between the two materials has improved osteoblast attachment and proliferation. Seyedjafari et al. (2010, 2011) has also observed similar results, and concluded that both HA and type I collagen are present in bone, therefore simultaneous coating of these materials will act as a perfect substrate for bone tissue engineering.

6.5 NEURAL GRAFTS

Repairing damaged neural tissue is one of the leading areas of research in regenerative medicine. Millions of people are suffering because of neural tissue dysfunction caused by traumatic brain injury, spinal cord injury, nerve injury, and neurodegenerative diseases like Parkinson's, Alzheimer's, and Huntington's diseases, and amyotrophic lateral sclerosis (Brandt, 2001; Fitch et al., 1999; McMurtrey, 2014; Park et al., 2009; Skovronsky et al., 2006; Vickers et al., 2009). As there is no curative therapy, the interdisciplinary field of neural tissue engineering has emerged as a possible solution for fabricating a biological substitute that can maintain, restore, or improve neural tissue function (Edalat et al., 2012; Gerardo-Nava et al., 2009; Orive et al., 2009). Scaffold design has pivotal role in nerve tissue engineering. Interest in employing electrospinning for scaffold fabrication is mainly due to the mechanical, biological, and kinetic properties of the scaffold being easily manipulated by altering the polymer composition and processing parameters (Greiner and Wendorff, 2007; Li et al., 2002; Teo and Ramakrishna, 2006). Multiple compositions of natural biomaterial-based nanofibers have been exerted to show significant effects on neuronal development and neurite formation, with positive results compared to clinically used autografts (Lee et al., 2010; Liu et al., 2010a,b; Scott et al., 2011; Stokols and Tuszynski, 2004; Stokols et al., 2006). Collagen-based biomaterials can also be used to develop innovative 3D tissue-engineered nervous system to promote axonal migration and myelination of neurons by Schwann cells through a connective tissue (Blais et al., 2009; Gingras et al., 2003, 2008). On the other hand, synthetic polymers blended with natural materials to enhance cellular attachment and neurite extension have also gained interest. Studies carried out by two different groups have shown that alignment of nanofibrous PCL/collagen scaffolds promoted proliferation and differentiation of human glioblastoma-astrocytoma epithelial-like cell line (Liu et al., 2010a,b; Schnell et al., 2007). In another study, it was proved that incorporation of ECM components, such as collagen with electrospun PCL scaffolds, improved axonal growth of human neuroblastoma cell line up to 600 μm (Gerardo-Nava et al., 2009). Prabhakaran et al. (2009) conducted in vitro studies to evaluate the potential of human bone marrow–derived mesenchymal stem cells (MSCs) for neuronal differentiation using poly(L-lactic acid)-*co*-poly-(3-caprolactone)/collagen nanofibrous scaffolds. These scaffolds were proven to be suitable substrates for the neuronal differentiation of MSCs, and promoted multipolar elongations, along with neurofilament protein and nestin expressions, which are typical characteristics of neuronal cells. Liu et al. (2010a,b) have reported the positive influence of phytochemical cross-linking on electrospun collagen over neural stem cell interactions. With the recent advancements in technology, some of the several natural product 3D scaffolds were successful in treating patients with neuronal disorders.

REFERENCES

Abegunde, D.O., Mathers, C.D., Adam, T., Ortegon, M., Strong, K., 2007. The burden and costs of chronic diseases in low-income and middle-income countries. Lancet 370, 1929–1938.

Ahmad, M., Benjakul, S., 2010. Extraction and characterisation of pepsin-solubilised collagen from the skin of unicorn leatherjacket (*Aluterus monocerous*). Food Chem. 120, 817–824.

Almeida, L.R., Martins, A.R., Fernandes, E.M., Oliveira, M.B., Correlo, V.M., Pashkuleva, I., Marques, A.P., Ribeiro, A.S., Durães, N.F., Silva, C.J., 2013. New biotextiles for tissue engineering: development, characterization and in vitro cellular viability. Acta Biomater. 9, 8167–8181.

Amit, M., Carpenter, M.K., Inokuma, M.S., 2000. Clonally derived human embryonic stem cell lines maintain pluripotency and proliferative potential for prolonged periods of culture. Dev. Biol. 227 (2), 271–278.

Ashmarin, I.P., Karazeeva, E.P., Lyapina, L.A., Samonina, G.E., 1998. The simplest proline-containing peptides PG, GP, PGP, and GPGG: regulatory activity and possible sources of biosynthesis. Biochemistry (Moscow) 63 (2), 119–124.

Athanasiou, K.A., Niederauer, G.G., Agrawal, C.M., 1996. Sterilization, toxicity, biocompatibility and clinical applications of polylactic acid/polyglycolic acid copolymers. Biomaterials 17, 93–102.

Athanasiou, K.A., Darling, E.M., Hu, J.C., 2009. Articular cartilage tissue engineering. In: Athanasiou, K.A., Leach, J.K. (Eds.), Synthesis Lectures on Tissue Engineering. first ed. Morgan & Claypool, San Rafael, CA, pp. 1–182.

Augustine, R., Kalarikkal, N., Thoma, S., 2014. Advancement of wound care from grafts to bioengineered smart skin substitutes. Prog. Biomater. 3, 103–113.

Avendano, M., Glymour, M.M., Banks, J., Mackenbach, J.P., 2009. Health disadvantage in US adults aged 50 to 74 years: a comparison of the health of rich and poor Americans with that of Europeans. Am. J. Pub. Health 99 (3), 540–548.

Bagheri, S., Bell, B., Khan, H., 2011. Tissue engineering. Elsevier Saunders, St. Louis, MO, (Chapter 9).

Bama, P., Vijayalakshimi, M., Jayasimman, R., Kalaichelvan, P.T., Deccaraman, M., Sankaranarayanan, S., 2010. Extraction of collagen from cat fish (*Tachysurus maculatus*) by pepsin digestion and preparation and characterization of collagen chitosan sheet. Int. J. Pharm. Pharm. Sci. 2 (4), 133–137.

Beachley, V., Wen, X., 2009. Effect of electrospinning parameters on the nanofiber diameter and length. Mater. Sci. Eng. C Mater. Biol. Appl. 29, 663–668.

Bhardwaj, N., Kundu, S.C., 2010. Electrospinning: a fascinating fiber fabrication technique. Biotechnol. Adv. 28, 325–347.

Birk, D.E., Bruckner, P., 2005. Collagen suprastructures. Top. Curr. Chem. 247, 185–205.

Blais, M., Grenier, M., Berthod, F., 2009. Improvement of nerve regeneration in tissue-engineered skin enriched with Schwann cells. J. Invest. Dermatol. 129, 2895–2900.

Böstman, O.M., Pihlajamäki, H.K., 2000. Adverse tissue reactions to bioabsorbable fixation devices. Clin. Orthop. Relat. Res. 371, 216–227.

Bosworth, L.A., Downes, S., 2012. Acetone, a sustainable solvent for electrospinning poly(e-caprolactone) fibres: effect of varying parameters and solution concentrations on fibre diameter. J. Polym. Environ. 20, 879.

Boudriot, U., Dersch, R., Greiner, A., Wendorff, J.H., 2006. Electrospinning approaches toward scaffold engineering—a brief overview. Artif. Organs 30, 785–792.

Brandt, R., 2001. Cytoskeletal mechanisms of neuronal degeneration. Cell Tissue Res. 305, 255–265.

Brogden, K.A., Ackermann, M., McCray, P.B., Tack, B.F., 2003. Antimicrobial peptides in animals and their role in host defences. Int. J. Antimicrob. Agents 22, 465–478.

Broughton, II, G., Janis, J.E., Attinger, C.E., 2006. The basic science of wound healing. Plast. Reconstr. Surg. 117 (7 Suppl.), 12S–34S.

Burke, J.F., Yannas, I.V., Quinby, Jr., W.C., Bondoc, C.C., Jung, W.K., 1981. Successful use of a physiologically acceptable artificial skin in the treatment of extensive burn injury. Ann Surg. 194 (4), 413–428.

Buttafoco, L., Kolkman, N.G., Engbers-Buijtenhuijs, P., Poot, A.A., Dijkstra, P.J., Vermes, I., Feijen, J., 2006. Electrospinning of collagen and elastin for tissue engineering applications. Biomaterials 27, 724–734.

Calori, G.M., Mazza, E., Colombo, M., Ripamonti, C., 2011. The use of bone-graft substitutes in large bone defects: any specific needs? Injury 42 (Suppl. 2), S56–S63.

Casper, C.L., 2004. Controlling surface morphology of electrospun polystyrene fibers: effect of humidity and molecular weight in the electrospinning process. Macromolecules 37, 573–578.

Cha, D.I., Kim, K.W., Chu, G.H., Kim, H.Y., Lee, K.H., Bhattarai, N., 2006. Mechanical behaviors and characterization of electrospun polysulfone/polyurethane blend nonwovens. Macromol. Res. 14, 331.

Chandrasekar, R., Zhang, L., Howe, J., Hedin, N., Zhang, Y., Fong, H., 2009. Fabrication and characterization of electrospun titania nanofibers. J. Mater. Sci. 44 (5), 1198.

Chen, J.P1., Li, S.F., Chiang, Y.P., 2010. Bioactive collagen-grafted poly-L-lactic acid nanofibrous membrane for cartilage tissue engineering. J. Nanosci. Nanotechnol. 10 (8), 5393–5398.

Chow, C.J., Yang, J.I., 2011. The effect of process variables for production of cobia (*Rachycentron canadum*) skin gelatine hydrolysates with antioxidant properties. J. Food Biochem. 35 (3), 715–734.

Chowdhury, M., Stylios, G., 2010. Effect of experimental parameters on the morphology of electrospun Nylon 6 fibres. Int. J. Basic Appl. Sci. 10 (06), 1609–1624.

Clark, A.F., 1996. The Molecular and Cellular Biology of Wound Repair, second ed. Plenum, New York, NY.

Cooley, J.F., 1992. Apparatus for electrically dispersing fluids. US Patent Specification 692,631.

Cui, H., Webber, M.J., Stupp, S.I., 2010. Self-assembly of peptide amphiphiles: from molecules to nanostructures to biomaterials. Biopolym. 94 (1), 1–18.

Dai, N.T., Williamson, M.R., Khammo, N., Adams, E.F., Coombes, A.G.A., 2004. Composite cell support membranes based on collagen and polycaprolactone for tissue engineering of skin. Biomaterials 25 (18), 4263–4271.

Damodaran, G., Collighan, R., Griffin, M., Pandit, A., 2009. Tethering a laminin peptide to a crosslinked collagen scaffold for biofunctionality. J. Biomed. Mater. Res. A 89A (4), 1001–1010.

Dan, Li, Younan, Xia, 2014. Electrospinning of nanofibers: reinventing the wheel? Adv. Mater. 16 (14), 1151–1170.

Daniel, J., Abe, K., McFetridge, P.S., 2005. Development of the human umbilical vein scaffold for cardiovascular tissue engineering applications. ASAIO J. 51, 252–261.

Dao, A.T., Jirsak, O., 2010. Roller electrospinning in various ambient parameters, Olomouc, Czech Republic, EU NANOCON, 12–14.

Deitzel, J.M., Kleinmeyer, J., Harris, D., Tan, N.C.B., 2001. The effect of processing variables on the morphology of electrospun nanofibers and textiles. Polymer 42, 261–272.

Ding, B., Kim, H.Y., Lee, S.C., Shao, C.L., Lee, D.R., Park, S.J., Kwag, G.B., Choi, K.J., 2002. Preparation and characterization of a nanoscale poly(vinyl alcohol) fiber aggregate produced by an electrospinning method. J. Polym. Sci. B Polym. Phys. 40 (13), 1261–1268.

Ding, W., Wei, S., Zhu, J., Chen, X., Rutman, D., Guo, Z., 2010. Manipulated electrospun PVA nanofibers with inexpensive salts. Macromol. Mater. Eng. 295, 958.

Djabourov, M., Bonnet, N., Kaplan, H., Favard, N., Favard, P., Lechaire, J.P., 1993. 3D analysis of gelatin gel networks from transmission electron microscopy imaging. J. Phys. II 3, 611–624.

Duan, R., Zhang, J., Du, X., Yao, X., Konno, K., 2009. Properties of collagen from skin, scale and bone of carp (*Cyprinus carpio*). Food Chem. 112, 702–706.

Edalat, F., Bae, H., Manoucheri, S., Cha, J.M., Khademhosseini, A., 2012. Engineering approaches toward deconstructing and controlling the stem cell environment. Annu. Biomed. Eng. 40, 1301–1315.

Erdmann, K., Cheung, B.W., Schroder, H., 2008. The possible roles of food-derived bioactive peptides in reducing the risk of cardiovascular disease. J. Nutr. Biochem. 19 (10), 643–654.

Evans, P., 1980. The healing process at cellular level: a review. Physiotherapy 66 (8), 256–259.

Fertala, A., Han, W.B., Ko, F.K., 2001. Mapping critical sites in collagen II for rational design of gene-engineered proteins for cell-supporting materials. J. Biomed. Mater. Res. 57, 48–58.

Fine, N.A., Mustoe, T.A., Wound, healing, 2005. Wound healing. In: Mulholland, M.W., Lillemore, K.D., Doherty, G.M., Maier, R.V. (Eds.), Greenfield's Surgery: Scientific Principles and Practice. fourth ed. Williams & Wilkins, Philadelphia, PA.

Fitch, M.T., Doller, C., Combs, C.K., Landreth, G.E., Silver, J., 1999. Cellular and molecular mechanisms of glial scarring and progressive cavitation: in vivo and in vitro analysis of inflammation-induced secondary injury after CNS trauma. J. Neurosci. 19, 8182–8198.

Fong, H., Chun, I., Reneker, D.H., 1999. Beaded nanofibers formed during electrospinning. Polymer 40, 4585.

Formals, A., 1934. Process and apparatus for preparing artificial threads. US Patent 1,975,704.

Fox, E., 1983. The Schoken Bible, vol. 1, Schoken, New York, NY.

Fraser, R.D., MacRae, T.P., Suzuki, E., 1979. Chain conformation in the collagen molecule. J. Mol. Biol. 129 (3), 463–481.

Fratzl, P., Gupta, H., Paschalis, E., Roschger, P., 2004. Structure and mechanical quality of the collagen-mineral nano-composite in bone. J. Mater. Chem. 14 (14), 2115–2123.

Frenot, A., Chronakis, I.S., 2003. Polymer nanofibres assembled by electrospinning. Curr. Opin. Colloid Interface Sci. 8, 64–75.

Fridrikh, S.V., Yu, J.H., Brenner, M.P., Rutledge, G.C., 2003. Controlling the fiber diameter during electrospinning. Phys. Rev. Lett. 90 (14), 1–4.

Fu, W., Liu, Z., Feng, B., Hu, R, He, X., Wang, H., Yin, M., Huang, H., Zhang, H., Wang, W., 2014. Electrospun gelatin/PCL and collagen/PLCL scaffolds for vascular tissue engineering. Int. J. Nanomed. 9, 2335–2344.

Gelse, K., Poschl, E., Aigner, T., 2003. Collagens-structure, function and biosynthesis. Adv. Drug Deliv. Rev. 55, 1531–1546.

Gerardo-Nava, J., Fuhrmann, T., Klinkhammer, K., Seiler, N., Mey, J., Klee, D., Moller, M., Dalton, P.D., Brook, G.A., 2009. Human neural cell interactions with orientated electrospun nanofibers in vitro. Nanomedicine (Lond.) 4, 11–30.

Gingras, M., Paradis, I., Berthod, F., 2003. Nerve regeneration in a collagen-chitosan tissue-engineered skin transplanted on nude mice. Biomaterials 24, 1653–1661.

Gingras, M., Beaulieu, M.M., Gagnon, V., Durham, H.D., Berthod, F., 2008. In vitro study of axonal migration and myelination of motor neurons in a three-dimensional tissue-engineered model. Glia 56, 354–364.

Gómez-Guillén, M.C., Turnay, J., Fernández-Díaz, M.D., Ulmo, N., Lizarbe, M.A., Montero, P., 2002. Structural and physical properties of gelatin extracted from different marine species: a comparative study. Food Hydrocoll. 16 (1), 25–34.

Gómez-Guillén, M.C., López-Caballero, M.E., López de Lacey, A., Alemán, A., Giménez, B., Montero, P., 2010. Antioxidant and antimicrobial peptide fractions from squid and tuna skin gelatin. In: Le Bihan, E. (Ed.), Sea By-Products as a Real Material: New Ways of Application. Transworld Research Network Signpost, Kerala, India, pp. 89–115, (Chapter 7).

Gómez-Guillén, M.C., Giménez, B., López-Caballero, M.E., Montero, M.P., 2011. Functional and bioactive properties of collagen and gelatine from alternative sources: a review. Food Hydrocoll. 25 (8), 1813–1827.

Gonçalves, F., Bentini, R., Burrows, M.C., Carreira, A.C.O., Kossugue, P.M., Sogayar, M.C., Catalani, L.H., 2015. Hybrid membranes of PLLA/collagen for bone tissue engineering: a comparative study of scaffold production techniques for optimal mechanical properties and osteoinduction ability. Materials 8, 408–423.

Greiner, A., Wendorff, J.H., 2007. Electrospinning: a fascinating method for the preparation of ultrathin fibers. Angew. Chem. Int. Ed. Engl. 46, 5670–5703.

Haghi, A.K., Akbari, M., 2007. Trends in electrospinning of natural nanofibers. Phys. Status Solidi A 204 (6), 1830–1834.

Haleem, A.M., Chu, C.R., 2010. Advances in tissue engineering techniques for articular cartilage repair. Oper. Tech. Orthop. 20, 76–89.

Hall, C.W., 1966. A hybrid system combining biological and synthetic materials. Trans Am. Soc. Artif. Intern. Organs 12, 340–345.

Harkness, R.D., 1966. Collagen. Sci. Prog. Oxf. 54, 257–274.

Harrison, R.G., 1907. Observations on the living developing nerve fiber. Proc. Soc. Exp. Biol. Med. 4, 140–143.

Hartgerink, J.D., Zubarev, E.R., Stupp, S.I., 2001. Supramolecular onedimensional objects. Curr. Opin. Solid State Mater. Sci. 5, 355–361.

He, W., Ma, Z., Teo, W.E., Dong, Y.X., Robless, P.A., Lim, T.C., Ramakrishna, S., 2009. Tubular nanofiber scaffolds for tissue engineered small-diameter vascular grafts. J. Biomed. Mater. Res. A 90 (1), 205–216.

He, X1., Fu, W., Feng, B., Wang, H., Liu, Z., Yin, M., Wang, W., Zheng, J., 2013. Electrospun collagen-poly(L-lactic acid-*co*-ε-caprolactone) membranes for cartilage tissue engineering. Regen. Med. 8 (4), 425–436.

Heino, J., Huhtala, H., Käpylä, J., Johnson, M.S., 2009. Evolution of collagen-based adhesion systems. Int. J. Biochem. Cell Biol. 41 (2), 341–348.

Huang, L., Nagapudi, K., Apkarian, R.P., Chaikof, E.L., 2001. Engineered collagen-PEO nanofibers and fabrics. J. Biomater. Sci. Polym. Ed. 12, 979.

Huang, Z.M., Zhang, Y.Z., Kotaki, M., Ramakrishna, S., 2003. A review on polymer nanofibers by electrospinning and their applications in nanocomposites. Compos. Sci. Technol. 63 (15), 2223–2253.

Huczko, A., 2000. Template-based synthesis of nanomaterials. Appl. Phys. A 70, 365–376.

Hulteen, J.C., Martin, C.R., 1997. A general template-based method for the preparation of nanomaterials. J. Mater. Chem. 7, 1075–1087.

Hunter, J., 1771. The Natural History of the Human Teeth. Johnson, London, UK.

Hunter, G., 1998. Specific soft tissue mobilisation in the management of soft tissue dysfunction. Man. Ther. 3 (1), 2–11.

Izadifar, Z., Chen, X., Kulyk, W., 2012. Strategic design and fabrication of engineered scaffolds for articular cartilage repair. J. Funct. Biomater 3, 799–838.

Janicki, P., Schmidmaier, G., 2011. What should be the characteristics of the ideal bone graft substitute? Combining scaffolds with growth factors and/or stem cells. Injury 42 (Suppl. 2), S77–S81.

Je, J., Qian, Z., Byun, H., Kim, S., 2007. Purification and characterization of an antioxidant peptide obtained from tuna backbone protein by enzymatic hydrolysis. Process Biochem. 42, 840–846.

Jeong, S.I., Kim, S.Y., Cho, S.K., Chong, M.S., Kim, K.S., Kim, H., 2007. Tissue-engineered vascular grafts composed of marine collagen and PLGA fibers using pulsatile perfusion bioreactors. Biomaterials 28, 1115–1122.

Ji, J., Ghosh, K., Li, B., Sokolov, J.C., Clark, R.A., Rafailovich, M.H., 2006. Dual-syringe reactive electrospinning of crosslinked hyaluronic acid hydrogel nanofiber for tissue engineering application. Macromol. Biosci. 6, 811–817.

Ji, G.L., Zhu, L.P., Zhu, B.K., Zhang, C.F., Xu, Y.Y., 2008. Structure formation and characterization of pvdf hollow fiber membrane prepared via tips with diluent mixture. J. Membr. Sci. 319, 264–270.

Jiang, H.L., Fang, D.F., Hsiao, B.S., Chu, B., Chen, W.L., 2004. Optimization and characterization of dextran membranes prepared by electrospinning. Biomacromolecules 5, 326.

Johnston Banks, F.A., 1990. Gelatin. In: Harris, P. (Ed.), Food Gels. Elsevier Applied Science, London, UK, pp. 233–289.

Jones, T.D., Chaffin, K.A., Bates, F.S., Annis, B.K., Hagaman, E.W., Kim, M.-H., Wignall, G.D., Fan, W.-H., Waymouth, R.M., 2002. Effect of tacticity on coil dimensions and thermodynamic properties of polypropylene. Macromolecules 35, 5061–5068.

Jongjareonrak, A., Benjakul, S., Visessanguan, W., Nagai, T., Tanaka, M., 2005. Isolation and characterisation of acid and pepsin-solubilised collagens from the skin of brownstripe red snapper (*Lutjanus vitta*). Food Chem. 93, 475–484.

Kanani, A.G., Bahrami, S.H., 2010. Review on electrospun nanofibers scaffold and biomedical applications. Trends Biomater. Artif. Organs 24 (2), 93–115.

Kannus, P., Parkkari, J., Järvinen, T.L., Järvinen, T.A., Järvinen, M., 2003. Basic science and clinical studies coincide: active treatment approach is needed after a sports injury. Scand. J. Med. Sci. Sports 13, 150–154.

Katta, P., Alessandro, M., Ramsier, R.D., Chase, G.G., 2004. Continuous electrospinning of aligned polymer nanofibers onto a wire drum collector. Nano Lett. 4 (11), 2215–2218.

Khantaphant, S., Benjakul, S., 2008. Comparative study on the proteases from fish pyloric caeca and the use for production of gelatine hydrolysate with antioxidative activity. Comp. Biochem. Physiol. 151, 410–419.

Ki, C.S., Baek, D.H., Gang, K.D., Lee, K.H., Um, I.C., Park, Y.H., 2005. Characterization of gelatin nanofiber prepared from gelatin–formic acid solution. Polymer 46 (14), 5094–5102.

Kim, S.K., Mendis, E., 2006. Bioactive compounds from marine processing byproducts—a review. Food Res. Int. 39 (4), 383–393.

Kim, S.K., Kim, Y.T., Byun, H.G., Nam, K.S., Joo, D.S, Shahidi, S.F., 2001. Isolation and characterization of antioxidative peptides from gelatine hydrolysate of Alaska. J. Agric. Food Chem. 49, 1984–1989.

Kim, K.-H., Jeong, L., Park, H.-N., Shin, S.-Y., Park, W.-H., Lee, S.-C., Kim, T.-I., Park, Y.-J., Seol, Y.-J., Lee, Y.-M., Ku, Y., Rhyu, I.-C., Han, S.-B., Chung, C.-P., 2005a. Biological efficacy of silk fibroin nanofiber membranes for guided bone regeneration. J. Biotechnol. 120 (3), 327–339.

Kim, B., Park, H., Lee, S.H., Sigmund, W.M., 2005b. Poly(acrylic acid) nanofibers by electrospinning. Mater. Lett. 59, 829.

Kim, R.Y., Fasi, A.C., Feinberg, S.E., 2014. Soft tissue engineering in craniomaxillofacial surgery. Ann. Maxillofac. Surg. 4 (1), 4–8.

Kittiphattanabawon, P., Benjakul, S., Visessanguan, W., Nagai, T., Tanaka, M., 2005. Characterization of acid-soluble collagen from skin and bone of bigeye snapper (*Priacanthus tayenus*). Food Chem. 89, 363–372.

Koski, A., Yim, K., Shivkumar, S., 2004. Effect of molecular weight on fibrous PVA produced by electrospinning. Mater. Lett. 58 (3–4), 493–497.

Kumar, N.S.S., Nazeer, R.A., 2012. Wound healing properties of type I collagen from the bone of marine fishes. Int. J. Pept. Res. Ther. 18 (3), 185–193.

Kumar, N.S.S., Nazeer, R.A., 2013. Characterization of acid and pepsin soluble collagen from skins of horse mackerels (*Magalaspis cordyla*) and croaker (*Otolithes ruber*). Int. J. Food Prop. 3 (16), 613–621.

Kwon, I.K., Kidoaki, S., Matsuda, T., 2005. Electrospun nano- to microfiber fabrics made of biodegradable copolyesters: structural characteristics, mechanical properties and cell adhesion potential. Biomaterials 26 (18), 3929–3939.

Ladd, M.R., Hill, T.K., Yoo, J.J., Lee, S.J., 2011. Electrospun nanofibers in tissue engineering. In: Lin, T. (Ed.), Nanofibers—Production, Properties and Functional Applications. InTech, Rijeka, Croatia.

Langer, R., Vacanti, J.P., 1993. Tissue engineering. Science 260, 920–926.

Larrondo, L., Manley, R.S.J., 1981. Electrostatic fibre spinning from polymer melts; electrostatic deformation of pendent drop of polymer melt. J. Polym. Sci. 19, 933.

Lavik, E., Langer, R., 2004. Tissue engineering: current state and perspectives. Appl. Microbiol. Biotechnol. 65, 1–8.

Lee, C.H., Singla, A., Lee, Y., 2001. Biomedical applications of collagen. Int. J. Pharm. 221, 1–22.

Lee, J.S., Choi, K.H., Ghim, H.D., Kim, S.S., Chun, D.H., Kim, H.Y., Lyoo, W.S., 2004. Role of molecular weight of atactic poly(vinyl alcohol) (PVA) in the structure and properties of PVA nanofabric prepared by electrospinning. J. Appl. Polym. Sci. 93 (4), 1638–1646.

Lee, J.Y., Bashur, C.A., Gomez, N., Goldstein, A.S., Schmidt, C.E., 2010. Enhanced polarization of embryonic hippocampal neurons on micron scale electrospun fibers. J. Biomed. Mater. Res. A 92, 1398–1406.

Leininger, R.I., Crowley, J.P., Falb, R.D., Grode, G.A., 1972. Three years' experience in vivo and in vitro with surfaces and devices treated by the heparin complex method. Trans. Am. Soc. Artif. Intern. Organs 18 (0), 312–315.

Li, D., Xia, Y.N., 2004. Electrospinning of nanofibers: reinventing the wheel? Adv. Mater. 16, 1151–1170.

Li, F., Zhao, Y., Song, Y., 2010. Core–shell nanofibers: nano channel and capsule by coaxial electrospinning. In: Kumar, A. (Ed.), Nanofibers. InTech, Rijeka, Croatia, (Chapter 22).
Li, W.J., Laurencin, C.T., Caterson, E.J., Tuan, R.S., Ko, F.K., 2002. Electrospun nanofibrous structure: a novel scaffold for tissue engineering. J. Biomed. Mater. Res. 60, 613–621.
Li, M., Mondrinos, M.J., Gandhi, M.R., Kob, F.K., Weiss, A.S., Lelkes, P.I., 2005. Electrospun protein fibers as matrices for tissue engineering. Biomaterials 26, 5999–6009.
Li, H., Xia, Y., Wu, J., He, Q., Zhou, X., Lu, G., Shang, L., Boey, F., Venkatraman, S.S., Zhang, H., 2012. Surface modification of smooth poly (L-lactic acid) films for gelatin immobilization. ACS Appl. Mater. Interfaces 4 (2), 687–693.
Liang, D., Hsiao, B.S., Chu, B., 2007. Functional electrospun nanofibrous scaffolds for biomedical applications. Adv. Drug Deliv. Rev. 59, 1392–1412.
Lin, T., Wang, H.X., Wang, H.M., Wang, X.G., 2004. The charge effect of cationic surfactants on the elimination of fibre beads in the electrospinning of polystyrene. Nanotechnology 15, 1375.
Lin, H.Y., Kuo, Y.J., Chang, S.H., Ni, T.S., 2013. Characterization of electrospun nanofiber matrices made of collagen blends as potential skin substitutes. Biomed. Mater. 8 (2), 025009.
Liu, L., Dzenis, Y.A., 2008. Analysis of the effects of the residual charge and gap size on electrospun nanofiber alignment in a gap method. Nanotechnology 2008 (19), 355307:1–:1355307.
Liu, X., Ma, P.X., 2004. Polymeric scaffolds for bone tissue engineering: on musculoskeletal bioengineering. Ann. Biomed. Eng. 32 (3), 477–486.
Liu, J., Song, H., zhang, L., Xu, H., Zhao, X., 2010a. Self-assembly-peptide hydrogels as tissue-engineering scaffolds for three-dimensional culture of chondrocytes in vitro. Macromol. Biosci. 10 (10), 1164–1170.
Liu, T., Teng, W.K., Chan, B.P., Chew, S.Y., 2010b. Photochemical crosslinked electrospun collagen nanofibers: synthesis, characterization and neural stem cell interactions. J. Biomed. Mater. Res. A 95, 276–282.
Loscertales, I.G., Barrero, A., Guerrero, I., Cortijo, R., Marquez, M., Ganan-Calvo, A.M., 2002. Micro/nano encapsulation via electrified coaxial liquid jets. Science 295, 1695–1698.
Lu, T., Li, Y., Chen, T., 2013. Techniques for fabrication and construction of three-dimensional scaffolds for tissue engineering. Int. J. Nanomed. 8, 337–350.
Martínez-Alvarez, O., Guimas, L., Delannoy, C., Fouchereau-Peron, M., 2007. Occurrence of a CGRP like molecule in siki (*Centroscymnus coelolepsis*) hydrolysate of industrial origin. J. Agric. Food Chem. 55 (14), 5469–5475.
Matthews, J.A., Wnek, G.E., Simpson, D.G., Bowlin, G.L., 2002. Electrospinning of collagen nanofibers. Biomacromolecules 3, 232.
Matthews, J.A., Boland, E.D., Wnek, G.E., Simpson, D.G., Bowlin, G.L., 2003. Electrospinning of collagen type II: a feasibility study. J. Bioact. Compat. Pol. 18, 125–134.
Mazoochi, T., Hamadanian, M., Ahmadi, M., Jabbari, V., 2012. Investigation of the morphological characteristics of nanofibrous membrane as electrospun in the different processing parameters. Int. J. Ind. Chem. 3, 2.
McGuire, K.S., Laxminarayan, A., Martula, D.S., Lloyd, D.R., 1996. Kinetics of droplet growth in liquid-liquid phase separation of polymer-diluent systems. Model development. J. Colloid Interface Sci. 182, 46–58.
McKee, M.G., Layman, J.M., Cashion, M.P., Long, T.E., 2006. Phospholipid nonwoven electrospun membranes. Science 311 (5759), 353–355.
McMurtrey, R.J., 2014. Patterned and functionalized nanofibers scaffolds in three-dimensional hydrogel constructs enhance neurite outgrowth and directional control. J. Neural Eng. 11 (066009), 15.
Mehta, R.H., Madsen, D.A., Kalika, D.S., 1995. Microporous membranes based on poly(ether-ether ketone) via thermally-induced phase separation. J. Membr. Sci. 107, 93–106.
Mendis, E., Rajapakse, N., Byun, H.G., Kim, S.K., 2005. Investigation of jumbo squid (*Dosidicus gigas*) skin gelatin peptides for their in vitro antioxidant effects. Life Sci. 77, 2166–2178.

Mikos, A.G., Herring, S.W., Ochareon, P., Elisseeff, J., Lu, H.H., Kandel, R., Schoen, F.J., Toner, M., Mooney, D., Atala, A., Van Dyke, M.E., Kaplan, D., Vunjak-Novakovic, G., 2006. Engineering complex tissues. Tissue Eng. 12, 3307–3339.

Mingyan, Y., Bafang, L., Xue, Z., 2009. Isolation and characterization of collagen from squid (*Ommastrephes bartrami*) skin. J. Ocean Univ. China 8 (2), 191–196.

Mitra, J., Tripathi, G., Sharma, A., Basu, B., 2013. Scaffolds for bone tissue engineering: role of surface patterning on osteoblast response. RSC Adv. 3 (28), 11073–11094.

Mit-Uppatham, C., Nithitanakul, M., Supaphol, P., 2004. Ultrafine electrospun polyamide-6 fibers: effect of solution conditions on morphology and average fiber diameter. Macromol. Chem. Phys. 205 (17), 2327–2338.

Mizuta, S., Takahide, T., Yoshinaka, R., 2003. Comparison of collagen types of arm and mantle muscles of the common Octopus (*Octopus vulgaris*). Food Chem. 81, 527–532.

Morais, J.M., Papadimitrakopoulos, F., Burgess, D.J., 2010. Biomaterials/tissue interactions: possible solutions to overcome foreign body response. AAPS J. 12 (2), 188–196.

Moriya, A., Maruyama, T., Ohmukai, Y., Sotani, T., Matsuyama, H., 2009. Preparation of poly(lactic acid) hollow fiber membranes via phase separation methods. J. Membr. Sci. 342 (1–2), 307–312.

Morton, W.J., 1902. Method of dispersing fluids. US Patent Specification 705,691.

Mostafa, N.Z., Talwar, R., Shahin, M., Unsworth, L.D., Major, P.W., Doschak, M.R., 2015. Cleft palate reconstruction using collagen and nanofiber scaffold incorporating bone morphogenetic protein in rats. Tissue Eng. Part A. 21 (1–2), 85–95.

Moutos, F.T., Guilak, F., 2008. Composite scaffolds for cartilage tissue engineering. J. Biorheol. 45, 501–512.

Moutos, F.T., Freed, L.E., Guilak, F., 2007. A biomimetic three-dimensional woven composite scaffold for functional tissue engineering of cartilage. Nat. Mater. 6 (2), 162–167.

Moutos, F.T., Estes, B.T., Guilak, F., 2010. Multifunctional hybrid three-dimensionally woven scaffolds for cartilage tissue engineering. Macromol. Biosci. 10 (11), 1355–1364.

Mow, V.C., Ratcliffe, A., Poole, A.R., 1992. Cartilage and diarthrodial joints as paradigms for hierarchical materials and structures. Biomaterials 13, 67–97.

Murray, P.E., Garcia-Godoy, F., Hargreaves, K.M., 2007. Regenerative endodontics: a review of current status and a call for action. J. Endod. 33 (4), 377–390.

Mustoe, T.A., 2006. Wound healing. In: Becker, J.M., Stucchi, A.F. (Eds.), Essentials of Surgery. Elsevier, Philadelphia, PA.

Mustoe, T.A., Cooter, R., Gold, M.H., et al., 2002. International clinical guidelines of scar management. Plast. Reconstr. Surg. 110, 560–572.

Nagai, Y., Unsworth, L.D., Koutsopoulos, S., Zhang, S., 2006. Slow release of molecules in self-assembling peptide nanofiber scaffold. J. Control. Release 115, 18–25.

Nain, A., Sitti, M., 2003. Micro- and nano-scale robotics. Proc. IEEE Nanotechnol. Conf. 2, 60–63.

Nain, A.S., Amon, F.C., Sitti, F M., 2005. Polymer micro/nanofiber fabrication using micro/nanopipettes, Proceedings of 2005 5th IEEE Conference on Nanotechnology, Nagoya, Japan.

Nakatani, S., Mano, H., Sampei, C., Shimizu, J., Wada, M., 2009. Chondroprotective effect of the bioactive peptide prolyl-hydroxyproline in mouse articular cartilage in vitro and in vivo. Osteoarthritis Cartilage 17 (12), 1620–1627.

Naveena, N., Venugopal, J., Rajeswari, R., Sundarrajan, S., Sridhar, R., Shayanti, M., Narayanan, S., Ramakrishna, S., 2012. Biomimetic composites and stem cells interaction for bone and cartilage tissue regeneration. J. Mater. Chem. 22 (12), 5239–5253.

Ngo, D.H., Ryu, B., Vo, T.S., Himaya, S.W.A., Wijesekara, I., Kim, S.K., 2011. Free radical scavenging and angiotensin-I converting enzyme inhibitory peptides from Pacific cod (*Gadus macrocephalus*) skin gelatine. Int. J. Biol. Macromol. 49 (5), 1110–1116.

Nichol, J.W., Khademhosseini, A., 2009. Modular tissue engineering: engineering biological tissues from the bottom up. Soft Matter 5 (7), 1312–1319.

Olde Damink, L.H.H., Dijkstra, P.J., Van Luyn, M.J.A., Van Wachem, P.B., Nieuwenhuis, P., Feijen, J., 1996. Cross-linking of dermal sheep collagen using a water-soluble carbodiimide. Biomaterials 17, 765–773.

Ondarcuhu, T., Joachim, C., 1998. Drawing a single nanofiber over hundreds of microns. Europhys. Lett. 42, 215–220.

Orive, G., Anitua, E., Pedraz, J.L., Emerich, D.F., 2009. Biomaterials for promoting brain protection, repair and regeneration. Nat. Rev. Neurosci. 10, 682–692.

Palmer, L.C., Stupp, S.I., 2008. Acc. Chem. Res. 41, 1674–1684.

Park, D.H., Eve, D.J., Borlongan, C.V., Klasko, S.K., Cruz, L.E., Sanberg, P.R., 2009. From the basics to application of cell therapy, a steppingstone to the conquest of neurodegeneration: a meeting report. Med. Sci. Monit. 15, RA23–RA31.

Pascual, J.I., Mendez, J., Gomez-Herrero, J., Baro, A.M., Garcia, N., Landman, U., Leudtke, W.D., Bogachek, E.N., Cheng, H.-P., 1995. Properties of metallic nanowires: from conductance quantization to localization. Science 267 (5205), 1793–1795.

Peck, M., Dusserre, N., McAllister, T.N., L'Heureux, N., 2011. Review: tissue engineering by self-assembly. Mater. Today 14 (5), 218–224.

Pelipenko, J, Kristl, J, Jankovic, B, Baumgartner, S, Kocbek, P., 2013. The impact of relative humidity during electrospinning on the morphology and mechanical properties of nanofibers. Int. J. Pharm. 456, 125–134.

Pham, Q.P., Sharma, U., Mikos, A.G., 2006a. Electrospun poly(e-caprolactone) microfiber and multilayer nanofiber/microfiber scaffolds: characterization of scaffolds and measurement of cellular infiltration. Biomacromolecules 7 (10), 2796–2805.

Pham, Q.P., Upma, S., Mikos, A.G., 2006b. Electrospinning of polymeric nanofibers for tissue engineering applications: a review. Tissue Eng. 12 (5), 1197–1211.

Piez, K.A., 1965. Characterization of a collagen from codfish skin containing three chromatographically different α chains. Biochemistry 4 (12), 2590–2596.

Place, E.S., George, J.H., Williams, C.K., Stevens, M.M., 2009. Synthetic polymer scaffolds for tissue engineering. Chem. Soc. Rev. 38 (4), 1139–1151.

Prabhakaran, M.P., Venugopal, J.R., Ramakrishna, S., 2009. Mesenchymal stem cell differentiation to neuronal cells on electrospun nanofibrous substrates for nerve tissue engineering. Biomaterials 30 (28), 4996–5003.

Raabe, O., Reich, C., Wenisch, S., Hild, A., Burg-Roderfeld, M., Siebert, H.C., Arnhold, S., 2010. Hydrolyzed fish collagen induced chondrogenic differentiation of equine adipose tissue-derived stromal cells. Histochem. Cell Biol. 134 (6), 545–554.

Ramakrishna, S., Fujihara, K., Teo, W., 2005. An Introduction to Electrospinning and Nanofibres. World Scientific Publishing Co Pte Ltd, Singapore.

Reneker, D.H., Yarin, A.L., 2008. Electrospinning jets and polymer nanofibers. Polymer 49, 2387–2425.

Rho, K.S., Jeong, L., Lee, G., Seo, B.M., Park, Y.J., Hong, S.D., Roh, S., Cho, J.J., Park, W.H., Min, B.M., 2006. Electrospinning of collagen nanofibers: effects on the behavior of normal human keratinocytes and early-stage wound healing. Biomaterials 27, 1452–1461.

Rous, P., Jones, F.S., 1916. A method for obtaining suspensions of living cells from the fixed tissues, and for the planting out of individual cells. J. Exp. Med. 23, 549–555.

Sahay, R., Kumar, P.S., Sridhar, R., Sundaramurthy, J., Venugopal, J., Mhaisalkar, S.G., Ramakrishna, S., 2012. Electrospun composite nanofibers and their multifaceted applications. J. Mater. Chem. 22, 12953–12971.

Sahoo, S., Ang, L.T., Goh, J.C., Toh, S.L., 2010. Growth factor delivery through electrospun nanofibers in scaffolds for tissue engineering applications. J. Biomed. Mater. Res. A 93 (4), 1539–1550.

Schnell, E., Klinkhammer, K., Balzer, S., Brook, G., Klee, D., Dalton, P., Mey, J., 2007. Guidance of glial cell migration and axonal growth on electrospun nanofibers of poly-epsilon-caprolactone and a collagen/poly-epsilon-caprolactone blend. Biomaterials 7 (28), 3012–3025.

Schofer, M.D., Boudriot, U., Wack, C., Leifeld, I., Grabedunkel, C., Dersch, R., Rudisile, M., Wendorff, J.H., Greiner, A., Paletta, J.R.J., Fuchs-Winkelmann, S., 2009. Effect of direct RGD incorporation in PLLA nanofibers on growth and osteogenic differentiation of human mesenchymal stem cells. J. Mater. Sci. Mater. Med. 20 (7), 1535–1540.

Scott, J.B., Afshari, M., Kotek, R., Saul, J.M., 2011. The promotion of axon extension in vitro using polymertemplated fibrin scaffolds. Biomaterials 32, 4830–4839.

Seyedjafari, E., Soleimani, M., Ghaemi, N., Shabani, I., 2010. Nanohydroxyapatite-coated electrospun poly(L-lactide) nanofibers enhance osteogenic differentiation of stem cells and induce ectopic bone formation. Biomacromolecules 11 (11), 3118–3125.

Seyedjafari, E., Soleimani, M., Ghaemi, N., Sarbolouki, M.N., 2011. Enhanced osteogenic differentiation of cord blood-derived unrestricted somatic stem cells on electrospun nanofibers. J. Mater. Sci. Mater. Med. Jan. 22 (1), 165–174.

Shanmugam, V., Ramasamy, P., Subhapradha, N., Sudharsan, S., Seedevi, P., Moovendhan, M., Krishnamoorthy, J., Shanmugam, A., Srinivasan, A., 2012. Extraction, structural and physical characterization of type I collagen from the outer skin of *Sepiella inermis* (Orbigny, 1848). Afr. J. Biotechnol. 11 (78), 14326–14337.

Shields, K.J., Beckman, M.J., Bowlin, G.L., Wayne, J.S., 2004. Mechanical properties and cellular proliferation of electrospun collagen type II. Tissue Eng. 10 (9), 1510–1517.

Shoulders, M.D., Raines, R.T., 2009. Collagen structure and stability. Annu. Rev. Biochemist. 78, 929–958.

Skovronsky, D.M., Lee, V.M., Trojanowski, J.Q., 2006. Neurodegenerative diseases: new concepts of pathogenesis and their therapeutic implications. Annu. Rev. Pathol. 1, 151–170.

Spira, M., Fissette, J.C., Hall, W.S., Hardy, S.B., Gerow, F., 1969. Evaluation of synthetic fabrics as artificial skin grafts to experimental burn wounds. J. Biomed. Mater. Res. 3 (2), 213–234.

Stainsby, G., 1987. Gelatin Gels. In: Pearson, A.M., Dutson, T.R., Bailey, A.J. (Eds.), Advances in Meat Research, Collagen as a Food. Van Nostrand Reinhold Company Inc, New York, NY, pp. 209–222.

Stevens, M.M., 2008. Biomaterials for bone tissue engineering. Mater. Today 11 (5), 18–25.

Stitzel, J., Liu, J.S., Lee, J., Komura, M., Berry, J., Soker, S., 2006. Controlled fabrication of a biological vascular substitute. Biomaterials 27, 1088–1094.

Stokols, S., Tuszynski, M.H., 2004. The fabrication and characterization of linearly oriented nerve guidance scaffolds for spinal cord injury. Biomaterials 25, 5839–5846.

Stokols, S., Sakamoto, J., Breckon, C., Holt, T., Weiss, J., Tuszynski, M.H., 2006. Templated agarose scaffolds support linear axonal regeneration. Tissue Eng. 12, 2777–2787.

Stupp, S.I., Pralle, M.U., Tew, G.N., Li, L.M., Sayar, M., Zubarev, E.R., 2000. Self-assembly of organic nano-objects into functional materials. MRS Bull. 25, 42–48.

Su, Y., Lu, B., Xie, Y., Ma, Z., Liu, L., Zhao, H., Zhang, J., Duan, H., Zhang, H., Li, J., Xiong, Y., Xie, E., 2011. Temperature effect on electrospinning of nanobelts: the case of hafnium oxide. Nanotechnology 22 (28), 285609.

Sui, G., Yang, X., Mei, F., Hu, X., Chen, G., Deng, X., Ryu, S., 2007. Poly-L-lactic acid/hydroxyapatite hybrid membrane for bone tissue regeneration. J. Biomed. Mater. Res. A 82 (2), 445–454.

Sukigara, S., Gandhi, M., Ayutsede, J., Micklus, M., Ko, F., 2003. Regeneration of Bombyx mori silk by electrospinning—part 1: processing parameters and geometric properties. Polymer 44 (19), 5721–5727.

Sung, H., Meredith, C., Johnson, C., Galis, Z.S., 2004. The effect of scaffold degradation rate on three-dimensional cell growth and angiogenesis. Biomaterials 25 (26), 5735–5742.

Supaphol, P., Mit-Uppatham, C.H., Nithitanakul, M., 2005. Ultrafine electrospun polyamide-6 fibers: effect of emitting electrode polarity on morphology and average fiber diameter. J. Polym. Sci. B Poly. Phys. 43, 3699–3712.

Takema, Y., Kimura, S., 1982. Two genetically distinct molecular species of Octopus muscle collagen. Biochem. Biophys. Acta 706, 123–128.

Tang, L., Chen, S., Sua, W., Wenga, W., Osako, K., Tanaka, M., 2015. Physicochemical properties and film-forming ability of fish skin collagen extracted from different freshwater species. Process Biochem. 50 (1), 148–155.

Teo, W.E., Ramakrishna, S., 2006. A review on electrospinning design and nanofibre assemblies. Nanotechnology 17, R89–R106.

Thanonkaew, A., Benjakul, S., Visessanguan, W., 2006. Chemical composition and thermal property of cuttlefish (*Sepia pharaonis*) muscle. J. Food Compos. Anal. 19, 127–133.

Theron, A., Zussman, E., Yarin, A.L., 2001. Electrostatic field-assisted alignment of electrospun nanofibers. Nanotechnol. J. 12 (3), 384–390.

Theron, S.A., Yarin, A.L., Zussman, E., Kroll, E., 2005. Multiple jets in electrospinning: experiment and modeling. Polymer 46, 2889–2899.

Thiersch, C., 1874. Ueber die feineren anatomischen Veränderungen bei Aufheilung von Haut auf Granulationen. Verh. dtsch. Ges. Chir. 3, 69–75.

Thomas, V., Dean, D.R., Jose, M.V., Mathew, B., Chowdhury, S., Vohra, Y.K., 2007. Nanostructured biocomposite scaffolds based on collagen coelectrospun with nanohydroxyapatite. Biomacromolecules 8 (2), 631–637.

Toro, C., Buriak, J.M., 2014. Template synthesis approach to nanomaterials: Charles Martin. Chem. Mater. 26 (17), 4889–4890.

Tu, R.S., Tirrell, M., 2004. Bottom-up design of biomimetic assemblies. Adv. Drug Deliv. Rev. 56, 1537–1563.

Tzaphlidou, M., 2004. The role of collagen and elastin in aged skin: an image processing approach. Micron 35, 173–177.

Venugopal, J., Ramakrishna, S., 2005. Biocompatible nanofiber matrices for the engineering of a dermal substitute for skin regeneration. Tissue Eng. 11 (5–6), 847–854.

Venugopal, J., Zhang, Y., Ramakrishna, S., 2006. In vitro culture of human dermal fibroblasts on electrospun polycaprolactone collagen nanofibrous membrane. Artif. Org. 30 (6), 440–446.

Vickers, J.C., King, A.E., Woodhouse, A., Kirkcaldie, M.T., Staal, J.A., McCormack, G.H., Blizzard, C.A., Musgrove, R.E., Mitew, S., Liu, Y., Chuckowree, J.A., Bibari, O., Dickson, T.C., 2009. Axonopathy and cytoskeletal disruption in degenerative diseases of the central nervous system. Brain Res. Bull. 80, 217–223.

Virchow, R., 1858. Die Cellular-Pathologie in ihrer Begründung auf Physiologische und Pathologische Gewebelehre. A. Hirschwald, Berlin, Germany.

Vrieze, S.D., Camp, T.V., Nelvig, A., Hagstrom, B., Westbroek, P., Clerck, K.D., 2009. The effect of temperature and humidity on electrospinning. J. Mater. Sci. 44, 1357–1362.

Wadea, T.L., Wegrowe, J.E., 2005. Template synthesis of nanomaterials. Eur. Phys. J. Appl. Phys. 29 (01), 3–22.

Walter, J.B., Israel, M.S., 1987. General Pathology. Churchill Livingstone.

Walker, A.A, 1998. Oldest glue discovered. Archaeology Institute of America, http://archive.archaeology.org/online/news/glue.html.

Watson, T., 2003. Soft tissue healing. In Touch 104, 2–9.

Watson, T., 2006. Tissue repair: the current state of art. J. Sportex Health 19, 8–12.

Wei, G., Ma, P.X., 2004. Structure and properties of nano-hydroxyapatite/polymer composite scaffolds for bone tissue engineering. Biomaterials 25 (19), 4749–4757.

Weiner, S., Wagner, H.D., 1998. The material bone: structure mechanical function relations. Annu. Rev. Mater. Sci. 28, 271–298.

Wnek, G.E., Carr, M.E., Simpson, D.G., Bowlin, G.L., 2003. Electrospinning of nanofiber fibrinogen structures. Nano Lett. 3 (2), 213–216.

Wollensak, G., Sporl, E., Pham, D.T., 2004. Biomechanical changes in the anterior lens capsule after trypan blue staining. J. Cataract Refract. Surg. 30 (7), 1526–1530.

Woodruff, M.A., Hutmacher, D.W., 2010. The return of a forgotten polymer—polycaprolactone in the 21st century. Prog. Polym. Sci. 35 (10), 1217–1256.

Xie, Z., Paras, C.B., Weng, H., Punnakitikashem, P., Su, L.C., Vu, K., Tang, L., Yang, J., Nguyen, K.T., 2013. Dual growth factor releasing multi-functional nanofibers for wound healing. Acta Biomater. 9 (12), 9351–9359.

Yang, Q., Li, Z., Hong, Y., Zhao, Y., Qiu, S., Wang, C., Wei, Y., 2004. Influence of solvents on the formation of ultrathin uniform poly(vinyl pyrrolidone) nanofibers with electrospinning. J. Polym. Sci. B Polym. Phys. 42 (20), 3721–3726.

Yang, J.I., Ho, H.Y., Chu, Y.J., Chow, C.J., 2008a. Characteristic and antioxidant activity of retorted gelatine hydrolysates from cobia (*Rachycentron canadum*) skin. Food Chem. 110, 128–136.

Yang, L., Fitie, C.F.C., van der Werf, K.O., Bennink, M.L., Dijkstra, P.J., Feijen, J., 2008b. Mechanical properties of native and cross-linked type I collagen fibrils. Biomaterials 29, 955.

Yuan, X., Zhang, Y., Dong, C., Sheng, J., 2004. Morphology of ultrafine polysulfone fibers prepared by electrospinning. Polym. Int. 53, 1704–1710.

Zargham, S., Bazgir, S., Tavakoli, A., Rashidi, A.S., Damerchely, R., 2012. The effect of flow rate on morphology and deposition area of electrospun nylon 6 nanofiber. J. Eng. Fibers Fabrics 7 (4), 42–49.

Zeugois, D.I., Panengad, P.P., Yew, E.S.Y., Sheppard, C., Phan, T.T., Raghunath, M., 2010. An in situ and in vitro investigation for the transglutaminase potential in tissue engineering. J. Biomed. Mater. Res. A 92A (1), 1310–1320.

Zeugolis, D.I., Khew, S.T., Yew, E.S.Y., Ekaputra, A.K., Tong, Y.W., Yung, L.Y.L., Hutmacher, D.W., Sheppard, C., Raghunath, M., 2008. Electro-spinning of pure collagen nano-fibres—just an expensive way to make gelatin? Biomaterials 29, 2293.

Zhang, Y., Chwee Teck, L., Ramakrishna, S, Huang, Z.M., 2005a. Recent development of polymer nanofibers for biomedical and biotechnological applications. J. Mater. Sci. Mater. Med. 16 (10), 933–946.

Zhang, Y.Z., Venugopal, J., Huang, Z.M., Lim, C.T., Ramakrishna, S., 2005b. Characterization of the surface biocompatibility of the electrospun PCL-collagen nanofibers using fibroblasts. Biomacromolecules 6 (5), 2583–2589.

Zhang, W.J., Liu, W., Cui, L., Cao, Y., 2007. Tissue engineering of blood vessel. J. Cell. Mol. Med. 11 (5), 945–957.

Zhang, M., Liu, W., Li, G., 2009. Isolation and characterisation of collagens from the skin of largefin longbarbel catfish (*Mystus macropterus*). Food Chem. 115, 826–831.

Zhang, F.X., Wang, A., Li, Z., He, S., Shao, L., 2011. Preparation and characterisation of collagen from freshwater fish scales. Food Nutr. Sci. 2, 818–823.

Zhao, Y.Y., Yang, Q.B., Lu, X.F., Wang, C., Wei, Y., 2005. Study on correlation of morphology of electrospun products of polyacrylamide with ultrahigh molecular weight. J. Polym. Sci. B Polym. Phys. 43 (16), 2190–2195.

Zhao, P., Jiang, H., Pan, H., Zhu, K., Chen, W., 2007. Biodegradable fibrous scaffolds composed of gelatin coated poly (e-caprolactone) prepared by coaxial electrospinning. J. Biomed. Mater. Res. A 83 (2), 372–382.

Zheng, W., Zhang, W., Jiang, X., 2010. Biomimetic collagen nanofibrous materials for bone tissue engineering. Adv. Eng. Mater. 12 (9), B451–B466.

Zhong, S.P., Teo, W.E., Zhu, X., Beuerman, R., Ramakrishna, S., Yung, L.Y.L., 2005. Formation of collagen-glycosaminoglycan blended nanofibrous scaffolds and their biological properties. Biomacromolecules 6 (6), 2998–3004.

Zhou, F.L., Gong, R.H., Porat, I., 2009. Three-jet electrospinning using a flat spinneret. J. Mater. Sci. 44, 5501–5508.

Zong, X.H., Kim, K., Fang, D.F., Ran, S.F., Hsiao, B.S., Chu, B., 2002. Structure and process relationship of electrospun bioabsorbable nanofiber membranes. Polymer 43, 4403.

Zuo, W.W., Zhu, M.F., Yang, W., Yu, H., Chen, Y.M., Zhang, Y., 2005. Experimental study on relationship between jet instability and formation of beaded fibers during electrospinning. Polym. Eng. Sci. 45, 704.

CHAPTER

12

NANOSTRUCTURED THERAPEUTIC SYSTEMS WITH BIOADHESIVE AND THERMORESPONSIVE PROPERTIES

Marcos L. Bruschi, Fernanda B. Borghi-Pangoni, Mariana V. Junqueira, Sabrina B. de Souza Ferreira

State University of Maringá, Maringá, Brazil

CHAPTER OUTLINE

1 INTRODUCTION

In recent decades, the development of the modified drug release systems has been well regarded. The aim is to promote a uniform release, decrease the number, and frequency of doses required to obtain the desired effect, increase the therapeutic effectiveness, and, at the same time, decrease side effects and increase patient adherence to therapy (Collet and Moreton, 2005; Nho et al., 2014).

In order to obtain the prolonged and/or controlled effect, it is necessary to use formulations containing "barriers" like chemical or physical bioadhesive systems (Collet and Moreton, 2005; Wise, 2000). These systems are composed by preparation of polymers that may exhibit the ability to adhere to biological substrates, and release modified form of the drug in the target site. When adhesion occurs in the mucosa is called mucoadhesive systems. Thus, these preparations may be administered in various pathways, such as skin and mucous membranes of the gastrointestinal tract, nasal, ocular, buccal, vaginal, rectal, oral, and periodontal (Andrews et al., 2009; Bruschi et al., 2015; Carvalho et al., 2010; Chang et al., 2002b; Shaikh et al., 2011).

The retention time of the formulation at the application site is related to the effectiveness of bioadhesive polymers, physicochemical characteristics of drug, and the pharmaceutical form (Andrews et al., 2009; Bruschi et al., 2006; Carvalho et al., 2010; Jones et al., 2009; Nho et al., 2014; Pereira, 2011; Shaikh et al., 2011). Thus, these properties can be obtained using bioadhesive materials, such as carbomers, polycarbophil, among others (Bruschi et al., 2007).

In addition, there are also the polymers that are responsive to the environment, when in aqueous dispersions, they can respond to body stimuli, such as temperature or pH. Besides this, polymer dispersion can increase the consistency in situ, making it very interesting in the drug-delivery field. In this sense, poloxamer 407, which is composed of ethylene oxide units and propylene oxide alternated, displays special interest (Carvalho et al., 2010; Kabanov et al., 2002; Kwon et al., 2015).

Therefore, the combination of polymers that are bioadhesive with the ones who are thermoresponsive enables the development of new drug delivery systems with superior contact capacity, better adhesion, prolonged residence time, and consequently, possible absorption at the desired application site and controlled within a drug delivery system (Andrews et al., 2009; Ferreira et al., 2015; Shaikh et al., 2011; Valenta, 2005).

2 BIOADHESIVE SYSTEMS

Adhesive materials have been extensively used throughout human history, from the first ointments of animal origin, used in prehistoric cave paintings within a high-strength adhesive dye, moreover, used in the manufacture of aircraft. On the other hand, adhesion processes play a very important role in nature with multiple implications on health and can be defined as "fixing" two surfaces together. When these occur in a biological environment, the phenomenon is called "bioadhesion" but if occur in a mucous membrane, named as "mucoadhesion" (Andrews et al., 2009).

Bioadhesion is defined as the state in which two materials, that at least one has a biological nature, are maintained together for a prolonged time by interfacial forces. The concept of mucoadhesion rose in the early 1980s as the attachment of biological or synthetic macromolecules to a mucosal epithelium (Carvalho et al., 2010; Hägerström et al., 2003).

Bioadhesive systems present the ability to get in close contact with the biological substrate; if the biological surface is the mucosal tissue, the system is called mucoadhesive (Bruschi et al., 2006; Carvalho et al., 2010; Pereira and Bruschi, 2012).

The mucous membrane behaves as a semipermeable barrier system where small molecules and ions are selected as gases, water, and nutrients are allowed to diffuse through (Khutoryanskiy, 2011). This epithelial layer is characterized by a surface that is covered by mucus, in which contains lipids, inorganic salts, glycoproteins, and 95% water by mass, producing a highly hydrated system. The most important glycoprotein of mucus is the mucin, responsible for its structuring, lubricating the epithelium, and other additional functions that depend on the epithelium covered.

The soluble mucin presents high-molecular weight (0.5–40 MDa) and is composed of 500 kDa sub-units linked together by intramolecular cystein-cystein disulfide bridges and peptide linkages. The thickness of mucus is approximately of 50–450 μm in the stomach to less than 1 μm in the oral cavity. The mucus gel is a dynamic system constantly converted through the secretion of mucins from the goblet cells (Bruschi et al., 2007; Khutoryanskiy, 2011).

Thermoresponsive systems display significant physicochemical changes in response to temperature modifications. The development of blends composed by bioadhesive thermoresponsive polymers provides new nanostructured therapeutic systems (Bruschi et al., 2006; Carvalho et al., 2010; Pereira and Bruschi, 2012).

The traditional dosage forms (e.g., liquid dispersions) have the inability to get the closeness contact with the mucosa during a great residence time to take place at the light application. This problem and the possible side effects of systemic administration of drugs have contributed interest in the mucoadhesive thermoresponsive drug delivery systems. These systems have been designed to release active agents to the different mucous (e.g., vaginal, ocular, buccal, and nasal) and their clinical, mechanical, and rheological evaluation have been established (Bruschi et al., 2006; Bruschi et al., 2015; Pereira et al., 2013). It is very significant that the mucoadhesive materials maintain the preparation at the action site and direct the active agent to a specific tissue or site (Bruschi et al., 2015).

It is a strategy adopted in many therapeutic systems with the aim to control the drug release, improve the drug bioavailability, reduce administration frequency, and drug targeting (Hägerström et al., 2003; Carvalho et al., 2010).

In this context, pharmaceutical polymer mucoadhesive systems can increase the availability and the contact time between these and the mucosa due to their ability to direct the location, and to control the release of the active agent in a particular region of the body (Andrews et al., 2009; Bruschi et al., 2007; Carvalho et al., 2010).

2.1 BIOADHESION THEORIES

In spite of, the chemical and physical mechanisms of bio/mucoadhesion are not hole understood yet; there are six theoretical classifications about many materials and polymer–polymer adhesion, which elucidate the phenomenon (Carvalho et al., 2010; Hägerström et al., 2003; Huang et al., 2000; Madsen et al., 2013; Mirza et al., 2013; Smart, 2005). These mechanisms must be considered as a complementary process involved in different stages of substrate-mucus interaction, instead of alternative or individual theories (Andrews et al., 2009; Zhu et al., 2013).

2.1.1 Electronic theory

This theory is based on the assumption that the bioadhesive hydrogel and target biological tissue possess opposite electrical charges. When these materials become in contact, the electronic transference forms an electronic double layer in the interface, where the attractive forces into the double layer determine the bioadhesive force (Andrews et al., 2009; Carvalho et al., 2010; Junginger et al., 2001; Sinko, 2005; Zhu et al., 2013).

2.1.2 Wettability theory

According to the wettability theory, the adhesion occurs as an incrustation process. The adhesive molecules permeate the surface irregularities, hardening, and then creating multiple binding sites (Sinko, 2005).

This mechanism is applicable to liquid or low viscosity bioadhesive systems, which demonstrate affinity for the surface to spread out on this, and characterize "spreadability" measure of drug delivery systems through the biological substrate. The affinity can be detected by contact angle techniques. Thereby, the smaller the contact angle, the greater affinity. If the contact angle is 0, the liquid spreads freely in the solid surface and moistens (Carvalho et al., 2010; Junginger et al., 2001; Zhu et al., 2013).

2.1.3 Adsorption theory

Adhesion is the result of several primary and secondary surface interactions between the adhesive polymer and mucous substrate. The primary interactions are made by ionic, covalent, metallic bonds, which are generally undesirable due to its permanence. On the other hand, the secondary bonds mainly arise due to Van der Walls, hydrophobic, and hydrogen interactions (Andrews et al., 2009; Zhu et al., 2013). Additionally, secondary interactions need less energy to "break", since they are considered weak individually. However, in large quantities can result in intense adhesion semipermanently. Consequently, these bonds are predominantly distributed in the interaction surface of mucoadhesion processes (Carvalho et al., 2010; Zhu et al., 2013).

2.1.4 Fracture theory

According to this theory, the force necessary to separate two surfaces is related to the adhesive bonds between the systems (Andrews et al., 2009; Zhu et al., 2013). The force (s_m) is calculated frequently in the rupture strength tests by ratio of maximum peel force (F_m) and total surface area (A_0) involved in the adhesive interactions, expressed by Eq. (12.1):

$$S_m = \frac{F_m}{A_0} \tag{12.1}$$

The fracture theory does not consider the interpenetration and the diffusion of polymer chains. Thus, it is appropriated for application in rigid and semirigid bioadhesive systems, where the polymeric chains do not penetrate the mucus chain (Carvalho et al., 2010). The fracture work will increase in that the polymer network chains are longer, and the degree of crosslinking is lower (Zhu et al., 2013).

2.1.5 Diffusion theory

This theory describes the interpenetration of mucoadhesive polymeric chains and mucin chains with a depth to create a semipermanent adhesive bond, in time-dependent mode (Carvalho et al., 2010; Zhu et al., 2013). There are several factors involved in these processes, such as molecular weight, cross-linking density, chain mobility/flexibility, and expansion capacity of both network. The temperature has been also related as an important environmental factor in this process, since the temperature increase, the polymeric dispersions become more fluid, such as carbomers. Then, the mucin chains interpenetrate easily. However, the increase of temperature may cause the decrease of cohesiveness, and become the polymeric systems weaker (Andrews et al., 2009; Zhu et al., 2013).

According to the literature, the interpenetration depth required to produce an efficient bioadhesive bond must be between 0.2–0.5 μm. This polymer interpenetration depth (l) and mucin chains can be estimated by Eq. (12.2):

$$l = (tD_b)^{1/2} \tag{12.2}$$

Where t is contact time, and D_b is a diffusion coefficient of the mucoadhesive material in mucus (Carvalho et al., 2010). Thereby, this theory represents a two-way diffusion process, where the polymers' permeability depends of diffusion coefficient of polymers that are realizing the interaction (Zhu et al., 2013).

Although, it is known that the longer polymeric chains can diffuse, interpenetrate and finally, entangled with a larger extension of mucous surface, it should be recognized that there is a critical length of chain up to 100,000 Da to obtain molecular entanglement and interpenetration. The polymeric chains that present length higher than 100,000 Da will act decreasing the polymeric mobility and then, interfacial penetration (Andrews et al., 2009; Zhu et al., 2013).

2.1.6 Mechanical theory

Mechanical theory considers the adhesion as an irregularities "filling" of an abrasive surface by a bioadhesive liquid. Hence, the presence of roughness increases the interfacial area available for interactions, contributing to dissipate energy (Carvalho et al., 2010).

2.2 BIOADHESIVE POLYMERS

Bioadhesive polymers may be designated as long molecules, composed of monomers (the same or similar units) interconnected by covalent bonds, which exhibit the ability to make adhesive interactions with biological substrates. Several advantages have been reported in the literature for these materials, such as residence time prolonged, specific sites, and modified drug release (Almeida et al., 2013b; Andrews et al., 2009; Carvalho et al., 2010; Gupta et al., 2013; Pereira, 2011; Wypych, 2009).

Their physicochemical characteristics, such as viscoelastic properties, presence of groups that makes hydrogen bonds, hydration and swelling degree, pH, charge, length, weight, and chain conformation make them different from other polymers (Almeida et al., 2013b; Andrews et al., 2009; Carvalho et al., 2014; Pliszczak et al., 2012; Wypych, 2009).

Bioadhesive bonds are formed by the interpenetration of polymer chains of substrates, followed by the formation of noncovalent interactions, such as Van der Waals connections, hydrophobic interactions, electrostatic forces, and hydrogen bonds. These interactions are due to the presence of hydrophilic groups in the polymer chains as carboxyl, hydroxyl, amide, and sulfate, which form the hydrogen bonds. In this way, the increased formation of the number of groups of hydrogen bonds, the bigger become the polymer chains enhancement and consequently. Consequently, there will be more interaction between polymer chains and the biological substrates (Andrews et al., 2009; Nho et al., 2014; Pliszczak et al., 2012; Reineke et al., 2013; Wypych, 2009; Zhu et al., 2013).

According to Roy and Prabhakar (2010) and Sudhakar et al. (2006), the ideal bio/mucoadhesive polymers should present some characteristics, such as:

1. Polymers must be easily available and the cost should not be high;
2. Polymers and their degradation products should be nontoxic, nonirritant, and free from impurities;

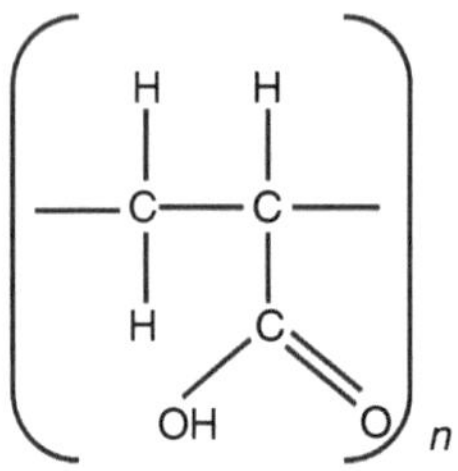

FIGURE 12.1 Acrylic Acid Monomer Unit

3. Adhere quickly and have strong interaction to the mucosa;
4. Display sufficient mechanical strength, good viscoelastic properties, and be pH biocompatible;
5. Good swelling, spreadability, wetting, solubility, and biodegrability properties;
6. Display a minimum impact on drug release and these release should be a unidirectional way through the mucosa;
7. Should facilitate the rate and extent of drug absorption;
8. Can be not affected by hydrodynamic condition and pH changes;
9. Should present bioadhesive properties in the both solid and liquid states;
10. Sufficiently crosslinked but not to the degree of suppression of bond forming groups;
11. Not cause any irritation or inconvenience to the patient and not interfere with the normal functions (talk, drink, eat, etc.);
12. Not promote the development of secondary infections.

Numerous conventional drug delivery systems that are already in use can become bioadhesive after including bioadhesive substances in their formulation (Bruschi et al., 2015).

2.2.1 Carbomers

The carbomer resins were initially produced and patented in 1957, and have shown great interest in several areas with various applications, mainly as thickener agents for drug delivery systems, flow behavior modifiers of polymer dispersions, as well as bioadhesive systems. The carbomers belong to a synthetic polymeric material class within high molecular weight containing acrylic acid crosslinked by chains of allyl sucrose or allyl ethers of pentaerythritol (Fig. 12.1) (Berginc et al., 2014; Bonacucina et al., 2004; Carvalho et al., 2013; Chawla and Saraf, 2012; Gómez-Carracedo et al., 2004; Guo, 2008; Hosmani et al., 2006; Sheet, 2002; Tamburic and Craig, 1995).

Carbomers are presented in a white-light powder, hygroscopic, with acid character and characteristic smell. They exhibit the ability to swell in water and increase 1000 times its original volume and 10 times its original size, in order to form a gel, mainly when exposed to lower pHs value between 4.0 and 6.0. The carboxyl groups presented in the acrylic acid chain are responsible for several of its features, such as gel formation due to the ionization of the chains and, when neutralized, present the phenomena of electrostatic repulsion of negatively charged groups. However, the carboxylic acid groups that are in low pH values remain unionized and retain the drug inside (Almeida et al., 2013b; Chawla and Saraf, 2012; Hosmani et al., 2006; Lu and Jun, 1998; Mirza et al., 2013; Sheet, 2002).

The carbomers are excellent bio/mucoadhesive systems due to its ability to absorb water, swell, and hydrate, keeping the crosslinked structure, and due to its complete insolubility in water. The mucoadhesion occur because of the hydrogen bonds between the carboxylic acid groups and carboxylic acid groups belonging to the sialic acid from the mucin glycoprotein (Almeida et al., 2013b; Berginc et al., 2014; Guo, 2008; Hosmani et al., 2006; Tamburic and Craig, 1995). The bioadhesive formulations containing carbomers can be used in the topical, nasal, oral, ophthalmic, vaginal, intestinal, and rectal routes (Bonacucina et al., 2004; Guo, 2008; Hosmani et al., 2006; Islam et al., 2004).

Some studies have assessed the toxicity of carbomers and demonstrated the low toxicity and even low potential for irritation. Thus, these polymers have wide acceptance for many bioadhesive applications (Carvalho et al., 2013; Chawla and Saraf, 2012; Hosmani et al., 2006).

Carbomers use on bioadhesive systems, when neutralized (pH 4.0–6.0), confer many advantages, such as safety and effectiveness in oral and topical applications, increased bioavailability, protection against the enzymes action, excellent bond strength, high viscosity at low concentrations, wide viscosity range and flow behavior characteristics, compatibility with multiple active ingredients, good thermal stability, excellent organoleptic characteristics, great acceptance by patient, and approval by many regulatory companies (Guo, 2008; Islam et al., 2004).

In addition, there are several classes of carbomers that differ by the cross-linking degree and manufacturing conditions, each class has its significance, and are used in different pharmaceutical formulations. The most studied carbomers are 934P (C934P), 971P (C971P), and 974P (C974P) (Bonacucina et al., 2004; Hosmani et al., 2006; Lubrizol, 2013a,c). The designation "P" of the Carbopol 934P refers to a pharmaceutical grade that presents a residual content of low- benzene, around 0.1%. When compared to common carbomers that present 0.5 %, these carbomers can be used orally (Sheet, 2002).

2.2.2 Polycarbophil

The polycarbophil or Noveon AA-1 (PCB) is a high molecular weight homopolymer, which contains acrylic acid cross-linked with polyalkenyl ethers or divinyl glycol (Fig. 12.2). The polycarbophil is known as one of the most potent bioadhesive agents, lightly cross-linked with chains of carboxylic groups (COOH).

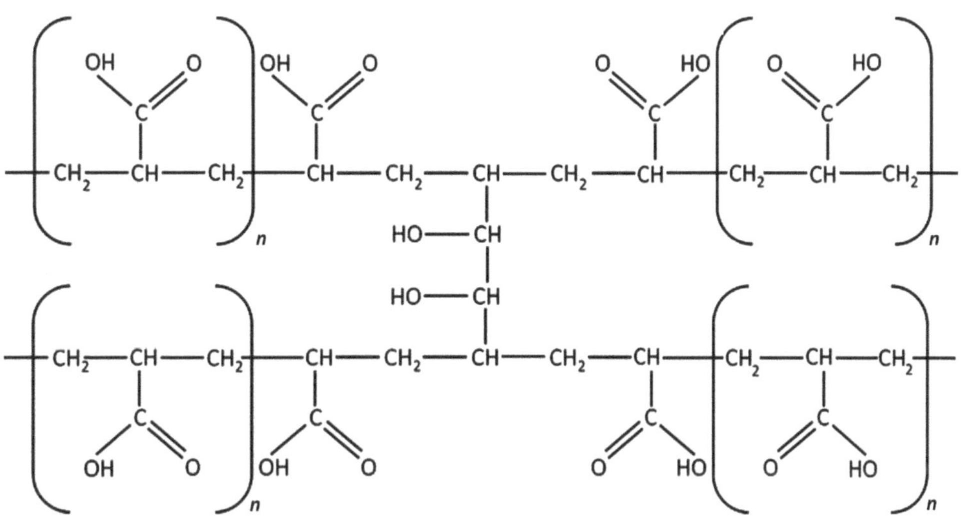

FIGURE 12.2 Structure of Polycarbophil

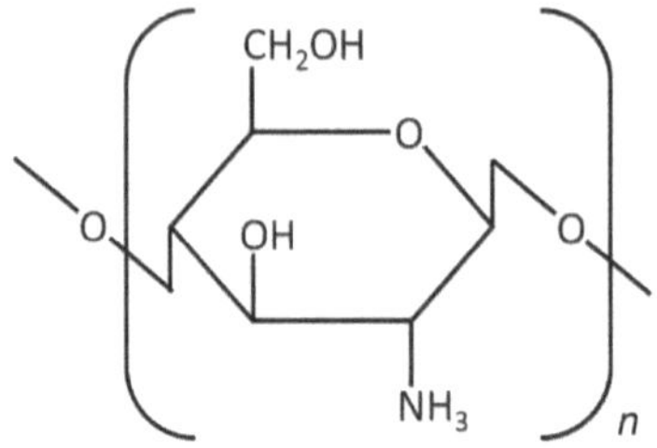

FIGURE 12.3 Structure of Chitosan

This polymer is used extensively to control the release of several drugs in mucosal membranes, which might be used in topical, buccal, nasal, vaginal, and ophthalmic routes, as well as carbomers. These are anionic polymers and may also be gelled at around pH 7.00, due to repulsion between the ionized chains, enabling the formation of high levels of entanglement between the polymer chains and the mucosal barrier, which results in a high mucoadhesive strength link (Almeida et al., 2013b; Carvalho et al., 2014; Gómez-Carracedo et al., 2004; Han et al., 2006; Ivarsson and Wahlgren, 2012; Kerec et al., 2002; Lubrizol, 2013b; Tamburic and Craig, 1995; Zhu et al., 2013).

2.2.3 Chitosan

Chitosan is the second most abundant polymer in nature, being a primary structural component of the outer skeletons of crustaceans, and also be found in each species of fungi, mollusks, and insects (George and Abraham, 2006).

Chitosan is a linear polysaccharide composed of randomly distributed β-(1,4)-linked 4)-linked N-acetyl-D-glucosamine units and D-glucosamine, being considered a seminatural bioadhesive polymer (Fig. 12.3). This presented as white or cream-white power or flakes, odorless, with hydrophilic and cationic charges characteristics and functional amino groups, being responsible for physicochemical properties (Bhattarai et al., 2010). This polysaccharide is insoluble at alkaline and neutral pH values, but is capable to form salts with organic and inorganic acids that when the amino functions are protonated resulting in a positively charged soluble polysaccharide (Illum et al., 2001).

Moreover, the polymer is nontoxic, biocompatible, and biodegradable. Almost all functional properties depend on the charge distribution, charge density, and chain length, but in general present a pseudoplastic and viscoelastic behavior (Ludwig, 2005; Rossi et al., 2001; Rowe et al., 2009).

Chitosan is well documented by the low toxicity, nonirritant material, biocompatibility, and biodegradation by human enzymes. These properties make chitosan an ideal candidate for controlled release formulations, being specially used to delivery of macromolecules, such as proteins, antigens, peptides, genes, and oligonucleotides (Bhattarai et al., 2010).

To development of hydrogels is necessary that chitosan network display two conditions, and how these interactions are physical, the gels can be reserved:

- The network should promote the access and residence of water molecules inside the network;
- The interaction between chains must be strong enough to form semipermeated junction points in the network.

The adhesive properties of chitosan display to persist well during continual contact with this substance and the substrate, which suggests that, in addition to hydration, many other mechanisms, such

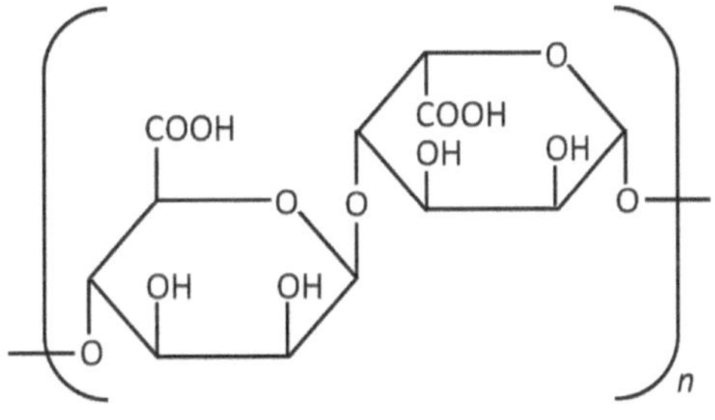

FIGURE 12.4 Structure of Alginate Composed by β-D-Mannurronic Acid and α-L-Guluronic Acid

as hydrogen bonding and ionic interaction might also be involved (George and Abraham, 2006). The ionic interactions occur thanks to interactions between cation amino groups of chitosan and negatively charged molecules and anions, as sialic acid residues of mucus glycoproteins or mucins, depending on environmental pH (Bhattarai et al., 2010; Roge et al., 2011).

2.2.4 Alginate

Alginate is extracted from brown algae (*Laminaria hyperboream*, *Ascophyllum nodosum*, and *Macrocytis pyrifera*) and bacterial (*Azotobacter vinelandii* and several *Pseudomonas* species), being classified as natural/seminatural polymers (Roge et al., 2011). This substance exists as a mixed salt of many cations as magnesium (Mg^{2+}), strontium (Sr^{2+}), barium (Ba^{2+}), and sodium (Na^{2+}), being this last the chiefly (Skjak-Braek et al., 1986). This structural is composed by two monomers: β-D-mannurronic acid residues and α-L-guluronic (Fig. 12.4), that alternating between then forming different geometries of mannurronic block regions, guluronic block regions, and irregular regions, and these diverse compositions and length determined the weight and consequently, the physical properties of this compound (Dupuy et al., 1994; Roge et al., 2011).

The two most important characteristics of salt alginate are water-soluble linear polysaccharide capable to absorb 200–300 times its own weight in water (Prajapati et al., 2012), and the ability to form gels when associated to divalent cations, as well as calcium (Ca^{2+}) (George and Abraham, 2006).

The sol–gel transition and the cross-linked occur mainly by the exchange of sodium ions from the guluronic with the bivalent cations, resulting stacking of the guluronic groups forming a structure called egg-box. Each egg-box can form junctions with many other chains structuring gel networks (Dupuy et al., 1994). The calcium content influence the interchains associations, for example, with low levels, these associations are temporary, giving rise to highly viscous and thixotropic solutions, but if the levels are higher, the precipitation, or gelation form permanent associations (George and Abraham, 2006).

The in situ formation of calcium cross-linked alginate gels was responsible for prolonged therapeutic effect, besides the mucoadhesive feature may aid in its utility as a potential delivery vehicle for drugs to mucosal (Gombotz and Wee, 2012). This adhesive behavior was due to the low surface tension of the alginate, which is lower than critical surface tension of mucin, resulting in good spreading and adhesion (Ludwig, 2005). Furthermore, it is known that polymers with charge density can be used as mucoadhesive agents, and how bigger the charge density better will be the adhesion. It was also been reported that polyanion polymers are more effective as bioadhesive than polycation or nonionic polymers (Chickering and Mathiowitz, 1995; George and Abraham, 2006).

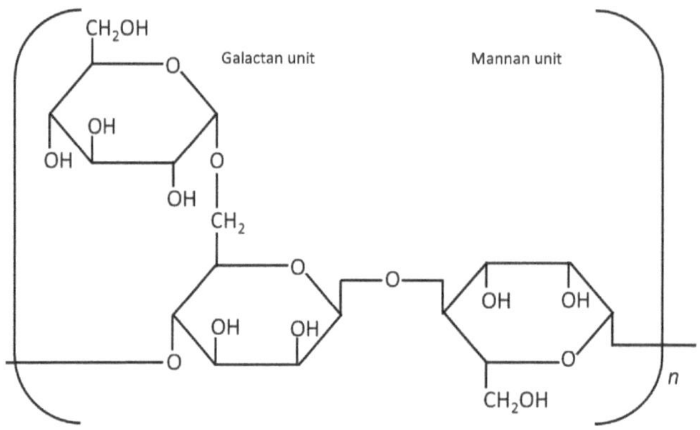

FIGURE 12.5 Structure of Guar Gum Composed by Ggalatan and Mannan Units

2.2.5 Cellulose derivatives

Cellulose is the most abundant organic compound in the world, which mostly produced by plants. In the market, it can be found with different mechanical and pharmaceutical properties, due to the range of shape, sizes, and degrees of crystallinity of their particles. Moreover, cellulose has been demonstrated many semisynthetic derivatives, as well cellulose ethers, and cellulose esters (Shokri and Adibkia, 2013). Some derivatives, such as ethyl cellulose, hydroxyethyl cellulose, hydroxypropyl cellulose, methylcellulose, carboxymethyl cellulose, hydroxypropylmethyl cellulose, and sodium carboxymethyl cellulose are used in bioadhesive preparations.

All these polymers have the ability to take up water from mucus contributing with adhesive characteristics (Ludwig, 2005). The cellulose derivatives have lesser dependency of adhesion time and adhesion force them to pH of medium when compared with polyacrylate and thiolated bioadhesive polymers (Shokri and Adibkia, 2013).

2.2.6 Guar gum

Guar gum is a natural compound obtained from endosperms of the seeds of *Cyamposis tetyragonolobus*. It is composed by galactan and mannan units combined through glycosides linkages and display a higher molecular weight (Fig. 12.5), these characteristics explain the high viscosity presented by this hydrocolloid polysaccharide. The optimum rate of hydration occurs between pH 7.5 and 9.0, but is stable in solution over pH range from 1.0 to 10.5 (Roge et al., 2011). The mucoadhesive property is due to electrostatic interaction between the positive charge of guar gum and negative charge present in the mucus (Singh et al., 2010).

2.2.7 Carrageenan

Carrageenan is an anionic polysaccharide extracted from some members of the class Rhodophyceae (red seaweed). It consists chiefly of potassium, magnesium, calcium, sodium, and ammonium sulfate esters of galactose and 3,6-anhydrogalactose copolymers (Fig. 12.6).

They are divided into three families according to the position of sulfate groups and the presence or absence of anhygalactose. The lambda-carrageenan is no gelling polymer with 35% ester sulfate by weight and no anhygalactose; the iota-carrageenan containing about 32% of ester sulfate and 30% of

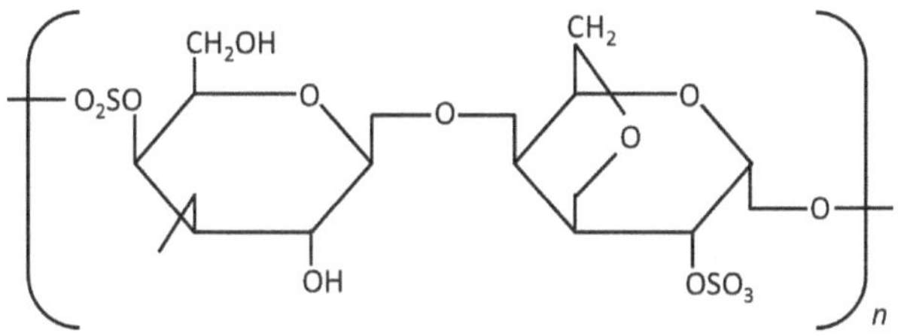

FIGURE 12.6 Structure of Carragenaan

3,6 anhydrogalactose, being able to produce elastic gels; and the kappa-carrageenan forms brittle gels, and containing about 20% of ester sulfate and 34% of 3,6 anhydrogalactose (Rowe et al., 2009).

The solubility depends of form and the type of carrageenan, that is, the sodium form of all three types is soluble in both cold and hot water. The other cations of iota and kappa are soluble only in hot water, and all cations of lambda in cold water. Besides, all carrageenan solutions are pseudoplastics with various degrees of yield value, and decrease of viscosity with the increase of temperature (Roge et al., 2011).

Studies on the effects on mucoadhesion of drugs to oropharyngeal area display that polymers had the greatest propensity for adhesion, and can be used in formulations for oral and buccal drug delivery (Rowe et al., 2009).

3 STIMULI-RESPONSIVE POLYMERS

Over many decades, the attention in stimuli-responsive polymers has increased and new smart polymers have been emerged due to the work committed to developing environmental sensitive materials (Priya James et al., 2014). These macromolecules are of great interest in biomedical applications, tissue engineering, cell encapsulation, and especially, controlled drug delivery (Bajpai et al., 2008; Priya James et al., 2014).

The stimuli-responsive polymers demonstrate the ability to respond to small changes in physical or chemical conditions with altering their physical properties (Almeida et al., 2013a; Bajpai et al., 2008; Hoffman, 2013; Hrubý et al., 2015; Priya James et al., 2014). These "smart," "intelligent," or "environmentally sensitive" materials have presented this designation due to the capability to change and return to their original shape, when an external trigger is added and, then removed respectively. Consequently, small stimulus can provide macroscopic adjustments in their building. The reversible fluctuations include physical state, shape, solubility, solvent interactions, hydrophilic, and lipophilic balance and conductivity changes, according to the stimulus (Bajpai et al., 2008; Priya James et al., 2014). In this sense, polymer–solvent and polymer–polymer interactions display readjustment in small changes, because of chain transition between extended and compacted coil states (Bajpai et al., 2008).

Thus, systems composed by these smart macromolecules are able to carry the bioactive molecules to the target tissue, cells, or even subcellular compartments and release in a triggered way. The use of smart polymers may act as a therapeutic modality itself or for targeting and controlled release/ activation of active molecule after reaching the destination (Hrubý et al., 2015). Because of these attributes, the development of systems composed by stimuli responsive polymers have demonstrated benefits, such as reduced frequency dose, ease of preparation, maintenance of desired therapeutic concentration with a single dose, prolonged release of incorporated drug reduced side effects, and improved stability (Priya James et al., 2014). Moreover, these systems offer potential economic alternatives to conventional separation processes for industrial applications (Bajpai et al., 2008).

Biodegradability is another concern that should be evaluated. Thus, these molecules should be eliminated after perform their job. However, depending on the target tissue, they can be removed with stool, or it is necessary to tailor the degradability in organism. So, this issue is considered a critical attribute (Hrubý et al., 2015).

Intelligent polymers may be present in formulations as dissolved in or phase-separated out of aqueous solutions adsorbed on or chemically-grafted aqueous–solid interfaces, or the smart polymers may be chemically cross-linked H-bonded and/or physically entangled in the form of hydrogels. These polymers can be combined with a wide range of bioactive molecules, such as proteins and peptides, nucleic acids, small organic molecules, and carbohydrates, by physical mixing, chemical conjugation, and complexation. Thus, these formulations can be applied in topical and mucosal surfaces, like as eye, nasal membrane, portions of gastrointestinal tract, anus, and vagina (Hoffman, 2013).

The stimuli that trigger the structural changes on smart polymers can be classified in three main groups: physical (temperature, ultrasound, light, or mechanical stress), chemical (pH and ionic strength) and biological (enzymes and other biomolecules). Besides, these signs are artificially controlled (magnetic or electric field, light, ultrasound) or naturally promoted by the body without external sign, such as pH of certain organs or related to a disease (cancer); temperature change or presence of enzymes) (Almeida et al., 2013b).

3.1 PHOTORESPONSIVE POLYMERS

The responsivity to light denotes a way to trigger an effect at the desired place by an external laser beam. Thus, these systems are useful in therapies of diseases where the lesion is readily available from outside of the body for light. Another interesting application may be for photodynamic therapy of skin cancer that consists in an application of photosensitizer followed by laser irradiation of tumor lesion (Hrubý et al., 2015).

3.2 GLUCOSE-RESPONSIVE POLYMERS

These macromolecules have the ability to mimic normal endogenous insulin secretion, which minimizes diabetic complications and can release the bioactive compound in a controlled manner. Glucose-responsive polymers are sugar-sensitive and show variability in response to the presence of glucose. Thereby, they have application in both glucose-sensing and insulin-delivery applications (Priya James et al., 2014).

3.3 pH-RESPONSIVE POLYMERS

The pH-responsive polymers show great interest, especially because of their many possibilities and applications in drug delivery systems (Peppas et al., 2000). These formulations are characterized by the modification of its structure by ionization of the polymeric groups with consequent electrostatic repulsion, with increased hydrophilic and swelling of the network polymer (Na and Bae, 2005; Qiu and Park, 2001).

Consequently, pH-responsivity is beneficial for site-specific delivery in the gastrointestinal tract, due to the wide range of pH values. Besides, these polymers are useful in drug delivery of solid tumors, since the differences in pH among health and cancer tissues are a prospect to develop pH-sensitive drug delivery systems (Hrubý et al., 2015).

These materials may have acid groups, such as sulfonic or carboxylic acids, or basic groups, such as ammonium salts. pH-responsive polymers with pKa between 4 and 8 are useful for biomedical applications (Hrubý et al., 2015). The anionic polymers are the most frequently used, tend to deprotonate, and swell when the external pH is greater than the pK groups that are in the polymer chain, like carbomer group. Cationic polymers tend to better protonate and ionize when external pH are lower than the pKb of the ionizable groups, such as poly(*N,N'*–diethylaminoethylmethacrylate) (Hrubý et al., 2015; Lin and Metters, 2006; Qiu and Park, 2001).

Despite of the wide range of application of these materials, some pH-responsive polymers, mainly cationic and with amphiphilic properties are cytotoxic. Moreover, the fact that the formulation should in many application sites modify their conformation with small changes in pH is sometimes hard to reach due to the protonation/desprotonation of the polymer proceeds within a broader range of pH than for low-molecular weight molecules, can be considered as a disadvantage (Hrubý et al., 2015).

3.4 FIELD-RESPONSIVE POLYMERS

These polymers react to the application of electric, magnetic, sonic, and electromagnetic fields. The magnetic field-responsive polymers are introduced by implementation of iron oxide nanoparticles, which can be guided to the target site by magnetic field. However, the magnetic guiding of these systems to organ or tumor deeply placed can cause difficulties in this application (Hrubý et al., 2015; Priya James et al., 2014).

On the other hand, the electric field-responsive polymers display large concentration of ions with able groups in their chain. Consequently, they also can be classified as pH-responsive polymers. Chitosan, alginate, and hyaluronic acid are natural polymers with electric field-responsive properties. These molecules convert electric energy into mechanical energy and have wide application in drug delivery. However, the selection of the electric current is a critical parameter, which can cause the drug release without stimulating the nerve ending in the surrounding tissue (Priya James et al., 2014).

3.5 THERMORESPONSIVE POLYMERS

Thermoresponsive polymers are the ones that exhibit conformational changes in response to temperature stimuli, especially near body temperature. They are the most commonly studied, have great use and present considerable interest in drug release systems (Hrubý et al., 2015; Kwon and Kim, 2005; Qiu and Park, 2001).

Moreover, these polymers have been demonstrated hydrophobic groups, such as methyl, ethyl, and propyl (Priya James et al., 2014). Thus, these systems exhibit changes above or below body temperature, and may be classified as positive or negative thermoresponsive polymers.

The positive thermoresponsive materials turn to gel above the upper critical solution temperature (USCT), which depends on the polymer structure, such as poloxamer, hydroxypropylcellulose, or methylcellulose. On the other side, the negative present the lower critical solution temperature (LCST) ranging from 25 to 32°C and are gelled above the freezing temperature and below LCST, for example the gelatin and carrageenan hydrogels (Kwon and Kim, 2005; Peppas et al., 2000; Qiu and Park, 2001).

The ideal thermoresponsive system is a solution that reports a free-flowing liquid at room temperature, and after body administration, turn into a gel (Eve Ruel-Gariépy, 2004; Matanović et al., 2014). The structure of thermoresponsive polymers evinces a very well balance between the hydrophilic and hydrophobic groups. As a result of this balance, a little change in the temperature of the polymer aqueous solution is capable of creating a new adjustment of the hydrophilic and hydrophobic interactions between the polymer blocks and water molecules (Bajpai et al., 2008; Matanović et al., 2014).

Besides, this phenomenon depends on the free energy of mixing or the entropy or the enthalpy of the system. Consequently, these systems can be used for forming injectable depots and may serve for sustained release of drugs (Hrubý et al., 2015; Priya James et al., 2014).

Many polymers display a thermoresponsive phase transition property. This polymer can be divided into two major classes, which are focused on their origin: synthetic materials and naturally occurring polymers. Some of them have been previously described. The natural polymers are chitosan, cellulose, gelatin, and xyloglucan, along with their derivatives. The synthetic polymers include poly(ethylene oxide)-*b*-poly(propyleneoxide)-*b*-poly(ethylene oxide) (PEO/PPO/PEO) block copolymers, poly-N-isopropylacrylamide(pNiPAAm), poly(ethylene oxide)-*b*-poly(D,L-lactic acid-*co*-glycolic acid)-*b*-poly(ethylene oxide) (PEO/PLGA/PEO) triblock copolymers, andamphiphilic triblock copolymers, composed of poly-caprolactone (PCL), and PEO (PEO/PCL/PEO) (Eve Ruel-Gariépy, 2004).

3.5.1 Poloxamer

The poloxamers, also known by the trademark Pluronic, Synperonic and Tetronic, were initially introduced between 1950 and have presented several pharmaceutical applications, as well as, excellent compatibility with other compounds (Fig. 12.7).

Studies showed some of the poloxamer's characteristics, especially thermoresponsiveness, high capacity to solubilize drugs, good drug release characteristics, and absence of toxicity in mucosal membranes, and thus widely recognized in the pharmaceutical area as a safe material (Chang et al., 2002b; Dumortier et al., 2006; Kikuchi and Okano, 2005; Liu et al., 2009; Mayol et al., 2008; Veyries et al., 1999).

Poloxamers are nonionic compounds that contains a large group of copolymers surfactants formed by chains of ethylene oxide block (EO) and propylene oxide (PO) (OEx–POy–OEx) (Almeida et al., 2013a; Asasutjarit et al., 2011; Bhowmik et al., 2013; Bruschi, 2006; Cleary et al., 2004; Dumortier et al., 2006). They are synthesized by polymerization of EO and PO units, in sequence, also in the presence of sodium hydroxide and potassium hydroxide. Poloxamers compounds have the chemical formula $HO[CH_2CH_2O]\,x[CH(CH_3)CH_2O]\,y[CH_2CH_2O]OH$, where y is greater than 14 (Dumortier et al., 2006; Kabanov et al., 2002).

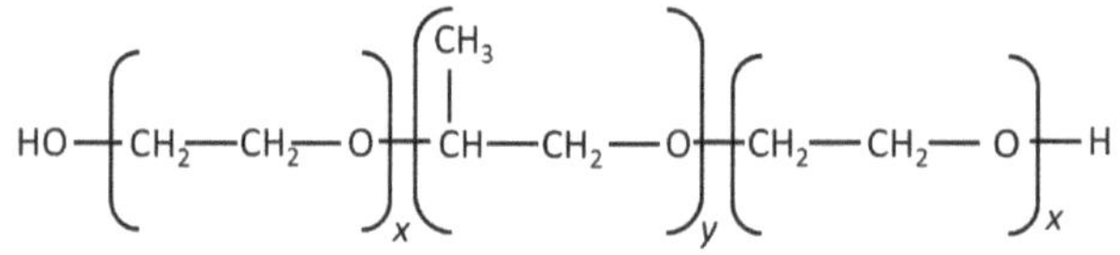

FIGURE 12.7 Structure of Poloxamer

According to the ratio between hydrophilic (EO) and lipophilic (PO) units, various copolymers block and size can be obtained within different molecular weights and physic-chemical properties (Almeida et al., 2013a; Asasutjarit et al., 2011; Kabanov et al., 2002; Veyries et al., 1999).

The poloxamer 407 or Pluronic F127 has particularly interest because of the thermoreversible properties, and can be useful in the optimization of drug delivery systems, and employed in many formulations like intravenous preparations, topical, ophthalmic, nasal, vaginal, and rectal, with no irritation or skin sensitivity (Bruschi et al., 2015; Yuan et al., 2012).

Poloxamer 407 aqueous solutions have the property of being a thermoresponsive system, which leads to a sol–gel transition due to temperature increase. When in aqueous dispersions, the individual molecules of the copolymer's block of P407 self-organize into micelles (micellization), when are in concentrations above the critical micelle concentration (CMC) in order to minimize the free energy from the solution (Almeida et al., 2013a; Dumortier et al., 2006). These micelles can be spherical, cylindrical, or lamellar, depending upon the length in the chain, containing EO and PB, the concentration of polymer and temperature, that leads to increased viscosity (Kabanov et al., 2002).

The advantages of poloxamers in liquid pharmaceutical forms are especially because that they allow a comfortable release at the action site, gelling at the site and may have modified release. The disadvantages of these polymers are weak mucoadhesive, and poor mechanical properties, short residence time due to easily dissolution at the action site (Asasutjarit et al., 2011; Dumortier et al., 2006; Jones et al., 2009; Mayol et al., 2008).

4 HYDROGELS

Since the first hydrogel containing hydroxyethyl methacrylate used in contact lenses, to the present day, there is a growing interest in hydrogels systems in different areas. In particular, for drug delivery systems, hydrogels are reported as an excellent proposes for bioadhesive devices or drugs release devices (Almeida et al., 2013b; Hoffman, 2002; Lefnaoui and Moulai-Mostefa, 2014; Lin and Metters, 2006; Peppas et al., 2000).

The hydrogels comprises polymeric networks containing three-dimensional hydrophilic homo or copolymers, crosslinked by bonds (crosslinks) which ensure the integrity and structure in order to prevent polymer chain's dissolution of the in an aqueous system. These connections can be chemical or physical. The chemical occurs through covalent bonds in polymers, such as hydroxyethyl methacrylate and the physical connections consisting hydrophobic interactions or hydrogen bonds as calcium alginate hydrogels (Almeida et al., 2013b; Chavda et al., 2013; Florence and Attwood, 2006; Hoare and Kohane, 2008; Hoffman, 2002; Jeong et al., 2005; Peppas et al., 2000).

The polymer chains are hydrophilic and this allows water absorption, mainly because of the presence of hydroxyl, carboxyl, and amide groups that permit the swelling of polymers and keep the polymer structure. This phenomenon occurs due to intra and intermolecular forces, such as covalent, ionic, Van der Waals forces, and hydrophilic interactions (Jeong et al., 2005; Sinko, 2005; Lefnaoui and Moulai-Mostefa, 2014).

Several advantages have been reported about hydrogels, such as biocompatibility, high porosity, biodegradable, and easily deformable. Hydrogels are most biocompatible with the body system because of the high concentration of water, similarity composition, and mechanical properties of the extracellular matrix. The highly porous structure allows the control of density of crosslinks, and the drug release in a dependent manner. Some disadvantages must be analyzed too. The drug quantity and homogeneity, in the hydrogel, must be considered, as well as the amount of water and the size of the pores in the polymer network. In addition, hydrogels can adopt the conformation of the local they are applied, thus being deformable (Hoffman, 2002; Hoare and Kohane, 2008).

5 MUCOADHESIVE THERMORESPONSIVE SYSTEMS

5.1 THE STATE OF THE ART

The mucoadhesive thermoresponsive drug delivery systems can be designed to release active agents to different mucous sites (e.g., vaginal, ocular, buccal, and nasal) for further evaluation (Bruschi et al., 2006, 2015). This kind of system can be used as a strategy to facilitate administration, maintain the preparation at the action site and direct the active agent to a specific tissue or site (Bruschi et al., 2015), protect the drug and control the drug release, improve drug bioavailability, reduce administration frequency, and drug targeting (Carvalho et al., 2010; Hägerström et al., 2003). In this sense, many researches have been conducted in order to study and characterize formulations for clinical possible adjuvant treatments.

Table 12.1 shows many materials used to prepare mucoadhesive and/or thermoresponsive systems used to study, characterize and/or investigate the behavior from the formulations. Most systems comprise semisolid polymeric formulations of gel. Hydrogels are also quite employed in research. There are few formulations using other dosage forms, such as microparticles or eye drops.

Among all polymers used to prepare mucoadhesive thermoresponsive formulations for characterization and/or clinical investigation in Table 12.1, it can be observed that poloxamer 407 and poly(acrylic acid) derivatives were the most common employed. Since, these systems represent a promising tool, because of their easy administration at liquid form, and improved adhesion on a biological surface for long periods in the gel state due to temperature stimulus, consequently, modifying and controlling the drug release (Baloglu et al., 2011).

Thus, when the temperature increases, the poloxamer 407 molecules in aqueous solutions aggregate into micelles, due to dehydratation of PPO chains in the core with an outer surface of hydrated PEO chains. Consequently, the hydrophilic chain of poly(acrylic acid) derivatives interact with the outer region of poloxamer 407 by hydrogen bonds, that can provide compatibility and miscibility (Almeida et al., 2013b; Bhowmik et al., 2013; Bruschi et al., 2007; Gratieri et al., 2010; Jones et al., 2010; Kehoe et al., 2012; Liu et al., 2009; Mayol et al., 2008; Zaki et al., 2007). The structure of the formulations composed by poloxamer 407 and poly(acrylic acid) derivatives is demonstrated in Fig. 12.8.

Due to the large development of bioadhesive thermoresponsive systems composed of poloxamer 407 and poly(acrylic acid) derivatives, Chen and Hoffman (2002) first developed the PAA-*g*-poloxamer in 2002. This system received the designation of Smart Hydrogels. It is known that polyacrylic acid (PAA) is a bioadhesive polymer capable to adhere to the hydrated mucosal of the nose, vagina, gastrointestinal tract, nose, and lungs. In contrast, the poloxamer is a thermoresponsive polymer composed of PO and EO.

The bonding between both polymers occurs by coupling monoamine-terminated poloxamer to PAA using dicyclohexyl carbodiimide, in a process considered complicated, and involving several intermediate purification steps. This platform is interesting for various ways of applications, already been tested for topical and vaginal administration (Chen and Hoffman, 2002; Matanović et al., 2014).

Table 12.1 Materials Used to Prepare a Mucoadhesive and/or Thermoresponsive Formulation for Characterization and/or Clinical Investigation

Dosage form	Polymer/Composition	Purpose	References
Gel	Hydroxyethylcellulose and polyvinylpyrrolidone	Treatment of periodontal diseases	Jones et al. (1996)
Gel	CMC, poly(acrylic acid) and sodium hyaluronate	Characterization of the system	Hägerström and Edsman (2001)
Gel	Poloxamers and polycarbophil	Vaginal disorders	Chang et al. (2002a)
Gel	Metolose	Treatment of pain and inflammation of the esophagus	Makó et al. (2009)
Gel	Poloxamer 407 and Carbopol 934P	Treatment of periodontal diseases	Bruschi et al. (2007)
Gel	Poloxamer 407, Carbopol 934P or gelatin	Treatment in the oral cavity	Hirata and Bruschi (2010)
Gel	Poloxamer 407 and chitosan		Gratieri et al. (2010)
Hydrogel	Hyperbranched block copolymers of ethylene oxide and propylene oxide (EG35 and EG 40) and Carbopol 974P		Bonacucina et al. (2004)
Hydrogel	Poloxamer 407, Carbopol 934P	Characterization of the system	Jones et al. (2009)
Hydrogel	N-isopropylacrylamide		Chang et al. (2010)
Hydrogel	Poloxamer 407 and poloxamer 188	Vulvovaginal candidiasis	Baloglu et al. (2011)
Hydrogel	Poloxamer 407 and Carbopol 934P	Dental pulp protection	Fabri et al. (2011)
Hydrogel	Carbopol 971P, Carbopol 974P, polycarbophil	Skin delivery systems	Carvalho et al. (2013)
Hydrogel	Poloxamer 407 and Carbopol 934P	Vaginal disorders	Pereira et al. (2013)
Hydrogel	Poloxamer, Carbopol 971P	Characterization of the system	Ferreira et al. (2015)
Hydrogel	Poloxamer, Carbopol 934P	Colorectal cancer	Borghi-Pangoni et al. (2015)
Hydrogel	Poloxamer, Carbopol 934P	Photodynamic therapy	Junqueira et al. (2016)
Hydrogel	Poloxamer, Carbopol 934P	Treatment of lesions caused by Herpes simplex	Mazia et al. (2016)
Eye drops	Elastin-like polypeptide (ELP)	Treatment for dry eyes	Wang et al. (2014)

5.2 ADVANTAGES AND DISADVANTAGES

The mucoadhesive drug delivery systems are an important route of administration. These systems are able to be flexible, elastic, and soft after administration favoring its use in emergency therapy case. Furthermore, they can be administrated in patients with trauma and/or unconscious. The absorption rates are maximized due to close contact between the system and absorbing membrane, besides, enables the use of drugs unstable in stomach acid, avoid the first-pass metabolism and increased the bioavailability.

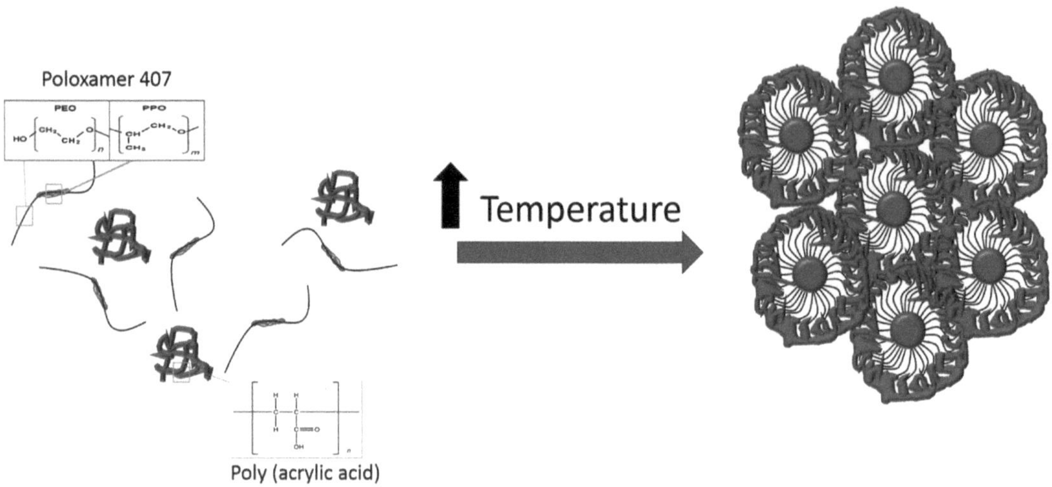

FIGURE 12.8 Schematic Representation of Bioadhesive Thermoresponsive System Composed by Poloxamer 407 and Poly(Acrylic Acid) Derivatives

However, it becomes a disadvantage because only drugs that require a small dose and with absorption by passive diffusion can be used, limiting the number of actives. Others limitations of this type of dosage form are also related to the drug, thus, active unstable in buccal pH, with unpleasant taste, and smell and/or irritation of the mucosa (Prajapati et al., 2012).

There are many bioadhesive polymers presenting advantages, such as specific sites, prolonged residence time, and modified drug release (Almeida et al., 2013b; Andrews et al., 2009; Carvalho et al., 2010; Gupta et al., 2013; Pereira, 2011; Wypych, 2009), such as carbomers, polycarbophil, chitosan, alginate, cellulose derivatives, guar gum, and carrageenan. This bioadhesive polymers can be used in the topical, nasal, oral, ophthalmic, vaginal, intestinal, and rectal routes (Almeida et al., 2013b; Carvalho et al., 2014; Gómez-Carracedo et al., 2004; Han et al., 2006; Ivarsson and Wahlgren, 2012; Kerec et al., 2002; Lubrizol, 2013b; Tamburic and Craig, 1995; Zhu et al., 2013).

The stimuli responsive polymers as a new smart, intelligent, or environmentally sensitive polymers and pH -responsive polymers are useful for biomedical applications (Hrubý et al., 2015) and site-specific delivery in a specific site. In this way, preparations composed by these polymers have demonstrated ease preparation, reduced frequency dose, improved stability, maintained of desired therapeutic concentration with a single dose and prolonged release of incorporated bioactive compounds (Matanović et al., 2014; Priya James et al., 2014). Specially, thermoresponsive polymers have been extensively investigated due to the simple application and low adverse effects on tissue compared to other stimuli (Matanović et al., 2014). Thus, these formulations present avoidance of toxic organic solvents in preparation, ability to deliver and solubilize hydrophilic and lipophilic drugs, reduced systemic effects, and sustained-release properties, and site-specific drug delivery (Priya James et al., 2014). On the other hand, these preparations display the concentration dependence of gelation temperature, when a highly diluted solution may have sol-gel transition temperature above body temperature while solutions that are more concentrated have gelation temperature between room and body temperature (Hrubý et al., 2015).

The hydrogels presents a hydrophilic polymer chains, which allows water absorption that permit the swelling of polymers keeping the polymer structure (Jeong et al., 2005; Lefnaoui and Moulai-Mostefa, 2014; Sinko, 2005). The hydrogels systems have many advantages, such as biocompatibility, high porosity, biodegradable, and easily deformable. They are most biocompatible with the body system because of the high concentration of water, similarity composition, and mechanical properties of the extracellular matrix. The highly porous structure allows the control of density of crosslinks, and the drug releases in a dependent manner. The mainly disadvantages is that the hydrogels is a deformable system, so can adopt the conformation of the local that they are (Hoare and Kohane, 2008; Hoffman, 2002).

Many times, the ability of the drug delivery system to be retained at the site of application is a key parameter. The bioadhesion and the rheological characteristics of the systems are properties, which can enhance the retention time of dosage forms at the site of application. The adhesion (intimate contact) between the polymeric system and the biological substrate will be facilitated by initial wetting of the dosage form on the site, by the interpenetration of the polymeric chains within the implant into the skin or mucus layer and the bond formation (e.g., van der Waals/hydrogen bonding).

Another strategy by which the retention time can be overcome involves the formulation of systems whose rheological properties are temperature dependent, rheological structuring occurring at body temperature, and which, in addition, offer bioadhesive properties. Therefore, the selection of the polymeric components plays an important role on the development of a bioadhesive thermoresponsive system, due to the importance of chemical structure on the bioadhesive interaction, as well as their rheological properties.

6 CHARACTERIZATION OF MUCOADHESIVE THERMORESPONSIVE SYSTEMS

During the development of bioadhesive thermoresponsive systems, several properties must be observed, such as easy removal of packing material, good spreadability, good bio/mucoadhesion over mucosal barriers, suitable gelation temperature, in order to contribute to the ease administration and adherence to treatment. Therefore, the obtaining of physicochemical properties, through mechanical and rheological analysis, allow to predict and control the optimal performance of the formulation (Baloglu et al., 2011; Bruschi et al., 2007; Jones et al., 1996; Pereira et al., 2013).

The mechanical properties are essential to the performance of polymer blends. Since they represent the systems answer to external mechanical influences expressed by the ability to develop reversible and irreversible deformation. The texture profile analysis is a simple and fast technique that can be applied in mechanical characterization of polymeric systems (Baloglu et al., 2011; Bruschi et al., 2007; Chang et al., 2002b; Jones et al., 1996; Pereira et al., 2013).

Moreover, the rheological characteristics need to be explored and understood. Since polymer blends exhibit complex flow behavior due to the interaction between the components, especially for thermoresponsive systems composed by bioadhesive polymers (Bruschi et al., 2007; Chang et al., 2002b; Jones et al., 1996).

6.1 RHEOLOGICAL ANALYSIS IN DEVELOPMENT AND CHARACTERIZATION OF MUCOADHESIVE THERMORESPONSIVE SYSTEMS

The rheological analysis have been employed in several studies with development and characterization of bioadhesive thermoresponsive systems, for ophthalmic, vaginal, and endodontic application (Asasutjarit et al., 2011; Bruschi et al., 2007; Chang et al., 2002a,b; Fabri et al., 2011; Gratieri et al., 2010; Han et al., 2006; Jones et al., 2009; Liu et al., 2009; Mayol et al., 2008; Pereira et al., 2013).

In this sense, the evaluation of rheological properties of these systems enables a greater understanding of the interaction between the mucoadhesive and thermoresponsive polymers, in order to select the formulations for particular application with better clinical performances. Furthermore, the assessing of the wide range of stresses that systems are subjected, during the manufacture, filling, storage, removing of the container before the application, as well as the in situ behavior, should be performed in order to prevent disruption of the system, for patient acceptability, and therapeutic efficacy (Baloglu et al., 2011; Carvalho et al., 2013; Jones et al., 2009; Lee et al., 2009).

Thus, a viscometer or rheometer apparatus can obtain the rheological characteristics of mucoadhesive thermoresponsive formulations. Both equipments are able to perform continuous shear rheometry (flow rheometry). However, some rheometer display controlled stress consequently; it is possible to evaluate the viscoelastic properties of formulations, gelation temperature, and interaction parameter, as well as, mucoadhesion strength.

Flow rheology describes the flow behavior of a formulation under a shear rate range and, then a shear rate versus shear stress can be obtained. Thereby, it is possible to classify (Aulton, 2005):

- Newtonian: shear rate is directly proportional to shear stress, and viscosity is constant with shear rate increases.
- Nonnewtonian substances: shear rate is not directly proportional to shear stress. Viscosity differs according to shear rate. Thus, they can be dilatant or shear-thinning.

Generally, mucoadhesive thermoresponsive systems display shear-thinning behavior, since it is a desirable and suitable property for pharmaceutical formulations. So, these systems display decrease of viscosity after the exposure to a shear stress, consequently the formulation flow readily, facilitating clinical administration uniformly over mucous tissue. However, under low shear conditions in application site, formulations return to original conformation (Bruschi, 2006; Carvalho et al., 2013; Fabri et al., 2011; Pereira et al., 2013). Beyond the flow behavior, it is essential to evaluate if the formulation demonstrate yield value that highlights the organization and interaction between the polymers, as well as the residence time of formulation (Almeida et al., 2013b; Curran et al., 2002; Schoenwald et al., 1978), while thixotropy and rheopexy also reveal the structuring of systems and a possible malleability or fragility, respectively (Carvalho et al., 2013; Fabri et al., 2011; Lee et al., 2009). Hence, this analysis evaluates the behavior of formulations in a range of conditions, since the manufacture until the application (Carvalho et al., 2013; Jones et al., 2009; Lee et al., 2009).

Under physiological and environmental conditions, the polymeric systems are exposed to shear stress that is insufficient to cause deformation and, thereby, it is important to examine the effects of these forces over rheological properties of formulations (Gratieri et al., 2010; Jones et al., 2009). So, the oscillatory rheology determines the viscoelastic properties of formulations by elastic (G') and viscous modulus (G"), and submits formulations to low shear stress, whereas keeps the structure intact and enables to simulate the formulations behavior under biological environment (Baloglu et al., 2011).

Therefore, the gelation temperature can be determined by oscillatory rheology, so frequency and shear stress are controlled, and formulation is submitted to a temperature scan. Temperature caused a drastic change in rheological behavior and elastic modulus is called gelation temperature. Before, gelation temperature, G' is low and formulation is an elastic liquid. In the gelation temperature, elastic modulus increases and the sample behave as a gel (Baloglu et al., 2011; Bruschi, 2006).

Moreover, rheological characterization is important to evaluate the behavior of formulations, since they are liquid at 5°C (during the storage), semisolid at 37°C in the application site (Ferreira et al., 2015).

6.2 TEXTURAL ANALYSIS

Throughout the development of the systems, containing mucoadhesive polymers should be evaluated the resistance of the blends to mucociliary clearance, which is the body's natural defense against the deposition impurities on the mucosal tissue and that can degrade the system on the application site (Carvalho et al., 2010). Under these conditions, the parameters in vivo can offer many variations in therapeutics effectiveness. For this reason, the texture profile analysis (TPA) allowed the evaluated of proprieties to ensure information of physical structure of the gels, furthermore, is considered a rapid, straightforward analytical technique to characterization of semisolid systems (Baloglu et al., 2011; Jones et al., 1996).

According to Szczesniak (1963), texture is defined as the characteristics disposition or connection of threads in a woven fabric or the disposition or manner of union of the particles of body or substance.

TPA is a penetrometric technique that has been extensively employed in the mechanical characterization of food materials and has emerged as a useful in the field of pharmaceutical characterizations. This is a versatile instrument, capable of measuring forces in extension or compression, being programmed with different hold time and analytical probe (Carvalho et al., 2013). From the resultant force-time curve, mechanical parameters can be calculated: hardness, compressibility, adhesiveness, elasticity, and cohesiveness (Jones et al., 1996).

Hardness is the force required to attain a given deformation and expresses the applicability of the gel to desired site. While compressibility is the force per unit time require deforming the system during the first compression cycle of the probe, and determines that removal of the prepared gel from the container and its spreadability on the application site. Both parameters depend of polymeric concentration and include the stress or work to remove the gel from the packaging material and administer on the application site. With these properties is possible quantify, the deformation of the sample on compression. Thus, both values should be lower to remove the gel from packaging material, to apply on the mucosal tissue and spread to epithelial tissue (Baloglu et al., 2011; Tuğcu-Demiröz et al., 2013).

On the other hand, adhesiveness represents a desire parameter in the development of bio/mucoadhesive systems, since the higher, the adhesion the better is the retention of the gel on the mucosal surface. This parameter can be defined as the work necessary to overcome the attractive forces between the surface of the product and the surface of the other materials with which the product comes in contact, being determined from the area under the force-distance curve in the graphic tension (Andrews et al., 2009; Baloglu et al., 2011). The adhesiveness depends of effects of adhesive and cohesive force, and this information can be related with the bioadhesion of the gels (Baloglu et al., 2011; Jones et al., 2010; Tuğcu-Demiröz et al., 2013).

Elasticity is described as the rate at which a deformed material goes back to its undeformed condition after the deforming force is removed, is determined as the ratio between the time when the maximum force occurs on the second compression by the time the maximum force of the first compression (Jones et al., 1996). Lower values of this parameter display better elasticity, because the ratio of the necessary times to the systems return the initial three-dimensional structure is smaller (Ferreira et al., 2015).

Cohesiveness allows an estimation of the extent of structural reformation following product administration, a factor, which will influence product performance. This parameter related to the structural reformation following successive shearing stress during application. The cohesiveness of the polymer is high if the attractive force of its own molecules is high in the gel. This parameter increases the performance of the product at the application site. The high value of cohesiveness provides full structural recovery after the gel application (Tuğcu-Demiröz et al., 2013).

6.3 MUCOADHESIVE ANALYSIS

The bioadhesive test is a complemented with rheological and mechanical (texture profile) characterization, techniques that can predict the flow and viscoelastic properties of semisolid systems (Carvalho et al., 2010). However, an analysis specially developed for mucoadhesion analysis does not exist. The majority of tests are adaptations of other tests and are useful to select the mucoadhesive materials and to elucidate the mechanisms of action (Carvalho et al., 2014).

The texture profile analysis (TPA) test is widely used by itself to characterize the bioadhesion, which is correlated with the work done to overcome the attractive forces between the sample and the surface of the probe (Carvalho et al., 2013).

The pharmaceutical dosage is a limiting factor for the choice of the method to be used, for example, tablets cannot be evaluated by rheological method (Woertz et al., 2013).

The most important methodology occurs in the texture analyzer. In this case, it is measure the force required to break the bonds between the model membrane and the mucoadhesive system. To model the membrane, mucin disc or mucosal membrane animal are used generally, which are fixed on to the probe and will stay in contact with the sample for a predetermined time (Carvalho et al., 2010, 2014). The results are obtained by force-time graphic, the area under curve is considered the mucoadhesive work, and the maximum pick is the maximum force obtains in shift (Carvalho et al., 2014; Woertz et al., 2013).

Another method uses a rheometer apparatus in oscillatory mode which is based in the macroscopy analysis of mucin-polymer interaction. The polymer is analyzed and the mucin are premixed for rheological test. Thus, the mucoadhesive force by monitoring of viscometric changes the system, when the result of the mixture is greater than the polymer and mucin results isolate, it is considered that there was a rheological synergism (Carvalho et al., 2010).

6.4 THE IN VITRO RELEASE ASSAYS

The in vitro release is a fundamental step for the development of a nanostructured drug delivery system. Generally, in vitro tests provide prior understanding of the formulation behavior and present good correlation with in vivo tests. However, it does not replace the need for clinical work, but has the advantage of previous knowledge of some factors of the drug release kinetics, a principle that occurs during the development of the formulation (Florence and Attwood, 2006).

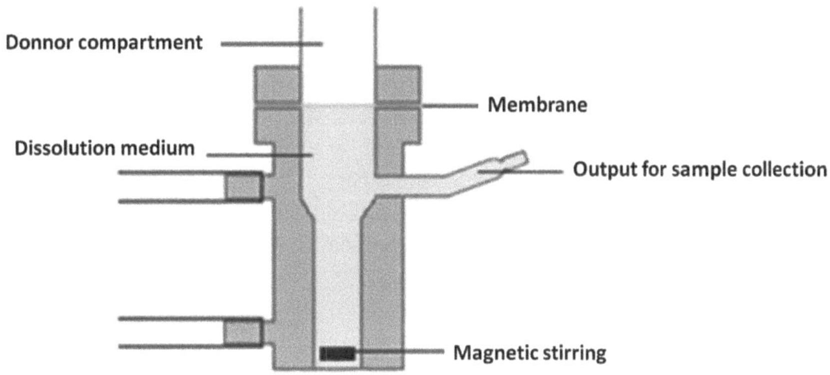

FIGURE 12.9 Scheme of the Release Apparatus

To investigate the mechanism of drug release, the data generated from these release studies can be fitted to the general release equation (Eq. 12.3) using logarithmic transformations and least-squares regression analysis (Bruschi et al., 2007; Fabri et al., 2011; Jones et al., 2009; Korsmeyer et al., 1983)

$$M_t / M_\infty = kt^n \tag{12.3}$$

Where:
M_t is the amount of drug released at time t,
$M\infty$ is the total drug content,
k is a constant incorporating structural and geometric characteristic of the device,
n is the release exponent, which may indicate the mechanism of drug release.

The mechanism of drug release that is under study can be analyzed by general equation, which describes the drug release from the polymeric systems matrix, with the use of analysis of the squares and logarithmic transformations. Thus $n = 0.5$ indicates controlled release by Fickian diffusion, $n = 1$ indicates a controlled release entirely by relaxation of polymer chains (transport case II) and intermediate values between 0.5 and 1 indicate an anomalous behavior (kinetic not Fickian not referred to the phenomena of diffusion and relaxation of polymer chains. In addition, values of $n > 1$ are linked to kinetic type Super Case II (Higuchi, 1963; Ritger and Peppas, 1987).

The Franz diffusion cell is one of the most used methods for in vitro permeation assay. The release profile can also be determined in the release apparatus by modified Franz diffusion cells. This apparatus consists of a glass circular cell, for a semisolid formulation study, with a volume capacity previously determined, with a temperature controller, by a thermostatic bath (Fig. 12.9).

6.5 PERMEATION TESTS

The advantages of the in vitro methods for the in vivo are that the first can be controlled on the study conditions, with no biological interfering, besides, are cheap, and easily performed (Allen et al., 2004).

Several studies containing experimental in vivo, in situ and in vitro methods have been developed to determine the permeability of drugs in the different systems, such as cultured cells, isolated tissues, and membranes (Balimane et al., 2000; Stewart et al., 1997). The animal tissue can be used in the in vitro studies using vertical or horizontal distribution chambers containing two compartments: a donor, and recipient.

The most commonly used test for permeation assay are the diffusion cells, especially the Franz diffusion cells, equipped with excised skin specimen, artificial skin, or different membranes, and isolated perfused tissue models (Wagner et al., 2001), separated by a membrane or tissue under study, which may be from intestinal tissue, rat skin, pig ear among others.

The intestinal rat segments are used frequently to establish a correlation between the drug permeability data with those observed in humans. The intestinal segments of rats are used frequently to establish a correlation of drug permeability data with those observed in humans (Ruan et al., 2006).

The permeation can be evaluated in normal mouse skin, that might be examined the penetration in the skin of hairless mice using different delivery vehicles (Boiy et al., 2007).

Cutaneous applications of drugs have been extensively studies in recent research. Some advantages of this route are the avoidance of first pass metabolism and the minimization of side effects. Some studies were carried out using skin of pig ears excised from the dorsal surface, with thickness normalized, also using in a Franz diffusion cell (1.57 cm^2 diffusion surface area), and maintained at 37°C by a circulating water bath (Lonni et al., 2015).

The rectal route is not the route of first choice, but it has many benefits to the patient, including local or systemic effects. In addition, systemic treatment using the rectal route is also a good alternative for treatment (Junqueira et al., 2016), this route can quickly reach systemic effect and is an effective route of administration for various compounds as analgesics, sedatives, anticonvulsants, antiinflammatory, antibiotics, antiepileptics (Van-Rijt and Sadler, 2009).

The second Fick's law is used generally in experiments where it employs a membrane separating two compartments, one of them with the highest concentration of the drug (donor compartment) and the other (receiver compartment) to zero concentration (sink conditions) thereby maintaining the gradient concentration, the cumulative amount of drug (m) passing through the membrane per unit area, and as a time function of until reaching the equilibrium state, and can be expressed in Eq. (12.4) below:

$$(\mathrm{dm/dt}) = DC_0K/h \tag{12.4}$$

Where:
C_0 is a constant drug concentration in the donor solution,
K is the partition coefficient of the solute between the vehicle and the membrane,
h is the membrane thickness.

When plotted on a graph over time of absorption (T), and cumulative amount absorbed (m) per unit of membrane area as a function of time, the balance is checked when the graph becomes linear, and this linear portion of the data extrapolated in order to obtain the intercept of the curve when $m = 0$ provide the lag time (time it takes to begin to occur at absorption) (Aulton, 2005).

7 CONCLUSIONS

Mucoadhesive thermoresponsive platforms constitute a potentially useful therapeutic modality of systems. They are generally composed by polymeric blends. Mucoadhesive materials allow the intimacy contact to specific sites and prolonged residence time. For this purpose, some polymers are used, such as poly(acrylic acid) derivatives, chitosan, alginate, cellulose derivatives, guar gum, and carrageenan.

Moreover, thermoresponsive systems are environmental-sensitive polymers, which can change their conformational structure in response to temperature, resulting in situ gelling systems. For this purpose, the poloxamers constitute the most commonly employed polymer class. Therefore, the association of these systems with bioadhesive materials provides an improved adhesion, easy administration, no irritancy, and can be used to delivery active agents to different mucous regions. Hydrogels are more frequently used, and are characterized by rheological, textural, mucoadhesive analysis, in vitro release assays, and permeation tests. In this context, the strategy to use bioadhesive and thermoresponsive polymers enables to produce versatile drug delivery systems. They have applications for many diseases, such as vulvovaginal candidiasis, periodontal disease, treatment of lesions caused by Herpes simplex virus, treatment of pain, and inflammation of esophagus and colorectal cancer as a promising tool to photodynamic therapy.

REFERENCES

Allen, Jr., L.V., Popovich, N.G., Ansel, H.C., 2004. Ansel's pharmaceutical dosage forms and drug delivery systems, eigth ed. Lippincott Williams & Wilkins, Philadelphia, 720 p.

Almeida, H., Amaral, M.H., Lobão, P., Sousa Lobo, J.M., 2013a. Applications of poloxamers in ophthalmic pharmaceutical formulations: an overview. Expert Opin. Drug Deliv. 10 (8), 1–15.

Almeida, H., Amaral, M.H., Lobão, P., Lobo, J.M.S., 2013b. In situ gelling systems: a strategy to improve the bioavailability of ophthalmic pharmaceutical formulations. Drug Discov. Today 19 (4), 400–412.

Andrews, G.P., Laverty, T.P., Jones, D.S., 2009. Mucoadhesive polymeric platforms for controlled drug delivery. Eur. J. Pharm. Biopharm. 71 (3), 505–518.

Asasutjarit, R., et al., 2011. Optimization and evaluation of thermoresponsive diclofenac sodium ophthalmic in situ gels. Int. J. Pharm. 411, 128–135.

Aulton, M.E., 2005. Pharmaceutics: The Science of Dosage Form Design, second ed. Churchill Livingstone, Edinburgh, 679 p.

Bajpai, A.K., et al., 2008. Responsive polymers in controlled drug delivery. Prog. Polym. Sci. 33 (11), 1088–1118.

Balimane, P.V., Chong, S., Morrisson, R.A., 2000. Current methodologies used for evaluation of intestinal permeability and absorption. J. Pharmacol. Toxicol. Methods 44, 301–312.

Baloglu, E., et al., 2011. Rheological and mechanical properties of poloxamer mixtures as a mucoadhesive gel base. Pharm. Dev. Technol. 16, 627–636.

Berginc, K., et al., 2014. Mucoadhesive liposomes as new formulation for vaginal delivery of curcumin. Eur. J. Pharm. Biopharm. 87, 40–46.

Bhattarai, N., Gunn, J., Zhang, M., 2010. Chitosan-based hydrogels for controlled, localized drug delivery. Adv. Drug Deliv. Rev. 62 (1), 83–99.

Bhowmik, M., et al., 2013. Effect of xanthan gum and guar gum on in situ gelling ophthalmic drug delivery system based on poloxamer-407. Int. J. Biol. Macromol. 62, 117–123.

Boiy, A., Roelandts, R., De Witte, P.M.A., 2007. Influence of application and formulation factors on the penetration of Hipericin in normal mouse skin and UV induced skin tumors. J. Photochem. Photobiol. B. Biol. 89 (2–3), 156–162.

Bonacucina, G., Martelli, S., Palmieri, G.F., 2004. Rheological, mucoadhesive and release properties of Carbopol gels in hydrophilic cosolvents. Int. J. Pharm. 282, 115–130.

Borghi-Pangoni, F.B., et al., 2015. Screening and in vitro evaluation of mucoadhesive thermoresponsive system containing methylene blue for local photodynamic therapy of colorectal cancer. Pharm. Res. 33 (3), 776–791.

Bruschi, M.L., 2006. Desenvolvimento e caracterização de sistemas de liberação de própolis intrabolsa periodontal. Doctoral Thesis at Faculdade de Ciências Farmacêuticas, Ribeirão Preto. p.318.

Bruschi, M.L., et al., 2006. Sistemas de liberação de fármaco intrabolsa periodontal. Revista Brasileira de Ciências Farmacêuticas 42 (1), 29–47, (In Portuguese).

Bruschi, M.L., et al., 2007. Semisolid systems containing propolis for the treatment of periodontal disease: in vitro release kinetics, syringeability, rheological, textural, and mucoadhesive properties. J. Pharm. Sci. 96 (8), 2074–2089.

Bruschi, M.L., et al., 2015. An overview of recent patents on composition of mucoadhesive drug delivery systems. Recent Pat. Drug. Deliv. Form. 9, 79–87.

Carvalho, F.C., et al., 2010. Mucoadhesive drug delivery systems. Braz. J. Pharm. Sci. 46 (1), 1–17.

Carvalho, F.C., et al., 2013. Rheological, mechanical, and bioadhesive behavior of hydrogels to optimize skin delivery systems. Drug Dev. Ind. Pharm. 39, 1750–1757.

Carvalho, F.C., et al., 2014. Plataformas bio(muco) adesivas poliméricas baseadas em nanotecnologia para liberação controlada de fármacos – propriedades, metodologias e aplicações. Polímeros 24 (2), 203–213, (In Portuguese).

Chang, J.Y., Oh, Y.-K., et al., 2002a. Prolonged antifungal effects of clotrimazole-containing mucoadhesive thermosensitive gels on vaginitis. J. Control. Rel. 82, 39–50.

Chang, J.Y., Oh, Y.K., et al., 2002b. Rheological evaluation of thermosensitive and mucoadhesive vaginal gels in physiological conditions. Int. J. Pharm. 241, 155–163.

Chang, Y., et al., 2010. Tunable bioadhesive copolymer hydrogels of thermoresponsive poly(N-isopropyl acrylamide) containing zwitterionic polysulfobetaine. Biomacromolecules 11 (4), 1101–1110.

Chavda, H., et al., 2013. Development of bioadhesive chitosan superporous hydrogel composite particles based intestinal drug delivery system. BioMed Res. Int., 1–10.

Chawla, V., Saraf, S.a., 2012. Rheological studies on solid lipid nanoparticle based carbopol gels of aceclofenac. Colloids Surf. B Biointerf. 92, 293–298.

Chen, G., Hoffman, A.S., 2002. Block and graft copolymers and methods relating thereto. US Patent 6,486,213B.

Chickering, D.E., Mathiowitz, E., 1995. Bioadhesive microspheres: I. A novel electrobalance-based method to study adhesive interactions between individual microspheres and intestinal mucosa. J. Control. Rel. 34 (3), 251–262.

Cleary, J., Bromberg, L., Magner, E., 2004. Adhesion of polyether-modified poly (acrylic acid) to mucin. Langmuir 20, 9755–9762.

Collet, J., Moreton, C., 2005. Modified-release peroral dosage forms. Delineamento de Formaspp. 298–313.

Curran, S.J., et al., 2002. Properties of carbopol solutions as models for yield-stress fluids. J. Food Sci. 67 (1), 176–180.

Dumortier, G., et al., 2006. A review of poloxamer 407 pharmaceutical and pharmacological characteristics. Pharm. Res. 23 (12), 2709–2728.

Dupuy, B., Arien, A., Perrot Minnot, A., 1994. FT-IR of membranes made with alginate/polylysine complexes. Variations with the mannuronic or guluronic content of the polysaccharides. Artif. Cells Blood Substit. Immobil. Biotechnol. 22 (1), 71–82.

Eve Ruel-Gariépy, J.-C.L., 2004. In situ-forming hydrogels—review of temperature-sensitive systems. Eur. J. Pharm. Biopharm. 58 (2), 409–426.

Fabri, F.V., et al., 2011. Preparation and characterization of bioadhesive systems containing propolis or sildenafil for dental pulp protection. Drug Dev. Ind. Pharm. 37, 1446–1454.

Ferreira, S.B., de, S., et al., 2015. Rheological, mucoadhesive and textural properties of thermoresponsive polymer blends for biomedical applications. J. Mech. Behav. Biomed. Mater. 55, 164–178.

Florence, A.T., Attwood, D., 2006. Physicochemical principles of pharmacy, fourth ed. Pharmaceutical Press, London, UK.

George, M., Abraham, T.E., 2006. Polyionic hydrocolloids for the intestinal delivery of protein drugs: alginate and chitosan—a review. J. Control. Rel. 114 (1), 1–14.

Gombotz, W.R., Wee, S.F., 2012. Protein release from alginate matrices. Adv. Drug Deliv. Rev. 64 (SUPPL.), 194–205.

Gómez-Carracedo, A., et al., 2004. Glass transitions and viscoelastic properties of Carbopol and Noveon compacts. Int. J. Pharm. 274, 233–243.

Gratieri, T., et al., 2010. A poloxamer/chitosan in situ forming gel with prolonged retention time for ocular delivery. Eur. J. Pharm. Biopharm. 75 (2), 186–193.

Guo, J., 2008. Carbopol polymers for pharmaceutical drug delivery applications. Drug Dev. Deliv. 3 (6), 1–3.

Gupta, N.V., et al., 2013. Bioadhesive vaginal tablets containing spray dried microspheres loaded with clotrimazole for treatment of vaginal Candidiasis. Acta Pharm. 63, 359–372.

Hägerström, H., Edsman, K., 2001. Interpretation of mucoadhesive properties of polymer gel preparations using a tensile strength method. J. Pharm. Pharmacol. 53, 1589–1599.

Hägerström, H., Edsman, K., Strømme, M., 2003. Low-frequency dielectric spectroscopy as a tool for studying the compatibility between pharmaceutical gels and mucous tissue. J. Pharm. Sci. 92 (9), 1869–1881.

Han, I.K., et al., 2006. Thermosensitive and mucoadhesive delivery systems of mucosal vaccines. Methods 38, 106–111.

Higuchi, T., 1963. Mechanism of sustained-action medication: theoretical analysis of rate of release of solid drugs dispersed in solid matrices. J. Pharm. Sci. 52 (12), 1145–1149.

Hirata, A.N., Bruschi, M.L., 2010. Development and characterisation of semisolid systems to deliver propolis in the oral cavity. Revista de Ciências Farmacêuticas Básica e Aplicada 31 (1), 1–8.

Hoare, T.R., Kohane, D.S., 2008. Hydrogels in drug delivery. Polymer 49, 1993–2007.

Hoffman, A.S., 2002. Hydrogels for biomedical applications. Adv. Drug Deliv. Rev. 43, 3–12.

Hoffman, A.S., 2013. Stimuli-responsive polymers: biomedical applications and challenges for clinical translation. Adv. Drug Deliv. Rev. 65 (1), 10–16.

Hosmani, A., Thorat, Y., Kasture, P., 2006. Carbopol and its pharmaceutical significance: a review. Pharmainfo. Net 4, 1–22.

Hrubý, M., Filippov, S.K., Štěpánek, P., 2015. Smart polymers in drug delivery systems on crossroads: which way deserves following? Eur. Polym. J. 65, 82–97.

Huang, Y., et al., 2000. Molecular aspects of muco- and bioadhesion. J. Control. Rel. 65 (1–2), 63–71.

Illum, L., et al., 2001. Chitosan as a novel nasal delivery system for vaccines. Adv. Drug Deliv. Rev. 51 (1–3), 81–96.

Islam, M.T., et al., 2004. Rheological characterization of topical carbomer gels neutralized to different pH. Pharm. Res. 21 (7), 1192–1199.

Ivarsson, D., Wahlgren, M., 2012. Comparison of in vitro methods of measuring mucoadhesion: ellipsometry, tensile strength and rheological measurements. Colloid Surf. B Biointerf. 92, 353–359.

Jeong, S.H., Fu, Y., Park, K., 2005. Hydrogels for oral administration. In: Kwon, G.S. (Ed.), Polymeric drug delivery systems. Taylor & Francis Group, Boca Raton, FL, pp. 195–214.

Jones, D.S., Woolfson, a D., Djokic, J., 1996. Texture profile analysis of bioadhesive polymeric semisolids: mechanical characterization and investigation of interactions between formulation components. J. Appl. Polym. Sci. 61 (12), 2229–2234.

Jones, D.S., et al., 2009. Rheological, mechanical and mucoadhesive properties of thermoresponsive, bioadhesive binary mixtures composed of poloxamer 407 and carbopol 974P designed as platforms for implantable drug delivery systems for use in the oral cavity. Int. J. Pharm. 372 (1–2), 49–58.

Jones, D.S., et al., 2010. Moisture-activated rheological struturing of nonaqueous poloxamine-poly(acrylic acid) systems designed as novel biomedical implants. J. Pharm. Sci. 99 (4), 1838–1854.

Junginger, H.E., Thanou, M., Verhoef, J.C., 2001. Mucoadhesive hydrogels in drug delivery. In: Swarbrick, J., Boylan, J.C. (Eds.), Encyclopedia of pharmaceutical technology. Marcel Dekker, New York, NY, pp. 169–192.

Junqueira, M.V., et al., 2016. Functional polymeric systems as delivery vehicles for methylene blue in photodynamic therapy. Langmuir 32 (1), 19–27.

Kabanov, A.V., Batrakova, E.V., Alakhov, V.Y., 2002. Pluronic block copolymers as novel polymer therapeutics for drug and gene delivery. J. Control. Rel. 82, 189–212.

Kehoe, S., et al., 2012. Characterization of PLGA based composite nerve guidance conduits: effect of F127 content on modulus over time in simulated physiological conditions. J. Mech. Behav. Biomed. Mater. 14, 180–185.

Kerec, M., et al., 2002. Mucoadhesion on pig vesical mucosa: Influence of polycarbophil/calcium interactions. Int. J. Pharm. 241, 135–143.

Khutoryanskiy, V.V., 2011. Advances in mucoadhesion and mucoadhesive polymers. Macromol. Biosci. 11 (6), 748–764.

Kikuchi, A., Okano, T., 2005. Hydrogels-stimuli sensitive hydrogels. In: Kwon, G.S. (Ed.), Polymeric drug delivery systems. Taylor & Francis Group, Boca Raton, FL, pp. 275–322.

Korsmeyer, R.W., Gurny, R., Doelker, E., Buri, P., Peppas, N.A., 1983. Mechanisms of solute release from porous hydrophilic polymers. Int. J. Pharm. 15, 25–35.

Kwon, Y.M., Kim, S.W., 2005. Thermosensitive biodegradable hydrogels for the delivery of therapeutic agents. In: Kwon, G.S. (Ed.), Polymeric drug delivery systems. Taylor & Francis Group, Boca Raton, FL, pp. 251–274.

Kwon, J., et al., 2015. A novel synthesis method for an open-cell microsponge polyimide for heat insulation. Polymer 56, 68–72.

Lee, C.H., Moturi, V., Lee, Y., 2009. Thixotropic property in pharmaceutical formulations. J. Control. Rel. 136 (2), 88–98.

Lefnaoui, S., Moulai-Mostefa, N., 2014. Investigation and optimization of formulation factors of a hydrogel network based on kappa carrageenan-pregelatinized starch blend using an experimental design. Colloid Surf. A Physicochem. Eng. Aspects 458, 117–125.

Lin, C.C., Metters, A.T., 2006. Hydrogels in controlled release formulations: network design and mathematical modeling. Adv. Drug Deliv. Rev. 58, 1379–1408.

Liu, Y., et al., 2009. Effect of carrageenan on poloxamer-based in situ gel for vaginal use: Improved in vitro and in vivo sustained-release properties. Eur. J. Pharm. Sci. 37 (3–4), 306–312.

Lonni, A.A.S.G., Munhoz, V.M., Lopes, G.C., Longhini, R., Borghi-Pangoni, F.B., Dos Santos, R.S., Junqueira, M.V., Natali, M.R., Leite-Mello, E., Guimaraes, F.B., Scarminio, I.S., Bruschi, M.L., Mello, J.C., 2015. Development and characterization of multiple emulsions for controlled release of *Trichilia catigua* (Catuaba) extract. Pharm. Dev. Tech. 11, 1083–1097.

Lu, G., Jun, H.W., 1998. Diffusion studies of methotrexate in Carbopol and Poloxamer gels. Int. J. Pharm. 160, 1–9.

Lubrizol, 2013a. Carbopol Polymers. Lubrizol Advanced Materials, Cleveland, OH.

Lubrizol, 2013b. Noveon AA-1 Polycarbophil. Lubrizol Advanced Materials, Cleveland, OH.

Lubrizol, 2013c. Pharmaceutical Polymers Typical Properties and Specifications. Lubrizol Advanced Materials, Cleveland, OH, 1–3.

Ludwig, A., 2005. The use of mucoadhesive polymers in ocular drug delivery. Adv. Drug Deliv. Rev. 57 (11), 1595–1639.

Madsen, K.D., et al., 2013. Development of an ex vivo retention model simulating bioadhesion in the oral cavity using human saliva and physiologically relevant irrigation media. Int. J. Pharm. 448 (2), 373–381.

Makó, Á., et al., 2009. Formulation of thermoresponsive and bioadhesive gel for treatment of oesophageal pain and inflammation. Eur. J. Pharm. Biopharm. 72 (1), 260–265.

Matanović, M.R., Kristl, J., Grabnar, P.A., 2014. Thermoresponsive polymers: Insights into decisive hydrogel characteristics, mechanisms of gelation, and promising biomedical applications. Int. J. Pharm. 472 (1–2), 262–275.

Mayol, L., et al., 2008. A novel poloxamers/hyaluronic acid in situ forming hydrogel for drug delivery: rheological, mucoadhesive and in vitro release properties. Euro. J. Pharm. Biopharm. 70, 199–206.

Mazia, R.S., et al., 2016. Formulation and evaluation of a mucoadhesive thermoresponsive system containing brazilian green propolis for the treatment of lesions caused by herpes simplex type I. J. Pharm. Sci. 105 (1), 113–121.

Mirza, M.A., et al., 2013. Development of a novel synergistic thermosensitive gel for vaginal candidiasis: An in vitro, in vivo evaluation. Colloid Surf. B Biointerf. 103, 275–282.
Na, K., Bae, Y.H., 2005. pH-sensitive polymers for drug delivery. In: Kwon, G.S. (Ed.), Polymeric drug delivery systems. Taylor & Francis Group, Boca Raton, FL, pp. 129–194.
Nho, Y.-C., Park, J.-S., Lim, Y.-M., 2014. Preparation of poly(acrylic acid) hydrogel by radiation crosslinking and its application for mucoadhesives. Polymers 6 (3), 890–898.
Peppas, N.a., et al., 2000. Hydrogels in pharmaceutical formulations. Eur. J. Pharm. Biopharm. 50 (1), 27–46.
Pereira, R.R.D.A., 2011. Desenvolvimento e caractterização de sistemas mucoadesivo termossensível contendo micropartículas de própolis para potencial tratamento de candidíase vulvovaginal. Master's Thesis at Universidade Estadual de Maringá, Paraná. p.171.
Pereira, R.R.D.A., Bruschi, M.L., 2012. Vaginal mucoadhesive drug delivery systems. Drug Dev. Ind. Pharm. 38 (6), 643–652.
Pereira, R.R.A., et al., 2013. Preparation and characterization of mucoadhesive thermoresponsive systems containing propolis for the treatment of vulvovaginal candidiasis. J. Pharm. Sci., 4215–4227.
Pliszczak, D., et al., 2012. Mucoadhesion evaluation of polysaccharide gels for vaginal application by using rheological and indentation measurements. Colloid Surf. B Biointerf. 92, 168–174.
Prajapati, V., Bansal, M., Sharma, P.K., 2012. Mucoadhesive buccal patches and use of natural polymer in its preparation – a review. Int. J. PharmTech Res. 4 (2), 582–589.
Priya James, H., et al., 2014. Smart polymers for the controlled delivery of drugs – a concise overview. Acta Pharm. Sinica B 4 (2), 120–127.
Qiu, Y., Park, K., 2001. Environment-sensitive hydrogels for drug delivery. Adv. Drug Deliv. Rev. 53, 321–339.
Reineke, J., et al., 2013. Can bioadhesive nanoparticles allow for more effective particle uptake from the small intestine? J. Control. Rel. 170 (3), 477–484.
Ritger, P.L., Peppas, N.A., 1987. A simple equation for description of solute release. II. Fickian and anomalous release from swellable devices. J. Control. Rel. 5, 37–42.
Roge, A.B., et al., 2011. Bioadhesive polymer: review. J. Pharm. Biom. Sci. 05 (05), 1–5.
Rossi, S., et al., 2001. Characterization of chitosan hydrochloride – mucin rheological interaction: influence of polymer concentration and polymer: mucin weight ratio. Eur. J. Pharm. Sci. 12, 479–485.
Rowe, R.C., Sheskey, P.J., Quinn, M.E., 2009. Handbook of pharmaceutical excipients, 6th ed. Pharmaceutical Press.
Roy, S.K., Prabhakar, B., 2010. Bioadhesive polymeric platforms for transmucosal drug delivery systems - a review. Tropical J. Pharm. Res. 9 (1), 91–104.
Ruan, L.P., Chen, S., Yu, B.Y., Zhu, D.N., Cordell, G.A., Qiu, S.X., 2006. Prediction of human absorption of natural compounds by the non-everted rat intestinal sac model. Eur. J. Med. Chem. 41, 605–610.
Schoenwald, R., et al., 1978. Influence of high-viscosity vehicles on miotic effect of pilocarpine. J. Pharm. Sci. 67 (9), 1280–1283.
Shaikh, R., et al., 2011. Mucoadhesive drug delivery systems. J. Pharm. Bioallied Sci. 3 (1), 89–100.
Sheet, T.D., 2002. Measurement and understanding of yield value in home care formulations. Lubrizol Advanced Materials, 1–3.
Shokri, J., Adibkia, K., 2013. Application of Cellulose and Cellulose Derivatives in Pharmaceutical Industries. In: van de Ven, T., Godbout, L. (Eds.), Cellulose - medical, pharmaceutical and electronic applications. InTech, Rijeka, Croatia, pp. 47–66.
Singh, M., Tiwary, A.K., Kaur, G., 2010. Investigations on interpolymer complexes of cationic guar gum and xanthan gum for formulation of bioadhesive films. Res. Pharm. Sci. 5 (2), 79–87.
Sinko, P.J., 2005. Martín- Físico-Farmácia e Ciências Farmacêuticas, 5th ed. Artmed, Porto Alegre, Rio Grande do Sul.
Skjak-Braek, G., Grasdalen, H., Larsen, B., 1986. Monomer sequence and acetylation pattern in osme bacterial alginates. Carbohydr. Res. 154, 239–250.

Smart, J.D., 2005. The basics and underlying mechanisms of mucoadhesion. Adv. Drug Deliv. Rev. 57, 1556–1568.

Stewart, B.H., Chan, O.H., Jezyk, N., Fleisher, D., 1997. Discrimination between drug candidates using models intestinal absorption for evaluation of intestinal absorption. Adv. Drug Deliv. Rev. 23, 27–45.

Sudhakar, Y., Kuotsu, K., Bandyopadhyay, A.K., 2006. Buccal bioadhesive drug delivery - a promising option for orally less efficient drugs. J. Control. Rel. 114 (1), 15–40.

Szczesniak, A.S., 1963. Classification of textural characteristics. J. Food Sci. 28 (4), 385–389.

Tamburic, S., Craig, D.Q.M., 1995. An investigation into the rheological, dielectrical and mucoadhesive properties of poly(acrylic acid) gel systems. J. Control. Rel. 37, 59–68.

Tuğcu-Demiröz, F., Acartürk, F., Erdog˘an, D., 2013. Development of long-acting bioadhesive vaginal gels of oxybutynin: Formulation, in vitro and in vivo evaluations. Int. J. Pharm. 457 (1), 25–39.

Valenta, C., 2005. The use of mucoadhesive polymers in vaginal delivery. Adv. Drug Deliv. Rev. 57 (11), 1692–1712.

Van-Rijt, S.H., Sadler, P.J., 2009. Current applications and future potential for bioinorganic chemistry in the development of anticancer drugs. Drug Discov. Today 14 (23–24), 1089–1097.

Veyries, M.L., et al., 1999. Controlled release of vancomycin from poloxamer gels. Int. J. Pharm. 192, 183–193.

Wagner, H., Kostka, K.H., Lehr, C.M., Schaefer, U.F., 2001. Interrelation of permeation and penetration parameters obtained from in vitro experiments with human skin and skin equivalents. J. Control. Rel. 75, 283–295.

Wang, H., et al., 2014. Hybrid elastin-like polypeptide-polyethylene glycol (ELP-PEG) hydrogels with improved transparency and independent control of matrix mechanics and cell ligand density. Biomacromolecules 15 (9), 3421–3428.

Wise, D.L., 2000. Handbook of pharmaceutical release technology. Marcel Dekker, New York, NY.

Woertz, C., et al., 2013. Assessment of test methods evaluating mucoadhesive polymers and dosage forms: An overview. Eur. J. Pharm. Biopharm. 85 (3), 843–853, Available at: http://dxdoiorg/10.1016/jejpb.2013.06.023.

Wypych, T.C., 2009. Desenvolvimento e avaliação de sistemas bucoadesivos contendo diclofenaco sódico. Dissertation at Universidade Federal do Paraná, Paraná.

Yuan, Y., et al., 2012. Thermosensitive and mucoadhesive in situ gel based on poloxamer as new carrier for rectal administration of nimesulide. Int. J. Pharm. 430, 114–119.

Zaki, N.M., et al., 2007. Enhanced bioavailability of metoclopramide HCl by intranasal administration of a mucoadhesive in situ gel with modulated rheological and mucociliary transport properties. Eur. J. Pharm. Sci. 32, 296–307.

Zhu, Z., et al., 2013. The development of polycarbophil as a bioadhesive material in pharmacy. Asian J. Pharm. Sci. 8 (4), 218–227.

CHAPTER

DESIGN CONSIDERATIONS IN THE DEVELOPMENT OF WOUND HEALING BIONANOMATERIALS

13

Parvez Alam
University of Edinburgh, Edinburgh, United Kingdom;
Abo Akademi University, Turku, Finland

CHAPTER OUTLINE

1 INTRODUCTION

Research into wound healing is of high significance due to the ubiquity and seriousness of wounds. Wounds are essentially breaks or lacerations to membranes causing damage to underlying tissues. Such breaks in the membrane may arise through, for example, violent impact, accidents, surgical procedures, or burns. The lack of proper treatment may result in sepsis, which in turn may instigate the need for amputation, or may lead to death. Poorly reformed membranes can result in serious levels of scarring and deformities that alter the properties and function of the final membrane structure. This is particularly evident with burn victims, where there are frequent problems with postclosure wound esthetics. There are several challenges when it comes to wound healing. Bacterial protection is critical since bacterial ingress may lead to infections and further complications in the healing process. Mechanical compatibility is also of considerable importance. Highly differential stresses developing at wound-biomaterial interfaces can cause the wound to rupture prematurely during the healing process that reinstigates problems associated with bacterial ingress and scarring. The elasticity of biomaterials should be designed in view of their compatibility with the skin from the host, which can itself differ in properties as a function of age and condition. Biomaterials must be able to breathe and as such, have a degree of semipermeability, yet they should be water retentive since wounds are less likely to rupture prematurely if the wound is kept moist. Wound healing biomaterials should further be hypoallergenic, nontoxic, and should not be immunogenic. Biocompatibility, ease of bioactivation, biodegradability, and durability are final fundamental elements of importance in the design of biomaterials. Apposite designs are therefore critical when it comes to the design and development of wound healing bionanomaterials. Several biological materials are excellent candidate wound healing materials. Many of them have an intrinsic biocompatibility, and are nontoxic to host or to their environment. Moreover, several of them are renewable, sustainable, biodegradable (RSB) materials, and fulfil current demands for environmentally responsible product development in several economically affluent countries. Nanomaterials are currently researched comprehensively as they will often have a higher-than-normal chemical affinity with respect to volume, and are very reactive at interfaces. Several biological materials are hierarchically structured and will have a nanoscale element to them. This chapter is concerned with the design and understanding of bionanomaterials with respect to their function in wound healing.

2 SKIN: PROPERTIES, GROWTH, AND REGENERATION

The skin is the largest organ of the human body and has several roles to fulfil. It acts as a barrier, offering protection against pathogens, germs, and radiation. It maintains and regulates the body temperature, synthesizes chemicals, and is the primary organ involved in the transfer of signals (pain, heat, pressure, etc.). It is also an organ that self-repairs through various cascade reactions, cuts, lesions, and burns. Finally, it has a waterproofing function yet, is able to absorb oils, which are vital to retaining sound mechanical elasticity. As mentioned before, mechanical compatibility and elasticity are vital to designing sounds biomaterials. Skin comprises three layers: the epidermis, the dermis, and the hypodermis (Pailler-Mattei et al., 2008), that is, a soft subcutaneous layer containing fat (Fig. 13.1) (Kong, 2011). The epidermis is very thin (approximately one tenth of a millimeter) and is itself made up of several layers. These include the stratum basalis, the stratum spinosum, the stratum granulosum, and the stratum corneum (Saxena, 2009). The dermis is considerably thicker (1–2 mm) and controls effectively, the mechanical resistance of the skin. The dermis is a composite comprised of two proteins; collagen and elastin, which provide strength and elasticity respectively to the body of material.

The skin has properties of a viscoelastic material (Barbenel and Evans, 1977) and both the dermis and hypodermis contribute to its mechanical response (Pailler-Mattei et al., 2008). Fig. 13.2 (from Pan et al., 1998) shows how the modulus of skin increases as a function of increasing strain, suggesting that there is a strain stiffening effect in skin, perhaps arising through collagen fiber alignment with the loading axis. Mechanical properties will also change with age, and this change is attributed directly

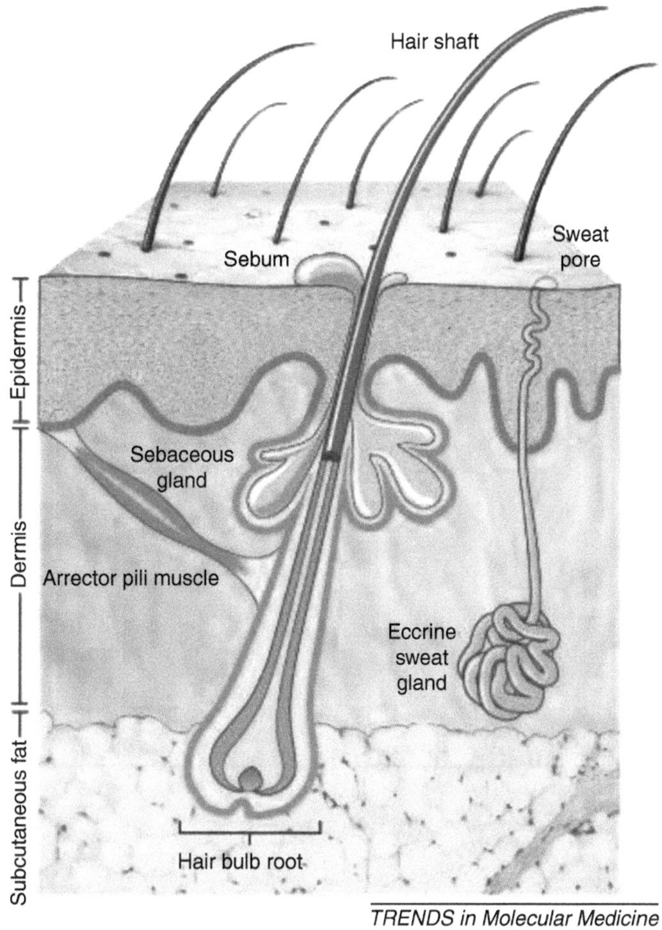

FIGURE 13.1 Representation of Human Skin

The topographical surface of human skin is irregular and consists of multiple skin appendageal structures, including hair follicles, sebaceous (oil) glands, and sweat glands and ducts. Each of the skin appendages provides a potential niche for skin bacteria. The external communication of the skin with the terrestrial environment and the heterogeneity of the appendageal invaginations contribute to the diversity of the skin microbiome.

From Kong, H.H., 2011. Skin microbiome: genomics-based insights into the diversity and role of skin microbes. Trends Mol. Med. 17, 320–328. Reproduced with permission from Elsevier.

to degradation of the elastin matrix around collagen fibers (Daly and Odland, 1979). Scarred skin has higher stiffness and lower elasticity as compared to healthy skin, and this tendency exacerbates with respect to the level of scarring (Clarke et al., 1996). Skin is, furthermore, an auxetic material (Evans and Anderson, 2000), in that it has a negative Poisson's ratio (widens as it is stretched), which in turn has several protective benefits, for example, in pregnancy. This is a difficult feature to guide and control in skin, and may relate to regeneration kinetics (Argyris, 1977) for wounded skin, which can be controlled via biomaterial use (Boucard et al., 2007) to achieve full skin reconstruction.

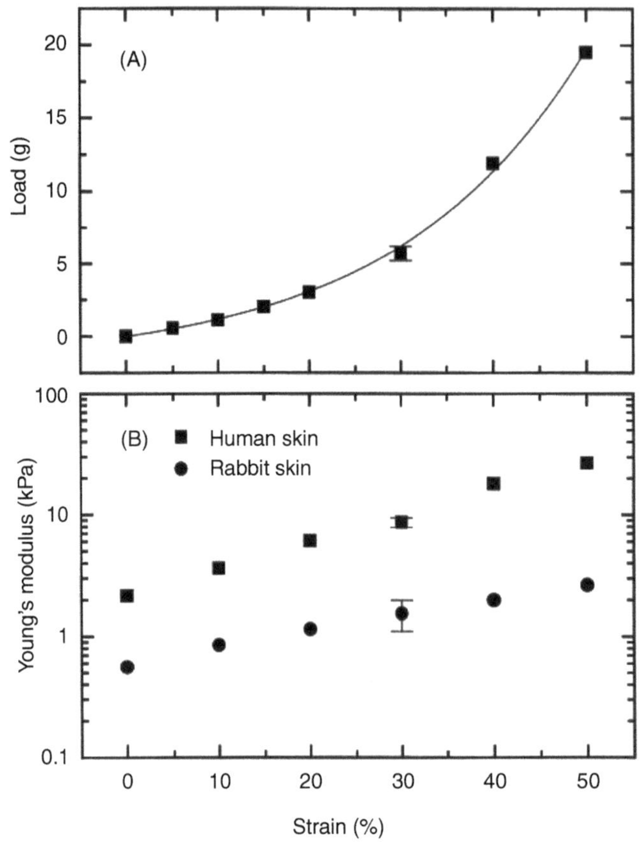

FIGURE 13.2

(A) Load as a function of strain from human skin. The squares are experimental measurements. (B) Young's modulus versus strain. Squares, From the same human tissue sample as in (A); circles, from rabbit skin. The error bars represent measurement uncertainties.

From Pan, L., Zan, L., Foster, F.S., 1998. Ultrasonic and viscoelastic properties of skin under transverse mechanical stress in vitro. Ultrasound Med. Biol. 24, 995–1007. Reproduced with permission from Elsevier.

The means by which the skin will heal naturally will depend on the geometry of the wound. If the wound edges are in close proximity, fibroblasts will occur swiftly throughout the wound, and there is essentially no wound contraction (termed: healing by primary intention, Webster et al., 2014). If, however, the wound is large and the edges are distant (termed: secondary intention), fibroblasts will form at the edges of the wound but will be disconnected from each other. This will cause contraction of the wound, which in turn increases the extent of scarring (Vermeulen et al., 2005). Birk and Trelstad (1996) reported that the final orientation of fibrous collagen in the scar tissue follows the orientations of the fibroblasts during the period of stroma synthesis. Both primary and secondary intentions are examples of skin repair, whereby wounds are closed, as opposed to returning it to a fully functional original state. Skin regeneration is somewhat different, in that regeneration includes a return of organ function, in this case the skin. When considering the use of bionanomaterials in wound healing, one must consider design principles, whereby skin regeneration is encouraged, as opposed to skin repair and closure, and how regeneration takes place. Most wounds will damage blood vessels. If the wound is sufficiently deep, hair follicles and sweat glands do not reform. Clots are a composite of platelets within a cross-linked network of fibrin fibers. Clots also contain several cytokines and growth factors that are released during platelet degranulation. The combination and concentration of growth factors instigates the process of wound closure by chemically inviting circulating inflammatory cells, which in turn initiates tissue movements and connective tissue contractions (Martin, 1997). Wound contraction is seen as an essential requisite to successful wound reepithelization. Yannas (2009) suggests that scar formation occurs through the coupled process of a tensile pull on collagen fibers from wound contraction. Thus excessive orientation of fibers may be a cause for scarred tissue, and by removing the tension from the wound site, theoretically scarring should be lessened. Thus, using a wound healing scaffold should reduce scar formation, and this is confirmed through the same author's earlier works (Yannas et al., 1981, 1982). Myofibroblasts are the critical cell types that permit wound contraction. Myofibroblastic differentiation is controlled by a transforming growth factor TGF-β1 (Desmouliere et al., 2005), which works in conjunction with mechanical tensile forces and extracellular matrix materials. Keratinocyte binding to laminin is a requisite for reepithelization. This is achieved through the expression of new integrins (fibronectin and vitronectin receptors), which allow collagen receptors to adhere to, and migrate over the wound matrix and wound dermis (Martin, 1997), leading eventually to complete coverage, and a new epidermis. Keratinocyte and fibroblast proliferation at wound edges is furthermore regulated by epidermal growth factors (EGFs), with heparin binding EGFs, and fibroblast growth factors (FGF) being a key regulator for proliferation of keratinocytes and fibroblasts respectively.

It can be now said that total regeneration of a wound has been dependent on its geometrical extent and depth. Recent work involving stem cell regeneration within wounds (Chen et al., 2008; Lee et al., 2009; Nakagawa et al., 2005; Wu et al., 2007) makes it progressively more realistic that total regeneration of wounds will be a future possibility, and that this technology will be integrated into biomaterials, especially at the nanoscale (Marmaras et al., 2012). Nanoscale biomaterials are an essential delivery and retention material for wound healing agents, whether for stem cells, or for chemical growth and proliferation factors. There are several benefits of using nanofeatured dressings and materials with respect to wound healing. One is in the available surface area that is in contact with the wound, and which can be functionalized for stem cell delivery or agents that stimulate myofibroblastic proliferation across the entire wound surface as opposed to just at the edges. Moreover, functionalized nanomaterials can quickly migrate to "hard-to-reach" areas within a wound, which increases the potential for a more complete wound regeneration. In the following section, we will touch upon different bionanomaterials that have been used, or show potential in wound healing technology.

3 BIONANOMATERIALS

This section introduces different classes of biological material of relevance to wound healing. The list includes both protein-based and polysaccharide-based materials, several of which are also hierarchically structured.

3.1 NANOCELLULOSES

Cellulose is a long chain polysaccharide of glucose molecules linked by β-1-4 glycosidic linkages (Fig. 13.3), and is the most dominant biopolymer on earth. These chains line up, and are held together by both intra- (Fig. 13.4) and intermolecular hydrogen bonding (Fig. 13.5), between hydroxyl groups and adjacent oxygen atoms, creating β-conform structures, and crystalline cellulose units. Nanocelluloses from trees comprise typically 95% crystalline cellulose, whereas those from bacteria (bacterial nanocellulose) can be up to 99% crystalline. Bacterial cellulose is the most "contaminant free" of the nanocelluloses in that lignin and hemicellulose residues are absent, unlike in plant based celluloses (Huang et al., 2014). Celluloses are biocompatible with cells and tissue (Barbie et al., 1990; Kino et al., 1998; Miyamoto et al., 1989), are nontoxic and moisture retentive (Klemm et al., 2006). The stiffness of nanocellulose is high (110–220 GPa, Moon et al., 2011), which would initially make it seem incompatible with skin, which has a much lower stiffness (3–70 MPa, Orgill, 1983). This can be altered and controlled by using cellulose in in combination with different materials to form composites (Ferras et al., 2012; Taokaew et al., 2012; Veliz et al., 2014; Wanna et al., 2013). There are several structural forms of nanocellulose. These include nanocrystals, nanofibers, nanowhiskers, nanorods, nanofibrils, and microbial (or bacterial) nanocellulose. Nanocellulose has unique properties and qualities that make it a suitable bionanomaterial for wound healing. These include a high specific surface area, unique rheological properties, liquid crystalline behavior, barrier properties, easy bioactivation, biocompatibility, renewability, sustainability, biodegradability, recyclability, and low toxicity.

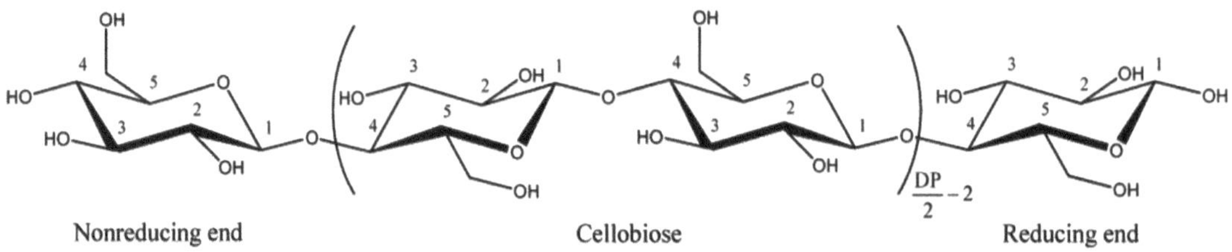

FIGURE 13.3 Chemical Structure of Cellulose

Reproduced with permission from Habibi, Y., Lucia, L.A., Rojas, O.J., 2010. Cellulose nanocrystals: self-assembly, and applications. Chem. Rev. 110, 3479–3500. Copyright of the American Chemical Society.

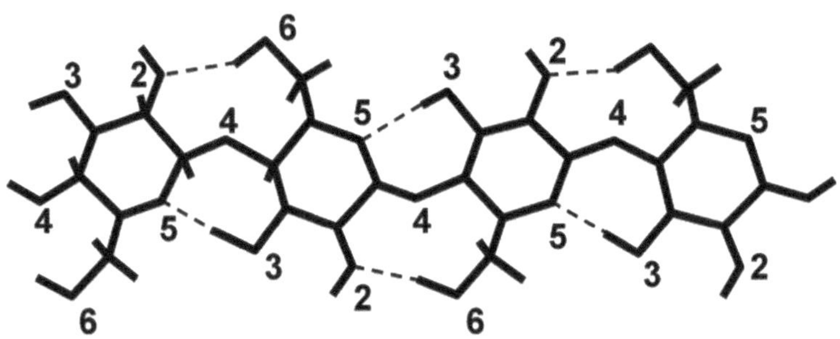

FIGURE 13.4 Intramolecular Hydrogen-Bonding Network in a Representative Cellulose Structure

Reproduced with permission from Habibi, Y., Lucia, L.A., Rojas, O.J., 2010. Cellulose nanocrystals: self-assembly, and applications. Chem. Rev. 110, 3479–3500. Copyright of the American Chemical Society.

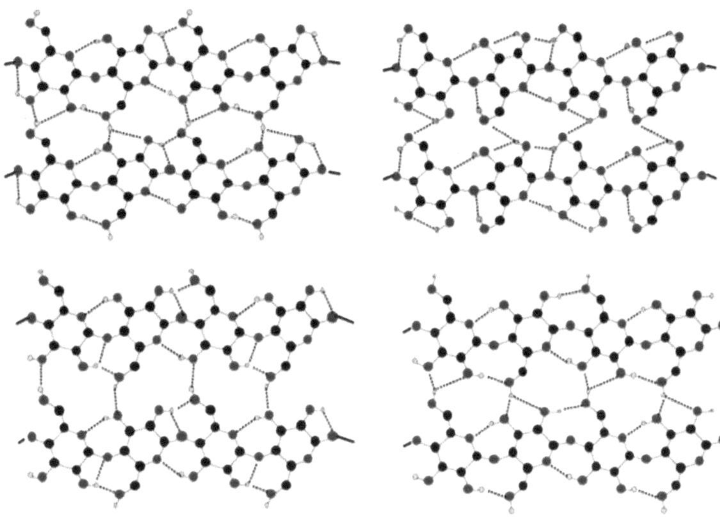

FIGURE 13.5 Hydrogen-Bond Networks in Cellulose 1α

Top left and right: the two alternative hydrogen-bond networks in cellulose 1α. Bottom: the dominant hydrogen-bond network in cellulose 1β: (left) chains at the origin of the unit cell; (right) chains at the center of the unit cell.

Reproduced with permission of Sturcova, A., His, I., Apperley, D.C., Sugiyama, J., Jarvis, M.C., 2004. Structural details of crystalline cellulose from higher plants. Biomacromolecules 5, 1333–1339. Copyright of the American Chemical Society.

There are several examples of cellulose based tissue regenerative research. For tissue engineering and wound healing, bacterial celluloses show currently greater potential than plant based cellulose nanofibers, foremost because they are of greater purity. Bacterial celluloses have been seeded with cells to instigate skin grafting mechanisms in severely damaged skin (Petersen and Gatenholm, 2011). Bacterial celluloses, in combination with FGF enhance biocompatibility (Lin et al., 2011). Lin and co-workers developed porous bacterial cellulose hydrogels that could act as scaffolds for not only FGF, but also collagen, hyaluronans, and elastin. One of the beneficial qualities of nanocellulose is that it when it forms a network, it has a very large specific surface area and topographical detail (roughness) that can be beneficial in the nucleation and proliferation of cells (Rambo et al., 2008). Often, the pores in nanocellulose films exhibit a form of nanoconfinement and water retention is high. High water retention is concurrently a function of the hydrophilic nature of cellulose itself and its ability to develop electrostatic interactions with water molecules. This is an important characteristic in nanocellulose bionanomaterials since water retention will prevent wound from prematurely drying up before scar tissue has had the chance to develop uniformly. The network structure in itself makes bacterial cellulose films or sheets ideal as temporary artificial skin substitutes (Huang et al., 2014). Bacterial cellulose sponges formed through freeze drying possesses larger pores (as well as nanoscale porosity) are also excellent templates for cell growth (Gao et al., 2011). Porosity is a crucial factor in cell growth and attachment. Variations of material shapes and sizes alter porosity and consequently the tortuosity (Alam et al., 2006) and permeability (Alam et al., 2009) of a material. If porosity is altered, bacterial cellulose films can facilitate considerably improved cell growth (Rosamond et al., 2008) though varying the porosity also has the effect of changing the surface structure, and the water retentive capabilities. Changing the surface structure

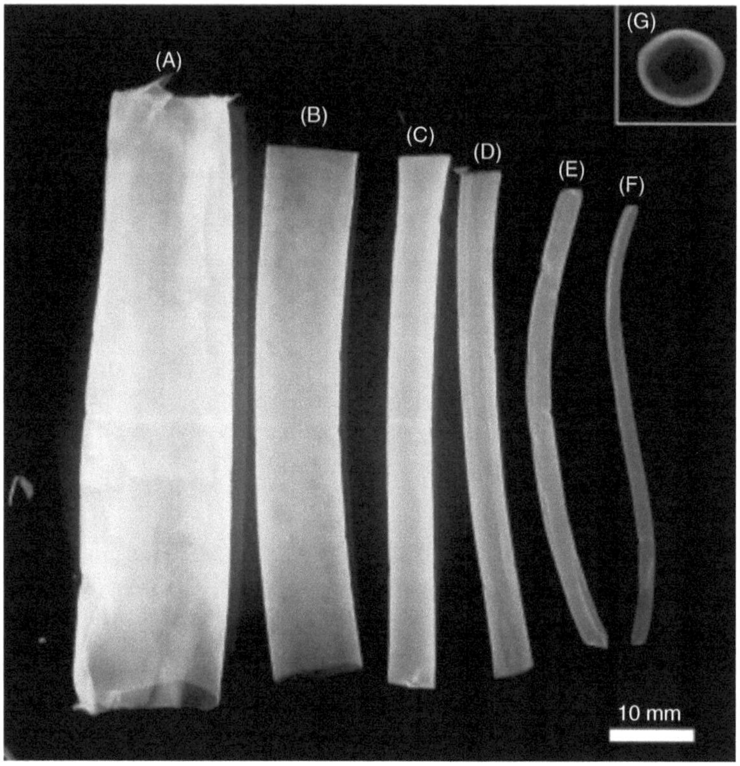

FIGURE 13.6 BC–TS Gels Cultured in the Silicone Tubes of Various Inner Diameters

(A–F) Lateral view of BC–TS with an outer diameter of 20, 10, 5, 4, 3, and 1 mm, respectively, and (F) top view of BC–TS with an outer diameter of 10 mm.

From Putra, A., Kakugo, A., Furukawa, H., Gong, J.P., Osada, Y., 2008. Tubular bacterial cellulose gel with oriented fibrils on the curved surface. Polymer 49, 1885–1891. Reproduced with permission from Elsevier.

is equivalent to changing the physics of blood–biomaterial interactions (Veliz et al., 2014); however, bacterial cellulose has been used to improve blood–biomaterial interactions by grafting to biomaterial surfaces (Bodin et al., 2007). Another benefit bacterial cellulose has over plant based nanocelluloses is that it can be grown in 3D shapes (Huang et al., 2014). Putra et al. (2008), for example, grew bacterial cellulose-based gels in the form of tubes. The tube diameters ranged from c. 1 to 10 mm and as such show great potential in vascular tissue applications (Fig. 13.6). An issue with bacterial cellulose is that it has low elongation compared to tissues of the human body. This problem can be circumvented by developing composites comprising both bacterial cellulose and an elastic blood clotting protein known as fibrin. The elongation of fibrin is over 300% (Liu et al., 2006) and by immersing wet bacterial cellulose sheets into a solution of fibrin together with a glutaraldehyde cross-linker, the properties of both can be combined (Brown et al., 2012).

Bacterial cellulose as described earlier, can promote blood–biomaterials interactions when coated to other biomaterials. By far the simplest and most utilized method of coating with bacterial cellulose is to grow the bacteria directly upon the biomaterial of interest (Meftahi et al., 2010). Moreover, bacterial cellulose can be functionalized in several ways. Silver infused with bacterial cellulose to form porous structural hybrids are considered apposite wound healing scaffolds with an additional antibacterial element (Berndt et al., 2013). Song et al. (2012) use silver nitrate to deposit silver nanoparticles upon nanocellulose fibers. The consequence of Song's technique is a relatively even coat of silver nanoparticles. This may or may not be beneficial. An even coating of silver ensures thorough antibacterial property, however, there remains the possibility that the silver nanoparticles concurrently mask effects of other compounds grafted to the cellulose to encourage, for example, cell proliferation and nucleation. Clays, such as montmorillonites when combined with bacterial cellulose also have antibacterial properties (Ul Islam et al., 2013). However, clays (kaolinites) have also been shown to have additional effects on blood clotting rates (Wanna et al., 2013) and blood attachment (Veliz et al., 2014). Gallic acid is an additional compound to cellulose based biomaterial hydrogels containing cyclodextrin (Pinho et al., 2014). Gallic acid is itself an antibacterial, but in the research of Pinho and coworkers, its antibacterial properties are found to continue functioning even when integrated within a hydrogel environment. Other natural polysaccharides, such as hyaluronic acid in combination with lysine based cationic surfactants coated to cellulose surfaces have convincing antibacterial effects (Bracic et al., 2014), though, potential masking of cellular proliferation compounds may be a problem if there is insufficient exposure. Cell (fibroblast) proliferation compounds may include organosilanes (Taokaew et al., 2015) which induce hydrophobic interactions with fibroblasts, in a similar fashion to that of hydrophobic amino acids attached by esterification to bacterial cellulose surfaces (Kalaskar et al., 2008). Aside from cell growth, attachment and antibacterial properties, there is immense potential benefit from being able to release antibiotics into the site of wounds in the cases where local infections spread and distribute about the site of the wound. Nanocellulose fibers are one means by which this may be possible since they have similar dimensions and certain similar characteristics to electrospun cellulose acetate, and other biopolymer component poly-ε-caprolactone and poly-L-lactic acid spun nanofibers (Buschle-Diller et al., 2007). Hydrogels made up of nanocellulose fibers may be used as release materials for antibiotics, antibacterial agents, and cell proliferation agents. However, hydrogels are noncovalently bonded and are prone to changes in mechanical properties with variations in the environment. This problem can be circumvented by cross-linking nanofibers which has been achieved through TEMPO oxidation of fiber surfaces followed by fibrillation and the attachment of carboxyl and aldehyde groups (Syverus et al., 2014) to the cellulose nanofiber surfaces. Cross-linking can then be secured with diamines.

For the most part, nanocellulose based wound healing materials have not yet established themselves and at present, research predominates the commercial utility of this material. There are several factors that can raise the price of utility of cellulose including energy resources, transportation, harvesting, purification, processing, and chemical modification. How ecofriendly the cellulose is postchemical modification is a conundrum of its own and great impetus ought to be placed on the development of low energy, degradable chemical modification of nanocellulose biomaterials.

3.2 CHITIN AND CHITOSAN

Chitin is perhaps the second most abundant biopolymer on earth after cellulose. It is a predominant biopolymer in the skeletal components of insects and crustaceans. It is a mucopolysaccharide consisting of 2-acetamido-2-deoxy-β-D-glucose linked through β-1-4 glycosidic bonds. *N*-deacetylation of chitin yields the derivative chitosan (Kumar, 2000) leaving *N*-acetylglucosamine and *N*-glucosamine units.

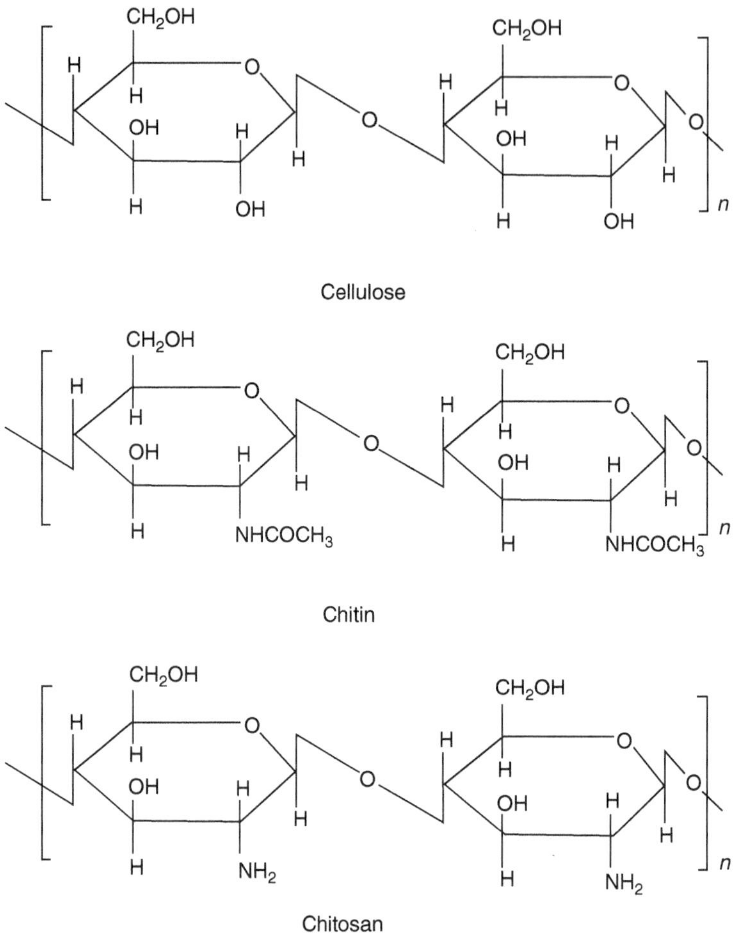

FIGURE 13.7 Structures of Cellulose, Chitin, and Chitosan

From Kumar, M.N.V.R., 2000. A review of chitin and chitosan applications. React. Funct. Polym. 46, 1–27. Reproduced by permission of Elsevier.

The structures of chitin and chitosan as compared with cellulose are shown in Fig. 13.7. Chitin exists in two main forms, α-chitin and β-chitin, each of which can be distinguished by their crystallographic structures (Rinuado, 2006), which are in turn related to the intermolecular hydrogen bond patterns (Figs. 13.8 and 13.9). Since chitosan has a primary amino group on C2 and an –OH group on C4, it is able to react under relatively mild conditions. Under acidic conditions, chitosan is soluble in water, and can be easily combined with other mild-acid soluble biopolymers.

Chitin and its derivative chitosan have been of interest in the areas of wound healing and skin replacement research because of biocompatibility, chemical compatibility, low toxicity, and biodegradability. It is utilized in many forms including powders, nonwovens, porous beads, gels, gauzes, fibers, and films. Chitosans comprise the *N*-acetylglucosamine unit, which has been linked to the metabolism

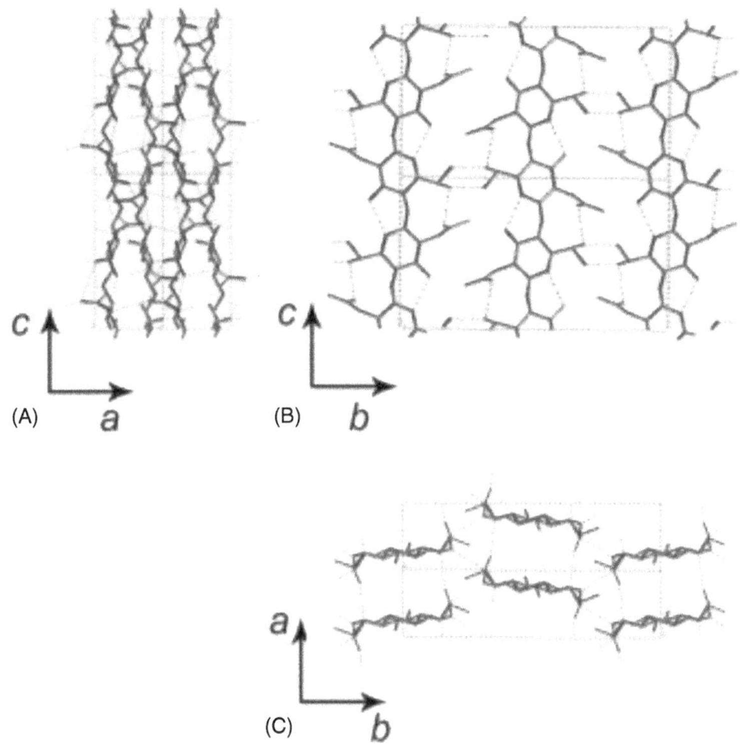

FIGURE 13.8 Structure of α-Chitin

(A) ac projection; (B) bc projection; (C) ab projection. The structure contains a statistical mixture of two conformations of the $-CH_2OH$ groups.

From Rinuado, M., 2006. Chitin and chitosan: properties and applications. Prog. Polym. Sci. 31, 603–632. Reproduced with permission from Elsevier.

of hexamines. This is presumed to instigate the cross-linking of collagen in the wound healing process. *N*-glucosamine units in chitosan may also play a role in wound healing. Glucosaminoglycans—chitin derivatives—are involved in granulation and creating structural integrity in collagen fibers. Since chitin is soluble under physiological conditions, chitin derivatives from locally solubilized chitin can manifest themselves in high concentrations at the site of the wound, thereby aiding collagen granulation (Singh and Ray, 2000). Granulation is preceded by the infiltration of inflammatory cells and chitin derivatives, such as chitosan accelerate the infiltration of such cells, most prominently polymorphonuclear leukocytes, macrophages, and foreign giant cells. Ueno et al. (2001) showed that this exacerbation of inflammatory cells is clear on the 3rd day, while granulation occurs on the 9th day of wound healing. By accelerating the rate of polymorphonuclear leukocytes, osteopontin is released in greater concentrations. Osteopontin is a glycosulated phosphoprotein which promotes cell adhesion for a variety of types of cells. The polymorphonuclear leukocytes accumulate osteopontin-mRNA expression, and this improves cell adhesion, and wound healing. Macrophages are important for local phagocytosis of foreign bodies and are beneficial to the process of wound healing. Chitosan *N*-acetylglucosamine binding

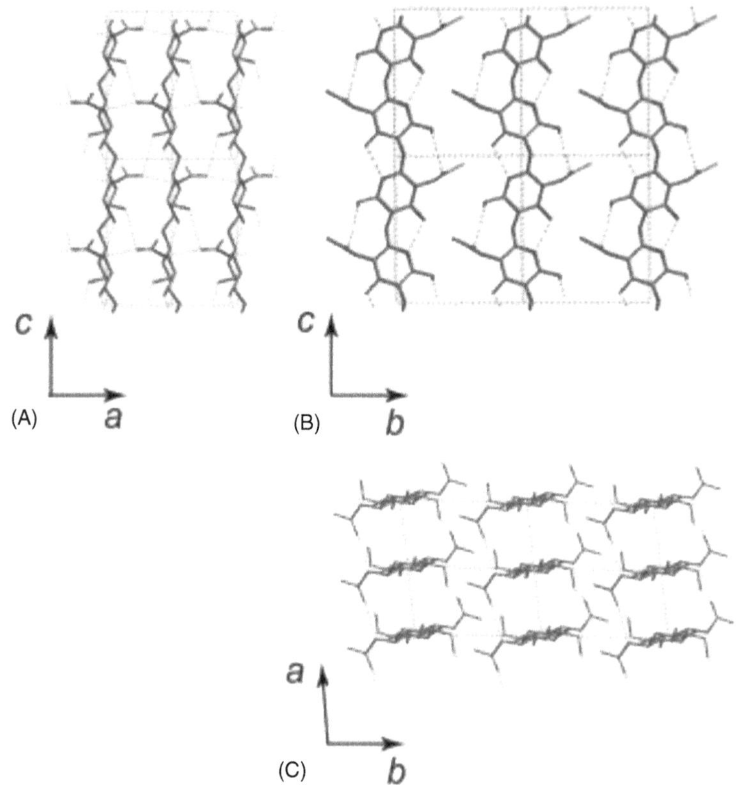

FIGURE 13.9 Structure of Anhydrous β-Chitin

(A) ac projection; (B) bc projection; (C) ab projection.

From Rinuado, M., 2006. Chitin and chitosan: properties and applications. Prog. Polym. Sci. 31, 603–632. Reproduced with permission from Elsevier.

to certain receptors is thought to mediate the release of the macrophages, though it still remains unclear as to the underlying mechanism. Fibroblastic proliferation is indirectly accelerated by chitin derivatives, such as chitosan in vivo by forming polyelectrolyte complexes with heparin. Heparin both stabilizes the FGF and enhances mitogenesis in combination with the FGF. These, alongside its anticoagulant and angiogenic properties, make heparin a wound healing agent that is effectively catalyzed by chitin derivatives (Mori et al., 1997). The efficiency of chitosan is tied to its molecular weight, and oligomers or monomers of chitosan accelerate wound healing more effectively than longer chain, higher molecular weight polymers (Minagawa et al., 2007), which clarifies that the *N*-acetylglucosamine parts that are most important to have present during wound healing.

Chitin derivatives require manufacturing into, for example, gauzes and fibrous mats for application to sites of wound healing. Fibrous mats made up of dibutyrylchitin (a water-soluble chitin derivative) have been used for burn wound and post traumatic wounds with relative success. Fibrous chitin based materials can be manufactured by electrospinning, phase separation, or by self-assemblage processes, the most popular of which is electrospinning. Carboxyethyl chitosan/poly(vinyl alcohol) solutions electrospun into nanofibrous materials are an example of material that works well in cell attachment and fibroblast proliferation in vitro. Degradation of chitosan electrospun mats under physiological conditions has a further benefit in that it prevents wound reopening during removal (Jayakumar et al., 2011).

3.3 SILKS

Silks are remarkably tough biopolymers with several intrinsic nanofeatures, each contributing to their ability to absorb mechanical energy. There are several types of silk, the most commonly researched with respect to bionanomaterials is from the silkworm, *Bombyx mori*. Spiders too have several types of silk that they extrude from spinnerets in a complex process that gives rise to ordered, disordered, and semiordered regions within each fiber. The combined interactions from each of these regions give silk a considerable capacity for energy absorption (Sintya and Alam, 2016). Silks are also electrically conductive, aiding in capture (Ortega-Jimenez and Dudley, 2013) and in thread preservation on impact (Pahlevan et al., 2014). Different silks comprise different proteins and are varied in their properties. Several silks have found biocompatible function as potential bionanomaterials. Moreover, they are degradable in the human body, nontoxic on degradation, easily functionalized, and several of their physical properties can be modified.

Silk can be manufactured into a number of architectures. The majority of researchers have worked on developing or artificially spinning fibers, predominantly through electrospinning, though several researchers have also considered self-assembling silk fibers from constituent proteins by effectively precipitating over several days from a potassium phosphate solubilized form (Slotta et al., 2007). A consequence of this type of precipitation is the development of folded nanocrystals, as opposed to sheared and stretched nanocrystals, as this is more typical during an extrusion from, for example, a spider spinneret. In terms of the intrinsic energies of folded predominantly polyalanine nanocrystals, there is little difference with respect to native silk, though the unfolding and failure mechanism of folded nanocrystals is likely to be different (Alam, 2014). Alcohols, such as methanol (Kim et al., 2005b) or ethanol (Jin et al., 2005) can be used to control the extent of beta-sheet development, and the mechanical properties of "precipitated self-assembled silks" can be better controlled. An important element of silk nanofiber precipitation is the drying strategy (Lu et al., 2012), which when coupled to dynamic molecular interactions in the presence of charged additives, cause nanofiber formation, crystallization being the concluding process. Ultrathin films used to cover other substrates are another architecture that can be used to coat other substrates for biomedical use. From a dissolved state, ultrathin (down to a few hundred nm) membranes can be manufactured by simple deposition and evaporation (Rockwood et al., 2011). Particles (Slotta et al., 2008) are a specialized form of architecture that may be beneficial as a delivery method of stem cells or myofibroblast proliferation stimulating compounds in deep wounds. These are made similarly to fibers, in that silk proteins are solvated in potassium phosphate solution, after which the proteins can nucleate to "seeds" introduced into the system. The particle sizes can be controlled by varying the temperatures of the mix. Electrospun fibers can be conjoined to form a mesh with porosity (Aznar-Cervantes et al., 2012). Advantages of porosity in view of wound healing include increased surface area for attachment and interactions, roughness which potentializes cellular interlocking with the bionanomaterial scaffold. Silk particles close to 2 μm in diameter can be manufactured with nanoscale surface "bulbs" and porosity using wet milling and spray drying methods. Particles diameters can reach sizes down to 200 nm, and the subsequent porosities and surface bulbs are then very small. The benefits of particulate production using milling techniques are in terms of the high

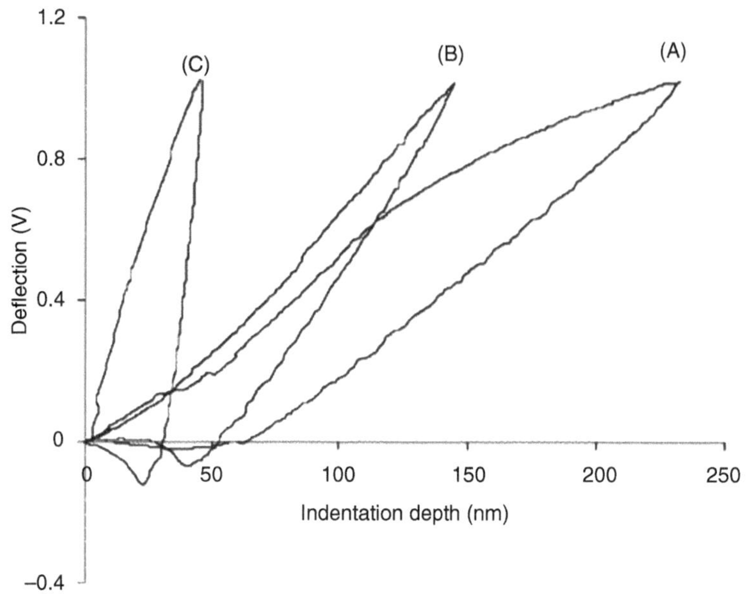

FIGURE 13.10 Representative Loading and Unloading Curves of Fibers

(A) As-spun fiber; (B) methanol-treated fiber; (C) water-extracted fiber.

Used with the permission of Wang, M., Jin, H.J., Kaplan, D.L., Rutledge, G.C., 2004. Mechanical properties of electrospun silk fibres. Macromolecules 37, 6856–6864. Copyright of the American Chemical Society.

volume of production with respect to time, and the use of little or no harmful chemicals. Using a top down approach (milling from silk fibers) requires considerable energy, and the process has drawbacks (Rajkhowa, 2014). Silk particles of this size with nanoporosity can be easily loaded with antibiotics or specific cellular growth factors (such as FGF), and integrated to surfaces or within biomaterials applied directly to wound sites.

The process of electrospinning is covered in a latter part of this chapter. Electrospun silk scaffolding is considered a highly promising scaffold structure for tissue regeneration and wound healing. This is because these are porous and rough materials with high specific surface areas (like bacterial cellulose networks), which in turn promote the nucleation and growth of cells. According to the research of Wang et al. (2004), electrospun silk fibers lose stiffness from native silks by as much as 18-fold, depending on the treatment method. Wang and coworkers spun *B. mori* silks from polyethylene oxide (PEO) solutions (A), from silk/PEO solutions and immersion into 90/10 by volume methanol/water solution (B), and from silk/PEO solutions, immersion into 90/10 by volume methanol/water solution, and then washing in water to remove PEO (C). The treatment affects significantly the final mechanical performance of electrospun silks with (C) yielding the highest stiffness, and (A) the highest energy absorptive capability (Fig. 13.10), as well as the diameters and morphology of the fibers (Fig. 13.11).

Fiber diameter can be taken to the nanoscale by using lowered concentrations of silk fibroin in solution prior to electrospinning. A study by Sukigara et al. (2003) electrospun silk fibers from fibroin/formic acid solutions, varying the concentration of the fibroin from 5% to 19.5% by weight. At 15% by

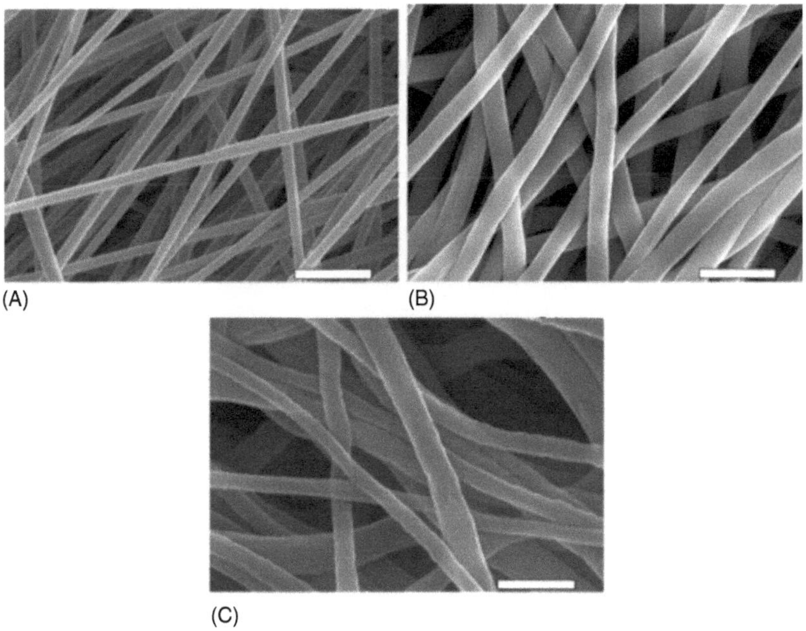

FIGURE 13.11 SEM Images of *B. mori* Silk/PEO Fibers

(A) As-electrospun fiber (scale bar, 5 μm); (B) methanol-treated electrospun fiber (scale bar, 2.5 μm); (C) water-extracted electrospun fiber (scale bar, 2 μm).

Used with the permission of Wang, M., Jin, H.J., Kaplan, D.L., Rutledge, G.C., 2004. Mechanical properties of electrospun silk fibres. Macromolecules 37, 6856–6864. Copyright of the American Chemical Society.

weight fibroin and below, and using 3 kV, nanofibers could be spun, however, when the concentration of fibroin is too high, the fibers enter the microscale. The critical transition from nano- to microfiber according to the research of Sukigara et al. (2003) seems very small (Fig. 13.12), and concentration is a vital design parameter with respect to geometry.

Electrospun silks once spun, may be functionalized for improved wound healing or regeneration efficiency. Schneider et al. (2009) incorporated epidermal growth factors within electrospun silk fiber mats by mixing and absorption, and found the release rate of the EGFs was highest in the first hours after which the release rate decreased as a function of time. Importantly though, Schneider and coworkers demonstrated that electrospun silk with epidermal growth factors incorporated significantly improved wound closure, reaching 100% closure at 48 h (as compared to <30% at 48 h for silk without EGFs) (Fig. 13.13).

Reduced water loss at a wound site is relevant since maintaining moist conditions is important in the healing process. Silk fibers are water absorbing, and their structures and properties can change somewhat as a function of water absorption. Surface grafting is one mean by which water absorption at interfaces can be reduced. Singha and Kapoor (2015) demonstrated this by using acrylonitrile grafted to the surfaces of electrospun silk fibers. Even at 100% relative humidity, the fibers absorbed less than half that of untreated electrospun silks. Hybrid silk-based materials show immense cell viability

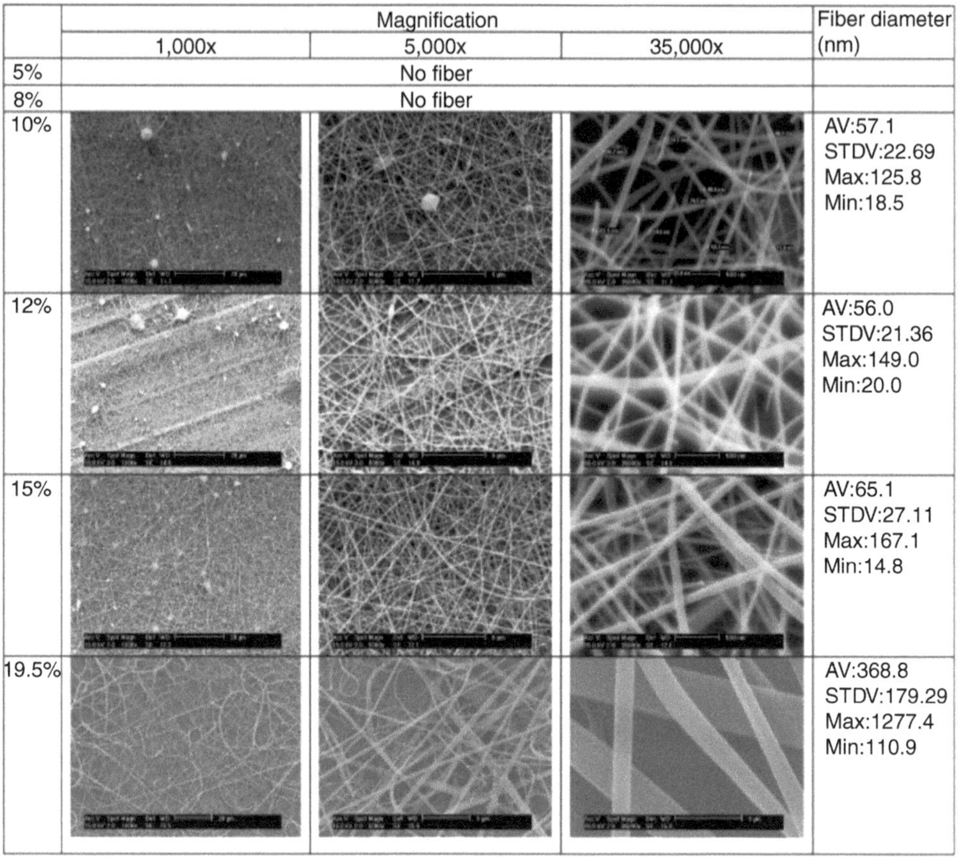

FIGURE 13.12 The Morphology of Fibers at Electric Field of 3 kV/cm at Concentrations From 5% to 19.5%, with a Constant Spinning Distance of 7 cm

The figure also shows the average, standard deviation, maximum and minimum values of the fiber diameter. Reproduced with permission from Elsevier.

From Sukigara, S., Gandhi, M., Ayutsede, J., Micklus, M., Ko, F., 2003. Regeneration of Bombyx mori *silk by electrospinning – part 1: processing parameters and geometric properties. Polymer 44, 5721–5727.*

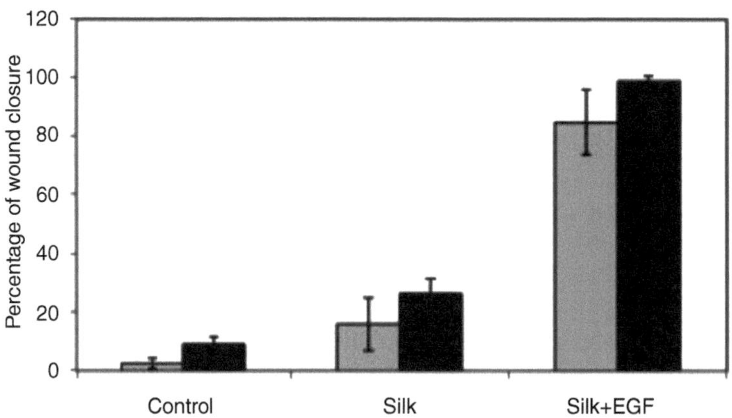

FIGURE 13.13 Percentage of Wound Closure After 24 *(Gray)* and 48 h *(Black)*

Data points are the average of $n = 7$ and error bars represent ± SD.

From Schneider, A., Wang, X.Y., Kaplan, D.L., Garlick, J.A., Egles, C., 2009. Biofunctionalised electrospun silk mats as a topical bioactive dressing for accelerated wound healing. Acta Biomater. 5, 2570–2578. Reproduced by permission of Elsevier.

potential. Chutipakdeevong et al. (2015) combined poly-ε-caprolactone with silk fibroin to electrospun wound dressing mats, and found respectable increases in cell viability when compared to neat poly-ε-caprolactone. Moreover, surface functionalization with modified silk fibroin (dip coated essentially from solution) yielded considerably greater improvements in cell viability tests as compared with neat poly-ε-caprolactone. One disadvantage of the silk fibroin coating as a fiber modification, is that it increases the fiber diameter somewhat, thus reducing surface area of contact. One way to circumvent this is to coat thinner fibers, however, that in itself can be challenging since very low concentrations of fibroin can result in the electrospinning of no fiber at all (Sukigara et al., 2003), refer to Fig. 13.12. Another electrospun hybrid material that functions well with silk fibroin as wound repair materials is gelatin (Shan et al., 2015). Generally, these fibers end up with larger diameters than poly-ε-caprolactone/fibroin fibers. However, they are able to increase marginally the rate of wound closure when loaded with astragaloside IV, a traditional Chinese medicinal plant-based compound. One important aspect of silk surface modification is that, provided the correct modification method is used with the right materials, nanofeatures can be induced upon the fibers themselves. This allows the surface area to increase considerably, and provides a greater number of nucleation points, to which cells (or loading compounds) can be attached. Sodium hydroxide treatments (Karahaliloglu et al., 2015) seem to have this effect and alter the surface energetics of the fibroins.

3.4 ALGINATE (ALGINIC ACID)

Alginate is a ubiquitous compound in brown algae and brown seaweeds. It consists of β-1-4 glycoside bonded β-D-mannuronic acid and α-L-glucuronic acid. A scheme of how these uronic acids are linked is provided in Fig. 13.14. Alginate found in nature is often a compound in combination with group 1 or 2 metals (such as sodium, potassium, calcium). Calcium alginate is one of the more popular materials

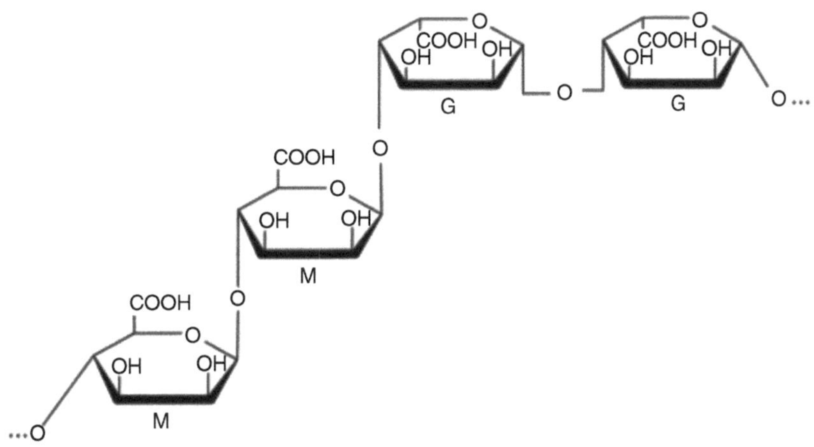

FIGURE 13.14 Scheme of an Alginate Segment With its two Monomers, D-Mannuronic (M) and L-Guluronic (G) Acids

From Paula, G.A., Benevides, N.M.B., Cunha, A.P., de Oliveira, A.V., Pinto, A.M.B., Morais, J.P.S., Azeredo, H.M.C., 2015. Development and characterization of edible films from mixtures of K-carrageenan, I-carrageenan and alginate. Food Hydrocoll. 47, 140–145. Reproduced with permission from Elsevier.

for cell immobilization (Smidsrod and Braek, 1990), and for that reason, has considerable potential benefit for cell adhesion and hence, proliferation in wounds. On contact with calcium, alginate becomes hardened, and is also finding use as drug-delivery nanoparticle encapsulates.

Alginate is available typically as freeze dried porous sheets or as fibers, though due to its high water retention, alginate turns into a gel on contact with a wet wound. This results from high levels of hydrophilic hydrogen bonding. The benefit of gelation on contact with a wound is effectively wound filling, which decreases the exudence of bodily fluids, and limits pathogenic and bacterial ingress into the wound. Gelation of alginate on contact with a wound is via an ionic exchange between the alginate and the blood. Cross-linking of the alginate occurs through calcium ions and calcium alginate is noted as having a secondary benefit in conjunction to its ability to encourage cell adhesion (Boateng et al., 2008). The benefit of applying alginate in a nanofibrous form is that it is able to more easily reach the smaller fissures of a wound prior to its gelation. Besides freeze drying, photolithography is a method by which nanofibrous alginate structures can be manufactured. A thermally degradable sacrificial stamp is made upon which the alginate can be layered and cross-linked (Shapira et al., 2014).

Complete alginate gelation inside a wound can be circumvented by creating nanofibers as a hybrid with another material. Hybrid materials have included calcium alginate with poly(vinyl alcohol) blended to a nanofibrous matrix and used in wound dressing (Tarun and Gobi, 2012). Shalumon et al. (2011) combined sodium alginate with poly(vinyl alcohol) alongside nanozinc oxide particles to form composites. Zinc oxide has an antibacterial effect, however, at higher concentrations it becomes cell toxic. The study by Shalumon et al. (2011), and balancing and optimizing the composite component concentrations is somewhat crucial in the development of multifunctional nanostructured dressings. Alginate/starch biocomposites loaded with nanosilver particles are another mean by which multifunctionality can be achieved, and these are well capable of inducing granulation as early as the 8th day after wound initiation (Arockianathan et al., 2012).

One of the main benefits of alginate-based wound dressings in a fibrous form is that when they become entrapped within a wound, they are still easily biodegradable. Alginate-based dressings are considered use worthy for moderate-to-heavily exuding wounds, and entrapped fibers can also be washed away using saline water. Removal is thus relatively painless and granulation tissue remains intact during the different processes of biodegradation, washing, and physical removal. One disadvantage in using alginate based dressing is that they cannot be used for dry wounds since they are so moisture absorbent that they may absorb vital fluids and prolong wound healing (Boateng et al., 2008).

3.5 HYALURONIC ACID

Hyaluronic acid is a prevalent polysaccharide in nature, the chemical structure of which is provided in Fig. 13.15. It is found in connective and epithelial tissues in several species, and is moreover secreted in large quantities from the epithelial layers of fish to form protective mucosal slime layers for their skin. Hyaluronic acid is an anionic nonsulfated glucosaminoglycan, and contributes to cell proliferation, and it is an obvious choice of biomaterial if cell proliferation within wounds is desired.

Under room temperature and humidity conditions, hyaluronic acid is in a gelated state and is not a nanomaterial. It has, however, found considerable use as an additive to nanomaterials to form hybrids, much like alginate as described previously. Eroglu et al. (2015), for example, hybridized hyaluronic acid with dipalmitoylphosphatidylcholine (DPPC) and resveratrol to form wound healing

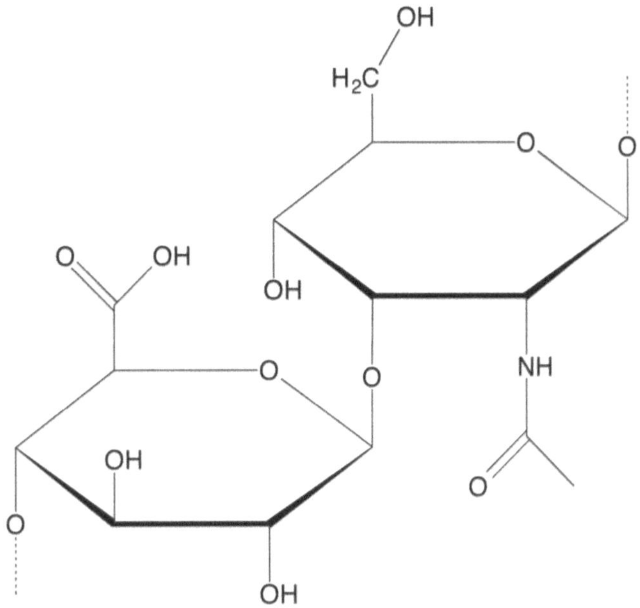

FIGURE 13.15 Hyaluronic Acid: Chemical Structure

microparticles. Microparticles are beneficial in that they can spread well and distribute well within a wound, however, nanoscale particles would do that more effectively and provide greater surface area with respect to the volume of particles added. Nanofibers coated by hyaluronic acid are reported in Croisier et al. (2014). Here they report electrospun poly-ε-caprolactone into nanofibrous form, and coated the fibers with either chitosan or hyaluronic acid. Such hybrid nanofibers formed into mats have the multiple benefits of improved wound healing, easy ingress into small wound fissures, and mechanical stability. Porous hyaluronic acid hydrogels have also been a matter of recent interest (Cam and Segura, 2014). In their work, they chemically sintered a network of poly methyl methacrylate microspheres, subsequently used for the formation of porous hydrogels. These materials encourage accelerated cellular proliferation within the wound as well as faster vascular growth. Chitosan-hyaluronic acid hydrogel combinations have also been used to coat poly(caprolactone) scaffolds for improved ligament regeneration (Deepthi et al., 2015). Though both microspheres and scaffolds are above the nanoscale, the hybrid chitosan-hyaluronic acid coatings are very thin, and can be applied at the nanoscale (depending on how prolonged coat-time is). As such, materials, such as hyaluronic acid that are gels under normal ambient conditions, can still be used, especially as hybrids, at the nanoscale in wound healing materials technology.

3.6 FUCOIDAN

Fucoidan is similar to alginate in that it is prevalent in brown algae. It is also similar to hyaluronic acid in that it is a sulfated polysaccharide; however, it contains fucose and has a backbond with a variety of glycosidic linkages (C1–3), (C1–4), or (C1–2). Fucoidan has several beneficial uses in wound healing

Table 13.1 Examples of Basic Nanostructures Found in Nature

Nanostructure	Examples of Structures in Nature	Function
Nanoporous nanocages	Biosilica of diatoms	Protective exoskeleton
Nanospatulae	Surfaces of gecko toes	Improved dry adhesion
Nanoflakes	Surfaces of butterfly wings	Iridescence and protection
Nanoparticles and nanoplates	Shells, corals and sponges	Structural support
Nanofilms	Waxy leaves	Protection and self-cleaning
Nanospikes	Cicada wing surfaces	Impaling and destruction of potential pathogens

From Alam, P., 2015. Mechanical properties of bio-nanostructured materials. In: Aliofkhazraei, M. (Ed.), Handbook of Mechanical Nanostructuring, vol. 1. Wiley-VCH, pp. 211–231 (Chapter 10). Reproduced with permission from Wiley-VCH.

and has found application, similarly to hyaluronic acid, in hybrid nanocoatings applied to biomaterials, for example, for dermal burn-based wounds (Sezer et al., 2007).

4 NANOSTRUCTURES

Nanostructures are, by definition, structures with a dimension in one direction that is lower than 100 nm. This may include a material with protuberances at a spacing distance of less than 100 nm, a porous material with pores lower than 100 nm in diameter, or a material with one dimension (e.g., a fiber or a particle) that is less than 100 nm.

4.1 BASIC STRUCTURES

Nanostructures in nature have different uses, shapes, and performance criteria. Table 13.1 (from Alam, 2015) describes a few of the more common basic nanostructure in nature by name, examples of where they may be found, and by their relative function.

4.2 SURFACE STRUCTURES

Several hydrophobic surfaces rely on natural nanofabrication. Hydrophobicity, as already discussed (Kalaskar et al., 2008), improves encourages fibroblasts in contact with biomaterials to develop correct extended structures. Hydrophobic nanosurfaces are ubiquitous in nature, and are found, for example, on the waxy leaves of water borne plants (Samaha et al., 2012), the wings of butterflies (Yan et al., 2007), and under the feet of water walking insects, such as pond skaters (Hu et al., 2003). Natural nanostructured surfaces, for example (Fig. 13.16), can yield design principles for bionanomaterials in view of increasing surface hydrophobicity to persuade superior growth and proliferation of fibroblasts. Biological materials, such as cellulose, are hydrophilic, and would have to be functionalized with hydrophobic chemicals to develop sound cell growth and proliferation. Other biological materials, such as silks, have alternating hydrophobic and hydrophilic regions (Romer and Scheibel, 2008), and are more agreeable than cellulose to direct interactions with cells involved in the wound healing process.

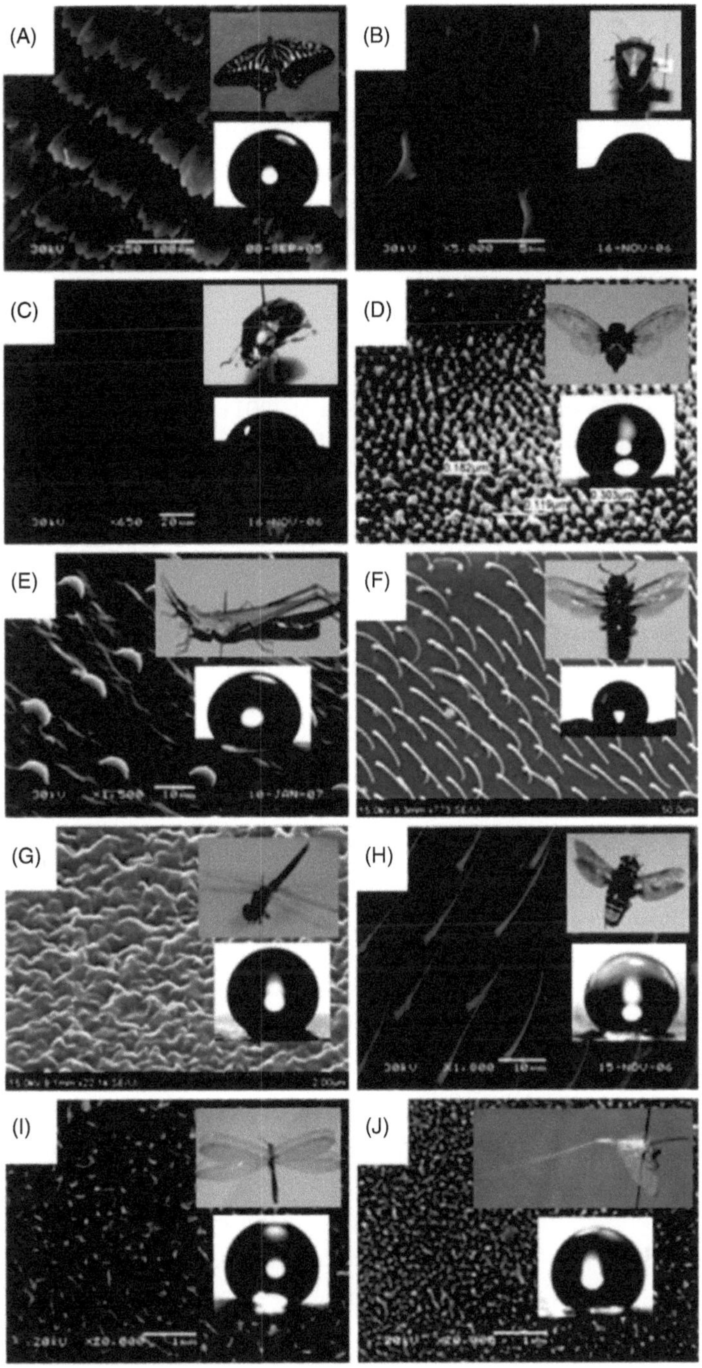

FIGURE 13.16 SEM and Contact Angle Analyses of Different Natural Nanostructured Wing Surfaces, Some are Wetting, Some are Nonwetting

(A) *Lepidoptera Papilio xuthus* (Linnaeus); (B) *Hemiptera Plautia stali* (Scott); (C) *Coleoptera Mimela testaceipes* (Motschulsky); (D) *Homoptera Meimuna opalifera* (Walker); (E) *Orthoptera Acrida cinerea cinerea* (Thunberg); (F) *Hymenoptera Vespa dybowskii* (Andre); (G) *Odonata Pantala flavescens*; (H) *Diptera Tabanus chrysurus* (Loew); (I) *Neuroptera Grocus bore* (Tjeder); (J) *Ephemeroptera Ephemera* sp.

From Byun, D., Hong, J., Saputra, Ko, J.H., Lee, Y.J., Park, H.C., Byun, B.K., Lukes, J.R., 2009. Wetting characteristics of insect wing surfaces. J. Bionic Eng. 6, 63–70. Reproduced with permission from Elsevier.

5 NANOFABRICATION

The following sections describe processing and manufacturing methods that can be used to both produce and improve the performance of bionanomaterials in view of wound healing.

5.1 ELECTROSPINNING

Electrospinning is a versatile process by which biopolymer nanofibers can be produced. Nanofibrous networks with high specific surface areas (Sehaqui et al., 2011) are a means by which bionanomaterial–cell interactions can be improved at interfaces. Electrospinning enables fibers of various forms to be made, including core shell and porous nanofibers (Khajavi and Abbasipour, 2012). At the heart of the process is a nozzle with a thin opening (diameter $< 100\ \mu m$). Polymer melt is pumped through this nozzle, which itself acts as an electrode. A high electric field is applied and 100–500 kV/m, and a counter electrode is used to then electrically attract the polymer melt. A mat for collecting fiber is attached usually to the counter electrode. The polymer melt is solvated typically, and the solvent evaporates as the polymer melt is drawn from the nozzle to the counter electrode. Evaporation of the solvent also causes the polymer to solidify and a nanofiber is formed with a diameter that is drawing rate dependent. In reality, the fiber dimensions are not equal from one end to another. Since the size is so small, simple factors, such as evaporation rate, changes in draw-velocity arising as a function of volume and distance, viscosity and variations in conductivity will all play a role and affect the final fiber dimensions. Fig. 13.17 (from Greiner and Wendorff, 2007) shows a basic laboratory electrospinning set up.

Since many biopolymers decompose as opposed to "melt," they must first be solvated. Several solvents are in use for particular biopolymers preelectrospinning. *B. mori* silks and spider silks are combined typically with poly ethylene oxides, after which water can be used as a solvent (Jin et al., 2002; Wang et al., 2004, 2006). Alternatively, formic acid is a good solvent for silk fibroin (Kim et al., 2003; Min et al., 2004), chitosan (Ohkawa et al., 2004), or combinations of silk fibroin and chitosan (Park et al., 2004). *N*-methylmorpholine/water or *N,N*-dimethylacetamide/lithium chloride solvents can be used for cellulose biopolymers (Kim et al., 2005a; Kulpinski, 2005). Hexafluoroisopropanol alcohol can has been used as a solvent for pure collagen (Matthews et al., 2002), which has considerable potential as a bionanomaterial in wound healing applications (Rho et al., 2006). In combination with polyethylene oxide, collagen can be solvated in hydrochloric acid (aqueous) or sodium chloride (aqueous) (Buttafoco et al., 2006; Huang et al., 2001).

5.2 SELF-ASSEMBLY

Self-assembly occurs naturally in biomaterials, such as silk (Sintya and Alam, 2016), and is also a cost effective means of functionalizing material surfaces by driving agglomerated structures (usually at the nano- or mesoscale), molecules and atoms into higher order configurations. Most commonly, the self-assemblage of nanostructured surfaces in materials science is achieved through molecular means. The process begins by the deposition of molecules as an initially disordered system, onto a surface. As molecules subsequently noncovalently interact with one another, they form ordered, patterned, and sometimes crystalline nanostructures. Self-assembly relies on an intrinsic molecular drive to lower Gibbs

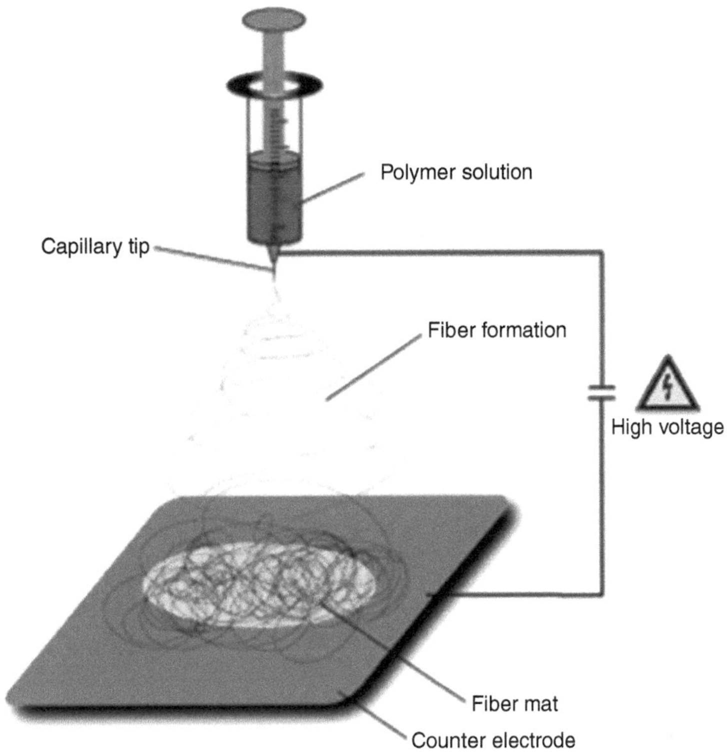

FIGURE 13.17 A Laboratory S for an Electrospinning Experiment with a Perpendicular Arrangement of the Electrodes

From Greiner, A., Wendorff, J.H., 2007. Electrospinning: a fascinating method for the preparation of ultrathin fibres. Angew. Chem. 46, 5670–5703. Reproduced with permission from Wiley-VCH.

free energy states. This in turn improves the thermodynamic stability of self-assembled matter over its disordered origin. What is noteworthy about the process of self-assembly is that it is typically governed by so-called "weak" secondary interactions, such as van der Waals forces, capillary forces, and electrostatic (H-bonding) forces. So called "strong" bonds, such as ionic, metallic, or covalent bonds play a lesser role, which is more "molecularly intraspecific." In essence, self-assembly is not a self-governed process. Typically, there is at least dependence on ambient conditions and upon surrounding materials in terms of chemistry, structure, order, and thermodynamics. Self-assembly in that sense is a cooperative form of assembly (Kogej, 2010; Uskokovic, 2008). As such, a desired surface pattern will more often than not require some form of surface functionalization alongside prepreparation of an appropriate environmental condition to coerce molecules into their "self-assembled" final state (Darling, 2007). A good way of illustrating this is by using the example of selective interfacial wetting in film. By coercing the wetting dynamics of films (using either physical or chemical methods), the growth morphology of the film layers can be controlled (Fig. 13.18).

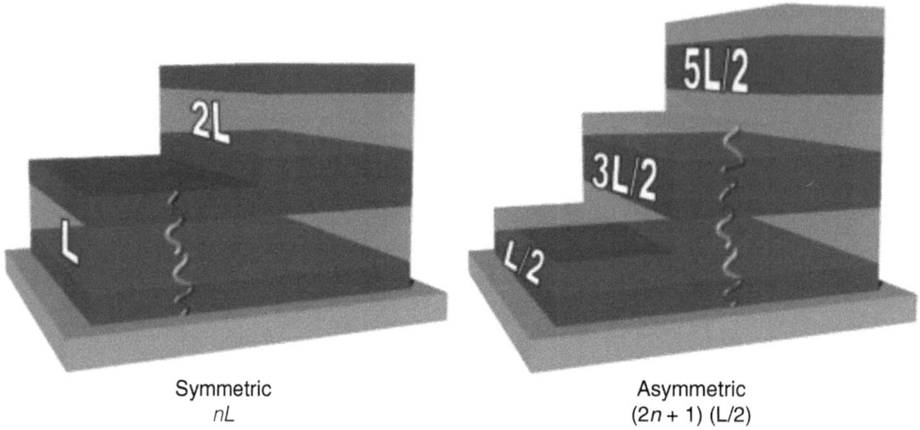

FIGURE 13.18 Schematics S the Effect of Selective Interfacial Wetting in BCP Thin Films

When both interfaces are wet by the same block (symmetric wetting), film thicknesses are quantized to integer multiples of L. In the case of asymmetric wetting, the quantization rule changes to $(2n + 1)(L/2)$. In either case, the selective wetting leads to a parallel orientation of the BCP domains with respect to the substrate plane.

From Darling, S.B., 2007. Directing the self-assembly of block copolymers. Prog. Polym. Sci. 32, 1152–1204. Reproduced with permission from Elsevier.

5.3 PRINTED NANOSTRUCTURES

3D printing is a new method by which microstructures and microfeatured surfaces can be developed. At present, it is difficult to produce nanofeatures due to several physical and engineering limitations. The process involves printing three dimensional solid objects from a single digital file. Individual layers are printed upon one another, making in essence, a laminated object. Depending on the type of polymer being printed, the speed at which printing takes place, and the temperature of the melt on-contact, these layers can in fact merge, thus removing the laminated arrangement. The most common methods for layer-by-layer printing include selective layer sintering (SLS) and fused deposition modelling (FDM). Input files are typically from computer aided designs (CAD) or from 3D scans. Stereolithography is the process used to deposit layers. Usually the resins used can be cured (very often UV curing is used), and a laser creates the pattern upon which resin deposition occurs. Though microscale protuberances are a current limitation, nanocomposites can be made by adding nanostructures (such as fibers or particles) into the resin melt (Campbell and Ivanova, 2013) (Fig. 13.19). Nevertheless, simple nanostructures can be printed by means of nanotransfer printing, Zaumseil et al. (2003). Though there may be some limitations in shape, we can reach the nanoscale using advanced printing methods is progress in itself.

6 CONCLUSIONS

Currently, there exists a concerted effort toward the development of wound healing bionanomaterials. Bionanomaterials often exhibit integrated qualities of biocompatibility with the human body and can furthermore be bioactivated to exacerbate cellular proliferation, or to control cellular morphology.

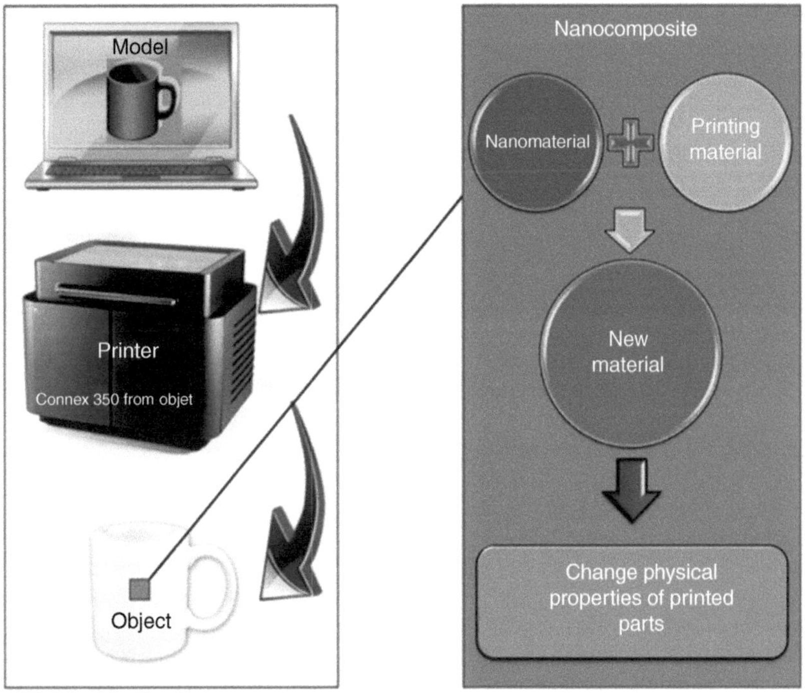

Nanocomposites made via 3D printing

FIGURE 13.19 Scheme Showing How Nanocomposites Can be Made Via 3D Printing Technology

From Campbell, T.A., Ivanova, O.S., 2013. 3D printing of multifunctional nanocomposites. Nano Today 8, 119–120. Reproduced with permission from Elsevier.

Bionanomaterial design for wound healing should take into consideration the type, function, compatibility, and ease of activation of biological materials. Physical and chemical attributes will guide cell proliferation success. Porosity, mass transfer, fluid retention, and scaffolding qualities all play a tremendous role in achieving uniform, uninterrupted cellular regeneration, and are critical considerations in design. Adhesion is vital in the creation of biointerfaces; however, excessive bond strengths and incompatible mechanical properties between the biomaterial and the skin may damage superficial healing tissue during kinematic motion. Bionanomaterials should thus be manufactured using apposite methods with an aim of fulfilling specific design requirements. Such design requirements must, in turn, address the importance of tissue variations, which occur with age and, indeed, local properties of wounded tissue will vary with respect to the means by which it is inflicted.

REFERENCES

Alam, P, 2014. Protein unfolding versus beta-sheet separation in spider-silk nanocrystals. Adv. Nat. Sci. Nanosci. Nanotechnol. 5 (1), 015015.

Alam, P., 2015. Mechanical properties of bio-nanostructured materials. Aliofkhazraei, M. (Ed.), Handbook of Mechanical Nanostructuring, 1, Wiley-VCH, Weinheim, Germany, pp. 211–231, (Chapter 10).

Alam, P., Byholm, T., Kniivila, J., Sinervo, L., Toivakka, M., 2009. Calculating the permeability of model paper coating structures comprising incongruent particle shapes and sizes. Micropor. Mesopor. Mater. 117, 685–688.

Alam, P., Byholm, T., Toivakka, M., 2006. Calculating tortuosity in quasi-random anisotropic packings. Nordic Pulp Paper Res. J. 21, 670–675.

Argyris, T.S., 1977. Kinetics of regression of epidermal hyperplasia in the skin of mice following abrasion. Am. J. Pathol. 88, 575–581.

Arockianathan, P.M., Sekar, S., Sankar, S., Kumaran, B., Sastry, T.P., 2012. Evaluation of biocomposites films containing alginate and sago starch impregnated with silver nano particles. Carbohydr. Polym. 90, 717–724.

Aznar-Cervantes, S., Roca, M.I., Martinez, J.G., Meseguer-Olmo, L., Cenis, J.L., Moraleda, J.M., Otero, T.F., 2012. Fabrication of conductive electrospun silk fibroin scaffolds by coating with polypyrrole for biomedical applications. Bioelectrochemistry 85, 36–43.

Barbenel, J.C., Evans, J.H., 1977. The time-dependent mechanical properties of skin. J. Investig. Dermatol. 69, 318–320.

Barbie, C., Chaveaux, D., Barthe, X., Baquey, C., Poustis, J., 1990. Biological behaviour of cellulosic material after bone implants: preliminary results. Clin. Mater. 5, 251–258.

Berndt, S., Wesarg, F., Wiegand, C., Kralisch, D., Muller, F.A., 2013. Antimicrobial porous hybrids consisting of bacterial nanocellulose and silver nanoparticles. Cellulose 20, 771–783.

Birk, D.E., Trelstad, R.L., 1996. Fibroblasts compartmentalise the extracellular space to regulate and facilitate collagen fibril bundle, and macro-aggregate formation. In: Reddi, A.H., Liss, A.A. (Eds.), Extracellular Matrix: Structure and Function. Wiley, New York, NY.

Boateng, J.S., Matthews, K.H., Stevens, H.N.E., Eccleston, G.M., 2008. Wound healing dressings and drug delivery systems: a review. J. Pharm. Sci. 97, 2892–2923.

Bodin, A., Bäckdahl, H., Fink, H., Gustafsson, L., Risberg, B., Gatenholm, P., 2007. Influence of cultivation conditions on mechanical and morphological properties of bacterial cellulose tubes. Biotechnol. Bioeng. 97, 425–434.

Boucard, N., Viton, C., Agay, D., Mari, D., Roger, T., Chancerelle, Y., Domard, A., 2007. The use of physical hydrogels of chitosan for skin regeneration following third degree burns. Biomaterials 28, 3478–3488.

Bracic, M., Perez, L., Martinez-Pardo, R.I., Kogej, K., Hribernik, S., Sauperl, O., Zemljic, L.F., 2014. A novel synergistic formulation between a cationic surfactant from lysine and hyaluronic acid as an antimicrobial coating for advanced cellulosic materials. Cellulose 21, 2647–2663.

Brown, E.E., Laborie M-PG, Zhang, J., 2012. Glutaraldehyde treatment of bacterial cellulose/fibrin composites: impact on morphology, tensile and viscoelastic properties. Cellulose 19, 127–137.

Buschle-Diller, G., Cooper, J., Xie, Z., Wu, Y., Waldrup, J., Ren, X., 2007. Release of antibiotics from electrospun biocomponent fibres. Cellulose 14, 553–562.

Buttafoco, L., Kolkman, N.G., Engbers-Buijtenhuijs, P., Poot, A.A., Dijkstra, P.J., Vermes, I., Feijen, J., 2006. Electrospinning of collagen and elastin for tissue engineering applications. Biomaterials 27, 724–734.

Cam, C., Segura, T., 2014. Chemical sintering generates uniform porous hyaluronic acid hydrogels. Acta Biomater. 10, 205–213.

Campbell, T.A., Ivanova, O.S., 2013. 3D printing of multifunctional nanocomposites. Nano Today 8, 119–120.

Chen, L., Tredget, E.E., Wu, P.Y.G., Wu, Y., 2008. Paracrine factors of mesenchymal stem cells recruit macrophages and endothelial lineage cells and enhance wound healing. PLoS One 3 (4), e1886.

Chutipakdeevong, J., Ruktanonchai, U., Supaphol, P., 2015. Hybrid biomimetic electrospun fibrous mats derived from poly-ε-caprolactone and silk fibroin protein for wound dressing application. J. Appl. Polym. Sci. 132, 41653.

Clarke, J.A., Cheng, J.C.Y., Leung, K.S., 1996. Mechanical properties of normal skin and hypertrophic scars. Burns 22, 443–446.

Croisier, F., Atanasova, G., Poumay, Y., Jerome, C., 2014. Polysaccharide-coated PCL nanofibres for wound dressing applications. Adv. Health. Mater. 3, 2032–2039.

Daly, C.H., Odland, G.F., 1979. Age-related changes in the mechanical properties of human skin. J. Investig. Dermatol. 73, 84–87.

Darling, S.B., 2007. Directing the self-assembly of block copolymers. Prog. Polym. Sci. 32, 1152–1204.

Deepthi, S., Sundaram, J.M.N., Chennazhi, K.P., Jayakumar, R., 2015. Chitosan-hyaluronic acid hydrogel coated poly(caprolactone) multiscale bilayer scaffold for ligament regeneration. Chem. Eng. J. 260, 478–485.

Desmouliere, A., Chaponnier, C., Gabbiani, G., 2005. Tissue repair, contraction and the myofibroblast. Wound Repair Regener. 13, 7–12.

Eroglu, I., Gokce, E.H., Tsapis, N., Tannverdi, S.T., Gokce, G., Fattal, E., Ozer, O., 2015. Evaluation of characteristics and in vitro antioxidant properties of RSV loaded hyaluronic acid-DPPC microparticles as a wound healing system. Colloid Surf. B Biointerf. 126, 50–57.

Evans, K.E., Anderson, K.L., 2000. Auxetic materials: the positive side of being negative. Eng. Sci. Ed. J. 9, 148–154.

Ferras, N., Stromme, M., Fellstrom, B., Pradhan, S., Nyholm, L., Mihranyan, A., 2012. In vitro and in vivo toxicity of rinsed and aged nanocellulose-polypyrrole composites. J. Biomed. Res. A 100A, 2128–2138.

Gao, C., Wan, Y., Yang, C., Dai, K., Tang, T., Luo, H., Wang, J., 2011. Preparation and characterization of bacterial cellulose sponge with hierarchical pore structure as tissue engineering scaffold. J. Porous Mater. 18, 139–145.

Greiner, A., Wendorff, J.H., 2007. Electrospinning: a fascinating method for the preparation of ultrathin fibres. Angew. Chem. 46, 5670–5703.

Hu, D.L., Chan, B., Bush, J.W.M., 2003. The hydrodynamics of water strider locomotion. Nature 424, 663–666.

Huang, L., Nagapudi, K., Apkarian, R.P., Chaikof, E.L., 2001. Engineered collagen-PEO nanofibres and fabrics. J. Biomater. Sci. 12, 979–993.

Huang, Y., Zhu, C., Yang, J., Nie, Y., Chen, C., Sun, D., 2014. Recent advances in bacterial cellulose. Cellulose 21, 1–30.

Jayakumar, R., Pabaharan, M., Kumar, P.T.S., Nair, S.V., Tamura, H., 2011. Biomaterials based on chitin and chitosan in wound dressing applications. Biotechnol. Adv. 29, 322–337.

Jin, H.J., Fridrikh, V.D., Rutledge, G.C., Kaplan, D.L., 2002. Electrospinning *Bombyx mori* silk with poly(ethylene oxide). Biomacromolecules 3, 1233–1239.

Jin, H.J., Park, J., Karageorgiou, V., Kin, U.J., Valluzzi, R., Cebe, P., et al., 2005. Water stable silk films with reduced beta-sheet content. Adv. Funct. Mater. 15, 1241–1247.

Kalaskar, D., Gough, J.E., Ulijn, R.V., Sampson, W.W., Scurr, D.J., Rutten, F.J.M., Alexander, M.R., Merry, C.L.R., Eichhorn, S.J., 2008. Controlling cell morphology on amino acid modified cellulose. Soft Matter 4, 1059–1065.

Karahaliloglu, Z., Ercan, B., Denkbas, E.B., Webster, T.J., 2015. Nanofeatured silk fibroin membranes for dermal wound healing applications. J. Biomed. Res. A 103A, 135–144.

Khajavi, R., Abbasipour, M., 2012. Electrospinning as a versatile method for fabricating coreshell hollow and porous nanofibres. Sci. Iran. 19, 2029–2034.

Kim, C.W., Frey, M.W., Marquez, M., Joo, Y.L., 2005a. Preparation of submicron-scale, electrospun cellulose fibres via direct dissolution. J. Polym. Sci. B Polym. Phys. 43, 1673–1683.

Kim, K.H., Jeong, L., Park, H.N., Shin, S.Y., Park, W.H., et al., 2005b. Biological efficacy of silk fibroin nanofibre membranes for guided bone regeneration. J. Biotechnol. 120, 327–339.

Kim, S.H., Nam, Y.S., Lee, T.S., Park, W.H., 2003. Silk fibroin nanofibre. Electrospinning, properties, and structure. Polym. J. 35, 185–190.

Kino, Y., Sawa, M., Kasai, S., Mito, M., 1998. Multiporous cellulose microcarrier for the development of hybrid artificial liver using isolated hepatocytes. J. Surg. Res. 79, 71–76.

Klemm, D., Schumann, D., Kramer, F., Hessler, N., Hornung, M., Schmauder, H.P., Marsch, S., 2006. Nanocellulose as innovative polymers in research and application. Polysaccharides II 205, 49–96.

Kogej, K., 2010. Association and structure formation in oppositely charged polyelectrolyte-surfactant mixtures. Adv. Colloid Interf. Sci. 158, 68–83.

Kong, H.H., 2011. Skin microbiome: genomics-based insights into the diversity and role of skin microbes. Trends Mol. Med. 17, 320–328.

Kulpinski, P., 2005. Cellulose nanofibres prepared by the *N*-methylmorpholine-*N*-oxide method. J. Appl. Polym. Sci. 98, 1855–1859.

Kumar, M.N.V.R., 2000. A review of chitin and chitosan applications. React. Funct. Polym. 46, 1–27.

Lee, E., Xia, Y., Kim, Y., Kim, W.-S., Kim, M.H., Kim, T.H., Park, B.-S., Sung, J.H., 2009. Hypoxia-enhanced wound-healing function of adipose-derived stem cells: Increase in stem cell proliferation and up-regulation of VEGF and bFGF. Wound Repair Regener. 17, 540–547.

Lin, Y.-K., Chen, K.-H., Ou, K.-L., Liu, M., 2011. Effects of different extracellular matrices and growth factor immobilization on biodegradability and biocompatibility of macroporous bacterial cellulose. J. Bioact. Compat. Polym. 26, 508–518.

Liu, W., Jawerth, L., Sparks, E., Falvo, M., Hantgan, R., Superfine, R., Lord, S., Guthold, M., 2006. Fibrin fibers have extraordinary extensibility and elasticity. Science 313, 634.

Lu, Q., Zhu, H., Zhang, C., Zhang, F., Zhang, B., Kaplan, D.L., 2012. Silk self-assembly mechanisms and control from thermodynamics to kinetics. Biomacromolecules 13, 826–832.

Marmaras, A., Lendenmann, T., Civenni, G., Franco, D., Poulikakos, D., Kurtcuoglu, V., Ferrari, A., 2012. Topography-mediated apical guidance in epidermal wound healing. Soft Matter 26, 6922–6930.

Martin, P., 1997. Wound healing—aiming for perfect skin regeneration. Science 276, 75–81.

Matthews, J.A., Wnek, G.E., Simpson, D.G., Bowlin, G.L., 2002. Electrospinning of collagen nanofibres. Biomacromolecules 3, 232–238.

Meftahi, A., Khajavi, R., Rashidi, A., Sattari, M., Yazdanshenas, M.E., Torabi, M., 2010. The effects of cotton gauze coating with microbial cellulose. Cellulose 17, 199–204.

Min, B.M., Lee, G., Kim, S.H., Nam, Y.S., Lee, T.S., Park, W.H., 2004. Electrospinning of silk fibroin nanofibres and its effect on the adhesion and spreading of normal human keratinocytes and fibroblasts in vitro. Biomaterials 25, 1289–1297.

Minagawa, T., Okamura, Y., Shigemasa, Y., Minami, S., Okamoto, Y., 2007. Effects of molecular weight and deacetylation degree of chitin/chitosan on wound healing. Carbohydr. Polym. 67, 640–644.

Miyamoto, T., Takahashi, S., Ito, H., Inagaki, H., 1989. Tissue compatibility to cellulose and its derivatives. J. Biomed. Mater. Res. 23, 125–133.

Moon, R.J., Martini, A., Nairn, J., Simonsen, J., Youngblood, J., 2011. Cellulose nanomaterials review: structure, properties and nanocomposites. Chem .Soc. Rev. 40, 3941–3994.

Mori, T., Okumura, M., Marsuura, M., Ueno, K., Tokura, S., Okamoto, Y., Minami, S., Fujinaga, T., 1997. Effects of chitin and its derivatives on the proliferation and cytokine production of fibroblasts in vitro. Biomaterials 18, 947–951.

Nakagawa, H., Akita, S., Fukui, M., Fujii, T., Akino, K., 2005. Human mesenchymal stem cells successfully improve skin-substitute wound healing. Br. J. Dermatol. 153, 29–36.

Ohkawa, K., Cha, D., Kim, H., Nishida, A., Yamamoto, H., 2004. Electrospinning of chitosan. Macromol. Rapid Comm. 25, 1600–1605.

Orgill, D.P., 1983. The effects of an artificial skin on scarring and contraction in open wounds. PhD Thesis, Harvard University.

Ortega-Jimenez, V.M., Dudley, R., 2013. Spiderweb deformation induced by electrostatically charged insects. Sci. Rep. 3, 1–4.

Pahlevan, M., Toivakk, M., Alam, P., 2014. Electrostatic charges instigate 'concertina-like' mechanisms of molecular toughening in MaSp1 (spider silk) proteins. Mater. Sci. Eng. C 41, 329–334.

Pailler-Mattei, C., Bec, S., Zahouani, H., 2008. In vivo measurements of the elastic mechanical properties of human skin by indentation tests. Med. Eng. Phys. 30, 599–606.

Pan, L., Zan, L., Foster, F.S., 1998. Ultrasonic and viscoelastic properties of skin under transverse mechanical stress in vitro. Ultrasound Med. Biol. 24, 995–1007.

Park, W.H., Jeong, L., Yoo, D.I., Hudson, S., 2004. Effect of chitosan on morphology and conformation of electrospun silk fibroin nanofibres. Polymer 45, 7151–7157.

Petersen, N., Gatenholm, P., 2011. Bacterial cellulose-based materials and medical devices: current state and perspectives. Appl. Microbiol. Biotechnol. 91, 1277–1286.

Pinho, E., Henriques, M., Soares, G., 2014. Cyclodextrin/cellulose hydrogel with gallic acid to prevent wound infection. Cellulose 21, 4519–4530.

Putra, A., Kakugo, A., Furukawa, H., Gong, J.P., Osada, Y., 2008. Tubular bacterial cellulose gel with oriented fibrils on the curved surface. Polymer 49, 1885–1891.

Rajkhowa, R., 2014. Silk powder for regenerative medicine. In: Kundu, S.C. (Ed.), Silk Biomaterials for Tissue Engineering and Regenerative Medicine. Woodhead Publishing Ltd, Sawston, UK.

Rambo, C.R., Recouvreux, D.O.S., Carminatti, C.A., Pitlovanciv, A.K., Antonio, R.V., Porto, L.M., 2008. Template assisted synthesis of porous nanofibrous cellulose membranes for tissue engineering. Mat. Sci. Eng. C 28, 549–554.

Rho, K.S., Jeong, L., Lee, G., Seo, B.M., Park, Y.J., Hong, S.D., Roh, S., Cho, J.J., Park, W.H., Min, B.M., 2006. Electrospinning of collagen nanofibers: effects on the behaviour of normal keratinocytes and early-stage wound healing. Biomaterials 27, 1452–1461.

Rinuado, M., 2006. Chitin and chitosan: properties and applications. Prog. Polym. Sci. 31, 603–632.

Rockwood, D.N., Preda, R.C., Yucel, T., Wang, X., Lovett, M.L., Kaplan, D.L., 2011. Materials fabrication from Bombyx mori silk fibroin. Nat. Protoc. 6, 1612–1631.

Romer, L., Scheibel, T., 2008. The elaborate structure of spider silk. Prion 2, 154–161.

Rosamond, W., Flegal, K., Furie, K., Go, A., Greenlund, K., Haase, N., Hailpern, S., Ho, M., Howard, V., Kissela, B., 2008. American heart association statistics committee and stroke statistics subcommittee. Heart disease and stroke statistics-2008 update: a report from the American Heart Association Statistics Committee and Stroke Statistics Subcommittee. Circulation 117 (4), e25–e146.

Samaha, M.A., Tafreshi, H.V., Gad-el-Hak, M., 2012. Superhydrophobic surfaces: from the lotus leaf to the submarine. Comptes Rendus Mecanique 340, 18–34.

Saxena, V., 2009. Biomechanics of skin. In: Orgill, D., Blanco, C. (Eds.), Biomaterials for Treating Skin Loss. Woodhead Publishing Ltd, Sawston, UK.

Schneider, A., Wang, X.Y., Kaplan, D.L., Garlick, J.A., Egles, C., 2009. Biofunctionalised electrospun silk mats as a topical bioactive dressing for accelerated wound healing. Acta Biomater. 5, 2570–2578.

Sehaqui, H., Zhou, Q., Ikkala, O., Berglund, L.A., 2011. Strong and tough nanopaper with high specific surface area and porosity. Biomacromolecules 12, 3638–3644.

Sezer, A.D., Hatipoglu, F., Cevher, E., Ogurtan, Z., Bas, A.L., Akbuga, J., 2007. Chitosan film containing Fucoidan as a wound dressing for dermal burn healing: preparation and in vitro/in vivo evaluation. AAPS PharmSciTech 8, E1–E8.

Shalumon, K.T., Anulekha, K.H., Nair, S.V., Nair, S.V., Chennazhi, K.P., Jayakumar, R., 2011. Sodium alginate/poly(vinyl alcohol)/nano ZnO composite nanofibres for antibacterial wound dressings. Int. J. Biol. Macromol. 49, 247–254.

Shan, Y.H., Peng, L.H., Liu, X., Chen, X., Xiong, J., Gao, J.Q., 2015. Silk fibroin/gelatin electrospun nanofibrous dressing functionalised with astragaloside IV induces healing and anti-scar effects on burn wound. Int. J. Pharm. 479, 291–301.

Shapira, A., Kim, D.H., Dvir, T., 2014. Advanced micro- and nanofabrication technologies for tissue engineering. Biofabrication 6, 020301.

Singh, D.K., Ray, A.R., 2000. Biomedical applications of chitin, chitosan, and their derivatives. J. Macromol. Sci. C 40, 69–83.

Singha, A.S., Kapoor, H., 2015. Evaluation of physico-chemical properties of modified silk fibre. J. Nat. Fibers 12, 604–616.

Sintya, E., Alam, P., 2016. Self-assembled semi-crystallinity at parallel beta-sheet nanocrystal interfaces in clustered MaSp1 (spider silk) proteins. Mater. Sci. Eng. C Mater. Biol. Appl. 58, 366–371.

Slotta, U., Hess, S., Spiess, K., Stromer, T., Serpel, L., Scheibel, T., 2007. Spider silk and amyloid fibrils: a structural comparison. Macromol. Biosci. 7, 183–188.

Slotta, U., Rammensee, S., Gorb, S., Scheibel, T., 2008. An engineered spider silk protein forms microspheres. Angew. Chem. Int. Ed. 47, 4592–4594.

Smidsrod, O., Braek, G.S., 1990. Alginate as immobilisation matrix for cells. Trends Biotechnol. 8, 71–78.

Song, J., Birbach, N.L., Hinestroza, J.P., 2012. Deposition of silver nanoparticles on cellulosic fibers via stabilization of carboxymethyl groups. Cellulose 19, 411–424.

Sukigara, S., Gandhi, M., Ayutsede, J., Micklus, M., Ko, F., 2003. Regeneration of *Bombyx mori* silk by electrospinning—part 1: processing parameters and geometric properties. Polymer 44, 5721–5727.

Syverus, K., Pettersen, S.R., Draget, K., Chinga-Carrasco, G., 2014. Controlling the elastic modulus of cellulose nanofibril hydrogels-scaffolds with potential in tissue engineering. Cellulose 22, 473–481.

Taokaew, S., Phisalaphong, M., Newby, B.Z., 2015. Modification of bacterial cellulose with organosilanes to improve attachment and spreading of human fibroblasts. Cellulose 22, 2311–2324.

Taokaew, S., Seetabhawang, S., Siripong, P., Phisalaphong, M., 2012. Biosynthesis and characterization of nanocellulose-gelatin films. Materials 6, 782–794.

Tarun, K., Gobi, N., 2012. Calcium alginate/PVA blended nano fibre matrix for wound dressing. Indian J. Fibre Text. Res. 37, 127–132.

Ueno, H., Mori, T., Fujinaga, T., 2001. Topical formulations and wound healing applications of chitosan. Adv. Drug Deliv. Rev. 52, 105–115.

Ul Islam, M., Khan, T., Khattak, W.A., Park, J.K., 2013. Bacterial cellulose-MMTs nanoreinforced composite films: novel wound dressing material with antibacterial properties. Cellulose 20, 589–596.

Uskokovic, V., 2008. Isn't self-assembly a misnomer? Multi-disciplinary arguments in favour of co-assembly. Adv. Colloid Interf. Sci. 141, 37–47.

Veliz, D.S., Alam, C., Toivola, D., Toivakka, M., Alam, P., 2014. On the nonlinear attachment characteristics of blood to bacterial cellulose/kaolin biomaterials. Colloids Surf. B Biointerf. 116, 176–182.

Vermeulen, H., Ubbink, D.T., Goossens, A., de Vos, R., Legemate, D.A., 2005. Systematic review of dressings and topical agents for surgical wounds by secondary intention. Br. J. Surg. 92, 665–672.

Wang, M., Jin, H.J., Kaplan, D.L., Rutledge, G.C., 2004. Mechanical properties of electrospun silk fibres. Macromolecules 37, 6856–6864.

Wang, Y., Blasioli, D.J., Kim, H.J., Kim, H.S., Kaplan, D.L., 2006. Cartilage tissue engineering with silk scaffolds and human chondrocytes. Biomaterials 27, 4434–4442.

Wanna, D., Alam, C., Toivola, D., Alam, P., 2013. Bacterial cellulose-kaolin nanocomposites for application as biomedical wound healing materials. Adv. Nat. Sci. Nanosci. Nanotechnol. 4 (4), 045002.

Webster, J., Scuffham, P., Stankiewicz, M., Chaboyer, W.P., 2014. Negative pressure wound therapy for skin grafts and surgical wounds healing by primary intention. Cochrane Database Syst. Rev. 10, 1–47, Art. No.: CD009261.

Wu, Y., Chen, L., Scott, P.G., Tredget, E.E., 2007. Mesenchymal stem cells enhance wound healing through differentiation and angiogenesis. Stem Cells 25, 2648–2659.

Yan, F., Gang, S., TongQing, W., Qian, C., LuQuan, R., 2007. Hydrophobicity mechanism of non-smooth pattern on surface of butterfly wing. Chinese Sci. Bull. 52, 711–716.

Yannas, I.V., 2009. Principles of skin regeneration. In: Orgill, D., Blanco, C. (Eds.), Biomaterials for Treating Skin Loss. Woodhead Publishing Ltd, Sawston, UK.
Yannas, I.V., Burke, J.F., Orgill, D.P., Skrabut, E.M., 1982. Regeneration of skin following closure of deep wounds with a biodegradeable template. Trans. Soc. Biomater. 5, 24–27.
Yannas, I.V., Burke, J.F., Warpehoski, M., Stasikelis, P., Skrabut, E.M., Orgill, D., Giard, D.J., 1981. Prompt long-term functional replacement of skin. Trans. Am. Soc. Artif. Intern. Org. 27, 19–22.
Zaumseil, J., Meitl, M.A., Hsu, J.W.P., Acharya, B.R., Baldwin, K.W., Loo, Y.L., Rogers, J.A., 2003. Three-dimensional and multilayer nanostructures formed by nanotransfer printing. Nano Lett. 3, 1223–1227.

CHAPTER

ROLE OF NANOSTRUCTURE MOLECULES IN ENHANCING THE BIOAVAILABILITY OF ORAL DRUGS

14

Muhammad A. Azmi*, Kaneez F. Shad**

**Isra University Karachi Campus, Karachi, Pakistan*

***University of Technology Sydney (UTS), Sydney, NSW, Australia*

CHAPTER OUTLINE

1 NANOTECHNOLOGY: SOME BASIC CONCEPTS

The term "nano" means small or very small or dwarf whereas "technology" clearly relates an emerging trend or design in certain systems. Putting these "two" together signifies as a technical knowledge in designing, creation, synthesis, fabrication and application of nanostructure molecules and nanomaterials at the nanometer scale. The concept of nanotechnology has been explained in many ways such as it is mainly concerned with the materials and systems that are significantly shared in the biological process on the basic of their nanoscale size. It also explains the fundamental relationships between physical properties, phenomena and the dimensions of materials used in the nanometer which are more correctly referred as Nanoscience. To describe this terminology, is the study of phenomena and categorization of materials on atomic, molecular and macromolecular scales of measurements whose properties differ significantly from other structures at a larger scale as reported by the Royal Society and Royal Academy of Engineering (2004).

Nanostructures for Novel Therapy

1.1 HISTORICAL BACKGROUND

The concept of nanotechnology is supported by many researchers in the field of nanotechnology. One of the researcher who clearly explained the basic concept of nanotechnology is the Nobel Prize Winner Richard Feynman in 1957 in his speech titled "There is plenty of room at the bottom" at the American Physiological Society's winter meeting. Feynman in his speech stated that the capabilities of assembling of atom-by-atom and nano-engineering could lead to new materials and pathways similar to biological system (Matija, 2004). He also presented a new technology vision of miniaturization of materials that are important in manipulating and regulating materials on a small scale, which he termed as Nanotechnology. Feynman in this sense visualized this technology by exploring a toolbox of nature to build nano-object as atom by atom and molecule by molecule (Majumder et al., 2007). The word nanoscale, nano-engineering, nanotechnology and nano-object according to Feynman idea are the basic terminology and tools used in modern technology.

The term "Nanotechnology" was first introduced in 1974 by Norio Taniguchi, a researcher at the University of Tokyo, who used this technology as an ability to assemble materials at nanoscale. In the 1980s, the development of nanotechnologies has been attributed by the invention of two important analytical tools that have revolutionized the imaging of individual atoms or molecules as well as their manipulation of surfaces at the nanoscale that provide significant importance in the field of nanotechnology (Cortie, 2004; Matija, 2004). These two tools are the scanning tunneling microscope (STM) and the atomic force microscope (AFM). Both these inventing tools are capable of imaging surfaces with atomic resolution. The STM was invented by Gerd Binning while AMF by Henrich Rohrer. STM was the first in realizing Feynman's vision of atom-by-atom fabrication. Besides these, Fullerence C_{60} was discovered in 1985 by Kroto's and his colleagues. In 1991, Saumio Lijima discovered carbon nanotubes and nanotechnology-based investments, which have also been launched for the progress of research and development in the field of nanotechnology (Matija, 2004; Roco, 2004).

Nanotechnology today focuses on the development, understanding and use of materials at the nanoscale or nanomaterials. Some biological systems contain nanosystems that provide some specific functions, such as locomotion, where actin moves along myosin and kinesin moves along microtubules (Eijkel and van den Berg, 2006). Also a DNA molecule can be seen as a self-assembly machine, which replicates itself and able to produce complex organisms in a right conditions. Ribosomes on the other hand construct protein molecules with the clear indications from DNA (Sheeparamatti et al., 2007). Research provide sufficient knowledge in recent years into the principles of natural systems would enhance the design and fabrication of man-made nanostructure that may reduce the activities of natural systems. For instance, biomolecules, such as proteins, peptides, DNA, lipids and carbohydrates can act as templates in such a way that their shape and chemical properties can be employed to arrange inorganic substances, such as metals on nanoscale (Bittner, 2005).

1.2 NANOMETER SCALE IN NANOSYSTEMS

In nanosystems, nanotechnology deals with small structures or small sized materials. The typical dimension extended from subnanometer to several hundred nanometers range. The nanometer scale is defined as 1–100 nm. One nanometer is one billionth of a meter. The size range is normally set to a minimum of 1 nm to avoid one single atom, or a very small group of atoms, and these are designated as "Nano-objects." The upper limit of nanoscale is 100 nm, but this is a "fluid" limit. Now, the question

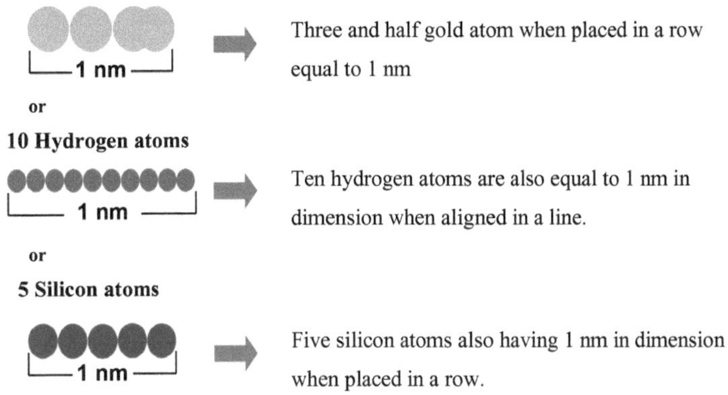

FIGURE 14.1 Different Atoms Showing Dimensions of 1 nm When Aligned in a Row

that arises here is why 100 nm and not 150 nm, or even why not 1–1000 nm; the reason is that materials exhibit unique properties at 1–100 nm. The change in properties is due to increase in surface area, and dominance of quantum effects associated with very small size and large surface area to volume ratio (Williams, 2008). There are several examples that explain this physical phenomenon. Copper is opaque at macroscale, but becomes transparent at nanoscale (Gao et al., 2003; Zong et al., 2005). Platinum is inert, but becomes a catalyst at nanoscale (Lau et al., 2005; Tian et al., 2007). Otherwise stable aluminum is combustible, while known insulator silicon becomes conductor at nanoscale (Heron et al., 2007; Hu et al., 2003; Patel-Predd, 2008; Shafirovich et al., 2006, 2007). The properties of gold at nanoscale are quite significant depending on the size of the nanoparticles. Here we are giving some examples of atoms regarding their nanoscale dimensions (Fig. 14.1).

Materials in the micrometer scale mostly exhibit physical properties similar to that of the bulk form. However, materials in the nanometer scale may exhibit physical properties distinctively different from that of bulk form. Any material in this size range exhibits remarkable specific properties, such as a transition from atoms or molecules to bulk form takes place in this range.

1.3 NANOMATERIAL

A nanomaterial is defined as an object that has at least one dimension in the nanometer scale, which is approximately 1–100 nm. Nanomaterials are categorized into different types on the basis of their dimensions. We are presenting here some important classes of nanostructured materials with their dimensions (Fig. 14.2).

1.3.1 Classification of nanomaterials

Nanomaterial can be classified into two main types:

1.3.2 Nonintentional produced nanomaterials

It represents to nanosized particles or materials that belong to the ingredients which are appropriate to the environment, that is, proteins, viruses produced by human activity without intention.

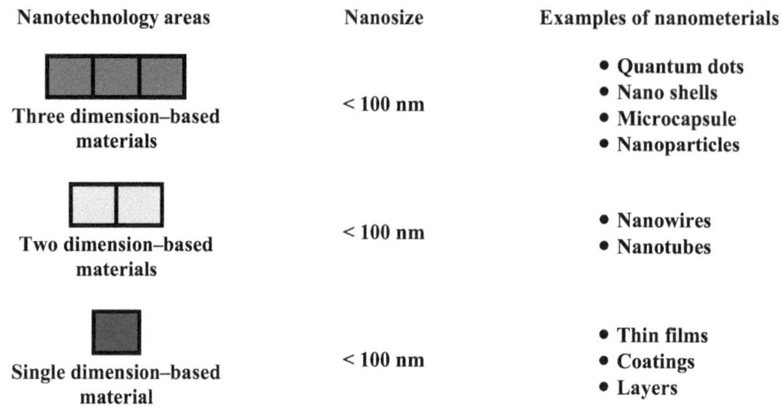

FIGURE 14.2 Dimension-Based Materials

1.3.3 Intentionally produced nanomaterials

This represents to nanomaterials produced deliberately through a well-designed fabrication process.

Intentionally produced nanomaterials are further classified into four subtypes:

- carbon-based materials
- metal-based materials
- dendrimers
- composites

1.3.3.1 Carbon-based materials

Nanomaterials are mostly composed of carbon, richly present in the hollow spheres, ellipsoids, or tubes. Spherical and ellipsoidal carbon nanomaterials are referred usually as fullerenes, while cylindrical ones are called nanotubes. These nanomaterial particles have many potential applications including films and coatings.

1.3.3.2 Metal-based materials

These nanomaterials include quantum dots (QDs), nanogold, nanosilver metal, and oxides, such as titanium oxide. A QD is a closely packed semiconductor crystal comprised of hundreds or thousands of atoms, and whose size is in few nanometers to a few hundred nanometers. This means that changing in the size of QDs changes their optical properties.

1.3.3.3 Dendrimers

These nanometerials are nanosized polymers built from branched units. The surface of a dendrimer has numerous chain ends, which can perform several chemical functions. This unique property could be useful for catalysts. In addition, this type of nanomaterial contains interior cavities into which other molecules could be adjusted or placed because of three-dimensional (3D) dendrimers, and this property may be useful for drug delivery.

1.3.3.4 Composites

This type of nanomaterial can combine with other nanoparticles or with larger bulk type materials. Nanoparticles like nanosized clays are already being added to products ranging from auto parts to packaging materials to enhance mechanical, thermal, barrier, and flame-retardant properties.

2 NANOSTRUCTURE MOLECULES OR NANOPARTICLES

A nanostructure is defined as any structure with one or more dimension, measuring in the nanometer scale range, that is, 10^{-9} m. Its structure is more clarified in this respect that a nanostructure should have a unique characteristic in terms of dimensions focusing between 1 nm and 100 nm, and this puts nanostructure molecules as an intermediate size between a molecule and a bacterium; nanoparticles are represented as any particles having a diameter smaller than 100 nm. In nanotechnology, a particle is represented as a small object, but behaves as a whole unit in relation to its transport and physical characteristics. The nanoparticle is categorized into various classes on the basis of diameter; in case of fine particles, that measure is between 100 and 2500 nm, whereas ultrafine particles are showing their size between 1 and 100 nm.

Nanoparticles behave with size-related properties that differ, by comparison with fine particles or bulk materials (Buzea et al., 2007). This shows that individual molecules are not called nanoparticles, but the collection of fine particles, when they fit into the structure molecule. In general, nanoparticles are particles having a diameter smaller than 100 nm and are potentially important in biomedical, optical, and in drug carrier systems. Another important characteristic is that nanoparticles have attractive forces in relation to covalent and noncovalent linkages, and this unique feature is based on their surface to mass ratio which is much large than other particles and materials, and this is important for increasing their ability to absorb and transport other compounds. Therefore, because of having large functional surface, nanoparticle is able to bind, absorb, and carry other compounds, such as drugs, proteins, etc. (Paul et al., 2004).

2.1 ROLE OF NANOPARTICLES IN BIOMEDICINE

The involvement of micro and nanoparticles in biomedicine particularly in drug-delivery systems has greater importance than conventional systems of drug delivery. Some of the major advantages of using these nanosized particles in drug delivery are:

- an ability of enhancing the product delivery
- having greater performance characteristics of the product
- extension of the bioactivity of drugs
- an ability to protect the product from environmental effects
- having more therapeutic characteristics with minimal side effects

Use of micro- and nanocolloidal drugs and other nanostructure molecules have crucial application in nanomedicine. Some of the important nanostructure molecules with their clinical applications are listed.

2.1.1 Liposomes

The term liposome is derived from Greek, where lipo means "fat" and soma means "body." It so called because its composition is mainly based of phospholipid (Fig. 14.3). These are small vesicular

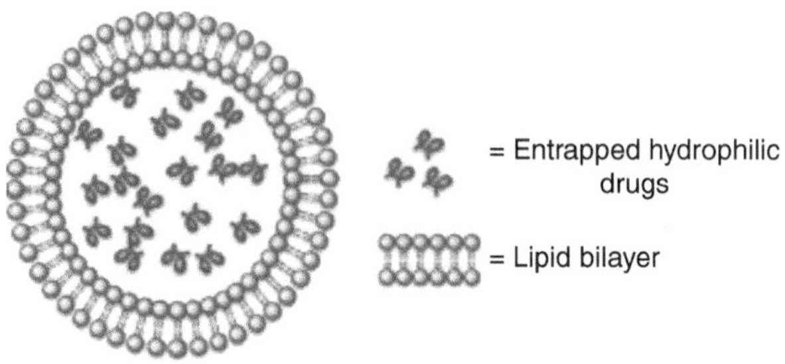

FIGURE 14.3 Liposomes

Reproduced with permission from Xu, Q., Kambhampati, S.P., Kannan, R.M., 2013. Nanotechnology approaches for ocular drug delivery. Middle East Afr. J. Ophthalmol. 20 (1), 26–37. Medknow Publications.

structures of 50–100 nm in dimension. The vesicles contain one or more aqueous compartments, which are completely enclosed by molecules with both hydrophilic and hydrophobic characteristics. The functional ability of the liposomes may vary in their composition, size, surface charge, and method of preparation (Yih and Al-Fandi, 2006). Their outline is composed of single or multiple bilayers. If they have only single lipid bilayer membrane, they are termed as unilamellar vesicles, or in some cases they are called large unilamellar vesicles, depending on their sizes. On the other hand, when they contain more than one bilayer membrane they are called multilamellar vesicles (Mozafari and Sahin, 2005). Liposomes are composite structure made of phospholipids and many other small molecules.

Liposomes that contain low or high pH can be constructed in such a way that the dissolved aqueous drugs could be charged in solution. As the pH neutralizes the drug within, the liposomes will also be neutralized, thus allowing it to pass freely through a membrane. These liposomes work to deliver drugs by a process of diffusion, rather than by direct cell fusion. Another way for drug delivery by liposomes is to target endocytosis. Liposomes exist in a particular size, and this is important in targeting natural macrophage–phagocytosis process. Such liposomes may be digested in macrophage, while phagosome releases the drugs. In addition, liposomes can also be constructed with opsonins and ligands, which are important in activating other cell types.

Application:

- models for artificial cells,
- carriers for the delivery of dyes to textiles,
- distribution of pesticides to plants, enzymes, and nutritional supplements to foods and cosmetics to the skin,
- microbubble contrast agents for enhanced ultrasound, and
- amphiphilic character, biocompatibility, and easy surface attachment of biotechnological drugs.

2.1.2 Polymeric micelles

These are supramolecular structures produced from self-assemblage of amphiphilic molecules (Fig. 14.4). They are present in colloidal range consisting of 50–200 nm, with an aggregation number defined as an average number of monomers to form micelles at a given time. The formation of micelles occurs when polymeric molecules combine in aqueous medium (solvent) forming core–shell structure or vesicles. These can be synthesized from the union of some amphiphilic copolymers, such

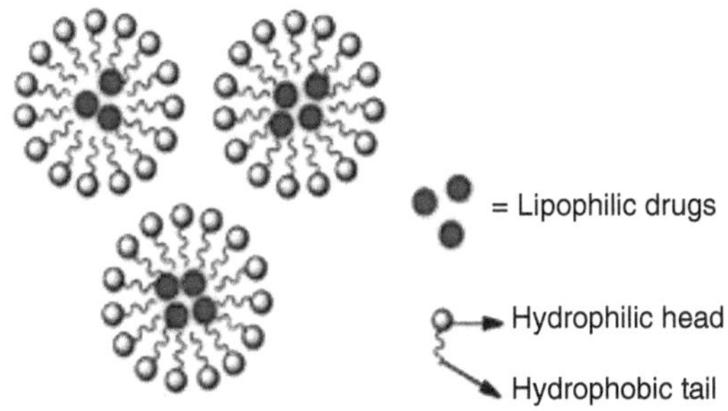

FIGURE 14.4 Polymeric Micelles

Reproduced with permission from Xu, Q., Kambhampati, S.P., Kannan, R.M., 2013. Nanotechnology approaches for ocular drug delivery. Middle East Afr. J. Ophthalmol. 20 (1), 26–37. Medknow Publications.

as polyethylene oxide, poly β-benzyl-L-aspartate, and poly *N*-isopropylacrylamide polystyrene. These structures are more stable than surfactant micelles in normal solutions (Jones and Leroux, 1999). The inner hydrophobic core is covered by a shell of hydrophilic polymers, for example, polyethylene glycol (Nishiyama and Kataoka, 2006). The hydrophobic core provides the poor incorporation of water-soluble and amphiphilic drugs, while its hydrophilic shell, which is less than 100 nm prolongs the circulation of the drug in the blood (Jones and Leroux, 1999). Polymeric micelles are formed when amphiphilic molecules combine with solvent, resulting in the self-association of amphiphilic monomers due to the decrease in free energy of the system. Self-association occurs due to the removal of hydrophobic fragments from aqueous surroundings by the formation of micelles with hydrophilic fragments, when exposed to water.

Factors affecting the micelles formation are:

- molecular weight of monomer
- aggregation number
- proportion and length of hydrophobic and hydrophilic chains

Types of polymeric micelles (based on intramolecular forces):

1. Conventional polymeric micelles: Result from hydrophilic interaction, for example, polyethylene oxide, polypropylene oxide.
2. Poly–ion complex micelles: Result from electrostatic interaction between apposite charged moieties, for example, polyelectrode.
3. Noncovalent connected polymeric micelles: A novel block copolymer technique obtained by self-assemblage of homopolymer, random copolymer, and graft copolymer in which the core and shells are noncovalently connected at their homopolymer chain by H-bonding. For example, poly(4-vinyl pyridine), a backbone, and carbonyl terminated polybutadiene as the graft.

Application:

- Involved in solubilization process, which leads to enhance solubility of water-insoluble molecules in water.
- Nanosized polymeric micelles elevate gastrointestinal uptake and enhance its bioavailability.
- Involved in targeted and intracellular delivery.

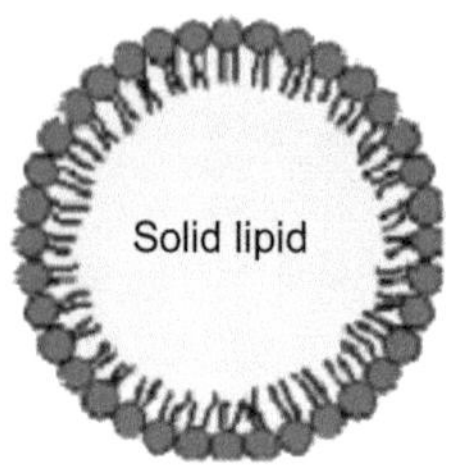

FIGURE 14.5 Solid Lipid Nanoparticle

Reproduced with permission from Xu, Q., Kambhampati, S.P., Kannan, R.M., 2013. Nanotechnology approaches for ocular drug delivery. Middle East Afr. J. Ophthalmol. 20 (1), 26–37. Medknow Publications.

2.1.3 Solid lipid nanoparticles

Typically spherical in shape, with an average diameter between 10 and 1000 nm (Muller et al., 2000) (Fig. 14.5). These nanoparticles possess a solid lipid core matrix that can solubilize lipophilic molecules. The lipid core is stabilized by surfactants which act as emulsifiers.

The solid lipid nanoparticles are made from solid lipids, such as triglycerides (e.g., tristearin), diglyceride (e.g., glycerol bahenate), monoglyceride (e.g., glycerol monostearate), fatty acids (e.g., stearic acid), steroids (e.g., cholesterol), and waxes (e.g., cetylpalmitate). The different classes of emulsifiers with respect to molecular weight and charge have been used to stabilize lipid dispersion (Mehnert et al., 2001). Solid lipid nanoparticles are of two types:

1. Nanospheres

   These are solid core spherical particulates, which contain drugs embedded within the matrix, or observed on the surface (e.g., matrix type) (Fig. 14.6).
2. Nanocapsules

   They are vesicular-like structure in which drugs are essentially encapsulated within the central core, surrounded by a polymeric sheath (e.g., reservoir type) (Fig. 14.7).

   Applications:

- Solid lipid nanoparticles (SLN) possess enhanced appropriateness and ability to produce large scale as compared to liposomes;

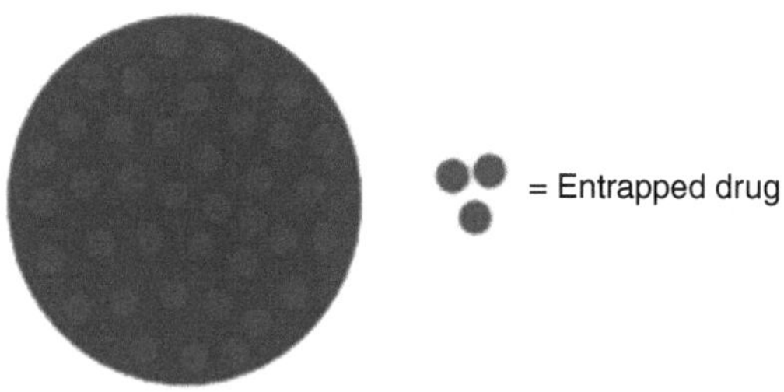

FIGURE 14.6 Nanospheres

Reproduced with permission from Xu, Q., Kambhampati, S.P., Kannan, R.M., 2013. Nanotechnology approaches for ocular drug delivery. Middle East Afr. J. Ophthalmol. 20 (1), 26–37. Medknow Publications.

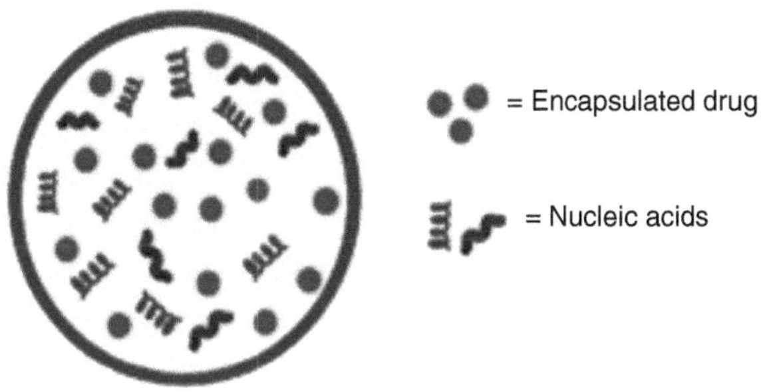

FIGURE 14.7 Nanocapsules

Reproduced with permission from Xu, Q., Kambhampati, S.P., Kannan, R.M., 2013. Nanotechnology approaches for ocular drug delivery. Middle East Afr. J. Ophthalmol. 20 (1), 26–37. Medknow Publications.

- These particles form the basis of colloidal drug-delivery system,
- Provide narrow and small size distribution of drugs to the specific site,
- Offer protection of incorporated drugs against chemical degradation.

2.1.4 Dendrimers

The term dendrimers comes from Greek word "Dendron" which translates as "tree." A dendrimer is a high-branched spherical structure having inner core of synthetic polymer with all functional groups facing away from the core just like the suction cup ball (Fig. 14.8).

It is a unique class of polymer whose particles ranges from 1 to 100 nm, though their size and shape can be precisely controlled. These are produced from macromolecules, such as polyamidoamine, polypropyleneimine, and polyaryl ether.

The unique property of dendrimer is based on their series of branches, multivalency, globular structure, and well defined molecular weight with controlled surface, which enhances their potential as carrier for drug delivery (Gillies and Frechét, 2005; Gupta et al., 2006). An important characteristic of

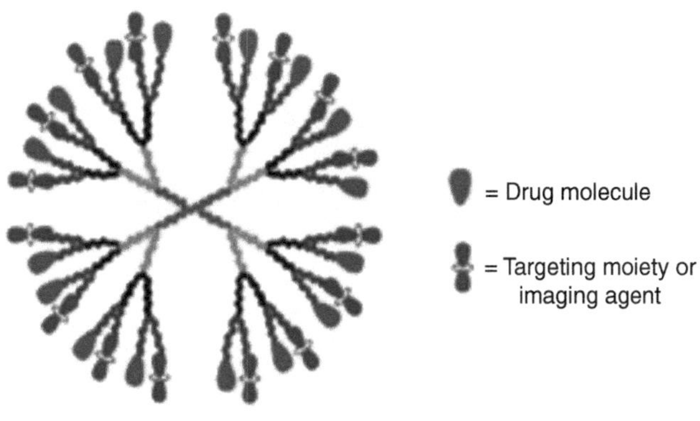

FIGURE 14.8 Dendrimers

Reproduced with permission from Xu, Q., Kambhampati, S.P., Kannan, R.M., 2013. Nanotechnology approaches for ocular drug delivery. Middle East Afr. J. Ophthalmol. 20 (1), 26–37. Medknow Publications.

dendrimers is that the uploaded drug can be incorporated both in the inner core and outer surface. Due to this ability, both hydrophobic and hydrophilic drugs can be incorporated into dendrimers at the same time.

The synthetic process of dendrimer is repeated, and larger dendrimers are created. This process is called "generation." The dendrimers are further classified into G1, G2, and G3 on the basis of molecular weight and number of surface groups. The presence of functional groups on the periphery of dendrimer acts as a handle for drug attachment and for other functional group attachment (Navath et al., 2010). Aspirin is one of the finest molecules to be encapsulated with the dendrimer structure. Polyamidoamine (PAMAM) is the most well-known dendrimer synthesized in two basic steps by using ethylenediamine and methylacrylate monomer to form a single generation dendrimer. These steps, when repeated, form the next generation of dendrimer.

Another dendritic effect is the "container property," which allows dendrimers to encapsulate pharmaceuticals, cosmetic ingredients, metal catalysts, and some other variety of small molecules (Zhang et al., 2015).

Application:

- Ability to encapsulate molecules and to carry very toxic medicines, such as cancer drugs directly to the cancerous cells without harming the normal cells,
- Act as time releasing agents for drugs,
- Vector to carry genetic materials into the cell nucleus,
- Can be used as catalyst,
- Act as nanocontainer for the incorporation of drugs with both hydrophilic and hydrophobic portions,
- Solubility enhancer for poorly soluble materials.

2.1.5 Nanopowder and nanocrystals

Nanopowders are composed of nanoparticles with an average diameter below 50 nm (Figs. 14.9 and 14.10). Most of the manufacturers of nanopowder produce micropowders that are assembled of nanoparticles, but the powder itself is rarely a nanopowder. Such compounds have two or three different cations (i.e., positively charged elements) in their chemical formula. The best example of chemical compound in this respect is calcium titanate ($CaTiO_3$).

FIGURE 14.9 Nanopowder

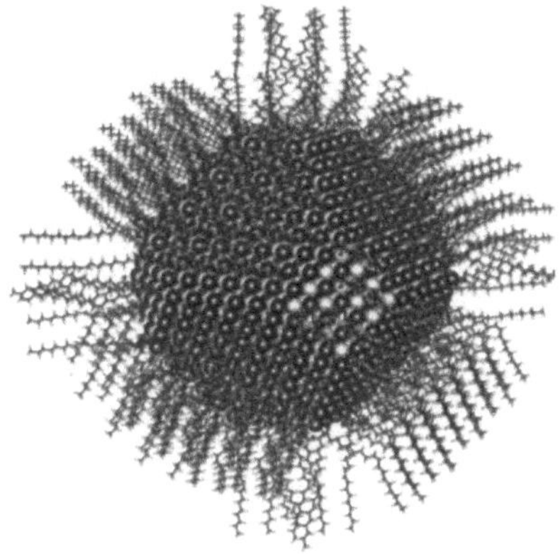

FIGURE 14.10 Nanocrystals

From https://en.wikipedia.org/wiki/Nanoparticle#/media/File:Colloidal_nanoparticle_of_lead_sulfide_(selenide)_with_complete_passivation.png

Nanocrystals are the aggregates of several hundreds or thousands of molecules that all combine in a crystalline form. They are composed of pure drugs with only a thin coating comprising of surfactant or combination of surfactants. These crystalline nanoparticles are between 2 and 100 nm in size. Nanocrystals are grouped in special manner, and the resulting drugs can be applied intravenously as nanosuspension, or bronchially through an inhaler. This small size nanoparticle enhances the surface–volume ratio and bioavailability of almost insoluble drugs. Poorly water-soluble drug compounds are grinded into nanocrystals that increases the surface area of these compounds. This results in an aqueous dispersion of the drug substance that behaves like a solution, which can be processed into final dosage forms for all routers of administration. It is also noted that the size of the particles is important for providing safe and effective passage through capillaries.

Application:

- Nanocrystal technology is important for evaluating new chemical entities that exhibit poor water solubility.
- These are also considered as valuable or effective tool for optimizing the performance of established drugs.
- These nanostructures are also important for moderately soluble drugs when a high concentration of drug is desired in a low volume of fluid.

2.1.6 Quantum dot

A QD is a crystal of semiconductor material whose diameter is in the order of several nanometers. The electronic properties of QDs are intermediate between the bulk semiconductors and discrete molecules, which are closely related to their size and shape (Brus, 2007; Murray et al., 2000; Norris, 1995). This property allows to tune the band gap, color emission, and absorption spectrum, which are considered to be highly significant, as the size of QDs can be controlled during fabrication process.

Application:

- QDs are particularly significant for optical importance due to their highly extinction coefficient property.

- Potential important in the treatment of cancer by involving targeted drug-delivery system.
- These are also useful to determine other biomolecules.

2.1.7 Synthesis of nanostructures

Basically, there are two approaches for the synthesis of nanomaterials and the fabrication of nanostructures: 'top–down approach" and "bottom–up approach." Milling is typically a top–down method in making nanoparticles, whereas colloidal dispersion is a good example of bottom–up approach in the synthesis of nanoparticles. Lithography may be considered as a hybrid approach (Majumder et al., 2007); since the growth of thin film is bottom–up, whereas etching is top–down (Guo, 2008), while nanolithography and nanomanipulation are commonly bottom–up approaches. Both approaches play very important role in industry, and also in nanotechnology, but there are some advantages and disadvantages to both approaches.

One of the major problems arises with top–down approach is the imperfection of the surface structure. It has been noted that the conventional top–down techniques, such as lithography can cause significant crystallographic damage in the pattern of processing (Das et al., 1993). An additional defects may also be introduced even during the etching techniques (Vieu et al., 2000). For example, a nanowire made by lithography is not smooth and may contain a lot of impurities and structural defects on their surfaces. Such imperfections would have a significant impact on physical properties and surface chemistry of nanostructure and nanomaterials. In connection to surface imperfections and other defects, top–down approaches may be introduced, and this will play an important role in the synthesis and fabrication of nanostructures and nanomaterials.

The bottom–up approach also provides a better chance in obtaining nanostructures with less defects, more homogenous chemical composition, and better short and long range order, while the top–down approach introduces internal stress in addition to surface defects and contamination.

3 ABSORPTION OF ORAL DRUGS IN THE GASTROINTENSTINAL TRACT

When any drug is transported from the lumen of the gut into the systemic circulation, it exerts its biological action, but the oral drug-delivery system has unfortunately many hurdles in order to penetrate the epithelial membrane. Although part of the absorption process occurs in the oral cavity and stomach due to the presence of salivary amylase and gastric protease (pepsin), while the main site of absorption is the small intestine. There are many pathways for the absorption of nutrients and other substances in the small intestine, such as transcellular pathway, active transport, facilitated diffusion, carrier-mediated transport process, paracellular pathway, and a fluid-phase endocytosis, which is called pinocytosis, but unfortunately some hurdles exist in the absorption of oral drugs to penetrate the epithelial membrane of the gastro intestinal tract when transported through the cells (transcellulary, but depending on the concentration gradient of the molecules across the intestinal membrane). The hydrophilic molecules, on the other hand, cannot diffuse easily through the intestinal mucosa because of low constituents of lipids (Daugherty and Mrsny, 1999). Thus, in the absence of particular and more specific membrane transported the only specified and more suitable route for the absorption of oral drugs is the paracellular pathway. However, in some cases, the oral drugs may be absorbed by a process of pinocytosis, which is a fluid-phase endocytic pathway where molecules can penetrate more easily inside the membrane vesicles.

As we know that route of administration of drug is safer and economical, and acceptance by the patients is important as well, the bioavailability of oral drugs is strongly faced by two factors, such as solubility and permeability (Gaucher et al., 2010). The bioavailability of drugs largely depends on their intrinsic properties and physiological conditions. Therefore, any drug when administered orally must face certain factors, such as the chemical and enzymatic nature of the fluid present in the GI tract, and the passage through mucus layer and epithelium before absorption (Deli, 2009; Schenk and Mueller, 2008). Concerning the intrinsic properties of drugs, certain factors, such as poor stability in the gastric environment, low mucosal permeability, and low absorption of drugs (Ensign et al., 2012). Similarly, different physiological factors including GI-transit time, pH, surface area, role in the absorption of drugs (Plapied et al., 2011).

The mechanisms involved in the enhancement of drug absorption are:

- Protection of the loaded drug from the harsh enhancement of gastro intestinal tract.
- Release of the loaded drug in a controlled manner at the target sites.
- Prolong period of time in the gut by mucoadhesion
- Inhibition of efflux pumps to improve drug accumulation.

The measurement of the amount of the drug in the plasma in a specific period of time indirectly indicates the rate and extent at which the active ingredient is absorbed from the drug product and becomes available at the site of action. Therefore, in order to achieve a better absorption and bioavailability of oral drugs, a particular drug should be stable at low gastric pH, and have excellent reproducible quality, as well as better dissolution profile and appropriate balance between hydrophilic and hydrophobic characteristics, so that the drug should be able to cross the intestinal epithelial membrane. To overcome all the restricted barriers and factors, several strategies should be considered, such as reduction of the size of drug particles (Chaumeil, 1998), salt formation (Serajuddin, 2007), and prodrug synthesis (Stella and Addae).

3.1 BIOAVAILABILITY ENHANCEMENT OF ORAL DRUGS

Bioavailability is one of the main pharmacokinetic properties of drugs, which is the fraction of an administered dose of unchanged drug that reaches the systemic circulation. When a drug is administered intravenously, its bioavailability is 100%. However, when a drug is given via other routes (such as oral), its bioavailability decreases due to the incomplete absorption, or first-pass metabolism, which indicates the intestinal and hepatic degradation or alteration of a drug or substance when taken by the mouth, and after absorption, removing some of the active substance from the blood before it enters into the general circulation. Clinically, first-pass metabolism is important when the fraction of the dose administered that escapes metabolism is small and variable. The liver is usually assumed to be the major site of first-pass metabolism of a drug administered orally, and besides the liver, other potential sites are the gastrointestinal tract (GIT), blood, vascular endothelium, and lungs. Thus, many drugs may be metabolized before adequate plasma concentrations are reached, resulting in poor bioavailability.

The extent of first-pass metabolism in the liver and intestinal wall depends on a number of physiological factors. The major factors are enzyme activity, plasma protein, and blood cell binding, as well as gastrointestinal motility. Many clinically important drugs undergo considerable first-pass metabolism after an oral dose. Drugs in this category include alprenolol, amitriptyline, 5-fluorouracil, hydralazine, isoprenaline, lignocaine, pethidine, morphine, neostigmine, nifedefine, pentazocine, and propanalol. One major therapeutic implication of extensive first-pass metabolism is that much larger oral doses are required than intravenous doses to achieve equivalent plasma concentrations.

Drugs given orally remain the preferred choice of patients because of simplicity, convenience, and acceptance (Gaucher et al., 2010; Stant et al., 2005). Development in the administration of drug still continues with respect to traditional oral delivery systems, but with nanoengineered-based improvements. There are some areas where nanoenhanced drugs could make big differences in increasing oral bioavailability and, on the other hand, could reduce undesirable side effects. By increasing bioavailability, nanoparticles can increase the yield in drug development and, more importantly, also help to treat previously untreatable conditions. In some cases, the oral route is not suitable for drugs that are poorly permeable or easily degradable in GIT. For example, delivery of proteins and peptides via oral route will be restricted by barriers, such as:

- epithelial cell lining
- mucus layer
- presence of proteolytic enzymes in the gut lumen (e.g., pepsin, trypsin and chymotrypsin)
- presence of proteolytic enzymes (i.e., endopeptidases) at the brush border membrane

However, drugs loaded in nanoparticles will be protected from the enzymatic degradation along the GIT, thus providing the potential benefit of enhanced absorption. Nanotechnology has some beneficial advantages in drug-delivery system, particularly for oral drugs, such as:

- The delivery of poorly water-soluble drugs.
- The targeting of drugs to the specific parts of GIT.
- The transcytosis of drugs across the tight intestinal barrier.
- The intracellular and transcellular delivery of large macromolecules.

4 NANOSCIENCE AND NANOTECHNOLOGY: FROM BASIC SCIENCE TO APPLICATIONS

Nanoscience offers the potential of dynamic revolutionary advances in many areas of science and technology, ranging from electronics and computing to biology and medicine, yet the realization of this promise depends essentially on the rational development of unique nanoscale structures whose properties and/or functions are controlled during materials synthesis.

Powerful semiconductor nanowires provide the capability for exploring new science and technology. First, we will be discussing "chemical" synthesis of complex modulated nanowires as a central material in nanoscience for enabling the bottom–up paradigm. Second, the concept of assembling a nanocomputer, introduced by Feynman in 1959, will be presented, and then the advances made of the past 15 years and more will be reviewed, with the culmination of recent breakthroughs of programmable nanoprocessors of unique complexity. In the end, we will present advances and opportunities at the boundaries between nanoelectronics and biology, with a focus on nanowire nanoelectronics for building interfaces to cells and tissues.

Feynman diagrams are pictorial representations of the mathematical expressions describing the behavior of subatomic particles (Fig. 14.11).

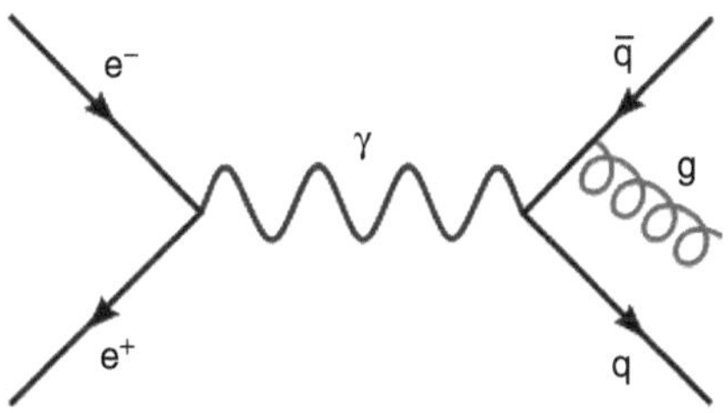

FIGURE 14.11 Example of Feynman Diagram

From https://en.wikipedia.org/wiki/Feynman_diagram

Quantum field theory: in this Feynman diagram, an electron and a positron crush, producing a photon (represented by the blue sine wave) that becomes a quark–antiquark pair, after which the antiquark radiates a gluon (represented by the green helix) (Fig. 14.12).

In this diagram, a kaon, made of an up and antistrange quark, decays both weakly and strongly into three pions, with intermediate steps involving a W boson and a gluon represented by the blue sine wave and green spiral, respectively (Feynman, 1992; Kaiser, 2005).

Nanoelectronics with cell networks in 3D to synthesize "cyborg" tissues (created from electrodes and wires combined on a nanoscale), once implanted, would sense and react to chemical and electrical changes. Direct electrical recording and stimulation of neural activity using microfabricated silicon and metal microwire probes have contributed extensively to basic neuroscience and therapeutic applications; however, the dimensional and mechanical mismatch of these probes with the brain tissue limits their stability in chronic implants, and decreases the neuron-device contact. Recently, a 3D macroporous nanoelectronic brain probe was developed that combines ultraflexibility and subcellular feature sizes to overcome these limitations (Xie et al., 2015). These ultraflexible probes were implanted frozen into rodent brains, and used to record multiplexed local field potentials and single-unit action potentials from the somatosensory cortex. Significantly, histology analysis revealed filling-in of neural tissue through the macroporous network and attractive neuron–probe interactions, consistent with long-term biocompatibility of the device.

Lately, the self-organized deposition of single wall carbon nanotubes (CNT) on prestructured electrode arrays with densities exceeding 106 CNT/cm^2 has been demonstrated. Between each electrode pair, a single CNT is arranged by the dielectrophoretic deposition process. All CNT are electrically active which provides the base for future applications in electronics and sensors.

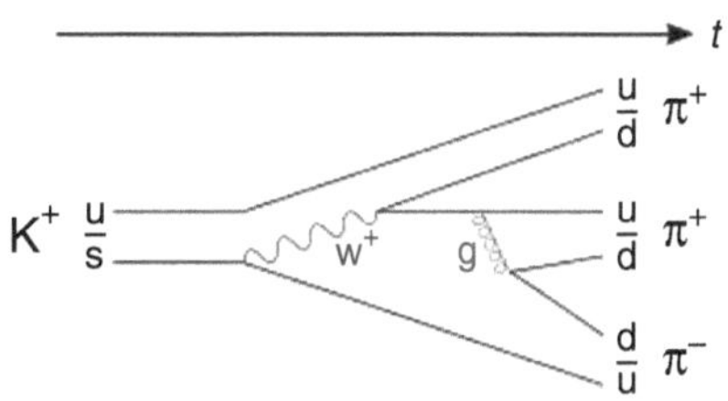

FIGURE 14.12 Example of a Kaon Diagram

From https://en.wikipedia.org/wiki/Feynman_diagram

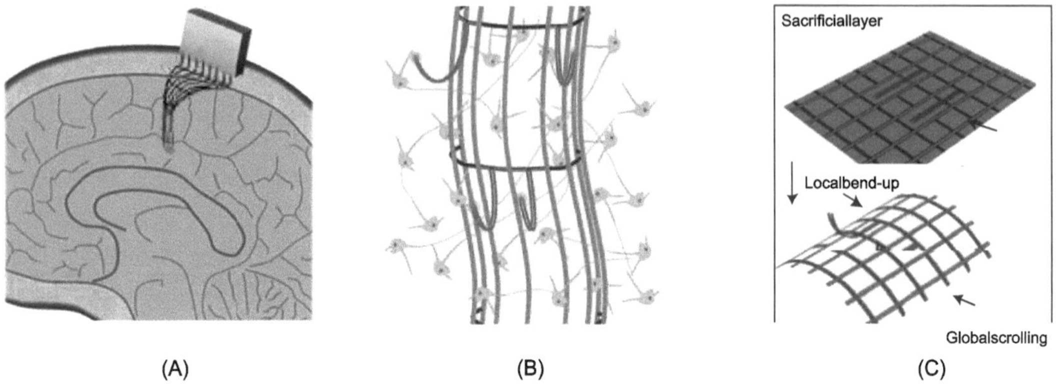

FIGURE 14.13 Nanoelectronic Probe and its Integration with Neural Tissue

Macro porous nano-electronic 3D neural probes. Schematic of the probe implanted in the brain. (A) The macro porous and flexible probe (yellow lines [black in print version]) is implanted in the brain and connected to the cranially mounted I/O connector. (B) Schematic of the microscopic interface of the macro porous probe (C) showing the steps of global scrolling.

Reproduced with permission from Xie, C., Liu, J., Fu, T.-M., Dai, X., Zhou, W., Lieber, C.M., 2015. Three-dimensional macroporous nanoelectronic networks as minimally invasive brain probes. Nat. Mater. 14, 1286–1292.

The reversible change of physical, mechanical and chemical properties of metallic nanostructures has been shown for nanoporous structures, and for thin films exposed to electrolytes. The tenability using an applied potential to control the change of properties is caused by the change of the electron density profile at the surface due to the electrochemical double layer formed in the electrolyte. Energy storage is one of the key aspects of mobility and the use of regenerative energy sources. Researchers developed strategies to meet these constraints focuses on implementing 3D macroporous nanoelectronic networks (Tian et al., 2012) where the macroporous nanoelectronic probe has a mesh-like structure designed to promote interpenetration and close integration with neural tissue (Fig. 14.13A–B). The mesh design is unique in having a two-dimensional (2D) open area of about 80%, feature sizes to sub-10 μm scale, and, importantly, high flexibility with an effective bending stiffness of $< 0.64 \times 10\ \mathrm{Nm}^2$ 4–7 orders of magnitude smaller than conventional Si (Lee et al., 2005), carbon fiber 14, and thin polyimide neural probes (Kim et al., 2013; Rousche et al., 2001). The exceptionally small bending stiffness yields mechanical interactions with tissue in the range of cellular forces. For instance, the force to deflect the two sensor-supporting arms by 10 μm (scale of a cell) is estimated to be about 10 nN, comparable to the single-cell migration force (du Roure et al., 2005).

4.1 ELECTRONICALLY TUNABLE NANOMATERIALS

The properties of materials are typically controlled in a static manner by the microstructure. This implies control of grain size, faulty concentration, structure, and metastability. As long as the microstructure does not change during the use of the material, the properties of the material are fixed, or irreversible. In contrast, in semiconducting materials, properties can be tuned by the application of an external field due to the space charge regions, which extend far from the interfaces. In metallic systems, this effect cannot be observed unless the dimensions of the structures are in the nanometer range. The reason for this different behavior is the small 3D dimension of the space charge regions due to the effective screening of the induced charges by the conduction electrons. In nanoporous metals and thin films exposed to appropriate electrolytes, it has been demonstrated that substantial changes of physical properties can be induced by the application of a potential between the nanostructured metal and a counter electrode. A simple model is proposed, based on the modification of electron density distribution at the interface of the metal and the electrolyte. Effectively, the corresponding change of the effective thickness of the sample is the major cause of the observed resistivity change. Additionally, a transparent conducting oxide, ITO, in a nanoparticulate form, has been prepared from a dispersion using spin coating. The observed resistivity changes, that is, the on/off ratio, can be as large as 2,000, that is, 200,000%, between the different values of the control potential. Moreover, the device exhibits field effect transistor behavior identical to a conventional semiconductor, but in this case observed in a material with a large charge carrier density exhibiting metallic conduction behavior. Additionally, the mobility is exceeding 30 cm^2/V s. The device can be used for printable electronics and transparent electronics. Solubility, the phenomenon of dissolution of solute in solvent to give a homogenous system, is one of the important parameters to achieve desired concentration of drug in systemic circulation for desired (anticipated) pharmacological response. Low aqueous solubility is the major problem encountered with formulation development of new chemical entities, as well as for the generic development. More than 40% NCEs (new chemical entities) developed in pharmaceutical industry are practically insoluble in water. Solubility is a major challenge for formulation scientist. Any drug to be absorbed must be present in the form of solution at the site of absorption. Various techniques are used for the enhancement of the solubility of poorly soluble drugs which include physical and chemical modifications of drug and other methods like particle size reduction, crystal engineering, salt formation, solid dispersion, use of surfactant, complexation, and so forth. Selection of solubility improving method depends on drug property, site of absorption, and required dosage form characteristics.

4.2 NANOSTRUCTURE METALS AND BIOAVAILABILITY

Improving a therapeutic profile of existing platinum (II) complexes, combination of 5,7-diterbutyl-1,2,4-triazolo[1,5-a]pyrimidine (DBTP) as N-donor ligand and two types of leaving groups: malonate (mal), as well as cyclobutane-1,1-dicarboxylate (CBDC) were used (Hoffmann et al., 2014).

Despite the extended use of porphyrin complexes in PDT (Agostinis et al., 2011; MacDonald and Dougherty, 2001), tetraplatinated porphyrins have so far not been studied for their anticancer properties upon light irradiation (Munakata et al., 2003). This is in contrast to ruthenated tetrapyridylporphyrin complexes (Pernot et al., 2012).Cisplatin is the first example of anticancer drug, which cross-links double-strand DNA and leads to DNA distortion and apoptosis of the tumor cell (Jamieson and Lippard, 1999). Derivatives or other platinum (Pt) compounds, such as carboplatin, oxaliplatin, and AMD473, were later synthesized and developed in order to reduce the side effects of cisplatin (Bakalova et al., 2008). The development of Pt complex as anticancer drug goes unrelenting over the years (Haxton and Burt, 2009). Pt-based antitumor therapy is complicated by severe side effects and intrinsic and acquired resistance of tumor cells. Implicated in cisplatin resistance is cyclooxygenase-2 (COX-2), a key enzyme in the biosynthesis of prostaglandins. COX-2 is overexpressed in many tumors, and plays a role in tumor initiation and progression (Ghosh et al., 2010). It is also associated with poor outcome in several types of cisplatin-treated cancer (Stewart, 2007). Thus, COX inhibitors are used

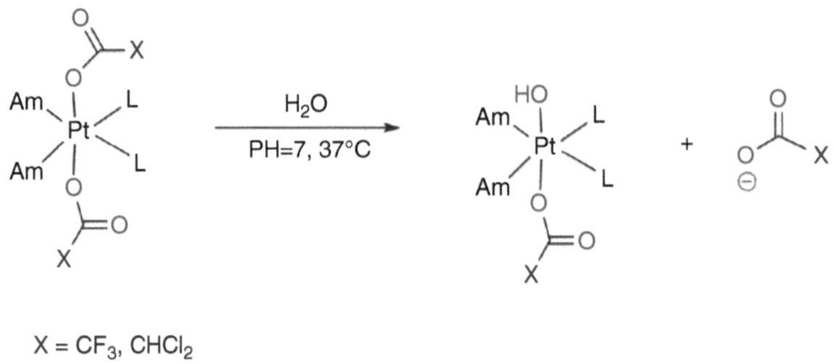

FIGURE 14.14 Pt(IV) Complexes with TFA or DCA Axial Ligands

Reproduced with permission from Metals in medicine: metal-related diseases. J. Biol. Inorg. Chem. (2014) 19 (Suppl. 2), S779–S814.

as chemopreventive and adjuvant chemotherapeutic agents. Clinical studies have shown synergistic effects when COX inhibitors are administered in combination with various antitumor agents, such as cisplatin (Hattori et al., 2001). Previous studies on the efficacy of COX inhibitors as antitumor agents results in potential discrepancies between clinical and cell culture studies. These differences were due to variance in pharmacokinetics and drugs delivery systems. To address this issue, scientists developed covalently linked conjugates of cisplatin with COX inhibitors (Neumann et al., 2014). These Pt(IV) complexes show extremely augmented cytotoxicity, compared to cisplatin, and overcome cisplatin-related resistance in tumor cells (Uddin et al., 2010). Literature indicates that Pt(IV) complexes with either trifluoroaceate TFA (Khokhar et al., 1997) or dichloroacetate DCA (Dhar and Lippard, 2009) axial ligands are potent anticancer agents. Recently, it was reported (as shown in Fig. 14.14) that Pt(IV) complexes with either TFA or DCA axial ligands can undergo rapid hydrolysis under physiological conditions (Wexselblatt et al., 2013).

Chemoresistance often occurs after successive treatment with single-agent chemotherapy. Combination therapy, the administration of a cocktail of different anticancer drugs, has been employed to overcome chemoresistance and improve the performance of single-agent chemotherapy.

However, differences in the pharmacokinetic profile of drugs often complicate treatment regimens. This can be overcome using nanomaterials, which allow simultaneous delivery of multiple drugs to the targeted site. In order to deliver the exact stoichiometric amount of cisplatin and doxorubicin, an inert Pt(IV) complex was designed for entrapment in tumor-targeting multiwalled carbon. Upon chemical reduction, equimolar of cisplatin and doxorubicin were released from the hydrophobic carrier, thereby achieving synchronous delivery of these two mechanistically complementary drugs (Chin et al., 2014). The use of light provides a unique mechanism for triggering drug release from delivery systems, or for activating drugs at a desired time and place. In this situation, photoactivated Pt(IV) prodrugs are very attractive compounds, since they can be selectively activated by visible light to become highly active species toward a range of cancer cell lines, including cisplatin-resistant cancer cells (Farrer et al., 2010).

Despite the potential of photoactivatable metallodrugs for cancer treatment, the application of the so-called targeted strategies to these complexes can be used to improve their pharmacological properties, such as aqueous solubility, cell uptake, and tumor selectivity, to further optimize their activity, and reduce the occurrence of unwanted adverse reactions (Barragàn et al., 2011, 2012; Grau-Campistany et al., 2013). New conjugates in which Pt(IV) complexes are covalently bound to tumor-targeting

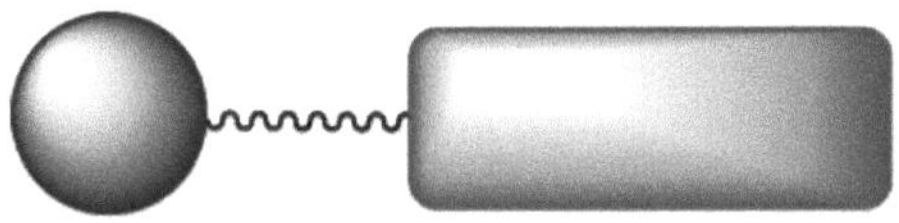

FIGURE 14.15 Photoactive Metallodrugs

Reproduced with permission from Metals in medicine: metal-related diseases. J. Biol. Inorg. Chem. (2014) 19 (Suppl. 2), S779–S814.

vectors based on peptides or glycosides, either nonmodified or caged with photoactivatable protecting groups. Following is an example of photoactive metallodrugs (Fig. 14.15).

Current research focus is to specifically target the tumor tissue, exploiting their unique properties (Galanski et al., 2011). A well-known strategy is albumin binding, since albumin is able to accumulate in the tumor tissue due to the enhanced permeability and retention (EPR) effect. Based on this knowledge, analbumin-targeting, bis-maleimide-functionalized Pt(IV) complex (1) along with an analogues succinimide-containing Pt (IV) complex, (2) which lacks the thiol affinity, was prepared. Both compounds showed potent anticancer activity, however with significant higher activity and even disease stabilization for 1 (Fig. 14.16).

Pt-based compounds are well-documented therapeutic agents used for the treatment of colorectal cancer (Kasparkova et al., 2008). The cytotoxicity of Pt drugs is attributed to multiple mechanisms, but primarily their ability to enter cells, hydrolyze, and covalently bind to DNA, causing the formation of DNA adducts. These events can lead to DNA damage responses and ultimately programmed cell death, apoptosis. The clinical efficacy of Pt drugs is limited by drawbacks, however, such as toxicity, but primarily by the high incidence of chemoresistance, intrinsic or acquired (Piccart et al., 2001). HSP70 is overexpressed in colorectal cancer, among other cancers, and is associated with cancer progression, chemotherapy resistance, and poor prognosis, as it is thought to provide cancer cells with a survival advantage (Murphy, 2013). HSP70 is therefore an exciting and legitimate anticancer target to develop novel platinum HSP70 inhibitor drug candidates as potential alternative treatments for colorectal cancer.

Platinum-containing therapeutics (e.g., cisplatin, oxaliplatin) have successfully been applied in medicine, however, have also shown undesirable side effects, and their use is connected with resistance of cancer cells (Alessio, 2011). The ongoing research in this field has proven that simple modification in the structure of platinum-based drugs does not guarantee desirable activity of the resulting compounds, even in the first panel of testing, that is, in vitro. For significant cytotoxicity, a suitable combination of

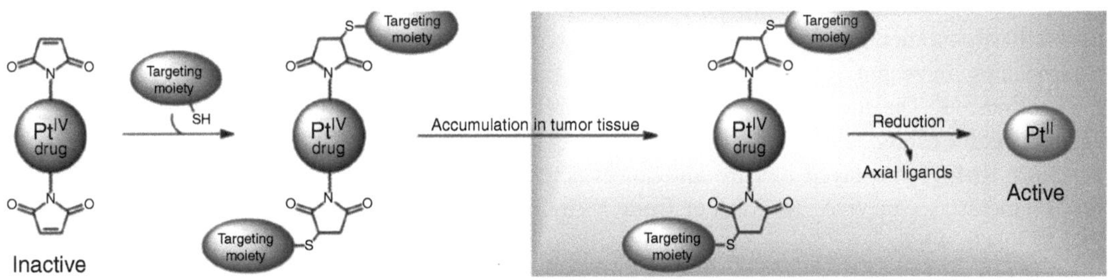

FIGURE 14.16 Pt(IV) Compounds

Reproduced with permission from Metals in medicine: metal-related diseases. J. Biol. Inorg. Chem. (2014) 19 (Suppl. 2), S779–S814.

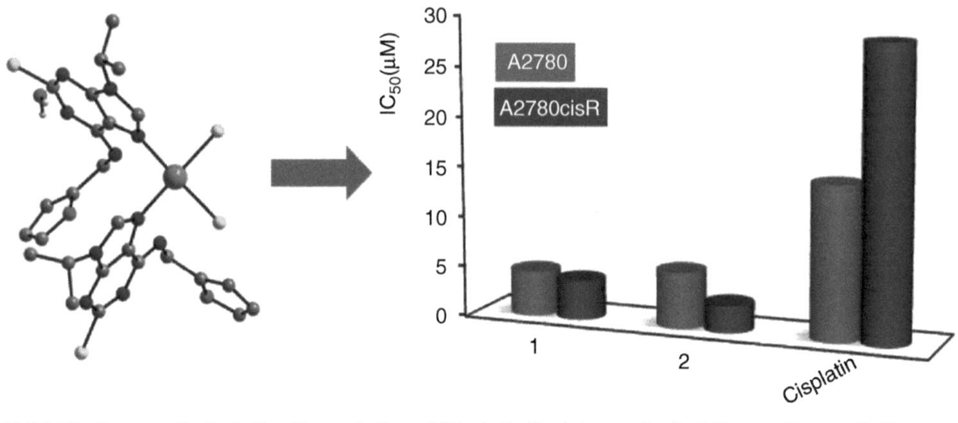

FIGURE 14.17 Comparatively Active Presentation of Cisplatin Resistance Against Human Cancer Cells

Complexes that are able to circumvent cisplatin resistance in A2780cisR (IC_{50} and 3 μM), while also being more cytotoxic against A2780 and HOS than cisplatin, and comparably active against MCF7 and G-361 human cancer cells.
Reproduced with permission from Metals in medicine: metal-related diseases. J. Biol. Inorg. Chem. (2014) 19 (Suppl. 2), S779–S814.

the leaving and carrier ligands in the studied molecule is crucial. In this work, N-donor carrier ligands (Ln) in the prepared dichloride, and oxalato complexes cis-[PtCl2(Ln)2] and [Pt(Ln)2(ox)] were based on a plant hormone kinetin (*N*6-furfuryladenine), which is nontoxic to human cells. Moreover, kinetin has been shown to be a beneficial compound for cells, as it has been applied in medicinal cosmetics ameliorating the signs of aging (Rattan and Clark, 1994) (Fig. 14.17).

The treatment of cancer cells with a spatial and temporal control is one of the most appealing features of photodynamic therapy PDT (Dolmans et al., 2003). This innovative medical technique exploits the synergistic action of light, oxygen, and a nontoxic photosensitizer (PS). Six novel Ru(II) polypyridyl complexes showed strong DNA binding affinity. 1 and 2 showed the best phototoxic effect upon irradiation at 420 nm, with a dark/light toxicity ratio of 150 and 40, respectively. The two complexes exhibited high cellular uptake, together with an outstanding nuclear accumulation (Fig. 14.18).

Thiophenolato-bridged highly cytotoxic p-cymene ruthenium complexes of [(g6-p-MeC6H4Pri) 2Ru2(SR)3] were recently synthesized against human ovarian cancer cells, with the IC_{50} values in the nanomolar range (Giannini et al., 2012). Their exact mechanism of action is still unclear, although the cytotoxicity of these complexes could be correlated to the lipophilicity of the corresponding thiophenol ligands (Giannini et al., 2013a,b). Fine-tuning of the complexes indicated the influence of arena ligands. Bioorganometallic chemistry provides an excellent platform to incorporate drug-like properties in molecules. In recent years, metal arene complexes have been extensively studied for the design of metallodrugs. The most successful examples are RAPTA-C and RM175 (Hartinger et al., 2012). Despite small difference in their structures, both exhibited contrasting biological activity and modes of action. RM175 displayed in vitro anticancer activity similar to that of cisplatin, while RAPTA-C inhibits metastasis in vivo. As evident from X-ray crystallographic experiments, RM175 prefers to bind to DNA, while RAPTA-C has high affinity to histone proteins of the nucleosome core particle (Adhireksan et al., 2014).

This demonstrates that a small structural variation can dramatically alter the reactivity of compounds with cellular targets and hence alter their biological activity.

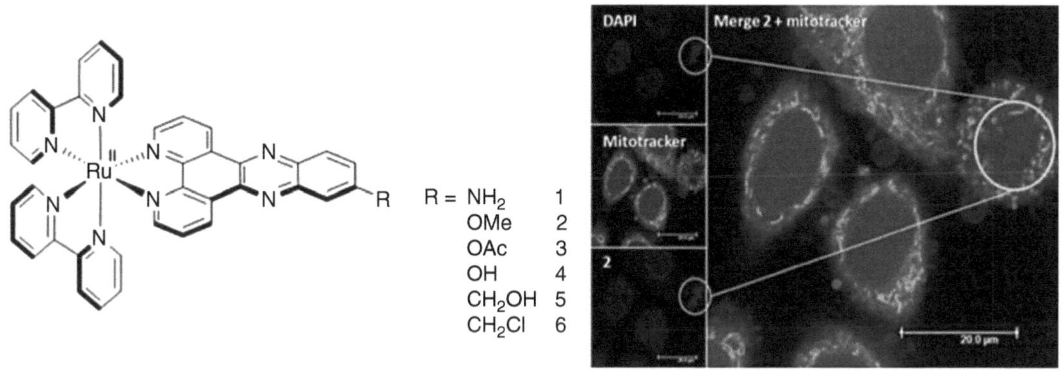

FIGURE 14.18 Photocleaving DNA

The ability to photocleave DNA at concentrations comparable with the IC_{50} values upon irradiation and the very efficient binding to DNA via intercalation (Kb × 10–6–10–7 M-1) suggested the involvement of DNA in the mechanism of phototoxic action of these compounds (Mari et al., 2014).

Reproduced with permission from Metals in medicine: metal-related diseases. J. Biol. Inorg. Chem. (2014) 19 (Suppl. 2), S779–S814.

Recently, the ability of a new complex, bis[N-(2-pyridyl)salicylideneimine-ONN] ruthenium(III) chloride, to bind CT DNA was designed against *Staphylococcus aureus* (MRSA), due the ability of ruthenium Ru(III) complex to bind DNA (Fig. 14.19).

This study helps to combat the challenges of treating cancer, as well as infection. Later pathology is getting difficult to treat, as bacterial strains rapidly develop resistance to existing drugs.

Gold(I) complexes are known for their luminescent properties, as well as for their potential as cytotoxic agents (Arcau et al., 2014; Hashmi, 2007; Ott, 2009). Recent work has also shown that compounds out of a series of gold(I) alkynylphosphane complexes showed antiangiogenic properties in zebrafish embryos, and inhibited thioredoxinreductase, a selenocysteine containing enzyme

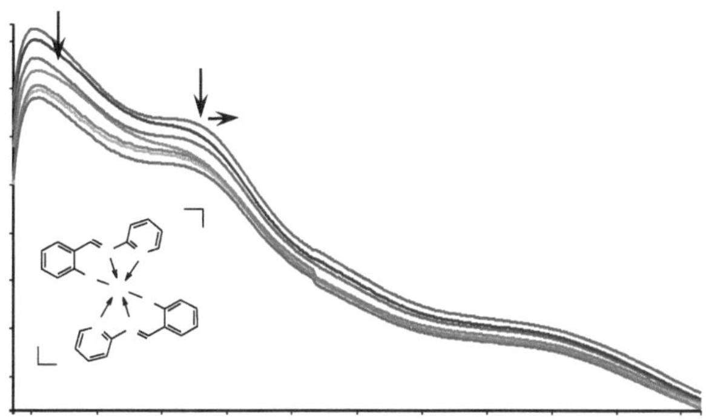

FIGURE 14.19 Ru(III) Compound Showing Significant Vancomycin Activity

Ru(III) compound has shown significant activity against community-acquired methicillin-resistant *Staphylococcus aureus* (CA-MRSA), hospital acquired methicillin-resistant *S. aureus* (HA-MRSA), methicillin-sensitive *S. aureus* (MSSA) reaching about 76 % of vancomycin activity, the reference antibiotic.

Reproduced with permission from Metals in medicine: metal-related diseases. J. Biol. Inorg. Chem. (2014) 19 (Suppl. 2), S779–S814.

that is crucial for cell proliferation, in the low-nanomolar range. Additionally, these complexes were cytotoxic against MCF-7 breast cancer, and HT-29 colon carcinoma cells, making them a very promising substance class for tumor therapy (Meyer et al., 2012). However, the downside of this type of organometallics is its poor solubility in aqueous medium, caused by the lipophilic triphenylphosphane ligand. Recent investigations present a new series of novel gold(I) alkynylphosphane complexes, for which we used the most promising alkynyl ligand of the former work, and modified the phosphane ligand to improve solubility.

Free radicals and other Reactive Oxygen Species (ROS) are important components in many physiological and pathological processes in the living cell (Winterbourn, 2008). Still, there is much unknown about the roles of ROS and oxidative damage in cellular processes. Elevated concentrations of ROS can cause oxidative stress, which severely impacts cellular function by damaging important cellular components (Dickinson et al., 2010). The goal of this project is to design novel catalytic tools for the detection and manipulation of ROS and oxidative stress in both healthy, as well as cancer cells, with a particular focus on the subcellular environment. This will give rise to new insights into the behavior of cancer cells, compared to healthy cells under oxidative stress. Ideally suited for this purpose are the bioinspired iron oxidation catalysts based on N4Py ligands (Li et al., 2010; Lubben et al., 1995; Roelfes et al., 2000).

There are pH-dependent equilibria between spin states and tetra- and pentadentate coordination of the ligand N4py to iron (Draksharapu et al., 2011) exhibiting three complexes, such that in acetonitrile the complexes retain their solid-state structure, but in water they exhibit immediate ligand exchange of the CH_3CN ligand(s) for hydroxide. It was also observed that in the pH range between 5 and 8, complexes 1 and 2 show equilibrium between two different species.

Recent advances in nanostructured materials gave rise to their natural evolution from the microscale to the nanoscale, and the capability for specific interactions with proteins, DNA, viruses, and other nanoscale biological structures.

Although no universally accepted nomenclature has been agreed upon, amorphous carbon with some tetrahydral (sp3) component is often referred to as diamond-like carbon (DLC) (Narayan et al., 1994; Sharma et al., 2000; Wei et al., 1999). These carbon solids exhibit atomic number densities greater than 3.19 g atom/cm^3 and may possess densities much closer to that of diamond, 3.51 g/cm^3, than that of graphite of 2.26 g/cm^3 (Moravec et al., 1980).

DLC has been used in a variety of vascular medical devices, including coronary artery stents, synthetic heart valves, left ventricular assist devices, and artificial hearts (Alanazi et al., 2000; Gutensohn et al., 2000; Tran et al., 1999; Yu et al., 2000). In vitro studies demonstrate platelet activation, and platelet adhesion and clot formation occur less often on DLC-coated surfaces than on titanium and stainless steel surfaces. Similarly, inorganic constituent of bone is made up of biological apatites, which provide strength to the skeleton, and act as a storehouse for calcium, phosphorus, sodium, and magnesium. These biological apatites are structurally similar, though not identical, to the mineral apatite hydroxyapatite [HAp, $Ca_{10}(PO_4)_6(OH)_2$] used as a carrier of plasmid DNA for nonviral gene delivery (Sfeir et al., 2003). Nano materials are more frequently used in ventricular assist devices for treating heart failing patients, especially with end-stage congestive heart failure (Mancini et al., 1998; Stevenson et al., 2004). There are six types of surface coatings for ventricular assist devices, that is, titanium nitride (TiN), DLC 2-methacryloyloxyethyl phosphorylcholine (MPC) polymer, heparin coatings, textured surfaces, and endothelial cell lining (Sin et al., 2009). Some of the advantages of the aforementioned six types are, for example, that TiN coating was developed (Harman et al., 1997) for improving the fretting resistance of the titanium alloy surface by increasing the surface hardness. DLC films (or coatings) can be deposited easily on many substrates of a wide area at room temperature (Alanazi et al., 2000), with excellent biocompatibility and hemocompatibility (Dion et al., 1993). Similarly, MPC polymer is very well matched with blood, due to its minimized protein adsorption and increased amount of free water (Nakabayashi and Williams, 2003).

Surfaces coated with heparin can prevent platelet adhesion and activation, leading to improved blood compatibility of artificial materials (Anderson and Clapper, 1998; Riesenfeld et al., 1995). Textured surface are capable to minimize the thromboembolic events, and reduce the need for systemic anticoagulation in patients with long-term left ventricular assist devices with a low risk of thromboembolism (Rose et al., 1994). Metabolically active endothelial lining plays a major role in preventing blood thrombosis, and are under investigation to improve blood compatibility of VADs (Nikolaychik et al., 1996; Scott-Burden et al., 1998).

"Activated bleomycin" (BLM) is an antitumor drug that effects the oxidative cleavage of DNA, and the oxidation of hydrocarbons. Stubbe and Kozarich (1996) cumulatively support its formulation as a low-spin hydroperoxidoiron (III) species (Sam et al., 1994). A model for BLM has been synthesized that mimics its reactivity, and gives rise to the same optical and EPR spectra; however, more detailed spectroscopic data is not yet available. In a related area, alkylperoxoiron (III) complexes have been implicated in hydrocarbon oxidations (Guajardo et al., 1993; Leising et al., 1993); several of them have been stabilized at low temperature, and spectroscopically characterized (Zang et al., 1993).

Up to this point, we discussed the unlimited advantages of using nanomaterials for the benefit of mankind by using them for making new medical equipment, drugs, and other relevant procedures. All natural and man-made materials (organic and inorganic) have both advantages and disadvantages, and so do nanomaterials. We are now going to talk about some obvious disadvantages of nanostructures. We have to see clearly the danger of using engineered nanoparticles, especially regarding the immune system. Understanding nanoimmuno-interactions is critical for the safe application of engineered nanoparticles in medicine (Fadeela, 2012).

Our first line of defense against microbial invasion is the innate immune system, which includes the recognition of pathogen-associated molecular patterns (PAMPs). The immune system also responds to damage-associated molecular patterns (DAMPs).

In this hypothetical diagram, the analogy is made between pathogens (microorganisms) that display PAMPs and damaged or stressed tissues that release DAMPs, such as HMGB1, a nonhistone nuclear protein that acts as a secreted alarmin, and engineered nanoparticles that are coated with a biocorona of proteins possibly unfolded, thereby revealing hidden epitopes, that may act as nanomaterial-associated molecular patterns (NAMPs). These molecular signatures are recognized by so-called pattern recognition receptors or PRRs—including the Toll-like receptors—on the surface of innate immune cells, or expressed within such cells. The activation of PRRs triggers inflammation and alerts the adaptive immune system to impending danger (Zitvogel et al., 2010).

There are so many factors involved in the aforementioned process affecting our immune system, including properties, such as size, shape, surface charge and solubility. Other causes can be the adsorption of biomolecules onto the surface of nanoparticles. For these reasons, we have to investigate the parameters determine particle clearance or biodegradation in a living system.

Nature already placed nanoscale phenomena in between macro and molecular creations, for example, the human immunodeficiency virus is 120 nm in diameter (Stephen, 2008). Moreover, humankind has been exposed to natural or anthropogenic nanoparticles for millennia (Buzea et al., 2007). For instance, when you light a candle on a dinner table, over 1.5 million nanodiamond particles form in the flame every second (Su et al., 2011). The innate immune system is the first line of defense against foreign intrusion, that is, microorganisms (and nanoorganisms, such as viruses) and particles, as well as cellular debris. We have to think carefully how we can modify engineered nanoparticles so they won't be considered as "danger" signals.

The nanoscale is not just an intermediate dimension between molecular and macroscopic states, but a passage for gathering, processing, and transmitting chemical-based information. Living cells are self-regulating and very composite chemical systems, which mainly run by nanoscale structures that are critical for its existence and survival. Human cells consist of molecular components (i.e., carbon and hydrogen), nanosubstances (such as enzymes, receptors, etc.), and bigger structures (e.g., microtubules, ribosomes). It is very clear from the present nanoscience studies that this research can lead to artificial life, but further research is needed to solve the dimensional constraints for the gathering, processing, and transmission of chemical-based information.

Nanoscience provides the interfaces between chemistry, physics, biology, mathematics, and medicinal sciences. More than three decades have passed since certain semiconductors (Brus, 1983; Rossetti et al., 1983) and iron oxides (Bell et al., 1984; Pierre et al., 1986) were shown to exhibit markedly different electronic and magnetic properties, respectively, compared with corresponding bulk materials, when limited to nanoscales. Later on, STM enabled individual atoms to be manipulated and positioned into quantum structures (Crommie et al., 1993; Eigler and Schweizer, 1990) leading to the discovery of nanocarbon (Iijima, 1991). These important advances validated the fact that confinement is the key for determining the intrinsic properties of structured matter. In general, we can say that particles, with constant volume but variable shape, differ significantly in their properties. Thus, work continued for the synthesis of nanoparticles in the form of rods (Yu et al., 1997), plates (Ni et al., 2002), and multipodal structures (Manna et al., 2000; Milliron and Scher, 2003), which have high-grade directional dependency.

Thus, structure, as well as size, plays a key role in determining the metallic/semiconducting properties of carbon nanotubes (Ebbesen et al., 1996) and the depletion of crystal defects, moderation of surface roughness, and selected expression of crystal faces (Alivisatos, 2000; Chen et al., 2005; Xiong and Xia, 2007) are additional parameters that contribute significantly to the continuing surge of activity in nanoscience today.

It is true that only some materials exhibit significant and useful changes in their inherent properties as a function of scale. But still nanoscience is progressing due to progressions in miniaturization and strengthening of supramolecular chemistry. Today, the advances in nanoscience are due to increasing capacity in nanoscale miniaturization of the macromolecules, with improvement in storage density, screening protocols, isolation of molecules and their conjugates, and maneuvers incorporation.

Furthermore, cooperative properties, such as superhydrophobicity (Li et al., 2007), plasmonic coupling (Lin et al., 2005), and magnetic/electronic behavior of devices (Talapin and Murray, 2005; Urban et al., 2007) can be modulated by miniaturization of structural components, and production of superlattice assemblies. And the ability to create multifaceted nanoscale substances with hybrid structure and composition is a major challenge and opportunity in the chemistry of systematized matter (Colfen et al., 2003).

Under these conditions, disadvantages of nanotechnology molecules cannot be ignored. There is evidence (Moyano et al., 2012) that nanoparticle hydrophobicity direct immune response target and mode of action. However, it may be difficult to preserve a hydrophobic material surface once exposed (Hubbell et al., 2009), as such surfaces are rapidly conceal through protein adsorption. Certainly, cells of the immune system may not recognize the nanoparticle surfaces, but there are indications that a long-lived protein corona is formed when nanoparticles are introduced into a biological environment (Mahon et al., 2012). When nanoparticles are presented to a physiological environment, such as human plasma, they selectively absorb biomolecules to form a so-called biomolecular corona on the surface of the nanoparticle (Mahon et al., 2012). It is possible that nanoparticle surfaces may be coated with bacterial endotoxin or lipopolysaccharide (LPS), a membrane component of all Gram-negative bacteria, especially if the nanoparticles are synthesized under nonsterile conditions. This could give rise to interference with the biological assessment of nanomaterials, for example, formation of maturing dendritic cells (Vallhov et al., 2006). Several studies have provided evidence for a long-lived corona of human plasma proteins on nanoparticles of different chemical composition (Casals et al., 2010; Cedervall et al., 2007; Lundqvist et al., 2011; Monopoli et al., 2011). The formation of the hard corona was shown to depend on both surface properties, such as degree of hydrophobicity and size (Lundqvist et al., 2008). The biocorona has been reported to impact on cellular uptake (Maiorano et al., 2010) and on cell signaling, including NF-κB-dependent cytokine production in human monocyte THP1 cell line (Deng et al., 2011). In an effort to meet the challenges of harmful cellular impacts of nanomaterial, a new drug-delivery platform was developed (Hu et al., 2011), consisting of biodegradable polymeric nanoparticles camouflaged with membrane lipids, and associated membrane proteins derived from erythrocytes. This study provides a new twist on the biocorona concept, whereby the nanoscale carriers are disguised as endogenous constituents derived from red blood cells (Fang et al., 2012). Similarly, immune cells cannot recognize the spores of airborne fungal pathogen *Aspergillus fumigatus* when surrounded by hydrophobic corona (Aimanianda et al., 2009). In another study, functional self-assembled protein coating on the porous silicon nanoparticles have shown a pronounced change in the degree of plasma protein adsorption in vitro, and altered biodistribution in vivo, when compared to uncoated nanoparticles (Sarparanta et al., 2012). This study provides further evidence that surreptitiousness can be engineered by manipulating the biocorona on nanoparticles. Careful assessment of the interactions of nanoparticles with the immune system is of the utmost importance for the safe development of the nanotechnologies, not least for biomedical use (Dobrovolskaia and McNeil, 2007). Important lessons to this regard can be learned from immunological principles for the recognition of pathogens (Shvedova et al., 2010). Thus, appreciation of the interface of nanomaterials with biological systems becomes the key for their safe and proficient application. It is progressively being accepted that the surface of nanomaterials would be covered by protein corona upon their arrival to the biological medium. The biological environment will then see the achieved modified surface of nanomaterials, and therefore further responses depend on the composition of corona. We may begin to understand the basic principles that will allow for the rational design of safe and useful nanomaterials.

The importance of the protein corona formed around nanoparticles upon entering a biological fluid has recently been highlighted. This corona is, when sufficiently long-lived, thought to govern the particles' biological fate. However, even this long-lived "hard" corona evolves and reequilibrates as particles pass from one biological fluid to another, and may be an important feature for long-term fate. Here, we observed the evolution of the protein corona as a result of transfer of nanoparticles from one biological fluid (plasma) into another (cytosolic fluid), a simple illustrative model for the uptake of nanoparticles into cells. While no direct comparison can be made to what would happen in, for example, the uptake pathway, the results confirm that significant evolution of the corona occurs in the second biological solution, but that the final corona contains a "fingerprint" of its history. This could be evolved to map the transport pathways utilized by nanoparticles, and eventually to predict nanoparticle fate and behavior.

In the end, we would like to discuss briefly the problems regarding the bioavailability of oral drugs. The phenomenon of dissolution of solute in solvent to give a homogenous system is one of the important parameters to achieve the desired concentration of drug in systemic circulation for expected pharmacological response. Low aqueous solubility is the major problem encountered with the formulation and the development of new chemical entities, as well as for the generic development of drugs. Any drug to be absorbed must be present in the form of solution at the site of absorption. Selection of solubility-improving methods depends on drug property, site of absorption, and required dosage. Due to ease of administration, high patient compliance, cost effectiveness, least sterility constraints, and flexibility, oral drugs are the most convenient and commonly employed route of drug delivery, and many drug companies are producing oral drug products. However, the major challenges of drug solubility, low permeability, and resultant poor bioavailability remain when designing oral drugs. High doses are required for the drugs that are water-soluble in order to reach therapeutic plasma concentrations. Low water solubility is the major problem faced by drug developers when formulating new chemical entities, as well as developing generic drugs, as drugs to be absorbed must be present in the form of an aqueous solution at the site of absorption. Most of the drugs are either weakly acidic or weakly basic, having poor aqueous solubility, with slow drug absorption leading to inadequate and variable bioavailability and gastrointestinal mucosal toxicity. The improvement of drug solubility by its oral bioavailability remains one of the most challenging aspects of the drug development process, especially for oral drug-delivery system. Nanosuspension tools have been developed as optimistic contenders for skillful distribution of hydrophobic drugs. This technology is applied to less soluble drugs that are unsolvable in both water and oils. An oral pharmacological nanosuspension is a biphasic system consisting of nanosized drug particles stabilized by surfactants for oral administration. The particle size distribution of the solid particles in nanosuspensions is usually less than one micron, with an average particle size ranging between 200 and 600 nm. Numerous methods are used for the preparation of nanosuspensions, including precipitation-system, media crushing, high-pressure homogenization in water, high-pressure homogenization in nonaqueous media, and a combination of precipitation and high-pressure homogenization. In future, we will be expecting more novel approaches of oral drug deliveries.

REFERENCES

Adhireksan, Z., Davey, G.E., Campomanes, P., Groessl, M., Clavel, C.M., Yu, H., et al., 2014. Ligand substitutions between ruthenium-cymene compounds can control protein versus DNA targeting and anticancer activity. Nat. Commun. 5, 3462.

Agostinis, P., Berg, K., Cengel, K.A., Foster, T.H., Girotti, A.W., Gollnick, S.O., et al., 2011. Photodynamic therapy of cancer: an update. CA Cancer J. Clin. 61, 250–281.

Aimanianda, V., Bayry, J., Bozza, S., Kniemeyer, O., Perruccio, K., Elluru, S.R., et al., 2009. Surface hydrophobin prevents immune recognition of airborne fungal spores. Nature 460, 1117–1121.

Alanazi, A., Nojirici, C., Kido, K., et al., 2000. Engineering analysis of diamond like carbon coated polymeric materials for biomedical applications. Artif. Organs 24 (8), 624–627.

Alessio, E., 2011. Bioinorganic Medicinal Chemistry. Wiley, Weinheim, Germany.
Alivisatos, P., 2000. Pure Appl. Chem. 72, 3.
Anderson, A.B., Clapper, D.L., 1998. Coatings for blood-contacting devices. In: Medical Plastics and Biomaterials.
Arcau, J., Andermark, V., Aguilo, E., Gandioso, A., Moro, A., Cetina, M., Lima, J.C., et al., 2014. Luminescent alkynyl-gold(I) coumarin derivatives and their biological activity. Dalton Trans. 43, 4426–4436.
Bakalova, A., Varbanov, H., Buyukliev, R., Momekov, G., Ferdinandov, D., Konstantinov, S., Ivanov, D., 2008. Synthesis, characterization and biological activity of Pt(II) and Pt(IV) complexes with 5-methyl-5(4-pyridyl)-2,4-imidazolidenedione. Eur. J. Med. Chem. 43, 958–965.
Barragàn, F., Lòpez-Senín, P., Salassa, L., Betanzos-Lara, S., Habtemariam, A., Moreno, V., et al., 2011. Photocontrolled DNA binding of a receptor-targeted organometallic ruthenium(II) complex. J. Am. Chem. Soc. 133, 14098–14108.
Barragàn, F., Carrion-Salipm, D., Gòmez-Pinto, I., Gonzàlez-Canto, A., Sadler, P.J., de Llorens, R., et al., 2012. Somatostatin Subtype-2 Receptor-Targeted Metal-Based Anticancer Complexes. Bioconjugate Chem. 23, 1838–1855.
Bell, S.H., Weir, M.P., Dickson, D.P.E., Gibson, J.E., Sharp, G.A, Peters, T.J., 1984. Mössbauer spectroscopic studies of human haemosiderin and ferritin. Biochim. Biophys. Acta 787, 227.
Bittner, A.M., 2005. Biomolecular rods and tubes in nanotechnology. Naturwissenschaften 92, 51–64.
Brus, L.E., 1983. A simple model for the ionization potential, electron affinity, and aqueous redox potentials of small semiconductor crystallites. J. Chem. Phys. 79, 5566.
Brus, L., 2007. Chemistry and physics of semiconductor nanocrystals. p. 1.
Buzea, C., Pacheco, I., Robbie, K., 2007. Nanomaterials and nanoparticles: sources and toxicity. Biointerphases 2, MR17–MR71.
Casals, E., Pfaller, T., Duschl, A., Oostingh, G.J., Puntes, V., 2010. Time evolution of the nanoparticle protein corona. ACS Nano 4, 3623–3632.
Cedervall, T., Lynch, I., Lindman, S., Berggård, T., Thulin, E., Nilsson, H., et al., 2007. Understanding the nanoparticle-protein corona using methods to quantify exchange rates and affinities of proteins for nanoparticles. Proc. Natl. Acad. Sci. USA 104, 2050–2055.
Chaumeil, J.C., 1998. Micronization: a method of improving bioavailability of poorly soluble drugs. Methods Find. Exp. Clin. Pharm. 20 (3), 211–215.
Chen, Y., Gu, K., Nie, C.G., Jiang, Z.Y., Xie, Z.X., Lin, C.J., 2005. Complex-shaped Metal Nanoparticles: Bottom-Up Syntheses. Chem. Commun. 4, 181.
Chin, C.F., Yap, S.Q., Li, J., Pastorin, G., Ang, W.H., 2014. Ratiometric delivery of cisplatin and doxorubicin using tumour-targeting carbon-nanotubes entrapping platinum(IV) prodrugs. Chem. Sci. 5, 2265–2270.
Colfen, H., Mann, S., 2003. Higher-order organization by mesoscale self-assembly and transformation of hybrid nanostructures. Angew. Chem. Int. Ed. Engl. 42, 2350.
Cortie, M.B., 2004. The weird world of nanoscale gold. Gold Bull. 37 (1), 12–19.
Crommie, M.F., Lutz, C.P., Eigler, D.M., 1993. Confinement of electrons to quantum corrals on a metal surface. Science 262, 218.
Das, B., Subramaniam, S., Melloch, M.R., 1993. Effects of electron-beam-induced damage on leakage currents in back-gated GaAs/AlGaAs devices. Semicond. Sci. Technol. 8, 1347.
Daugherty, A.L., Mrsny, R.J., 1999. Regulation of intestinal epithelial paracellular barrier. Pharmaccutic Sci. Techn. Today 2 (7), 281–287.
Deli, M.A., 2009. Potential use of tight junction modulators to reversibly open membranous barriers and improve drug delivery. Biochim. Biophys. Acta 1788 (4), 892–910.
Deng, Z.J., Liang, M., Monteiro, M., Toth, I., Minchin, R.F., 2011. Nanoparticle-induced unfolding of fibrinogen promotes Mac-1 receptor activation and inflammation. Nat. Nanotechnol. 6, 39–44.
Dhar, S., Lippard, S.J., 2009. Mitaplatin, a potent fusion of cisplatin and the orphan drug dichloroacetate. Proc. Natl. Acad. Sci. USA 106, 22199–22204.

Dickinson, B.C., Sriikun, D., Chang, C.J., 2010. Mitochondrial-targeted fluorescent probes for reactive oxygen species. Curr. Opin. Chem. Biol. 14, 50.
Dion, I., Roques, X., Baquey, C., et al., 1993. Hemocompatibility of diamond-link carbon coating. Biomed. Mater. Eng. 3 (1), 51–55.
Dobrovolskaia, M.A., McNeil, S.E., 2007. Immunological properties of engineered nanomaterials. Nat. Nanotechnol. 2, 469–478.
Dolmans, D., et al., 2003. Photodynamic therapy for cancer. Nature Rev. Cancer. 3, 380–387.
Sin, D., Kei, H.L., Miao, X., 2009. Surface coatings for ventricular assist devices. J. Expert Rev. Med. Devices 6 (1), 51–60.
Draksharapu, A., Qian, L., Logtenberg, H., Tieme, A., et al., 2011. Ligand exchange and spin state equilibria of Fe^{II}(N4Py) and related complexes in aqueous media. Inorg. Chem. 51 (R), 900–913.
du Roure, O., et al., 2005. Force mapping in epithelial cell migration. Proc. Natl. Acad. Sci. USA 102, 2390–2395.
Ebbesen, T.W., Lezec, H.J., Hiura, H., Bennett, J.W., Ghaemi, H.F., Thio, T., 1996. Electrical conductivity of individual carbon nanotubes. Nature 382, 54.
Eigler, D.M., Schweizer, E.K., 1990. Positioning single atoms with a scanning tunnelling microscope. Nature 344, 524.
Eijkel, J.C.T., van den Berg, A., 2006. The promise of nanotechnology for separation devices from a top–down approach to nature- inspired separation devices. Electrophoresis 27, 677–685.
Ensign, L.M., Cone, R., Hanes, J., 2012. Oral drug delivery with polymeric nanoparticles: "the gastrointestinal barriers". Adv. Drug Deliv. Rev. 64 (6), 557–570.
Fadeela, B., 2012. Clear and present danger? Engineered nanoparticles and the immune system. Swiss Med. Wkly. 142, 13609.
Fang, R.H., Hu, C.M., Zhang, L., 2012. Nanoparticles disguised as red blood cells to evade the immune system. Expert Opin. Biol. Ther. 12, 385–389.
Farrer, N.J., Woods, J.A., Salassa, L., Zhao, Y., Robinson, K.S., Clarkson, G., et al., 2010. A potent trans-diimine platinum anticancer complex photoactivated by visible light. Angew. Chem. Int. Ed. 49, 8905–8908.
Feynman, R.P., 1992. There's plenty of room at the bottom (data storage). J. Microelectromech. Syst. 1 (1), 60–66.
Galanski, M., Keppler, B.K., Kratz, F., Senter, P., Steinhagen, H., 2011. Drug Delivery in Oncology. Wiley-VCH Verlag GmbH, Weinheim, Germany, pp. 1605–1629.
Gao, S., Zhao, y., Gou, P., Chen, N., Xie, y., 2003. Preparation of Cu $A10_2$ nanocrystalline transparent thin films with high conductivity. Nanotechnology 14, 538–541.
Gaucher, G., Satturwar, P., Jones, M.C., Furtos, A., Leroux, J.C., 2010. Polymeric micelles for oral drug delivery. Eur. J. Pharm. Biopharm. 76 (2), 147–158.
Ghosh, N., Chaki, R., Mandal, V., Mandal, S.C., 2010. Pharmacol. Rep. 62, 233.
Giannini, F., Furrer, J., Ibao, A.F., Süss-Fink, G., Therrien, B., Zava, O., et al., 2012. J. Biol. Inorg. Chem. 17, 951–960.
Giannini, F., Furrer, J., Süss-Fink, G., Clavel, C.M., Dyson, P.J., 2013a. Crystal structure of bis[μ-(4-methoxyphenyl)methanethiolato-κ2 *S*:*S*]bis[chlorido(η6-1-isopropyl-4-methylbenzene)ruthenium(II)] chloroform disolvate. J Organomet Chem. 744, 41–48.
Giannini, F., Paul, L.E.H., Furrer, J., Therrien, B., Süss-Fink, G., 2013b. Crystal structure of bis[-(4-methoxyphenyl)methanethiolato-2*S*:*S*]bis[chlorido(6-1-isopropyl-4-methylbenzene)ruthenium(II)] chloroform disolvate. New J. Chem. 37, 3503–3511.
Gillies, E.R., Frechét, J.M.J., 2005. Dendrimer and dendritic polymer in drug delivery. Drug Discov. Today 10 (1), 35–43.
Grau-Campistany, A., Massaguer, A., Carrion-Salip, D., Barragàn, F., Artigas, G., Lòpez- Senìn, P., 2013. Conjugation of a Ru(II) arene complex to neomycin or to guanidinoneomycin leads to compounds with differential cytotoxicities and accumulation between cancer and normal cells. Mol. Pharm. 10, 1964–1976.
Guajardo, R.J., Hudson, S.E., Brown, J., Mascharak, P.K., 1993. Models for non-heme iron containing oxidation enzymes. J. Am. Chem. Soc 115, 7971–7977.

Guo, P., 2008. Preface: a special issue on bionanotechnology. J. Nanosci. Nanotechnol. 5 (12), 04–12.

Gupta, U., Agashe, H.B., Asthana, A., Jain, N.K., 2006. A review of in vitro – in vivo investigations on dedrimers: the novel nanoscopic drug carriers. Nanomedicine 2, 66–73.

Gutensohn, K., Beythien, C., Bau, J., Fenner, T., Grewe, P., Koester, R., Padmanaban, K., Kuehnl, P., 2000. In vitro analyses of diamond-like carbon coated stents. Thromb. Res. 99 (6), 577–585.

Harman, M.K., Banks, S.A., Hodge, W.A., 1997. Analysis of a retrieved hip implant with titanium nitride coating. Arthroplasty 12 (8), 938–944.

Hartinger, C.G., Nolte, M.N., Dyson, P.J., 2012. Organometallics 31, 5677–5685.

Hashmi, A.S.K., 2007. Chem. Rev. 107, 3180–3211.

Hattori, K., Matsushita, R., Kimura, K., Abe, Y., et al., 2001. Biol. Pharm. Bull. 24, 1214.

Haxton, K.J., Burt, H.M., 2009. J. Pharm. Sci. 98, 2299–2316.

Heron, J.S., Fournier, T., Bourgeois, O., 2007. Surface effect on the pholon transport of silicon nanowire. J. Phys. 1 (Conferences Series 92), 1–4.

Hoffmann, K., Maj, E., Wojtezak, A., Wietrzyk, J., Lakomska, I., 2014. Lipophilic platinum(II) complexes with dicarboxylates: in vitro antiproliferative activity as well as the mechanism of the suppression of tumour cells growth. J. Biol. Inorg. Chem. 19 (Suppl 2), S779–S814.

Hu, X., Guo, T., Fu, X., Hu, X., 2003. Nanoscale oxide nanostructure induced by dynamic electric field on Si with AFM. Appl. Surf. Sci. 217 (1-4), 34–38.

Hu, C.M., Zhang, L., Aryal, S., Cheung, C., Fang, R.H., Zhang, L., 2011. Erythrocyte membrane—camouflaged polymeric nanoparticles as a biomimetic delivery platform. Proc. Natl. Acad. Sci. USA 108, 10980–10985.

Hubbell, J.A., Thomas, S.N., Swartz, M.A., 2009. Materials engineering for immunomodulation. Nature 462, 449–460.

Iijima, S., 1991. Helical microtubules of graphitic carbon. Nature 354, 56.

Jamieson, E.R., Lippard, S.J., 1999. Chem. Rev. 99, 2467–2498.

Jones, M.C., Leroux, J.C., 1999. Polymeric micelles—a new generation of colloidal drug carriers. Eur. J. Pharm. Biopharm. 356, 333–344.

Kaiser, D., 2005. Drawing Theories Apart: the Dispersion of Feynman Diagrams in Postwar Physics. University of Chicago Press, Chicago, IL.

Kasparkova, J., Vojtiskova, M., Natile, G., Brabec, V., 2008. Chem. Eur. J. 14, 1300.

Khokhar, A.R., AlBaker, S., Shamsuddin, S., Siddik, Z.H., 1997. J. Med. Chem. 40, 112–116.

Kim, T.I., et al., 2013. Injectable, cellular-scale optoelectronics with applications for wireless optogenetics. Science 340, 211–216.

Lau, J., Maye, M.M., Kariuki, N.N., Wang, L., Njoki, P., Lin, y., et al., 2005. Electrocatalytic oxidation of methanol: carbon-supported gold-platinum nanoparticles catalysts prepared by two – phase protocol. Catalysis Today 99 (3–4), 291–297.

Lee, H., Bellamkonda, R.V., Sun, W., Levenston, M.E., 2005. Biomechanical analysis of silicon microelectrode-induced strain in the brain. J. Neural Eng. 2, 81–89.

Leising, R.A., Kim, J., Pérez, M.A., Que, Jr., L., 1993. Alkane functionalization at (.mu.-oxo)diiron(III) centers. J. Am. Chem. Soc. 115, 9524–9530.

Li, X.M., Reinhoudt, M., Crego-Calama, M., 2007. What do we need for a superhydrophobic surface? A review on the recent progress in the preparation of superhydrophobic surfaces. Chem. Soc. Rev. 36, 1350.

Li, Q., van den Berg, T.A., Feringa, B.L., Roelfes, J.G., 2010. Mononuclear Fe(II)-N4Py complexes in oxidative DNA cleavage: structure, activity and mechanism. Dalton Trans. 39, 8012.

Lin, S., Li, M., Dujardin, E., Girard, C., Mann, S., 2005. One-dimensional plasmon coupling by facile self-assembly of gold nanoparticles into branched chain networks. Adv. Mater. 17, 2553.

Lubben, M., Meetsma, A., Wilkinson, E.C., Feringa, B.L., Que, Jr., L., 1995. Angew. Chem. Int. Ed. 34, 1512.

Lundqvist, M., Stigler, J., Elia, G., Lynch, I., Cedervall, T., Dawson, K.A., 2008. Nanoparticle size and surface properties determine the protein corona with possible implications for biological impacts. Proc. Natl. Acad. Sci. USA 105, 14265–14270.

Lundqvist, M., Stigler, J., Cedervall, T., Berggård, T., Flanagan, M.B., Lynch, I., et al., 2011. The evolution of the protein corona around nanoparticles: a test study. ACS Nano 5, 7503–7509.
MacDonald, I.J., Dougherty, T.J., 2001. Basic principles of photodynamic therapy. J. Porphyrins Phthalocyanines 5, 105–129.
Mahon, E., Salvati, A., Baldelli-Bombelli, F., Lynch, I., Dawson, K.A., 2012. Designing the nanoparticle-biomolecule interface for targeting and therapeutic delivery. J. Control. Rel. 161, 164–174.
Maiorano, G., Sabella, S., Sorce, B., Brunetti, V., Malvindi, M.A., Cingolani, R., et al., 2010. Effects of cell culture media on the dynamic formation of protein-nanoparticle complexes and influence on the cellular response. ACS Nano 4, 7481–7491.
Majumder, D.D., Banerjee, R., Ulrichs, C.H., Mewis, I., 2007. Nanomaterials: science of bottom–up and top–down. IETE Tech. Rev. 24 (1), 9–25.
Mancini, D.M., Beniaminovitz, A., Levin, H., et al., 1998. Low incidence of myocardial recovery after left ventricular assist device implantation in patients with chronic heart failure. Circulation 98, 2383–2389.
Manna, L., Scher, E.C., Alivisatos, A.P., 2000. J. Am. Chem. Soc 122, 12700.
Mari, C., et al., 2014. DNA Intercalating RuII Polypyridyl Complexes as Effective Photosensitizers in Photodynamic Therapy. Chem. Eur. J. 20 (44), 14421–14436.
Matija, L., 2004. Reviewing paper: nanotechnology: artificial versus natural self-assembly. FME Trans. 32, 1–14.
Mehnert et al., 2001.Solid lipid nanoparticles, Wikipedia, 1–3.
Meyer, A., Bagowski, C.P., Kokoschka, M., Stefanopoulou, M., Alborzinia, H., Can, S., et al., 2012. On the biological properties of alkynyl phosphine gold(I) complexes. Angew. Chem. Int. Ed. 51, 8895–8899.
Milliron, D.J., Scher, E.C., 2003. Nat. Mater. 2, 382.
Monopoli, M.P., Walczyk, D., Campbell, A., Elia, G., Lynch, I., Bombelli, F.B., et al., 2011. Physical-chemical aspects of protein corona: relevance to in vitro and in vivo biological impacts of nanoparticles. J. Am. Chem. Soc. 133, 2525–2534.
Moravec, T.J., Schmidt, P., Spencer, E.G., 1980. Am. Ceram. Soc. Bull. 59 (3), 329.
Moyano, D.F., Goldsmith, M., Solfiell, D.J., Landesman-Milo, D., 2012. Nanoparticle hydrophobicity dictates immune response. J. Am. Chem. Soc. 134, 3965–3967.
Mozafari, M.R., Sahin, N.O., 2005. Manufacturing methods and mechanism of formation of lipid vesicles in nanoliposomes: from fundamentals to recent developments. Trafford Publishing Ltd, Oxford, UK, pp. 39–48.
Muller, R.H., Mader, K., Gohla, S., 2000. Solid lipid nanoparticles (SLN) for controlled drug delivery—a review of the state of art. Eur. J. Pharm. Biopharm. 50 (1), 161–177.
Munakata, H., Imai, H., Nakagawa, S., Osada, A., Uemori, Y., 2003. Chem. Pharm. Bull. 51, 614–615.
Murphy, M.E., 2013. The HSP70 family and cancer. Carcinogenesis 34, 1181.
Murray, C.B., Kagan, C.R., Bowendi, M.G., 2000. Synthesis and characterization of monodispersenanocrystals and close-packed nanocrystals assemblies. Annu. Rev. Mater. Res. 30 (1), 545–555.
Nakabayashi, N., Williams, D.F., 2003. Preparation of non-thrombogenic materials using 2-methacryloyloxyethylphosphorylcholine. Biomaterials 24 (13), 2431–2435.
Royal Society and Royal Academy of Engineering, 2004. Nanosciences and nanotechnologies: opportunities and uncertainties. Royal Society and Royal Academy of Engineering, London, UK, http://www.nanotech.org.uk/finalreport.html.
Narayan, J., et al., 1994. Diamond, diamond-like and titanium nitride biocompatible coatings for human-body parts. Mater. Sci. Eng. B 25, 5–10.
Navath, R.S., Menjoge, A.R., Wang, B., Romero, R., Kannan, S., Kannan, R.M., 2010. Amino acid-functionalized dendrimers with heterobifunctional chemoselective peripheral groups for drug delivery applications. Biomacromolecules 11 (6), 1544–1563.
Neumann, W., Crews, B.C., Marnett, L.J., Hey-Hawkins, E., 2014. Chem. Med. Chem. 9 (6), 1150–1153.
Ni, Y.H., Ge, X.W., Zhang, Z.C., Ye, Q., 2002. Fabrication and characterization of the plate-shaped γ-Fe2O3 nanocrystals. Chem. Mater. 14, 1048.

Nikolaychik, V.V., Wankowski, D.M., Samet, M.M., Lelkes, P.I., 1996. In vitro testing of endothelial cell monolayers under dynamic conditions inside a beating ventricular prosthesis. ASAIO J. 42 (5), M487–M494.

Nishiyama, N., Kataoka, K., 2006. Current state achievements and future prospects of polymeric micelles as nanocarriers for drug and gene delivery. Pharmacol. Ther. 112, 630–648.

Norris, D.J., 1995. Measurement and assignment of the size-dependent optical spectrum in cadmium Selinide (CdSe) Quantum dots, PhD thesis.

Ott, I., 2009. On the medicinal chemistry of gold complexes as anticancer drugs. Coord. Chem. Rev. 253, 1670–1681.

Patel-Predd, P., 2008. Turning waste heat in to power: research shows that silicon is as efficient as pricier materials. Technol. Rev. 14, 1–2.

Paul, J., Borm, M., Kreyling, W., 2004. Toxicological hazards of inhaled nanoparticles potential implications for drug delivery. J. Nanosci. Nanotechnol. 4 (6), 1–11.

Pernot, M., Bastogne, T., Barry, N.P.E., Therrien, B., Koellensperger, G., Hann, S., et al., 2012. Systems biology approach for in vivo photodynamic therapy optimization of ruthenium-porphyrin compounds. J. Photochem. Photobiol. B 117, 80–89.

Piccart, M.J., Lamb, H., Vermorken, J.B., 2001. Current and future potential roles of the platinum drugs in the treatment of ovarian cancer. Ann. Oncol. 12, 1195.

Pierre, T.G., Bell, S.H., Dickson, D.P.E., Mann, S., Webb, J., Moore, G.R., Williams, R.J.P., 1986. Mössbauer spectroscopic studies of the cores of human, limpet and bacterial ferritins. Biochim. Biophys. Acta 870, 127.

Plapied, L., Duhem, N., des Rieux, A., Preat, V., 2011. Fate of polymeric nanocarriers for oral drug delivery. Curr. Opin.Colloid Interface Sci. 16 (3), 228–237.

Rattan, S.I.S., Clark, B.F.C., 1994. Biochem. Biophys. Res. Commun. 201, 665–672.

Riesenfeld, J., Olsson, P., Sanchez, J., Mollnes, T., 1995. Surface modification with functionally active heparin. Med. Dev. Technol. 6 (2), 24–31.

Roco, M.C., 2004. The US national nanotechnology initiative after 3 years (2000-2003). J. Nanopart. Res. 6, 1–10.

Roelfes, J.G., Branum, M.E., Wang, L., Que, L., Feringa, B.L., 2000. J. Am. Chem. Soc. 122, 11517.

Rose, E.A., Levin, H.R., Oz, M.C., et al., 1994. Artificial circulatory support with textured interior surfaces. A counterintuitive approach to minimizing thromboembolism. Circulation 90 (5), 1187–1191.

Rossetti, R., Nakahara, S., Brus, L.E., 1983. Quantum size effects in the redox potentials, resonance Raman spectra, and electronic spectra of CdS crystallites in aqueous solution. J. Chem. Phys. 79, 1086.

Rousche, P.J., et al., 2001. Flexible polyimide-based intracortical electrode arrays with bioactive capability. IEEE Trans. Biomed. Eng. 48, 361–371.

Sam, J.W., Tang, X.J., Peisach, J., 1994. Am. Chem. Soc. 116, 5250–5256.

Sarparanta, M., Bimbo, L.M., Rytkönen, J., Mäkilä, E., Laaksonen, T.J., Laaksonen, P., et al., 2012. Intravenous delivery of hydrophobin-functionalized porous silicon nanoparticles: stability, plasma protein adsorption and bio distribution. Mol. Pharm. 9, 654–663.

Schenk, M., Mueller, C., 2008. The mucosal immune system at the gastrointestinal barrier. Best Pract. Res. Clin. Gastroenterol. 22 (3), 391–409.

Scott-Burden, T., Tock, C., Bosely, J., et al., 1998. Nonthrombogenic, adhesive cellular lining for left ventricular assist devices. Circulation 10 (98), II339–II345.

Serajuddin, A.T.M., 2007. Salt formation to improve drug solubility. Adv. Drug Deliv. Rev. 59 (7), 603–616.

Sfeir, C., Bennett, J., Kumta, P., 2003. Mol. Ther. 7 (5), 225.

Shafiovich, E., Diakov, V., Verma, A., 2007. Combustion-assisted hydrolysis of sodium borohydride for hydrogen generation. Int. J. Hydrogen Energy 32 (2), 207–211.

Shafirovich, E., Diakov, V., Verma, A., 2006. Combustion of novel chemical mixture of hydrogen generation. Combust. Flame 144 (1-2), 415–418.

Sharma, A.K., et al., 2000. Mater. Sci. Eng. B 77 (2), 139–143.

Sheeparamatti, B.G., Sheeparamatti, R.B., Kadadevaramath, J.S., 2007. Inspiration from nature. IETE Techn. Rev. 24 (1), 5–8.
Shvedova, A.A., Kagan, V.E., Fadeel, B., 2010. Close encounters of the small kind: adverse effects of man-made materials interfacing with the nanocosmos of biological systems. Annu. Rev. Pharmacol. Toxicol. 50, 63–88.
Stant, V.P., Smith, D., Leroux, J.C., 2005. Enhancement of oral bioavailability of poorly water- soluble drugs by polyethylene glycol self-assemblies. J. Control. Rel. 104 (2), 289–300.
Stella, V.J., Addae, K.W., 2007. Prodrugstratigies to overcome poor water solubility. Adv. Drug Deliv. Rev. 59 (7), 677–694.
Stephen, M., 2008. Life as a nanoscale phenomenon. Angew. Chem. Int. Ed. 47 (29), 5306–5320.
Stevenson, L.W., Miller, L.W., Desvigne-Nickens, P., et al., 2004. Left ventricular assist device as destination for patients undergoing intravenous inotropic therapy: a subset analysis from REMATCH (Randomized Evaluation of Mechanical Assistance in Treatment of Chronic Heart Failure). Circulation 110, 975–981.
Stewart, D.J., 2007. Crit. Rev. Oncol. Hematol. 63, 12.
Stubbe, J., Kozarich, J.W., 1996. Mechanisms of bleomycin-induced DNA degradation. Chem. Rev. 87, 1107–1136.
Su, Z., Zhou, W., Zhang, Y., 2011. New insight into the soot nanoparticles in a candle flame. Chem. Commun. 47, 4700–4702.
Talapin, D.V., Murray, C.B., 2005. PbSe nanocrystal solids for n- and p-channel thin film field-effect transistors. Science 310, 86.
Tian, N., Zhou, Z.y., Sun, S.G., Ding, y., Wang, Z.L., 2007. Synthesis of tetrahexahedral platinum nanocrystal with high index facets and high electro-oxidation activity. Science 316, 732–735.
Tian, B., et al., 2012. Macroporous nanowire nanoelectronic scaffolds for synthetic tissues. Nat. Mater. 11, 986–994.
Tran, H.S., Puc, M.M., Hewitt, C.W., Soll, D.B., Marra, S.W., Simonetti, V.A., Cilley, J.H., DelRossi, A.J., 1999. Diamond-like carbon coating and plasma or glow discharge treatment of mechanical heart valves. J. Invest. Surg. 12 (3), 133–140.
Uddin, M.J., Crews, B.C., Blobaum, A.L., Kingsley, P.J., Gorden, D.L., McIntyre, J.O., et al., 2010. Cancer Res. 70, 3618.
Urban, J.J., Talapin, D.V., Shevchenko, E.V., Kagan, C.R., Murray, C.B., 2007. Nat. Mater. 6, 115.
Vallhov, H., Qin, J., Johansson, S.M., Ahlborg, N., Muhammed, M.A., Scheynius, A., et al., 2006. The importance of an endotoxin-free environment during the production of nanoparticles used in medical applications. Nano Lett. 6, 1682–1686.
Vieu, C., Carcenac, F., Pepin, A., Chen, y., Mejias, M., Lebib, L., et al., 2000. Appl. Surf. Sci. 164, 111.
Wei, Q., et al., 1999. Preparation and mechanical properties of composite diamond-like carbon thin films. J. Vac. Sci. Technol. A 17 (6), 3406–3414.
Wexselblatt, E., Yavin, E., Gibson, D., 2013. Platinum(IV) prodrugs with haloacetato ligands in the axial positions can undergo hydrolysis under biologically relevant conditions. Angew. Chem. Int. Ed. 52, 6059–6062.
Williams, D., 2008. The relationship between biomaterials and nanotechnology. Biomaterials.
Winterbourn, C.C., 2008. Reconciling the chemistry and biology of reactive oxygen species. Nat. Chem. Biol. 4, 278.
Xie, C., Liu, J., Fu, T.M., Dai, X., Zhou, W., Charles, M., 2015. Three-dimensional macroporous nanoelectronic networks as minimally invasive brain probes. Nat. Mater. 14, 1286–1293.
Xiong, Y., Xia, Y., 2007. Shape-controlled synthesis of metal nanostructures: the case of palladium. Adv. Mater. 19, 3385.
Yih, T.C., Al-Fandi, M., 2006. Engineered nanoparticles as precise drug delivery systems. J. Cell. Biochem. 97, 1184–1190.
Yu, Y., Chang, S., Lee, J., Wang, C.R., 1997. Gold nanorods: electrochemical synthesis and optical properties. J. Phys. Chem. B 101, 6661.

Yu, L.J., Tay, B.K., Sheeja, D., Fu, Y.Q., Miao, J.M., Zhang, B.K., et al., 2000. A comparative study between pure and Al-containing amorphous carbon films. Surf. Coat. Technol. 128, 484–488.

Zang, Y., Elgren, T.E., Dong, Y.L., Que, Jr., J., 1993. Am. Chem. Soc 115, 811–813.

Zhang, H., Ma, Y., Xie, Y., An, Y., Huang, Y., Zhu, Z., Yang, C.J., 2015. A controllable aptamer-based self-assembled DNA dendrimer for high affinity targeting, bioimaging and drug delivery. Sci. Res. 5, 10099.

Zitvogel, L., Kepp, O., Kroemer, G., 2010. Decoding cell death signals in inflammation and immunity. Cell 140, 798–804.

Zong, R.L., Zhou, J., Li, B., Fu, M., Shi, S.K., Li, L.T., 2005. Optical properties of transparent copper nanorod and nanowise arrays embedded in anodic alumina oxide. J. Chem. Phys. 123, 094710–094715.

CHAPTER

IMPROVEMENT STEPS OF PLASTIC SURGERY TO TISSUE ENGINEERING BY NANOTECHNOLOGY

15

Ulvan Ozad*, Anı Cinpolat, Gamze Bektas**, Zumreta Rizvanovic****

**Near East University Hospital, Nicosia, North Cyprus*

***Private Clinic, Istanbul, Turkey*

CHAPTER OUTLINE

1 INTRODUCTION

Nanotechnology is the modification of matter in the scale of one billion times smaller than a meter. Manipulation of the atoms and molecules in such a small scale could not be observed with the naked human eye or optical microscopes, and could only be visualized by special imaging systems that enable nanosized visualization (Drexler, 1986, 1992). When high technology imaging systems were developed and the definition of the "nano" size became a topic for debate, the National Nanotechnology Initiative has brought a clear definition to the dimensions of matter that could be studied in nanoscale: 1–100 nm (Coombs and Robinson, 1996).

Plastic and reconstructive surgery is the specialty that probably benefits the most from nanotechnology. Understanding the development of nanotechnology into nanomedicine and the characteristics of nanomaterials are important for appreciating its uses in therapeutics, medical diagnostics, and tissue engineering.

2 DEVELOPMENT OF NANOTECHNOLOGY INTO NANOMEDICINE

Nanomedicine is one of the most extensively studied fields of medicine. As the research in nanotechnology developed, new mechanisms and materials were also invented; this resulted in extending the applications of nanotechnology to all fields of science, such as engineering, medicine, and materials science. Increasing the use of nanotechnology into the daily lives of people also brought the discussions about the environmental sustainability of nanomaterial consumption (Buzea et al., 2007). This created the foundations of nanotechnology research, and initiated the entry of new regulations to the nanotechnology field.

Although nanotechnology is seen and accepted as a field emerging in the past decade, the nanotechnology concept was first mentioned in 1959, where the possibility of atom manipulation in synthesis process was discussed by Richard Feynman (Drexler, 2009). Eric Drexler, a follower of Feynman, explained the idea of nanoscale atomic control in the book he published in 1986, as well as cofounding an institute for increasing the understanding about the importance of nanotechnology: the Foresight Institute (Drexler, 1986; Foresight Institute, 2015).

Drexler's intensive work opened the doors to the acknowledgement of nanotechnology as a field. However, two inventions taking place in the same timeframe not only shaped the research, but also increased the understanding and interest in nanotechnology: scanning tunneling microscope (STM) and fullerenes. The invention of STM invention enabled the imaging of atoms, as well as enabling manipulation at an atomic level. The ability to image and visualize nanoscaled particles attracted attention to the nanotechnology field (Binnig and Rohrer, 1986). Fullerenes, the second invention, were discovered as graphene tubes and, although the original intention was not use in nanotechnology field, fullerene discovery provided great potential for nanotechnology application (Kroto et al., 1985).

Nanotechnology is one of the main interests of the current research, in all scientific fields, ranging from medicine to engineering. It is the study of particles in the nanoscale, 1×10^{-9} m. Especially after the millennium age, interest of the medical field in nanotechnology has greatly increased, and a new field, nanomedicine, has been created. Nanomedicine investigates all medical aspects of nanotechnology, such as cell repair and drug delivery. Developments in nanotechnology are changing the future of therapeutics, especially in drug delivery and imaging (Jaffer and Weissleder, 2004; Taton et al., 2001; Tyagi and Kramer, 1996).

Nanotechnology has been used increasingly in the medical field, to the extent that a new specialty is named after it. The application of nanotechnology to medicine is called nanomedicine (Freitas, 1999). This field has a wide range of applications, including nanomaterials, nanoimaging, nanosensors, and nanobiologics. Also, nanotechnology is used in a wide range of diagnostic and therapeutic applications in medicine, such as drug delivery, cancer imaging, tumor diagnosis, and analytic systems.

Currently, nanosized materials have countless uses. Nanotubes, nanorods, nanoparticles, nanopillars, nanopores, nanodots, and many more nanomaterials have been discovered and found to be useful in the medical field.

Nanofibers were one of the first nanomaterials that found wide use in an extensive range of products, from housing to aeronautical industry. Nanofiber use has developed very quickly, without realization of the possible side-effects to health. It was discovered afterward that inhalation of nanofibers through the air led to a range of occupational lung diseases, especially to lung fibrosis (Byrne and Baugh, 2008). Also, nanoparticles were found to cause an increase in inflammation, and increase the aging of the skin (Elder, 2006; Jonaitis et al., 2010; Wu et al., 2009). Following these, many studies took place to demonstrate the harmful effects of nanotechnology. After these studies on the negative implications of nanotechnologically created substances on health, handling of these materials attracted extra care and attention. The regulation of nanotechnology products became even more serious and tighter (Rollins, 2009).

Plastic and reconstructive surgery is a specialty concerned with restoring image. It is interested in cancer, degenerative conditions, trauma, and many other diseases. In order to understand the influence of nanotechnology on the field of plastic and reconstructive surgery, it is essential to understand the nanomaterials used in therapeutics, use of nanotechnology in diagnostics, drug-delivery mechanisms, and the use of nanotechnology in tissue engineering.

3 NANOMATERIALS USED IN MEDICAL THERAPEUTICS

Understanding the nanomaterials and nanodevices used in the medical field is vital for understanding the therapeutic applications of nanotechnology.

Nanopores are the most basic nanomaterials, originally created by the automotive industry, but found a great use in the medical field (Desai et al., 1998). In medicine, it was the possibility of creating nanopores appropriately large to enable nutrient and oxygen delivery, but small enough to prevent immune complex and pathogen delivery; this is how the possible implications of such a material on restoring health was realized. In this system, nutrients (such as glucose), and hormones (such as insulin), exchange gases (such as oxygen and carbon dioxide), and are able to move through the nanopores; substances that could cause disease (such as viruses, bacteria, or immune complexes) are larger, and they could not move through the nanopores. Therefore this was acknowledged as a great way of facilitating certain therapeutic methods. Glucose mechanism, perhaps, is one of the best examples for the use of nanopores. This system could be useful for neurodegenerative conditions, such as Parkinson's disease or Alzheimer's disease. New encapsulated cells, encapsulated neurons, could be supplied to the brain, accompanied by electric stimulation in order to facilitate the release of neurotransmitters (Tao and Desai, 2003).

External flow regulation through nanopores is also possible through electrical charging (Lee and Martin, 2002). For example, positively charged ions would be repelled, whereas negatively charged ions would be attracted to a positively charged nanopore. In the same way, when a membrane made up of nanopores receives a negative signal and becomes negatively charged, only positively charged particles could pass through. This mechanism enables controlling ion transport by not only the size, but also according to the charge. Therefore, this is found to be potentially useful for separating drugs (Lee et al., 2002; Mitchell et al., 2002).

After the discovery of numerous nanopore uses, it was found that nanopores are also functional in DNA studies because nanopores could be used to differentiate the dissimilar nucleotide bases of the DNA: purine, and pyrimidine. Through this discovery, the functional use of nanopores in DNA sequencing has been initiated (Meller et al., 2000; Storm et al., 2003).

Fluorescence has been increasingly used in medical diagnosis. The use of fluorescent dyes is very complicated since specific dyes, molecules, and lasers are required, depending on colors. In order to overcome this problem, nanocrystals have been developed and called quantum dots. The color of the molecule could be varied by changing the composition of the nanocrystal, or, simply by changing the size. This has provided an easier and more economical way of imaging and diagnosis.

Quantum dots could be used in various fields, such as cancer detection, DNA microarrays, and drug screening (Gao et al., 2002; Gerion et al., 2002; Hanaki et al., 2003; Medintz et al., 2003; Wu et al., 2003).

The increased use of nanotechnology in drug delivery guided scientists to conduct research in the discovery of new novel nanosized materials to be used in drug-delivery mechanisms. After the success of carbon fullerenes, nanoshells were developed (Sershen et al., 2000; West and Halas, 2000). These gold-coated nanospheres could be ingested in the body of the cancer patient, and transported to the site of the cancer. Afterward, an infrared laser could be used to provide heat and release the medication in the nanoshell by melting the capsule. This mechanism provides a target-specific therapy method in cancer. Similarly, nanoshells could be used for diabetic patients, where heating would release insulin (Rice University, 2003).

Quantum dots could be effectively used in labeling, since both electrical and optical methods would be available for this process. It is found to be successful in vitro, and partially successful in vivo (Ballou et al., 2004; Chan et al., 2002; Dubertert et al., 2002; Gao et al., 2004; Jaiswal et al., 2003). Quantum dots could also be used, combined by fluorescent magnetic resonance imaging, to detect numerous conditions, such as cancer, arthritis, stroke, or atherosclerosis (Dardzinski et al., 2001; Dousset et al., 1999; Harisinghani et al., 2003; Kooi et al., 2003; Perez et al., 2004; Ruehm et al., 2001; Saleh et al., 2004).

Nanosized particles, although small in size, have a very high surface area to volume ratio. This has led researchers to question the possibility of making use of the high surface area in creating artificial binding sites, and achieve molecular imprinting (Shi and Ratner, 2000; Zimmerman et al., 2002). Targeted nanoparticles, such as targeted radioimmunotherapeutics are approved by the US Food and Drug Agency (FDA) for use in cancer treatment (Dixon, 2003).

Nanotechnology research improved with great velocity over a very short period of time. The commercialization of nanotechnology drug-delivery applications started in 2005, when nanocrystals started to be used in brands, such as TriCor and Emend. From 2006 to 2008, the FDA reviewed these products, and clinical testing of phase three trials took place for few years. Later, many other products, such as biological nanoparticles and nanocomposites have been developed, and went into clinical and preclinical testing. In 2018, self-assembling nanocubes are planned to go through early stage development (NHI, 2006).

In cell studies, microparticles with magnetic features have been increasingly used in separation of cells or cell structures. With developments in nanotechnology, it was questioned whether nanoparticles could be used for removal of infectious pathogens or certain toxins from the blood. This has been successful in the animal models, and is currently a promising research for dialysis treatment of renal patients (Herrmann et al., 2013a; Kang et al., 2014).

Nanomagnets, due to large surface area to volume ratio and their nanosize, have been found to be more useful than hemoperfusion, and the possibility of using nanotechnology to purify the blood, especially by direct removal of the infectious pathogens in sepsis, or by removal of toxins in infections, is still under preclinical research (Herrmann et al., 2009, 2013b; Lee et al., 2014; Schumacher et al., 2013; Yung et al., 2009).

There are countless types of nanomaterials that are invented, patented, and used in nanomedicine; ranging from one dimensional (1D) films to three-dimensional (3D) structures. Fig. 15.1 summarizes the most common nanomaterials used in nanomedicine.

FIGURE 15.1 The Most Common Materials Used in Nanomedicine

One-dimensional nanothin films; two-dimensional nanowires, nanofibers, and nanotubes; three-dimensional (3D) nanopowders, nanoparticles, nanopores, nanocapsules, nanosponges, quantum dots, fullerenes, and dendrimers are commonly used in nanomedicine.

4 NANOTECHNOLOGY IN DIAGNOSTICS

As the developments in the medical field improve, more effective and efficient methods of treatments to diseases are searched. The success of the treatment is influenced greatly by diagnostics, especially by imaging.

Nanotechnology research provides promising developments for tumor imaging. The nanoparticles produced from metals, such as magnesium oxide, are coated with antibodies of cancer cells and, when delivered, they locate the cancerous tissue by traveling to the cancer site. Magnetic resonance imaging performed afterward would enable the precise anatomical location of the cancer tissue.

Cadmium selenide quantum dots are used in a similar manner. These particles become fluorescent in ultraviolet light, and are used as a guide by surgeons in the removal of cancerous tissue. Numerous cancer types are in the interest of plastic surgery; therefore, improvements in cancer diagnosis and treatment has also created a benefit to the plastic surgery field.

Nanotechnology is also effectively applied to genetic studies. For example, nanosized gold particles are used for determination of the genetic sequence by DNA tagging. Intercellular structures, such as nucleic acids, could be studied by nanopores. In histology, nanotechnology is not only used in intercellular structures; microorganisms could also be detected by the similar technology. The availability of cancer detection by using this technology have led to minimally invasive diagnosis and surgery of cancer with better healing time (Hall, 2005; Zheng et al., 2005).

5 NANOTECHNOLOGY AND DRUG DELIVERY

In the past decades, there was a strong shift in the delivery methods of therapeutics and new treatments from the generalized, nonspecific methods, such as formulations and systemic delivery techniques, to intelligent methods, such as cell-targeted therapies, and individually engineered delivery systems.

Development of nanotechnology has influenced greatly the application of therapeutic materials; mainly by providing the opportunity to perform a precise control mechanism on the dimensions, interfaces, and morphologies of materials.

In nanomedicine, drug delivery is one of the most important fields of nanotechnology application. Nanoparticles are used to aid the delivery of medications to specific targets, such as cells. Furthermore, the total amount of drug administered and the drug side-effects could be significantly lowered by locating the precise amount of active medication particles onto the live region. This approach would not only minimize the side-effects of the therapeutics used, due to decreased amount of usage, but would also decrease treatment expenses. By this method, bioavailability of the medication increases significantly, and faster, sensitive drug delivery would be achieved (Boisseau and Loubaton, 2011). However, the efficacy in this drug-delivery method requires several factors to be satisfied. Encapsulation of the medications should be performed in an effective way, the medication should be delivered to the target in the human body; furthermore, the medication should be released into the body in an appropriate way. Numerous materials are used in the nanoscale characterization of the systems used in drug delivery. The most commonly used materials are lipids and polymers. One other advantage of drug-delivery systems is the chance of improving the solubility by addition of hydrophilic components (Nagy et al., 2011). Nanomaterial functionality could be improved through biological molecules as the nanosize of materials is consistent with size of the biological structures.

One of the greatest applications of nanotechnology in the medical field is development of nanoparticles to aid delivery of therapeutic substances to the specific targets, such as cells. Therapeutic applications could be medications or other therapy methods, such as heat. Particles are specifically engineered to go to targeted cells, to enable treatment of the desired cells or tissue. This method is mostly mentioned in cancer treatment where carcinogenic cells are targeted by anticancer therapeutics. Numerous nanocarriers have been created for delivery of anticancer medications to the carcinogenic tissue (Ghaz-Jahanian et al., 2015; Nguyen-Ngoc and Raymond, 2015; Pérez-Herrero and Fernández-Medarde, 2015; Wang et al., 2012). This enables preservation and protection of healthy cells. Currently, chemotherapy delivery to carcinogenic cells is one of the most important research projects in nanomedicine. Vaccines and insulin release mechanisms are designed with nanotechnology drug-delivery mechanisms.

Although there is extensive research taking place on nanomedical drug-delivery systems, only two trials have been successful in animal experiments, so far. The first mechanism involves use of nanosized gold shells in cancer treatment (Soni et al., 2013). The second mechanism involves use of liposomes for the drug delivery in vaccinations (Mozafari, 2006). These two methods require clinical trials in the human population in order to receive approval for use. One use of nanotechnology in medicine is the creation of nanosized electromechanic devices in order to release drugs actively, which would be a great development for implantable treatments, such as for infections. Drug release in nanoscaled drug-delivery systems could be possibly used in infection treatments in order to reduce drug overdosing, as well as to prevent antibiotic resistance (Banoee et al., 2010; Borzabadi-Farahani et al., 2013).

As explained, nanosized drug-delivery systems are still being developed and tested; however, medications created by using nanotechnology are present and available for commercial use. Abraxane, Cornell dots, Doxil, and Minicell are some examples of nanotechnologically created drugs. Abraxane is the active anticancer agent paclitaxel, binding to albumin in nanoscale; it is used for the treatment of specific cancer types, such as breast cancer and pancreatic cancer (FDA, 2012, 2013). Cornell dots (C dots) are silica particles with dyes that cause fluorescence. These nanoparticles are particularly useful for location of tumors and carcinogenic tissue (Gibney, 2013). Doxil is a liposome-derived medication that was developed for use in Kaposi's sarcoma. However, the range of use of this medication extended to cover multiple myeloma, as well as certain cancer types, such as ovarian cancer (Martis et al., 2012). Minicell is a complex nanotechnology product especially designed for terminal cancer patients, in order to minimize the cancer treatment side-effects. Phase one trials of this medication have been successful (Fierce Drug Delivery, 2015).

One great advantage of nanosized particles is the increased surface area, compared to volume. This provides a greater area for attachment, and helps attachment to the cancer tissue. Also, the small size of the particles helps transportation and accumulation at cancer sites (Shuming et al., 2007). Current conventional anticancer treatments have important limitations, such as insolubility, and development of drug resistance. New generation nanomedicines are being designed to overcome these important limitations and offer a more successful cancer treatment (Chidambaram et al., 2011).

6 USE OF NANOTECHNOLOGY IN TISSUE ENGINEERING

Tissue engineering is an emerging subject that involves both medical and engineering studies. As the life expectancy of the general population is increasing, the amount of patients suffering from degenerative conditions increases. With improving technology, the amount and impact of trauma cases are also increasing. This has led to a search for regenerating tissues, in order to replace the damage. Cartilage, nerve, and cardiac tissues have limited regeneration capacity; therefore, regeneration is now accepted as the only solution for completely treating the pathological, degenerative, or traumatic diseases of these tissues. However, the tissue-engineering field is not limited to the tissues with limited regeneration capacity; bone and skin are two tissue types which are able to regenerate but are still in the interest of the tissue-engineering field. This is because bone loss in various conditions could not be completely replaced by the human body. Traffic accidents and bone tumors are two examples of this condition. In the skin, although open wounds could be closed by the body itself, scarring tissue would remain. Therefore, these tissue types are also extensively studied in the tissue-engineering field.

Modern lifestyle in developed countries and the increased use of processed, carbohydrate-rich foods result in obesity. Developments in medicine and therapeutics as well as improvements in living conditions have created the resulting effect of increased life expectancy. These two factors, obesity and increased life expectancy, lead to an increase in cartilage damage, especially in weight-bearing joints. In the human body, cartilage regeneration is limited. This results in the total effect of destruction in cartilage leading to loss of mobility and pain. Cartilage tissue engineering has been developed in the past decades to search for a permanent treatment to cartilage degeneration caused by trauma and disease. There are four constituents that are vital in cartilage tissue engineering: the cell source, extracellular matrix, scaffold, and bioreactors. This is demonstrated in Fig. 15.2.

The creation of a scaffold with great precision in the desired characteristics has several requirements. Numerous methods have been created to produce nanopatterns that are required for cell regeneration. Nanofeatures could be found in many different forms, such as nanodots, nanopits, and nanopillars (Hanawa et al., 1997; Knabe et al., 2004).

In nanomaterials, the arrangement pattern is also prepared in nanoscale. There are numerous lithography techniques, such as imprint lithography, dip pen lithography, photolithography, and electron beam lithography that achieve patterning on the nanosurface. However, a successful method of copying the surface density and integrin epitope arrangement has not been achieved yet.

Scaffold production techniques have changed to become more precise with the introduction of nanotechnology to the engineering field, and numerous new scaffold production techniques have been

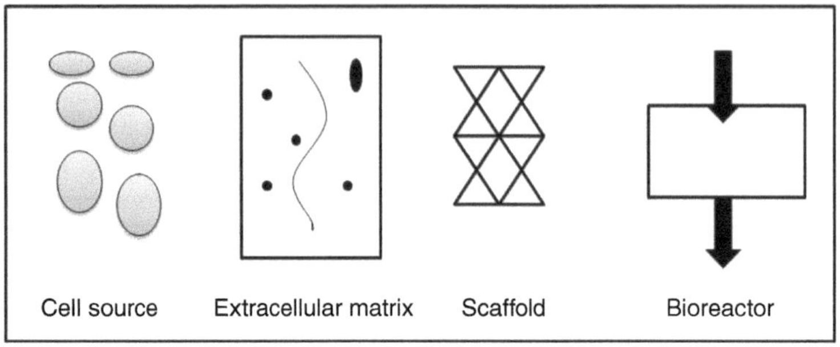

FIGURE 15.2 Requirements for Tissue Regeneration

A cell source that new cells could be regenerated from, extracellular matrix, a 3D scaffold to allow cell adhesion and expansion, and a bioreactor for providing the optimum medium; these are the requirements of tissue regeneration.

developed, such as self-assembly, phase separation, and electrospinning, in order to create nanofibrous scaffolds.

The self-assembly method involves the organization of molecules without any external intervention or force applied. Van der Waals, hydrogen bunds, and electrostatic forces have a great influence in the molecule arrangement of self-assembly technique. With this technique, nanofibers as thin as 10 nm could be formed. A great advantage of this method is providing biological compatibility in numerous arrangements. However, this is a complex procedure, and production could be very complicated, which is unfavorable for mass cartilage tissue regeneration (Lim et al., 2014).

The phase separation method is preferred in nanoporous scaffolds. In this technique, the polymer biomaterial is dissolved in the solvent to create a solution, with the aid of high temperatures. After the dissolving process, the temperature is decreased to create a phase separation and, by this method, numerous shapes, such as pores and pits, could be created. The phase separation aims to differentiate polymer-rich and solvent-rich components of the polymer solution. The remaining separated solutions could be cooled, followed by removal of the solvent through various methods, such as drying or freezing. Fibers as thin as 50 nm, and porosity as high as 98% could be achieved by this technique. Scaffolds created by the phase separation technique were found to be effective in many stages of cartilage regeneration, like cell proliferation and cell differentiation. This method is also suitable for mass production (Lim et al., 2014).

Nanotechnology is used increasingly in the production of precise scaffolds for extracellular matrix growth, cell regeneration, adhesion, and cell differentiation in cartilage tissue engineering (Kay et al., 2002; Oseni et al., 2011). Among all nanoscaled cartilage scaffold engineering methods, the most successful technique is electrospinning. In this technique, a charged polymer fluid is poured onto a surface loaded with the opposite charge. Fiber diameter and scaffold topology can be controlled precisely. Solution properties, such as viscosity and conductivity, can also be controlled, as well as the parameters requiring precision, such as needle tip design, and general factors, such as air temperature and humidity. Fiber diameters can be precisely controlled to be as thin as few nanometers by using the electrospinning method. This method works by creating a field of high electrical gradient between the spinneret and the collector device. The electrical force generated is redirected onto the solution that contains the polymer in order to overcome the charges that are carried in the solution, leading to solvent evaporation. The polymer that is left after the evaporation process continues to lengthen toward the collector. The collector device could be rotated to achieve various fiber alignments. This method is effective and economical. This technique is very successful in achieving the anatomical shapes of many tissues, such as the dermal tissue; therefore, it is one of the most promising scaffold production techniques in nanomedicine.

In tissue engineering, eloctrospinning is used to create a 3D nanofibrous networks made up of collagen. In the scaffold, collagen fibers are created not only to enable the anchoring and adhesion of the cells, but also to allow proliferation and cell differentiation (Tran et al., 2006). The advantage of using electrospinning for the creation of nanofiber scaffolds is having the ability to control the parameters tightly, for example, porosity and pore diameters. With this method, membranes could be produced to regenerate the dermis layer of the skin (Khil et al., 2003). This engineered tissue holds the ability to control the gas transport and water loss as well as demonstrating antimicrobial protection.

Bioreactors are devices that are used to provide optimal conditions for tissue growth. The aims are to provide gas exchange by carrying oxygen and removing carbon dioxide, provide nutrients (such as glucose), and remove waste. They also perform uniform distribution of these substances in order to preserve tissue viability. Numerous bioreactors have been developed in order to satisfy all these parameters; however, the research for the optimum parameter still continues (Eniwumide et al., 2009; McMahon et al., 2010; Sengers et al., 2005).

Cartilage has an avascular nature leading to a very limited regeneration capacity. This is due to low cellularity and minimal blood supply to the tissue. Osteoarthritis and rheumatoid arthritis are the two most common conditions leading to destruction in cartilage. Other than these conditions, congenital anomalies, tumors, trauma, and numerous degenerative diseases also cause destruction in the cartilage tissue. Currently, these conditions are managed temporarily to improve the range of motion and to limit the pain.

The cartilage is very important for the mobility of people because it is the tissue covering the joints at the ends of the bones. Loading and trauma could cause damage onto the cartilage, with long-term consequences. As the regeneration capacity of the cartilage is extremely low, a regenerative approach was found to be necessary in order to treat and heal degenerative cartilage conditions completely (Lopez-Ruiz et al., 2013).

One of the critical issues in cartilage engineering is the regeneration of chondrocytes in accurate morphology. In weight-bearing joints, the cartilage is arranged in layers where the chondrocyte morphology of the superficial layer is different compared to that of the deep layer. This is one of the main limitations that could not be overcome in artificial cartilage regeneration.

In vitro replication of cartilage cells have the drawback of losing the redifferentiation ability and phenotype formation through cell replication. Countless factors ranging from the growth factor use to creation of a 3D scaffolds to enable growth have an influential role in cartilage regeneration.

The immature cartilage has a greater ability to regenerate in comparison to the mature cartilage (Hunter, 1742; Namba et al., 1998). Low regeneration and repair capacity of the hyaline cartilage lead to a search for the creation of a scaffold with the accurate characteristics to enable tissue growth. Intracellular signaling is also vital for the scaffold structure since the intracellular pathways influence greatly the gene expression as well as cell anchoring. The use of nanotechnology to form scaffolds made up of nanosized fibers improve effectively the anabolic mechanisms in cell regeneration.

In tissue engineering, cartilage regeneration is achieved through several stages. The first stage is culturing the stem cells (Barry et al., 2001). This is followed by seeding the cells onto a 3D scaffold

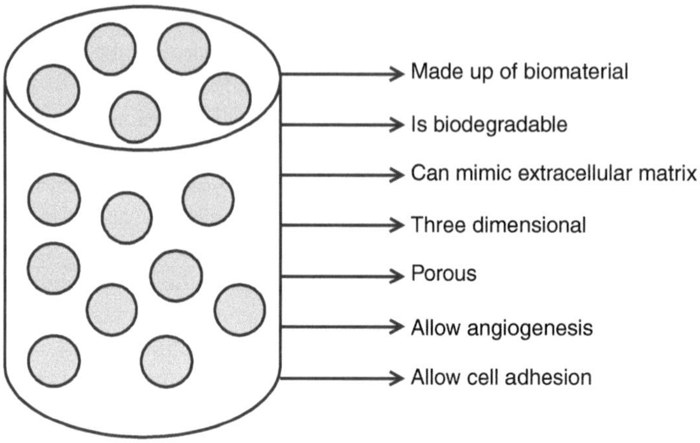

FIGURE 15.3 Important Properties of Scaffolds

Scaffolds should possess numerous qualities in order to be efficient in tissue regeneration. The use of biomaterials that are biodegradable, having a 3D setting, the ability to mimic the extracellular matrix, possessing the appropriate porosity characteristics that would effectively allow angiogenesis and cell adhesion are some of the vital scaffold requirements.

created from biomaterials. The choice of scaffold is important as this structure would provide a 3D matrix for tissue growth, imitate the biomechanical properties of extracellular matrix, and allow cell adhesion (Sams and Nixon, 1995; Silver and Pins, 1992). Afterward, bioreactors are used to provide optimum growth medium (Plunkett and O'Brien, 2010).

The skin is the barrier of the human body and it acts as a protective layer against external dangers. Due to its function, the skin could receive numerous trauma attacks, such as burns, ulcers, wounds, or lacerations. The skin is made up of two layers: epidermis and dermis. Both layers have different healing durations and mechanisms. In a great majority of trauma cases, skin heals with scarring tissue. Nanotechnology research has generated a potential for regenerating skin and restoring the physiological appearance. Scaffold formation is essential for skin regeneration, similar to all tissue-engineering fields. Scaffolds need to possess numerous characteristics. They need to be biocompatible, so that an adverse reaction would not be regenerated after application; they need to enable cell attachment in order to enable proliferation, and scaffolds need to mimic the extracellular matrix of the tissue that is requested to be regenerated as much as possible (Huang and Fu, 2010; Mohamed and Xing, 2012). Some of the important properties required to be present in an efficient scaffold are summarized in Fig. 15.3.

Mimicking the physical properties of the skin is an important goal to achieve since the mechanical barrier function is required for protection of the human being. Currently, the greatest challenge encountered in skin tissue engineering is achieving the adhesion of the skin graft to the site of application (Liu and Cao, 2007; Stynes et al., 2008).

Similarly, wound healing benefits from nanotechnology as well. Creation of nanofibrous scaffolds made from gelatin and poly(L-lactic acid)-*co*-poly(E-caprolactone) demonstrated to be successful in fibroblast growth and proliferation (Venugopal et al., 2006).

The application of growth factors onto the wounds that have become chronic over the time, by electrospinning the fibers including fibroblast growth factors, was found to improve the recovery rate, together with achieving angiogenesis and epithelization in mice experiments (Yang et al., 2012).

As the animal experiments demonstrated success, the effective use of nanotechnology in skin graft applications to humans is still under development and experimentation.

The bone, due to congenital conditions, trauma, degeneration, or cancer, is one of the greatest interests of transplant medicine (Arrington et al., 1996; Giannoudis et al., 2005; Miyazaki et al., 2009). Nanotechnology is also used in bone tissue engineering. Nanobiomaterials, such as carbon nanotube and graphene are used to create scaffolds with precise diameters and porosity to allow optimum bone regeneration in bone tissue engineering (Lalwani et al., 2013a,b).

Autologous and allogenic bone grafts have been commonly used to restore the bone defects; however, these methods demonstrated numerous drawbacks, especially due to the surgery itself. Therefore, new methods of bone regeneration are still researched (Kneser et al., 2006).

Electrospinning, in bone tissue engineering, is successfully used in the creation of nanofibrous scaffolds made up of synthetic and natural polymers which induce the differentiation of mesenchymal stem cells as well as the formation of the minerals that are abundant in the bone (Frohbergh et al., 2012; Kim et al., 2005; Nguyen et al., 2012; Yoshimoto et al., 2003).

The introduction of new materials onto nanofibers, such as nanohydroxyapatite, demonstrated satisfying cell attachment, proliferation, and differentiation functions on stem cells derived from the adipose tissue (Ravichandran et al., 2012).

Apart from the nanofibers, nanotube structures were found to be useful in bone regeneration. Self-assembled nanotubes mimic the collagen structure that is abundant in the bone. In this way, nanohydroxyapatite is used to facilitate cell adhesion and bone regeneration (Sun et al., 2012). Hydroxyapatite is one of the materials that is abundant in the bone matrix, and has been successfully used in bone tissue engineering. These developments are promising for the use of nanotechnology in a clinical setting for bone reconstruction.

Another tissue with very low regeneration capacity is the nervous tissue. Trauma or degenerative nerve conditions could lead to major negative results, such as sensory and motor function loss. Autologous grafting is used currently as a treatment method; however, this means the sensory and motor function of another site in the body is sacrificed; only certain lengths and regions of nervous tissue could be grafted (Peran et al., 2013). The greatest challenge in the regeneration of many tissues, especially in cartilage, was achieving cell differentiation; however, in nervous tissue, the greatest challenge is mimicking the extracellular matrix. Nanofiber polymers are thought to address this problem (Tan et al., 2012).

In nerve studies, electrical conduction is as crucial as the regeneration of the cells because the main function of the nerve cells is conducting signals between sensory or motor stimulus and the central nervous system (Ibrahim et al., 2012; Subramaniam et al., 2012). Therefore, the newly regenerated tissue has to possess adequate conduction characteristics in order to be functional.

Cardiac tissue engineering also benefits from the use of nanotechnology. The cardiac tissue, unlike the skeletal muscle, is made up of a special type of muscle called the smooth muscle. It is not able to regenerate or work voluntarily. As the eating habits of the developed countries changed to include fast foods that are high in saturated fats, and the physical activity performed by the working population has decreased significantly, heart disease started to create a great burden on health, and became one of the biggest causes of mortality and morbidity. Ischemic heart disease, stroke, and valvular heart disease are the most frequently encountered cardiovascular conditions. Myocardial infarction, also known as heart attack, is simply caused by the blockage of one or more vessels that carry oxygen and glucose to the cardiac muscles to perform respiration. Due to the blockage, cardiac muscles become ischemic and die. Since these cells could not regenerate efficiently, a fibrous scar tissue forms over the ischemic site. This causes a significant decrease in the function of the heart. The heart is a muscle tissue that actively responds to the hormonal or electrochemical changes in body signaling. Therefore, this presents a great challenge in cardiac muscle regeneration studies. The conductive and elastic properties of the cardiac tissue need to be regenerated in order to prevent malfunctioning or dysfunctioning (Kocher et al., 2001; Prabhakaran et al., 2011; Tandon et al., 2013). Injecting autologous cells into the infarct area had some advantages and improvement in tissue function, especially in the left ventricular function (Fuchs et al., 2006; Penn et al., 2002; Peran et al., 2013; van Ramshorst et al., 2009; Tse et al., 2007).

Tissue-engineering technology has enabled the in vitro creation of engineered heart muscle with appropriate properties and contractility. Also, heart valves which have a critical role in the circulation of blood have been engineered using nanomaterials. In particular, nanomaterial injections were found to be useful in restoring the function of heart valves that are not working properly.

The use of scaffolds made up of nanofibers, using the electrospinning method, is a feasible technique for regenerating the myocardial tissue. A critical issue is the determination of the material to be used. Several materials and compounds have been tried in cardiomyocyte regeneration (Kai et al., 2011; Mukherjee et al., 2011; Shin et al., 2004).

In the regeneration of tissue-engineered heart valves, efficiency could be improved by two approaches. Bioreactors are used to grow the cells in optimum conditions, and this approach was found to improve the regeneration success (Aleksieva et al., 2012; Martin et al., 2004; Peran et al., 2013). As explained, nanoparticles are extensively used in drug delivery. The other method is using nanoparticles for drug-delivery mechanisms in order to decrease thrombus or embolus formation after cardiac surgery (Winter et al., 2006, 2008; Sharma et al., 2010; Peters et al., 2009; Ji et al., 2012).

7 CONCLUSIONS

Nanomedicine involves all aspects of the medical field, such as control, repair, and monitoring of one certain or general human systems at cellular and molecular level, using nanoscaled structures and devices (Sahoo, 2005). The use of nanotechnology in restoring and monitoring human health is very promising, especially in imaging, diagnostics, tissue engineering, and drug delivery. Plastic surgery is extensively interested in all these subjects, especially in cancer treatment and tissue engineering. The diagnosis and treatment of cancer, which is the backbone of the regenerative part of the field, has been completely changed by the introduction of nanotechnology. Drug delivery, infection control, genetic diagnostics are all important for the field. The introduction of nanotechnology to the tissue-engineering field has created new opportunities in the regeneration field. Numerous products are already on the market, while countless ideas and products are still in the process of testing and approval. With these nanotechnological developments, tissue regeneration has become one of the most extensive research fields in science, and new inventions and developments for achieving effective organ and tissue regenerations are extremely promising.

REFERENCES

Aleksieva, G., Hollweck, T., Thierfelder, N., Haas, U., Koenig, F., Fano, C., Dauner, M., Wintermantel, E., Reichart, B., Schmitz, C., Akra, B., 2012. Use of a special bioreactor for the cultivation of a new flexible polyurethane scaffold for aortic valve tissue engineering. Biomed. Eng. Online 11, 92.

Arrington, E.D., Smith, W.J., Chambers, H.G., Bucknell, A.L., Davino, N.A., 1996. Complications of iliac crest bone graft harvesting. Clin. Orthop. Relat. Res. 329, 300–309.

Ballou, B., Lagerholm, B.C., Ernst, L.A., Bruchez, M.P., Waggoner, A.S., 2004. Noninvasive imaging of quantum dots in mice. Bioconjug. Chem. 15, 79–86.

Banoee, M., Seif, S., Nazari, Z.E., Jafari-Fesharaki, P., Shahverdi, H.R., Moballegh, A., Moghaddam, K.M., Shahverdi, A.R., 2010. ZnO nanoparticles enhanced antibacterial activity of ciprofloxacin against *Staphylococcus aureus* and *Escherichia coli*. J. Biomed. Mater. Res. B Appl. Biomater. 93 (2), 557–561.

Barry, F., Boynton, R.E., Liu, B., Murphy, J.M., 2001. Chondrogenic differentiation of mesenchymal stem cells from bone marrow: differentiation-dependent gene expression of matrix components. Exp. Cell Res. 268, 189–200.

Binnig, G., Rohrer, H., 1986. Scanning tunneling microscopy. IBM J. Res. Dev. 30 (4), 355–369.

Boisseau, P., Loubaton, B., 2011. Nanomedicine, nanotechnology in medicine. C. R. Phys. 12 (7), 620.

Borzabadi-Farahani, A., Borzabadi, E., Lynch, E., 2013. Nanoparticles in orthodontics, a review of antimicrobial and anti-caries applications. Acta Odontol. Scand. 72 (6), 413–417.

Buzea, C., Pacheco, I.I., Robbie, K., 2007. Nanomaterials and nanoparticles: sources and toxicity. Biointerphases 2 (4), MR17–MR71.

Byrne, J.D., Baugh, J.A., 2008. The significance of nanoparticles in particle-induced pulmonary fibrosis. McGill J. Med. 11 (1), 43–50.

Chan, W.C., Maxwell, D.J., Gao, X., Bailey, R.E., Han, M., Nie, S., 2002. Luminescent quantum dots for multiplexed biological detection and imaging. Curr. Opin. Biotechnol. 13, 40–46.

Chidambaram, M., Manavalan, R., Kathiresan, K., 2011. Nanotherapeutics to overcome conventional cancer chemotherapy limitations. J. Pharm. Pharm. Sci. 14 (1), 67–77.

Coombs, R.R.H., Robinson, D.W., 1996. Nanotechnology in Medicine and the Biosciences. Gordon-Breach, Amsterdam, ISBN 2-88449-080-9.

Dardzinski, B.J., Schmithorst, V.J., Holland, S.K., Boivin, G.P., Imagawa, T., Watanabe, S., Lewis, J.M., Hirsch, R., 2001. MR imaging of murine arthritis using ultrasmall superparamagnetic iron oxide particles. Magn. Reson. Imaging 19, 1209–1216.

Desai, T.A., Chu, W.H., Tu, J.K., Beattie, G.M., Hayek, A., Ferrari, M., 1998. Microfabricated immunoisolating biocapsules. Biotechnol. Bioeng. 57, 118.

Dixon, K.L., 2003. The radiation biology of radioimmunotherapy. Nucl. Med. Commun. 24, 951.

Dousset, V., Ballarino, L., Delalande, C., Coussemacq, M., Canioni, P., Petry, K.G., Caille, J.M., 1999. Comparison of ultrasmall particles of iron oxide (USIOP) -enhanced T2-weighted, conventional T2-weighted, and gadolinium enhanced T1-weighted MR images in rats with experimental autoimmune encephalomyelitis. Am. J. Neuroradiol. 20, 223–227.

Drexler, K.E., 1986. The Coming Era of Nanotechnology. Anchor Books/Doubleday, New York, NY.

Drexler, K.E., 1992. Nanosystems: Molecular Machinery, Manufacturing and Computation. John Wiley & Sons, New York, NY.

Drexler, K.E., 2009. There's plenty of room at the bottom. Available from: http://metamodern.com/2009/12/29/theres-plenty-of-room-at-the-bottom%E2%80%9D-feynman-1959/

Dubertert, B., Skourides, P., Norris, D.J., Noireaux, V., Brivanlou, A.H., Libchaber, A., 2002. In vivo imaging of quantum dots encapsulated in phospholipid micelles. Science 298, 1759–1762.

Elder, A., 2006. Tiny inhaled particles take easy route from nose to brain. Available from: https://www.urmc.rochester.edu/news/story/1191/tiny-inhaled-particles-take-easy-route-from-nose-to-brain.aspx

Eniwumide, J.O., Lee, D.A., Bader, D.L., 2009. The development of a bioreactor to perfuse radially-confined hydrogel constructs: design and characterization of mass transport properties. Biorheology 46 (5), 417–437.

FDA, 2012. Highlights of prescribing information, abraxane for injectable suspension. Press announcement.

FDA, 2013. FDA approves Abraxane for late-stage pancreatic cancer. Press announcement.

Fierce Drug Delivery, First trial in humans of 'minicells': a completely new way of delivering anti-cancer drugs. Available from: http://www.fiercepharma.com/r-d/first-trial-humans-of-minicells-a-completely-new-way-of-delivering-anti-cancer-drugs

Foresight Institute, 2015. Available from: http://www.foresight.org

Freitas, Jr., R.A., 1999. Nanomedicine, Volume I: Basic Capabilities. CRC Press, Boca Raton, FL.

Frohbergh, M.E., Katsman, A., Botta, G.P., Lazarovici, P., Schauer, C.L., Wegst, U.G., Lelkes, P.I., 2012. Electrospun hydroxyapatite-containing chitosan nanofibers crosslinked with genipin for bone tissue engineering. Biomaterials 33, 9167–9178.

Fuchs, S., Kornowski, R., Weisz, G., Satler, L.F., Smits, P.C., Okubagzi, P., Baffour, R., Aggarwal, A., Weissman, N.J., Cerqueira, M., et al., 2006. Safety and feasibility of transendocardial autologous bone marrow cell transplantation in patients with advanced heart disease. Am. J. Cardiol. 97, 823–829.

Gao, X., Chan, W.C., Nie, S., 2002. Quantum-dot nanocrystals for ultrasensitive biological labeling and multicolor optical encoding. J. Biomed. Opt. 7, 532.

Gao, X., Cui, Y., Levenson, R.M., Chung, L.W.K., Nie, S., 2004. In vivo cancer targeting and imaging with semiconductor quantum dots. Nat. Biotechnol. 22, 969–976.

Gerion, D., Parak, W.J., Williams, S.C., Zanchet, D., Micheel, C.M., Alivisatos, A.P., 2002. Sorting fluorescent nanocrystals with DNA. J. Am. Chem. Soc. 124, 7070.

Ghaz-Jahanian, M.A., Abbaspour-Aghdam, F., Anarjan, N., Berenjian, A., Jafarizadeh-Malmiri, H., 2015. Application of chitosan-based nanocarriers in tumor-targeted drug delivery. Mol. Biotechnol. 57 (3), 201–218.

Giannoudis, P.V., Dinopoulos, H., Tsiridis, E., 2005. Bone substitutes: an update. Injury 36, S20–S27.

Gibney, M, 2013. Cornell nanosized 'dots' for diagnostics may now deliver drugs. Available from: http://www.fiercepharma.com/r-d/cornell-nanosized-dots-for-diagnostics-may-now-deliver-drugs

Hall, J.S., 2005. Nanofuture: What's Next for Nanotechnology. Prometheus Books, Amherst, NY.

Hanaki, K., Momo, A., Oku, T., Komoto, A., Maenosono, S., Yamaguchi, Y., Yamamoto, K., 2003. Semiconductor quantum dot/albumin complex is a long-life and highly photostable endosome marker. Biochem. Biophys. Res. Commun. 302 (3), 496.

Hanawa, T., Kon, M., Ukai, H., Murakami, K., Miyamoto, Y., Asaoka, K., 1997. Surface modifications of titanium in calcium-ion-containing solutions. J. Biomed. Mater. Res. 34, 273–278.

Harisinghani, M.G., Barentsz, J., Hahn, P.F., Deserno, W.M., Tabatabaei, S., Hulsbergen van de Kaa, C., de la Rosette, J., Weissleder, R., 2003. Noninvasive detection of clinically occult lymph-node metastases in prostate cancer. N. Engl. J. Med. 348, 2491–2499.

Herrmann, I.K., Grass, R.N., Stark, W.J., 2009. High-strength metal nanomagnets for diagnostics and medicine: carbon shells allow long-term stability and reliable linker chemistry. Nanomedicine 4, 787–798.

Herrmann, I.K., Schlegel, A., Graf, R., Schumacher, C.M., Senn, N., Hasler, M., Gschwind, S., Hirt, A.M., Gunther, D., Clavien, P.A., Stark, W.J., Beck-Schimmer, B., 2013a. Nanomagnet-based removal of lead and digoxin from living rats. Nanoscale 5, 8718–8723.

Herrmann, I.K., Urner, M., Graf, S., Schumacher, C.M., Roth-Z'graggen, B., Hasler, M., Stark, W.J., Beck-Schimmer, B., 2013b. Endotoxin removal by magnetic separation-based blood purification. Adv. Healthc. Mater. 2, 829–835.

Huang, S., Fu, X., 2010. Naturally derived materials-based cell and drug delivery systems in skin regeneration. J. Control. Release 142, 149–159.

Hunter, W., 1742. Of the structure and diseases of articulating cartilages. Phil. Trans. 42, 514–521, (Clin. Ortop. Relat. Res. 317, 3–6).

Ibrahim, A., Ohshima, H., Allison, S.A., Cottet, H., 2012. Determination of effective charge of small ions, polyelectrolytes and nanoparticles by capillary electrophoresis. J. Chromatogr. 1247, 154–164.

Jaffer, F.A., Weissleder, R., 2004. Seeing within—molecular imaging of the cardiovascular system. Circ. Res. 94 (4), 433–445.

Jaiswal, J.K., Mattoussi, H., Mauro, J.M., Simon, S.M., 2003. Long-term multiple color imaging of live cells using quantum dot bioconjugates. Nat. Biotechnol. 21, 47–51.

Ji, J., Ji, S.Y., Yang, J.A., He, X., Yang, X.H., Ling, W.P., Chen, X.L., 2012. Ultrasound-targeted transfection of tissue-type plasminogen activator gene carried by albumin nanoparticles to dog myocardium to prevent thrombosis after heart mechanical valve replacement. Int. J. Nanomed. 7, 2911–2919.

Jonaitis, T.S., Card, J.W., Magnuson, B., 2010. Concerns regarding nano-sized titanium dioxide dermal penetration and toxicity study. Toxicol. Lett. 192 (2), 268–269.

Kai, D., Prabhakaran, M.P., Jin, G., Ramakrishna, S., 2011. Guided orientation of cardiomyocytes on electrospun aligned nanofibers for cardiac tissue engineering. J. Biomed. Mater. Res. B Appl. Biomater. 98B, 379–386.

Kang, J.H., Super, M., Yung, C.W., et al., 2014. An extracorporeal blood-cleansing device for sepsis therapy. Nat. Med. 20, 1211–1216.

Kay, S., Thapa, A., Haberstroh, K.M., Webster, T.J., 2002. Nanostructured polymer/nanophase ceramic composites enhance osteoblast and chondrocyte adhesion. Tissue Eng. 8, 753–761.

Khil, M.S., Cha, D.I., Kim, H.Y., Kim, I.S., Bhattarai, N., 2003. Electrospun nanofibrous polyurethane membrane as wound dressing. J. Biomed. Mater. Res. B Appl. Biomater. 67, 675–679.

Kim, K.H., Jeong, L., Park, H.N., Shin, S.Y., Park, W.H., Lee, S.C., Kim, T.I., Park, Y.J., Seol, Y.J., Lee, Y.M., et al., 2005. Biological efficacy of silk fibroin nanofiber membranes for guided bone regeneration. J. Biotechnol. 120, 327–339.

Knabe, C., Berger, G., Gildenhaar, R., Klar, F., Zreiqat, H., 2004. The modulation of osteogenesis in vitro by calcium titanium phosphate coatings. Biomaterials 25, 4911–4919.

Kneser, U., Schaefer, D.J., Polykandriotis, E., Horch, R.E., 2006. Tissue engineering of bone: the reconstructive surgeon's point of view. J. Cell. Mol. Med. 10, 7–19.

Kocher, A.A., Schuster, M.D., Szabolcs, M.J., Takuma, S., Burkhoff, D., Wang, J., Homma, S., Edwards, N.M., Itescu, S., 2001. Neovascularization of ischemic myocardium by human bone-marrow-derived angioblasts prevents cardiomyocyte apoptosis, reduces remodeling and improves cardiac function. Nat. Med. 7, 430–436.

Kooi, M.E., Cappendijk, V.C., Cleutjens, K.B.J.M., Kessels, A.G.H., Kitslaar, P.J.E.H.M., Borgers, M., Frederik, P.M., Daemen, M.J.A.P., van Engelshoven, J.M.A., 2003. Accumulation of ultrasmall superparamagnetic particles of iron oxide in human atherosclerotic plaques can be detected by in vivo magnetic resonance imaging. Circulation 107, 2453–2458.

Kroto, H.W., Heath, J.R., O'Brien, S.C., Curl, R.F., Smalley, R.E., 1985. C_{60}: Buckminsterfullerene. Nature 318 (6042), 162.

Lalwani, G., Henslee, A.M., Farshid, B., Parmar, P., Lin, L., Qin, Y.X., Kasper, F.K., Mikos, A.G., Sitharaman, B., 2013a. Tungsten disulfide nanotubes reinforced biodegradable polymers for bone tissue engineering. Acta Biomater. 9 (9), 8365–8373.

Lalwani, G., Henslee, A.M., Farshid, B., Lin, L., Kasper, K.F., Qin, Y.X., Mikos, A.G., Sitharaman, B., 2013b. Two-dimensional nanostructure- reinforced biodegradable polymeric nanocomposites for bone tissue engineering. Biomacromolecules 14 (3), 900–909.

Lee, J.J., Jeong, K.J., Hashimoto, M., Kwon, A.H., Rwei, A., Shankarappa, S.A., Tsui, J.H., Kohane, D.S., 2014. Synthetic ligand-coated magnetic nanoparticles for microfluidic bacterial separation from blood. Nano Lett. 14, 1–5.

Lee, S.B., Martin, C.R., 2002. Electromodulated molecular transport in gold-nanotube membranes. J. Am. Chem. Soc. 124, 11850.

Lee, S.B., Mitchell, D.T., Trofin, L., Nevanen, T.K., Soderlund, H., Martin, C.R., 2002. Antibody- based bio-nanotube membranes for enantiomeric drug separations. Science 296, 2198.

Lim, E.H., Sardinha, J.P., Myers, S., 2014. Nanotechnology biomimetic cartilage regenerative scaffolds. Arch. Plast. Surg. 41 (3), 231–240.
Liu, W., Cao, Y.L., 2007. Application of scaffold materials in tissue reconstruction in immunocompetent mammals: our experience and future requirements. Biomaterials 28, 5078–5086.
Lopez-Ruiz, E., Peran, M., Cobo-Molinos, J., Jimenez, G., Picon, M., Bustamante, M., Arrebola, F., Hernandez-Lamas, M.C., Delgado-Martinez, A.D., Montanez, E., Marchal, J.A., 2013. Chondrocytes extract from patients with osteoarthritis induces chondrogenesis in infrapatellar fat pad-derived stem cells. Osteoarthr. Cartil. 21, 246–258.
Martin, I., Wendt, D., Heberer, M., 2004. The role of bioreactors in tissue engineering. Trends Biotechnol. 22, 80–86.
Martis, E.A., Badve, R.R., Degwekar, M.D., 2012. Nanotechnology based devices and applications in medicine: an overview. Chron. Young Sci. 3 (1), 68–73.
McMahon, L.A., Barron, V., Prina-Mello, A., Prendergast, P.J., 2004. The state-of-the-art in cartilage bioreactors. In: Predergast, P.J., McHugh, P.E. (Eds.), Topics in Biomechanical Engineering. TCBE & NCBES, Dublin, Ireland, (Chapter 4).
Medintz, I.L., Clapp, A.R., Mattoussi, H., Goldman, E.R., Fisher, B., Mauro, J.M., 2003. Self-assembled nanoscale biosensors based on quantum dot FRET donors. Nat. Mater. 2, 630.
Meller, A., Nivon, L., Brandin, E., Golovchenko, J., Branton, D., 2000. Rapid nanopore discrimination between single polynucleotide molecules. Proc. Natl. Acad. Sci. USA 97, 1079.
Mitchell, D.T., Lee, S.B., Trofin, L., Li, N., Nevanen, T.K., Soderlund, H., Martin, C.R., 2002. Smart nanotubes for bioseparations and biocatalysis. J. Am. Chem. Soc. 124, 11864.
Miyazaki, M., Tsumura, H., Wang, J.C., Alanay, A., 2009. An update on bone substitutes for spinal fusion. Eur. Spine J. 18, 783–799.
Mohamed, A., Xing, M.M., 2012. Nanomaterials and nanotechnology for skin tissue engineering. Int. J. Burns Trauma 2, 29–41.
Mozafari, M.R. (Ed.), 2006. Nanocarrier Technologies: Frontiers of Nanotherapy, Springer pp. 10–11, 25–34 (Chapters 1 and 2).
Mukherjee, S., Gualandi, C., Focarete, M.L., Ravichandran, R., Venugopal, J.R., Raghunath, M., Ramakrishna, S., 2011. Elastomeric electrospun scaffolds of poly (L-lactide-*co*-trimethylene carbonate) for myocardial tissue engineering. J. Mater. Sci. Mater. Med. 22, 1689–1699.
Nagy, Z.K., Zsombor, K., Balogh, A., Vajna, B., Farkas, A., Patyi, G., Kramarics, A., Marosi, G., 2011. Comparison of electrospun and extruded soluplus-based solid dosage forms of improved dissolution. J. Pharm. Sci. 101, 1.
Namba, R.S., Meuli, M., Sullivan, K.M., Le, A.X., Adzick, N.S., 1998. Spontaneous repair of superficial defects in articular cartilage in a fetal lamb model. J Bone Joint Surg. Am. 80 (1), 4–10.
National Health İnformation (NHI), LLC, 2006. Nanomedicine. Device and Diagnostics Report, Nanobiotech News Data.
Nguyen, L.T., Liao, S., Chan, C.K., Ramakrishna, S., 2012. Enhanced osteogenic differentiation with 3D electrospun nanofibrous scaffolds. Nanomedicine 7, 1561–1575.
Nguyen-Ngoc, T., Raymond, E., 2015. Reinvention of chemotherapy: drug conjugates and nanoparticles. Curr. Opin. Oncol. 27, 232–242.
Oseni, A.O., Crowley, C., Boland, M.Z., Butler, P.E., Seifalian, A.M., 2011. Cartilage tissue engineering: the application of nanomaterials and stem cell technology. In: Eberli, D. (Ed.), Tissue Engineering for Tissue and Organ Regeneration. InTech, Rijeka, Croatia, Available from: http://www.intechopen.com/books/tissue-engineering-for-tissue-and-organ-regeneration/cartilage-tissue-engineering-the-application-of-nanomaterials-and-stem-cell-technology.
Penn, M.S., Francis, G.S., Ellis, S.G., Young, J.B., McCarthy, P.M., Topol, E.J., 2002. Autologous cell transplantation for the treatment of damaged myocardium. Prog. Cardiovasc. Dis. 45, 21–32.

Peran, M., Garcia, M.A., Lopez-Ruiz, E., Jimenez, G., Marchal, J.A., 2013. How can nanotechnology help to repair the body? Advances in cardiac, skin, bone, cartilage and nerve tissue regeneration. Materials 6, 1333–1359.

Perez, J.M., Josephson, L., Weissleder, R., 2004. Use of magnetic nanoparticles as nanosensors to probe for molecular interactions. Chem. Bio. Chem. 5, 261–264.

Pérez-Herrero, E., Fernández-Medarde, A., 2015. Advanced targeted therapies in cancer: drug nanocarriers, the future of chemotherapy. Eur. J. Pharm. Biopharm. 93, 52–79.

Peters, D., Kastantin, M., Kotamraju, V.R., Karmali, P.P., Gujraty, K., Tirrell, M., Ruoslahti, E., 2009. Targeting atherosclerosis by using modular, multifunctional micelles. Proc. Natl. Acad. Sci. USA 106, 9815–9819.

Plunkett, N.A., O'Brien, F.J., 2010. 4. 3. Bioreactors in tissue engineering. Stud. Health Technol. Inform. 152, 214–230.

Prabhakaran, M.P., Kai, D., Ghasemi-Mobarakeh, L., Ramakrishna, S., 2011. Electrospun biocomposite nanofibrous patch for cardiac tissue engineering. Biomed. Mater. 6, 055001:1–:155001.

Ravichandran, R., Venugopal, J.R., Sundarrajan, S., Mukherjee, S., Ramakrishna, S., 2012. Precipitation of nanohydroxyapatite on PLLA/PBLG/Collagen nanofibrous structures for the differentiation of adipose derived stem cells to osteogenic lineage. Biomaterials 33, 846–855.

Rice University Press, 2003. Halas Wins Innovator Award in Fight Against Breast Cancer. Available from: http://www.eurekalert.org/pub_releases/2003-07/ru-hwi070903.php

Rollins, K., 2009. Nanobiotechnology regulation: a proposal for self-regulation with limited oversight. Nanotechnol. Law Bus. 6 (2), 221–225.

Ruehm, S.G., Corot, C., Vogt, P., Kolb, S., Debatin, J.F., 2001. Magnetic resonance imaging of atherosclerotic plaque with ultrasmall superparamagnetic particles of iron oxide in hyperlipidemic rabbits. Circulation 103, 415–422.

Sahoo, S.K., 2005. Applications of nanomedicine. ABPN 9 (20), 1048–1050.

Saleh, A., Schroeter, M., Jinkmanns, C., Hartung, H.P., Modder, U., Jander, S., 2004. In vivo MRI of brain inflammation in human ischaemic stroke. Brain 127, 1670–1677.

Sams, A.E., Nixon, A.J., 1995. Chondrocyte-laden collagen scaffolds for resurfacing extensive articular cartilage defects. Osteoarthr. Cartil. 3 (1), 47–59.

Schumacher, C.M., Herrmann, I.K., Bubenhofer, S.B., Gschwind, S., Hirt, A.M., Beck-Schimmer, B., Günther, D., Stark, W.J., 2013. Quantitative recovery of magnetic nanoparticles from flowing blood: trace analysis and the role of magnetization. Adv. Funct. Mater. 23, 4888–4896.

Sengers, B.G., Heywood, H.K., Lee, D.A., Oomens, C.W., Bader, D.L., 2005. Nutrient utilization by bovine articular chondrocytes: a combined experimental and theoretical approach. J. Biomech. Eng. 127 (5), 758.

Sershen, S.R., Westcott, S.L., Halas, N.J., West, J.L., 2000. Temperature-sensitive polymer–nanoshell composites for photothermally modulated drug delivery. J. Biomed. Mater. Res. 51, 293.

Sharma, G., She, Z.G., Valenta, D.T., Stallcup, W.B., Smith, J.W., 2010. Targeting of macrophage foam cells in atherosclerotic plaque using oligonucleotide-functionalized nanoparticles. Nano Life 1, 207–214.

Shi, H., Ratner, B.D., 2000. Template recognition of protein-imprinted polymer surfaces. J. Biomed. Mater. Res. 49, 1.

Shin, M., Ishii, O., Sueda, T., Vacanti, J.P., 2004. Contractile cardiac grafts using a novel nanofibrous mesh. Biomaterials 25, 3717–3723.

Shuming, N., Xing, Y., Kim, G.J., Simmons, J.W., 2007. Nanotechnology applications in cancer. Annu. Rev. Biomed. Eng. 9 (1), 257–288.

Silver, F.H., Pins, G., 1992. Cell growth on collagen: a review of tissue engineering using scaffolds containing extracellular matrix. J. Long Term Effect Med. Impl. 2 (1), 67–81.

Soni, S., Tyagi, H., Taylor, R.A., Kumar, A., 2013. Role of optical coefficients and healthy tissue-sparing characteristics in gold nanorod-assisted thermal therapy. Int. J. Hypertherm. 29 (1), 87–97.

Storm, A.J., Chen, J.H., Ling, X.S., Zandbergen, H.W., Dekker, C., 2003. Fabrication of solid-state nanopores with single-nanometre precision. Nat. Mater. 2, 537.

Stynes, G., Kiroff, G.K., Morrison, W.A., Kirkland, M.A., 2008. Tissue compatibility of biomaterials: Benefits and problems of skin biointegration. ANZ J. Surg. 78, 654–659.

Subramaniam, R., Astot, C., Juhlin, L., Nilsson, C., Ostin, A., 2012. Determination of S-2-(*N*,*N*-diisopropylaminoethyl)- and S-2-(*N*,*N*-diethylaminoethyl) methylphosphonothiolate, nerve agent markers, in water samples using strong anion-exchange disk extraction, in vial trimethylsilylation, and gas chromatography-mass spectrometry analysis. J. Chromatogr. 1229, 86–94.

Sun, L., Zhang, L., Hemraz, U.D., Fenniri, H., Webster, T.J., 2012. Bioactive rosette nanotube-hydroxyapatite nanocomposites improve osteoblast functions. Tissue Eng. 18, 1741–1750.

Tan, A., Rajadas, J., Seifalian, A.M., 2012. Biochemical engineering nerve conduits using peptide amphiphiles. J. Control. Release 163, 342–352.

Tandon, V., Zhang, B., Radisic, M., Murthy, S.K., 2013. Generation of tissue constructs for cardiovascular regenerative medicine: from cell procurement to scaffold design. Biotechnol. Adv. 31, 722–735.

Tao, S.L., Desai, T.A., 2003. Microfabricated drug delivery systems: from particles to pores. Adv. Drug Deliv. Rev. 55, 315.

Taton, T.A., Lu, G., Mirkin, C.A., 2001. Two-color labeling of oligonucleotides array via size-selective scattering of nanoparticle probes. J. Am. Chem. Soc. 123, 5164–5165.

Tran, P.A., Zhang, L., Webster, T.J., 2006. Carbon nanofibers and carbon nanotubes in regenerative medicine. Adv. Drug Deliv. Rev. 61, 1097–1114.

Tse, H.F., Thambar, S., Kwong, Y.L., Rowlings, P., Bellamy, G., McCrohon, J., Thomas, P., Bastian, B., Chan, J.K., Lo, G., et al., 2007. Prospective randomized trial of direct endomyocardial implantation of bone marrow cells for treatment of severe coronary artery diseases (PROTECT-CAD trial). Eur. Heart J. 28, 2998–3005.

Tyagi, S., Kramer, F.R., 1996. Molecular beacons: probes that fluoresce upon hybridization. Nat. Biotechnol. 14, 303–308.

van Ramshorst, J., Bax, J.J., Beeres, S.L., Dibbets-Schneider, P., Roes, S.D., Stokkel, M.P., de Roos, A., Fibbe, W.E., Zwaginga, J.J., Boersma, E., et al., 2009. Intramyocardial bone marrow cell injection for chronic myocardial ischemia: a randomized controlled trial. JAMA 301, 1997–2004.

Venugopal, J.R., Zhang, Y., Ramakrishna, S., 2006. In vitro culture of human dermal fibroblasts on electrospun polycaprolactone collagen nanofibrous membrane. Artif. Organs 30, 440–446.

Wang, A.Z., Langer, R., Farokhzad, O.C., 2012. Nanoparticle delivery of cancer drugs. Annu. Rev. Med. 63, 185–198.

West, J.L., Halas, N.J., 2000. Applications of nanotechnology to biotechnology—commentary. Curr. Opin. Biotechnol. 11, 215.

Winter, P.M., Caruthers, S.D., Zhang, H., Williams, T.A., Wickline, S.A., Lanza, G.M., 2008. Antiangiogenic synergism of integrin-targeted fumagillin nanoparticles and atorvastatin in atherosclerosis. JACC Cardiovasc. Imaging 1, 624–634.

Winter, P.M., Neubauer, A.M., Caruthers, S.D., Harris, T.D., Robertson, J.D., Williams, T.A., Schmieder, A.H., Hu, G., Allen, J.S., Lacy, E.K., et al., 2006. Endothelial alpha(v)beta3 integrin targeted fumagillin nanoparticles inhibit angiogenesis in atherosclerosis. Arterioscler. Thromb. Vasc. Biol. 26, 2103–2109.

Wu, J., Liu, W., Xue, C., Zhou, S., Lan, F., Bi, L., Xu, H., Yang, X., Zeng, F.D., 2009. Toxicity and penetration of TiO2 nanoparticles in hairless mice and porcine skin after subchronic dermal exposure. Toxicol. Lett. 191 (1), 1–8.

Wu, X., Liu, H., Liu, J., Haley, K.N., Treadway, J.A., Larson, J.P., Ge, N., Peale, F., Bruchez, M.P., 2003. Immunofluorescent labeling of cancer marker Her2 and other cellular targets with semiconductor quantum dots. Nat. Biotechnol. 21, 41.

Yang, Y., Xia, T., Chen, F., Wei, W., Liu, C., He, S., Li, X., 2012. Electrospun fibers with plasmid bFGF polyplex loadings promote skin wound healing in diabetic rats. Mol. Pharm. 9, 48–58.

Yoshimoto, H., Shin, Y.M., Terai, H., Vacanti, J.P., 2003. A biodegradable nanofiber scaffold by electrospinning and its potential for bone tissue engineering. Biomaterials 24, 2077–2082.

Yung, C.W., Fiering, J., Mueller, A.J., Ingber, D.E., 2009. Micromagnetic–microfluidic blood cleansing device. Lab. Chip. 9, 1171.

Zheng, G., Patolsky, F., Cui, Y., Wang, W.U., Lieber, C.M., 2005. Multiplexed electrical detection of cancer markers with nanowire sensor arrays. Nat. Biotechnol. 23 (10), 1294–1301.

Zimmerman, S.C., Wendland, M.S., Rakow, N.A., Zharov, I., Suslick, K.S., 2002. Synthetic hosts by monomolecular imprinting inside dendrimers. Nature 418, 399.

CHAPTER

16

DENDRIMERS AND DENDRONIZED MATERIALS AS NANOCARRIERS

Eliana D. Farias, Lydia M. Bouchet, Verónica Brunetti, Miriam C. Strumia
National University of Cordoba, Córdoba, Argentina

CHAPTER OUTLINE

1 INTRODUCTION

Dendrimers are synthetic molecules that have attracted attention because of their very particular physical structure and chemical properties. Their monomeric units are designed in such manner that the number of terminal groups grows geometrically (Tomalia et al., 1990) as the radius of the polymer

Nanostructures for Novel Therapy

increases linearly thus investing them with a radial gradient of properties (Fig. 16.1A). These polymers exhibit structural integrity and homogeneity, a well-defined internal cavity available to encapsulate or "catch" another molecule, a controlled composition, and multiple homogeneous active terminal groups for given reactions (Astruc, 2003). Dendrons (Fig. 16.1B), the wedged-shaped sections of a dendrimer, are monodisperse, and can be used for chemical modification of a substrate, maintaining the desired dendritic effect with cost-effective synthesis. Despite their less controlled structure, compared to dendrimers, random hyperbranched polymers (HB) are nevertheless capable of exercising dendritic effect, and have the advantage that they can be synthesized more easily (Wurm and Frey, 2012) (Fig. 16.1C). Dendrigrafts lie structurally speaking midway between the high structure control of dendrimers and random HB, and use their polymeric chains as building blocks to generate numerous branch units (Fig. 16.1D). The multiple dendron decoration of the backbone conformation of a lineal polymer creates a further architectural dendritic subclass called dendronized polymer (DP) (Fig. 16.1E). Finally, a new class of dendritic polymer called hyperbranched polydendron (HP), bearing two untethered ends on each primary chain of the cross-linked backbone architecture, with one chain end having an ideally branched dendron (Fig. 16.1F).

Dendrimers are also suitable supra-molecular nanoscale modules for creating the controlled assembly of more complex nanostructures leading to the appearance of megamers. These supramacromolecular structures are formed by two or more randomly assembled dendrimers and structure-controlled

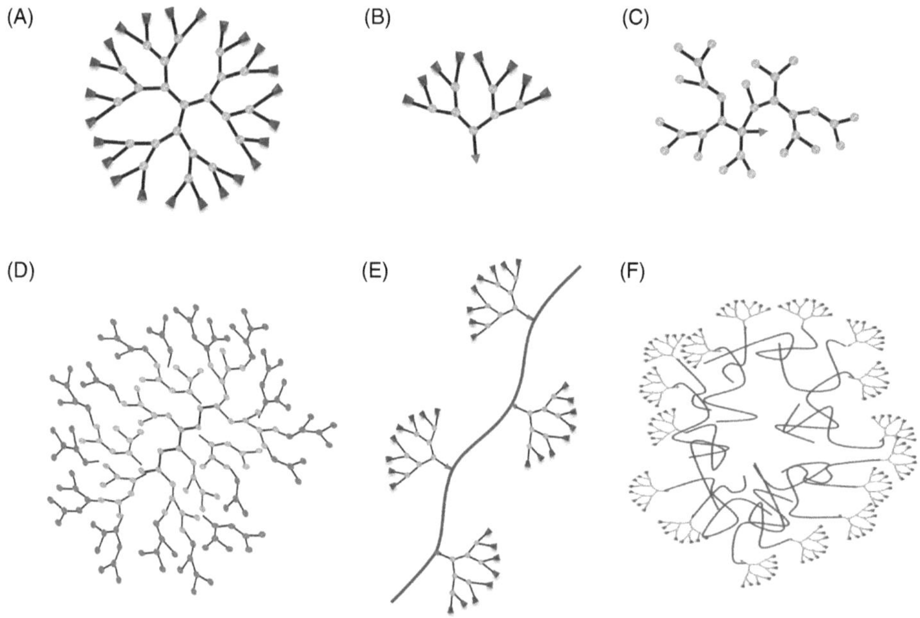

FIGURE 16.1 Scheme of Supramolecular Dendritic Structures

(A) dendrimer, (B) dendron, (C) hyperbranched polymer, (D) dendrigraft, (E) dendronized polymer (DP), and (F) hyperbranched polydendrons (HP).

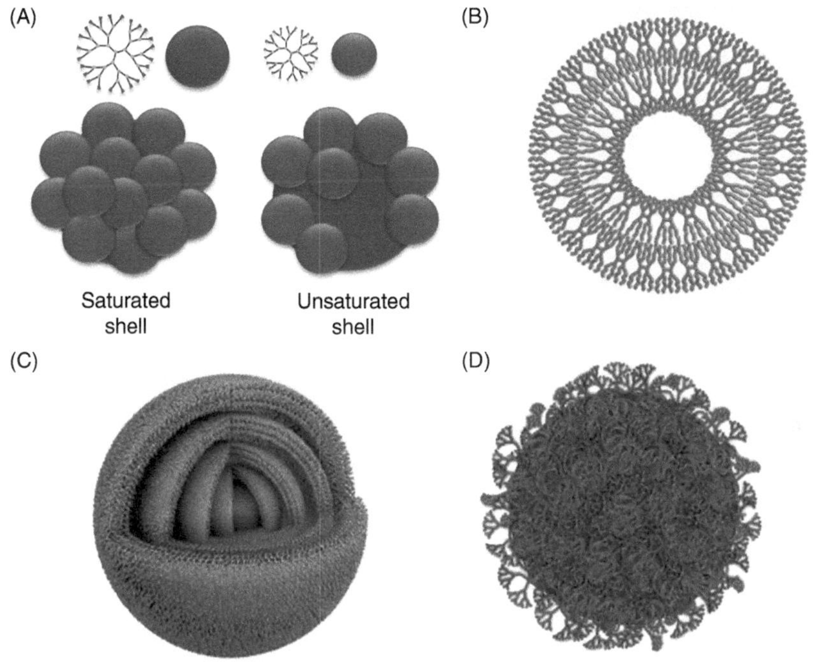

FIGURE 16.2 Cartoon Pictures of Supramacromolecular Structures

(A) megamers, (B) dendrimersome, adapted from reference (Sun et al., 2015), with permission of The Royal Society of Chemistry (C) onion-like dendrimersome.

Adapted from Gawande et al. (2015), with permission of The Royal Society of Chemistry, (d) HP-based nanoparticles, reprinted from reference (Hatton et al., 2014) published by The Royal Society of Chemistry.

clusters known as core–shell tecto(dendrimers) (Fig. 16.2A). Amphiphilic janus dendrimers self-assembly into a subclass of megamers called dendrimersomes (Fig. 16.2B) to mimic single-bilayer biological membranes. Recently, Percec et al. (2010) discovered six primary structures of single–single amphiphilic janus dendrimers that self-assemble into uniform onion-like dendrimersomes by simple injection in water and buffers (Zhang et al., 2014) (Fig. 16.2C). In addition, the self-assembly of HP to controllably form highly monodisperse, stable nanoparticles with control of size and chemistry was recently reported by Hatton et al. (2014), exhibiting a highly surprisingly uniform behavior from such polydisperse materials (Fig. 16.2D).

Another class of interesting structures is formed by the combination of dendritic molecules and inorganic nanoparticles (NPs), forming hybrid-dendrimers. When trapped or encapsulated in dendritic pockets, these NPs are called dendrimer-encapsulated NPs (DENPs), and the branched interior of the dendrimer prevents aggregation of the NPs (Fig. 16.3A). NPs can also be structural units of the hybrid structure in the form of a core surrounded by dendrimers (similar to megamers) which interact with the surface through their terminal functional groups (Fig. 16.3B), and are called dendrimer-stabilized NPs

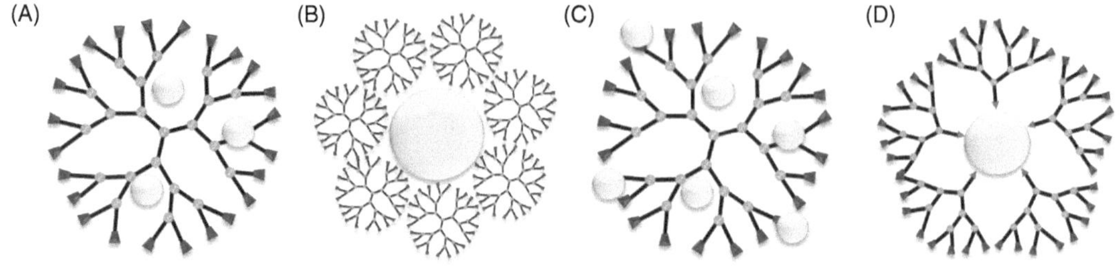

FIGURE 16.3 Scheme of Different Dendritic Hybrid Structures

(A) dendrimer-encapsulated nanoparticles (DENPs); (B) dendrimer-stabilized nanoparticles (DSNPs); (C) DENPs + DSNPs; and (D) nanoparticle-cored dendrimers (NCDs).

(DSNPs). An inorganic core radially functionalized with dendrons by specific interactions and impeding NP aggregation (Fig. 16.3C), is called a nanoparticle-cored dendrimer (NCD). NPs can be metallic (Au, Ag, Pd, Pt), magnetic (iron oxide), or semiconductors (CdSe, CdS).

2 SYNTHETIC APPROACH

The first methodologies for dendritic synthesis involved divergent and the convergent approaches that even today remain the standard procedure for the synthesis of dendrons and dendrimers. In the divergent approach developed by Tomalia (Lothian-Tomalia et al., 1997), Newkome (Newkome et al., 1992) and Vögtle (Mekelburger et al., 1992), the dendritic growth is carried out from the multi-functionalized core to the periphery through a series of repetitive coupling and activation steps (Fig. 16.4A). The first step consists in the coupling of the functional groups of the core with the monomer that introduces a branch amplification region in each coupling site and thus increases the number of terminal groups (Grayson and Fréchet, 2001). The reaction then continues with the activation of peripheral groups, which may involve a deprotection reaction, an activation by coupling with another species, or transformation of a functional group. This additional activation step is performed to avoid the presence of reactive groups in the coupling step that could otherwise give rise to uncontrolled polymerization processes. The successive repetition of the coupling and activation reactions leads to the growth of the dendritic structure.

The convergent approach was first developed by Welsh and Smith (2011). In this methodology, the growth starts from the periphery of the molecule, and proceeds inwards via the coupling of the end group at each branch of monomer (Fig. 16.4B). The next step also involves the deprotection of the new dendritic unit, which can then react again with another monomer. The successive repetition of these steps leads to the growth of a dendron structure that eventually reacts with the central core to form the dendrimer. This route gives greater structural control, and also enables the generation of well-defined asymmetric dendrimers combining dendrons with different peripheral groups (Lee et al., 2008; Liu et al., 2014).

The synthesis of hyperbranched polymers consists of a one-pot method that is carried out in a single-step, via reactions of polycondensation, polyaddition or radical polymerization, using multifunctional monomers (Fig. 16.4C) (Vögtle et al., 2009). The advantage of this methodology, compared with the synthesis of dendrimers, is that its simplicity enables the preparation of HB polymers on a

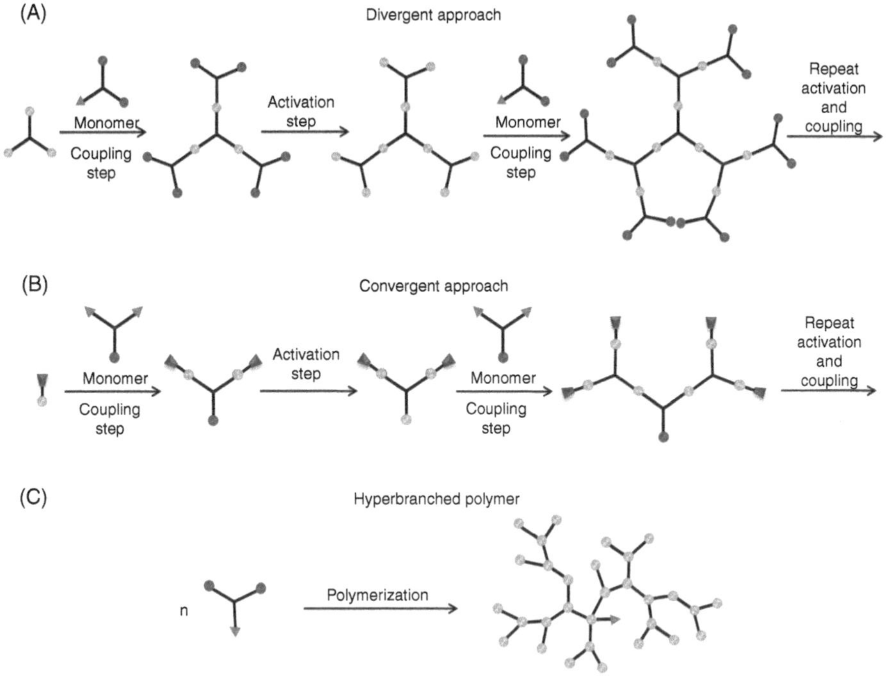

FIGURE 16.4 Scheme of Different Synthetic Approaches of (A and B) Dendrimers, and (C) Hyperbranched polymers

larger-scale. However, this "one-pot" synthesis leads to the formation of heterogeneous products with different molar mass and branching units (Voit and Lederer, 2009).

Dendronized polymers can be synthesized by three main pathways: the graft-to, the graft-from, and the macromonomer approaches. The graft-to approach consists in the attachment of the desired generation of dendrons to a polymer with the appropriate functionalization (Fig. 16.5A) (Paez et al., 2012). In the graft-from approach, a dendron of first generation is attached to the functionalized polymer, and from there, the higher generations of the dendrons are grown (Fig. 16.5B). Finally, the macromonomer methodology, which is the most common synthetic route, uses monomeric building blocks with dendrons attached to undergo polymerization (Fig. 16.5C) (Schluter, 2003).

Megamers can be synthesized using two different approaches, saturated-shell-architecture and unsaturated-shell-architecture. The first involves the combination of a limited amount of dendrimer with amine functionalization in its periphery as a core reagent, and an excess of dendrimer with carboxylic acid functionalities as a shell reagent. In a first reaction step, these two reagents self-assemble into an electrostatically driven supramolecular architecture. After an equilibration time, a carbodiimide reagent is added to induce the covalent bond formation. This methodology allows the formation of monodisperse saturated-shell megamers. The second methodology involves the direct covalent bond formation of partially filled megamers. As in the first case, a limited amount of a nucleophilic terminated dendrimer reacts with an electrophilic dendrimer, acting as a shell reagent, to obtain a semicontrolled partially filled structure (Tomalia, 2005).

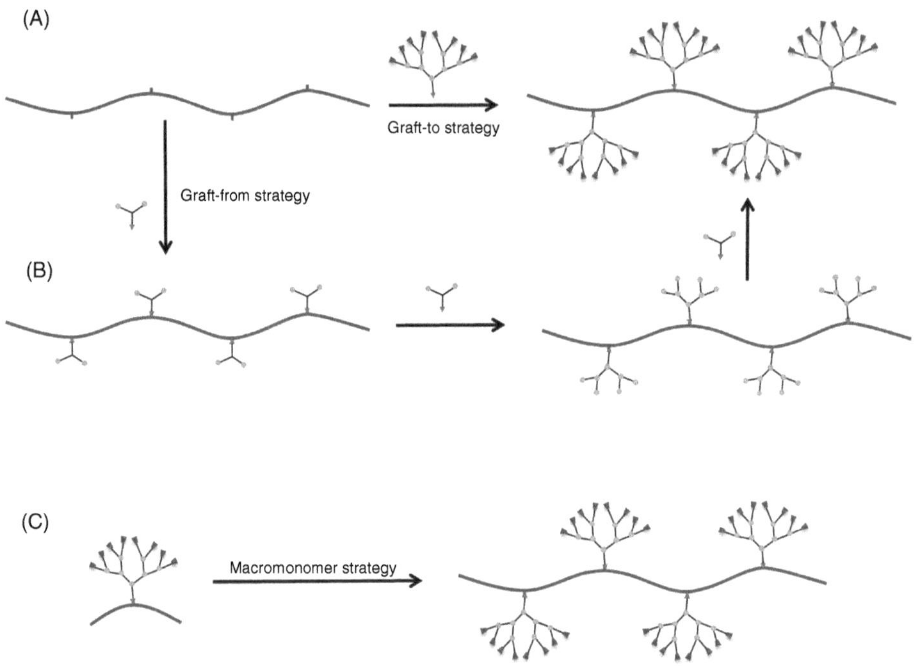

FIGURE 16.5 Scheme of the Three General Synthetic Strategies Toward Dendronized Polymers

(A) graft-to; (B) graft-from; and (C) macromonomer route.

Dendrimersomes are synthesized by self-assembled amphiphilic janus dendrimers having hydrophobic and hydrophilic structures. Percec et al. (2010) reported a methodology whereby the self-assembly reaction takes place in a water-soluble solvent in water or buffer, and produces uniform, impermeable, and stable vesicles with excellent mechanical properties (Zhang et al., 2014). Not only single vesicles but also onion-like vesicles can be synthesized using this approach. The onion-like vesicles consist of concentric alternating bilayers of janus dendrimers, and can be obtained merely by varying the concentration of dendrimer (Zhang et al., 2014).

Hyperbranched polydendron-nanoparticles are interesting novel materials mainly based on their feasible and rapid synthesis, which allow them to maintain all the benefits of dendrimers. Rannard and colleagues combined aspects of dendrimer and branched vinyl polymer strategies to simultaneously create linear-dendritic hybrid copolymers and introduce interchain branching to produce very high-molecular weight complex unimolecular architectures in a one-pot controlled radical polymerization (Hatton et al., 2014). According to the authors, the self-assembled nanoparticles have probably been formed by a nucleation/aggregation mechanism. One addition of hexane to acetone solutions of the branched polymers showed first a collapse of the HP forming an increasingly dense structure. These nuclei grow as they aggregate to generate larger structures that appear to be somewhat swollen within the solvent environment. Then, HP-nanoparticles

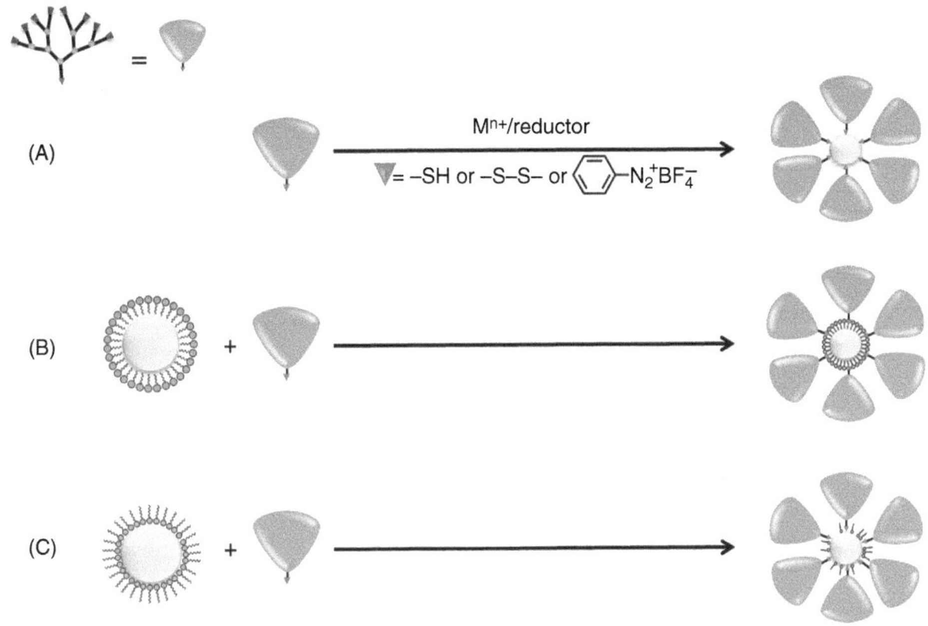

FIGURE 16.6 Schematic Illustration of the Three General Synthetic Strategies Toward NCDs

(A) direct method; (B) convergent approach; and (C) ligand exchange method.

decrease in size to form more dense NPs in the progressively poorer solvent environment prior to precipitation.

For the synthesis of hybrid dendritic materials such as nanoparticle-cored dendrimers (NCD), three different methods have been applied: the direct method, the convergent approach, and the ligand-exchange method. The first involves one-step growth of the NCD and reduction of a metallic cation in the presence of dendrons with appropriate functionalities at their focal point (Fig. 16.6A) (Brunetti et al., 2015). The moieties most commonly used are thiols, disulfides (Love et al., 2006), and diazonium salts (Kumar and Gopidas, 2010), to obtain NCD with metal-sulfur bonds in the first two cases, and with metal-carbon bonds in the latter case. The convergent approach is an indirect method involving the attachment of the dendrons through a reaction between the moieties of the ligand of the nanoparticle and the functionality present in the focal point of the dendron (Fig. 16.6B). The third approach, ligand exchange method (Fig. 16.6C), is also indirect and consists in the partial replacement of the molecules present in the nanoparticle by dendrons. The advantage of the last two methods is that the core size of the nanoparticle remains unchanged during the reaction.

One innovative way of synthesizing dendritic nanoparticles (NPs) was recently developed by Krishnakumar and Gopidas (2014). Fréchet-type dendron NPs, have been synthesized by a simple one-pot reaction by etching off the gold core in a first-generation gold nanoparticle-cored dendrimer (Fig. 16.7). Dissolution of the gold core leads to the generation of a small volume of closely packed dendron radicals which undergo very rapid coupling and addition reactions to form the Fréchet-type dendron NPs.

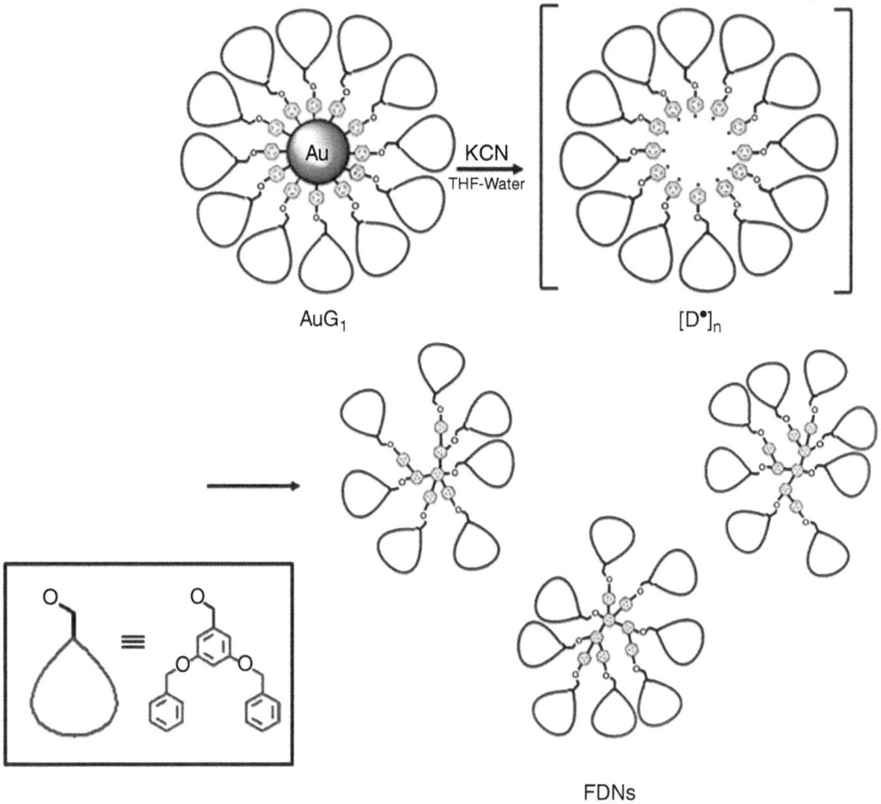

FIGURE 16.7 Scheme of the Synthesis of Fréchet-type Dendron NPs

Reprinted from Krishnakumar and Gopidas (2014), with kind permission of The Royal Society of Chemistry.

3 PHYSICO-CHEMICAL PROPERTIES OF DENDRITIC MOLECULES USEFUL FOR NANOCARRIERS

In general, nanocarriers are multifunctional systems having at least one dimension ranging between 1–100 nm that can transport the bioactive agent to the targeted site with improved therapeutic activity over the "free" form of a bioactive deliverer or conventional drugs, such as vaccines (Calderon et al., 2010). Based on the overall size and shape, their flexibility and their three architectural components, dendrimers and dendritic molecules have turned out to be interesting nanocarriers (Fig. 16.8). First, the interior void space delimited by the branched amplification region influences the endoreceptor capability of the carrier. Second, the inner composition determines the nature of guest-host properties. Lastly, the tailored surface end-groups have control over their exoreceptor attributes, providing the key stage for drug conjugation or acting as "intelligent gates" that manage the entry or exit of drugs from the interior. Relevant aspects in drug-delivery, such as solubility, in vivo stability, biodistribution, and biocompatibility may be optimized by tuning the multiple peripheral-groups (Duncan et al., 2010; Patel et al., 2013). For instance, PAMAM, a dendrimer bearing amine groups, can be protonated under

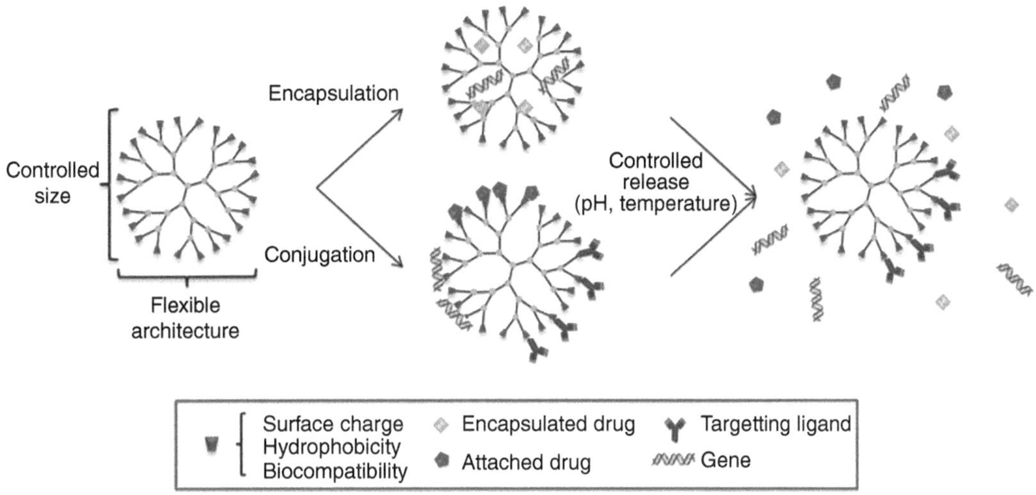

FIGURE 16.8 Cartoon Depictions of Some Possibilities for Dendritic Polymers as Nanocarrier

physiological or acidic conditions to generate a positively charged density on the surface. It is well known that cationic dendrimers induce electrostatic interaction with negatively charged components of cell membranes, and thus facilitate paracellular permeability, but few studies have reported cytotoxic effects by using positively charged surface functionalized nanocarriers (Froehlich, 2012). Both attributes, assisting cellular uptake and inducing toxicity, can easily be exploited either as a drug delivery nanoplatform in line with the requirements of a particular application or as an active positive control of toxicology, respectively. Another common modification on peripheral-dendritic surfaces is the conjugation of polyethylene glycol (PEG) chains to improve biocompatibility. Recently, a review covering all the interesting aspects of PEGylated dendrimer-mediated drug delivery such as reticuloendothelial system uptake, drug leakage, immunogenicity, hemolytic toxicity, cytotoxicity, hydrophobicity, solubilization enhancement, DNA transfection, siRNA delivery, transepithelial transport, and tumor targeting, was published by Thakur et al. (2015).

One of the key functions of a nanocarrier is to protect the bioactive agent from the environment, for example, from degradation and deactivation, while directing the drugs to the specific site of action and releasing them when they finally reach their target. After encapsulation, it is not only the guest that is protected from its environment, but also the system to which the medicine is applied. The access of the spatial-control over the carrier-drug conjugation arising from the dendritic architecture fortunately enhances the targeting capability of the nanocarrier. End-groups are optimal for the bioconjugation of drugs or targeting moieties and the small molecular guest can be encapsulated into the well-defined void space of the dendritic-host. The simple synthetic tuning of dendritic architecture in terms of the chemical functional group and the extent of the dendritic branching is able to exercise considerable control over the final nanocarrier properties (Brunetti et al., 2015). For instance, carriers based on hybrid-dendrimers including metal NPs, quantum dots or magnetic NPs have the additional ability of themselves becoming imaging agents able to monitor the distribution of a drug in the body in real time, thus becoming in a "theranostic" system (Xie et al., 2010, 2011).

To evaluate the optimal size of nanocarriers, two factors should be considered. Renal excretion through the kidneys determines the minimum size since the threshold of renal clearance is about 45 kDa (hydrodynamic diameter of approximately 4–6 nm) (Soo Choi et al., 2007). The other important factor is the capture of large particles by the immune system. In this context, it is desirable to develop nanocarriers ranging between 10–30 nm in order to maintain equilibrium between blood circulation and the extracellular space, increasing blood circulation times to achieve therapeutic effect (Brunetti et al., 2015). Additionally, the (bio)degradability of the carrier may prevent accumulation in different organs, avoiding inflammation or consequences of toxicity. Thus, the introduction of tailored groups that can be cleaved by a particular trigger, is a good strategy for achieving controlled release.

4 DRUG–DENDRITIC POLYMER INTERACTIONS

There are two main interaction mechanisms between dendrimers or dendritic polymers and therapeutic drugs: simple encapsulation and conjugation. The simple encapsulation comprises physical entrapment of guest molecules in the inner cavities of the dendritic polymers and/or multiple noncovalent chemical interactions, like hydrogen bonding, between the drug and the host molecule. The use of these highly branched molecules as molecular containers was proposed for the first time by Maciejewski (1982). Since then, numerous reports based on different guest-dendritic host systems were published. The internal void of dendrimers is mostly hydrophobic in nature, allowing interaction with poorly soluble drugs through hydrophobic interactions. The size and shape of the guest and cavity establish the number of guests entrapped in the dendritic box, while the architecture of the dendritic molecule is important as well. Interestingly, a shape-selective liberation can be accomplished by removing the shell either partially or completely, using finely tuned hydrolytic conditions (Jansen et al., 1995).

Due to its multiple surface functionalities, a multivalent dendritic molecule is capable of carrying multiple therapeutic molecules to construct dendritic polymer–drug conjugates. By employing different generations or by modifying coupling conditions, the number of drug molecules per conjugate can vary. One way of achieving dendritic nanoconjugates involves electrostatic interactions. The high density of functional groups on the periphery, like amine groups and carboxyl groups in PAMAM, helps to enhance the solubility of hydrophobic drugs by means of the charge density of the surface. Diflunisal, ibuprofen, ketoprofen, indomethacin, and naproxen, among other nonsteroidal antiinflammatory drugs with carboxyl groups, have been complexed with dendritic polymers by electrostatic interactions (Tripathy and Das, 2013). On the other hand, covalent conjugation to the chain ends of the dendritic molecule can be achieved to selectively conjugate several targeting groups, solubilizing groups or different drug molecules. Usually, ester or amide bonds are employed that can be hydrolyzed inside the cell by endosomal or lysosomal enzymes (Menjoge et al., 2010). Liu and Fréchet (1999) explored the attachment of three model drugs, each with a different functional group (hydroxyl, amine, or carboxyl groups) by using a variety of potentially cleavable bonds (carbonate, carbamate, and ester), and reported that even after the attachment of five cholesterol molecules, the resulting conjugates were still water-soluble.

In dendrimers with a steric crowding at the surface, the controlled release from polymer–drug conjugates plays a critical role in carrier efficacy. The drug release characteristics of ibuprofen-G4 PAMAM dendrimer conjugates with ester, amide, and peptide linkers were explored by Kannan and coworkers, in order to gain a deeper insight into steric effects and the choice of linkage to achieve effective drug delivery formulations (Kurtoglu et al., 2010). These authors reported that the hydrolysis of ester conjugates showed clear pH-dependent rate, whereas amide ones were stable at all pH buffers. In addition, conjugates without linkers were found to be stable against enzymatic hydrolysis due to steric effects. Since esterase and protease enzymes are both comparable in size to dendrimers, steric hindrance becomes crucial in enzymatic cleavage of drugs from dendrimer conjugates. Moreover, a novel pH-sensitive amphiphilic dendritic polyrotaxane drug-polymer conjugate was synthesized via a pH-responsible hydrazone bond with 1.84 wt.% of doxorubicin (Kang et al., 2015). In this report, the authors demonstrated that drug-polymer conjugates self-assembled to form micelles in an aqueous solution showing a significantly faster doxorubicin release at mildly acidic pH of 6.0 and 5.0, and almost no burst release at the physiological pH of 7.4; thus, they could efficiently deliver doxorubicin to the nuclei of the tumor cells. Heparin drug conjugates presented a similar behavior as pH-responsive drug delivery carriers also for doxorubicin (She et al., 2013). Another pH-sensitive polymeric carrier was developed for delivery of the anticancer drug bortezomib (BTZ) to cancer cells. Messersmith and coworkers reported a novel strategy based on facile conjugation of BTZ to catechol-conjugated poly(ethylene glycol) carriers designed to be taken up selectively by cancer cells through cell surface receptor-mediated mechanisms (Su et al., 2011). According to the authors, this polymer was chosen for its ability to reduce nonspecific interactions with proteins and cells, and the catechol moiety was exploited for its ability to bind and release borate-containing therapeutics such as BTZ in a pH-dependent manner. They demonstrated that these cancer-targeting drug-polymer conjugates dramatically enhanced cellular uptake, proteasome inhibition, and cytotoxicity toward breast carcinoma cells in comparison with nontargeting drug-polymer conjugates (Su et al., 2011).

5 TYPES OF TARGETING BY NANODELIVERY SYSTEM

Targeted drug delivery comprises a method of delivery therapeutics preferentially to the right place in a body, at the precise time, the appropriate dose, and for the suitable period of time. Targeted drug delivery is decisive in cancer therapy to reduce the systemic toxicity resulting from nonspecific drug delivery, and to increase drug delivery efficiency. The two general approaches of targeting include active targeting and passive targeting.

An active targeting may be achieved by conjugation of nanocarriers with a drug to a desired target site. This strategy is essentially based on specific interactions, such as antibody–antigen or ligand-receptor, and usually involves linking the therapeutic agent to a targeting moiety that distinguishes and acts together with receptors expressed particularly by the target.

A broad range of targeting ligands has been explored for dendritic polymer-based nanocarriers. The folate receptor (FR) is a recognized tumor biomarker, known to be stable, inexpensive, and nonimmunogenic, as compared to proteins such as monoclonal antibodies; moreover, folic acid binds to the FR at cell surfaces with very high affinity ($K_d = 10^{-1}$ M), and is internalized by receptor-mediated endocytosis (Vllasaliu et al., 2013). In 1997, the first report of Gadolinium complexes of folate-conjugated dendrimer-chelates showed an increase in the longitudinal relaxation rate of tumor cells expressing the high-affinity FR, demonstrating that folate-conjugated magnetic resonance imaging contrast agents represent a promising approach to tumor targeting (Wiener et al., 1997). Four years later, Wiener and coworkers reported a generation four polyamidoamine (PAMAM) folate-dendrimers that specifically target the high affinity FR overexpressed in more than 80% of ovarian tumors (Konda et al., 2001).

Since then, numerous researchers have explored the advantages of dendrimer-folate targeting to address novel carriers for cancer treatment. For instance, a device based on an G5-ethylenediamine core PAMAM dendrimer containing folic acid, fluorescein, and methotrexate covalently attached to the surface to provide targeting, imaging, and intracellular drug delivery capabilities showed that targeted delivery improved the cytotoxic response of the cells to methotrexate 100-fold over free drug (Quintana et al., 2002). The expression of the FR is amplified in a variety of human tumors and can serve also as a molecular target for boron neutron capture therapy (BNCT). FR was successfully targeted in cancer cells using folic acid conjugates of boronated PEG containing G3–PAMAM dendrimers to obtain ^{10}B concentrations necessary for BNCT, reducing the uptake of these conjugates by the reticuloendothelial system (Shukla et al., 2003). The folate-poly(ethylene glycol) conjugates of G 3.5 PAMAM with different degrees of substitution and loaded with indomethacin showed to be the ideal choice for targeted delivery of antiarthritic drugs to inflammation with reduced gastric-related side-effects and higher targeting efficiency (Chandrasekar et al., 2007). The hydrophobic drug curcumin was also conjugated to the surface of folic acid-PEG-PAMAM dendrimer and the targeting capability of these conjugates was observed in the human hepatocellular carcinoma cell line (Gao et al., 2013). An amphiphilic dendrimer-like star polymer (DLSP) comprised of a hydrophobic star Poly-L-lactide acid core and a hydrophilic 2,2-bis(hydroxymethyl)propionic acid-based polyester dendron shell was synthesized, and the surface functional groups were modified to afford multiple carboxylic acid groups to conjugate amine-functionalized folic acids yielding a folate-DLSP hybrid as a carrier for targeted drug delivery that showed an enhanced cellular uptake against KB cells (a human epidermal carcinoma cell line) higher than that of the neat DLSP (Cao et al., 2010). The folate-poly(ethylene glycol)-G 3.5 PAMAM-functionalized was found to encapsulate and solubilize luminescent CdSe/ZnS quantum dots (QDs) through direct ligand-exchange reactions showing in HeLa cells that cellular uptake of folate-QDs carriers was more significant than that of nonfolate QDs in vivo imaging experiment, key for optical detection of tumor cells (Zhao et al., 2010). A human epidermoid cancer cell line was targeted with folated Ag-dendrimer nanocomposites, and the labeled cancer cells were subsequently destroyed by the microbubbles generated by the enhanced energy uptake of the Ag Nps from the laser irradiation, as compared to unlabeled cells, developing an interesting method to selectively destroy cancer cells by combining biochemical targeting with topical laser irradiation (Tse et al., 2011). The folic acid conjugated to poly(ethylene glycol)-modified G 3.5 PAMAM with doxorubicin and superparamagnetic magnetite was successfully constructed as a tumor targeted and a pH-responsive drug release system, enhancing the effect of magnetic resonance imaging (MRI) contrast and cancer therapy (Chang et al., 2012). Baker and coworkers reported the design and synthesis of a photo-releasable FR-targeting methotrexate-tethered dendrimer-based nanoconjugate; they demonstrated methotrexate release from the conjugate through a light-controlled mechanism and evaluated its cytotoxic activity in KB cells following exposure to UV light (Choi et al., 2012). Folate-targeted G4 PAMAM dendrimers were encapsulated into poly(ethylene glycol)-*b*-poly(D, L-lactide) NPs, using a double emulsion method, and selectively interacted with FR overexpressing KB cells upon release in a temporally controlled manner, providing evidence that selective cellular interactions can be kinetically controlled by the nanohybrid design (Sunoqrot et al., 2012). A bioprobe based on folate–copper G5 PAMAM conjugates was developed to recognize cancer cells by synchrotron X-ray fluorescence analysis (Zhang et al., 2013).

Monoclonal antibodies comprise other attractive targeting ligands for selective delivery of chemotherapeutic drugs while reducing undesirable side effects caused by drug accumulation in healthy tissues. Trastuzumab—that binds to human epidermal growth factor receptor 2 (HEGFR2)—was used by

Sun and coworkers as a targeting agent in a Trastuzumab-PEG-PAMAM conjugate carrying paclitaxel specifically to cells overexpressing HEGFR2. The antibody-dendrimer conjugate was located in the cytoplasm, showing an in vitro cytotoxicity lower than free paclitaxel and in vivo targeting accumulated in the breast cancer model more efficiently than nonconjugated PAMAM (Ma et al., 2015). An effective dendritic polyglycerol-based nanocarrier to specifically target and destroy cancer cells was provided with a dendritic core as a multifunctional anchoring point, doxorubicin coupled through a pH-sensitive linker, a fluorescence marker, poly(ethylene glycol), as solubilizing and shielding moiety, and a single-chain variable fragment (scFv) antibody conjugated (Hussain et al., 2013).

Some peptides and aptamers may appear as ligands with overexpressed biological uptake systems in cancerous tissue. A fluorescein isothiocyanate-G5 PAMAM dendrimer-entrapped Au nanoparticle modified with alpha-tocopheryl succinate and arginine-glycine-aspartic acid peptide was recently developed for targeted chemotherapy, and computed tomography imaging of cancer cells (Zhu et al., 2015). This theranostic platform for cancer is capable of targeting cancer cells, overexpressing $\alpha v\beta 3$ integrin (a receptor for vitronectin), and specifically restraining the growth of cancer cells due to the role played by peptide-mediated targeting. A further case involves Chlorin e6 (Ce6) and arginine-glicine-aspartic acid peptide, which were conjugated to G7 PAMAM dendrimer for targeted photodynamic therapy of solid tumors. Targeted nanoconjugates demonstrated approximately 16-fold enhancement in receptor-specific cellular delivery of Ce6 into integrin-expressing A375 cells, compared to free Ce6, thus being able to cause massive cell killing at low nanomolar concentrations under photoirradiation (Yuan et al., 2015). In addition, multivalent aptamer-tethered dendritic nanostructures were developed as multifunctional vehicles for targeted cancer cell imaging and drug delivery. A multifunctional DNA dendrimer constructed from Y-shaped building blocks with predesigned base-pairing hybridization including fluorophores, targeting DNA aptamers and intercalated doxorubicin, was recently developed by Yang and coworkers, and exerted a potent toxicity for target cancer cells (human T cell acute lymphoblastic leukemia cell line) with low side effects for nontarget cells (human Burkitt's lymphoma cell line) (Zhang et al., 2015).

Unlike active targeting, a passive one uses the inherent ability of conjugated-dendritic molecules to extravasate and selectively accumulate in the tumor tissues based on the enhanced permeability and retention (EPR) effect (Maeda et al., 2009). The passive transport of drugs is based on the passage by convection or passive diffusion of nanocarriers through the tumor capillaries into the interstitial space and cancerous cells. As a consequence of a rapid process of angiogenesis are capillaries with abnormalities in their structure (small pores), depending on the type of tumor. In addition, the interstitial pressure is higher in the center of the cancerous tissue than in the tumor periphery because the tumors lack a well-defined lymphatic system. Drugs and nanoparticles that enter into the interstitial space remain retained more time inside than in normal tissue. The combination of these porous capillaries and deficient lymphatic drainage effect leads to EPR. Thus, dendrimers may achieve passive-mediated targeting to a cancer cell simply by control of their size and physicochemical properties. In practice, several of the targeted dendrimers can act as active and passive targeting moieties. For example, the partially acetylated G5-PAMAM dendrimer was conjugated with the active targeting moiety avidin and labeled with 99mTc, exhibiting much higher cellular uptake into Hela cells than those without avidin conjugate (Xu et al., 2011). In another study, the arginine-glycine-aspartic acid peptide-PEG–PAMAM-doxorubicin conjugate showed in vitro and in vivo targeting to both tumor neovascular endothelial cells and tumor cells (Zhu et al., 2011). In this case, the arginine-glycine-aspartic acid peptide was used to actively target the drug delivery system to tumor neovascular endothelial cells and tumor cells, while the PEG-PAMAM dendritic carrier allowed passive targeting to the tumor site by enhanced permeability and retention effect (Zhu et al., 2011).

6 ROUTES OF ADMINISTRATION OF DENDRITIC MOLECULES

There are several routes of administration of dendritic molecules for drug delivery in pharmaceutical applications, among them we can highlight the oral, ocular, intravenous, topical, transdermal, and pulmonary routes. In all cases, these systems are designed to enhance the properties of drug formulation.

6.1 INTRAVENOUS DRUG DELIVERY

Lam et al. reported the development of a series of linear-dendritic block copolymers, referred to as telodendrimer, containing PEG and dendritic oligocholic acids (CA). These telodendrimers are able to self-assemble into stable micelles in aqueous solution and to encapsulate doxorubicin (DOX). These micelles were effectively used for the delivery of DOX in the treatment of B-cell lymphoma. The comparison between the micelles formed with the different telodendrimers synthesized showed that the one containing a PEG chain of 2 kDa and 4 CA subunits in the telodendrimer (PEG2k-CA4) was the one that had higher stability in physiological conditions, higher capacity to load DOX and a more persistent release profile. In vitro studies also showed that DOX-loaded micelles were able to internalize in Raji lymphoma cells and release DOX. Finally, DOX-PEG2k-CA4 micelles not only enhanced the anticancer efficacy in vivo after being injected into female BALB/c mice via tail vein, they also found that it significantly reduced the cardiotoxicity associated with DOX in comparison with free DOX and a commercial formulation of PEGylated liposomal DOX (Doxil), at equivalent doses (Xiao et al., 2011). Another interesting example of intravenous drug delivery is the one reported by Jain and coworkers. In this work, dendritic PEG-lysine nanoparticles conjugated with chondroitin sulfate A in their surface were loaded with artemether (ART). ART is a potent antimalaria drug that has been included in the World Health Organization list of essential drugs for this disease; however, it has the disadvantage of being only oil-soluble and having a low stability (a half-life of 3–5 h). The dendritic nanoparticles with ART were tested both in vivo and in vitro, and showed that they were able to release ART and sustain its blood concentration up to 10 h at steady state levels of bioactives in albino rats, acting as a suitable drug delivery system for this kind of drug (Bhadra et al., 2005).

6.2 ORAL DRUG DELIVERY

The oral administration of silybin and silybin-PAMAM complex was studied by Huang et al. (2011). Silybin is a natural product used in the treatment of hepatitis and cirrhosis; yet, its clinical use is restricted by its exceptionally low solubility in water and oil, and by its reduced absorption in the gastrointestinal tract. In this work, they found that PAMAM of different generations considerably improved the solubility of the drug; the complex also showed a sustained release of silybin in vitro and enhanced its oral bioavailability in vivo (Huang et al., 2011). There are also other interesting examples of oral administration of PAMAM-drug complexes. Ke et al. (2008) used doxorubicin as a drug model for loading into PAMAM, and they reported that this complex enhanced the cellular uptake of DOX in

Caco-2 cells, as compared with the free drug. After oral administration, the bioavailability led by the PAMAM-DOX complex was 200 times higher in comparison with free DOX (Ke et al., 2008).

6.3 TOPICAL DRUG DELIVERY

Winnicka et al. (2012) have developed and evaluated a hydrogel containing PAMAM dendrimers and ketoconazole (KET). KET is a well-known antifungal, highly lipophilic, and insoluble in water; therefore, its use is difficult in clinical applications. In this work, the authors developed a PAMAM dendrimer of second and third generation that contains an amino group in its surface, and can strongly enhance the solubility of KET. In addition they found that this formulation increased the antifungal activity of KET against Candida strains in comparison with pure KET hydrogel and its commercial available formulation (Nizoral) (Winnicka et al., 2012)

6.4 TRANSDERMAL DRUG DELIVERY

Nonsteroidal antiinflammatory drugs are one of the most normally used drugs due to their antiinflammatory, analgesic, and antipyretic properties. This kind of drugs has been commonly used in chronic musculoskeletal conditions, chronic rheumatoid- and osteoarthritis. One of the most common routes of administration for this kind of compounds is the oral one. However, it has the disadvantage of producing disturbance in the gastrointestinal tract. Taking this into account, Yiyun et al. (2007) studied the formation of PAMAM-ketoprofen and diflunisal complex for in vitro and in vivo transdermal delivery, using the ability of PAMAM dendrimers to facilitate this process. In vitro permeation studies showed that the amount of drug that penetrates rat skin was significantly higher for the complexes, and in vivo assays demonstrated that this formulation enhances the bioavailability of the drug (Yiyun et al., 2007).

6.5 PULMONARY DRUG DELIVERY

Kannan and coworkers reported the conjugation of methylprednisolone (MP), a well-known drug used in the treatment of lung inflammation associated with asthma, with PAMAM dendrimers, and they tested the ability of this conjugate to improve the airway delivery of MP in vivo. The MP-PAMAM conjugates improved the ability of MP to decrease the ovalbumin allergen-induced inflammation with a lower concentration of MP, in comparison with free MP and nonconjugated PAMAM, without producing undesired side effects such as nonspecific inflammation. These results are in good agreement with those showed in the considerably longer residence time in the lungs of the conjugates in comparison with free MP (Inapagolla et al., 2010).

6.6 OCULAR DRUG DELIVERY

Jain and coworkers investigated the loading of acetazolamide in poly(propyleneimine) dendrimers, and evaluated the loading efficiency, toxicity, ocular irritation, and intraocular pressure reduction in vivo by using New Zealand albino rabbits. At the concentration range used, the drug-PPI formulation showed to be weakly irritant. They also found that this formulations leads to an enhancement of the acetazolamide residence in the eye, and an improvement in the intraocular pressure lowering effect in vivo. As a consequence, PPI dendrimer-based formulation appears to be a promising ophthalmic drug delivery system for the treatment of glaucoma (Mishra and Jain, 2014).

7 SUCCESSFUL EXAMPLES OF DENDRITIC MOLECULES AS NANOCARRIERS

One of the first complete reviews on dendrimer-based nanovectors for tumor-targeted drug and gene delivery was published by Fréchet (1994) with an emphasis on the micelle-like behavior of dendrimers with hydrophobic core and hydrophilic periphery. A decade later, Svenson and Tomalia illustrated clearly the potential of this fourth architectural class of polymers and substantiated the high optimism for the future of dendrimers in drug delivery systems in their review of biomedical applications of dendrimers (Svenson and Tomalia, 2005). Despite the promising future foreseen in these studies, after 40 years of research no more than a few dendrimers have been translated into clinical practice. As far as we know, only two dendrimers have been developed for in vivo medical applications: a poly(lysine) dendrimer functionalized with 32 naphthalene disulfonate units on its surface called VivaGel (Starpharma Holdings Ltd) that is at the moment in phase three clinical evaluation for the prevention of bacterial vaginosis (VivaGel®, 2015); and a hydrogel-dendrimer liquid ocular bandage called OcuSeal (BeaverVisitec International) designed to promote healing by protecting the corneal incision from eyelid trauma (OcuSeal®, 2015). Recently, a critical evaluation of the status of dendrimers as drug carriers, carried out by Svenson (2015), postulated that instead of concentrating on large numbers of surface functional groups, dendrimer researchers should be formulating the question of how many groups are really needed. Svenson also commented on the need to introduce internal functional groups to improve the carrier-to-active ratio, and sustain the solubility and interaction behavior between nanocarriers, and components of the mononuclear phagocyte system regardless of the amount of drug loading (Svenson, 2015). Taking into account this reasoning, other dendritic polymers should be evaluated for delivery applications. For instance, hybrid nanocomposites such as nanoparticle-cored dendrimers combine the advantage of stimuli-responsive dendritic coating to regulate the drug release behavior under different conditions with improving the biocompatibility and in vivo half-time circulation of the inorganic nanoparticles. Recent examples of dendrimers and other dendritic molecules designed as vehicles in drug delivery are described later.

7.1 DENDRIMERS

In cancer treatment, the success of chemotherapy relies on the nature of the drug, its dosage, and how effectively it is delivered to the target cell. Owing to their nonspecific action, even under optimal conditions several anticancer drugs induce toxicity to normal tissue as well. It is therefore necessary to be able to precision-target anticancer drugs to cancer cells only. Saraswathy et al. (2015) designed a complex nanocarrier based on the polyamidoamine dendrimer (PAMAM) that has a unique nanoscale architecture and multifunctionality for drug delivery (Saraswathy et al., 2015). PAMAM was conjugated with PEG to reduce cytotoxicity and increase the circulation time in the blood stream. To enhance its inherent encapsulation via H-bonding capability, it was also conjugated with cyclodextrins (βCDs) to provide another hydrophobic microenvironment capable of loading suitably sized hydrophobic drug molecules via hydrophobic-hydrophobic interaction. Doxorubicin (Dox), a model hydrophobic anticancer drug, was effectively encapsulated with a loading level of 16.8 wt.%. Finally, to enhance the tumor accumulation of nanocarriers in cancer cells, cRGD (ciclyc arginine-glicine-aspartic acid sequences) used as a tumor-targeting ligand was selectively conjugated onto the distal ends of the PEG arms. The resulting βCDs–PAMAM–PEG–cRGD polymer was able to form stable and uniform nanoparticles (NPs) in aqueous solution. The precise, controlled, and smart design of these NPs, showing a high capacity to encapsulate small hydrophobic drugs and also a pH-sensitive drug release behavior (62% at pH 5.3, and 19% at pH 7.4) converted them into a successful nanoplatform for controlled release. These NPs possess less off-target threats to healthy cells and are suitable for intravenous injection. Saraswathy et al. (2015) also showed that cRGD-conjugated NPs exhibit higher cellular uptake in U87MG human glioblastoma cells overexpressing αvβ3 receptors, than in nontargeted NPs, resulting in higher and specific cytotoxicity, and a superior antiproliferative effect than that of free drug and nontargeted NPs (Saraswathy et al., 2015).

Another antiangiogenic agent for cancer therapy currently under investigation is the poly-L-lysine (PLL) dendrimer. Jain et al. (2014) recently reported folate conjugated 5.0 G poly-L-lysine dendrimers (FPLL) as an efficient nanocarrier for doxorubicin hydrochloride (Dox), and showed that the formulation was selectively taken up by the tumor cells rather than free drug (Jain et al., 2014). Based on the fact that these Dox-conjugates showed initial rapid in vitro release followed by gradual slow release, with a greater release at acidic pH, the authors also reported that this formulation caused a significant delay in tumor growth with a significantly prolonged survival time.

Dendrimer-based therapeutic molecules have been successfully developed for the treatment of ischemia-induced neonatal white matter injury. Kannan and colleagues demonstrated Cy5-labeled G-4(OH) PAMAM dendrimer uptake in cells involved in ischemic injury, and in the ongoing inflammation leading to secondary injury that is consistent with the region-specific inflammatory response following ligation (Nance et al., 2015). According to these authors, the uptake was time and site-specific dependent, with early dendrimer localization in astrocytes, the immediate responders to injury, and later dendrimer localization in microglia, which respond 2–3 days after injury.

In the case of infectious diseases such as leishmaniasis, salmonellosis, candidiasis, etc. the targeted delivery of antiparasitic drugs to macrophages is essential since the parasites reside and multiply within host macrophages. Recently, Jain et al. (2015) reported a promising antiparasitic drug using mannose anchored dendrimers for targeted delivery of Amphotericin B (AmB) to macrophages for treatment of visceral leishmaniasis as a model infectious disease. AmB was loaded into mannose conjugated 5.0G poly(propylene imine) dendrimers (MPPIA) and studied for its in vitro and in vivo intracellular macrophage targeting potential and antileishmanial activity (Jain et al., 2015). In their results, Jain and coworkers found that MPPIA show enhanced cellular uptake in J774A.1 macrophage cells without toxicity, and significantly improve inhibitory activity against intracellular *Leishmania donovani amastigotes* and in vivo infection. On the one hand, cell uptake was significantly higher for MPPI than for 5.0G PPI dendrimers, possibly attributable to the mannose receptor specific targeting of ligand-conjugated dendrimers. On the other hand, the pharmacological activity and toxicity of AmB depends on its aggregation state and the reduction in its cytotoxicity, when incorporated into mannose-anchored dendrimers, might be due to the release of the drug in a monomeric state (Jain et al., 2015).

7.2 HYPERBRANCHED POLYMERS

An exhaustive examination of numerous examples of HB-polyethers (mainly HB-polyglycerol), HB-polyesters (functionalized Boltorn), HB-polyethylene and HB-polystyrene, among others, used as vehicles for drug delivery applications, was recently reported by Kurniasih et al. (2015) According to these authors, the tailored flexibilities and composition of HB, together with its ease of synthesis and scale-up, indicate a bright future in drug delivery applications.

One interesting and novel HB-based nanocarrier for paclitaxel, a known chemotherapeutic agent with low aqueous solubility, was recently developed by Zarrabi and Vossoughi (2015). In this report, conjugation of β-cyclodextrin to HB-polyglycerol (β-CD-*g*-PG) was used to load paclitaxel by including it in the hydrophobic cavity of cyclodextrin as guest. The hybrid nanocarrier shows a number of advantages over cyclodextrin alone: it has more hydrophilic surface with several vacancies for the attachment of targeting and tracing agents, and also has several hydrophobic vacancies between its branches which could lead to higher drug loading (Zarrabi and Vossoughi, 2015). Thus, the paclitaxel/β-CD-*g*-PG inclusion complex proves to be a promising alternative for the injectable paclitaxel formulation with minimum side effects.

Another successful example of HB-based nanocarriers is a unique mechanism of drug release based on amphiphilic-to-hydrophilic transition. Yan and coworkers synthesized an amphiphilic HB polymer with alternative hydrophobic selenide groups and hydrophilic phosphate segments in the dendritic backbone of developed H_2O_2-responsive nanocarriers (Liu et al., 2013). Based on the fact that the hydrophobic selenide groups turn into hydrophilic selenone after oxidation, this smart drug-vehicle can be dissembled under the exclusively oxidative microenvironment within cancer cells (Fig. 16.9).

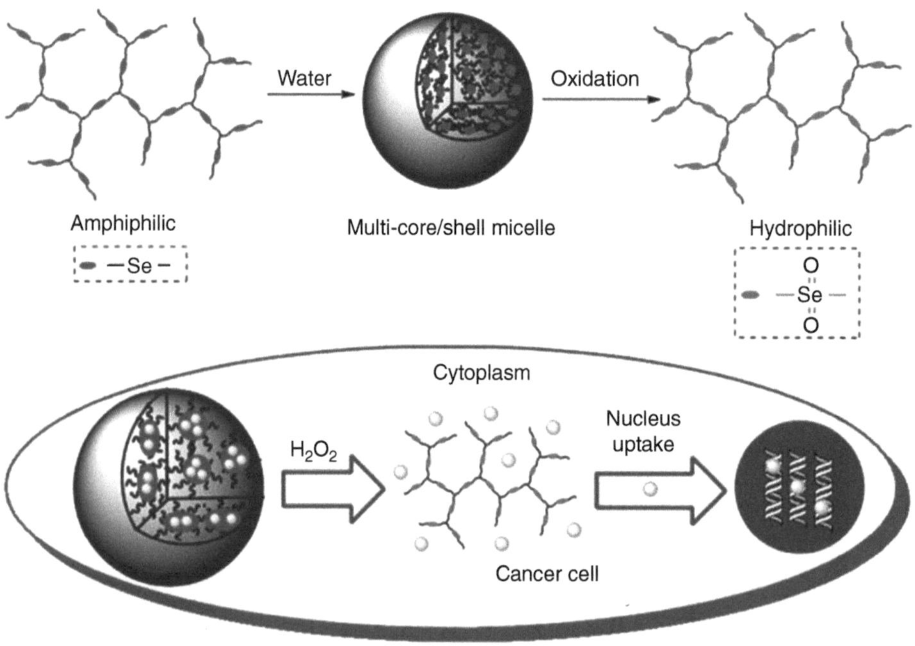

FIGURE 16.9 Cartoon Depiction of Self-Assembled Micelles and the Oxidation-Triggered Disassembly

Reprinted from Liu et al. (2013), with kind permission of The American Chemical Society.

7.3 DENDRONIZED POLYMERS

Dendrimers of third and fifth generation dendrimers are suitable as nanocarriers owing to their sufficiently open structure for the loading and subsequent opportune release of drug molecules. However, cationic PAMAM dendrimers, the most used dendrimers for low-aqueous-soluble drugs, cause a certain degree of toxicity toward cells. One successful way of dealing with this undesired effect is the chemical combination of PAMAM with chitosan (CS) to generate the corresponding DPs. Zhou et al. recently reported that dendronized chitosan derivatives CS-*g*-PAMAM ($G = 2,3$) caused a low level of morphological change and membrane disruption of rat red blood cells, indicating an excellent degree of hemocompatibility with these cells (Zhou et al., 2015). In fact, due to its nontoxic, biodegradable, biocompatible and nonimmunogenic properties, chitosan has been the subject of considerable interest in terms of its controlled release capacity, and is currently a promising constructive block of DP-based vehicles for therapeutic drugs. Mukhopadhyay et al. (2014) recently reported a PAMAM (G2)-conjugated chitosan through the N-carboxyethyl chitosan methyl ester intermediate for oral insulin delivery. In this case, the self-assembled nanoparticles with insulin loading were prepared through a mild process from insulin and dendronized chitosan, and revealed an almost doubly-enhanced insulin bioavailability, compared to native CS-NPs.

7.4 DENDRIMERSOMES AND ONION-LIKE DENDRIMERSOMES

The self-assembly of janus dendrimers into uniform dendrimersomes serves to mimic biological membranes and enables the delivery of drugs and genes. Percec et al. (2010) proposed that dendrimersomes "marry" the stability and mechanical strength obtainable from polymers with the biological function of stabilized phospholipid liposomes, plus superior uniformity of size, ease of formation, and chemical functionalization. Dendrimersomes formed from amphiphilic janus dendrimers composed of TEG-functionalized gallic acid as the hydrophilic block and G3 photodegradable dendrons, recently developed by Nazemi and Gillies (2014), have been shown to exhibit triggered release of both hydrophilic (fluorescein) and hydrophobic (Nile red) payloads upon irradiation with UV light. These results suggest the promise of these systems for controlled encapsulation and release applications.

Hammond et al. reported the self-assembly of linear-dendritic polymer (LDP) micelles functionalized with folate on 10% of the surface for cell targeting (Poon et al., 2011). These micelles were loaded with paclitaxel (PTX), one of the most effective anticancer drugs, and the efficiency of this system was evaluated in tumor targeting. The synthesized micelles have a high degree of colloidal stability, which enhances the circulation time in blood and the bioavailability of PTX. Moreover, the presence of folate on the surface of the micelle dramatically increases tumor targeting, leading to greater drug accumulation over a 5-day period (Fig. 16.10); upgrading the therapy, for instance a low PTX dose of 2.5 mg/kg administered once in 4–5 days, is sufficient to achieve significant tumor regression.

In 2014, Zhang et al. reported the self-assembly of amphiphilic janus dendrimers into uniform onion-like vesicles with predictable size and number of bilayers by simple injection of a solution of amphiphilic janus dendrimer with specific primary structure into water or buffer (Zhang et al., 2014). The authors proposed the use of these dendrimersomes as a tool for the construction of promising time-dependent delivery devices containing multiple water-insoluble and water-soluble payloads incorporated in each bilayer of these onion-like architectures.

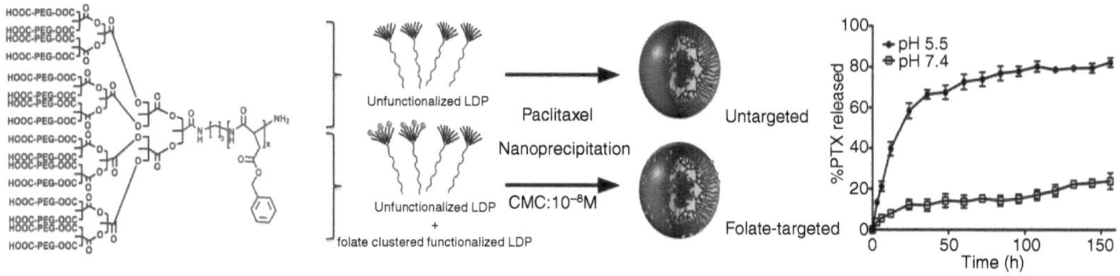

FIGURE 16.10 Chemical Structure of the LDP and Scheme Showing the Preparation of PTX-Encapsulated LDP Micelles that either Present or Not Present Folate Clusters for Enhanced Cell Targeting

In vitro release of PTX at 37°C in fetal calf serum shows that PTX will be stable in circulation, but acidic endosomes will trigger release of PTX.

Adapted from Poon et al. (2011), with permission from Elsevier.

7.5 HYPERBRANCHED POLYDENDRON-BASED NANOPARTICLES

The hyperbranched polydendron-based nanoparticles (HP-NPs) are recently developed uniform nanostructures that have undergone preliminary pharmacological evaluation to establish low cytotoxicity and enhanced permeation through model intestinal epithelial cells (Hatton et al., 2015). According to the authors, the careful design of these materials allows tailoring permeation through gut epithelium models and accumulating specially in macrophages suggesting potential for oral dosing. They synthesized seven HP (one HP bearing a G2 dendron and copolymerization of 2-hydroxypropyl methacrylate (HPMA) and ethylene glycol dimethacrylate (EGDMA), five HP also with varying PEG content and one entirely PEG initiated HP). Afterward, these HPs were subjected to nanoprecipitation under varying condition of solvents to form the corresponding HP-NPs. Cytotoxicity of these seven aqueous HP-nanoprecipitates was evaluated against the human epithelial colorectal adenocarcinoma (Caco-2) cell line and no appreciable cytotoxicity was observed at achievable concentrations in assays assessing either adenosine triphosphate or 3-(4,5-dimethylthiazol-2-yl)2,5-diphenyltetrazolium bromide turnover (Hatton et al., 2015). Moreover, Nile Red (NR) fluorescence monitoring was used to determine HP passage from the apical compartment (A) of the transwell plate to the basolateral well (B) and vice versa. The P ratio, calculated as the ratio of (A > B)/(B > A), provides a relative indication of apparent oral absorption in the presence of active transport proteins. The A > B permeation of these HP-NPs with encapsulated NR was approximately an order of magnitude higher than observed for B > A permeation in all cases and P ratios across the materials were all considerably higher than those of the aqueous/DMSO solution of NR (Hatton et al., 2015). These results are good enough to suggest prospect of oral administration.

7.6 DENDRIMER-ENCAPSULATED NANOPARTICLES (DENPs)

Magnetic nanoparticles incorporated into nanocarriers are of interest because of their ease of recovery after use. Thus, encapsulation of magnetic NPs in dendritic molecules is a versatile strategy for developing dendrimer-based nanodevices with many surface functional groups available for covalent immobilization of biomolecules (enzymes, e.g.) providing a means of facilitating product separation.

Lin and coworkers covalently immobilized a bacterial γ-glutamyltranspeptidase (BLGGT) on the hyperbranched poly(amido acid)s (HBPAA) which previously encapsulated γ-Fe_2O_3 NPs and proved that the presence of magnetic NPs allowed recovery and reuse of the grafted catalyst for at least 10 cycles with 32% of the initial activity (Juang et al., 2014).

7.7 NANOCORED DENDRIMERS (NCDs)

In recent years, nanocored dendrimers (NCDs) have emerged as innovative alternatives for use as nanocarriers owing to their versatility, exhibiting greater colloidal stability than nondendronized NPs, the ability to encapsulate a large range of drugs and molecules, and also having tunable membrane properties (Brunetti et al., 2015). One of the first examples of this appears in a study performed by Gopidas et al. (2003), who show that water soluble Au NCDs are capable of encapsulating pinacyanol chloride, proving this material to be a good candidate as a nanocarrier.

One important feature of NCDs of relevance to their use as nanocarriers is that they possess a high degree of surface coverage, leaving the inside pockets for drug upload. Forni and colleagues reported the synthesis of physiologically compatible nanoparticles of polyester poly(D,L-lactic-coglycolic acid) (PLGA) using Lin's amine dendrons functionalized with peptide and β-D-glucose (Costantino et al., 2006). This method allows shifting the amount of peptide bound to the surface of the NP, a very useful attribute opening up the possibility of obtaining NPs with different degrees of coverage on the surface. Furthermore, these NPs with dendritic architecture were able to cross the blood–brain barrier when tested in in vivo experiments. Both of these features are important for the potential application of NCDs in drug delivery. Another important characteristic for application purposes is good stability in biological media. Daniel et al. (2011) reported the preparation of Au-NCD using poly(propyleneimine) dendron as coating. This type of NPs shows remarkable stability in a wide range of conditions and pH levels. These authors also found that at pH around 4.5 the internal tertiary amines of the dendron are protonated, enabling lysosomal escape, a requisite for optimal drug delivery (Daniel et al., 2011). Another example of NDCs with excellent stability in biologically relevant test media containing proteins and electrolytes is that reported by Cho et al. (2014). This investigation showed the development of Au-NDCs using positively charged PEGylated dendrons as coating. These hybrid nanomaterials were evaluated both in vivo and in vitro and showed an apparent lack of toxicity under the conditions tested, properties rendering NCDs interesting candidates as vehicles for drug delivery. A successful example is that reported by He et al. (2012), where superparamagnetic iron oxide nanoparticles (SPIONs) coated with a water soluble dendritic-linear-brush-like triblock copolymer were used for delivery of doxorubicin (Dox), a well-known anticancer drug. One of the advantages of this system is the low cytotoxicity, compared with individual polymers and also the low hemoglobin release (4%) in vitro, indicating that NCDs are not harmful to erythrocytes. This amount of hemolysis is comparable with dextran and PBS and significantly lower than comparable concentrations of PEI. These in vitro characteristics make this NDC a highly biocompatible system. With regard to drug delivery using Hela cells as a model, NCDs showed a higher IC_{50} compared with free Dox because the more compact polymeric structure of the NCDs results in a lower diffusion rate of the drug. The advantage of this hybrid system, even though the amount of Dox required is higher, is that side effects on normal cells are lower, making these nanoparticles good candidates for controlled drug release.

An alternative application of NPs as nanocarriers is their use for gene delivery. This kind of therapy offers great promise for cancer and genetic disorders, though its successful application requires the effective transport of nucleic acid into cells. In this respect, NPs constitute interesting transfection agents. Gold NPs in particular provide a good alternative since they are bioinert, nontoxic, and easy to synthesize, and functionalize. Ghosh et al. (2008) reported the use of lysine dendron-functionalized gold nanoparticles as DNA packing agents for use as a noncytotoxic transfection vector. Quantification of the efficiency of DNA delivery showed these NCDs to be significantly better than NPs coated with lysine and free polylysine. Subsequent studies, also performed by Kim et al. (2012), reported that gold nanoparticles coated with dendritic cationic ligands have been successfully used to condense siRNA for delivery. These dendronized NCDs, fully complexed with siRNA, suppressed β-gal expression by ~50%, similar to the values obtained with Lipofectamine (Invitrogen), a commercially available vehicle with validated protocol. Furthermore, it was found that in the absence of AuNPs, siRNA, or the correct sequence of siRNA, the β-gal was not suppressed to any significant degree.

Another interesting example is the use of dendronized magnetic NPs as vectors. Gunduz and colleagues have reported the immobilization polyinosinic-polycytidylic acid (poly(I:C)), a synthetic double-stranded RNA with significant toxicity for tumor cells, onto magnetic nanoparticles coated with different generations of PAMAM dendron. One of the advantages of the use of these nanoparticles is that they can be targeted to the tumor site using a magnetic field, resulting in the accumulation of the poly(I:C) on an exact target. The use of PAMAM as a coating is also very advantageous for two main reasons: first, PAMAM is able to prolong the period of time that NPs can circulate in blood; and second, because this dendrimer has an amino group on its surface that can be modified with different functionalities or linked with different biomolecules. The authors also demonstrated that nanoparticles coated with higher generations of PAMAM (G7, G6, and G5) were more appropriate for delivering poly(I:C), since they have more functional groups at the surface; the authors call for further in vivo and in vitro research to come up with novel strategies for the selective targeting and killing of tumor cells (Khodadust et al., 2013). Xu et al. (2014) reported the modification of quantum dots with low generation of peptide dendrons to generate a multifunctional supramolecular hybrid dendrimer capable of gene delivery and fluorescence imaging. These NCDs have the ability to bind DNA through noncovalent interactions, and assemble into compact nanoclusters. This supramolecular cluster greatly enhanced the transfection efficiency of pGL3-Luc, compared to a single peptide dendron, and also enhanced the luciferase gene transfection with respect to polyethylenimine (PEI), using a culture medium containing 10% fetal bovine serum (Fig. 16.11). This kind of hybrid also showed low cytotoxicity. Thus, Xu and colleagues concluded that this hybrid strategy offers some unique features for intracellular tracking, protein expression monitoring and in vivo bioimaging, all of which contribute to the better understanding of gene delivery pathways, and to improved delivery systems (Xu et al., 2014).

8 CONCLUSIONS

The significant advantages of dendritic polymer architecture over nondendritic nanocarriers reside mainly in the multivalency of dendritic molecules, which are useful for encapsulating or conjugating multiple drug molecules while allowing the controlled addition of targeting onto the nanocarriers. In addition, the synergy between their multifunctionality and nanoscale size gives rise to a chemical smartness along their molecular scaffolding yielding tunable properties that facilitate adaptation to the environment (e.g., solubility, biocompatibility, biodistribution, etc.), and which are not frequently found in conventional polymers or self-assembled nanostructures like liposomes. Based on the number

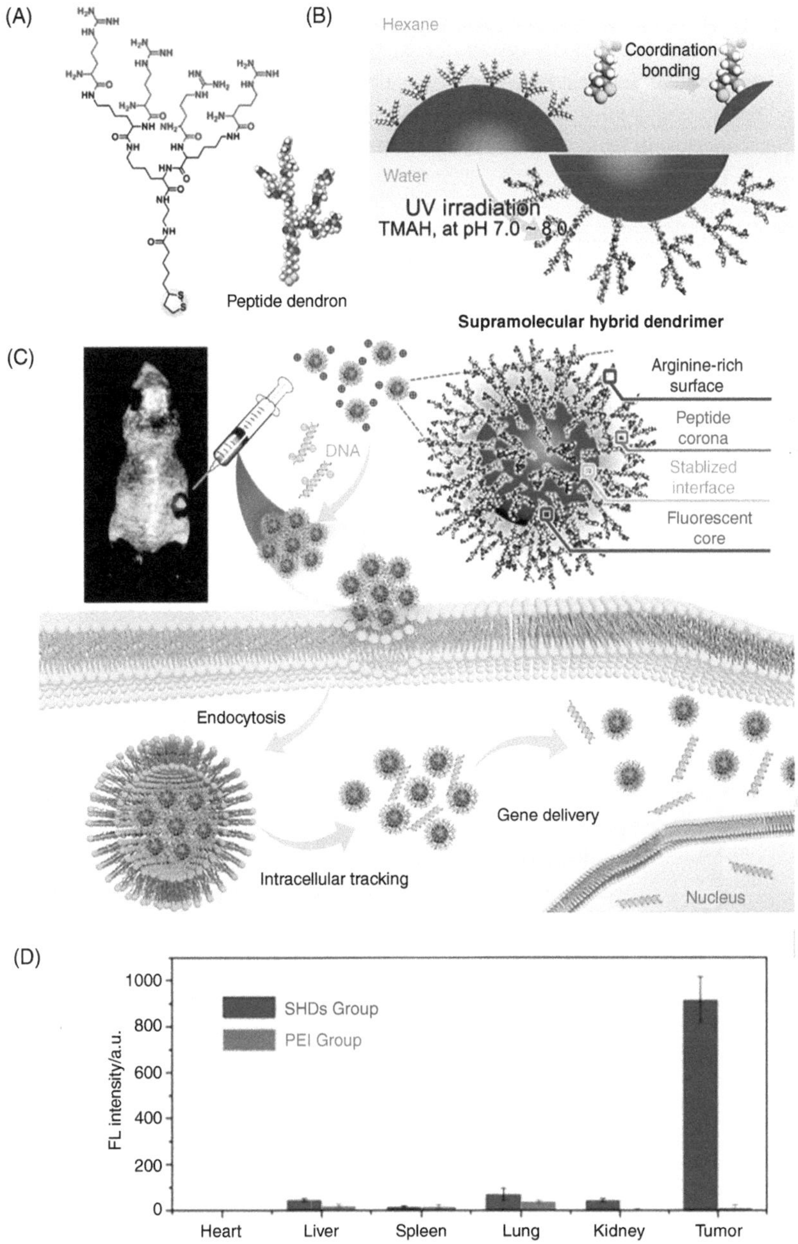

FIGURE 16.11 Schematic Illustrations for Self-Assembly and Biomedical Applications of SHDs

(A) Chemical structure of the dualfunctionalized PDs; (B) self-assembly of PDs onto quantum dots via coordination interactions; (C) SHDs with hierarchical nanostructures for gene delivery and biological tracking in vitro and in vivo; (D) ex vivo fluorescence intensity of organs after intratumoral injection for 48 h.

Reprinted from Xu, X., Jian, Y., Li, Y., Zhang, X., Tu, Z., Gu, Z., 2014. Bio-inspired supramolecular hybrid dendrimers self-assembled from low-generation peptide dendrons for highly efficient gene delivery and biological tracking. ACS Nano, 8, 9255–9264. With kind permission of The American Chemical Society.

of reports published in the past 40 years, it is clear that a substantial part of dendritic-based nanocarrier research has been focused on the utilization of dendrimers and hyperbranched polymers for the development of biomedical devices. In recent years, novel dendritic polymers, such as DPs, dendrimersome, HPs, and nanocored dendrimers, are emerging for delivery applications. In particular, nanocored dendrimers offer superior performance due to their ability to build theranostic devices that can monitor the distribution of a payload in the system in real time. Undoubtedly, developments in the design and application of novel dendronized materials, such as nanocarriers, augur well for progress in the pharmaceutical and nanomedicine fields.

ACKNOWLEDGMENTS

Financial support from ANPCyT (2011-0654), CONICET (112-20110101029), SECYT-UNC and CYTED (214RT0482) is gratefully acknowledged. EDF and LMB acknowledge CONICET for the receipt of a fellowship.

REFERENCES

Astruc, D., 2003. Dendrimers and nanosciences. Comptes Rendus Chimie 6, 709–711.

Bhadra, D., Bhadra, S., Jain, N.K., 2005. PEGylated peptide-based dendritic nanoparticulate systems for delivery of artemether. J. Drug Deliv. Sci. Technol. 15, 65–73.

Brunetti, V., Bouchet, L.M., Strumia, M.C., 2015. Nanoparticle-cored dendrimers: functional hybrid nanocomposites as a new platform for drug delivery systems. Nanoscale 7, 3808–3816.

Calderon, M., Quadir, M.A., Sharma, S.K., Haag, R., 2010. Dendritic polyglycerols for biomedical applications. Adv. Mater. 22, 190–218.

Cao, W., Zhou, J., Wang, Y., Zhu, L., 2010. Synthesis and in vitro cancer cell targeting of folate-functionalized biodegradable amphiphilic dendrimer-like star polymers. Biomacromolecules 11, 3680–3687.

Chandrasekar, D., Sistla, R., Ahmad, F.J., Khar, R.K., Diwan, P.V., 2007. Folate coupled poly(ethyleneglycol) conjugates of anionic poly(amidoamine) dendrimer for inflammatory tissue specific drug delivery. J. Biomed. Mater. Res. A 82, 92–103.

Chang, Y., Liu, N., Chen, L., Meng, X., Liu, Y., Li, Y., Wang, J., 2012. Synthesis and characterization of DOX-conjugated dendrimer-modified magnetic iron oxide conjugates for magnetic resonance imaging, targeting, and drug delivery. J. Mater. Chem. 22, 9594–9601.

Cho, T.J., Maccuspie, R.I., Gigault, J., Gorham, J.M., Elliott, J.T., Hackley, V.A., 2014. Highly stable positively charged dendron-encapsulated gold nanoparticles. Langmuir 30, 3883–3893.

Choi, S.K., Thomas, T.P., Li, M.H., Desai, A., Kotlyar, A., Baker, J.R., 2012. Photochemical release of methotrexate from folate receptor-targeting PAMAM dendrimer nanoconjugate. Photochem. Photobiol. Sci. 11, 653–660.

Costantino, L., Gandolfi, F., Bossy-Nobs, L., Tosi, G., Gurny, R., Rivasi, F., Angela Vandelli, M., Forni, F., 2006. Nanoparticulate drug carriers based on hybrid poly(D,L-lactide-*co*-glycolide)-dendron structures. Biomaterials 27, 4635–4645.

Daniel, M.C., Grow, M.E., Pan, H., Bednarek, M., Ghann, W.E., Zabetakis, K., Cornish, J., 2011. Gold nanoparticle-cored poly(propyleneimine) dendrimers as a new platform for multifunctional drug delivery systems. New J. Chem. 35, 2366–2374.

Duncan, B., Kim, C., Rotello, V.M., 2010. Gold nanoparticle platforms as drug and biomacromolecule delivery systems. J. Control. Release 148, 122–127.

Fréchet, J.M.J., 1994. Functional polymers and dendrimers: reactivity, molecular architecture, and interfacial energy. Science 263, 1710–1715.

Froehlich, E., 2012. The role of surface charge in cellular uptake and cytotoxicity of medical nanoparticles. Int. J. Nanomed. 7, 5577–5591.

Gao, X.L., Zhao, Y.J., Wang, Z., 2013. Synthesis and performance study of folate-conjugated curcumin prodrug dendrimer derivative. J. Int. Pharm. Res. 40, 359–364.

Gawande, M.B., Zboril, R., Malgras, V., Yamauchi, Y., 2015. Integrated nanocatalysts: a unique class of heterogeneous catalysts. J. Mater. Chem. A 3, 8241–8245.

Ghosh, P.S., Kim, C.K., Han, G., Forbes, N.S., Rotello, V.M., 2008. Efficient gene delivery vectors by tuning the surface charge density of amino acid-functionalized gold nanoparticles. ACS Nano 2, 2213–2218.

Gopidas, K.R., Whitesell, J.K., Fox, M.A., 2003. Metal-core-organic shell dendrimers as unimolecular micelles. J. Am. Chem. Soc. 125, 14168–14180.

Grayson, S.M., Fréchet, J.M.J., 2001. Convergent dendrons and dendrimers: from synthesis to applications. Chem. Rev. 101, 3819–3867.

Hatton, F.L., Chambon, P., Mcdonald, T.O., Owen, A., Rannard, S.P., 2014. Hyperbranched polydendrons: a new controlled macromolecular architecture with self-assembly in water and organic solvents. Chem. Sci. 5, 1844–1853.

Hatton, F.L., Tatham, L.M., Tidbury, L.R., Chambon, P., He, T., Owen, A., Rannard, S.P., 2015. Hyperbranched polydendrons: a new nanomaterials platform with tuneable permeation through model gut epithelium. Chem. Sci. 6, 326–334.

He, X., Wu, X., Cai, X., Lin, S., Xie, M., Zhu, X., Yan, D., 2012. Functionalization of magnetic nanoparticles with dendritic-linear-brush-like triblock copolymers and their drug release properties. Langmuir 28, 11929–11938.

Huang, X., Wu, Z., Gao, W., Chen, Q., Yu, B., 2011. Polyamidoamine dendrimers as potential drug carriers for enhanced aqueous solubility and oral bioavailability of silybin. Drug Dev. Ind. Pharm. 37, 419–427.

Hussain, A.F., Krüger, H.R., Kampmeier, F., Weissbach, T., Licha, K., Kratz, F., Haag, R., Calderón, M., Barth, S., 2013. Targeted delivery of dendritic polyglycerol-doxorubicin conjugates by scFv-SNAP fusion protein suppresses EGFR+ cancer cell growth. Biomacromolecules 14, 2510–2520.

Inapagolla, R., Guru, B.R., Kurtoglu, Y.E., Gao, X., Lieh-Lai, M., Bassett, D.J.P., Kannan, R.M., 2010. In vivo efficacy of dendrimer-methylprednisolone conjugate formulation for the treatment of lung inflammation. Int. J. Pharm. 399, 140–147.

Jain, K., Gupta, U., Jain, N.K., 2014. Dendronized nanoconjugates of lysine and folate for treatment of cancer. Eur. J. Pharm. Biopharm. 87, 500–509.

Jain, K., Verma, A.K., Mishra, P.R., Jain, N.K., 2015. Surface-engineered dendrimeric nanoconjugates for macrophage-targeted delivery of amphotericin B: formulation development and in vitro and in vivo evaluation. Antimicrob. Agents Chemother. 59, 2479–2487.

Jansen, J.F.G.A., Meijer, E.W., De Brabander-Van Den Berg, E.M.M., 1995. The dendritic box: shape-selective liberation of encapsulated guests. J. Am. Chem. Soc. 117, 4417–4418.

Juang, T.Y., Kan, S.J., Chen, Y.Y., Tsai, Y.L., Lin, M.G., Lin, L.L., 2014. Surface-functionalized hyperbranched poly(amido acid) magnetic nanocarriers for covalent immobilization of a bacterial L-glutamyltranspeptidase. Molecules 19, 4997–5012.

Kang, Y., Zhang, X.M., Zhang, S., Ding, L.S., Li, B.J., 2015. PH-responsive dendritic polyrotaxane drug-polymer conjugates forming nanoparticles as efficient drug delivery system for cancer therapy. Polym. Chem. 6, 2098–2107.

Ke, W., Zhao, Y., Huang, R., Jiang, C., Pei, Y., 2008. Enhanced oral bioavailability of doxorubicin in a dendrimer drug delivery system. J. Pharm. Sci. 97, 2208–2216.

Khodadust, R., Mutlu, P., Yalcin, S., Unsoy, G., Gunduz, U., 2013. Polyinosinic:polycytidylic acid loading onto different generations of PAMAM dendrimer-coated magnetic nanoparticles. J. Nanopart. Res. 15, 1860–1871.

Kim, S.T., Chompoosor, A., Yeh, Y.C., Agasti, S.S., Solfiell, D.J., Rotello, V.M., 2012. Dendronized gold nanoparticles for siRNA delivery. Small 8, 3253–3256.

Konda, S.D., Aref, M., Wang, S., Brechbiel, M., Wiener, E.C., 2001. Specific targeting of folate-dendrimer MRI contrast agents to the high affinity folate receptor expressed in ovarian tumor xenografts. Magn. Reson. Mater. Phys. Biol. Med. 12, 104–113.

Krishnakumar, S., Gopidas, K.R., 2014. Organic nanoparticles composed of Fréchet-type dendrons: synthesis, characterization, self-assembly and reversible guest encapsulation. J. Mater. Chem. B 2, 5576–5584.

Kumar, V.K.R., Gopidas, K.R., 2010. Synthesis and characterization of gold-nanoparticle-cored dendrimers stabilized by metal-carbon bonds. Chem. Asian J. 5, 887–896.

Kurniasih, I.N., Keilitz, J., Haag, R., 2015. Dendritic nanocarriers based on hyperbranched polymers. Chem. Soc. Rev. 44, 4145–4164.

Kurtoglu, Y.E., Mishra, M.K., Kannan, S., Kannan, R.M., 2010. Drug release characteristics of PAMAM dendrimer–drug conjugates with different linkers. Int. J. Pharm. 384, 189–194.

Lee, H.I., Lee, J.A., Poon, Z., Hammond, P.T., 2008. Temperature-triggered reversible micellar self-assembly of linear-dendritic block copolymers. Chem. Commun., 3726–3728.

Liu, M., Fréchet, J.M.J., 1999. Designing dendrimers for drug delivery. Pharm. Sci. Technol. Today 2, 393–401.

Liu, J., Pang, Y., Zhu, Z., Wang, D., Li, C., Huang, W., Zhu, X., Yan, D., 2013. Therapeutic nanocarriers with hydrogen peroxide-triggered drug release for cancer treatment. Biomacromolecules 14, 1627–1636.

Liu, L., Li, W., Yan, J., Zhang, A., 2014. Thermoresponsive dendronized polymeric sensors. J. Polym. Sci. A Polym. Chem. 52, 1706–1713.

Lothian-Tomalia, M.K., Hedstrand, D.M., Tomalia, D.A., Padias, A.B., Hall, Jr., H.K., 1997. A contemporary survey of covalent connectivity and complexity. The divergent synthesis of poly(thioether) dendrimers. Amplified, genealogically directed synthesis leading to the de Gennes dense packed state. Tetrahedron 53, 15495–15513.

Love, C.S., Ashworth, I., Brennan, C., Chechik, V., Smith, D.K., 2006. Dendron-protected Au nanoparticles-effect of dendritic structure on chemical stability. J. Colloid Interf. Sci. 302, 178–186.

Ma, P., Zhang, X., Ni, L., Li, J., Zhang, F., Wang, Z., Lian, S., Sun, K., 2015. Targeted delivery of polyamidoamine-paclitaxel conjugate functionalized with anti-human epidermal growth factor receptor 2 trastuzumab. Int. J. Nanomed. 10, 2173–2190.

Maciejewski, M., 1982. Concepts of trapping topologically by shell molecules. J. Macromol. Sci. Chem. A 17, 689–703.

Maeda, H., Bharate, G.Y., Daruwalla, J., 2009. Polymeric drugs for efficient tumor-targeted drug delivery based on EPR-effect. Eur. J. Pharm. Biopharm. 71, 409–419.

Mekelburger, H.B., Jaworek, W., Vögtle, F., 1992. Dendrimers, arborols, and cascade molecules: breakthrough into generations of new materials. Angew. Chem. Int. Ed. 31, 1571–1576.

Menjoge, A.R., Kannan, R.M., Tomalia, D.A., 2010. Dendrimer-based drug and imaging conjugates: design considerations for nanomedical applications. Drug Discov. Today 15, 171–185.

Mishra, V., Jain, N.K., 2014. Acetazolamide encapsulated dendritic nano-architectures for effective glaucoma management in rabbits. Int. J. Pharm. 461, 380–390.

Mukhopadhyay, P., Sarkar, K., Bhattacharya, S., Mishra, R., Kundu, P.P., 2014. Efficient oral insulin delivery by dendronized chitosan: in vitro and in vivo studies. RSC Adv. 4, 43890–43902.

Nance, E., Porambo, M., Zhang, F., Mishra, M.K., Buelow, M., Getzenberg, R., Johnston, M., Kannan, R.M., Fatemi, A., Kannan, S., 2015. Systemic dendrimer-drug treatment of ischemia-induced neonatal white matter injury. J. Control. Release 214, 112–120.

Nazemi, A., Gillies, E.R., 2014. Dendrimersomes with photodegradable membranes for triggered release of hydrophilic and hydrophobic cargo. Chem. Commun. 50, 11122–11125.

Newkome, G.R., Nayak, A., Behera, R.K., Moorefield, C.N., Baker, G.R., 1992. Cascade polymers: synthesis and characterization of four-directional spherical dendritic macromolecules based on adamantane. J. Org. Chem. 57, 358–362.

OcuSeal®, 2015. Available from: http://www.beaver-visitec.com/products/ocuseal-ovd-saline-solutions.cfm

Paez, J.I., Martinelli, M., Brunetti, V., Strumia, M.C., 2012. Dendronization: a useful synthetic strategy to prepare multifunctional materials. Polymers 4, 355–395.

Patel, P.V., Soni, T.G., Thakkar, V.T., Gandhi, T.R., 2013. Nanoparticle as an emerging tool in pulmonary drug delivery system. Micro Nanosyst. 5, 288–302.

Percec, V., Wilson, D.A., Leowanawat, P., Wilson, C.J., Hughes, A.D., Kaucher, M.S., Hammer, D.A., Levine, D.H., Kim, A.J., Bates, F.S., Davis, K.P., Lodge, T.P., Klein, M.L., Devane, R.H., Aqad, E., Rosen, B.M., Argintaru, A.O., Sienkowska, M.J., Rissanen, K., Nummelin, S., Ropponen, J., 2010. Self-assembly of janus dendrimers into uniform dendrimersomes and other complex architectures. Science 328, 1009–1014.

Poon, Z., Lee, J.A., Huang, S., Prevost, R.J., Hammond, P.T., 2011. Highly stable, ligand-clustered "patchy" micelle nanocarriers for systemic tumor targeting. Nanomed. Nanotechnol. Biol. Med. 7, 201–209.

Quintana, A., Raczka, E., Piehler, L., Lee, I., Myc, A., Majoros, I., Patri, A.K., Thomas, T., Mulé, J., Baker, Jr., J.R., 2002. Design and function of a dendrimer-based therapeutic nanodevice targeted to tumor cells through the folate receptor. Pharm. Res. 19, 1310–1316.

Saraswathy, M., Knight, G.T., Pilla, S., Ashton, R.S., Gong, S., 2015. Multifunctional drug nanocarriers formed by cRGD-conjugated βCD-PAMAM-PEG for targeted cancer therapy. Colloid. Surf. B Biointerf. 126, 590–597.

Schluter, A.D., 2003. The macromonomer route to dendronized polymers. Comptes Rendus Chimie 6, 843–851.

She, W., Li, N., Luo, K., Guo, C., Wang, G., Geng, Y., Gu, Z., 2013. Dendronized heparin-doxorubicin conjugate based nanoparticle as pH-responsive drug delivery system for cancer therapy. Biomaterials 34, 2252–2264.

Shukla, S., Wu, G., Chatterjee, M., Yang, W., Sekido, M., Diop, L.A., Müller, R., Sudimack, J.J., Lee, R.J., Barth, R.F., Tjarks, W., 2003. Synthesis and biological evaluation of folate receptor-targeted boronated PAMAM dendrimers as potential agents for neutron capture therapy. Bioconjugate Chem. 14, 158–167.

Soo Choi, H., Liu, W., Misra, P., Tanaka, E., Zimmer, J.P., Itty Ipe, B., Bawendi, M.G., Frangioni, J.V., 2007. Renal clearance of quantum dots. Nat. Biotechnol. 25, 1165–1170.

Su, J., Chen, F., Cryns, V.L., Messersmith, P.B., 2011. Catechol polymers for pH-responsive, targeted drug delivery to cancer cells. J. Am. Chem. Soc. 133, 11850–11853.

Sun, H.-J., Zhang, S., Percec, V., 2015. From structure to function via complex supramolecular dendrimer systems. Chem. Soc. Rev. 44, 3900–3923.

Sunoqrot, S., Bae, J.W., Pearson, R.M., Shyu, K., Liu, Y., Kim, D.H., Hong, S., 2012. Temporal control over cellular targeting through hybridization of folate-targeted dendrimers and PEG-PLA nanoparticles. Biomacromolecules 13, 1223–1230.

Svenson, S., 2015. The dendrimer paradox - high medical expectations but poor clinical translation. Chem. Soc. Rev. 44, 4131–4144.

Svenson, S., Tomalia, D.A., 2005. Dendrimers in biomedical applications - reflections on the field. Adv. Drug Deliv. Rev. 57, 2106–2129.

Thakur, S., Kesharwani, P., Tekade, R.K., Jain, N.K., 2015. Impact of pegylation on biopharmaceutical properties of dendrimers. Polymer 59, 67–92.

Tomalia, D.A., 2005. The dendritic state. Mater. Today 8, 34–46.

Tomalia, D.A., Naylor, A.M., Goddard, III, W.A., 1990. Starbust dendrimers: molecular-level control of size, shape, surface chemistry, topology, and flexibility from atoms to macroscopic matter. Angew. Chem. Int. Ed. 29, 138–175.

Tripathy, S., Das, M.K., 2013. Dendrimers and their applications as novel drug delivery carriers. J. Appl. Pharm. Sci. 3, 142–149.

Tse, C., Zohdy, M.J., Ye, J.Y., O'Donnell, M., Lesniak, W., Balogh, L., 2011. Enhanced optical breakdown in KB cells labeled with folate-targeted silver-dendrimer composite nanodevices. Nanomed. Nanotechnol. Biol. Med. 7, 97–106.

VivaGel®, (2015). Available from: http://www.starpharma.com/vivagel/vivagel_clinical_trials

Vllasaliu, D., Casettari, L., Bonacucina, G., 2013. Folic acid conjugated chitosan nanoparticles for tumor targeting of therapeutic and imaging agents. Pharm. Nanotechnol. 1, 184–203.

Vögtle, F., Richardt, G., Werner, N., 2009. Dendrimer chemistry: concepts, syntheses, properties, application. Wiley-VCH, Weinheim, Germany.

Voit, B.I., Lederer, A., 2009. Hyperbranched and highly branched polymer architectures-synthetic strategies and major characterization aspects. Chem. Rev. 109, 5924–5973.

Welsh, D.J., Smith, D.K., 2011. Comparing dendritic and self-assembly strategies to multivalency - RGD peptide-integrin interactions. Org. Biomol. Chem. 9, 4795–4801.

Wiener, E.C., Konda, S., Shadron, A., Brechbiel, M., Gansow, O., 1997. Targeting dendrimer-chelates to tumors and tumor cells expressing the high-affinity folate receptor. Invest. Radiol. 32, 748–754.

Winnicka, K., Wroblewska, M., Wieczorek, P., Sacha, P.T., Tryniszewska, E., 2012. Hydrogel of ketoconazole and PAMAM dendrimers: formulation and antifungal activity. Molecules 17, 4612–4624.

Wurm, F., Frey, H. 2012. Hyperbranched Polymers: Synthetic Methodology, Properties, and Complex Polymer Architectures. Matyjaszewski, K., Möller, M., (Eds.), Polymer science: a comprehensive reference, 10 volume set. Elsevier Science, Netherlands.

Xiao, K., Luo, J., Li, Y., Lee, J.S., Fung, G., Lam, K.S., 2011. PEG-oligocholic acid telodendrimer micelles for the targeted delivery of doxorubicin to B-cell lymphoma. J. Control. Release 155, 272–281.

Xie, J., Lee, S., Chen, X., 2010. Nanoparticle-based theranostic agents. Adv. Drug Deliv. Rev. 62, 1064–1079.

Xie, J., Liu, G., Eden, H.S., Ai, H., Chen, X., 2011. Surface-engineered magnetic nanoparticle platforms for cancer imaging and therapy. Acc. Chem. Res. 44, 883–892.

Xu, X., Zhang, Y., Shen, Y., 2011. Targeting Hela cancer cells with avidin-dendrimer. He Jishu/Nucl. Techn. 34, 293–298.

Xu, X., Jian, Y., Li, Y., Zhang, X., Tu, Z., Gu, Z., 2014. Bio-inspired supramolecular hybrid dendrimers self-assembled from low-generation peptide dendrons for highly efficient gene delivery and biological tracking. ACS Nano 8, 9255–9264.

Yiyun, C., Na, M., Tongwen, X., Rongqiang, F., Xueyuan, W., Xiaomin, W., Longping, W., 2007. Transdermal delivery of nonsteroidal anti-inflammatory drugs mediated by polyamidoamine (PAMAM) dendrimers. J. Pharm. Sci. 96, 595–602.

Yuan, A., Yang, B., Wu, J., Hu, Y., Ming, X., 2015. Dendritic nanoconjugates of photosensitizer for targeted photodynamic therapy. Acta Biomater. 21, 63–73.

Zarrabi, A., Vossoughi, M., 2015. Paclitaxel/β-CD-*g*-PG inclusion complex: an insight into complexation thermodynamics and guest solubility. J. Mol. Liquids 208, 145–150.

Zhang, Y., Xu, X., Wang, L., Lin, J., Zhu, Y., Guo, Z., Sun, Y., Wang, H., Zhao, Y., Tai, R., Yu, X., Fan, C., Huang, Q., 2013. Dendrimer-folate-copper conjugates as bioprobes for synchrotron X-ray fluorescence imaging. Chem. Commun. 49, 10388–10390.

Zhang, S., Sun, H.-J., Hughes, A.D., Moussodia, R.-O., Bertin, A., Chen, Y., Pochan, D.J., Heiney, P.A., Klein, M.L., Percec, V., 2014. Self-assembly of amphiphilic Janus dendrimers into uniform onion-like dendrimersomes with predictable size and number of bilayers. Proceedings of the National Academy of Sciences, 111, 9058–9063.

Zhang, H., Ma, Y., Xie, Y., An, Y., Huang, Y., Zhu, Z., Yang, C.J., 2015. A controllable aptamer-based self-assembled DNA dendrimer for high affinity targeting, bioimaging and drug delivery. Sci. Rep., 5.

Zhao, Y., Liu, S., Li, Y., Jiang, W., Chang, Y., Pan, S., Fang, X., Wang, Y.A., Wang, J., 2010. Synthesis and grafting of folate-PEG-PAMAM conjugates onto quantum dots for selective targeting of folate-receptor-positive tumor cells. J. Colloid Interf. Sci. 350, 44–50.

Zhou, Y., Li, J., Lu, F., Deng, J., Zhang, J., Fang, P., Peng, X., Zhou, S.F., 2015. A study on the hemocompatibility of dendronized chitosan derivatives in red blood cells. Drug Design Dev. Ther. 9, 2635–2645.

Zhu, S., Qian, L., Hong, M., Zhang, L., Pei, Y., Jiang, Y., 2011. RGD-modified PEG-PAMAM-DOX conjugate: in vitro and in vivo targeting to both tumor neovascular endothelial cells and tumor cells. Adv. Mater. 23, H84–H89.

Zhu, J., Fu, F., Xiong, Z., Shen, M., Shi, X., 2015. Dendrimer-entrapped gold nanoparticles modified with RGD peptide and alpha-tocopheryl succinate enable targeted theranostics of cancer cells. Colloid. Surf. B Biointerf. 133, 36–42.

CHAPTER

GOLD NANOSTRUCTURES: PREPARATION, PROPERTIES, APPLICATION IN SERS, AND BIOPHOTONICS

17

Petar A. Atanasov

Bulgarian Academy of Sciences, Sofia, Bulgaria

CHAPTER OUTLINE

1 INTRODUCTION

The synthesis, study and application of nanoparticles (NPs) and nanostructures (NSs) of different compositions and characteristics expressed great interest because of its novel and unexpected unique properties, compared to these of the bulk materials (Edelstein and Cammarata, 1996; Gudiksen et al., 2002; Noguez, 2007; Shi et al., 2010). As a result of the progress in nanotechnologies in the past two decades, nanosystems find nowadays applications in many areas, such as chemistry, optics, biology, medicine, microelectronics, etc.

Recently, the development of reliable and low cost methods for fabrication of metal NPs, NP arrays, and NSs becomes an important branch of nanotechnology (Amoruso et al., 2005a; Atanasov et al., 2014; Edelstein and Cammarata, 1996; Guowei, 2012). The interest is driven by the unique properties of the metal structures having nanosized dimensions. The most interesting characteristics of the metal NPs are their optical properties. They are governed by the excitation of collective oscillations of the free electrons (plasmons) into the structures. These oscillations are characterized by a resonant wavelength, which falls usually in the ultraviolet (UV) or visible (VIS) spectral range for the most of metals. Especially for the noble metals, that is, gold (Au) or silver (Ag), and copper (Cu), the resonance wavelength is in the near UV and VIS spectral ranges, where most of the commercial lasers operate, and makes them suitable for efficient applications (Kreibig and Vollmer, 1995; Noguez, 2007).

The excitation of plasmons in the metal NSs results in significant enhancement of their extinction coefficient (Tanaka et al., 2010). Thus, the noble metallic NSs are used in development of systems for imaging, diagnostics, therapy and catalysis (Jain et al., 2007; Shi et al., 2010). Specifically, Au NPs have attracted intensive interest because of peculiarities of their absorption spectra in the region of the surface plasmon resonance and the specific and very important applications in nanobiotechnology.

Different chemical and physical techniques are developed for creation of NPs and NSs. The chemical methods are wet technologies, which make use of precursors and NPs produced are inside a solution. Generally, they are more productive for creation of nanomaterials compared to the physical methods. However, they are sensitive with respect to composition, crystallinity and especially presence of contamination traces (Masala and Seshadri, 2004).

The use of lasers to build up nanomaterials and NSs—one of the most employed physical methods—has a continuously growing interest. In particular, pulsed laser ablation of solid targets in vacuum, background gas or liquid, is a relatively easy and commonly used method for generating of NPs or build-up of NPs-assembled materials (Besner and Meunier, 2010). For instance, laser ablation via ns-pulses in ambient gas can be used for quenching the ablated plume. Thus, a homogeneous nucleation of the ablated species occurs. This technique is found to be a rather versatile for synthesis of NPs. However, several drawbacks are presented, connected with the optimization of the background gas type and its pressure, as well as the complexity of the expansion of the plume.

Recently, laser ablation of solid or thin film targets in liquids is also found as a simple and effective method for the preparation of colloidal NPs suspensions (Guowei, 2012; Nikolov et al., 2011; Nikov et al., 2012, 2013a; Perez et al., 2008; Zhang et al., 2009). The type and characteristics of the surrounding solvent significantly influence on the NPs growth and their properties. In some cases surfactants are added in order to prevent the aggregation of the produced NPs.

The excitation spectrum of noble metallic subwavelength structures is determined by its surface plasmon resonances. The energies of plasmon resonances depend strongly on the shape and composition of the nanostructure. The tunability of the plasmon resonances of noble metallic NPs can be exploited to position the optical resonances at specific wavelength regions of interest, and lead to a wide range of applications. The strong local electromagnetic field enhancement accompanied with the surface plasmon resonances is also used to manipulate light-matter interactions (Tanaka et al., 2010; Terakawa et al., 2012; and references quoted in these works), and noble metallic subwavelength structures are widely applied in surface-enhanced Raman spectroscopy (SERS). In addition to their fundamental importance, plasmonic nanostructures are receiving a great deal of attention for their potential applications in areas, such as subwavelength waveguides, optical nanoantennas, photovoltaic technology for efficient light coupling into solar cells, metamaterials, chemical, and biological sensing, medicine.

The electromagnetic field in the near field zone of metal NSs possesses specific features, which are attributed to the localization and enhancement of the field intensity. It is demonstrated that the field can be enhanced by few orders of magnitude compared to the incident one, and has an evanescent behavior, as it decreases rapidly with the distance from the metal surface. Hence, the metal NSs give the possibility of localization of enhanced electromagnetic fields in a zone whose spatial characteristics are defined by the structure size, but not by the incident wavelength (Kawata, 2001; Tanaka et al., 2010). This property breaks the diffraction limit of the far field optics, and is used in development of subwavelength optical devices and applications.

The great enhancement of the electromagnetic field intensity in close vicinity of the NPs surface is the basis of one of the most developed applications of the noble metallic nanostructures: SERS (Baker and Moore, 2005; Moskovits, 2005). It is a technique used for detection and analysis of broad range of materials as drugs, explosives, soil, water and food pollutions, viruses and bacteria, and DNA (Fan et al., 2011). The high sensitivity of SERS allows detection of even single molecule (Kneipp et al., 1997). The unique characteristics of the near field around the metal NPs are also used for enhancing the conversion efficiency in the photovoltaic devices (Pillai et al., 2007; Schaadt et al., 2005). However, the strong dependences of the near field properties on the parameters of the NS (size and shape of the particles, interparticle distance), incident irradiation and environment, often cause difficulties in fabrication of reliable and reproducible systems for such applications.

Fabrication of NP arrays with desirable parameters on different substrates is still a significant challenge. The fabrication of metal NSs by electron and ion beam lithography is usually slow and highly cost processes, where the chemistry-based methods require multisteps and toxic reagents. For some applications as plasmonic solar cells and bionanochips fabrication, where NP monolayer on conductive electrodes are required this will leads to additional steps of deposition. Furthermore, the need of low cost, reliable, quick, and easy fabrication of metal nanostructures on different substrates is still not completely satisfied.

The interaction of intense, nanosecond (ns) or femtosecond (fs) laser pulses with solid-state bulk targets or thin films is very interesting from the point of view of the variety of possibilities offered both for basic science and technology. Specifically, the use of fs-laser pulses opens up the opportunity to decouple the initial photoexcitation of the target from the subsequent dynamics of the removed material, that is, plume, in contrast to the case of longer ones: ns-pulses. In this case the pulse width is less than the typical electron-lattice thermalization time (typically ≈ few picoseconds) and finally, the irradiated material is driven into an extremely excited state followed by a rapid quenching, resulting in a material blow-off composed of fast ions, atoms, clusters, and nanoaggregates of the target material. Fs-laser ablation of metals and semiconductors in a vacuum environment is investigated extensively. Recent research has in particular focused on the formation mechanism of NPs created by the irradiation of a solid metal or thin film targets with intense fs-laser pulses of different duration in vacuum (Amoruso et al., 2007a,b, 2011, 2014; Eliezer et al., 2004; Perez and Lewis, 2002).

In the following, article studies and results are presented on the development of several methods for generating Au NPs and NPs arrays, and some examples of their application, that is, in SERS and biophotonics. Ns and fs-laser ablation of bulk Au targets or thin films in vacuum is studied and used, respectively, to manipulate the production of NSs on different substrates: dielectrics, metal, ceramics, plastics. Moreover, results on laser annealing of single or bimetal thin films for fabrication of alloyed nanostructures is also presented. Tuning of the optical properties of Au NSs, fabricated on flexible plastics substrates, is also shown. The properties of the structures produced are experimentally characterized by a variety of analytical methods, such as fast photography, optical emission spectroscopy, FE SEM, AFM, TEM, Raman spectroscopy. Moreover, theoretical methods as molecular dynamic (MD) simulation, finite difference time domain (FDTD), and methods based on generalized multiparticle Mie theory (GMMT) are applied, and the results are compared with the experimental ones. Two possible applications of the Au NPs or NSs are envisaged, that is, in SERS and biophotonics: photothermal cancer cell therapy, and plasmonic nanometric optical tweezers.

2 EXPERIMENTAL

Three laser systems are used in the experiments: excimer, solid-state ns- and fs-lasers. XeCl laser system ($\lambda = 308$ nm) delivers pulses with duration of $\tau_p = 30$ ns, and repetition rate of 5 Hz. The second experimental set up uses a ns-Nd:YAG laser operating on basic (1064 nm), SHG (532 nm), THG (355 nm), or FHG (266 nm), respectively, with $\tau_p = 15$ ns, and repetition rate of 10 Hz. The fs-laser system consisted of a Nd:Glass laser lasing at $\lambda = 1055$ nm ($\tau_p \approx 0.9$ ps and repetition rate of 33 Hz), SHG of $\lambda = 527$ nm, and FHG of $\lambda = 263$ nm ($\tau_p \approx 300$ fs), respectively.

The morphology of the deposited NPs or NSs is analyzed by AFM or SEM analyses, and the shape of the NPs is viewed by TEM. XRD and EDS analyses are performed to check the elemental properties of the NPs. The experimental study of the plume and its evolution in time in vacuum conditions are described in details by Amoruso et al. (2007a,b, 2011, 2014) and Nedyalkov et al. (2007a). Fast photography is also applied to follow the vacuum expansion dynamics of the ablated species.

3 THEORETICAL

3.1 MOLECULAR DYNAMIC SIMULATIONS

The interaction of the fs-laser pulses with Au solid targets and its decomposition is described by the classical MD technique described by Allen and Tildesley (1987) together with two-temperature (2T) modeling. Details are given by Amoruso et al. (2005b, 2007a,b) and Nedyalkov et al. (2007a). Since the electrons are not included explicitly in the MD model, their role, and the heating of the material, are taken into account by using a one-dimensional, two-temperature diffusion model (1D-TTM) described by Anisimov et al., (1974a, 1974b) and Amoruso et al. (2014). The temperature into the material predicted by the 2T model is implemented in the MD by a scaling of the atoms' velocities. As a result, the MD simulations predict plume propagation and size distribution of the generated NPs, in the case of the Au solid and thin film targets. The time evolution of the system consisting of n Au atoms, which interacts with a potential U is described by the set of equations of motion:

$$m_i \frac{\partial^2 r_i}{\partial t^2} = -\nabla_i U(r_1, r_2, \cdots r_N) \tag{17.1}$$

where m_i, mass of the Au atom and $F_i = -\nabla_i U(r_1, r_2, \ldots r_n)$, force.

The initial positions of the Au atoms arranged in a cubic face-centered crystal lattice, at initial temperature, the atom velocities, and the interatomic potential are the input parameters.

The interaction between the atoms is governed by the Morse potential, and the specific parameters for Au are taken from Girifalco and Weizer (1959), and Mohammed et al. (1984). The algorithm proposed by Allen and Tildesley (1987) is used to integrate the equation of motion. In order to simulate an infinite material in lateral directions, periodic boundary conditions are imposed in the *x–y* directions. Specific boundary conditions are applied on the bottom of the computational cell in order to prevent the reflection of shock waves (Zhigilei and Garrison, 1999a,b). The absorption of laser energy and redistribution of the heat into the metal are governed initially by the free electrons.

By solving the equation of motion (17.1), one can obtain the trajectories of the atoms, the temporal and spatial evolution and distribution of the energy, temperature, and pressure of the system. In this way, different processes as phase changes, deformations (shockwaves), material growth, and realization of different ablation mechanisms can be investigated.

3.2 FINITE DIFFERENCE TIME DOMAIN (FDTD) METHOD

The electromagnetic field distribution and the optical near field enhancement in the vicinity of Au NPs, when they interact with electromagnetic field, are calculated by FDTD method (Amoruso et al., 2014; Atanasov et al., 2014; Hirano et al., 2013; Nedyalkov et al., 2007a; Tanaka et al., 2010). The interaction of light with the metal nanoobjects is simulated by FDTD approach according to Taflove and Hagness (2000), Sullivan (2000), and Taflove (2005). It has recently become a start-of-the-art method for solving Maxwell's equations in complex geometries and inhomogeneous systems, consisting of variety of materials and environments. Being a direct time and space solution, it offers the user a unique insight into all types of problems in electromagnetism and photonics. FDTD algorithm is used when feature sizes, that is, NPs, are on the order of a wavelength.

The simulated system is divided into elementary cells, where the electric and magnetic field components are calculated at each time step by the scheme of Taflove and Hagness (2000). The dielectric function of the Au particles is described by Drude model, as the input parameters are taken from Johnson and Christy (1972) and Vial et al. (2005). Simulations are made for a system that consists of Au NPs placed on different substrates. The structure of the NP array is taken from its SEM images. The incident irradiation is a plane wave at specific wavelength of 785 or 800 nm. The dielectric function of the substrate is taken from Palik (1998), and the electric field intensity (an input parameter for FDTD simulation) is assumed to be 1 V/m^2.

3.3 GENERALIZED MULTIPARTICLE MIE MODELING

Near-field properties around metallic particles can basically be explained by Mie scattering theory (Liu et al., 2005). Mie theory, also called Lorenz-Mie theory, or Lorenz-Mie-Debye theory, is an analytic solution of Maxwell's equations for the scattering of the electromagnetic radiation by spherical particles (also called Mie scattering). Mie theory provides rigorous analytic solutions for light scattering by an isotropic sphere embedded in a homogeneous medium (Xu and Khlebtsov, 2003; Xu, 2003). The optical properties of two-dminesional (2D) or three-dimensional (3D) Au NPs arrays are studied in the far-field zone on the basis of the GMM model using FORTRAN codes

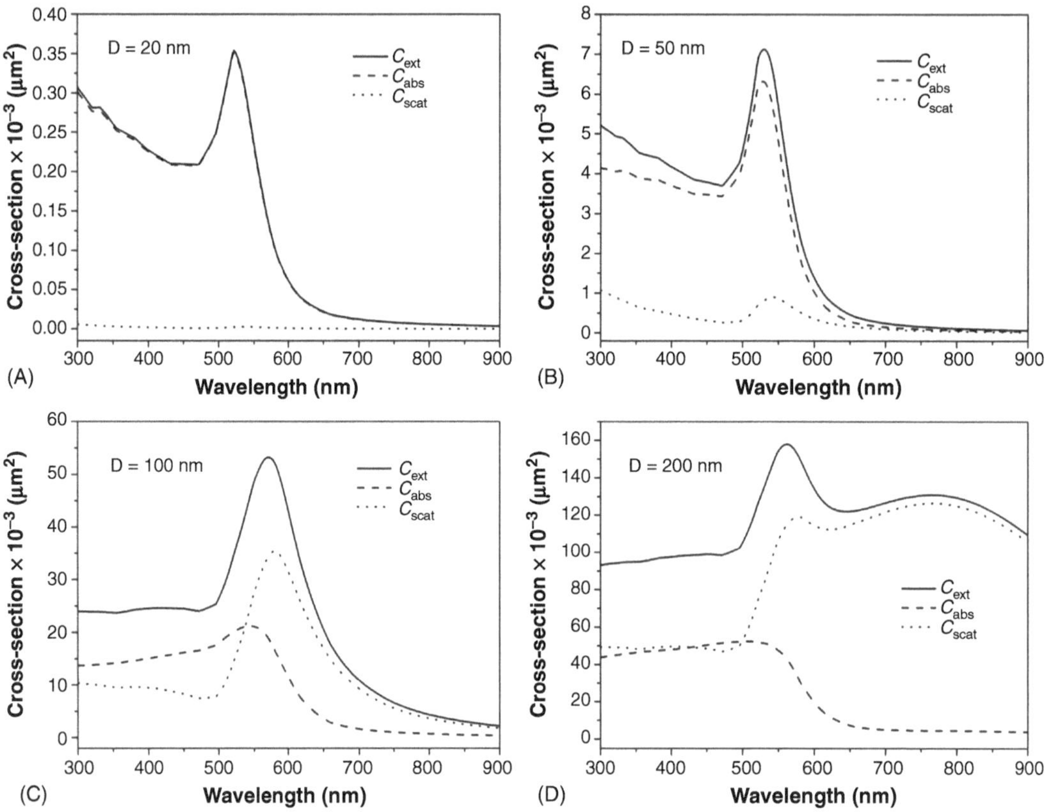

FIGURE 17.1 The Extinction C_{ext}, Scattering C_{scat}, and Absorption C_{abs} Cross-Sections' Spectra of Au NPs

Diameters of: (A) 20 nm, (B) 50 nm, (C) 100 nm, and (D) 200 nm.

available on the web Xu, Y.-L.-ftp://astro.ufl.edu/pub/xu/gmm01ff. The code is modified in order to calculate extinction (scattering and absorption) cross-sections and particles' efficiency in the UV-VIS range (Dimitrov et al., 2010). The input data are the coordinates of the each particle, its radius, and the complex refractive index for the bulk Au, taken from Johnson and Christy (1972). All calculations are performed in water or in air at 20°C as surrounding medium. The results indicate that the optical properties depend strongly on the irradiation conditions (light polarization), array structure (configuration, number of Au NPs, and interparticle distance), dielectric properties of the surrounding media, and particle characteristics.

Fig. 17.1 reports the dependence of the calculated optical extinction C_{ext}, absorption C_{abs}, and scattering C_{scat} cross-sections of spherical Au NPs with diameters 20, 50, 100, and 200 nm as a function of the wavelength. As seen, the scattering increases significantly with the NPs diameter contrary to the absorption, which has opposite tendency.

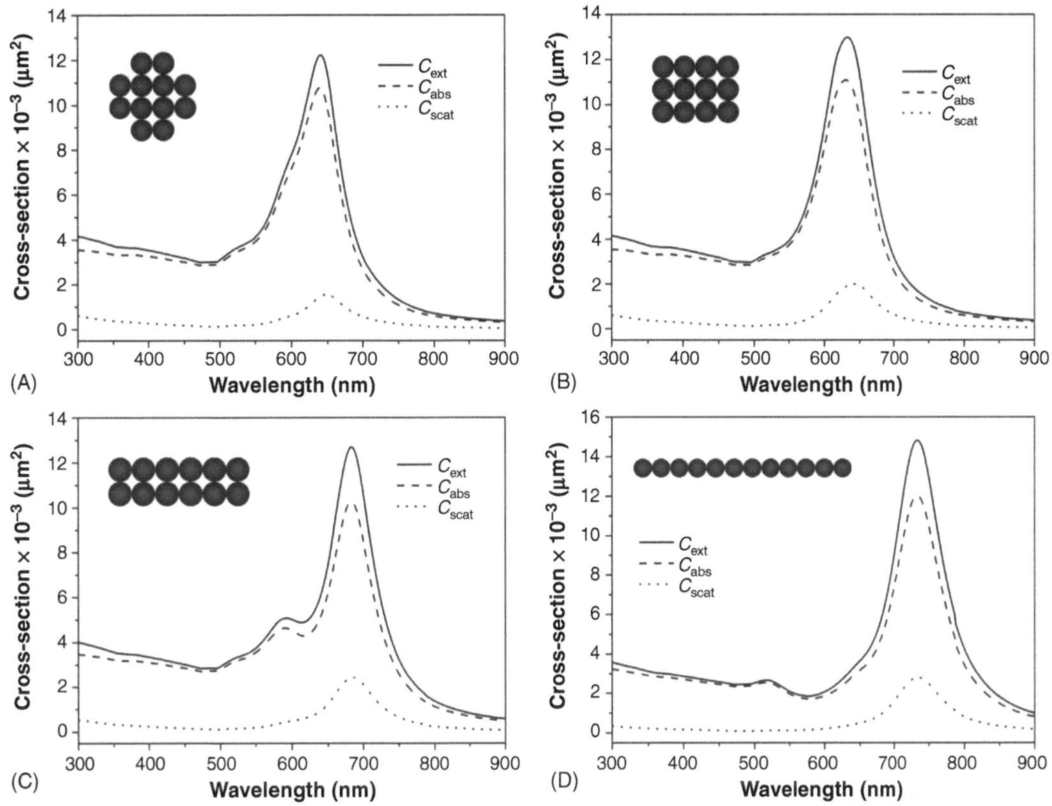

FIGURE 17.2 The Extinction C_{ext}, Scattering C_{scat}, and Absorption C_{abs} Cross-Sections' Spectra of Different 2D Configurations of 12 Au NPs with Diameter of 20 nm

It is worth noting that second peak appears in the scattering and the extinction spectra (Fig. 17.1D). The dipole resonance is much wider, and shifted toward longer wavelengths (the maximum is around 770 nm), when the quadruple resonance has a maximum at around 570 nm.

Fig. 17.2 depicts C_{ext}, C_{abs}, and C_{scat} cross-sections' spectra of different combination of 12 Au NPs with diameter of 20 nm. As seen, the cross-sections depend quite significantly on the arrangement of the particles; the maximums are red, shifted when NPs are linearly arranged.

3D clusters are composed by different number and combination of 20 nm diameter Au NPs and Fig. 17.3 presents the spectra of C_{scat} and C_{abs} of different 3D clusters. As it can be seen, the only one peak of scattering (Fig. 17.3A) is expressed, when in case of the absorption, two peaks are evident with different positions depending on the configuration of the Au NPs.

The results reported are given as examples and the method developed is used in defining optimal properties of the NPs arrays used in photothermal cancer therapy and bioimaging.

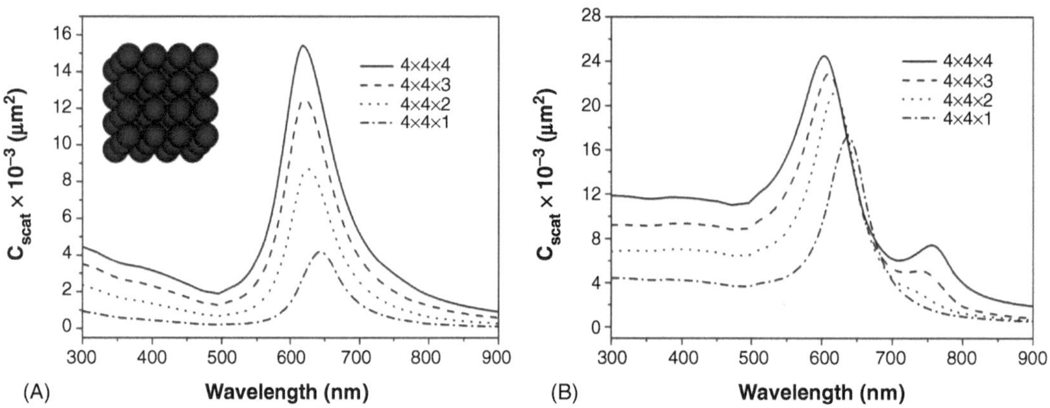

FIGURE 17.3 The scattering Cscat (A) and absorption Cabs (B) Coss-Sections' spectra of different 3D configurations of various numbers of Au NPs with diameter of 20 nm

3.4 LOCAL HEATING OF Au NPs BY Ns- AND Fs-LASER PULSES

As mentioned before, the electromagnetic field distribution and the optical near-field enhancement in the vicinity of NPs, when they interact with the electromagnetic field, can be calculated by FDTD method.

In order to estimate the temperature of the Au NP, when irradiated by ns-laser pulses, a model combining the heat diffusion equation and NP optical properties calculation is applied. In this model, the temperature evolution of a particle and the surrounding medium (assumed water) during and after the interaction with the laser pulse is traced by 2D heat diffusion model (Nedyalkov et al., 2010a,b; Pustovalov, 2005):

$$C(T)\frac{\partial T}{\partial t} = \nabla(k(T)T) + S \tag{17.2}$$

where $S = IC_{abs}/V_p$ and S and I are the laser energy source and laser intensity, respectively. For example, the maximal temperature in Au NPs with diameters of 80 and 200 nm as a function of the applied ns-laser fluence is presented in Fig. 17.4.

The temperature evolution of the particle during and after the interaction with the fs-laser pulse is described by 1D 2T model:

$$C_e\frac{\partial T_e}{\partial t} = -\gamma(T_e - T_i) + \frac{IC_{abs}}{V_p} \tag{17.3}$$

$$C_i\frac{\partial T_i}{\partial t} = \gamma(T_e - T_i) \tag{17.4}$$

where T_e and T_i, temperatures of the electron and lattice systems, respectively; I, laser intensity and $V_p = 4/3\pi R^3$ is the NP volume. C_{abs} is the absorption cross-section of the particle, C_i and C_e, are the lattice and electron system heat capacities, and γ is the electron-lattice coupling constant, respectively. The spatial distribution in the particle heating is neglected and it is assumed to be homogeneous.

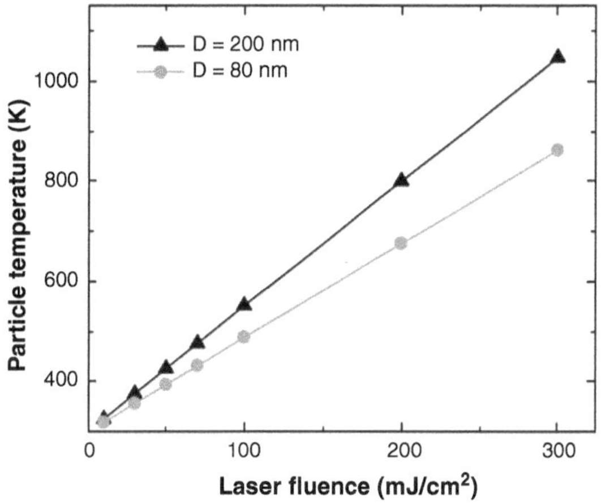

FIGURE 17.4 Dependence of the Maximal Temperature Reached in Nanoparticles with Diameters of 40 and 100 nm on the Incident Laser Fluence

Ns-laser pulse with duration of 15 ns at wavelength of 532 nm is considered.

Reprinted from Nedyalkov N.N., Imamova, S.E., Atanasov, P.A., Obara, M., 2010a. Gold nanoparticles as nanoheaters and nanolenses in the processing of different substrate surfaces. J. Phys. Conf. Series 223, 012035.

The system of Eqs. (17.3) and (17.4) is solved using a classical finite difference scheme, and taking into account the temperature dependence of the electron heat capacity, according to the relation $C_e = A_e T_e$, where $A_e = 70\ \mathrm{J/m^3K^2}$, and the temperature dependence of C_i (Chowdhury and Xu, 2003). The value of the coupling coefficient used is $\gamma = 3 \times 10^{16}\ \mathrm{W/m^3K}$.

Since the optical properties of nanoparticles, generally depend on their properties, the absorption cross-section C_{abs} is obtained on the base of Mie scattering theory using the equation:

$$C_{abs} = C_{ext} - C_{sca} \tag{17.5}$$

where the extinction C_{ext} and scattering C_{sca} cross-sections are taken from Papavassiliou (1979). The optical parameters for Au used in the calculation are taken from Johnson and Christy (1972).

For example, the maximal temperature in Au NPs as a function of the applied laser fluence is presented in Fig. 17.5.

As it is clearly expected from the comparison in both cases that no matter of difference in Au NPs diameters (Figs. 17.4 and 17.5), the maximal temperatures and its growth rate with fluence are much bigger in case of ns-laser pulse irradiation compared to this for fs-laser pulse.

4 METHODS FOR PREPARATION OF GOLD NANOPARTICLES AND NANOSTRUCTURES. APPLICATION IN SERS

As mentioned, SERS is one of the most advanced techniques for the detection of molecules at extremely low concentrations (Kneipp et al., 1997; Xu et al., 1999), and thus has very high potential for chemical and biosensing applications (Shanmukh et al., 2006). In order to develop a SERS-based sensor, an appropriate and efficient substrate is required to ensure high enhancement factors, on the one

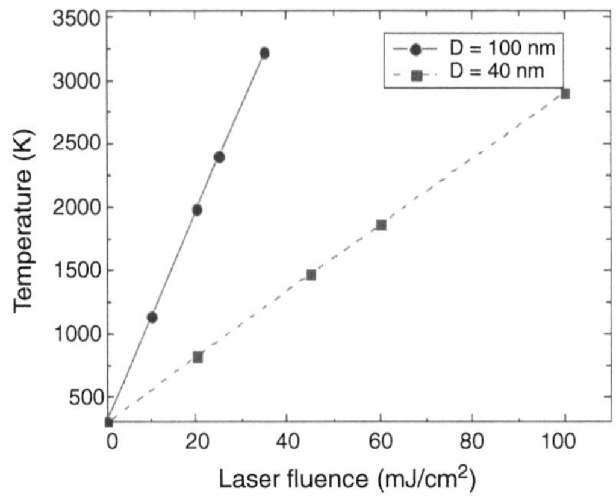

FIGURE 17.5 Calculated Dependence of the Maximal Temperature Reached in Au NPs with Diameter of 80 and 200 nm as a Function of the Applied Laser Fluence

Fs-laser pulse with duration of 100 fs at wavelength of 800 nm is considered.

Reprinted from Nedyalkov N.N., Imamova, S.E., Atanasov, P.A., Obara, M., 2010b. Gold nanoparticles as nanoheaters and nanolenses in the processing of different substrate surfaces. J. Phys. Conf. Series 223, 012035.

hand, and which is also to be stable, easy, and relatively inexpensive, on the other (Beermann et al., 2009; Deniz and Lad, 2011; Suzuki et al., 2007).

4.1 Ns-PULSED LASER DEPOSITION OF Au THIN FILMS AND LASER NANOSTRUCTURING

Au films are produced by pulsed laser deposition (PLD) technique on quartz, with XeCl excimer laser (Imamova et al., 2010). The films are deposited by laser fluence of 1.5 J/cm^2, in ambient pressure of 4×10^{-3} Pa, and at room or 250°C temperature of the substrate. The estimated deposition rate is ~15 nm/min, and the film thicknesses vary with the deposition time. The same excimer laser radiation is used for annealing at the same environmental conditions and laser fluencies between 100 and 280 mJ/cm^2 and 20 subsequent pulses. Micro Raman system operated at 785 nm is used to characterize the SERS properties of the fabricated Au NSs. SERS measurements are carried out for Rodamine 6G (R6G) dissolved in ethanol.

The sequence of SEM images on Fig. 17.6 presents these surfaces' nanomodification.

As seen, the annealed surfaces consist of nanosized spherical particles, as the particles' diameters depend on the incident laser fluence and deposition conditions. Moreover, its diameter is proportional to the film thicknesses. These parameters can be used as an efficient factor to control particle density and size distribution. The larger particles are produced at the lowest fluence applied (see Fig. 17.6B). The fluence 100 mJ/cm^2 is found to be the threshold for thin film modification, and its increase causes fragmentation of the droplets to NPs with smaller sizes (Fig. 17.6B and D). Moreover, the NPs produced at room temperature (Fig. 17.6B) are spherical with sizes in the range of 10–40 nm, while these deposited at 250°C are nonspherical, with sizes in the range of 20–60 nm (Fig. 17.6D). The difference in the observed structures can be related to the different degree of crystallinity, influenced by the substrate temperature.

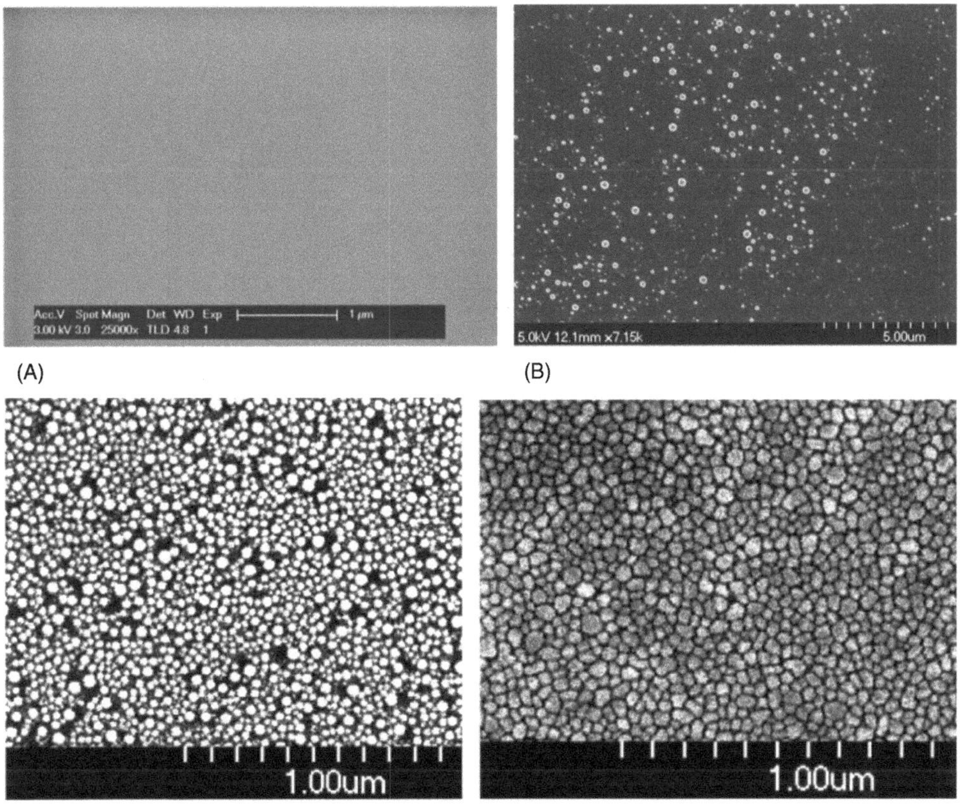

(C) (D)

FIGURE 17.6 SEM Images of Annealed ~60 nm Au Films

(A) As-deposited at room temperature; (B) after annealing with 20 pulses and fluence of 100 mJ/cm^2; (C) after annealing with 20 pulses and fluence of 130 mJ/cm^2; (D) film deposited at 250°C and annealed with 20 pulses at fluence of 130 mJ/cm^2.

Reprinted in part from Imamova, S., Dikovska, A., Nedyalkov, N., Atanasov, P., Sawczak, M., Jendrzejewski, R., Śliwinski, G., Obara, M., 2010. Laser nanostructuring of thin Au films for application in surface enhanced Raman spectroscopy. J. Optoelectron. Adv. Mater. 12, 500–504.

Fig. 17.7 reports Raman spectra of R6G deposited on quartz, as in line (NR), and nanostructured substrates produced at different conditions, as in lines (a) and (b). A direct comparison between spectra of the NR and SERS spectra (a) and (b) on Fig. 17.6 shows the magnification factors of the enhanced signal.

As is seen from Fig. 17.8, Au NSs having junctions between NPs are producing larger enhancement of the electromagnetic field, which confirms the comparison of the Raman signals in Fig. 17.7A and B. Higher SERS enhancement is related to the higher particle density and existence of junctions between Au NPs, that is, these structures are better for SERS applications according to Zhang and Noguez (2008). Finally, the estimated enhancement factor obtained in this study is in the range of 10^6–10^7.

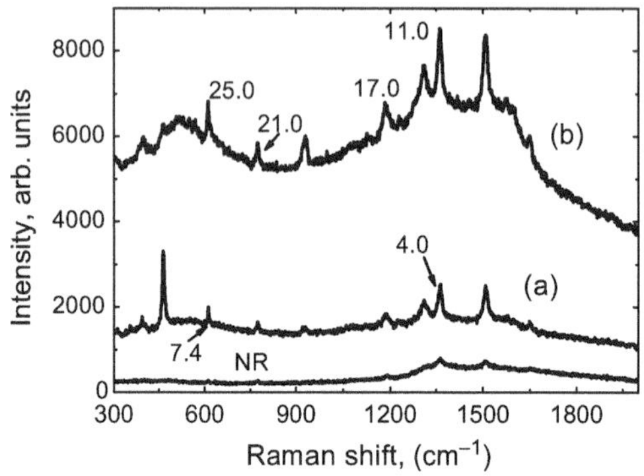

FIGURE 17.7 A Comparison Between the SERS Spectra of R6G

As-deposited on: quartz substrate, line *NR*; structure presented in Fig. 17.6C, line *(a)*; structure given in Fig. 17.6D, line *(b)*.

Reprinted from Imamova, S., Dikovska, A., Nedyalkov, N., Atanasov, P., Sawczak, M., Jendrzejewski, R., Śliwinski, G., Obara, M., 2010. Laser nanostructuring of thin Au films for application in surface enhanced Raman spectroscopy. J. Optoelectron. Adv. Mater. 12, 500–504.

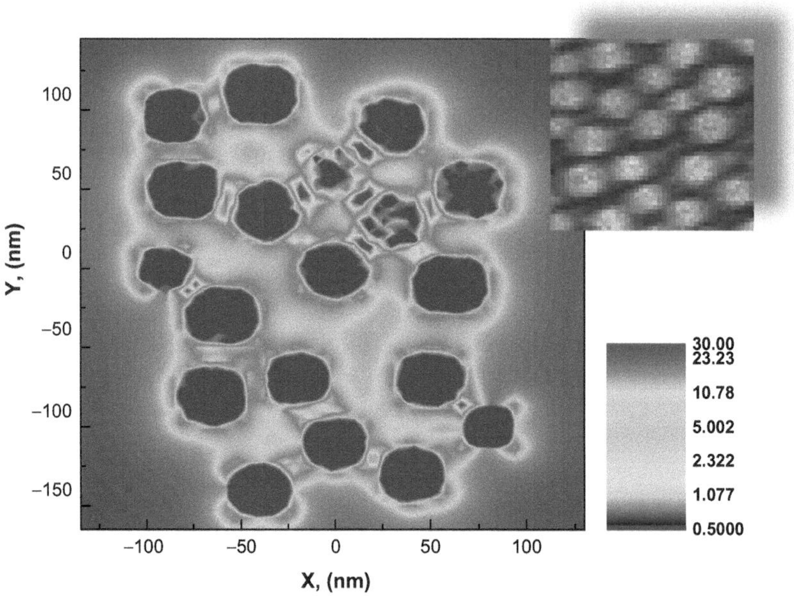

FIGURE 17.8 Near-Field Intensity Distribution in the Vicinity of Au NP Array, Calculated by FDTD

The insert presents SEM image of the experimentally produced structure, used for construction of the simulated system. The area contains 19 spherical Au NPs with sizes between 30 and 40 nm, arranged on quartz substrate at distances of 5–20 nm. The color scale indicates the ratio between the obtained and the incident field intensity.

Reprinted from Imamova, S., Dikovska, A., Nedyalkov, N., Atanasov, P., Sawczak, M., Jendrzejewski, R., Śliwinski, G., Obara, M., 2010. Laser nanostructuring of thin Au films for application in surface enhanced Raman spectroscopy. J. Optoelectron. Adv. Mater. 12, 500–504.

4.2 BIMETAL NPs AND SERS APPLICATION

In this section, the effort is directed toward description of fabrication and study of the properties of Au-based alloys. Such NSs possess some extended new characteristics, compared to the single Au particle (Major et al., 2009; Salem et al., 2003; Sun et al., 2000). Special interest is paid to the noble metal/noble metal and noble metal/ferromagnetic NPs, which may possibly be applied in catalysis, biophotonics, and magneto-optics. Au/Ni, Ag/Ni, and Ag/Co NPs have significant combined plasmon and magnetic properties (Gaudry et al., 2003). In the case of Au/Ag NP alloy, by changing the ratio between both metals, the plasmon resonance wavelength can be tuned precisely. Link et al. (1999) reported on a linear dependence of the plasmon absorption maximum position on the composition of such nanoparticles. This effect can increase the efficiency of the SERS and the effect of such structures in photothermal therapy of cancer cells or for bioimaging.

THG of the ns-Nd:YAG laser (λ = 355 nm) is applied in order to produce bimetal (Au/Ag and Au/Ni) NPs in vacuum. The concentrations of the both metals into the deposited thin films are controlled by changing the areas of the sections on the targets (Nikov et al., 2013b). The fluence during deposition of the films is 1.5 J/cm^2 and the annealing is made at the energy density of 64 or 132 mJ/cm^2 and different number of the applied laser pulses.

As it is clear from Fig. 17.9, the laser parameters do not affect notably the morphology of the NSs, nor their optical properties.

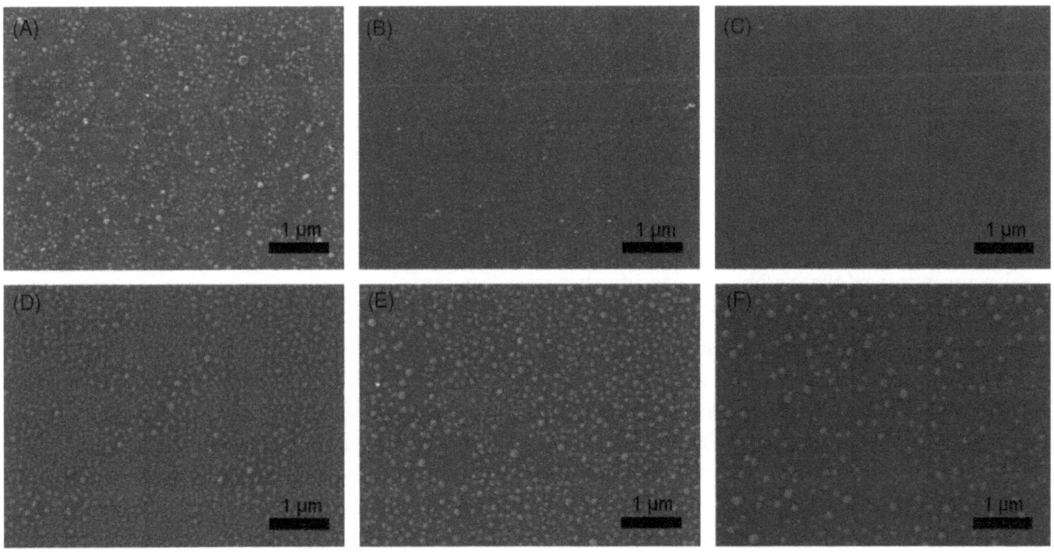

FIGURE 17.9 SEM Images of Au/Ag (Upper Sequence) and Au/Ni (Down Sequence) NSs

Fabricated at: fluence 64 mJ/cm^2 and 5 pulses (A, D); fluence 132 mJ/cm^2 and 5 pulses (B, E); fluence 132 mJ/cm and 20 pulses (C, F). Au/Ag and Au/Ni systems have same ratio for both metals.

Reprinted from Nikov Ru.G., Nedyalkov N.N., Atanasov P.A., Grochowska K., Iwulska A., Sliwinski G., 2013b. Laser nanostructuring of Au/Ag and Au/Ni films for application in SERS. Proc. SPIE 8770, 87700D, 1–7.

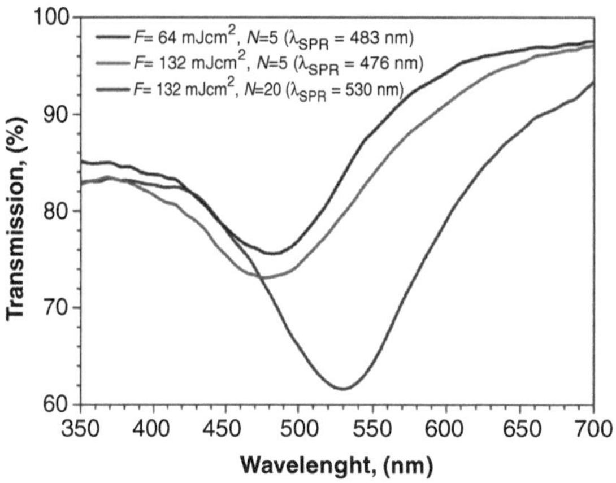

FIGURE 17.10 Transmission Spectra of the Au/Ag NSs Given in Fig. 17.6A–C

The wavelength of the surface plasmon resonance (SPR) in case of the respective nanostructure is indicated in the brackets.

Reprinted from Nikov Ru.G., Nedyalkov N.N., Atanasov P.A., Grochowska K., Iwulska A., Sliwinski G., 2013b. Laser nanostructuring of Au/Ag and Au/Ni films for application in SERS. Proc. SPIE 8770, 87700D, 1–7.

The optical transmission spectra of the Au/Ag nanostructures are shown in Fig. 17.10.

The position of surface plasmon resonance (SPR) is red shifted when higher laser fluence is applied for annealing, and the increase of the number of the laser pulses cause and increase the extinction. However, there is no significant influence of the laser fluence on the optical properties at equal number of pulses. In addition, the characteristics of the structures do not express significant influence on their composition.

EDX analysis reveals that the structures of as-deposited thin film is preserved into the laser annealed structures, that is, the fabricated NSs are composed by bimetallic NPs. It is well known that Au/Ag alloy phase is stable, and it exists in solid or liquid state. On the contrary, equilibrium phase diagram of the Au–Ni system shows a miscibility gap with a maximal temperature of 1083 K at 70.6 At% Ni (Wittig and Liu, 2005), and below this temperature the alloy phase is not stable. Hence, the precise formation of the Au/Ni NSs need further investigations.

The samples produced are tested with R6G (Fig. 17.11). Fig. 17.11A shows SERS spectra of R6G deposit on the Au/Ag films, nanostructured by 20 subsequent laser pulses with fluence of 132 mJ/cm^2.

The enhancement factor of the samples is estimated and compared to the reference signal from R6G, deposited on the film. In all cases the concentration of the analyte is the same. The greatest enhancement (enhancement factor up to 20) of the most characteristic Raman peaks is observed in the NS of the highest Ag concentration (Au/Ag = 40/60). Here, the enhancement of the SERS signal is due to the strong localization of the near field intensity as a result of formation of nonregularities of position of the NPs. Similar feature and enhancement by the same order of magnitude is already demonstrated by Fan et al. (2011). It is worth noting that resonance wavelength of 521 nm in the Raman spectrum of the NS having highest Ag concentration is very near to the excitation wavelength of the Raman spectrometer (514 nm). The SERS spectra of R6G, deposited on Au/Ni NSs with different compositions are shown in Fig. 17.11B. It is evident that the enhancement of this structure is weaker compared to the Au/Ag NSs, produced at the same conditions and laser parameters. The largest enhancement factors of 6.5–9 for the

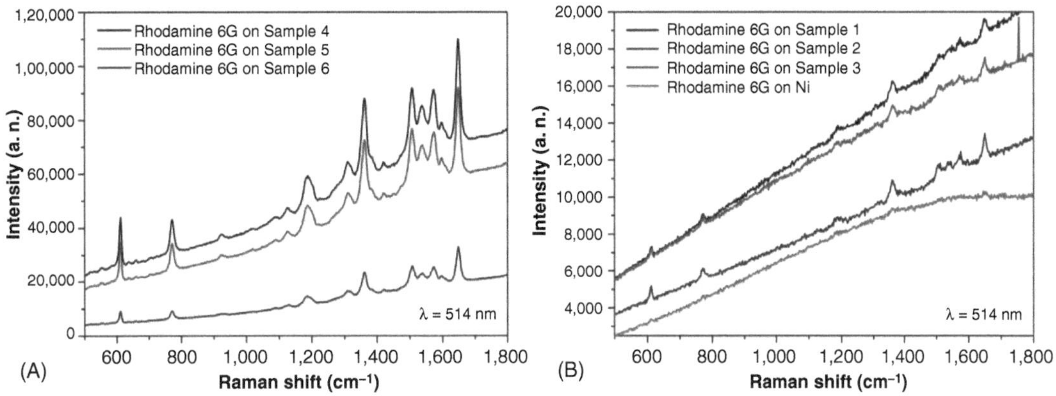

FIGURE 17.11 SERS Spectra of R6G, Adsorbed on NSs with Different Structure Composition

(A) Au/Ag samples: 4, 40/60; 5, 50/50; and 6, 75/25; (B) Au/Ni samples: 1, 50/50; 2, 80/20, and 3, 60/40. Excitation by 514 nm radiation is used in all measurements.

Reprinted from Nikov Ru.G., Nedyalkov N.N., Atanasov P.A., Grochowska K., Iwulska A., Sliwinski G., 2013b. Laser nanostructuring of Au/Ag and Au/Ni films for application in SERS. Proc. SPIE 8770, 87700D, 1–7.

main peaks in the Raman spectra are obtained in the case of Au/Ni NS with the highest Au concentration (Au/Ni = 80/20). The much weaker SERS signal is produced in the case of a pure Ni NS (Fig. 17.11B).

The properties of the Au or bimetal NSs depend on the production environmental conditions (Fig. 17.12). The structuring is accomplished in air or in vacuum. SEM images of the Au/Ni (75/25) films, annealed in air and in vacuum by 10 pulses at fluence of 470 mJ/cm^2, are presented in Fig. 17.12A and B, respectively.

When annealing is made in air (Fig. 17.12A), the thin Au film is transformed to closely situated NPs, where in case of vacuum (Fig. 17.12B) relatively larger droplets (100–300 nm) are formed, and some of them are connected by thin Au nanowires. The differences in both structures are probably related to the different surface energies of the metal/environment and substrate/environment boundaries. Moreover, in case of annealing in air, the interaction of laser radiation may cause an oxidation and change of the balance of the surface energy at the boundaries.

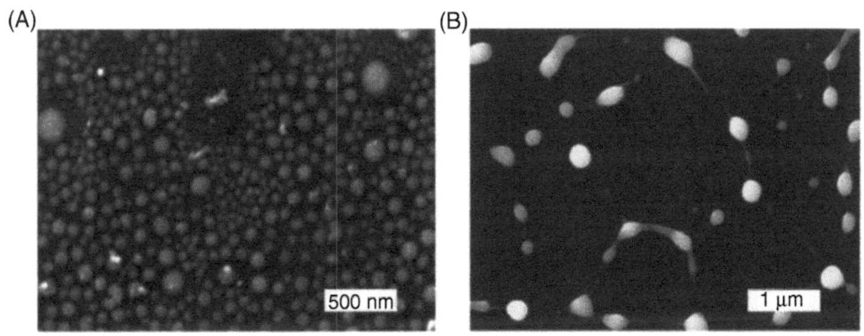

FIGURE 17.12 SEM images of the Au/Ni (75/25) films, annealed in air (A) and in vacuum (B) the annealing is produced by 10 pulses at fluence of 470 mJ/cm^2.

4.3 DEPOSITION ON DIELECTRICS, METALS, AND POLYMERS

A simple and efficient method is proposed for fabrication of Au NPs on Al_2O_3 ceramic surface (Nedyalkov et al., 2014) or metal substrate (Nikov et al., 2015). In case of metal substrates, Al and Ta are used. The method is based on laser-assisted dewetting of Au film deposited on the surface of the substrates. The ceramic material being composed of crystallites has a porous structure with large surface area. In case of pure Al, a film of oxide is formed at room temperature and, subsequently, the situation is similar to that of the ceramics. The homogeneous cover of NPs on such surface, combined with the high mechanical strength, the chemical inertness, and the low cost, makes the proposed structure an efficient candidate in catalysis applications and sensor devices. In the case of Ta, no clear explanation for the results can be given at this stage.

In all cases, the Au films are deposited by PLD method using 355 nm pulses delivered from ns-Nd:YAG laser. The deposition of films lasts 2 or 3 min, respectively, by equal laser fluence of 1.5 J/cm^2. Depending on the deposition time, the thickness of the films is ~60 or 80 nm, respectively. Laser annealing—the second step of the technology—is accomplished with the same laser wavelength at fluences between 0.06 and 0.3 J/cm^2 and by one or five pulses.

The Au NPs are well separated and distributed over all alumina crystallite surfaces (Nikov et al., 2015). SEM images also reveal that the particles' characteristics and distribution do not depend on the crystalline surface orientation, with respect to the laser beam direction (perpendicular to the ceramic sample). The histograms indicate that when single pulse irradiation is applied, the NPs have a broad size distribution ranged from 35 to 265 nm, with a standard deviation of ~45 nm with respect to the mean particles size. In the case of five laser pulses annealing, 96% of the counted particles are in the size range from 30 to 75 nm, with a standard deviation of about three times lower than when single pulse irradiation is applied. The uncompleted decomposition of the Au thin film observed at low fluence and single pulse irradiation results in a complex structure that ensures formation of large number of irregularities, that is, hot spots, where the near-field intensity is strongly enhanced. Such structure expresses a high efficiency for SERS application. R6G dissolved in methanol at concentration of 200 nM is used for checking the SERS properties of such structures (Fig. 17.13) (Nedyalkov et al., 2014).

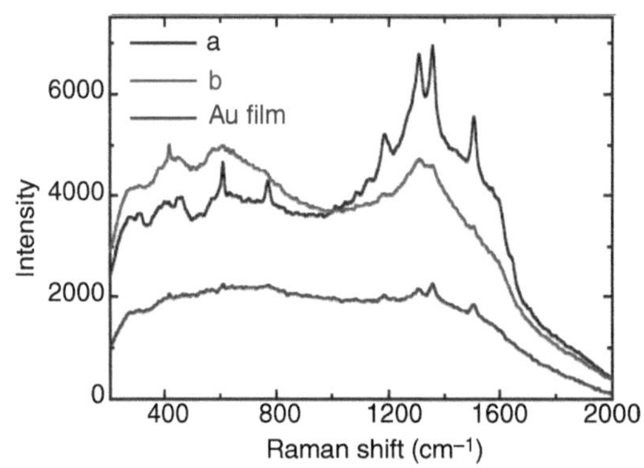

FIGURE 17.13 Raman Spectra of R6G Deposited on Different Structures on Alumina Ceramic Substrate

The black spectra is obtained using the structure produced by 1 pulse annealing, the red produced by 5 pulses, and the blue shows spectrum from unprocessed Au film on alumina. Annealing is produces by irradiation with fluence of 0.3 J/cm^2.

Reprinted from Nedyalkov, N., Nikov, Ru., Koleva, M., Atanasov, P.A., Constantinescu, C., Delaporte, Ph., Grojo, D., 2014. Nanoparticle-decorated ceramic as substrate in surface enhanced Raman spectroscopy. Appl. Surf. Sci. 336, 16–20.

The SEM images and size distributions of the Au NPs fabricated on Al and Ta are depicted in Fig. 17.14 (Nikov et al., 2015).

When annealing of Au film deposited on Ta, raising the laser fluence from 130 to 210 mJ/cm^2 leads to an increase of the mean size of the Au NPs from 44 to 53 nm, whereas in the case of Au NPs on Al substrate, a decrease of the mean particle size (from 44 to 34 nm) and the width of the size distribution (from 17 to 8 nm) occur. The optical extinction spectra of the Au NSs possess surface plasmon resonance, as the position of the resonance wavelength depends on the processing conditions via the characteristics of the formed structures. The estimation on the basis of FDTD simulation shows that the intensity of the near-field of the produced NSs is enhanced about three orders of magnitude compared to that of the incident one.

The obtained enhancement is verified experimentally as the fabricated substrates are used for SERS analysis (Fig. 17.15).

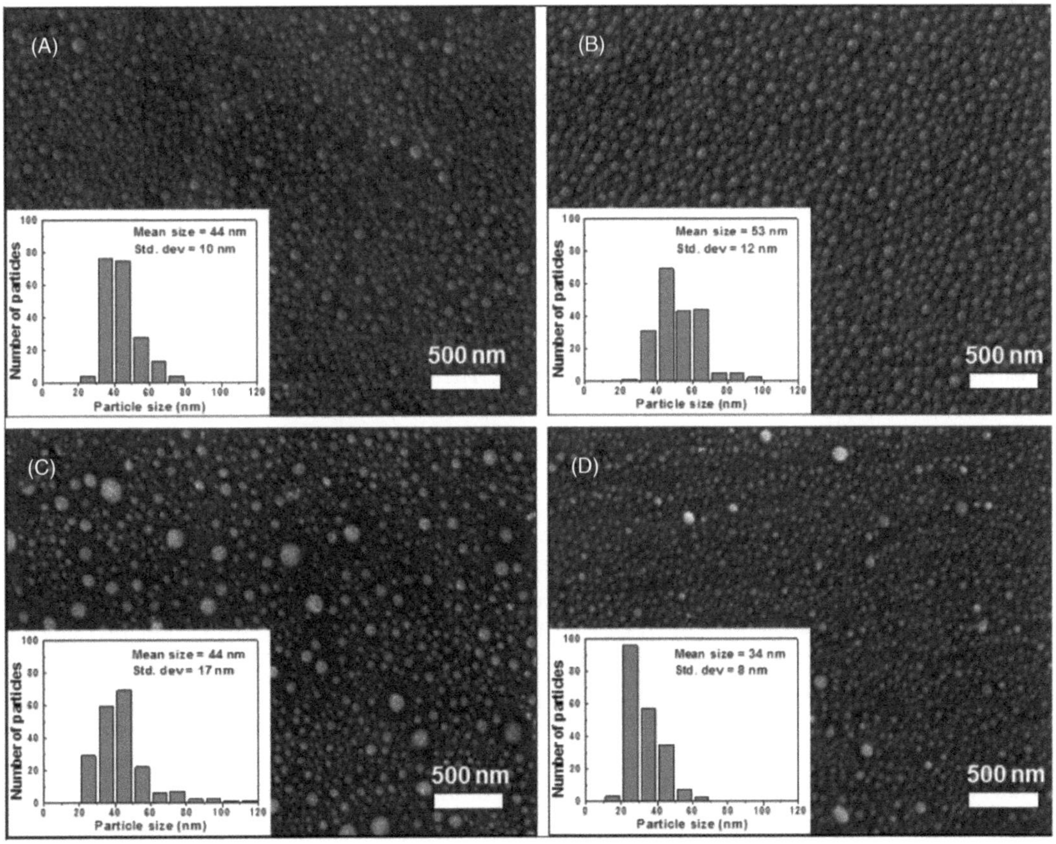

FIGURE 17.14 SEM Images of Laser Processed Au Films (60 nm thickness)

Films on: Ta (A) and (B); Al (C) and (D). The films are irradiated by 5 pulses at fluence of 0.13 J/cm^2 (A) and (C); 0.21 J/cm^2 (B) and (D). The insets indicate particle size distributions.

Reprinted from Nikov, Ru.G., Nedyalkov, N.N., Atanasov, P.A., Delaporte, Ph., Grojo D., 2015. Fabrication and characterization of metal nanostructures on metal substrates. Proc. SPIE, 9447, 94470K1–94470K7.

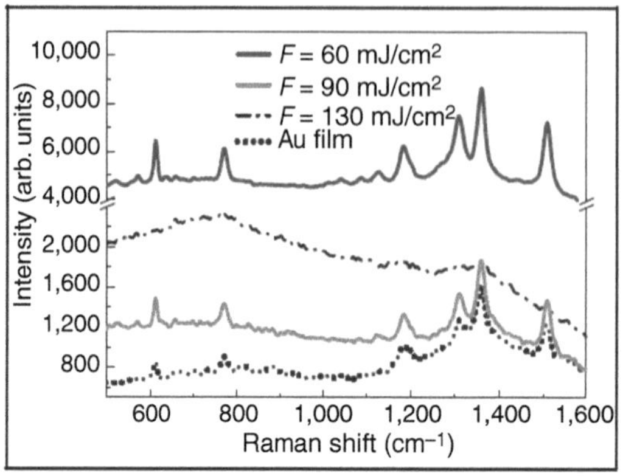

FIGURE 17.15 Raman Spectra of 10 μM R6G Adsorbed by Au NSs Fabricated by on Ta Substrate 5 Laser Pulses at Different Fluence

The Raman signal from the same quantity of R6G absorbed on as-deposited Au film is presented for comparison.

Reprinted from Nikov, Ru.G., Nedyalkov, N.N., Atanasov, P.A., Delaporte, Ph., Grojo D., 2015. Fabrication and characterization of metal nanostructures on metal substrates. Proc. SPIE, 9447, 94470K1–94470K7.

As seen, the most efficient structure, with respect to the value of the intensity of the Raman peaks, is that produced by laser fluence of 0.06 mJ/cm^2. This structure is composed by closely separated NPs with interparticle distance of few nanometers that ensures maximal enhancement of the near-field intensity (Moskovits, 2005).

The already described two-step technology is successfully applied to produce Au NPs on polymeric substrate (Atanasov et al., 2014; Nikov et al., 2013c). The challenge here is connected with the low degradation's temperature of the most of such materials, so no heating can be applied.

100 nm Au thin films are deposited on PVC and PMMA substrates in vacuum by a 355 nm ns-Nd:YAG laser. The energy density for ablation of solid Au target is 1.5 J/cm^2. The as-deposited films are annealed by different numbers of laser pulses (1, 2, 5, or 10) in the fluence range from 0.05 to 0.17 J/cm^2. Fig. 17.16 depicts SEM images of the Au NSs fabricated on PMMA and PVC substrates, respectively (Nikov et al., 2013c).

The optical extinction spectra of Au NSs produced on PMMA as function of annealing fluence and numbers of pulses are shown in Fig. 17.17A and B, respectively. In all cases, the plasmon resonance is clearly expressed, as its position is shifted toward longer wavelengths with increasing of the annealing fluence and do not depend on the number of laser pulses applied.

Fig. 17.18 reports Raman spectra of R6G spread on Au thin film deposited on PVC, and Au NSs produced after pulsed laser annealing, respectively. It is clearly seen that the Raman spectrum caused by Au NSs have quite strong peaks, compared to these of the nonprocessed Au film where peaks characterized R6G cannot be distinguished.

An excellent advantage of the Au NSs produced on polymeric flexible substrates is possibility to tuning smoothly of the plasmon position by mechanical bending of the substrate (Atanasov et al., 2014; Nikov et al., 2013c). It is found that by bending the structure at 1.5 mm, a smooth change of the position of the plasmon resonance by about 30 nm is achieved.

FIGURE 17.16 SEM Images of Au Nanostructures Fabricated at F = 0.09 mJ/cm² with Single L Pulse on Different Substrates

(A) PMMA and (B) PVC. The insert in (A) shows the native surface of the gold film on PMMA after deposition and in (B), more detailed view of the fabricated structure.

Reprinted from Nikov, Ru., Nedyalkov, N., Atanasov, P.A., Terakawa, M., Shimizu, H., Obara, M., 2013b. Tuning the optical properties of gold nanostructures fabricated on flexible substrates. Appl. Surf. Sci. 264, 779–782.

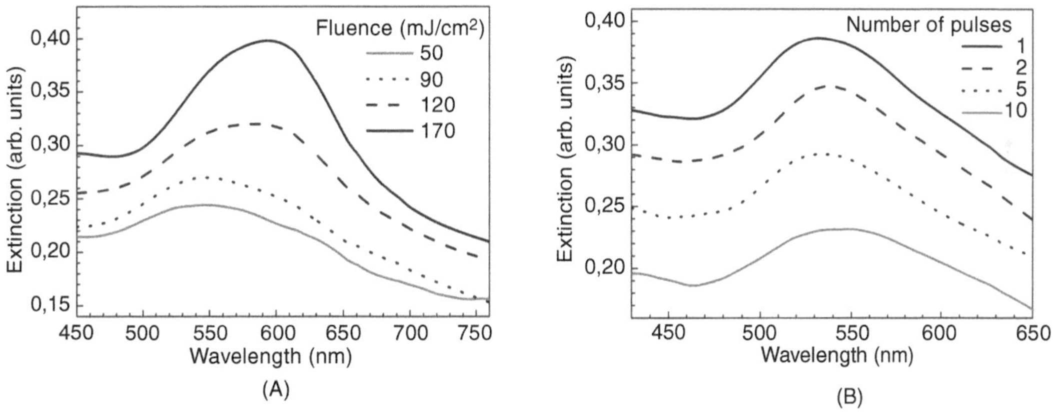

FIGURE 17.17 Optical Extinction Spectra of Au NSs Produced on PMMA

(A) Laser annealing of the Au films with thickness of ~100 nm by 1 pulse and different laser fluences and (B) dependence on number of pulses at 90 mJ/cm^2.

4.4 PREPARATION OF Au NANORODS FOR SERS APPLICATION

Pulsed laser deposition (PLD) technique at glancing angle (GLAD) modification is applied by Dikovska et al. (2013) and Atanasov et al. (2014) in order to produce Au substrate with a controllable and desired morphology for SERS application. During the GLAD process, the angle between the axis of the Au plume and the substrate surface is varied between ≈5 and 15 degree. The main mechanisms controlling the growth of the structure are the shadowing effect combined with the surface diffusion of the species

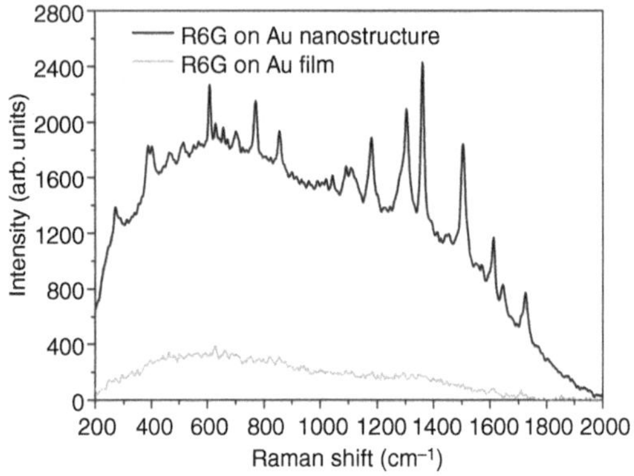

FIGURE 17.18 Comparison of the Raman Spectra of R6G on Au Thin Film (Gray Curve) and Au NSs (Black Curve) Deposited on PVC Substrate

Reprinted from Nikov, Ru., Nedyalkov, N., Atanasov, P.A., Terakawa, M., Shimizu, H., Obara, M., 2013b. Tuning the optical properties of gold nanostructures fabricated on flexible substrates. Appl. Surf. Sci. 264, 779–782.

over the substrate. By changing the incident angle, laser fluence, substrate to target distance, and/or substrate rotation speed, the size, shape, and arrangement of the metal nanorods can be tailored.

The THG of a ns-Nd:YAG laser ($\lambda = 355$ nm) is used for ablation of Au solid target with fluence of ~1–3 J/cm^2. The substrate—quartz—is situated at a distance of 4 or 5 cm from the target. The experiments are carried out at room or 100°C substrate temperatures in vacuum and varying the deposition time. The Au vapor will be deposited onto the shadow of the nucleus created by the first particles reaching substrate and a columnar structure develops. AFM image of Au nanorod arrays is depicted in Fig. 17.19. In this case, the Au nanorods are aligned parallel to each other and tilted at 60–70° with respect to the normal of the substrate's surface. The average length of the nanorods is 50–70 nm and increases monotonically with deposition time.

It is worth noting that Au nanorods with shorter length are also observed, probably due to the competition between neighboring nanorods. The measured diameter of the Au nanorods is in the range of 10–20 nm, and the average gap between two adjacent rods is estimated to be ~33 nm, resulting in a density of approximately 640 rods/μm^2. This is higher by a factor of 20–50 to these reported by other methods (Shanmukh et al., 2006). At shorter target/substrate distance, that is, 4 cm, the average gap between the nanorods further decreases, and on further shortening of the distance, the neighboring nanorods are joining together.

The as-produced Au nanorod arrays are analyzed by XRD technique (Fig. 17.20). It indicates that Au nanorods are polycrystalline, with average crystallite sizes of ~8 nm. The face-centered cubic lattice has parameter $a = 4.072(8)$ Å, which practically coincides with the usually reported value of $a = 4.072$ Å (ICDD Entry # 00-065-8601). The XRD pole (the inset in Fig. 17.20) reveals that the Au nanorods grow at direction with respect to the substrate whose plane is parallel to the plane of the SiO_2 substrate. This means that the growth direction of nanorods is probably along the same lines as the laser plasma direction.

SERS measurements of the produced structures are applied by using 1×10^6 R6G molecules spread and dried over an area of 1×1 cm^2 (~10^8 μm^2). SERS spectrum of R6G on Au nanorods substrate

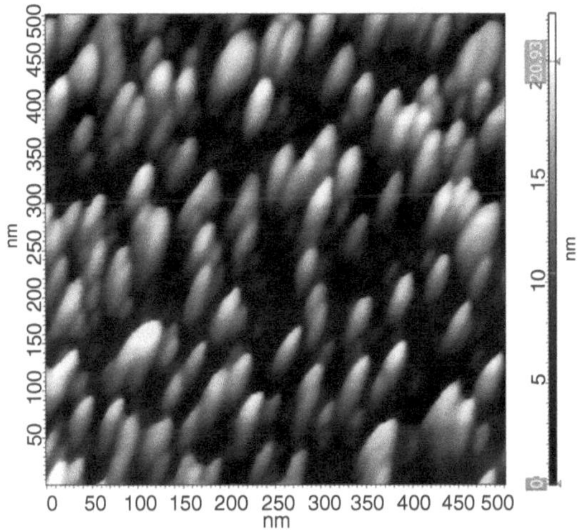

FIGURE 17.19 AFM Image of Au Nanorods, Prepared by GLAD Technique

The experimental conditions are: laser fluence 1 J/cm^2; target/substrate distance 5 cm; room temperature and in vacuum.

Reprinted from Dikovska, A.Og., Avdeev, G.V., Nedyalkov, N.N., Koleva, M.E., Atanasov, P.A., 2013. Preparation of metal nanorods substrates for SERS application. Proc. SPIE, 8770, 877006–877011.

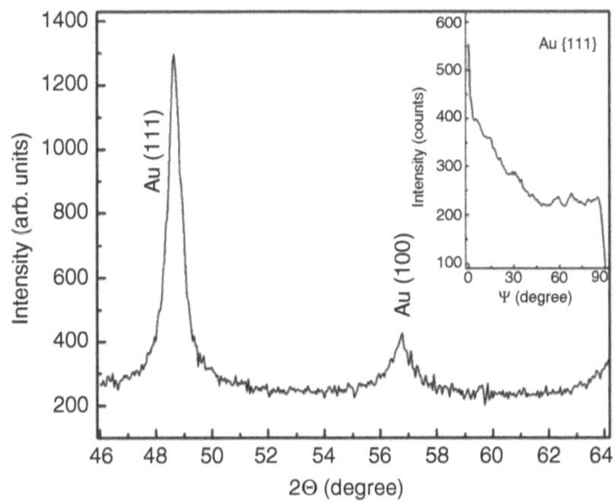

FIGURE 17.20 XRD Spectrum of Au Nanorod Arrays

Rods prepared at the same experimental conditions as in Fig. 13.19. The inset indicates XRD pole figure of Au nanorods.

Reprinted from Dikovska, A.Og., Avdeev, G.V., Nedyalkov, N.N., Koleva, M.E., Atanasov, P.A., 2013. Preparation of metal nanorods substrates for SERS application. Proc. SPIE, 8770, 877006–877011.

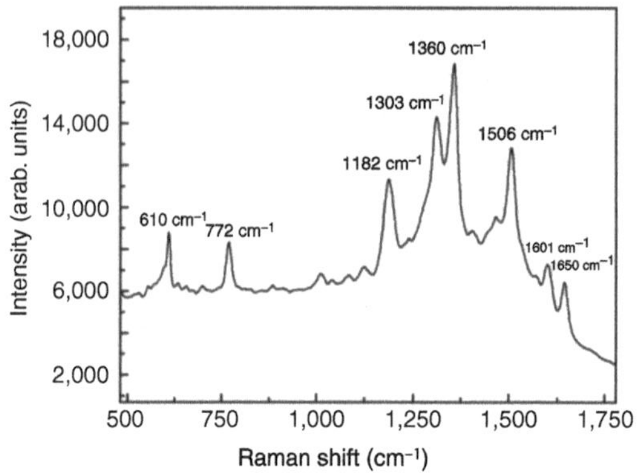

FIGURE 17.21 SERS Spectrum of Au Nanorode Substrate Fabricated at Laser Fluence of 3 J/cm^2

Reprinted from Dikovska, A.Og., Avdeev, G.V., Nedyalkov, N.N., Koleva, M.E., Atanasov, P.A., 2013. Preparation of metal nanorods substrates for SERS application. Proc. SPIE, 8770, 877006–877011.

prepared at laser fluence of 3 J/cm^2 is presented in Fig. 17.21. All indicated strong peaks are attributed to the Raman spectrum of the R6G molecules. At the same time, no signal is registered from nonprocessed Au thin film.

The enhancement factor of the Au nanostructures is evaluated to be up to 10^5, depending on the experimental conditions. The SERS enhancement factor strongly depends on the aspect ratio and the spatial arrangement of the nanorods. Moreover, the intensity of the Raman signal increases with deposition time.

4.5 GOLD NPs PRODUCED BY Fs-LASER PULSE ABLATION OF THIN FILMS

Fs-laser pulse ablation in vacuum is a viable route for the direct synthesis of NPs (Amoruso et al., 2005a; Eliezer et al., 2004; Perrière et al., 2007), and is attractive for several reasons: (1) the interaction between the laser pulses having very short durations and the plume (ablated material) is negligible, contrary to the case of the longer pulse durations; (2) the vacuum environment allows avoiding the complications by the presence of a background gas. Along these lines, when the laser pulse duration is much shorter than electron-phonon coupling time (typically ≈10 ps for most of the metals), the target material undergoes a fast quasi-isochoric heating, and thus becomes of almost solid-density condition, with a temperature and pressure above the critical point. The subsequent relaxation from this extremely excited state occurs through mechanisms of fragmentation, homogeneous nucleation and vaporization, which take place simultaneously at different ratios (Lewis and Perez, 2010).

The exact mechanisms of ablation and the size properties of the different species composing the plume depend on the local deposited energy density and the initial thermodynamic state of the specific layer in depth of the target from which they originate (Amoruso et al., 2007a; Lewis and Perez, 2010; Perez and Lewis, 2002). The local temperature inside the irradiated material can become, in some points, higher than several thousand Kelvin, and decreases exponentially in depth of the target—the deeper the layer is in the material, the lower its initial temperature. Consequently, the NPs of different sizes are generated by the different internal layers into the target, which becomes parts of the plume, and then they fly ahead at different velocities, which depend on their sizes—the smaller, the faster. Such behavior is confirmed by various experimental results (Amoruso et al., 2007c). Hence, by influencing the temperature profile in the depth of the target, one can control of the size and properties of the produced NPs.

In the case of Au film, the situation is different, and a rather uniform heating in depth of the film is observed due to the weak electron-phonon coupling of Au, and the ballistic energy transport by excited electrons (Ivanov and Zhigilei, 2003).

The absorption of laser energy and redistribution of the heat into the metals is governed initially by the free electrons. Since the electrons are not included explicitly in a MD model, their role, and the heating of the material, are taken into account by using one dimensional two-temperature diffusion model (1D-TTM) as proposed by Anisimov et al. (1974b). 1D-TTM-MD is applied with the assumption that the laser beam diameter is much larger than the energy penetration depth (Amoruso et al., 2007a, 2014). The values of the parameters for Au used are taken from Chowdhury and Xu (2003); Anisimov et al. (1974a,b), and Cheng and Xu (2005), and the decomposition of the Au material is simulated by the classical MD technique according to Allen and Tildesley (1987), and Amoruso et al. (2011).

The 1D-TTM-MD algorithm is applied to simulate fs-laser ablation decomposition of the both different Au systems: a bulk target and a thin film. In the former case, fs-laser ablation process of a bulk Au target is studied by applying specific boundary conditions at the bottom of the computational cell in order to prevent the reflection of shock waves. In case of thin Au film on glass substrate, it was proved that during the short time of the energy deposition into the lattice, the heat conduction into a substrate with low conductivity does not affect the heating of Au film.

In case of bulk Au, the heating (temperature) is very inhomogeneous, varying almost exponentially with depth and spanning over several thousand K over a distance of about 1 μm.

The structure and composition of the fs-laser ablation plume predicted by the MD simulations are in a good agreement with experimental observations addressing the spatial segregation of the species produced during fs-laser ablation of metals, and the presence of two main components, atoms and NPs, characterized by different expansion dynamics (Amoruso et al., 2005b, 2007a, 2011).

The thermodynamic trajectories of the Au bulk target indicate that at 40 ps, after the laser pulse onset, the surface layer reaches a maximal temperature of ≈5000 K. This fast quasi-isochoric heating is then followed by a rapid expansion, and at about 60 ps the thermodynamic trajectory of this slice of material enters the liquid-vapor metastable region enclosed by the spinodal curve. At this point, a progressive separation of gas and liquid phases takes place, that is, the decomposition mechanisms associated to phase explosion (Cheng and Xu, 2005; Perez et al., 2008; Perez and Lewis, 2002, 2003). The region at depth of 20 nm follows a rather similar behavior. The deeper part of the ablated material is isochorically heated to a lower temperature of ≈3500 K, close to the Au boiling temperature (3127 K) within 40 ps after the laser pulse onset. The target material initially located at different depths undergoes different thermodynamic relaxation and decomposition resulting in the different structure and size of the product produced.

A more homogeneous heating of the irradiated material is achieved when the depth distribution of the absorbed energy is restricted by the material dimensions, as in the case of a thin film target (Amoruso et al., 2011, 2014). This is proved by the temperature spatial profile, as calculated by

1D-TTM applied to a 100 nm thick Au film, at 40 ps after the laser pulse onset, when thermal equilibrium between both electron and lattice subsystems is established. It is observed that the system temperature is almost uniform within the entire depth (100 nm) of the Au thin film, differing by only of about ≈50 K between the front and the rear film surfaces. In this case, the NP sizes seem to not vary notably with *z*, thus proving that fs-laser ablation of thin film results in the generation of NPs characterized by a more homogeneous size distribution with respect to that obtained by fs-laser ablation of a bulk target, in similar conditions. In particular, the NPs size generated during fs-laser ablation of a thin Au film looks very similar to those present in the front layers of the fs-laser plume produced from the bulk target.

Theoretical and experimental studies of fs-laser ablation have demonstrated a correlation between the expected mean fragment size, *d*, and the material strain rate η ($d \propto \eta^{-2/3}$), as proposed by Perez and Lewis (2002), Glover (2003), and Amoruso et al. (2007c). This explains the smaller dimensions of the NPs produced in fs-laser ablation of a thin film, as well as the fact that their sizes are rather similar to those of the NPs generated in the decomposition of the uppermost layers of the bulk target. In this respect, the NPs produced from a thin film, and the largest nanoscale chunks present in the deeper layers of the fs-laser ablation plume from the bulk target, indicate a 2–3 times reduction of the larger NPs size, in fairly good agreement with the prediction based on the different strain rates, as well as it is confirmed by the experiments (Amoruso et al., 2014). Finally, this indicates that a more uniform and narrower NPs size distribution is produced in the case of fs-laser ablation of thin film with respect to the bulk counterpart, thus suggesting it as a viable approach toward a better control on NPs dimensions. The comparison of the size distributions of Au NPs produced by fs-laser ablation from solid and thin film targets are depicted in Fig. 17.22 (Amoruso et al., 2014; Atanasov et al., 2014).

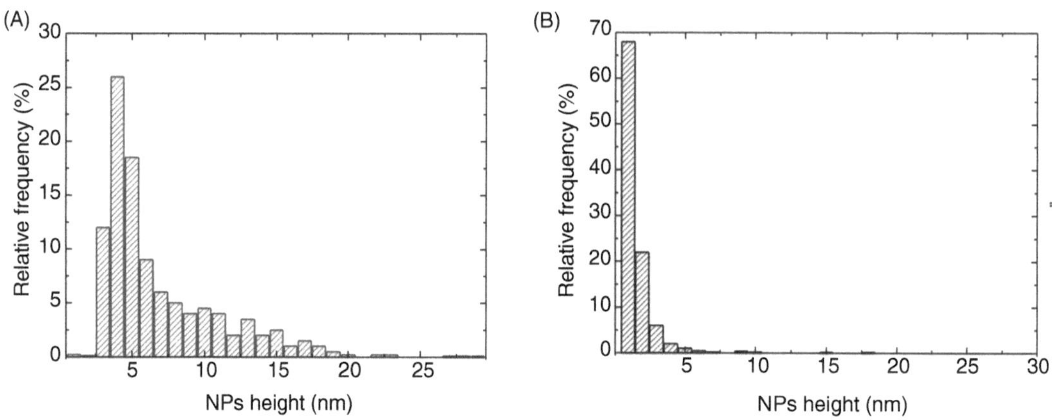

FIGURE 17.22 Size Histograms of the Au NPs Height Produced by Fs-Laser Ablation of (A) Au Bulk Target, and (B) Au 100 nm Thin Film

Reprinted from Amoruso, S., Nedyalkov, N.N., Wang, X., Ausanio, G., Bruzzese, R., Atanasov, P.A., 2014. Ultrashort-pulse laser ablation of gold thin film targets: Theory and experiment. Thin Solid Films 550, 190–198.

4.6 Au NPs PRODUCED IN LIQUIDS

Gold spherical NPs produced by pulsed laser ablation of solid targets in liquid are reported by Turkevich et al. (1951) and Nikov et al. (2012). In the past decade, different forms of the Au NPs, such as nanowires, nanocubes, nanoprisms, nanoplates, or tapes have been generated.

In brief, the biggest difference between laser ablation of solid target in vacuum or gas environment and that happening in liquid is that the latter suppresses the development of the plasma plume, that is, it influences the thermodynamics and kinetic evolution of the plasma, phase transformation, and formation of the NPs. Moreover, a new type of chemistry appears because of interaction of the particles with the liquid confinement. Finally, the cooling of the liquid is quite important, inducing the creation of NPs of smaller sizes.

A possible simple experimental set up of laser generation of NPs in distilled water is shown in Fig. 17.23.

Nd:YAG laser generating on four different wavelengths is used in the experiments (Nikov et al., 2011). TEM pictures of the Au NPs produced by four wavelengths of the laser are exposed in Fig. 17.24. Here, values of low fluence denote the energy density at which the laser ablation of the target starts, when high-energy density and optical breakdown occur at given wavelength in the experimental configuration (Nikov et al., 2011).

Several features can be summarized: (1) at low fluencies, with the exception of the use of NIR laser light, the NPs are smaller, and the size distributions are narrower with respect to the cases at high fluencies, and the Au NPs are quite small, at ~4 nm; (2) in the case presented in Fig. 17.24F, the big NPs are connected by Au nanowires, a phenomenon that is still unclear; (3) most of the Au NPs are spherical.

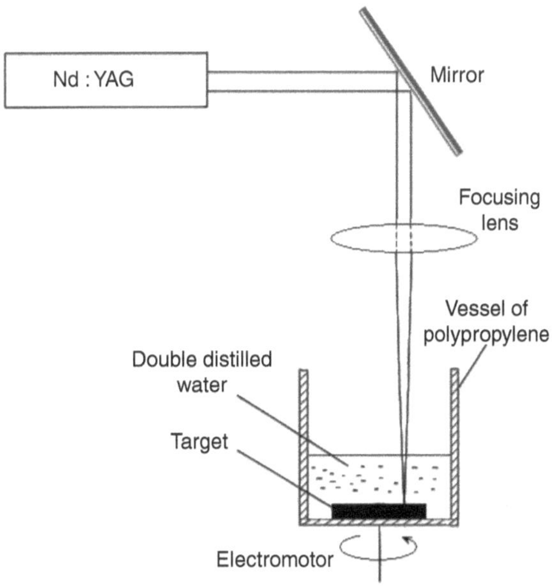

FIGURE 17.23 Experimental Setup for Laser Ablation of Solid Target in Water, Used for Production of Au NPs

Reprinted from Nikov, R.G., Nikolov, A.S., Atanasov, P.A., 2011. Preparation of gold and silver nanoparticles by pulsed laser ablation of solid target in water. Proc. SPIE 7747, 774708.

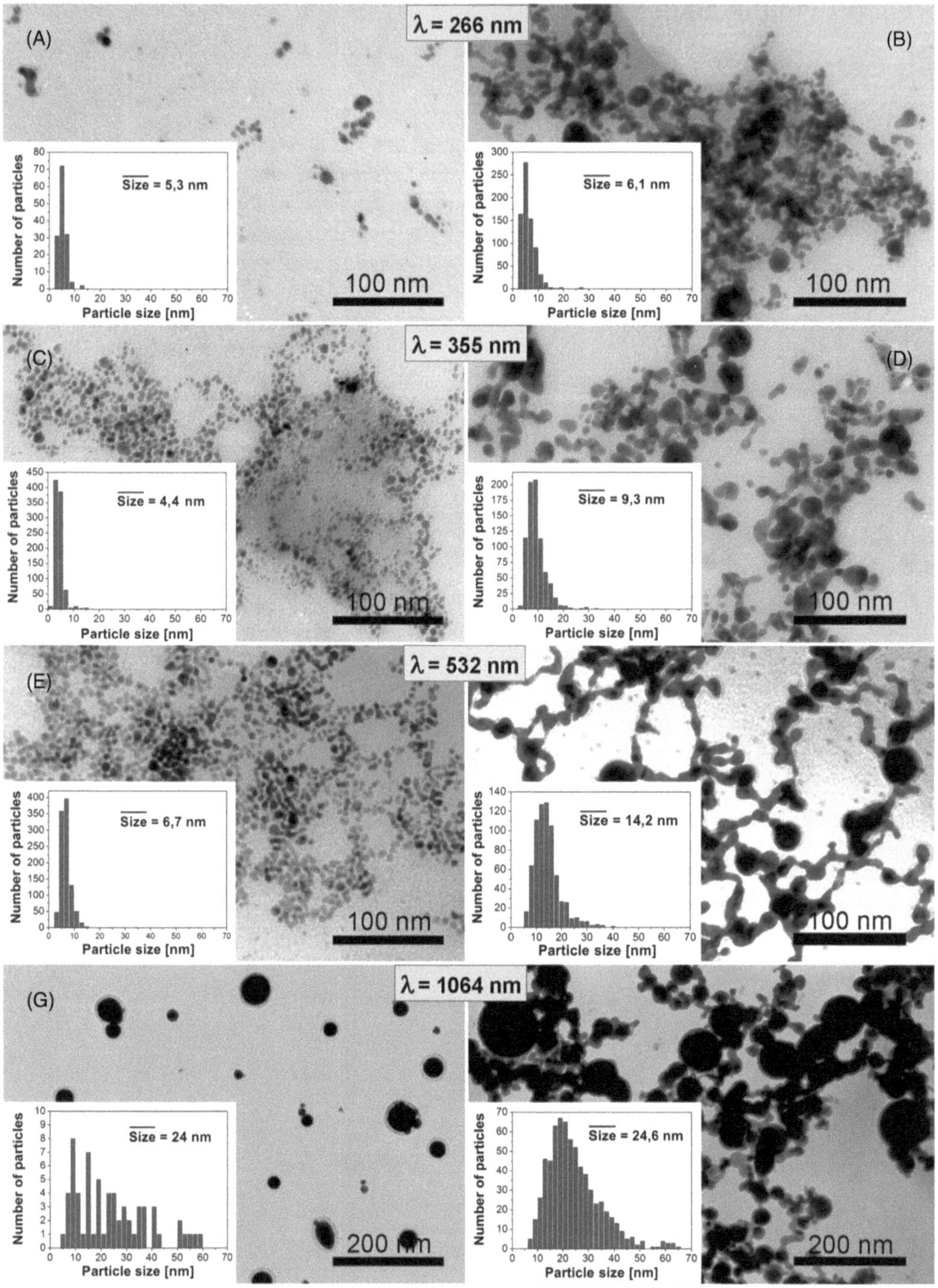

FIGURE 17.24 TEM Pictures of Au NPs

Pictures produced by four laser wavelengths at two fluencies: low in left column, and high in the right, respectively. The insets depict size distributions of the Au NPs.

5 APPLICATION OF Au NANOSTRUCTURES IN BIOPHOTONICS

5.1 LASER HEATING OF Au NPs: PHOTOTHERMAL THERAPY OF CANCER CELLS

As mentioned before, efficient plasmon excitation results in drastic enhancement of the coupling efficiency of the particle with the electromagnetic field. This is expressed by an enhancement of the absorption and scattering cross-section at the resonance wavelength. This enhancement may ensure absorption that is orders of magnitude higher than the conventional organic dyes used as absorbers in biophotonics applications (Jain et al., 2007). In addition, Au NPs are chemically stable, can be easily functionalized and are not toxic. These properties make them a promising alternative in photothermal therapy of cancer cells, as proposed by Jain et al. (2007) and Pissuwan et al. (2006). The plasmon resonance wavelength can be efficiently tuned in a wide spectral range by changing the particles size, shape, structure, and local environment dielectric properties (Noguez, 2007). In this way, working in tissue transparency window is possible when in vitro applications are designed.

The theoretical and experimental works (Jain et al., 2007; Pissuwan et al., 2006; Pustovalov, 2005; Zharov et al., 2005) devoted to applications of Au NPs in biophotonics show that particle heating up to temperatures high enough to induce cell damage can be achieved at laser fluences lower than the established safety standards. According to Volkov et al. (2007) and Letfullin et al. (2006), the cell killing mechanism can be related to irreversible biophysical and/or chemical effects, bubble formation, and subsequent generation of strong stress waves that modify cell membrane or organelles, fragmentation of NPs that produces fragments with high energy. The efficient and reliable application of metal NPs in the field of biophotonics also needs information about particle response to the laser radiation.

Different theoretical and experimental studies were performed to investigate particle melting, reshaping, and fragmentation when exposed to laser radiation (Hartland et al., 2003; Inasawa et al., 2005; Sassaroli et al., 2009; Takami et al., 1999). Since these modifications will influence the electromagnetic field-particle interaction, they will also influence cell-processing efficiency (Letfullin et al., 2006).

Here, some experimental results of Au NPs-assisted photothermal therapy of HeLa (Nedyalkov et al., 2010a, 2011, 2012a,b), HEp-2, and 8MGBA human cancer cells are presented. The influence of particle dimensions and incident laser parameters on cell viability is studied.

In order to make experiments in vitro, the cells, after a specific preparation procedure, are mixed with colloid of Au NPs, and after 24 h are exposed to laser treatment. TEM analyses are applied before laser treatment in order to get information whether the NPs with diameter of 40 or 100 nm are entering inside the cells (through its membrane) and coupling to the cells' nuclei. Fig. 17.25 depicts TEM pictures of the NPs ensembles inside the HeLa cells. The insets present, respectively, the optical properties of these specific NPs ensembles inside the cells, calculated by GMM.

Fig. 17.26 shows HeLa cell viability 6 h after 5 s laser treatment. In the cases of 100 and 200 nm particles, laser treatment leads to visible decrease of cell viability, contrary to the case when using 40 nm particles, where no visible effect is observed. The low cytotoxic efficiency of 40 nm particles can be attributed to less effective bubble formation, since the threshold for this process is inversely proportional to the particle's dimension (Zharov et al., 2005). Furthermore, the cooling time of NPs is scaled as r^2 (r is particle radius), defining a longer time for cell heating when big particles are considered.

Dependence of the HeLa cells viability on laser fluence applied is presented in Fig. 17.27. Clear tendency of decrease of the viability with the increase of the applied laser fluence is seen, as at the highest fluence (100 mJ/cm^2) applied it reaches 65%.

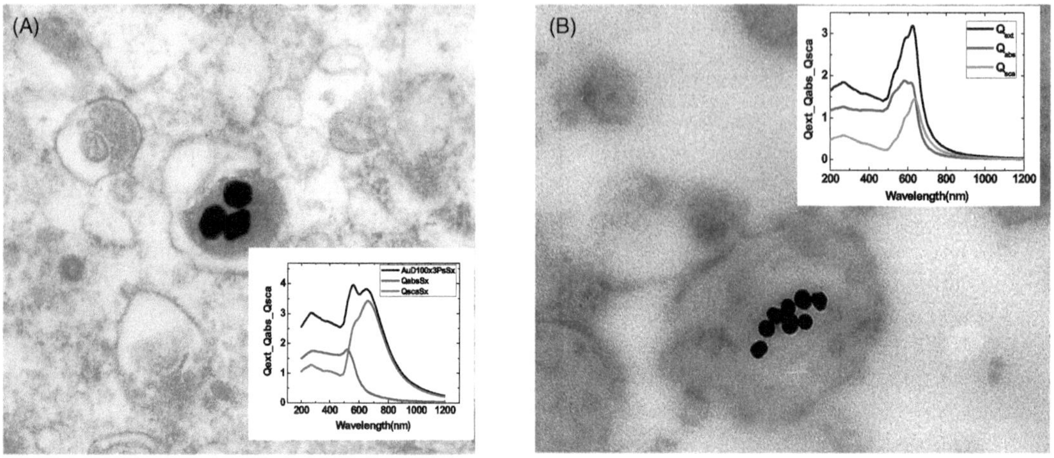

FIGURE 17.25 TEM Views of HeLa Cells Mixed with Colloid of Au NPs With Diameter of: (A) 40 nm, and (B) 100 nm

The insets represent optical parameters of the Au NPs calculated by GMM.

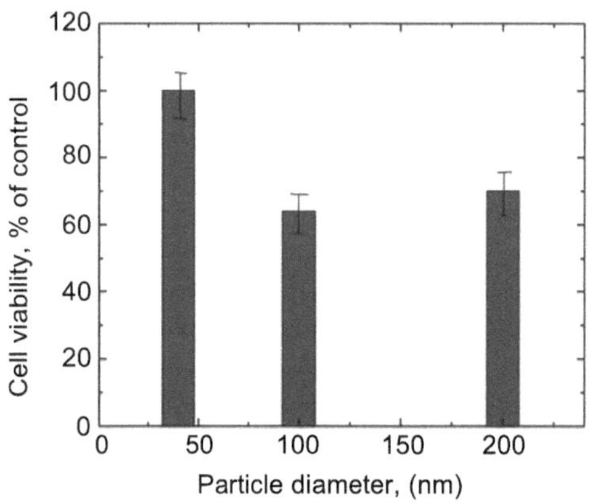

FIGURE 17.26 HeLa cell viability after Au NPs-assisted photothermal treatment

The ns-laser ($\lambda = 532$ nm) fluence is 90 mJ/cm^2; irradiation time is 5 s.

Reprinted from Nedyalkov, N.N., Atanasov, P.A., Toshkova, R.A., Gardeva, E.G., Yossifova, L.S., Alexandrov, M.T., Karashanova, D., 2012a. Laser heating of gold nanoparticles: photothermal cancer cell therapy. Proc. SPIE, 8427, 84272, P1-P7.

Fig. 17.28 represents the experimental data for cytotoxic effect of Au NPs with diameter of 40 and 100 nm, irradiated by laser pulses with fluence of 25 and 80 mJ/cm^2 on HeLa, HEp-2, and 8MGBA human cancer cells, respectively. The laser fluences are chosen to ensure realization of two different mechanisms of laser pulse/particle interaction. At fluence of 25 mJ/cm^2, the Au particles with diameter of 100 nm are melted, but not evaporated. For particles with diameter of 40 nm, this fluence will drive the temperature up to about 1000 K. The second regime (F = 80 mJ/cm^2) ensures complete evaporation of the NPs with diameter of 100 nm, and melting of those of 40 nm.

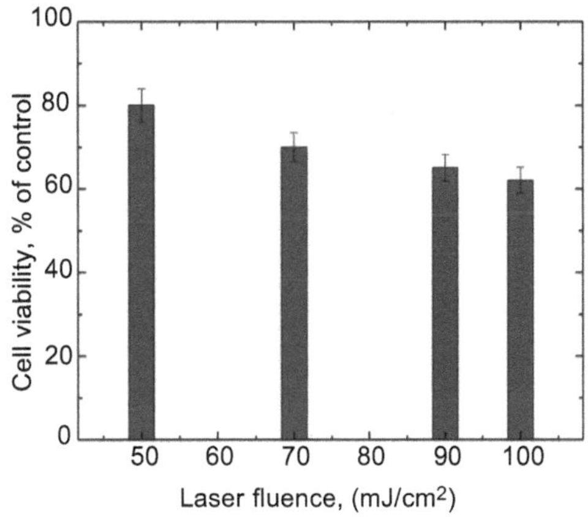

FIGURE 17.27 Cell Viability as a Function of the Applied Laser Fluence

The NPs diameter is 100 nm and the irradiation time is 5 s.

Reprinted from Nedyalkov, N.N., Atanasov, P.A., Toshkova, R.A., Gardeva, E.G., Yossifova, L.S., Alexandrov, M.T., Karashanova, D., 2012a. Laser heating of gold nanoparticles: photothermal cancer cell therapy. Proc. SPIE, 8427, 84272, P1-P7.

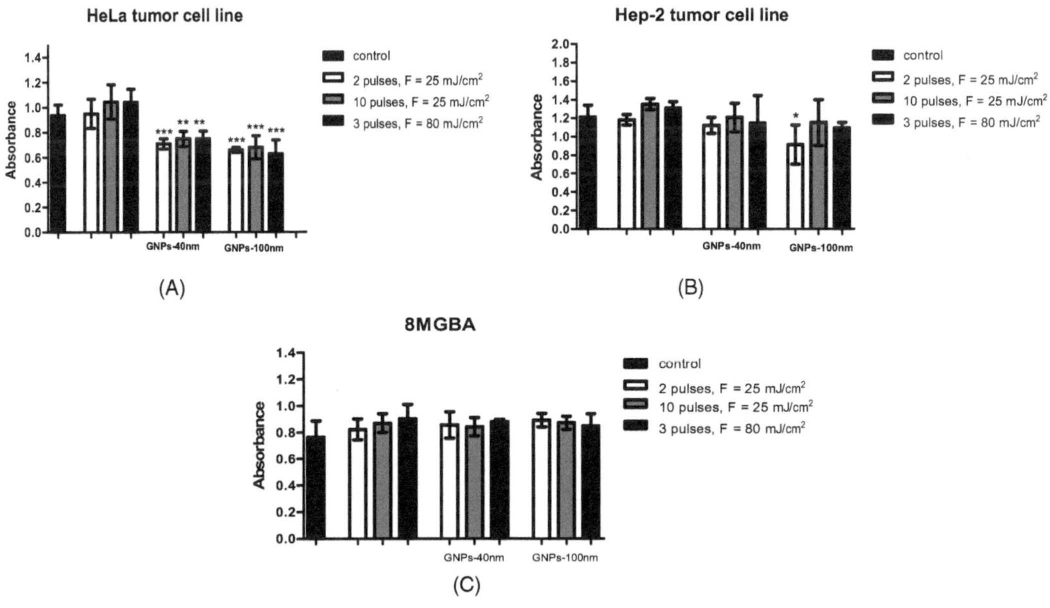

FIGURE 17.28 Viability of HeLa (A), HEp-2 (B), and 8MGBA (C) Tumor Cells, Estimated by MTT Assay

The results are taken 24 h after laser treatment.

Reprinted in part from Nedyalkov N.N., Imamova, S.E., Atanasov, P.A., Toshkova, R.A., Gardeva, R.G., Yossifova, L.S., Alexandrov, M.T., 2010. Nanosecond laser heating of gold nanoparticles. Application in photothermal cancer cell therapy, Compt. Rend. de l'Acad. Bulg. Des. Sci. 63, 5, 767–774.

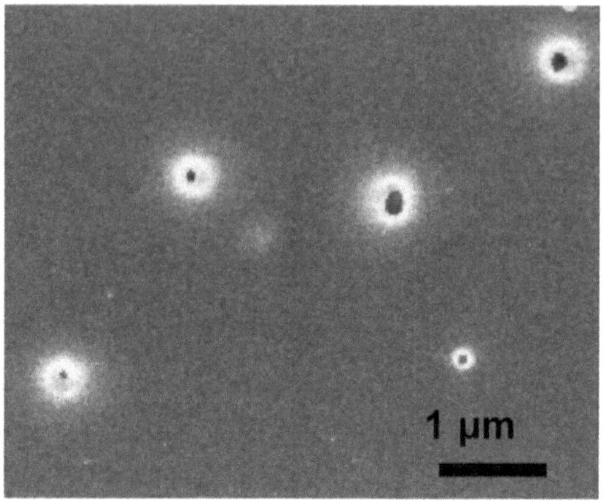

FIGURE 17.29 SEM Image of Gelatin Surface

Gelatin surface after deposition of gold nanoparticle with diameter of 100 nm irradiated by a single pulse at fluence of 25 mJ/cm^2.

Reprinted from Nedyalkov, N.N., Atanasov, P.A., Toshkova, R.A., Gardeva, E.G., Yossifova, L.S., Alexandrov, M.T., Karashanova, D., 2012a. Laser heating of gold nanoparticles: photothermal cancer cell therapy. Proc. SPIE, 8427, 84272, P1–P7.

The red bar in Fig. 17.28 represents the behavior of the cells that are not treated by laser radiation or mixed with Au colloid. The first group of the three bars represents the behavior of the cells, when treated by laser radiation without presence of Au NPs. It is seen that at the presented conditions the laser treatment itself has no toxic effect on the cells. The next two groups of bars represent the results of laser treatment of cells, when mixed with Au colloid with diameters of 40 nm (GNPs 40 nm) and 100 nm (GNPs 100 nm) of the Au NPs. In both cases, the cytotoxic effect of the laser heated Au NPs is clearly expressed. The obtained results indicate that the application of Au NPs with diameter of 100 nm is more efficient, compared to those of 40 nm. This result is in agreement with the theoretical predictions (Nedyalkov et al., 2010a, 2012a) for more efficient laser heating of particles with diameter of 100 nm. The results for application of laser fluence of 80 mJ/cm^2, which will lead to decomposition of the 100 nm particles also indicates that this mechanism can be even more efficient than in the case of particle heating, without its evaporation. Furthermore, as seen in Fig. 17.28, a statistically positive reliable effect is obtained in case of HeLa cells, and in case of HEp-2 only, when Au NPs have 100 nm diameter, and two subsequent laser pulses with energy density of 25 mJ/cm^2 are applied.

The photothermal effect is confirmed by applying similar treatment of low-temperature melting materials, such as gelatin or paraffin. Fig. 17.29 presents a SEM picture of laser-treated 100 nm Au NPs, deposited on a gelatin.

Finally, some in vivo experiments on photothermal therapy of hamsters, infected by experimental myeloma tumor of Graffi are performed. The tumors are injected hypodermically and when tumors grow up to 0.5–1.0 cm in diameter, 40 or 100 nm Au NPs are injected into the tumors. Then, the tumors are irradiated by 100 laser pulses with energy density of 25 or 80 J/cm^2, respectively. An image of the experiment is shown in Fig. 17.30.

The change of the tumor volume is inspected, and the survival time of the different groups of hamsters is investigated. Fig. 17.31 represents the survival time of different groups of the animals, compared to the untreated (control) group, or that only treated by laser radiation.

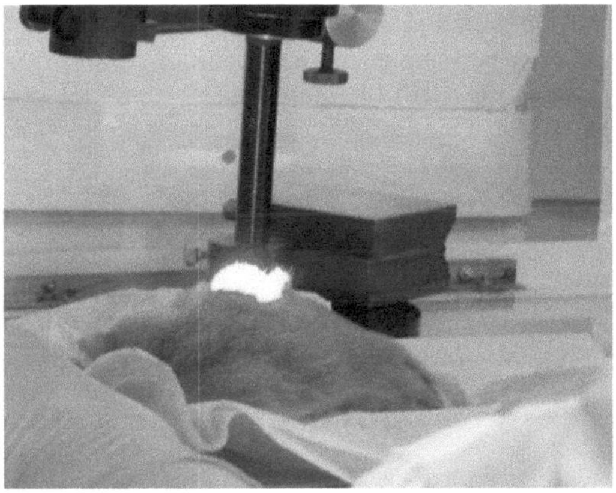

FIGURE 17.30 Picture of the In Vivo Laser Treatment of Myeloma Tumor of Graffi in Hamster

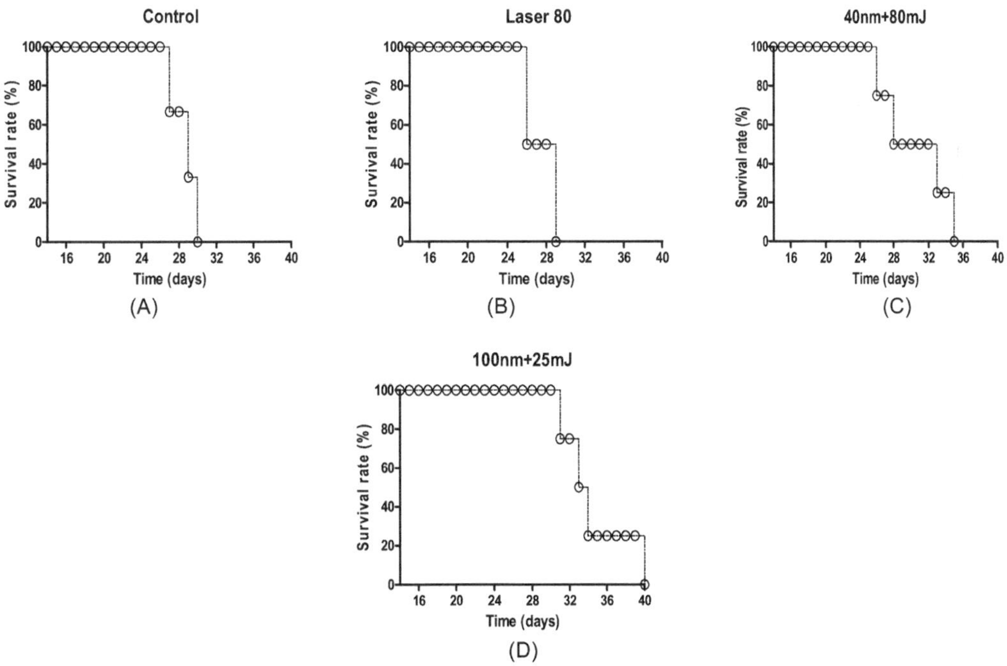

FIGURE 17.31 Viability of Three Different Groups of Hamsters Having Myeloma tumor of Graffi

(A) nontreated animals (control); (B) only laser-treated at fluence of 80 mJ/cm^2 and repetition rate of 10 Hz; (C) 40 nm Au NPs are injected into the tumor, and laser treatment by 80 mJ/cm^2 is applied; (D) 100 nm Au NPs are injected into the tumor, and laser treatment by 25 mJ/cm^2 is applied.

The average lifetime of the nontreated animals (Fig. 17.31A) is 30 days. There is no effect on statistical surviving when only laser treatment with fluence of 80 J/cm^2 is applied (Fig. 17.31B). As seen, the best results are obtained when 100 nm Au NPs are applied together with laser treatment with energy density of 25 mJ/cm^2: the lifetime is even increased up to 40 days (Fig. 17.31D).

In the frame of this study, in vivo experiments on the transplacental clastogenic and epigenetic effects of Au NPs (40 and 100 nm) on mice are studied (Balansky et al., 2013). In brief, the data obtained demonstrate for the first time the transplacental size-dependent clastogenic and epigenetic effects of Au NPs in mouse fetus, thus highlighting new aspects concerning the putative genotoxicity of Au NPs during a vulnerable period of life.

As a conclusion, the obtained results demonstrate the ability of the method and show some optimal conditions for tumor cells treatment.

5.2 PLASMONIC OPTICAL TWEEZERS IN AN ASYMMETRIC SPACE OF Au NANOSTRUCTURED SUBSTRATES

The manipulation of nanoscale objects is one very important and fast growing application field. Optical tweezers (trapping) is a state-of-the-art technology. Far-field optical tweezers are reported by Ashkin et al. (1986). Nowadays, Au NP pairs are widely used structures giving a high optical field enhancement factor in a nanogap space. The spectral shift of plasmon resonance and dependence of interparticle distance (Rechberger et al., 2003; Yang et al., 2010), as well as performance in heterodimer systems (Slaughter et al., 2010) are studied. Plasmonic sensing of viruses is proposed by Fan et al. (2011) and Dantham et al. (2012), and the viruses are trapped and may optically be killed in a 3D nanospace with an unfocused 800 nm fs-laser radiation. The proposed system may potentially be used for plasmonic dialysis of viruses ex vivo in the blood flow of patients suffering from sepsis and/or an infectious disease, such as kidney dialysis. Water has low optical absorption at 800 nm and large heat capacity, makeing the thermal effect negligible. The merit of using not cw-laser, but fs-laser, is to give trapping force digitally by the laser pulse number. Commercially available 80–100 MHz repetition rate fs-laser of 1 W average output power can easily be used to trap viruses in an enhanced plasmonic potential, and simultaneously the trapped viruses may be deactivated by the enhanced pulsed optical field. By reducing the intensity of fs-laser pulses, one can make nondestructive trapping of single viruses and immobilize them. The average optical intensity of the unfocused fs-laser is quite low, and the thermal effect may be negligible.

Near-field trapping characteristics of systems consisting of Au NPs pair placed on substrates as SiO_2, Si, and Au in water, and pumped by 800 nm fs-laser, are used for near-field optical trapping (Hirano et al., 2013). The dependence of the enhanced optical near-field distribution on the substrate material is calculated using a 3D FDTD method, which gives good agreement with the experiments for the calculation of the enhanced optical field distribution of near-field light (Miyanishi et al., 2009; Terakawa et al., 2012).

Fig. 17.32 depicts a schematic of the simulation model. The model consists of a pair of Au NPs (diameter of d, and gap distance of g) placed on a substrate in water. A y-polarized plane wave source of 800 nm is set 1.2 μm above the substrate. More details are described by Hirano et al. (2013) and Atanasov et al. (2014). The enhancement factor is defined as $|E|^2/|E_0|^2$ and the incident optical electrical field is set to 1 V/m.

Fig. 17.33 reports the dependence of maximal enhancement factor on the gap distance. The dashed line shows a maximal enhancement factor obtained with a single Au NP (the interparticle distance is

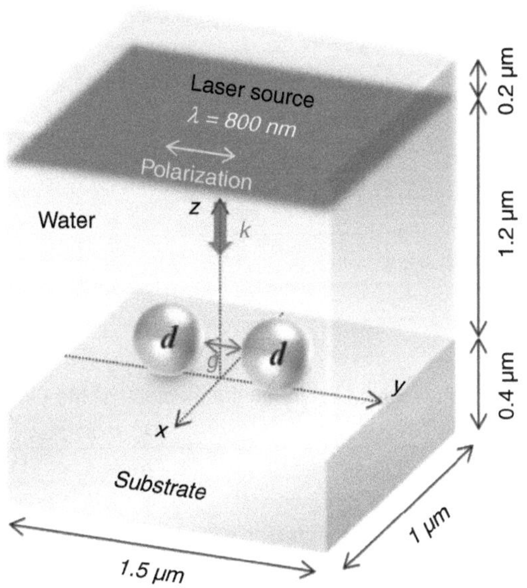

FIGURE 17.32 Schematic of the Simulation Model for Au NS pair Placed on a Substrate

Reprinted from Hirano, K., Shimizu, H., Enami, T., Terakawa, M., Obara, M., Nedyalkov, N.N., Atanasov, P.A., 2013. Plasmonic nanometric optical tweezers in an asymmetric space of gold nanostructured substrates. J. Nanotechnol. Diagn. Treat. 1, 2–10.

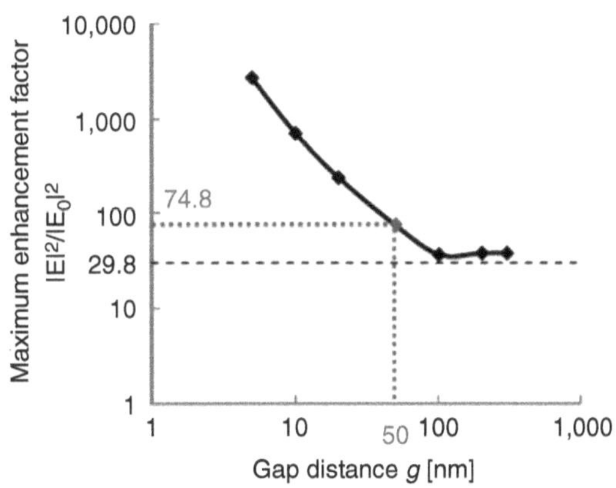

FIGURE 17.33 Dependence of the Maximal Enhancement Factor on the Gap Distance

The diameter of Au NS is 200 nm. The substrate material is SiO_2. A black dashed line shows the maximal enhancement factor achieved with a single Au NP on SiO_2.

Reprinted from Hirano, K., Shimizu, H., Enami, T., Terakawa, M., Obara, M., Nedyalkov, N.N., Atanasov, P.A., 2013. Plasmonic nanometric optical tweezers in an asymmetric space of gold nanostructured substrates. J. Nanotechnol. Diagn. Treat. 1, 2–10.

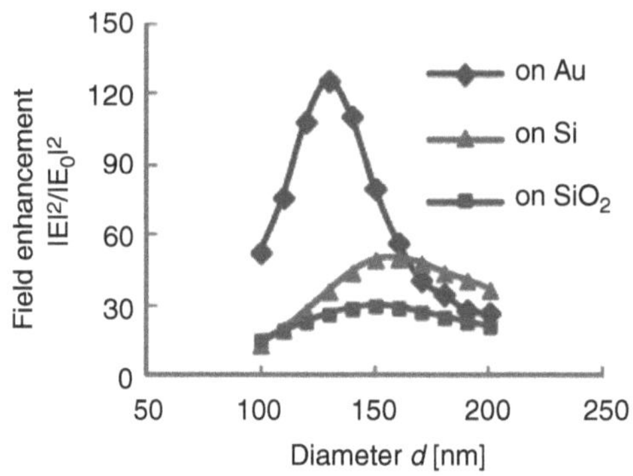

FIGURE 17.34 Dependence of the Enhancement Factor on the Diameter at the Gap Center for the Three Different Substrate Materials, SiO_2, Si, and Au

The gap distance is 50 nm.

Reprinted from Hirano, K., Shimizu, H., Enami, T., Terakawa, M., Obara, M., Nedyalkov, N.N., Atanasov, P.A., 2013. Plasmonic nanometric optical tweezers in an asymmetric space of gold nanostructured substrates. J. Nanotechnol. Diagn. Treat. 1, 2–10.

infinite). The results indicate that the maximal enhancement factor decreases on increasing the gap distance g, and approaching that of the single Au NP for $g > 100$ nm.

It is evident that there is still sufficiently high enhancement factor for a pair of Au NPs with a gap distance of 50 nm (the red dotted line in Fig. 17.33). This dimension is enough smaller for relatively small viruses as yellow fever virus, norovirus, hepatitis B or C virus, and so on.

The dependence of the field enhancement factor on the Au NP diameter at a gap distance of 50 nm for different substrates as SiO_2, Si and Au are presented in Fig. 17.34. As it can be seen, the largest field enhancement factor is achieved in the case of Au substrate.

Fig. 17.35 shows the enhancement factor distributions for the three substrates. Each interparticle distance giving respective highest factor is used based on the data presented in Fig. 17.33.

Enhanced field zones are generated under the Au NPs for Si and Au substrates, but not in the case of SiO_2 substrates. The enhanced zone at the contact point is due to the contribution of the image charge in the high refractive-index and metal substrates (Miyanishi et al., 2009; Nedyalkov et al., 2007b; Terakawa et al., 2012). The amount of image charge q' inside dielectric material can be expressed as

$$q' = -\frac{\varepsilon_{\rm m} - \varepsilon_{\rm sub}}{\varepsilon_{\rm sub} + \varepsilon_{\rm m}} q$$

where q is the amount of the charge, $\varepsilon_{\rm m}$ and $\varepsilon_{\rm sub}$ are permittivity of the surrounding medium and the substrate, respectively.

The highest charge is produced in the case of Au substrate, and the lowest one in case of SiO_2 substrate. This is because the plasmon coupling between Au NPs and the Au substrate is stronger and the enhancement factor is also increased in the gap of Au NP pairs.

M13 bacteriophages are inactivated at a threshold intensity of about 60 $MW \cdot cm^{-2}$, while the tobacco mosaic virus is inactivated at a threshold intensity of about 250 $MW \cdot cm^{-2}$ (Tsen et al., 2008). Therefore, these data provide strong support that target viruses have an inactivation threshold of the order of a few hundred $MW \cdot cm^{-2}$ or lower, when illuminated with a near-IR 80 fs-laser. The HIV samples are

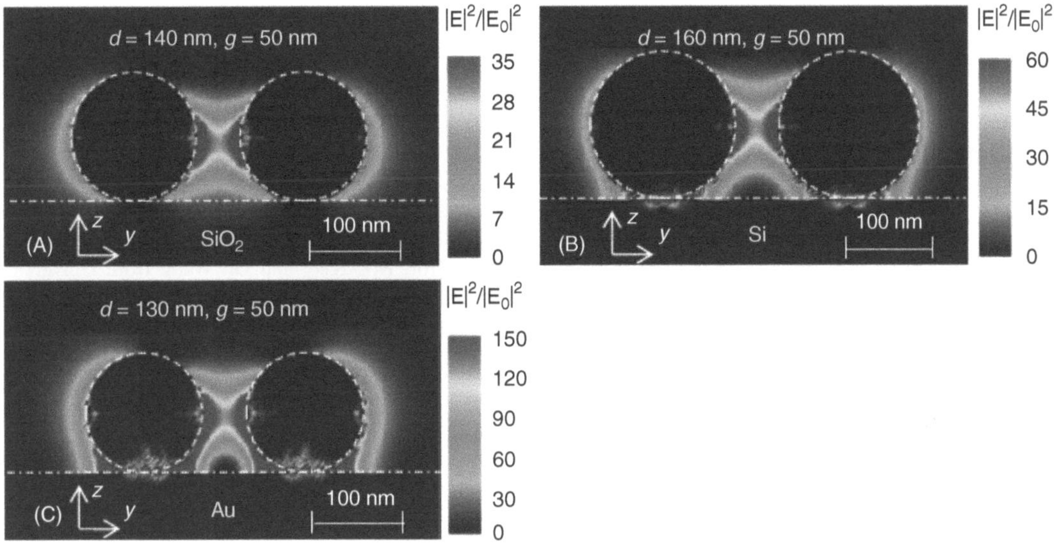

FIGURE 17.35 Enhanced Field Intensity Distributions at the Gap Center for the Three Different Substrate Materials, SiO_2, Si, and Au

The gap distance is 50 nm, while respective optimal diameter is different. Note that each intensity scale is different.

Reprinted from Hirano, K., Shimizu, H., Enami, T., Terakawa, M., Obara, M., Nedyalkov, N.N., Atanasov, P.A., 2013. Plasmonic nanometric optical tweezers in an asymmetric space of gold nanostructured substrates. J. Nanotechnol. Diagn. Treat. 1, 2–10.

irradiated with 776 nm, 500 fs pulse (Tsen et al., 2008). It was found that HIV in vitro remains infectious when irradiation laser intensities are lower than ~500 MW·cm^{-2}. However, as the laser intensity exceeds 1.1 GW·cm^{-2}, HIV loses its infectivity. Since in case of Au substrate the increase of the optical intensity in the gap is about 150, this highly enhanced intensity can inactivate such trapped viruses with 800 nm fs-laser at incident intensities of tens of MW·cm^{-2}, which is easily obtainable with commercially available fs-laser systems.

The proposed system may potentially be used for plasmonic dialysis of viruses ex vivo in the blood flow of patients suffering from sepsis and/or an infectious disease, such as kidney dialysis. With low intensity fs-laser pulses one can make nondestructive trapping of single viruses, that is interesting to immobilize them and inspect them via, for example, multiphoton excitation.

6 CONCLUSIONS

In this chapter studies and results are presented on the development of several advanced methods for the fabrication of Au NPs, NSs, and NPs arrays, and examples of their application in SERS and biophotonics. Ns- and fs-laser systems are used, and methods for controlling and manipulating the properties of the structures are also presented. The properties of the structures produced are experimentally characterized by a variety of advanced techniques, and theoretically studied by very modern methods, and the results are compared.

Some of the scientific studies and applications presented here can be considered as a first step toward much deeper and multilateral investigations in the field of nanoscience and nanotechnologies.

ACKNOWLEDGMENTS

The author would like to express his deep gratitude to his coauthors: Prof. Obara and his former PhD students from Keio University, Japan; Prof. Nedyalkov, Dr. Dikovska, Dr. Ru. Nikov, Dr. Nikov, Dr. Nikolov, and Mrs. Imamova from his former laboratory, IE-BAS, Bulgaria; Prof. Amoruso, Prof. Bruzzese, and Dr. Wang from University Federico II, Napoli, Italy; Prof. Sliwinski and his collaborators from IF-FM, Gdansk, Poland; Prof. Delaporte and Dr. Grojo from LP3, UMR 7341 CNRS/Aix-Marseille Univ., Marseille, France; Prof. Toschkova and her students from IEMPA-BAS, and Prof. Balansky from NCO, Sofia, Bulgaria.

REFERENCES

Allen, M.P., Tildesley, D.J., 1987. Computer simulation of liquids. Clarendon Press, Oxford, UK.

Amoruso, S., Ausanio, G., Bruzzese, R., Vitiello, M., Wang, X., 2005a. Femtosecond laser pulse irradiation of solid targets as a general route to nanoparticle formation in a vacuum. Phys. Rev. B 71, 033406.

Amoruso, S., Bruzzese, R., Vitiello, M., Nedialkov, N.N., Atanasov, P.A., 2005b. Experimental and theoretical investigation of femtosecond laser ablation of Al in vacuum. J. Appl. Phys. 98 (4), 044907.

Amoruso, S., Bruzzese, R., Wang, X., Nedialkov, N.N., Atanasov, P.A., 2007a. Femtosecond laser ablation of nickel in vacuum. J. Phys. D. Appl. Phys. 40, 331–340.

Amoruso, S., Bruzzese, R., Wang, X., Nedialkov, N.N., Atanasov, P.A., 2007b. An analysis of the dependence on photon energy of the process of nanoparticles generation by femtosecond laser ablation in a vacuum. Nanotechnology 18, 145612.

Amoruso, S., Bruzzese, R., Wang, X., Ausanio, G., Lanotte, L., 2007c. Laser-induced modification of the size distribution of nanoparticles produced during ultrashort laser ablation of solid targets in vacuum. J. Phys. B: At. Mol. Opt. Phys. 40 (6), 1253.

Amoruso, S., Nedyalkov, N.N., Wang, X., Ausanio, G., Bruzzese, R., Atanasov, P.A., 2011. Ultrafast laser ablation of gold thin film targets. J. Appl. Phys. 110 (124303), 1–4.

Amoruso, S., Nedyalkov, N.N., Wang, X., Ausanio, G., Bruzzese, R., Atanasov, P.A., 2014. Ultrashort-pulse laser ablation of gold thin film targets: theory and experiment. Thin Solid Films 550, 190–198.

Anisimov, S.I., Kapelovich, B.L., Perel'man, T.L., 1974a. Electron emission from metal surfaces exposed to ultrashort laser pulses. JETP 39, 375–377.

Anisimov, S.I., Kapelovich, B.L., Perel'man, T.L., 1974b. Emission of electrons from the surface of metals induced by ultrashort laser pulses. Zh. Eksp. Teor. Fiz. 66, 776–781.

Ashkin, A., Driedzic, J.M., Bjorkholm, J.E., Chu, S., 1986. Observation of a single-beam gradient force optical trap for dielectric particles. Opt. Lett. 11 (5), 288–290.

Atanasov, P.A., Nedyalkov, N.N., Dikovska, A.O.g., Nikov, R.u., Amoruso, S., Wang, X., Bruzzese, R., Hirano, K., Shimizu, H., Terakawa, M., Obara, M., 2014. Noble metallic nanostructures: preparation, properties, applications. J. Phys. Conf. Ser. 514, 012024.

Baker, G.A., Moore, D.S., 2005. Progress in plasmonic engineering of surface-enhanced Raman-scattering substrates toward ultra-trace analysis. Anal. Bioanal. Chem. 382 (8), 1751–1770.

Balansky, R., Longobardi, M., Ganchev, G., Iltcheva, M., Nedyalkov, N., Atanasov, P., Toshkova, R., De Flora, S., Izzotti, A., 2013. Transplacental clastogenic and epigenetic effects of gold nanoparticles in mice. Mutat. Res. 751, 42–48.

Beermann, J., Novikov, S.M., Leosson, K., Bozhevolnyi, S.I., 2009. Surface enhanced Raman microscopy with metal nanoparticle arrays. J. Opt. A Pure Appl. Opt. 11, 075004.

Besner, S., Meunier, M., 2010. Laser synthesis of nanomaterials. In: Sugioka, K., Meunier, M., Piqué, A. (Eds.), Laser Precision Microfabrication. Springer-Verlag, Berlin, Germany.

Cheng, Ch., Xu, X., 2005. Mechanisms of decomposition of metal during femtosecond laser ablation. Phys. Rev. B 72, 165415.

Chowdhury, I.H., Xu, X., 2003. Heat transfer in femtosecond laser processing of metal. Numer. Heat Transfer A 44, 219–232.

Dantham, V.R., Holler, S., Kolchenko, V., Zan, W., Arnold, S., 2012. Taking whispering gallery-mode single virus detection and sizing to the limit. Appl. Phys. Lett. 101 (4), 043704.

Deniz, D., Lad, R.J., 2011. Temperature threshold for nanorod structuring of metal and oxide films grown by glancing angle deposition. J. Vac. Sci. Technol. A 29, 011020.

Dikovska, A.O.., Avdeev, G.V., Nedyalkov, N.N., Koleva, M.E., Atanasov, P.A., 2013. Preparation of metal nanorods substrates for SERS application. Proc. SPIE 8770, 877006–877011.

Dimitrov, I.G., Nedyalkov, N.N., Atanasov, P.A., 2010. Optical properties of gold nanoparticles arrays. In: Proceedings of IE BAS-Keio GCOE Workshop on Nanophotonics. Institute of Electronics of the Bulgarian Academy of Sciences, Sofia, Bulgaria, pp. 84–92.

Edelstein, A.S., Cammarata, R.C., 1996. Nanomaterials: synthesis, properties and applications. Institute of Physics Publishing, Bristol, UK.

Eliezer, S., Eliaz, N., Grossman, E., Fisher, D., Gouzman, I., Henis, Z., Pecker, S., Horovitz, Y., Fraenkel, M., Maman, S., Lereah, Y., 2004. Synthesis of nanoparticles with femtosecond laser pulses. Phys. Rev. B 69, 144119.

Fan, M., Andrade, G.F.S., Brolo, Al.G., 2011. A review on the fabrication of substrates for surface enhanced Raman spectroscopy and their applications in analytical chemistry. Anal. Chimica Acta 693 (1-2), 7–25.

Gaudry, M., Cottancin, E., Pellarin, M., Lermé, J., Arnaud, L., Huntzinger, J.R., Vialle, J.L., Broyer, M., Rousset, J.L., Treilleux, M., Mélinon, P., 2003. Size and composition dependence in the optical properties of mixed (transition metal/noble metal) embedded clusters. Phys. Rev. B 67 (15), 155409.

Girifalco, L.A., Weizer, V.G., 1959. Application of the Morse potential function to cubic metals. Phys. Rev. 114 (3), 687.

Glover, T.E., 2003. Hydrodynamics of particle formation following femtosecond laser ablation. J. Opt. Soc. Am. B 20, 125–131.

Gudiksen, M.S., Laudon, L.J., Wang, J., Smith, D.C., Lieber, C.M., 2002. Growth of nanowire superlattice structures for nanoscale photonics and electronics. Nature 415, 617–620.

Guowei, Y. (Ed.), 2012. Laser Ablation in Liquids: Principles and Applications in the Preparation of Nanomaterials. Pan Stanford Publishing, Singapore.

Hartland, G.V., Min, H., John, E.S., 2003. Softening of the symmetric breathing mode in gold particles by laser-induced heating. J. Phys. Chem. B 107, 7472–7478.

Hirano, K., Shimizu, H., Enami, T., Terakawa, M., Obara, M., Nedyalkov, N.N., Atanasov, P.A., 2013. Plasmonic nanometric optical tweezers in an asymmetric space of gold nanostructured substrates. J. Nanotechnol. Diagn. Treat. 1, 2–10.

Imamova, S., Dikovska, A., Nedyalkov, N., Atanasov, P., Sawczak, M., Jendrzejewski, R., Śliwinski, G., Obara, M., 2010. Laser nanostructuring of thin Au films for application in surface enhanced Raman spectroscopy. J. Optoelectron. Adv. Mater. 12, 500–504.

Inasawa, S., Sugiyama, M., Yamaguchi, Y., 2005. Laser-induced shape transformation of gold nanoparticles the melting point: the effect of surface melting. J. Phys. Chem. B 109, 3104–3111.

Ivanov, D.S., Zhigilei, L.V., 2003. Combined atomistic-continuum modeling of short-pulse laser melting and disintegration of metal films. Phys. Rev. B 68, 064114.

Jain, P.K., El-Sayed, I.H., El-Sayed, M.A., 2007. Au nanoparticles target cancer. Nano Today 2 (1), 18–29.

Johnson, P.B., Christy, R.W., 1972. Optical constants of the noble metals. Phys. Rev. B 6 (12), 4370.
Kawata, S., 2001. Near-Field Optics and Surface Plasmon Polaritons. Springer, Berlin, Germany.
Kneipp, K., Wang, Y., Kneipp, H., Perelman, L.T., Itzkan, I., Dasari, R.R., Feld, M.S., 1997. Single molecule detection using surface-enhanced Raman scattering (SERS). Phys. Rev. Lett. 78 (9), 1667–1670.
Kreibig, U., Vollmer, M., 1995. Optical Properties of Metal Clusters. Springer, Berlin, Germany.
Letfullin, R.R., Joenathan, C.h., George, T.h.F., Zharov, V.P., 2006. Laser-induced explosion of gold nanoparticles: potential role for nanophotothermolysis of cancer. Nanomedicine 1, 473–480.
Lewis, L.J., Perez, D., 2010. Theory and simulation of laser ablation—from basic mechanisms to applications. In: Sugioka, K., Meunier, M., Piqué, A. (Eds.), Laser Precision Microfabrication. Springer-Verlag, Berlin, Germany.
Link, S., Wang, Z.L., El-Sayed, M.A., 1999. Alloy formation of gold-silver nanoparticles and the dependence of the plasmon absorption on their composition. J. Phys. Chem. B 103, 3529–3533.
Liu, Z., Wei, Q., Zhang, X., 2005. Surface plasmon interference nanolithography. Nano Lett. 5 (5), 957–961.
Major, K.J., De, C., Obare, S.O., 2009. Recent advances in the synthesis of plasmonic bimetallic nanoparticles. Plasmonics 4, 61–78.
Masala, O., Seshadri, R., 2004. Synthesis routes for large volumes of nanoparticles. Ann. Rev. Mater. Res. 34, 41–81.
Miyanishi, T., Sakai, T., Nedyalkov, N.N., Obara, M., 2009. Femtosecond-laser nanofabrication onto silicon surface with near-field localization generated by plasmon polaritons in gold nanoparticles with oblique irradiation. Appl. Phys. A 96 (4), 843–850.
Mohammed, K., Shukla, M.M., Milstein, F., Merz, J.L., 1984. Lattice dynamics of face-centered-cubic metals using the ionic Morse potential immersed in the sea of free-electron gas. Phys. Rev. B 29, 3117.
Moskovits, M., 2005. Surface-enhanced Raman spectroscopy: a brief retrospective. J. Raman Spectrosc. 36 (6–7), 485–496.
Nedyalkov, N.N., Atanasov, P.A., Amoruso, S., Bruzzese, R., Wang, X., 2007a. Laser ablation of metals by femtosecond pulses: theoretical and experimental study. Appl. Surf. Sci. 253 (19), 7761–7766.
Nedyalkov, N.N., Atanasov, P.A., Obara, M., 2007b. Near-field properties of a gold nanoparticle array on different substrates excited by a femtosecond laser. Nanotechnology 18 (30), 305703–305707.
Nedyalkov, N.N., Imamova, S.E., Atanasov, P.A., Toshkova, R.A., Gardeva, R.G., Yossifova, L.S., Alexandrov, M.T., 2010a. Nanosecond laser heating of gold nanoparticles. Application in photothermal cancer cell therapy. Compt. Rend. Acad. Bulg. Sci. 63 (5), 767–774.
Nedyalkov, N.N., Imamova, S.E., Atanasov, P.A., Obara, M., 2010b. Gold nanoparticles as nanoheaters and nanolenses in the processing of different substrate surfaces. J. Phys. Conf. Ser. 223, 012035.
Nedyalkov, N.N., Imamova, S.E., Atanasov, P.A., Toshkova, R.A., Gardeva, E.G., Yossifova, L.S., Alexandrov, M.T., Obara, M., 2011. Interaction of gold nanoparticles with nanosecond laser pulses: Nanoparticle heating. Appl. Surf. Sci. 275, 5456–5459.
Nedyalkov, N.N., Atanasov, P.A., Toshkova, R.A., Gardeva, E.G., Yossifova, L.S., Alexandrov, M.T., Karashanova, D., 2012a. Laser heating of gold nanoparticles: photothermal cancer cell therapy. Proc. SPIE 8427, 1–7.
Nedyalkov, N.N., Dikovska, A., Dimitrov, I., Nikov, Ru., Atanasov, P.A., Toshkova, R.A., Gardeva, E.G., Yossifova, L.S., Alexandrov, M.T., 2012b. Far- and near-field optical properties of gold nanoparticle ensembles. Quant. Electron. 42 (12), 1123–1127.
Nedyalkov, N., Nikov, Ru., Koleva, M., Atanasov, P.A., Constantinescu, C., Delaporte, Ph., Grojo, D., 2014. Nanoparticle-decorated ceramic as substrate in surface enhanced Raman spectroscopy. Appl. Surf. Sci. 336, 16–20.
Nikolov, A.S., Nedyalkov, N.N., Nikov, R.G., Atanasov, P.A., Alexandrov, M.T., 2011. Characterization of Ag and Au nanoparticles created by nanosecond pulsed laser ablation in double distilled water. Appl. Surf. Sci. 257, 5278–5282.

Nikov, R.G., Nikolov, A.S., Atanasov, P.A., 2011. Preparation of gold and silver nanoparticles by pulsed laser ablation of solid target in water. Proc. SPIE 7747, 774708.

Nikov, R.G., Nikolov, A.S., Nedyalkov, N.N., Dimitrov, I.G., Atanasov, P.A., Alexandrov, M.T., 2012. Stability of contamination-free gold and silver nanoparticles produced by nanosecond laser ablation of solid targets in water. Appl. Surf. Sci. 258 (23), 9318–9322.

Nikov, R.G., Nikolov, A.S., Nedyalkov, N.N., Atanasov, P.A., Alexandrov, M.T., Karashanova, D.B., 2013a. Processing condition influence on the characteristics of gold nanoparticles produced by pulsed laser ablation in liquids. Appl. Surf. Sci. 274, 105–109.

Nikov, Ru.G., Nedyalkov, N.N., Atanasov, P.A., Grochowska, K., Iwulska, A., Sliwinski, G., 2013b. Laser nanostructuring of Au/Ag and Au/Ni films for application in SERS. Proc. SPIE 8770 (87700D), 1–7.

Nikov, Ru., Nedyalkov, N., Atanasov, P.A., Terakawa, M., Shimizu, H., Obara, M., 2013c. Tuning the optical properties of gold nanostructures fabricated on flexible substrates. Appl. Surf. Sci. 264, 779–782.

Nikov, Ru.G., Nedyalkov, N.N., Atanasov, P.A., Delaporte, Ph., Grojo, D., 2015. Fabrication and characterization of metal nanostructures on metal substrates. Proc. SPIE 9447, 94470K1–194470K.

Noguez, C., 2007. Surface plasmons on metal nanoparticles: the influence of shape and physical environment. J. Phys. Chem. C 111, 3806–3819.

Palik, E.D., 1998. Handbook of Optical Constants of Solids. Academic Press, San Diego, CA.

Papavassiliou, G.C., 1979. Optical properties of small inorganic and organic metal particles. Prog. Solid St. Chem. 12 (3), 185–271.

Perez, D., Lewis, L.J., 2002. Ablation of solids under femtosecond laser pulses. Phys. Rev. Lett. 89, 255504.

Perez, D., Lewis, L.J., 2003. Molecular-dynamics study of ablation of solids under femtosecond laser pulses. Phys. Rev. B 67, 184102.

Perez, D., Béland, L.K., Deryng, D., Lewis, L.J., Meunier, M., 2008. Numerical study of the thermal ablation of wet solids by ultrashort laser pulses. Phys. Rev. B 77, 014108.

Perrière, J., Boulmer-Leborgne, C., Benzerga, R., Tricot, S., 2007. Nanoparticle formation by femtosecond laser ablation. J. Phys. D Appl. Phys. 40, 7069.

Pillai, S., Catchpole, K.R., Trupke, T., Green, M.A., 2007. Surface plasmon enhanced silicon solar cells. J. Appl. Phys. 101 (9), 093105.

Pissuwan, D., Valenzuela, S.M., Cortie, M.B., 2006. Therapeutic possibilities of plasmonically heated gold nanoparticles. Trends Biotech. 24, 62–67.

Pustovalov, V.K., 2005. Theoretical study of heating of spherical nanoparticle in media by short laser pulses. Chem. Phys. 308, 103–108.

Rechberger, W., Hohenau, A., Leitner, A., Krenn, J.R., Lamprecht, B., Aussenegg, F.R., 2003. Optical properties of two interacting gold nanoparticles. Opt. Commun. 220 (1–3), 137–141.

Salem, A.K., Searson, P.C., Leong, K.W., 2003. Multifunctional nanorods for gene delivery. Nat. Mater. 2, 668–671.

Sassaroli, E., Li, K.C.P., O'Neill, B.E., 2009. Numerical investigation of heating of a gold nanoparticle and the surrounding microenvironment by nanosecond laser pulses for nanomedicine applications. Phys. Med. Biol. 54, 5541–5550.

Schaadt, D.M., Feng, B., Yu, E.T., 2005. Enhanced semiconductor optical absorption via surface plasmon excitation in metal nanoparticles. Appl. Phys. Lett. 86 (6), 063106.

Shanmukh, S., Jones, L., Driskell, J., Zhao, Y., Dluhy, R., Tripp, R.A., 2006. Rapid and sensitive detection of respiratory virus molecular signatures using a silver nanorods array SERS substrate. Nano Lett. 6, 2630–2636.

Shi, J., Votruba, A.R., Farokhzad, O.C., Langer, R., 2010. Nanotechnology in drug delivery and tissue engineering: from discovery to applications. Nano Lett. 10, 3223–3230.

Slaughter, L.S., Wu, Y., Willingham, B.A., Nordlander, P., Link, S., 2010. Effects of symmetry breaking and conductive contact on the plasmon coupling in gold nanorod dimmers. ACS Nano 4 (8), 4657–4666.

Sullivan, D.M., 2000. Electromagnetic Simulation Using the FDTD Method. IEEE Press Series, New York, NY.

Sun, S.H., Murray, C.B., Weller, D., Folks, L., Moser, A., 2000. Nanocrystal superlattices monodisperse FePt nanoparticles and ferromagnetic FePt. Science 287, 1989–1992.

Suzuki, M., Nakajima, K., Kimura, K., Fukuoka, T., Mori, Y., 2007. Au nanorod arrays tailored for surface-enhanced Raman spectroscopy. Anal. Sci. 23, 829–832.

Taflove, A., 2005. Computational electromagnetics: the finite-difference time-domain method. Artech House, Boston, MA.

Taflove, A., Hagness, S.C., 2000. Computational electrodynamics: the finite-difference time-domain method. Artech House, Boston.

Takami, A., Kurita, H., Koda, S., 1999. Laser-induced size reduction of noble metal particles'. J. Phys. Chem. B 103, 1226–1232.

Tanaka, Y., Terakawa, M., Obara, M., Nedyalkov, N.N., Atanasov, P.A., 2010. Plasmonic nanopatterning of the material surface mediated with gold nanoparticles excited by a femtosecond laser pulse. In: Chow, P.E. (Ed.), Gold Nanoparticles: Properties, Characterization and Fabrication. NOVA Science Publishers, New York, NY, pp. 173–204.

Terakawa, M., Takeda, S., Tanaka, Y., Obara, G., Miyanishi, T., Sakai, T., Sumiyoshi, T., Sekita, H., Hasegawa, M., Viktorovitch, P., Obara, M., 2012. Enhanced localized near field and scattered far field for surface nanophotonics applications. Prog. Quantum Electron. 36 (1), 194–271.

Tsen, K.T., Tsen, S.W.D., Hung, C.F., Wu, T.C., Kiang, J.G., 2008. Selective inactivation of human immunodeficiency virus with subpicosecond near-infrared laser pulses. J. Phys. Condens. Matter 20 (25), 252205–252214.

Turkevich, J., Stevenson, P.C., Hillier, J., 1951. A study of the nucleation and growth processes in the synthesis of colloidal gold. Discuss. Faraday Soc. 11, 55–75.

Vial, A., Grimault, A.S., Macias, D., Barchiesi, D., Chapelle, M.I., 2005. Improved analytical fit of gold dispersion: application to the modeling of extinction spectra with a finite-difference time-domain method. Phys. Rev. B 71 (8), 085416.

Volkov, A.N., Sevilla, C., Zhigilei, L.V., 2007. Numerical modelling of short pulse laser interaction with Au nanoparticle surrounded by water. Appl. Surf. Sci. 253, 6394–6399.

Wittig, M.X., Liu, X., 2005. Ultrathin Metal Films: Magnetic and Structural Properties. Springer.

Xu, Y.-L., 2003. Scattering Mueller matrix of an ensemble of variously shaped small particles. J. Opt. Soc. Am. A 20 (11), 2093–2105.

Xu, Y.-L., Khlebtsov, N.G., 2003. Orientation-averaged radiative properties of an arbitrary configuration of scatterers. J. Quant. Spectrosc. Radiat. Tranf. 79, 1121–1137.

Xu, H., Bjerneld, J., Kall, M., Borjesson, L., 1999. Spectroscopy of single hemoglobin molecules by surface enhanced Raman scattering. Phys. Rev. Lett. 83, 4357–4360.

Yang, S.C., Kobori, H., He, C.L., Lin, M.H., Chen, H.Y., Li, C., et al., 2010. Plasmon hybridization in individual gold nanocrystal dimers: direct observation of bright and dark modes. Nano Lett. 10 (2), 632–637.

Zhang, J.Z., Noguez, C., 2008. Plasmonic optical properties and applications of metal nanostructures. Plasmonics 3, 127–150.

Zhang, Q., Tan, Y.N., Xie, J., Lee, J.Y., 2009. Colloidal synthesis of plasmonic metallic nanoparticles. Plasmonics 4 (1), 9–22.

Zharov, V.P., Galitovskaya, E.N., Johnson, C., Kelly, Th., 2005. Synergistic enhancement of selective nanophotothermolysis with gold nanoclusters: potential for cancer therapy. Lasers Surg. Med. 37, 219–226.

Zhigilei, L.V., Garrison, B.J., 1999a. Molecular dynamics simulation study of the fluence dependence of particle yield and plume composition in laser desorption and ablation of organic solids. Appl. Phys. Lett. 74 (9), 1341–1343.

Zhigilei, L.V., Garrison, B.J., 1999b. Multiscale modeling of materials. Mat. Res. Soc. Symp. Proc. 538, 491.

CHAPTER

QUANTUM DOTS FOR BIOIMAGING AND THERAPEUTIC APPLICATIONS

18

George Vlasceanu*, Alexandru Mihai Grumezescu*, Irina Gheorghe, Mariana C. Chifiriuc**, Alina M. Holban****

**University Politehnica of Bucharest, Bucharest, Romania*

***Research Institute of the University of Bucharest (ICUB), Bucharest, Romania*

CHAPTER OUTLINE

1 INTRODUCTION

Quantum dots (QDs), also known as nanoscaled semiconductors, are in line with the quantum mechanics principle of quantum confinement. Featuring an energy band gap according to which the spectrum of radiation absorption and emission is wavelength-wise displayed, their essential properties are size-dependent (Mendonca and De Azevedo, 2016). Their energy band gap increases inversely with the QDs' size, thus their intrinsic features and the conjugation processing play a key role in the optimization of their optical behavior (He et al., 2016; Mendonca and De Azevedo, 2016).

Even though QDs are usually used in electronic devices, in recent years, an interdisciplinary domain of bioassaying and bioimaging has raised more interest in the science community. Offering more advantages compared to the standard chemical (organic fluorophores and proteins with fluorescent emission) staining techniques, QD-mediated imaging has become one of the most explored approaches when it comes to molecular, cellular, and intracellular labeling and biodetection (Onoshima et al., 2015).

Generally biologically unsafe, QDs are often subjected to surface modifications for biocompatibility enhancement and toxicity decrease. Their rich surface chemistries make it easier to engineer conjugation with organic materials with or without targeted biological activity or other chemical compounds meant to model their behavior after the implantation in the organism (Liu et al., 2015a). The fluorescent/phosphorescent emissions are to be factored in according to the nature of the medium in which QDs are employed for specific detection. Environments with strong fluorescence resulting from the features of biological fluids need phosphorescent sensors which can avoid the recording of interferences with the biomatrices due to prolonged delay time (Bi et al., 2014). Less common than fluorescence assays, phosphorescent-relied techniques have the advantage of light scattering eluding and also of a wider gap between excitation and emission spectra (Ertas and Kara, 2015).

Widely used for (Finetti et al., 2016), the QDs wide clinical implementations are still a desideratum as concerns related to their heavy metal content and often unforeseeable interactions with the living structures have been raised (Liu et al., 2015a). Factors taken into account for tuning in order to decrease the cytotoxicity cover aspects like QDs' core composition, existing coatings, and the size of the particles (Liu et al., 2015a). Sometimes, the size cannot be controlled when synthesizing QDs according to a certain technique (Mir et al., 2016), but core compositions and specific coating are easily adapted when designed (Liu et al., 2015a). Nonetheless, green synthesis methods were developed for QDs (Li et al., 2016; Uddandarao and Mohan, 2016; Xu et al., 2016).

For superior mimicking of the natural recognition sites or entities, QDs can be embedded into microbeads which improve the sensing activities due to a larger superficial area and multiplexing ability. Molecularly imprinted polymers are specially designed to target molecules according to size, shape, and superficial functional groups, which can reproduce bioreceptors and antibodies, supplementing the affinity of the sensor they are included in (Liu et al., 2016f). Capping and conjugations with various compounds were reported: para-methoxy-*N*-methylamphetamine (PMMA), Ru organocomplex (Huang et al., 2016), heparin (Maguire et al., 2014), L-cysteine (Ertas and Kara, 2015), and also metal/nonmetal ions doping (Sadeghi et al., 2016; Xue et al., 2016) for enhanced properties and behavior. The development of Core/shell/shell QDs also revealed more adequate behavior when those materials were exposed to biological environment (Durgadas et al., 2012; Ji et al., 2014; Li et al., 2016).

2 QDs FOR BIOIMAGING APPLICATIONS

Manganese (Mn^{2+}) doped QDs could act like ratiometric pH sensors (as they count on the coupling of only two fluorophores), with simple sensing and broadened applicability. The CdSSe QDs were synthesized and doped with manganese ions according to a recently described protocol. The MnQDs display two emission bands, for green (510 nm) and red (610 nm), depending on the dopant emission. The reversibility of protonation and deprotonation of the carboxylic acid groups on the QDs surface leads to the registering of a pH value. High stability and sensitivity were obtained in the range of 4.6–8.5, corresponding to the pH values of most human internal media, making it suitable for both cellular and interstice measurement (Pratiwi et al., 2016).

Manganese doped ZnS QDs could be used for the detection of urea in biological fluids. The phosphorescence properties of this QD ensures data registration without the need of deoxidants and inducers; moreover, the detection happens without interfering with autofluorescence or the scattering light originated in the matrix of mercaptopropionic acid. To overcome most disadvantages related with previously synthesized sensors (CdSe/ZnS QDs as indicators, urease as catalyst), a new approach developed the possibility of room-temperature measurement of this new system (Bi et al., 2014). The nanocomplex was able to detect quercetin. Aluminum ions addition determined the amplification of the phosphorescent signal once they determined the QDs aggregation as a result of the interaction with the superficial carboxyl groups (Zhang et al., 2015a). Nonchemically obtained PMMA-CdSe/ZnS QDs are also granted with superior optical properties, high stability, and emission property preservation after being water solvated. Used for the manufacturing of composite polymeric films, this structure could also be used for the monitoring of cellular activity (such as adherence and proliferation).

A novel optical sensor based on modified polymers, containing Mn^{2+} doped ZnS QDs was developed for the sensitive recognition of florfenicol. Results revealed that the fluorescence intensity of Mn:ZnS QDs increases linearly with the florfenicol concentration (Sadeghi et al., 2016).

The intracellular interaction between idarubicin and DNA was assessed by Mn-doped ZnS QDs capped with L-cysteine. The in vitro study aimed at detecting the drug–DNA interaction at room temperature. The drug adsorption on the surface of the QD impressively quenched the room-temperature phosphorescence, as a result of a combined mechanism of static and dynamic quenching. The addition of DNA determined the formation of a new drug–macromolecule complex, as the idarubicin desorbed. This simple, low-cost assay has another advantage as it avoids biochemical interferences (Ertas and Kara, 2015). Manganese doping photoluminescence reduction was reported for the multinary $CuInS_2$/ZnS core/shell QDs (Hua et al., 2014).

2.1 PbSe QDs

The synthesis routes for QDs used for the nanofabrication of semiconductors used in bioimaging are numerous, but in the medical field, bioinspired methods are more desirable. Though the field is still under development, PbSe QDs were synthesized in a cost-effective, ecofriendly manner from fungal biomass of *Aspergillus terreus*. The stress Pb and Se manifest toward the fungal cell activates a metal detoxification. This mechanism, connected with the phytochelatins, metallothioneins, and super oxide dismutases release grant the fungi the ability of biosynthesizing PbSe QDs, with absorption and emission peaks at 303/425 nm (Jacob et al., 2016).

2.2 Zn-CONTAINING QDs

Highly specific ZnO QDs conjugated with antimicrobial peptides $UBI_{29\text{–}41}$ and MPA is both able to detect and manifest cytotoxicity toward *Staphylococcus aureus*, *Escherichia coli*, and *Bacillus subtilis*. The affinity assay showed that the ZnO QDs uptake in *S. aureus* and *E. coli* was superior, according to the fluorescence intensity. Bacterial detection was evaluated both in vitro and in vivo, on mouse models; the infected areas were easily imaged indicating excellent bacteria-pointing ability of the complexed QDs (Chen et al., 2015). Ultrasonically synthesized ZnO QDs, under strictly controlled temperature and time, exhibited exceptional UV absorption abilities, which could be exploited for various biosystems sensitive to UV exposure (Yang et al., 2016).

Optical techniques for biomolecular imaging are being intensively researched. Various other combinations of QDs have been synthesized and tested with the purpose of labeling and target highlighting (Adegoke et al., 2016; Yang et al., 2015b).

InP/ZnS QDs were in vivo assessed for toxicity by using BALB/c mice models. After 84 days of evaluation, it was concluded that even under extremely high doses (25 mg/kg) the nanostructures did not lead to detectable side-effects related to either of hematology, biodistribution, blood chemistry, organ histology, nor body weight. The theory further formulated revolved around the lack of cadmium ion in the QDs, thus their future use as nanocarriers of optical markers could be possible (Lin et al., 2015a).

Aspergillus flavus was proven to be able to synthesize ZnS QDs after metal absorption in Zn enriched environment, in consonance with the internal mechanism reported for the other fungal species (Uddandarao and Mohan, 2016).

2.3 Ag-CONTAINING GDs

Ag_2S QDs displayed noteworthy in vivo and in vitro bioimaging properties in NIR-II second near-infrared interval (1000–1400 nm). With a maximum emission positioned at 1045 nm, stability, low levels of cytotoxicity, Ag_2S QDs could be used in bioimaging application as it might offer deeper penetration into the tissues, enhanced image contrast, low phototoxicity, and photobleaching (Zhao and Song, 2014). Considering that many organic substances emit in the infrared domain, a large number of authors described QDs with directed labeling properties, which can offer satisfying images of biological matter. Such QD complex has been recently investigated. The research team synthesized Ag_2Te QDs for targeted cell imaging featuring tunable emission in near infrared (NIR) (1000–1700). Two types of Ag_2Te QDs were revealed and evaluated, one simple and another Ag_2Te core/SiO_2 shell. L02, A549, HeLa, HepG2 cell lines were used for the cancer marking properties investigation (Jin et al., 2016).

Near infrared emitting polymer coated $AgInS_2$ QDs were used for a preliminary in vivo study, with promising result (Tan et al., 2015a). In vivo biodetection of tumor cells was achieved and reprised with CdHgTe QDs, conjugated with antivascular endothelial growth factor in mice models. The water soluble QDs were injected in the tail vein of mice, and their motility was fluorescent imaged at numerous time intervals. The aim of the study was to assess the SCG-7901 cells (gastric cancer cells) biosensing behavior of the functionalized QDs. A comparative study of bare and functionalized QDs study concluded that the lack of superficial conjugated antibodies makes it impossible to detect the location of the QDs. Last but not least, it was observed that the obtained complex is metabolized and excreted through urine (Pang et al., 2016).

2.4 Cd-CONTAINING CDs

Cadmium ions could be efficiently detected by employing SnO_2 QDs platforms. The low temperature hydrothermally synthesized crystalline QDs, with a size averaging to 3 nm and a sensitivity of 77.5 10^2 nA ppm/cm^2, respectively a lower detection limit of 0.5 ppm, can effectively mediate the detection of Cd ions from solutions in situ (Bhanjana et al., 2015). Cu^{2+}, as well as L-cysteine, can be rapidly and label-free detected by a fluorescent nanosystem of graphene QDs decorated with europium. The strong affinity of copper ions and L-cysteine to the complex resulted in an excellent linear relationship in 0.1–10 μM and 0.5–50 μM for Cu^{2+} and L-cysteine, respectively (Lin et al., 2015b). Recently, a nanohybrid of graphene QDs, polypirrole and Prussian blue was successfully used for L-cysteine amperometric determination (Wang et al., 2016c).

CdTe tumor-targeting QDs were used in vivo on a mouse model as imaging and chemotherapeutic agents. The cancer-targeting was achieved by encapsulating the QDs and doxorubicin in mixed nanopolymersomes polyethyleneglycol/poly(lactic-*co*-glycolic acid) (PEG/PLGA) and activated by folate conjugation for strong binding with the receptors from the surface of the malignant cells. After intravenous injection, the intricate system was monitored on the 4T1 breast adenocarcinoma cells in mice. Fluorescence microscopy imaging revealed that 6 h after administration, the QDs were accumulating at the tumor sites. In time, the synergistic antitumoral effect of QDs and doxorubicin was highlighted, with low cytotoxicity and agglomeration in the healthy tissues, according to the histopathological and physiological tests (Alibolandi et al., 2016).

Regenerative medicine might benefit from CdTe and CdSe QDs properties, proven to be efficient on labeling stem cells for fluorescent imaging in vivo, with higher sensitivity. Stem cell labeling could either be accomplished by standard conjugation or by QDs transduction into the stem cells, the latter being questioned for the cytotoxicity complications that could occur. On the other hand, the chemical modifications (with low concentrations of cationic liposomes, peptides rich in arginine and lysine sequences, nanocarriers) are, so far, considered the only safe approach as such molecules have been clinically used for DNA and proteins transductions into cells (Onoshima et al., 2015).

For the detection of ochratoxin A, a metabolite with teratogenous and immunosuppressive effects, a system of enhanced electrochemiluminescence of RuSi nanoparticles by energy transfer with CdTe QDs in a polysaccharide matrix was developed by Wang and coworkers. Ochratoxin A is produced generally by *Aspergillus* and *Penicillium* (Wang et al., 2016d). Aflatoxins are secondary metabolites produced by *A. flavus* and *Aspergillus parasiticus* that can be the cause of acute and chronic toxicity in humans. Self-assembled star shaped ZnS QDs coated with thioglycolic acid provide excellent toxin detection properties, and high sensitivity, as showed by the employed photoluminescence techniques (Bhardwaj et al., 2016).

Highly selective DNA biosensors, consisting of a ruthenium-single color QDs could prove useful in clinical diagnosis and forensic investigations. The CdTe complex was tested on 1% human serum samples, with feasible results and state of the art targeting ability against one-, two- and three-base mismatched sequences (Su et al., 2015). The applicability of the CdTe QDs fluorescence was further detailed and justified for various biosystems, immunoassays, and biosensors (Estupinan-Lopez et al., 2016). Nucleic acids can be detected with graphene QDs whose activity relies on the catalytic activity of G-quadruplex DNAzyme (Wang et al., 2016a). The description of CdTe QDs synthesis via electrochemical cavity cell approach was highlighted and their emission spectrum and respective applications were discussed. The nanocrystaline structures, with light emission from 520 to 576 nm (green to orange) were stabilized with mercaptopropionic acid, 2-thiosalicylic acid, and cysteamine, and since the fabrication route did not involve toxic compounds and residues, bioapplicability is considered for these systems (Passos et al., 2016).

Though the biocompatibility of QDs is acceptable generally, studies show that CdTe QDs might also have in vivo side-effects, such as influencing lipid metabolism in certain organs (i.e., *Caenorhabditis elegans* intestine) (Wu et al., 2016b), or the specific accumulation in tissues (cadmium ions or CdSe QDs in zebrafish larvae) (Zarco-Fernandez et al., 2016).

Breakthroughs in molecular staining and imaging are being made on daily basis. Recent studies revealed hydrogel incorporated (Sachdev et al., 2016) or bare QDs mediated sensing and quantified various lung cancer biomarkers (Liu et al., 2016c), pH sensitive CdTe loaded nanogels for both imaging and targeted drug delivery (Li et al., 2015), antibody conjugated CdSe/ZnS QDs immunohistochemical analysis, and cancer biomarkers labeling, qualitative and quantitative assay of postdigestion compounds in blood by MoS_2 QDs (Vasilescu et al., 2016).

A biosensor consisting of CdTe QDs was used to detect herring DNA. At the "turn off" the QDs' fluorescence could be quenched by safranin T (ST) via an electron transfer phenomenon. The ST could be embedded into the DNA due to its high affinity to it, forming a stable complex. As the ST attached to the QDs merged with the macromolecule, the "turn on" fluorescence was restored to the initial parameters. The selectivity and sensitivity of the sensor are remarkable, the QDs system being able to detect DNA amounts of 10.8 ng/mL, with the lack of interference from organic compounds, metal ions etc.; thus, it could be presumed a significant step forward in the DNA monitoring (Wang et al., 2016b).

Sensors for hepatitis B and hepatitis C viruses were revealed, based on the properties of electrochemiluminescence manifested by multicolor QDs and gold nanoparticles. The DNA sensors, consisting of CdTe QDs linked on a glass carbon electrode by grapheme nanosheets, were able to selectively capture the viral DNA. The electrochemiluminescence behavior derives from the interaction of the QDs with $K_2S_2O_8$, chosen as coreactant, where SO^{*-}_4 radicals react with the CdTe QDs, leading them to an excited state via electron transfer annihilation. The sensitive, reproducible, and stable complex sensor was further optimized in terms of pH, functional groups, intensity, and target DNA concentration (Liu et al., 2016b).

Room temperature prepared CdSe QDs were tested for the sensing ability of ascorbic acid, oxalic acid, citric acid, and hydrogen peroxide. Comparatively, hot injection synthesized QDs were investigated for the same properties. Nonetheless, differences of size were also easily observed: the "greener" room temperature route lead to one size QDs (3.3 nm) while the particles obtained under hot injection approach sized from 2.5 to 6.3 nm (Mir et al., 2016). Excellent monitoring of Fe^{3+} ions can be achieved by the employment of economically, green synthesized graphene QDs composites (graphene + lignin-sulfonate-based carbon materials). Considering its cost-effectiveness and the abundance of raw materials, it is believed that for the practical applications of these kind of nanocomposite the development of similar compounds will gain more interest (Xu et al., 2016). Comparable results were reported for graphene QDs sensitive to iron ions and pH value. Apart from the limit of detection of 0.26 μM and the ability to detect exclusively iron ion from metal ion mixtures, the reported nanocompound also manifests sensitivity to the pH values (2–12), suggesting the probability of developing optical pH sensors; additionally, its good biocompatibility was highlighted (Zhang et al., 2016).

Golgi protein-73 can be biosensed by employing an immunoassay based on carboxylic-functionalized-Mn-modified CdTe/CdS QDs system. The anti-GP73 antibody capped QD are able to specifically combine with GP73, when the fluorescent intensity quenches, with an inferior detection limit of 10 ng/mL (Liu et al., 2016e).

A label-free fluorescent sensor for biomolecule detection (adenosine, dopamine, 17-β-estradiol) was developed with QDs and the Ru complex $[Ru(bpy)_2dppz]^{2+}$. The CdTe QDs system features rapid analysis and detection, straightforward targeting, and extremely low detection limits: 101, 19, and 37 nM, respectively (Huang et al., 2016).

An immunoassay based on QDs was developed to concurrently detect H5 and H5 subtypes of Influenza A virus. Water-soluble, carboxyl-functionalized CdSe QDs were specifically designed to bind the antiviral antibodies via an amide group. One QD was able to bind one H5 and one H9 strain, resulting in a sandwich complex, with detection limits of 0.016 and respectively 0.25 hemagglutinating units. Under UV excitation at 365 nm, the complexes emit a bright fluorescent light, which was detected in real time with a lateral flow immunoassay system (Wu et al., 2016a).

Bovine serum albumin (BSA) was used as QDs precursors (Cd^{2+}) sequestering protein for an in situ synthesis of CdSe QDs. The ratio protein and cadmium ions, as well as the chelating time, temperature, and pH were pointed out as crucial parameters that could influence the QDs quality and activity. After the thermogravimetric analysis, the results pointed to the existence of conjugating bonds between the CdSe QDs and the –OH, –NH, and –SH functional groups in the BSA. The fluorescence tests ensured the opportunity to assess the labeling properties on *E. coli* (Wang et al., 2011).

Cationic QDs could adhere to bacterial surfaces and their interaction was assessed by Yang et al. (2015a) for *E. coli* DH5α, *B. subtilis* B168, *Listeria monocytogenes* CMCC(B) 54002, *Pseudomonas aeruginosa* ATCC 27853, *Staphylococcus epidermidis* CMCC(B) 269069, *Proteus vulgaris* CMCC(B) 49027, *Salmonella paratyphi* A CMCC(B) 50093, *S. aureus*, *Bacillus cereus*, and *Serratia sp.* Cysteamine-stabilized CdTe QDs' adherence was investigated by fluorescence spectrometry and microscopy. The evaluation revealed discriminatory results, which were interpreted considering that Gram-negative bacteria benefits from extra protection ensured by the second lipid layer (Yang et al., 2015a). In a different study, strains of *P. aeruginosa* was labeled with three types (COOH—, COOH—PEG—, NH_2—PEG—) of CdSe/ZnS QDs after direct incubation, with more convenient results for the latter. The low toxicity of this type of QDs could endorse their use for clinical treatment (Kumar et al., 2016). Recently, a paper subjected a mechanism of preserving the fluorescent properties of CdTe QDs-based sensors and biosensors, by embedding them between montmorillonite layers; as a plus, the exhibited luminescence is superior to the one sourced in bare CdTe QDs (Guo et al., 2016).

Labeling tumor cells could be eased by QDs employment. CdSe-core ZnS-shell QDs, with functionalized polymer coatings were proven to exhibit superior luminescence and photostability compared with flourophores of organic origin for glioma cells. The possibility of cancerous cell motility and metastasis monitoring using QDs is being researched (Onoshima et al., 2015). Another strategy of cancer diagnosis by bioimaging the tumor presence and growth was presented by Sukharova et al. (2016). The complex consists of CdSe core/shell ZnS QDs and specific antibodies (single domain antibodies). These nanoprobes, when excited with at least two photons, provide fluorescent images in the near IR domain (Sukharova et al., 2016).

Different means of tumor detection by QDs rests on the pH response as it targets the hypoxic cancer cell. The strategy aims at siRNA delivery to the tumor site, which can be difficult due to the lack of angiogenesis and the presence of efflux transporters. The designed system to overcome these shortcomings dwells on a CdTe QD core and a shell of 2-deoxyglucose (DG)-polyethylene glycol (PEG) connected with the compound of lipoic acid, lysine and 9-poly-D-arginine (LA–Lys–9R) by a hydrazone bond for irreversible and site-specific delivery of siHIF-1α. With an undoubtful biocompatibility, the QD system was tested on EA.hy926 human umbilical vein cells, HepG2 human hepatocellular carcinoma cells, U87MG human glioma cells, SGC-7901 gastric cancer cells, and MCF-7 human breast adenocarcinoma cells and its cytotoxicity was MTT assayed (Zhu et al., 2015).

Chaotic proliferation of commonly spread colorectal cancer cells could be visualized with paramagnetic hydrophobic $CdSe/CdS^{2ML}/Cd_{0.75}Zn_{0.25}S/Cd_{0.5}Zn_{0.5}S/Cd_{0.25}Zn_{0.25}S/ZnS^{2ML}$ core/shell QDs according to a recent "imaging-biopsy" strategy. Their coating consists in paramagnetic pentetic acid-Gd coupled BSA and a superficial antibody, anti-Glut1. In vitro, in vivo, and ex vivo tests were performed with encouraging results (Xing et al., 2015).

Au nanorods decorated CdSe/CdS/ZnS QDs silicon coated structure help cancer imaging due to bimodal fluorescent plasmonic effects (Sneha et al., 2016). A 549 cancer cell line was selectively marked after transferrin functionalization. The test subjected cells were able to internalize this complex. To avoid interference between the fluorescent and plasmonic signals, each layer was specially designed: the gold nanorods would emit at 625 nm, and the nanostructure as a whole, at values between 550 and

750 nm (Wu et al., 2015). Gd-dopes $CuInS_2$ have similar properties that favor their used for bimodal imaging as well (Zheng et al., 2016).

Pathogenic entities labeling has also been remarkably achieved in the past years by employing various QDs and tagging strategies.

Vijian et al. (2016) described a diagnosis test based on QDs that could be able to detect pathogenic entities that could infect human organism. The PbS, CdS, and ZnS QDs functionalized with specific DNA probes from *Vibrio cholerae*, *Salmonella sp.*, and *Shigella sp.* were targeted toward sequences of nonprotein coding RNA of the mentioned species. The high sensitivity the assay (60, 59, and 53 attomolar, respectively) recommends it as a viable instrument of detecting pathogens from clinical samples, food, or other sources (Vijian et al., 2016).

Specific DNA and superficial antigens of *Leishmania* species (*L. tropica, L. major, L. aethiopica, L. mexicana, L. braziliensis braziliensis, L. braziliensis panamensis, L. braziliensis guyanensis, L. donovani, L. infantum*, and *L. chagasican*) were detected and quantified by using a combination of previously functionalized magnetic beads and CdSe QDs. With the limit of detection of 3125 nm/μL and 10^3 cell/mL for DNA and protein, the approach could be considered for clinical implementation, particularly in resource-poor areas where the infections are frequent (Andreadou et al., 2016).

Viral (AcMNPV) infected *Spodoptera frugiperda* was easily visualized by a fluorescent QDs assay (Zhang et al., 2013), CdTe QDs, and CdTe/ZnS core/shell (Yang et al., 2013) and carbon nanostructures (Weng et al., 2015). QDs labeled *E. coli* in vitro, silica-coated CdTe/CdS QDs proved efficient for HeLa and A549 cell labeling (Ma et al., 2014), AgInS2-ZnS QDs for MCF7 cancer cells staining (Fahmi and Chang, 2016), AgInS2-ZnSe QDs as alternative for Cd-/Pb- based QDs for bioimaging (Che et al., 2016), while CdZnSe/ZnS QDs could be selectively tuned for PC12, C17.2, and HUVEC cells labeling (Soenen et al., 2014), whereas nanoemulsion containing functionalized core–shell InGaP/ZnS QDs can selectively image phagocytic and nonphagocytic immune cells (Lim et al., 2010). In the cancer field, imaging could be enriched by the employment of CuInZnS/ZnS QDs (Vinayagam et al., 2016), microfluidic fabrication of CdSe QDs with fatty amines used as ligands for an alleged superior dispersion behavior (Tian et al., 2016).

Bifunctional oligonucleotides conjugated QDs were synthesized for the first time and their activity was monitored and described after immunological and cell imaging assays. Characterized by a strong 560 nm green emission fluorescent signal, the cytosine-phosphate-guanine functionalized CdTe QDs' features endorse them as strong immunostimulatory agents (Tao et al., 2015). Antitumor antibody conjugated multishell CdSe/ZnS/CdS/ZnS QDs fluorescent effect can aid disseminating cancer cells from healthy tissues (Linkov et al., 2016).

Techniques for fluorescence, phosphorescence, sensitivity, and biocompatibility enhancement have also been developed for a better biological behavior, in terms of parameter optimization: thickness of self-assembled (InAs/GaAs) QDs (Mohanta et al., 2016), surface modification of CdS QDs (Bastos-Arrieta et al., 2016), carbon nanotube coating of CdTe QDs for MCF-7 cell labeling, and possible bioactive compounds targeted delivery (Baslak et al., 2015), dispersion factors establishing for CdSSe/ZnS QDs colloids seizing (Vanengen Spivey, 2016), carbon quantum dots coupling with thermoresponsive polymers for increased metabolic degradation resistance (Kavitha et al., 2016), the capping agent influence on the essential properties (structural, photocatalytic and optical) of ZnS QDs (Rajabi and Farsi, 2016), overcoming the toxicity by increasing the impact/effect on the biological systems (Fan et al., 2016), enhancing the in situ doping with active metal particles resulted from the metabolism process, as biofluids sanitation approach after metal implants metal ions emission (Danscher and

Stoltenberg, 2006), metal ions influence on QDs nanocomposites ($CdTe:Cu^{2+}/TiO_2$ nanotubes) (Chen et al., 2016), (Ni^{12+} irradiation on PbS QDs) (Chowdhury et al., 2005), crystalline size tuning for emission shifting (e.g., Co-ferrite QDs) (El-Sayed and Agami, 2015), superficial postprocessing on $CuInS_2$ QDs, prior to being integrated in a biosensor (Akdas et al., 2015), ferromagnetic QDs for possible eased cancer targeting and treatment (Ge QDS with Mn doping) (Ma and Lou, 2012; Thota et al., 2016), increasing the sensitivity to various diseases specific markers (β-HCG and AFP of Down Syndrome) (Liu et al., 2013), etc.

Regenerative medicine might benefit from CdTe and CdSe QDs properties, proven to be efficient on labeling stem cells for fluorescent imaging in vivo, with higher sensitivity (Onoshima et al., 2015).

2.5 GRAPHENE-BASED QDs

The early detection of the cardiac marker antigen troponin 1 (cTn1) in blood vessels could prevent a myocardial infarction in initial stages. With this aim, amine functionalized anti-cTn1 conjugated graphene QDs sensors based on fluorescence resonance energy transfer were synthesized. The analysis involved photoluminescence tests to evaluate the interaction of the targeted cTn1 with the specific antibody. Exhibiting superior specificity, a limit of detection of 0.192 pg/mL and low response to various nonspecific antigens, the methods could benefit from validation for clinical use (Bhatnagar et al., 2016). According to a study, graphene QDs could be tuned in terms of density defects and functional groups by gamma-ray irradiation. It was estimated that the decrease of defects could exponentially improve the sensitivity, stability, and the detection limits of the sensors QDs (Lu et al., 2016). Retinal and choroidal vasculature development can be prevented by angiotensin receptor blocker conjugated QDs which could be tuned to behave as "nanodressings" for ocular lesions too (Henning et al., 2015). Renal diseases could be similarly prevented and also detected by a composite system of graphene QDs and Fe_3O_4 nanoparticles anti-IgG conjugated for urine entities analysis. Highly sensitive and specific, with a wide linear range, this novel system represents a model for a near-future fluorescent assay for clinically investigating renal disease (Jiang et al., 2016).

Biocompatible carbon dots doped with sulfur and nitrogen were simple and cost-effective hydrothermally synthesized for biomedical applications, such as doxycycline probing and multicolor imaging in living cells. Cytotoxicity tests were performed on HepG2 and fluorescent quenching effect of the drug was registered. As the feature small sizes, good photostability, and biocompatibility, bioimaging assays were also carried out. The QDs complex-stained cells, under 405, 488, 514 nm exhibited blue, green, and red coloration, thus proving the systems manifest advanced optical properties, with an ample excitation wavelength (Xue et al., 2016). Hydrothermally modified in dopamine and cysteine solution, fluorescent carbon dots exhibit cytocompatible properties and are efficient for cell (L929 fibroblast-like cells) probing. It was shown that living cells are selectively recognized, labeled and imaged, while the dead cells are left aside. The incubation of the QDs (40%) with cells results in an overwhelming viability of around 140%. The cell bioimaging reveals the intracellular environment, as well as the cell wall and membrane. The adhesion depends on the average QDs size (Jana et al., 2016).

Liu et al. reported a simple method for nitrogen and sulfur codoping of graphene QDs, aiming to obtain a compound able to monitor the pyrophosphate ions in synovial fluid of arthritis patients. Pyrophosphate ion ($P_2O_7^{4-}$) is involved in DNA polymerization, ATP hydrolysis, and other metabolic processes, including the pathological process of arthritis, thus its monitoring raises interest. The fabrication method of QDs includes a combination of pyrolysis and dialysis treatment of the raw materials.

N-S graphene QDs can be quenched by the addition of Fe^{3+}; however, the addition of pyrophosphate ions results in the fluorescence recovery (Liu et al., 2016g). Similarly, green synthesized nitrogen doped carbon dots are able to detect ferric ions in water. With blue fluorescence emission at 411 nm after being excited at 310 nm, and the possibility of extending the applicability to bioimaging, due to their low cytotoxicity upon MDA-MB-231cells, this low cost versatile biosensor could be used in the future for standard assays which today are being performed with more toxic compounds (Edison et al., 2016).

Glutathione detection was perfected by a sensor based on graphene QDs doped with nitrogen and coated with AgNPs. The nanocomposite is able to catalyze the oxidation of 3,3'5,5'-tetramethylbenzidine to oxTMB in H_2O_2 rich environments accompanied by a color change (bright blue, 652 nm). Glutathione is a reducing agent for oxTMB; a strong relationship is illustrated between the UV-VIS absorption intensity and the glutathione concentration. This sensor has an important role in future blood serum analysis (Ju et al., 2016).

Gadolinium (III) chelates functionalized carbon QDs feature, represents an ideal bioimaging fluorescent properties, magnetic resonance, which could lead to their employment in dual-mode comprehensive visual investigation of biological samples. Being able to penetrate cell membranes, solvating in water and manifesting extremely low cytotoxicity, the new hybrid nanosystem could become in the future a strong ally in the intracellular labeling and bioimaging (Shi et al., 2015b).

Carbon QDs were also obtained for support of AgNPs for the electroanalytical evaluation of H_2O_2 in fetal bovine serum. The sensor was based on carbon QDs hydrothermally synthesized from a biopolymer, embedded with silver under UV irradiation, (Jahanbakhshi and Habibi, 2016). Heparin can be effectively detected by a phosphorescent mercaptopropionic acid capped Mn-doped ZnS QDs biosensor, at room temperature (Zhang et al., 2015b). The complex proved to be effective for phosphopeptides sensing in the presence of zirconium (IV) as a quenching agent (Gong and Fan, 2015). Moreover, heparin conjugated CdTe QDs are suited for the in vitro imaging of a considerable cell lines spectrum: THP-1, A549, and Caco-2 (Maguire et al., 2014).

To understand the viral infection, Wen et al. theorized that the internal nucleocapsid of enveloped viruses could be labeled with QDs. In their experiment, the baculovirus strain was marked with straptavidin-conjugated QDs and subjected to a series of tests (Western blot assay, TEM, FTIR) to visualize the viral entities as they were infecting the Sf9 cell culture. Comparative one-step growth curves tests showed that the tagging functionalized QDs did not interfere with the infectivity of the virus. The focus of the study, tracking the strains on its infection route was successfully accomplished, as the optical and analytical results that were obtained could provide opportunities of more thorough evaluations (Wen et al., 2014).

Intracellular ionic detection was approached by Liu et al. in a recent study. The authors created a graphene QDs chemsensor aimed at identifying and imaging Hg^{2+}, which has always raised considerable interest due to the health problems and environmental issues that cascade after mercury contamination. The QD sensor, based on fluorescence resonance energy transfer, the graphene supplies the electrons while the rhodamine derivatives bonded to it stands for the energy acceptor when mercury ions are encountered. Water soluble, biocompatible, highly selective, and nontoxic, the sensor can be practically used in intracellular and water samples mercury detection (Liu et al., 2015b). Green synthesized nitrogen-doped graphene QDs were found to be suitable for pH-sensitive photoluminescence and mercury ions detection. The nitrogen dopage provides the graphene QDs with fluorescence, independent from wavelength excitation (an unique maximum emission at 445 nm). Furthermore, it was assessed that metal ions can have an efficient quenching effect as a result of nonradiative electron transfer, thus, making it convenient for strikingly sensitive for Hg^{2+}detection, at concentrations as low as 8.6 nM (Shi et al., 2015a).

Peroxidase-like magnetic $ZnFe_2O_4$/graphene QDs as a mimic enzymatic label could be used to sense DNA through an electrochemical approach. The electron transfer rate was improved by modifying immobilized graphene and Pd nanowires on a glass carbon electrode. The peroxidase-like catalytic activity of the nanohybrid resides on the synergistic activity of the graphene QDs and of the magnetic compound. Under favorable conditions, the biosensor had a detection limit of 6.2×10^{-17} M, and a liner range between 10^{-16} and 5×10^{-9} M. Apart from the robust properties, this system is also environmentally friendly since it avoids the use of enzyme labels and presents itself with high stability (Liu et al., 2014).

High quantum yield carbon QDs can be used for labeling the albumin contained in bovine serum with the benefits of extending the fluorescent biomarking to other biomolecules (Liu et al., 2016d). Nitrogen-doped carbon dots derived from BSA and formic acid could also be used for live cells (Tca-8113 tongue squamous cancer line) nuclear bioimaging due to their one- or two-photon fluorescent behavior (Tan et al., 2015b).

Gram-positive rods could be better assayed with vancomycin-modified (amine and carboxyl condensation reaction) carbon dots. It was shown that this rapid and convenient strategy, with a detection limit of 9.40×10^4 cfu/mL could directly target bacterial cultures, without any additional treatment. The vancomycin amount is a crucial factor that was considered for the optimization, as an excess could influence the morphology and the size of the carbon dots (Zhong et al., 2015). 3 nm fluorescent core mannose/folic acid carbon QDs are also suited for *E. coli* labeling, according to a recent study. Further, it was demonstrated that the 365, 450 nm, respectively emission maxima QDs, could detect the pathogen at low concentrations of 100 colony forming units/mL. The targeting properties of the carbon QDs also apply to folate receptor-overexpressing tumor cells (Lai et al., 2016).

Graphitic QDs UV-assisted synthesized upon graphite treatment, also manifest ferromagnetic properties (Swain et al., 2013).

Ecofriendly microwave assisted graphitic carbon nitride QDs with green fluorescent emission was found suited for in vitro bioimaging of HeLa cells. 488 nm laser pulse excitement revealed notable fluorescence inside the tagged cells after 24 h of cultivation in a medium with QDs addition. Giving its biocompatibility associated with low cytotoxicity, this system is promising for future bioimaging and biosensing in clinical field (Li et al., 2016).

Even though fluorescence imaging techniques for in vivo QDs-complexed stem cells monitoring have reached unparalleled peaks, with the development of ultrasensitive cameras, scattering, absorption, and autoflourescence attenuation methods (i.e., by emitting in the infrared "Biological Optical Window" of 700–900 nm), not all drawbacks have been overcome. Most complications are derived from the doubtful safety and high functionality of QDs. National or regional regulations are also corroborating in the deterring of stem cell regenerative applications (Onoshima et al., 2015).

The binding strategies are not always efficient, so further investigations are being required (Liu et al., 2016a). The biologically toxic cadmium released from cadmium containing QDs could agglomerate in the tissues of the tested lab animals. It was shown that QDs exposed zebra fish larvae exhibit cadmium bioaccumulations even at low dosages administration. Precautions are to be taken, considering that the metal ions is among the class I carcinogens listed by the International Agency of Research of Cancer and their adverse effects due to their small dimensions have not yet fully assessed for longer time periods (Zarco-Fernandez et al., 2016). Unwanted interaction with biological fluid proteins can be an important factor that unbalances the organic medium, on the one hand, and generates glitches, inaccuracies or impotence of QDs sensors (Hossinzadeh et al., 2016). Another drawback could be caused by narrow emission spectra of the QDs, that also need additional research (Grijseels et al., 2016).

3 QDs FOR THERAPY

3.1 ANTIBACTERIAL THERAPY

QDs, as part of nanocomposites, could improve the antibacterial activity of the more studied particles and compounds, some of which are already clinically available (Habiba et al., 2015). A comparative test on Gram-positive and Gram-negative bacteria on silver nanoparticles (Ag NPs), graphene QDs (GQDs), and Ag-GQDs embedded in polyethylene-glycol matrices revealed that the growth inhibition of *P. aeruginosa* and *S. aureus* is superior as a consequence of the synergistic effect of the dispersed nanocompounds; hence, the inhibitory concentration of Ag-GQDs is lower compared to those of AgNPs' and GQDs'. Moreover, normal mammalian cells incubated with Ag-GQDs maintained maximum viability as GQDs decorating the AgNPs decreased the citotoxicity as a result of their biocompatibility (Habiba et al., 2015).

A new nanocomposite with biomedical and possible antimicrobial activity was proposed by Ding et al. (2016), consisting of carbon QDs dispersed in a matrix of ZnO foam, under UV and VIS light irradiation conditions, it showed excellent photocatalytic activity in terms of rhodamine, methyl orange, and methylene blue degradation. This behavior can be explained through a series of factors, like the properties of transferring photo-induces electrons and up-converted photoluminescence QDs manifest, the light-trap effect ZnO features (Ding et al., 2016). Other oxide QDs have gained more interest in recent years; Solanki et al. (2015) reported the synthesis of rutile TiO_2 QDs via ECR sputtering at convenient conditions (room temperature). Depending on the nanostructure size, superparamagnetic behavior or ferromagnetic behavior was observed, being suggested that it could be used for localized targeting of bionetworks (Solanki et al., 2015).

3.2 CANCER THERAPY

CdTe tumor-targeting QDs were efficiently used in vivo on mouse model to act as both imaging and chemotherapeutic agents. The cancer targeting was achieved by encapsulating the QDs and doxorubicin in mixed nanopolymersomes (PEG/PLGA) and activated by folate conjugation for strong binding with the receptors from the surface of the malignant cells (Alibolandi et al., 2016). The citotoxicity of CdSe, CdTe, and InP QDs toward two human cancer cell lines (gastric adenocarcinoma, BGC-832, and neuroblastoma, SH-SY5Y) were thoroughly discussed in a paper (Liu et al., 2015a). The CdTe and InP QDs were obtained and assessed as core–shells as well (CdTe/ZnS, InP/ZnS) and the CdSe as core/shell/shell, namely, CdSe/CdS/ZnS. Before performing the bioassays, their surface suffered physicochemical modifications—PEGylation, Pluronic F127 encapsulation, and mercaptopropionic acid modifications—to facilitate their dispersion in aqueous media. Mercaptopropionic acid modified CdSe formula exhibited the most stable and convenient photoluminescence activity. The cancer cell viability tests and imaging revealed that the cell lines are selectively sensitive to certain QDs formulations, event al lower dosages (Liu et al., 2015a). Upon UV exposure bright blue emitting CuSe/ZnS/silica core/shell/shell QDs were extensively tested for their specific behavior during blood interaction for prospective in vivo tests. The methotrexate conjugate nanostructures were then tested for cytotoxicity on C_6 glioma and HepG2 cancer cell lines, with good results. As a future application it was theorized that QDs could serve as dual modal imaging probes for optical investigations and as MRI contrast agents, due to their paramagnetic copper core (Durgadas et al., 2012).

4 CONCLUSIONS

In this chapter, both QDs synthesis methods and bioimaging techniques were reviewed, with the focus on the latter. Recent studies revealed that bioimaging emerged as a relatively recent field, as the integration of chemistry, physics, and biology in an intricate overlapping. Commonly used for electronic devices, QDs have been investigated for the most explored approaches when it comes to molecular, cellular, and intracellular labeling, as a more effective approach compared to standard chemical-staining techniques. The broad absorption/narrow emission spectrum, durable fluorescence, high quantum yields, size dependent photoluminescence, outstanding photostability, and photobleaching-resistant behavior, endorse QDs for a broad spectrum of sensor-mediated analyses. Their intrinsic features and the conjugation processing are essential in the optimization of their optical behavior, in terms of fluorescent/phosphorescent emissions, according to the targeted entities to be labeled and the nature of their medium. Regenerative and palliative medicine can also benefit from the properties QDs feature: nanocarrier behavior, anticancer, and antipathogenic activities.

QDs can manifest distinctive degrees of toxicity, but strategies for their decreasing have been theorized, studied, and applied. Cappings, coatings, ion doping, or organic compounds conjugation were shown to broaden the spectrum of their applicability. Disadvantages still remain and are to be addressed, especially for in vivo applications. The exposure of biosystems to QDs must be done carefully as some are classified as carcinogens. Moreover, ecology-based concerns were also raised. Uncapped bare QDs are more difficult to control and identify in living organisms, but exceptions exist and binding strategies require further evaluations.

REFERENCES

Adegoke, Q., Mashazi, P., Nykong, T., Forbes, P.B.C., 2016. Fluorescence properties of alloyed ZnSeS quantum dots overcoated with ZnTe and ZnTe/ZnS shells. Opt. Mater. 54, 104–110.

Akdas, T., Distaso, M., Kuhri, S., Winter, B., Birajdar, B., Spiecker, E., Guldi, D.M., Peukert, W., 2015. The effects of post-processing on the surface and the optical properties of copper indium sulfide quantum dots. J. Colloid Interf. Sci. 445, 337–347.

Alibolandi, M., Abnous, K., Sadeghi, F., Hosseinkhani, H., Ramezani, M., Hadizadeh, F., 2016. Folate receptor-targeted multimodal polymersomes for delivery of quantum dots and doxorubicin to breast adenocarcinoma: in vitro and in vivo evaluation. Int. J. Pharm. 500, 162–178.

Andreadou, M., Liandris, E., Gazouali, M., Mataragka, A., Tachtsidis, I., Goutas, N., Vlachodimitropoulos, D., Ikonomopoulos, J., 2016. Detection of Leishmania-specific DNA and surface antigens using a combination of functionalized magnetic beads and cadmium selenite quantum dots. J. Microbiol. Methods 123, 62–67.

Baslak, C., Kars, M.D., Karaman, M., Kus, M., Cengeloglu, Y., Ersoz, M., 2015. Biocompatible multi-walled carbon nanotube–CdTe quantum dot–polymer hybrids for medical applications. J. Lumin. 160, 9–15.

Bastos-Arrieta, J., Munoz, J., Stenbock-Fermor, A., Munoz, M., Maraviev, D.N., Cespedes, F., Tsarkova, L.A., Baeza, M., 2016. Intermatrix Synthesis as a rapid, inexpensive and reproducible methodology for the in situ functionalization of nanostructured surfaces with quantum dots. Appl. Surf. Sci. 368, 417–426.

Bhanjana, G., Dilbaghi, N., Kumar, R., Umar, A., Kumar, S., 2015. Sno2 quantum dots as novel platform for electrochemical sensing of cadmium. Anal. Chim. Acta 891, 261–268.
Bhardwaj, H., Singh, C., Pandey, M.K., Sumana, G., 2016. Star shaped zinc sulphide quantum dots self-assembled monolayers: preparation and applications in food toxin detection. Sens. Actuat. B Chem. 231, 624–633.
Bhatnagar, D., Kumar, V., Kumar, A., Kaur, I., 2016. Graphene quantum dots FRET based sensor for early detection of heart attack in human. Biosens. Bioelectron. 79, 495–499.
Bi, L., Dong, X.T., You, Y.H., 2014. Room-temperature phosphorescence sensor based on manganese doped zinc sulfide quantum dots for detection of urea. J. Lumin. 153, 356–360.
Che, D., Zhu, X., Wang, H., Duan, Y., Zhang, Q., Li, Y., 2016. Aqueous synthesis of high bright and tunable near-infrared AgInSe2–ZnSe quantum dots for bioimaging. J. Colloid Interf. Sci. 463, 1–7.
Chen, H., Zhang, M., Li, B., Chen, D., Dong, X., Wang, Y., Gu, Y., 2015. Versatile antimicrobial peptide-based ZnO quantum dots for in vivo bacteria diagnosis and treatment with high specificity. Biomaterials 53, 532–544.
Chen, Q., Geng, Y., Zhou, C., Song, J., Zhou, L., 2016. Influence of Cu^{2+} dopant in optical property of CdTe quantum dots and photoelectrochemical performance of CdTe:Cu^{2+}/TiO2 nanotube arrays. Phys. E 79, 59–66.
Chowdhury, S., Mohanta, D., Ahmed, G.A., Dolui, S.K., Avasthi, D.K., Choudhury, A., 2005. J. Lumin. 114, 95–100.
Danscher, G., Stoltenberg, M., 2006. Silver enhancement of quantum dots resulting from (1) metabolism of toxic metals in animals and humans, (2) in vivo, in vitro and immersion created zinc–sulphur/zinc–selenium nanocrystals, (3) metal ions liberated from metal implants and particles. Prog. Histochem. Cytochem. 41, 57–139.
Ding, D., Lan, W., Yang, Z., Zhao, X., Chen, Y., Wang, J., Zhang, X., Zhang, Y., Su, Q., Xie, E., 2016. A simple method for preparing ZnO foam/carbon quantum dots nanocomposite and their photocatalytic applications. Mater. Sci. Semiconduct. Process. 47, 25–31.
Durgadas, C.V., Sreenivasan, K., Sharma, C.P., 2012. Bright blue emitting CuSe/ZnS/silica core/shell/shell quantum dots and their biocompatibility. Biomaterials 33, 6420–6429.
Edison, T.N.J.I., Atchudan, R., Shim, J.J., Kalimuthu, S., Ahn, B.C., Lee, Y.R., 2016. Turn-off fluorescence sensor for the detection of ferric ion in water using green synthesized N-doped carbon dots and its bio-imaging. J. Photochem. Photobiol. B Biol. 158, 235–242.
El-Sayed, H.M., Agami, W.R., 2015. Controlling of optical energy gap of Co-ferrite quantum dots in poly (methyl methacrylate) matrix. Superlatt. Microstruct. 83, 651–658.
Ertas, N., Kara, H.E.S., 2015. L-Cysteine capped Mn-doped ZnS quantum dots as a room temperature phosphorescence sensor for in-vitro binding assay of idarubicin and DNA. Biosens. Bioelectron. 70, 345–350.
Estupinan-Lopez, C., Tolentino Dominguez, C., Cabral Filho, P.E., Santos, B.S., Fontes, A., De Araujo, R.E., 2016. A pH dependence study of CdTe quantum dots fluorescence quantum yields using eclipsing thermal lens spectroscopy. J. Lumin. 174, 17–21.
Fahmi, M.Z., Chang, J.Y., 2016. Potential application of oleylamine-encapsulated AgInS2-ZnS quantum dots for cancer cell labeling. Procedia Chem. 18, 112–121.
Fan, J., Sun, Y., Wang, S., Li, Y., Zeng, X., Cao, Z., Yang, P., Song, P., Wang, Z., Xian, Z., Gao, H., Chen, Q., Cui, D., Ju, D., 2016. Inhibition of autophagy overcomes the nanotoxicity elicited by cadmium-based quantum dots. Biomaterials 78, 102–114.
Finetti, C., Plavisch, L., Chiari, M., 2016. Use of quantum dots as mass and fluorescence labels in microarray biosensing. Talanta 147, 397–401.
Gong, Y., Fan, Z., 2015. Highly selective manganese-doped zinc sulfide quantum dots based label free phosphorescent sensor for phosphopeptides in presence of zirconium (IV). Biosens. Bioelectron. 66, 533–538.
Grijseels, S.C.M., Van Bree, J., Koenraad, P.M., Toropov, A.A., Klimko, G.V., Ivanov, S.V., Pryor, C.E., Silov, A.Y., 2016. Radiative lifetimes and linewidth broadening of single InAs quantum dots in an AlxGa(1 − x) As matrix. J. Lumin. 176, 95–99.

Guo, Y., Ge, X., Guan, J., Wu, L., Zhao, F., Li, H., Mu, X., Jiang, Y., Chen, A., 2016. A novel method for fabricating hybrid biobased nanocomposites film with stable luorescence containing CdTe quantum dots and montmorillonite-chitosan nanosheets. Carbohydr. Polym. 145, 13–19.

Habiba, K., Bracho-Rincon, D.P., Gonzales-Feliciano, J.A., Villalobos-Santos, J.C., Makarov, V.I., Ortiz, D., Avalos, J.A., Gonzales, C.I., Weiner, B.R., Morell, G., 2015. Synergistic antibacterial activity of PEGylated silver–graphene quantum dots nanocomposites. Appl. Mater. Today 1, 80–87.

He, L., Mei, S., Chen, Q., Zhang, W., Zhang, J., Zhu, J., Chen, G., Guo, R., 2016. Two-step synthesis of highly emissive C/ZnO hybridized quantum dots with a broad visible photoluminescence. Appl. Surf. Sci. 364, 710–717.

Henning, R., Ohlmann, A., Staffel, J., Pollinger, K., Haunberger, A., Breunig, M., Schweda, F., Tamm, E.R., Goepferich, A., 2015. Multivalent nanoparticles bind the retinal and choroidal vasculature. J. Control. Release 220, 265–274.

Hossinzadeh, G., Maghari, A., Sabouri, A.A., Moosavi-Movahedi, A.A., 2016. Unfolding of insulin at the surface of ZnO quantum dots. Int. J. Biol. Macromol. 86, 169–176.

Hua, J., Zhang, Y., Yuan, X., Cheng, H., Meng, X., Zhao, J., Li, H., 2014. Photoluminescence properties of Cu–Mn–In–S/ZnS core/shell quantum dots. Superlat. Microstruct. 73, 214–223.

Huang, H., Shi, S., Gao, X., Gao, R., Zhu, Y., Wu, X., Zang, R., Yao, T., 2016. A universal label-free fluorescent aptasensor based on Ru complex and quantum dots for adenosine, dopamine and 17β-estradiol detection. Biosens. Bioelectron. 79, 198–204.

Jacob, J.M., Sharma, S., Balakrishnan, R.M., 2016. Exploring the fungal protein cadre in the biosynthesis of PbSe quantum dots. J. Hazard. Mater., pii: S0304-3894.(15)30317-4. doi: 10.1016/j.jhazmat.2015.12.056

Jahanbakhshi, M., Habibi, B., 2016. A novel and facile synthesis of carbon quantum dots via salep hydrothermal treatment as the silver nanoparticles support: application to electroanalytical determination of H_2O_2 in fetal bovine serum. Biosens. Bioelectron. 81, 143–150.

Jana, J., Ganguly, M., Das, B., Dhara, S., Negishi, Y., Pal, T., 2016. One pot synthesis of intriguing fluorescent carbon dots for sensing and live cell imaging. Talanta 150, 253–264.

Ji, X., Peng, F., Zhong, Y., Su, Y., He, Y., 2014. Fluorescent quantum dots: synthesis, biomedical optical imaging, and biosafety assessment. Colloid Surf. B Biointerf. 124, 132–139.

Jiang, D., Ni, D., Liu, F., Zhang, L., Liu, L., Pu, X., 2016. A fluorescent imaging assay of cast in renal disease based on grapheme quantum dots and Fe_3O_4 nanoparticles. Clin. Chim. Acta 454, 94–101.

Jin, H., Gui, R., Sun, J., Wang, Y., 2016. Glycerol-regulated facile synthesis and targeted cell imaging of highly luminescent Ag2Te quantum dots with tunable near-infrared emission. Colloid Surf. B Biointerf. 143, 118–123.

Ju, J., Zhang, R., Chen, W., 2016. Photochemical deposition of surface-clean silver nanoparticles on nitrogen-doped graphene quantum dots for sensitive colorimctric detection of glutathione. Sens. Actuat. B 228, 66–73.

Kavitha, T., Kim, J.O., Jang, S., Kim, D.P., Kang, I.K., Park, S.Y., 2016. Multifaceted thermoresponsive poly(*N*-vinylcaprolactam) coupled with carbon dots for biomedical applications. Mater. Sci. Eng. C 61, 492–498.

Kumar, N., Wiraja, C., Palnisamy, K., Marsili, E., Xu, C., 2016. Heat shock mediated labelling of *Pseudomonas aeruginosa* with quantum dots. Colloid. Surf. B Biointerf. 142, 259–265.

Lai, I.P.J., Harroun, S.G., Chen, S.Y., Unnikrihsnan, B., Li, Y.J., Huang, C.C., 2016. Solid-state synthesis of self-functional carbon quantum dots for detection of bacteria and tumor cells. Sens. Actuat. B 228, 465–470.

Li, H., Shao, F.Q., Huang, H., Feng, J.J., Wang, A.J., 2016. Eco-friendly and rapid microwave synthesis of green fluorescent graphitic carbon nitride quantum dots for vitro bioimaging. Sens. Actuat. B 226, 506–511.

Li, Z., Xu, W., Wang, Y., Shah, B.R., Zhang, C., Chen, Y., Li, Y., Li, B., 2015. Quantum dots loaded nanogels for low cytotoxicity, pH-sensitive fluorescence, cell imaging and drug delivery. Carbohydr. Polym. 121, 477–485.

Lim, Y.T., Cho, M.Y., Kang, J.H., Noh, Y.W., Cho, J.H., Kwan, S.H., Chung, J.W., Chung, B.H., 2010. Perfluorodecalin/[InGaP/ZnS quantum dots] nanoemulsions as ^{19}F MR/optical imaging nanoprobes for the labeling of phagocytic and nonphagocytic immune cells. Biomaterials 31, 4964–4971.

Lin, G., Ouyang, Q., Hu, R., Ding, Z., Tian, J., Yin, F., Xu, G., Chen, Q., Wang, X., Yong, K.T., 2015a. In vivo toxicity assessment of non-cadmium quantum dots in BALB/c mice. Nanomed. Nanotechnol. Biol. Med. 11, 341–350.

Lin, L., Song, X., Chen, Y., Rong, M., Wang, Y., Zhao, L., Zhao, T., Chen, X., 2015b. Europium-decorated graphene quantum dots as a fluorescent probe for label-free, rapid and sensitive detection of Cu^{2+} and L-cysteine. Anal. Chim. Acta 891, 261–268.

Linkov, P., Krinvekov, V., Nabiev, I., Samokhvalov, P., 2016. High quantum yield CdSe/ZnS/CdS/ZnS multishell quantum dots for biosensing and optoelectronic applications. Mater. Today Proc. 3, 104–108.

Liu, D., Wu, F., Zhou, C., Shen, H., Yuan, H., Du, Z., Ma, L., Li, L.S., 2013. Multiplexed immunoassay biosensor for the detection of serum biomarkers—β-HCG and AFP of Down Syndrome based on photoluminescent water-soluble CdSe/ZnS quantum dots. Sens. Actuat. B Chem. 186, 235–243.

Liu, J., Hu, R., Liu, J., Zhang, B., Wang, Y., Liu, X., Law, W.C., Liu, L., Ye, L., Yong, K.T., 2015a. Cytotoxicity assessment of functionalized CdSe, CdTe and InP quantum dots in two human cancer cell models. Mater. Sci. Eng. C 57, 222–231.

Liu, J.L., Walper, S.A., Turner, K.B., Brozozog Lee, A., Medintz, I.L., Susumu, K., Oh, E., Zabetakis, D., Goldman, E.R., Anderson, G.P., 2016a. Conjugation of biotin-coated luminescent quantum dots with single domain antibody-rhizavidin fusions. Biotechnol. Rep. 10, 56–65.

Liu, L., Wang, X., Ma, Q., Lin, Z., Chen, S., Li, Y., Lu, L., Qu, H., Su, X., 2016b. Multiplex electrochemiluminescence DNA sensor for determination of hepatitis B virus and hepatitis C virus based on multicolor quantum dots and Au nanoparticles. Anal. Chim. Acta 916, 92–101.

Liu, L., Wu, S., Jing, F., Zhou, H., Jia, C., Li, G., Cong, H., Jin, Q., Zhao, J., 2016c. Bead-based microarray immunoassay for lung cancer biomarkers using quantum dots as labels. Biosens. Bioelectron. 80, 300–306.

Liu, M., Liu, T., Li, Y., Xu, H., Zheng, B., Wang, D., Du, J., Xiao, D., 2015b. A FRET chemsensor based on graphene quantum dots for detecting and intracellular imaging of Hg2+. Talanta 143, 442–449.

Liu, P., Zhang, C., Liu, X., Cui, P., 2016d. Preparation of carbon quantum dots with a high quantum yield and the application in labeling bovine serum albumin. Appl. Surf. Sci. 368, 122–128.

Liu, W., Yang, H., Ma, C., Ding, Y.N., Ge, S., Yu, J., Yan, M., 2014. Graphene–palladium nanowires based electrochemical sensor using $ZnFe_2O_4$–graphene quantum dots as an effective peroxidase mimic. Anal. Chim. Acta 852, 181–188.

Liu, W., Zhang, A., Xu, G., Wei, F., Yang, J., Hu, Q., 2016e. Manganese modified CdTe/CdS quantum dots as an immunoassay biosensor for the detection of Golgi protein-73. J. Pharm. Biomed. Anal. 117, 18–25.

Liu, Y., Liu, L., He, Y., He, Q., Ma, H., 2016f. Quantum-dots-encoded-microbeads based molecularly imprinted polymer. Biosens. Bioelectron. 77, 886–893.

Liu, Z., Xiao, J., Wu, X., Lin, L., Weng, S., Chen, M., Cai, X., Lin, X., 2016g. Switch-on fluorescent strategy based on N and S co-doped grapheme quantum dots (N-S/GQDs) for monitoring pyrophosphate ions in synovial fluid of arthritis patients. Sens. Actuat. B 229, 217–224.

Lu, S., Liao, F., Wang, T., Zhu, L., Shao, M., 2016. Tuning surface properties of graphene oxide quantum dots by gamma-ray irradiation. J. Lumin. 175, 88–93.

Ma, X., Lou, C., 2012. The ferromagnetic properties of Ge magnetic quantum dots doped with Mn. Appl. Surf. Sci. 258, 2906–2909.

Ma, Y., Li, Y., Zhong, X., 2014. Silica coating of luminescent quantum dots prepared in aqueous media for cellular labeling. Mater. Res. Bull. 60, 543–551.

Maguire, C.M., Mahfoud, O.K., Rakovich, T., Gerard, V.A., Prina-Mello, A., Gun'ko, Y., Volkov, Y., 2014. Heparin conjugated quantum dots for in vitro imaging applications. Nanomed. Nanotechnol. Biol. Med. 10, 1853–1861.

Mendonca, L.T.B., De Azevedo, W.M., 2016. A fast bottom-up route for preparing CdS quantum dots using laser ablation in a liquid environment. J. Lumin. 171, 79–84.

Mir, I.A., Das, K., Rawat, K., Bohidar, H.B., 2016. Hot injection versus room temperature synthesis of CdSe quantum dots: a differential spectroscopic and bioanalyte sensing efficacy evaluation. Colloid. Surf. A Physicochem. Eng. Aspects 494, 162–169.

Mohanta, A., Jang, D.J., Wang, F.Y., Wang, J.S., 2016. Effect of GaAs spacer layer thickness on optical properties of multi-stacked InAs/GaAs quantum dots. J. Lumin. 175, 16–20.
Onoshima, D., Yukawa, H., Baba, Y., 2015. Multifunctional quantum dots-based cancer diagnostics and stem cell -therapeutics for regenerative medicine. Adv. Drug Deliv. Rev. 95, 2–14.
Pang, L., Cui, H., Liu, Y., Zhong, W., 2016. Anti-VEGF antibody conjugated CdHgTe quantum dots as a fluorescent probe for imaging in living mouse. J. Lumin. 173, 274–278.
Passos, S.G.B., Freitas, D., Dias, J.M.M., Neto, E.T., Navarro, M., 2016. One-pot electrochemical synthesis of CdTe quantum dots in cavity cell. Electrochim. Acta 190, 689–694.
Pratiwi, F.W., Hsia, C.H., Kuo, C.W., Yang, S.M., Hwu, Y.K., Chen, P., 2016. Construction of single fluorophore ratiometric pH sensors using dual emission Mn^{2+}-doped quantum dots. Biosens. Bioelectron. 84, 133–140.
Rajabi, H.R., Farsi, M., 2016. Study of capping agent effect on the structural, optical and photocatalytic properties of zinc sulfide quantum dots. Mater. Sci. Semiconduct. Process. 48, 14–22.
Sachdev, A., Matai, I., Gopinath, P., 2016. Carbon dots incorporated polymeric hydrogels as multifunctional platform for imaging and induction of apoptosis in lung cancer cells. Colloid Surf. B Biointerf. 141, 242–252.
Sadeghi, S., Jahani, M., Foroogh, B., 2016. The development of a new optical sensor based on the Mn doped ZnS quantum dots modified with the molecularly imprinted polymers for sensitive recognition of florfenicol. Spectrochim. Acta A Mol. Biomol. Spectr. 159, 83–89.
Shi, B., Zhang, L., Lan, C., Zhao, C., Zhao, J., Su, Y., Zhao, S., 2015a. One-pot green synthesis of oxygen-rich nitrogen-doped grapheme quantum dots and their potential application in pH-sensitive photoluminescence and detection of mercury(II) ions. Talanta 142, 131–139.
Shi, Y., Pan, Y., Zhong, J., Yang, J., Zheng, J., Cheng, J., Song, R., Yi, C., 2015b. Facile synthesis of gadolinium (III) chelates functionalized carbon quantum dots for fluorescence and magnetic resonance dual-modal bioimaging. Carbon 93, 742–750.
Sneha, M., Oluwatobi, S.O., Sandile, P.S., Soney, C.G., Patrice, M., Dider, R., Nandakumar, K., Sabu, T., 2016. Green synthesis of yellow emitting PMMA–CdSe/ZnS quantum dots nanophosphors. Mater. Sci. Semiconduct. Process. 39, 587–595.
Soenen, S.J., Manshian, B.B., Himmelreich, U., Demeester, J., Braeckmans, K., De Smedt, S.C., 2014. The performance of gradient alloy quantum dots in cell labeling. Biomaterials 35, 7249–7258.
Solanki, V., Mishra, I., Joshi, S.R., Mishra, P., Dash, P., Mishra, N.C., Kanjilal, D., Varman, S., 2015. Room temperature superparamagnetism in rutile TiO2 quantum dots produced via ECR sputtering. Nucl. Instrum. Methods Phys. Res. B 365, 82–85.
Su, C., Liu, Y., Ye, T., Xiang, X., Ji, X., He, Z., 2015. Rolling cycle amplification based single-color quantum dots–ruthenium complex assembling dyads for homogeneous and highly selective detection of DNA. Anal. Chim. Acta 853, 495–500.
Sukharova, A., Hafian, H., Turini, M., Chames, P., Baty, D., Plout, M., Cohen, J.H.M., Millot, J.M., Nabiev, I., 2016. Multiphoton imaging of tumor biomarkers in situ using singledomain antibodies conjugated with quantum dots in a set orientation. Mater. Today Proc. 3, 523–526.
Swain, A.K., Li, D., Bahadur, D., 2013. UV-assisted production of ferromagnetic graphitic quantum dots from graphite. Carbon 57, 346–353.
Tan, L., Liu, S., Li, X., Chronakis, I.S., Shen, Y., 2015a. A new strategy for synthesizing AgInS2 quantum dots emitting brightly in near-infrared window for in vivo imaging. Colloid. Surf. B Biointerf. 125, 222–229.
Tan, M., Li, X., Wu, H., Wang, B., Wu, J., 2015b. N-doped carbon dots derived from bovine serum albumin and formic acid with one- and two-photon fluorescence for live cell nuclear imaging. Colloid. Surf. B Biointerf. 136, 141–149.
Tao, Y., Zhang, Y., Ju, E., Ren, J., Qu, X., 2015. One-step synthesized immunostimulatory oligonucleotides-functionalized quantum dots for simultaneous enhanced immunogenicity and cell imaging. Colloid. Surf. B Biointerf. 126, 585–589.
Thota, S., Kashyap, S.C., Sharma, S.K., Reddy, V.R., 2016. Micro Raman, Mossbauer and magnetic studies of manganese substituted zinc ferrite nanoparticles: role of Mn. J. Phys. Chem. Solids 91, 136–144.

Tian, Z.H., Xu, J.H., Wang, Y.J., Luo, G.S., 2016. Microfluidic synthesis of monodispersed CdSe quantum dots nanocrystals by using mixed fatty amines as ligands. Chem. Eng. J. 285, 20–26.

Uddandarao, P., Mohan, R., 2016. ZnS semiconductor quantum dots production by an endophytic fungus *Aspergillus flavus*. Mater. Sci. Eng. B 207, 26–32.

Vanengen Spivey, A.G., 2016. Group velocity dispersion of CdSSe/ZnS core–shell colloidal quantum dots measured with white light interferometry. Opt. Commun. 363, 31–36.

Vasilescu, I., Eremia, S.A.V., Kusko, M., Radoi, A., Vasile, E., Radu, G.L., 2016. Molybdenum disulphide and graphene quantum dots as electrode modifiers for laccase biosensor. Biosens. Bioelectron. 75, 232–237.

Vijian, D., Chinni, S.V., Yin, L.S., Lertanantawong, B., Surareungchai, W., 2016. Non-protein coding RNA-based genosensor with quantum dots as electrochemical labels for attomolar detection of multiple pathogens. Biosens. Bioelectron. 77, 805–811.

Vinayagam, J., Chen, G.R., Huang, T.Y., Ho, J.H., Ling, Y.C., Ou, K.L., Chang, J.Y., 2016. Aqueous synthesis of CuInZnS/ZnS quantum dots by using dual stabilizers: a targeting nanoprobe for cell imaging. Mater. Lett. 173, 242–247.

Wang, G.L., Fang, X., Wu, X.M., Hu, X.L., Li, Z.J., 2016a. Label-free and ratiometric detection of nuclei acids based on grapheme quantum dots utilizing cascade amplification by nicking endonuclease and catalytic G-quadruplex DNAzyme. Biosens. Bioelectron. 81, 214–220.

Wang, L., Liu, S., Hao, C., Zhang, X., Wang, C., He, Y., 2016b. A dual-fluorescence biosensor assembled by quantum dots and phenazinium dyes: a comparative study for DNA detection. Sens. Actuat. B 229, 145–154.

Wang, L., Tricard, S., Yue, P., Zhao, J., Fang, J., Shen, W., 2016c. Polypyrrole and graphene quantum dots @ Prussian Blue hybrid film on graphite felt electrodes: Application for amperometric determination of L-cysteine. Biosens. Bioelectron. 77, 1112–1118.

Wang, Q., Chen, M., Zhang, H., Wen, W., Zhang, X., Wang, S., 2016d. Enhanced electrochemiluminescence of RuSi nanoparticles for ultrasensitive detection of ochratoxin A by energy transfer with CdTe quantum dots. Biosens. Bioelectron. 79, 561–567.

Wang, Q., Ye, F., Fang, T., Niu, W., Liu, P., Min, X., Li, X., 2011. Bovine serum albumin-directed synthesis of biocompatible CdSe quantum dots and bacteria labeling. J. Colloid Interf. Sci. 355, 9–14.

Wen, L., Lin, Y., Zheng, Z.H., Zhang, Z.L., Zhang, L.J., Wang, L.Y., Wang, H.Z., Pang, D.W., 2014. Labeling the nucleocapsid of enveloped baculovirus with quantum dots for single-virus tracking. Biomaterials 35, 2295–2301.

Weng, C.I., Chang, H.T., Lin, C.H., Shen, Y.W., Unnikrishnan, B., Li, Y.J., Huang, C.C., 2015. One-step synthesis of biofunctional carbon quantum dots for bacterial labeling. Biosens. Bioelectron. 68, 1–6.

Wu, F., Youan, H., Zhou, C., Mao, M., Liu, Q., Shen, H., Cen, Y., Qin, Z., Ma, L., Li, L.S., 2016a. Multiplexed detection of influenza A virus subtype H5 and H9 via quantum dot-based immunoassay. Biosens. Bioelectron. 77, 464–470.

Wu, Q., Chen, L., Huang, L., Wang, J., Liu, J., Hu, C., Han, H., 2015. Quantum dots decorated gold nanorod as fluorescent-plasmonic dual-modal contrasts agent for cancer imaging. Biosens. Bioelectron. 74, 16–23.

Wu, Q., Zhi, L., Qu, Y., Wang, D., 2016b. Quantum dots increased fat storage in intestine of *Caenorhabditis elegans* by influencing molecular basis for fatty acid metabolism. Nanomed. Nanotechnol. Biol. Med. 12, 1175–1184.

Xu, L., Mao, W., Huang, J., Li, S., Huang, K., Li, M., Xia, J., Chen, Q., 2016. Economical, green route to highly fluorescence intensity carbon materials based on ligninsulfonate/graphene quantum dots composites: application as excellent fluorescent sensing platform for detection of Fe3+ ions. Sens. Actuat. B 230, 54–60.

Xing, X., Zhang, B., Wang, X., Liu, F., Shi, D., Cheng, Y., 2015. An "imaging-biopsy" strategy for colorectal tumor reconfirmation by multipurpose paramagnetic quantum dots. Biomaterials 48, 16–25.

Xue, M., Zhang, L., Zhan, Z., Zou, M., Huang, Y., Zhao, S., 2016. Sulfur and nitrogen binary doped carbon dots derived from ammonium thiocyanate for selective probing doxycycline in living cells and multicolor cell imaging. Talanta 150, 324–330.

Yang, C., Xie, H., Li, Y., Zhang, J.K., Su, B.L., 2013. Direct and rapid quantum dots labelling of *Escherichia coli* cells. J. Colloid Interf. Sci. 393, 438–444.

Yang, C., Xie, H., Li, Q.C., Sun, E.J., Su, B.L., 2015a. Adherence and interaction of cationic quantum dots on bacterial surfaces. J. Colloid Interf. Sci. 450, 388–395.

Yang, D., Wang, L., Lv, W.B., Hao, Z.B., Luo, Y., 2015b. Growth and characterization of phosphor-free white light-emitting diodes based on InGaN blue quantum wells and green–yellow quantum dots. Superlatti. Microstruct. 82, 26–32.

Yang, W., Zhang, B., Ding, N., Ding, W., Wang, L., You, M., Zhang, Q., 2016. Fast synthesize ZnO quantum dots via ultrasonic method. Ultrason. Sonochem. 30, 103–112.

Zarco-Fernandez, S., Coto-Garcia, A.M., Munoz-Olivas, R., Sanz-Landaluze, J., Rainieri, S., Camara, C., 2016. Bioconcentration of ionic cadmium and cadmium selenide quantum dots in zebrafish larvae. Chemosphere 148, 328–335.

Zhang, C., Cui, Y., Song, L., Liu, X., Hu, Z., 2016. Microwave assisted one-pot synthesis of graphene quantum dots as highly sensitive fluorescent probes for detection of iron ions and pH value. Talanta 150, 54–60.

Zhang, F., Zheng, Z., Liu, S.L., Lu, W., Zhang, Z., Zhang, C., Zhou, P., Zhang, Y., Long, G., He, Z., Pang, D.W., Hu, Q., Wang, H., 2013. Self-biotinylation and site-specific double labeling of baculovirus using quantum dots for single-virus in-situ tracking. Biomaterials 34, 7506–7518.

Zhang, Z., Miao, Y., Lain, L., Yan, G., 2015a. Detection of quercetin based on Al^{3+}-amplified phosphorescence signals of manganese-doped ZnS quantum dots. Anal. Biochem. 489, 17–24.

Zhang, Z., Miao, Y., Zhang, Q., Lian, L., Yan, G., 2015b. Selective room temperature phosphorescence detection of heparin based on manganese-doped zinc sulfide quantum dots/polybrene self-assembled nanosensor. Biosens. Bioelectron. 68, 556–562.

Zhao, Y., Song, Z., 2014. Phase transfer-based synthesis of highly stable, biocompatible and the second near-infrared-emitting silver sulfide quantum dots. Mater. Lett. 126, 78–80.

Zheng, Y., Zou, Y., Jiang, J., 2016. Synthesis of Gd-doped CuInS2 quantum dots exhibiting photoluminescence and high longitudinal relaxivity. Mater. Lett. 168, 86–89.

Zhong, D., Zhuo, Y., Feng, Y., Yang, X., 2015. Employing carbon dots modified with vancomycin for assaying Gram-positive bacteria like *Staphylococcus aureus*. Biosens. Bioelectron. 74, 546–553.

Zhu, H.Y., Zhang, S.Y., Ling, Y., Meng, G.L., Yang, Y., Zhang, W., 2015. pH-responsive hybrid quantum dots for targeting hypoxic tumor siRNA delivery. J. Control. Release 220, 529–544.

CHAPTER

LIPID-BASED SYNTHETIC GENE CARRIERS

19

Nily Dan

Drexel University, Philadelphia, PA, United States

CHAPTER OUTLINE

1 INTRODUCTION

The idea to nucleic-acid based therapies utilizing DNA or siRNA as drugs to modify malfunctioning cellular activities was proposed first in the 1970s (Friedmann and Roblin, 1972). Indeed, nucleic acids have been studies for the relief or reversal of symptoms of diseases, such as cystic fibrosis, or different cancers (Abou-El-Enein et al., 2014; Caplen et al., 1995; Carnio et al., 2014; Dahlman et al., 2014; Forsythe et al., 1996; Fujiwara et al., 1994; Merdan et al., 2002; Pensado et al., 2014; Qasim and Thrasher, 2014; Sahay et al., 2013; Schroeder et al., 2010; Whitehead et al., 2009; Xue et al., 2014).

Therapeutic activity of nucleic acids requires that they be delivered into the target cells. Unlike more traditional drug delivery vehicles, however, whose role is to target specific cell populations and/or enable drug release of prolonged periods of time, nucleic acid carriers must address several functions. First is the encapsulation of the nucleic acid, to protect it from degradation by environmental agents, such as the immune system. As a result, carriers must be stable under in vivo circulation conditions and throughout the stages of cellular uptake, transport, and release from endosomes. Once the nucleic acid is delivered in the cell interior, therapeutic activity can occur only if the carrier destabilizes and releases the nucleic acid into the appropriate cell location. This is achieved by designing environmentally sensitive carriers that respond to the different conditions appropriately (Choi et al., 2014; Dan and Danino, 2014; Ramamoorth and Narvekar, 2015; Yang et al., 2014).

The large size of nucleic acids hinders cellular uptake. Their condensation into smaller, endocytosis-relevant dimensions in the carriers facilitates uptake efficiency. In addition, carrier design can allow selective binding to specific cell populations (Abou-El-Enein et al., 2014; Caplen et al., 1995; Carnio et al., 2014; Dahlman et al., 2014; Forsythe et al., 1996; Fujiwara et al., 1994; Merdan et al., 2002; Pensado et al., 2014; Qasim and Thrasher, 2014; Sahay et al., 2013; Schroeder et al., 2010; Whitehead et al., 2009; Xue et al., 2014).

Several types of nucleic acid carriers have been examined as potential delivery agents, with different advantages and disadvantages: Viral carriers are highly efficient, but require complex processing, and their use in vivo can sometimes lead to harmful side effects (Hastie and Samulski, 2015; Laakkonen and Yla-Herttuala, 2015; Li and Huang, 2000; Pierce and Bennett, 2015; Yang et al., 2014). Polymer-based synthetic carriers, polyplexes, are benign and highly effective as condensation agents (Chen et al., 2014; Mintzer and Simanek, 2009; Namvar et al., 2015; Neuberg and Kichler, 2014; Osada, 2014; Wagner, 2014), but their high stability can hinder the necessary gene unpackaging stage (Schaffer et al., 2000).

Lipid-based carriers, or lipoplexes, are arguably the best synthetic alternative to viral carriers (Caracciolo and Amenitsch, 2012; Evans et al., 2003; Jafari et al., 2012; Junquera and Aicart, 2014; Koltover et al., 1998; Merkel et al., 2009; Radler et al., 1997; Simberg et al., 2001; Tros de Ilarduya and Duezguenes, 2013; Xiong et al., 2011). DNA-based lipoplexes for gene delivery have been studied extensively for several decades, while recent investigations focused on small interfering (si)RNA carriers. Although differing in the nature of nucleic acid and the therapeutic functions they are designed to fill, both types of carriers are similar in structure and properties (Akinc et al., 2009; Dahlman et al., 2014; de Fougerolles et al., 2007; Desigaux et al., 2007; Juliano et al., 2008).

Lipoplex characteristics are defined on two different length-scales: local and large scale. On the local level of order 10 nm is the *equilibrium* structure, dominated by the lipid properties. The large majority of lipids assemble into bilayers, forming vesicles (or liposomes) in vitro (Cullis and de Kruijff, 1979). Few lipid species form hexagonal phases (Fig. 19.1), while other types of aggregate geometries, such as spherical micelles requires mixing lipids with small-molecule amphiphiles, cholesterol, polymers, or proteins (Ackerman and Feigenson, 2015; Feigenson, 2009; Goni et al., 2008; Koynova and Tenchov, 2001).

Lipoplexes based on bilayer-favoring lipids form multilamellar assemblies with alternating layers of lipid bilayers and ordered nucleic acid layers. The latter are characterized by a unique spacing between molecules, set by the system composition. Lipoplexes based on lipids that favor hexagonal arrays are assemble into structures with an underlying hexagonal structure, as sketched in Fig. 19.1.

In addition to the type of lipids, local lipoplex structure is set by a balance between the electrostatic attraction between the cationic lipids and the anionic nucleic acid, and the repulsive interactions between the similarly charged nucleic acids (Bruinsma, 1998; Dan, 1996, 1997, 1998; Harries et al., 1998; May and Ben-Shaul, 1997). Lipoplex equilibrium microstructure has been studied widely and is well understood. Systematic investigations established a consistent phase diagrams that links the lipid properties, the cationic lipid to DNA ratio (L/D), and other system parameters to the geometry and packing density of the nucleic acid chains (Kapoor et al., 2012; Ma et al., 2007; Safinya et al., 2011; Tros de Ilarduya et al., 2010; Wasungu and Hoekstra, 2006).

Lipoplexes can also be characterized on the aggregate macro-scale, namely on length-scales of the order of 100 nm–1 μm. Unlike the equilibrium properties, these are not as well studied or understood because they are generally the result of irreversible, kinetic processes rather than equilibrium ones, such as flocculation (Simberg et al., 2001, 2003). Sensitive to both global system features (e.g., the type and size of the lipid vesicles) and methodologies (the order and manner of mixing), lipoplex features on these length-scales are heterogeneous and cannot be linked to a well-defined phase diagram (Barreleiro and Lindman, 2003; Dan and Danino, 2014; Kennedy et al., 2000; Pozharski and Macdonald, 2007). Furthermore, as will be discussed in detail later, lipoplexes may get trapped in states that require unusually long timescales to equilibrate (Dan and Danino, 2014).

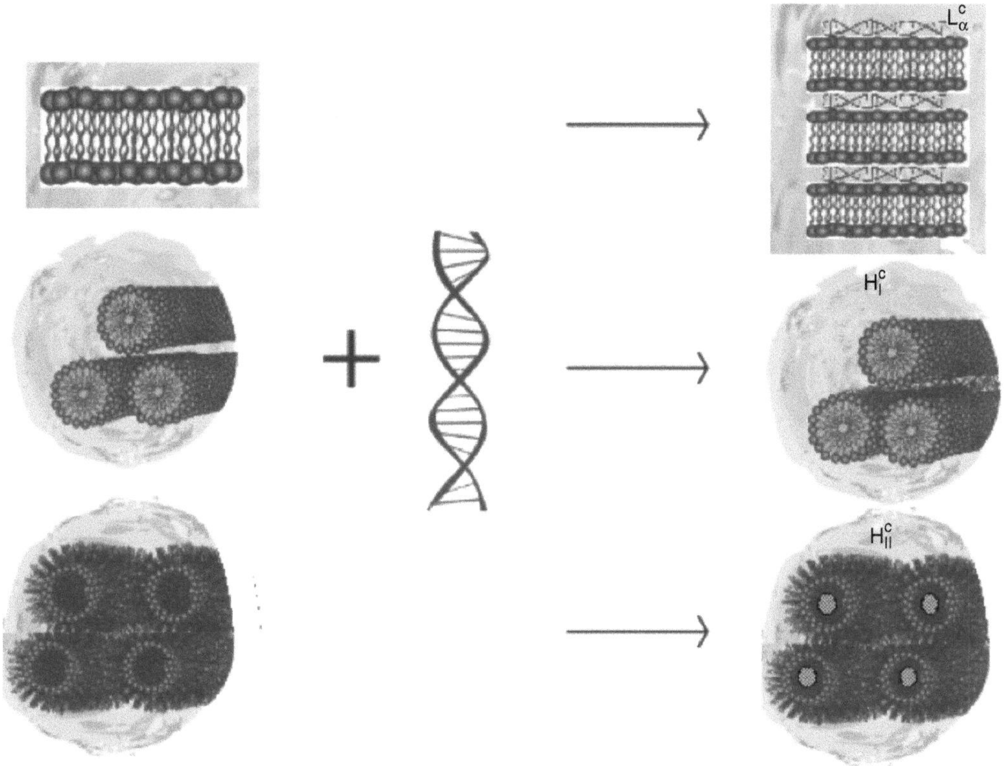

FIGURE 19.1 Complexation Between Nucleic Acids and Lipid Phases Yields Lipoplexes Whose Local Structure is Set by the Underlying Lipid Phase

Top: lipoplexes based on lamellar phases (L_{α}) form multilamellar structures where the nucleic acid forms flat layers intercalated with bilayer layers, $L_{\alpha}C$. Complexes between hexagonal phases (H_{I}, center and H_{II}, bottom) yield lipoplexes with similar structures.

The majority of lipoplex studies focus on their structure and the correlation between structure and biological function (Kapoor et al., 2012; Ma et al., 2007; Safinya et al., 2011; Tros de Ilarduya et al., 2010; Wasungu and Hoekstra, 2006), Yet, understanding the kinetics of lipoplex self-assembly is essential both for their utilization, and as a model for multistage, multitime scale, and multilength-scale self-assembly process (Dan and Danino, 2014). This chapter summarizes the results of lipoplex studies investigating the equilibrium structure, and the kinetics of their formation.

2 EQUILIBRIUM LIPOPLEX STRUCTURE ON THE NANOSCALE

The complex formation between nucleic acids and cationic lipid phases is driven by electrostatics, resembling polyelectrolyte adsorption on oppositely charged surfaces Netz and Andelman (2003). The macroions, or nucleic acids, trap a large number of counterions, as do the oppositely charged surfaces or lipid assemblies. This gives rise to an entropic penalty that is relieved when complexation between the macroion and the surface releases these counterions (Gregory and Barany, 2011; Netz and Andelman, 2003).

Polyelectrolyte adsorption is controlled by a combination of the polymer MW and charge density. Due to the high MW of the chain, a large number of segments attach to the surface. Therefore, even a low adsorption energy per individual segment (which accounts for the counterion release) can still lead to a large gain per molecule (Fleer et al., 1993). The typically low binding energy of individual polymer segments means that they can detach and reattach through thermal fluctuations. However, the adsorption of an entire polymer chain is irreversible, since the probability that all the segments detach at the same time is exceedingly low. The segmental binding varies with the solution ionic strength, as in any electrostatic-driven system (Sukhishvili and Granick, 1998).

Lipoplex formation is qualitatively similar. Although the binding energy between a nucleotide charge and the cationic lipid layer is of order the thermal energy (~kT, where k is the Boltzmann constant and T temperature) or less, multiple site adsorption leads to irreversible binding, while local rearrangement can still take place via thermal fluctuations. In both systems, the entropy gain caused by counterion release compensates for the associated loss of due to polymer or nucleic acid configurational entropy loss associated with chain immobilization on the surface (Fig. 19.2). Theoretical models that predict lipoplex assembly in terms of counterion release (see, for example, Dan, 1996) were later confirmed by calorimetry analysis of DNA/lipid complexes (Kennedy et al., 2000; Pozharski and Macdonald, 2002, 2003).

Although qualitatively similar, nucleic acid complexation with lipid phases and polyelectrolyte adsorption on oppositely charged colloidal surfaces differ in two, significant, ways. First is the type of adsorbing macroion: the persistence length of DNA is both large (of order 50 nm)—much larger than the bilayer thickness, and insensitive to ionic strength (Nelson, 2013). Indeed, the persistence length of DNA is comparable in some cases to the vesicle size. In contrast, the typical rigidity of polyelectrolytes is of order 1 nm—much smaller than the size of any colloidal particle, and highly sensitive to the solution conditions (Netz and Andelman, 2003) (Fig. 19.3).

As a result, polyelectrolyte adsorbed layers display "loops and trains," and the layer thickness is highly sensitive to the ionic strength (Sukhishvili and Granick, 1998), as sketched in Fig. 19.3. In the nucleic acid, which cannot form loops and trains on the relevant length-scales, the binding of even one segment means a flat layer whose thickness is of order the nucleic acid diameter, which is therefore insensitive to the solution ionic strength (Koltover et al., 1998; Safinya et al., 2011). Also, although theoretically possible, detachment of any specific segment is unlikely.

The second significant difference between nucleic acid complexes and polyelectrolyte adsorption is the substrate characteristics. Colloids are rigid, so that their surfaces (and surface charges) are immobile. Lipid assemblies can rearrange in response to external forces, such as the adsorption of nucleic acids. This leads to a responsive process where adsorption of the nucleic acid in a specific lipid location causes disruption and rearrangement in the lipid surface, for example, changing the local lipid packing.

As noted, although most lipids are bilayer forming in aqueous solutions, some common species favor hexagonal or bicontinuous geometries (Nelson, 2013). Thermodynamic analysis of DNA lipoplexes suggested that the complex structure would be set by the preferred lipid phase geometry (Bruinsma, 1998; Dan, 1996, 1997, 1998; Harries et al., 1998; May and Ben-Shaul, 1997). However, DNA adsorption was predicted to affect the local lipid packing and cause local "buckling" (Dan, 1996). In contrast, the large persistence length of nucleic acids, and its insensitivity to the environment,

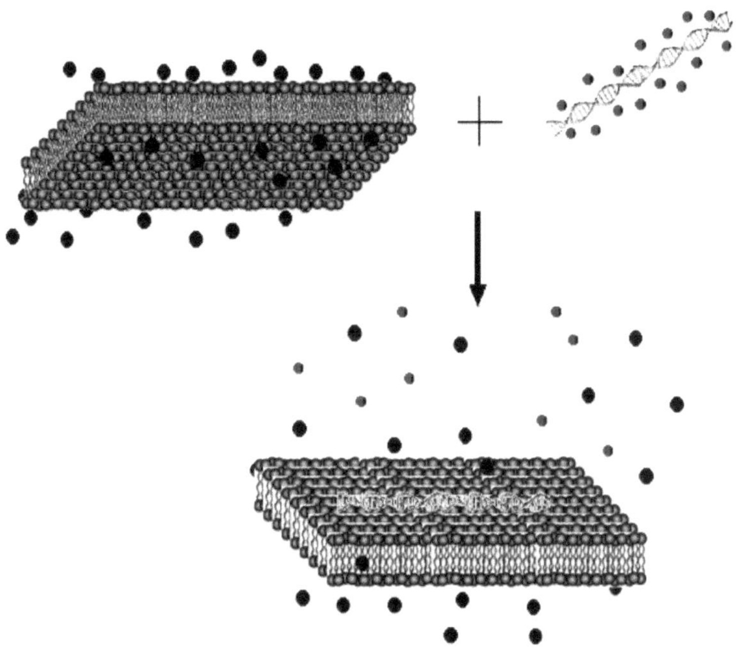

FIGURE 19.2 Counterion Release During Complexation Between Macroions-Like Nucleic Acids, and Charged Substrates, Such as the Cationic Liposomes

Both species trap a fraction of their counterions in their vicinity. Complexation allows the release of the counterions, resulting in a large gain in entropic energy.

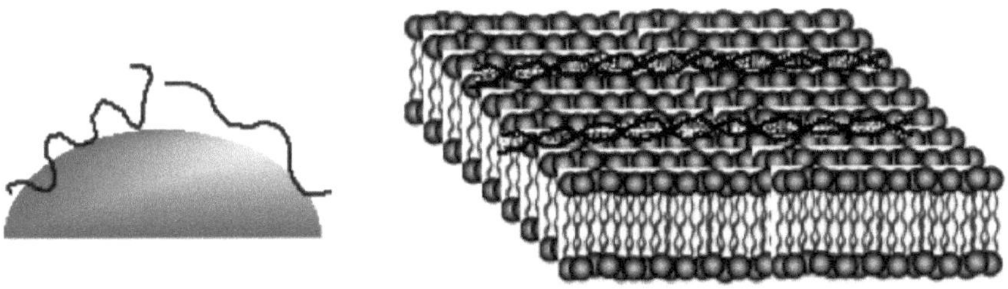

FIGURE 19.3 Comparison of Polyelectrolyte Adsorption on Colloidal Particles (Left) Versus Nucleic Acid Adsorption on Liposomes (Right)

In the case of polyelectrolytes, the polymer persistence length is small relative to the colloid size, leading to the formation of an adsorbed layer characterized by loops and tails. In the case of nucleic acid adsorption, the high persistence length leads to adsorption as a flat layer.

suggests that the nucleic acid characteristics (lambda-phage DNA or circular plasmid DNA) would not affect the local structure of lipoplexes, as indeed verified experimentally (Koltover et al., 1998; Safinya et al., 2011).

As expected, single component, bilayer-favoring lipids were shown to form bilayer-type lipoplexes (Koltover et al., 1998; Safinya et al., 2011), with an L_α^C structure composed of DNA layers intercalated between lipid bilayers (Radler et al., 1997). Similar arrays were found also in mixtures of cationic and nonionic, bilayer favoring lipids (such as dioleoyl-trimethylammonium propane–DOTAP mixtures with the neutral dioleoyl-phosphadtidycholin, DOPC): Lamellar L_α^C phase (Koltover et al., 1999; Radler et al., 1997), with the DNA strands within a layer regularly spaced, organized in a two-dimensional (2D) liquid-crystal smectic array (Ahsan et al., 1998; Bruinsma, 1998; Dan, 1997; Harries et al., 1998; Hirsch-Lerner and Barenholz, 1999; Kennedy et al., 2000; May and Ben-Shaul, 1997; Radler et al., 1997; Safinya et al., 2011). Characteristic dimensions for L_α^C lipoplexes are membrane thickness of ≈40Å, and water gap accommodating a DNA monolayer between lipid bilayers of order 25Å. The characteristic spacing between DNA molecules in a monolayer ranged from 25 to 50Å for isoelectric complexes, a function of membrane charge density, and the ratio between the cationic lipid and DNA charges, L/D (Radler et al., 1997; Safinya et al., 2011).

Mixtures of lipid species that may favor different phases led to other types of lipoplex geometries, with the order and packing of the nucleic acid varying as a function of parameters, such as the overall charge ratio, the mixture composition, and charge density (Koltover et al., 1998). Complexes containing DOTAP, DOPC and the nonionic, inverted hexagonal forming dioleoyl-phosphadtidylethanolamine (DOPE) formed L_α^C lipoplexes when the DOTAP content was high, and an inverted hexagonal phase, H_{II}^C (honeycomb) in systems with high DOPE fractions. In this complex geometry, an inverse lipid monolayer coats the DNA in a form similar to that of cylindrical micelles that aggregate into a hexagonal lattice (Tilcock et al., 1982). The diameter of the inverse micelles in H_{II}^C lipoplexes is 28Å (Koltover et al., 1998; Safinya et al., 2011), similar to the characteristic dimension in the bilayer assemblies.

Both the equilibrium L_α^C and H_{II}^C microstructures were affected by the lipoplex environment. Osmotic stress was shown to induce a reversible transition between lamellar and hexagonal arrays in isoelectric complexes (Danino et al., 2009; Scarzello et al., 2005), while temperature increases caused conversion of *N*-methyl-4-(dioleyl)methylpyridinium (SAINT) and DOPE complexes from lamellar to H_{II} structures.

Although lamellar and hexagonal phases are the most common lipoplex geometry, other structures were observed, in particular in systems with multivalent headgroups. Lipoplexes containing the cationic MVLBG2 (a molecule with a dendrimer headgroup with +16 charges) and DOPC mixtures exhibited typical lamellar L_α^C phase at low MVLBG2 fraction, but at higher MVLBG2 content formed a dual-lattice structure composed of cylindrical lipid micelles intercalated with DNA in a hexagonal symmetry (H_I^C) substructure (Ewert et al., 2006), or a distorted hexagonal H_I^C phase (Zidovska et al., 2009). Gemini surfactant-based complexes transitioned from lamellae to H_I^C arrays upon acidification (Wasungu and Hoekstra, 2006; Wasungu et al., 2006). Bicontinuous cubic phases (defined as Q_{II} or V_{II}) were also found, in a narrow composition window in lecithin and dodecyltrimethylammonium (DTA)-based lipoplexes (Bilalov et al., 2009), as were cubic bicontinuous phases in complexes between short DNA and DOTAP mixtures with a nonionic lipid monoolein (GMO), which has been previously found to form nonlamellar hexagonal and bicontinuous cubic phases (Larsson, 1983). Temperature changes were found to induce a reversible transition from bicontinuous to inverted hexagonal phases (Leal et al., 2013).

The equilibrium properties of lipoplexes are also set by the composition (namely the ratio between nucleic acid, cationic, and nonionic lipids). Theoretical analysis predicted that, since the density of cationic charges in the lipid assembly sets the density of binding sites, increasing the fraction of cationic lipids (i.e., the L/D ratio) would reduce the average spacing between nucleic acids, in the limit of low DNA fraction (Bruinsma, 1998; Dan, 1996, 1997, 1998; Harries et al., 1998; May and Ben-Shaul, 1997). However, as the DNA concentration on the surface increases, a balance between attractive and repulsive DNA-DNA interactions was expected to lead to coexistence between closely packed and more dilute DNA regions (Dan, 1996), a phenomenon that was later verified experimentally (Farago et al., 2008).

RNA interference (RNAi) as a pathway for posttranscriptional gene-silencing, which was identified by Fire et al. (1998), led to interest in utilizing lipid-based small-interfering RNA (siRNA) complexes as a potential cancer therapy, as well as a therapeutic route for a number of chronic diseases (Davidson and Paulson, 2004; Ponnappa, 2009; Takeshita and Ochiya, 2006; Whitehead et al., 2009). Like DNA-based lipoplexes, the characteristics of siRNA-lipid complexes are determined by the lipid composition and the L/D charge ratio. For example, mixtures of custom-made aminoglycoside derivatives and siRNA (Desigaux et al., 2007) formed lamellar aggregates with either concentric multilamellar or ordered microdomains, similar to DNA complexes with those same lipid components. Multilamellar siRNA lipid complexes were also found in mixtures of bis(11-ferrocenylundecyl) dimethylammonium bromide (Aytar et al., 2013), or an $L_{\alpha}^{\mathrm{siRNA}}$ in low GMO-DOTAP mixtures with low GMO content (Safinya et al., 2011). At higher GMO concentrations a hexagonal array of inverse cylindrical micelles ($\mathrm{H}_{\mathrm{II}}^{\mathrm{siRNA}}$) was observed (Leal et al., 2010, 2011; Safinya et al., 2011). In addition, as in the case of DNA-based complexes, the inclusion of multivalent components led to the formation of other types of assemblies, such as mixtures of pentavalent cationic lipid with excess GMO and siRNA that displayed a double gyroid cubic phase, with the siRNA embedded in the water channels (Leal et al., 2011).

Yet, DNA and siRNA complexes show some key differences (Schroeder et al., 2010). siRNA is orders of magnitude shorter than DNA, close to the dimensions of the bilayer thickness. Therefore, although DNA complexes are insensitive to the nucleic acid length, in siRNA complex properties were found to vary with the length, indeed, the biological activity of siRNA complexes occurs when they are below 100 nm. siRNA complexes are also more sensitive to enzymatic degradation, requiring additional stabilization modifications that are not necessary in DNA-based systems (De Paula et al., 2007; Juliano et al., 2008; Kim and Rossi, 2007), and biomedical function that require targeting of specific tissues led to the inclusion of targeting ligands, PEG moieties for steric stabilization, and pH-sensitive groups to induce siRNA release (extensively reviewed in Mintzer and Simanek, 2009).

3 THE KINETICS OF LIPOPLEX FORMATION

Studying the kinetics of lipoplex formation is complicated by the different time scales and length-scales involved; structures are characterized by length-scales ranging from 10 nm (lipid and nucleic acid spacing) to 1 μm (aggregate dimensions), evolving over several timescales. Better understanding may be achieved by comparing it to two similar processes: cluster nucleation and growth, and polyelectrolyte adsorption. Like lipoplex formation, these structures evolve through similar kinetic steps that include initial, rapid, complex formation followed by slower stages of growth.

Nanoclusters nucleation occurs via a rapid initial nucleation step, from a supersaturated solution. The rate—regardless of the type of nucleation (homogeneous or heterogeneously)—is set by thermodynamic equilibrium properties (Edelstein and Cammarata, 1996)

$$J = J_0 \exp\left\{-\frac{\Delta G^*}{kT}\right\} \tag{19.1}$$

where J, formation (number, per unit volume per time); k Boltzmann constant; and T, temperature. The energy barrier for the formation of a critical cluster is defined by ΔG^*, which is higher for homogeneous than heterogeneous nucleation. The preexponential rate J_0 depends on parameters, such as temperature, atomic vibration frequency, or the critical cluster size (Gregory and Barany, 2011). It should be noted that because of the implicit dependence of J_0 on temperature, there is an optimal T value at which the overall rate of nucleation is maximal.

While the critical cluster formation step is roughly universal, cluster growth stages display a range of different mechanisms. Most common is that of flocculation, where cluster collision leads to their fusion into larger aggregates. The rate of flocculation depends on the collision frequency and the probability of cluster adhesion (Gregory and Barany, 2011). The former is insensitive to specific system characteristics, increasing with cluster concentration and mobility, while the probability of cluster-cluster adhesion depends on short-range cluster-cluster interactions (Gregory and Barany, 2011). Another mechanism for cluster growth is one where individual molecules continuously combine with existing clusters, leading to their growth. The Johnson-Mehl-Avrami-Kolmogorov (JMAK) equation, and other similar approaches predict the effect of system parameters on the addition rate of growth, with time scales that are slower than initial nucleation rates (Edelstein and Cammarata, 1996).

The adsorption of charged polymer chains, polyelectrolytes, on oppositely charged surfaces (usually colloidal particles) follows similar kinetic steps. The initial adsorption, where the polymer chains comes into contact with the surface, follows a rate constant similar to that of Eq. (19.1): the driving force for adsorption, manifested in the value of ΔG^*, is due to the trapped polymer and surface counterions, that are released when adsorption occurs. As a result, ΔG^* varies with the solution ionic strength, as well as the surface and polyelectrolyte charge density. Unlike nucleation processes, though, increasing surface coverage decreases the driving force, so that ΔG^* and rate of polymer adsorption decrease with time (Gregory and Barany, 2011).

As in nucleation, the rate prefactor J_0 increases with increasing concentration of the components. However, J_0 in polymeric systems differs from that of nucleation, in its dependence on the polymer chain length: as the chain MW increases, the characteristic adsorption time decreases (Filippova, 1998). This is due to the fact that once one segment is adsorbed on the surface, chain connectivity brings neighboring ones closer to the surface, where they are more likely to bind. The resulting cooperativity can increase the rate by an order of magnitude (Filippova, 1998). More detailed analysis of polyelectrolyte adsorption than the somewhat qualitative Eq. (19.1) combines collision probability with an associated rate constant based on the Smoluchowski expression (Gregory and Barany, 2011), an approach validated by experiments (Enarsson and Wagberg, 2008; Iruthayaraj et al., 2005; Mcfarlane et al., 2008; Sedeva et al., 2009; van Heiningen and Hill, 2011).

Charged colloidal particles are stabilized against flocculation by electrostatic repulsion. The adsorption of oppositely charged polyelectrolytes can neutralize the particle charge or even cause charge inversion (Barany and Szepesszentgyorgyi, 2004), thereby leading to flocculation (Popa et al., 2009, 2010). In addition to flocculation driven by electrostatics, polymer chains can also bridge neighboring particles—thereby binding them into a larger aggregate (Dickinson and Eriksson, 1991). The kinetics of these processes depends on several parameters. In dilute systems, the rate is slow (~minutes) because the probability of species collision is low. In systems with high polymer concentrations, flocculation is rapid (Gregory and Barany, 2011) (Fig. 19.4).

The second kinetic stage of flocculate formation is followed, in both clusters and polyelectrolyte adsorption, by a reorganization and equilibration step. In polyelectrolytes, the strength of electrostatic

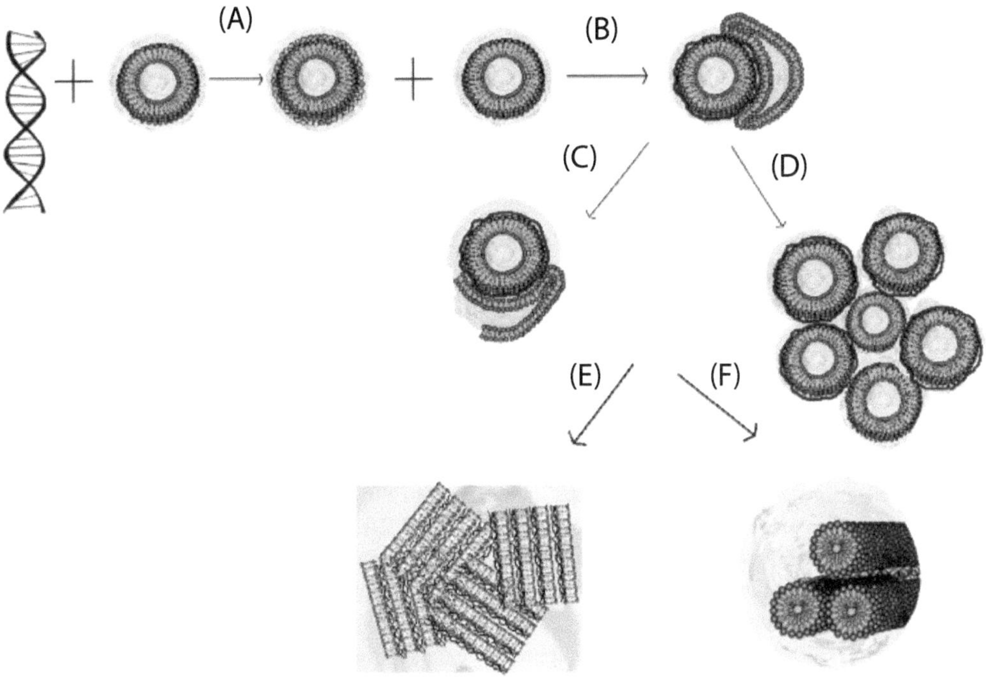

FIGURE 19.4 Suggested Routes to Lipoplex Formation

(A) Initially, nucleic acid strands adsorb onto the surface of cationic liposomes. This process is rapid, occurring over time scales the order ms or sec; (B) partially or fully coated liposomes interact with uncoated liposomes, in a first flocculation stage; (C) if the liposomes can rupture easily, liposome disintegration leads to the formation of multilayered or coated structures; (D) if the liposomes are rigid, floccs form where the underlying liposome structure remains. This process is characterized by time scales of order minutes to hour. The structures formed in this stage continue to rearrange internally over time scales the order of hours, until reaching the optimal local structure (E, F), similar to those sketched in Fig. 19.1.

interactions between the oppositely charged species leads to trapping of polyelectrolyte chains in meta-stable states during the initial and intermediate adsorption stages. The equilibration of these metastable arrangements requires a long-time (Gregory and Barany, 2011). In clusters, the flocculation stage also gives rise to meta-stable aggregates that require a long time to reach an equilibrated internal structure (Lin et al., 1989).

In both systems, the internal equilibration of aggregates is not manifested by changes in global properties, such as aggregate size, but by the evolution in internal order on length-scales that are of order the molecular dimensions, and requires long time scales of order hours (Edelstein and Cammarata, 1996; Lin et al., 1989), or even days (Gregory and Barany, 2011).

Lipoplex formation follows the same three kinetic stages as shown in Fig. 19.4: rapid complexation between the nucleic acid and the lipid phase, followed by an intermediate consolidation/flocculation step, with a final, long-time stage of internal rearrangement and equilibration.

Experimental studies of lipoplex kinetics tend to focus on detailed examination of L_{α}^{C}–DNA complexes. However, it is expected that the process be similar, qualitatively, for all types of lipoplex geometries. Furthermore, some of the kinetic features are independent of the lipoplex underlying structure, so that the results may apply quantitatively as well.

The initial step of lipoplex formation—the "adsorption" of nucleic acid on cationic lipid aggregates—is driven by the attraction between oppositely charged macro-ions (the nucleic acid and the lipid assembly). As a result, it is expected to resemble the same step in polyelectrolyte adsorption, and is insensitive to the lipid's preferred geometry or the lipid aggregate characteristics. Indeed, comparing the initial rate of complex formation between extruded vesicles that are uniform and small and large, polydispersed vortexed ones did not show significant differences (Pozharski and Macdonald, 2007). Lipoplex formation in systems of monovalent cationic lipids (1-[2-(9-(Z)-octadecenoyloxy)ethyl]-2-(8-(Z)-heptadecenyl)-3-(hydroxyethyl)imidazolinium chloride (DOTIM), and 1,2-dimyristoyl-sn-glycero-3-ethylphosphocholine (EDMPC) mixed with two neutral lipids (cholesterol and diphytanoyl phosphatidylethanolamine, DipPE) was studied by Lai and van Zanten (2002) using time-resolved multiangle laser light scattering (TR-MALLS). They find that the L/D ratio is the primary factor setting the rate in the initial lipoplex formation steps, insensitive to the type of cationic or nonionic lipid. Interestingly, the rates were not symmetrical, that is, the rate at L/D = 2:1 differed from that of L/D= 1:2. Another study found that when the DNA was in excess (0.24 and 0.75 charge ratio), the time constant was nearly twice as much as that systems with identical lipid composition and excess lipids (1.5, 3, and 5) (Barreleiro and Lindman, 2003). The effect of the L/D ratio is attributed to the smaller number of complexes in excess DNA systems when compared to excess lipid systems (Lai and van Zanten, 2002).

The initial, rapid binding stage creates lipid aggregates (usually vesicles) coated by adsorbed nucleic acids. This step is followed by a flocculation stage, where the complexes aggregate to form larger clusters. Since the experimental methodologies follow this stage by measuring the average aggregate size, "steady state" is defined by a constant aggregate size that stays constant with time.

The effect of temperature on the initial aggregation step was studied by Braun et al. (2005) in dimethyldioctadecylammonium bromide (DDAB) and DOTAP systems. They found that the initial rate constant decreased with temperature in both as expected from a process that follows Eq. (19.1), although the DOTAP activation energy was estimated to be approximately 5 times higher than the DDAB's activation energy.

In the DOTIM/EDMPC–DNA complexes, Lai and van Zanten found that complex size, mass and density increased with concentration, and were maximal is systems where L/D approached 1 (Lai and van Zanten, 2002). The effect of the charge ratio on the rate at the flocculation stage was significant: solutions with excess of either component reached a "steady-state" in complex growth quickly, while mixtures with an L/D of order 1 continued growing over longer periods, of order 10 min following the initial step (Lai and van Zanten, 2001). A similar observation was made by an earlier study of DOTAP/DNA systems, where complexes in excess lipid or excess DNA (L/D of 0.5 and 1.5) reached their final size within seconds, while those with an L/D = 1 required 40 min to reach that stage (Gershon et al., 1993). In addition to the effect of the L/D ratio, it was found that lipoplexes formed in DMEM/F-12 (a cell growth medium that contains divalent Ca^{2+} and Mg^{2+} ions) where electrostatic interactions are highly screened were larger than ones formed in dextrose solutions.

Analyzing these observations clearly suggests that the second stage of aggregate formation, where lipoplexes grow (Lai and van Zanten, 2001), is highly dependent on electrostatic interactions between the initially formed complexes. In systems with highly asymmetric charge ratios, complexes become saturated with one charge or another (e.g., if DNA is in excess, it coats all aggregates until they carry a net anionic charge), which inhibits further flocculation, and leads to smaller aggregates relatively rapidly. However, in systems where the charge ratio is near 1, nucleic acid adsorption on the complexes neutralizes the stabilizing electrostatic repulsion, and allows the formation of large floccs that evolve over long periods of time. The presence of divalent ions in solution screens electrostatic forces (attractive or repulsive) and suppresses flocculation.

While the initial rate is insensitive to the properties of the lipid aggregates, the kinetics of the flocculation stage were found to vary depending on the underlying vesicle characteristics (Pozharski and Macdonald, 2007). Systems with an L/D = 1 ratio of DNA with dioleoyl-O-ethylphosphocholine (EDOPC) based on extruded vesicles with uniform, relatively small size were stable over long periods of time that ranged in hours. In contrast, complexes that formed from large, nonuniform vortexed vesicles displayed rates of order minutes. When the lipid was in excess (L/D = 2), the stability of the coated extruded vesicles decreased. Similar results were found in DOTAP-DOPE and DOTAP-cholesterol mixtures (Zhang et al., 2003), although the effect of charge ratio was weaker in cholesterol-containing systems than in those containing DOPE, suggesting that uncharged lipoplex components affect the complexation kinetics.

Temperature increase was found to decrease the time scales for the growth stage in both DDAB than for DOTAP, although (contrary to the observation for the initial rate) the activation energy higher for DDAB than for DOTAP (Braun et al., 2005).

Lindman (Barreleiro and Lindman, 2003; Barreleiro et al., 2003) was the first to suggest that the apparent steady state in the growth stage where the aggregates reach a constant in size, is followed by an additional step. In this step, which cannot be probed using techniques that follow global aggregate characteristics, such as size, changes in the DNA conformations take place leading to rearrangement of complexes into the multilamellar or hexagonal equilibrium structures over longer time scales (of order minutes to hours). Since this stage requires internal rearrangement, it is expected to be sensitive to the molecular properties of the constituent components. Stopped-flow turbidity and fluorescence show that this stage depends on the L/D charge ratio and lipid composition, which may be attributed to an associated decrease in bilayer fluidity (Barreleiro and Lindman, 2003). The presence of the third step was supported by neutron scattering experiments that found a third time scale for multilamellar reorganization of order ~10 min (Barreleiro et al., 2003). Solid state and diffusion NMR studies found that in systems containing a mix of cationic and zwitterion lipid (i.e., lipids with no net charge that do not bind to the nucleic acid), the zwitterions displayed diffusion coefficients typical for lipids in bilayers (Leal et al., 2008). In contrast, the cationic component displayed a diffusion rate that was an order of magnitude slower, while the nucleic acid was relatively static (Leal et al., 2008). These results suggest that (1) the complexation between cationic lipids and nucleic acid reduces their mobility and (2) that there is local segregation where regions of DNA-cationic lipid complexes coexist with regions of pure lipid composed of the noncationic component.

4 DISCUSSION

The equilibrium structure of lipoplexes is defined on local, molecular length-scales of order 10 nm. Set by thermodynamics, it is well characterized and understood. Bilayer-favoring lipids form L_α^C lipoplexes with DNA layers intercalated between lipid bilayers, either in single cationic lipid systems or in cationic/nonionic ones (Koltover et al., 1998; Safinya et al., 2011). The DNA strands are regularly spaced in a 2D liquid-crystal smectic array whose dimensions depend on the L/D ratio (Ahsan et al., 1998; Bruinsma, 1998; Dan, 1997; Harries et al., 1998; Hirsch-Lerner and Barenholz, 1999; Kennedy et al., 2000; May and Ben-Shaul, 1997; Radler et al., 1997; Safinya et al., 2011). Lipoplexes based on lipids that favor nonlamellar phases form geometries, such as inverted hexagonal L_{α}^{C} or H_{II}^{C} honeycombs (Tilcock et al., 1982; Koltover et al., 1998; Safinya et al., 2011; Tilcock et al., 1982), or bicontinuous cubic phases (Bilalov et al., 2009; Larsson, 1983).

siRNA complex characteristics are also determined by the lipid composition and the L/D charge ratio (Aytar et al., 2013; Desigaux et al., 2007; Leal et al., 2010, 2011; Safinya et al., 2011), although siRNA complexes depend on the nucleic acid length, while DNA ones do not.

In contrast, the large-scale features of lipoplexes, such as their size, are set by the kinetics route. As a result, they are more complex and heterogeneous than the equilibrium characteristics. The kinetics of lipoplex formation are divided into three stages: initial adsorption, flocculation and growth, and equilibration. These differ in their characteristic timescales and their dependence on system parameters.

The initial step in lipoplexes formation consists of a rapid, electrostatically driven binding between the anionic nucleic acid and cationic lipid aggregate-typically vesicles. The process resembles critical nucleation or the adsorption of polyelectrolytes on oppositely charged surfaces, and follows Arrhenius kinetics [Eq. (19.1)] with an activation energy that depends on lipid type (Braun et al., 2005). Since binding requires collision between the nucleic acid and lipid aggregate, increasing the concentration of components (for fixed L/D ratio) increases linearly the rate (Lai and van Zanten, 2002). However, binding takes place between the nucleic acid charges—proportional to the nucleic acid concentration—and the lipid charges available on the outer surface of the vesicles. As a result, for a given overall lipid concentration, increasing the concentration of nucleic acid should increase the rate of binding by a factor of 2, when compared to increasing the cationic charges by the same amount—as indeed observed. "Diluting" the cationic lipids by mixing with a nonionic component reduces the membrane charge density, that is, the density of potential adsorption site, and thus reduces the probability and rate of adsorption (Simberg et al., 2001; Zuidam and Barenholz, 1997, 1998). Similarly to polyelectrolyte adsorption, the rate varies nonmonotonically with the charge ratio, reaching a maximum in systems with charge ratio or order unity.

The irreversibility and rapid nature of the initial aggregate formation means that the mixing methodology used to form the aggregates significantly affects their properties: if the two solutions (one with lipid vesicles, one with nucleic acid) are directly mixed, the average, local concentrations of the complexes is similar to those of the overall mixture—namely, mixing lipids with nucleic acids at a 2:1 ratio will yield a local ratio of 2:1 in the complex. However, if the complexes are formed by dropwise addition of one solution to the other—which is a common practice—the environment in which the first complexes form differs from that of the last ones at the end of the process, thereby giving rise to heterogeneity: for example, if nucleic acid is added dropwise to a solution of lipids, initially the ratio of L to D will be extremely high, and will represent the local structure of those complexes that are very different from those of the final steps, as observed in the similar process of polyelectrolyte adsorption on colloidal particles (Szilagyi et al., 2014).

Regardless of system specifics, this initial step is rapid and irreversible, occurring on a timescale of less than seconds.

The next stage of lipoplex formation involves complexes aggregation and growth, in a process that is similar to that occurring in particles carrying adsorbed polyelectrolyte layers. The charged chains (polymer or nucleic acid) bind in the initial step in a random and irreversible manner to the surface (colloidal particle or lipid aggregate). The adsorption neutralizes some, or all of the electrostatic repulsion between the surfaces, so that flocculation can occur. In addition, exposed binding sites on both components (lipid aggregates and nucleic acid, or particles and polyelectrolytes) can cause bridging. This process evolves over periods of time in the order minutes, until the larger aggregates, or floccs, reach a "steady state" size.

As noted, the preparation route affects the properties of the initial aggregates, leading to differences in the initial aggregate heterogeneity. This, in turn, affects the kinetics and end-results of the growth stage: in highly heterogeneous systems, flocculation is more likely, and rapid, due to differences in complex charge. In contrast, homogeneous complexes are likely to repel each other, and thus grow slowly or not at all. A similar effect is observed depending on the type of vesicles used—extruded, homogeneous ones versus vortexed, polydisperse ones (Pozharski and Macdonald, 2007). Yet, surprisingly, this issue is underemphasized in the literature, in the context of lipoplexes.

In addition, the molecular properties of the vesicles also affect the growth process. Rigid vesicles that resist rearrangement and rupture will follow the same flocculation kinetics as colloid/polyelectrolyte complex to form heterogeneous, large complexes. However, if the cohesion of the lipid bilayer is low, so that it ruptures under moderate forces, the strains induced by the adsorbed nucleic acid, and the flocculation events, will induce rearrangement that leads to more ordered and uniform structures on relatively shorter time scales.

In addition, the two systems differ in the nature of the chain. Polyelectrolytes are highly flexible, when compared to nucleic acids. As a result, they tend to form loops and trains, while nucleic acids are more likely to either bind flat to the surface (thereby forming multiple, sequential bonds), or to "beads on a string" type morphology where vesicles adhere to the chain in only one location. In either case, nucleic acid rigidity will lead to flocculation occurring by DNA binding (flat) to two vesicles, as indeed shown experimentally. These will lead to different interactions between the aggregates, and thus different growth rates.

Once the macroscale aggregate structure has been stabilized, so that no more vesicles or nucleic acids are added to the lipoplex, the final kinetic step takes place, where the local order of the complex develops. As noted, the initial and growth stages in the adsorption of any chain macroion (polyelectrolyte or nucleic acid) leads to locally trapped, nonequilibrium arrangements, due to the overall irreversible binding. These relax toward equilibrium states, over periods of time that are orders of magnitude longer than the first kinetic stages. The internal structure and, specifically, the morphology of the nucleic acid domains, evolves until reaching the equilibrium structure. During this period, the macrofeatures of the aggregates remain unchanged (namely, number of lipid or nucleic acid molecules, size), so that determining the kinetics of this stage require techniques that can probe the internal, molecular-scale configurations. This stage requires time scales that are orders of magnitude longer than the previous steps (the order of hours). Such equilibration occurs also in systems of polyelectrolyte adsorption on solid (colloidal) surfaces. Although qualitatively similar, though, there are significant differences in the mechanism of equilibration between polyelectrolyte/colloidal assemblies and lipoplexes.

As noted, the rigidity of nucleic acids is orders of magnitude higher than that of polyelectrolytes. Therefore, the nucleic acid adsorbs in a flat conformation on the surface, forming a large number of sequential bonds. The flexibility of polyelectrolytes allows them to form loops and trains, so that there may be a significant length of chain that is unbound between bound segments. Since the binding energy per segment is typically lower than kT, each segment in the polyelectrolyte can detach and reattach, thereby leading to equilibration. However, in the case of nucleic acid, rearrangement would require a coordinated detachment of several segments, which is highly improbable.

The differences between the two systems are also manifested in the nature of the substrate on which the charged chain adsorbed. Clearly, colloidal particles cannot rearrange their internal structure, while lipid-based assemblies do. As a result, one of the critical parameters affecting the rate is the lipid bilayer rigidity and fluidity. Rearrangement occurs through a 2D flow of the lipids until the equilibrium order is obtained. Studies show that the mobility of cationic lipids bound to the nucleic acid is lower than that of unbound lipids, but both types are usually mobile to some degree.

Thus, while the final, equilibration step is dominated by the polymer chain kinetics in polyelectrolyte adsorption, in lipoplexes it is set by the mobility of the lipids.

As noted, studies tend to focus on the kinetics of lamellar lipoplexes. The first two stages are largely independent of the specific lipid aggregate structure, more sensitive issues, such as the L/D ratio, and the mixing procedure, and thus should be similar for lamellar and nonlamellar systems. However, the last, equilibration stage requires the evolution of local equilibrium structure, and is expected to be highly sensitive to the local properties and structure of the complexes. For example, the formation of local bilayer/DNA alternating lamellae does not require the breakup of the bilayers when transitioning from the DNA adsorbed on the surface of a spherical vesicle to the equilibrium array, while the formation of a hexagonally arrayed or bicontinuous lipoplexes is based on complete rearrangement of the lipid assemblies.

The molecular properties of the nucleic acid are not expected to affect lipoplex kinetics significantly: although the diffusion rate of long chain-like molecules depends on their MW, the initial complexation is more sensitive to the concentration of components and the mixing protocol. The second stage, of complex flocculation, is mostly dependent on the mobility of the complexes than the specific properties of the nucleic acid. In the last stage, where internal rearrangement takes place, higher MW chains have lower mobility, but as noted earlier, the dominant mechanism for the evolution of the equilibrium structure is due to lipid rearrangement so that the nucleic acid characteristics should play only a minor role is setting the rate.

The efficacy of lipoplex transfection has been linked to both their local and large-scale structure (Caracciolo and Amenitsch, 2012; Junquera and Aicart, 2014; Ma et al., 2007; Martin et al., 2005; Safinya, 2001; Tros de Ilarduya et al., 2010; Wasungu and Hoekstra, 2006; Xiong et al., 2011; Yang et al., 2014; Zhang et al., 2012). In general, transfection experiments focus on the correlation between carrier characteristics and performance. However, the cell type plays a crucial role, as do physiological conditions (see, for example, Audouy et al., 2000; Balasubramaniam et al., 1996; Chan et al., 2014; Pollard et al., 1998). As a result, a lipoplex that performs effectively in one cell population or under specific conditions may perform poorly when transfecting a different type of cell. Furthermore, the underlying chemistry of the lipoplex (i.e., specific nature of the lipids), with only a minor effect of the typical lipoplex structure, may significantly affect interactions with cell components.

Transfection is a multistep, complex process (Elouahabi and Ruysschaert, 2005). As may be expected, the charge ratio of the lipoplexes is the most important feature in determining their efficacy, since the net charge will set their interactions with the various cell components (Bouxsein et al., 2007; Candiani et al., 2010; Koumbi et al., 2006; Masotti et al., 2009; Ramezani et al., 2009; Romoren et al., 2004; Ruozi et al., 2003; Salvati et al., 2006; Zhao et al., 1997). Indeed, effective lipoplexes carry a net cationic charge, which enhances their interactions with the anionic biological membranes (Koumbi et al., 2006).

Correlating global lipoplex properties to transfection efficiency is more complicated. Several studies show that the efficiency of delivery was optimized for a specific size range (Jaaskelainen et al., 2000; Ross and Hui, 1999; Templeton et al., 1997). However, the optimal sizes varied between experiments. Other studies do not observe a simple size effect (Koumbi et al., 2006; Stegmann and Legendre, 1997; Xu and Anchordoquy, 2008). The lack of clear correlation between complex size and transfection efficiency is most likely due to the fact that complex size is not an equilibrium-unique feature of these complexes. As noted, the overall complex size is set by the kinetic route, independent of the underlying equilibrium structure. Therefore, complexes with the same size may have very different L:D ratio, or may be lamellar or hexagonal in their local packing (Ma et al., 2007).

5 CONCLUSIONS

Modifying cellular functions by transfection, either through the insertion of genes or siRNA, promises to be an effective route to new therapies that address a range of diseases. Of the different types of synthetic nucleic acid carriers, lipid-based ones could become effective delivery vehicles for both in vitro and in vivo applications. However, optimizing their efficiency requires detailed understanding of the relationship between properties and performance.

The local, equilibrium properties of lipoplexes are well understood, and can be directly correlated to such parameters as the L/D ratio, or lipid molecular characteristics. Studies of lipoplexes focused, as a rule on their equilibrium features and the correlation between the local structure and transfection performance. However, results were not always consistent, since global features—set by the kinetics of complex formation—also play a role in the transfection process.

Compilation of studies examining the kinetics of lipoplexes reveal three stages: a rapid initial step where the nucleic acid adsorbs onto the cationic lipid aggregate (typically vesicle). This stage depends mostly on the component concentration and mixing process, and takes place over periods of seconds. Once the initial complexes form, they flocculate into larger aggregates, over intermediate time scales whose rate depends on the L/D charge, the structure complexes formed in the initial stage, and lipid characteristics. Although the time scales for this stage are of order minutes, they can vary widely and the resulting structures may be highly heterogeneous. The end of this step is determined when the aggregates reach an apparent steady state, where their size and composition remains unchanged with time. The last stage is that of equilibration, where the local internal structure evolves to the equilibrium structure. The dominant equilibration mechanism is that of lipid rearrangement, so that the rate is determined by the lipid mobility.

The large range in time scales and length-scales during lipoplex formation mean that most experimental measurements miss one or other of the three stages, due to the constraints of the specific technique: identification of the first, initial stage requires time resolution in the order of seconds. Methods that measure the size (or mass) of the aggregates are optimal for examining the flocculation and growth stage, but usually cannot observed the rapid initial complexation step, or probe the internal rearrangement occurring during the equilibration step. Indeed, such measurements tend to suggest that steady state is obtained when the aggregates stop growing, although this is an indicator of the end of the second step. Probes for molecular-level properties will yield typically data relevant to the last, equilibration stage (and potentially the initial binding step if they has the relevant temporal resolution), but do not provide information regarding the second flocculation step.

The efficacy of lipoplex-based transfection depends on their local and global properties. Understanding lipoplex kinetics will enable the design of more efficient carriers for nucleic acids that would perform better throughout the various transfection stages. In addition, understanding this complex, multi time and length-scale system will enable better understanding of other types of complexation processes.

CONFLICT OF INTERESTS

The author declares that there is no conflict of interests regarding the publication of this paper.

REFERENCES

Abou-El-Enein, M., Bauer, G., Reinke, P., Renner, M., Schneider, C.K., 2014. A roadmap toward clinical translation of genetically-modified stem cells for treatment of HIV. Trends Mol. Med. 20, 632–642.

Ackerman, D.G., Feigenson, G.W., 2015. Lipid bilayers: clusters, domains and phases. In: Parmryd, I. (Ed.), Essays in biochemistry: Membrane Nanodomains. Portland Press, London, UK, pp. 33–42.

Ahsan, A., Rudnick, J., Bruinsma, R., 1998. Elasticity theory of the B-DNA to S-DNA transition. Biophys. J. 74, 132–137.

Akinc, A., Goldberg, M., Qin, J., Dorkin, J.R., Gamba-Vitalo, C., Maier, M., Jayaprakash, K.N., Jayaraman, M., Rajeev, K.G., Manoharan, M., Koteliansky, V., Rohl, I., Leshchiner, E.S., Langer, R., Anderson, D.G., 2009. Development of lipidoid-siRNA formulations for systemic delivery to the liver. Mol. Ther. 17, 872–879.

Audouy, S., Molema, G., De Leij, L., Hoekstra, D., 2000. Serum as a modulator of lipoplex-mediated gene transfection: dependence of amphiphile, cell type and complex stability. J. Gene Med. 2, 465–476.

Aytar, B.S., Muller, J.P., Kondo, Y., Talmon, Y., Abbott, N.L., Lynn, D.M., 2013. Redox-based control of the transformation and activation of siRNA complexes in extracellular environments using ferrocenyl lipids. J. Am. Chem. Soc. 135, 9111–9120.

Balasubramaniam, R.P., Bennett, M.J., Aberle, A.M., Malone, J.G., Nantz, M.H., Malone, R.W., 1996. Structural and functional analysis of cationic transfection lipids: the hydrophobic domain. Gene Ther. 3, 163–172.

Barany, S., Szepesszentgyorgyi, A., 2004. Flocculation of cellular suspensions by polyelectrolytes. Adv. Colloid Interf. Sci. 111, 117–129.

Barreleiro, P.C.A., Lindman, B., 2003. The kinetics of DNA-cationic vesicle complex formation. J. Phys. Chem. B 107, 6208–6213.

Barreleiro, P.C., May, R.P., Lindman, B., 2003. Mechanism of formation of DNA-cationic vesicle complexes. Faraday Discuss. 122, 191–201, discussion 269–282.

Bilalov, A., Olsson, U., Lindman, B., 2009. A cubic DNA-lipid complex. Soft Matter 5, 3827–3830.

Bouxsein, N.F., Mcallister, C.S., Ewert, K.K., Samuel, C.E., Safinya, C.R., 2007. Structure and gene silencing activities of monovalent and pentavalent cationic lipid vectors complexed with siRNA. Biochemistry 46, 4785–4792.

Braun, C.S., Fisher, M.T., Tomalia, D.A., Koe, G.S., Koe, J.G., Middaugh, C.R., 2005. A stopped-flow kinetic study of the assembly of nonviral gene delivery complexes. Biophys. J. 88, 4146–4158.

Bruinsma, R., 1998. Electrostatics of DNA cationic lipid complexes: isoelectric instability. Eur. Phys. J. B 4, 75–88.

Candiani, G., Pezzoli, D., Ciani, L., Chiesa, R., Ristori, S., 2010. Bioreducible Liposomes for gene delivery: from the formulation to the mechanism of action. PLoS One 5 (10), e13430.

Caplen, N.J., Alton, E., Middleton, P.G., Dorin, J.R., Stevenson, B.J., Gao, X., Durham, S.R., Jeffery, P.K., Hodson, M.E., Coutelle, C., Huang, L., Porteous, D.J., Williamson, R., Geddes, D.M., 1995. Liposome-mediated CFTR gene-transfer to the nasal epithelium of patients with cystic fibrosis. Nat. Med. 1, 39–46.

Caracciolo, G., Amenitsch, H., 2012. Cationic liposome/DNA complexes: from structure to interactions with cellular membranes. Eur. Biophys. J. Biophys. Lett. 41, 815–829.

Carnio, S., Novello, S., Bironzo, P., Scagliotti, G.V., 2014. Moving from histological subtyping to molecular characterization: new treatment opportunities in advanced non-small-cell lung cancer. Expert Rev. Anticancer Ther. 14, 1495–1513.

Chan, C.-L., Ewert, K.K., Majzoub, R.N., Hwu, Y.-K., Liang, K.S., Leal, C., Safinya, C.R., 2014. Optimizing cationic and neutral lipids for efficient gene delivery at high serum content. J. Gene Med. 16, 84–96.

Chen, W., Meng, F., Cheng, R., Deng, C., Feijen, J., Zhong, Z., 2014. Advanced drug and gene delivery systems based on functional biodegradable polycarbonates and copolymers. J. Control. Release 190, 398–414.

Choi, Y.S., Lee, M.Y., David, A.E., Park, Y.S., 2014. Nanoparticles for gene delivery: therapeutic and toxic effects. Mol. Cell. Toxicol. 10, 1–8.

Cullis, P.R., de Kruijff, B., 1979. Lipid polymorphism and the functional roles of lipids in biological membranes. Biochim. Biophys. Acta 559, 399–420.

Dahlman, J.E., Barnes, C., Khan, O.F., Thiriot, A., Jhunjunwala, S., Shaw, T.E., Xing, Y., Sager, H.B., Sahay, G., Speciner, L., Bader, A., Bogorad, R.L., Yin, H., Racie, T., Dong, Y., Jiang, S., Seedorf, D., Dave, A., Sandhu, K.S., Webber, M.J., Novobrantseva, T., Ruda, V.M., Lytton-Jean, A.K.R., Levins, C.G., Kalish, B., Mudge, D.K., Perez, M., Abezgauz, L., Dutta, P., Smith, L., Charisse, K., Kieran, M.W., Fitzgerald, K., Nahrendorf, M., Danino, D., Tuder, R.M., Von Andrian, U.H., Akinc, A., Panigrahy, D., Schroeder, A., Koteliansky, V., Langer, R., Anderson, D.G., 2014. In vivo endothelial siRNA delivery using polymeric nanoparticles with low molecular weight. Nat. Nanotechnol. 9, 648–655.

Dan, N., 1996. Formation of ordered domains in membrane-bound DNA. Biophys. J. 71, 1267–1272.

Dan, N.L., 1997. Multilamellar structures of DNA complexes with cationic liposomes. Biophys. J. 73, 1842–1846.

Dan, N., 1998. The structure of DNA complexes with cationic liposomes—cylindrical or flat bilayers? Biochim, Biophys. Acta Biomembr. 1369, 34–38.

Dan, N., Danino, D., 2014. Structure and kinetics of lipid-nucleic acid complexes. Adv. Colloid Interf. Sci. 205, 230–239.

Danino, D., Kesselman, E., Saper, G., Petrache, H.I., Harries, D., 2009. Osmotically induced reversible transitions in lipid-DNA mesophases. Biophys. J. 96, L43–L45.

Davidson, B.L., Paulson, H.L., 2004. Molecular medicine for the brain: silencing of disease genes with RNA interference. Lancet Neurol. 3, 145–149.

de Fougerolles, A., Vornlocher, H.P., Maraganore, J., Lieberman, J., 2007. Interfering with disease: a progress report on siRNA-based therapeutics. Nat. Rev. Drug. Discov. 6, 443–453.

De Paula, D., Bentley, M.V., Mahato, R.I., 2007. Hydrophobization and bioconjugation for enhanced siRNA delivery and targeting. RNA 13, 431–456.

Desigaux, L., Sainlos, M., Lambert, O., Chevre, R., Letrou-Bonneval, E., Vigneron, J.P., Lehn, P., Lehn, J.M., Pitard, B., 2007. Self-assembled lamellar complexes of siRNA with lipidic aminoglycoside derivatives promote efficient siRNA delivery and interference. Proc. Natl. Acad. Sci. USA 104, 16534–16539.

Dickinson, E., Eriksson, L., 1991. Particle flocculation by adsorbing polymers. Adv. Colloid Interf. Sci. 34, 1–29.

Edelstein, S., Cammarata, R.C., 1996. Nanomaterials: Synthesis, Properties and Applications. Inst. of Physics Publishing, London, UK.

Elouahabi, A., Ruysschaert, J.-M., 2005. Formation and intracellular trafficking of lipoplexes and polyplexes. Mol. Ther. 11, 336–347.

Enarsson, L.-E., Wagberg, L., 2008. Adsorption kinetics of cationic polyelectrolytes studied with stagnation point adsorption reflectometry and quartz crystal microgravimetry. Langmuir 24, 7329–7337.

Evans, H.M., Ahmad, A., Ewert, K., Pfohl, T., Martin-Herranz, A., Bruinsma, R.F., Safinya, C.R., 2003. Structural polymorphism of DNA-dendrimer complexes. Phys. Rev. Lett. 91, 075501.

Ewert, K.K., Evans, H.M., Zidovska, A., Bouxsein, N.F., Ahmad, A., Safinya, C.R., 2006. A columnar phase of dendritic lipid-based cationic liposome-DNA complexes for gene delivery: hexagonally ordered cylindrical micelles embedded in a DNA honeycomb lattice. J. Am. Chem. Soc. 128, 3998–4006.

Farago, O., Ewert, K., Ahmad, A., Evans, H.M., Gronbech-Jensen, N., Safinya, C.R., 2008. Transitions between distinct compaction regimes in complexes of multivalent cationic lipids and DNA. Biophys. J. 95, 836–846.

Feigenson, G.W., 2009. Phase diagrams and lipid domains in multicomponent lipid bilayer mixtures. Biochim. Biophys. Acta Biomembr. 1788, 47–52.

Filippova, N.L., 1998. Adsorption and desorption kinetics of polyelectrolytes on planar surfaces. Langmuir 14, 1162–1176.

Fire, A., Xu, S., Montgomery, M.K., Kostas, S.A., Driver, S.E., Mello, C.C., 1998. Potent and specific genetic interference by double-stranded RNA in *Caenorhabditis elegans*. Nature 391, 806–811.

Fleer, G., Cohen Stuart, M.A., Scheutjens, J.M.H.M., Cosgrove, T., Vincent, B., 1993. Polymers at Interfaces. Springer, Berlin, Germany.

Forsythe, J.A., Jiang, B.H., Iyer, N.V., Agani, F., Leung, S.W., Koos, R.D., Semenza, G.L., 1996. Activation of vascular endothelial growth factor gene transcription by hypoxia-inducible factor 1. Mol. Cell. Biol. 16, 4604–4613.

Friedmann, T., Roblin, R., 1972. Gene therapy for human genetic disease? Science 175, 949–955.

Fujiwara, T., Grimm, E.A., Mukhopadhyay, T., Zhang, W.W., Owenschaub, L.B., Roth, J.A., 1994. Induction of chemosensitivity in human lung cancer cells in-vivo by adenovirus-mediated transfer of the wild type P53 gene. Cancer Res. 54, 2287–2291.

Gershon, H., Ghirlando, R., Guttman, S.B., Minsky, A., 1993. Mode of formation and structural features of DNA-cationic liposome complexes used for transfection. Biochemistry 32, 7143–7151.

Goni, F.M., Alonso, A., Bagatolli, L.A., Brown, R.E., Marsh, D., Prieto, M., Thewalt, J.L., 2008. Phase diagrams of lipid mixtures relevant to the study of membrane rafts. Biochim. Biophys. Mol. Cell Biol. Lipids 1781, 665–684.

Gregory, J., Barany, S., 2011. Adsorption and flocculation by polymers and polymer mixtures. Adv. Colloid Interf. Sci. 169, 1–12.

Harries, D., May, S., Gelbart, W.M., Ben-Shaul, A., 1998. Structure, stability, and thermodynamics of lamellar DNA-lipid complexes. Biophys. J. 75, 159–173.

Hastie, E., Samulski, R.J., 2015. Recombinant adeno-associated virus vectors in the treatment of rare diseases. Exp. Opin. Orphan Drugs 3, 675–689.

Hirsch-Lerner, D., Barenholz, Y., 1999. Hydration of lipoplexes commonly used in gene delivery: follow-up by laurdan fluorescence changes and quantification by differential scanning calorimetry. Biochim. Biophys. Acta Biomembr. 1461, 47–57.

Iruthayaraj, J., Poptoshev, E., Vareikis, A.V., Makuska, R., Van Der Wal, A., Claesson, P.M., 2005. Adsorption of low charge density polyelectrolyte containing poly(ethylene oxide) side chains on silica: effects of ionic strength and pH. Macromolecules 38, 6152–6160.

Jaaskelainen, I., Peltola, S., Honkakoski, P., Monkkonen, J., Urtti, A., 2000. A lipid carrier with a membrane active component and a small complex size are required for efficient cellular delivery of anti-sense phosphorothioate oligonucleotides. Eur. J. Pharm. Sci. 10, 187–193.

Jafari, M., Soltani, M., Naahidi, S., Karunaratne, D.N., Chen, P., 2012. Nonviral approach for targeted nucleic acid delivery. Curr. Med. Chem. 19, 197–208.

Juliano, R., Alam, M.R., Dixit, V., Kang, H., 2008. Mechanisms and strategies for effective delivery of antisense and siRNA oligonucleotides. Nucleic Acids Res. 36, 4158–4171.

Junquera, E., Aicart, E., 2014. Cationic lipids as transfecting agents of DNA in gene therapy. Curr. Top. Med. Chem. 14, 649–663.

Kapoor, M., Burgess, D.J., Patil, S.D., 2012. Physicochemical characterization techniques for lipid based delivery systems for siRNA. Int. J. Pharm. 427, 35–57.

Kennedy, M.T., Pozharski, E.V., Rakhmanova, V.A., Macdonald, R.C., 2000. Factors governing the assembly of cationic phospholipid-DNA complexes. Biophys. J. 78, 1620–1633.

Kim, D.H., Rossi, J.J., 2007. Strategies for silencing human disease using RNA interference. Nat. Rev. Genet. 8, 173–184.

Koltover, I., Salditt, T., Radler, J.O., Safinya, C.R., 1998. An inverted hexagonal phase of cationic liposome-DNA complexes related to DNA release and delivery. Science 281, 78–81.

Koltover, I., Salditt, T., Safinya, C.R., 1999. Phase diagram, stability, and overcharging of lamellar cationic lipid-DNA self-assembled complexes. Biophys. J. 77, 915–924.

Koumbi, D., Clement, J.-C., Sideratou, Z., Yaouanc, J.-J., Loukopoulos, D., Kollia, P., 2006. Factors mediating lipofection potency of a series of cationic phosphonolipids in human cell lines. Biochim. Biophys. Acta 1760, 1151–1159.

Koynova, R., Tenchov, B., 2001. Interactions of surfactants and fatty acids with lipids. Curr. Opin. Colloid Interf. Sci. 6, 277–286.

Laakkonen, J.P., Yla-Herttuala, S., 2015. Recent advancements in cardiovascular gene therapy and vascular biology. Hum. Gene Ther. 26, 518–524.

Lai, E., van Zanten, J.H., 2001. Monitoring DNA/poly-L-lysine polyplex formation with time-resolved multiangle laser light scattering. Biophys. J. 80, 864–873.

Lai, E., van Zanten, J.H., 2002. Real time monitoring of lipoplex molar mass, size and density. J. Control. Release 82, 149–158.

Larsson, K., 1983. Two cubic phases in monoolein–water system. Nature 304, 664.

Leal, C., Sandstrom, D., Nevsten, P., Topgaard, D., 2008. Local and translational dynamics in DNA-lipid assemblies monitored by solid-state and diffusion NMR. Biochim. Biophys. Acta Biomembr. 1778, 214–228.

Leal, C., Bouxsein, N.F., Ewert, K.K., Safinya, C.R., 2010. Highly efficient gene silencing activity of siRNA embedded in a nanostructured gyroid cubic lipid matrix. J. Am. Chem. Soc. 132, 16841–16847.

Leal, C., Ewert, K.K., Shirazi, R.S., Bouxsein, N.F., Safinya, C.R., 2011. Nanogyroids incorporating multivalent lipids: enhanced membrane charge density and pore forming ability for gene silencing. Langmuir 27, 7691–7697.

Leal, C., Ewert, K.K., Bouxsein, N.F., Shirazi, R.S., Li, Y., Safinya, C.R., 2013. Stacking of short DNA induces the gyroid cubic-to-inverted hexagonal phase transition in lipid-DNA complexes. Soft Matter 9, 795–804.

Li, S., Huang, L., 2000. Nonviral gene therapy: promises and challenges. Gene Ther. 7, 31–34.

Lin, M.Y., Lindsay, H.M., Weitz, D.A., Ball, R.C., Klein, R., Meakin, P., 1989. Universality in colloid aggregation. Nature 339, 360–362.

Ma, B., Zhang, S., Jiang, H., Zhao, B., Lv, H., 2007. Lipoplex morphologies and their influences on transfection efficiency in gene delivery. J. Control. Release 123, 184–194.

Martin, B., Sainlos, M., Aissaoui, A., Oudrhiri, N., Hauchecorne, M., Vigneron, J.P., Lehn, J.M., Lehn, P., 2005. The design of cationic lipids for gene delivery. Curr. Pharm. Design 11, 375–394.

Masotti, A., Mossa, G., Cametti, C., Ortaggi, G., Bianco, A., Del Grosso, N., Malizia, D., Esposito, C., 2009. Comparison of different commercially available cationic liposome-DNA lipoplexes: parameters influencing toxicity and transfection efficiency. Colloid Surf. B Biointerf. 68, 136–144.

May, S., Ben-Shaul, A., 1997. DNA-lipid complexes: stability of honeycomb-like and spaghetti-like structures. Biophys. J. 73, 2427–2440.

Mcfarlane, A., Yeap, K.Y., Bremmell, K., Addai-Mensah, J., 2008. The influence of flocculant adsorption kinetics on the dewaterability of kaolinite and smectite clay mineral dispersions. Colloid Surf. A Physicochem. Eng. Aspects 317, 39–48.

Merdan, T., Kopecek, J., Kissel, T., 2002. Prospects for cationic polymers in gene and oligonucleotide therapy against cancer. Adv. Drug Deliv. Rev. 54, 715–758.

Merkel, O.M., Mintzer, M.A., Sitterberg, J., Bakowsky, U., Simanek, E.E., Kissel, T., 2009. Triazine dendrimers as nonviral gene delivery systems: effects of molecular structure on biological activity. Bioconjug. Chem. 20, 1799–1806.

Mintzer, M.A., Simanek, E.E., 2009. Nonviral vectors for gene delivery. Chem. Rev. 109, 259–302.

Namvar, A., Bolhassani, A., Khairkhah, N., Motevalli, F., 2015. Physicochemical properties of polymers: an important system to overcome the cell barriers in gene transfection. Biopolymers 103, 363–375.

Nelson, P., 2013. Biological Physics. Freeman, New York, NY.

Netz, R.R., Andelman, D., 2003. Neutral and charged polymers at interfaces. Phys.Rep.Rev.Sect.Phys.Lett. 380, 1–95.

Neuberg, P., Kichler, A., 2014. Recent Developments in Nucleic Acid Delivery with Polyethylenimines. In: Huang, L., Liu, D., Wagner, E. (Eds.), Nonviral Vectors for Gene Therapy Lipid- and Polymer-Based Gene Transfer. Elsevier, Amsterdam, The Netherlands.

Osada, K., 2014. Development of functional polyplex micelles for systemic gene therapy. Polym. J. 46, 469–475.

Pensado, A., Seijo, B., Sanchez, A., 2014. Current strategies for DNA therapy based on lipid nanocarriers. Exp. Opin. Drug Deliv. 11, 1721–1731.

Pierce, E.A., Bennett, J., 2015. The status of RPE65 gene therapy trials: safety and efficacy. Cold Spring Harbor Persp. Med. 5 (9), a017285.

Pollard, H., Remy, J.S., Loussouarn, G., Demolombe, S., Behr, J.P., Escande, D., 1998. Polyethylenimine but not cationic lipids promotes transgene delivery to the nucleus in mammalian cells. J. Biol. Chem. 273, 7507–7511.

Ponnappa, B.C., 2009. siRNA for inflammatory diseases. Curr. Opin. Investig. Drugs 10, 418–424.

Popa, I., Gillies, G., Papastavrou, G., Borkovec, M., 2009. Attractive electrostatic forces between identical colloidal particles induced by adsorbed polyelectrolytes. J. Phys. Chem. B 113, 8458–8461.

Popa, I., Gillies, G., Papastavrou, G., Borkovec, M., 2010. Attractive and repulsive electrostatic forces between positively charged latex particles in the presence of anionic linear polyelectrolytes. J. Phys. Chem. B 114, 3170–3177.

Pozharski, E., Macdonald, R.C., 2002. Thermodynamics of cationic lipid-DNA complex formation as studied by isothermal titration calorimetry. Biophys. J. 83, 556–565.

Pozharski, E., Macdonald, R.C., 2003. Lipoplex thermodynamics: determination of DNA-cationic lipoid interaction energies. Biophys. J. 85, 3969–3978.

Pozharski, E.V., Macdonald, R.C., 2007. Single lipoplex study of cationic lipoid-DNA, self-assembled complexes. Mol. Pharm. 4, 962–974.

Qasim, W., Thrasher, A.J., 2014. Progress and prospects for engineered T cell therapies. Br. J. Haematol. 166, 818–829.

Radler, J.O., Koltover, I., Salditt, T., Safinya, C.R., 1997. Structure of DNA-cationic liposome complexes: DNA intercalation in multilamellar membranes in distinct interhelical packing regimes. Science 275, 810–814.

Ramamoorth, M., Narvekar, A., 2015. Non viral vectors in gene therapy—an overview. J. Clin. Diagn. Res. 9, GE01–GE06.

Ramezani, M., Khoshhamdam, M., Dehshahri, A., Malaekeh-Nikouei, B., 2009. The influence of size, lipid composition and bilayer fluidity of cationic liposomes on the transfection efficiency of nanolipoplexes. Colloids Surf. B Biointerf. 72, 1–5.

Romoren, K., Thu, B.J., Bols, N.C., Evensen, O., 2004. Transfection efficiency and cytotoxicity of cationic liposomes in salmonid cell lines of hepatocyte and macrophage origin. Biochim. Biophys. Acta Biomembr. 1663, 127–134.

Ross, P.C., Hui, S.W., 1999. Lipoplex size is a major determinant of in vitro lipofection efficiency. Gene Ther. 6, 651–659.

Ruozi, B., Forni, F., Battini, R., Vandelli, M.A., 2003. Cationic liposomes for gene transfection. J. Drug Target. 11, 407–414.

Safinya, C.R., 2001. Structures of lipid-DNA complexes: supramolecular assembly and gene delivery. Curr. Opin. Struct. Biol. 11, 440–448.

Safinya, C.R., Ewert, K.K., Leal, C., 2011. Cationic liposome-nucleic acid complexes: liquid crystal phases with applications in gene therapy. Liq. Cryst. 38, 1715–1723.

Sahay, G., Querbes, W., Alabi, C., Eltoukhy, A., Sarkar, S., Zurenko, C., Karagiannis, E., Love, K., Chen, D., Zoncu, R., Buganim, Y., Schroeder, A., Langer, R., Anderson, D.G., 2013. Efficiency of siRNA delivery by lipid nanoparticles is limited by endocytic recycling. Nat. Biotechnol. 31 (7), 653–658.

Salvati, A., Ciani, L., Ristori, S., Martini, G., Masi, A., Arcangeli, A., 2006. Physico-chemical characterization and transfection efficacy of cationic liposomes containing the pEGFP plasmid. Biophys. Chem. 121, 21–29.

Scarzello, M., Chupin, V., Wagenaar, A., Stuart, M.C., Engberts, J.B., Hulst, R., 2005. Polymorphism of pyridinium amphiphiles for gene delivery: influence of ionic strength, helper lipid content, and plasmid DNA complexation. Biophys. J. 88, 2104–2113.

Schaffer, D.V., Fidelman, N.A., Dan, N., Lauffenburger, D.A., 2000. Vector unpacking as a potential barrier for receptor-mediated polyplex gene delivery. Biotechnol. Bioeng. 67, 598–606.

Schroeder, A., Levins, C.G., Cortez, C., Langer, R., Anderson, D.G., 2010. Lipid-based nanotherapeutics for siRNA delivery. J. Intern. Med. 267, 9–21.

Sedeva, I.G., Fornasiero, D., Ralston, J., Beattie, D.A., 2009. The influence of surface hydrophobicity on polyacrylamide adsorption. Langmuir 25, 4514–4521.

Simberg, D., Danino, D., Talmon, Y., Minsky, A., Ferrari, M.E., Wheeler, C.J., Barenholz, Y., 2001. Phase behavior, DNA ordering, and size instability of cationic lipoplexes - Relevance to optimal transfection activity. J. Biol. Chem. 276, 47453–47459.

Simberg, D., Danino, D., Talmon, Y., Minsky, A., Ferrari, M.E., Wheeler, C.J., Barenholz, Y., 2003. Phase behavior, DNA ordering and size instability of cationic lipoplexes: relevance to optimal transfection activity. J. Lipos. Res. 13, 86–87.

Stegmann, T., Legendre, J.Y., 1997. Gene transfer mediated by cationic lipids: lack of a correlation between lipid mixing and transfection. Biochim. Biophys. Acta Biomembr. 1325, 71–79.

Sukhishvili, S.A., Granick, S., 1998. Polyelectrolyte adsorption onto an initially-bare solid surface of opposite electrical charge. J. Chem. Phys. 109, 6861–6868.

Szilagyi, I., Trefalt, G., Tiraferri, A., Maroni, P., Borkovec, M., 2014. Polyelectrolyte adsorption, interparticle forces, and colloidal aggregation. Soft Matter 10, 2479–2502.

Takeshita, F., Ochiya, T., 2006. Therapeutic potential of RNA interference against cancer. Cancer Sci. 97, 689–696.

Templeton, N.S., Lasic, D.D., Frederik, P.M., Strey, H.H., Roberts, D.D., Pavlakis, G.N., 1997. Improved DNA: liposome complexes for increased systemic delivery and gene expression. Nature Biotechnol. 15, 647–652.

Tilcock, C.P., Bally, M.B., Farren, S.B., Cullis, P.R., 1982. Influence of cholesterol on the structural preferences of dioleoylphosphatidylethanolamine-dioleoylphosphatidylcholine systems: a phosphorus-31 and deuterium nuclear magnetic resonance study. Biochemistry 21, 4596–4601.

Tros de Ilarduya, C., Duezguenes, N., 2013. Delivery of therapeutic nucleic acids via transferrin and transferrin receptors: lipoplexes and other carriers. Exp. Opin. Drug Deliv. 10, 1583–1591.

Tros de Ilarduya, C., Sun, Y., Duezguenes, N., 2010. Gene delivery by lipoplexes and polyplexes. Eur. J. Pharm. Sci. 40, 159–170.

van Heiningen, J.A., Hill, R.J., 2011. Polymer adsorption onto a micro-sphere from optical tweezers electrophoresis. Lab Chip 11, 152–162.

Wagner, E., 2014. Polymers for Nucleic Acid Transfer - an Overview. In: Huang, L., Liu, D., Wagner, E. (Eds.), Nonviral vectors for gene therapy lipid- and polymer-based gene transfer. Elsevier, Amsterdam, The Netherlands.

Wasungu, L., Hoekstra, D., 2006. Cationic lipids, lipoplexes and intracellular delivery of genes. J. Control. Release 116, 255–264.

Wasungu, L., Stuart, M.C., Scarzello, M., Engberts, J.B., Hoekstra, D., 2006. Lipoplexes formed from sugar-based gemini surfactants undergo a lamellar-to-micellar phase transition at acidic pH. Evidence for a non-inverted membrane-destabilizing hexagonal phase of lipoplexes. Biochim. Biophys. Acta 1758, 1677–1684.

Whitehead, K.A., Langer, R., Anderson, D.G., 2009. Knocking down barriers: advances in siRNA delivery. Nat. Rev. Drug Discov. 8, 129–138.

Xiong, F., Mi, Z., Gu, N., 2011. Cationic liposomes as gene delivery system: transfection efficiency and new application. Pharmazie 66, 158–164.

Xu, L., Anchordoquy, T.J., 2008. Cholesterol domains in cationic lipid/DNA complexes improve transfection. Biochim. Biophys. Acta Biomembr. 1778, 2177–2181.

Xue, W., Dahlman, J.E., Tammela, T., Khan, O.F., Sood, S., Dave, A., Cai, W., Chirino, L.M., Yang, G.R., Bronson, R., Crowley, D.G., Sahay, G., Schroeder, A., Langer, R., Anderson, D.G., Jacks, T., 2014. Small RNA combination therapy for lung cancer. Proc. Natl. Acad. Sci. USA 111, E3553–E3561.

Yang, J., Liu, H., Zhang, X., 2014. Design, preparation and application of nucleic acid delivery carriers. Biotechnol. Adv. 32, 804–817.

Zhang, Y., Garzon-Rodriguez, W., Manning, M.C., Anchordoquy, T.J., 2003. The use of fluorescence resonance energy transfer to monitor dynamic changes of lipid-DNA interactions during lipoplex formation. Biochim. Biophys. Acta. 1614, 182–192.

Zhang, S., Zhi, D., Huang, L., 2012. Lipid-based vectors for siRNA delivery. J. Drug Target. 20, 724–735.

Zhao, D.D., Watarai, S., Lee, J.T., Kouchi, S., Ohmori, H., Yasuda, T., 1997. Gene transfection by cationic liposomes: comparison of the transfection efficiency of liposomes prepared from various positively charged lipids. Acta Med. Okayama 51, 149–154.

Zidovska, A., Evans, H.M., Ewert, K.K., Quispe, J., Carragher, B., Potter, C.S., Safinya, C.R., 2009. Liquid crystalline phases of dendritic lipid-DNA self-assemblies: lamellar, hexagonal, and DNA bundles. J. Phys. Chem. B 113, 3694–3703.

Zuidam, N.J., Barenholz, Y., 1997. Electrostatic parameters of cationic liposomes commonly used for gene delivery as determined by 4-heptadecyl-7-hydroxycoumarin. Biochim. Biophys. Acta 1329, 211–222.

Zuidam, N.J., Barenholz, Y., 1998. Electrostatic and structural properties of complexes involving plasmid DNA and cationic lipids commonly used for gene delivery. Biochim. Biophys. Acta 1368, 115–128.

CHAPTER

USING MICROSENSORS TO PROMOTE THE DEVELOPMENT OF INNOVATIVE THERAPEUTIC NANOSTRUCTURES

20

Corinne Dejous*, Hamida Hallil*, Vincent Raimbault*,, Rajagopalan Rukkumani†, Jatinder Vir Yakhmi‡**

**University of Bordeaux, Bordeaux, France*

***LAAS-CNRS, Toulouse, France*

†Pondicherry University, Kalapet, India

‡Homi Bhabha National Institute, Anushaktinagar, India

CHAPTER OUTLINE

1 INTRODUCTION

Innovative therapeutic nanostructures are assuming increasing significance due to their potential in the design of nanodrugs and their delivery (Nikalje, 2015), notwithstanding the major issues related to the efficiency and toxicity of these therapeutic nanostructures. These issues and the design of nanostructures are also strongly linked to the mechanisms involved in the drug delivery process, and their interaction with biological barriers has to be considered with caution (Blanco et al., 2015).

Micro- and nanotechnologies have been increasingly employed in conjunction with conventional characterization tools, which have been adapted to evaluate the relevant characteristics and properties of therapeutic nanostructures for use in nanodrugs. The aim of this chapter is to present the most recent advances in micro- and nanofluidics that may allow to overcome the drawbacks of conventional characterization, coupled with advantages, such as low volumes and in-through assays. The field of microsensors can be utilized to promote the development of therapeutic nanostructures. Recognition strategies and materials elaborated for therapeutic purposes by targeting specific cells have been investigated for specific detection of diseases at early stage or even during the follow-up of treatment for evaluating effectiveness and efficiency of drugs. Ample literature exists on the use of microsensors, but it is mostly in vitro or ex vivo detection. However, currently, there is enormous interest on potentialities for in-vivo sensing and integration of a whole implantable microsystem combining drug delivery and wireless communication with remote system, and we wish to highlight this in the present Chapter. A major attraction of microsensors is their use as convenient and complementary tools, *vis-a-vis* the modern analytic labs requiring large instruments and skilled users, although the latter do deliver more informative signal on complex sample composition. Not much effort has been spent to study the potentialities of microsensors to characterize the properties of therapeutic nanostructures, including the drug itself, or the drug carrier and delivery system, and their influence on living materials. Besides, the microsensors offer numerous advantages towards dynamic monitoring of interactions with living materials, with conditions closer to in vivo experiments when integrating microfluidics, in addition to capabilities for efficient high-throughput screening due to in-flow and array abilities with low sample volumes. Use of microsensors can greatly facilitate studies of the efficiency or toxicity of therapeutic agents, or the resistance of cells to them, providing both scientific and economic advantages.

This chapter is divided in three main parts. In Section 1, we describe nano-objects in relation with the therapeutic field, their main properties to be studied, and the in vitro and in vivo issues involved. In Section 2, we present micromachined functions potentially relevant for fluid handling, namely, flow control and cell exposure system with liquid or air interface, generally named lab-on-a-chip (LOC), and the very recently developed multi-organ chip (MOC). In Section 3, we describe measurements techniques and the principles of main microsensors, while underlining the benefits, which would accrue by the use of advanced materials and wireless multisensors. The association of both conventional and microsensing techniques with fluidic chips is particularly emphasized through relevant examples.

2 NANO-OBJECTS DEFINITION AND PROPERTIES

Engineered nanomaterials, objects with at least one dimension below 100 nm, are becoming rapidly a part of our daily life in the form of cosmetics, for medical imaging, new drug delivery technologies, biosensors, therapeutics, and various industrial products. Due to their unique characteristics including superparamagnetic or fluorescent properties, and small size comparable to biomolecules, nanostructured materials have emerged as novel bioimaging, diagnostic, and therapeutic agents. The first FDA-approved application of a nanotherapeutic drug was made in 1995 (Barenholz, 2012), which has been followed by more than 50 nanotherapeutic drug approvals. The forms and physicochemical properties of nanomaterials can, however, be extremely varied and need to be ascertained to get the best therapeutic benefits from them. Especially, the combinations of various nanostructured materials with different properties can offer synergetic multifunctional nanomedical platforms, which make it possible to accomplish multimodal imaging, and simultaneous diagnosis, and therapy. Moreover, the conjugation of targeting moieties on the surface of these multifunctional nanomaterials gives them specific targeted imaging and therapeutic properties (Barenholz, 2012).

There are two major application areas for nanoobjects in medicine: (1) contrast agents for magnetic resonance imaging (MRI) and (2) vector for chemotherapeutic agents. Other applications include biodetection, photothermal therapy (PTT), using plasmonic properties, and drug delivery.

One of the current challenges prompting considerable debate in nano (eco) toxicology and health impact of these nanoparticles is the issue of concentration of nanoparticles used in experimental studies to characterize the potential hazards of these materials. In particular, these debates relate to the use of "realistic" doses representative of actual exposures of humans or the environment studied. This is partly due to insufficient exposure information, but also due to the limits of some models of study that do not permit to take into account the effect of chronic exposure to low doses, which are repeated over several weeks or months. The characterization of the behavior of nanomaterials in the environment and the exposure levels is important. Continued research in nanotoxicology has not yet thrown up clear answers, or complete and exhaustive regulation. The concept of a specific dose, which is a basic concept in (eco) toxicology, received special attention in 2014. Then, there is the problem of an effective dose, the one that is actually delivered to the test system, and the metric to use. Recent studies have advocated the decreasing of doses administered, taking into account the potential transformation of nanomaterials during testing.

The use of mathematical models to determine effective doses is very beneficial for both in vivo and in vitro studies. For in vitro studies, a mathematical computer model proposed by Teeguarden and coworkers has been employed to predict the delivered dose—in vitro sedimentation, diffusion model, and dosimetry (Hinderliter et al., 2010)—in conjunction with the experimentally measured characteristics of nanoparticles (Cohen et al., 2014). This integrative approach involves measuring the effective density of the agglomerates by volumetric Harvard centrifugation method (VLM) (Deloid et al., 2014). The estimate of the delivered dose could be validated experimentally for 20 varieties of commonly studied nanoparticles to know the dose delivered over a period of time in the required metric [mass, surface or number of particles, (Cohen et al., 2014)]. These functions can be used to determine the effective dose in vitro. However, this integrative strategy is impractical to nanomaterials with a high form factor (called HARN for high aspect ratio nanomaterial), such as carbon nanotubes.

2.1 NANO-OBJECTS ENCOUNTERED IN NANOTHERAPEUTICS

Although the field of nanotechnology is relatively recent, there is already a vast range of nanoobjects with different characteristics, such as, their shape, surface rugosity, chemical composition (organic or inorganic), size (1 nm to > 100 nm) or surface charge that can strongly affect their properties. Fig. 20.1 depicts the most common shapes of nano-objects, usually called nanoparticles (NPs, also named nanocarriers in the context of drug delivery) that are encountered in the field of nanotherapeutics.

The size of the NPs has direct relation with the long circulation time in the body, typically in blood stream or respiratory tract, and better permeability during drug delivery. In the case of cancer therapy, an enhanced permeability and retention (EPR) effect, helps to accumulate the NPs preferably in the tumor tissues rather than in normal tissues.

Cytotoxicity studies of nanostructures may be performed for studying both therapeutic efficacy as well as nanotoxicity. However, depending on the objective of the study, the experimental data may be accordingly generated and interpreted. While in case of nanotoxicity studies, the effect of nanostructures may be studied on normal cell lines and tissues, in the case of therapeutic efficacy studies, on the other hand, similar experimental study may be done on diseased/abnormal cells and tissues.

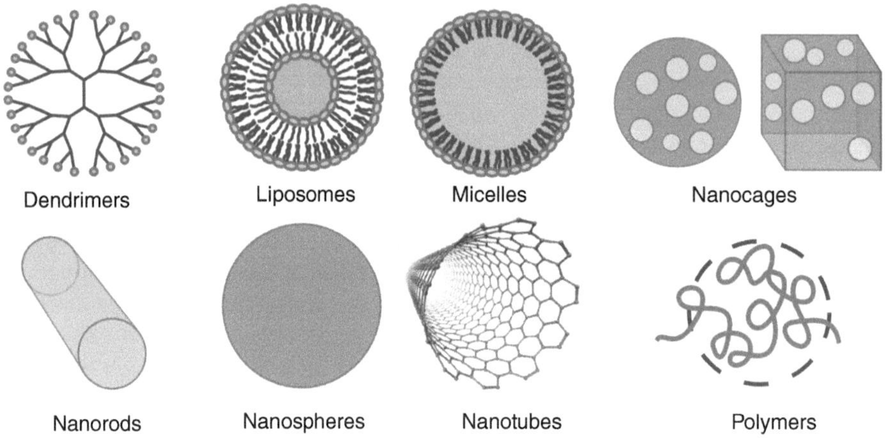

FIGURE 20.1 Illustration of the Most Common Shapes of Nanocarriers Used in Nanotherapeutics

Adapted from Wong, P.T., Choi, S.K., 2015. Mechanisms of drug release in nanotherapeutic delivery systems. Chem. Rev., 115(9), 3388–3432.

3 INTEGRATED ANALYTE HANDLING

3.1 MICROFLUIDICS FUNCTIONS

Microfluidics aims at manipulating fluid flows at the microscale. The advent of soft lithography through the work of Whitesides (Xia and Whitesides, 1998), has brought forth microfluidics as a cheaper, faster, and easier method for handling small amount of fluids than the previous microelectronic-based fabrication processes, offering new paradigms in life sciences thanks to dimensions that are now comparable to the objects that are manipulated.

We now describe microfluidic systems that are already used, or could be used in a near future with great potentialities, in the field of nanotherapeutics. These systems can be sorted in two categories:

- Microfluidics systems for the handling of body fluids (for example, blood, saliva, urine) and their analysis.
- Cell-culture based microfluidic systems, used to mimic the behavior of organs and evaluate their reaction to nano-objects.

These microfluidic systems can be used for the development of new nano-carriers, for direct therapy, or to assess their toxicity.

3.1.1 Body fluids handling and analysis using microfluidics

The manipulation of body fluids by the mean of microfluidics attracts a lot of interest due to interesting potentialities, offering faster, simpler, and cheaper methods. Kong et al. (2014) give an insight on microfluidic emerging schemes that addresses challenges in blood filtering. These challenges concern two different types of actions: (1) the removal of unwanted pathogens and/or toxins from whole blood; (2) the isolation of rare cells that possess interesting therapeutic properties.

The removal of pathogens in blood has been demonstrated by using a microfluidic based separation device to clean blood for sepsis therapy (Kang et al., 2014). The authors engineered magnetic nanobeads with opsonins (molecules which facilitates phagocytosis), in the form of mannose-binding lectin (MBL). Building on the ability of MBL to bind with a vast range of pathogenic microorganisms, like *Staphylococcus aureus* or *Escherichia coli*, the functionalized nanobeads can be separated by using an applied magnetic field, in a modified H-filter microfluidic chip, inspired by the living spleen. This H-filter is made of two adjacent channels, one containing flowing blood with flow rates as high as 1.25 l/h, and the other perfused by saline solution at a much lower intermittent flow rates, of the order of 0.00 to 0.01 l/h. A series of microchannels connect the two aforementioned channels, thus allowing a direct interface between blood and saline solution. By using an appropriate blood flow rate, and an applied magnetic field using permanent magnets, the engineered superparamagnetic nanobeads are extracted from the blood through the small microchannels, and then discarded into the saline solution. This approach shows promising results for sepsis treatment by the removal of pathogen and endotoxins, but could also be used for the capture of bacterial cells for further analysis. This could be an efficient way of identifying the infection origin. While the work presented by Kang et al. (2014) did not aim at any specificity toward a certain type of pathogenic microorganisms, the same technique could be applied by using a different type of functionalization to bring selectivity to the system.

Lin et al. (2015), have reported an alternative microfluidic method for the capture, isolation and concentration of endothelial colony-forming cells (ECFCs). Of great therapeutic potential, these ECFCs can be found circulating in the peripheral blood, but in very small proportion, compared to other circulating cells, making their capture, and isolation particularly challenging. The conventional method to do so is based on density gradient centrifugation from peripheral blood samples to separate mononuclear cells, followed by erythrocyte lysis and several cell washes, which involves number of preprocessing steps, is time-consuming, and is known to lead to a high cell loss. Compared to this conventional centrifugation-based method (Fig. 20.2), the microfluidic device offers the same performance in terms of processing time and captured ECFCs, but offers the ability to process without blood pretreatment, meaning without the density gradient centrifugation and erythrocyte lysis, thus facilitating the processing of large volumes, and adding the possibility of a continuous blood processing. The ability to perform continuous capture of ECFCs, before recirculating the fluid back into the patient, could greatly facilitate the processing of large peripheral blood volumes, and in fact may favor a better collection of ECFCs, even when present at very low concentrations.

Once the cells of interest have been isolated, their deliverance is another challenge to overcome. The cell delivery step is known to represent a strong limitation to the efficiency of cell therapy. Use of a delivery vehicle is an efficient mean of enhancing cell therapy efficiency. By using 3D microstructured gelatin microcryogels, Li et al. (2014) have developed such vehicles in the form of 3D microniches. The Gelatin Microcryogels (GMs) are fabricated by means of a PMMA microstencil array. The microstencil is composed of 600 circular microwells, each microwell being 400 μm in radius. The structuration of the PMMA is achieved by laser prototyping. After etching, the microstencil is treated using a plasma O_2 cleaner to increase its hydrophobicity. Gelatin is used as the retaining matrix for the cells, as it offers adequate cell adhesion sites combined with enzymatic degradability properties. Moreover, gelatin is inexpensive. A gelatin precursor is dissolved in deionized water and then incubated on ice. Glutaraldehyde is added to the solution, and then pipetted on top of the microstencil array chip. A cover slide is used to scrape the prepolymer on the microstencil array, and once the microstencil array is filled with the gelatin solution, it is placed for cryogelation, and then lyophilized. The GMs are then ejected

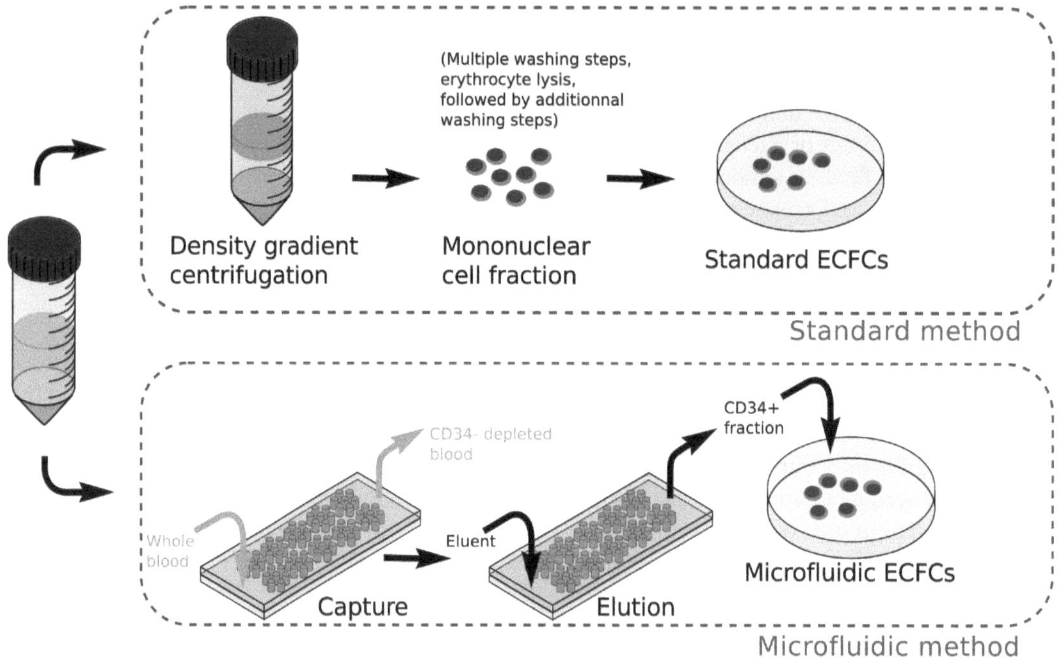

FIGURE 20.2 Comparison of a Standard and a Microfluidic Method for ECFCs Capture

The ability of the microfluidic method to be performed without blood pretreatment offers the opportunity to perform continuous blood processing, thus facilitating the process of large volumes.

Adapted from Lin, R.-Z., Hatch, A., Antontsev, V.G., Murthy, S.K., Melero-Martin, J.M., 2015. Microfluidic capture of endothelial colony-forming cells from human adult peripheral blood: phenotypic and functional validation in vivo. Tissue Eng. Part C: Methods 21(3), 274–283.

from the mold, and thoroughly washed. The GMs undergo sterilization, and are then exposed to an isolated and cultured human adipose-derived mesenchymal stem cells (hMSCs). The cells are absorbed into the GMs due to their porosity and dehydrated state. The GMs are then incubated in a humidified chamber for 2 h at 37°C.

These microniches enable a great enhancement of the extracellular matrix and cell–cell interactions, thus mimicking the tissue structure. The result is a better retention of the cells, and also a better survival, compared to free cell suspensions.

3.1.2 Mimicking the human body: organs-on-a-chip

As opposed to conventional drug based treatments that relies on specific chemicals or molecules, the size and physico-chemical properties of nanoparticles implies a different behavior while exposed to a biological environment. The study of interactions between nanoparticles and specific cells is, by and large, performed in vitro, generally in petri dishes or cell culture plates, and their effect on real organs extrapolated, using various strategies. These standard methods have inherent drawbacks, especially due to the lack of any regulation of the flow conditions, although these have a strong influence on the cell growth and their interactions with the products under investigation.

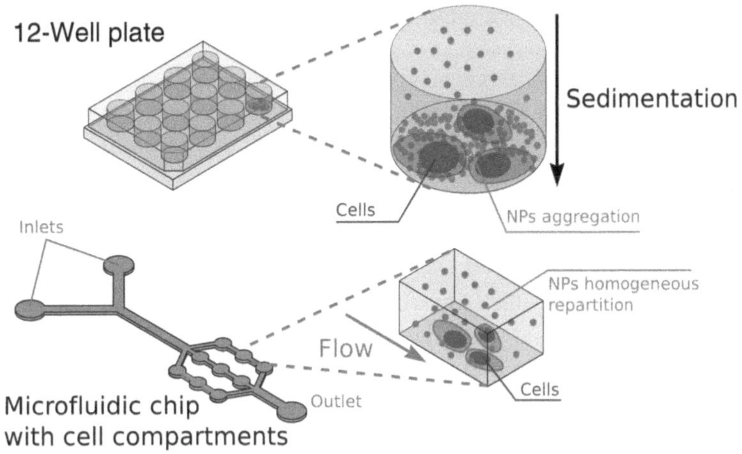

FIGURE 20.3 Illustration of Difference in Kinetics in a Classic In Vitro Assay Performed in a Multiwell Culture Plate and in a Microfluidic Device

Classical cell cultures in petri dish suffer from sedimentation (top), while a microfluidic flow cell facilitates an even distribution as in the nonstatic surrounding media of living organism (bottom).

Adapted from Mahto, S.K., Yoon, T.H., Rhee, S.W., 2010. Cytotoxic effects of surface-modified quantum dots on neuron-like pc12 cells cultured inside microfluidic devices. Biochip J., 4(1), 82–88.

Understanding, analyzing, and more generally, observing nano-objects interactions in vivo is not an easy task. As the research and development on nano-objects continues, more and more different types of nanoparticles are susceptible to be ingested. This ingestion can be either voluntary, in the case of a therapy based on nanoparticles, or involuntary, such as, from air pollution, self-cleaning clothes, degradation. Once ingested, the observation of nanoparticles is obviously very limited in situ due to the vast network of tissues that prevent noninvasive techniques to be efficient.

Due to their size and peculiar properties, the propagation and diffusion of nanoparticles is very different from what can be expected from small molecules. Depending on their mass, nanoparticles can rapidly get sedimented, which, in the frame of cell culture in plates or petri dish, lead to unrealistic exposition of cultured cells. This is illustrated in the top portion of Fig. 20.3, where one can observe a strong gradient of concentration oriented toward the bottom of the well. This situation is barely plausible in a living organism, where the surrounding media is generally nonstatic, thus preventing sedimentation. As a comparison, the bottom portion of Fig. 20.3 shows the distribution of nanoparticles in a microfluidic flow cell, where the concentration is more evenly distributed inside the channel. The concentration of nanoparticles to which the cells are exposed in this scenario is much smaller than in the case of static analysis, and also more representative of a typical exposition in the body, keeping in mind that the flow conditions has to reflect the flow conditions encountered in vivo.

During the past decade, a lot of different organs or tissues have been implemented as organs-on-chips, including liver, kidney, intestine, lung, heart, smooth and striated muscle, fat, bone, marrow, cornea, skin, blood vessels, nerve or blood–brain barrier (BBB), and a comprehensive list of references can be found in the review by Bhatia and Ingber (2014). Although less frequently done due to their complexity, some authors have also studied the case of multiple organs or tissues on a single chip, either to mimic the interactions between multiple organs, or to expose several organs separately to the same conditions. These microfluidic systems are generally referred to as body-on-a-chip, or human-on-a-chip, and are targeted to a more systemic approach of the drugs/organs interactions.

In a 2013 article in the *Lab on a Chip* journal, Wagner and colleagues presented a microfluidic chip that has the size of a standard microscope slide, but still including a peristaltic micropump, media reservoirs, and tissue culture spaces separated by microfluidc channels. This chip, referred by the authors as a multiorgan-chip (MOC) is developed for the field of cosmetics. As the skin is the target tissue of this industry, skin cells are cultured on chip, as well as liver cells, liver being regarded as the prime target for toxicity evaluations of drugs (Wagner et al., 2013). The authors reported successful use of Transwell permeable membranes, which are widely available for traditional multiwell plates assays, allowing the presence of an air–liquid (AIL) interface favorable to a better stability of the skin biopsies cultures. This AILinterface also allow dermal exposure experiments, which considering the numerous cosmetic products containing nanoparticles could translate into a better way to study the effects of nano-objects exposure.

The use of skin biopsies compared to skin equivalents offers better fidelity in reproducing the in vivo behavior in this microfluidic setup. This refined microfluidic setup has allowed a long-term stability of the cell culture, up to 28 days with the Transwell membranes. This long-term stability is of crucial interest to study long-term effects of exposure to nanoparticles.

Maschmeyer et al. (2015) have enriched the MOC and reported a four-organ chip, exhibiting the same 28 days viability as in their previous article. They have successfully implemented intestine, liver, skin and kidney tissue cultures in four distinct compartments. The MOC then features a surrogate blood circuit and an excretory circuit. These additional features allow the authors to perform adsorption, distribution, metabolism, and excretion profiling, also known as ADME, a method which is widely used in pharmacology to assess the performance of drugs.

The presence of an excretory circuit is very interesting in the scope of nanoparticles study, as it adds the ability to estimate the amount of nanoparticles not uptaken in the various cell compartments. By quantifying this amount, one can obtain a better estimate of the effective dose applied to the tissues, which is of paramount importance in nanotherapeutics studies. The authors rely currently on external characterization techniques to assess the various parameters. The tissue viability is monitored by a measurement of the lactate dehydrogenase (LDH) levels in the various cell compartments. Thanks to fluidic access ports, it is possible to sample the circulating fluids for external analysis, in that case allowing absorbance measurements for LDH activity.

3.2 AIR–LIQUID INTERFACE

3.2.1 Air–liquid interface issues

Therapeutic applications of nanodelivery systems for the treatment of airway diseases are considered a promising solution for public health and a suitable route for pharmaceutical industry. Beyond that, it can be noted an increasing interest for aerosols in systemic drug delivery, gene therapy, and vaccination (Laube, 2014). Asthma, diabetes, thrombosis, parathyroid disease, vaccination, mutative defects in lung, lung cancers, infections, and pulmonary hypertension are targets for topical inhalation treatment.

The possibility of noninvasive administration of drug by inhalation aerosols, makes the lung an attractive target for medication administration. Indeed, this noninvasive method avoids first pass metabolism and promotes the direct delivery of nanoparticles to the target site of action. This provides a large area for locally administered treatment against respiratory diseases (Mansour et al., 2009).

Even though, the field of drug delivery is growing fast for therapy of respiratory illnesses, there are concerns about their biodegradability, potential toxicity, and elimination. In fact, the nanoparticles are deposited at all surfaces of the respiratory system. This nanoparticles distribution leads to damaging effects on biological tissues, such as oxidative stress, prothrombotic effects, proinflammatory, and consequently of DNA damage or the development of fibrosis or pulmonary emphysema. More generally, environmental exposure to nanomaterials is inevitable as nanomaterials become part of our daily life, and consequently, nanotoxicity research is gaining attention, to better control the dosimetry. Several epidemiologic studies have shown that their side effects can result in morbidity and mortality in susceptible parts of the population (Newby et al., 2014). Safety characterization of nanoparticles becomes more and more important. However, studies on the toxicity of nanoparticles in animals are very expensive and repeated/long duration exposures are often not possible to achieve, as in other experimental models due to limited duration of cell cultures, harmful effects of installations realized in vivo, low number of assessments, time consuming, and ethically controversial methodologies (Steinritz et al., 2013).

A compelling way of studying the effects on human biology without animal or human beings is to perform in vitro experiments using explants or isolated cells from human organs or biopsies, as well as cell lines. Ex vivo experiments are performed typically, under immersed conditions.

Several cell culture models of the respiratory tract are described in the literature using both continuous and primary cell cultures. In both such cellular models, it is important that epithelial cells form a tight monolayer in order to represent the natural epithelial barrier. Monolayer tightness and integrity are assessed classically by measuring transendothelial/transepithelial electrical resistance (TEER) and potential difference across the monolayer. Monolayers of lung epithelial cells allow the characterisation of drug transport and assessment of potential drug, and formulation toxicity (Fernandes and Vanbever, 2009). For example, the study by Schlinkert et al. (2015) focused on the oxidative potential of differently charged silver and gold nanoparticles coated with either sodium citrate or chitosan on three human lung epithelial cell types. TEER measurements and immunofluorescence staining of tight junctions were performed to test the growth characteristics of the cells. Cytotoxicity was measured by means of the CellTiter-Blue that provides a homogeneous, fluorescent method for monitoring cell viability, and with lactate dehydrogenase assay. Cellular and cell-free reactive oxygen species (ROS) production was measured using the DCFH-DA assay. Different growth characteristics as well as NP-induced cytotoxicity effects and ROS production were shown in the three cell types used. Cell-free ROS measurements showed a significant increase in ROS production with an increase in chitosan coating.

Furthermore, the phenomena of agglomeration and sedimentation are strongly related to the test medium used and studied nanoparticle or the employed exposure protocol that could affect the particle-induced biological response (Lenz et al., 2009). In addition, experimental models also lead to issues related to maintaining exposure concentrations and physicochemical properties of nanoparticles during testing. These questions on dosimetry also concern the evaluation of the real applied dose during the experiment, as previously presented in fluids flows. The local concentration in contact with the cells can, therefore, increase with the density and state of aggregation/agglomeration of the nanoparticles. For example, physical methods (that is. ultrasonication) or chemical (covalent functionalization of the surface or addition of dispersing agents) used to better disperse and stabilize the nanoparticle suspensions may modify their surface and, therefore, artificially modulate their toxicity. It is, therefore, desirable to use as dispersant, molecules that can really be encountered by the nanoparticles in the environment (natural organic matter, such as humic acids, for example) or in the target organ studied.

Additionally, many in vitro experiments at low doses and single exposure seem to show no damaging effect on cells. This necessitates a caution before making any conclusions, since long-term studies or on repeated exposure, do show often deleterious effects. Using air–liquid interface tools, with experiment conditions closer to physiological behaviors, allows to improve the relevance of conclusions about pharmacological and toxicological effects.

Oral application is a common noninvasive method of drug delivery and inhalation therapy shows promise not only for treatment of respiratory diseases, but also for drug delivery to the systemic circulation.

3.2.2 Air and aerosol exposure systems

In order to mimic realistic transport and toxicity studies, air–liquid interface culture should be combined with application of aerosols. Culture in the AILinterface (ALI) is used in more advanced in vitro exposure models where cells are cultured on semipermeable membranes of a Transwell insert. The insert is arranged into a culture well, the medium is fed from the original site and cells are exposed to an aerosol on the apical part.

Transwell cultures were first used for permeability studies of gastrointestinal cells and later adapted to other cell types (Hidalgo et al., 1989). Fröhlich et al. (2013) present several systems commercially available to expose Transwell cultures to aerosols and compared two in vitro systems to assess cellular effects of nanoparticles-containing aerosols. The study evaluated the utility of a Vitrocell system compared with a conventional MicroSprayer (Fröhlich and Meindl, 2015; Meindl et al., 2015). A549 cells were cultured in the AILinterface, exposed to nanoparticle aerosols and characterized by measurement of TEER and staining for tight junction proteins. Deposition and distribution rates of polystyrene particles and of carbon nanotubes on the cells were assessed. The study demonstrated that cytotoxicity of amine-functionalized polystyrene nanoparticles was significantly higher when applied as an aerosol on cells cultured in AILinterface culture compared with nanoparticle suspensions tested in submersed culture. It was concluded that the Vitrocell system was better to assess the efficacy of aerosolized nanoparticles at therapeutic doses because it mimics better the low flow velocities in the alveoli.

Similar to other commercialized system, ALICE (air liquid interface exposure) system is based on the nebulization by a vibrating membrane of nanoparticles or dissolved compounds. Cloud settling combined with sedimentation of single particles ensure valuable delivery to cells cultured at the air liquid interface, with exposure durations kept within about 10 min, a good repeatability better than 12% and low-stress. The exposure chamber combined with an incubation chamber is connected to a flow system providing humidity and temperature conditions suitable for cell cultivation, as well as a quartz crystal microbalance (QCM, described in Section 4.2.4; detection limit 0.09 mg/cm^2, 5 MHz) for real-time determination of the cell deposited dose material (Lenz et al., 2009). This advanced in vitro testing strategy was used to investigate the risk of HARN (high aspect ratio nanoparticles) from perspective of inhalation exposure, as presented by Clift et al. (2014). Cellulose nanocrystals (CNCs) with differing aspect ratios (9 and 80) were employed as model HARN samples. Specifically, well-dispersed, and characterized CNC suspensions were aerosolized using an ALICE system. The biological impact (cytotoxicity, oxidative stress levels, and pro-inflammatory effects) of each HARN sample was assessed.

Furthermore, this experienced method was validated using both long fiber amosite asbestos (LFA) and crystalline quartz as a positive control in the ALICE system. These promising results demonstrated a successful approach towards investigation of the hazard of HARN aerosols.

In the same manner, Lenz et al. (2014) introduced the very recent progress in the ALICE-CLOUD technology, as rapid and dosimetrically accurate tool for aerosolized liquid drug screening with pulmonary cells cultured under realistic air liquid interface (ALI) setup. In this work, they confirmed the antiinflammatory effect of aerosolized bortezomib on human epithelial lung cells (A549), with faster uptake kinetics than under nonaerosolized submerged conditions, thus demonstrating its interest for inhalation therapy.

The new, fully integrated, and transportable deposition chamber NACIVT provides a promising tool for reliable, acute, and sub-acute dose-response studies of (nano)particles in air-exposed tissues cultured at the AILinterface. NACIVT studies have been developed and fully characterized for its performance by Jeannet et al. (2014). The system features preventing of cellular stress during exposure, then allowing long term exposures, with conditioning of the temperature and humidity which are computer-controlled. High-throughput screening is allowed by using a continuous airflow with airborne NPs that are then deposited simultaneously on Transwell inserts of up to 24 cell cultures. Polystyrene and silver particles were deposited uniformly on the 24 permeable membranes. Results of interaction particle cell comparison confirmed that particles could reach the cell surface and were taken up by cells. As demonstrated in control experiments, there was no evidence for any adverse effects on human bronchial epithelial cells (BEAS-2B) due to the exposure treatment in NACIVT.

3.3 CONCLUSIONS

The techniques discussed above may represent outstanding innovative routes to promote the development of therapeutic nanostructures, regardless of the mode of drug delivery (injectable, ingestable, inhalable, implantable, etc.). Indeed, they can offer biologically more relevant experimental conditions, compared to current submerged cell culture systems. More generally, emerging biotechnology with nanoscience and especially, in the field of therapeutic nanostructures will fully benefit from recent advanced characterization and fabrication techniques to generate new smart sensors to solve environmental and biological issues.

4 MEASUREMENTS TECHNIQUES

4.1 CONVENTIONAL CHARACTERIZATION TECHNIQUES

Characterization of nanostructures involves mainly the fields of nanometrology and bioanalyses. Size, structure, such as specific surface area (SSA) and particle size distribution (PSD), Zeta potential, chemical composition (speciation, concentration, chemical sensitivity, spatial resolution) are among key parameters to be measured when analyzing NPs. Main details of selected techniques are described by Bensebaa (2013a), extracted from a report on the emerging field of nanometrology (Lojkowski et al., 2006), which gives a comprehensive survey on nanometrology techniques, characterisation of particle size in nanopowders, and bulk nanocrystalline materials as well as nanometrology in the nanometre range of thin film systems and surfaces. Now we briefly describe principles of conventional techniques mainly used in nanometrology and bioanalyses for therapeutic nanostructures.

4.1.1 Imaging techniques

4.1.1.1 Atomic force microscopy and electron microscopy

Atomic force microscopy (AFM)-based techniques are used widely to study the mechanical behavior of the cellular structure and its interaction with the extracellular matrix or other cell-to-cell interactions. The basic contact mode is well-adapted for solid samples, with Young's modulus between 10^9 and 10^{12} Pa, whereas biological samples with soft structure can be characterized using Tapping Mode. The Young's modulus of biological materials is in the range of 10^3–10^5 Pa (live cells) and 10^8–10^9 Pa (native proteins). Both these modes are in the repulsion interaction regime between the cantilever and the studied surface. The noncontact mode is the attractive interaction regime, with a possibility of fast imaging (< 1 s) (Endo and Nagamune, 2010; Ikai, 2010).

Atomic force microscopy is one of the most preferred instrumentation to study surface smoothness as well as tunnel size and linearity (Mosier et al., 2014) of microchips. AFM images can give information on the presence and size of the grooves and crevices present in the tunnel of microchip. A large-size groove may hinder the flow of the analyte, thus affecting the time period of sample flow, resulting in erroneous analysis. Similarly, crevices present may result in entrapping and holding back a significant amount of sample, generating erroneous results that are nonreproducible. Defined channels of specified dimension play a critical role in sample tunneling. The external pressure applied and the amount of sample inserted is precisely specified for a microchip, therefore any error in tunnel size and width, while designing the microchip, may not give reproducible results as calculated theoretically. Scanning and transmission electron microscopy (SEM, TEM) can confirm the dimensional analysis done by AFM, thus affirming the fabrication process.

4.1.1.2 Fluorescent imaging techniques

Unlike AFM and EM, fluorescent imaging techniques have an advantage of studying the behavior of nanostructure in real time (Mustafi et al., 2014). However, the nanostructure needs to have an intrinsic fluorescence property or they may be tagged with certain fluorescent dyes, biomolecules, or a ligand. The intensity of the fluorescence may correspond to the biodistribution of the nanostrustures in an in vivo model. Drug delivery systems are often known to be studied by fluorescence imaging (Endo and Nagamune, 2010). Confocal fluorescent imaging techniques, based on the principle of merging multiple complementary fluorescent images captured simultaneously, such that real time location of a nanostructure may be followed, provide insight into the interaction and uptake of nanostructures by intracellular organelles within a live cell or a tissue within an organ (Moraru et al., 2014). At lower resolution, fluorescent imaging techniques are also known to be used to study the morphologies of cells, and their nuclei and the distribution of nanotherapeutics within Yang et al. (2015).

Fluorescence resonance energy transfer (FRET) is another technique that has been recently used in nanostructure-organellar membrane interaction studies, thus, giving us an insight into the uptake of nanostructures at subcellular level. FRET works on the principle of transfer of resonance energy from donor photochrome to acceptor photochrome when the two come in contact with each other. Therefore, to study NP uptake at cellular and subcellular level, both these surfaces must either have inherent fluorescence property or may be tagged by certain photochromes that actively couple and recognize each other for significant energy transfer (Yamada and Harashima, 2014).

4.1.1.3 Other imaging techniques

A number of highly sensitive imaging techniques have been developed recently based on interaction of subatomic particles with external spectral rays. Techniques, such as MRI and positron-emitting techniques could be highly sensitive in studying biodistribution of nanostructures even in clinical models (Hong et al., 2013).

4.1.2 Spectral analysis techniques

Fourier transform infrared spectroscopy (FTIR), and surface-enhanced Raman spectroscopy (SERS) are some of the commonly used techniques in studying the drug-polymer interactions, which are essential in drug-delivery designing. A number of studies have reported the importance of FTIR in studying the influence of functional groups on drug-loading and entrapment efficiency (Huang et al., 2008). However, not all nanostructures respond actively to FTIR, though many nanostructures are known to be more sensitive to Raman spectroscopy. In addition to studying the manipulation of functional groups in nanostructures, which can optimize the controlled drug-releasing capabilities, Raman spectroscopy has also been frequently used in studying interaction of nanostructures to the cell membranes, and thus cellular uptake (Goodhead et al., 2015).

4.1.3 Bioanalytical techniques

Most of the bioanalytical techniques are either used to study the behavior of nanostrustures on various in vitro models, such as 2D and 3D cell cultures, or the biochemical and pharmaceutical response of nanostructures in the in vivo models. Liquid chromatography coupled with other sensitive techniques, such as mass spectroscopy can provide real time data on the go (Wang et al., 2014). Biodistribution studies of nanostructures are often performed using LC–MS/MS and HPLC. However, ELISA is more commonly used to detect biochemical regulation of important biomolecules, in response to the exposure of nanostructures, in both in vivo and in vitro models. It is based on quantitative tagging of the "bait" molecule to the biomolecule under study, which is in turn tagged with a fluorescent molecule or an enzyme which results in quantitative study of target biomolecule (Ritchie et al., 2015). Other techniques, such as polymerase chain reaction (PCR) may also be used to study the effect of nanostructures on gene regulation in in vitro and in vivo models (Jiang et al., 2015). Of late, techniques, such as flow cytometry have been used to study biostructural and bioanalytical parameters. Apoptosis and cell cycle arrest are crucial in studying toxicity of nanostructures in in vitro and in vivo models. Flow cytometry can be used to study both these with precision (Verschoor et al., 2015). Flow cytometry is also used to study interaction of nanostructures with cell surfaces and uptake of nanostructures by the cells, using side scatter data (Gharagozloo et al., 2015).

4.1.4 Cytotoxicity studies

Cytotoxicity studies are used to screen potential therapeutic nanostructures, under which in vitro assays are conducted, such as MTT and XTT assays for screening the therapeutic nanostructures. Both the MTT and XTT are colorimetric dyes, which are converted into intense color-generating formazan crystals in live cells. However, in dead cells these dyes fail to get converted into farmazan crystals thus resulting in linear decrease in colorimetric absorbance, in dose- or time-dependent manner. It may be noted that in the case of inorganic nanostructures, such cytotoxicity assays may generate false data because of interference in crystal formation (Tournebize et al., 2013). Cytotoxicity studies of in vivo models are generally studied with a technique of much higher sensitivity, namely, terminal deoxynucleotidyl transferase dUTP nick end labeling (TUNEL) assay (Segura et al., 2015). TUNEL assay can be used for a precise study of the DNA fragments within a cell, which are one of the most significant stages in triggering of apoptosis and cell death. Thus, effect of nanostructures on the DNA fragmentation may be studied in both in vivo and in vitro models for cytotoxicity.

Each of the above tests has some advantages and drawbacks. Conventional in vitro tests can give a wide range of parameters, but they suffer from lengthy and cumbersome procedures and large equipment. Furthermore, they are generally endpoint assays, as cells are treated and characterized at a fixed time. In vivo tests on animals suffer from ethical issues, high cost, and a lack of reproducibility and correlation with clinical effects.

4.1.5 Integration of microfluidics with conventional characterization techniques, and applications

Integrating microfluidics functions with conventional tools can have many benefits. Miniaturization and adaptable modules lend themselves to the use of conventional techniques towards high throughput screening. For example, the parallel artificial membrane assay (PAMPA) from Pharmacelsus (2015) allows the evaluation of passive transport properties of compounds and can be used for the determination of the intestinal permeability as a key parameter for the selection of compounds in early drug discovery. Permeation through the artificial membrane is measured photometrically or by LC–MS/MS. As another example, microfluidic channels used by Kim et al. (2011) combined with colorimetric MTT assay allowed a study of shear stress effect on the cytotoxicity of NPs.

4.2 BIOSENSING PRINCIPLES

As mentioned, recognition strategies and materials, especially based on nanostructures, such as NPs and other materials of interest for therapeutic purposes, can be involved for diagnostic and imaging. They can also be used for a better understanding of how cells work, to monitor efficiency of therapy, or for screening of the efficiency and toxicity of therapeutic nanostructures. Microsensors integrate recognition and electronic elements, leading to devices that can fulfil conveniently such functions, as described in this chapter. Furthermore, they allow the monitoring of dynamic response, resulting in a continuous monitoring of biological activity. In a typical biosensor, target analytes are immobilized on NPs or more generally biofunctional compounds, grafted on the transducer that converts this surface layer changes into an output detection signal. A comprehensive survey of microsensing technologies can be found for example in Bensebaa (2013b). A selection of main transduction principles and examples of biosensors based on each one of them are presented below.

4.2.1 Electrochemical

Electrochemical sensors are among the most studied microsensors. A review on their principles and architectures can be found for example in Grieshaber et al. (2008). They are typically based on a redox reaction involving the target analyte in the electrolyte at a working electrode, resulting in the variation of an electrical signal. Well-known examples are the blood glucose sensor, or the respiratory carbon dioxide sensor. When one measures a current or a difference in electrodes potentials, the method is called amperometry and potentiometry, respectively. Potentiometry is typically used for ion-selective electrodes (ISE) measurements (pH electrode and other ion-selective electrodes). Widely used among these electroanalytical methods, voltammetry, or voltamperometry is a subclass of amperometry, measuring the current as the applied potential is varied. Cyclic voltammetry consists in applying an up and down linearly varying potential between the working and the reference electrodes, and representing the current generated externally from the counter electrode to the working electrode. The characteristic resulting curve is called a cyclic or CV.

Similarly, in electrochemical (or more generally electrical) impedance spectroscopy (EIS) or impedimetry, an alternative voltage is applied, with a frequency varying typically in the range from 10^{-3} to 10^5 Hz, and current at the same frequency (linear system) is measured. The resulting impedance is generally represented on a Nyquist diagram, with imaginary part as a function of the real part. This allows for example to monitor the function of membrane proteins, which are electrogenic, for example, they can translocate a net charge across the membrane, or the effect of chemical species on whole microorganisms (Tekaya et al., 2013), or even the seeding of endothelial cells on biomaterial surfaces, such as collagen or fibronectin, which then makes it challenging to enhance hemocompatible surfaces or characterize tumoral cell adhesion on host tissues (Zanina et al., 2013). An example of EIS data is given in Fig. 20.4D.

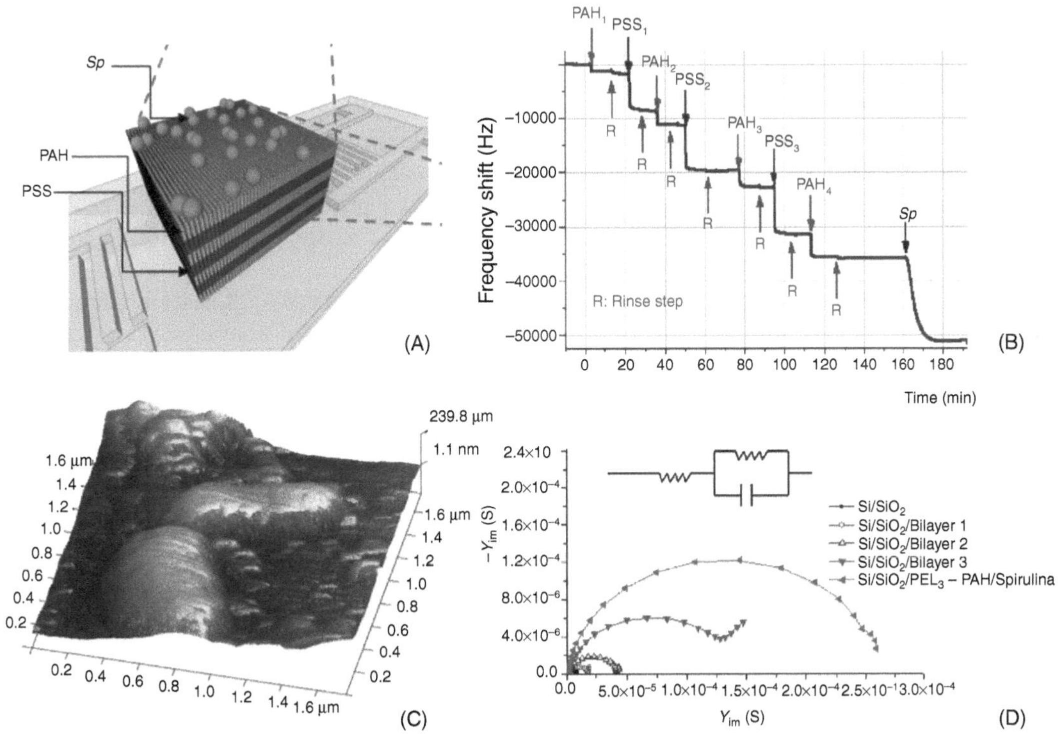

FIGURE 20.4 Saw Sensor, Real-Time Frequency, AFM Image, and Results of EIS Measurements

(A) Schematic of a SAW sensor with hybrid biofilm of polyelectrolyte multilayer/microorganisms; (B) real-time frequency during in-flow deposit of the biofilm; (C) AFM image of the biofilm surface; (D) results obtained with static EIS measurements after each step.

From Tekaya, N., Gammoudi, I., Braiek, M., Tarbague, H., Moroté, F., Raimbault, V., Jaffrezic Renault, N., 2013. Acoustic, electrochemical and microscopic characterization of interaction of arthrospira platensis biofilm and heavy metal ions. J. Environ. Chem. Eng. 1(3), 609–619, with permission from Elsevier.

Details of electrochemical detection can be found in Lefrou et al. (2012). These electroanalytical methods generally require a three electrodes setup (working, counter, reference) with a potentiostat. Screen-printed electrodes (SPEs) allowed recently some improvements in terms of miniaturization (device, sample volume), reproducibility and low cost (Dropsens). In order to overcome a lack of selectivity, electrochemical detection is often coupled to separation techniques, such as chromatography or electrophoresis, or used in a detection strategy combining a different principle, mainly optical, such as luminescence (electrochemiluminescence ECL), fluorescence, or SERS (Eliaz, 2012).

A review of latest advances in the field of electrochemical biosensors, which are much used transducing technologies, with various strategies to employ new developments in nanotechnology, can be found in Zhu et al. (2015). Micro-/nanoelectrode arrays (or MEA) can be used during conductometric sensing. Examples of in vivo applications in the field of neurobiology, namely, detection of dopamine and of adrenaline release have been cited in Lefrou et al. (2012). In the review by Jaffrezic-Renault and Dzyadevych (2008), there are presented numerous advantages of printed interdigitated electrodes (IDEs) for conductometric enzyme biosensors, without a need for a reference electrode.

4.2.2 Optical

Optical devices are based on emission and/or reception of light. The pulse oximeter that measures the absorption of light through blood is a typical example of in vivo application. Fluorescence and chemiluminescence are typical ways used in conventional imaging techniques, and can be involved in biosensing transduction. DNA biochip microarray is a popular example of such a device with application to genome characterization. Similar devices can be developed and applied in the screening of new therapeutic compounds, such as platform for cell cultures used for cytotoxicity studies (Lee et al., 2014), or as previously seen with MTT assays.

Among optical label-free techniques mostly adaptable for drug screening, surface plasmon resonance (SPR) allows to monitor binding events between molecules ranging from ions to viruses. This technique also offers other advantages, such as affinity and specificity, binding kinetics from transient part of the real-time response (association/dissociation), concentration from steady-state response (before/after binding or dissociation, down to the nanomolar range), and low sample volume required down to a few microliters. A useful animation on the SPR principle is proposed by GE Healthcare Life Sciences (2015). Basically, the sensor chip is composed of a thin layer of metal (gold) on dielectric prism (glass). Target analyte is captured specifically from the sample in microfluidic continuous flow, with biorecognition ligand immobilized on the gold surface of the removable sensor chip. Ligand/analyte interaction alters the refractive index close to the surface, modifying the angle of minimum intensity of light reflected at the specific angle of total internal reflection (TIR) from the glass side of the sensor surface. The change is related to the mass of material bound on the surface.

A limitation of SPR technique is the local sensitivity, as plasmon resonance occurring at TIR angle is confined within a few nanometers to about 200 nm from the metal/dielectric interface. The need for integration of optical setup (incident light, photodetection) is also an issue, but technological progress tends to overcome these limits. A big plus point is that SPR measurement principle is adaptable to array devices.

4.2.3 Electrical

Electrical microsensors are based on electronic components, typically passive dipole, such as resistance and capacitance (electric dipoles), or active device, such as field effect transistor (FET). In FET (3-pole component), chemical modification of the gate polarization due to target species immobilization or presence, influences the electron conductance between the drain and the source. Among them, ion sensitive FET (ISFET) are based on ionic sensitive membrane, in the same way as ISE. Interdigitated Electrode (IDE) are involved typically in capacitive sensing, and are of particular interest in the detection of electroactive species. Similarly, MEA can be considered among electrical microsensors, sometimes based on electrochemical reactions. This massively parallelizable technology is used widely to access electrophysiological signals in vivo or in vitro: namely, neural circuit connectivity, physiology, and pathology. Applications in electrical stimulation, cardiovascular, stem cell, or retina research, drug, and toxicology tests were also claimed at the biennial International Meeting on Substrate-Integrated Microelectrode Arrays (MEA Meeting, 2014) and in reviews on MEAs (Gross, 2011; Johnstone et al., 2010; McConnell et al., 2012; Stett et al., 2003). MEA and IDE principles can be combined in interdigitated array microelectrodes (IDA), such as those proposed by MicruX Technologies (2015). Combining MEA with active complementary metal oxide semiconductor (CMOS) microelectronic technology allows higher integration (several thousands of electrodes), yielding higher resolution with integrated amplification and real-time processing. The general principle has been explained in 3Brain (2015). The potential of such devices extends to in vitro studies on models of disease or the response of cellular networks to pharmacological compounds related to pathologies, such as cardiac arrhythmias, memory impairment due to Alzheimer's disease, or vision impairment caused by ganglion cell degeneration in the retina (Jones et al., 2011).

4.2.4 Mechanical

Microsensors based on mechanical transduction are among the most versatile ones, as mechanical parameters are always modified when target compounds are immobilized on bioreceptive film. Indeed, commonly assessed mass-effect usually modifies resonant frequencies of a mechanical device, depending upon materials and architecture of the transducer. Most known and commercialized mechanical sensors are based on quartz crystal microbalance (QCM), made typically of a piezoelectric disk of quartz with electrodes. Application of alternating potential at synchronous frequency induces mechanical vibration of the disk. Mounted in closed-loop circuit, the devices oscillates at a resonant frequency that decreases linearly as rigidly linked mass increases. Such devices are used to monitor the thickness of coating, as well as any other change due to mass or visco-elastic effect within the surface layer mechanically moved by the wave. An additional information can be derived from measurement with QCM-D system. In this case indeed, a pulse instead of a continuously alternating potential is applied periodically, and both frequency and wave decay are monitored.

A widely used technique in the field of biosensing, QCM is more and more supplanted by surface acoustic wave (SAW) devices, when even lower detection threshold are needed. SAW devices can also be cost effective if collectively fabricated with planar microtechnologies. Typical SAW devices are based on generating/receiving IDE(s) deposited on a piezoelectric substrate, such as represented on Fig. 20.4A. The crystallographic orientation determines the type of waves that can be generated and that can propagate into the structure. Commonly used devices are based on Rayleigh waves, but Love waves devices are particularly adapted to liquid applications as they involve shear horizontal waves (Rocha-Gaso et al., 2013). Such devices are known to allow the real-time monitoring of mechanical parameters of films on the surface, such as illustrated in Fig. 20.4B (Gammoudi et al., 2014a,b). SAW devices provide a label-free technique, with easily accessible surface, compatible with microfluidic chips, they offer a high sensitivity as well as a great versatility that allow them to target many applications related to the development of therapeutic nanostructures.

Microcantilevers or other mobile microstructures, involved in physical microsensors, such as microaccelerometers and other microsystems, have also been proposed as (bio)chemical microsensors. Such devices can be used in two manners. In the first one, the whole microstructure resonates, in a similar manner as acoustic devices but suffer from issues related to high viscous attenuation and mechanical fragility. A second operation mode is employed in static state, with stress at the interface (rigid link) between the main cantilever part and the coated sensitive film when ad-/absorbing target species. This operation mode requires an optical read-out system or the integration of a strain gauge to measure the cantilever deflection.

Other mechanical-based devices can respond to ultrasound probes, by emitting an ultrasound wave and analyzing the ultrasonic signal back. These technologies are compared in Qiu et al. (2015) and among them piezoelectric micromachined ultrasound transducer (PMUT) arrays appear as promising for devices gathering integrated sensing, actuation, and imaging, included potentially in disposable eye lenses for intraocular pressure control, as a major factor outlined in glaucoma (Mareschal et al., 2010). Besides, exploring routes offered by technological means, it gives rise to innovative structures, such as a scalable SAW-based cantilevered device for miniaturized bio-chemical sensor described in De Simoni et al. (2015). In this device, the presumed change in the residual surface stress of the beam body induced by biotin–streptavidin interaction can be measured as the frequency shift of an integrated SAW device, instead of a nonintegrable optical deflection readout. The observed sensitivity and resolution ($\sim$10 nM) are far better than what would be available from a typical measurement with QCM-D. Furthermore, such integration of a SAW device leads to high frequencies, namely 2.8 GHz, for a wireless readout, as presented later.

4.2.5 Contributions made by latest developments in advanced materials

When targeting specific compounds, or in order to increase the effective surface area, microsensors are required for sensitive layers whose properties are modified by the immobilization or the presence of targeted compounds. Polymeric materials have been widely used with numerous advantages as receiving matrices, but of late, nanomaterials and nanostructured surfaces are fast emerging as candidates, among them carbon nanotubes (CNTs) (Mehra et al., 2014) and graphene-like materials, for their unique properties, as demonstrated with SAW devices resulting in outstanding sensitivity for vapor detection (Nikolaou et al., 2015). A description of sensors with CNTs, associated with polymeric materials, can be found in Wang and Yeow (2009), reviewing gas-sensing applications. Applications in the biomedical field with CNTs and various functionalization strategies, in association with FETs and electrochemical transducers, are reviewed in Jacobs et al. (2010). Some examples in the detection of neurotransmitters, proteins, small molecules, such as glucose and DNA are available. Association of carbon-based materials with polymers or other nanomaterials, such as chitosan as a biopolymer-derived material well-known for its properties in therapeutic systems (Zargar et al., 2015), results in hybrid materials with interesting applications, as sensitive film on FET or optical devices, for drug and drug delivery systems (Artiles et al., 2011). Optical properties of graphene-like materials are useful (Ping et al., 2015), for example, with graphene oxide as a generic quencher for numerous fluorophores, including organic dyes, conjugated polymers, and quantum dots.

4.2.6 Impact of wireless compatibility

In most cases, electrochemical and electrical microsensors are used under direct current (DC) conditions, or alternating current (AC) at frequencies lower than a few hundreds of kilohertz. Besides, high frequency resonant devices are more and more studied and developed, as an attractive way of microsensing with outstanding sensitivity, and compatibility with wireless readout. Indeed, they can be designed for passive operation mode, with antennas as transceiver that can excite them and be sensitive to any resonant frequency change due to a modification related to interaction with the sample adjacent to the device itself. Acoustic devices, among them SAW ones, are generally well adapted as possibly passive components. An example of wireless acoustic (Love) wave biosensor is presented in Haekwan et al. (2014), targeting originally simultaneous detection of multiple bio-molecules in a single sensor. Recent developments on microsensors offer interesting devices based on resonant components and inexpensive printed technologies, such as antennas on flexible substrates with applications to detections of vapors (Trafton, 2014), which may be further developed for liquid media.

Examples of technologies for wireless biosystems can be found in Jow et al. (2014), illustrating the use of transcutaneous electromagnetic interactions in near field with inductively powered device for neuroprostheses, such as cochlear and retinal implants stimulation (forward transmission) or neural recording (back telemetry). Such systems and devices confer interesting possibilities for in vivo behavioral experiments.

The systems based on electromagnetic terahertz waves appear as other promising wireless detection technologies. Indeed, safe for living organisms, in contrast with X-ray radiation, THz waves can be transmitted through many materials and analyze internal structure, as chemical bonds vibrate at a characteristic frequency when exposed (Advantest, 2013).

4.3 APPLICATIONS OF MICROSENSORS INTEGRATED WITH ANALYTE HANDLING FUNCTIONS

In nanotherapeutics, microsensors combined with microfluidic functions can offer good opportunities for several different scenarios at different scales, with advantages, such as label-free and time-resolved measurements. Among them, the first function of microfluidics is to handle the analyte and control the flow and distribution of compounds inside at the near interface with the sensing surface, in an optimized manner. Furthermore, microfluidics allows to extract, sort, and immobilize a specific type of cell to observe, for example, permeability regarding different kinds of nanocarriers. In the case of cell sorting with microfluidics, the cells of interest can be either "wanted" or "unwanted." Selecting unwanted cell (infected cells per example), can be interesting for counting, or to observe lysis in reaction to the drug delivered by the nanocarrier. Selecting "wanted" cells are desirable when one wants to isolate rare cells possessing remarkable properties. In the case of small population of cells, having miniaturized sensors, dimensions of which are compatible with the small sample volume is of great interest. A huge potential for single cell experiments (Shah et al., 2014) is to obtain useful data to implement better models, and enhancing more precise models to transpose in vitro results to in vivo.

Among the combinations of microfluidics and microsensors, those involving controlled fluid-flow with integrated electrodes, such as EIS up to microelectrode arrays (MEA), as presented earlier in the text on electrical and electrochemical microsensors, may appear as one of the simplest ones to study cell behavior. Many other combinations of microfluidics and microsensors are investigated. Furthermore, conventional characterization techniques and new ones based on microsensors are complementary, and can be mixed in a single set-up to enrich the knowledge on phenomena involved in the interaction between compounds or cells.

4.3.1 Real-time monitoring of biofilms, formation, and interaction with species

Works described in Gammoudi et al. (2014b) and Tekaya et al. (2013) are an example of the real-time monitoring of the formation of a biofilm of PEM/microorganisms (*E. Coli* bacteria or *Arthrospira plantensis* microalgae) on SAW device combined with microfluidic chip. This allowed the optimization of the deposition process of this biofilm, as sensitive film for heavy metals (HM) detection. The toxicity induced by HM on the microorganisms have been characterized, with different effects depending upon the microorganism (*E. Coli* or microalgae), resulting in a real-time detection threshold of Cd^{2+} and Hg^{2+} better than 10^{-12} M. Enhancement of the biofilm grafting and cells viability, thereby the sensor lifetime, by adding thin film of titanium was also demonstrated (Gammoudi et al., 2014a). Results and mechanisms involved were confirmed with static EIS and AFM measurements (before/after HM), with extra details on the mechanical shape and behavior of the cells. Typical results are illustrated in Fig. 20.4. Similar real-time monitoring experiments have been investigated by us with in-flow deposit of other biofilms, such as collagen or fibronectin (unpublished results), known to allow studies of the influence on tumoral cells adhesion properties, as shown by static EIS measurements in Zanina et al. (2013).

Another such combination of complementary technologies is proposed commercially by Biacore, with surface plasmon resonance and matrix-assisted laser desorption/ionization mass spectrometry system (SPR-MALDI MS) for interaction analysis with low femtomole amounts of the target material. As an illustration (Zhukov et al., 2004) developed a fully automated procedure for Biacore 3000 that performs all steps of ligand capture (on SPR chip, from analyte in microfluidic flow), elution, deposition onto a MALDI target and sample preparation for MALDI MS analysis, as an intact protein as well as after on target digestion. Calmodulin was used as a brain's protein model, selectively captured from biological extract by an interaction partner immobilized on the sensor chip. Low volumes and automatic sample processing increase the throughput of SPR MS analysis.

EIS-based devices are available as commercial systems with array capabilities, for example in real-time cell analysis (RCTA) from xCELLigence (Otero-González et al., 2012; Teng et al., 2013). Some authors (Nguyen et al., 2013) refer to EIS as electrical cell-substrate impedance sensing (ECIS), and have claimed the integration of ECIS in a Boyden chamber design to study cell migration, which is one hallmark of malignant tumor progression.

4.3.2 Evaluation of the risk of toxicity

Impedance measurements with interdigitated electrodes or three-electrode system are of particular interest to perform real-time monitoring of cellular responses under various culture conditions (Lei, 2014). As an example, a setup based on impedance spectroscopy using an Autolab potentiostat/galvanostat as reported in Douville et al. (2010) and integrated Ag/AgCl electrodes is applied to the real time monitoring of TEER, commonly used for cytotoxicity studies as previously presented (see Section 3.2). As another example, 3D-cultures in microfluidic devices with real time monitoring by EIS are presented as a convenient way to study multicellular spheroids (MCSs) as a model for investigating cytotoxic effects of anticancer drugs in Luongo et al. (2013). Networks based on MEAs can be used as even more effective and rapid screening platforms in pharmacology, toxicology, for drug testing, and tissue-based biosensors (Gross, 2011), with increased integration by using CMOS-based MEA (MEA Meeting, 2014).

4.3.3 Real-time monitoring of cancer cell metabolism for drug testing

Among the typical applications of microsystems integrating microsensors and analyte handling, Alborzinia (2015) described the real-time and label-free monitoring and analysis of living cancer cells and drug-induced effects, by using an online cell biosensor chip system. The silicon biosensor chip serves as a platform for a monolayer of cells to be measured, and integrates three types of microsensors: ISFET, Clark-type, and IDE structures, for the measurement of physiological parameters, respectively: pH for extracellular acidification (glycolysis) and O_2 for oxygen consumption (respiration rate), providing information on the cellular energy metabolism, and cellular impedance as an image of cellular adhesion, confluence, and morphology. It is combined with automatic fluidics system from (Bionas, 2012) which supplies the cells at programmable times with solution containing drug to be tested.

4.3.4 Blood–brain barrier microsystem

Booth and Kim (2012) reported the detailed design, fabrication, and characterization of the developed in vitro dynamic thin-membrane micro blood–brain barrier (μBBB) system including multi-layered polymer fabrication, cell culturing procedure, and validation of the developed models through three most common methods: optical imaging (cell imaging to observe structure and morphology), static and transient permeability tests (to evaluate barrier selectivity), and TEER measurements under different concentrations of various tracers (to evaluate cell confluence and tight junction integrity). This study demonstrates that μBBB effectively mimics the dynamic cerebro-vascular environment with fluid shear stress, and the results from this characterization study indicate that the model expresses sufficient key characteristics of a BBB model. Real-time TEER response was shown to be feasible through measurement of transient effects histamine testing. Permeability assays were demonstrated in the system, with a selective permeability over a wide range of tracer sizes. These characteristics indicate that the μBBB system is a useful and enabling tool for further studies of BBB function and delivery. It can be used to monitor changes in barrier function in response to various environmental stimuli, such as barrier-enhancing or barrier-opening drugs. Finally, through permeability assays the system can be used to predict the rate of delivery of new drugs across the BBB. Many researchers actually investigate BBB models by using microsystem integrating microfluidic and microsensing (MEA) functions, mixed with conventional techniques (Schultz et al., 2015).

4.3.5 In vitro toxicology, air–liquid interface

Recently, the technology of nanoaerosols has been developed, and drug delivery by this route has reduced significantly the required dose of inhaled drugs by several orders of magnitude, compared to delivery by the oral route. A major challenge with nanoaerosolized drugs arises in measuring and controlling the amount of inhaled doses, such as, the total amount of nanoaerosolized material deposited in the lungs and their toxicity. Yet, another challenge is the relatively low rate of administration due to the rapid coagulation of nanoaerosol at high concentrations. Therefore, special devices and techniques are required to estimate the delivered dose (Morozov et al., 2014) as aimed by the association of nebulization systems to generate nanoaerosol from biological and therapeutic substances and a micro sensor (QCM) device to estimate the doses obtained by mice exposed to these nanoaerosolized substances.

Chortarea et al. (2015) aimed to mimic inhalation of multiwalled CNTs (MWCNTs) in vitro as realistically as possible, by producing MWCNTs aerosols via an ALICE system previously presented, combined with a 3D epithelial airway barrier model cultivated at the air–liquid interface (ALI). The deposited MWCNT mass was quantified by the QCM which was placed inside the exposure chamber. At the end of each aerosolization process, the QCM was taken out of the chamber and left to air–dry. The deposition of MWCNTs was quantified using the online QCM during the entire nebulization process. The QCM results combined with the TEM images showed that the ALICE revealed a dose dependent, reproducible, and controlled deposition of MWCNTs. Although in both repeated exposure scenarios MWCNTs were found to interact with the different cell types, they did not induce any cytotoxicity or alterations in cell morphology, nor did they elucidate any significant increase in proinflammatory markers, compared to control cultures.

The development of physiologically relevant test methods to analyze the potential irritant effects in the respiratory tract caused by e-cigarette aerosols is widely studied in the literature. A recent study reported the development of a protocol to evaluate the acute irritant potential of e-cigarette aerosols (and formulations) and cigarette smoke exposed at the ALI, using quartz crystal microbalance (QCM) technology to support exposure scenarios and a commercially available human 3D airway tissue model, EpiAirway (MatTek Corporation) (Neilson et al., 2015). QCM were installed into a Vitrocell Systems GmbH combined with a Victrocell VC 10 Smoking Robot (VC10/141209), and were used in an offline capacity to support exposure scenarios. This study confirmed and strengthened the applicability of QCM technology to act as a qualitative and quantitative tool to support in vitro exposure scenarios/experimentation.

4.3.6 Miscellaneous applications

We discussed in the preceding sections, some applications in which microsensors and integrated analyte handling systems are used. But there can be several other applications, such as, those proposed by Richter et al. (2011) based on a lab-on-a-chip capable of monitoring the collagen produced by human fibroplast cells (NHDF), responsible for the maintenance of soft tissue integrity, by using contactless dielectric microsensors based on passivated IDEs, comprising high-density interdigitated capacitors. They show that collagen production can assess cellular stress response to various stimuli, including exposure to NPs. Liu et al. (2013) have used an electrofluidic Wheatstone bridge for pressure measurements that can monitor the localized pressure in a microfluidic environment, and have proposed that both the shear stress and hydrostatic pressure are critical for the growth of human umbilical vein endothelial cells (HUVECs). Mohammed and Desmulliez (2014) have presented a polymeric autonomous capillary microfluidic system, with planar lenses and biosensing elements embedded on chip for improved troponin I cardiac biomarker detection, with 24 pg/mL limit of detection within a few minutes, which is a first step toward point-of-care instrumentation.

4.3.7 Innovative high throughput characterization techniques and standard issues

All these developments and products discussed above aim at high throughput analysis, with fluids (vapor or liquid) handling systems which can simulate as well as study real in vivo conditions. Compared to the conventional measurements mainly based on nanometrology and bioassays, microsensors offer alternative convenient techniques, which can be integrated with microfluidics, and thereby, offering automatic and real-time monitoring of dynamic interactions between cells or other tissues and therapeutic nanostructures. Such systems need to be increasingly used by laboratories working on engineered nanomaterials. On the other side, teams working on biosensors should take into account the existing guidelines and regulations in this field. For example, the society Biacore claims seamless transition into GxP-regulated environment (General Electric Company, GE Healthcare), with GxP the collective abbreviation for Good x Practice quality guidelines and regulations for the bio/pharmaceutical industry, where "x" specifies the stage of drug development or production (laboratory, clinical, manufacturing, storage, distribution, review), as established by the United States Food and Drug Administration (FDA) and published in the Code of Federal Regulations (Nanjwade, 2015). Similar guidelines are published by the European Medicines Agency (EMA) or the Organization for Economic Cooperation and Development (OECD).

5 CONCLUSIONS

A major challenge for microsensors centers on simultaneous detection of multiple targets in complex samples, combining good sensitivity, specificity and stability with dynamical ability and fast response, as well as reproducibility, reliability, portability, and low cost. These properties rely on both the transducer itself, as described above, as well as on the recognition material, which acts as a specific amplifier for the target compound, and also determines the microsensor reproducibility. This implies a synergy of multifunctional materials and microarrays sensing devices, with innovative detection strategies. Among them, label-free methods with electrical readout are preferred, as compatible with technological integration, coupled with a reduced vulnerability to contamination, plus simpler sample handling. A recent review on progress of new label-free techniques for biosensors (Sang et al., 2015) emphasized FET biosensors with carbon-based materials and silicon nanowires, as well as other miscellaneous types of biosensors based on nanotechnologies, among them are magnetoelastic, optical-based, and surface stress-based biosensors. Embedded-system engineering can ensure convenient data processing with cointegration of electronics, informatics, and communication features.

Finally, this chapter tried to point to the considerable opportunities offered by the use of biosensors combined with analyte handling microsystems, as complementary devices, compared to conventional assays used within the development of therapeutic nanostructures. In particular, label-free and time-resolved capabilities, as well as similarity to in vivo conditions, open new routes for efficient and high throughput screening of new drugs.

Apart from the very useful applications of microsensors in the field of nanostructures for therapy, one has to assess the environmental impacts of these nanostructures and the fate of pharmaceuticals, in general. There is a growing awareness and concern about the detrimental effects of the nanostructured materials as environmental toxic pollutants, and therefore, there is a crucial need to develop monitoring networks of nanoparticles in the air we breathe or of emissions of particles of therapeutic origin in wastewater.

REFERENCES

Alborzinia, H., 2015. Real-time monitoring of cancer cell metabolism for drug testing. Spektrum Springer, Berlin, Germany.

Artiles, M.S., Rout, C.S., Fisher, T.S., 2011. Graphene-based hybrid materials and devices for biosensing. Adv. Drug Deliv. Rev. 63 (14–15), 1352–1360.

Barenholz, Y., 2012. Doxil—the first FDA-approved nano-drug: lessons learned. J. Control. Release 160 (2), 117–134.

Bensebaa, F., 2013a. Nanoparticle Technologies - From Lab to Market. Interface science and technology, 19, Elsevier, Amsterdam, The Netherlands, pp. 1–84.

Bensebaa, F., 2013b. Nanoparticle Technologies—From Lab to Market. Interface science and technology, 19, Elsevier, Amsterdam, The Netherlands, pp. 185–277.

Bhatia, S.N., Ingber, D.E., 2014. Microfluidic organs-on-chips. Nat. Biotechnol. 32 (8), 760–772.

Blanco, E., Shen, H., Ferrari, M., 2015. Principles of nanoparticle design for overcoming biological barriers to drug delivery. Nat. Biotechnol. 33 (9), 941–951.

Booth, R., Kim, H., 2012. Characterization of a microfluidic in vitro model of the blood–brain barrier (BBB). Lab Chip 12 (10), 1784–1792.

Chortarea, S., Clift, M.J.D., Vanhecke, D., Endes, C., Wick, P., Petri-Fink, A., Rothen-Rutishauser, B., 2015. Repeated exposure to carbon nanotube-based aerosols does not affect the functional properties of a 3D human epithelial airway model. Nanotoxicology, 1–11..

Clift, M.J.D., Endes, C., Vanhecke, D., Wick, P., Gehr, P., Schins, R.P.F., Rothen-Rutishauser, B., et al., 2014. A comparative study of different in vitro lung cell culture systems to assess the most beneficial tool for screening the potential adverse effects of carbon nanotubes. Toxicol. Sci. 137 (1), 55–64.

Cohen, J.M., Teeguarden, J.G., Demokritou, P., 2014. An integrated approach for the in vitro dosimetry of engineered nanomaterials. Parti. Fibre Toxicol. 11 (1), 20.

De Simoni, G., Signore, G., Agostini, M., Beltram, F., Piazza, V., 2015. A surface-acoustic-wave-based cantilever bio-sensor. Biosens. Bioelectron. 68, 570–576.

Deloid, G., Cohen, J.M., Darrah, T., Derk, R., Rojanasakul, L., Pyrgiotakis, G., Wohlleben, W., Demokritou, P., 2014. Estimating the effective density of engineered nanomaterials for in vitro dosimetry. Nat. Commun. 5, 3514.

Douville, N.J., Tung, Y.C., Li, R., Wang, J.D., El-Sayed, M.E.H., Takayama, S., 2010. Fabrication of two-layered channel system with embedded electrodes to measure resistance across epithelial and endothelial barriers. Anal. Chem. 82 (6), 2505–2511.

Eliaz, N. (Ed.), 2012. Applications of Electrochemistry and Nanotechnology in Biology and Medicine II, Modern Aspects of Electrochemistry. Springer US, Boston, MA.

Endo, I., Nagamune, T., 2010. Nano/Micro Biotechnology. Endo, I., Nagamune, T. (Eds.), Advances in biochemical engineering/biotechnology, 119, Springer, Berlin Heidelberg, Germany.

Fernandes, C.A., Vanbever, R., 2009. Preclinical models for pulmonary drug delivery. Expert Opin. Drug Deliv. 6 (11), 1231–1245.

Fröhlich, E., Meindl, C., 2015. In Vitro Assessment of Chronic Nanoparticle Effects on Respiratory Cells. In: Soloneski, S., Larramendy, M.L. (Eds.), Nanomaterials - toxicity and risk assessment. InTech, Rijeka, Croatia.

Fröhlich, E., Bonstingl, G., Höfler, A., Meindl, C., Leitinger, G., Pieber, T.R., Roblegg, E., 2013. Comparison of two in vitro systems to assess cellular effects of nanoparticles-containing aerosols. Toxicol. In Vitro 27-360 (1), 409–417.

Gammoudi, I., Blanc, L., Moroté, F., Grauby-Heywang, C., Boissière, C., Kalfat, R., Dejous, C., et al., 2014a. High sensitive mesoporous TiO2-coated love wave device for heavy metal detection. Biosens. Bioelectron. 57, 162–170.

Gammoudi, I., Raimbault, V., Tarbague, H., Moroté, F., Grauby-Heywang, C., Othmane, A., Cohen-Bouhacina, T., et al., 2014b. Enhanced bio-inspired microsensor based on microfluidic/bacteria/love wave hybrid structure for continuous control of heavy metals toxicity in liquid medium. Sens. Actuat. B Chem. 198, 278–284.

Gharagozloo, M., Rafiee, A., Chen, D.W., Foldvari, M., 2015. A flow cytometric approach to study the mechanism of gene delivery to cells by gemini-lipid nanoparticles: an implication for cell membrane nanoporation. J. Nanobiotechnol. 13 (1), 62.

Goodhead, R.M., Moger, J., Galloway, T.S., Tyler, C.R., 2015. Tracing engineered nanomaterials in biological tissues using coherent anti-Stokes Raman scattering (CARS) microscopy – a critical review. Nanotoxicology 9 (7), 928–939.

Grieshaber, D., MacKenzie, R., Vörös, J., Reimhult, E., 2008. Electrochemical biosensors—sensor principles and architectures. Sensors 8 (3), 1400–1458.

Gross, G.W., 2011. Multielectrode arrays. Scholarpedia 6 (3), 5749.

Haekwan, O., Chen, F., Kunnyun, K., Keekeun, L., 2014. Wireless and simultaneous detections of multiple bio-molecules in a single sensor using Love wave biosensor. Sensors 14 (11), 21660–21675.

Hidalgo, I.J., Raub, T.J., Borchardt, R.T., 1989. Characterization of the human colon carcinoma cell line (Caco-2) as a model system for intestinal epithelial permeability. Gastroenterology 96 (3), 736–749.

Hinderliter, P.M., Minard, K.R., Orr, G., Chrisler, W.B., Thrall, B.D., Pounds, J.G., Teeguarden, J.G., 2010. ISDD: a computational model of particle sedimentation, diffusion and target cell dosimetry for in vitro toxicity studies. Particle Fibre Toxicol. 7 (1), 36.

Hong, H., Chen, F., Cai, W., 2013. Pharmacokinetic issues of imaging with nanoparticles: focusing on carbon nanotubes and quantum dots. Mol. Imaging Biol. 15 (5), 507–520.

Huang, J., Wigent, R.J., Schwartz, J.B., 2008. Drug-polymer interaction and its significance on the physical stability of nifedipine amorphous dispersion in microparticles of an ammonio methacrylate copolymer and ethylcellulose binary blend. J. Pharm. Sci. 97 (1), 251–262.

Ikai, A., 2010. A review on: atomic force microscopy applied to nano-mechanics of the cell. Adv. Biochem. Eng. Biotechnol. 119, 47–61.

Jacobs, C.B., Peairs, M.J., Venton, B.J., 2010. Review: carbon nanotube based electrochemical sensors for biomolecules. Anal. Chim. Acta 662 (2), 105–127.

Jaffrezic-Renault, N., Dzyadevych, S.V., 2008. Conductometric microbiosensors for environmental monitoring. Sensors 8 (4), 2569–2588.

Jeannet, N., Fierz, M., Kalberer, M., Geiser, M., Burtscher, H., 2014. Nano aerosol chamber for in-vitro toxicity (NACIVT) studies. Nanotoxicology 5390, 1–9.

Jiang, S., Yang, J., Shao, J., Liao, X., Lu, Z., Jiang, H., 2015. Real-time polymerase chain reaction for diagnosing infectious mononucleosis in pediatric patients: a systematic review and meta-analysis. J. Med. Virol..

Johnstone, A.F.M., Gross, G.W., Weiss, D.G., Schroeder, O.H.U., Gramowski, A., Shafer, T.J., 2010. Microelectrode arrays: a physiologically based neurotoxicity testing platform for the 21st century. Neurotoxicology 31 (4), 331–350.

Jones, I.L., Livi, P., Lewandowska, M.K., Fiscella, M., Roscic, B., Hierlemann, A., 2011. The potential of microelectrode arrays and microelectronics for biomedical research and diagnostics. Anal. Bioanal. Chem. 399, 2313–2329.

Jow, U.-M., McMenamin, P., Kiani, M., Manns, J.R., Ghovanloo, M., 2014. EnerCage: a smart experimental arena with scalable architecture for behavioral experiments. IEEE Trans. BioMed. Eng. 61 (1), 139–148.

Kang, J.H., Super, M., Yung, C.W., Cooper, R.M., Domansky, K., Graveline, A.R., Ingber, D.E., et al., 2014. An extracorporeal blood-cleansing device for sepsis therapy. Nat. Med. 20 (10).

Kim, D., Lin, Y., Haynes, C.L., 2011. On-chip evaluation of shear stress effect on cytotoxicity of mesoporous silica nanoparticles. Anal. Chem. 83 (22), 8377–8382.

Kong, J.E., Kahkeshani, S., Pushkarsky, I., Di Carlo, D., 2014. Research highlights: micro-engineered therapies. Lab Chip 14 (24), 4585–4589.

Laube, B.L., 2014. The expanding role of aerosols in systemic drug delivery, gene therapy and vaccination: an update. Translat. Resp. Med. 2, 3.

Lee, D.W., Lee, M.Y., Ku, B., Yi, S.H., Ryu, J.H., Jeon, R., Yang, M., 2014. Application of the DataChip/MetaChip technology for the evaluation of ajoene toxicity in vitro. Arch. Toxicol. 88 (2), 283–290.

Lefrou, C., Fabry, P., Poignet, J.-C., 2012. Electrochemistry: the basics, with examples. Springer-Verlag, Berlin Heidelberg.

Lei, K.F., 2014. Review on impedance detection of cellular responses in micro/nano environment. Micromachines 5 (1), 1–12.

Lenz, A.G., Karg, E., Lentner, B., Dittrich, V., Brandenberger, C., Rothen-Rutishauser, B., Schmid, O., et al., 2009. A dose-controlled system for air-liquid interface cell exposure and application to zinc oxide nanoparticles. Part. Fibre Toxicol. 6, 32.

Lenz, A.-G., Stoeger, T., Cei, D., Schmidmeir, M., Semren, N., Burgstaller, G., Schmid, O., et al., 2014. Efficient bioactive delivery of aerosolized drugs to human pulmonary epithelial cells cultured in air–liquid interface conditions. Am. J. Resp. Cell Mol. Biol. 51 (4), 526–535.

Li, Y., Liu, W., Liu, F., Zeng, Y., Zuo, S., Feng, S., Du, Y., et al., 2014. Primed 3D injectable microniches enabling low-dosage cell therapy for critical limb ischemia. Proc. Natl. Acad. Sci. USA 111 (37), 13511–13516.

Lin, R.-Z., Hatch, A., Antontsev, V.G., Murthy, S.K., Melero-Martin, J.M., 2015. Microfluidic capture of endothelial colony-forming cells from human adult peripheral blood: phenotypic and functional validation in vivo. Tissue Eng. C Methods 21 (3), 274–283.

Liu, M.-C., Shih, H.-C., Wu, J.-G., Weng, T.-W., Wu, C.-Y., Lu, J.-C., Tung, Y.-C., 2013. Electrofluidic pressure sensor embedded microfluidic device: a study of endothelial cells under hydrostatic pressure and shear stress combinations. Lab Chip 13 (9), 1743–1753.

Lojkowski, W., Turan, R., Proykova, A., Daniszewska, A. (2006). Nanometrology: Nanoforum Thematic Network Report. Available from: http://nanoparticles.org/pdf/nanometrology.pdf

Luongo, K., Holton, A., Kaushik, A., Spence, P., Ng, B., Deschenes, R., Bhansali, S., et al., 2013. Microfluidic device for trapping and monitoring three dimensional multicell spheroids using electrical impedance spectroscopy. Biomicrofluidics 7 (3), 34108.

Mansour, H.M., Rhee, Y.S., Wu, X., 2009. Nanomedicine in pulmonary delivery. Int. J. Nanomed. 4, 299–319.

Mareschal, O., Loiseau, S., Fougerat, A., Valbin, L., Lissorgues, G., Saez, S., Poullain, G., et al., 2010. Piezoelectric aluminum nitride resonator for oscillator. IEEE Trans. Ultrason. Ferroelectr. Freq. Control 57 (3), 513–517.

Maschmeyer, I., Lorenz, A.K., Schimek, K., Hasenberg, T., Ramme, A.P., Hübner, J., Marx, U., et al., 2015. A four-organ-chip for interconnected long-term co-culture of human intestine, liver, skin and kidney equivalents. Lab Chip 15, 2688–2699.

McConnell, E.R., McClain, M.A., Ross, J., Lefew, W.R., Shafer, T.J., 2012. Evaluation of multi-well microelectrode arrays for neurotoxicity screening using a chemical training set. Neurotoxicology 33 (5), 1048–1057.

Mehra, N.K., Mishra, V., Jain, N.K., 2014. A review of ligand tethered surface engineered carbon nanotubes. Biomaterials 35 (4), 1267–1283.

Meindl, C., Stranzinger, S., Dzidic, N., Salar-Behzadi, S., Mohr, S., Zimmer, A., Fröhlich, E., 2015. Permeation of therapeutic drugs in different formulations across the airway epithelium in vitro. PLoS One 10 (8), e0135690.

Mohammed, M.I., Desmulliez, M.P.Y., 2014. Autonomous capillary microfluidic system with embedded optics for improved troponin I cardiac biomarker detection. Biosens. Bioelectron. 61, 478–484.

Moraru, A.D., Costuleanu, M., Sava, A., Costin, D., Peptu, C., Popa, M., Chiselita, D., 2014. Intraocular biodistribution of intravitreal injected chitosan/gelatin nanoparticles. Rom. J. Morphol. Embriol. 55 (3), 869–875.

Morozov, V.N., Kanev, I.L., Mikheev, A.Y., Shlyapnikova, E.A., Shlyapnikov, Y.M., Nikitin, M.P., van Hoek, M.L., et al., 2014. Generation and delivery of nanoaerosols from biological and biologically active substances. J. Aerosol Sci. 69, 48–61.

Mosier, A.P., Peters, S.B., Larsen, M., Cady, N.C., 2014. Microfluidic platform for the elastic characterization of mouse submandibular glands by atomic force microscopy. Biosensors 4, 18–27.

Mustafi, N., Grünberger, A., Mahr, R., Helfrich, S., Nöh, K., Blombach, B., Frunzke, J., 2014. application of a genetically encoded biosensor for live cell imaging of l-valine production in pyruvate dehydrogenase complex-deficient corynebacterium glutamicum strains. PLoS One 9 (1), e85731.

Neilson, L., Mankus, C., Thorne, D., Jackson, G., DeBay, J., Meredith, C., 2015. Development of an in vitro cytotoxicity model for aerosol exposure using 3D reconstructed human airway tissue; application for assessment of e-cigarette aerosol. Toxicol. In Vitro 29 (7), 1952–1962.

Newby, D.E., Mannucci, P.M, Tell, G.S., Baccarelli, A.A., Brook, R.D., Donaldson, K., Storey, R.F., et al., 2014. Expert position paper on air pollution and cardiovascular disease. Eur. Heart J. 7, 83–93.

Nguyen, T.A., Yin, T.I., Reyes, D., Urban, G.A., 2013. Microfluidic chip with integrated electrical cell-impedance sensing for monitoring single cancer cell migration in three-dimensional matrixes. Anal. Chem. 85 (22), 11068–11076.

Nikalje, A.P., 2015. Nanotechnology and its applications in medicine. Med. Chem. 5, 081–089.

Nikolaou, I., Hallil, H., Deligeorgis, G., Conédéra, V., Dejous, C., Rebière, D., 2015. Printed graphene layer by layer on SAW devices for gas detection applications. Sensors, 2015 IEEE. IEEE, Busan, South Korea.

Otero-González, L., Sierra-Alvarez, R., Boitano, S., Field, J.A., 2012. Application and validation of an impedance-based real time cell analyzer to measure the toxicity of nanoparticles impacting human bronchial epithelial cells. Environ. Sci. Technol. 46 (18), 10271–10278.

Ping, J., Zhou, Y., Wu, Y., Papper, V., Boujday, S., Marks, R.S., Steele, T.W.J., 2015. Recent advances in aptasensors based on graphene and graphene-like nanomaterials. Biosens. Bioelectron. 64, 373–385.

Qiu, Y., Gigliotti, J., Wallace, M., Griggio, F., Demore, C., Cochran, S., Trolier-McKinstry, S., 2015. Piezoelectric micromachined ultrasound transducer (PMUT) arrays for integrated sensing. Actuat. Imaging Sensors 15 (4), 8020–8041.

Richter, L., Charwat, V., Jungreuthmayer, C., Bellutti, F., Brueckl, H., Ertl, P., 2011. Monitoring cellular stress responses to nanoparticles using a lab-on-a-chip. Lab Chip 11 (15), 2551–2560.

Ritchie, B.M., Connors, J.M., Sylvester, K.W., 2015. Comparison of an IgG-epecific enzyme-linked immunosorbent assay cutoff of 0.4 versus 0. 8 and 1. 0 optical density units for heparin-induced thrombocytopenia. Clin. Appl. Thromb. Hemost.doi: 10.1177/1076029615606532.

Rocha-Gaso, M.-I., Jimenez, Y., Francis, L.A., Arnau, A., 2013. Love Wave Biosensors: A Review. In: Rinken, T. (Ed.), State of the art in biosensors - general aspects. InTech, Rijeka, Croatia, pp. 277–310.

Sang, S., Wang, Y., Feng, Q., Wei, Y., Ji, J., Zhang, W., 2015. Progress of new label-free techniques for biosensors: a review. Crit. Rev. Biotechnol., 1–17.

Schlinkert, P., Casals, E., Boyles, M., Tischler, U., Hornig, E., Tran, N., Duschl, A., et al., 2015. The oxidative potential of differently charged silver and gold nanoparticles on three human lung epithelial cell types. J. Nanobiotechnol. 13 (1).

Schultz, L., Zurich, M.-G., Culot, M., da Costa, A., Landry, C., Bellwon, P., Bal-Price, A., et al., 2015. Evaluation of drug-induced neurotoxicity based on metabolomics, proteomics and electrical activity measurements in complementary CNS in vitro models. Toxicol. In Vitro.

Segura, S., Efthimiadi, L., Porcher, C., Courtes, S., Coronas, V., Krantic, S., Moyse, E., 2015. Leptin-dependent neurotoxicity via induction of apoptosis in adult rat neurogenic cells. Front. Cell. Neurosci. 9, 350.

Shah, P., Kaushik, A., Zhu, X., Zhang, C., Li, C.-Z., 2014. Chip based single cell analysis for nanotoxicity assessment. Analyst 139 (9), 2088–2098.

Steinritz, D., Möhle, N., Pohl, C., Papritz, M., Stenger, B., Schmidt, A., Aufderheide, M., et al., 2013. Use of the Cultex® Radial Flow System as an in vitro exposure method to assess acute pulmonary toxicity of fine dusts and nanoparticles with special focus on the intra- and inter-laboratory reproducibility. Chem. Biol. Interact. 206 (3), 479–490.

Stett, A., Egert, U., Guenther, E., Hofmann, F., Meyer, T., Nisch, W., Haemmerle, H., 2003. Biological application of microelectrode arrays in drug discovery and basic research. Anal. Bioanal. Chem. 377 (3), 486–495.

Tekaya, N., Gammoudi, I., Braiek, M., Tarbague, H., Moroté, F., Raimbault, V., Jaffrezic Renault, N., et al., 2013. Acoustic, electrochemical and microscopic characterization of interaction of Arthrospira platensis biofilm and heavy metal ions. J. Environ. Chem. Eng. 1 (3), 609–619.

Teng, Z., Kuang, X., Wang, J., Zhang, X., 2013. Real-time cell analysis—a new method for dynamic, quantitative measurement of infectious viruses and antiserum neutralizing activity. J. Virol. Methods 193 (2), 364–370.

Tournebize, J., Sapin-Minet, A., Bartosz, G., Leroy, P., Boudier, A., 2013. Pitfalls of assays devoted to evaluation of oxidative stress induced by inorganic nanoparticles. Talanta 116, 753–763.

Verschoor, C.P., Lelic, A., Bramson, J.L., Bowdish, D.M.E., 2015. An introduction to automated flow cytometry gating tools and their implementation. Front. Immunol. 6, 380.

Wagner, I., Materne, E.-M., Brincker, S., Süssbier, U., Frädrich, C., Busek, M., Marx, U., et al., 2013. A dynamic multi-organ-chip for long-term cultivation and substance testing proven by 3D human liver and skin tissue co-culture. Lab Chip 13 (18), 3538–3547.

Wang, Y., Yeow, J.T.W., 2009. A review of carbon nanotubes-based gas sensors. J. Sensors 2009 (ID 493904), 1–24.

Wang, X.-L., Zhu, Y., Fang, Q., 2014. Coupling liquid chromatography/mass spectrometry detection with microfluidic droplet array for label-free enzyme inhibition assay. Analyst 139 (1), 191–197.

Xia, Y., Whitesides, G.M., 1998. Soft lithography. Angew. Chem. Int. Ed. 37 (5), 550–575.

Yamada, Y., Harashima, H., 2014. A method for screening mitochondrial fusogenic envelopes for use in mitochondrial drug delivery. Methods Mol. Biol. 1141, 57–66.

Yang, M., Jiang, P., Hoffman, R.M., 2015. Early reporting of apoptosis by real-time imaging of cancer cells labeled with green fluorescent protein in the nucleus and red fluorescent protein in the cytoplasm. Anticancer Res. 35 (5), 2539–2543.

Zanina, N., Mora, L., Othmane, A., Bénard, M., Duncan, A., Jouenne, T., Souiri, M., et al., 2013. Differences in Caco-2 cell attachment, migration on collagen and fibronectin coated polyelectrolyte surfaces. Biotechnol. Bioproc. Eng. 18 (1), 144–154.

Zargar, V., Asghari, M., Dashti, A., 2015. A review on chitin and chitosan polymers: structure, chemistry, solubility, derivatives, and applications. Chem. Bio. Eng. Rev. 2 (3), 204–226.

Zhu, C., Yang, G., Li, H., Du, D., Lin, Y., 2015. Electrochemical sensors and biosensors based on nanomaterials and nanostructures. Anal. Chem. 87 (1), 230–249.

Zhukov, A., Schürenberg, M., Jansson, Ö., Areskoug, D., Buijs, J., 2004. Integration of surface plasmon resonance with mass spectrometry: automated ligand fishing and sample preparation for MALDI MS using a biacore 3000 biosensor. J. Biomol. Techn. 15 (2), 112–119.

WEB BIBLIOGRAPHY

3Brain, 2015. Available from: http://www.3brain.com/how-it-works

Advantest, 2013. Available from: https://www.advantest.com/products/terahertz-spectroscopic-imaging-systems/about-terahertz-waves

Bionas, 2012. Available form: http://www.bionas-discovery.com/prodservices/instruments/system2500/

Bruker Corporation, 2015. Products, Available from: https://www.bruker.com/products.html

Dropsens, Products, 2015. Available from: http://www.dropsens.com/en/home.html

GE Healthcare Life Sciences, 2015. Principle of Surface Plasmon resonance (SPR) used in Biacore systems. Video presentation Available from: https://www.youtube.com/watch?t=89&v=o8d46ueAwXI

General Electric Company, GE Healthcare, Biacore 3000, System Information. Available from: https://www.biacore.com/lifesciences/products/systems_overview/3000/system_information/index.html and parent links.

MEA Meeting, 2014. 9th Int. Meeting on Substrate-Integrated Microelectrode Arrays. Available from: http://www.nmi.de/fileadmin/PDF/Broschueren/MEA2014_Proceedings_web.pdf

MicruX Technologies, Electrochemical Sensors, 2015. Available from: http://www.micruxfluidic.com/products-electrochemical_sensors.html

Nanjwade B.K., 2015. GxP and cGxP in Bio/Pharmaceutical Industry. Available from: http://fr.slideshare.net/bknanjwade/gxp-and-cgxp-in-biopharmaceutical-industry

Pharmacelsus, 2015. Available from: http://www.pharmacelsus.de

Trafton, A., 2014. Detecting gases wirelessly and cheaply, MIT News Office. Available from: http://news.mit.edu/2014/wireless-chemical-sensor-for-smartphone-1208

CHAPTER

THE EFFECT OF NANOSTRUCTURED SURFACES ON STEM CELL FATE

21

Dorothee Schipper, Patrick Babczyk, Fatma Elsayed, Stephanie E. Klein, Margit Schulze, Edda Tobiasch

Bonn-Rhine-Sieg University of Applied Sciences, Rheinbach, Germany

CHAPTER OUTLINE

1 INTRODUCTION

The current demographic trend of longevity that comes along with an aging population results in a rapidly growing demand for donor organs worldwide. However, organ donation rates are low, leading to a wide shortfall between demand for, and supply of, transplantable organs. Several organ transplant scandals in the past years discouraged potential donors in addition. Also political regulations and systems influence the donation rates. Some countries use an opt-in system, where only those who have given explicit consent will donate, others an opt-out system. However, Europe, the USA, as well as Asia have the same problem of donor organ shortage. This has forced researchers to search for alternatives. Stem cells are a promising approach due to their unique features of self-renewal and differentiation capacities as shown by the hematopoietic stem cell transplantation that was already established 50 years ago (Thomas et al., 1957), and is nowadays considered the gold standard to treat numerous malignant and nonmalignant blood diseases. However, this is a cellular therapy and the application of stem cells gets tremendously more difficult if tissues are addressed. Great efforts in research and clinical studies have been made until in 2008 the world's first transplantation of a donated trachea seeded with the patient's own stem cells was conducted (Macchiarini et al., 2008).

Tissues are much simpler structures than whole organs because they are highly complex, and consist of a multitude of cell types whose concerted actions must be understood.

The in vivo environment provides various clues at a nanolevel that decisively influence the cells behavior. To meet the requirements in tissue engineering, a common approach is to mimic the native cellular and tissue environments of the organ in question as much as possible (Peloso et al., 2015).

The discovery of fullerenes and carbon nanotubes stimulated an enormous progress in the development of novel nanomaterials and their surface modification techniques to be used in various biomedical applications including stem cell based tissue engineering and scaffold fabrication (Nasibulin et al., 2007). Within the past two decades, a variety of novel micro- and nanoprocessing techniques have been developed to generate scaffolds of tailored three-dimensional (3D) structure and surface topography. In particular, novel artificial scaffold surfaces are designed that mimic the native surface topography of the extracellular matrix (EMC), including the corresponding polarity and long-range length scale to improve cell adhesion, differentiation, proliferation, and growth. In addition to the chemical nature of the surfaces, the 3D structured scaffolds have to show similar mechanical properties and stiffness as the natural substrates (Discher et al., 2005). Thus, a number of in vitro and in vivo studies could clearly demonstrate, that the interaction of stem cells, in particular mesenchymal stem cells, with nanostructured surfaces plays a key role in stem cell fate (Bettinger et al., 2009). Both scaffold properties and cell characteristics, such as cell type, size, and geometry influence the strength of those interactions resulting in differences in local migration, cell polarization, and signaling processes. Natural and artificial nanosurfaces are able to induce and even control changes in cell morphology and alignment mainly depending on the intensity of cell adhesion (Grotheer et al., 2015; Jensen et al., 2015). In the following, nanostructured surfaces, their fabrication, and their use in stem cell research are explained in detail.

2 NANOSTRUCTURED SURFACES

The following section provides an overview of the most important nanomaterials used in tissue engineering. In addition, the main surface modification methods are discussed to generate appropriate surface topographies including chemical and physical vapor deposition, 3D printing, self-assembly techniques, electrospinning, and other rapid prototyping methods.

2.1 NANOMATERIAL FABRICATION

Cellular behavior can be influenced and even determined in a controlled manner by topographically patterned surfaces. In particular, the surface pattern symmetry and its regularity may result in significant differences in cell responses (Curtis and Wilkinson, 1997). Surface roughness in micro- and nanoscale is known to influence biocompatibility of materials used for tissue engineering applications. Furthermore, cell adhesion and alignment strongly depend on micro- and nanotopographical features. Nanostructured materials, such as carbon nanotubes, exhibit exceptional properties including robust mechanical and thermal stability, which made them interesting candidates for many different applications including electronics and biomedicine. So nanospheres are used as reinforcements to improve especially the mechanical behavior and stiffness of biomaterials (Martins-Junior et al., 2013). In the last decade, a variety of nanostructured biomaterials was developed and fabricated for tissue regeneration including cardiac, skin, bone, cartilage, and nerve (Peran et al., 2013). Structures designed in nanoscale include various architectures, such as nanospheres and nanoparticles, nanofibers, nanotubes, and nanodendrimers, respectively (Grotheer et al., 2015; Kim et al., 2011; Oliveira et al., 2011). Regarding their chemical composition they can be classified into three main groups: (1) inorganic glasses and ceramics (Vallet-Regí and Salinas, 2015); (2) organic and polymeric nanomaterials (Jensen et al., 2015); and (3) nanocomposites or hybrid materials (Namgung et al., 2011).

Synthetic polymers exhibit an essential advantage since their molecular weight, copolymerization ratio, and polydispersity can easily be adjusted to control the degradation rate, making them very attractive materials for tissue engineering applications. However, it becomes more and more important to adapt novel nanotechniques and methods to tailor 3D structure and topography to those materials that were developed specifically for tissue regeneration. Both nanotechniques and nanomaterials must show a sufficient compatibility in particular regarding processing parameters (temperature, pressure, irradiation). Today, a broad variety of natural and synthetic polymers is under investigation, and both groups do show specific advantages and disadvantages. In particular, the potential of natural proteins is limited regarding their stability during nanoprocessing whereas synthetic polymers are much easier in handling but often limited regarding their biocompatibility. However, physical and mechanical properties of synthetic polymers and even their biodegradation kinetics can be fine-tuned via variation of the monomer composition. Thus, synthetic polymers and corresponding polymer fiber composites (hybrid materials) represent the most attractive group of scaffolds for tissue engineering applications (Bettinger et al., 2009). Today, material scientists have managed to fabricate a series of functional nanostructures that vary in type and form, for example, collagen-based composites for bone regeneration including various combinations of collagen fibers with nanostructured hydroxyapatites (HA), polysaccharides, polyethylene glycol, cellulose, and hyaluronate derivatives (Van Vlierberghe et al., 2011). Enhancement of cell-based therapeutic angiogenesis was reported using a novel type of injectable scaffold based on HA/polymer nanocomposite microspheres and collagen nanofibers (Mima et al., 2012). Composites of hydroxyapatite-silica nanoparticles reinforcing poly(lactic acid) scaffolds were synthesized and showed positive effects regarding their influence on scaffold bioactivity (Guerzoni et al., 2014). Poly(lactic-*co*-glycolic) acid nanoparticles were used as drug release polymers and loaded onto chitosan bioactive glasses to generate a combined scaffold-release system for bone defect therapy (Nazemi et al., 2014). Another approach combines the controlled synthesis of colloidal nanoparticles with freeze-drying technique for bone tissue engineering applications: porous nanocomposite scaffolds based on poly(vinyl alcohol) and colloidal HA nanoparticles were prepared. Poursamar et al. (2011) performed in vitro experiments with osteoblasts and found positive effects of those well-defined porous scaffolds onto cell migration and growth. The regulation of human endothelial progenitor cell maturation by artificial polyurethane-based nanocomposites was studied by Hung et al. (2014).

2.2 SURFACE MODIFICATION TECHNIQUES

From a chemical point of view, surface characteristics include hydrophilicity versus hydrophobicity, polar versus nonpolar functionalities, neutral versus charged surfaces, all defined by nature and amount of the corresponding functional groups. Depending on electron negativity strength of the atoms nonpolar or polar structures can be created, hydrophilicity can be changed in a controlled manner that in turn significantly influences cell adhesion processes. The design of tailor-made surfaces can be achieved using different chemical surface modification techniques including physical, mechanical, and chemical adhesion (physiosorption, chemisorption). A much stronger connection to the surface is realized via chemical grafting resulting in covalent bonds of higher bond strength. According to Dupré, the adhesion power of any substance on a surface depends on the surface energies of the substance to be adhered and their interface energy values (De Gennes et al., 2004).

Whitesides et al. (2001) first described the patterning of proteins and cells using the so-called "soft stereolithography." In general, stereolithography is an important technology based on the solidification of a photopolymer with ultraviolet laser irradiation. After solidification the generated two-dimensional (2D) patterns are packed up to obtain the required 3D structure. A huge number of studies are published regarding surface modification in biomedical research, comprehensively reviewed by several groups including discussion of the most relevant aspects influencing (stem) cell adhesion on scaffold surfaces (Burdick and Vunjak-Novakovic, 2009; Schulze and Tobiasch, 2012). In 2011, Chew and Low reported a review that focusses the influence of physical and biochemical effects of the scaffold nature and topography on stem cell adhesion and differentiation, in particular for neural and cardiovascular systems (Chew and Low, 2011). Today, various fabrication methods are available for creating and tailoring nanostructured surfaces in short and long-range order, schematically summarized in Fig. 21.1.

2.2.1 CVD, PVD

Chemical and physical vapor depositions are conventional techniques for surface functionalization and modification known for a long time (Kwaśny, 2009). Very recently, Park reported a study using the so-called "initiated chemical vapor deposition" (iCVD) process for the preparation of glycidyl methacrylate-based nanolayers of variable thickness up to 200 nm that cover titanium implants and show improved cell adhesion and attachment (Park et al., 2015).

2.2.2 3D printing

Nanomaterials are intensively studied particularly to develop bone grafts. The performance of an implant mainly depends on the combined interaction of bone-forming cells, substrate (scaffold) and bioactivating substances (growth factors). To achieve bone formation, artificial 3D implants have to mimic the native surroundings, such as extracellular matrix and bone tissue composition. Here, the new technology of the so-called *organ-tissue-printing* enables a precise 3D organization of the components important for bone formation. In addition, it addresses aspects, such as graft porosity and vascularization. This "bone printing" is a very interesting approach due to the fact that it combines rapid prototyping technologies to produce precisely structured 3D scaffolds of the required size, shape, and porosity for bone regeneration (Fedorovich et al., 2011). The preparation and characterization of a 3D printed scaffold based on functionalized polyesters was reported by Seyednejad and coworkers. Porous scaffolds were prepared based on a hydroxyl functionalized polycaprolactone via 3D plotting that enables the formation of a highly porous scaffold with an interconnected pore structure. The cell attachment properties and differentiation behavior of human mesenchymal stem cells seeded onto the scaffolds were evaluated. Results could demonstrate that cells filled the pores of the scaffold within one week, displayed increased metabolic activity, and supported osteogenic differentiation (Seyednejad et al., 2011).

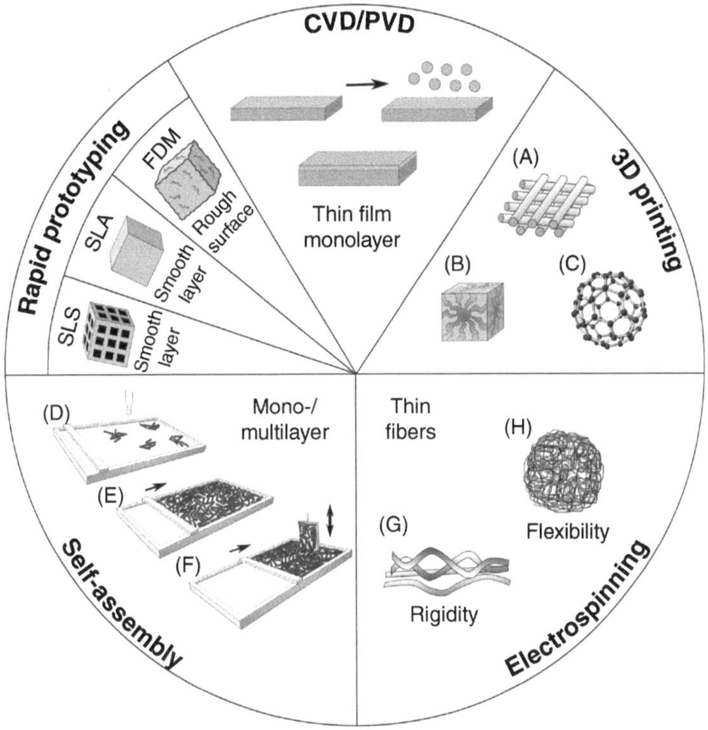

FIGURE 21.1 Surface Modification Techniques

These techniques include rapid prototyping (RP) methods, such as selective laser sintering (SLS), selective laser ablation (SLA), fused deposition modeling (FDM), chemical and physical vapor deposition (CVD, PVD), 3D printing methods resulting in tailor-made layered (A), cubic (B) and spherical (C) structures; various self-assembly methods, that is, Langmuir-Blodgett technique for monolayer formation including (D) spreading of polymer solution, (E) compression to single monolayer, and (F) film transformation onto substrates; and electrospinning of rigid (G) and flexible (H) polymers.

2.2.3 Self-assembly techniques

In additi-on to functionalization via chemical reaction, solid substrates (scaffolds) can be covered by ultrathin films, single monolayers or multilayers using different methods: (1) molecular self-assembling, (2) the co-called layer-by-layer (LbL) method, or (3) the Langmuir–Blodgett (LB) technique. Via film-coating, both surface chemistry and topography can be varied in a very controlled manner (Whitesides and Grzybowski, 2002). Picart and coworkers studied new methods for the anchoring of biomolecules onto scaffold surfaces using multiple functionalities of polyelectrolyte multilayer films (Boudou et al., 2010). Moby and coworkers reported a study using polyelectrolyte multilayers that were designed for vascular tissue engineering applications. Human mesenchymal stem cell differentiation into endothelial-like cells could be observed on surfaces coated with polyelectrolyte multilayer films (Moby et al., 2011). In between, a variety of depositing methods have been developed for LbL formation including dip coating, spin coating, and spraying. For future use and industrial applications of LbL films, the total time required for mono and multilayer film preparation, and the transfer of the layer to an appropriate substrate, are important limitations. Therefore, rapid methods, such as spraying are under investigation. In addition, the adhesion and anchorage of the layers to the substrate was improved, in particular for poly(ethylene) and poly(tetrafluoro ethylene), that usually require priming methods due to their very hydrophobic surfaces. Marchenko et al. (2010) reported a so-called bottom-up approach, using assembled polyelectrolyte multilayers and studied their effects on the self-assembly behavior of particles, including nanoparticles, and carbon nanotubes. For many biomaterial applications, relatively large surface areas are required—they are beyond the limits of traditional lithography. Lenhert and coworkers used a recently developed method called "Langmuir-Blodgett lithography" to pattern polystyrene surfaces and investigate osteoblast alignment, elongation, and migration. This "nanoimprinting" allows the fabrication of nanostructured surface areas for a broad variety of natural and artificial biomaterials. Lehnert could achieve spaced grooves of different depths (50 up to 150 nm) on silicon substrates that show a very high degree of periodicity. These topographies were transferred into polystyrene surfaces via nanoimprinting and osteoblasts were cultured on the structured polymer surfaces. The authors could show that the osteoblasts are able to align, elongate, and migrate parallel to the patterns. In addition, the osteoblasts show a significant anisotropic behavior to the surfaces resulting in enhanced cell adhesion/attachment (Lenhert et al., 2005). Another approach to generate patterned surfaces is the so-called "protein-mediated molecular self-assembly technique," a powerful method used for the fabrication of biomimetic biomaterials with controlled size and shape. Very recently, Wang and coworkers developed a molecular template made of two proteins, silk fibroin and albumin. The template is able to both promote and inhibit hydroxyapatite formation, respectively, to be used to synthesize HA nanoparticles of different controlled shape. As such, nanospheres were generated using the combined albumin/silk fibroin template, whereas nanorods were obtained using solely the silk fibroin template. The authors could show that the nanoparticle shape has a crucial impact on the cellular attachment. Thus, hydroxyapatite nanospheres were absorbed by mesenchymal stem cells at a faster rate. Regarding cell proliferation and migration, both nanospheres and nanorods showed no significant influence. The osteoblastic differentiation of MSCs was significantly supported by HA nanospheres, compared to HA nanorods. In summary, those templates based on two proteins, silk fibroin and albumin, are combinations of a promoter and inhibitor protein that enable the preparation of differently shaped nanoparticles (Wang et al., 2015). Uniform calcium phosphate/polydopamine nanolayers were generated on the surface of β-tricalciumphosphate (β-TCP) bioceramics drenched with tris-dopamine solution. The dopamine concentration, reaction temperature, and reaction time are three key factors inducing the formation of a uniform Ca-P/polydopamine composite nanolayer. The surface roughness and hydrophilicity of β-TCP ceramics was significantly enhanced by those self-assembled Ca-P/polydopamine composite nanolayers. In addition, they could stimulate the adhesion, attachment and proliferation of human bone marrow stromal cells as well as their alkaline phosphate activity and bone-related gene expression (i.e., Runx2, COL1, ALP, OCN). The results suggest that those self-assembled Ca-P/polydopamine composite nanolayers represent an appropriate system for guided surface modification of biomaterials for bone regeneration application due to significant improvement of physicochemical surface properties and bioactivity (Wu et al., 2014).

2.2.4 Electrospinning

This method provides fibrous scaffolds that are able to mimic the 3D dimension and topology of extracellular matrix fibers, comprehensively reviewed by Wei and Ma (2008). In several studies, it was shown that the nanoscaled collagen fibrillary structure with a diameter of 50–500 nm enhances the cell–matrix interactions. Filaments of different composition and morphologies can be formed on the nanometer scale that are already used as medical membranes and scaffold materials. Lim et al. (2011)

studied the micropatterning of electrospun nanofiber scaffolds based on poly(caprolactone) and gelatin using a femtosecond laser ablation technique. In another approach, Fan used patterning of polymer composites via electrospinning to study the role of nanostructured mesoporous silicon in discriminating calcification (Fan et al., 2011). Hassan and coworkers studied an electrospinning system, were the fiber diameter of polycaprolactone showed entanglements that could be varied via polymer concentration. They could demonstrate that at low polymer concentration the resulting scaffold exhibit a bead structure whereas smooth fiber structures are formed at higher concentrations. Furthermore, the diameter of the fibers could be increased via addition of nanohydroxyapatite. As a result, those HA/PCL nanocomposites showed improved bioactivity (up to 7 days), compared to conventional polycaprolactone fibers. In addition, the nanocomposites exhibit enhanced hydrophilicity (studied via contact angle measurements and water uptake) finally resulting in improved osteoconductive activity (Hassan et al., 2014).

2.2.5 Rapid prototyping methods

Material surfaces can be micro- and nanostructured using so-called solid freeform or rapid prototyping techniques (Faid et al., 2005; Wei and Ma, 2008). The processes do show differences in resolution and all are characterized by advantages and certain limitations. Thus, highest resolutions up to 70–250 μm can be realized via selective laser ablation (SLA). However, SLA requires appropriate liquid photopolymers that are still limited in availability, whereas 3D print is a rather fast process but often characterized by weak bonding between powder particles (Peltola et al., 2008). The so-called fused deposition method (FDM) has attracted interests due to its ability to form 3D structures by LBL deposition. FDM technology allows easy and flexible material implementation for scaffold fabrication. In this case, polycaprolactone (PCL) calcium phosphate and tricalcium phosphate (TCP) may also be used for bone tissue engineering (Abbah et al., 2009). In the selective laser sintering (SLS) method, powders of polymers, ceramics, and metals are sintered with high energy. The surface of the powder bed is sintered to create a 2D cross-sectional pattern of a 3D structure. The new powder is applied layer by layer until the desired 3D structure is reached. The SLS technology involves the handling of inorganic ceramics that facilitate important structural properties for bone regeneration (Duan et al., 2010). With the SLS technology controlled, related, and highly porous microstructures using poly(caprolactone) and poly(lactic acid) could be produced (Williams et al., 2005). Differences in the size of nanotopographic features significantly influence the cell behavior including mechanical aspects, in particular substrate stiffness. Gerecht et al. (2007) could demonstrate that nanotopographic structured ECM alters the morphology and proliferation of human embryonic stem cells through cytoskeletal-mediated mechanisms. They synthesized poly(dimethylsiloxane) scaffolds with gratings up to 600 nm that are able to induce a well-defined alignment and elongation of embryonic stem cells. In addition, they could show that nanotopographic patterns altered the organization of various cytoskeletal components (i.e., F-actin, vimentin, tubulin). Changes in proliferation and morphology were repealed by the effect of actin-disrupting agents (Gerecht et al., 2007). In addition to tailored topography, nanobased technologies enable the production of scaffolds with a controllable interconnected pore network, allowing an improved cell migration and nutrient exchange. A comprehensive review with focus on the effects of order and symmetry onto cell differentiation and proliferation was published by Langer and his group, including numerous in vitro and in vivo studies (Bettinger et al., 2009). A variety of nanolithography techniques is discussed including conventional photolithography, electronbeam photolithography, and interference lithography. The advantage of lithographic methods is their ability to manufacture scaffolds of very precise 3D structure. However, a rather complex equipment is required, in addition the process is time-consuming and rather expensive. In conclusion they emphasize that there are significant differences in cell behavior on nanostructured scaffolds, compared to conventional scaffolds even if they were prepared with a roughness of a similar length scale. In addition to those techniques discussed here, there are various alternatives to generate nanostructured scaffold surfaces including synthetic methods based on di- and multiblock copolymers of well-defined macromolecular morphology (Ruiz et al., 2007).

2.3 NANOSTRUCTURED SURFACES IN STEM CELL RESEARCH

In stem cell based bone regeneration, nanomaterial applications are mainly focused to improve mechanical properties of scaffold materials. Most promising approaches are represented by nanocomposites, comprehensively reviewed by Peran et al. (2013). For rat bone marrow stromal cells, Kim et al. (2011) could show that the early osteogenic signal expression depends on two factors: the hydroxyapatite nanoparticle content and the initial cell seeding density. Sitharaman introduced a novel nanoparticle-enhanced biophysical stimulus based on the so-called photoacoustic effect. Results could show that this photoacoustic effect influences differentiation of bone marrow-derived stromal cells grown on poly(lactic-*co*-glycolic acid) polymer films toward osteoblasts. Osteodifferentiation of the cells due to photoacoustic stimulation is significantly enhanced by the presence of single-walled carbon nanotubes in the polymer (Sitharaman et al., 2011). Peng et al. (2012) studied nanofibrous composites of HA and chitosan to be used as biomimetic scaffolds for guided stem cell differentiation and induced bone regeneration. To do so, they varied the content of osteogenic supporting molecules, such as ascorbic acid or glycerol phosphate. Chitosan nanofibers were generated via electrospinning and studied regarding their effects on stem cell adhesion, differentiation, and growth. In conclusion, the HA-chitosan nanocomposites showed improved cell adhesion and proliferation of mesenchymal stem cells, compared to conventional chitosan-based scaffolds. In addition, the expression of osteogenic genes (collagen, Runx2, alkalinephosphate, osteocalcin) was clearly upregulated. Thus, the authors emphasized the significant effects on osteogenic commitment of MSCs (Peng et al., 2012). Brammer and coworkers studied hydrophobic micro- versus nanopillar type fabrication techniques to explore the interaction of mesenchymal stem cells with the physical environment. In particular, the nanopillars are able to induce the formation of 3D cell aggregates that show enhanced osteogenic activity and controlled differentiation. It was found that the nanopillar surface had significant impact on cell adhesion and growth (Brammer et al., 2011). Lai and coworkers studied the influence of surface nanostructured titanium substrates on the fate of mesenchymal stem cells, in particular, proliferation and growth. Here, a technique was used called "surface mechanical attrition treatment" (SMAT) to produce titanium surface nanostructures. In vitro studies at cellular and molecular levels were performed using mesenchymal stem cells. The results suggest that nanostructured surfaces significantly supported the growth of MSCs, including adhesion, filament orientation, proliferation, and gene expression. This SMAT approach will further be exploited for the generation of high performance titanium-based implants (Lai et al., 2012). Jensen and coworkers studied the osteogenic potential of dental pulp stem cells on three different HA and β-TCP functionalized polycaprolactone scaffolds produced via rapid prototyping and fused deposition methods. They found that dental pulp stem cells showed the ability to differentiate into an osteogenic lineage. Functionalized scaffolds promoted Ca-ion deposition after 3 weeks of cultivation. Thus caprolactone-based scaffolds can be used to stimulate osteogenic differentiation, and support cell migration to be used for large bone defects (Jensen et al., 2015). Very recently, Magalhães et al. found that the chondrogenic differentiation of bone marrow-derived mesenchymal stem cells was influenced by the composition and morphology of chitosan-coated poly(lactic acid) based scaffolds; a hydrogel-like coating supported the formation of heterogeneous tissues, with well-defined areas of inefficient cell condensation or matrix maturation, with improved chondrogenic phenotype of the MSCs; however a fiber-like coating could induce a spatial distribution that strongly supports interactions of the scaffold and cells resulting in a more stable hyaline-like cartilage tissue (Magalhães et al., 2015). Salinas and his group developed the synthesis of novel mesoporous bioactive glasses based on SiO_2–CaO–P_2O_5 glasses with substitutions including various oxides (i.e., Ga_2O_3, Ce_2O_3, ZnO) for bone tissue engineering applications. These mesoporous bioactive glasses are obtained via rapid prototyping methods resulting in a 3D structure including interconnected pores of well-defined size. The final network supports cell adsorption and simultaneously, the release of bone inducing molecules. In particular, the structure and size (at atomic scale) of mesoporous bioactive glasses can be varied in a broad range via introduction of different inorganic oxides (Vallet-Regí and Salinas, 2015).

In summary, the exploitation of nanotopographic techniques to generate appropriate nanostructures to be used for scaffold design in tissue engineering still represents a major challenge. In particular, novel nanofabrication techniques and recently developed nanomaterials have to be adapted to each other to improve the cell adhesion process and optimize the interaction of cells at the nanostructured scaffold surface. It should be emphasized, that the nanofabrication techniques discussed in this chapter represent only a limited number of selected examples; further methods, such as "two-photon polymerization" are recently reported to generate 3D scaffold structures (Song et al., 2015).

3 STEM CELLS WITHIN THE HUMAN BODY AND THEIR USE IN REGENERATIVE MEDICINE

The term stem cell was first introduced in the late 19th century, when researchers all over the world started to investigate the development of different tissues in plants and animals. The first cells known to arise from a single ancestral cell type have been blood cells. Accordingly, in the middle of the 20th century, first experiments with bone marrow transplantations were made, and it became clear that these cells have the potential to differentiate into the hematopoietic lineages. In 1980s, the first embryonic stem cells was established, and by the end of the last century, many sources of stem cells had been found in the adult body as well. Finally, a new approach was conducted with the development of induced pluripotent stem cells (iPS) from terminally differentiated cells in 2006 (Fig. 21.2A).

The following section will give an overview of the history of stem cell research and their use in regenerative medicine.

Blood (hematopoietic) stem cells derived from bone marrow were the only stem cells to be known until 1950s. Consequently, it is not surprising that the transplantation of bone marrow had occupied the minds of scientists and physicians for decades, and was finally conducted in 1957 (Thomas et al., 1957). In the 1970s, it became clear that bone marrow also harbors another type of stem cells that were called bone-marrow stromal cell (BMSC) (Friedenstein, 1990). As a consequence, the theory of a stem cell niche was postulated, suggesting that hematopoietic stem cells are regulated by their cellular environment, such as the BMSCs in the bone marrow (Schofield, 1978).

At about the same time (1978), cancer stem cells were described. For the first time, the progression of some cancerous diseases that could not be explained by the two-step carcinogenesis model could be explained. Today, it is broadly accepted that both models are represented in the various maligned diseases depending on the cancer type (Bonnet and Dick, 1997).

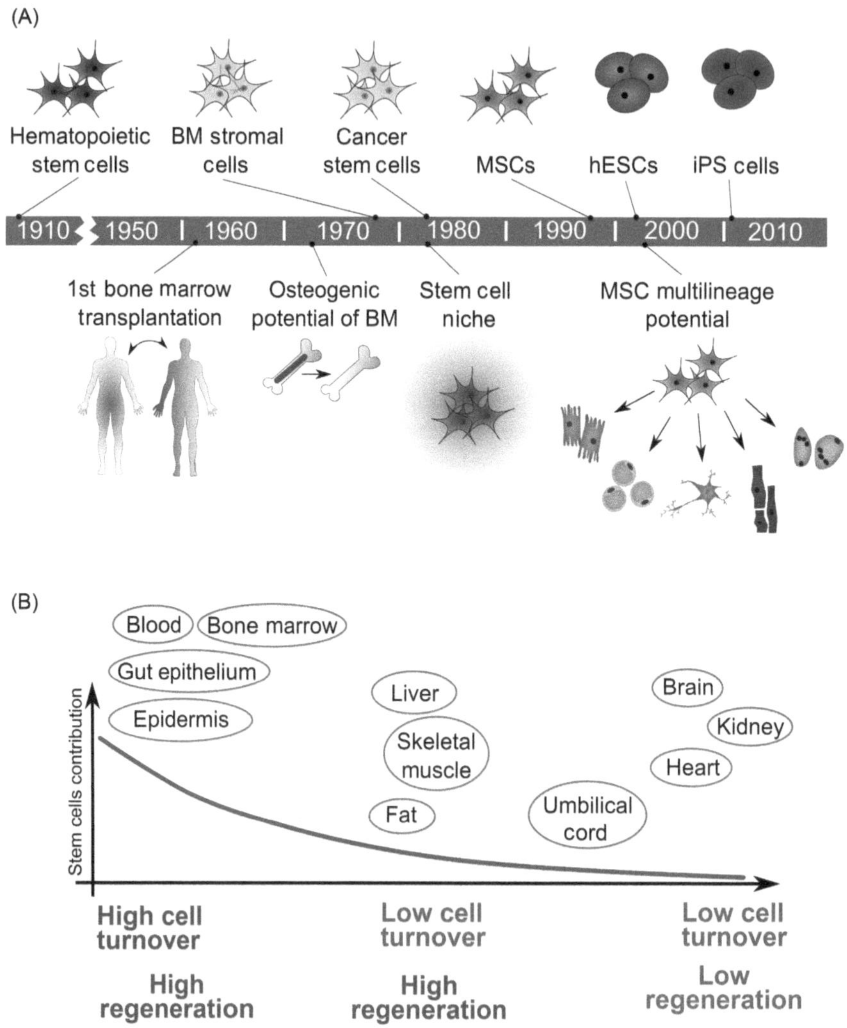

FIGURE 21.2 Milestones of Stem Cell Research

(A) The timeline represents key events in the history of stem cell research. The first stem cells discovered were blood stem cells. This resulted in the first bone marrow transplantation in the late 1950s, and the discovery of bone marrow stromal cells. 1978 cancer stem cells were found before, in the early 1980s, the discovery of embryonic stem cells. Ten years later, the term mesenchymal stem cell was coined. Another milestone in stem cell research was the discovery of the induced pluripotent stem cells in 2006. (B) For a long time, MSCs were thought to be found only in tissues with a high cell turnover and a high regeneration rate, such as blood, bone marrow, and gut. Within the recent decades, MSCs could also be found in tissues with a low cell turnover and high regeneration, such as fat and the skeletal muscle and even in tissues with low regeneration, such as brain and the heart. The amount of stem cells is depending on the cell turnover.

A new milestone was reached by the generation of embryonic stem cells (ESCs) in 1981. Evans and Martin were the first researchers to generate ESCs from embryos and 17 years later, human embryonic stem cells from blastocytes were successfully derived (Martin, 1981; Thomson et al., 1998). However, this type of stem cells was and still is controversially discussed due to ethical concerns since human embryos have to be dismantled to gain human ESCs and due to the risk of teratoma formation. Nevertheless, this is the most potent type of natural stem cells.

At that time, it was also generally expected that stem cells were only present within tissues that have a high cell turnover and a high regeneration capacity, such as blood (Prockop, 1997) (Fig. 21.2B). Therefore, the concept of nonhematopoietic stem cells did not resonate with many. But bone marrow, gut epithelium, the epidermis, or other tissues also feature these attributes and consequently adult stem cells were found within, and Caplan coined the term mesenchymal stem cell (MSC) (Caplan, 1991).

Driven by the idea that every tissue with the capacity of maintenance and repair must have progenitor cells from which adult somatic cells are derived, researchers began to investigate tissues also with a low turnover and a high regeneration capacity, such as liver, skeletal muscle, and adipose tissue.

During the past decade adult stem cells were even found in tissues with a low cell turnover and a low regeneration capacity, such as brain, lung, and heart (Jiang et al., 2002a; Messina et al., 2004; Sabatini et al., 2005). However, it is obvious that adult stem cells are limited in their capacity to repair large defects, such as a heart attack. Therefore, the idea of proliferating them in vitro for replacement material was obvious.

An interesting source for such repair or replacement strategies are MSCs due to their multipotential. Today, many sources for MSCs are known and, regarding their high number, the International Society for Cellular Therapy has proposed minimal criteria for their definition: MSCs must adhere to plastic, in vitro differentiate toward the osteogenic, chondrogenic, and adipogenic lineage, and they must be positive for three specific markers, and lack expression of three other markers and surface molecules (Dominici et al., 2006).

The plasticity of MSCs, meaning their transgermal differentiation potential, has been shown in many publications (Babczyk et al., 2014; Zippel et al., 2010). Although the expectation was that their potential is limited to the mesenchymal lineage, it was shown that multipotent adult progenitor cells copurified with MSCs could be differentiated into cells of the visceral mesoderm, neuroectoderm (neurons), and the endoderm (liver, lung, bladder) (Jiang et al., 2002b). Many scientists continue to contribute to this so-called “transdifferentiation,” but this increased potential is still controversially discussed.

The potential of MSCs is not their only feature with importance to regenerative medicinc. It was discovered that they also have the capacity for migrating from their stem cell niche to damaged areas (Ferrari et al., 1998). Another important function of MSCs was uncovered in 2006, by showing that these cells secrete a variety of important growth factors (Fraser et al., 2006). It was revealed that MSCs differentiating toward the adipogenic lineage secrete 420 soluble factors that influence other cells greatly (Zhong et al., 2010).

This and their availability, the fewer ethical concerns than embryonic stem cells, and their low immunogenicity prompted scientists to use MSCs for cellular therapies with and without scaffolds to restore lost tissue in vivo. Surprisingly, the clinical studies showed that the cells were lost after some weeks and failed supply the distinctive cell types needed to restore function. Nevertheless, patients recovered better with than without the stem cell application. Today, scientists presume that the benefits there are due to positive paracrine effects on the tissue via the secreted soluble MSC factors.

In 2006, a breakthrough in stem cell research was reached by creating the first artificial stem cells, named induced pluripotent stem cells (iPS), due to their similarity in potency to ESCs. These cells derive from adult and terminally differentiated cells, and are reprogrammed by transcription factors (Takahashi and Yamanaka, 2006). While Yamanaka initially used four transcription factors (c-Myc, Klf-4, Oct-4, Sox-2) to achieve pluripotency, research has progressed since then so that only a decreased number of factors is necessary now depending on the differentiation status of the initially targeted cells (Yamanaka, 2008). Since these cells can be generated from a patient's own tissue, iPS are moving into focus for regenerative approaches very fast. Consequently, the Nobel prize was given for this achievement to Shinya Yamanaka in 2012. However, recent studies showed that the epigenetic pattern of these cells does not match the one of ESCs but still refers to the one of the targeted original cells. This, and the risk of teratoma formation, shared with ESCs, move their use in regenerative medicine in the far future.

A better understanding of the respective stem cell type is necessary for an effective implementation of stem cells with and without scaffolds in regenerative medicine. Thus, the effect of the stem cell environment is of major interest since it encompasses chemical and physical factors influencing stem cell fate.

4 INTERACTIONS OF NANOSTRUCTURED SURFACES AND STEM CELLS

The stem cell niche is not only the microenvironment that houses stem cells, but is also a functional dimension that protects the cells from differentiating, and maintains their capacity to self-renew. Several factors, such as physical, mechanical, cellular, or chemical interactions influence the stem cell niche and, therefore, characterize each individual microenvironment. Nanostructured surfaces are nowadays used in tissue engineering to mimic this environment perfectly, and direct stem cell fate into the desired lineage. This section is defining the stem cell niche first before an overview about the factors influencing the niche is given. It is completed by a paragraph on stem cells on nanostructured surfaces that focusses on soft and hard tissue engineering.

4.1 THE STEM CELL NICHE

The stem cell niche is a spatial location or microenvironment within an organ that houses stem cell populations, and their secreted products. The definition also encompasses a functional dimension, namely protecting the stem cells from depletion by preventing their differentiation, and supporting their self-renewal. Hence, the stem cell niche regulates the stem cells' participation in maintenance, repair, and generation of tissues, while preventing the host from excessive stem cell proliferation (Scadden, 2006). The factors that shape the stem cell niche are summarized in Fig. 21.3, and discussed in detail in the following sections.

4.1.1 Mechanical and physical factors of the stem cell niche

Mechanical triggers encompass mainly the extracellular matrix (ECM) that is a key player of the stem cell niche, although its interactions with stem cells as well as its composition vary largely. The ECM forms the structural environment and consequently determines matrix geometry, elasticity, stiffness, and topography. Those factors can vary from tissue to tissue and during aging or injury. In vitro, the substrate elasticity has been shown to directly influence the stem cell fate (Chitteti et al., 2015; Engler et al., 2006). External mechanical forces, such as the compression of the surrounding cells or the stiffness of the ECM need to be balanced by the cells and is regulated by a complex mechanosensing machinery.

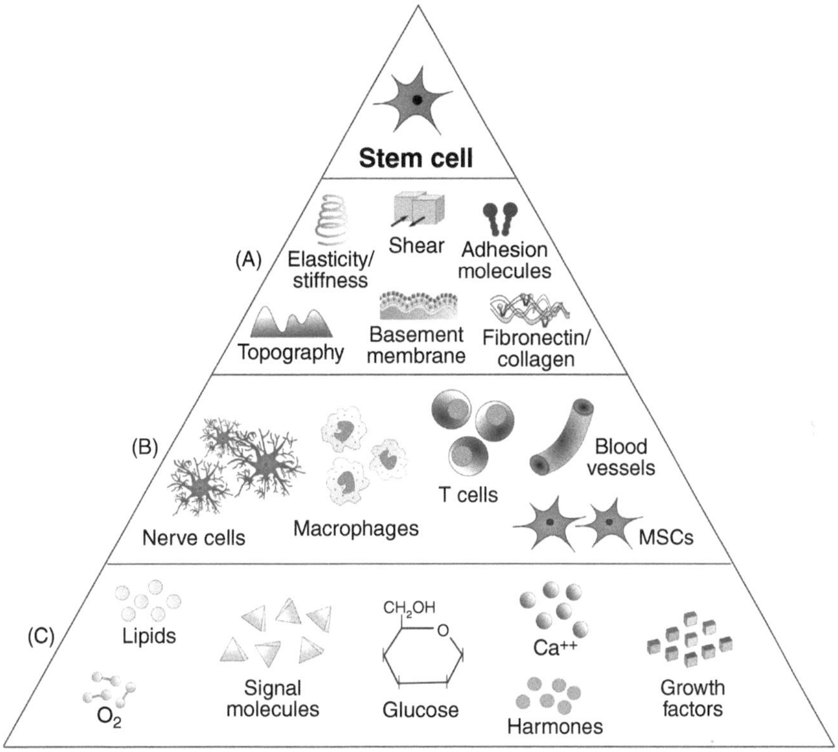

FIGURE 21.3 The Stem Cell Niche

Factors influencing the stem cell niche can be roughly categorized in three groups: physical and mechanical factors, such as (A) shear forces, elasticity, and topography, (B) cellular issues, such as immune and nerve cells, nearby blood vessels, and neighboring stem cells; and (C) soluble factors, such as oxygen, glucose, hormones, growth factors, or signaling molecules.

In their natural niche, stem cells also interact with the matrix topography. Nanostructured surfaces providing topographical clues can be used in vitro to differentiate stem cells. They might offer a way of augmenting chemical induction or even provide a better alternative (Seidlits et al., 2008).

Physical triggers, such as oxygen tension, ionic strength, and pH can influence the fate of stem cells as well. Hypoxic environments are common for several cell populations such as cardiac progenitors and HSCs (Kimura and Sadek, 2012). In vitro, hypoxic conditions can be used to promote survival, proliferation, and function after engraftment (Muscari et al., 2013). Next to oxygen, calcium is one of the chemical elements studied mainly with respect to the HSC niche. G-coupled calcium-sensing receptors expressed by HSCs seem to influence the localization and function of the cells in bone (Adams et al., 2006). The pH is known to influence cell proliferation and even small variations in pH substantially affect the cell performance. A reduction in the pH of 0.5 (7.1 instead of 7.6) increased the cloning efficiency of progenitor cells and a pH between 7.2 and 7.4 was shown to be optimal for differentiation in vitro (McAdams et al., 1997).

Besides mechanical and physical factors, the stem cell niche is formed by others cells that influence each niche in an individual way.

4.1.2 Cells influencing the stem cell niche

Cell–cell interactions play a major role in controlling stem cell fate (Schulze and Tobiasch, 2012). The stem cell niche of many adult tissues is composed of a multitude of cell types—each with a distinct function. Besides their neighboring cells, the stem cell pool itself and its direct progeny are important players in regulating the stem cell niche. Whereas cell–ECM interactions are mainly concerted via adhesion receptors, such as integrins that bind to ECM proteins, cell–cell interactions encompass a complex paracrine system (Tanentzapf et al., 2007). An example of cellular interaction is the Notch signaling that has been associated with the regulation of cellular fate, the differentiation of progenitor cells, and the self-renewal of adult stem cells. The Notch pathway was also shown to regulate cell survival and proliferation (Dontu et al., 2004). Besides Notch, Wnt signaling plays an important role in major developmental processes in the embryonic state, and is responsible for regulating differentiation, maintenance, and self-renewal of adult mammalian tissue stem cells (Van Camp et al., 2014). Another signaling molecule that should be mentioned in connection with cell–cell interactions is Hedgehog. As a key mediator involved in several cellular and embryonic developmental processes, Hedgehog signaling is among others responsible for adult stem cell self-renewal (Mooney et al., 2015).

4.1.3 Soluble factors stimulating stem cells

Stem cells and their niche communicate indirectly via secreted factors. Some of these factors are already widely used in clinics to modulate stem cell fate. For example, cytokines, such as granulocyte colony-stimulating factor, are applied to support the treatment of bone marrow failure, hematological malignancies, and rare genetic disorders (Lane et al., 2014). Moreover, interleukines, such as IL-11 or IL-6 are well-known key players in regulating stem cell fate (Notara et al., 2010). Other metabolic factors broadly affecting stem cell fate are purinergic derivates. Besides its role as the main energy source of the cell, ATP also acts as a substrate in signal transduction pathways and is a major factor that influences the stem cell niche. It was shown to promote self-renewal (Xi and Xie, 2005), and maintenance of multipotency of MSCs during ischemic stress (Mylotte et al., 2008). ATP, UTP, and their breakdown products also act as ligands for purinergic receptors, which can be divided into P1 and P2 receptors (Kaebisch et al., 2015). The latter were shown to be involved in mesenchymal stem cell proliferation and differentiation toward the adipogenic and osteogenic lineage, as well as toward endothelial and smooth muscle cells under microgravity (Zhang et al., 2014; Zippel et al., 2012).

Growth factors play a key role in regulating various cellular functions, such as proliferation and differentiation. Specific peptide sequences, such as arginine–glycine–aspartic acid (RGD) that are present in ECM proteins, such as fibronectin, were shown to regulate MSC function and can be added to natural and synthetic biomaterials to increase their functionality (Griffin et al., 2015). Other well-known growth factors with an influence on stem cell fate are members of the transforming growth factor-β family that are, for example, involved in neuronal/glial differentiation and survival, and neural stem cell maintenance (Kandasamy et al., 2014).

4.2 STEM CELLS ON NANOSTRUCTURED SCAFFOLDS FOR SOFT TISSUES

Soft tissues, including muscle, skin, fat, nerves, blood vessels, fibrous tissue, and others, are connecting, supporting, and surrounding structures, and organs of the body. Despite active research in this field, soft tissue engineering still lacks the transition from bench to bedside. Until now, only some of these simple structures can be engineered. The simplicity and the urgent clinical need for better and more biocompatible blood vessel replacements explain why vascular tissue engineering is a major field in soft tissue engineering.

A vascular conduit that effectively mimics the native original has to include its nonthrombogenic properties and long-term patency. Especially in vascular tissue engineering, nanotechnologies can be used to optimize the inner surface of grafts, providing a basis for cellularization, promoting mobilization of surrounding cells, and directing the differentiation of stem cells incorporated within the graft (Hashi et al., 2007). Often used materials, such as nanofibers and nanostructured polymer surfaces can imitate the organization of the natural vascular extracellular matrix and thus improve their performance as a graft (Schulze and Tobiasch, 2012).

One approach is the electrospinning of polymers or naturally occurring materials into nanofibers. These fibers simulate the dimension and structure of native collagen and elastin fibers very well, and can form scaffolds with high porosity and adequate surface area-to-volume ratio (Ravi and Chaikof, 2010). Mesenchymal stem cells can proliferate on electrospun nanofibers of different materials, such as PLLA/Collagen (1:1) or L-lactide-*co*-trimethylene carbonate (Dargaville et al., 2013; Jia et al., 2013; Wang et al., 2014). In an artery bypass experiment, small-diameter nanofibrous scaffolds showed as well organized layers as their native counterpart (Hashi et al., 2007). Mesenchymal stem cells grown on electrospun fiber meshes with patterned nano- and microgrooves or geometric shapes imprinted in different tones at physiological temperatures, can change morphology and cytoskeletal organization.

In addition, nanoparticles can be used to functionalize biomaterial in a way that promotes their suitability as vascular graft. Polymer-based nanoparticles made of poly(carbonate-urea)urethane and polyhedral oligomeric silsesquioxane as fillers for biomaterials have been reported to show heparin-like behavior at the blood-material interphase, and make nanocomposites an attractive scaffold for vascular tissue engineering (Kannan et al., 2006a,b).

Especially for larger scaffolds (>200–500 μm), host angiogenesis is needed to ensure sufficient nutrient supply and waste removal in the new tissue. Addressing this problem, peptide-based self-assembling nanofibrous hydrogels have been developed. Cell-mediated degradation and proangiogenic moieties that are entrapped in the hydrogel promote angiogenesis by providing rapid infiltration of hematopoietic and mesenchymal cells that quickly form a mature vascular network (Kumar et al., 2015).

4.3 STEM CELLS ON NANOSTRUCTURED SCAFFOLDS FOR HARD TISSUES

Hard tissues are biological tissues that embed minerals into soft matrices, forming a protective shield and providing structural support. Bone, cartilage, tooth enamel, and dentin are some examples of structures where mineralized tissues are found (Espinosa et al., 2009). The repair and regeneration of hard tissues, such as bone and teeth is an active research field in modern biomedical engineering and clinical practice (Han et al., 2013).

Bone tissue (as an example for hard tissues) has excellent regenerative ability. Though, if the injury exceeds a critical size, bone formation is impaired and surgical intervention is required (Murphy et al., 2013). Despite the usage of bone graft techniques for decades to reconstruct large bone defects, obstacles can still diminish their applications in therapy. For instance autographs are limited in supply and induce additional donor site morbidity. On the other hand, allografts and xenografts could cause pathogen transmission and rejection by the recipient's body (Baghaban Eslaminejad and Faghihi, 2011). These facts have encouraged scientists to look for appropriate alternatives, leading to bone construct manufacturing (Faghihi and Eslaminejad, 2014).

Bone tissue engineering involves three main components, an artificial extracellular matrix scaffold, stem cells with the potential to differentiate into osteoblasts, and osteoinductive growth factors to direct the growth and differentiation of cells toward osteogenic lineage within the scaffold (Rauh et al., 2011).

Successful bone regeneration requires multiple critical considerations. Natural bone is composed of 70% inorganic calcium phosphate crystals, and 30% w/v organic collagen fibers. This structure has been used as a model to imitate bone on nanoscale level (Walmsley et al., 2015; Zietz et al., 2013). The dominant nanocrystalline component in bone is hydroxyapatite with a thickness of 2–5 nm (Kaplan et al., 1994). Nanometer-size proteins assemble together, forming the nanostructured ECM that influences osteoblasts and mesenchymal stem cells in their adhesion, proliferation, and differentiation (Zhang and Webster, 2009). Advances in nanotherapeutic approaches lead to further manipulation of the artificial ECM providing a more appropriate surface for angiogenesis and cellular proliferation. Furthermore, the spatial and temporal release of signaling molecules should be controlled as they guide the survival and differentiation of cells. Last, regarding biocompatibility, artificial nanomaterials ought to stay inert or resorb in a foreseeable and controlled manner for remodeling (Walmsley et al., 2015). In recent years, tissue engineers used various modifications, such as addition of bioactive molecules or nanoparticles, to enhance attachment and proliferation of stem cells on the scaffold (Zeng et al., 2012).

An optimal scaffold should be biocompatible and biodegradable, allowing for a substitution with functional tissue. Many scaffolds have been designed to provide support and tensile strength, and to serve as a connected network for cell adhesion, movement, and tissue ingrowth, in order to mimic the ECM (Kleinman et al., 2003). The usage of nanoparticle-modified composite scaffolds can facilitate bone regeneration significantly, compared to biodegradable polymeric materials or ceramics. The integration of nanoscale organic and inorganic materials into polymeric scaffolds may provide the appropriate surface and mechanical properties necessary for support, as well as cellular adhesion, differentiation, and incorporation with the surrounding milieu (Kim and Fisher, 2007).

To enhance the attachment of cells within scaffolds, the RGD cell adhesion ligand has been incorporated. This improved the ability for scaffolds to emulate the extracellular matrix and guide behavior and development of implanted cells (Shakesheff et al., 1998). The attachment and differentiation of MSCs to osteoprogenitors has been shown to increase upon seeding onto manufactured nanocomposites poly(L-lactide) (PLA)-based scaffolds incorporated with carbon nanotubes, and microhydroxyapatite (HA) particles (Ciapetti et al., 2012). It was also found that the incorporation of a biomimetic HA surface could change the microenvironment of adherent preosteoblasts, and allow for survival and extension of cellular projections along textured surfaces (Chou et al., 2005).

In the recent years, combining different biomaterials into a single scaffold has been approached as a means of balancing the advantages and disadvantages of each material to create the optimal scaffold for stem cell-based therapy. As an example of this approach, integrating PLA, HA, and collagen into a single scaffold improved the viability of engrafted hMSCs. It also augmented osteoinduction, which increased osteogenic gene expression, such as significantly elevated levels of osteocalcin, and enhanced matrix mineralization, as shown by elevated calcium and Alizarin Red staining (Balaji Raghavendran et al., 2014).

It should be considered that bones are considerably complex organs due to a combination of hard tissues (compact and trabecular bone) and soft tissues (bone marrow) in a unique hierarchical structure (Polini et al., 2013). Regarding this fact the processing of nanometer-sized fibers to produce highly porous scaffolds will promote easy cell migration and nutrient diffusion. Besides the availability of large surface area for cell attachment, they also allow the delivery of drugs and growth factors due to their increased porosity. All these criteria allow the interaction between cells and extracellular matrix as seen during normal tissue development (Sharma et al., 2014).

5 FUTURE PERSPECTIVE

The global expand of the population combined with its increased aging will rise the clinical need for tissue regeneration or replacement and even whole organs. The application of nanotechnology for scaffolds coupled with stem cells is a promising approach that might help to meet the future clinical needs for replacement and regeneration of tissues. So far, the majority of successful tissue engineering methods resulted only in simple tissues, playing a relatively small role in clinical treatment of patients. Currently, supplemental bladders, small arteries, skin grafts, cartilage, and trachea have been implanted in patients (Baiquera et al., 2010), but the procedures are still in their infancies and very costly. However, when it comes to increasing levels of complexity, as it is the case for whole organs with a multitude of cell types involved, stem cells on nanostructured surfaces are currently pushed to their limits. Complex organs, such as the kidney would require a multitude of intricate structures, specific cell types, and an extensive network of blood vessels and nerve cells in order to build up a functional organ. Tremendous scientific and technical challenges need to be overcome to ensure such a structural hierarchy and complex functionality of engineered tissues and organs in the future.

ACKNOWLEDGMENTS

This work was supported by BMBF, AdiPaD, [1720X06]; BMBF, FHprofUnt, [03FH012PB2]; FH-Extra, "Europäischer Fonds für regionale Entwicklung", "Europa – Investition in unsere Zukunft", EFRE co-financed NRW Ziel 2: "Regionale Wettbewerbsfähigkeit und Beschäftigung" [z1112fh012]; BMBF, IngenieurNachwuchs, Persolmplant, [03FH019IX5].

REFERENCES

Abbah, S.A., Lam, C.X.L., Hutmacher, D.W., Goh, J.C.H., Wong, H.-K., 2009. Biological performance of a polycaprolactone-based scaffold used fusion cage device in a large animal model of spinal reconstructive surgery. Biomaterials 30, 5086–5093.

Adams, G.B., Chabner, K.T., Alley, I.R., Olson, D.P., Szczepiorkowski, Z.M., Poznansky, M.C., Kos, C.H., Pollak, M.R., Brown, E.M., Scadden, D.T., 2006. Stem cell engraftment at the endosteal niche is specified by the calcium-sensing receptor. Nature 439, 599–603.

Babczyk, P., Conzendorf, C., Klose, J., Schulze, M., Tobiasch, E., 2014. Stem cells on biomaterials for synthetic grafts to promote vascular healing. J. Clin. Med. 3, 39–87.

Baghaban Eslaminejad, M.R., Faghihi, F., 2011. Mesenchymal stem cell based bone engineering for bone regeneration. In: Eberli, D. (Ed.), Regenerative Medicine and Tissue Engineering - Cells and Biomaterials. InTech publishers, Rijeka, Croatia, pp. 57–82.

Baiquera, S., Birchall, M.A., Macchiarini, P., 2010. Tissue-engineered tracheal transplantation. Transplantation 85, 485–491.

Balaji Raghavendran, H.R., Puvaneswary, S., Talebian, S., Raman Murali, M., Vasudeveraj Naveen, S., Krishnamurithy, G., 2014. A comparative study on in vitro osteogenic priming potential of electron spun scaffold PLLA/HA/ Col, PLLA/HA, and PLLA/Col for tissue engineering application. PLoS One, 1–16.

Bettinger, C.J., Langer, R., Borenstein, J.T., 2009. Engineering substrate topography at the micro- and nanoscale to control cell function. Angew. Chem. Int. Ed. 48, 5406–5415.

Bonnet, D., Dick, J.E., 1997. Human acute myeloid leukemia is organized as a hierarchy that originates from a primitive hematopoietic cell. Nat. Med. 3, 730–737.

Boudou, T., Crouzier, T., Ren, K., 2010. Multiple functionalities of polyelectrolyte multilayer films: new biomedical applications. Adv. Mater. 22, 441–467.

Brammer, K.S., Choi, C., Frandsen, J., Oh, S., Jin, S., 2011. Hydrophobic nanopillars mesenchymal stem cell aggregation and osteo-differentiation. Acta Biomater. 7, 683–690.

Burdick, J.A., Vunjak-Novakovic, G., 2009. Engineered microenvironments for controlled stem cell differentiation. Tissue Eng. A 15, 205–219.

Caplan, A.I., 1991. Mesenchymal stem cells. J. Orthop. Res. 9, 641–650.

Chew, S.Y., Low, W.C., 2011. Scaffold-based approach to direct stem cell neural and cardiovascular differentiation: an analysis of physical and biochemical effects. J. Biomed. Mater. Res. A 29, 355–374.

Chitteti, B.R., Kacena, M.A., Voytik-Harbin, S.L., Srour, E.F., 2015. Modulation of hematopoietic progenitor cell fate in vitro by varying collagen oligomer matrix stiffness in the presence or absence of osteoblasts. J. Immunol. Methods 425, 108–113.

Chou, Y.F., Huang, W., Dunn, J.C., Miller, T.A., Wu, B.M., 2005. The effect of biomimetic apatite structure on osteoblast viability, proliferation, and gene expression. Biomaterials, 285–295.

Ciapetti, G., Granchi, D., Devescovi, V., Baglio, S.R., Leonardi, E., Martini, D., 2012. Enhancing osteoconduction of PLLA-based nanocomposite scaffolds for bone regeneration using different biomimetic signals to MSCs. Int. J. Mol. Sci. 13, 2439–2458.

Curtis, A., Wilkinson, C., 1997. Topographical control of cells. Biomaterials 18, 1573–1583.

Dargaville, B.L., Vaquette, C., Rasoul, F., Cooper-White, J.J., Campbell, J.H., Whittaker, A.K., 2013. Electrospinning and crosslinking of low-molecular-weight ppoly(trimethylene carbonate-co-(L)-lactide) as an elastomeric scaffold for vascular engineering. Acta Biomater. 9, 6885–6897.

De Gennes, P.G., Brochard-Wyart, F., Quéré, D., 2004. Capillary and wetting phenomena—drops, bubbles, pearls, waves. Springer Verlag, New York, NY.

Discher, D.E., Janmey, P., Wang, Y.-l., 2005. Tissue cells feel and respond to the stiffness of their substrate title. Science 310, 1139–1143.

Dominici, M., Le Blanc, K., Mueller, I., Slaper-Cortenbach, I., Marini, F., Krause, D., Deans, R., Keating, A., Prockop, Dj., Horwitz, E., 2006. Minimal criteria for defining multipotent mesenchymal stromal cells. The International Society for Cellular Therapy position statement. Cytotherapy 4, 315–317.

Dontu, G., Jackson, K.W., McNicholas, E., Kawamura, M.J., Abdallah, W.M., Wicha, M.S., 2004. Role of Notch signaling in cell-fate determination of human mammary stem/progenitor cells. Breast Cancer Res. 6, R605–R615.

Duan, B., Wang, M., Zhou, W.Y., Cheung, W.L., Li, Z.Y., Lu, W.W., 2010. Three-dimensional nanocomposite scaffolds fabricated via selective laser sintering for bone tissue engineering. Acta Biomater. 6, 4495–4505.

Engler, S.J., Sen, S., Sweeney, H.L., Discher, D.D., 2006. Matrix elasticity directs stem cell lineage specification. Cell 126, 677–689.

Espinosa, H.D., Rim, J.E., Barthelat, F., Buehler, M.J., 2009. Merger of structure and material in nacre and bone – perspectives on de novo biomimetic materials. Prog. Mater. Sci. 54, 1059–1100.

Faghihi, F., Eslaminejad, M., 2014. The effect of nano-scale topography on osteogenic differentiation of mesenchymal stem cells. Biomed. Pap. Med. Fac. Univ. Palacky Olomouc Czech Repub. 158, 5–16.

Faid, K., Voicu, R., Bani-Yaghoub, M., 2005. Rapid fabrication and chemical patterning of polymer microstructures and their applications as a platform for cell cultures. Biomed. Microdev. 7, 179–184.

Fan, D.M., Akkaraju, G.R., Couch, E.F., 2011. The role of nanostructured mesoporous silicon in discriminating in vitro calcification for electrospun composite tissue engineering scaffolds. Nanoscale 3, 354–361.
Fedorovich, N.E., Alblas, J., Hennink, W.E., Öner, F.C., Dhert, W.J.A., 2011. Organ printing: the future of bone regeneration? Trends Biotechnol. 29, 601–606.
Ferrari, G., Cusella-De Angelis, G., Coletta, M., Paolucci, E., Stornaiuolo, A., Cossu, G., Mavilio, F., 1998. Muscle regeneration by bone marrow-derived myogenic progenitors. Science 279, 1528–1530.
Fraser, J.K., Wulur, I., Alfonso, Z., Hedrick, M.H., 2006. Fat tissue: an underappreciated source of stem cells for biotechnology. Trends Biotechnol. 24, 150–154.
Friedenstein, A.J., 1990. Osteogenic stem cells in bone marrow. In: Heersche, J.N.M., Kanis, J.A. (Eds.), Bone and Mineral Research. Elsevier, Amsterdam, The Netherlands, pp. 243–272.
Gerecht, S., Vunjak-Novakovic, G., Langer, R., 2007. Engineering biomaterials for vascular differentiation and regeneration. Circulation 116, II235.
Griffin, M.F., Butler, P.E., Seifalian, A.M., Kalaskar, D.M., 2015. Control of stem cell fate by engineering their micro and nanoenvironment. World J. Stem Cells 7, 37–50.
Grotheer, V., Schulze, M., Tobiasch, E., 2015. Trends in Bone Tissue Engineering: Proteins for Osteogenic Differentiation and the Respective Scaffolding. Protein Purification—Principles and Trends. iConcept Press Ltd, Hong Kong.
Guerzoni, S., Deplaine, H., El Haskouri, J., 2014. Combination of silica nanoparticles with hydroxyapatite reinforces poly (L-lactide acid) scaffolds without loss of bioactivity. J. Bioact. Compat. Polym. 29, 15–31.
Han, Z., Rider, A., Ishaq, M., Kumar, S., Kondyurin, A., Bilek, M., Levchenko, I., Ostrikov, K., 2013. Carbon nanostructures for hard tissue engineering. RSC Adv. 3, 11058–11072.
Hashi, C.K., Zhu, Y., Yang, G.Y., Young, W.L., Hsiao, B.S., Wang, K., Chu, B., Li, S., 2007. Antithrombogenic property of bone marrow mesenchymal stem cells in nanofibrous vascular grafts. Proc. Natl. Acad. Sci. USA 104, 11915–11920.
Hassan, M.I., Sultana, N., Hamdan, S., 2014. Bioactivity assessment of poly((-caprolactone)/hydroxyapatite electrospun fibers for bone tissue engineering application. J. Nanomater. 2014, 1–6.
Hung, H.S., Yang, Y.C., Lin, Y.C., Lin, S.Z., Kao, W.C., Hsieh, H.H., Chu, M.Y., Fu, R.H., Hsu, S.H., 2014. Regulation of human endothelial progenitor cell maturation by polyurethane nanocomposites. Biomaterials 35, 6810–6821.
Jensen, J., Christian, D., Kraft, E., Lysdahl, H., Foldager, C.B., Chen, M., Kristiansen, A.A., Rölfing, J.H.D., Bünger, C.E., 2015. Functionalization of polycaprolactone scaffolds with hyaluronic acid and β-TCP facilitates migration and osteogenic differentiation of human dental pulp stem cells in vitro. Tissue Eng. A 21, 729–740.
Jia, L., Prabhakaran, M.P., Qin, X., Ramakrishna, S., 2013. Stem cell differentiation on electrospun nanofibrous substrates for vascular tissue engineering. Mater. Sci. Eng. C 33, 4640–4650.
Jiang, Y., Jahagirdar, B.N., Reinhardt, R.L., Schwartz, R.E., Keene, C.D., Ortiz-Gonzalez, X.R., Reyes, M., Lenvik, T., Lund, T., Blackstad, M., Du, J., Aldrich, S., Lisberg, A., Low, W.C., Largaespada, D.A., Verfaillie, C.M., 2002a. Pluripotency of mesenchymal stem cells derived from adult marrow. Nature 418, 41–49.
Jiang, Y., Vaessen, B., Lenvik, T., Blackstad, M., Reyes, M., Verfaillie, C.M., 2002b. Multipotent progenitor cells can be isolated from postnatal murine bone marrow, muscle, and brain. Exp. Hematol. 30, 896–904.
Kaebisch, C., Schipper, D., Babczyk, P., Tobiasch, E., 2015. The role of purinergic receptors in stem cell differentiation. Comput. Struct. Biotechnol. J. 13, 75–84.
Kandasamy, M., Lehner, B., Kraus, S., Ramm Sander, P., Marschallinger, J., Rivera, F.J., Trümbach, D., Ueberham, U., Reitsamer, H.A., Strauss, O., Bogdahn, U., Couillard-Depres, S., Aigner, L., 2014. TGF-beta signalling in the adult neurogenic niche promotes stem cell quiescence as well as generation of new neurons. J. Cell. Mol. Med. 18, 1444–1459.
Kannan, R.Y., Salacinski, H.J., De Groot, J., Clatworthy, I., Bozec, L., Horton, M., Butler, P.E., Seifalian, A.M., 2006a. The antithrombogenic potential of a polyhedral oligomeric silsesquioxane (POSS) nanocomposite. Biomacromolecules 7, 215–223.

Kannan, R.Y., Salacinski, H.J., Sales, K.M., Butler, P.E., Seifalian, A.M., 2006b. The endothelialization of polyhedral oligomeric silsesquioxane nanocomposites: an in vitro study. Cell. Biochem. Biophys. 45, 129–136.

Kaplan, F.S., Hayes, W.C., Keaveny, T.M., Boskey, A., Einhorn, T.A., Iannotti, J.P., 1994. Biomaterials. In: Simon, S.P. (Ed.), Orthopedic Basic Science. American Academy of Orthopaedic Surgeons, Columbus, OH, pp. 127–145.

Kim, K., Fisher, J.P., 2007. Nanoparticle technology in bone tissue engineering. J. Drug Target 15, 241–252.

Kim, K., Dean, D., Lu, A., Mikos, A.G., Fisher, J.P., 2011. Early osteogenic signal expression of rat bone marrow stromal cells is influenced by both hydroxyapatite nanoparticles content and initial cell seeding density in biodegradable nanocomposites scaffolds. Acta Biomater. 7, 1249–1264.

Kimura, W., Sadek, H.A., 2012. The cardiac hypoxic niche: emerging role of hypoxic microenvironment in cardiac progenitors. Cardiovasc. Diagn. Ther. 2, 278–289.

Kleinman, H.K., Philp, D., Hoffman, M.P., 2003. Role of the extracellular matrix in morphogenesis. Curr. Opin. Biotechnol. 14, 526–532.

Kumar, V.A., Taylor, N.L., Shi, S., Wang, B.K., Jalan, A.A., Kang, M.K., Wickremasinghe, N.C., Hartgerink, J.D., 2015. Highly angiogenic peptide nanofibers. ACS Nano 9, 860–868.

Kwaśny, W., 2009. Predicting properties of PVD and CVD coatings based on fractal quantities describing their surface. J. Achiev. Mater. Manufact. Eng. 37, 125–192.

Lai, M., Cai, K., Hu, Y., Yang, X., Liu, Q., 2012. Regulation of the behaviors of mesenchymal stem cells by surface nanostructured titanium. Coll. Surf. B Biointerf. 97, 211–220.

Lane, S.W., William, D.A., Watt, F.M., 2014. Modulating the stem cell niche for tissue regeneration. Nat. Biotechnol. 32, 795–803.

Lenhert, S., Meier, M.B., Meyer, U., 2005. Osteoblast alignment, elongation and migration on grooved polystyrene surface patterned by Langmuir-Blodgett lithography. Biomaterials 26, 563–570.

Lim, Y.C., Johnson, J., Fei, Z.Z., 2011. Micropatterning and characterization of electrospun poly(epsilon-caprolactone)/gelatin nanofiber tissue scaffolds by femtosecond laser ablation for tissue engineering applications. Biotechnol. Bioeng. 108, 116–126.

Macchiarini, P., Jungebluth, P., Go, T., Asnaghi, M.A., Rees, L.E., Cogan, T.A., Dodson, A., Martorell, J., Bellini, S., Parnigotto, P.P., Dickinson, S.C., Hollander, A.P., Mantero, S., Conconi, M.T., Birchall, M.A., 2008. Clinical transplantation of a tissue-engineered airway. Lancet 372 (9655), 2023–2030.

Magalhães, J., Lebourg, M., Deplaine, H., Ribelles, J.L.G., Blanco, F.J., 2015. Effect of the physicochemical properties of pure or chitosan-coated poly(L-lactic acid) scaffolds on the chrodrogenic differentiation of mesenchymal stem cells from osteoarthritic patients. Tissue Eng. A 21, 716–728.

Marchenko, I., Yashchenok, A., German, S., 2010. Polyelectrolytes: influence on evaporative self-assembly of particles and assembly of multilayers. Polymers 2, 690–708.

Martin, G.R., 1981. Isolation of a pluripotent cell line from early mouse embryos cultured in medium conditioned by teratocarcinoma stem cells. Proc. Natl. Acad. Sci. USA 78, 7634–7638.

Martins-Junior, P.A., Alcantara, C.E., Resende, R.R., 2013. Carbon nanotubes: directions and perspectives in oral regenerative medicine. J. Dent. Res. 92, 575–583.

McAdams, T.A., Miller, W.M., Papoutsakis, E.T., 1997. Variations in culture pH affect the cloning efficiency and differentiation of progenitor cells in ex vivo haemopoiesis. Br. J. Haematol. 97, 889–895.

Messina, E., De Angelis, L., Frati, G., Morrone, S., Chimenti, S., Fiordaliso, F., Salio, M., Battaglia, M., Latronico, M.V., Coletta, M., Vivarelli, E., Frati, L., Cossu, G., Giacomello, A., 2004. Isolation and expansion of adult cardiac stem cells from human and murine heart. Circ. Res. 95 (9), 911–921.

Mima, Y., Fukumolo, S., Koyama, H., 2012. Enhancement of cell-based therapeutic angiogenesis using a novel type of injectable scaffolds of hydroxyapatite-polymer nanocomposite microspheres. PLoS One 7, e35199.

Moby, V., Labrude, P., Kadi, A., 2011. Polyelctrolyte multilayer film and human mesenchymal stem cells: An attractive alternative in vascular engineering approaches. J. Biomed. Mater. Res. 96A, 313–319.

Mooney, C.J., Hakimjavadi, R., Fitzpatrick, E., Kennedy, E., Walls, D., Morrow, D., Redmond, E.M., Cahill, P.A., 2015. Hedgehog and resident vascular stem cell fate. Stem Cell. Int. 2015, 1–16.

Murphy, C.M., O'Brien, F.J., Little, D.G., Schindeler, A., 2013. Cell-scaffold interactions in the bone tissue engineering triad. Eur. Cell. Mater. 26, 120–132.

Muscari, C., Giordano, E., Bonafè, F., Govoni, M., Pasini, A., Guarnieri, C., 2013. Priming adult stem cells by hypoxic pretreatments for applications in regenerative medicine. J. Biomed. Sci. 20, 63.

Mylotte, L.A., Duffy, A.M., Murphy, M., O'Brien, T., Samali, A., Barry, F., Szegezdi, E., 2008. Metabolic flexibility permits mesenchymal stem cell survival in an ischemic environment. Stem Cells 5, 1325–1336.

Namgung, S., Kim, T., Baik, K.Y., Lee, M., Nam, J.-M., Hong, S., 2011. Fibronectin–carbon-nanotube hybrid nanostructures for controlled cell growth. Small 7, 56–61.

Nasibulin, A.G., Anisimov, A.S., Pikhitsa, P.V., 2007. Investigations of NanoBud formation. Chem. Phys. Lett. 446, 109–114.

Nazemi, K., Moztarzadeh, F., Jalali, N., Asgari, S., Mozafari, M., 2014. Synthesis and characterization of poly(lactic-co-glycolic) acid nanoparticles-loaded chitosan/bioactive glass scaffolds as a localized delivery system in the bone defects. Biomed. Res. Int. 2014, 898930.

Notara, M., Shortt, A.J., Galatowicz, G., Calder, V., Daniels, J.T., 2010. IL6 and the human limbal stem cell niche: a mediator of epithelial-stromal interaction. Stem Cell Res. 3, 188–200.

Oliveira, J.M., Sousa, R.A., Malafaya, P.B., 2011. In vivo study of dendron-like nanoparticles for stem cells "tune-up": from nano to tissues. Nanomedicine 7, 914–924.

Park, S.W., Lee, D., Lee, H.R., Moon, H.-J., Lee, B.R., Ko, W.-K., Song, S.-J., Lee, S.J., Shin, K., Jang, W., Yi, J.-K., Im, S.G., Kwon, I.K., 2015. Generation of functionalized polymer nanolayer or implant surface via initiated chemical vapor deposition (iCVD). J. Colloid Interf. Sci. 439, 34–41.

Peloso, A., Dhal, A., Zambon, J.P., Li, P., Orlando, G., Atala, A., Soker, S., 2015. Current achievements and future perspectives in whole-organ bioengineering. Stem Cell. Res. Ther. 6, 107.

Peltola, S., Sanna, M., Melchels, F., 2008. A review of rapid prototyping techniques for tissue engineering purposes. Ann. Med. 40, 268–280.

Peng, H., Yin, Z., Liu, H., Chen, X., Feng, B., Yuan, H., Su, B., Ouyang, H., Zhang, Y., 2012. Electrospun biomimetic scaffold of hydroxyapatite/chitosan supports enhanced osteogenic differentiation of mMSCs. Nanotechnology 23, 485102.

Peran, M., Angel-Garcia, M., Lopez-Ruiz, E., 2013. How can nanotechnology help to repair the body? Advances in cardiac, skin, bone, cartilage and nerve. Tissue Regen. Mater. 6, 1333–1359.

Polini, A., Bai, H., Tomsia, A.P., 2013. Dental applications of nanostructured bioactive glass and its composites. Wiley Interdiscip. Rev. Nanomed. Nanobiotechnol. 5, 399–410.

Poursamar, S.A., Azami, M., Mozafari, M., 2011. Controllable synthesis and characterization of porous polyvinyl alcohol/hydroxyapatite nanocomposite scaffolds via an in situ colloidal technique. Colloid. Surf. B Biointerf. 84, 310–316.

Prockop, J.D., 1997. Marrow stromal cells as stem cells for nonhematopoietic tissues. Science 276 (5309), 71–74.

Rauh, J., Milan, F., Gunther, K.P., Stiehler, M., 2011. Bioreactor systems for bone tissue engineering. Tissue Eng. B Rev. 4, 263–280.

Ravi, S., Chaikof, E.L., 2010. Biomaterials for vascular tissue engineering. Regen. Med. 5, 107.

Ruiz, R., Sandstrom, R.L., Black, C.T., 2007. Induced orientational order in symmetric diblock copolymer thin films. Adv. Mater. 19, 587–591.

Sabatini, F., Petecchia, L., Tavian, M., Jodon de Villeroché, V., Rossi, G.A., Brouty-Boyé, D., 2005. Human bronchial fibroblasts exhibit a mesenchymal stem cell phenotype and multilineage differentiating potentialities. Lab. Invest. 85 (8), 962–971.

Scadden, D.T., 2006. The stem-cell niche as an entity of action. Nature 441, 1075–1079.

Schofield, R., 1978. The relationship between the spleen colony-forming cell and the haemopoietic stem cells. Blood Cells 4, 7–25.

Schulze, M., Tobiasch, E., 2012. Artificial scaffolds and mesenchymal stem cells for hard tissues. In: Scheper, T. (Ed.), Tissue Engineering III: Cell - Surface Interactions for Tissue Culture. Springer Verlag, Berlin Heidelberg, Germany.

Seidlits, S.K., Lee, J.Y., Schmidt, C.E., 2008. Nanostructured scaffolds for neural applications. Nanomedicine 3, 183–199.

Seyednejad, H., Gawlitta, D., Dhert, W.J., 2011. Preparation and characterization of a three-dimensional printed scaffold based on a functionalized polyester for bone tissue engineering applications. Acta Biomater. 7, 1999–2006.

Shakesheff, K., Cannizzaro, S., Langer, R., 1998. Creating biomimetic microenvironments with synthetic polymer–peptide hybrid molecules. J. Biomater. Sci. Polym. Ed. 9 (5), 507–518.

Sharma, S., Srivastava, D., Grover, S., Sharma, V., 2014. Biomaterials in tooth tissue engineering: a review. J. Clin. Diagn. Res. 8, 309–315.

Sitharaman, B., Avti, P.K., Schaefer, K., 2011. A novel nanoparticle-enhanced photoacustic stimulus for bone tissue engineering. Tissue Eng. A 13–14, 1851–1858.

Song, S.W., Jeong, Y., Kwon, S., 2015. Photocurable polymer nanocomposites for magnetic, optical, and biological applications. IEEE Sel. Topics Quant. Electron. 21, 4800212.

Takahashi, K., Yamanaka, S., 2006. Induction of pluripotent stem cells from mouse embryonic and adult fibroblast cultures by defined factors. Cell 126, 663–676.

Tanentzapf, G., Devenport, D., Godt, D., Brown, N.H., 2007. Integrin-dependent anchoring of a stem-cell niche. Nat. Cell Biol. 9, 1413–1418.

Thomas, E.D., Lochte, Jr., H.L., Lu, W.C., Ferrbee, J.W., 1957. Intravenous infusion of bone marrow in patients receiving radiation and chemotherapy. N. Engl. J. Med. 257 (11), 491–496.

Thomson, J.A., Itskovitz-Eldor, J., Shapiro, S.S., Waknitz, M.A., Swiergiel, J.J., Marshall, V.S., Jones, J.M., 1998. Embryonic stem cell lines derived from human blastocysts. Science 282, 1145–1147.

Vallet-Regí, M., Salinas, A.J., 2015. Glasses in bone regeneration: a multiscale issue. J. Non-Crystall. Solids 432, 9–14.

Van Camp, J.K., Beckers, S., Zegers, D., Van Hul, W., 2014. Wnt signaling and the control of human stem cell fate. Stem Cell Rev. 10, 207–229.

Van Vlierberghe, S., Dubruel, P., Schacht, E., 2011. Biopolymer-based hydrogels as scaffolds for tissue engineering applications: a review. Biomacromolecules 12, 1387–1408.

Walmsley, G.G., McArdle, A., Tevlin, R., Momeni, A., Atashroo, D., Hu, M.S., Feroze, A.H., Wong, V.W., Lorenz, P.H., Longaker, M.T., Wan, D.C., 2015. Nanotechnology in bone tissue engineering. Nanomedicine 11 (5), 1253–1263.

Wang, Y., Hu, J., Jiao, J., Liu, Z., Zhou, Z., Zhao, C., Chang, L.J., Chen, Y.E., Ma, P.X., Yang, B., 2014. Engineering vascular tissue with functional smooth muscle cells derived from human iPS cells and nanofibrous scaffolds. Biomaterials 35, 8960–8969.

Wang, J., Yang, G., Wang, Y., Du, Y., Liu, H., Zhu, Y., Mao, C., Zhang, S., 2015. Chimeric protein template-induced shape control of bone mineral nanoparticles and its impact on mesenchymal stem cell fate. Biomacromolecules 16, 1987–1999.

Wei, G., Ma, P.X., 2008. Nanostructured biomaterials for regeneration, nano-scaled drug release systems incorporated into nanostructured biomaterials represents a novel and promising strategy to tissue regeneration. Adv. Funct. Mater. 18, 3568–3582.

Whitesides, G.M., Grzybowski, B., 2002. Self-assembly at all scales. Science 295, 2418–2421.

Whitesides, G.M., Ostuni, E., Takayama, S., 2001. Soft lithography in biology and biochemistry. Ann. Rev. Biomed. Eng. 3, 335–373.

Williams, J.M., Adewunmi, A., Schek, R., Flanagan, C.L., Krebsbach, P.H., Feinberg, S.E., Hollister, S.J., Das, S., 2005. Bone tissue engineering using polycaprolactone scaffolds fabricated via selective laser sintering. Biomaterials 26, 4817–4827.

Wu, C., Han, P., Liu, X., Xu, M., Tian, T., Chang, J., Xiao, Y., 2014. Mussel-inspired bioceramic with self-assembled Ca-P/polydopamine composite nanolayer: preparation, formation mechanics, improved cellular bioactivity and osteogenic differentiation of bone marrow stromal cells. Acta Biomater. 10, 428–438.

Xi, R., Xie, T., 2005. Stem cell-renewal controlled by chromatin remodelling factors. Science 310, 1487–1489.

Yamanaka, S., 2008. Induction of pluripotent stem cells from mouse fibroblasts by four transcription factors. Cell Prolif. 41 (Suppl 1), 51–56.

Zeng, X.B., Hu, H., Xie, L.Q., Lan, F., Jiang, W., Wu, Y., Gu, Z.W., 2012. Magnetic responsive hydroxyapatite composite scaffolds construction for bone defect reparation. Int. J. Nanomed. 7, 3365–3378.

Zhang, L., Webster, T.J., 2009. Nanotechnology and nanomaterials: promises for improved tissue regeneration. Nano Today 4, 66–80.

Zhang, Y., Lau, P., Pansky, A., Kassack, M., Hemmersbach, R., Tobiasch, E., 2014. The influence of simulated microgravity on purinergic signaling is different between individual culture and endothelial and smooth muscle cell coculture. Biomed. Res. Int. 2014, 1–11.

Zhong, J., Krawczyk, S.A., Chaerkady, R., Huang, H., Goel, R., Bader, J.S., Wong, G.W., Corkey, B.E., Pandey, A., 2010. Temporal profiling of the secretome during adipogenesis in humans. J. Proteome Res. 9, 5228–5238.

Zietz, C., Bergschmidt, P., Lange, R., Mittelmeier, W., Bader, R., 2013. Third-body abrasive wear of tibial polyethylene inserts combined with metallic and ceramic femoral components in a knee simulator study. Int. J. Artif. Organs 36, 47–55.

Zippel, N., Schulze, M., Tobiasch, E., 2010. Biomaterials and mesenchymal stem cells for regenerative medicine. Recent Pat. Biotechnol. 4, 1–22.

Zippel, N., Limbach, C.A., Ratajski, N., Urban, C., Luparello, C., Pansky, A., Kassack, M.U., Tobiasch, E., 2012. Purinergic receptors influence the differentiation of human mesenchymal stem cells. Stem Cell Dev. 21, 884–900.

CHAPTER

22

APPLICATIONS OF APTAMERS FOR THE DIAGNOSIS AND THERAPY OF DIFFERENT DISEASES

Sepideh Khoee, Somayeh Khezrian

University of Tehran, Tehran, Iran

CHAPTER OUTLINE

1 INTRODUCTION

More than 20 years ago Gold and Ellington announced independently a new class of targeting ligands, referred to as aptamers (Ellington and Szostak, 1990; Tuerk and Gold, 1990). As of June 2015, there were more than 5000 published articles in the PubMed database that included the term "aptamer"

(Sun and Zu, 2015). Aptamers are functional single-stranded DNA or RNA oligonucleotides, or peptides that can bind to target molecules with high affinity and specificity. The dissociation constant (K_d) of aptamers typically are within the picomolar to nanomolar range. The typical structural patterns in aptamers contain stems, internal loops, purine-rich bulges, hairpin structures, hairpins, pseudoknots, kissing complexes, and G-quadruplex structures (Feigon et al., 1996; Patel et al., 1997). DNA and RNA aptamer libraries typically contain 70–90 nucleotides in length, but the whole length of aptamers is not necessary for direct interaction with target, or for folding into the configuration to help target-binding. Owing to the nucleic acid nature of aptamers, they interact with complementary strands of DNA sequences, as well as their molecule targets, and can also be used in diagnostic applications or as antidotes in in vivo testing. By folding into well-defined secondary and tertiary structures and shapes, aptamers can recognize with high specificity and affinity to a wide array of targets with distinctive molecular binding motifs. In addition, aptamers have been integrated with biomolecular systems, and created optical, electrochemical, or other aptasensors. The word aptamer originates from the Latin word *aptus*, meaning "to fit," and the Greek word *meros*, meaning "region." In general, their high affinity and specificity aptamers have been referred to antibody-antigen assay to capture and screen target molecules. Structurally, aptamers functional groups are similar to antibodies; as such, they mimic and extend many properties of antibodies, consequently, the term "chemical antibodies" has been applied to these synthesized oligonucleotides (Fang and Tan, 2010). In comparison, aptamers reveal some characteristic features or advantages over antibodies, and they have attracted more attention to substitute antibodies in the development of many innovative bioapplications, such as biosensing, diagnostic, as well as biomedical and imaging assays. However, in spite of clinical therapy, aptamers are used in many other applications, including detection of bacterial and viral infections, detection of tumor biomarkers, detection of proteins in Western blot, as in vivo imaging agents, as target-specific ligands in liquid chromatography, surface plasmon resonance (SPR) assays, and biosensors (optical, electrochemical). Despite of antibodies which are generally produced by in vivo process, preparation of aptamers is easy and inexpensive, with high reproducibility by in vitro selection procedure, without concerning animal destruction or challenging the animal system to select the sites on the target proteins, as necessary for antibody production. Aptamers have less immunogenicity, greater specificity, and higher affinity to their target than antibodies. By folding into distinct tertiary structures, aptamers can recognize any given targets, including small chemical compounds to large multidomain proteins or polysaccharides, including organic dyes, amino acids, drugs, antibodies, as well as biological live cancer cells and tissues; however, antibodies only bind to larger molecules and are limited to immunogenic proteins. In many ways, aptamers are strong chemicals rather than being biological; furthermore, they have natural abilities to tolerate rigid conditions, including heat, pH, ion concentration, and organic solvents, so they can be denatured and renatured multiple times, without losing activity, because they can recover their natural conformation after chemical or thermal condition. The other advantages of aptamers in comparison to antibodies lie mainly in their rapid intracellular penetration, due to the smaller size and molecular weight of them (~1–2 nm, ~5–15 kDa), compared to antibodies (~10 nm, and ~155 kDa). Besides long-term stability, aptamers also possess many advantages, involving high surface density and low steric hindrance, unwanted immunogenicity, and a nontoxic and nonproteinaceous nature. Moreover, whereas site-specific modification of antibodies is difficult, the chemical nature of aptamers facilitates their simple modification of by different functional moieties (such as nanomaterials, optical, electrochemical reporter, radioisotopes, and small-interfering RNA—siRNA) at different sites (5′ or 3′ end or on the aptamer backbone, or on the bases), and can be used for clinical and diagnostic and therapeutic purposes in much the same way as monoclonal antibodies (Sun and Zu, 2016).

2 SELECTION OF APTAMERS

2.1 SYSTEMATIC EVOLUTION OF LIGANDS BY EXPONENTIAL ENRICHMENT

High-affinity nucleic acid ligands for a protein were isolated by a procedure that depends on alternate cycles of ligand selection from pools of variant sequences, and amplification of the bound species. Multiple rounds exponentially enrich the population for the highest affinity species that can be clonally isolated and characterized (Tuerk and Gold, 1990). The synthetic oligonucleotides can be manufactured and screened in parallel for a specific functionality, such as ligand-specific (aptamers) or catalysis-specific requirements (DNAzyme or aptazyme). The starting point of selection and purification of an aptamer is evolved with a primary library that is often referred to as "a pool" of random RNA or DNA nucleic acid sequences ($\sim 10^{13}$–10^{15} different individual sequences in a combinatorial library oligonucleotides, with 20–70 bp random sequences), using a simple procedure called "systematic evolution of ligands by exponential enrichment" (SELEX) (Ellington and Szostak, 1992; Tuerk and Gold, 1990). SELEX process has been used against various types of targets for aptamer selection including, inorganic and small organic molecules, proteins, peptides, carbohydrates, and antibiotics, as well as complex target mixtures, such as whole cells and organisms. SELEX technology is a main and extensively used method in pharmaceutical, molecular biology, and therapeutic research, which is mostly modified to select the aptamer with more efficient and less time consuming as well as with characteristic binding qualities. The basic principle of SELEX is in vitro selection, which can be traced back to 1990 (Tuerk and Gold, 1990), when Tuerk and Gold had been studying the interaction between the bacteriophage T4 DNA polymerase (gp43) and the ribosome binding site of the mRNA that encodes it. They developed a novel method for simply selecting preferred binding sequences from a population of random sequences, which they called "systematic evolution of ligands by exponential enrichment" (SELEX). Conceptually, a successful SELEX is needed to retain the specific binding sequences and release the nonspecific binding sequences during these processes. Preparation of a large pool of DNA sequences through SELEX procedure is easier, compared to other libraries, much the same aptamers like nucleic acids are composed of just four types of nucleotides, so it is easier to calculate the all variety of nucleic acid libraries. A nucleic acid pool that has "n" randomized possible situations owns 4^n diverse sequences. with any one of them having a unique structure. For example, a nucleic acid pool with 20 randomized positions contains 4^{20} different sequences. Because of the mutations happen during the amplification method, at the end the total number of selected species enhances. A typical SELEX process is shown schematically in Fig. 22.1.

In brief, the SELEX procedure is described by replication cycles and enzymatic amplification which includes five main continuous steps, involving:

1. *Binding*: mixing the library and target molecules.
2. *Partitioning (rinsing)*: separation of target-bound oligonucleotides from nonbinding ones by several stringent washing steps.
3. *Elution* of target-bound oligonucleotides.
4. *Amplification and mutation of the selected nucleic acids*: generation of a new enriched pool single stranded DNA aptamers by PCR (DNA SELEX), and single-stranded RNA aptamers from the double-stranded DNA PCR products, via in vitro transcription by RT-PCR (RNA SELEX).
5. *Separation*: conditioning and sequencing the new enriched pool of selected oligonucleotides after a number of rounds of selection.

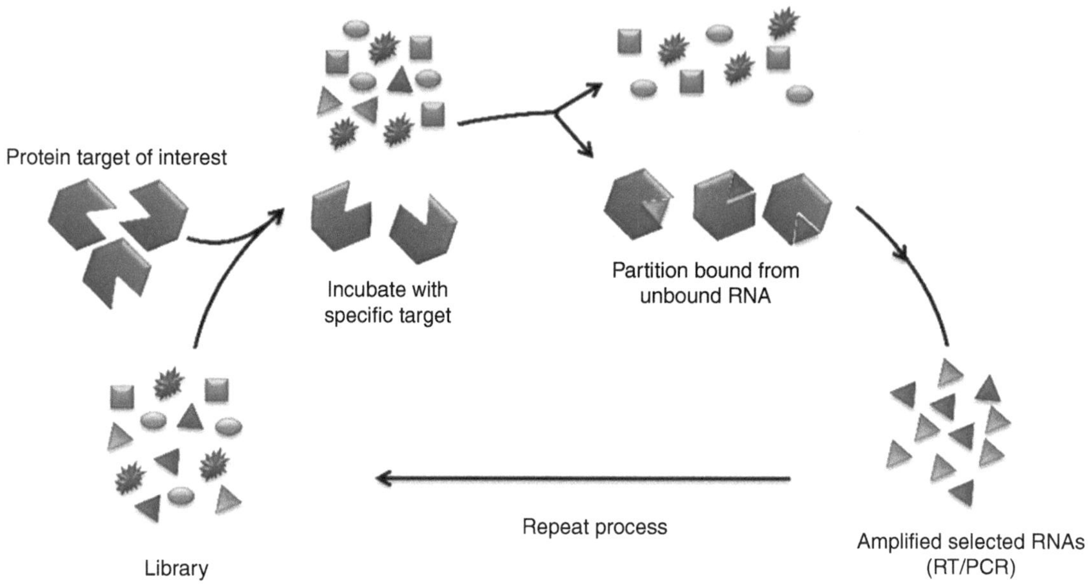

FIGURE 22.1 In Vitro Selection of RNA Aptamers by SELEX

Oligonucleotides of a randomized library are subjected to selection.

Numerous cycles of selection and amplification are needed in competition with each of the binding specific sequences of oligonucleotides and chosen aptamers with maximum affinity, released from a population. In general, the number of selection cycles needed preferentially to amplify the particular oligonucleotide or a set of specific oligonucleotides from 6 to 20 cycles is common. After the population has been separated to those sequences that bind the ligand with high affinity, the discrete sequences are cloned into bacteria using a standard procedure, and individuals are sequenced and characterized to isolate aptamers with the desired properties, including affinities and specificities. The whole process can last from several days to several months, but an average of 2–4 weeks is enough. Mutation and truncation experiments can be mainly performed in the amplification procedure to limit the minimum-binding region within the aptamer sequence, and so increase the total number of selected species. The full-length of aptamer generated in a typical SELEX procedure contains 75–120 nucleotides that include fixed sequence regions at both ends, separated by a variable domain of 25–60 nucleotides. Finally, most of the aptamers are exposed to some post-SELEX processes in order to improve the stability of the aptamers by modification of nucleotides, or to use the aptamers in analytical experiments by attachment of linker molecules.

2.2 CELL-BASED SELEX

The classic SELEX process is often highly time- and labor-consuming. Selection of aptamers with desired properties in this methodology is only done by using a single purified form of proteins as targets. Moreover, the conformation of aptamers selected through in vitro classic SELEX techniques are usually different from their conformation under in vivo selection, thus, this traditional SELEX is not permanently effective for in vivo applications or under the biological situations. Aptamers also give more precise information on diseases and have excellent clinical application for many different purposes in biomedical and biotechnological fields. In recent years the selection of aptamers from complex mixtures like red blood cell membranes and whole cells has been performed. A new procedure has been developed that aims at increasing the use of live cells as targets for in vitro selection of aptamer by a process termed cell-based SELEX or cell-SELEX (Shangguan et al., 2006). Cell-SELEX refers to the procedure whereby live cells are used to elect aptamers against target recognition. The length of randomized area for cell-SELEX is often in the limited area of 20–80 nt. The cell tissue membrane is a complex structure, and has multiple molecules on its surface. Cell-SELEX against the whole live cells upon treatment of a gene, or a virus particle, could select special aptamer sequences to recognize several molecules on the membrane of live cells. Consequently, many aptamers have been synthesized from the multiplication of biological cells or unknown molecules. Especially, cell-SELEX enables the targeting of a disease state without previous knowledge of molecular changes concerning the new condition, for example, the number and type of proteins on the cell membrane surface. Besides, in this method the cellular proteins keep their natural conformations. Cell-SELEX also determines the differences between two types of cells that are mentioned, including "normal versus tumor cells," as well as "different types of tumor cells or pathological cells at different stages and even among different patients." This current technology allows simple and cost-effective aptamer production through slight batch-to-batch variation, and with facility for site-specific modifications either during or after synthesis, aimed at improvement of chemical properties, especially for biological and therapeutic application. Based on these advantages, current cell-based SELEX is a successful procedure for multiple applications, such as the study of diagnostic and cancer therapy. Several aptamers were developed via cell-SELEX to improve, differentiate, and subclassify pathological or healthy cells, using unknown biomarkers as targets. A series of therapeutic aptamers are currently selected for particular illness therapies in clinical trials, including prostate-specific membrane antigen (PMSA) (Kavosi et al., 2014), allergy (Khezrian et al., 2013; Salimi et al., 2014), platelet-derived growth factor (PDGF) (Ye et al., 2016), liver cancer (Chen et al., 2016), acute lymphocytic leukemia (Taghdisi et al., 2010), gastric carcinoma cancer (Ding et al., 2015), malignant melanomas (Herrmann et al., 2014), tenascin-C (Li et al., 2013).

3 APTAMERS AS THERAPEUTICS

3.1 PEGAPTANIB

However, the use of aptamers as therapeutics was approved in 1990, and, in December 2004, the United States Food and Drug Administration (US FDA) was able to successfully approve the first aptamer as therapeutic agent in humans, under the name of Macugen (Gragoudas et al., 2004). Pegaptanib was first discovered in a 2′-ribo purine/2′-fluoro pyrimidine transcript library by SELEX. Pegaptanib sodium (Macugen) is an antivascular endothelial growth factor (anti-VEGF) RNA aptamer for the treatment of all types of neovascular age-related macular degeneration (AMD). Neovascularization is caused by growth of the concentration of vascular endothelial growth factor (VEGF). Macugen can be used as two major agents, anti-VEGF agents and an anticancer agent (Fig. 22.2A).

3.2 AS1411

AS1411, formerly AGRO001, is the first-in-human anticancer aptamer, and the first-in-class drug, and also is the first G-quadruplex DNA aptamer (G-quadruplex is a particular G-rich phosphodiester oligonucleotide that consists in four guanine by guanine Hoogsteen hydrogen bonds) (Fig. 22.3). The target-binding

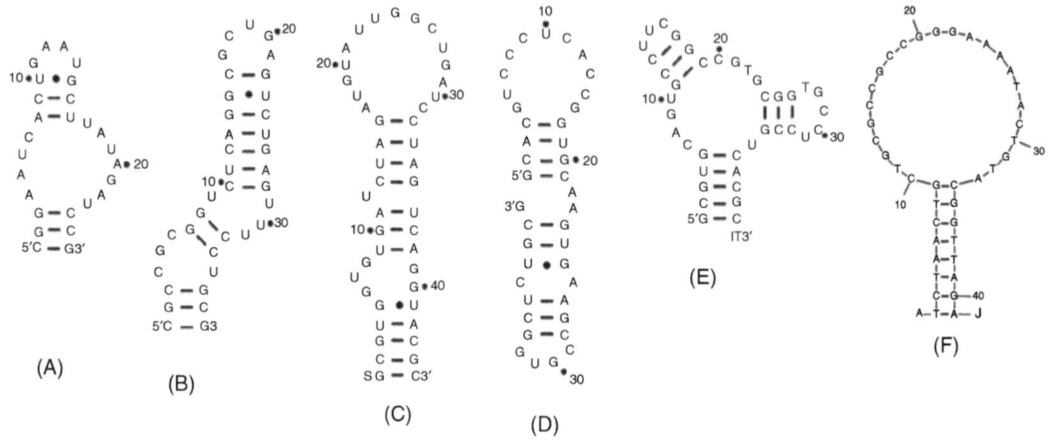

FIGURE 22.2

Secondary structure of Pegaptanib (A), ARC1905 (B), NOX-A12 (C), NOX-E36 (D), ARC1779 (E), and Sgc8c (F) (Padma Sundaram et al., 2013).

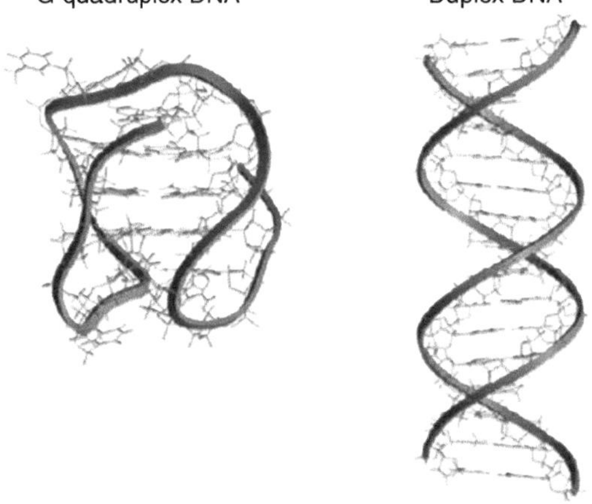

FIGURE 22.3 Structures of Quadruplex and Double-Stranded (Duplex) DNA

typically results in a change of conformation of G-quadruplex structures, so G-quadruplexes have more compressed structures that possess distinct binding sites for small molecules.

More recently, AS1411 has been tested in more than 80 human cell lines, and it was discovered that AS1411 had antiproliferative activity in almost every cancer cell types (lung cancer, breast cancer, prostate cancer, colon cancer, pancreatic cancer, renal cell carcinoma, ovarian cancer, cervical cancer, leukemia and lymphoma, melanoma glioblastoma, neuroblastoma, sarcoma, and gastric cancer), holding multiple therapeutic possibilities. AS1411 is a guanosine-rich 26-mer sequence (5′-GGTGGTGGTGGTTGTG-GTGGTGGTGG-3′) that can induce apoptosis, and thus inhibit the growth of malignant cells, is now under phase two trials for the cure of renal cancer and acute myeloid leukemia (Agrawal, 2015). AS1411 inhibits activation of nuclear factor-kappaB (NF-κB) by creating a complex with NF-κB essential modulator (NEMO) and nucleolin. Nucleolin, as the cancer-associated protein, is a versatile protein that migrates at ~110 kDa and placed first in the nucleolus and also present on the cytoplasm and cell membrane. AS1411 specifically binds to the external domain of the membrane protein nucleolin on malignant cells, cancer cells have high levels of surface nucleolin, compared to normal cells, causing pleiotropic antiproliferative effects because it interferes with some of the normal functions of nucleolin, which plays a role in many cancer-associated pathways (Bates et al., 2009).

3.3 ARC1905

ARC1905 (Ophthotech Corporation) is a 39 nucleotide, PEGylated, stabilized aptamer capped at the 3′-terminus with an inverted nucleotide to reduce 3′-exonuclease-mediated degradation and to reduce the renal filtration. ARC1905 as a therapeutic aptamer selectively binds to complement component 5 (C5) for inhibiting activation of C5-mediated inflammatory and the membrane attack complex (MAC) activities, and can achieve in both dry and wet age-related macular degeneration (AMD) while sparing the immunoprotective functions of the complement system (Fig. 22.2B).

3.4 NOX-A12

NOX-A12 is a spiegelmer aptamer that is a 45 nucleotide-long linked to 40 kDa PEG. Spiegelmers are based on DNA or RNA that are intended to bind and inhibit pharmacologically related target molecules. Compared to aptamers synthesized by naturally occurring chemistry (D-stereoisomers), spiegelmers are built on a backbone of "mirror-image" RNA or DNA oligonucleotides-L-stereoisomers (Keefe et al., 2010). They are biostable as regards the use of the nonnatural L-nucleotides in spiegelmers and provide a combine the benefits of small molecule drugs and biopharmaceuticals. Like aptamers, spiegelmers bind to their targets with high affinity and selectivity, but they are resistant against enzymatic degradation by nucleases. And thereupon, these properties make spiegelmers remarkable substances for drug development. NOX-A12 is being developed for use with autologous hematopoietic stem cell transplants and has therapeutic abilities to multiple myelomas and non-Hodgkin's lymphoma (Fig. 22.2C)

3.5 NOX-E36

NOX-E36 (NOXXON Pharma) is also a Spiegelmer aptamer that binds and neutralizes the human chemokine CCL2 (C-C chemokine ligand 2 is a small cytokine that belongs to the CC chemokine family), also binds with high affinity to the proinflammatory chemokine the Monocyte Chemoattractant Protein 1 [MCP-1 (CCL2)]. NOX-E36 is a 40 nucleotide-long L-RNA linked to 40 kDa PEG. This aptamer has been developed for the treatment of complications arising from type II diabetes mellitus, nephropathy, and lupus nephritis (Fig. 22.2D). Based on preclinical stage work it is believed that the neutralization of this chemokine will prevent infiltration of proinflammatory cells into the kidney, thereby allowing existing inflammation to resolve over time. The expected downstream effects include preservation of podocyte numbers, as well as conservation of renal structure and function.

3.6 ARC1779

ARC1779 is a therapeutic aptamer, which binds to the A1 domain of von Willebrand factor (vWF, a blood glycoprotein involved in hemostasis), and inhibits its interaction with the glycoprotein Ib receptor

(GPIb) on platelet membranes, and potentially blocking pathological thrombosis. ARC1779 is an ideal candidate for patients with acute myocardial infarction (AMI) or thrombotic microangiopathy. vWF is a large multimeric glycoprotein that is a carrier protein for one of the clotting factors. As a systemic prothrombotic condition, AMI is evidenced by increased the vWF, in the elderly (Fig. 22.2E)

3.7 SGC8C

Sgc8c aptamer, 5′-ATC TAA CTG CTG CGC CGC CGG GAA AAT ACT GTA CGG TTA GA-3′ has only a single loop–stem structure. Sgc8c is selected by cell-SELEX and has been confirmed to be a specific aptamer for the endosomes of CCRF-CEM (human T cell lymphoblast-like cell line) cells. Sgc8c aptamer targets leukemia biomarker protein tyrosine kinase-7 (Fig. 22.2F)

4 CHALLENGES OF APTAMERS

Aptamers are isolated and chemically synthesized in vitro against any given molecule targets, without needing cell lines or concerning animal destruction. Aptamers, because of their unprecedented properties as molecular probes with high specificity and selectivity, are successful candidates for usage in targeted therapy when their targets possess special biological activities in disease development. Moreover, aptamer therapeutics have been performed by intravenous (IV), subcutaneous (SC), or intravitreal (IVT) methods. Of course, aptamers are not without disadvantages (Bouchard et al., 2010). Despite the relatively easy selection of aptamers for therapeutic applications, aptamers have some intrinsic disadvantages for in vivo applicability that put the development of aptamer therapeutics at risk. Depending on the therapeutic and in vivo use of the aptamer, most targets are cell-surface proteins present in the blood or interstitial fluid compartments, such as the surface of the vasculature. For most aptamers, it is sensible to remain the aptamer in blood or interstitial fluid compartments for a long time, but in these media, aptamers are severely subject to nuclease-mediated degradation, fast renal filtration, and rapid biodistribution from the plasma compartment into the tissues. These problems have been largely overcome and under these limitations, the aptamers could not really play a more robust and efficient act for in vivo systems.

4.1 NUCLEASE RESISTANCE

A wide type of DNA and RNA aptamers are readily produced using SELEX. The aptamers compositions as nucleic acids are susceptible to degradation by the nucleases present in biological fluids, to a lesser amount physical and chemical degradation. Regrettably, this issue can be more and more notable for RNA, compared with DNA aptamers because RNA molecules are much less resistant against hydrolysis in blood or cells, so the half-lives of many RNA molecules in blood are just a few minutes or less, while the half-lives of DNA are of up to hours (Griffin et al., 1993). Accordingly, it is important to know that in vitro selection technologies have been performed with RNA libraries, but also carried out with DNA libraries. Generally, unmodified aptamers have metabolic instability in biological fluids (blood and tissue) therefore nonspecific delivery occurs at the surface of aptamers due to the short period of action. To overcome these limitations, the sequence of aptamers and other therapeutic oligonucleotide types must be chemically modified in order to increase nuclease-resistance (thus resist exo- and endonuclease degradation), target affinity and specificity for in vivo applications, including therapeutics, molecular imaging, or drug delivery. Not all chemical modifications can recommend the SELEX technology, thus these modified nucleotides cannot be applied to each section of SELEX process. Indeed, additional improvements can be achieved for the selection of the initial library randomized pool or for amplification of the ultimate aptamer in SELEX process. The situations that stay unmodified at the end of the in vitro selection method can be modified by post-SELEX for subsequent optimization of the aptamers. Therefore, modification of the pools should be suitable for the molecular biology steps to be performed in the SELEX procedure, so this approach depends on DNA or RNA polymerases that can allow modified nucleotide triphosphates. For example, the mutant T7 RNA polymerase can modify 2'-fluoro-pyrimidine and 2'-amino-pyrimidine nucleotide triphosphates at aptamer, either before or after SELEX process. As mentioned earlier, the "modified aptamers" are meant as a different kind of modifications to evaluate interactions between aptamers and their molecule targets, easy detection, resistance, and stability. The resistance of the aptamer to nuclease degradation can be introduced by systematic chemical modification of backbone or side chains, combining artificial nucleotides, and capping aptamer ends, thus allowing oligos to reach their intracellular targets. Moreover, application of these modifications will prepare various extra advantages, including increasing aptamer affinity, enhanced library diversity, and improving the result rate of SELEX. Modified nucleotide production by introducing the functional groups must follow two purposes, for example, the modified sequences must be more chemically varied like proteins, and must stay simple amplifiers and mutagens, like nucleotides. Structurally, it is the chemical substitution of 2′-OH or 2′-H group from RNA or DNA moieties with different chemical groups (e.g., 2′-O-methyl ribose purines and pyrimidines, 2′-fluoro pyrimidines and 2′-amino pyrimidines), modifying the phosphate (e.g., phosphorothioate, phosphoramidate, morpholino), modifying the nucleoside (e.g., α anomer, bicyclic sugar), modifying the 3′-end-capping, and unexpectedly the use of polyamide backbone (peptide nucleic acid) that are the most usual strategies to improve resistance of these molecules to enzymes, such as nucleases. Some techniques perform both chemical modifications simultaneously, consisting of phosphorodithioate (PS2) and 2′-O-methyl (2′-OMe) on one nucleotide (Wu et al., 2014). An effective system for selection of nuclease-resistant aptamers is spiegelmer knowledge. Spiegelmers (Bilik et al., 2007) are biologically stable mirror-image oligonucleotide ligands in which D-ribose replaces L-ribose, in other words, all of the sugars are the mirror image of natural nucleic acids sugars. As expected, L-enantiomeric RNA or DNA ligands have shown good resistance against enzymatic degradation by nucleases.

4.2 FAST RENAL FILTRATION

Aptamers, as well as all oligonucleotide classes, due to their small size and low molecular weight, are subject to degradation by nuclease and, as such, renal clearance, without any uptake by the liver and spleen, as reticuloendothelial system (RES) organs. To resolve these limitations, oligonucleotide sequences should be distributed with the blood stream into RES organs, or the size of the aptamer should be enhanced by conjugating them with chemical functions, for example, cholesterol, polyethylene glycol (PEG) polymer, 2′-O-methyl nucleotides, and nanomaterials. For example, conjugation of aptamers with high molecular weight polymers, such as PEG, can enhance its in vivo biocompatibility and can also greatly increase the blood cyclic lifetime by at least one day by prolonged tissue residence. One example of a polyethylene glycol-containing aptamer is pegaptanib sodium (Macugen) which has been modified with 40 kDa PEG to reduce degradation by nuclease and rapid renal clearance; Macugen contains 2′-hydroxy, 2′-fluoro, and 2′-O-methyl nucleotides.

4.3 MASS PRODUCTION

The other main challenge in selection of therapeutic aptamers is to join the economical manufacture procedure and also confirming the quality of the end-product by taking the necessary parameters including raw material availability, production yield, cost, purity, reproducibility, confirmation, and consistency during scaling up the process. Purification of the raw material (DNA and RNA phosphoramidites) usually produced in reactors by large-scale column chromatography and also the liquid reagent used in each step of the solid phase synthesis cycle and each step of the solid phase synthesis cycle depends on the liquid reagent. Another critical hurdle for therapeutic oligonucleotides is to apply the modifications throughout the preparation process to improve the productive quality and yield without side products. As mention formerly, the postsynthesis conjugation can be applied whenever the modification group is not amenable to or is unstable under the deprotection conditions. Functional groups including thiol-, amine-, and aldehyde can be used for the modification of oligonucleotides. For this conjugation, pericyclic reactions grow in mild conditions, with high yield and no side products (McKenna, 2012).

5 NANOTECHNOLOGY AND APTAMERS

Nanotechnology is the science that produces a wide range of new materials with a size of 1–100 nm range in one, two, or three dimensions. Furthermore, nanotechnology deals with the processes that happen at molecular system or nanoscale sizes. In recent years, nanotechnology has provided an enormous application in bioanalysis and biomedicine. Pharmaceutical nanotechnology has focused on the application of nanoscience in pharmaceutical domains in the form of nanomaterials, such as magnetic nanoparticles (iron oxide), gold, and silver nanoparticles (AgNPs), or as nanodevices for drug delivery, diagnostic on a molecular level, imaging and bioimaging, therapies, active implants, biosensors. Nanoparticles (NPs) are currently used in many medical applications and due to their excellent potential advantages; many more of them have been proposed for patient and medical services. The conjugation of NPs with receptor molecules on the cell surface can increase the uptake of NPs in different types of cell lines. For this purpose, various receptor molecules, such as antibodies, peptides, and DNA/RNA aptamers have been used to conjugate with NPs (Lu and Liu, 2006). Potential advantages of nanomaterials are well accepted in science, and some interpreters argue that nanotechnology commitments exceed the influence of the Industrial Revolution, planning to become a $1 trillion market by 2015 (Jain et al., 2012). Because of the unique characteristics of aptamers in conjugation with other receptor molecules, various sensing biomedicine systems have been improved based on the conjugation of various aptamers with novel nanomaterials. As a result, these combinations improve the inherent properties of aptamers, including nuclease digestion, conformational flexibility, connection characterization, and aptamer density on nanomaterial surface. Meanwhile, the aptamer-conjugated NPs have led to advances in fine-tuned diagnostics focused on detecting the antigen associated with diseases (e.g., diabetes mellitus, infectious, neurodegenerative disease, cancer), as well as detecting the microorganisms and viruses associated with infections, and also promotion in drug delivery and imaging in vivo. This study domain has already seen remarkable promotion in recent years. We will mainly focus on recent progress in the development of aptamer-based nanomaterials for cancer therapy and imaging. In this specific field, the progress has been remarkable.

5.1 APTAMER-ASSEMBLED METALLIC NANOMATERIALS

For centuries, metal NPs have attracted the attention of scientists due to their unique properties including ultrafine size and good dispersion, high stability, the unique electric, magnetic, optical, and catalytic properties, nontoxic and biocompatibility, antibacterial nature, abundant surface chemistry, and easy surface functionalization. Meanwhile, metal NPs (e.g., gold, silver, and copper nanoparticles), due to their attractive size- and shape-dependent optoelectronic and physicochemical properties, exhibit large nonlinear optical effects, nonlinear electronic properties and rapid response time, according to their surface plasmon resonance (SPR) scattering (Thakkar et al., 2010). The intensity of light scattering is based on the size of the particle. SPR allows the molecule specified imaging and diagnosis of numerous diseases for instance cancer. Among metal NPs, gold NPs (AuNPs) are more popular. The ranges of sizes in AuNPs are from 1 nm to several hundred nanometers. The key factors for the sustainable development of the AuNPs are strong light-scattering properties, oxidation stability, good biocompatibility, high intracellular resistance, and high DNA-loading capacity. The interaction of biomolecules with AuNPs can be carried out by directly adsorption or through the thiol–gold interaction. As well as, functionalization of AuNPs with alkyl thiols terminated with —OH, —COOH, or —NH_2 group in order to improve their performance is the other way to connect the biomolecules to the surface of AuNPs upon activation of functional groups (Mariani et al., 2015). These advantages make AuNPs an effective tool for highly sensitive molecular detection.

Many targeted metallic NPs modified with aptamers have been reported that can be accumulated on the surface of a particular kind of cancer cell via the recognition of the aptamer by its target on the surface of cell. For example, Kim and coworkers have reported aptamer functionalized AuNPs to inhibit discoidin domain receptors 2 (DDR2)-mediated carcinogenesis that identified as a novel therapeutic target in lung cancer. They used an AuNP-Apt-based protein delivery system for transporting peptides, containing the hexahistidine (His)- or glutathione S-transferase (GST)-tagged transmembrane juxtamembrane 1/2 (TM-JM1/2), into a lung cancer cell line. By interaction of AuNP-Apt based TM-JM1/2 peptides with DDR2, the activation of DDR2 led to decrease in cell proliferation and invasion (Fig. 22.4). This innovative aptamer-based AuNP system allows the effective, local, and targeted delivery of functional proteins for both in vitro and in vivo approaches. Therefore, this system can be used for the loading and delivery of proteins for the development of protein-based biomedical applications, including recombinant protein therapeutics and bioimaging technologies (Kim et al., 2015; Ryou et al., 2014).

By using an anticancer aptamer, Malik and coworkers developed AS1411 (a novel nucleolin-targeted DNA aptamer)-linked gold nanospheres (AS1411-GNS) for breast cancer therapy (Malik et al., 2015). In this study, GNS were first functionalized with AS1411 aptamer and then delivered to target tumor cells. The result showed the same activity for AS1411—5-nm gold nanospheres for in vitro and in vivo studies. Compared to unconjugated AS1411 or GNS linked to control oligonucleotides, AS1411-GNS had superior cellular uptake and markedly increased antiproliferative/cytotoxic effects. Moreover, in a mouse model of breast cancer, systemic administration of AS1411-GNS could completely inhibit tumor growth with no signs of toxicity. The reported results establish the potential of AS1411-GNS as a future nanomedicine and highlight the need for additional research—especially for in vivo studies—to facilitate translation to the clinic.

In another example, Danesh and coworkers have reported sgc8 aptamer-wrapped AuNPs for targeting and controlled release delivery of daunorubicin (Dau) to the Molt-4 cells (human acute lymphoblastic leukemia T-cell, target). Dau was released from Apt–Dau–AuNPs complex in a pH-dependent

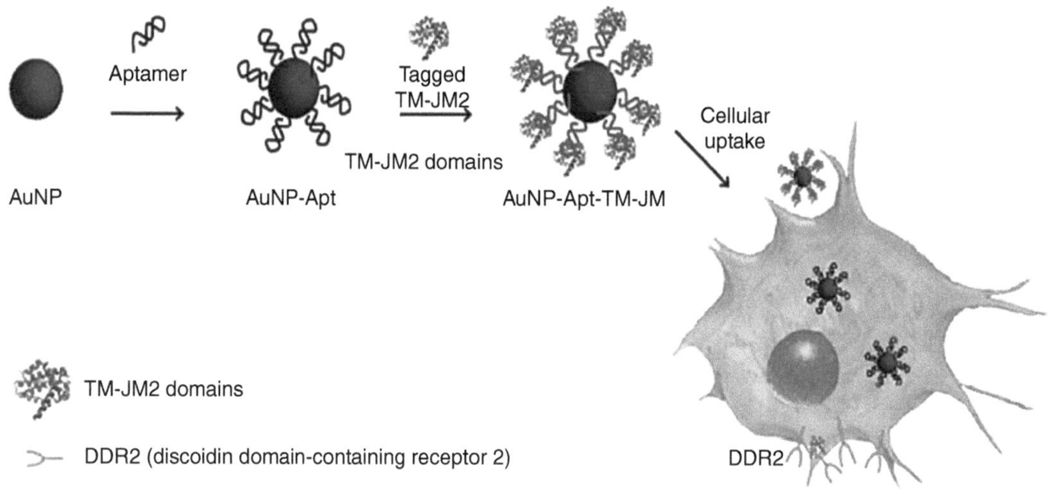

FIGURE 22.4 Schematic Diagram of the Generation of TM-JM1/2 Peptides, Including the Aptamer-Conjugated AuNP Delivery System (Kim et al., 2015)

manner, and the rate of drug release increased about fourfold at pH 5.5, in comparison to the pH 7.4. The results of flow cytometry analysis and confocal imaging showed that the complex was effectively internalized into Molt-4 cells, but not into U266 cells (B lymphocyte human myeloma, nontarget). Therefore, this system could reduce cytotoxic effects of Dau by targeted delivery and controlled release of this drug to tumor cells. Moreover, this complex has characteristics, which make it an applicable and ideal delivery system for clinical use (Danesh et al., 2015).

Recent studies have found that fluorescence spectroscopy is an effective technique to detect biomolecules, and it is also widely used for pharmaceutical and biological applications due to its high temporal and spatial resolution. AuNPs, AgNPs, and quantum dots (QDs) have also been known to use in different kinds of fluorescent assays to detect cancer.

Li and coworkers developed Sgc8 linked Fluorescein isothiocyanate (FAM) aptamer-functionalized AgNPs for apoptosis-based cancer therapy and fluorescence-enhanced cell imaging (Li et al., 2015a). The surface of AgNPs was coated with Sgc8 or TDO5 aptamers and then Ag-Sgc8, Ag-TDO5 bind to CCRF-CEM (CCL-119, T-cell line, human acute lymphoblastic leukemia) and Ramos cells (CRL-1596, B-cell line, human Burkitt's lymphoma) particularly. However, Ag-Sgc8 and Ag-TDO5 accomplished the specific apoptosis of target cells. As shown in Fig. 22.5, based on electrostatic adherence principle, Ag-Sgc8 could enter into Ramos cells and induce vitro cytotoxicity on Ramos cells. As resulted the apoptosis of cells depended on the concentration of aptamer–silver conjugates, as well as the incubation time between cells and aptamer–silver conjugated. Thus, Ramos cells were more easily going into the apoptotic process in low concentration of Ag-TDO5 (0.37 nM); moreover, the apoptotic and necrotic cells population increased with the incubation time between Ag-TDO5 and Ramos cells.

In addition, star-shaped gold nanoparticles (AuNPs) exhibit several advantages including tunable plasmon bands, high two-photon action cross section and optical properties tunable in the near-infrared (NIR) region. AuNPs with their attractive properties have been exploited for the hyperthermic demolition of tumor cells and have been used as drugs in (PTT) photothermal therapy (Nehl et al., 2006).

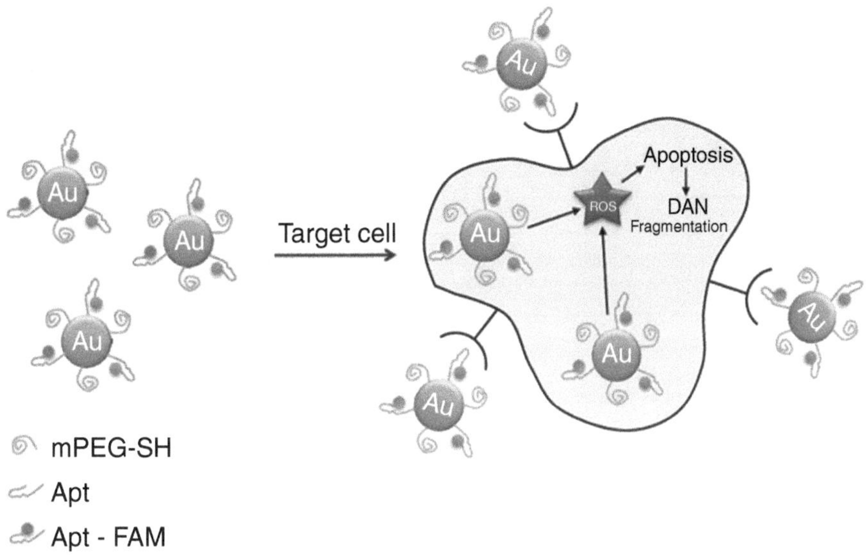

FIGURE 22.5 Aptamer-Silver Conjugates Induced Apoptosis for Specific Cancer Therapy and Fluorescence-Enhanced Cell Imaging

Compare to chemotherapy and radiation therapy, noninvasive photothermal therapy is more efficiently to destroy target cells and diminish unpleasant damage to surrounding cells. For designing a powerful photothermal probe that can selectively and effectively target both types of prostate cancer cells, Jo and coworkers have developed the first ultra-effective photothermal agent, namely, dual aptamer-modified gold nanostars, for targeted therapy of prostate cancers, namely PSMA(+) and PSMA(–) cells, without any impact on the cell viability (Jo et al., 2014). The most important aim of Jo and coworkers was overcome various problems, such as short half-life of the particles in the circulation, low permeability of the particles, and instability of aptamers in the blood.

As illustrated in Fig. 22.6, AuNPs were freshly synthesized by a seed-mediated method and then bare-AuNPs were swathed by mercaptopolyethylene glycol monomethyl ether (PEG-AuNS). Thiol-modified A10 and cysteine modified DUP-1 aptamers were introduced into PEG-AuNPs via disulfide bonds at a ratio of 1:2, which can give rise to discrepancies in the photothermolysis ability for PSMA(+) and PSMA(–) cells. Dual-AuNPs were found only in the cytoplasmic part and not in intranuclear region because the comparatively large Dual-AuNPs (~62 nm) were not unable to pass through the nuclear pores (~50 nm). The results clearly showed that Dual-AuNPs could simultaneously target PSMA(+) and PSMA(–) cells with great selectivity.

In addition, core–shell and alloyed nanomaterial (e.g., Au/Ag core shell NPs, and gold nanorods or GNRs) with higher optical, and electronic properties have been used in surface-enhanced Raman scattering (SERS) compared to pure AuNPs and AgNPs as SERS substrate. For example, Bamrungsap and coworkers have reported the use of dual mode nanotags for targeting and imaging of specific cancer markers expressed on the membrane of cervical cancer cells (Bamrungsap et al., 2015). The nanotags include of three principal compartments, including 4-aminothiophenol (4-ATP) labeled Au–Ag NRs designed for SERS signal generation, aptamers that act as the targeting ligands and fluorophore labeled

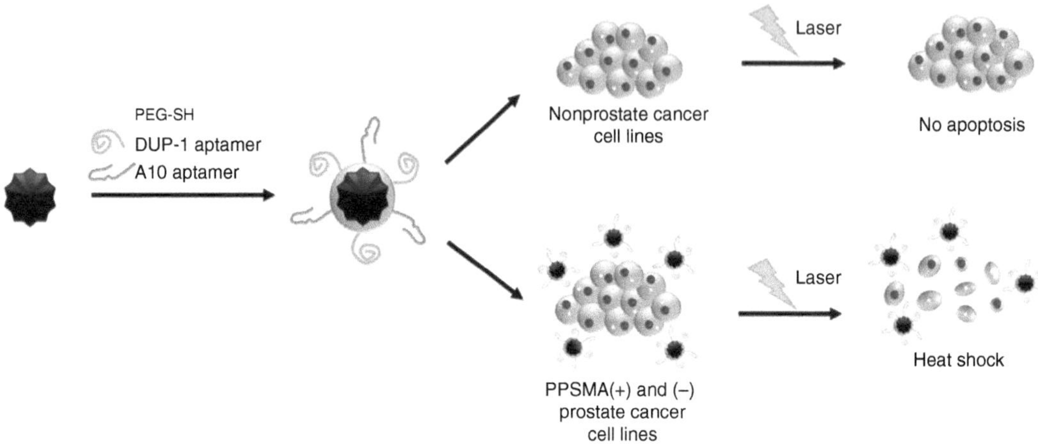

FIGURE 22.6 Functionalization of AuNPs was Immobilized and Selective Photothermolysis for Only PSMA(+) and PSMA(−) Prostate Cancer Cells by Dual-AuNPs

aptamer for fluorescence imaging. Therefore, nanotags suitable for surface enhanced Raman scattering (SERS) and fluorescence detection and imaging. In this study, nanotags were manufactured on Au–Ag NRs surface followed, by layer-by-layer process, using the conjugation of fluorescently-labeled aptamers specific to the target cancer cells to form clusters, and then inducing the coupling of electromagnetic fields. Likewise, strong detection signal derived from both high affinities of the aptamers conjugated NRs toward their target cells and the collective electromagnetic fields ascribed to the strong longitudinal surface plasmon oscillation of Au–Ag NRs coupled with the overlapping with the excitation wavelength (Fig. 22.7).

Fluorescent semiconductor nanocrystals (also known as Quantum dots or QDs) are light-emitting particles with a few nanometers in diameter whose size and shape can be precisely controlled by the duration, temperature, and ligand molecules used in the synthesis. Owing to the considerable advantages of QDs against customary fluorophores, including low toxicity, greater photostability, size-tunable photoluminescence spectra, wide absorption with narrow photoluminescence spectra, low photobleaching, high quantum yields, resistance to chemical degradation, and steady absorption spectra covering the ultraviolet to NIR ranges, they have been increasingly utilized as biological imaging and labeling probes (Michalet et al., 2005). Surface-modified QDs conjugated with aptamers, peptides, or small molecules bound to antigens present on target cells have been applied to study targeted imaging, therapy, and drug delivery in cancer. As well as, by modifying the QDs with aptamers, sensitive sensing methods have been resulted for simultaneous imaging and therapeutic applications and medical applications (Pandurangan and Mounika, 2012). For instance, by using two different aptamer-conjugated QDs, Lee and coworkers have been developed a sensitive method for intracellular localization or diagnosis of cancer biomarkers, and multiplex imaging of different cancer cell lines (Lee et al., 2015). In this study, two different QDs, QD605 and AD655, were conjugated to tenascin-C aptamer (TTA-1) and AS1411 aptamer, respectively. The formation of QD-aptamer conjugates was approved using transmission electron microscopy (TEM) imaging and electrophoretic mobility shift assay (EMSA). To appraise the cancer-targeting specificity of QD-TTA-1 and QD-AS1411, three cancer cell lines containing DU145

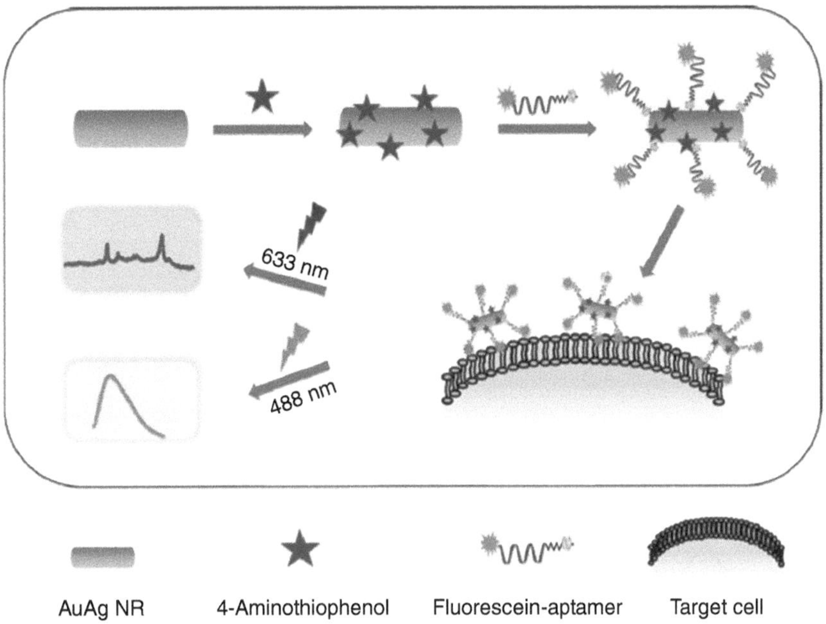

FIGURE 22.7 Schematic Image of the Dual-Mode Nanotag Preparation and Target Cell Detection (Bamrungsap et al., 2015)

cells (a human prostate cancer cell line), U-87 MG cells (a human glioblastoma-astrocytoma cell line), and A549 cells (a human lung carcinoma cell line) were treated and incubated with QD-TTA-1 and QD-AS1411. Fluorescence emission spectra showed that QD-TTA-1 and QD-AS1411 were stable in all three cancer cell lines. QD-TTA-1 exhibited a high level of fluorescence intensity in all three cancer cell lines but the fluorescence intensity of QD-AS1411 was highest in U-87 MG cells and lowest in A549 cells. These results indicated that there were different expression profiles of tenascin-C and nucleolin in different cancers.

As regards the near-infrared (NIR) region between 650 and 900 nm, recently the development of lesser toxic and more environmentally friendly of near-infrared emission quantum dots (NIR QDs) in biomedical imaging and detection has been attracted more attention. The combination of anticancer drugs to water-soluble NIR CuInS2 QDs can comfort higher drug payload, lesser systemic toxicity, prolong in vivo drug retention time, and improved drug solubility and targeting. Subsequently, Lin and coworkers have been reported a novel daunorubicin (DNR)-loaded MUC1 aptamer-near infrared (NIR) CuInS2 quantum dot (DNR–MUC1–QDs) conjugates which can image and sense the delivery of drugs to target prostate cancer cells (Lin et al., 2014). In this work, the MUC1 aptamer has been coated onto the surface of CuInS QDs with the (CGA)7 to achieve a novel aptamer carrier. The effect of incorporation of numerous CG sequences within the stem of the MUC1 aptamer on the capacity of the aptamer to DNR loading and delivering has been evaluated. DNR is an anticancer drugs used to treat acute leukemias with the ability to inhibit the proliferation of cancer cells via intercalating into the DNA structure in cell nuclei. In this study, carboxyl CuInS2 QDs was activated in the presence of EDC and NHS and then the resulting *N*-hydroxy-succinimide-activated QDs solutions were added to

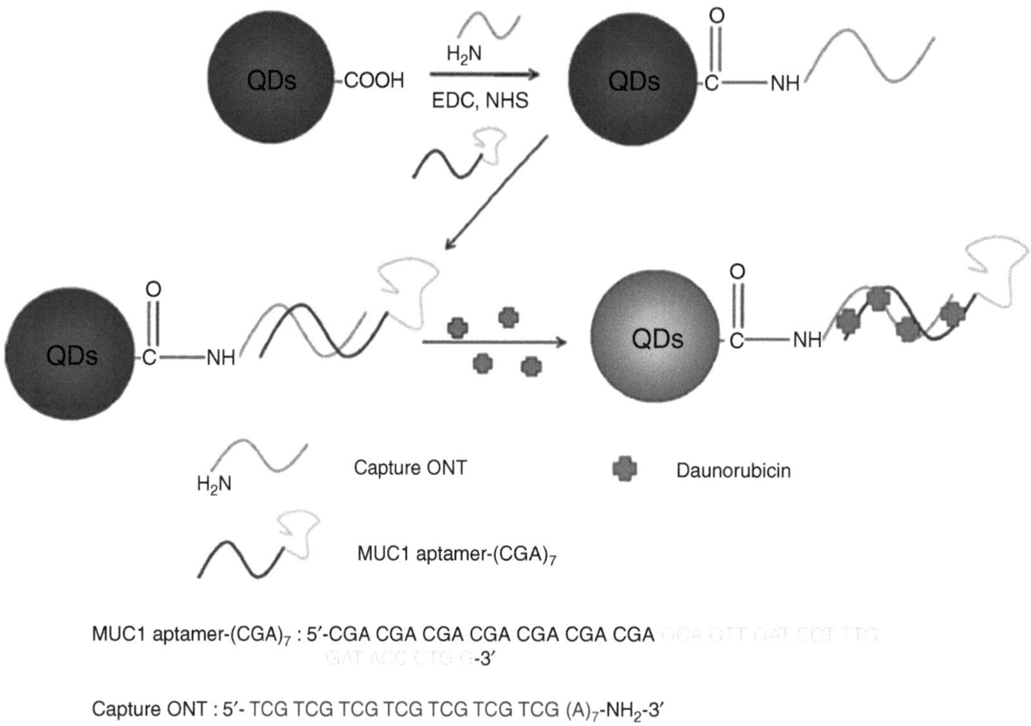

FIGURE 22.8 Schematic Illustration of the Method for Preparing DNR–MUC1–QDs (Li et al., 2014)

NH_2-modified capture ONTs. The MUC1 aptamer was hybridized to capture ONT-coated QDs. The amount of loaded DNR was determined by fluorescence emission of CuInS2 QDs at 655 nm, with excitation at 620 nm (Fig. 22.8). It was found that the drug loading of DNR on the MUC1–QDs system was more favorable in alkaline medium than in acidic and on the other hand, the release of the DNR under acid condition was more favorable. Herein, DNR acted as an anthraquinone derivative, so, compared with the PL intensity of QDs and MUC1–QDs conjugates, there was significant decrease in the PL intensity of the DNR–MUC1–QDs (Li et al., 2014).

Another important variety of multifunctional NPs is superparamagnetic iron oxide nanoparticles (SPIONs). Owing to their intrinsic superparamagnetism, SPIONs have been widely used for magnetic resonance imaging (MRI) contrast enhancement, magnetically induced hyperthermia, tissue-specific release of therapeutic agents, hyperthermia, and magnetic field assisted radionuclide therapy. Moreover, the surface of SPIONs can be easily functionalized with targeting ligands to improve the accumulation of therapeutics in their target region (Mahmoudi et al., 2011).

Wang and coworkers have reported an A10 aptamer conjugated thermally cross-linked superparamagnetic iron oxide nanoparticles (TCL-SPIONs), which can both detect prostate cancer (PCa), and deliver targeted chemotherapeutic agents directly to the PCa cells, without significant loss in cytotoxicity (Wang et al., 2008). In this study, they conjugated A10 aptamers to TCL-SPIONs, using standard coupling chemistry to give the TCL-SPION–Apt bioconjugate formulation which followed by loading

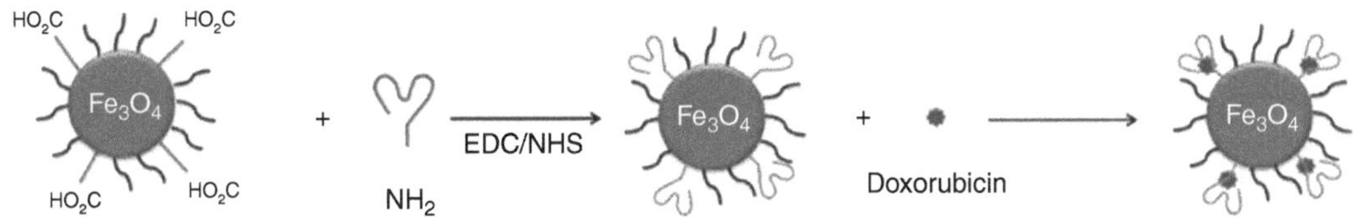

FIGURE 22.9 Schematic Illustration of the TCL-SPION–Apt Bioconjugate System

Dox to the conjugate via intercalation (Fig. 22.9). Incubation of TCL-SPION–Apt bioconjugates with non-PSMA expressing PCa cells (PC3 cells), and TCL-SPION with PSMA-expressing PCa cells (LNCaP) and PC3 cells, did not show intracellular uptake of the NPs. PSMA is a well-established marker for PCa cells and, under these conditions, normal prostate, kidney, brain, and small intestine tissue are expressed at relatively low levels. On the other hand, Wang et al. (2008) investigated the potential of TCL-SPION–Apt bioconjugates as a therapeutic carrier. The result of spectrofluorophotometer confirmed that TCL-SPION–Apt bioconjugates can be used as therapeutic carriers for the delivery of Dox, leading to selective delivery to PSMA-expressing cells, without significant loss in cytotoxicity. Consequently, TCL-SPION–Apt bioconjugates could differentially target PSMA-expressing PCa cells, and the cytotoxicity of Dox-loaded TCL-SPION–Apt bioconjugates was nearly as same as free Dox.

5.2 APTAMER-ASSEMBLED CARBON BASED NANOSTRUCTURES

Carbon-based nanostructures, such as carbon nanotubes, fullerene, graphene/graphene oxide, nanodiamond, and carbon NPs are novel manufactured materials which have been the focus of interest in a wide range of biological, electronic, optical, environmental, and medical research projects. The unique structure and considerable properties of these materials (such as electrical, optoelectronic, mechanophysical, chemical, and biocompatibility) are caused by their exceedingly small dimensions and large surface-to-volume ratio.

In the field of nanomedicine, carbon nanotubes (CNTs) and graphene (GN) are most widely used as versatile carriers to transport a variety of bioactive molecules containing drugs, proteins, siRNA, oligonucleotides, and aptamers through endocytosis. They can also act as biomarkers, a substance for the detection of cancers at their primary stage, with high selectivity and sensitivity. Aptamers are a special class of nucleic acid molecules with extra potential for carbon nanotube modification. The aptamer-conjugated CNTs may be formed via covalent and noncovalent interactions, and thus can stabilize the aptamers against nuclease-mediated degradation, and improve their solubility and binding affinity (Mehra et al., 2014).

Taghdisi and coworkers have reported sgc8 aptamer-wrapped SWNTs for reversible targeting and controlled release delivery of daunorubicin (Dau) to cancer cells (Taghdisi et al., 2011). The sgc8c DNA aptamer was used as targeting ligand to leukemia biomarker protein tyrosine kinase-7 (PTK7), and was recognized with high affinity ($k_d = 1$ nM) among normal human bone marrow aspirates. As SWCNTs have highly delocalized π electron-rich structure, their surface could easily be functionalized via π–π interaction. Sgc8c aptamer noncovalently binds to SWNTs. The π stacking interactions between the nucleotide bases and the sidewalls of SWNT played an important role in forming a stable complex. Dau-aptamer-SWNTs tertiary complex delivery system is able to specifically deliver and internalize Dau to Molt-4 cells. Dau was released from Dau-aptamer-SWNTs tertiary complex in a pH-dependent manner, and the rate of drug release increased about sixfold at pH 5.5. Flow cytometric analysis showed that the tertiary complex was internalized effectively to Molt-4 cells (target), but not to U266 cells (nontarget). Cytotoxicity of Dau-aptamer-SWNTs tertiary complex also confirmed internalization data. Dau-aptamer-SWNTs tertiary complex was less cytotoxic in U266 cells, when compared to Dau alone. No significant change in viability between Dau and complex-treated Molt-4 cells was observed. Moreover, application of antisense oligonucleotides against the aptamer could almost prevent delivery of Dau to Molt-4 cells. Therefore, this system could reverse targeting and controlled release of Dau to cancer cells while, at the same time, reducing the drug's cytotoxic effects. Metastases are the cause of 90% of cancer-related deaths. Metastasis occurs when tumor cells migrate from the primary tumor site and penetrate to blood vessels, recognized as circulating tumor cells (CTCs). Viraka Nellore and coworkers have used 3D graphene oxide foam as the macroscaffold for the attachment of aptamers to porous garphene oxide membranes. Their results show that aptamer-conjugated graphene oxide membranes not only could specifically capture SKBR3 breast cancer cells, LNCaP prostate cancer cells, and SW-948 colon cancer cells separately and simultaneously from infected blood, but could also accurately identify the captured cancer cells via multicolor fluorescence imaging (Viraka Nellore et al., 2015). In this study, a 2D graphene oxide sheet has been produced using a modified Hummers method, and was then modified with different types of aptamers via carboxylic acid groups of 2D graphene oxide. For accurate identification, Alexa Fluor 488 dye modified YJ-1 aptamer was used that can specifically capture SW-948 colon cancer cells from infected blood, and can be identified from blue fluorescence image. Cy2-modified S6 aptamer was applied to specifically bind SK-BRr3 breast cancer cells, and was identified from red color fluorescence image. Similarly, Cy5-modified A9 aptamer was for specific capturing of LNCaP prostate cancer cells from blood sample, and recognized by red color fluorescence image. 3D graphene oxide foam was developed from aptamer-attached 2D graphene oxide using PEG as a cross-linking agent. Based on the fact that graphene oxide is an ultra-high efficient quencher for fluorescent assay, when dye-conjugated aptamers were attached to graphene oxide, fluorescence from the dye is totally quenched by graphene oxide. In presence of targeted cancer cells, by interaction of the cancer cells with dye-conjugated aptamer, the distance between dye and graphene oxide increases and, as a result, fluorescence persists.

5.3 APTAMER-CONJUGATED POLYMERIC NANOPARTICLES

Another successful carrier system involves polymeric NPs, composed of biocompatible and biodegradable polymers with a significant therapeutic potential in various stages of preclinical and clinical development. The polymers utilized to prepare these NPs are either of natural or of synthetic origin. Drug targeting for effective accumulation in the body is often hampered by the rapid recognition of the carrier system by the RES, and by the following kidney and/or hepatic elimination. Meanwhile, for modulated drug delivery to solid tumors, which locate outside the blood compartment, the carrier is required to exhibit a sufficient half-life in the blood compartment, and also the capability of extravasation at the tumor site. Recent developments led to the design of drug carriers with prolonged circulation in the vascular system (Otsuka et al., 2012). Polymeric NPs have been extensively acted as particulate carriers for almost every branch of medicine, including oncology, cardiology, immunology, pulmonary, neurology, endocrinology, ophthalmology, dentistry, and orthopedics. These NPs as therapeutics have many advantages including, subcellular size, keep safe nucleic acids from degradation, capability to encapsulate and especially controlled- and sustained-release drug value. Polymeric NP platforms are categorized by their unique physicochemical structures; they contain solid polymeric NP, polymeric micelle, polymer conjugate, dendrimer, polymersome, polyplex, and polymer–lipid hybrid system.

5.3.1 Polymeric nanoparticles

Many polymers, such as poly(D,L lactide-*co*-glycolide) (PLGA), polyethylene glycol (PEG), polyglycolide, polylactide (PLA), and polycaprolactone (PCL), have been utilized in a diverse set of medical applications. Additionally, PEG-functionalized PLGA NPs are mostly useful because pegylated polymeric NPs have considerably decreased systemic clearance, compared with the same particles lacking PEG (Dhar et al., 2008). For example, by using salinomycin-loaded PEGylated poly(lactic-*co*-glycolic acid) nanoparticles (SAL-NP) conjugated with CD133 aptamers (Ap–SAL–NP), Ni and coworkers developed a robust drug-delivery system for specifically targeting $CD133^+$ osteosarcoma CSCs. CD133 aptamer is regarded as a cancer stem cells (CSCs) marker of osteosarcoma, which is the most common primary bone malignancy in childhood and adolescence. In addition, Salinomycin is a polyether ionophore antibiotic which has been used to kill various CSCs, like osteosarcoma CSCs. In this study, SAL–NP was prepared from PLGA (17 k)-PEG (3 k)-COOH and salinomycin at weight ratio of 25:1 via an emulsion/solvent evaporation method. For preparation of Ap–SAL–NP, CD133 aptamers conjugated to the NPs via an EDC/NHS technique. The particle size and morphological examination of NPs were analyzed by dynamic light scattering (DLS) and transmission electron microscopy (TEM), respectively. The NPs were of desired particle size ($\sim$150 nm), drug encapsulation efficiency ($\sim$50%), and drug release profile. Significantly, Ap–SAL–NP could selectively kill $CD133^+$ osteosarcoma CSCs both in vitro and in vivo, as reflected by the tumorsphere formation and proportion of Saos-2 CD133+ cells (Ni et al., 2015).

Chitosan is a natural polysaccharide biopolymer prepared by the *N*-deacetylation of chitin a principal component of the crustacean exoskeleton. Chitosan has excellent biological and chemical properties, such as biocompatibility, biodegradability, polycationicity, antibacterial, and antifungal properties, high film-forming ability, good water permeability, and adhesion. Thus chitosan has been widely used in food and bioengineering industries, including the encapsulation of active food ingredients, wound dressing, soft tissue engineering, tissue repair, and local delivery of a variety of therapeutics, such as proteins, drugs, cells. As a drug carrier, chitosan helps overcome some characteristics of drugs, such as insolubility and hydrophobicity (Wang et al., 2011). Additionally, because of the presence of reactive amino and hydroxyl functional groups in the structure of chitosan, and susceptibility to chemical modifications, it has been used as an immobilizing support for biofabrication to build the biology-device interface (Kavosi et al., 2015; Khezrian et al., 2013). Numerous studies have demonstrated the increased clinical potential for chitosan NPs, compared with chitosan alone. Mucin 1 (MUC1) protein is a major cell-surface-associated glycoprotein that contains a hydrophobic membrane-spanning domain of 31 amino acids, a cytoplasmic domain of 69 amino acids, and an extracellular domain consisting of a region of nearly identical repeats of 20 amino acids per repeat. MUC1 is overexpressed in various epithelial carcinomas along with an alteration in the glycosylation pattern that is specific for each epithelial carcinoma. In addition, MUC1 is a famous cancer marker present in a variety of malignant cancers and may facilitate the vascular spread of carcinoma cells. Therefore, MUC1 is commonly expressed in a variety of malignant tumors containing lung, pancreatic, colon, breast, stomach, colorectal, ovarian, bladder carcinomas, and pancreas (Hu et al., 2014). Because of increasing the expression of MUC1 in cancer cells, serum assays for MUC1 are potentially helpful in tumor detection. Varnamkhasti and coworkers have been developed a surface modified of core–shell chitosan nanoparticulate system (Varnamkhasti et al., 2015). Chitosan as core and HA–SN-38 conjugates as shell decorated with MUC1 aptamer in order to target the mucin glycoprotein producing colon cancer cells. Hyaluronan (HA) is a bacterial polymer with high mucoadhesive property which has recently been employed with chitosan-based NPs, and was broadly used in drug-delivery systems. HA has a main role in cell proliferation and has the ability to bind specifically to HA receptors CD44, which are overexpressed in colon cancer (Wielenga et al., 1993). The SN-38 (7-ethyl-lo-hydroxycamptothecin) metabolite inhibits DNA topoisomerase I, and therefore used for promoting irinotecan anticancer activity and its side effects. Due to the extreme low solubility of SN-38 in biological milieu, Varnamkhasti group formed the SN-38 conjugated HA with an appropriate linker via its 20-hydroxyl group. The SN-38 conjugated HA has several advantages, such as increasing SN-38 stability by protecting its lactone ring that is critical for antitumor activity, and increasing the solubility of the drug. Subsequently, the negative charge HA–SN-38 conjugate was electrostatically linked to the positive chitosan NPs, and formed a core–shell NPs with appropriate size (125 nm) and charge. Afterward, in order to target the mucin glycoprotein producing colon cancer cells the NPs were modified with MUC1 aptamer (Fig. 22.10). In vitro release study proved that SN-38 is cleaved twice more in the acidic medium simulating cancer cell microenvironment, than in physiologic pH—which is important about target therapy, and minimizing related side-effects in off-target organs. By using flow cytometry and confocal microscopy, the targeting efficiency was evaluated, and it was demonstrated that MUC1 decoration increased the uptake of NPs by cancerous cell (HT29), while CHO as MUC1 negative cell line showed no improved uptake of decorated NPs. Investigating the effect of protein corona on NP cytotoxicity revealed that the adsorption of serum proteins does not exert a significant difference between the cytotoxicity of the unmodified and aptamer-modified NPs. However, nonprotein coated aptamer-modified NPs enhanced the cellular cytotoxic effect of aptamer-modified NPs because the protein corona can block the accessibility of targeting aptamer ligands, and the interference in the appropriate ligand–receptor attachment will decrease.

5.3.2 Hydrogel

Hydrogel is a three-dimensional (3D) polymer network, in which the hydrophilic polymer chains (e.g., ether, amine, hydroxyl, sulfate, and carboxyl) are connected by the physical or chemical bond therefore it can absorb large amount of water or biological fluids. Hydrogel possess several interesting properties, such as good biocompatibility, no side-effects, good biodegradability performance, and tissue compatibility arising from the high water content, and soft and rubbery consistency. Furthermore, hydrogels can be chemically modified with numerous ligands for targeted drug delivery. It is reported that, in order to slow drug release and achieve pulsatile release, the protein drugs were released with the aid of complementary oligonucleotide. Three different strategies can be used to produce aptamer-based hydrogels. The first method includes cross-linking of nucleic acid unit components by hybridization or self-assembly to form the hydrogel (Chang et al., 2012). The second method involves polymerization of organic component and cross-linker to form polymer networks with entrapped functional nucleic acids or FNAs (Liu et al., 2015). The third method involves the modification of hydrophilic polymer chains by nucleic acid aptamers as chain branches, and next, the formation of the nucleic acid aptamer hydrogel by cross-linking the branches (Chu et al., 2015; Li et al., 2015). For instance, Wang and coworkers have used the aptamer-cross-linked hydrogel as a drug carrier by encapsulating the cancer drug doxorubicin (Dox) to test the release properties (Wang et al., 2015). In this study, for visualizing the gelling formation and transition, AuNPs were employed as indicating agents. AuNPs were modified

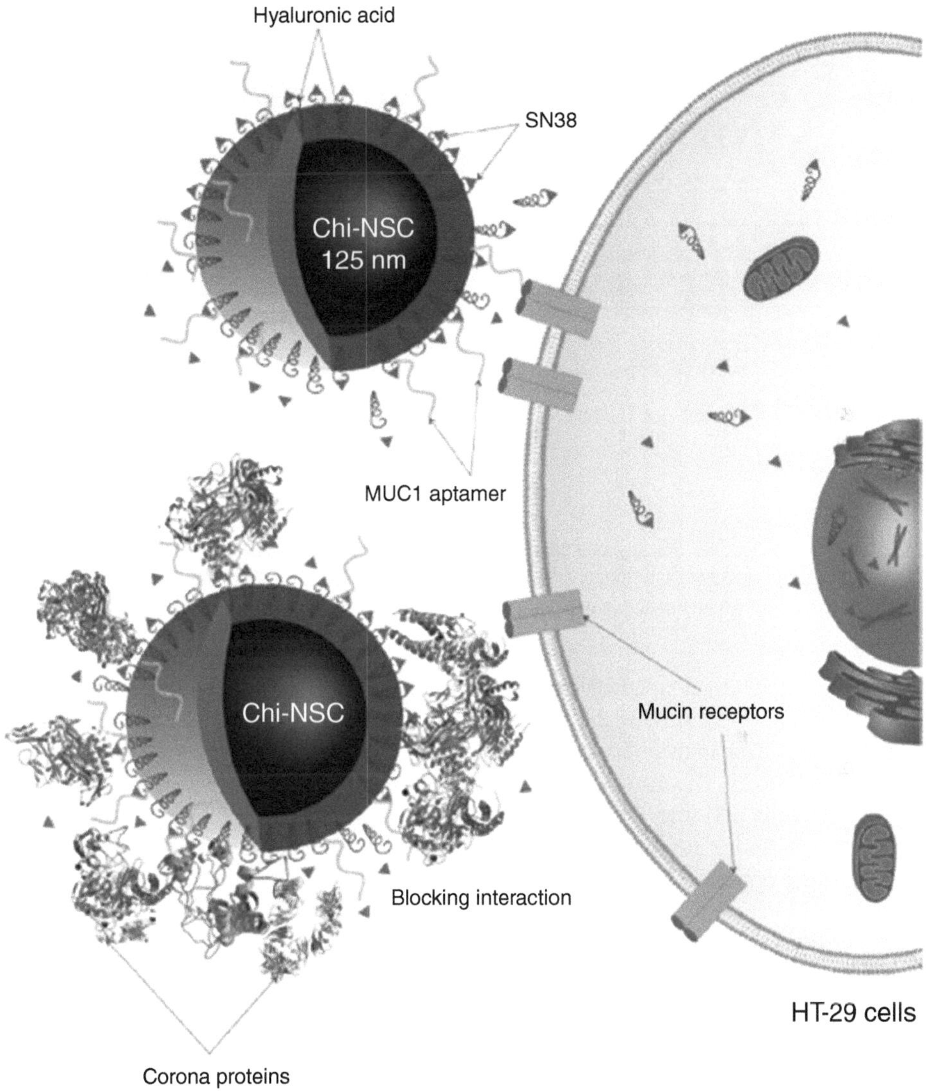

FIGURE 22.10 The Interaction of Bare MUC1 Aptamer-Conjugated Nanoparticles and Protein-Coated MUC1 Aptamer-Conjugated Nanoparticles With MUC 1 Positive Cell (Varnamkhasti et al., 2015)

with bovine serum albumin (BSA) to avoid aggregation, and then mixed well with polymer strand A (PS-A) and polymer strand B (PS-B) before adding AS1411 (Fig. 22.11B). Finally, by introduction of aptamers, homogenous red hydrogel was formed, which AuNPs trapped inside and distributed well. As the hybridization proceeded, the cross-linking ratio of polyacrylamide increased, resulting in the increase of viscosity of the polymer solution (Fig. 22.11A). Upon introduction of a target protein nucleolin, AS1411 interacted with it, and the gel dissolved as a result of reducing the cross-linking

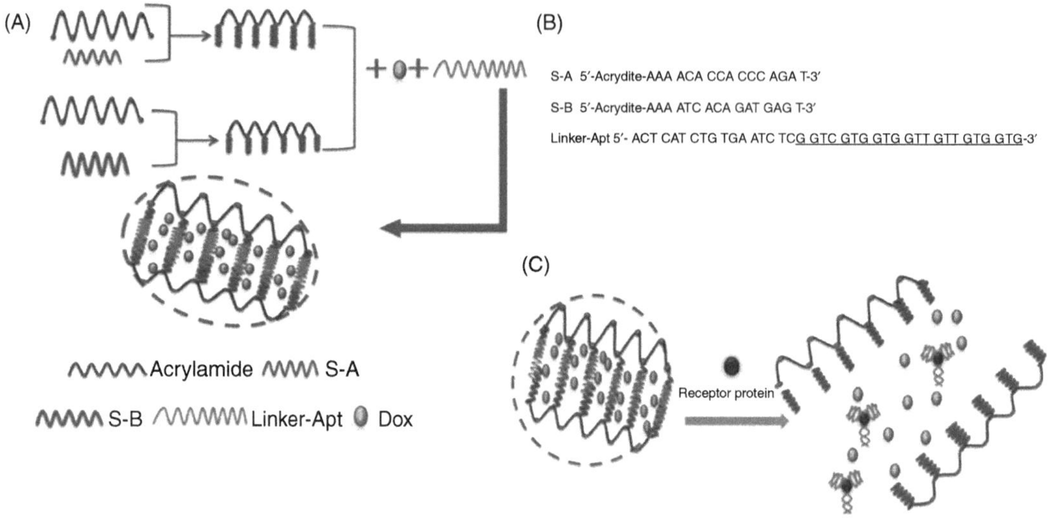

FIGURE 22.11

The synthesis process of aptamer-functionalized hydrogel (A), sequences of oligonucleotides (AS1411 aptamer with underline) (B), and working principle of aptamer-functionalized hydrogel for target drug delivery (C) (Wang et al., 2015)

density by competitive target–aptamer binding. Then, the Dox encapsulated in the hydrogels was released (Fig. 22.11C). By increasing the concentration of nucleolin, the density of the cross-linker decreased, thereby resulting in the enhancement of hydrogel dissociation, and consequently increasing in the fluorescence intensity. The results indicated that the release of Dox was much quicker at pH 6.3 (the surface of cancer cells pH) than that at pH 7.4 (normal physiological pH). Meanwhile, the biological activity of the cross-linker AS1411 was significantly higher under acidic condition. As such, pH 6.3 was expected as the optimum pH for drug release.

5.3.3 Dendrimers

Dendrimers are almost monodisperse macromolecules with a highly arranged structure consisting of tree-like arms or branches that are built around a small molecule or a linear polymer core. Dendrimers are nanosized, homogeneous, multivalent, and also easily functionalized particles. The unique potential to control size, the molecular weight, the charge density, high in vivo stability, and structural properties make these nanomaterials ideal candidates for anticancer therapies and diagnostic imaging. Polyamidoamine (PAMAM) dendrimers are a particular class of cationics dendritic polymers. The PAMAM form PAMAM dendrimer/DNA complexes via the electrostatic interactions with negatively charged DNA, without damaging and deactivating the DNA (Zhang et al., 2015a). As drug-delivery systems, these PAMAM dendrimer/DNA complexes show high solubility, sustained release, decrease systemic toxicity, and selective accumulation in solid tumors. To overcome the drawbacks of conventional cancer therapies, Lee and coworkers have established a novel drug-delivery system against prostate cancer based on a dendrimer and a single-strand DNA-A9 PSMA (prostate-specific membrane antigen) RNA aptamer hybrid (Lee et al., 2011). In their design, PAMAM dendrimer was conjugated with an amine and an adenine linker (5′-NH2-A10-(TCG) 7-3′), then an elongated version of A9 RNA aptamer specific

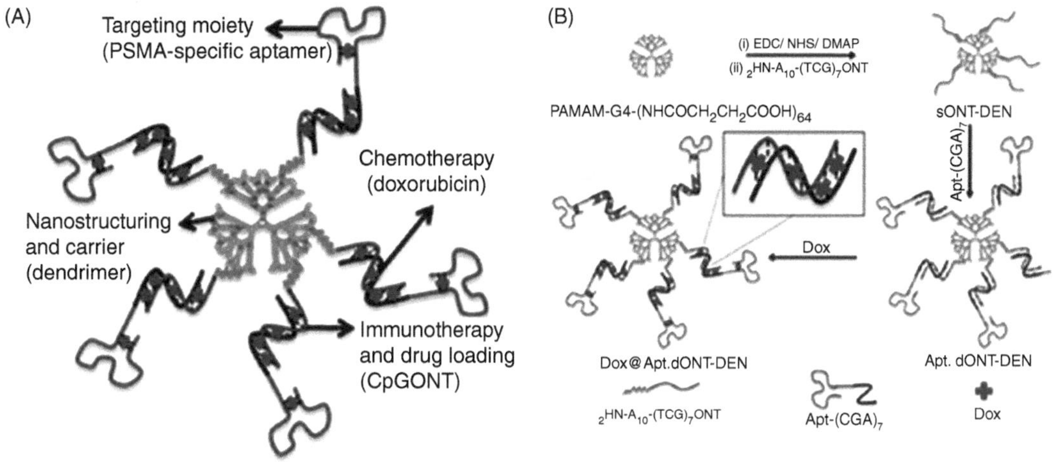

FIGURE 22.12 Schematic Bioconjugation Representation of the Dox@Apt·dONT-DEN for Targeted Chemoimmunotherapy (Lee et al., 2011)

for prostate-specific membrane antigen (PSMA) hybridized with the sONT–DEN (single-stranded oligonucleotide–dendrimer conjugated) to create a double-stranded Apt·dONT–DEN conjugate. Dox was selected as a chemotherapeutic agent because it forms a stable complex with duplex oligonucleotides (dONTs) by intercalating into base pairs, preferentially into consecutive CG pairs. In the last step, Dox intercalated into the double-stranded dONT site to yield Dox-loaded Apt·dONT–DEN conjugates (Dox@Apt·dONT–DEN) (Fig. 22.12).

The antitumor efficacy of Dox@Apt·dONT–DEN was evaluated by a 22RV1 xenograft tumor model. The results showed that the tumor growth profiles following intravenous injection (4 times treatment) of each therapeutic modality. Measurement of tumor size on day 36 announced that Dox@Apt·dONT–DEN (dose: 1 mg Dox/kg) demonstrated excellent antitumor efficacy, reducing tumor volume by 78%, when compared to the saline-treated control group (619 ± 52 mm^3 vs. 2726 ± 411 mm^3). In contrast, their results suggest that Dox@Apt·dONT–DEN complexes had the much greater antitumor efficacy, compared to the same dose of free-Dox or aptamer-free dendrimer conjugate (Dox@dONT–DEN)

5.3.4 Polymeric micelles

Polymeric micelles are nanosized core/shell structures formed by amphiphilic di- or tri-block copolymers. Polymeric micelles with the inherent and modifiable properties have been demonstrated beneficial for site-specific delivery of antitumor drugs to cancer. Polymeric micelles contain a core–shell structure. The hydrophobic part of the block copolymer forms the core, which encapsulates the poor water-soluble and hydrophobic anticancer drugs; whereas, the hydrophilic block of the copolymer forms the outer shell which decreases the interactions of drug with the aqueous environment (Otsuka et al., 2012). Cancer chemotherapy may cause severe side effects, leaving patients under extreme distress. To overcome this problem, an interest has been raised in the application of block copolymer micelles as novel carrier systems for anticancer agents because of the high drug-loading capacity of the inner core as well as of the unique disposition characteristics in the body. Pluronic F127 is an amphiphilic polymer with the terminal hydroxyl groups. Although Pluronic F127 has been widely used in

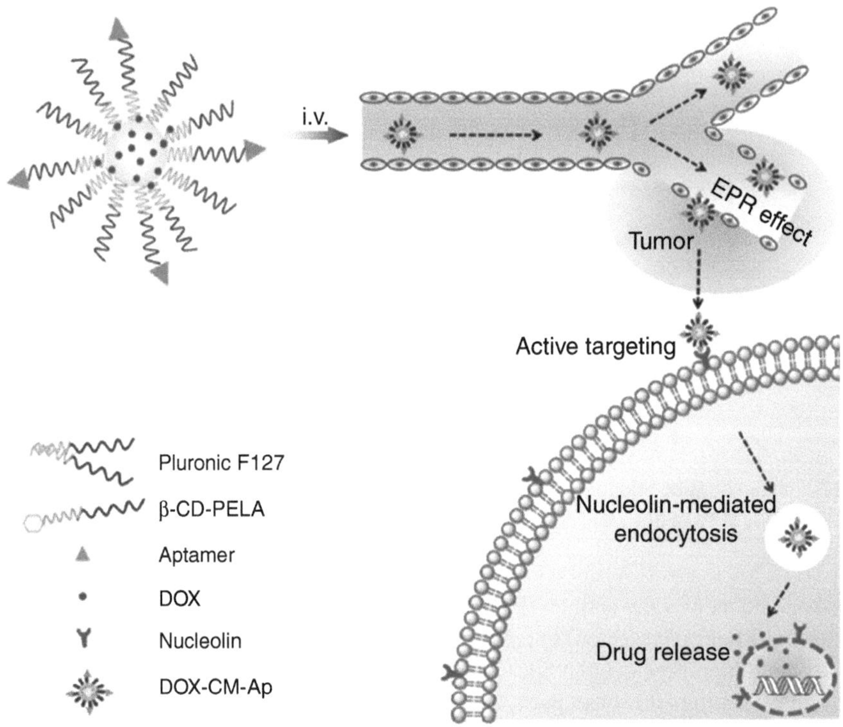

FIGURE 22.13 Schematic Diagram of the Multifunctional Composite Micelles Modified by Aptamer AS1411 and the Mechanism of Interaction Between Nucleolin Receptors and Aptamer AS1411 Ligands (Li et al., 2015b)

nanodrug-delivery systems, but it regrettably suffers from the low loading drug capacity and physical resistance of its micelle formulation. To overcome this shortcoming, binary mixed systems containing Pluronic F127 and other block copolymers have been developed. Interestingly, Li and coworkers designed a hybrid system containing Pluronic F127 and beta-cyclodextrin-linked poly(ethylene glycol)-*b*-polylactide (β-CD-PELA), which could form self-assembled micelles with a downward shift in the critical micelle concentration (CMC) to improve micelle stability and Dox-loading capacity (Li et al., 2015b). As shown in Fig. 22.13, Li and coworkers have constructed a binary composite micelles (CM) composed of AS1411-modified Pluronic F127 and β-CD-PELA to load Dox to accomplish targeted therapy of human breast cancer. The pharmacokinetic study confirmed that CM-Ap could significantly prolong the blood circulation time of Dox compared to free Dox. Moreover, in vivo real-time imaging study revealed that aptamer-conjugated micelles had better tumor targeting potency than nontargeted micelles. Based on a receptor-mediated endocytosis pathway and controllable cellular release of Dox, Dox-CM-Ap was validated to enhance the in vivo antitumor efficacy of Dox and prevent acute cardiotoxicity.

5.3.5 Liposome

Another promising polymeric NPs involves liposome. A liposome is a bilayer structure in nature which originally consisted in naturally occurring and/or synthetic phospholipids. According to the presence of a hydrophilic core and a hydrophilic bilayer, liposomes can use for delivery of

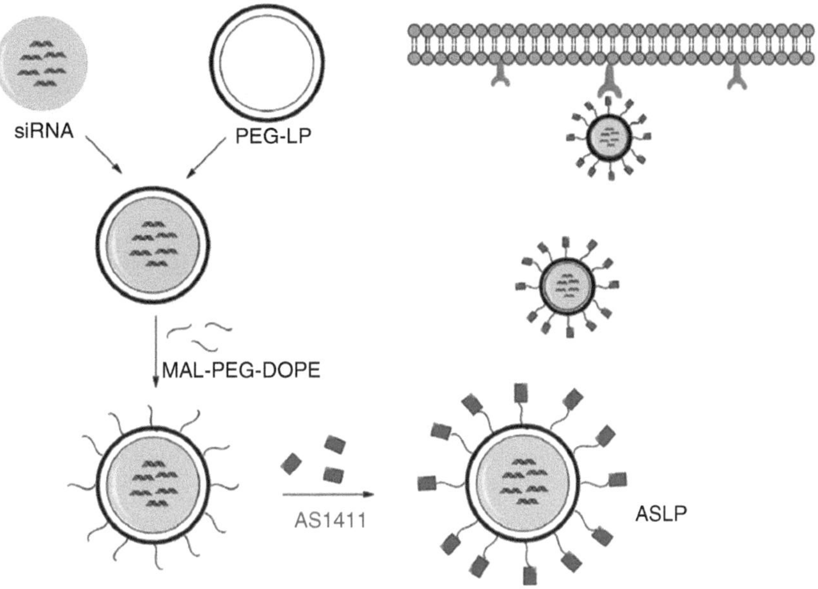

FIGURE 22.14 Schematic Illustration of Delivery of siRNA Using AS1411-Conjugated ASLPs (Li et al., 2014)

hydrophilic molecules. Liposome based system are created a new field in drug delivery and cancer therapy (Torchilin, 2005; Vemuri and Rhodes,1995). Liposomes can be loaded with highly toxic or poorly soluble drug (Zhang et al., 2015b), and surface modification with polyethylene glycol (PEG) (Xu et al., 2015) may prolong their circulation time and enhance preferential accumulation in tumor sites. Likewise, by conjugation of the cationic liposomes with the negatively charged nucleic acid aptamers, the residence time of aptamers in plasma are increased. Furthermore, cytotoxicity caused by cationic liposomes will be also greatly decreased. Li and coworkers have been reported a new nucleolin-targeting liposome to deliver anti-BRAF siRNA (siBraf) with high specificity and efficiency to melanoma cell A375 (Li et al., 2014). Melanoma is a complex genetic disease which is the chief cause of death in patients with skin cancer. RNA interference (RNAi) is a classical gene-silencing mechanism that employs an evolutionarily conserved process through which double-stranded RNA (dsRNA) induces the silencing of cognate genes. dsRNA is converted to siRNAs that direct RNAi. In other words, siRNAs are synthetic mediators of RNAi. The defining features of small-silencing RNAs are their short lengths (between 20 and 23 base pairs), and their association with members of the ribonuclease proteins called Argonautes, which they guide to the efficient silencing of genes, In addition, the delivery of siRNA into cells is a promising strategy for gene therapy (Ghildiyal and Zamore, 2009). The silencing BRAF gene by RNAi is a substitute approach for melanoma therapy by the induction of apoptosis and cancer cell growth inhibition. In this study, at first liposomes composed of 3b-[*N*-(*N*′, *N*′-dimethylaminoethane) carbamoyl] cholesterol/dioleoylphosphatidylethanolamine (Dc-Chol/DOPE) and (maleimide-poly(ethylene glycol)-succinimidyl valerate) MAL–PEG–DOPE were prepared. Then siRNA at different concentrations mixed with the naked liposomes gently to form siRNA/liposome complexes. Conjugation of the aptamer AS1411 to the cationic liposomes was done by a PEG linker. Finally, the ability of AS1411-PEG-liposome (ASLP) to deliver functional siRNA (siBraf) confirmed by the inhibited BRAF protein expression in A375 cells (Fig. 22.14). As a result, ASLP/siBraf complex exhibited significant silencing activity and inhibition of human growth in vitro. The complexes of ASLP/siBraf showed a successful downregulation of human BRAF mRNA expression within a short period of treatment in vivo.

6 CONCLUSIONS

Aptamers are functional single-stranded DNA or RNA oligonucleotides that can bind to target molecules with high affinity and specificity. Aptamers with commercial potential have been rapidly used in a broad range of applications, including molecular medicine. Bioconjugation of aptamers with a diverse number of nanomaterials have a potentially powerful tool to enhance novel diagnostic and therapeutic approaches, including targeted drug-delivery agents, and a myriad of cancers diagnostics and detection. Nanomaterials have different electrochemical, mechanical, optical, and magnetic properties, medical diagnostic and drug-delivery agents with diverse characteristics. NPs facilitate targeting and delivery of the drugs to the cellular tissues, while avoiding removal from the blood, and gathering in tissues involved in the RES, for example, in the liver, and spleen and lymph tissue.

REFERENCES

Agrawal, P., 2015. Recent advances in anticancer drugs development: G-quadruplex as new drug target. J. Pharmacovigil. 3, 1–2.

Bamrungsap, S., Treetong, A., Apiwat, C., Wuttikhun, T., Dharakul, T., 2015. SERS-fluorescence dual mode nanotags for cervical cancer detection using aptamers conjugated to gold-silver nanorods. Microchim. Acta 183, 249–256.

Bates, P.J., Laber, D.A., Miller, D.M., Thomas, S.D., Trent, J.O., 2009. Discovery and development of the G-rich oligonucleotide AS1411 as a novel treatment for cancer. Exp. Mol. Pathol. 86, 151–164.

Bilik, K.U., Ergüven, E., Klussmann, S., Jarosch, F., Wielinga, P.Y., Lutz, T.A., Riediger, T., 2007. In-vitro and in-vivo antagonistic action of an anti-amylin Spiegelmer. Neuroreport 18, 1855–1859.

Bouchard, P.R., Hutabarat, R.M., Thompson, K.M., 2010. Discovery and development of therapeutic aptamers. Annu. Rev. Pharmacol. Toxicol. 50, 237–257.

Chang, D., Lei, J., Cui, H., Lu, N., Sun, Y., Zhang, X., Gao, C., Zheng, H., Yin, Y., 2012. Disulfide cross-linked nanospheres from sodium alginate derivative for inflammatory bowel disease: preparation, characterization, and in vitro drug release behavior. Carbohydr. Polym. 88, 663–669.

Chen, X., Pan, Y., Liu, H., Bai, X., Wang, N., Zhang, N., 2016. Label-free detection of liver cancer cells by aptamer-based microcantilever biosensor. Biosens. Bioelectron. 79, 353–358.

Chu, T-W., Feng, J., Yang, J., Kopecˇek, J., 2015. Hybrid polymeric hydrogels via peptide nucleic acid (PNA)/DNA complexation. J. Control. Release 15, 30135–30138.

Danesh, N., Lavaee, P., Ramezani, M., Abnous, K., Taghdisi, S.M., 2015. Targeted and controlled release delivery of daunorubicin to T-cell acute lymphoblastic leukemia by aptamer-modified gold nanoparticles. Int. J. Pharm. 489, 311–317.

Dhar, S., Gu, F.X., Langer, R., Farokhzad, O.C., Lippard, S.J., 2008. Targeted delivery of cisplatin to prostate cancer cells by aptamer functionalized Pt(IV) prodrug-PLGA–PEG nanoparticles. Proc. Natl. Acad. Sci. USA 105, 17356–17361.

Ding, F., Guo, S., Xie, M., Luo, W., Yuan, C., Huang, W., Zhou, Y., Zhang, X.L., Zhou, X., 2015. Diagnostic applications of gastric carcinoma cell aptamers in vitro and in vivo. Talanta 134, 30–36.

Ellington, A.D., Szostak, J.W., 1990. In vitro selection of RNA molecules that bind specific ligands. Nature 346, 818–822.

Ellington, A.D., Szostak, J.W., 1992. Selection in vitro of single-stranded DNA molecules that fold into specific ligand-binding structures. Nature 355, 850–852.

Fang, X., Tan, W., 2010. Aptamers generated from cell-SELEX for molecular medicine: a chemical biology approach. Acc. Chem. Res. 43, 48–57.

Feigon, J., Dieckmann, T., Smith, F.W., 1996. Aptamer structures from A to zeta. Chem. Biol. 3, 611–617.

Ghildiyal, M., Zamore, P.D., 2009. Small silencing RNAs: an expanding universe. Nat. Rev. Genet. 10, 94–108.

Gragoudas, E.S., Adamis, A.P., Cunningham, E.T., Feinsod, M., Guyer, D.R., 2004. Pegaptanib for neovascular age-related macular degeneration. N. Engl. J. Med. 351, 2805–2816.

Griffin, L.C., Tidmarsh, G.F., Bock, L.C., Toole, J.J., Leung, L.L., 1993. In vivo anticoagulant properties of a novel nucleotide-based thrombin inhibitor and demonstration of regional anticoagulation in extracorporeal circuits. Blood 81, 3271–3276.

Herrmann, A., Priceman, S., Swiderski, P., Kujawski, M., Xin, H., Cherryholmes, G.A., Zhang, W., Zhang, C., Lahtz, C., Kowolik, C., Forman, S.J., Kortylewski, M., Yu, H., 2014. CTLA4 aptamer delivers STAT3 siRNA to tumor-associated and malignant T cells. J. Clin. Invest. 124, 2977–2987.

Hu, R., Wen, W., Wang, Q., Xiong, H., Zhang, X., Gu, H., Wang, S., 2014. Novel electrochemical aptamer biosensor based on an enzyme–gold nanoparticle dual label for the ultrasensitive detection of epithelial tumour marker MUC1. Biosens. Bioelectron. 53, 384–389.

Jain, S., Hirst, D.G., O'Sullivan, J.M., 2012. Gold nanoparticles as novel agents for cancer therapy. Br. J. Radiol. 85, 101–113.

Jo, H., Youn, H., Lee, S., Ban, C., 2014. Ultra-effective photothermal therapy for prostate cancer cells using dual aptamer-modified gold nanostars. J. Mater. Chem. B 2, 4862–4867.

Kavosi, B., Salimi, A., Hallaj, R., Amani, K., 2014. A highly sensitive prostate-specific antigen immunosensor based on gold nanoparticles/PAMAM dendrimer loaded on MWCNTS/chitosan/ionic liquid nanocomposite. Biosens. Bioelectron. 52, 20–28.

Kavosi, B., Salimi, A., Hallaj, R., Moradi, F., 2015. Ultrasensitive electrochemical immunosensor for PSA biomarker detection in prostate cancer cells using gold nanoparticles/PAMAM dendrimer loaded with enzyme linked aptamer as integrated triple signal amplification strategy. Biosens. Bioelectron. 74, 915–923.

Keefe, A.D., Pai, S., Ellington, A., 2010. Aptamers as therapeutics. Nat. Rev. Drug Discov. 9, 537–550.

Khezrian, S., Salimi, A., Teymourian, H., Hallaj, R., 2013. Label-free electrochemical IgEaptasensor based on covalent attachment of aptamer onto multiwalled carbon nanotubes/ionic liquid/chitosan nanocomposite modified electrode. Biosens. Bioelectron. 43, 218–225.

Kim, D., Yeom, J.H., Lee, B., Lee, K., Bae, J., Rhee, S., 2015. Inhibition of discoidin domain receptor 2-mediated cancer progression by gold nanoparticle-aptamer-assisted delivery of peptides containing transmembrane juxtamembrane 1/2 domain. Biochem. Biophys. Res. Commun. 464, 392–395.

Lee, I., An, S., Yu, M.K., Kwon, H.K., Im, S.H., Jon, S., 2011. Targeted chemoimmunotherapy using drug-loaded aptamer–dendrimer bioconjugates. J. Control. Release 155, 435–441.

Lee, J., Kang, H.J., Jang, H., Lee, Y.J., Lee, Y.S., Ali, B.A., Al-Khedhairy, A.A., Kim, S., 2015. Simultaneous imaging of two different cancer biomarkers using aptamer-conjugated quantum dots. Sensors 15, 8595–8604.

Li, Y., Qiao, H., Yan, W., Zhang, J., Xing, C., Wang, H., Zhang, B., Tang, J., 2013. Molecular recognition force spectroscopy study of the dynamic interaction between aptamer GBI-10 and extracellular matrix protein tenascin-C on human glioblastoma cell. J. Mol. Recognit. 26, 46–50.

Li, L., Hou, J., Liu, X., Guo, Y., Wu, Y., Zhang, L., Yang, Z., 2014. Nucleolin-targeting liposomes guided by aptamer AS1411 for the delivery of siRNA for the treatment of malignant melanomas. Biomaterials 35, 3840–3850.

Li, Y., Maciel, D., Rodrigues, J., Shi, X., Tomás, H., 2015. Biodegradable polymer nanogels for drug/nucleic acid delivery. Chem. Rev. 16, 8564–8608.

Li, H., Hu, H., Zhao, Y., Chen, X., Li, W., Qiang, W., Xu, D., 2015a. multifunctional aptamer-silver conjugates as theragnostic agents for specific cancer cell therapy and fluorescence-enhanced cell imaging. Anal. Chem. 87, 3736–3745.

Li, X., Yu, Y., Ji, Q., Qiu, L., 2015b. Targeted delivery of anticancer drugs by aptamer AS1411 mediated Pluronic F127/cyclodextrin-linked polymer composite micelles. Nanomedicine 11, 175–184.
Lin, Z., Ma, Q., Fei, X., Zhang, H., Su, X., 2014. A novel aptamer functionalized CuInS2 quantum dots probe for daunorubicin sensing and near infrared imaging of prostate cancer cells. Anal. Chim. Acta 818, 54–60.
Liu, R., Huang, Y., Ma, Y., Jia, S., Gao, M., Li, J., Zhang, H., Xu, D., Wu, M., Chen, Y., Zhu, Z., Yang, C., 2015. Design and synthesis of target-responsive aptamer-cross-linked hydrogel for visual quantitative detection of ochratoxin A. ACS Appl. Mater. Interf. 7, 6982–6990.
Lu, Y., Liu, J., 2006. Functional DNA nanotechnology: emerging applications of DNAzymes and aptamers. Curr. Opin. Biotechnol. 17, 580–588.
Mahmoudi, M., Sant, S., Wang, B., Laurent, S., Sen, T., 2011. Superparamagnetic iron oxide nanoparticles (SPIONs): development, surface modification and applications in chemotherapy. Adv. Drug. Deliv. Rev. 63, 24–46.
Malik, M.T., O'Toole, M.G., Casson, L.K., Thomas, S.D., Bardi, G.T., Reyes-Reyes, E.M., Ng, C.K., Kang, K., Bates, P.J., 2015. AS1411-conjugated gold nanospheres and their potential for breast cancer therapy. Oncotarget 6, 22270–22281.
Mariani, S., Scarano, S., Ermini, M.L., Bonini, M., Minunni, M., 2015. Investigating nanoparticle properties in plasmonic nanoarchitectures with DNA by surface plasmon resonance imaging. Chem. Commun. (Camb.) 51, 6587–6590.
McKenna, N., 2012. Redefining oligonucleotide production. Genetic Eng. Biotechnol. News 32, 37–38.
Mehra, N.K., Mishra, V., Jain, N.K., 2014. A review of ligand tethered surface engineered carbon nanotubes. Biomaterials 35, 1267–1283.
Michalet, X., Pinaud, F.F., Bentolila, L.A., Tsay, J.M., Doose, S., Li, J.J., Sundaresan, G., Wu, A.M., Gambhir, S.S., Weiss, S., 2005. Quantum dots for live cells, in vivo imaging, and diagnostics. Science 307, 538–544.
Nehl, C.L., Liao, H., Hafner, J.H., 2006. Optical properties of star-shaped gold nanoparticles. Nano Lett. 6, 683–688.
Ni, M., Xiong, M., Zhang, X., Cai, G., Chen, H., Zeng, Q., Yu, Z., 2015. Poly(lactic-*co*-glycolic acid) nanoparticles conjugated with CD133 aptamers for targeted salinomycin delivery to CD133+ osteosarcoma cancer stem cells. Int. J. Nanomed. 10, 2537–2554.
Otsuka, H., Nagasaki, Y., Kataoka, K., 2012. PEGylated nanoparticles for biological and pharmaceutical applications. Adv. Drug. Deliv. Rev. 64, 246–255.
Padma Sundaram, H.K., Byrne, Mark E., Wower, Jacek, 2013. Therapeutic RNA aptamers in clinical trials. Eur. J. Pharm. Sci. 48, 259–271.
Pandurangan, D.K., Mounika, S.K., 2012. Quantum dot aptamers-an emerging technology with wide scope in pharmacy. Int. J. Pharm. Pharm. Sci. 5, 24–31.
Patel, D.J., Suri, A.K., Jiang, F., Jiang, L., Fan, P., Kumar, R.A., Nonin, S., 1997. Structure, recognition and adaptive binding in RNA aptamer complexes. J. Mol. Biol. 272, 645–664.
Ryou, S.M., Yeom, J.H., Kang, H.J., Won, M., Kim, J.S., Lee, B., Seong, M.J., Ha, N.C., Bae, J., Lee, K., 2014. Gold nanoparticle–DNA aptamer composites as a universal carrier for in vivo delivery of biologically functional proteins. J. Control Release 196, 287–294.
Salimi, A., Khezrian, S., Hallaj, R., Vaziry, A., 2014. Highly sensitive electrochemical aptasensor for immunoglobulin E detection based on sandwich assay using enzyme-linked aptamer. Anal. Biochem. 466, 89–97.
Shangguan, D., Li, Y., Tang, Z., Cao, Z.C., Chen, H.W., Mallikaratchy, P., Sefah, K., Yang, C.J., Tan, W., 2006. Aptamers evolved from live cells as effective molecular probes for cancer study. Proc. Natl. Acad. Sci. USA 103, 11838–11843.
Sun, S., Zu, Y., 2015. A highlight of recent advances in aptamer technology and its application. Molecules 20, 11959–11980.
Sun, W.T., Zu, Y., 2016. Aptamers: versatile molecular recognition probes for cancer detection. Analyst 41, 403–415.

Taghdisi, S.M., Lavaee, P., Ramezani, M., Abnous, K., 2011. Reversible targeting and controlled release delivery of daunorubicin to cancer cells by aptamer-wrapped carbon nanotubes. Eur. J. Pharm. Biopharm. 77, 200–206.

Taghdisi, S.M., Abnous, K, Mosaffa, F., Behravan, J, 2010. Targeted delivery of daunorubicin to T-cell acute lymphoblastic leukemia by aptamer. J. Drug Target 18, 277–281.

Thakkar, K.N., Mhatre, S.S., Parikh, R.Y., 2010. Biological synthesis of metallic nanoparticles. Nanomedicine 6, 257–262.

Torchilin, V.P., 2005. Recent advances with liposomes as pharmaceutical carriers. Nat. Rev. 4, 145–160.

Tuerk, C., Gold, L., 1990. systematic evolution of ligands by exponential enrichment: RNA ligands to bacteriophage T4 DNA polymerase. Science 249, 505–510.

Varnamkhasti, B.S., Hosseinzadeh, H., Azhdarzadeh, M., Vafaei, S.Y., Esfandyari-Manesh, M., Mirzaie, Z.H., Amini, M., Ostad, S.N., Atyabi, F., Dinarvand, R., 2015. Protein corona hampers targeting potential of MUC1 aptamer functionalized SN-38 core–shell nanoparticles. Int. J. Pharm. 494, 430–444.

Vemuri, D., Rhodes, C.T., 1995. Preparation and characterization of liposomes as therapeutic delivery systems: a review. Pharm. Acta Helv. 70, 95–111.

Viraka Nellore, B.P., Kanchanapally, R., Pramanik, A., Sinha, S.S., Chavva, S.R., Hammell, A., Ray, P.C., 2015. Aptamer conjugated graphene oxide membranes for highly efficient capture and accurate identification of multiple types of circulating tumor cells. Bioconjug. Chem. 2, 235–242.

Wang, A.Z., Bagalkot, V., Vasilliou, C.C., Gu, F., Alexis, F., Zhang, L., Shaikh, M., Yuet, K., Cima, M.J., Langer, R., Kantoff, P.W., Bander, N.H., Jon, S., Farokhzad, O.C., 2008. Superparamagnetic iron oxide nanoparticle–aptamer bioconjugates for combined prostate cancer imaging and therapy. Chem. Med. Chem. 3, 1311–1315.

Wang, J.J., Zeng, Z.W, Xiao, R.Z., Xie, T., Zhou, G.L., Zhan, X.R., Wang, S.L, 2011. Recent advances of chitosan nanoparticles as drug carriers. Int. J. Nanomed. 6, 765–774.

Wang, Z., Xia, J., Cai, F., Zhang, F., Yang, M., Bi, S., Gui, R., Li, Y., Xia, Y., 2015. Aptamer-functionalized hydrogel as effective anti-cancer drugs delivery agents. Colloid Surf. B Biointerf. 134, 40–46.

Wielenga, V.J., Heider, K.H., Offerhaus, G.J., Adolf, G.R., Van Den Berg, F.M., Ponta, H., Herrlich, P., Pals, S.T., 1993. Expression of CD44 variant proteins in human colorectal cancer is related to tumor progression. Cancer Res. 53, 4754–4756.

Wu, S.Y., Yang, X., Gharpure, K.M., Hatakeyama, H., Egli, M4., Mcguire, M.H., Nagaraja, A.S., Miyake, T.M., Rupaimoole, R., Pecot, C.V., Taylor, M., Pradeep, S., Sierant, M., Rodriguez-Aguayo, C., Choi, H.J., Previs, R.A., Armaiz-Pena, G.N., Huang, L., Martinez, C., Hassell, T., Ivan, C., Sehgal, V., Singhania, R., Han, H.D., Su, C., Kim, J.H., Dalton, H., Kovvali, C., Keyomarsi, K., Mcmillan, N.A., Overwijk, W.W., Liu, J., Lee, J.S., Baggerly, K.A., Lopez-Berestein, G., Ram, P.T., Nawrot, B., Sood, A.K., 2014. 20-OMe-phosphorodithioate-modified siRNAs show increased loading into the RISC complex and enhanced anti-tumour activity. Nat. Commun. 4459, 1–12.

Xu, H., Hu, M., Yu, X., Li, Y., Fu, Y., Zhou, X., Zhang, D., Li, J., 2015. Design and evaluation of pH-sensitive liposomes constructed by poly(2-ethyl-2-oxazoline)-cholesterol hemisuccinate for doxorubicin delivery. Eur. J. Pharm. Biopharm. 91, 66–74.

Ye, S., Zhai, X., Wu, Y., Kuang, S., 2016. Dual-primer self-generation SERS signal amplification assay for PDGF-BB using label-free aptamer. Biosens. Bioelectron. 79, 130–135.

Zhang, H., Ma, Y., Xie, Y., An, Y1., Huang, Y., Zhu, Z., Yang, C.J., 2015a. A controllable aptamer-based self-assembled DNA dendrimer for high affinity targeting, bioimaging and drug delivery. Sci. Rep. 5, 1–8.

Zhang, W., Wang, G., Falconer, J.R., Baguley, B.C., Shaw, J.P., Liu, J., Xu, H., See, E., Sun, J., Aa, J., Wu, Z., 2015b. Strategies to maximize liposomal drug loading for a poorly water-soluble anticancer drug. Pharm. Res. 4, 1451–1461.

CHAPTER

RECENT PROGRESS IN THERAPEUTIC DIAGNOSIS USING PHOTONIC CRYSTAL NANOSTRUCTURES

23

Saeed Olyaee, Alieh Naraghi
Shahid Rajaee Teacher Training University, Tehran, Iran

CHAPTER OUTLINE

1 INTRODUCTION TO PHOTONIC CRYSTAL NANOSTRUCTURES

Photonic crystals (PCs) originally introduced by Yablonovitch and John in 1987 are engineered nanostructures with a periodic arrangement of materials having different refractive indices (RI) (John, 1987; Yablonovitch, 1987). They could have period in one (1D), two (2D), or three dimensions (3D). Fig. 23.1 illustrates three configurations of PCs with different periodicities. The interference of reflections and refractions of photons within these materials, and particularly at the boundaries between them, provide constructive or destructive patterns, produce a phenomenon referred to as photonic bands

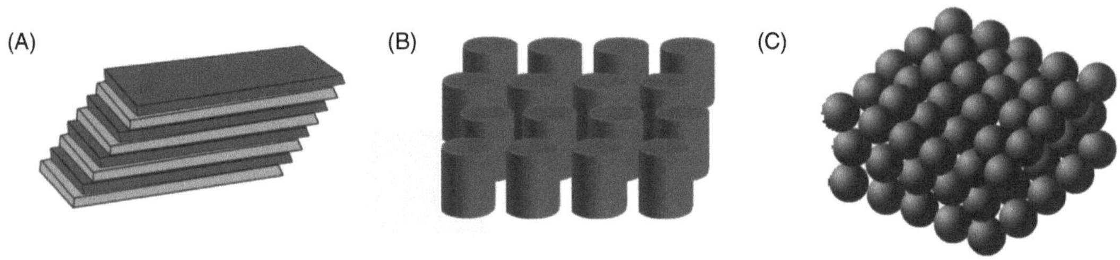

FIGURE 23.1

Different configurations of PCs with (A) 1D, (B) 2D, and (C) 3D periodicities.

which represent a continuous range of light frequencies (of a particular polarization, in a certain direction) with the ability to pass through the PC. On the other hand, the range of frequencies of light that cannot pass through a PC are called photonic bandgaps (PBGs). Photons with bandgap frequencies can only penetrate the PC evanescently and are reflected by the material. Extensive studies have resulted in progress for the development of many photonic devices based on PCs capable of confining, controlling, and routing the light in a photonic chip. In addition, PCs are also suitable candidates for the development of chemical and biological sensors with the ability to significantly enhance the light–matter interaction through the slow-light effect, leading to enhanced interaction in terms of time and intensity (Lai et al., 2012, 2013; Li et al., 2014). PCs are, therefore, used as sensitive devices for detecting molecular interactions because of this ability combined with the high dependence on their spectral response with the surrounding medium.

2 ORIGIN OF PHOTONIC BANDGAP

Maxwell's equations can be used as a starting point for explaining many optical phenomena, including photonic band structure. The following set of equations represents Maxwell's equations in SI units:

$$\begin{aligned} \nabla \cdot B &= 0 \qquad & \nabla \times E &= -\frac{\partial B}{\partial t} \\ \nabla \cdot D &= \rho \qquad & \nabla \times H &= -\frac{\partial D}{\partial t} + J \end{aligned} \tag{23.1}$$

where, ρ and J represent the free charge and current densities, and E, D, B, and H are the electric displacement, magnetic induction, and magnetic fields, respectively. All six variables are time- and position-dependent. For dielectric media that are nonmagnetic, nondispersive, linear, and isotropic, the electric fields E and D and magnetic fields H and B are related through the constitutive relations:

$$\begin{aligned} D &= \varepsilon_0 \varepsilon E \\ B &= \mu_0 H \end{aligned} \tag{23.2}$$

where, ε is the dielectric function (of position and time). The permittivity and permeability constants of free space are respectively denoted by ε_0 and μ_0. It should be noted that under these assumptions, the refractive index is approximately equal to the square root of the dielectric function, $n \cong \sqrt{\varepsilon}$. The defining equations can be obtained by combining Eqs. (23.1) and (23.2), using the assumption of harmonic time-dependence at any given frequency:

$$\nabla\times\left(\frac{1}{\varepsilon(r)}\nabla\times H(r)\right)=\left(\frac{\omega}{c}\right)^2 H(r) \tag{23.3a}$$

$$E(r)=\frac{i}{\omega\varepsilon_0\varepsilon(r)}\nabla\times H(r) \tag{23.3b}$$

where, ω is the angular frequency of propagating light. The solution to the Eq. (23.3a) can entirely determine steady-state distribution of magnetic fields within a PC, which is specified by the dielectric function $\varepsilon(r)$. Then, Eq. (23.3b) can be used to calculate electric fields. The photonic band structure described earlier can be obtained from the solutions to Eq. (23.3a). This eigenvalue equation may have solutions in the form of discrete functions or bands of solution depending on $\varepsilon(r)$. A photonic band is obtained for cases where Eq. (23.3a) holds over a continuous range of frequencies. On the other hand, a PBG will be obtained where no solution is found for Eq. (23.3a) over a continuous range of frequencies [for the given $\varepsilon(r)$]. A discrete solution to Eq. (23.3a), corresponding to a specific isolated frequency, describes a confined optical cavity state at that frequency within a PBG. Such discrete solutions are the result of certain geometric defects in the lattice periodicity. The problem, however, is that Eq. (23.3a) cannot be solved analytically for most practical PC geometries, leading researchers to use numerical modeling when predictions are needed about how the various dielectric configurations will behave. However, perturbation theory can be used in relation to an electric-field equivalent of Eq. (23.3a) in an attempt to find a useful relation when considering the small changes in refractive index that often occur in PC sensors:

$$\frac{\Delta\omega}{\omega}\cong-\frac{1}{2}\frac{\int d^3r\Delta\varepsilon(r)|E(r)|^2}{\int d^3r\varepsilon(r)|E(r)|^2}\cong-\frac{\Delta n}{n}.(\text{fraction of}\int\varepsilon|E|^2\text{ in perturbed region}) \tag{23.4}$$

The aforementioned equation shows a linear relationship between the fractional change in the frequency of an optical mode of the PC and the fractional change in refractive index as well as the fractional portion of electric-field energy in the region of refractive index perturbation (Joannopoulos et al., 2008).

3 PHOTONIC CRYSTAL SLABS

Although not physically realizable, the simplest form of a 2D PC could be made using a periodic dielectric crystal pattern over a two-dimensional plane which limits photon propagation. In reality, the dielectric materials extend homogeneously into the third dimension. In this respect, two rather different implementations of 2D PCs exist: photonic crystal fibers (PCFs) and photonic crystal slabs (Olyaee et al., 2011; Olyaee and Taghipour, 2012a,b). PCFs are in turn divided into two main groups. The first one is index-guiding PCFs that consist of cladding including air holes that go all along the fiber. These holes allow for the interaction of light with analytes located in the cladding holes through the evanescent field (Begum and Namihira, 2012; Naraghi et al., 2013; Olyaee et al., 2014, 2015; Olyaee and Naraghi, 2013; Otupiri et al., 2014; Yu et al., 2010). Another type of PCFs are PBG fibers that include large hollow core and guide light via PBG mechanism (Chow et al., 2014; Otupiri et al., 2014). While the use of 2D PC fibers for physical and biochemical sensing has been reported, we will not focus on those structures here and instead refer readers to Begum and Namihira (2012), Chow et al. (2014), Dinish et al. (2014), Naraghi et al. (2013), Olyaee et al. (2014), Otupiri et al. (2014), Padmanabhan et al. (2010), and Yu et al. (2010). Slab-PCs are better candidates for more accurate approximation of the zero-thickness case of 2D PCs. The order of thickness of the dielectric materials in the third dimension here is often determined by the crystal lattice spacing, and light confinement in that direction results from total internal reflection (TIR). Given the possibility of integration with other nanocomponents on the chip, we focus on slab PCs in this chapter.

There are two kinds of 2D PCs slabs: waveguides and cavities. A cavity is created by introducing a local defect, for example, by removing one lattice point. Point defect is the smallest cavity in the PC lattice that has dimensions about the wavelength of light and could confine light strongly and is sufficient for lots of applications, such as nanofilters and low threshold nanolasers and photonic chips (Tokushima et al., 2000). This effect doesn't arise just through lattice point removal; in addition, such resonation of localized optical modes in the altered region can be produced by perturbing the dimension or location or the material of one or more lattice points (Fig. 23.2). Sharp transmission peaks (or dips) at the corresponding frequencies can be observed in the optical transmission spectrum as far as there is sufficient spectral separation between supported defect modes in the bandgap. Changes in refractive index within the optical cavity may affect the frequency of these localized optical resonances and therefore highly site-specific material interactions can be identified by monitoring the frequencies of such resonances. Various designs can be proposed for these point-like defects in which many periods of unaltered crystal lattices contain a region of lattice alterations. A selection of such designs and implementations in sensors is covered in Section 5.6.2.

Getting to gather multiple cavities, light could couple from one cavity to another. In this way a path for guiding light is created. In other words, a PC waveguide is created. This type of defect is used to strategically guide light between two locations within a slab-PC. The ability to make low-loss large angle bends at a single lattice point is considered one of the unique features of PC waveguides (Song et al., 2005; Tokushima et al., 2000); however, almost all PC waveguides used in analytical sensors provide a collinear connection that links two ridge waveguides at opposite edges of a slab-PC.

Biosensing constitutes a particular application of linear waveguide in a PC slab. In these applications a waveguide is used to guide a given bandwidth of light in an approach similar to a conventional core/cladding waveguide structure where total internal reflection of the slab and the bandgap of the 2D PC structure are used to confine the light in vertical and lateral direction, respectively.

Three conditions for waveguides must hold for having optimal performance (Johnson et al., 2000): (1) supporting true guided modes by waveguide, that is, no leaky modes; (2) having single-mode

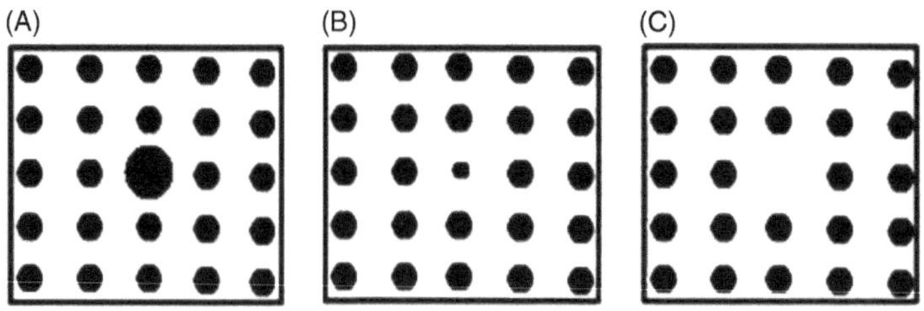

FIGURE 23.2

Photonic crystal point defect cavities created by increasing (A), reducing (B), and removing (C) the hole.

waveguide in the frequency range of interest; and (3) prohibiting radiation losses by confining the guided mode to the bandgap. The basis for using the third criterion in biosensors is PC waveguides. The changes in the refractive index on the surface or at the sidewalls of the holes of the PC slab considerably affect the bandgap (Najafgholinezhad and Olyaee, 2014; Olyaee and Bahabady, 2014a; Olyaee and Najafgholinezhad, 2013a). For example, by increasing the refractive index, the bandgap shifts toward longer wavelengths (blue shifted), this in turn will result in guiding leaky modes and blue shift of cut-off wavelength. The refractive index on the surface, changed by binding of molecules, can be determined through measuring this cut-off wavelength (Threm et al., 2012).

If a cavity comes close to a PC waveguide (Fig. 23.3), light could couple this cavity and waveguide, directly or reversely. However, only lights with resonance frequency could couple sufficiently. In this way certain wavelengths can come out from the waveguide. The cavity resonates only if its length is the multiple of wavelength of light. This means that a larger resonator is able to support more wavelengths.

Instead of cavity coupled to the waveguide from the side it can be embedded in the waveguide and used as a wavelength selection transmission filter. Fig. 23.4 shows an example of this configuration.

There is another kind of defect called nanoring resonator. Nanoring resonators consist of a ring waveguide defect adjacent to a bus waveguide defect in crystal lattice. Light from a laser travels in bus waveguide and evanescently couples into the ring resonator and the light of resonant wavelength will be coupled into drop waveguide eventually. Equality of the optical path length around the ring

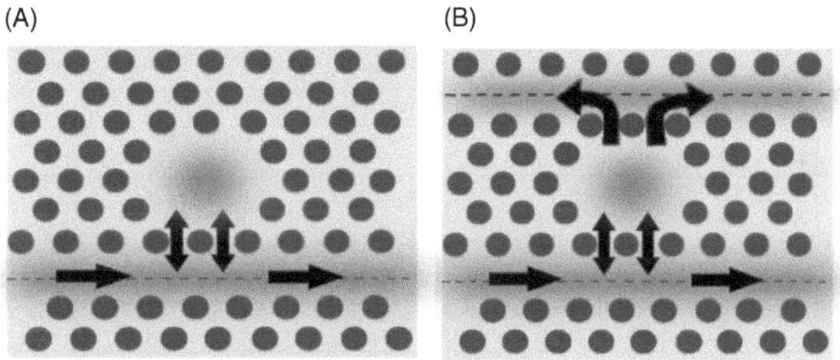

FIGURE 23.3 Interactions Between PC Cavity and Waveguide

Coupling light between a cavity and a waveguide (A) and coupling light between a cavity and two waveguides (B).

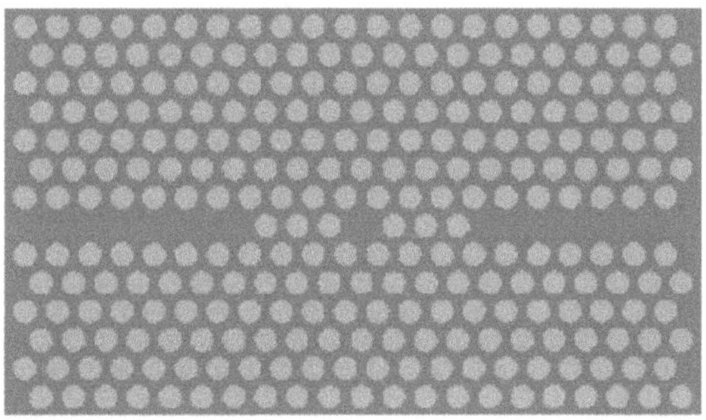

FIGURE 23.4 Wavelength Selection Transmission Filter

This structure transmits the resonance wavelength.

Reproduced from Lončar, M., Scherer, A., Qiu, Y., 2003. Photonic crystal laser sources for chemical detection. Appl. Phys. Lett. 82 (26), 4648–4650.

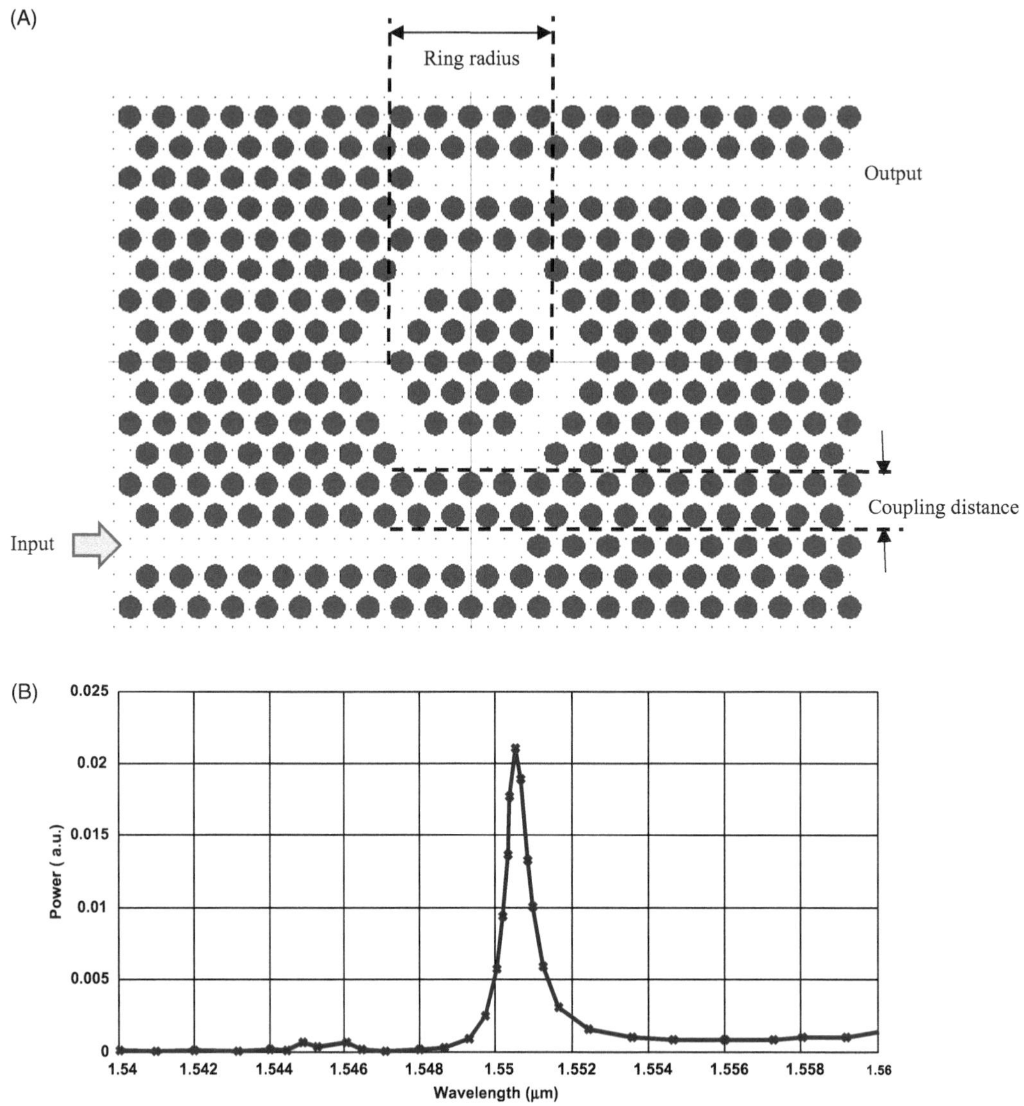

FIGURE 23.5

(A) Sketch of the PC-Based Nanoring Resonator and (B) Output Spectra of the PCs Based Nanoring Resonator in (A).

Modified from Hsiao, F.L., Lee, C., 2010. Computational study of photonic crystals nano-ring resonator for biochemical sensing. Sens. J. IEEE 10 (7), 1185–1191.

with an exact multiple of the wavelength of the excitation light realizes the conditions for constructive interference and the ring is said to be on-resonance. Thus, output spectrum of drop port reveals a resonant peak. As seen in Fig. 23.5A, the bottom-terminal waveguide known as the bus waveguide provides for excitation at the resonant mode of the nanoring resonator in the light coming from the left side of the waveguide (indicated here by yellow arrow). Then the resonant light is coupled to the upper-terminal waveguide, often referred to as the drop waveguide. Resonant wavelength and output spectra are changed with the variation of two important parameters indicated in Fig. 23.5A: ring radius and coupling distance. Fig. 23.5B shows the output spectra of a PC-based nanoring resonator (Hsiao and Lee, 2010). More examples of PC-based nanoring resonators could be found in Section 5.6.3.

4 MATERIALS AND FABRICATION METHODS

Depending on the intended application, numerous configurations of materials can be used to implement PC structures for sensing. In addition to the periodic structure distribution, the spectral response of the PC will also depend, to a large extent, on the refractive index contrast between its materials. There are different configurations of materials used for the development of PC structures that indicate PBGs are wide enough for sensing applications, and such structures can be implemented using aluminum gallium arsenide (AlGaAs, $n_{AlGaAs} = 3.37$) on gallium arsenide (GaAs, $n_{GaAs} = 2.89$), indium gallium arsenide phosphate (InGaAsP, $n_{InGaAsP} = 3.4$) on gallium arsenide, silicon nitride (Si_3N_4, $n_{Si3N4} = 1.98$) on silicon oxide (SiO_2, $n_{SiO2} = 1.45$), and silicon (Si, $n_{Si} = 3.45$) on silicon oxide. Clearly, the last configuration (silicon on silicon oxide, i.e., silicon-on-insulator) is widely used by researchers and developers, and hence more frequently reported in the literature.

Electron-beam lithography and reactive ion etching (RIE) are among the methods commonly employed for fabricating 2D PC structures, such as PC slabs, microcavities, and waveguide structures. A number of cost-effective methods have been proposed for fabrication of PC slabs for biosensor applications (Block et al., 2007). Nanostructures can be produced using nanoimprint lithography and injection molding. Unlike the previous techniques, this alternative method needs no etching process in light exposure for the lithography step. This can result in a smaller number of fabrication steps and a low-cost and high-throughput PC fabrication method. Nanoimprint lithography was first proposed by Chou et al. (1996). This type of lithography consists of two important techniques: thermoplastic and photonanoimprint lithography. For this purpose, a mold produced through conventional methods, with the desired nanostructure, is pressed in a thermoplastic polymer under a certain pressure and heat, or into a photocurable liquid resist with an ensuing UV-light exposure. As a standard technique, injection molding is used in a variety of products. It often involves injection of a thermoplastic or thermosetting plastic material into a mold cavity, and then cooling and hardening it to the desired shape. Using this technique, parts with macroscopic and microscopic features, for example, a petri dish containing a nanostructured surface, can be fabricated in one step. A high index layer on top of these nanostructures is needed for creating PCs. This needs to be coated in a further step (Block et al., 2008; Cho et al., 2011; Kouba et al., 2006).

5 INTRODUCTION TO BIOSENSORS

Today, the interest in monitoring and adjusting many parameters in various cases related to nanotechnology, food industry, clinical diagnostics, biomedical research, health, medicine, environmental protection, and forensics is growing. Biotechnology is a broad term covering a wide range from industrial biotechnology to environmental biotechnology. On the other hand, medical biotechnology has a defined role that includes technologies used in the pharmaceutical industry and diagnostics. Biomolecular interactions of molecules are very important in drug identification. Significant information about the molecules can be obtained by closely observing these interactions in real time. Furthermore, label-free biosensing can greatly improve the diagnostics of diseases as indicated, for example, by detection of biomarkers that have improved the diagnostics of cancer (Cunningham, 2010; Homola, 2008; Pineda et al., 2009; Ymeti et al., 2007; Zheng et al., 2005). Here we introduce the basics of biosensors.

5.1 BASICS OF BIOSENSORS

A biosensor is a system designed to recognize biochemical molecules, referred to as analytes, based on their binding on a biochemical layer; the information is then converted by the biosensor into a measurable and analyzable signal (Mohanty and Kougianos, 2006; Olyaee et al., 2013; Olyaee and Najafgholinezhad, 2013b). The system usually consists of three parts: the biological layer that detects and recognizes analyte(s) of interest; the transducer which transforms the chemical information provided by the biological layer into measurable mechanical, optical, or electrical signal; and the read-out section which is capable of visualization and interpretation of the sensing information. However, sensing area is considerably reduced in size in small biosensors, requiring particular attention that should be paid to the delivery of the target analyte(s). Thus, a photonic biosensing system can be regarded as a system basically consisting of a sample delivery system. Fig. 23.6 shows a simplified schematic of a biosensor with the four parts noted earlier.

The sensor uses its recognition layer to selectively detect the presence of the target analytes. This part often consists of a specific coating of proper receptors with the ability to detect the target molecules and distinguish them from other particles present in the sample. The transducer then converts the chemical/physical signal provided by the recognition layer into a measurable signal that can be quantified in the read-out system.

5.2 SURFACE CHEMISTRY

Sensors based on transient wave can measure the changes in the refractive index in the surroundings of the structure and then transform it into a measurable signal. This mechanism is used in optical sensors

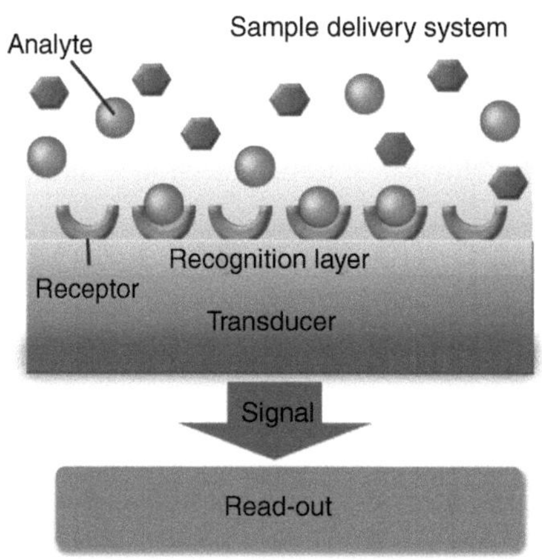

FIGURE 23.6 Schematic Representation of a Biosensor Including Biological Layer, Sample Delivery System, Transducer, and Read-Out of the Output Signal

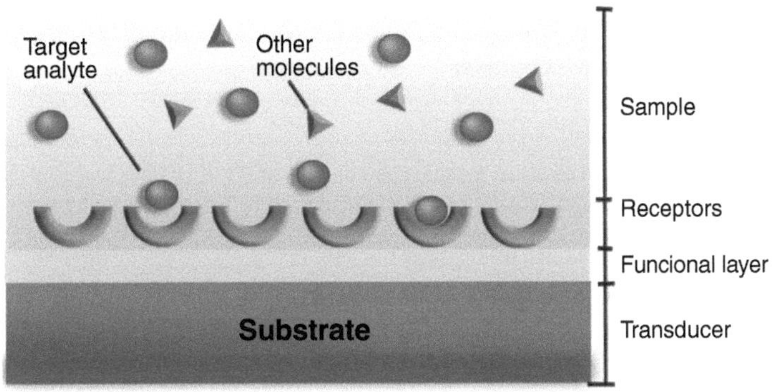

FIGURE 23.7 The Parts Involved in the Detection Process for an Affinity-Based Biosensor

to detect the presence of particles on the structure's surface. However, optical sensors are not able to specifically detect the presence of particular biological molecules and distinguish them from the background molecules. An additional layer is often used to coat the sensing area of a photonic sensor in order to give it the ability for such differentiation. This coating layer allows for attachment to the surface of proper receptors specific to the target molecules to be detected as well as to simultaneously repel the binding of the nondesirable particles of the analyzed sample (Rogers, 2000). The basic parts (sample, receptors, functional layer, and transducer) involved in the detection process are shown in Fig. 23.7. The additional layer in an affinity biosensor is responsible for promoting the binding of an organic molecule (i.e., the receptor) to an inorganic substrate or surface.

Biochemical functionalization (or biofunctionalization) of the sensing structure is a technique used to develop the recognition layer which consists of the molecular receptors. Biofunctionalization is an extremely important process for fabricating an acceptable biosensor because the quality of the biochemical coating determines several intrinsic parameters, such as sensitivity, detection limit, long-term stability, or uniformity. Receptors can be attached to the sensing structure through various mechanisms including physical attachment, chemical attachment, or molecular recognition. Since the recognition layer is contained in the receptors, different types of molecules can be used for this purpose; these include enzymes, antibodies, proteins, oligonucleotides, DNA, cells, or aminoacids. Receptors often used for the development of affinity biosensors are listed in Table 23.1 (Wang and Liu, 2011).

In some cases it is very difficult to directly detect the target analyte because of its intrinsic physical and biological properties. In such cases, indirect detection is achieved by attaching a label to the target analyte and measuring this attached element instead. These labels can be easily detected for their particular fluorescence, radioactivity, or metallic nature; applications that involve gene sequencing, gene expression microarrays, protein–protein interaction screening, and cell imaging often require assays through the detection of fluorescent proteins, chemical tags, and nanoparticles. Fluorescence-based assays for achieving detection sensitivity in the <10 pg/mL range (Gonzalez et al., 2008a,b), subdiffraction limit imaging (Eid et al., 2009; Hell, 2003), and throughput of up to millions of assays per chip (Schena et al., 1995) rely on various materials for detection tools, fluorescent tags, surface chemistries, and light sources. Despite the great potential of fluorescence-based detection, in addition to difficulty of controlling, the labeling approaches can interfere with receptor affinity, can change protein dynamics, and cannot reveal the real-time dynamics. On the other hand, label-free assay methods are preferred in

Table 23.1 Specific Biomolecules Widely Employed for the Functionalization of Biosensing Devices and Their Recognition Mechanism

Bioreceptor Component	Recognition Mechanism
Enzymes, proteins	They can catalyze specific chemical reactions. It can be measured by conversion of the analyte into detectable signal. The detection of analyte can act as an inhibitor or activator of the reaction.
Antibodies and antigens	The antigens trigger the immune response of the organism to produce antibodies. Therefore, the antibody specifically recognizes the antigen in the sample.
DNA strands	The recognition process is based on the hybridization of complementary DNA strands through double helix.
Cellular structures	The whole microorganism can be used as a recognition element, such as noncatalytic receptor.
Biomimetic	The recognition is achieved using genetically engineered molecules: for instance, artificial membrane and molecularly imprinted polymers (MIP) imitate the shape of the bioreceptor molecules in order to attract the target molecules.

From Wang, P., Liu, Q., 2011. Biomedical Sensors and Measurement. Springer Science & Business Media. 2011 ed.

some cases because of their capability in studying biomolecular binding events or cellular processes based on physical property of the analytes. Label-free methods are often easier to implement compared to tag-based methods because the former transduce biological phenomena instantly to a measurable output; however, these methods need some type of sensor surface on which assays must take place. In label-free assay, the receptor is attached on the surface. Next, the detection is carried out by flowing the complementary analyte diluted in a buffer. The refractive index will change on the surface since target molecules will bind to the immobilized probes via affinity interaction.

Optical transducers employ the interaction of the light with the target analytes to convert the biological information into a measurable signal. In the majority of optical sensors, the sensing mechanism relies on the variation of the effective index of the propagating mode that occurs when a variation of the chemical or physical parameters is produced on the surface of the structure.

Effective index variations can be transformed into a measurable signal (e.g., intensity or frequency variations), by making use of different configurations of photonic structures [e.g., Mach–Zehnder interferometer (MZI), ring resonators (RR), PCs]. In the following subsection, we will present advantages of using PC biosensors.

5.3 ADVANTAGES OF USING PC NANOSTRUCTURES FOR BIOSENSING

Nanophotonic biosensors can be employed in the diagnosis of infectious diseases and avoid the viral spreads, sensing of cancer biomarkets and Alzheimer's detection, and in pharmacology (drug/vaccine discovery). Here, a clear difficulty is to develop, implement, and test nanostructures that can localize light below subdiffraction limit and enhance the field, promote light–molecule interaction, and lead to eventual compact and portable devices for point-of-care diagnostics. Handheld diagnostic capabilities with a "lab-on-a-chip" can be achieved by the means of miniaturization and integration of devices based on micro/nanoarrays of sources, detectors, and active or passive biosensing surfaces. In addition, complicated intra and intercellular biochemical processes can be better observed and understood by making use of real-time multimodal sensing technologies. Owning periodic nanostructures, PCs allow high confinement of the optical field in the defect regions designed for sensing purposes, as well as the enhancement in the light–matter interaction provoked by the slow-light effect.

5.4 IMPORTANT PARAMETERS FOR PC BIOSENSORS

In the following list, we have selected a number of parameters, which are usually considered in evaluating the performance of a biosensor (photonic or not). It should be noted that in some areas of application, some requirements will be very strict while other parameters can vary more freely.

1. *Full-width at half-maximum (FWHM)*: The sensitivity and sensor quality factor is improved at the low values of full-width half maximum of the resonance.
2. *Quality factor*: Quality factor is one of the most important parameters in the study of PC sensors and is calculated using Eq. (23.5). By definition, the quality factor is resonance wavelength shift toward the spectral FWHM of the resonance.

$$Q = -\frac{\lambda_0}{\Delta\lambda_{\mathrm{FWHM}}} \tag{23.5}$$

   In this equation λ_0 is the center resonance frequency and $\Delta\lambda_{\mathrm{FWHM}}$ is the spectral FWHM of the resonance. High quality factor is desirable for PC slab biosensors that measure the shift of resonance wavelength, and low quality factor is good when the change of intensity is used.
3. *Bulk sensitivity*: This parameter shows the extent to which changes in the sensor's output depend on the homogeneous change of the medium; for example, bulk sensitivity of an optical sensor depends on the homogeneous changes in the refractive index. In optical biosensors, this parameter is directly dependent on the efficiency of the interaction between the light and the matter.
4. *Sensitivity*: This is obtained by measuring the variation of the output signal caused by a certain chemical/physical variation in an experiment and the amount of the analyte. This parameter depends on the sensitivity of the transducer as well as some other factors including the recognition layer, the affinity of the molecules, the buffer conditions, or the fluidic architecture.
5. *System noise*: This parameter indicates changes in the signal in a static measurement. Resolution can be calculated using the value for the noise of the system obtained in a steady-state condition without any intentional variation of the sensor. The signal noise is typically calculated in terms of the standard deviation (σ) of the stabilized signal.
6. *Detection limit*: This shows the lowest detectable concentration or amount of analytes. The parameter can be determined in different units, such as in refractive index units (RIU), for optical sensors to characterize the bulk performance of an optical sensor, in surface mass density, in total mass, or in analyte concentration, $\mathrm{DL} = \frac{\sigma}{S}$, where σ is the system noise and S represents the sensitivity.
7. *Selectivity*: Selectivity depends on the chemical modification of the transducer to a very large extent and indicates the capability to distinguish the target analyte from other substances in the sample.
8. *Dynamic range*: This indicates the range between the largest and the smallest signal values that can be directly related with sensing events.

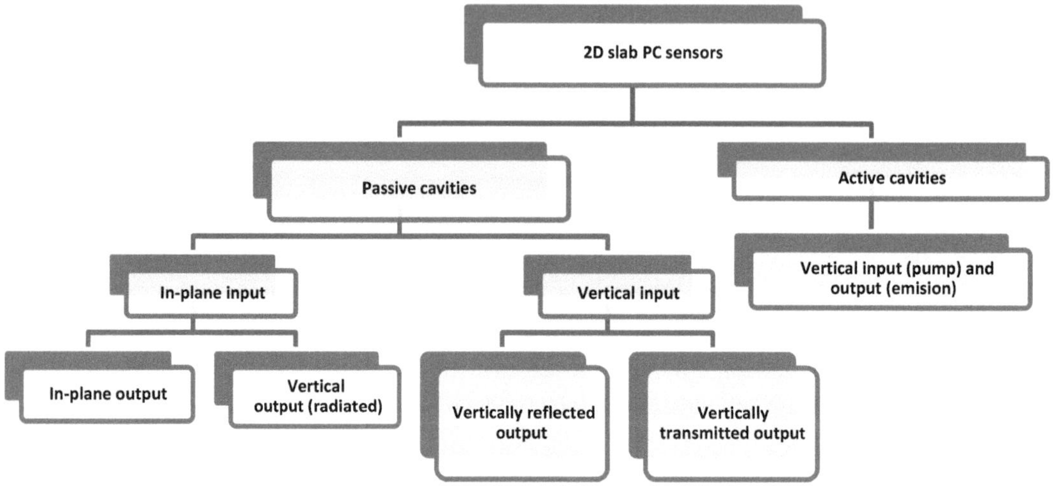

FIGURE 23.8 Measurement Methods Relating to the Input and Output Light in 2D Slab-PCs

Reproduced from Baker, J.E., Sriram, R., Miller, B.L., 2015. Two-dimensional photonic crystals for sensitive microscale chemical and biochemical sensing. Lab Chip 15(4), 971–990, with permission from The Royal Society of Chemistry, Copyright 2015.

Cost, throughput or multiplexing level, portability, sample volume, and energy consumption, are also important parameters in sensing performance.

5.5 METHODS OF MEASUREMENT

According to Baker et al. (2015), we can classify measurement schemes of slab biosensors as shown in Fig. 23.8. According to Fig. 23.8, we could consider two main kinds of PC slab biosensors, PCs with passive cavities and with active cavities. The strong optical resonances that can be achieved in slab-PC cavities have made them useful when implemented in both kinds of sensors. For biosensors with passive cavity we could introduce light in-plane or perpendicular to the surface of the sensor. For perpendicular input we could detect reflected or transmitted output. And for in-plane input we could detect in-plane output (waveguided) or reflected ones (radiated).

Output intensities have been used as an indicator of changes while in other applications the changes of resonance wavelength have been utilized for this purpose.

In-plane waves are more dominant tools for measurement compared to waves perpendicular to the surface. However, illuminating a passive 2D PC slab surface, followed by observing the reflected or transmitted light radiating perpendicularly from the slab, has also been used as a measurement technique. Here the analysis usually relies on the spectrum of reflected or transmitted light. In addition, changes in overall intensity of a broad reflection peak (Baker et al., 2015; Chakravarty et al., 2005; Dündar et al., 2010; Endo et al., 2012b; Hashimoto et al., 2014; Terada et al., 2014) have been observed for this purpose in one study while another study examined changes in the reflectance spectrum that were monitored (Zecca et al., 2014). Huang et al. (2009) also monitored shifts in the wavelengths of spectral features, but for transmitted light.

PC architectures using in-plane-wave can be designed using a variety of structures, as discussed in the next subsection.

5.6 EXAMPLES OF PC SLAB BIOSENSORS

For convenience, various PC geometries will be discussed according to the following groupings: PC with waveguide-defect, with cavity-defect, nanoring resonator defect, and combination of these defects, and biosensors without defects.

5.6.1 Biosensor-based PC waveguides

PC waveguides can be used as a platform for biosensors. Cut-off wavelength of waveguide is used as an indicator to detect changes in the refractive index. Compared with other PC biosensors, waveguide-based PC biosensors have larger sensing area. This large area requires a larger sample volume consequently.

Buswell et al. (2008) demonstrated a specific detection of streptavidin to biotin using these kinds of sensors. Waveguide was formed by filling the entire holes of a row. The binding of streptavidin to immobilized biotin causes a red-shift of 0.86 nm of the cut-off wavelength for a 2.5-nm streptavidin film. The bulk sensitivity of the shown waveguide biosensor is 88 nm/RIU. To improve the sensitivity, they reduced the dimension of entire holes of a row instead of filling them. With this device, design of the cut-off wavelength is not modified compared to the previous design. Hence, the response is similar except that the modes are shifted to higher frequencies inside the bandgap. However, the advantage of this design is the increased surface area available for sensing in the central high-field region. This modification results in an improvement of the sensitivity up to 120 nm/RIU, which corresponds to an improvement of 40% in comparison to the sensor with a regular waveguide. Figs. 23.9 and 23.10 show the waveguide with filling holes and reducing the dimension of holes, respectively.

Using slotted PC–WGs is another technique to design and implement PC-WG-based sensing structures. This technique was originally introduced by the group of Krauss (Falco, 2008; Scullion et al., 2011). As seen in Fig. 23.11, in this approach a narrow trench is created in the middle of a PC–WG. The approach was an attempt to combine the advantages of PC-based sensing structures with the extremely high confinement and intensity of the optical field in slot waveguides (Almeida et al., 2004).

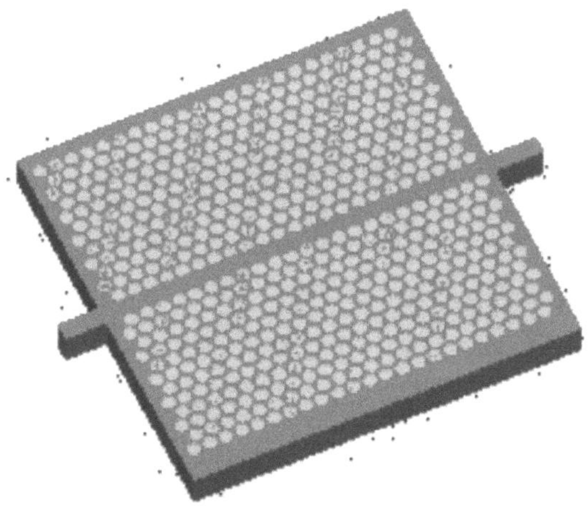

FIGURE 23.9 PC Waveguide Biosensor Constructed With Filling the Entire Holes of a Row

Modified from Buswell, S.C., Wright, V.A., Buriak, J.M., Van, V., Evoy, S., 2008. Specific detection of proteins using photonic crystal waveguides. Opt. Express 16 (20), 15949–15957.

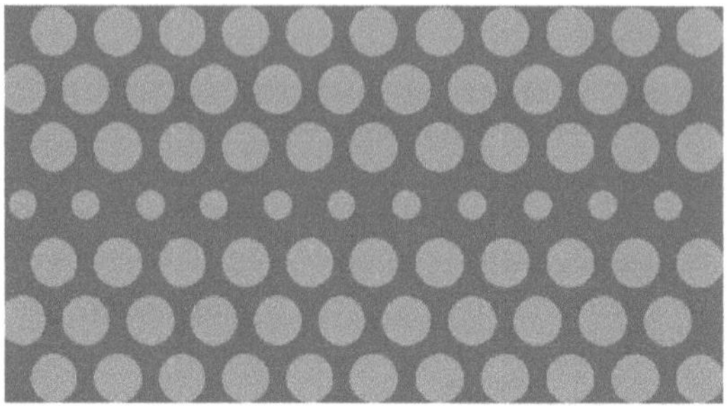

FIGURE 23.10 PC Waveguide Biosensor Constructed With Reducing the Dimension of the Entire Holes of a Row

Reproduced from Buswell, S.C., Wright, V.A., Buriak, J.M., Van, V., Evoy, S., 2008. Specific detection of proteins using photonic crystal waveguides. Opt. Express 16 (20), 15949–15957.

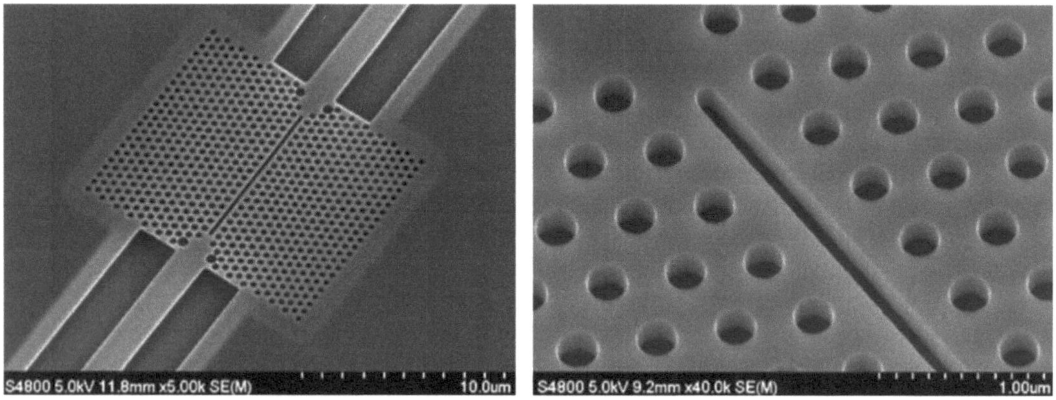

FIGURE 23.11 Scanning Electron Microscope (SEM) Images of Slotted PC–WG

From Scullion, M.G., Di Falco, A., Krauss, T.F., 2011. Slotted photonic crystal cavities with integrated microfluidics for biosensing applications. Biosens. Bioelectron. 27(1), 101–105, with permission from Elsevier, Copyright 2011.

This approach resulted in a sensitivity of 800 nm/RIU for bulk refractive index sensing (Falco, 2008) while PDMS-based microfluidic delivery system (Scullion et al., 2011) was successfully employed to detect concentrations of avidin as low as 15 nM. In addition, the small size of this structure ($\approx 2.2\ \mu m^2$) allowed to achieve a minimum detectable mass of 100 ag.

5.6.2 Biosensor-based PC cavity

Biosensors with PC waveguide can be easily integrated into an optical chip; however, the readout of the cut-off wavelength is not precise enough for sensitive detection. On the other hand, a PC nanocavity creates resonances with a high quality factor that provides higher level of precision for sensitive detection. Now there are a variety of PC cavity based biosensors. These biosensors typically comprise cavities side-coupled to the PC waveguide, in which cavity is the sensing area and waveguide is used as a light path in and out of the crystal and coupling of light into the cavity.

Lx cavities are formed by successive removal of adjacent holes. Here x represents the number of removing holes; in other words, x represents the length of cavity. For example, L3 cavity is formed by removing three adjacent holes from a lattice hole. So far L3, 4, 7, 13, cavities are demonstrated for

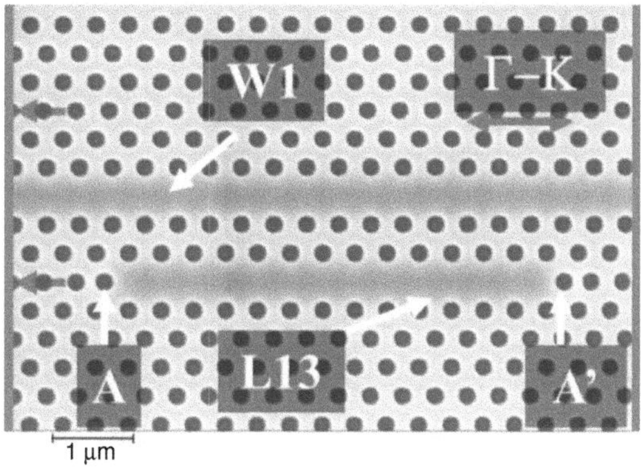

FIGURE 23.12 SEM Image of Sensor Architectures With a Point-Like Defect Placed Within the Vicinity of a Bus W1 Waveguide for Evanescent Side Coupling From the Waveguide to the Cavity

From Chakravarty, S., Zou, Y., Lai, W., Chen, R. T., 2012. Slow light engineering for high Q high sensitivity photonic crystal microcavity biosensors in silicon. Biosens. Bioelectron. 38, 170–176, with permission from Elsevier, Copyright 2012.

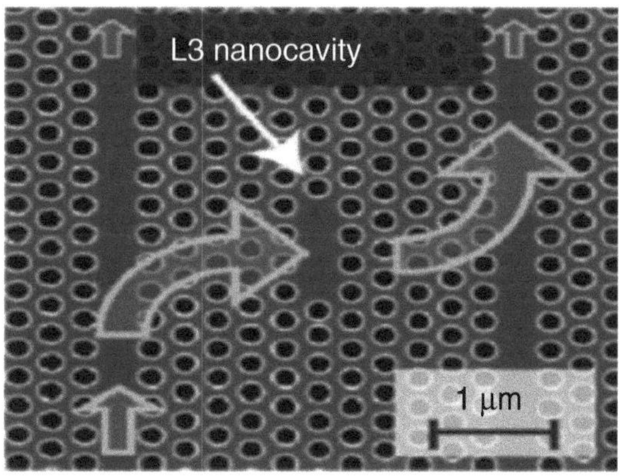

FIGURE 23.13 SEM Image of the PhC Structure Showing the L3 Nanocavity in the Immediate Vicinity of the PhC–WG

From Dorfner, D., Zabel, T., Hürlimann, T., Hauke, N., Frandsen, L., Rant, U., Finley, J., 2009. Photonic crystal nanostructures for optical biosensing applications, Biosens. Bioelectron. 24(12), 3688–3692, with permission from Elsevier, Copyright 2009.

biosensing application. Figs. 23.12 and 23.13 show these cavities (Chakravarty et al., 2012; Dorfner et al., 2009).

Liu and Salemink demonstrated using of chip L7 cavity for biosensing application. This structure could overcome the limitations of light overlap (Liu and Salemink, 2012). L7 cavity is formed by filling of seven adjacent holes on the Γ–*K* direction of lattice. They could achieve more light confinement and *Q* factor compared to previous works. For improving their structure, they introduced some small holes with certain distance in the cavity area. Fig. 23.14 shows the SEM image of the improved structures. By penetrating the solution of water and ethanol to the sensor through a fluid microchannel, they achieved the sensitivity of 460 nm/RIU.

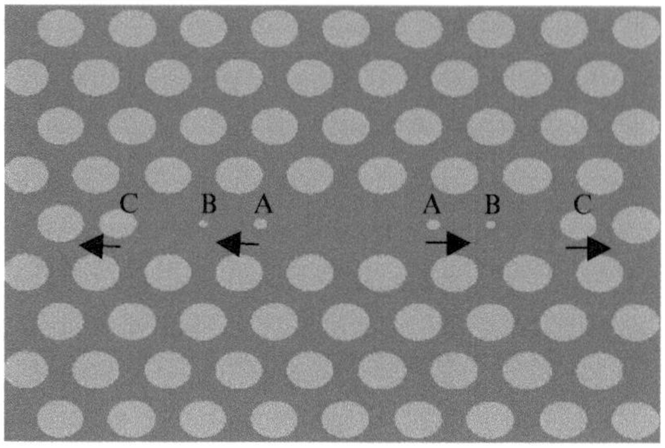

FIGURE 23.14 SEM Image of Modified L7 Cavity PC Nanostructures

From Liu, Y., Salemink, H.W.M., 2012. Photonic crystal-based all-optical on-chip sensor. Opt. Express 20(18), 19912–19920.

Most studies on surface modification procedures and biosensing have reported relatively simple architectures. However, the approach recently proposed by Chakravarty et al. (2013) for developing a more complicated biomolecular detection using cell lysates presented a considerable advancement in this area based on employing an array of four multiplexed cavities of type L13 (Figs. 23.12 and 23.13) with a high Q (~26,000) used to detect a protein associated with lung cancer, that indicated a sensitivity as low as 2 cells/μL for a sensing surface area of 11 μm^2. In addition, nonspecific binding with control antibodies in this study demonstrated the selectivity of the sensing device and its capability in biosensing (Yang et al., 2015).

Sensors with the ability of recording concentrations of analytes over a wide dynamic range of at least 5 orders of magnitude are required for some areas of biosensing, including those involving drug discovery and therapeutic drug monitoring (Scullion et al., 2011). Taking into account that typical L-type PC microcavities have a dynamic range of approximately 3–4 orders of magnitude in biosensing (experiments conducted by Yang et al. (2015) revealed that multiplexing of PC sensors with different defect geometry is one possible way to obtain a wide dynamic range covering 6 orders of magnitude. In this approach, four distinct cavities denoted by L3, L13H, L55, and L21 (L13H indicates that defect holes smaller than the radius of the holes in the bulk PC lattice) are implemented on the four arms of a 1×4 multimode interference (MMI) power splitter and are immobilized with the same probe (biotin). Avidin can be used here as a target protein while bovine serum albumin (BSA) plays the role of a blocking agent.

5.6.3 PC based on nanoring resonator defect

Although there are many biosensors based on a ring resonator with high sensitivity and Q factors, further increasing of Q factor and enlarging the free space range need reduction in ring radius (nanometric). However, this leads to a sharp increase in bending losses of conventional dielectric microring waveguides (Krioukov et al., 2002; Kwon and Steier, 2008).

Using PBG properties of PC nanostructures provide an opportunity to create very sharp bends without bending losses. Binding events along the surface of the nanoring increase the refractive index in the evanescent field area, effectively optically prolong the ring and causing the resonant picks to red shift to longer wavelengths.

Hsiao and Lee (2010) demonstrated a PC nanoring resonator for biochemical sensing. Using direct-write dip-pen nanolithography (DPN) to perform immobilization of protein and DNA in nanometer scale patterns and holes (Haaheim et al., 2005; Hsiao and Lee, 2011) gave them an opportunity of applying a single hole of micro/nanocavity of the PC resonators as the sensing element. They discussed the effect of ring radius and coupling space (as indicated in Fig. 23.5A), and position of sensing holes on Q factor of biosensor. The nanoring resonator is assumed to be packaged inside a fluidic channel. Sensing holes are particular holes at the boundary of the resonator near the bus waveguide fully occupied with biomolecules, such as proteins and DNAs, and changing effective index inside these holes due to the binding are investigated. Fig. 23.15 marks these holes as I1 to I3, and O1 to O5, where "I" refers to inner and "O" refers to outer boundaries. Fig. 23.16 indicates the shift of resonant peaks in terms of different sensing

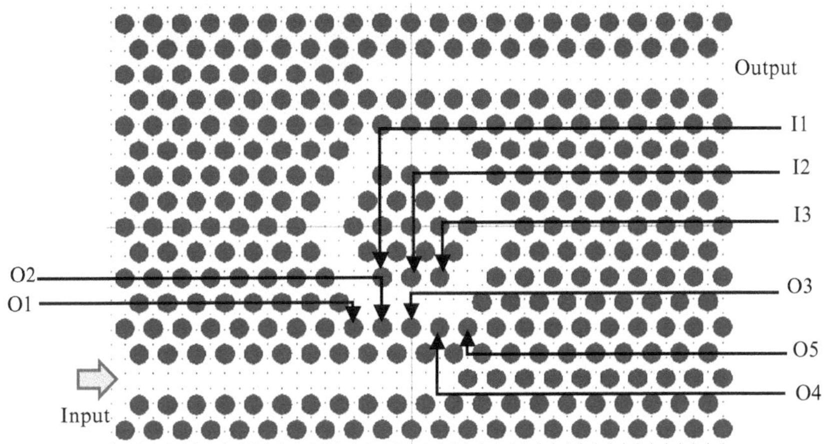

FIGURE 23.15 Sketch of Nanoring Resonator PC

Reproduced from Hsiao, F.L., Lee, C., 2010. Computational study of photonic crystals nano-ring resonator for biochemical sensing. Sens. J. IEEE 10 (7), 1185–1191.

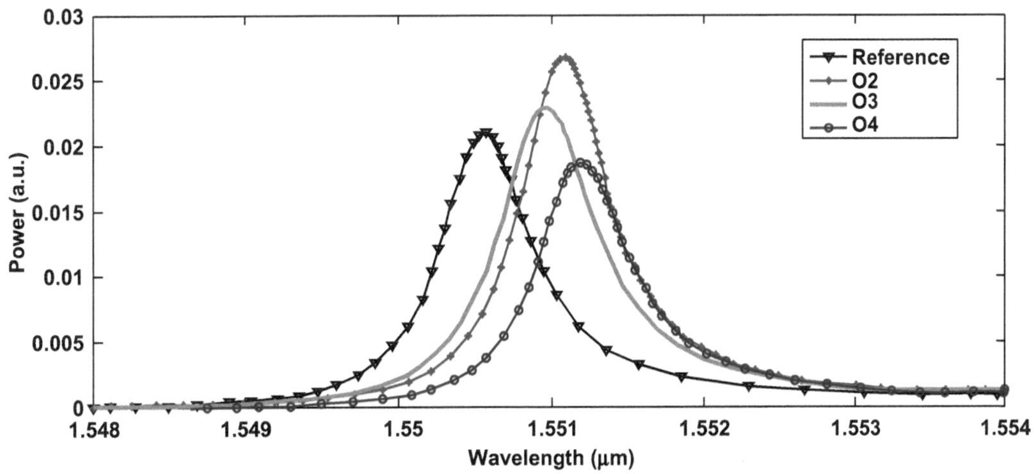

FIGURE 23.16 Output Spectra of Reference Case and Cases of Biomolecules Captured in O2, O3, and O4

Data taken from Hsiao, F.L., Lee, C., 2010. Computational study of photonic crystals nano-ring resonator for biochemical sensing. Sens. J. IEEE 10 (7), 1185–1191.

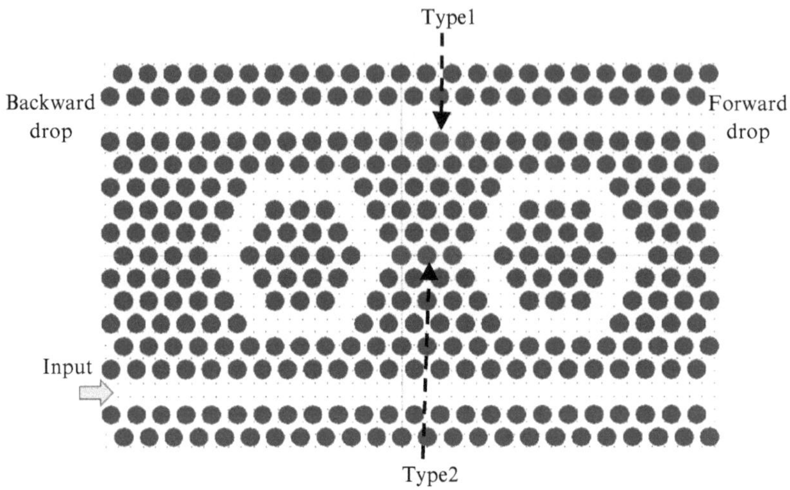

FIGURE 23.17 Sketch of Dual Ring Resonator on PC Including the Position of Sensing Hole

From Hsiao, F.-L., Lee, C., 2011. Nanophotonic biosensors using hexagonal nanoring resonators: computational study. J. Micro Nanolithogr. MEMS MOEMS 10(1), 013001.

holes with respect to reference condition (black curve) that assumes water as analyte. The minimum detectable biomolecule weight in a sensing hole for a nanoring resonator of two-hole and three-hole coupling distance is derived as 0.23 and 0.2 fg, respectively and the quality factors of the nanoring resonators of two-hole and three-hole coupling distances are derived as 2400 and 3200, respectively. It shows promising applications which require detection limit in the order of a single copy of DNA.

Another structure with dual hexagonal ring resonator was designed by Hsiao and Lee (2011) as shown in Fig. 23.17. They selected two positions for sensing holes that are marked in Fig. 23.17 as type 1 and type 2. The drop waveguide has two outputs. A forward drop is referred to the case of the propagated dropped light is parallel, and a backward drop is referred to the case of the propagated dropped light is antiparallel to the input light. There are three holes in between the nearest vertexes of these two nanorings. Forward and backward spectra of both type 1 and 2 are shown in Figs. 23.18 and 23.19, respectively.

As seen in each case, only one port (backward drop or forward drop) reacts to binding of biomolecule.

Variety of geometric based on PC nanoring resonators can be used as a basis for improving the sensitivity and Q factor and enhance detection limit (Bahabady and Olyaee, 2015; Olyaee and Bahabady, 2014a,b, 2015).

Our research group demonstrated a nanoring resonator biosensor for detection of glucose concentration in the urine sample and salinity of seawater (Olyaee and Bahabady, 2014b). The ring resonator is formed by two mixed curvatures and sandwiched by two waveguides in the PC nanostructures with lattice constant of 410 nm and air holes radius of 120 nm, as shown in Fig. 23.20.

The results reveal that for a unit change in the refractive index, the intensity of transmission spectrum was decreased or increased by about 12.5. Also, regression coefficients for detection of glucose concentration and detection of seawater salinity were, respectively, obtained as 0.9945 and 0.9996, which showed a good linearity.

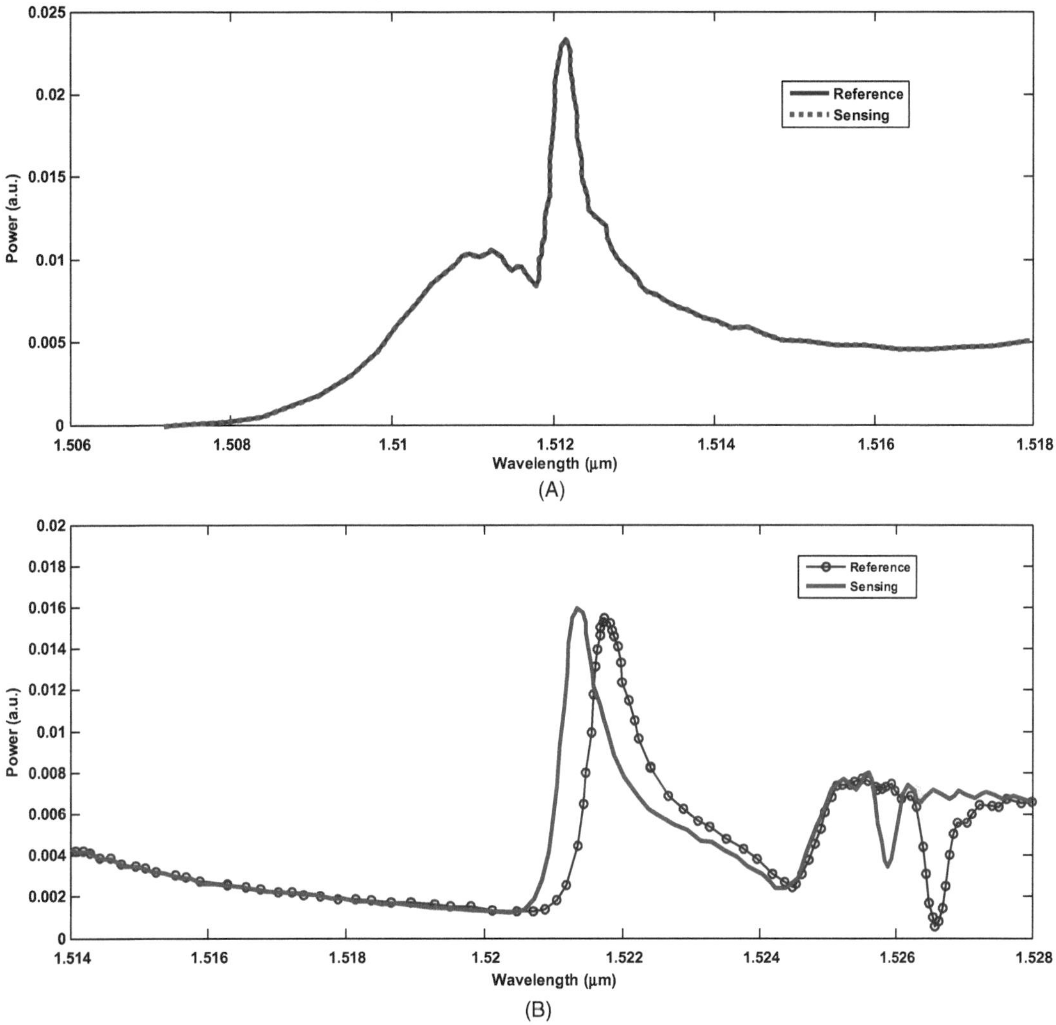

FIGURE 23.18

(A) Forward and (B) backward drop spectra of reference and sensing cases of type 1.

Data taken from Hsiao, F.L., Lee, C., 2011. Nanophotonic biosensors using hexagonal nanoring resonators: computational study. J. Micro Nanolithogr. MEMS MOEMS 10 (1), 013001.

In other work, our research group used two consecutive curves as nanoring resonators sandwiched by two waveguides in silicon-based PC nanostructures with 220-nm thickness on isolator (Olyaee and Bahabady, 2014a). The wave enters through the bottom waveguide, the resonance occurs in the ring resonator, and the wave is transmitted from bottom to top waveguides. Finally, wave leaves the top waveguide and is detected by the time monitor. The preliminary layout of the biosensor based on PC nanoring resonator is shown in Fig. 23.21. The vertical, horizontal, and coupling distances are marked on this figure. These parameters are studied for the optimal design of the structure. The horizontal distance was equal to the number of rows of air holes between the two curves. The vertical distance

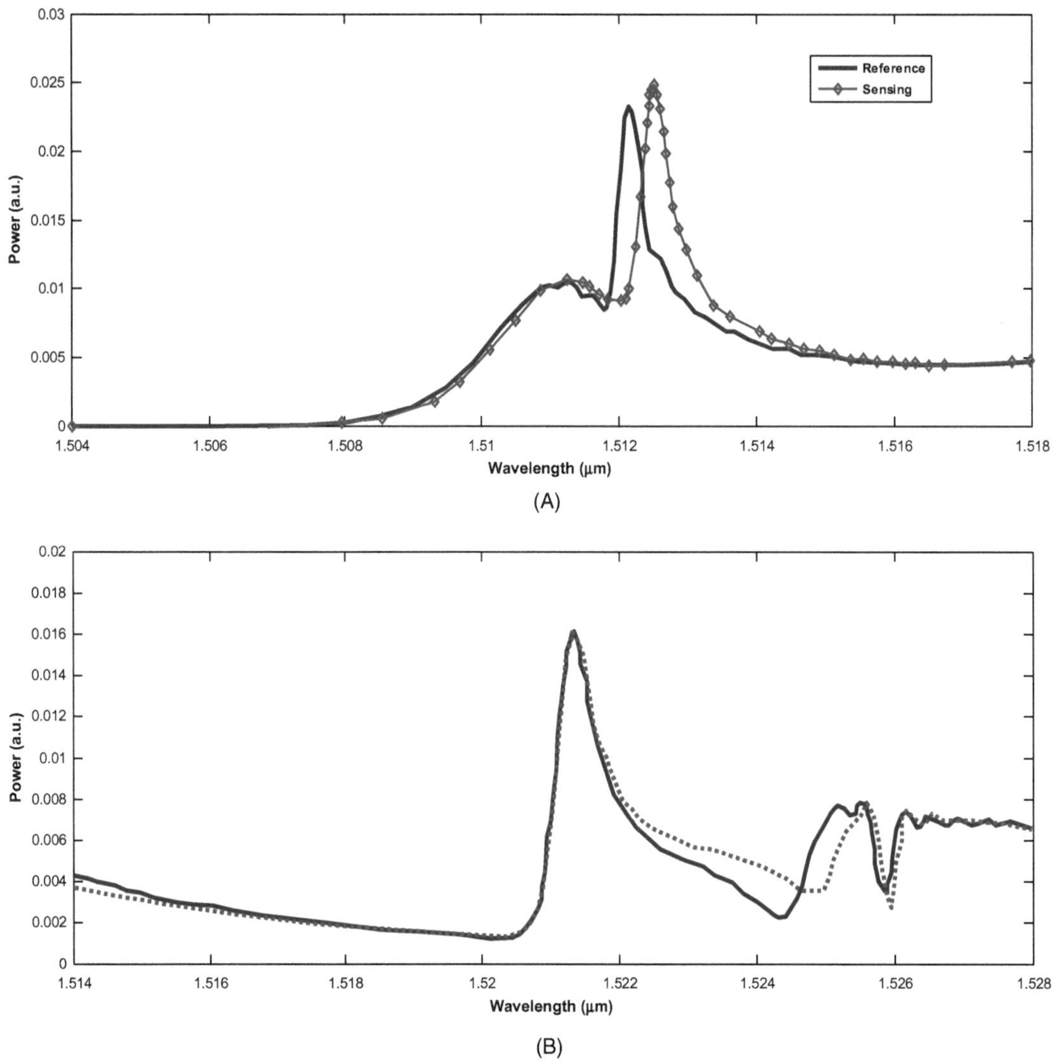

FIGURE 23.19

(A) Forward and (B) backward drop spectra of reference and sensing cases of type 2.

Data taken from Hsiao, F.L., Lee, C., 2011. Nanophotonic biosensors using hexagonal nanoring resonators: computational study. J. Micro Nanolithogr. MEMS MOEMS 10 (1), 013001.

meant that the right curve (curve 2) was several rows higher than the left one (curve 1). Also, coupling distance was the number of rows of holes between the waveguides and ring resonator.

The PBG of this structure for TM gap extended from 0.258 to 0.340 eV. The corresponding wavelength range is equal to the wavelengths between 1205 and 1640 nm. Changing the refractive index of the sensing hole by binding the target analyte, the intensity of transmission spectrum is shifted to a higher value. Therefore, for a unit change in the refractive index of the biosensor, intensity of

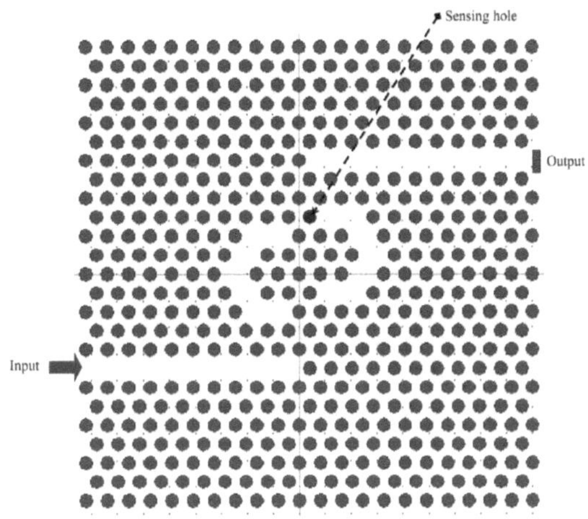

FIGURE 23.20 Sketch of the Biosensor Based on PC Nanoring Resonator

Reproduced from Olyaee, S., Bahabady, A.M., 2014b. Two-curve-shaped biosensor using photonic crystal nano-ring resonators. J. Nanostruct. 4 (3), 303–308.

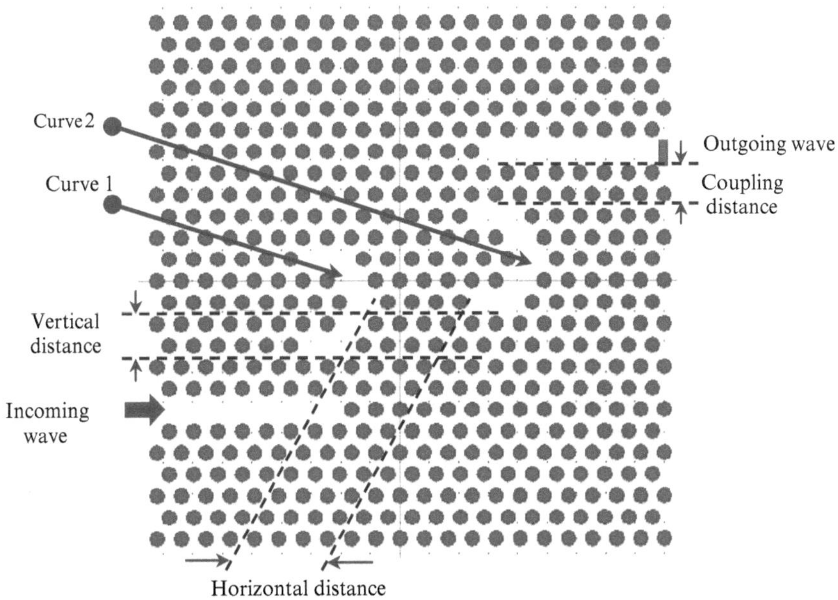

FIGURE 23.21 The Preliminary Layout of the Biosensor Based on PC Nanoring Resonator

From Olyaee, S., Bahabady, A.M., 2014a. Designing a novel photonic crystal nano-ring resonator for biosensor application. Opt. Quant. Electron. 47 (7), 1881–1888.

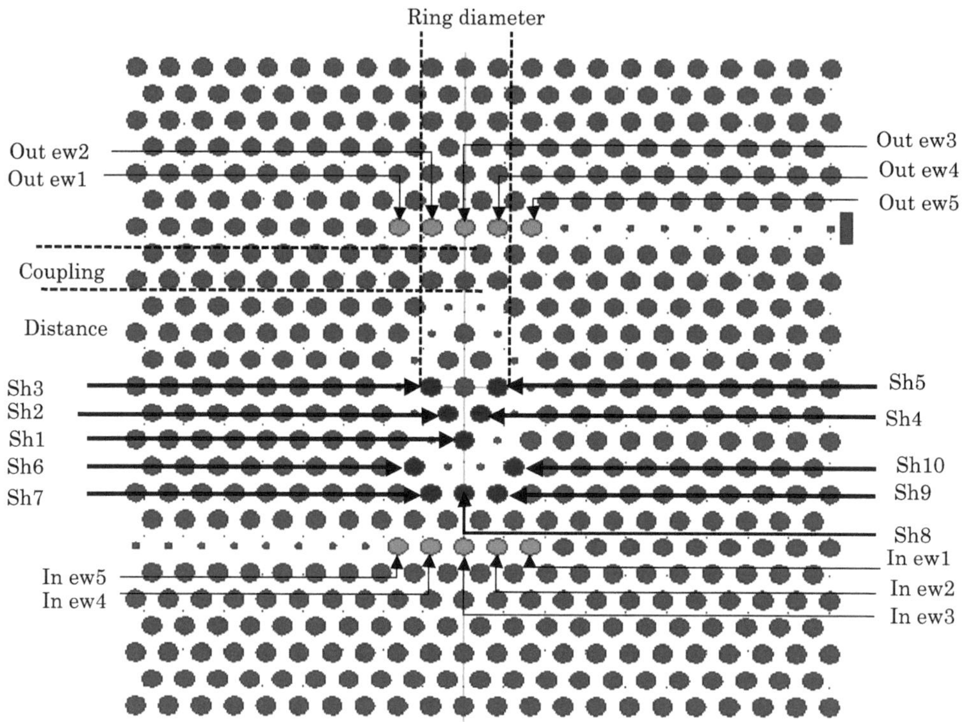

FIGURE 23.22 Sketch of the Biosensor Based on PC Nanoring Resonator

Beneath the waveguide is the input and top of the waveguide is the output.

From Olyaee, S., Bahabady, A.M., 2015. Design and optimization of diamond-shaped biosensor using photonic crystal nano-ring resonator. Optik Int. J. Light Electron Opt. 126(20), 2560–2564, with permission from Elsevier, Copyright 2010.

transmission spectrum has been reduced by 14.26 units. The results reveal that the regression coefficient of the biosensor was about 0.99967.

To enhance the sensitivity and Q factor and to improve detection limit, our research group proposed using diamond-shaped nanoring resonator on the PC structure (Olyaee and Bahabady, 2015). There are several parameters that play important roles in the performance of the diamond-shaped nanoring resonator including: (1) the coupling distances between the ring resonator and the waveguides, (2) the hole size of the ring resonator and waveguides, (3) the end-holes of the bus and drop waveguides, and (4) selecting the best sensing hole for binding an analyte. The resonant wavelength and output spectra are changed by these parameters. Fig. 23.22 shows the sketch of the proposed biosensor and its important parameters. Refractive index of air and silicon are considered as 1 and 2.825, respectively. The lattice constant of the PC structures and radius of air holes (R_a) are respectively equal to 410 and 120 nm. The size of the diamond nanoring resonator diameter is equal to three rows of holes, that is, as small as 1.1 μm. The PBG of the structure is calculated by the plane wave expansion (PWE) method. This hexagonal lattice does not exhibit bandgap for TE gap and the PBG for TM gap extends from 0.258 to 0.340. The corresponding wavelength ranges from 1205 to 1640 nm.

According to the results, the resonant wavelength shift in the range of 1.33–1.54 is linearly proportional to the refractive index variations. The quality factor and the sensitivity of biosensor are 3700 and

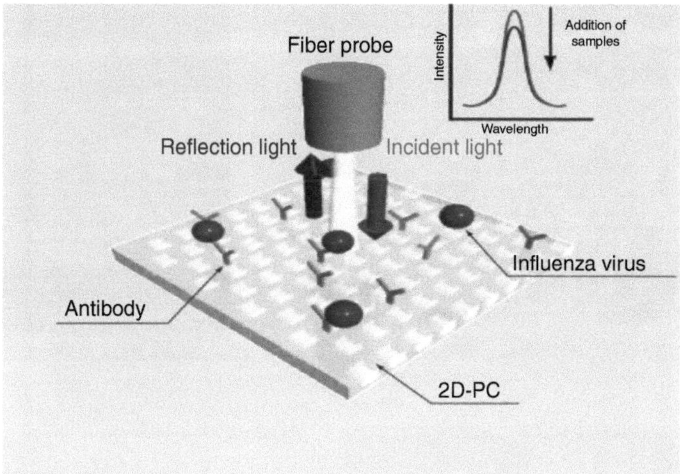

FIGURE 23.23 Principle of Reflectometric Detection for Influenza Virus in Human Saliva on the Basis of NIL-Based 2D PC

From Endo, T., Ozawa, S., Okuda, N., Yanagida, Y., Tanaka, S., Hatsuzawa, T., 2010. Reflectometric detection of influenza virus in human saliva using nanoimprint lithography-based flexible two-dimensional photonic crystal biosensor. Sens. Actuat. B 148(1), 269–276, with permission from Elsevier, Copyright 2010.

3.4 nm/fg, respectively. The minimum detectable biomolecule weight in a sensing hole for a diamond-shaped nanoring resonator is derived as 0.029 fg.

5.6.4 PC slab biosensors without defects

A review of some PC nanostructures without any defect is presented in this subsection. The structural color of these PC nanostructures can be used to realize on-site biosensor. Devices with a wide range of applications, particularly for therapeutic diagnosis, can be developed based on an entirely visual biosensing method that requires no power supply, is easy to use, and noninvasive. For instance, lack of power supply is an essential challenge in outdoor applications. Examples include water tests and blood tests among other applications. Furthermore, the low cost of these methods make them proper options for third world countries. Another advantage is lowered T threshold of adopting such devices and improvement in the diagnostics of diseases. Structural color of PC nanostructures could be used to realize real-time noninvasive biosensor.

As noted earlier, incident light will diffract according to Bragg's law when reaching PC. The reflected light in the visible spectrum is often referred to as the structural color. Endo et al. demonstrated reflectometric detection of influenza virus in human saliva with the detection limit of 1 ng/mL, using a flexible 2D-PC biosensor by employing nanoimprint lithography (Alexeev et al., 2003; Ben-Moshe et al., 2006; Endo et al., 2010; Lee and Asher, 2000; Sharma et al., 2004). Nanoimprint lithography (NIL) can be employed to produce 2D PCs with different periodicities (100, 150, 200, and 250 nm) and hole diameters, onto a cycloolefin polymer (COP) film (Kameoka et al., 2001; Laib and MacCraith, 2007). Fig. 23.23 illustrates the detection principle when using 2D PC. The authors used a dual-functional polymeric self-assembled monolayer coated onto the 2D-PC surface for blocking procedure to restrict the bounding of nonspecific contaminant proteins onto the antibody-immobilized 2D-PC surface (Avtushenko et al., 1996; Moldoveanu et al., 1995).

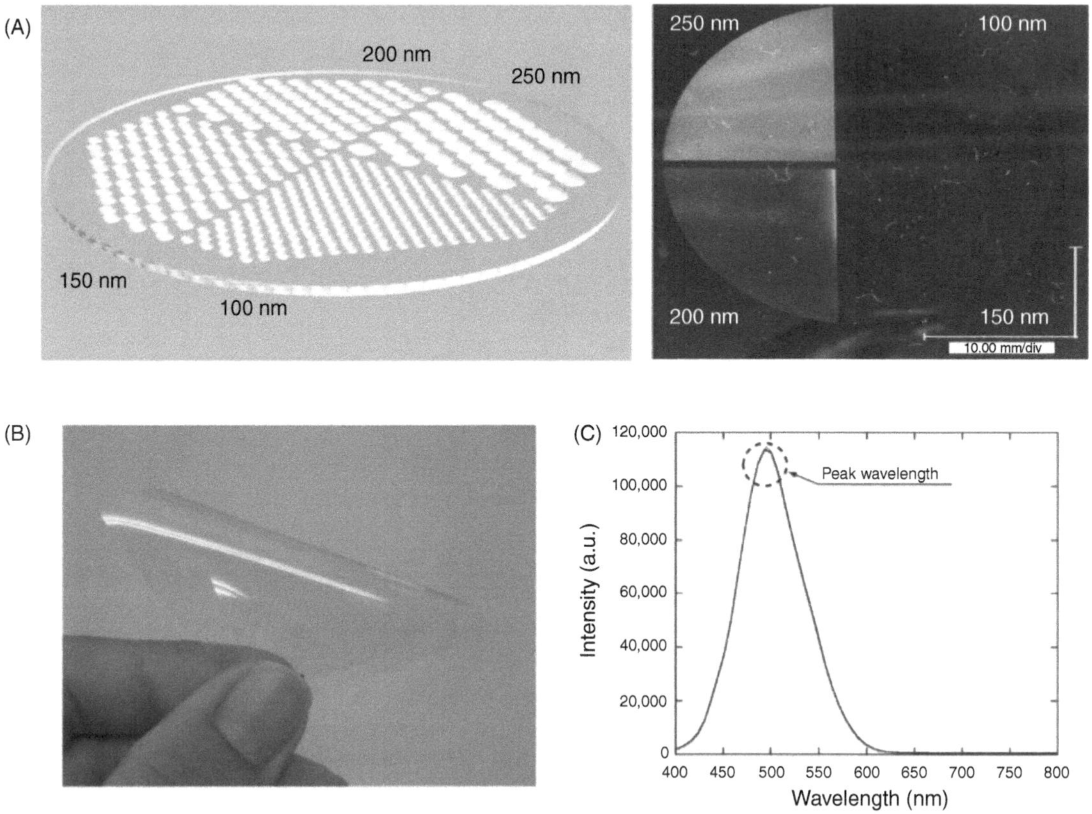

FIGURE 23.24

(A) The structural color of NIL-based 2D PC with different periodicities, (B) photograph of NIL-based 2D-PC using COP film, and (C) reflection spectrum of the 250-nm 2D-PC and photograph of experimental setup for optical characterization.

From Endo, T., Ozawa, S., Okuda, N., Yanagida, Y., Tanaka, S., Hatsuzawa, T., 2010. Reflectometric detection of influenza virus in human saliva using nanoimprint lithography-based flexible two-dimensional photonic crystal biosensor. Sens. Actuat. B 148(1), 269–276, with permission from Elsevier, Copyright 2010.

Fig. 23.24 depicts a 2D-PC fabricated using NIL. According to Bragg's law, different structural color will be found in a 2D-PC with a different periodicity (Fig. 23.24A). High flexibility can be achieved with the structural color (Fig. 23.24B). In addition, in order to identify influenza virus in human saliva based on the structural color from 2D-PC in visible region, the authors utilized hole diameter and periodicity of 250 nm so the structural color is green, and the wavelength of reflection peak is 508 nm (Fig. 23.24C). Furthermore, other hole diameters and periodicities (100, 150, and 200 nm) cannot be used to detect particular reflection peaks in visible region.

Calibration characterization of the influenza virus that was dissolved in 20 mmol/L phosphate buffer (pH 7.4) is carried out as an approach to assessment of sensitivity and specificity of antibody immobilized 2D PC. In this way, one can monitor the variations in reflection intensities of the antibody-immobilized 2D PC, until 10 pg/mL of influenza virus, depending on the influenza virus concentrations (Fig. 23.25).

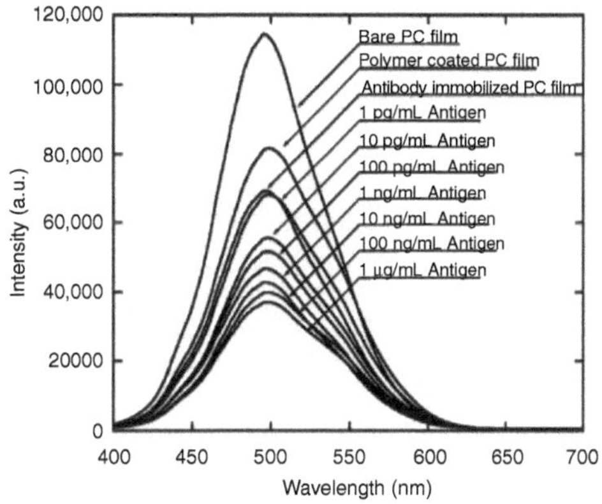

FIGURE 23.25 Wavelength With Respect to the Intensity for Different Concentrations of Influenza Virus Dissolved in 20 mmol/L Phosphate Buffer (pH 7.4)

From Endo, T., Ozawa, S., Okuda, N., Yanagida, Y., Tanaka, S., Hatsuzawa, T., 2010. Reflectometric detection of influenza virus in human saliva using nanoimprint lithography-based flexible two-dimensional photonic crystal biosensor. Sens. Actuat. B 148(1), 269–276, with permission from Elsevier, Copyright 2010.

Approximation shows that about 1.8×10^5 virions are present in the area irradiated by the optical fiber bundle. The authors used the AFM images, and in particular made use of antibody-immobilized NIL-based 2D PC, to detect the influenza virus.

In another study conducted by the same researchers, reflectometric detection of insulin was demonstrated using the same structure, which resulted in the detection limit 0.5 μg/mL of insulin. The structure claimed to provide versatile analytical systems that can be utilized in quick diagnosis of serious diseases, such as diabetes, cancer, and infection diseases (Endo et al., 2011, 2012a).

6 PC LASER BIOSENSORS AND LAB-ON-CHIP DEVICES

In this section after a brief review of using PC lasers in medical applications, we introduce lab-on-chip (LOC) devices and consequently discuss some demonstration integrated photonic structures for point of care (POC) applications.

6.1 MOTIVATIONS FOR USING PC LASERS

PC technology is expected to make a breakthrough in optoelectronics in the next decades. It will greatly enhance the performance of current technologies and will open up completely new application areas. Due to changes in society, such as an ageing population, unhealthy nutrition habits, and growing number of obese people, the market for consumer medical products is booming. PC technology combined with laser technology can provide the ingredient needed to develop low-cost consumer diagnostic products that diagnostic equipment manufacturers have been waiting for. PC laser could be used as high sensitive miniature biosensors or as ultra-fast low-threshold with high output power lasers used in service to therapeutic diagnosis. Here we just discuss PC laser biosensors and refer the reader to Altug and

Vučković (2005), Buckley et al. (2014), Diao et al. (2013), Fushman et al. (2008), Haque et al. (2013), Kita et al. (2008), and Yamilov et al. (2006) for PC lasers as light sources.

6.2 PC LASER AS BIOSENSORS

So far we discussed passive resonators and found that by providing narrower linewidth and high *Q*-factor cavities we could resolve the limitation of detection of smaller wavelength shifts associated with the analytes at low concentration or biomolecules with low molecular weight, such as drug compounds (Vollmer et al., 2002; White and Fan, 2008). Although high *Q*-factor passive resonators can considerably enhance detection resolution; but, this will generally lower the sensitivity and dynamic range of the system, however, there have been some cases of using passive resonators where high *Q* factor and high sensitivity were achieved simultaneously (White et al., 2006). Furthermore, a high precision alignment for evanescent light in and out coupling is often needed for obtaining high *Q*-factor optical resonators. This can limit the practical application of such resonators. Due to their ability in creating narrow linewidth stimulated emission, simplified instrumentation, and no need for the strict control of evanescent coupling to waveguides or tapered optical fibers, active resonator sensors, such as laser-based optical biosensors have been the focus of numerous studies (Lončar et al., 2003; Lu et al., 2008).

A gain medium is optically pumped into active cavities using light, and then laser emission from the cavity is measured. For such cavities, direction of the input (pump) and output (emission) light, is for most cases, perpendicular to the plane of the slab (as shown in Fig. 23.8). The same measurement scheme is used for almost all PC slab biosensors with active cavities use. Gain medium is pumped into layer(s) of alloys of III–V semiconductor quantum structures contained in the PC, and then the output emission is analyzed. The emission wavelength is related to the RI of the low-RI regions of the PC, and therefore changes in this wavelength can be interpreted as material changes within the PC. Such applications have involved planar quantum well layers (Kim et al., 2009; Kita et al., 2008, 2011; Lu et al., 2010; Shota et al., 2012) as well as embedded quantum dots (Chakravarty et al., 2005; Dündar et al., 2010) in measurements.

According to a technique proposed by Kita et al., an array of PC nanolaser structures can be used along with a bandpass filter and InGaAs camera in order to increase the number of measurements and eliminate the need for a spectrometer (Baker et al., 2015). This method results in an emission shift across the filter edge wavelength induced by target binding, and when the illumination passes from the nanolaser to the camera, one can simultaneously monitor a large number of binding sensors.

Watanabe et al. demonstrated the detection of refractive index and surface charges simultaneously through nanolaser biosensors comprising of an air-bridge PC slab, asymmetric H0 nanocavity, and a nanoslot (NS) fabricated in undoped GaInAsP quantum-well (QW) epiwafer (Watanabe et al., 2015). Since the wavelength and intensity behave independently (Fig. 23.26B), simply acquiring the laser spectrum allows the refractive index and surface charges to be simultaneously sensed. The GaInAsP PC nanolaser is immersed in a solution. Changing in PH of this solution alter the emission intensity of the PC nanolaser. This fact is the result of change in the redox potential, which in turn cause nonradiative surface recombination. Using this design, they demonstrated the simultaneous detection based on the adsorption of charged polyelectrolytes and the hybridization of deoxyribonucleic acids (DNAs), alternately. Fig. 23.26 depicts a schematic of the proposed nanolaser and its behavior with respect to adsorption of biomolecules.

Abe et al. (2011, 2012) used an array of PC nanolasers as the base of an imaging technique. By attaching the live cell to the surface of the PC array, each nanolaser works as a pixel and they could

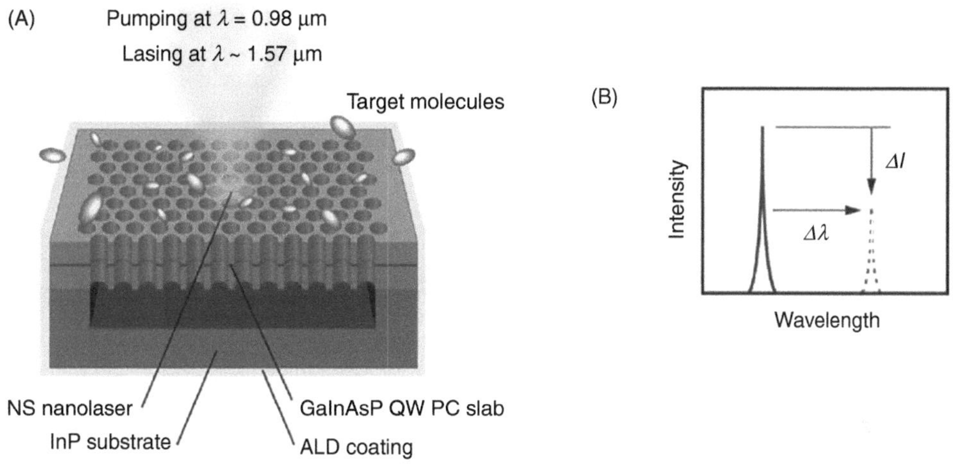

FIGURE 23.26 Sensing Biomolecules Using a PC Nanolaser

(A) Schematic of a nanolaser sensor and (B) schematic behavior of an emission spectrum. The behavior before and after the adsorption of biomolecules are, respectively, shown by the *solid line* and *dashed line* peaks.

Reprinted with permission from Watanabe, K., Kishi, Y., Hachuda, S., Watanabe, T., Sakemoto, M., Nishijima, Y., Baba, T., 2015. Simultaneous detection of refractive index and surface charges in nanolaser biosensors. Appl. Phys. Lett. 106(2), 021106. Copyright 2015, AIP Publishing LLC.

provide images of live cell behavior, gathering information of pixels and continuously mapping all pixels. Theoretical modal volume of each nanolaser is $0.15(\lambda/n)^3$, where λ is the laser wavelength and n is the refractive index. Therefore, many devices can be densely integrated on a chip. Due to its narrow spectral linewidth and cavity exposed to air, it shows a high index resolution of 10^{-5} RIU order. Fig. 23.27 shows the sketch of PC nanolaser array.

They experimentally demonstrated 5-μm pitch 6 × 6 array of GaInAsP semiconductor H0-type PC nanolasers and attached HeLa cells to the surface of the device. By pumping the small spot of beam

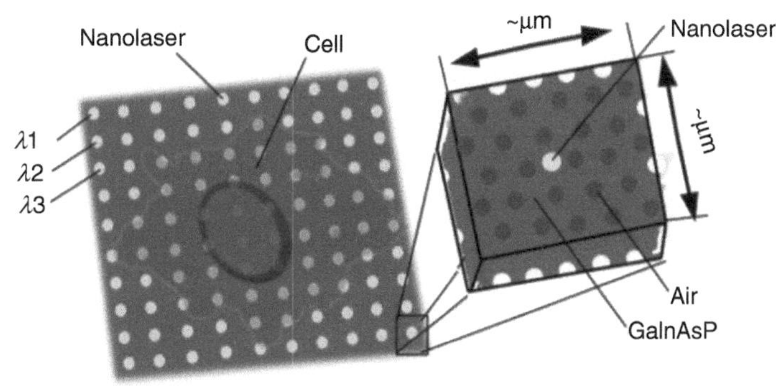

FIGURE 23.27 Sketch of Nanolaser Array for Using in Cell Imaging Applications

Reproduced from Abe, H., Narimatsu, M., Kita, S., Tomitaka, A., Takemura, Y., Baba, T., 2011. Live cell imaging using photonic crystal nanolaser array, 15th International Conference on Miniaturized Systems for Chemistry and Life Sciences October 2–6, 2011, Seattle, Washington, USA, Micro-TAS, 593, pp. 1433–1435.

to each nanolaser, they found the relation between its wavelength and position, then measured all the wavelengths repeatedly using a large spot of pump beam every 1 min. They injected nocodazole to give rise to some reactions. By injecting nocodazoles the dimension of cells gradually reduced and they were finally removed from the surface of array. After that actino-mycin D was injected, and the measurement was continued until its volatilization led to a simple wavelength shift for all nanolasers.

6.3 LAB-ON-CHIP DEVICES

LOC, refers to miniature devices that can perform the tasks of regular laboratories including filtering, mixing, and detecting the target analytes. In addition to reducing the size, these devices have a lot of advantages over other current analysis technologies. Using LOC devices causes reduction of sample volume requirements in order of μL/nL instead of mL as for laboratory procedures, and consequently reduction of cost. The analysis can be performed much faster due to reduction of processing structures (Duval et al., 2012; Washburn and Bailey, 2011). Since miniaturization and integration of devices based on microarrays of sources, detectors, and active or passive biosensing surfaces are available via LOC, detection of extremely low concentrations of analytes are possible (Lee et al., 2011). Moreover, the development of real-time multimodal sensing technologies will enable significant advances in our ability to observe and understand complex intra- and intercellular biochemical processes. Finally, since novel high-sensitivity micro/nanosensors are integrated within these LOC devices, they will allow for the detection of extremely low concentrations of the target analytes.

Probably the main objective of biosensors and LOC devices is for early therapeutic diagnosis using so-called POC devices which are used to provide in situ, rapid, reliable, minimally invasive, and early information about the patient's health status in order to determine the most appropriate treatment for him/her.

PCs that allow for control over light in very small dimensions and using novel fabrication processes including nanoimprint lithography and nanoreplica molding and injection molding, offer a potential to integrate with microfluidic on-chip for biosensing applications in planar geometries (Mortensen et al., 2008). Small volumes of analyte can be transported to the sensing region by microfluidic channels.

There are some challenging issues in the case of on-chip biosensors, such as the ability for integration of nanophotonic biosensors and microfluidic or multiplex sensing of analytes on a chip. Concerning the former case, Choi and Cunningham (2007) integrated microfluidic network and PC biosensors into a 96-well microplate through the nanoreplica molding process. Using this method one can generate patterns and structures on the nanometer scale.

They demonstrated detection of protein–protein interactions. Fig. 23.28 shows an SEM image of an open microfluidic channel network embedded with the PC biosensor and a photograph of the fabricated polymer microfluidic network with integrated PC structures attached on the bottom of a 96-well microplate (Choi and Cunningham, 2007). As shown in Fig. 23.28, 88 parallel flow channels with PC structure and 8 common inlets/outlets are integrated on the bottom of a 96-well microplate. Only 220 nL are necessary to fill 11 flow channels. The authors suggest that this system can be used with 11 different ligands, which are subsequently exposed to a single analyte (Choi and Cunningham, 2007).

Scullion et al. used functionalized slotted PC cavities with integrated microfluidics to detect dissolved avidin concentrations as low as 15 nM or 1 μg/mL. They were able to detect a surface mass coverage of 60 pg/mm^2 with a cavity sensing area of 2.2 μm^2. Using slotted waveguide cavity geometry causes strong modal overlap with the analyte, and consequently the high sensitivity over an extremely small area is achieved (Scullion et al., 2011).

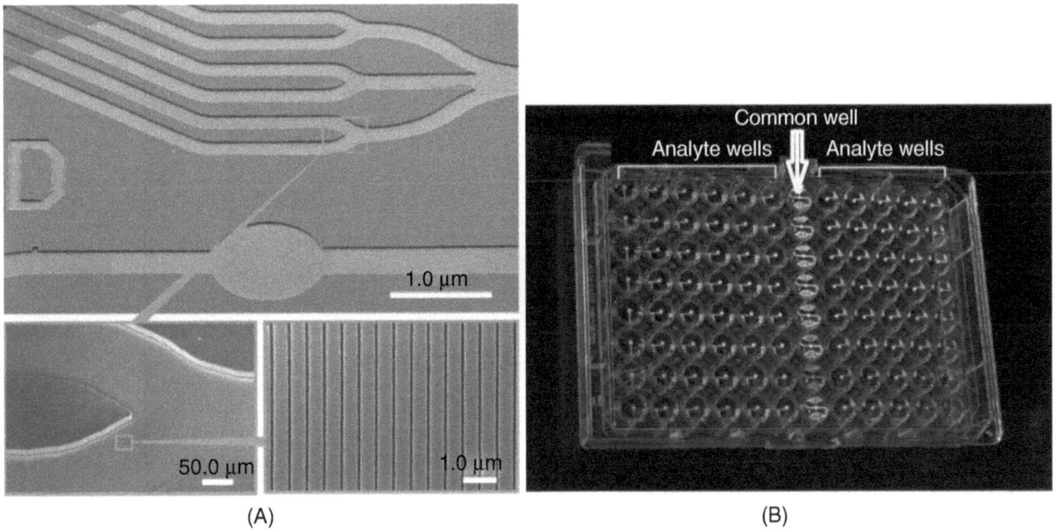

FIGURE 23.28

(A) SEM image of the microfluidic channels and PC structures molded in the polymer replica, (B) photograph of the fabricated polymer microfluidic network with integrated PC structures attached on the bottom of a 96-well microplate.

Reproduced from Choi, C.J., Cunningham, B.T., 2007. A 96-well microplate incorporating a replica molded microfluidic network integrated with photonic crystal biosensors for high throughput kinetic biomolecular interaction analysis. Lab Chip 7(5), 550–556, with permission from The Royal Society of Chemistry, Copyright 2007

Huang et al. (2009) proposed a novel approach by combining nanophotonics and nanofluidics. In this approach they used free-standing silicon nitride PC. The nanohole arrays in the PC were used as sensing structures as well as nanofluidic channels. The analyte directed to the PC surface goes through the nanoscale hole arrays and flows to the bottom channel. This method overcomes the mass transport limitations and therefore improves the light and analyte interaction. Both simulated and experimental measurements showed a sensitivity of 510 nm/RIU when the stream delivery was performed through the nanohole openings of the PC sensing structure instead of using the laminar flows (Fig. 23.29).

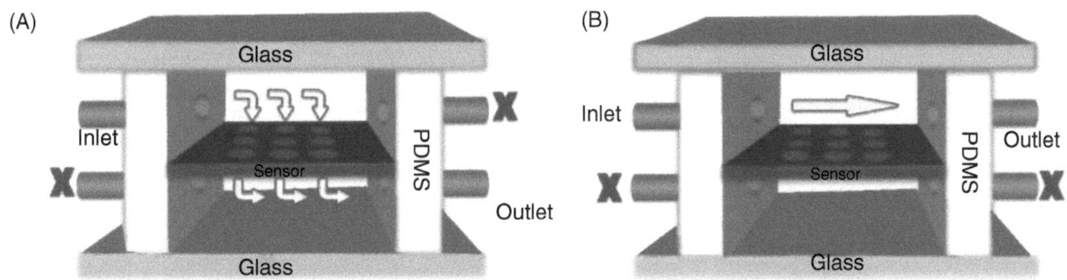

FIGURE 23.29 Schematic Diagram of the Actively Controlled Flow Process

(A) The solution flows through the hole array and (B) conventional flowing scheme.

Reproduced from Huang, M., Yanik, A.A., Chang, T.Y., Altug, H., 2009. Sub-wavelength nanofluidics in photonic crystal sensors. Opt. Express 17 (26), 24224–24233.

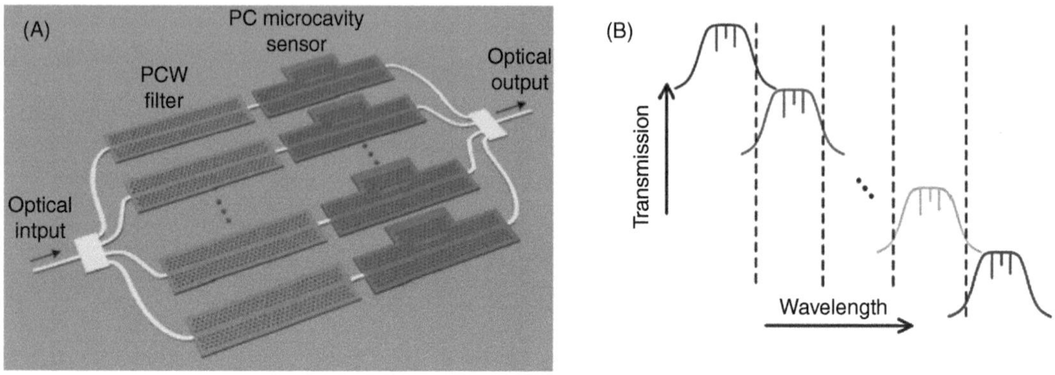

FIGURE 23.30

(A) Proposed single-input single-output PC biosensor array and (B) the corresponding transmission spectra from each channel.

Reprinted with permission from Yan, H., Zou, Y., Chakravarty, S., Yang, C.J., Wang, Z., Tang, N., Chen, R.T., 2015. Silicon on-chip bandpass filters for the multiplexing of high sensitivity photonic crystal microcavity biosensors. Appl. Phys. Lett. 106(12), 121103. Copyright 2015, AIP Publishing LLC.

Concerning multiplex sensing of analytes on-chip, Yan et al. (2015) demonstrated silicon on-chip bandpass filters for the multiplexing of high-sensitivity PC microcavity biosensors. They used L55 PC cavities as biosensors in series with PC waveguides as bandpass filters. They chose L55 PC cavities due to their ultra-low detection limit of 50 femto-molar, and added PCW to filter a narrow band containing resonances of the PC biosensors to avoid resonance overlap of PC biosensors that were connected in parallel. Fig. 23.30 shows the proposed single-input single-output PC biosensor array and the corresponding transmission spectra from each channel.

The proposed PCW bandpass filter can also serve as an optofluidic spectral filter (Measor et al., 2011; Schmidt and Hawkins, 2011). To allow both localized control and high refractive index modulation, the passband can be designed easily by engineering PC geometry. The design is also highly reconfigurable when integrated with microfluidics (Psaltis et al., 2006). This type of optofluidic filter is a promising component in future LOC systems. As discussed in Section 5.6.2 using the parallel combination of four cavities, with different geometrical arrangements, Yang et al. experimentally demonstrated multiplexing sensing with a wide dynamic range covering 6 orders of magnitude.

Lee et al. (2011) demonstrated compact, parallel sensor architecture using PC nanostructures assembled with vertical cavity surface emitting lasers (VCSELs) and GaAs photodetectors which enable multiplexed, cost-effective on-chip biosensing.

7 CONCLUSIONS

An important objective that should be pursued in the area of modern bioanalytics is to develop fast, sensitive, noninvasive, and specific means of detecting molecular biomarkers to identify normal biologic processes, pathogenic processes, or pharmacologic responses to a therapeutic intervention. Designing, fabricating, and testing nanostructures for creating changes in light on-chip in order to localize light below subdiffraction limit present an essential challenge in this area along with the attempts that should be made to enhance the field, increase light–molecule interaction, and create compact and portable devices that can be used for diagnosis at POC. Using PCs, we have tremendous opportunities for on-chip sensing of analytes related to medical diagnostics, such as cancer markers, allergy markers, heart attack markers, antibodies, drugs, and hormones.

In this chapter we tried to introduce PC biosensors concerning the possibility of integration on-chip. Two basic principles are applied to almost all label-free optical biosensors. The first one is to create a surface coated with molecules capable of capturing a specific targeted molecule. Unlike label or tag-based detection which requires a multilevel chemical reaction for adding a radioactive or fluorescent molecule to the target molecules, the capture process here is typically carried out in a single step, making it more time efficient. The second principle is to optically design a surface capable of concentrating light, which is emitted from an excitation source, at the settlement area of molecules. For this purpose, the optical response should be sensitive to changes occurring near the capture area. Capture events are often observed in the form of scattering, shifts in resonance frequency, or a dip in transmission or reflection. Therefore, label-free biosensors are essentially mechanisms consisting of two characteristics: selectivity which is provided through a chemistry-based process and sensitivity that is provided by an optical transducer. The focus of this chapter was on the optical transducer and here we discuss the applications of PC transducers. PCs are periodic nanostructures that could have period in 1D, 2D, or 3D. PC structures can confine photons to nanoscale dimensions, much less than a wavelength, resulting in dramatic modifications or enhancements of a variety of light–matter interactions. Although there are several reported works in the case of 1D and 3D PC biosensors, the focus was on the special 2D type called slab PCs. To evaluate the performance of a biosensor element several parameters could be considered, such as quality factor, sensitivity, detection limit, dynamic range, and multiplexing level. Three kinds of slab PCs including waveguides, nanocavities, nanoring resonators, and a combination of them as well as the novel theoretical and experimental works about using them as biosensing elements were introduced. In PC waveguides, cut-off wavelength is used as an indicator to detect changes in the refractive index. Waveguides are also used as light terminals in combination with nanocavities and nanoring resonators.

Although PC waveguides can be easily integrated into an optical chip, the readout of the cut-off wavelength is not precise enough for sensitive detection. On the other hand, a PC nanocavity creates resonances with a high quality factor that provides higher level of precision for sensitive detection. By reducing the size of the ring resonator, in order of nanometers, that is sandwiched with two waveguides, we could achieve high Q biosensors. Another kind of PC biosensor is a slab without any defects. Recently researchers' work on using the structural color of PC slabs for reflectometric detection of target molecules, such as DNA, proteins, and virus.

Although high Q-factor passive resonators can considerably enhance detection resolution, but this will generally lower the sensitivity and dynamic range of the system. Due to their ability in creating narrow linewidth stimulated emission, simplified instrumentation, and no requirements for high precision evanescent coupling to waveguides or tapered optical fibers, active resonator sensors, such as laser-based optical biosensors have been the focus of numerous studies. In addition to its use as a biosensor, PC laser could also be used as ultra-fast low-threshold with high-output power lasers used in service to therapeutic diagnosis. In the case of active cavities, gain medium is pumped into layer(s) of III–V semiconductor quantum structures contained in the PC, and then the output emission is analyzed. The emission wavelength is related to the RI of the low-RI regions of the PC, and therefore changes in this wavelength can be interpreted as material changes within the PC. After the introduction of some

PC lasers demonstration we had a brief review of LOC devices that referred to some kind of miniature devices use as an ordinary laboratory, including light sources, filters, and detectors with the main task of being used in therapeutic diagnosis. Reviewing some demonstration of PC and microfluidic integration, the potential of PC nanostructures for on-chip POC diagnosis is proved.

REFERENCES

Abe, H., Furumoto, T., Narimatsu, M., Kita, S., Nakamura, K., Takemura, Y., Baba, T., 2012. Direct live cell imaging using large-scale nanolaser array. IEEE Sens. Conf. 1565, 1–4.

Abe, H., Narimatsu, M., Kita, S., Tomitaka, A., Takemura, Y., Baba, T., 2011. Live cell imaging using photonic crystal nanolaser array, 15th International Conference on Miniaturized Systems for Chemistry and Life Sciences October 2–6, 2011, Seattle, Washington, USA, Micro-TAS, 593, pp. 1433–1435.

Alexeev, V.L., Sharma, A.C., Goponenko, A.V., Das, S., Lednev, I.K., Wilcox, C.S., Asher, S.A., 2003. High ionic strength glucose-sensing photonic crystal. Anal. Chem. 75 (10), 2316–2323.

Almeida, V.R., Xu, Q., Barrios, C.A., Lipson, M., 2004. Guiding and confining light in void nanostructure. Opt. Lett. 29 (11), 1209–1211.

Altug, H., Vučković, J., 2005. Photonic crystal nanocavity array laser. Opt. Express 13 (22), 8819–8828.

Avtushenko, S.S., Sorokin, E.M., Zoschenkova, N.Y., Zacharova, N.G., Naichin, A.N., 1996. Clinical and immunological characteristics of the emulsion form of inactivated influenza vaccine delivered by oral immunization. J. Biotechnol. 44 (1), 21–28.

Bahabady, A.M., Olyaee, S., 2015. Two-curve-shaped biosensor for detecting glucose concentration and salinity of seawater based on photonic crystal nano-ring resonator. Sens. Lett. 13 (9), 774–777.

Baker, J.E., Sriram, R., Miller, B.L., 2015. Two-dimensional photonic crystals for sensitive microscale chemical and biochemical sensing. Lab Chip 15 (4), 971–990.

Begum, F., Namihira, Y., 2012. Photonic Crystal Fiber for Medical Applications. INTECH Open Access Publisher, Rijeka, Croatia, pp. 229–246.

Ben-Moshe, M., Alexeev, V.L., Asher, S.A., 2006. Fast responsive crystalline colloidal array photonic crystal glucose sensors. Anal. Chem. 78 (14), 5149–5157.

Block, I.D., Chan, L.L., Cunningham, B.T., 2007. Large-area submicron replica molding of porous low-k dielectric films and application to photonic crystal biosensor fabrication. Microelectron. Eng. 84 (4), 603–608.

Block, I.D., Pineda, M., Choi, C.J., Cunningham, B.T., 2008. High sensitivity plastic-substrate photonic crystal biosensor. Sens. J. IEEE 8 (9), 1546–1547.

Buckley, S., Radulaski, M., Zhang, J.L., Petykiewicz, J., Biermann, K., Vučković, J., 2014. Multimode nanobeam cavities for nonlinear optics: high quality resonances separated by an octave. Opt. Express 22 (22), 26498–26509.

Buswell, S.C., Wright, V.A., Buriak, J.M., Van, V., Evoy, S., 2008. Specific detection of proteins using photonic crystal waveguides. Opt. Express 16 (20), 15949–15957.

Chakravarty, S., Lai, W.C., Zou, Y., Drabkin, H.A., Gemmill, R.M., Simon, G.R., Chen, R.T., 2013. Multiplexed specific label-free detection of NCI-H358 lung cancer cell line lysates with silicon based photonic crystal microcavity biosensors. Biosens. Bioelectron. 43, 50–55.

Chakravarty, S., Topol'Ancik, J., Bhattacharya, P., Chakrabarti, S., Kang, Y., Meyerhoff, M.E., 2005. Ion detection with photonic crystal microcavities. Opt. Lett. 30 (19), 2578–2580.

Chakravarty, S., Zou, Y., Lai, W., Chen, R.T., 2012. Slow light engineering for high Q high sensitivity photonic crystal microcavity biosensors in silicon. Biosens. Bioelectron. 38, 170–176.

Cho, E., Kim, B., Choi, S., Han, J., Jin, J., Lim, J., Kang, S., 2011. Design and fabrication of label-free biochip using a guided mode resonance filter with nano grating structures by injection molding process. J. Nanosci. Nanotechnol. 11 (1), 417–421.

Choi, C.J., Cunningham, B.T., 2007. A 96-well microplate incorporating a replica molded microfluidic network integrated with photonic crystal biosensors for high throughput kinetic biomolecular interaction analysis. Lab Chip 7 (5), 550–556.

Chou, S.Y., Krauss, P.R., Renstrom, P.J., 1996. 25-nanometer resolution. Science 272, 85–87.

Chow, K.K., Short, M., Lam, S., McWilliams, A., Zeng, H., 2014. A Raman cell based on hollow core photonic crystal fiber for human breath analysis. Med. Phys. 41 (9), 092701.

Cunningham, B.T., 2010. Photonic crystal surfaces as a general purpose platform for label-free and fluorescent assays. J. Lab. Autom. 15 (2), 120–135.

Diao, Z., Bonzon, C., Scalari, G., Beck, M., Faist, J., Houdré, R., 2013. Continuous-wave vertically emitting photonic crystal terahertz laser. Laser Photon. Rev. 7 (5), L45–L50.

Dinish, U.S., Balasundaram, G., Chang, Y.T., Olivo, M., 2014. Sensitive multiplex detection of serological liver cancer biomarkers using SERS-active photonic crystal fiber probe. J. Biophoton. 7, 956–965.

Dorfner, D., Zabel, T., Hürlimann, T., Hauke, N., Frandsen, L., Rant, U., Finley, J., 2009. Photonic crystal nanostructures for optical biosensing applications. Biosens. Bioelectron. 24 (12), 3688–3692.

Dündar, M.A., Ryckebosch, E.C., Nötzel, R., Karouta, F., van IJzendoorn, L.J., Van der Heijden, R.W., 2010. Sensitivities of InGaAsP photonic crystal membrane nanocavities to hole refractive index. Opt. Express 18 (5), 4049–4056.

Duval, D., González-Guerrero, A.B., Dante, S., Osmond, J., Monge, R., Fernández, L.J., Zinoviev, K.E., Domínguez, C., Lechuga, L.M., 2012. Nanophotonic lab-on-a-chip platforms including novel bimodal interferometers, microfluidics and grating couplers. Lab Chip 12 (11), 1987–1994.

Eid, J., Fehr, A., Gray, J., Luong, K., Lyle, J., Otto, G., Zhao, P., 2009. Real-time DNA sequencing from single polymerase molecules. Science 323 (5910), 133–138.

Endo, T., Ozawa, S., Okuda, N., Yanagida, Y., Tanaka, S., Hatsuzawa, T., 2010. Reflectometric detection of influenza virus in human saliva using nanoimprint lithography-based flexible two-dimensional photonic crystal biosensor. Sens. Actuat. B 148 (1), 269–276.

Endo, T., Henares, B.M., Hisamoto, H., 2012a. Development of label-free biosensor for the detection of adenosine diphosphate as a universal kinase/atpase assay using nanoimprinted flexible two-dimensional photonic crystal, 16th International Conference on Miniaturized Systems for Chemistry and Life Sciences, Okinawa, Japan, pp. 950–952.

Endo, T., Okuda, N., Tanaka S., Hisamoto, H., 2011. Nanoimprinted flexible photonic crystal for single-step label-free biosensor, 15th International Conference on Miniaturized Systems for Chemistry and Life Sciences, October 2–6, Seattle, Washington, USA, pp. 571–573.

Endo, T., Sato, M., Kajita, H., Okuda, N., Tanaka, S., Hisamoto, H., 2012b. Printed two-dimensional photonic crystals for single-step label-free biosensing of insulin under wet conditions. Lab Chip 12 (11), 1995–1999.

Falco, D., 2008. Slotted photonic crystal waveguides and cavities for slow light and sensing applications, 5th IEEE International Conference on Group IV Photonics. pp. 228–230.

Fushman, I., Englund, D., Altug, H., Ellis, B., Faraon, A., Vučković, J., 2008. Ultrafast photonic crystal nanocavity lasers and optical switches, Integrated Optoelectronic Devices 2008, International Society for Optics and Photonics, pp. 688910–688910.

Gonzalez, R.M., Seurynck-Servoss, S.L., Crowley, S.A., Brown, M., Omenn, G.S., Hayes, D.F., Zangar, R.C., 2008a. Development and validation of sandwich ELISA microarrays with minimal assay interference. J. Proteome Res. 7 (6), 2406–2414.

Gonzalez, R.M., Varnum, S.M., Zangar, R.C., 2008b. Sandwich ELISA microarrays: generating reliable and reproducible assays for high-throughput screens. Biomarker Methods in Drug Discovery and Development, Humana Press, Totowa, New Jersey, USA, pp. 273–290.

Haaheim, J., Eby, R., Nelson, M., Fragala, J., Rosner, B., Zhang, H., Athas, G., 2005. Dip Pen Nanolithography (DPN): process and instrument performance with NanoInk's NSCRIPTOR system. Ultramicroscopy 103 (2), 117–132.

Haque, M., Zacharia, N.S., Ho, S., Herman, P.R., 2013. Laser-written photonic crystal optofluidics for electrochromatography and spectroscopy on a chip. Biomed. Opt. Express 4 (8), 1472–1485.

Hashimoto, W., Endo, T., Sueyoshi, K., Hisamoto, H., 2014. Development of novel protease assay device using a nanoimprinted two-dimensional photonic crystal. Chem. Lett. 43 (11), 1728–1730.
Hell, S.W., 2003. Toward fluorescence nanoscopy. Nat. Biotechnol. 21 (11), 1347–1355.
Homola, J., 2008. Surface plasmon resonance sensors for detection of chemical and biological species. Chem. Rev. 108 (2), 462–493.
Hsiao, F.L., Lee, C., 2010. Computational study of photonic crystals nano-ring resonator for biochemical sensing. Sens. J. IEEE 10 (7), 1185–1191.
Hsiao, F.L., Lee, C., 2011. Nanophotonic biosensors using hexagonal nanoring resonators: computational study. J. Micro Nanolithogr. MEMS MOEMS 10 (1), 013001.
Huang, M., Yanik, A.A., Chang, T.Y., Altug, H., 2009. Sub-wavelength nanofluidics in photonic crystal sensors. Opt. Express 17 (26), 24224–24233.
Joannopoulos, J.D., Johnson, S.G., Winn, J.N., Meade, R.D., 2008. Photonic Crystals: Molding the Flow of Light. Princeton University Press, New Jersey, USA, 18.
John, S., 1987. Strong localization of photons in certain disordered dielectric superlattices. Phys. Rev. Lett. 58 (23), 2486.
Johnson, S.G., Villeneuve, P.R., Fan, S., Joannopoulos, J.D., 2000. Linear waveguides in photonic-crystal slabs. Phys. Rev. B 62 (12), 8212.
Kameoka, J., Craighead, H.G., Zhang, H., Henion, J., 2001. A polymeric microfluidic chip for CE/MS determination of small molecules. Anal. Chem. 73 (9), 1935–1941.
Kim, S., Lee, J., Jeon, H., Kim, H.J., 2009. Fiber-coupled surface-emitting photonic crystal band edge laser for biochemical sensor applications. Appl. Phys. Lett. 94 (13), 3503.
Kita, S., Hachuda, S., Otsuka, S., Endo, T., Imai, Y., Nishijima, Y., Baba, T., 2011. Super-sensitivity in label-free protein sensing using a nanoslot nanolaser. Opt. Express 19 (18), 17683–17690.
Kita, S., Nozaki, K., Baba, T., 2008. Refractive index sensing utilizing a cw photonic crystal nanolaser and its array configuration. Opt. Express 16 (11), 8174–8180.
Kouba, J., Kubenz, M., Mai, A., Ropers, G., Eberhardt, W., Loechel, B., 2006. Fabrication of nanoimprint stamps for photonic crystals. J. Phys. 34 (1), 897–903.
Krioukov, E., Klunder, D.J.W., Driessen, A., Greve, J., Otto, C., 2002. Sensor based on an integrated optical microcavity. Opt. Lett. 27 (7), 512–514.
Kwon, M.S., Steier, W.H., 2008. Microring-resonator-based sensor measuring both the concentration and temperature of a solution. Opt. Express 16 (13), 9372–9377.
Lai, W.C., Chakravarty, S., Zou, Y., Chen, R.T., 2012. Silicon nano-membrane based photonic crystal microcavities for high sensitivity bio-sensing. Opt. Lett. 37 (7), 1208–1210.
Lai, W.C., Chakravarty, S., Zou, Y., Guo, Y., Chen, R.T., 2013. Slow light enhanced sensitivity of resonance modes in photonic crystal biosensors. Appl. Phys. Lett. 102 (4), 041111.
Laib, S., MacCraith, B.D., 2007. Immobilization of biomolecules on cycloolefin polymer supports. Anal. Chem. 79 (16), 6264–6270.
Lee, K., Asher, S.A., 2000. Photonic crystal chemical sensors: pH and ionic strength. J. Am. Chem. Soc. 122 (39), 9534–9537.
Lee, M.M., O'Sullivan, T.D., Cerruto, A., Liu, V., Zhang, J., Levi, O., Harris, J.S., 2011. Integrated photonic structures for parallel fluorescence and refractive index biosensing, SPIE Defense, Security, and Sensing, International Society for Optics and Photonics, pp. 803406–803406.
Li, T., Li, L., Zhang, G., 2014. A nano-scaled force sensor based on a photonic crystal nanocavity resonator and a microcantilever. ECS J. Solid State Sci. Technol. 3 (7), Q146–Q151.
Liu, Y., Salemink, H.W.M., 2012. Photonic crystal-based all-optical on-chip sensor. Opt. Express 20 (18), 19912–19920.
Lončar, M., Scherer, A., Qiu, Y., 2003. Photonic crystal laser sources for chemical detection. Appl. Phys. Lett. 82 (26), 4648–4650.

Lu, M., Choi, S.S., Wagner, C.J., Eden, J.G., Cunningham, B.T., 2008. Label free biosensor incorporating a replica-molded, vertically emitting distributed feedback laser. Appl. Phys. Lett. 92 (26), 261502.

Lu, T.W., Lin, P.T., Sio, K.U., Lee, P.T., 2010. Optical sensing of square lattice photonic crystal point-shifted nanocavity for protein adsorption detection. Appl. Phys. Lett. 96 (21), 213702.

Measor, P., Phillips, B.S., Chen, A., Hawkins, A.R., Schmidt, H., 2011. Tailorable integrated optofluidic filters for biomolecular detection. Lab Chip 11 (5), 899–904.

Mohanty, S.P., Kougianos, E., 2006. Biosensors: a tutorial review. IEEE Potentials 25 (2), 35–40.

Moldoveanu, Z., Clements, M.L., Prince, S.J., Murphy, B.R., Mestecky, J., 1995. Human immune responses to influenza virus vaccines administered by systemic or mucosal routes. Vaccine 13 (11), 1006–1012.

Mortensen, N.A., Xiao, S., Pedersen, J., 2008. Liquid-infiltrated photonic crystals: enhanced light-matter interactions for lab-on-a-chip applications. Microfluid. Nanofluid. 4 (1–2), 117–127.

Najafgholinezhad, S., Olyaee, S., 2014. A photonic crystal biosensor with temperature dependency investigation of micro-cavity resonator. Optik Int. J. Light Electron Opt. 125 (21), 6562–6565.

Naraghi, A., Olyaee, S., Najibi, A., Leitgeb, E., 2013. Photonic crystal fiber gas sensor for using in optical network protection systems. Network and Optical Communications (NOC), 18th European Conference on and Optical Cabling and Infrastructure, pp. 175–180.

Olyaee, S., Arman, H., Naraghi, A., 2015. Design, simulation, and optimization of acetylene gas sensor using hollow-core photonic bandgap fiber. Sens. Lett. 13 (5), 387–392.

Olyaee, S., Bahabady, A.M., 2014a. Designing a novel photonic crystal nano-ring resonator for biosensor application. Opt. Quant. Electron. 47 (7), 1881–1888.

Olyaee, S., Bahabady, A.M., 2015. Design and optimization of diamond-shaped biosensor using photonic crystal nano-ring resonator. Optik Int. J. Light Electron Opt. 126 (20), 2560–2564.

Olyaee, S., Bahabady, A.M., 2014b. Two-curve-shaped biosensor using photonic crystal nano-ring resonators. J. Nanostruct. 4 (3), 303–308.

Olyaee, S., Najafgholinezhad, S., 2013a. Computational study of a label-free biosensor based on a photonic crystal nanocavity resonator. Appl. Opt. 52 (29), 7206–7213.

Olyaee, S., Najafgholinezhad, S., 2013b. A high quality factor and wide measurement range biosensor based on photonic crystal nanocavity resonator. Sens. Lett. 11 (3), 483–488.

Olyaee, S., Naraghi, A., 2013. Design and optimization of index-guiding photonic crystal fiber gas sensor. Photon. Sens. 3 (2), 131–136.

Olyaee, S., Taghipour, F., 2012a. Ultra-flattened dispersion hexagonal photonic crystal fibre with low confinement loss and large effective area. IET Optoelectron. 6 (2), 82–87.

Olyaee, S., Taghipour, F., 2012b. Doped-core octagonal photonic crystal fiber with ultra-flattened nearly zero dispersion and low confinement loss in a wide wavelength range. Fiber Integrated Opt. 31 (3), 178–185.

Olyaee, S., Najafgholinezhad, S., Banaei, H.A., 2013. Four-channel label-free photonic crystal biosensor using nanocavity resonators. Photon. Sens. 3 (3), 231–236.

Olyaee, S., Naraghi, A., Ahmadi, V., 2014. High sensitivity evanescent-field gas sensor based on modified photonic crystal fiber for gas condensate and air pollution monitoring. Optik Int. J. Light Electron Opt. 125 (1), 596–600.

Olyaee, S., Taghipour, F., Izadpanah, M., 2011. Nearly zero-dispersion, low confinement loss, and small effective mode area index-guiding PCF at 1.55 μm wavelength. Front. Optoelectron. China 4 (4), 420–425.

Otupiri, R., Akowuah, E.K., Haxha, S., Ademgil, H., AbdelMalek, F., Aggoun, A., 2014. A novel birefringent photonic crystal fiber surface plasmon resonance biosensor. Photon. J. IEEE 6 (4), 1–11.

Padmanabhan, S., Shinoj, V.K., Murukeshan, V.M., Padmanabhan, P., 2010. Highly sensitive optical detection of specific protein in breast cancer cells using microstructured fiber in extremely low sample volume. J. Biomed. Opt. 15 (1), 017005–117005.

Pineda, M.F., Chan, L.L.Y., Kuhlenschmidt, T., Choi, C.J., Kuhlenschmidt, M., Cunningham, B.T., 2009. Rapid specific and label-free detection of porcine rotavirus using photonic crystal biosensors. Sens. J. IEEE 9 (4), 470–477.

Psaltis, D., Quake, S.R., Yang, C., 2006. Developing optofluidic technology through the fusion of microfluidics and optics. Nature 442 (7101), 381–386.
Rogers, K.R., 2000. Principles of affinity-based biosensors. Mol. Biotechnol. 14 (2), 109–129.
Schena, M., Shalon, D., Davis, R.W., Brown, P.O., 1995. Quantitative monitoring of gene expression patterns with a complementary DNA microarray. Science 270 (5235), 467–470.
Schmidt, H., Hawkins, A.R., 2011. The photonic integration of non-solid media using optofluidics. Nat. Photon. 5 (10), 598–604.
Scullion, M.G., Di Falco, A., Krauss, T.F., 2011. Slotted photonic crystal cavities with integrated microfluidics for biosensing applications. Biosens. Bioelectron. 27 (1), 101–105.
Sharma, A.C., Jana, T., Kesavamoorthy, R., Shi, L., Virji, M.A., Finegold, D.N., Asher, S.A., 2004. A general photonic crystal sensing motif: creatinine in bodily fluids. J. Am. Chem. Soc. 126 (9), 2971–2977.
Shota, K.I.T.A., Otsuka, S., Hachuda, S., Tatsuro, E.N.D.O., Yasunori, I.M.A.I., Nishijima, Y., Toshihiko, B.A.B.A., 2012. Photonic crystal nanolaser biosensors. IEICE Trans. Electron. 95 (2), 188–198.
Song, B.S., Noda, S., Asano, T., Akahane, Y., 2005. Ultra-high-Q photonic double-heterostructurenanocavity. Nat. Mater. 4 (3), 207–210.
Terada, Y., Hashimoto, W., Endo, T., Seto, H., Murakami, T., Hisamoto, H., Miura, Y., 2014. Signal amplified two-dimensional photonic crystal biosensor immobilized with glyco-nanoparticles. J. Mater. Chem. B 2 (21), 3324–3332.
Threm, D., Nazirizadeh, Y., Gerken, M., 2012. Photonic crystal biosensors towards on-chip integration. J. Biophoton. 5 (8–9), 601–616.
Tokushima, M., Kosaka, H., Tomita, A., Yamada, H., 2000. Lightwave propagation through a 120 sharply bent single-line-defect photonic crystal waveguide. Appl. Phys. Lett. 76 (8), 952–954.
Vollmer, F., Braun, D., Libchaber, A., Khoshsima, M., Teraoka, I., Arnold, S., 2002. Protein detection by optical shift of a resonant microcavity. Appl. Phys. Lett. 80 (21), 4057–4059.
Wang, P., Liu, Q., 2011. Biomedical Sensors and Measurement. Springer Science & Business Media, Zhejiang university press, China.
Washburn, A.L., Bailey, R.C., 2011. Photonics-on-a-chip: recent advances in integrated waveguides as enabling detection elements for real-world, lab-on-a-chip biosensing applications. Analyst 136 (2), 227–236.
Watanabe, K., Kishi, Y., Hachuda, S., Watanabe, T., Sakemoto, M., Nishijima, Y., Baba, T., 2015. Simultaneous detection of refractive index and surface charges in nanolaser biosensors. Appl. Phys. Lett. 106 (2), 021106.
White, I.M., Fan, X., 2008. On the performance quantification of resonant refractive index sensors. Opt. Express 16 (2), 1020–1028.
White, I.M., Oveys, H., Fan, X., 2006. Liquid-core optical ring-resonator sensors. Opt. Lett. 31 (9), 1319–1321.
Yablonovitch, E., 1987. Inhibited spontaneous emission in solid-state physics and electronics. Phys. Rev. Lett. 58 (20), 2059.
Yamilov, A., Wu, X., Liu, X., Chang, R.P.H., Cao, H., 2006. Self-optimization of optical confinement in an ultraviolet photonic crystal slab laser. Phys. Rev. Lett. 96 (8), 083905.
Yan, H., Zou, Y., Chakravarty, S., Yang, C.J., Wang, Z., Tang, N., Chen, R.T., 2015. Silicon on-chip bandpass filters for the multiplexing of high sensitivity photonic crystal microcavity biosensors. Appl. Phys. Lett. 106 (12), 121103.
Yang, C.J., Yan, H., Zou, Y., Chakravarty, S., Tang, N., Wang, Z., Chen, R.T., 2015. Wide dynamic range sensing in photonic crystal microcavity biosensors, SPIE BiOS, International Society for Optics and Photonics, pp. 931713–931713.
Ymeti, A., Greve, J., Lambeck, P.V., Wink, T., van Hövell, S.W., Beumer, T.A., Kanger, J.S., 2007. Fast, ultrasensitive virus detection using a Young interferometer sensor. Nano Lett. 7 (2), 394–397.
Yu, X., Zhang, Y., Kwok, Y.C., Shum, P., 2010. Highly sensitive photonic crystal fiber based absorption spectroscopy. Sens. Actuat. B 145 (1), 110–113.
Zecca, D., Qualtieri, A., Magno, G., Grande, M., Petruzzelli, V., Prieto-Simon, B., Stomeo, T., 2014. Label-free photonic crystal based immunosensors for diagnostic applications. Photon. J. IEEE 6 (6), 1–7.
Zheng, G., Patolsky, F., Cui, Y., Wang, W.U., Lieber, C.M., 2005. Multiplexed electrical detection of cancer markers with nanowire sensor arrays. Nat. Biotechnol. 23 (10), 1294–1301.

CHAPTER

NOVEL CARRIERS AND APPROACHES: INSIGHT FOR PSORIASIS MANAGEMENT

24

Surbhi Dubey, Rajeev Sharma, Nishi Mody, Suresh P. Vyas
Dr. Hari Singh Gour Central University, Sagar, India

CHAPTER OUTLINE

1 INTRODUCTION

Psoriasis is an autoimmune T-cell mediated disorder of the skin accompanied by constrained, red, thickened plaques with an overlying silver-white scale and relapsing episodes of inflammation and hyperkeratosis. The chief cause of psoriasis is unknown. Now-a-days, various treatment approaches with different antipsoriatic drugs having different modes of action and routes of administration, are established. Many conventional treatments focus on suppressing symptoms associated with psoriasis and have significant side effects (Traub and Marshall, 2007). Currently, topical therapy, phototherapy, and systemic therapy are used for the treatment but not any of these approaches is found to promote complete cure for the disease. These ailment strategies are unsafe, ineffective, and are associated with both unfortunate cosmetic appearance and associated toxicities leading to poor patient compliance in long-term use. Major issue for ineffective therapy is the lack of an ideal carter for antipsoriatic drug delivery. Hence researchers are putting efforts for using colloidal carriers, such as ethosomes, liposome, niosomes, transferosome, solid lipid nanoparticles, microspheres, micelles, dendrimers, etc. for achieving better therapeutic regimens as the distinctive drug cargoes.

2 TYPES OF PSORIASIS

Psoriasis is classified based on degree of inflammatory process, localization of rashes, cute/chronic patient conditions, and additional clinical parameters into four types: chronic plaque, guttate, pustular, and erythroderma (Galimova et al., 2008). The characteristic features applied for classifying psoriasis are mentioned in the Table 24.1.

Plaque psoriasis/psoriasis vulgaris is the most common. Lesion emerges from a small papule, which then aggregates to form plaques (Jong De, 1997). Guttae psoriasis arises due to augmentation of immune response by streptococcal infection. It initiates from small pink papules that becomes scaly periodically. Erythrodermic psoriasis is linked with excessive protein loss, fluid loss, and imbalanced homeostasis. It is a severe progression arising as a result of worsening of prevalent psoriasis. It needs hospitalization and aggressive treatment since condition gets complicated by increased susceptibilty to staphylococcal infections, pustulosis, arthopathy, and growth retardation (Peter et al., 2000). Another severe form, Pustular psoriasis characterized by external postulation of the lesions is commonly localized to the palms and soles but may be generalized. The lesions have little systemic effect but may be hard to treat. Generalized pustular psoriasis is often associated with fever and malaise, and there may be fluid and electrolyte disturbances and infection. This form of psoriasis has a high rate of relapse.

Table 24.1 Characteristics Features for Psoriasis Classification

Feature	Description
Age of onset	Type I: begins before the age of 40 years account for more than 75% of cases Type II: begins after the age of 40 years
Measure of severity	
• Based on body surface area affected	<5%: Mild psoriasis 5%–10%: Moderate psoriasis >10%: Severe psoriasis
• Based on number of sites affected	One site, many site, and generalized
Pattern of distribution	Extensory, inverse, seborrhoeic, widespread
Anatomical site	Nail, hair, anal, scalp, genital
Morphological characters	Pustular, nonpustular but also plaque, nummular, guttate, gyrate, rupioid, elephantine, ostraceous, erythrodermic
Degree of inflammation	Inflammatory versus mainly hyperkeratotic
Velocity of progression	Stable, unstable, eruptive

3 PATHOGENESIS OF PSORIASIS

Physiological appearance of psoriasis is due to extensively altered differentiation of keratinocytes, which leads to formation of scaly plaques. New cells formed in the stratum basale of the epidermis drift toward the stratum corneum accumulating keratin along with the disappearance of organelles. The cells get completely filled with the keratin while they reach keratin, which offers protection to the skin. Keratinocytes in the basal layer divide approximately once every 2 weeks and the process of exfoliation and substitution by new cells known as epidermal cycle takes about 4 weeks.

This cell cycle occurs at a faster rate in psoriatic skin wit cell division occurring at every 1.5 days and movement of keratinocytes toward startum corneum take place in just 4 days, which leads to improper differentiation and maturation of the cells. The stratum corneum consists of partially keratinized and abnormal cells giving scaly appearance with 3–5 times thicker epidermis than normal. Blood vessels in the papillary layer of the dermis dilate in psoriasis, and inflammatory cells, such as neutrophils, infiltrate the epidermis.

Other distinctive histological markers of psoriasis:

- Small foci in the stratum corneum containing neutrophils and considerable mononuclear leukocytes (T cells and dendritic cells, DCs) in the epidermis following the dermis.
- Marked infiltration of mononuclear leukocytes (T cells and DCs) into elongated/hyperplastic blood vessels in the papillary dermal region.
- Redness of the psoriatic lesion due to marked dilation of the blood vessels.
- Endothelial cells acquire characteristics of high endothelial venules usually present in lymph nodes, due to adherence of lymphocytes, monocytes, and neutrophils to the endothelial cells.
- Endothelial cells are activated in psoriatic lesions, as is indicated by staining for intracellular adhesion molecule-1 (ICAM-1, also known as CD54), vascular cell adhesion molecule-1 (VCAM-1, or CD106) and E-selectin (CD62E).

3.1 MEDICAL PATHOGENESIS OF PSORIASIS

Medical pathogenesis of psoriasis involves three events, which are discussed in the following sections.

3.1.1 T lymphocytes activation

This process involves recognition and interaction of antigen by the antigen-presenting cells (APCs) present in the epidermis and dermis and is performed through following steps (Fig. 24.1).

1. APC which is attacked by the antigen undergoes surface modification by generation of antigen-Major Histocompatibilty Complex (MHC) on its surface. Such surface modified APC is then drifted to the lymph node.
2. In the lymph node, binding of antigen to receptor on T lymphocytes by the MHC molecule resulting in the activation of T lymphocyte.
3. Costimulation is the second signal required for the activation of T lymphocytes otherwise the T lymphocytes become unresponsive and undergoes apoptosis. Costimulation involves nonantigen/cell-cell interaction, such as pairing of receptor with ligand on the T cell; these pairs include lymphocyte functional antigen (LFA)-3 interacting with CD2, B7 interacting with CD28, and ICAM-1 interacting with LFA (Lebwohl, 2003).

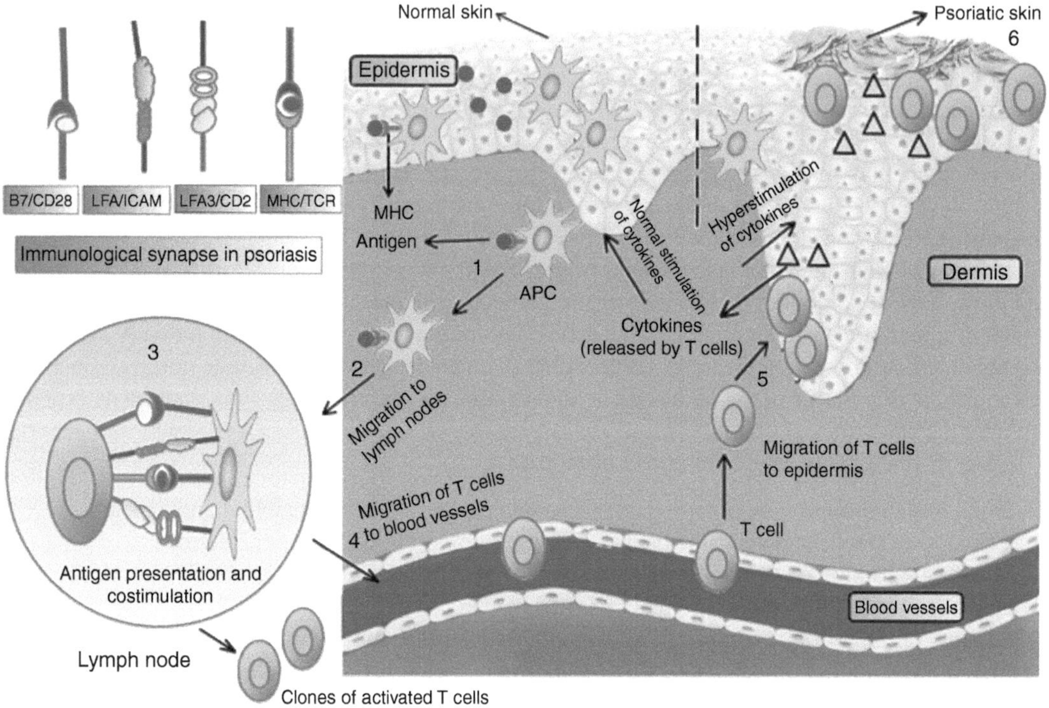

FIGURE 24.1 Different Events in the Pathogenesis of Psoriasis

(1) Binding of antigen to MHC of antigen presenting cells (MHC); (2) migration of antigen carrying APC to the lymph nodes; (3) antigen presentation by APC to T cell and costimulation; (4) migration of clones of activated T cells to blood vessels; (5) migration of activated T cell to epidermis and release of cytokines; and (6) formation of psoriatic skin by the influence of cytokines.

Adapted with permission from Pradhan, M., Singh, D., Singh, M.R., 2013. Novel colloidal carriers for psoriasis: current issues, mechanistic insight and novel delivery approaches. J. Control. Release 170, 380–395.

3.1.2 Migration into the skin

The activated lymphocytes either proliferate into memory-effector cell or moves into the systemic circulation where they interact with endothelial cells of the blood vessel through cell–cell interaction and transit to the inflamed skin (Lebwohl, 2003).

4 FUNCTION OF CYTOKINES

Cytokines play a key role in succession of psoriasis. Tumor necrosis factor alpha (TNF-α), interleukin-23 (IL-23), and IL-17 are the major cytokines which stimulate in the production of other proinflammatory cytokines and psoriasis lesions formation (Pietrzak et al., 2008). Other potentially important cytokines in psoriasis with their roles have been discussed in Table 24.2.

Table 24.2 Potentially Important Cytokines in Psoriasis

S. No.	Cytokine	Role in Psoriasis	References
1	IL-1	IL-1 acts through activated T cells and triggers the production of IL-2 and IFN-γ. Neutrophils, monocytes, eosinophils, and basophils are activated under the influence of IL-1. IL-1 also stimualates macrophages to synthesize TNF, IL-6, and IL-8.	Dinarello (2002)
2	IL-2	IL-2 triggers B-cells differentiation, as well as production and action of NK cells, monocytes, and macrophages. Il-2 encourages synthesis of IFN-g, TNF, IL-6, and IL-2R, as well as participates in self-production of IL-2.	Kemmett et al. (1990)
3	IL-4	This interleukin transforms the T-helper cell phenotype into Th2 cells thus participates in direct suppression of Th1-mediated inflammation. IL-4 also initiates synthesis of IL-6, extracellular matrix proteins, and ICAM-1.	Ghoreschi et al. (2003), Asadullah et al. (2002)
4	IL-6	IL-6 participates in the cellular development and differentiation of dermis and epidermis. It may directly trigger movement of T-cell into the epidermis as they contribute as chemotactic factors for T cells.	Hirano et al. (1990)
5	IL-7	Participates in development and differentiation of specifically T-cells (TCRγδ T cells) and B cells. IL-7 induces production of IFN-γ by combined effort of IL-2 and IL-12.	Spitz and Oppenheim (2000)
6	IL-8	Involved in formation of new blood vessels and encourage chemotaxis and degranulation of neutrophils.	Iizasa et al. (2000)
7	IL-10	It exerts negative effect on naive T-cells differentiation thus obstructs production of type 1 cytokines in Th1 cells, keratinocytes, monocytes, and macrophages. Synthesis of type 2 cytokines and IL-1 receptor antagonists is triggered by the action of IL-10.	Asadullah et al. (1998)
8	IL-11	IL-11 exerts negative effect on the synthesis of proinflammatory cytokines, such as IL-1β, IL-12 p40, TNF, and IL-6. It is involved in the reduction or inhibition of IFN-γ production and ICAM-1 expression. It induces Th2-type response by exerting positive effect on the synthesis of IL-4 and IL-5.	Godic (2004)
9	IL-15	It regulates the activation, proliferation, and endurance of NK cells. This interleukin stimulates formation of new blood vessels and T cells expression of IL-17.	Berard et al. (2003), Elder (2007)
10	IL-17	Induces the synthesis of IL-6 and IL-8 from epithelial, endothelial, and fibroblastic cells. Production of interleukins, such as TNF, IL-1β, IL-6, IL-10, and IL-12 from macrophages is also induced by IL-17	Witowski et al. (2004)
11	IL-18	The synthesis of IFN-γ in many immunocompetent cells, such as NK cells, T helper, and cytotoxic cells is stimulated by the combined effect of IL-18 and IL-12. IL-18 participates in synthesis of some interleukins, such as IL-4 and IL-13 from T cells.	Muhl and Pfeilschifter (2004)

(Continued)

Table 24.2 Potentially Important Cytokines in Psoriasis (*cont.*)

S. No.	Cytokine	Role in Psoriasis	References
12	IL-19	IL-19 positively regulates the synthesis of IL-4 while negatively regulates production of IFN-γ.	Gallagher et al. (2004)
13	IL-20	Exerts excessive proliferative effect on keratinocytes and switch on the IL-20 receptor to alter inflammation of skin. Participate and encourage CD8+ T cells–based expression of IL-6 and TNF.	Wang et al. (2003)
14	IL-23	Exerts stimulatory effect on memory T cells to evoke production and proliferation of IFN-γ. Participates in the inflammatory process of peripheral tissues and triggers a specific group of T-cell to produce IL-17 as a chief effector interleukin.	Aggarwal et al. (2003)

Adapted with permission from Pradhan, M., Singh, D., Singh, M.R., 2013. Novel colloidal carriers for psoriasis: current issues, mechanistic insight and novel delivery approaches. J. Control. Release 170, 380–395.

5 TREATMENT APPROACHES FOR PSORIASIS

Conventional treatment of psoriasis is based on the degree of severity. Presently three modes of treatment are established: (1) topical therapy, (2) phototherapy, and (3) systemic therapy.

Mild and limited psoriasis treatment includes topical corticosteroids, tars, anthralin, calcipotriene (a vitamin D3 analog), and tazarotene (a retinoid). For psoriasis treatment topical therapies are considered first line treatment. Phototherapy is suggested in case of noneffectiveness of topical therapy or relentlessness of the psoriasis condition, which is followed by systemic medications. Detailed account on available antipsoriatic drugs including their mechanism of action, pharmacokinetics, side effects, and marketed formulations has been discussed in Table 24.3.

5.1 DISADVANTAGES OF CONVENTIONAL TREATMENT

- Phototherapy and systemic agents exhibit numerous ill effects, such as hepatotoxicity, renal toxicity, hypertension, hyperlipidemia, and skin cancer (Pradhan et al., 2013).
- Topical application of drugs causes side effects, such as dermatitis, irritation, skin discoloration, or coloration.
- Photosensitivity and increases risk of cancer for patients receiving phototherapy.
- Systemic delivery of these agents causes nausea, vomiting, constipation, and GI upset.
- Systemic drugs are also prone to cause hepatotoxicity, renal toxicity, hypertension, and cancer.
- Conventional therapy requires frequent dosing of 3 or 4 times a day therefore present poor patient compliance and increased side effects due to overdose.
- Not appropriate for long-term use.

5.2 CHALLENGES FOR PSORIASIS THERAPY

Although several treatment options are available for psoriasis therapy, the cure of ailment is still remains a challenging task because of the following two main reasons.

Table 24.3 Available Treatment Options for Psoriasis

S. No.	Drug	Mechanism of Action	Pharmacokinetics	Side Effect	References
[I]	**Topical**				
(i)	Calcineurin inhibitors				
a.	Tacrolimus	Tacrolimus acts by inhibiting the activity of calcineurin phosphatase (cytoplasmicenzyme). Calcineurin phosphatase inhibitsthe translocation of NFAT (nuclear factor of activated T cells) in the T cells.	Bioavailability—20% less after eating food rich in fat, protein binding 75%–99%, metabolism—hepatic, half-life—11.3 h, excretion—mostly fecal, dosing—0.05%–0.1%.	Initial applications of this drug cause burning or itching sensation. Flu-like symptoms, headache, cough, and burning eyes are rare side effects.	Cather et al. (2001)
b.	Pimecrolimus	Its mode of action is very similar to tacrolimus but acts more selectively with no effect on dendritic (Langerhans) cells and low skin permeation as compared with topically applied tacrolimus.	Bioavailability—low systemicabsorption, protein binding—74%–87%, metabolism hepatic, dosing—0.05%–0.1%.	Burning, soreness of treated skin,swelling of lips, face, tongue, and throat.	Grassberger and Meingassner (2001)
(ii)	Dithranol	Dithranol impairs energy supply to the cell by accumulating inside the mitochondria which may be probably occurring by free radicals released from oxidation of dithranol. Dithranol interfere with the replication of DNA and slows down the extreme cell division that occurs in psoriatic plaques.	Bioavailability—0%, protein binding—0%, metabolism—absorbed and oxidized within the skin, dosing—begin with 0.5% preparation for long-term therapy or 1% for short-contact therapy.	Skin irritation such burning, itching, brown discoloration of diseased as and surrounding skin.	Mendonc and Burden (2003)
(iii)	Tars	Elements of coal tar are absorbed percutaneously. It causes suppression of the hyperplastic skin occurring in some proliferative disorders.	Adult dose—0.100%–4.000%, pedriatic dose—up to 0.100%.	Staining, irritation and carcinogenicrisk.	Roelofzen et al. (2010)
(iiv)	Vitamin D3 analog				
a.	Calcipotriol	It acts by slowing down the production of proinflammatory cytokines, such as interleukin (IL)-8 which is an important player in inflammatory process of psoriasis. It also triggers the production of antiinflammatory cytokines, such as IL-4 and IL-10.	Bioavailability—5%–6%, metabolism hepatic, excretion—biliary.	Pruritus, burning, sharp pain, erythema, and rare hypercalcemia.	Kragballe (1992)

(Continued)

Table 24.3 Available Treatment Options for Psoriasis (*cont.*)

S. No.	Drug	Mechanism of Action	Pharmacokinetics	Side Effect	References
b.	Calcitriol	It acts by favoring the calcium absorption in the intestines and retention at the kidneys leading to increased levels of calcium in the serum, decreased level of serum phosphatase levels and parathyroid hormone, decreased bone resorption, increased resorption of renal tubule phosphate. The exact mechanism of action of calcitriol in psoriasis treatment has not been established.	Onset of action: Oral—~2–6 h, absorption—rapid oral, protein binding—99.9%, half-life elimination—children ~27 h, normal adults—5–8 h, excretion primarily feces, urine.	Anorexia, headache, thirst, sweating, polyuria, and hypercalcemia.	Kragballe (1992)
(v)	Keratolytic agents				
a.	Salicylic acid	Salicylic acid acts by reducing intercellular bonding of corneocyte. It causes decrease in the pH of the stratum corneum leading to hydration and swelling of corneocyte which further leads to desquamation.	It is readily absorbed from the skin, Protein binding—50%–80%, apparent volume of distribution—0.1–0.2 L/kg, half-life—2–4.5 h, metabolism—80% in liver, excretion—mainly by kidney	Skin irritation, stinging in the area where applied and serious side effects include dizziness, headache, fast breathing, ringing, or buzzing in ears.	Federman and Fraelish (1999)
b.	Urea	Intercellular matrix gets dissolved by the action of urea which leads to loosening of horny skin layer and shedding scaly skin at regular intervals	Pharmacokinetic data for topically applied urea is not yet known.	Severe skin irritation, mild stinging, itching and skin rash.	Fluhr et al. (2008)
c.	Omega-3 fatty acids	It acts by favoring the inhibition of lymphoproliferation, antigen presentation, and adhesion molecule presentation. Omega-3 fatty acid also participates in the inhibition of Th1 and Th2 generated responses and prevents production of proinflammatory cytokine.	Peak omega-3 fatty acid plasma levels—2–8 h after absorption, optimal dose—3–6 g daily.	Rash, itching, wheezing, difficult breathing, or swallowing.	McCusker and Grant-Kels (2010)
d.	Cod liver oil	It acts as substrate for cyclooxygenase and lipoxygenase activities leading to increased leukotriene B5 and prostaglandin 3 productions.	Presystemic metabolism is about 50%.	Hypervitaminosis, hypersensitivity reaction, and topical irritation.	Suvarna (2008)

e.	Glycolic acid	Glycolic acid acts by disrupting the adhesion of corneocytes present in the lower layers of stratum corneum, leading to disconnection of the bulky upper layer from the newly synthesized lower layer.	Glycolic acid has p*K*a of 3.83,metabolism—liver.	Inflammation and burning sensation on the applied area.	Erardesca et al. (1998)
(vi)	Corticosteroid				
a.	Class I—superpotent Super potent: clobetasol propionate, diflorasone diacetate, halobetasol propionate, betamethasone dipropionate	They perform their activity by controlling theactivities of cellular proteins, which leads to broad range of antiinflammatory, immunesuppressant, and vasoconstrictionactions. Control over these activities is achieved both by modifying the expression of gene and through nongenomic mechanisms.	When these topical corticosteroids are absorbed through skin they follow the same pharmacokinetic pathways as that of systemic corticosteroids, metabolism—in liver, excretion—by kidneys.	Skin infections, perioral dermatitis, skin atrophy, hypertrichosis and striae.	Fenton and Plosker (2004), Maccaroni et al. (2009), Mohandas et al. (2009)
b.	Class II—potent Amcinonide, mometasone furoate, fluocinolone acetonide	They perform their activity by controlling the activities of cellular proteins which leads to broad range of antiinflammatory, immunosuppressant, and vasoconstrictionactions. Control over these activities is achieved both by modifying the expression of gene and through nongenomic mechanisms.	Dosage—topical as 0.1% cream/lotion/ointment, apply 2–3 times/day.	Skin thinning, striae, skin atrophy, telangiectasias, purpura, acneiform eruptions, burning, itching, irritation, dryness, olliculitis, hypertrichosis,and secondary infection.	Crim et al. (2001), Fowler et al. (2010), Laws and Young (2010)
c.	Class III – upper mid strength Triamcinolone acetonide, Betamethasone valerate	They exert their stimulatory action on phospholipase A2 inhibitory proteins, collectively called lipocortins. These proteinsinhibit the discharge of arachidonic acid andregulate the biosynthesis of inflammationcausing substances, such as prostaglandinsand leukotrienes.	Proteins binding—comparatively low and remains consistent over a wideconcentration range (0.03–3.2 ng/mL), metabolism—very rapid, excretion renal.	Irritation, dryness, folliculitis,hypertrichosis, acneiform eruptions, hypopigmentation, perioral dermatitis,allergic contact dermatitis andinfection.	Fowler et al. (2010), Naldi et al. (2011)
d.	Class IV—mid strength Flurandrenolide,hydrocortisone valerate	They exert their stimulatory action on phospholipase A2 inhibitory proteins, collectively called lipocortins. These proteins inhibit the discharge of arachidonic acid and regulate the biosynthesis of inflammation causing substances, such as prostaglandins and leukotrienes.	When these topical corticosteroids are absorbed through skin they follow the same pharmacokinetic pathways as that of systemic corticosteroids, metabolism—in liver, excretion—by kidneys.	Polyuria, polydipsia, polyphagia, weight gain, increased susceptibility to infection, gastro intestinal ulceration, pancreatitis, osteoporosis, hyperglycemia, steroid myopathy.	Rosso and Friedlander (2005)

(Continued)

Table 24.3 Available Treatment Options for Psoriasis (*cont.*)

S. No.	Drug	Mechanism of Action	Pharmacokinetics	Side Effect	References
e.	Class V—lower mid strength Prednicarbate	The chief mechanism of action involves the down regulation of IL-1α, which further inhibits the cytokine in keratinocytes. The inhibition of IL-1α in keratinocytes leads to antiinflammatory effects.	When these topical corticosteroids are absorbed through skin they follow the same pharmacokinetic pathways as that of systemic corticosteroids, metabolism—in liver, excretion by kidneys.	Burning sensation, itching, rash, folliculitis, reddening of skin, weeping skin lesion may occur.	Gupta and Chow (2004)
f.	Class VI—mild Betamethasone valerate	They exert their stimulatory action on phospholipase A2 inhibitory proteins, collectively called lipocortins. These proteins inhibit the discharge of arachidonic acid and regulate the biosynthesis of inflammation causing substances, such as prostaglandins and leukotrienes.	Metabolism—hepatic, half-life—5.6 h, excretion—renal.	Irritation, dryness, folliculitis, hypertrichosis, acneiform eruptions, hypopigmentation, perioral dermatitis, allergic contact dermatitis, secondary infection, and striae.	Naldi et al. (2011)
(vii)	Retinoids				
a.	Tazarotene	Tazarotene works by undergoing hydrolysis in the skin into its active metabolite tazarotene acids in presence of esterases enzyme. This further interects with the nuclear retinoic acid receptors RAR-beta and RAR-gamma and thus allows the proliferation and differentiation of epidermis. Moreover, there is decreased appearance of the inflammation causing substances in the epidermis and dermis.	Systemic absorption—following nonocclusive topical application is less than 1%, protein binding 99%, half-life—18 h, dosing—0.05–0.1%.	Pruritus, burning sensation of skin,erythema, irritation, inflamed, and dry skin.	Weinstein et al. (1997)
b.	Acitretin	Acitretin acts by modulating the mitotic activity and differentiation of the keratinocytes and slowing down intraepidermal movement of neutrophilic granulocytes.	Absorption—36–95% in the intestine, half-life—approximately 2 days, dosage—0.5–0.8 mg/kg daily.	Vitamin A toxicity (cheilitis, xerosis, nose bleeds, alopecia, increased skin fragility).	Rozenbli et al. (2003)
[II]	Systemic				

(i)	Biologics				
a.	Efalizumab	Efalizumab exerts its inhibitory action and prevents interaction of T cell to the endothelial cells, movement of T-cell from the blood vessels into the tissue, interaction of T cells to keratinocytes, as well as activation of T cell.	Elimination half-life—4–5 days, dosing—1 mg/kg weekly subcutaneous (s.c.), first dose—0.7 mg/kg.	Flu-like reactions, leukocytosis and lymphocytosis, rebound, exacerbation and arthralgia.	Kruger (2002), Sullivan and Preda (2009)
b.	Etanercept	TNFα is a Etanercept interacts with soluble TNFα and blocks the cascade of inflammation stimulated by TNFα which is a chief pro inflammatory cytokine contributing major role in the development and regulation of inflammatory processes	Bioavailability—56%–76% s.c., half-life—70–132 h.	Local reactions, infections, sore throat, headache, dizziness, fatigue, hair loss, and rash.	Papp et al. (2005), Saini et al. (2005)
c.	Infliximab	It exerts its action by interacting with soluble and receptor bound TNFα. It may act by two ways, one by binding with soluble TNFα and thus neutralizing pro inflammatory activity. Another way is, binding to cell bound TNF α and ultimately leading to removal of the affected cells.	Plasma half-life—8–9.5 days, dose—5 mg/kg as single dose for every alternate week followed by regular dose after 8 weeks	Infusion reactions, severe infections, progression of heart failure, and rare liver failure.	Saini et al. (2005), Winterfield and Menter (2004)
d.	Alefacept	It interacts with CD2 on T lymphocytes and prevents the costimulatory interaction between LFA-3/CD2 thus interferes with T-lymphocyte activation.	Bioavilability—63% intramuscular (i.m.), half-life—270 h approx.	Decrease in CD4+ count, itching, nausea, muscle aches, malignancies and allergies risk of infections	Korman and Moul (2005), Saini et al. (2005), Winterfield and Menter (2004)
e.	Adalimumab	It acts by nullifying the activity of TNF-α.	Bioavilability—64% s.c., half-life—10–20 days, dose—80 mg s.c. loading dose, then 40 mg s.c. every other week.	Swelling, redness, pain, and itching at the site of application, shortness of breath and swelling of ankles or feet.	Menter et al. (1980), Strober et al. (2008), Uslu (2007)
(ii)	Miscellaneous				
a.	Cyclosporine	Cyclosporine inhibits the activity of phosphatase of the calcium–calmodulin–calcineurin complex and promotes the translocation of NFAT following the production of NFAT-dependent cytokine.	Dosing—interval therapy is0.5 mg/kg every 14 days.	Renal failure, increase of blood pressure, liver failure, nausea, anorexia, vomiting, diarrhea, hypertrichosis,gingival hyperplasia, and tremor.	Berth (2005)

(*Continued*)

Table 24.3 Available Treatment Options for Psoriasis (*cont.*)

S. No.	Drug	Mechanism of Action	Pharmacokinetics	Side Effect	References
b.	Methotrexate	It acts as folic acid antagonist, which causes competitive inhibition of the enzyme dihydrofolate reductase. Methotrexate prevents the conversion of dihydrofolic acid to tetrahydrofolic acid as methotrexate high affinity to dihydrofolate reductase as compared with dihydrofolic acid.	Initial dose—5–15 mg/ week,maintenance dose—5–22.5 mg/week dependingon effect.	Liver fibrosis/cirrhosis, pneumonia/alveolitis, bone marrow depression, renal damage, necrosis of soft tissue and bone.	Kalariya and Misra (2004)
(iii)	Phototherapy				
a.	Ultraviolet B therapy	Ultraviolet B radiation acts by binding with the nuclear DNA. Interaction between UVB and DNA causes generation of pyrimidine dimers and other photoproducts leading to inhibition of DNA synthesis. Ultimately, in addition to inhibition of proliferating lymphocytes there is decreased proliferating keratinocytes.	Not defined.	Risk of skin cancer and erythema.	Lebwohl et al. (2001)
b.	Psoralen plus ultraviolet A therapy (PUVA)	Psoralen activated by UV radiation combines with DNA to generate tight binding between the two strands of the double helical DNA and inhibits DNA synthesis. Thus PUVA prevents the aberrant cell proliferation inpsoriasis plaques by avoiding the cells to duplicate their DNA.	Not defined.	Long-term PUVA therapy causes skincancer, most notably squamous cell carcinoma (SCC) and to lesser extentbasal cell carcinoma and malignant melanoma and premature skin aging.	Lebwohl et al. (2001)
c.	Methoxsalen	Methoxsalen activated by UV radiation binds with DNA and forms firm associations between the strands of the double helical DNA structure inhibiting the DNA synthesis.	Half-life—approx 2 h, dosage—thin layer of 0.15% preparation onto the affected areas 15 min before UVA exposure.	Pruritis, mild transient erythema,odema, headache, dizziness, achneiform eruption and severeskin pain.	Wozel (2008)

Adapted with permission from Pradhan, M., Singh, D., Singh, M.R., 2013. Novel colloidal carriers for psoriasis: current issues, mechanistic insight and novel delivery approaches. J. Control. Release 170, 380–395.

5.2.1 Lack of efficient drug delivery carrier

It is now clearly understood that presence of effective drug alone cannot afford effective and successful therapy. It is very essential for the drug to reach the desired site of action. The reasons for failure of therapy is mainly due to inherent inability of drug to reach site of action due to poor absorption, rapid metabolism, and elimination, drug distribution to other tissues combined with high drug toxicity, inappropriate integration in conventional dosage form and high fluctuation of plasma level due to unpredictable bioavailability after per oral administration, including the influence of food on plasma levels (Mehnert and Mader, 2012). Novel drug-delivery system offers promising strategies to overcome the limitations of conventional drug therapy and to obtain localized delivery and controlled delivery of the active drug according to the specific need of the therapy. Other merits of novel carriers include protection of drug against external and internal environment, controlled drug release, longer duration of action, passive and active targeting can be achieved, improvement of drug therapeutic efficacy and effective internalization (Pradhan et al., 2013).

5.2.2 Lack of animal models

This threatening disease occurs only in human beings which leads to lack of appropriate animal models which could entirely imitate the histopathopathological and immonogenity of psoriasis as in human. Number of immunogenetic models have been established but none of them have possesed all characteristic features and limitations of the disease. Following three models are commonly used but they also suffer from certain limitations.

5.2.2.1 Spontaneous mutation model

In this model mutation is generated by errors in replication of DNA, formation of spontaneous lesions, and transposable genetic elements. The model presents characteristics, such as hyerkeratosis and scaly formation but is deficient in T cell and neutrophil responses (Mizutani et al., 2003). So, it does not resemble psoriasis mimic completely. Examples: homozygousasebia (Scd1ab/Scd1ab), flaky skin mice (Ttcfsn/Ttcfsn), and spontaneous chronic proliferative dermatitis mutation (Sharpincpdm/Sharpincpdm).

5.2.2.2 Xenotransplantation model

In this process xenotransplantation is used to develop psoriasis. The process involves grafting of section of psoriatic skin on genetically tailored mice. The drawback of this model is presence of histological characters, which are absent in psoriasis, such as presence of granular layer in arthymic nude mice presence of immature killer (NK) cells in AGR 129 model and prevalence of mature NK cell in SCID mice. Examples: athymic nude mice (Fraki et al., 1983), severe combined immunodeficient micc (SCID) (Nickoloff et al., 1995), and spontaneous AGR129 model (Boyman et al., 2004).

5.2.2.3 Transgenic model

This model is developed by modification of gene but the modification is done to at a single gene level. This model is also an inappropriate because psoriasis is a multigenetic disorder and hence lacks number of psoriatic factors.

6 NOVEL COLLOIDAL CARRIERS FOR PSORIASIS TREATMENT

Historically development of new drug has been observed in need of an effective and safe therapy. But it is a time-consuming development requiring lot of efforts and huge amount of money for investigation and trials. It has been found that safety and efficacy of drug are lost at the expense of improper distribution of the drug within the biological system and its deviation from the desired site of action, that is, the target site. The objective of site-specific delivery can be accomplished by using novel carrier systems to deliver the molecules to specific receptor sites without afflicting the normal tissues and organs of the body. The basis lies in skillfully designed carrier, which brings desirable alteration in the surrounding microenvironment where the drug delivery is to be accomplished. It involves the modification(s) in physicochemical characteristics of the active moiety and alterations in barrier properties of the biological membranes at various locations, which directs the improved transportation of drug toward diseased location (Katare et al., 2010). Table 24.4 Gives the details of various treatment options available, their mechanism of action and their limitations.

6.1 LIPID-BASED COLLOIDAL CARRIERS

Lipid-based colloidal carriers are well tolerated since they are composed of physiological lipids which are usually nontoxic and degrade in the body to nontoxic components. Lipoidal carrier advantage of controlled release, stability, targetability, high drug load, biodegradability, economical, and certain lipid carriers, such as SLN and NLC can carry both lipophilic and hydrophilic drugs. It also possesses unique advantage of being identical to many similar or physiologically occurring compounds (Rawat et al., 2006).

Many subclasses of lipid-based carriers are devised as a potential candidate for the delivery of antiporiatic drug topically, as well as systemically which include: lipidic vesicular carriers (tansferosome, niosomes, liposomes, ethosomes), microemulsions, and lipidic particulate carriers (solid lipid nanoparticles, nanostuctured lipids).

6.1.1 Liposomes

Liposomes are microscopic phospholipid bilayered vesicles comprising of one or more concentric lipid bilayer enclosing an aqueous chamber. Cholesterol is included for providing integrity to the bilayer by increasing microviscosity and reducing permeability of the membrane to water-soluble molecules, thereby providing rigidity to the membrane and stabilizing the vesicles.

Liposomes are most preferred carriers for the topical formulation and are superior to conventional topical formulation because (1) they are capable of carrying drugs with opposing solubilities, that is, lipophilic and hydrophilic and (2) the major structural component of liposomal carrier, phospholipid, gets easily integrated with the skin lipids and assures the desired hydration conditions to get better penetration and localization of drug in the layers of the skin (Moghimi and Patel, 1993; Schmid and Korting, 1996). Liposome-based formulations are used for the treatment of variety of dermatological conditions including psoriasis. Liposomal formulations have also been found to be promising for providing improved therapeutic effect and reduced side effect of the topical antipsoriatic drugs. For example, liposomal encapsulation of retenoids, such as vitamin A acid or tretinoin reduced the local irritation. Skin deposition of liposomal formulation depends on lipid composition, method of preparation, and thermodynamic state of the bilayers of liposomes (Bouwstra and Honeywell-Nguyen, 2002; El Maghraby et al., 2006).

6.1.1.1 Mode of action of conventional liposomes

Liposomes offer enhanced the topical therapeutic by number of mechanisms which are: (1) intact vesicular skin penetration, (2) the penetration enhancing effect, (3) the adsorption effect, and (4) the penetration of liposomes through the transappendageal route.

Table 24.4 Novel Colloidal Carrier for Psoriasis

S No	Novel Carrier	Drug	Advantage	Method of Preparation	References
1.	Liposome	Tacrolimus	Facilitated entry into the tough barrier consisting of stratum corneum. Better solubility of drug.	Thin film hydration method	Patel et al. (2010)
		Anthralin	Increase the stability of drug. Increases the penetration of drug through skin.	Thin film hydration method	Saraswat et al. (2007)
		5-aminolevulinic acid	Enhance drug uptake	—	Casas and Batlle (2006)
		Temoporin	Higher rate of drug transfer across skin	Thin film hydration method	Dragicevic-Curic et al. (2009a)
2.	Microemulsion	Methotrexate	Sixfold higher permeation flux than that from the corresponding solutions in mice skin.	—	Trotta et al. (1996)
		Psoralens	Enhanced drug uptake and provide photostability.	—	Baroli et al. (2000)
		Cyclosporine	Increase the stability of drug	—	Umezawa and Ozawa (2007)
3.	Niosome	Methotrexate	Better efficacy, tolerance, and patient compliance, when compared to a marketed formulation.	Thin film hydration	Lakshmi et al. (2007)
		Dithranol	Enhance stability of drug. Augmented in vitro permeation using mouse abdominal skin (permeation rate constant for dithranol loaded formulation was 7.78 $\mu g/cm^2/h$) as compared to the cream base	Thin film hydration	Agrawal et al. (2010)
4.	Solid lipid nanoparticulate (SLN)	Methotrexate	Improved drug accumulation in human cadaver skin.		Misra et al. (2004)
			Significantly improved therapeutic index in terms of average percent improvement in healing (APIH) of lesions and reduction in average score of degree of erythema and scaling		Shah et al. (2007)
		Tretinoin	Significant improvement in photostability and decreased irritant effect as compared to methanolic tretinoin solution.	Emulsification–solvent diffusion (ESD) method	Jensen et al. (2010)
		Betamethasone 17	Improved permeation and retention of drug in both barrier-impaired and intact skin	Hot high pressure homogenization technique	

(*Continued*)

Table 24.4 Novel Colloidal Carrier for Psoriasis (*cont.*)

S No	Novel Carrier	Drug	Advantage	Method of Preparation	References
5.	Nanostructured lipid carrier	Tacrolimus	Provide good occlusion property and solubility. Enhanced encapsulation efficiency and skin localization.	High pressure homogenization	Pople and Singh (2011)
		Vitamin D analogs	Enhanced skin permeation and negligible skin irritation.	—	Lin et al. (2010)
		Acitretin	No major systemic side effects, the main topical side effect was irritation	Solvent diffusion method	Agrawal et al. (2010)
		Corticosteroids	PEG-NLCs, high entrapment efficiency and improved drug stability.	Modified microemulsion method	Doktorovova et al. (2010)
		Methotrexate	Higher skin penetration when compared to free methotrexate.	Hot homogenization combined with ultrasonication techniques	Pinto et al. (2014)
6.	Ethosomes	5-aminolevulinic acid	Better skin permeation and targeting deeper skin layer.	Thin-film hydration method	Fang et al. (2009)
		Tacrolimus	Higher entrapment efficiency and localization of tacrolimus in dermis by ethosomes in contrast to classic liposomes.	Mechanical dispersion method	Li et al. (2012)
		Methotrexate	Greater retention of MTX drug into the skin for a longer period of time for better therapeutic effect	Mechanical-dispersion method	Dubey et al. (2008)
7.	Invasomes	Temoporfin	Enhanced drug uptake	Membrane extrusion method and conventional film forming method	Dragicevic-Curic et al. (2009b)
8.	Transferosomes	Methotrexate	Better permeation and retention in comparison to aqueous solution and conventional liposomes	Reverse-phase evaporation technique	Trotta et al. (2004)
		Corticosteroids	Improved activity of entrapped drug even at doses lower than that used in conventional formulations for treatment of dermal disorders due to better drug retention in the target organ	Thin film hydration and proprietary manufacturing technique	Cevc et al. (1997)

Mezei and Gulasekharam (1982) were first to report the concept of intact vesicular skin penetration and suggested that liposomes act as reservoir at the site of action and large size of liposomes prevents their entry into the capillary circulation. They also reported that the liposomes composed of dipalmitoylphosphatidylcholine (DPPC) and cholesterol (CH) (1.1:0.5, molar ratio) reduced the percutaneous absorption along with the 4–5 times increase in the concentration of triamcinolone acetonide in the epidermis and dermis as compared to standard ointment.

A second possible mode of action, the penetration enhancing effect was supported by the finding that a fluorescent lipid bilayer marker showed improved delivery into deep stratum corneum after skin treatment with empty vesicles which was parallel to that obtained from carriers encapsulating the marker. This suggested that improved delivery was not due to intact vesicular penetration, supporting a second possible mode of action, the penetration enhancing effect (Kirjavainen et al., 1996). Liposomes have the potentiality to act as a penetration enhancer by causing loosening of the lipidic stratum corneum overcoming the barrier property of the skin with the subsequent partitioning of the drug in skin. Liposome forming lipid adheres to the surface of the skin followed by the destabilization and fusion or mixing with the lipid matrix of stratum corneum leading to penetration of liposome through skin. Therefore it is suggested that the liposomes made of lipid similar to skin lipids (like ceramides, cholesterol, free fatty acids, and cholesteryl sulfate) are more effective for dermal delivery than the phosholipid vesicles (Downing et al., 1993). For example, skin-lipid liposomes are prepared to improve corticosteroid dermal delivery and hence, their therapeutic effectiveness, which provided highest drug disposition within the deeper skin layers, that is, in the epidermis and dermis (Fresta and Puglisi, 1997). It has been found that pretreatment of the skin with liposomes did not promote the absorption of hydrophobic or hydrophilic drugs concluding that liposomes stratum corneum interaction hypothesis was invalid. Therefore, it is has been suggested that liposomes must be applied concurrently with the drug or the drug must be encapsulated within them.

6.1.2 Transferosomes

Transferosomes are highly elastic lipid vesicles capable of penetrating intact skin and are used for noninvasive or needle-free delivery of therapeutic agent. The conventional liposomes rarely provide transdermal effect and mainly elicit localizing effects while transferosomes are several orders of magnitudes more elastic than the standard liposomes capable of carrying therapeutic effect to the deeper layer of skin. These are self-aggregate vesicles having high flexiblilty and deformability, which permits them to overcome the mechanical barrier generated by the stratum corneum when applied to nonocclusive skin. Transferosomes are metastable vesicles, which makes them capable of permeating intact skin by squeezing through the intracellular sealing lipid or the pores present in the stratum corneum. They squeeze through the skin by the process of diffusion by following the natural water gradient across the epidermis thereby minimizing the possibility of vesicle rupture in the skin. They are capable of shrinking to one-tenth of the original size and therefore, vesicles upto the size of 200–300 nm can penetrate easily (Gupta et al., 2012). They are composed of phospholipids (soya phosphatidylcholine, egg phosphatidylcholine, dipalmityl phosphatidylcholine, etc.) as the main component and 10%–25% single chain surfactant as an edge activator (sodium cholate, tween 80, span-80), alcohol as solvent (3%–10%), and saline phosphate buffer (pH 6.5–7) (Pirvu et al., 2010).

6.1.3 Mechanism of penetration of transfersomes

The high permeability of these vesicles is attributed to the edge activator or surfactant which destabilizes the lipid bilayer of the vesicles rendering them flexibility and deformability for the diffusion through skin (Bouwstra and Honeywell-Nguyen, 2002; Cevc, 1996). A number of mechanisms have been proposed by which deformable vesicles could improve skin delivery of drugs. Following two mechanisms are being accepted predominantly.

6.1.3.1 Mechanism I

Transferosomes resembles drug carrier system, maintaining integrity of vesicles while entering the skin. This permits the vesicles entrapped drug to penetrate into the skin. This mechanism was proposed by Cevc et al. (Cevc and Blume, 1992, 2001; Cevc et al., 2002) and they reported that high flux of vesicles into the intact skin occurred due to xerophobia (the tendency to avoid dry surroundings). It is the naturally occurring "transdermal osmotic gradients" is generated to prevent water loss from the skin and sustains a water gradient between epidermis (75%) and dermis (15%) thereby acts as the driving force for vesicles to penetrate the skin (Schatzlein and Cevc, 1995). Energetically favorable interaction occurs between the hydrophilic residues of lipid and water present in their proximity. This interaction causes every polar lipid to draw water and thus prevents induction of dehydration spontaneously (Bain et al., 1997; Panchagnula, 1993). As a result, all polar lipid-based vesicles progress from dry locality to the sites of adequate water concentration. Consequently when the suspension of transferosomes is applied to the partly dehydrated skin surface by water evaporation loss, the vesicles experience osmotic gradient and thereby moves along the gradient. (Schatzlein et al., 1995).

6.1.3.2 Mechanism II

According to this concept transferosomes work as permeation enhancers. These bilayer vesicles enters the stratum corneum by transforming and disrupting the highly organize the intercellular lipid.

Transferosomes are reported to be widely used and are more effective, safe, and specific than conventional topical formulation for skin diseases, such as psoriasis. Deformable liposomes of methrotrexate were prepared for psoriasis, which show better permeation in comparison to aqueous solution and conventional liposomes. Tranfersomes of corticosteroids were prepared by Cevc et al. (1997) retained drug in the target organ and showed better activity at doses lower than that used in conventional formulations for skin disorders.

Hydrocortisone and dexamethasone containing deformable liposomes were prepared by Cevc and Blume. As compared to conventional cream it produced three- to fivefold increase in potency of entrapped drug with fourfold increase in duration of action (Cevc and Blume, 2004).

6.2 NIOSOME

Requirement of cryogenic atmosphere for the handling of phospholipid vesicles and stability problem during storage owing to oxidation of lipid have incited the research for the use of nonionic surfactants–based vesicular drug-delivery systems known as niosomes. Niosomes are microscopic vesicles made from nonionic surfactant (chiefly alkyl or dialkyl polyglycerol ether class) and cholesterol following hydration which may be unilamellar or multilamellar (Madhav and Saini, 2011). These vesicles consist of polar head region oriented inside and at the periphery where hydrophilic drug can be entrapped and nonpolar region formed by the bilayered hydrophobic tail where hydrophobic drug will be entrapped. Thus, niosomes serve as amphiphilic carrier (Khan et al., 2011). Various types of on ionic surfactant used for the preparation of niosomes include polyglycerol alkyl ethers, crown ethers, ester-linked surfactants, glucosyldialkyl ethers, polyoxyethylene alkyl ethers, Brij, Tweens, and Spans. These are relatively nontoxic, mild, and carry no charge (Azeem et al., 2009).

Although niosomes and liposomes are amphiphilic, have similar function, similar physical composition, both are used for targeted and sustained drug-delivery system and the property of both depend on their composition and preparation niosomes offer several advantages over liposomes. Niosomes possess high chemical stability, intrinsic skin permeation and lower cost of production, which makes them more attractive for industrial manufacturing (Hu and Rhodes, 2000). Furthermore, during the preparation and storage no special conditions, such as low temperature or inert environment is needed (Verma and Bindal, 2012).

6.2.1 Mechanism of niosomes penetration

Following mechanisms are proposed for the mechanism by which niosomes enter the skin:

1. Niosomes act as permeation enhancers and directly fuse with the stratum corneum and diffuse as a whole (Marianeccia et al., 2014).
2. In certain part of stratum corneum where water content is high, reformation of large niosomes takes place to form new small niosomal vesicles. This is attributed to smaller diameter of lipid lamellar spaces of the stratum corneum than noisome vesicles makes this mechanism more meaningful.
3. Noisomes interact with stratum corneum with aggregation, fusion, and adhesion to the cell surface, which causes a high thermodynamic activity gradient of the drug at the vesicle-stratum corneum surface, which is the driving force for the penetration of lipophilic drugs across the stratum corneum.

Several niosomal formulations are developed for the treatment of psoriasis and are found to be effective than conventional dosage forms. Dithranol loaded nisomes were formulated by thin film hydration method. Niosomal formulation showed enhanced in vitro permeation using mouse abdominal skin (permeation rate constant for dithranol loaded formulation was 7.78 $\mu g/cm^2/h$) as compared to the cream base (permeation rate constant was 4.10 $\mu g/cm^2/h$) (Katare et al., 2001). Niosomal gel was prepared by Lakshmi and associates consisting of methotrexate-loaded niosomal vesicles in chitosan gel. They performed double-blind placebo controlled study on healthy human volunteers and psoriasis patients and compared its efficacy with marketed methotrexate gel. The results revealed that Psoriasis Area Severity Index (PASI) score reduced to 3 times after 12 weeks of niosomal methotrexate gel suggesting its superior clinical usefulness, acceptability and patient compliance (Lakshmi et al., 2007).

6.2.2 Ethosomes

Ethosomes are recently developed noninvasive novel colloidal carriers for topical application and enable drugs to reach the deep skin layers and/or the systemic circulation. They are made of phospholipids, ethanol, and water. Main constituent of ethosome which renders easy accessibility to the skin is presence of high concentration of ethanol which creates disruption of the lipid present in skin making pathway for penetration of vesicles into the skin (Dubey et al., 2007). Although the exact process of drug delivery by ethosomes remains a matter of speculation (Dayan and Touitou, 2000), most likely, a combination of processes contributes to the enhancing effect (Touitou et al., 2000). However, previous studies (Dayan and Touitou, 2000; Touitou et al., 2000) that compared permeation enhancement of drugs from ethosomal systems versus hydroethanolic solutions showed that permeation enhancement from ethosomes was much greater than would be expected from ethanol alone. A synergistic mechanism was suggested between ethanol, vesicles, and skin lipids (Touitou et al., 2000). The drug absorption probably occurs in following two phases—ethanol effect and ethosomes effect. The mechanism of its penetration enhancing effect is well known and is dependent on the following effects.

Effects of ethanol: Ethanol may provide the vesicles with soft flexible characteristics, which allow them to more easily penetrate into deeper layers of the skin. Ethanol acts as a penetration enhancer through the skin, penetrates into intercellular lipids, increases the fluidity of cell membrane lipids, and decreases the density of lipid multilayer of cell membrane. (Verma and Fahr, 2004).

Ethosome effects: Increased cell membrane lipid fluidity caused by the ethanol of ethosomes results increased skin permeability. So, the ethosomes permeates very easily inside the deep skin layers, where it gets fused with skin lipids and releases the drugs into deep layer of skin (Touitou et al., 2000).

It was also proposed that phospholipid vesicles with ethanol may penetrate into the skin and influence the bilayer structure of the stratum corneum and this may lead to enhancement of drug penetration (Elsayed et al., 2006).

7 LIPID-BASED PARTICULATE CARRIERS

In these carrier systems the drug is either dissolved in the lipid matrix. The lipid restricts the mobility of drug preventing its rapid release from the system and thus, provides the controlled and sustained release. The drug release properties depend on the type of lipid employed and hence can be modified by selecting suitable lipid of choice. These particulate carriers are of two types.

7.1 SOLID LIPID NANOPARTICLES AND NANOSTUCTURED LIPID CARRIERS

Solid Lipid Nanoparticles (SLNs): SLNs are submicron sized (50–1000 nm) lipid particles comprising of lipid core made from high melting lipid and coated by surfactant. Physiological and nontoxic lipids are basically used for the preparation of SLN which makes them widely acceptable for topical drug delivery. SLNs are formed by dissolving or dispersing the hydrophobic drug into the lipid matrix or by dispersing the drug into the outer coating around the solid lipid core.

Despite of the aforementioned advantages SLN suffers from the potential limitations of low drug loading capability and discharge of drug from the matrix over the period of time during storage (Hirlekar et al., 2011). The reason for such behavior is the transformation of less orders arrangement of the lipid in the particle matrix to the well-ordered β-modification during the storage after production. Such ordered arrangement consists of perfect and compact crystal lattice with little space for the accommodation of the drug and hence drug leaches out.

7.2 NANOSTRUCTURED LIPID CARRIERS

NLC matrix is made up of the mixture of spatially different lipid molecules, which includes mixture of solid and liquid lipid and an aqueous phase containing a surfactant or a mixture of surfactants. Usually, the solid and liquid lipid are mixed in the ratio of 70:30 upto 99.9:0.1 and amount of surfactant included may range from 1.5% to 5% (w/v).

Unstructured solid and liquid lipids matrix creates more imperfection in the matrix to house larger number of drug molecules than SLN. Regardless of the presence of liquid lipid, NLC matrix is solid at room/body temperature (Chen et al., 2010). Imperfections in NLC results in loosely arranged crystal lattice than SLN thereby increasing the drug loading capacity and minimizing drug expulsion during storage. Drug-release profile can be easily modulated by varying the lipid matrix composition (Muller et al., 2002).

7.3 MECHANISM OF SKIN PERMEATION OF LIPID PARTICULATE

Lipid-based nanoparticulate systems and vesicular colloidal carriers have been explored not only to enhance percutaneous absorption but also for drug targeting to the skin or even to its substructure (Pardeike et al., 2009). Lipid nanoparticulates promote skin permeation through adhesion, occlusion, and hydration effects on the skin. On application to the skin particulate form a monolayer film by adhesion, which check the water loss due to evaporation from skin owing to the occlusive nature of hydrophobic monolayer film (Souto et al., 2007). This property keeps the skin hydrated consequently causing loosening of corneocytes packing followed by the opening of intercorneocyte gap junctions and increased and deeper skin permeation (Wissing and Muller, 2003). Occlusion effect is dependent on the concentration of lipid, sample volume applied, particle size, and crystallinity of particles (Teeranachaideekul et al. 2008). Higher barrier is created by the small-sized particles for the water evaporation while bigger particles causes more water loss (Souto and Muller, 2008). The rigid and intact particles are unable to cross the atratum corneum the lipid-based vesicles can pass through in intact state (Cevc and Blume, 2004). The lipid-based particulate system tends to follow the transappendageal pathway as high lipophilicity promotes follicular deposition (Desai et al., 2010).

7.4 ADVANTAGES OF LIPID NANOPARTICULATE CARRIERS

- Large surface area due to small size allows wider and better interaction with the stratum corneum increasing the amount of drug penetrating the skin.
- Negligible irritation and biocompatible carriers as it incorporates physiologically tolerated, nonirritative, and nontoxic lipid (Prow et al., 2011; Singh et al., 2013).
- SLN protected the drug from degradation and offers sustained and controlled-release profile avoiding drug dosing, toxicity, and dose fluctuations.
- Biphasic drug release can be achieved. It involves initial burst of the drug present in the shell followed by sustained and prolonged release from solid core.
- SLN possesses occlusive properties upon topical application which prevents water loss and favors drug penetration into the skin.
- Production of SLN can be carried out organic solvent free and can be easily scaled up, for example, by high-pressure homogenization.

8 EMULSION-BASED CARRIERS

Microemulsions are clear, homogeneous, transparent, thermodynamically stable dispersions of water and oil, stabilized by a surfactant, usually in combination with a cosurfactant (typically a short-chain alcohol). They are of two types oil-in-water (O/W) or water-in-oil (W/O) with the size ranging from 5 to 100 nm (Pradhan et al., 2013). These are exploited as one of the efficient carriers for transdermal drug delivery for the treatment of various skin diseases including psoriasis. They offer following advantages:

- Both hydrophilic and hydrophobic drug can be delivered through microemulsion.
- They are thermodynamic stable thereby rendering (long shelf life).
- They can be easily formulated since they carry zero interfacial and hence forms spontaneously without the requirement of extra efforts to manage interfacial tension.
- They are less viscose and exhibit Newtonian behavior.

- Very small droplet size offers high surface area and high solubilization capacity.
- The small droplets size provide better adherence to skin surface and enhanced transport of bioactive molecules in a more controlled fashion.
- Microemulsion can be easily sterilized by filtration.
- Microemulsion are capable of penetrating the skin easily due to penetration enhancing capabilities of its amphiphilic components and surfactants and free mobility of the drug in the vehicle. Drug release form these system is governed by the interaction of the drug and surfactant along with the partitioning of drug in oil and water phase (Shaji and Reddy, 2004).

As compared to the conventional emulsion the permeation rates from microemulsion are significantly higher. The drug mobility between internal and the external phase in the conventional emulsion formulations, such as cream and lotions is limited by the strong interaction occurring between surfactants at the interfacial membrane film. In addition emulsions are not stable during storage causing fluctuations in bioavailability. In microemulsion such problems are overcome by the presence of co-surfactant, which forms flexible interfacial film and facilitates the lowering of interfacial tension of the surfactant film at the interphase. Cosurfactant promotes spontaneous formation of the microemulsion and provides thermodynamic stability (Kreilgaard et al., 2000). The drug undergoes thermodynamic process of drug diffusion between the two phases through the elastic interfacial surfactant film increases partitioning and drug diffusion into the stratum corneum (Sintov and Shapiro, 2004).

9 CONCLUSIONS

Despite the availability of number of treatment options and the significant advancement in the understanding of pathophysiology of psoriasis the search for the optimum approaches for the ailment still remains a major task. Available treatment options are mainly concerned with the suppression of disease rather than its cure and modification. Such challenges require the remedy consisting of combination of drug regimen and novel carriers. A variety of novel lipid-based colloidal drug-delivery system are investigated, such as liposomes, microemulsion, nanoemulsions, emulsomes, niosomes, transferosomes, nanostructured lipid carriers, solid lipid nanoparticles clearly present a tool to overcome the many challenges associated with topical antipsoriatic drug therapy. These novel carriers are widely explored and have revealed the better therapeutic profile owing enhanced rate and extent of drug delivery to the skin. The future for the psoriatic therapy demands for further upliftment and investigation to be carried out with antipsoriatic drugs to establish the clinical utility and industrial scale-up of techniques for manufacturing these potential novel carriers.

REFERENCES

Aggarwal, S., Ghilardi, N., Xie, M.H., Sauvage de, F.J., Gurney, A.L., 2003. Interleukin-23 promotes a distinct CD4 T cell activation state characterized by the production of interleukin-17. J. Biol. Chem. 278, 1910–1914.

Agrawal, Y., Petkar, K.C., Sawant, K.K., 2010. Development, evaluation and clinical studies of acitretin loaded nanostructured lipid carriers for topical treatment of psoriasis. Int. J. Pharm. 401, 93–102.

Asadullah, K., Sterry, W., Stephanek, K., 1998. IL-10 is a key cytokine in psoriasis. J. Clin. Invest. 101, 783–794.

Asadullah, K., Sterry, W., Trefzer, U., 2002. Cytokines: interleukin and interferon therapyin dermatology. Clin. Exp. Dermatol. 27, 578–584.

Azeem, A., Anwer, M.K., Talegaonkar, S., 2009. Niosomes in sustained and targeted drug delivery: some recent advances. J. Drug Target. 17 (9), 671–689.
Bain, K.R., Hadgkraft, A.J., James, W.J., Water, K.A., 1997. Prediction of Percutaneous Penetration. STS Publishing, Cardiff, Vol. 3b, pp. 226–234.
Baroli, B., Lopez-Quintela, M.A., Delgado-Charro, M.B., Fadda, A.M., Blanco- Méndez, J., 2000. Microemulsion for topical delivery of 8-methoxsalen. J. Control Rel. 69, 209–218.
Berard, M., Brandt, K., Bulfone-Paus, S., Tough, D.F., 2003. IL-15 promotes the survival of naïve and memory phenotype CD8+ T cells. J. Immunol. 170, 5018–5026.
Berth, J.J., 2005. The use of ciclosporin in psoriasis. J. Dermatol. Treat. 16, 258–277.
Bouwstra, J.A., Honeywell-Nguyen, P.L., 2002. Skin structure and mode of action of vesicles. Adv. Drug Deliv. Rev. 54, S41–S55.
Boyman, O., Hefti, H.P., Conrad, C., Nickoloff, B.J., Suter, M., Nestle, F.O., 2004. Spontaneous development of psoriasis in a new animal model shows an essential role for resident T cells and tumor necrosis factor-alpha. J. Exp. Med. 199, 731–736.
Casas, A., Batlle, A., 2006. Aminolevulinic acid derivatives and,liposome delivery as strategies for improving 5-aminolevulinic acid-mediated photodynamic therapy. Curr. Med. Chem. 13, 1157–1168.
Cather, J.C., Abramovits, W., Menter, A., 2001. Cyclosporine and tacrolimus in dermatology. Dermatol. Clin. 19, 119–137.
Cevc, G., 1996. Transferosomes, liposomes and other lipid suspensions on the skin: permeation enhancement, vesicle penetration, and transdermal drug delivery. Crit. Rev. Ther. Drug Carrier Syst. 13, 257–388.
Cevc, G., Blume, G., 1992. Lipid vesicles penetrate into intact skin owing to the transdermal osmotic gradients and hydration force. Biochim. Biophys. Acta. 1104, 226–232.
Cevc, G., Blume, G., 2001. New, highly efficient formulation of diclofenac for the topical, transdermal administration in ultradeformable drug carriers. Transferosomes. Biochim. Biophys. Acta 1514, 191–205.
Cevc, G., Blume, G., 2004. Hydrocortisone and dexamethasone in very deformable drug carriers have increased biological potency, prolonged effect, and reduced therapeutic dosage. Biochim. Biophys. Acta 1663, 61–73.
Cevc, G., Blume, G., Schatzlein, A., 1997. Transfersomes-mediated transepidermal delivery improves the regio-1 specificity and biological activity of corticosteroids in vivo. J. Control. Release 45, 211–226.
Cevc, G., Schatzlein, A., Richardsen, H., 2002. Ultradeformable lipid vesicles can penetrate the skin and other semi-permeable barriers unfragmented. Evidence from double label CLSM experiments and direct size measurements. Biochim. Biophys. Acta. 1564, 21–30.
Chen, C.C., Tsai, T.H., Huang, Z.R., Fang, J.Y., 2010. Effects of lipophilic emulsifiers on the oral administration of lovastatin from nanostructured lipid carriers: physicochemical characterization and pharmacokinetics. Eur. J. Pharm. Biopharm. 74, 474–482.
Crim, C., Pierre, L.N., Daley-Yates, P.T., 2001. A review of the pharmacology and pharmaco-kinetics of inhaled fluticasone propionate and mometasonefuroate. Clin. Ther. 23, 1339–1354.
Dayan, N., Touitou, E., 2000. Carriers for skin delivery of trihexyphenidyl HCl: ethosomes vs. liposomes. Biomaterials 21, 1879–1885.
Desai, P., Patlolla, R.R., Singh, M., 2010. Interaction of nanoparticles and cell penetrating peptides with skin for transdermal drug delivery. Mol. Membr. Biol. 27 (7), 247–259.
Dinarello, C.A., 2002. The IL-1 family and inflammatory diseases. Clin. Exp. Rheumatol. 20, S1–S13.
Doktorovova, S., Araujo, J., Garcia, M.L., Rakovsky, E., Souto, E.B., 2010. Formulating fluticasone propionate in novel PEG containing nanostructured lipid carriers. Coll. Surf. B Biointerf. 75, 538–542.
Downing, D.T., Abraham, W., Wegner, B.K., Willman, K.W., Marshall, J.L., 1993. Partition of dodecyl sulfate into stratum corneum lipid liposomes. Arch. Dermatol. Res. 285, 151–157.
Dragicevic-Curic, N., Scheglmann, D., Albrecht, V., Fahr, A., 2009a. Development of liposomes containing ethanol for skin delivery of temoporfin: characterization and in vitro penetration studies. Coll. Surf. B Biointerf. 74, 114–122.

Dragicevic-Curic, N., Scheglmann, D., Albrecht, V., Fahr, A., 2009b. Development of different temoporfin-loaded invasomes-novel nanocarriers of temoporfin: characterization, stability and in vitro skin penetration studies. Coll. Surf. B Biointerf. 70, 198–206.

Dubey, V., Mishra, D., Jain, N.K., Dutta, T., Nahar, M., Saraf, D.K., 2007. Transdermal drug delivery of antipsoriatic agent via ethanolic liposomes. J. Control. Release 123, 148–154.

Dubey, V., Mishra, D., Jain, M., Kumar, N., 2008. Novel methothrexate-based topical therapies for effective treatment of psoriasis. Clin. Dermatol. 24, 124–130.

El Maghraby, G.M., Williams, A.C., Barry, B.W., 2006. Can drug-bearing liposomes penetrate intact skin? J. Pharm. Pharmacol. 58, 415–429.

Elder, J.T., 2007. IL-15 and psoriasis: another genetic link to Th17. J. Invest. Dermatol. 127, 2495–2497.

Elsayed, M.M.A., Abdallah, O.Y., Naggar, V.F., Khalafallah, N.M., 2006. Deformable liposomes and ethosomes: mechanism of enhanced skin delivery. Int. J. Pharm. 322, 60–66.

Erardesca, B., Vignoli, P., Istante, D., Ona, R., 1998. Effects of glycolic acid on psoriasis. Clin. Exp. Dermatol. 23, 190–191.

Fang, Y.P., Huang, Y.B., Wu, P.C., Tsai, Y.H., 2009. Topical delivery of 5-amino-levulinic acid-encapsulated ethosomes in a hyperproliferative skin animal model using the CLSM technique to evaluate the penetration behavior. Eur. J. Pharm. Biopharm. 73, 391–398.

Federman, D., Fraelish, C., 1999. Topical psoriasis therapy. Am. Fam. Physician 59, 957–962.

Fenton, C., Plosker, G.L., 2004. Calcipotriol / betamethasone dipropionate: a review of itsuse in the treatment of psoriasis vulgaris. Am. J. Clin. Dermatol. 5, 463–478.

Fluhr, J.W., Cavallotti, C., Berardesca, E., 2008. Emollients, moisturizers, and keratolytic agents in psoriasis. Clin. Dermatol. 26, 380–386.

Fowler, J., Faad, M.D., Fowler, L., 2010. Physician and patient assessment of triamcinoloneacetonide spray for steroid responsive dermatoses. J. Clin. Aesthet. Dermatol. 3, 27–31.

Fraki, J.E., Briggaman, R.A., Lazarus, G.S., 1983. Transplantation of psoriatic skin onto nude mice. J. Invest. Dermatol. 80, 31–35.

Fresta, M., Puglisi, Gi., 1997. Corticosteroid dermal delivery with skin-lipid liposomes. J. Control. Release 44 (2–3), 141–151.

Galimova, E.S., Akhmetova, V.L., Khusnutdinova, E.K., 2008. Molecular genetic basis of susceptibility to psoriasis. Russ. J. Genet. 44, 513–522.

Gallagher, G., Eskdale, J., Jordan, W., 2004. Human interleukin-19 and its receptor: a potential role in the induction of Th2 responses. Int. Immunopharmacol. 4, 615–626.

Ghoreschi, K., Thomas, P., Breit, S., 2003. Interleukin-4 therapy of psoriasis induces Th2 responses and improves human autoimmune disease. Nat. Med. 9, 40–46.

Godic, 2004. New approaches to psoriasis treatment: a review. Acta Dermatovenerol. 3, 50–57.

Grassberger Stuetz, M., Meingassner, J.G., 2001. Pimecrolimus (Elidel, SDZ ASM 981)—preclinical pharmacologic profile and skin selectivity. Semin. Cutan. Med. Surg. 20, 233–241.

Gupta, A.K., Chow, M., 2004. Prednicarbate (Dermatop®): profile of a corticosteroid. J. Cutan. Med. Surg. 8, 244–247.

Gupta, A., Aggarwal, G., Singla, S., Arora, R., 2012. Transfersomes: a novel vesicular carrier for enhanced transdermal delivery of sertraline: development, characterization, and performance. Eval. Sci. Pharm. 80 (4), 1061–1080.

Hirano, T., Akira, S., Taga, T., Kishimoto, T., 1990. Biological and clinical aspects of interleukin 6. Immunol. Today 11, 443–449.

Hirlekar, R., Garse, H., Kadam, Vilasrao, 2011. Solid lipid nanoparticles and nanostructured lipid carriers: a review. Curr. Drug Ther. 6 (4), 240–250.

Hu, C., Rhodes, D.G., 2000. Proniosomes: a novel drug carrier preparation. Int. J. Pharm. 206, 110–122.

Iizasa, H., Matsushima, K., Oppenheim, J.J., 2000. IL-8. Cytokine Reference. Academic Press, San Diego, pp. 1061–1067.

Jensen, L.B., Magnussson, E., Gunnarsson, L., Vermehren, C., Nielsen, H.M., Petersson, K., 2010. Corticosteroid solubility and lipid polarity control release from solid lipid nanoparticles. Int. J. Pharm. 390, 53–60.

Jong De, E.M., 1997. The course of psoriasis. Clin Dermatol. 15, 687–692.

Kalariya, M., Misra, A.N., 2004. Methotrexate-loaded solid lipid nanoparticles for topicaltreatment of psoriasis: formulation and clinical implications. AAPS Pharm. Sci. Technol. 4, 225–241.

Katare, O.P., Agarwal, R., Vyas, S.P., 2001. Preparation and in vitro evaluation of liposomal / niosomal delivery systems for antipsoriatic drug dithranol. Int. J. Pharm. 228, 43–52.

Katare, O.P., Kaisar, R., Singh, B., Dogra, S., 2010. Novel drug delivery systems in topical treatment of psoriasis: rigors and vigors. Indian J. Dermatol. Venereol. Leprol. 76 (6), 612–621.

Kemmett, D., Symons, J.A., Colver, G.B., Duff, G.W., 1990. Serum-soluble interleukin 2receptor in psoriasis. Failure to reflect clinical improvement. Acta Derm. Venereol. 70, 264–267.

Khan, A., Sharma, P.K., Visht, S., Malviya, R., 2011. Niosomes as colloidal drug delivery system: a review. J. Chronother. Drug Deliv. 2 (1), 15–21.

Kirjavainen, M., Urtti, A., Jaaskelainen, I., Suhonen, T.M., Paronen, P., ValjakkaKoskela, R., Kiesvaara, J., Monkkonen, J., 1996. Interaction of liposomes with human skin in vitro-the influence of lipid composition and structure. Biochim. Biophys. Acta 1304, 179–189.

Korman, N.J., Moul, D.K., 2005. Alefacept for the treatment of psoriasis: a review of thecurrent literature and practical suggestions for everyday clinical use. Semin. Cutan. Med. Surg. 24, 10–18.

Kragballe, K., 1992. Treatment of psoriasis with calcipotriol and other vitamin D analogues. J. Am. Acad. Dermatol. 27, 1001–1008.

Kreilgaard, M., Pedersen, E.J., Jaroszewski, J.W., 2000. NMR characterisation and transdermal drug delivery potential of microemulsion systems. J. Control. Release 69, 421–433.

Kruger, J.G., 2002. The immunologic basis for the treatment of psoriasis with new biologic agents. J. Am. Acad. Dermatol. 46, 1–23.

Lakshmi, P.K., Devi, G.S., Bhaskaran, S., Sacchidanand, S., 2007. Niosomal methotrexate gel in the treatment of localized psoriasis: phase I and phase II studies. Ind. J. Dermatol. Venereol. Leprol. 73, 157–162.

Laws, P.M., Young, H.S., 2010. Topical treatment of psoriasis. Expert. Opin. Pharmacother. 11, 1999–2009.

Lebwohl, M., 2003. Psoriasis. Lancet 361, 1197–1204.

Lebwohl, M., Feldman, S.R., Walther, R., Morgan, P., Gutkin, S.W., 2001. Clinical management of psoriasis: principles and practice. Cutis 67, 1–15.

Li, G., Fan, C., Li, X., Fan, Y., Wang, X., Li, M., Liu, Y., 2012. Preparation and in vitro evaluation of tacrolimus-loaded ethosomes. Sci. World J. 2012, 1–6.

Lin, Y.K., Huang, Z.R., Zhuo, R.Z., Fang, J.Y., 2010. Combination of calcipotriol and methotrexate in nanostructured lipid carriers for topical delivery. Int. J. Nanomed. 5, 117–128.

Maccaroni, E., Giovenzana, G.B., Palmisano, G., Botta, D., Volante, P., Masciocchi, N., 2009. Structures from powders: diflorasonediacetate. Steroids 74, 102–111.

Madhav, N.V.S., Saini, A., 2011. Niosomes: a novel drug delivery system. IJRPC 1, 498–511.

Marianeccia, C., Marziob, L.D., Rinaldia, F., Celiab, C., 2014. Niosomes from 80s to present: the state of the art. Science 205, 187–206.

McCusker, M.M., Grant-Kels, J.M., 2010. Healing fats of the skin: the structural and immunologic roles of omega-6 and omega-3 fatty acids. Clin. Dermatol. 28, 440–451.

Mehnert, W., Mader, K., 2012. Solid lipid nanoparticles: production, characterization and applications. Adv. Drug Deliv. Rev. 64, 83–101.

Mendonc, C.O., Burden, A.D., 2003. Current concepts in psoriasis and its treatment. Pharmacol. Therapeut. 99, 133–147.

Menter, A., Tyring, S.K., Gordon, K., Kimball, A.B., Leonardi, C.L., Langley, R.G., Mezei, M., Gulasekharam, V., 1980. Liposomes-a selective drug delivery system for the topical route of administration. Lotion dosage form. Life Sci. 26, 1473–1477.

Mezei, M., Gulasekharam, V., 1982. Liposomes-a selective drug delivery system for the topical route of administration: gel dosage form. J. Pharm. Pharmacol. 34, 473–474.
Misra, A.K., Padhi, B.K., Chougule, M., 2004. Methotrexate-loaded solid lipid nanoparticles for topical treatment of psoriasis: formulation & clinical implications. Drug Del. Tech. 4 (8.).
Mizutani, H., Yamanaka, K., Konishi, H., Murakami, T., 2003. Animal models of psoriasis and pustular psoriasis. Arch. Dermatol. Res. 295, 567–568.
Moghimi, S.M., Patel, H.M., 1993. Current progress and future prospects of liposomes in dermal drug delivery. J. Microencapsul. 10, 155–162.
Mohandas, S., Rai, R., Srinivas, C.R., 2009. Halobetasol versus clobetasol: a study of potency. Indian J. Dermatol. Venereol. Leprol. 75, 186–187.
Muhl, H., Pfeilschifter, J., 2004. Interleukin-18 bioactivity: a novel target for immunopharmacological anti-inflammatory intervention. Eur. J. Pharmacol. 500, 63–71.
Muller, R.H., Radtke, M., Wissing, S.A., 2002. Nanostructured lipid matrices for improved microencapsulation of drugs. Int. J. Pharm. 242, 121–128.
Naldi, L., Yawalkar, N., Andrzej, K., Jean-Paul, O., Paolo, M., Stefano, R., Giuseppe, M., 2011. Efficacy and Safety of the betamethasone valerate 0.1% plaster in mild-to-moderate chronic plaque psoriasis: a randomized, parallel-group, active-controlled, phase III study. Am. J. Clin. Dermatol. 12, 191–201.
Nickoloff, B.J., Kunkel, S.L., Burdick, M., Strieter, R.M., 1995. Severe combined immunodeficiency mouse and human psoriatic skin chimeras: validation of a new animal model. Am. J. Pathol. 146, 580–588.
Panchagnula, R., 1993. Transdermal delivery of drugs. Ind J. Pharmacol. 29, 140–156.
Papp, K.A., Tyring, S., Lahfa, M., Prinz, J., Griffiths, C.E., Nakanishi, A.M., Zitnik, R., Kerkhof, P.C.V., Melvin, L., 2005. A global phase III randomized controlled trial of etanercept in psoriasis: safety, efficacy, and effect of dose reduction. Br. J. Dermatol. 152, 1304–1312.
Pardeike, J., Hommoss, A., Muller, R.H., 2009. Lipid nanoparticles (SLN, NLC) in cosmetic and pharmaceutical dermal products. Int. J. Pharm. 366, 170–184.
Patel, S.S., Patel, M.S., Salampure, B., Vishwanath, B., Patel, N.M., 2010. Development and evaluation of liposomes for topical delivery of Tacrolimus (Fk-506). J. Sci. Res. 2 (3), 587–598.
Peter, B.P., Weissman, F.G., Gill, M.A., 2000. Pathophysiology and treatment of psoriasis. Am. J. Health Syst. Pharm. 57, 645–659.
Pietrzak, A.T., Zalewska, A., Chodorowska, G., Krasowska, D., Michalak-Stoma, A., Nockowski, P., Osemlak, P., Paszkowski, T., Roli?ski, J.M., 2008. Cytokines and anticytokines in psoriasis. Clin. Chim. Acta 394, 7–21.
Pinto, M.F., Moura, C.C., Nunes, C., Segundo, M.A., Costa Lima, S.A., Reis, S., 2014. A new topical formulation for psoriasis: Development of methotrexate-loaded nanostructured lipid carriers. Int. J. Pharm. 477 (1–2), 519–526.
Pirvu, C.D., Hlevca, C., Ortan, A., Prisada, R.A., 2010. Elastic vesicles as drugs carriers through the skin. Farmacia 58, 2.
Pople, P.V., Singh, K.K., 2011. Development and evaluation of colloidal modified nanolipid carrier: application to topical delivery of tacrolimus. Eur. J. Pharm. Biopharm. 79, 82–94.
Pradhan, M., Singh, D., Singh, M.R., 2013. Novel colloidal carriers for psoriasis: current issues, mechanistic insight and novel delivery approaches. J. Control. Release 170, 380–395.
Prow, T.W., Grice, J.E., Lin, L.L., Faye, R., Butler, M., Becker, W., Wurm, E.M.T., Yoong, C., Robertson, T.A., Soyer, H.P., Roberts, M.S., 2011. Nanoparticles and microparticles for skin drug delivery. Adv. Drug Deliv. Rev. 63, 470–491.
Rawat, M., Singh, D., Saraf, S., Saraf, S., 2006. Nanocarriers: promising vehicle for bioactive drugs. Biol. Pharm. Bull. 29, 1790–1798.
Roelofzen, J.H.J., Aben, K.K.H., Oldenhof, U.T.H., Coenraads, P.J., Alkemade, H.A., Van de Kerkhof, P.C.M., Van der Valk, P.G.M., Kiemeney, L.A.L.M., 2010. No increased risk of cancer after coal tar treatment in patients with psoriasis and eczema. J. Invest. Dermatol. 130, 953–961.

Rosso, J.D., Friedlander, S.F., 2005. Corticosteroids: options in the era of steroid-sparing therapy. J. Am. Acad. Dermatol. 53, S50–S58.
Rozenbli, M., Lebwohl, M., Spuls, P.I., 2003. Retrospective study of the efficacy of narrowband UVB and acitretin. J. Dermatol. Treat. 14, 17–20.
Saini, R., Tutrone, W.D., Weinberg, J.M., 2005. Advances in therapy for psoriasis: an overview of infliximab, etanercept, efalizumab, alefacept, adalimumab, tazarotene, and pimecrolimus. Curr. Pharm. Des. 11, 273–280.
Saraswat, A., Agrawal, R., Katare, O.P., Kaur, I., Kumar, B., 2007. A randomized, double-blind, vehicle-controlled study of a novel liposomal dithranol formulation in psoriasis. J. Dermatol. Treat. 18, 40–45.
Schatzlein, A., Cevc, G., 1995. Skin penetration by phospholipids vesicles, transfersomes as visualized by means of the confocal scanning laser microscopy, In: Phospholipids: characterization, metabolisim, and novel biological applications. Proceedings of the 6th International Colloquium, pp. 191–209.
Schmid, M.H., Korting, H.C., 1996. Therapeutic progress with topical liposome drugs for skin disease. Adv. Drug Deliv. Rev. 18, 335–342.
Shah, A.K., Date, A.A., Joshi, M.D., Patravale, V.B., 2007. Solid lipid nanoparticles of tretenoin: potential in topical delivery. Int. J. Pharm. 345, 163–171.
Shaji, J., Reddy, M.S., 2004. Microemulsion as drug delivery system. Pharma. Times 36, 17–24.
Singh, D., Dubey, P., Pradhan, M., Singh, M.R., 2013. Ceramic nanocarriers: versatile nanosystem for protein and peptide delivery. Expert Opin. Drug Deliv. 10, 241–259.
Sintov, A.C., Shapiro, L., 2004. J. Control. Release 95, 173.
Souto, E.B., Muller, R.H., 2008. Cosmetic features and applications of lipid nanoparticles (SLN, NLC). Int. J. Cosmet. Sci. 30, 157–165.
Souto, E.B., Almeida, A.J., Muller, R.H., 2007. Nanoparticles (SLN®, NLC®) for cutaneous drug delivery:structure, protection and skin effects. J. Biomed. Nanotechnol. 3, 317–331.
Spitz, H., Oppenheim, J.J., 2000. IL-7. Cytokine Reference. Academic Press, San Diego, 137–153.
Strober, B.E., Kaul, M., Gu, Y., Okun, M., Papp, K., 2008. Adalimumab therapy for moderate to severe psoriasis: a randomized, controlled phase III trial. J. Am. Acad. Dermatol. 58, 106–115.
Sullivan, J.R., Preda, V.A., 2009. Treatments for severe psoriasis. Aust. Prescr. 32, 14–18.
Suvarna, B.S., 2008. Secrets of omega-3 oil. Med. J. 6, 406–411.
Teeranachaideekul, V., Boonme, P., Souto, E.B., Muller, R.H., Junyaprasert, V.B., 2008. Influence of oil content on physicochemical properties and skin distribution of Nile red-loaded NLC. J. Control. Release 128, 134–141.
Touitou, E., Dayan, N., Bergelson, L., Godin, B., Eliaz, M., 2000. Ethosomes-novel vesicular carriers for enhanced delivery: characterization and skin penetration properties. J. Control. Release 65, 403–418.
Traub, M., Marshall, K.M.S., 2007. Psoriasis—pathophysiology, conventional, and alternative approaches to treatment. Altern. Med. Rev. 12 (4), 319–330.
Trotta, M., Pattarino, F., Gasco, M.R., 1996. Influence of counter ions on the skin permeation of methotrexate from water–oil microemulsions. Pharm. Acta Helv. 71, 135–140.
Trotta, M., Peira, E., Carlotti, M.E., Gallarate, M., 2004. Deformable liposomes for dermal administration of methotrexate. Int. J. Pharm. 270, 119–125.
Umezawa, Y., Ozawa, A., 2007. Optimal time for therapeutic drug monitoring of cyclosporine microemulsion in patients with psoriasis. Int. J. Dermatol. 46, 763–766.
Uslu, M., 2007. Adalimumab. Anti Inflamm. Anti Allergy Agents Med. Chem. 6, 217–227.
Verma, A.K., Bindal, M., 2012. A review on niosomes: an ultimate controlled and novel drug delivery carrier. Int. J. Nanoparticles 5 (1), 73–87.
Verma, D.D., Fahr, A., 2004. Synergistic penetration enhancement. Effect of ethanol and phospholipids on the topical delivery of cyclosporin A. J. Control. Release 97 (1), 55–66.
Wang, Y.C., Li, H.H., Chen, P.C., Hsue, J.H., Yang, J.J., Chang, M.S., 2003. IL-20: promoter analysis and characterization of biological function. Eur. Cytokine New. 14, 174.

Weinstein, G.D., Krueger, J.G., Lowe, N.J., Duvic, M., Friedman, D.J., Jegasothy, B.V., Jorizzo, J.L., Shmunes, E., Tschen, E.H., Lew-Kaya, D.A., Lue, J.C., Sefton, J., Gibson, J.R., Chandraratna, R.A., 1997. Tazarotene gel, a new retinoid, for topical therapy of psoriasis: vehicle-controlled study of safety, efficacy, and duration of therapeutic effect. J. Am. Acad. Dermatol. 37, 85–92.
Winterfield, L., Menter, A., 2004. Infliximab. Dermatol. Ther. 17, 409–426.
Wissing, S.A., Muller, R.H., 2003. Cosmetic applications for solid lipid nanoparticles (SLN). Int. J. Pharm. 254, 65–68.
Witowski, J., Ksi?zek, K., Jorres, A., 2004. Interleukin-17: a mediator of inflammatoryresponses. Cell. Mol. Life Sci. 61, 567–579.
Wozel, G., 2008. Psoriasis treatment in difficult locations: scalp, nails, and intertriginous areas. Clinics Dermatol. 26, 448–459.

CHAPTER

USE OF NANOSTRUCTURES BASED ON NOBLE METALS IN NANOBIOMEDICINE

25

Jalpa Soni

Indian Institute of Science Education and Research, Kolkata, India

CHAPTER OUTLINE

1 INTRODUCTION

While bulk noble metals have many favorable properties, such as high thermal and electrical conductivity, specific mechanical properties and high reflectivity of incident radiation, a whole new range of their potential applications have transpired with the advancements of nanotechnology. The newfound ability to fabricate the structures of nanoscale dimensions would have led to many dramatic and useful properties; most interesting of those being their optical response (Maier, 2007). These intriguing optical properties of nanomaterials are governed by the collective oscillations of free/conduction electrons, which are in resonance with the incident electromagnetic (EM) field, known as the surface plasmon resonance (SPR). Although SPRs have gained immense research interest in recent times, owing to both their potential applications, as well as for fundamental understanding of light-matter interaction; metal nanoparticles were used ages ago for giving rich colors to stained glass. Opening up the field of plasmonics, these nanomaterials have allowed the manipulation and control of light–matter interactions at nanoscale dimensions providing compelling possibilities of applications and fundamental studies (Maier, 2007). SPRs have found numerous practical applications in science and technology, such as ultra-high sensitive chemical and biomedical sensing, biomolecular manipulation, labeling/detection, contrast enhancement in biomedical imaging, surface enhanced spectroscopy (Raman and fluorescence), new generation optical devices including plasmonic wave guiding nanodevices, optical information processing, and data storage among many others (Anker et al., 2008; Kern et al., 2012;

Kneipp et al., 1997; Lal et al., 2007; Maier et al., 2003, 2001; Nie and Emory, 1997; Ohtsu et al., 2002; Ozbay, 2006; Zijlstra et al., 2009).

With the solutions of Maxwell's equations for spherical particles, derived by Mie (1908), calculating the wavelength-dependent scattering and absorption/extinction cross-sections of solid spherical particles of subwavelength dimensions became a realistic possibility. Mie theory-based calculations correctly predict the extinction peaks and their size-dependent wavelength shifts for nano-sized spherical scatterers, and help to explain their intriguing optical properties. Additionally, there have been various other approaches to solve the scattering problem for nonspherical particles for which the Mie theory is not applicable. These include various numerical models, such as the discrete dipole approximation (DDA), the T-matrix method for axially symmetric nonspherical scatterers, finite difference time domain (FDTD), and finite element method (FEM) for complex nanostructures and the multiple multipole method. Moreover, a variety of experimental tools have been used to measure the resulting optical properties of the nanoparticles/structures, including, Rayleigh scattering measurements in dark field microscope, confocal microscopy, near field scanning optical microscopy (NSOM), a variety of nonlinear scattering measurements, and time-resolved measurements among many others (Beversluis et al., 2003; Bouhelier et al., 2003; Ditlbacher et al., 2005; Du et al., 2009; Klar et al., 1998; Klein et al., 2006; Lindfors et al., 2004; Nelayah et al., 2007; Schuck et al., 2005; Sönnichsen et al., 2000; Stockman et al., 2004). These techniques have been used for spectroscopic and imaging studies on various types of nanoparticles (e.g., single or assembly of multiple nanoparticles of any shape) and complex nanostructures.

Additionally, recent developments of powerful characterization tools and refined synthetic methodologies have greatly helped to achieve unprecedented control over the dimensions, morphology, and optical properties of metal nanostructures. This greater degree of control on tuning parameters of plasmon resonances has, in turn, expanded the range of applications of nanostructures, especially in biomedical applications. For example, it is possible to utilize the unique optical properties of metal nanostructures in biological environments by tuning the plasmon resonances into the near infrared (NIR) region where absorption from water and hemoglobin is low (Weissleder, 2001), or to tune them to desirable wavelengths where they can affect the specific targets only (Kelly et al., 2003). However, certain critical issues of using nanostructures for biomedical applications, such as, bioactivity, compatibility, toxic effects, and nanobio interfacial properties, must be addressed before clinical applications. One of the biggest challenges in using the advent of nanomaterials in biomedicine is to achieve high biodegradability while minimizing the toxicity. Owing to the great potential of novel applications, the field of plasmonics is rapidly expanding and there have been numerous studies exploring various effects and applications of nanostructures. In this chapter, however, I will focus on the noble metal-based nanoparticles/nanostructures and summarize the advances of their potential use in the field of biomedicine.

This chapter is organized as follows: an overview of the optical properties of metal nanostructures is given in Section 2. The biomedical applications of gold and silver nanoparticles are given in Section 3, followed by the chapter summary in Section 4.

2 OPTICAL PROPERTIES OF METAL NANOSTRUCTURES

The intriguing optical properties of metal nanoparticles are due to their unique interaction with light. This interesting optical response of metal nanostructures results from the resonant collective oscillations of free electrons with the incident EM radiation, known as the SPR, which can be propagating along a metal-dielectric surface or localized (LSPR) in a metal nanoparticle. The SPR oscillations suffer either radioactive decay, resulting in light scattering, or nonradiative decay by converting the absorbed light to heat and the scattering, absorption and extinction cross-sections are strongly enhanced at the resonance frequencies (Bohren and Huffman, 2008; Maier, 2007). This section provides a brief overview on the plasmon resonance properties of metal nanoparticles/nanostructures.

2.1 BASIC PRINCIPLE OF PLASMON RESONANCE

The plasmon resonance properties of metals are described by the dielectric function, which can be defined using the Drude–Sommerfeld model as

$$\in(\omega) = \in_\infty - \frac{\omega_p^2}{\omega^2 + i\gamma\omega}; \omega_p = \sqrt{\frac{Ne^2}{\in_0 m_e}} \tag{25.1}$$

where, ω is the angular frequency of light, γ is the electron collision frequency in the bulk, and ω_p is the bulk plasma frequency of the free electrons, which is determined by the free electron density N and the effective mass m_e of the electrons. The quantity $\in_\infty$ has the contribution from bound electrons to the dielectric function. This simple model well-predicts the many properties of real metals, such as the optical properties as described by the frequency dependent dielectric function $\in(\omega)$. However, additional collisions of conduction electrons occurring due to the surface scattering become prominent in case of nanoparticles because of high surface to volume ratio. Although, the SPR can be composed of different orders (from lowest order dipolar to multipoles) depending on the size of the nanoparticle (with respect to the wavelength of light); the electron oscillations are predominantly dipolar in nature when the particle radius is much smaller than the wavelength of light ($r \ll \lambda$). Within this limit, the dipolar polarizability α and scattering and absorption coefficients for spherical scatterers can be defined as follows (Bohren and Huffman, 2008)

$$\alpha = \in_0 3V \frac{\in_r - 1}{\in_r + 2} \tag{25.2}$$

$$C_{sca} = \frac{\kappa^4}{6\pi}(3V)^2 \frac{(\in_r' - 1)^2 + \in_r''^2}{(2 + \in_r')^2 + \in_r''^2} \tag{25.3}$$

$$C_{abs} = k3V \frac{3\in''}{(2 + \in_r')^2 + \in_r''^2} \tag{25.4}$$

where, $k = 2\pi/\lambda$, $\in_r$ is the relative dielectric constant ($\in_r = \in_r' + i\in_r''$) and V is the volume of the metal sphere. It is clear from Eq. (25.2) that the condition for plasmon resonance to occur (for spherical scatterers) is to have the real part of the dielectric constant $\in_r' = -2$.

2.2 NONSPHERICAL NANOPARTICLES AND OTHER NANOSTRUCTURES

For nonspherical particles, however, the relation between the polarizability and the dielectric function becomes not so straightforward resulting in the modified resonance conditions. Irrespective of the particle shape though, the real part of the metal dielectric function (ϵ_r') governs the resonance frequency and the imaginary part of the dielectric function (ϵ_r'') takes care of the broadening and absorptive dissipation occurring due to damping and dephasing of the plasmon oscillations (Bohren and Huffman, 2008).

Among noble metals, the resonance frequencies of gold, silver, and copper for spherical nanoparticles fall in the visible region of the EM spectrum (Kelly et al., 2003; Maier, 2007; Mie, 1908), from which gold and silver are most widely explored for optical applications since copper is easily oxidized. For example, the LSPR for a gold nanosphere in the 10–20 nm size range peaks around 520 nm in water and around 380 nm for a silver nanospheres (Fig. 25.1), and shows red shift with increase in the nanosphere size (Kelly et al., 2003). The tunability of the resonances can be achieved by changing the size parameter of the nanoparticles, as well as by changing their shape. For instance, rod-shaped particles have two resonances arising from the plasmon oscillation along the short axis and the long axis (Link et al., 1999; Murphy et al., 2005). Strongly dependent on the aspect ratio (defined as length-to-width ratio) of the nanorod, these are called the transverse and longitudinal resonances respectively. The longitudinal LSPR experiences red shift with the increase in the nanorod aspect ratio and the peak position moves toward NIR from visible (Link et al., 1999).

Another way to tune the LSPR is by using a metal nanoshell, where the shell thickness is the tuning parameter for resonance position. The near-exponential decrease in the LSPR frequency by decreasing the shell thickness-to-core radius ratio was shown recently (Jain and El-Sayed, 2007). This trend was shown to be universal and independent of the various parameters of a nanoshell, such as size, core material, shell metal, and/or surrounding medium, which makes the design of nanoshells with desired

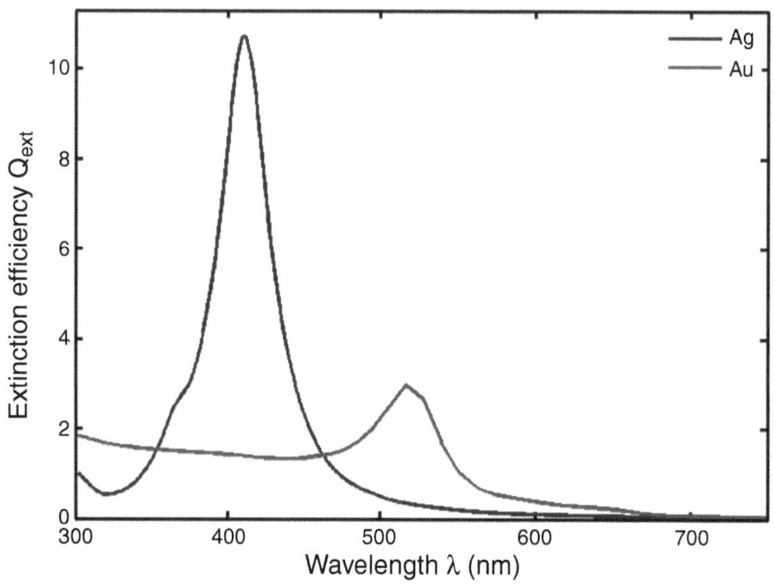

FIGURE 25.1 Extinction Coefficient (Calculated Using Mie Theory) for Silver and Gold Nanospheres of Radius 20 nm Showing the Plasmon Resonance Peaks Around 380 and 520 nm for Silver and Gold, Respectively (Surrounding Medium is Water)

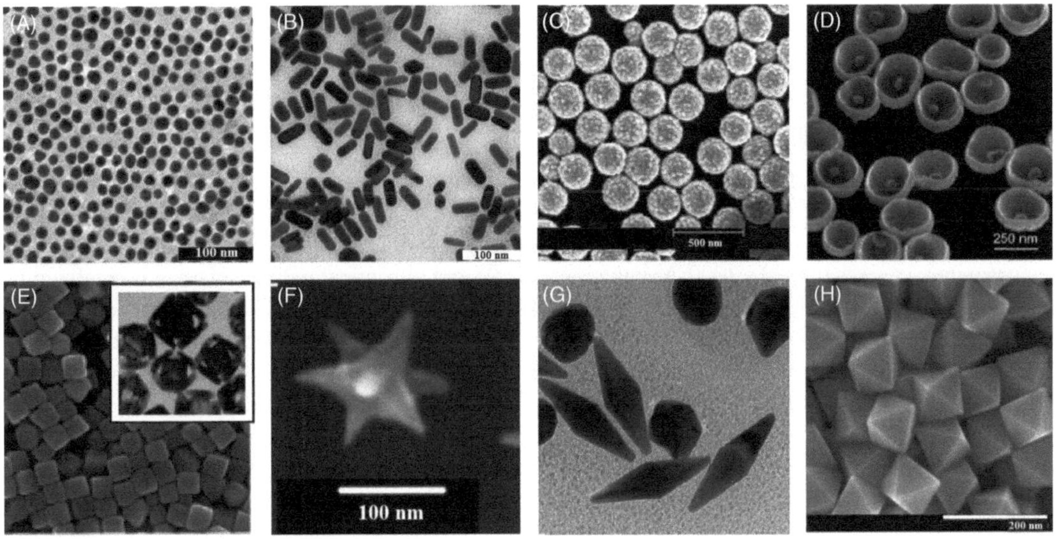

FIGURE 25.2

TEM images of (A) 15 nm colloidal gold, (B) 15 × 50 nm nanorods, (C) 160/17 nm (silica/gold) nanoshells, (D) 250 nm Au nanobowls with 55 nm Au seeds, (E) silver cubes and gold nanocages (insert), (F) nanostars, (G) bipyramids, and (H) octahedral.

Reprinted with permission from Khlebtsov, N.G., Dykman, L.A., 2010. Optical properties and biomedical applications of plasmonic nanoparticles. J. Quant. Spectrosc. Radiat. Transfer 111, 1–35. © 2009 Elsevier Ltd.

optical resonance to be easy for specific applications. Similarly, hollow nanocages and nanocubes have also been shown to show similar visible-NIR tunability with the wall thickness and void size of the particle being the control parameters (Hu et al., 2006a; Sun et al., 2002). Several examples of various gold nanoparticles of different shapes have been shown in Fig. 25.2.

3 APPLICATIONS IN BIOLOGICAL SYSTEMS AND BIOMEDICINE

Metal nanoparticles have expanded the use of nanomaterials in biological and biomedical applications owing to their exceptional optical properties. Their applications include biomedical imaging, biosensing, drug delivery, and disease-related studies spanning from detection and monitoring to diagnosis and therapy. However, it is important to understand and consider their toxic effects on human health and environment before using them as clinical tools since such toxic effects can be much stronger in case of nanomaterials due to their larger surface area as they can get in direct contact with cells causing more effective interactions (Lok et al., 2007). Understanding the toxic nature of metal structures also help to determine the choice of metal for specific applications, for example, silver nanoparticles are particularly used in many consumer products (e.g., detergents, food storage containers) due to their antibacterial properties while gold structures have a greater biocompatibility for various in vivo applications. Practically, simpler control-parameters to optimize the nanostructures and the ability to direct them to specific locations in the body through targeting is particularly advantageous for many disease treatments, such as cancer, in which case, the broad side effects can be reduced by focused treatments.

In this section, I will discuss the range of biological applications that have been proposed and demonstrated with the use of metal (gold and silver) nanoparticles.

3.1 GOLD (Au) NANOPARTICLES

Gold nanoparticles are most widely used for in vivo studies owing to their better biocompatibility. Additionally, new functionality to gold nanostructures is added by the gold—thiolate bond. For instance, temperature dependent drug delivery can be arranged by using thermo-sensitive polymer bound gold nanostructures (Sershen et al., 2000; Yavuz et al., 2009) and all-optical pH sensors can be realized by conjugation of pH sensitive molecules with gold nanostructures (Qiu and Park, 2012). Direct conjugation of certain therapeutics to the surface of the nanostructures using the stealth-like properties of PEG coatings is also possible, which can transport highly toxic drugs to the tumor site with minimum side effects (Paciotti et al., 2004).

It has been shown that while free CTAB molecules are detrimental to human cells, bound CTAB molecules are not, and the fact that they can be separated by configuration or dialysis membrane makes the gold nanoparticles to be quite safe for in vivo and clinical studies. Also, the surface of gold nanoparticles can be modified such that CTAB is either replaced or covered by compatible molecules. Takahashi et al. (2006) have shown that the cytotoxicity is greatly reduced for phosphatidylcholine (PC)-modified Au-nanorods as compared to the twice-centrifuged CTAB capped Au-nanorods. Studies on surface chemistry related effects on cellular uptake, toxicity, and gene expression have shown that the PDADMAC-coated [poly(diallyldimethylammonium chloride)] or PSS-coated [poly(4-styrenesulfonic acid)] Au-nanorods have negligible toxic effects on cell function (Hauck et al., 2008). In another case, the ligand exchange with polyethyleneglycol (PEG) polymers by mixing the mPEG-SH with nanorod solution for hours not only reduces the toxicity of gold nanorods, but also increases their biocompatibility (Niidome et al., 2006).

The mainly exploited optical properties are scattering and absorption occurring in the plasmon resonance bands of the nanoparticles. The different optical responses resulting from the strong optical properties of nanoparticles have been used for various applications as shown in Scheme 25.1. For example, scattered light is particularly used in imaging techniques for contrast enhancement with the particles of large scattering cross-sections, while absorption is used in a variety of techniques including drug delivery, photoacoustic imaging and photothermal treatments (Jain et al., 2008; Lal et al., 2008; Skrabalak et al., 2007; Song et al., 2009; Yavuz et al., 2009). Some of these techniques will be described in more detail in the following sections.

3.1.1 Bioimaging and diagnostics

Colloidal gold nanoparticles induced transmission electron microscopy (TEM) has been predominantly used for detection of biointeractions for a long time (Hayat, 1989). Modern instruments have led to the development of high-resolution transmission electron microscope (HRTEM) along with digital recording units and image processing systems bettering the conventional methods of TEM. Modern biomedicinal studies utilize the immune electron spectroscopy mainly to identify the causative agents of infectious diseases and their surface antigens (Bunin et al., 2005). For the same purpose the scanning probe microscopy, scanning electron microscopy (SEM), and fluorescence microscopy are also frequently used (Drygin et al., 2009). Additionally, nonspherical particles, such as nanorods, nanoshells, nanocages, nanostars, and other types of particles (Fig. 25.2) have also been used in recent times along with the conventional spherical nanoparticles (Khlebtsov and Dykman, 2010).

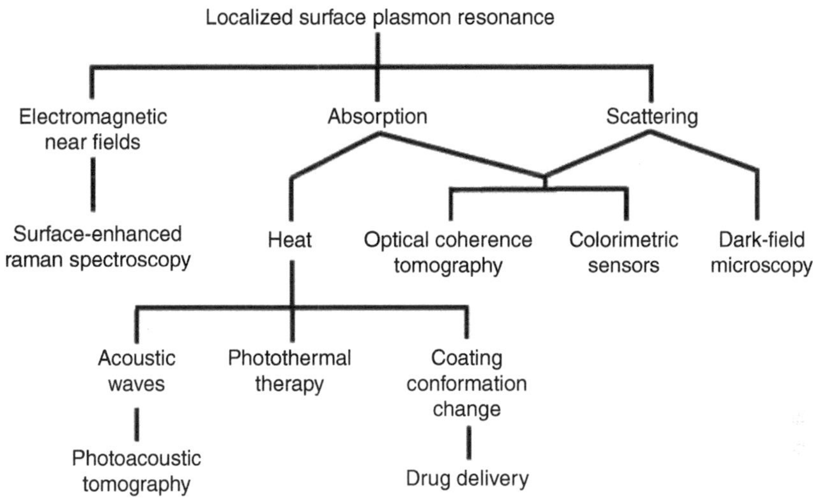

SCHEME 25.1 Localized Surface Plasmon Resonance (LSPR) is the Origin of a Number of Interesting Phenomena Observed with Gold Nanostructures

These properties have enabled a number of applications.

Reproduced with permission from Cobley, C.M., Chen, J., Cho, E.C., Wang, L.V., Xia, Y.,2011. Gold nanostructures: a class of multifunctional materials for biomedical applications. Chem. Soc. Rev. 40, 44–56. The Royal Society of Chemistry 2011.

Confocal microscopy is another technique that collects the light coming from the objects located only in the focal plane to detect microobjects. With this method, 3D images of samples can be generated by scanning the sample along their height and making superposition of scanograms. The use of gold nanoparticles and their conjugates with antigens in confocal microscopy is being increasingly used in biomedical applications (Wang et al., 2010). This makes the real-time detection of gold penetrating into living cells (e.g., cancer cells) possible at the single particle level along with the estimate of their amount (Klein et al., 2010). Confocal images can be obtained by confocal fluorescence microscopy, resonance scattering confocal microscopy and/or two-photon luminescence confocal microscopy. The two-photon luminescence method has a major advantage in terms of strong reduction of background signal, thus enhancing the contrast by a large degree. The use of gold nanoparticles for two-photon luminescence for the visualization of oncomarkers on the surface or inside a cell in the presence of other objects has been demonstrated successfully (Maiorano et al., 2010).

One of the most popular methods in bioimaging is the dark-field microscopy, which beats the resolution limit of a conventional light microscope. Similar to the Tyndall effect, the signal collected in dark field microscope has contribution only from the object-scattered light and the direct light goes out of the field-of-view of the collection lens. Thus, the scattering object shines brightly against the dark background in the image obtained from a dark field microscope. Using gold nanoparticles for biospecific interactions offer more advantages in dark field microscopy owing to their larger scattering cross-sections than the fluorescence labels (by 3–5 orders of magnitude) (Khlebtsov and Dykman, 2010). A simple and reliable scheme for oncologic diseases diagnostics has been demonstrated using this very principle; following which, other shapes, such as nanorods, nanoshells, nanostars, and nanocages were also used

(Aaron et al., 2008; Au et al., 2010; El-Sayed et al., 2005; Huang et al., 2006; Loo et al., 2005a; Schultz, 2003). Many other applications of gold nanoparticles in resonance dark field microscopy including the detection of microbial cells and metabolites, surface-receptors of cells, bioimaging of tumor cells, and study of endocytosis have also been demonstrated (Dykman and Khlebtsov, 2011). The sensitivity of plasmon resonance in the local dielectric environments is more enhanced for nonspherical particle shapes compared to the spheres of equivalent volumes owing to the fact that local EM fields are much stronger near sharp regions of the particles or in the subnanometer gaps between two nanoparticles. This allows the nonspherical nanoparticles with unconventional shapes to have unprecedented potential for in vivo biomedical imaging and diagnostics applications (Dykman and Khlebtsov, 2011; Hutter et al., 2010; Schultz, 2003).

Other than the confocal and dark field microscopy, photoacoustic microscopy, fluorescence correlation microscopy, optical coherence tomography, and X-ray and magnetoresonance tomography methods have also been used for various biological diagnostic applications using gold nanoparticles (Dykman and Khlebtsov, 2011).

3.1.2 Biomedical sensing

Gold has also been exploited for healthcare applications since many centuries, became less popular in 18th century, but still continued to be used in 19th century; by which time it was used to treat syphilis, alcoholism, and tuberculosis. Gold is still being used as an active ingredient in the drug Aurofin for the treatment of rheumatoid arthritis. Lange used a variation of the Zsigmondy flocculation test to identify the altered proteins in cerebrospinal fluid of patients with syphilis is credited to be the first verifiable use of gold nanoparticles in diagnosis (Lange, 1912).

With modern fabrication/synthesis methods, gold nanoparticles can be prepared in a range of size parameters with desirable precision and accuracy, and for the fact that they are stable and can be easily functionalized with various recognition molecules, such as antibodies, antigens, oligonucleotides, and so on, they have been gaining importance in high sensitive biomedical sensing/labeling applications. It has been well understood that the increase in diameter of gold nanospheres leads to increase in the extinction cross-section, accompanied by a red shift in the plasmon band, and various studies involving the effect of size parameter on biomedical applications of gold nanospheres have concluded that particles with around 80 nm diameter are most visible to the human eye and have the extinction cross section of around $5 \times 10^{10}\ M^{-1}\ cm^{-1}$ making them a better choice over organic dyes whose extinction cross-sections are typically four to five orders of magnitudes lower. Basically, the favorable characteristics of gold nanoparticles, such as low cost, ease of conjugation, and unusually unique optical properties make them ideal candidates for detection and diagnostics applications.

3.1.2.1 Distance-dependent detection schemes

Typically, the distance between nanoparticles in a solution is greater than 1000 nm. Although the dipole–dipole interaction between neighboring particles is not significant at this distance, it becomes prominent when this distance is less than the particles' diameter. Such overlap between dipoles of neighboring particles shifts the plasmon resonance band inducing color changes, which were first used for detection by Leuvering et al. (1980). Although the shift of the resonance band is more prominent for high concentration of target molecules in terms of color change, for lower concentrations the detection of such changes requires instruments, such as spectrophotometer. The sensitivity of such measurements has indeed been improved with more advanced detection schemes, such as hyper-Rayleigh scattering

(HSS) and differential light-scattering spectroscopy (DLSS) (Khlebtsov et al., 2004; Zhang et al., 2003). This approach of using the sol particle immunoassays (SPIA's) has also been extended to monovalent molecules (Aslan et al., 2004; Leuvering et al., 1983). The sensitivity of SPIA's in undiluted serum or urine samples was limited by the background signal fluctuations for a long time (Gribnau et al., 1986), which was later understood to be induced by the changes in refractive index (caused by nonspecific binding) that can be compensated by proper optimization schemes (Englebienne et al., 2003). Distance dependent immunoassays have also been developed with gold nanoshells (silica as core material), where, the ratio of the core-to-shell dimensions acts as a tuning parameter to systematically vary the extinction maxima from 700 to 1300 nm where the interference from absorbing molecules is minimal (Hirsch et al., 2003). An earlier version of this method was applied to detect chorionic gonadotropin in the urine of pregnant women (Leuvering et al., 1983).

On the other hand, the nucleic acids are more responsive to the distance dependent changes compared to the antigens, which are too large to produce significant changes when sandwiched between gold nanoparticles. The familiar colorimetric scheme for nucleic acid, which exploits the same principle, was first developed by Mirkin's group (Elghanian et al., 1997). This method uses single stranded target sequences or linker molecules, sandwiched between gold nanoparticles which are conjugated to thiolated reporter oligonucleotides, and the reversible color change from red to purple on temperature rise accompanies the detection. The length of the target sequence determines the rate and extent of this color change (Storhoff et al., 2000) and the Northwestern spot test can be performed to record the reverse color change (Storhoff et al., 1998). The temperature at which the reverse color change happens (melting temperature) for a given sequence depends primarily on the salt concentration with influence from the interparticle distance and density of attached oligonucleotides. This melting temperature is always higher than that of the corresponding duplex without the attached gold nanoparticles. Although some reports have suggested that the melting temperatures induced by this method are narrower than those of the conventional assays (Murphy and Redmond, 2005; Murphy et al., 2004; Taton et al., 2001), several studies involving fluorescent dyes via high resolution melting point curves have shown that this is not true in general applies to only certain cases (Liew et al., 2004). Later, Murphy and Redmond (2005) developed a compact instrument for melting curve analysis of gold conjugates in small volumes. Further, Li et al. (2005) developed a distance dependent detection scheme for single base polymorphisms (SNPs), which are the most abundant form of genetic mutation and their identification is crucial in many ways (e.g., diagnostics, disease prevention, and prognosis). In this method, genomic target sequences sandwiched between gold nanoparticles are conjugated to oligonucleotides and treated with a high fidelity DNA ligase (Fig. 25.3). The sol remains purple on heating when the target and capture sequences are a perfect match for each other keeping the gold nanoparticles linked together, and changes to red when there is a mismatch. Recently the gel electrophoresis was combined with the gold nanoparticles conjugated to single oligonucleotides to quantify SNPs (Qin and Yung, 2007).

However, the concentrations of nucleic acids in biological matrices are much lower limiting the sensitivity in many cases, which can be further improved by amplification techniques involving either polymerase chain reaction (PCR) or silver enhancement. Although PCR is widely used for amplification in molecular biology, attempts to combine it with gold nanoparticles have not been successful for various reasons. One of the major obstacles in this regard is the thermal instability that leads to dissociation of monothiolated oligonucleotides from nanoparticles (Nicewarner Peña et al., 2002). The thermal stability can be enhanced by using more than one anchoring thiol for conjugation, but the performance of such conjugates under PCR conditions is yet to be understood completely

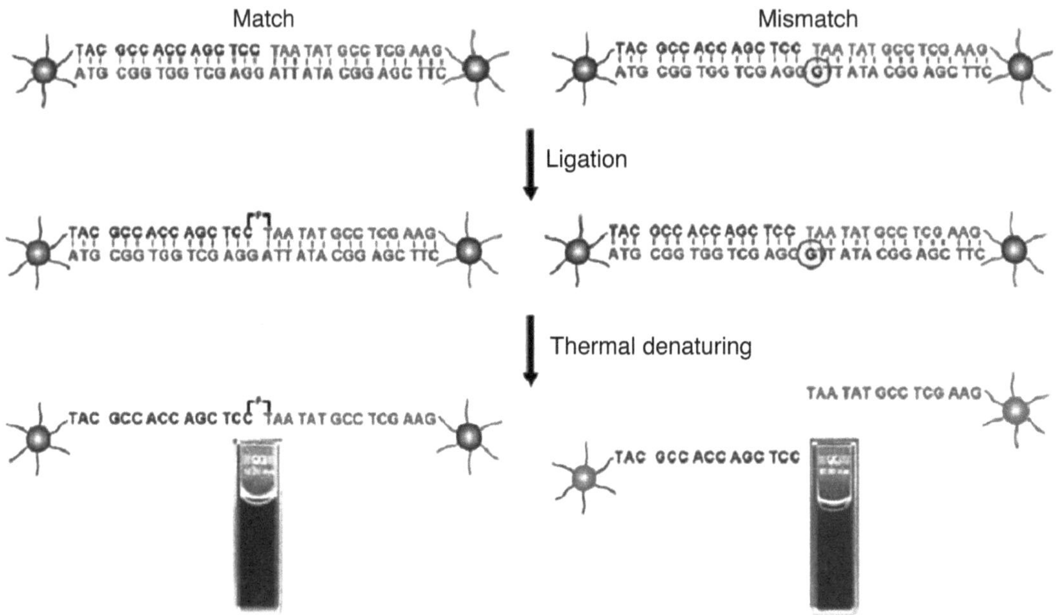

FIGURE 25.3 Detection of Single Base Imperfections in DNA Using Distance Dependent Detection Technique

At low temperatures, the extinction spectra of gold nanoparticles linked by matched and mismatched sequences show red shift. The red shift of the single base mismatch [encircled (yellow circle in the web version)] in the target sequences can be reversed by heating, as they are not joined by the enzyme.

Reproduced with permission from Wilson F R., 2008. The use of gold nanoparticles in diagnostics and detection. Chem. Soc. Rev. 37, 2028–2045. The Royal Society of Chemistry 2008.

(Chan and Fox, 1999; Letsinger et al., 2000). Several isothermal amplification methods have been proposed as an alternative solution, many of which were later found to generate unsuitable products due to contributions from different enzymes (Chan and Fox, 1999). The only successful example of an isothermal nucleic acid amplification technique for distance dependent detection with gold nanoparticles is shown in Fig. 25.4 (Tan et al., 2005). Here an exponentially amplified trigger sequence at 55°C is reformatted onto a linker oligonucleotide, which agglutinates nanoparticles. The maximum color change is assured by the appropriate choice of length and sequence of the linker, achieving more than 10^6-fold amplification within 5 min. This protocol allows the target concentrations, as small as 100 fM to be detected with the naked eye. Recently the trigger sequence was excised from genomic DNA that was derived from herpes simplex virus (Tan et al., 2007).

Another set of useful nucleic acid sequences are aptamers, which recognize molecules in similar ways as antibodies recognize antigens. Producing aptamers is easier than antibodies; they are more stable in harsh conditions and perform better than antibodies for target molecules having low molecular weights. The first demonstration of detecting thrombin by sandwiching it between gold nanoparticles conjugated to thiolated aptamers was given in 2004 (Pavlov et al., 2004). In this method there was no color change, instead the change in optical density was observed that was proportional to the amount of thrombin in the sample. A similar approach was used by Huang et al. (2005) to detect platelet-derived

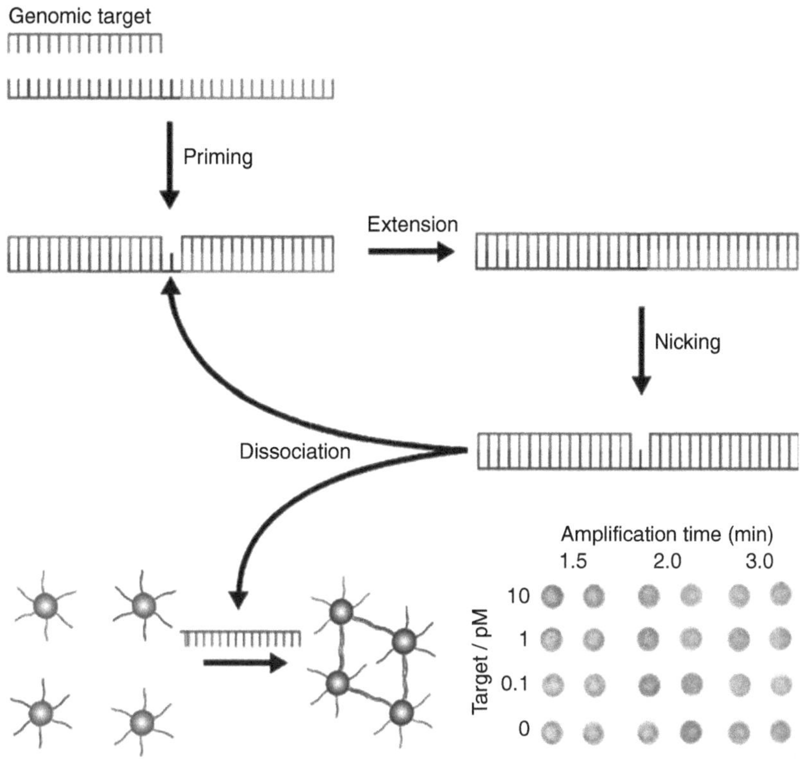

FIGURE 25.4 Simplified Scheme of the Isothermal Amplification Method (EXPAR—Exponential Amplification Reaction)

In this method, the genomic target sequences are reformatted onto short linker sequences which are then amplified at a constant temperature of 55°C. The effects of time and target concentration on the color of the gold nanoparticles can be deduced from the Northwestern spot tests (lower right hand corner).

Reproduced with permission from Wilson F R., 2008. The use of gold nanoparticles in diagnostics and detection. Chem. Soc. Rev. 37, 2028–2045. The Royal Society of Chemistry 2008.

growth factor (PDGF), which resulted in a color change. The limitation of aptamer-based approach, that it can only detect the targets with minimum of two aptamer binding sites, was lifted by a more general approach that can detect low molecular weight molecules having only one binding site (Liu and Lu, 2005).

The gold nanoparticles conjugated to thiolated oligonucleotides were later found to precipitate at high NaCl concentrations in the presence of complementary target sequences. Sato et al. (2003, 2006) showed that a single terminal base mismatch cannot only prevent such precipitation, but is also useful in detecting SNPs. Such distance dependent detection schemes have also been realized for detection of enzymic reactions by using substrates leading to aggregation or dissociation of gold nanoparticles (Wilson, 2008).

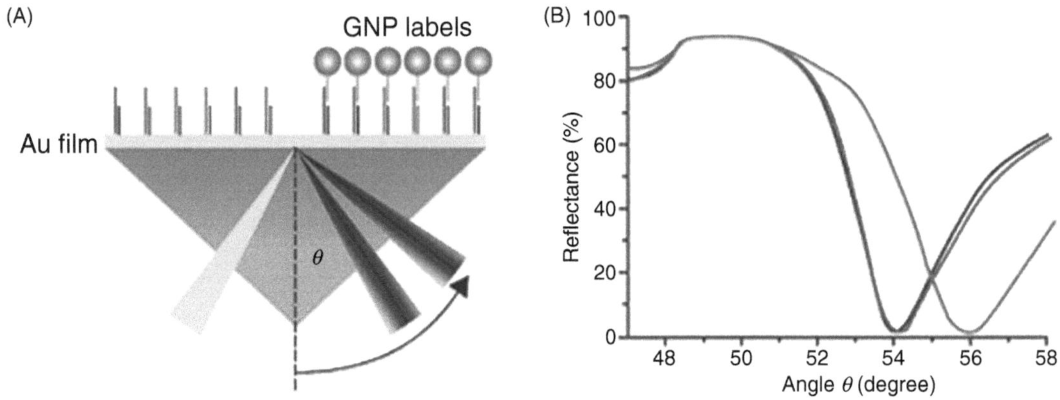

FIGURE 25.5

(A) A conventional surface plasmon resonance (SPR) biosensor composed of a thin film (~50 nm) of gold located at the interface between a prism and the sample. Small changes in the deflection angle θ are introduced by the low molecular weight molecules such as oligonucleotides upon their binding to the sensing surface [between black and dark gray (green in the web version) lines in the graph (B)], whereas much larger changes in the deflection angle are observed when the target molecules are sandwiched between the surface and gold nanoparticle labels, [between dark gray (green in the web version) and light gray (red in the web version) lines in graph (B)]. Reproduced with permission from Wilson F R., 2008. The use of gold nanoparticles in diagnostics and detection. Chem. Soc. Rev. 37, 2028–2045. The Royal Society of Chemistry 2008.

3.1.2.2 Other detection schemes

The growth and formation of gold nanoparticles can also be used to detect several enzymes that produce reducing agents. Small gold nanoparticles were shown to be enlarged as a result of enzyme catalysis, which can be exploited to detect glucose in micromolar concentrations (Baron et al., 2005). Such size dependent detection scheme was also extended for detection of neurotransmitters and inhibitors of acetylcholine esterase (Baron et al., 2005). Another sensing parameter is the refractive index that affects the resonance responses. This principle was used for the first time in 1990s to detect the biomolecular interactions at planar macroscopic gold surfaces (Malmqvist, 1999). When a gold-coated glass substrate (with a gold thin film of ~50 nm) is kept at the prism-sample interface (Fig. 25.5), there is a sharp drop in the intensity of the reflected light from the thin film when it couples with conducting electrons of the metal generating a strong evanescent wave. The changes in the refractive index of the region strongly influence the wavelength at which the intensity drops maximally and can be monitored by tracking the angle (θ). Generally, the capture molecules are coated to the sample to accumulate specific target molecules in the evanescent wave's affected region. Since the exponential nature of decay for evanescent waves ensure a very narrow region of their effect (only about 200 nm), this method can be used to monitor specific targets in bulk also. Additionally, the effect of most biological molecules on the SPR is proportional to their mass only as they have similar refractive indices. This limits the sensitivity of the SPR biosensors to around 1 pg of molecules per square millimeter of the sensing surface.

The widely popular immunochromatographic test systems of today were launched around 10 years ago by several companies for instrument-free diagnostics. These tests found wide range of applications, such as the detection of narcotic agents, toxins, early diagnostics of pregnancy, and screening

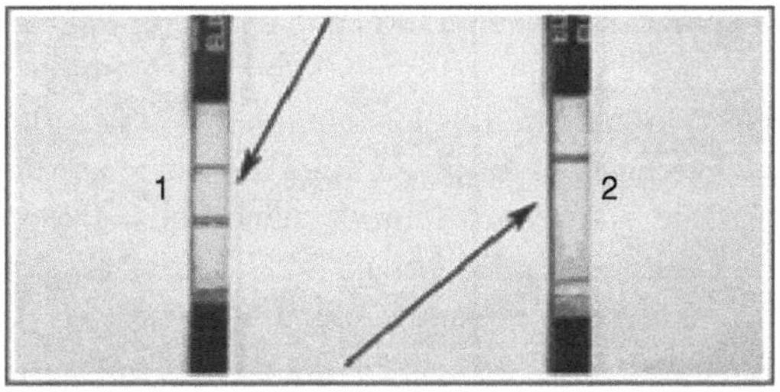

FIGURE 25.6

Positive (1) and negative (2) results of an immunochromatographyc assay.

Adapted from Wilson F R., 2008. The use of gold nanoparticles in diagnostics and detection. Chem. Soc. Rev. 37, 2028–2045; Dykman, L.A., Khlebtsov, N.G., 2011. Gold nanoparticles in biology and medicine: recent advances and prospects. Acta Nat. 3, 34.

of extremely dangerous and urogenital infections owing to the high sensitivity and specificity of the immunoassays. Although various enzymes, stained latexes, and quantum dots are used as labels in these systems, the use of gold nanoparticles still dominates the choice of labels to a huge extent (Aveyard et al., 2008; Wilson, 2008). This test method exploits the eluent motion along the membrane (lateral diffusion) that results in the formation of specific immune complexes, which are detected as stained bands on different membrane regions showing positive or negative results (Fig. 25.6). Very simple and convenient to use with visual assessment of the results, these tests have been demonstrated to be highly stable and allows reproducible results with reliable accuracy.

3.1.3 Therapeutic applications

3.1.3.1 Photothermal therapy

The photothermal cell damage technique has shown considerable promise in recent times for the treatment some of the infectious diseases (Velev and Kaler, 1999). The principle of this method is simple yet ingenious: gold nanoparticles become hot when irradiated at the light wavelength corresponding to their plasmon resonance wavelength, killing the target cells if they are located inside or around them. The location of these gold nanoparticles near the target cells can be achieved by their conjugation with antibodies or other molecules.

The thermal therapy for cancer dates back to the 18th century, where microwave, ultrasound, and radio radiation were used for either local heating or the entire organism (Brunelle, 2001). While the later takes around 1 h at 41–47°C, the local heating can be done within 3–4 min at 70°C. However, the use of both these methods has considerable restrictions in terms of healthy tissues getting damaged in the process. These restrictions were significantly lifted by the use of laser radiation that allows the controlled and direct damage of the tumor tissues (Möller et al., 2001). The drawback of the laser therapy is the requirement of using high power lasers for efficient tumor cell death restricting the selectivity. This was further improved when the gold nanoparticles were used for the first time in photothermal therapy (Urban et al., 2003). Later termed as the plasmonic photothermal therapy (PPTT), the target cells can be selectively damaged in this method using 20–30 nm gold nanospheres radiated with 20 ns pulse laser (532 nm) (Kim et al., 2000). The most promising direction for the use of gold nanoparticles in PPTT is for the types of cancers that are chemotherapy-resistant. The use of PPTT for clinical trials depends on finding the solution of several problems, such as the optimal optical properties of nanoparticles, contrast-enhancement in a tumor, reduction of overall toxicity, and improving the delivery of optical radiation to targets.

Theoretically, the effects of size, shape, structure, and degree of aggregation of gold nanoparticles on the photothermal efficiency have been extensively studied (Huang et al., 2007; Khlebtsov et al., 2006). For example, the intensity of radiation that is required for destruction of cancer cells is reduced by a factor of 20 when small aggregates of 30 nm particles are used instead of individual particles. The PPTT has also been demonstrated with gold nanorods, nanoshells, and composite gold–silver nanocages (Huang et al., 2006; Hu et al., 2006b; Huff et al., 2007; Loo et al., 2005b). Further discussion on the parameter-optimization of nanoparticles can be found elsewhere (Dykman and Khlebtsov, 2011; Lukianova-Hleb et al., 2010).

Targeted delivery of nanoparticles into the tumor requires the contrast-enhancement in the desired biotarget and reduction of side effects due to the accumulation of nanoparticles in other organs. The gold nanoparticles conjugated with polythylenglycol (PEG), or with antibodies to certain marker, proteins of tumor cells can be used as delivery strategies to overcome this problem. The use of PEG enhances the bioavailability and stability of nanoparticles providing more time for their circulation in blood flow, and prevents self-aggregation. The PEG-induced delivery method is called passive delivery, while the one that uses antibodies is called the active delivery. The active delivery method using antibodies to tumor markers is a more reliable and efficient method, an example of which is shown in Fig. 25.7 (Lukianova-Hleb et al., 2010). In addition to antibodies, active delivery method can also be performed with folic acid, ligand of numerous folate receptors of tumor cells and hormones (Huff et al., 2007; Lu et al., 2009).

Another important aspect of the PPTT methods is the efficient radiation-delivery to specific biotargets. Although the tissue penetration depth can be improved with IR radiation because of more absorption of biotissue chromophores at these wavelengths, it remains limited to 5–10 mm (Tuchin, 2010) making the search for alternative solutions a necessity. The possible solutions to overcome this problem have been suggested in the form of nanosecond pulse laser for increasing the irradiation-intensity, use of fiber-optic devices for target delivery inside the tissue and using radio radiation for deeper penetration. A detailed account of PPTT methods using gold nanoparticles can be found in this review (Kennedy et al., 2011).

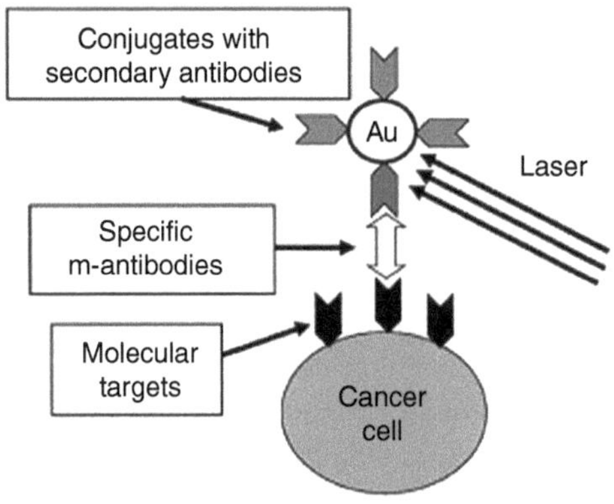

FIGURE 25.7 Scheme of Plasmonic Photothermal Therapy (PPTT) with Active Delivery of Au Nanoparticles to the Cancer Cells

Reprinted with permission from Khlebtsov, N.G., Dykman, L.A., 2010. Optical properties and biomedical applications of plasmonic nanoparticles. J. Quant. Spectrosc. Radiat. Transfer 111, 1–35. 2009 Elsevier Ltd.

3.1.3.2 Photodynamic therapy

The photodynamic method uses light-sensitive agents, such as photosensitizers with typically visible light for therapy of oncological diseases (Wilson, 2010). In general, the sensitizers are sent into the organism either intraversaly or administered applicatively, which selectively accumulate in the target tissues/cells and kill them when irradiated with the light corresponding to their (i.e., sensitizers') absorption maximum. Along with heat mechanism, the photochemical generation of singlet oxygen with subsequent formation of highly active radicals is also important for photodynamic therapy (PDT) in order to induce necrosis and apoptosis of tumor cells. Although the damage inflicted on the microvessels of the tumor results in its death, the patient's tissue remain highly sensitive to light even after the treatment as the photosynthesizers remain in the organism for a long time, which is a major drawback of the PDT.

In order to enhance the efficacy of PDT methods, the use of metal nanoparticles has been demonstrated. Several studies have reported the successful drug-delivery in polyelectrolyte capsules on gold nanoparticles, which disintegrate under laser radiation delivering the therapeutic agent to the targets (Sershen et al., 2000; Thomas and Kamat, 2003). Additionally, photoactive agents and peptides within conjugates have also been used for intracellular penetration (Liu et al., 2010; Thomas and Kamat, 2003). Composite nanoparticles including nanoshells, magnetic particles, photodynamic dyes along with PEG, and antibodies have also been proposed for PDT methods (Bardhan et al., 2010).

Although gold nanostructures have shown their potential in plasmon-assisted treatments (PPTT and PDT) in initial studies, a number of aspects, such as stability, biocompatibility, chemical interaction between nanoparticle conjugates in physiological environments, blood circulation time, penetration into the tumor, interaction with the immune system, excretion of nanoparticles, and so on, require further studies for such applications. With proper optimization of technical parameters, selective PPTT and PDT methods are soon expected to be upgraded for the clinical trials (Kennedy et al., 2011).

3.1.3.3 As therapeutic agents

In addition to diagnostics and cell photothermolysis applications, gold nanoparticles are also being used as therapeutic agents. Colloidal gold for rheumatoid arthritis was successfully applied for the first time in 1997, followed by numerous studies on its many aspects (Abraham et al., 1997; Abraham, 2008). Studies with intraarticular and subcutaneous introduction of colloidal gold into rats with collagen- and pristan-induced arthritis showed promising results (Brown et al., 2008; Tsai et al., 2007). The positive effects in these studies were attributed to the increased antiangiogenic activity that resulted from the binding between gold nanoparticles and the vascular endothelial growth factor, thus decreasing the macrophage infiltration and inflammation.

A colloidal gold vector to deliver the TNF into solid tumors in mice has been demonstrated, which accumulated specifically in tumor cells, and was not detected in any other healthy organs (Paciotti et al., 2004; Paciotti et al., 2006). As shown in Fig. 25.8, the color of the tumor was changed to red/purple color corresponding to the colloidal gold and tumor-specific activity of the TNF. Since the required doses of the drug are small for colloidal gold-TNF vector as compared to the native TNF for the antitumor reaction, the toxicity is reduced while the efficacy is enhanced. Following this proof-of-principle demonstration of tumor-treatment, the proposed method has been studied in vast details, reaching to a point where the intravenous administration based on a gold nanoparticle–TNF conjugate (named AurImmune) has already passed the second stage of clinical trials.

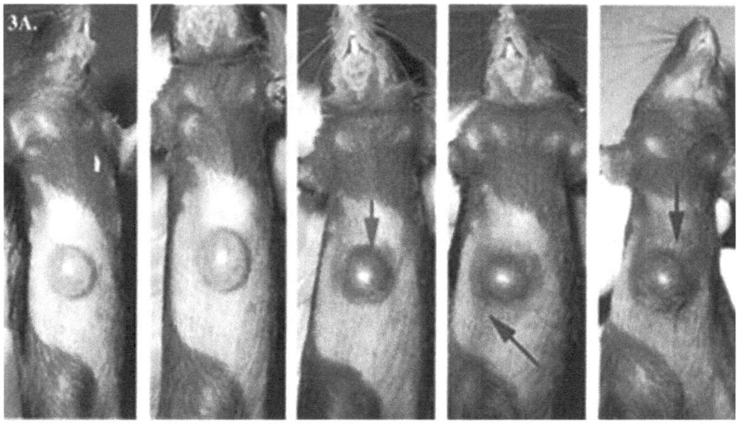

FIGURE 25.8 Accumulation of the PT-cAu-TNF Vector in MC-38 Tumors Over Time

A MC-38 tumor-burdened C57/BL6 mouse was intravenously injected with 15 μg of the PT–cAu–TNF vector. The ventral surface of the animal was photographed at the indicated times, showing the color changes of the tumor over 5 h. Red arrows show tumor uptake of the vector; blue arrows show accumulation of the vector in the tissues surrounding the tumor.

Reprinted with permission from Paciotti, G.F., Myer, L., Weinreich, D., Goia, D., Pavel, N., Mclaughlin, R.E., Tamarkin, L.,2004, Colloidal gold: a novel nanoparticle vector for tumor directed drug delivery. Drug Delivery 11, 169–183. The American Association for Cancer Research.

In an extensive study, the antiangiogenic properties of gold nanoparticles have been explored in vitro and in vivo (Mukherjee et al., 2005). The gold nanoparticles were found to be interacting with heparin-binding glycoproteins, which mediate angiogenesis (the formation of new blood vessels in organs or tissues) even in tumor tissues, making the gold nanoparticles to inhibit their activity. Since one of the main aspects of tumor is the uncontrolled cell-growth governed by the intensive angiogenesis; the antiangiogenic properties of gold nanoparticles are expected to be effective for tumor therapy. Several reports have also demonstrated the enhanced-apoptosis of the chronic lymphocytic leukemia cells using gold nanoparticles (Bhattacharya et al., 2007; Mukherjee et al., 2007).

3.1.4 Drug-delivery

Targeted delivery of drugs with the help of gold nanoparticles is a continuously developing field where antitumor agents and antibiotics have shown most promising results (Pissuwan et al., 2011). There are various antitumor agents with which the gold nanoparticles have been conjugated, such as, paclitaxel, methotrexate, daunorubicine, hemcytabin, 6-mercaptopurine, dodecylcysteine, sulfonamide, 5-fluorouracil, platinum complexes, kahalalide, tamoxifen, herceptin, doxorubicin, prospidin, and so on (Chen et al., 2007). Both in vitro models using tumor cell cultures and in vivo methods (in mice with induced tumors) have been used to assess the effect of conjugates in targeted drug delivery. Additionally target molecules, such as cetuximab, which improve anchoring and penetration into the target cells have also been used along with active substances. Multimodal delivery systems have also been proposed in which a gold nanoparticle is loaded simultaneously with several therapeutic and auxiliary agents (Kim et al., 2009a).

Antibacterial agents can also be delivered with the help of gold nanoparticles. Colloidal gold conjugated with vancomycin has been successfully demonstrated to be effective on enteropathogenic strains of *Escherichia coli*, *Enterococcus faecium*, and *Enterococcus faecalis* (Gu et al., 2003). Similarly, a complex of ciprofloxacin with gold nanoshells was used for high antibacterial activity toward *E. coli* (Rosemary et al., 2006), while a conjugate of the antileukemia drug 5-fluorouracil with colloidal gold, showed a noticeable antibacterial and antifungal effect against arrange of bacteria/fungi (Selvaraj and Alagar, 2007). On the other hand, the stable complexes with gold nanoparticles could not be achieved for some antibiotics, such as ampicillin, streptomycin, kanamycin, hentamycin, neomycin, ciprofloxacin, gatifloxacin, and norfloxacin, which are active against various bacterial families (Grace and Pandian, 2007a,b; Saha et al., 2007). Although the activity of these antibiotics can be increased by 12%–40% when mixed with colloidal gold, the underlying mechanisms for the enhancement of antibacterial effect of drugs is yet to be completely understood. For example, it has been shown that the antibacterial activity, remain significantly unchanged for gentamicin drug when used with and without its conjugation with gold nanoparticles (Burygin et al., 2009). The increased antioxidant properties of tocoferol complex and the therapeutic effects of the antirheumatic drug etanercept, conjugated with gold nanoparticles and their potential advantages have also been studied (Chamberland et al., 2008; Nie et al., 2007).

Gold nanoparticles have also been exploited for gene therapy for genetics and acquired diseases. In gene therapy, the genetic structures are launched into cells and organisms for therapeutic motives (Dykman and Khlebtsov, 2011; Miller, 1992). The inserted gene can give the desired result by its own expression or can suppress the damaged or over-expressed gene (partially or completely). In this regard, the structure and function of the affected gene can be readjusted with the help of gold nanoparticles. For example, gold nanoparticles were used in one study to efficiently deliver the genetic material into the cytoplasm and cell nucleus (Patel et al., 2010).

3.1.5 Other applications

The immunological properties of colloidal gold have also generated immense interest in last few decades. This is governed by the strong dependence of immunogenicity and antigenic specificity on the physiochemical properties of the compounds. This was first postulated by J. Bordet and is more commonly known as the physiochemical immunity theory. It has been shown in various studies that a rigorous antigen along with colloidal metals triggers the production of antibodies upon its introduction. The various effects of colloidal gold on nonspecific immune reactions have been studied in quite detail (Pacheco, 1925). The agents for antibody synthesis are generally chosen from the proteins, polysaccharides, and certain synthetic polymers for their higher immune response, whereas many biological compounds, such as vitamins, hormones, antibiotics, narcotics, and so on, are not considered to for antibody synthesis owing to their low immune response as a result of their low molecular weight.

Colloidal gold particles as a carrier to produce antibodies to glutamic acid were first demonstrated in 1986. Subsequently, this method was applied and further developed to produce antibodies to various haptens and rigorous antigens, including, amino acids, platelet-activating factor, quinolinic acid, biotin, recombinant peptides, lysophosphatide acid, endostatin, peptides of viral capsid of B and C hepatitis, influenza, murrain, α-amidated peptides, actin, antibiotics, azobenzene, Aβ-peptide, clenbuterol, surface *Yersinia* antigens, transmissible gastroenteritis virus, and tuberculin (Dykman and Khlebtsov, 2011; Pow and Crook, 1993; Wilson, 2008).

Gold colloids were experimentally used as the carriers of protein antigen in antiviral vaccine for tick-borne encephalitis virus in 1996 with better results than its commercial analogues (Demenev et al., 1996).

Nanoparticle-DNA carriers are most popularly produced using colloidal gold and protein encoded gene construction sequences for DNA vaccines using gold nanoparticles have been extensively studied (Zhao et al., 2003). Despite numerous studies proving the immunological properties of gold nanoparticles and many experimental evidences exploiting their usefulness, the underlying mechanism that leads to these properties of gold particles is not understood completely. However, the somewhat reliable discussion on some aspects of this topic can be found in (Pow and Crook, 1993; Villiers et al., 2010).

3.2 SILVER (Ag) NANOPARTICLES

Silver has been used for medical applications since 19th century. In general, silver ions and silver-based compounds show high toxicity to microorganisms including 16 major species of bacteria. The well-known toxic properties of silver (which are significantly enhanced in case of nanoparticles due to increased surface area) have made it an excellent choice of metal for many applications, such as water and air purification, food production, cosmetics, and numerous household products along with many others. The recent advancements of nanotechnology have further extended the range of such applications, making silver nanoparticles to be the most commonly used nanomaterial in consumer products (Marambio-Jones and Hoek, 2010). Clothing, respirators, contraconceptives, antibacterial sprays, detergent, dietary supplements, cutting boards, socks, shoes, cell phones, laptop keyboards, and children's toys are just a smattering of the numerous applications that exploit the antimicrobial properties of silver.

Along with textile engineering and electronics, the major applications of silver nanoparticles include their use as catalysts, as optical sensors of zeptomole (10^{-21}) concentrations and most importantly as a bactericidal and as a therapeutic agent in biomedicine. Different forms of silver nanomaterials, such as metallic silver nanoparticles, silver chloride particles, silver-impregnated zeolite powders, and activated carbon materials, dendrimer–silver complexes and composites, polymer-silver nanoparticle composites, silver–titanium dioxide composite nanopowders show antimicrobial properties (Marambio-Jones and Hoek, 2010). Furthermore, silver nanoparticles may exhibit additional antimicrobial capabilities that are not seen in bulk or ionic silver metal (Chen and Schluesener, 2008). Several specific microbes against which silver nanoparticles have been shown as effective biocides are: various bacterial families of *E. coli*, *Staphylococcus aureus*, *Staphylococcus epidermis*, *Leuconostoc mesenteroides*, *Bacillus subtilis*, *Klebsiella mobilis*, and *Klebsiella pneumonia*, and so on; various fungi such as *Aspergillus niger*, *Candida albicans*, *Saccharomyces cerevisia*, *Trichophyton mentagrophytes*, and *Penicillium citrinum* and various virii, such as Hepatitis B, HIV-1, syncytial virus, and so on (Marambio-Jones and Hoek, 2010). Despite numerous applications, the action of silver on microbes is yet to be completely understood and though there have been several studies exploring this issue, the underlying mechanism is not fully understood and/or agreed upon. The commonly proposed mechanism for the antibacterial properties of silver nanoparticles is the release of silver ions along with the generation of reactive oxygen species (ROS) and subsequent cell membrane damage (Hwang et al., 2008; Kim et al., 2007). However, several other contradictory views based on experimental findings are also described in the literature regarding this mechanism (Marambio-Jones and Hoek, 2010).

Other than the various antimicrobial properties, silver nanoparticles are also toxic to higher cell lines, such as zebra fish, clams, rats, and humans (Stebounova et al., 2011; Wijnhoven et al., 2009). It has been found that silver nanoparticles can accumulate and damage certain tissues like liver, lungs, and olfactory bulbs in rodents and can also penetrate the blood–brain barrier. These evidences suggest the potential impacts of silver nanoparticles for human and ecosystem related applications, which can be further extended with the more widespread use of silver nanomaterials. For example, silver nanoparticles released to the environment at the point of chemical or other hazardous production units may help to reduce environmental pollution.

3.2.1 Antibacterial activity

The toxicity, that is, detrimental effects on a living organism, of silver is an important property in context to many sterilization related applications. Antibacterial properties of silver have been studied and exploited for such applications for quite some time now.

The first report showing minimum inhibitory concentration (MIC) against *E. coli* (below 6.6 nM) was reported in 2007, using13.4 nm silver nanoparticles prepared by reduction of silver nitrate with sodium borohydride (Kim et al., 2007). The same nanoparticles were also shown to have MIC against *S. aureus* above 33 nM. Following this, silver nanoparticles 16 nm diameter, were also demonstrated to completely inhibit colony forming units (CFU) ability of *E. coli* (60 μg/mL) (Raffi et al., 2008). Commercially available silver nanopowders have also been shown to reduce the colony formation in *E. coli* and *S. aureus* if used in proper concentrations. Furthermore, the MIC ranging between 13.5 and 1.69 μg/mL has been obtained upon exposure of silver nanoparticles (26 nm size) for various bacterial strains including those isolated from different human clinical materials (Kvitek et al., 2008). Silver nanoparticles with a size distribution from 14 to 710 nm in an aerosol form also produced 76% reduction in the CFUs of *B. subtilis*.

Nanocomposites of dendrimer–silver have also been shown as effective antibacterials. For example, composites of poly(amidoamine) dendrimer–silver have been reported to be effective against *S. aureus*, *P. aeruginosa*, *E. coli*, *B. subtilis*, and *K. mobilis* (Balogh et al., 2001). Recently 7-log reduction, in CFUs of *E. coli* was demonstrated with the help of 333.3 μg/mL of Ag-loaded zeolite (Inoue et al., 2002). Further, MICs of 78 μg Ag^{+}/mL for *E. coli* and 39 μg Ag^{+}/mL for *S. aureus* and *P. aeruginosa* along with antibacterial activity against *Listeria monocytogenes* has also been demonstrated (Cowan et al., 2003).

3.2.2 Antimicrobial activity to other microorganisms

Silver nanoparticles are also effective against various fungi, virii, and algae. For example, silver nanoparticles restrain the growth of *A. niger* when stabilized with hyperbranched polymers (size 1.4–7.1 nm) and/or Myramistin (size 10 nm) (Vertelov et al., 2008). Further, the Myramistin stabilized silver particles with different sizes also showed toxic effects for other bacteria such as *S. cerevisiae*, *T. mentagrophytes*, and *C. albanicans* (Kim et al., 2007). The silver nanoparticles have also been used to disrupt the cell membrane structure that restrains the reproduction process, thus showing the antifungal activity against *C. albanicans* (Kim et al., 2009b,c).

Kim et al. (2007) first showed the ability of suppressing the yeast growth using silver nanoparticles stabilized with hyperbranched polymers. Subsequently, HPAMAMN$(CH_3)_2$/AgNPs composites were used to restrict the growth of *P. citrinum*. Additionally, the toxic effects of silver nanoparticles against algae were also explored, where the photosynthetic yield of *Chlamydomonas reinhardtii* was reduced with the help of silver nanoparticles (Navarro et al., 2008).

Antiviral effects of silver nanoparticles are also explored by various research groups. For example, the replications properties of hepatitis B virus can be inhibited by silver nanoparticles of 10 nm size (Ravindran et al., 2013). Further, the PVP-coated silver nanoparticles (sizes 1–10 nm) restrains the attachment of the HIV-1 virus to host cells. The similar PVP-coated silver nanoparticles were found to reduce the respiratory syncytial virus infection by 44% in another study. Silver nanoparticles of different sizes (1–70 nm) lead to enhance virus removal when impregnated to polysulfone ultrafiltration membranes (Ravindran et al., 2013). This particular ability of silver nanoparticles helps to improve water disinfection by low-pressure membrane filtration.

3.2.3 Toxicity to mammalian cells

The toxicity of silver nanoparticles against higher organisms has been studied quite extensively in the last decade. The positive results of these studies have indeed proven the significant potential in this direction. The silver nanoparticles have been shown to be toxic to zebra fish, *Drosophila melanogaster*, a fruit fly, different cell lines of mice, rats, and also humans (Greulich et al., 2009; Ravindran et al., 2013). I will provide few examples of silver nanoparticle toxicity for mammalian cells in this chapter. A detailed and focused analysis on this particular topic can be found in the literature (Greulich et al., 2009; Panyala et al., 2008).

The evidence of silver nanoparticle toxicity to mammalian cells was first found from the in vitro studies. In these studies, the silver nanoparticles were found to be able to reduce the mitochondrial function along with increased membrane leakage of mouse spermatogonial stem cell and rat liver cells (Braydich-Stolle et al., 2005). Later studies on human fibrosarcoma and human skin/carcinoma cells found that the cellular levels of glutathione are reduced in the presence of silver nanoparticles when used in a topical antimicrobial agent (Ravindran et al., 2013). The resultant cellular damage and lipid peroxidation was attributed to the oxidative stress as indicated by the reduced glutathione levels. It was also found in this study that for both the cell types, the required dose for apoptosis (0.78–1.56 μg/mL) was much smaller that required for necrosis (12.5 μg/mL). The in vivo studies further confirmed the possibility of defining a safe range for the use of silver nanoparticles as a topical microbial agent for such applications. In another study, similar differences of required doses of silver nanoparticles for apoptosis and/or necrosis in mouse fibroblasts and liver cells were found. It was also suggested that the cellular antioxidant mechanisms limits the oxidative stress even if the silver nanoparticles enter into the cells.

The in vivo studies to find the toxic effects of silver nanoparticles were also performed. In the study by Sung et al. (2008) silver nanoparticles reduced lung function with production of inflammatory lesions in the lungs during a 90-day inhalation in rats (Sung et al., 2008). On the other hand, Kim et al. (2008) showed that the silver nanoparticles tend to accumulate in the olfactory bulbs of Sprague-Dawley rats and also accumulated in the brain.

Hsin et al. (2008) performed the mechanistic studies to understand the toxic effects of silver nanoparticles in mouse fibroblast and human colon cancer cells. The dose of silver nanoparticle required to induce apoptosis to 43.4% of fibroblast cells was found to be 50 μg/mL, while a larger dose of 100 μg/mL was required to produce necrosis to 40.2% of the cancer cells. A mitochondrial-mediated pathway along with generation of ROS in the cell was found to be the mechanism for apoptosis in fibrosblast cells. This leads to the activation of apoptosis regulators, namely JNK and p53 proteins, which induces the migration of protein Bax to the surface of the mitochondria resulting in the release of cytochrome C from mitochondria and cleavage of PARP.

A possible mechanism of silver nanoparticle toxicity to human cells was proposed by Asharani et al. (2008). It was suggested that the silver nanoparticles affect the mitochondrial respiratory chain that causes ROS generation and also affects the ATP production, subsequently leading to DNA damage. The authors of this study also found that the silver nanoparticles are potentially toxic even in small doses. They further concluded that the silver nanoparticles are cytotoxic, genotoxic, and antiprolofera-tive, and therefore are equally toxic for human glioblastoma and normal human fibroblasts cells.

4 SUMMARY

The use of gold and silver nanoparticles and their composites has generated immense interest for a broad range of biomedical applications over a last few decades. Overall, the optical properties of gold and silver nanoparticles are indeed intriguing and have shown definite promise for the development of a new generation of vaccines, therapeutics, and medicines. Whether in biology for determination of nucleic acids, proteins, and metabolites, or in medicine for drug screening, analysis of antibodies and antigens, diagnostics of infectious diseases; and even in chemistry for environmental express monitor-ing, quantitative analysis of solutions, and dispersed systems; the metal nanoparticles have shown great importance in numerous applications. The important properties that fundamentally affect the optical re-sponse of nanoparticles and/or nanostructures are the particle size and shape, interparticle distance, and the optical properties of the local environment, acting as control parameters for tuning of resonances according to the desired application.

However, understanding the mechanisms behind the toxicity of metal nanoparticles is also of great interest to widen the range of their applications, as well as to control the specific applications.

On the other hand, there is also a growing need to evaluate the potential human and ecosystem haz-ards associated with silver or gold since a wide range of products containing these metal nanoparticles (especially silver) are getting utilized worldwide. This, in turn, makes the additional research directed toward safe designs and proper disposal of products containing metal nanomaterials a necessity to pre-vent creating new risks to humans or the environment. Thus, in vivo, in vitro, and environmental stud-ies in future should be directed toward systematic understanding of various side-effects of nanoscaled metal, their safe transport, and toxicity.

REFERENCES

Aaron, J., De la Rosa, E., Travis, K., Harrison, N., Burt, J., José-Yacamán, M., Sokolov, K., 2008. Polarization microscopy with stellated gold nanoparticles for robust monitoring of molecular assemblies and single biomolecules. Opt. Express 16, 2153–2167.

Abraham, G.E., 2008. Clinical applications of gold and silver nanocolloids. Orig. Internist 15, 132–157.

Abraham, E., Peter, B., Himmel, G.U.Y., 1997. Management of rheumatoid arthritis: rationale for the use of colloidal metallic gold. J. Nutr. Environ. Med. 7, 295–305.

Anker, J.N., Hall, W.P., Lyandres, O., Shah, N.C., Zhao, J., Van Duyne, R.P., 2008. Biosensing with plasmonic nanosensors. Nat. Mater. 7, 442–453.

Asharani, P.V., Low Kah Mun, G., Hande, M.P., Valiyaveettil, S., 2008. Cytotoxicity and genotoxicity of silver nanoparticles in human cells. ACS Nano 3, 279–290.

Aslan, K., Lakowicz, J.R., Geddes, C.D., 2004. Nanogold-plasmon-resonance-based glucose sensing. Anal. Biochem. 330, 145–155.

Au, L., Zhang, Q., Cobley, C.M., Gidding, M., Schwartz, A.G., Chen, J., Xia, Y., 2010. Quantifying the cellular uptake of antibody-conjugated au nanocages by two-photon microscopy and inductively coupled plasma mass spectrometry. ACS Nano 4, 35.

Aveyard, J., Nolan, P., Wilson, R., 2008. Improving the sensitivity of immunoassays by tuning gold nanoparticles to the tipping point. Anal. Chem. 80, 6001–6005.
Balogh, L., Swanson, D.R., Tomalia, D.A., Hagnauer, G.L., Mcmanus, A.T., 2001. Dendrimer-silver complexes and nanocomposites as antimicrobial agents. Nano Lett. 1, 18–21.
Bardhan, R., Chen, W., Bartels, M., Perez-Torres, C., Botero, M.F., Mcaninch, R.W., Contreras, A., Schiff, R., Pautler, R.G., Halas, N.J., 2010. Tracking of multimodal therapeutic nanocomplexes targeting breast cancer in vivo. Nano Lett. 10, 4920–4928.
Baron, R., Zayats, M., Willner, I., 2005. Dopamine-, L-DOPA-, adrenaline-, and noradrenaline-induced growth of Au nanoparticles: assays for the detection of neurotransmitters and of tyrosinase activity. Anal. Chem. 77, 1566–1571.
Beversluis, M.R., Bouhelier, A., Novotny, L., 2003. Continuum generation from single gold nanostructures through near-field mediated intraband transitions. Phys. Rev. B 68, 115433.
Bhattacharya, R., Patra, C.R., Verma, R., Kumar, S., Greipp, P.R., Mukherjee, P., 2007. Gold nanoparticles inhibit the proliferation of multiple myeloma cells. Adv. Mater. 19, 711–716.
Bohren, C.F., Huffman, D.R., 2008. Absorption and scattering of light by small particles. John Wiley & Sons.
Bouhelier, A., Beversluis, M.R., Novotny, L., 2003. Characterization of nanoplasmonic structures by locally excited photoluminescence. Appl. Phys. Lett. 83, 5041–5043.
Braydich-Stolle, L., Hussain, S., Schlager, J.J., Hofmann, M.-C., 2005. In vitro cytotoxicity of nanoparticles in mammalian germline stem cells. Toxicol. Sci. 88, 412–419.
Brown, C.L., Whitehouse, M.W., Tiekink, E.R.T., Bushell, G.R., 2008. Colloidal metallic gold is not bio-inert. Inflammopharmacology 16, 133–137.
Brunelle, S., 2001. Electroimmunoassay technology for foodborne pathogen detection. IVD Technol. 7, 55–66.
Bunin, V.D., Ignatov, O.V., Guliy, O.I., Voloshin, A.G., Dykman, L.A., O'Neil, D., Ivnitskii, D., 2005. Study of the electrophysical properties of listeria monocytoge nes cells during interaction with monoclonal antibodies. Biophysics 50, 299–302.
Burygin, G.L., Khlebtsov, B.N., Shantrokha, A.N., Dykman, L.A., Bogatyrev, V.A., Khlebtsov, N.G., 2009. On the enhanced antibacterial activity of antibiotics mixed with gold nanoparticles. Nanoscale Res. Lett. 4, 794–801.
Chamberland, D.L., Agarwal, A., Kotov, N., Fowlkes, J.B., Carson, P.L., Wang, X., 2008. Photoacoustic tomography of joints aided by an Etanercept-conjugated gold nanoparticle contrast agent—an ex vivo preliminary rat study. Nanotechnology 19, 095101.
Chan, A.B., Fox, J.D., 1999. NASBA and other transcription-based amplification methods for research and diagnostic microbiology. Rev. Med. Microbiol. 10, 185–196.
Chen, X., Schluesener, H.J., 2008. Nanosilver: a nanoproduct in medical application. Toxicol. Lett. 176, 1–12.
Chen, Y.-H., Tsai, C.-Y., Huang, P.-Y., Chang, M.-Y., Cheng, P.-C., Chou, C.-H., Chen, D.-H., Wang, C.-R., Shiau, A.-L., Wu, C.-L., 2007. Methotrexate conjugated to gold nanoparticles inhibits tumor growth in a syngeneic lung tumor model. Mol. Pharma. 4, 713–722.
Cowan, M.M., Abshire, K.Z., Houk, S.L., Evans, S.M., 2003. Antimicrobial efficacy of a silver-zeolite matrix coating on stainless steel. J. Ind. Microbiol. Biotechnol. 30, 102–106.
Demenev, V.A., Shchinova, M.A., Ivanov, L.I., Vorob'Eva, R.N., Zdanovskaya, N.I., Nebaykina, N.V., 1996. Perfection of methodical approaches to designing vaccines against tick-borne encephalitis (Originally published in Russian). Vopr. Virusol. 41, 107–110.
Ditlbacher, H., Hohenau, A., Wagner, D., Kreibig, U., Rogers, M., Hofer, F., Aussenegg, F.R., Krenn, J.R., 2005. Silver nanowires as surface plasmon resonators. Phys. Rev. Lett. 95, 257403.
Drygin, Y.F., Blintsov, A.N., Osipov, A.P., Grigorenko, V.G., Andreeva, I.P., Uskov, A.I., Varitsev, Y.A., Anisimov, B.V., Novikov, V.K., Atabekov, J.G., 2009. High-sensitivity express immunochromatographic method for detection of plant infection by tobacco mosaic virus. Biochemistry (Moscow) 74, 986–993.

Du, C., You, Y., Zhang, X., Johnson, K., Shen, Z., 2009. Polarization-dependent confocal imaging of individual Ag nanorods and nanoparticles. Plasmonics 4, 217–222.

Dykman, L.A., Khlebtsov, N.G., 2011. Gold nanoparticles in biology and medicine: recent advances and prospects. Acta Nat. 3, 34.

Elghanian, R., Storhoff, J.J., Mucic, R.C., Letsinger, R.L., Mirkin, C.A., 1997. Selective colorimetric detection of polynucleotides based on the distance-dependent optical properties of gold nanoparticles. Science 277, 1078–1081.

El-Sayed, I.H., Huang, X., El-Sayed, M.A., 2005. Surface plasmon resonance scattering and absorption of anti-EGFR antibody conjugated gold nanoparticles in cancer diagnostics: applications in oral cancer. Nano Lett. 5, 829–834.

Englebienne, P., Hoonacker, A.V., Verhas, M., 2003. Surface plasmon resonance: principles, methods and applications in biomedical sciences. J. Spec. 17, 255–273.

Grace, A.N., Pandian, K., 2007a. Antibacterial efficacy of aminoglycosidic antibiotics protected gold nanoparticles—a brief study. Colloids Surf. A 297, 63–70.

Grace, A.N., Pandian, K., 2007b. Quinolone antibiotic-capped gold nanoparticles and their antibacterial efficacy against gram positive and gram negative organisms. J. Bionanosci. 1, 96–105.

Greulich, C., Kittler, S., Epple, M., Muhr, G., Köller, M., 2009. Studies on the biocompatibility and the interaction of silver nanoparticles with human mesenchymal stem cells (hMSCs). Langenbeck's Arch, Surg. 394, 495–502.

Gribnau, T.C.J., Leuvering, J.H.W., Van Hell, H., 1986. Particle-labelled immunoassays: a review. J. Chromatogr. B 376, 175–189.

Gu, H., Ho, P.L., Tong, E., Wang, L., Xu, B., 2003. Presenting vancomycin on nanoparticles to enhance antimicrobial activities. Nano Lett. 3, 1261–1263.

Hauck, T.S., Ghazani, A.A., Chan, W.C.W., 2008. Assessing the effect of surface chemistry on gold nanorod uptake, toxicity, and gene expression in mammalian cells. Small 4, 153–159.

Hayat, M.A., 1989. Colloidal gold: principles, methods and applications. Academic Press, San Diego.

Hirsch, L.R., Jackson, J.B., Lee, A., Halas, N.J., West, J.L., 2003. A whole blood immunoassay using gold nanoshells. Anal. Chem. 75, 2377–2381.

Hsin, Y.-H., Chen, C.-F., Huang, S., Shih, T.-S., Lai, P.-S., Chueh, P.J., 2008. The apoptotic effect of nanosilver is mediated by a ROS-and JNK-dependent mechanism involving the mitochondrial pathway in NIH3T3 cells. Toxicol. Lett. 179, 130–139.

Hu, M., Chen, J., Li, Z.-Y., Au, L., Hartland, G.V., Li, X., Marquez, M., Xia, Y., 2006a. Gold nanostructures: engineering their plasmonic properties for biomedical applications. Chem. Soc. Rev. 35, 1084–1094.

Hu, M., Petrova, H., Chen, J., Mclellan, J.M., Siekkinen, A.R., Marquez, M., Li, X., Xia, Y., Hartland, G.V., 2006b. Ultrafast laser studies of the photothermal properties of gold nanocages. J. Phys. Chem. B 110, 1520–1524.

Huang, C.-C., Huang, Y.-F., Cao, Z., Tan, W., Chang, H.-T., 2005. Aptamer-modified gold nanoparticles for colorimetric determination of platelet-derived growth factors and their receptors. Anal. Chem. 77, 5735–5741.

Huang, X., El-Sayed, I.H., Qian, W., El-Sayed, M.A., 2006. Cancer cell imaging and photothermal therapy in the near-infrared region by using gold nanorods. J. Am. Chem. Soc. 128, 2115–2120.

Huang, X., Qian, W., El-Sayed, I.H., El-Sayed, M.A., 2007. The potential use of the enhanced nonlinear properties of gold nanospheres in photothermal cancer therapy. Laser. Surg. Med. 39, 747–753.

Huff, T.B., Tong, L., Zhao, Y., Hansen, M.N., Cheng, J.X., Wei, A., 2007. Hyperthermic effects of gold nanorods on tumor cells. Nanomedicine 2, 125–132.

Hutter, E., Boridy, S., Labrecque, S., Lalancette-Hébert, M., Kriz, J., Winnik, F.M., Maysinger, D., 2010. Microglial response to gold nanoparticles. ACS Nano 4, 2595–2606.

Hwang, E.T., Lee, J.H., Chae, Y.J., Kim, Y.S., Kim, B.C., Sang, B.I., Gu, M.B., 2008. Analysis of the toxic mode of action of silver nanoparticles using stress-specific bioluminescent bacteria. Small 4, 746–750.

Inoue, Y., Hoshino, M., Takahashi, H., Noguchi, T., Murata, T., Kanzaki, Y., Hamashima, H., Sasatsu, M., 2002. Bactericidal activity of Ag–zeolite mediated by reactive oxygen species under aerated conditions. J. Inorg. Biochem. 92, 37–42.

Jain, P.K., El-Sayed, M.A., 2007. Universal scaling of plasmon coupling in metal nanostructures: extension from particle pairs to nanoshells. Nano Lett. 7, 2854–2858.

Jain, P.K., Huang, X., El-Sayed, I.H., El-Sayed, M.A., 2008. Noble metals on the nanoscale: optical and photothermal properties and some applications in imaging, sensing, biology, and medicine. Acc. Chem. Res. 41, 1578–1586.

Kelly, K.L., Coronado, E., Zhao, L.L., Schatz, G.C., 2003. The optical properties of metal nanoparticles: the influence of size, shape, and dielectric environment. J. Phys. Chem. B 107, 668–677.

Kennedy, L.C., Bickford, L.R., Lewinski, N.A., Coughlin, A.J., Hu, Y., Day, E.S., West, J.L., Drezek, R.A., 2011. A new era for cancer treatment: gold-nanoparticle-mediated thermal therapies. Small 7, 169–183.

Kern, A.M., Meixner, A.J., Martin, O.J.F., 2012. Molecule-dependent plasmonic enhancement of fluorescence and Raman scattering near realistic nanostructures. ACS Nano 6, 9828–9836.

Khlebtsov, N.G., Dykman, L.A., 2010. Optical properties and biomedical applications of plasmonic nanoparticles. J. Quant. Spectrosc. Radiat. Transfer 111, 1–35.

Khlebtsov, N.G., Bogatyrev, V.A., Melnikov, A.G., Dykman, L.A., Khlebtsov, B.N., Krasnov, Y.M., 2004. Differential light-scattering spectroscopy: a new approach to studying of colloidal gold nanosensors. J. Quant. Spectrosc. Radiat. Transfer 89, 133–142.

Khlebtsov, B., Melnikov, A., Zharov, V., Khlebtsov, N., 2006. Absorption and scattering of light by a dimer of metal nanospheres: comparison of dipole and multipole approaches. Nanotechnology 17, 1437.

Kim, J.-H., Cho, J.-H., Cha, G.S., Lee, C.-W., Kim, H.-B., Paek, S.-H., 2000. Conductimetric membrane strip immunosensor with polyaniline-bound gold colloids as signal generator. Biosens. Bioelectron. 14, 907–915.

Kim, J.S., Kuk, E., Yu, K.N., Kim, J.-H., Park, S.J., Lee, H.J., Kim, S.H., Park, Y.K., Park, Y.H., Hwang, C.-Y., 2007. Antimicrobial effects of silver nanoparticles. Nanomed. Nanotechnol. Biol. Med. 3, 95–101.

Kim, Y.S., Kim, J.S., Cho, H.S., Rha, D.S., Kim, J.M., Park, J.D., Choi, B.S., Lim, R., Chang, H.K., Chung, Y.H., 2008. Twenty-eight-day oral toxicity, genotoxicity, and gender-related tissue distribution of silver nanoparticles in Sprague-Dawley rats. Inhal. Toxicol. 20, 575–583.

Kim, C.-K., Ghosh, P., Rotello, V.M., 2009a. Multimodal drug delivery using gold nanoparticles. Nanoscale 1, 61–67.

Kim, J., Lee, J., Kwon, S., Jeong, S., 2009b. Preparation of biodegradable polymer/silver nanoparticles composite and its antibacterial efficacy. J. Nanosci. Nanotechnol. 9, 1098–1102.

Kim, K.-J., Sung, W.S., Suh, B.K., Moon, S.-K., Choi, J.-S., Kim, J.G., Lee, D.G., 2009c. Antifungal activity and mode of action of silver nano-particles on *Candida albicans*. Biometals 22, 235–242.

Klar, T., Perner, M., Grosse, S., Von Plessen, G., Spirkl, W., Feldmann, J., 1998. Surface-plasmon resonances in single metallic nanoparticles. Phys. Rev. Lett. 80, 4249.

Klein, M.W., Enkrich, C., Wegener, M., Linden, S., 2006. Second-harmonic generation from magnetic metamaterials. Sci. 313, 502–504.

Klein, S., Petersen, S., Taylor, U., Rath, D., Barcikowski, S., 2010. Quantitative visualization of colloidal and intracellular gold nanoparticles by confocal microscopy. J. Biomed. Opt. 15, 036015–136015.

Kneipp, K., Wang, Y., Kneipp, H., Perelman, L.T., Itzkan, I., Dasari, R.R., Feld, M.S., 1997. Single molecule detection using surface-enhanced Raman scattering (SERS). Phys. Rev. Lett. 78, 1667.

Kvitek, L., Panáček, A., Soukupova, J., Kolar, M., Vecerova, R., Prucek, R., Holecová, M., Zboril, R., 2008. Effect of surfactants and polymers on stability and antibacterial activity of silver nanoparticles (NPs). J. Phys. Chem. C 112, 5825–5834.

Lal, S., Link, S., Halas, N.J., 2007. Nano-optics from sensing to waveguiding. Nat. Photon. 1, 641–648.

Lal, S., Clare, S.E., Halas, N.J., 2008. Nanoshell-enabled photothermal cancer therapy: impending clinical impact. Acc. Chem. Res. 41, 1842–1851.

Lange, C., 1912. Über die Ausflockung von Goldsol durch Liquor cerebrospinalis, In: Hirschwald, A (Ed.). BKW, Berlin.

Letsinger, R.L., Elghanian, R., Viswanadham, G., Mirkin, C.A., 2000. Use of a steroid cyclic disulfide anchor in constructing gold nanoparticle-oligonucleotide conjugates. Bioconjugate Chem. 11, 289–291.

Leuvering, J.H.W., Thal, P., Waart, M.V., Schuurs, A., 1980. Sol particle agglutination immunoassay for human chorionic gonadotrophin. Fresenius' J. Anal. Chem. 301, 132–1132.

Leuvering, J.H.W., Thal, P.J.H.M., White, D.D., Schuurs, A.H.W.M., 1983. A homogeneous sol particle immunoassay for total oestrogens in urine and serum samples. J. Immunol. Methods 62, 163–174.

Li, J., Chu, X., Liu, Y., Jiang, J.-H., He, Z., Zhang, Z., Shen, G., Yu, R.-Q., 2005. A colorimetric method for point mutation detection using high-fidelity DNA ligase. Nucleic Acids Res. 33, e168–e1168.

Liew, M., Pryor, R., Palais, R., Meadows, C., Erali, M., Lyon, E., Wittwer, C., 2004. Genotyping of single-nucleotide polymorphisms by high-resolution melting of small amplicons. Clin. Chem. 50, 1156–1164.

Lindfors, K., Kalkbrenner, T., Stoller, P., Sandoghdar, V., 2004. Detection and spectroscopy of gold nanoparticles using supercontinuum white light confocal microscopy. Phys. Rev. Lett. 93, 037401.

Link, S., Mohamed, M.B., El-Sayed, M.A., 1999. Simulation of the optical absorption spectra of gold nanorods as a function of their aspect ratio and the effect of the medium dielectric constant. J. Phys. Chem. B 103, 3073–3077.

Liu, J., Lu, Y., 2005. Fast colorimetric sensing of adenosine and cocaine based on a general sensor design involving aptamers and nanoparticles. Angew. Chem. 45, 90.

Liu, S.-Y., Liang, Z.-S., Gao, F., Luo, S.-F., Lu, G.-Q., 2010. In vitro photothermal study of gold nanoshells functionalized with small targeting peptides to liver cancer cells. J. Mater. Sci. 21, 665–674.

Lok, C.-N., Ho, C.-M., Chen, R., He, Q.-Y., Yu, W.-Y., Sun, H., Tam, P.K.-H., Chiu, J.-F., Che, C.-M., 2007. Silver nanoparticles: partial oxidation and antibacterial activities. JBIC J. Biol. Inorg. Chem. 12, 527–534.

Loo, C., Hirsch, L., Lee, M.-H., Chang, E., West, J., Halas, N., Drezek, R., 2005a. Gold nanoshell bioconjugates for molecular imaging in living cells. Optics Lett. 30, 1012–1014.

Loo, C., Lowery, A., Halas, N., West, J., Drezek, R., 2005b. Immunotargeted nanoshells for integrated cancer imaging and therapy. Nano Lett. 5, 709–711.

Lu, W., Xiong, C., Zhang, G., Huang, Q., Zhang, R., Zhang, J.Z., Li, C., 2009. Targeted photothermal ablation of murine melanomas with melanocyte-stimulating hormone analog—conjugated hollow gold nanospheres. Clin. Cancer Res. 15, 876–886.

Lukianova-Hleb, E.Y., Anderson, L.J.E., Lee, S., Hafner, J.H., Lapotko, D.O., 2010. Hot plasmonic interactions: a new look at the photothermal efficacy of gold nanoparticles. Phys. Chem. Chem. Phys. 12, 12237–12244.

Maier, S.A., 2007. Plasmonics: fundamentals and applications. Springer Science & Business Media, NY, USA.

Maier, S.A., Brongersma, M.L., Kik, P.G., Meltzer, S., Requicha, A.A.G., Atwater, H.A., 2001. Plasmonics—a route to nanoscale optical devices. Adv. Mater. 13, 1501–1505.

Maier, S.A., Kik, P.G., Atwater, H.A., Meltzer, S., Harel, E., Koel, B.E., Requicha, A.A.G., 2003. Local detection of electromagnetic energy transport below the diffraction limit in metal nanoparticle plasmon waveguides. Nat. Mater. 2, 229–232.

Maiorano, G., Sabella, S., Sorce, B., Brunetti, V., Malvindi, M.A., Cingolani, R., Pompa, P.P., 2010. Effects of cell culture media on the dynamic formation of protein—nanoparticle complexes and influence on the cellular response. ACS Nano 4, 7481–7491.

Malmqvist, M., 1999. Biacore: an affinity biosensor system for characterization of biomolecular interactions. Biochem. Soc. Trans. 27, 335–340.

Marambio-Jones, C., Hoek, E.M.V., 2010. A review of the antibacterial effects of silver nanomaterials and potential implications for human health and the environment. J. Nanopart. Res. 12, 1531–1551.

Mie, G., 1908. Beiträge zur optik trüber medien, speziell kolloidaler metallösungen. Annalen der Physik 330, 377–445.

Miller, A.D., 1992. Human gene therapy comes of age. Nature 357, 455–460.

Möller, R., Csaki, A., Köhler, J.M., Fritzsche, W., 2001. Electrical classification of the concentration of bioconjugated metal colloids after surface adsorption and silver enhancement. Langmuir 17, 5426–5430.

Mukherjee, P., Bhattacharya, R., Wang, P., Wang, L., Basu, S., Nagy, J.A., Atala, A., Mukhopadhyay, D., Soker, S., 2005. Antiangiogenic properties of gold nanoparticles. Clin. Cancer Res. 11, 3530–3534.

Mukherjee, P., Bhattacharya, R., Bone, N., Lee, Y.K., Patra, C.R., Wang, S., Lu, L., Secreto, C., Banerjee, P.C., Yaszemski, M.J., 2007. Potential therapeutic application of gold nanoparticles in B-chronic lymphocytic leukemia (BCLL): enhancing apoptosis. J Nanobiotechnol. 5, 1–13.

Murphy, D., Redmond, G., 2005. Optical detection and discrimination of cystic fibrosis-related genetic mutations using oligonucleotide–nanoparticle conjugates. Anal. Bioanal. Chem. 381, 1122–1129.

Murphy, D., O'Brien, P., Redmond, G., 2004. Sub-picomole colorimetric single nucleotide polymorphism discrimination using oligonucleotide–nanoparticle conjugates. Analyst 129, 970–974.

Murphy, C.J., Sau, T.K., Gole, A.M., Orendorff, C.J., Gao, J., Gou, L., Hunyadi, S.E., Li, T., 2005. Anisotropic metal nanoparticles: synthesis, assembly, and optical applications. J. Phys. Chem. B 109, 13857–13870.

Navarro, E., Piccapietra, F., Wagner, B., Marconi, F., Kaegi, R., Odzak, N., Sigg, L., Behra, R., 2008. Toxicity of silver nanoparticles to *Chlamydomonas reinhardtii*. Environ. Sci. Technol. 42, 8959–8964.

Nelayah, J., Kociak, M., Stéphan, O., De Abajo, F.J.G., Tencé, M., Henrard, L., Taverna, D., Pastoriza-Santos, I., Liz-Marzán, L.M., Colliex, C., 2007. Mapping surface plasmons on a single metallic nanoparticle. Nat. Phys. 3, 348–353.

Nicewarner Peña, S.R., Raina, S., Goodrich, G.P., Fedoroff, N.V., Keating, C.D., 2002. Hybridization and enzymatic extension of Au nanoparticle-bound oligonucleotides. J. Am. Chem. Soc. 124, 7314–7323.

Nie, S., Emory, S.R., 1997. Probing single molecules and single nanoparticles by surface-enhanced Raman scattering. Science 275, 1102–1106.

Nie, Z., Liu, K.J., Zhong, C.-J., Wang, L.-F., Yang, Y., Tian, Q., Liu, Y., 2007. Enhanced radical scavenging activity by antioxidant-functionalized gold nanoparticles: a novel inspiration for development of new artificial antioxidants. Free Radic. Biol. Med. 43, 1243–1254.

Niidome, T., Yamagata, M., Okamoto, Y., Akiyama, Y., Takahashi, H., Kawano, T., Katayama, Y., Niidome, Y., 2006. PEG-modified gold nanorods with a stealth character for in vivo applications. J. Control. Rel. 114, 343–347.

Ohtsu, M., Kobayashi, K., Kawazoe, T., Sangu, S., Yatsui, T., 2002. Nanophotonics: design, fabrication, and operation of nanometric devices using optical near fields. IEEE J. Sel. Top. Quant. Electron. 8 (4), 839–862.

Ozbay, E., 2006. Plasmonics: merging photonics and electronics at nanoscale dimensions. Science 311, 189–193.

Pacheco, G., 1925. Studies on the Action of Metallic Colloids on Immunisation (vol. 18). Memórias do Instituto Oswaldo Cruz. pp. 81–149.

Paciotti, G.F., Myer, L., Weinreich, D., Goia, D., Pavel, N., Mclaughlin, R.E., Tamarkin, L., 2004. Colloidal gold: a novel nanoparticle vector for tumor directed drug delivery. Drug Delivery 11, 169–183.

Paciotti, G.F., Kingston, D.G.I., Tamarkin, L., 2006. Colloidal gold nanoparticles: a novel nanoparticle platform for developing multifunctional tumor-targeted drug delivery vectors. Drug Dev. Res. 67, 47–54.

Panyala, N.R., Peña-Méndez, E.M., Havel, J., 2008. Silver or silver nanoparticles: a hazardous threat to the environment and human health. J. Appl. Biomed. 6, 117–129.

Patel, P.C., Giljohann, D.A., Daniel, W.L., Zheng, D., Prigodich, A.E., Mirkin, C.A., 2010. Scavenger receptors mediate cellular uptake of polyvalent oligonucleotide-functionalized gold nanoparticles. Bioconjugate Chem. 21, 2250–2256.

Pavlov, V., Xiao, Y., Shlyahovsky, B., Willner, I., 2004. Aptamer-functionalized Au nanoparticles for the amplified optical detection of thrombin. J. Am. Chem. Soc. 126, 11768–11769.

Pissuwan, D., Niidome, T., Cortie, M.B., 2011. The forthcoming applications of gold nanoparticles in drug and gene delivery systems. J. Control. Rel. 149, 65–71.

Pow, D.V., Crook, D.K., 1993. Extremely high titre polyclonal antisera against small neurotransmitter molecules: rapid production, characterisation and use in lightand electron-microscopic immunocytochemistry. J. Neurosci. Methods 48, 51–63.

Qin, W.J., Yung, L.Y.L., 2007. Nanoparticle-based detection and quantification of DNA with single nucleotide polymorphism (SNP) discrimination selectivity. Nucleic Acids Res. 35, e111.

Qiu, Y., Park, K., 2012. Environment-sensitive hydrogels for drug delivery. Adv. Drug Delivery Rev. 64, 49–60.

Raffi, M., Hussain, F., Bhatti, T.M., Akhter, J.I., Hameed, A., Hasan, M.M., 2008. Antibacterial characterization of silver nanoparticles against *E. coli* ATCC-15224. J. Mater. Sci. Technol. 24, 192–196.

Ravindran, A., Chandran, P., Khan, S.S., 2013. Biofunctionalized silver nanoparticles: advances and prospects. Colloids Surf. B 105, 342–352.

Rosemary, M.J., Maclaren, I., Pradeep, T., 2006. Investigations of the antibacterial properties of ciprofloxacin@ SiO_2. Langmuir 22, 10125–10129.

Saha, B., Bhattacharya, J., Mukherjee, A., Ghosh, A.K., Santra, C.R., Dasgupta, A.K., Karmakar, P., 2007. In vitro structural and functional evaluation of gold nanoparticles conjugated antibiotics. Nanoscale Res. Lett. 2, 614–622.

Sato, K., Hosokawa, K., Maeda, M., 2003. Rapid aggregation of gold nanoparticles induced by non-cross-linking DNA hybridization. J. Am. Chem. Soc. 125, 8102–8103.

Sato, K., Onoguchi, M., Sato, Y., Hosokawa, K., Maeda, M., 2006. Non-cross-linking gold nanoparticle aggregation for sensitive detection of single-nucleotide polymorphisms: optimization of the particle diameter. Anal. Biochem. 350, 162–164.

Schuck, P.J., Fromm, D.P., Sundaramurthy, A., Kino, G.S., Moerner, W.E., 2005. Improving the mismatch between light and nanoscale objects with gold bowtie nanoantennas. Phys. Rev. Lett. 94, 017402.

Schultz, D.A., 2003. Plasmon resonant particles for biological detection. Curr. Opin. Biotechnol. 14, 13–22.

Selvaraj, V., Alagar, M., 2007. Analytical detection and biological assay of antileukemic drug 5-fluorouracil using gold nanoparticles as probe. Int. J. Pharm. 337, 275–281.

Sershen, S.R., Westcott, S.L., Halas, N.J., West, J.L., 2000. Temperature-sensitive polymer–nanoshell composites for photothermally modulated drug delivery. J. Biomed. Mater. Res. 51, 293–298.

Skrabalak, S.E., Chen, J., Au, L., Lu, X., Li, X., Xia, Y., 2007. Gold nanocages for biomedical applications. Adv. Mater. 19, 3177.

Song, K.H., Kim, C., Cobley, C.M., Xia, Y., Wang, L.V., 2009. Near-infrared gold nanocages as a new class of tracers for photoacoustic sentinel lymph node mapping on a rat model. Nano Lett. 9, 183.

Sönnichsen, C., Geier, S., Hecker, N.E., Von Plessen, G., Feldmann, J., Ditlbacher, H., Lamprecht, B., Krenn, J.R., Aussenegg, F.R., Chan, V.Z., 2000. Spectroscopy of single metallic nanoparticles using total internal reflection microscopy. Appl. Phys. Lett. 77, 2949–2951.

Stebounova, L.V., Guio, E., Grassian, V.H., 2011. Silver nanoparticles in simulated biological media: a study of aggregation, sedimentation, and dissolution. J. Nanopart. Res. 13, 233–244.

Stockman, M.I., Bergman, D.J., Anceau, C., Brasselet, S., Zyss, J., 2004. Enhanced second-harmonic generation by metal surfaces with nanoscale roughness: nanoscale dephasing, depolarization, and correlations. Phys. Rev. Lett. 92, 057402.

Storhoff, J.J., Elghanian, R., Mucic, R.C., Mirkin, C.A., Letsinger, R.L., 1998. One-pot colorimetric differentiation of polynucleotides with single base imperfections using gold nanoparticle probes. J. Am. Chem. Soc. 120, 1959–1964.

Storhoff, J.J., Lazarides, A.A., Mucic, R.C., Mirkin, C.A., Letsinger, R.L., Schatz, G.C., 2000. What controls the optical properties of DNA-linked gold nanoparticle assemblies? J. Am. Chem. Soc. 122, 4640–4650.

Sun, Y., Mayers, B.T., Xia, Y., 2002. Template-engaged replacement reaction: a one-step approach to the large-scale synthesis of metal nanostructures with hollow interiors. Nano Lett. 2, 481–485.

Sung, J.H., Ji, J.H., Yoon, J.U., Kim, D.S., Song, M.Y., Jeong, J., Han, B.S., Han, J.H., Chung, Y.H., Kim, J., 2008. Lung function changes in Sprague-Dawley rats after prolonged inhalation exposure to silver nanoparticles. Inhal. Toxicol. 20, 567–574.

Takahashi, H., Niidome, Y., Niidome, T., Kaneko, K., Kawasaki, H., Yamada, S., 2006. Modification of gold nanorods using phosphatidylcholine to reduce cytotoxicity. Langmuir 22, 2–5.

Tan, E., Wong, J., Nguyen, D., Zhang, Y., Erwin, B., Van Ness, L.K., Baker, S.M., Galas, D.J., Niemz, A., 2005. Isothermal DNA amplification coupled with DNA nanosphere-based colorimetric detection. Anal. Chem. 77, 7984–7992.

Tan, E., Erwin, B., Dames, S., Voelkerding, K., Niemz, A., 2007. Isothermal DNA amplification with gold nanosphere-based visual colorimetric readout for herpes simplex virus detection. Clin. Chem. 53, 2017–2020.

Taton, T.A., Lu, G., Mirkin, C.A., 2001. Two-color labeling of oligonucleotide arrays via size-selective scattering of nanoparticle probes. J. Am. Chem. Soc. 123, 5164–5165.

Thomas, K.G., Kamat, P.V., 2003. Chromophore-functionalized gold nanoparticles. Acc. Chem. Res. 36, 888–898.

Tsai, C.Y., Shiau, A.L., Chen, S.Y., Chen, Y.H., Cheng, P.C., Chang, M.Y., Chen, D.H., Chou, C.H., Wang, C.R., Wu, C.L., 2007. Amelioration of collagen-induced arthritis in rats by nanogold. Arthritis Rheum. 56, 544–554.

Tuchin, V.V., 2010. Handbook of photonics for biomedical science. CRC Press, NY, USA.

Urban, M., Möller, R., Fritzsche, W., 2003. A paralleled readout system for an electrical DNA-hybridization assay based on a microstructured electrode array. Rev. Sci. Instrum. 74, 1077–1081.

Velev, O.D., Kaler, E.W., 1999. In situ assembly of colloidal particles into miniaturized biosensors. Langmuir 15, 3693–3698.

Vertelov, G.K., Krutyakov, Y.A., Efremenkova, O.V., Olenin, A.Y., Lisichkin, G.V., 2008. A versatile synthesis of highly bactericidal Myramistin® stabilized silver nanoparticles. Nanotechnology 19, 355707.

Villiers, C.L., Freitas, H., Couderc, R., Villiers, M.-B., Marche, P.N., 2010. Analysis of the toxicity of gold nano particles on the immune system: effect on dendritic cell functions. J. Nanopart. Res. 12, 55–60.

Wang, G., Stender, A.S., Sun, W., Fang, N., 2010. Optical imaging of non-fluorescent nanoparticle probes in live cells. Analyst 135, 215–221.

Weissleder, R., 2001. A clearer vision for in vivo imaging. Nat. Biotechnol. 19, 316–1316.

Wijnhoven, S.W.P., Peijnenburg, W.J.G.M., Herberts, C.A., Hagens, W.I., Oomen, A.G., Heugens, E.H.W., Roszek, B., Bisschops, J., Gosens, I., Van De Meent, D.I.K., 2009. Nano-silver—a review of available data and knowledge gaps in human and environmental risk assessment. Nanotoxicology 3, 109–138.

Wilson, R., 2008. The use of gold nanoparticles in diagnostics and detection. Chem. Soc. Rev. 37, 2028–2045.

Wilson, B.C., 2010. Handbook of Photonics for Biomedical Science/ed. CRC Press, VV tuchin. Boca raton.

Yavuz, M.S., Cheng, Y., Chen, J., Cobley, C.M., Zhang, Q., Rycenga, M., Xie, J., Kim, C., Song, K.H., Schwartz, A.G., 2009. Gold nanocages covered by smart polymers for controlled release with near-infrared light. Nat. Mater. 8, 935–939.

Zhang, C.X., Zhang, Y., Wang, X., Tang, Z.M., 2003. Hyper-Rayleigh scattering of protein-modified gold nanoparticles. Anal. Biochem. 320, 136–140.

Zhao, Z., Wakita, T., Yasui, K., 2003. Inoculation of plasmids encoding Japanese encephalitis virus PrM-E proteins with colloidal gold elicits a protective immune response in BALB/c mice. J. Virol. 77, 4248–4260.

Zijlstra, P., Chon, J.W.M., Gu, M., 2009. Five-dimensional optical recording mediated by surface plasmons in gold nanorods. Nature 459, 410–413.

CHAPTER

INNOVATIVE NONVIRAL VECTORS FOR SMALL-INTERFERING RNA DELIVERY AND THERAPY

26

Danielle C. Arruda, Céline Hoffmann*, Christine Charrueau*, Pascal Bigey*, Virginie Escriou***

**CNRS UMR 8258; INSERM UMR-S 1022; Paris Descartes University; Chimie ParisTech; Paris, France*

***Minas Gerais Federal University, Belo Horizonte, Brasil*

CHAPTER OUTLINE

1 INTRODUCTION

RNAi is a sequence-specific gene-silencing phenomenon mediated by double-stranded RNA. RNAi is made possible through the activity of evolutionarily conserved RNA-protein machinery that is used by microRNAs (miRNAs), endogenously produced noncoding RNAs. This interference effect is mediated by double-stranded RNA (dsRNA), 21- to 25-nucleotide RNA cleavage products of longer dsRNA. Introduction of synthetic versions of these 21-nucleotide RNA duplexes with perfect homology to their target sequences could recapitulate the RNAi effect in mammalian cells and were termed siRNA (small-interfering RNAs). The transfection of RNA duplexes (siRNAs) can induce specific gene silencing in mammalian cells by loading of the duplexes onto a multiprotein complex known as the RNA-induced silencing complex (RISC). Only 6 years have passed between RNAi discovery in 1998 and the first RNAi-based clinical trial in 2004. This extremely rapid development was achieved thanks to a technology well advanced for the manipulation of nucleic acids that arise from the fields of gene therapy and the therapeutic application of antisense oligonucleotides. However, it soon became clear that effective and nontoxic delivery systems are necessary for the effective application of siRNA in a number of therapeutic applications. We present here the various barriers to siRNA administration as drugs and the newly developed strategies of chemical modification and delivery system to pass these barriers. We conclude with a review of the various siRNA-based clinical trials that have already been conducted or are ongoing.

2 OBSTACLES TO IN VIVO siRNA DELIVERY: THE BIOLOGICAL BARRIERS AND THE IMMUNOLOGICAL RESPONSE

An intracellular delivery is necessary for siRNA molecules to trigger RNA interference (Kanasty et al., 2012). In order to reach their intracytoplasmic target, siRNAs have to overcome a number of limiting factors (Scomparin et al., 2015). These factors include several biological barriers at the systemic level and at the cellular level, as well as the immunological response to their administration. In addition, the obstacles to be overcome may vary according to the administration route.

2.1 THE BIOLOGICAL BARRIERS

The main biological barriers encountered in vivo are: (1) siRNA stability, (2) cellular uptake and endosomal escape, and (3) pharmacokinetic features (Juliano et al., 2009; Whitehead et al., 2009).

2.1.1 siRNA stability

SiRNAs are hydrophilic polyanionic macromolecules, relatively small compared to many other biopolymers (size <10 nm, MW ~ 13 kDa), but relatively large compared to classical small molecule drugs (Gallas et al., 2013). SiRNAs have to be stable in the environments to which they are exposed. However, as exogenous nucleic acids, unmodified siRNAs are extremely sensitive to enzymatic degradation by exonucleases and endonucleases in biological fluids. Chemical modifications are useful to increase siRNA stability provided that RNA interference is preserved (Lorenzer et al., 2015).

2.1.2 Cellular uptake and endosomal escape

Effective cellular uptake is one of the most significant barriers for the targeted delivery of siRNAs (Gallas et al., 2013). Due to their physicochemical characteristics, naked siRNAs are unable to cross membranes. The diffusion of these hydrophilic and negatively charged molecules through the lipophilic and anionic cellular membrane is impossible (Scomparin et al., 2015). While the exposure of cultured cells to millimolar concentrations of siRNAs may allow cellular uptake to some extent, such high concentrations cannot be used in vivo due to toxicity issues. Therefore, the biological barrier of cellular uptake has to be overcome thanks to modifications of siRNAs, either chemical modifications or inclusion in adapted delivery systems. Hence, by enhancing siRNA lipophilicity, cholesterol conjugation can increase affinity to cellular membranes (Soutschek et al., 2004). Formulation of siRNAs can efficiently promote their cellular uptake. The endocytosis pathway is then the main mechanism of formulated siRNA cellular uptake (Lorenzer et al., 2015). Whatever the endocytosis pathway, caveolar-dependent uptake, clathrin- and caveolin-independent pathways, or micropinocytosis (Doherty and McMahon, 2009; Huotari and Helenius, 2011; Juliano et al., 2012; Sandvig et al., 2011), siRNAs are finally included in an endosomal vesicle. The next step essential to provide efficient RNA interference into the cytosol is the endosomal escape of siRNAs before their degradation in lysosomes. This step represents a crucial issue yet not completely elucidated and controlled. Hence, the intracellular endosome trafficking is complex, with a great extent of recycling and return of the vesicles to the cell surface, compared to maturation into late endosomes and lysosomes, or delivery of their content to the cytosol (Lorenzer et al., 2015). Enhanced endosomal escape of siRNAs remains to be achieved thanks to adequate delivery systems.

Interestingly, beside endocytosis, another pathway involving possible direct fusion of lipoplexes with the cell membrane has been evidenced. Although minor compared to endocytosis, this pathway allowed functional siRNA delivery since gene silencing remained efficient even after blocking endocytosis (Lu et al., 2009).

2.1.3 Pharmacokinetics

Following IV injection, siRNAs are exposed to serum proteins, nucleases, and cells of the innate immune system (Kanasty et al., 2012). Interactions between siRNAs and serum proteins may be either beneficial or detrimental. On the one hand, delivery outcomes of protein-associated siRNAs can be improved through increased circulation time and enhanced uptake into target tissue. On the other hand, interactions between siRNAs and proteins can lead to the formation of aggregates that are prone to opsonization, or may embolize small capillaries. More commonly, adsorption of the complement system components on siRNAs promotes rapid phagocytosis of the resulting particles by macrophages (Kanasty et al., 2012).

SiRNAs are eliminated by renal filtration whose limits for macromolecules are 2–4 nm in size, that is, 20–40 kDa in weight for nucleic acids (Lorenzer et al., 2015). The phenomenon is rapid with reported serum half-lives of less than 5 min (Gao et al., 2009; Soutschek et al., 2004; Weitzer and Martinez, 2007). This elimination can be efficiently reduced by increasing siRNA molecular size thanks to attachment of ligands, incorporation into larger delivery systems, or significant binding to plasma proteins (Dirin and Winkler, 2013). Besides kidney filtration, the hepatobiliary clearance has been shown to be an important pathway for siRNA elimination (Huang et al., 2011).

Using larger particles to reduce renal elimination has detrimental consequences, since above 200 nm, particles are rapidly cleared by hepatic and splenic phagocytotic cells. This phenomenon can be limited by the surface functionalization of the particles with amphiphilic PEG chains (Lorenzer et al., 2015).

Distribution into organs is determined by the endothelial barrier whose gap junctions vary from extremely tight in blood–brain barrier, to continuously small in standard arteries and capillaries in heart, lungs and muscle, and fenestrated in digestive mucosa, exocrine/endocrine glands and kidney glomerulus (Kanasty et al., 2012). As for the liver, it possesses sinusoidal capillaries characterized by a discontinuous fenestrated endothelium with higher permeability (Lorenzer et al., 2015). This characteristic, along with the well-perfused nature of the liver, may explain the preferential hepatic uptake of siRNA formulations through passive targeting. Though access to hepatocytes is favored by the structure of the hepatic endothelium, phagocytosis by Küpffer cells also accounts for important siRNA uptake by the liver (Kanasty et al., 2012).

To reach the other organs and tissues, an active process has to be implemented, requiring functionalization of the delivery systems with ligands able to bind specifically and with high affinity to receptors of the desired target (Gallas et al., 2013).

Besides healthy tissues, malignant ones can have increased vascular permeability, thereby allowing an easier access of siRNA delivery systems (Kanasty et al., 2012).

2.2 THE IMMUNOLOGICAL RESPONSE

As nonself nucleic acids, siRNAs can be recognized by the innate immune system and induce its response characterized by the production of small signaling molecules, the cytokines (Kanasty et al., 2012).

2.2.1 Recognition of siRNA molecules

Recognition of siRNAs occurs through pattern recognition receptors (PRRs); two general classes of siRNA-recognizing PRRs include Toll-like receptors (TLRs) that react to siRNAs in a sequence-dependent manner, and cytoplasmic receptors reacting to siRNAs in a sequence-independent fashion (Takeuchi and Akira, 2010).

Among Toll-like receptors that recognize structurally conserved regions of foreign pathogens, at least three have relevance to siRNA delivery: TLR3 which recognizes doubled-stranded RNAs, and TLR7 and TLR8 which recognize single-stranded RNAs. Their location is endosomal and on the cell surface of selected cell populations for TLR3, while TLR7 and TLR8 are located exclusively in the intracellular vesicles including endosomes, lysosomes, and the endoplasmic reticulum of plasmacytoid dendritic cells, B cells, and myeloid cells (Marques et al., 2006).

In addition, several cytoplasmic receptors mediate immune responses to siRNAs. They include the protein kinase PKR (Marques et al., 2006), and the cytoplasmic PRR RIG-I, expressed in fibroblasts and dendritic cells (Kato et al., 2005), both causing an interferon response.

2.2.2 Activation of the immunological response

Activation of the previously cited receptors triggers an immune signaling cascade that leads to an increased transcription of mRNA that encodes type I interferons and inflammatory cytokines. It is difficult for siRNAs to escape immunostimulation. However, the nature and degree of immunostimulation can be modulated by the siRNA sequence, the siRNA structure and chemistry, and the composition of the delivery system (Kanasty et al., 2012). Hence, sequences rich in guanosine and uridine provoke stronger immunostimulation than sequences substituted with adenosine (Heil et al., 2004). The chemical structure of siRNA duplex, as well as the nucleotide structure can also increase or reduce, or even eliminate the innate immune response. Finally, siRNA delivery vehicles play a role in siRNA-mediated immune activity. By promoting siRNA cellular uptake through endocytosis, the most frequently employed materials for siRNA delivery, that is, cationic and lipid materials, increase siRNA exposure to TLRs of the endosomal compartments, and to RIG-I and PKR cytoplasmic receptors. Therefore, they may raise immunostimulation. The intensity and nature of the siRNA-induced immune response may also be altered by the size, charge, and biodistribution of the delivery systems (Kanasty et al., 2012).

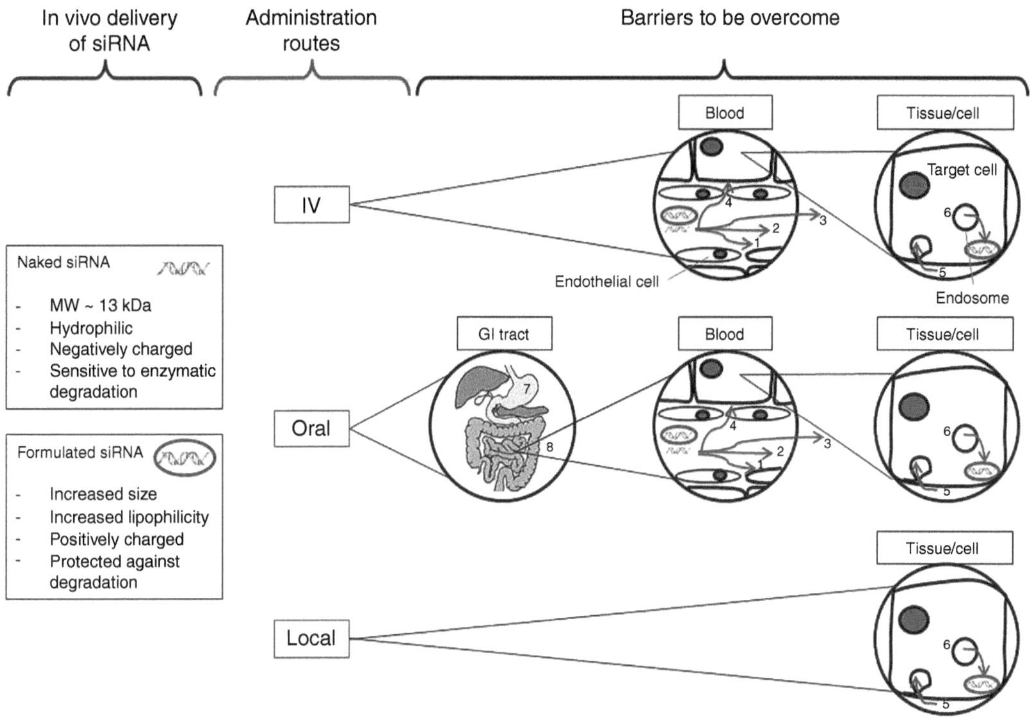

FIGURE 26.1 The Challenge of siRNA Delivery According to the Administration Route: Crossing Successive Biological Barriers and Overcoming the Immunological Response to Attain the Intracytoplasmic RNAi Machinery and to Trigger the Sequence-Specific mRNA Degradation

1, Degradation by nucleases; 2, immunostimulation; 3, renal elimination; 4, endothelial barrier; 5, cellular uptake; 6, endosomal escape; 7, degradation in GI fluids; and 8, gut absorption.

2.3 THE BARRIERS TO BE OVERCOME ACCORDING TO THE ADMINISTRATION ROUTE

Tissue targeting represents a significant challenge for therapeutic applications of siRNA formulations. Obstacles to siRNA delivery vary according to the administration route (Fig. 26.1). The systemic administration by the IV route currently represents the simplest delivery strategy (Borna et al., 2015). It first exposes siRNAs to blood and then to tissues/cells. If adequately formulated, siRNAs may overcome the cellular uptake and endosomal escape barriers to provide efficient cell delivery. However, besides passive targeting of the liver following IV injection that allows perspectives in the treatment of hepatic diseases, expansion of the target space to other organs and tissues is needed (Gallas et al., 2013). The oral administration route offers several advantages over the IV route, including ease of administration due to noninvasive access, along with the versatility to treat local diseases as well as systemic ones (O'Neill et al., 2011). However, this route of administration adds further barriers to siRNA delivery compared to the IV route: (1) exposure to the acidic gastric environment and to the enzyme-rich gastrointestinal tract responsible for rapid degradation; (2) exposure to mucus layer and glycocalyx that must be crossed prior to reaching the underlying epithelium; and (3) absorption through the gut mucosa

(O'Neill et al., 2011). Once absorbed, the subsequent barriers meet those encountered following IV administration. Finally, the direct administration into the site of interest by local routes, such as intradermal, subcutaneous, pulmonary, intraocular routes, for example, is currently the most straightforward way to reach the target tissue. Barriers to siRNA delivery are reduced when local administration routes are used. Hence, neither gastrointestinal challenge nor systemic circulation are involved. In addition, compared to systemic delivery methods that need high amounts of siRNAs, localized delivery required lower amounts that may not impact the whole organism (Pardridge, 2007).

To conclude this first part, the clinical use of siRNAs necessitates the improvement of their stability, and their inclusion in adequate delivery systems capable of protecting them from degradation, to promote their cellular uptake and their endosomal escape, and to overcome immune response stimulation, in order to enable safe and efficient RNA interference in vivo (Scomparin et al., 2015). Formulation strategies shall take into account the constraints linked to the administration route envisaged.

3 CHEMICAL MODIFICATIONS OF siRNA

The global efficacy of a siRNA depends on a number of factors, including stability, bioavailability, specificity, and potency on one side and adverse effects like off-target downregulation or triggering of an immune response on the other side. Naked siRNAs are cleared by kidneys in less than 1 h and degraded by nucleases in human serum within minutes. Furthermore, their negative charges prevent them from efficiently entering the cell. Once into a cell, they can last for weeks (unless the cells divide), and as catalysts, only a few hundred copies will result in a long-lasting gene knockdown. Then, their intrinsic properties only will determine their efficacy. Some of these barriers may be overcome as already described, but working on the chemistry of the siRNA seems, however, necessary to overcome others (Engels, 2013).

Most of the siRNAs used are of synthetic origin. As the chemical synthesis of such compounds is easy and well handled, it allows the incorporation of a variety of modified nucleosides at the desired position (Fig. 26.2). The challenge then is to improve the in vivo properties of stability (nuclease degradation), potency, and specificity, and to diminish the off-target and induced inflammation effects, without interfering with the RISC-induced silencing of gene expression. Modifications can occur on the sugar moiety, the nucleobase, the backbone, or at the ends as conjugates. Conjugates usually aim at improving targeting and delivery. In this case, modifications on the passenger strand are preferred since the 5′-phosphate on the guide strand is necessary for the siRNA activity (Nykänen et al., 2001), and modification at the 3′ end is controversial since it sometimes impairs the siRNA function. The chemical modifications should not disturb the structure of the double helix, which is necessarily of A-type.

3.1 STABILITY

The easiest way of protecting siRNAs against nucleases is to modify the internucleotide phosphodiester linkage. The most common modification is phosphorothioate PS which has been known for a long time to bring strong nuclease resistance. However, fully modified phosphorothioate siRNAs are not used since a few groups reported a decreased activity and some toxicity. However, phosphorothioate modification is still widely used, particularly at the two overhang positions of each strand. Stability is also

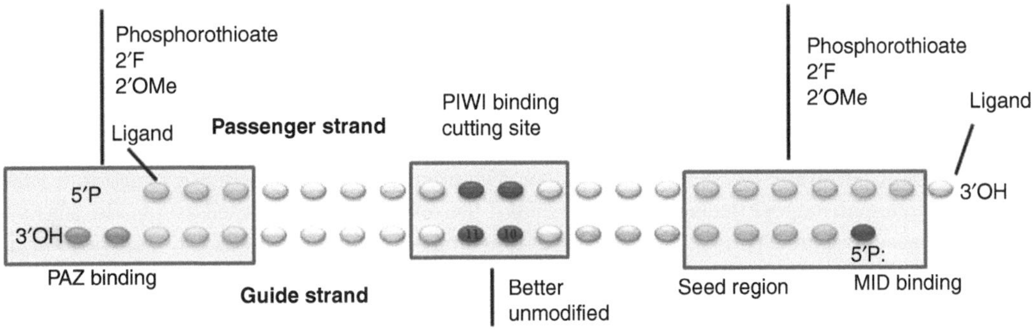

FIGURE 26.2 Chemical Modifications of siRNA

The main chemical modifications and their sites are shown. The cleavage site is preferred unmodified. 5′P, 5′ phosphate end; phosphorothioate, phosphorothioate bond; 2′-OMe, OMethyl substitution in position 2′ of the ribose moiety; 2′-F, fluoro substitution in position 2′ of the ribose moiety; ligand, end modification is preferred for a targeting ligand. PIWI, MID, and PAZ refer to the binding domains of the human AGO2 nuclease of the RISC complex.

improved by modification on the 2′ position of the sugar, since 2′-OH is not required for RISC-induced activity. A number of chemical modifications have been suggested, but most of the siRNA entering clinical trials contain 2′OMe and 2′F motifs. 2′OMe not only increase nuclease stability, but also the binding affinity for the complementary strand (Deleavey and Damha, 2012). However, fully 2′OMe-modified siRNAs have been shown to be inactive, particularly in the guide strand because it disrupts the type A helix. Therefore, it is of great interest to alternate 2′OMe nucleosides with 2′F nucleosides. Although 2′F modification does not bring nuclease resistance by itself, mixing the two mentioned modifications results in an improved resistance without decreasing the siRNA activity. Both modifications can be incorporated in both strands, although modification at the cleavage site of the passenger strand is not recommended, and are already FDA-approved (found in Macugen, which is an approved anti-VEGF aptamer used to treat wet AMD, Ng et al., 2006).

3.2 POTENCY, SPECIFICITY, AND OFF-TARGET EFFECTS

In addition to the knockdown of the intended gene, siRNAs might have adverse effects called "off-target" effects. Two kinds of off-target silencing have been described (Vaish et al., 2011). One is due to incorporation of the passenger strand onto RISC and results in the cleavage of the sequence complementary to the normal target. Therefore, correct loading of the guide strand is critical. The other one is an miRNA-like downregulation and is caused by the partial complementarity of the guide strand to a nontargeted mRNA, particularly in the seed region. It can be caused by either strand of the siRNA.

It has been demonstrated that the strand containing the least stable 5′-end is preferentially utilized as the guide strand in RISC. Then modifying the relative stabilities of the two ends of the siRNA duplex may be a way to bias a correct strand loading. The simplest way is destabilizing the 5′ end of the guide by introducing a mismatch. SiRNAs having a blunt end at the 3′ end of the passenger and a 2 bases 3′-overhang at the 3′-end of the guide also allow correct loading on RISC (Rose et al., 2005). Furthermore, the blunt end may also increase the knockdown potency.

Finally, some chemical modifications of siRNA strand-end favor a correct loading by incorporation of stabilizing modifications in the 5′-end of the passenger or the 3′-end of the guide, or by incorporation of destabilizing modifications in the 3′ end of the passenger (Bramsen et al., 2009).

Chemical modifications of siRNA overhangs can also be utilized to favor guide selection during RISC loading irrespective of the thermodynamic asymmetry of the duplex. Particularly, LNA near the 5′ end of the sense strand has been shown to favor incorporation of the antisense strand into RISC, thereby reducing off-target effects that would result if the sense strand were used as a guide. LNAs, or locked nucleic acids, are chemically modified RNA analogs containing a 2′-OH C4′ methylene bridge which gives a conformation that makes hybridization with the complementary strand more energetically favorable, and confers a high stability to the duplex (5–6°C per modified base).

A correct loading of the guide strand is important since it avoids saturation of the RNAi machinery and increases the potency of the siRNA and subsequently its specificity. It also allows using lower dose of the compound, thereby decreasing the miRNA like off-target effect.

This latter effect can be abolished by using 2′OMe modification in the siRNA. Remarkably, a single 2′OMe modification at position 2 of the guide strand strongly reduced off-target silencing without affecting the inhibition of the intended target (Jackson et al., 2006).

3.3 IMMUNOGENICITY

Another adverse effect of siRNAs is the activation of an innate immune response mainly through TLR7 recognition and production of interferon alpha and proinflammatory cytokines IL6 or TNF alpha (Judge et al., 2005), particularly when they are formulated with lipids. If possible, the immunostimulatory motifs should be eliminated in the sequence design of the siRNA. Also, in addition to increasing the nuclease stability, 2′-OMe modification on one strand has been extensively used to generate fully functional siRNAs with reduced immunogenicity by preventing the TLR recognition (Morrissey et al., 2005).

3.4 CONJUGATES FOR TARGETING

Covalently conjugating a siRNA to a ligand offers several advantages: mainly, adding resistance to endonucleases and allowing targeting to a specific organ, thereby reducing off-target effects by the use of lower doses and limiting dissemination to unwanted cell types. Furthermore, a unique chemically synthetized compound might be easier to handle from a regulatory point of view than a multicomponent mixture. Finally, conjugation should not interfere with the siRNA-silencing efficacy since it should be possible to modify only the passenger strand, either at the 5′ or the 3′ position. SiRNAs have been conjugated to RNA aptamers, antibodies, receptor ligands, and small molecules (cholesterol, folic acid, galactose) (see Wittrup and Lieberman, 2015 for reviews). Conjugation of the 3′ end of the passenger to cholesterol allowed a prolongation of the siRNA half-life in the serum by incorporation into circulating lipoproteins, which bring resistance to nucleases and prevent kidney elimination. More than 50% downregulation of ApoB mRNA in the liver was observed, as well as a 70% drop of apolipoprotein B in circulation (Soutschek et al., 2004).

Recently, the Alnylam company developed a chemically modified siRNA (2′-OMe, 2′-F, and phosphorothioate) conjugated to a trivalent *N*-acetylgalactosamine (GalNac), which is a high-affinity ligand of the asialoglycoprotein. Conjugation was performed at the 3′ end of the passenger strand by solid phase synthesis. The compound was stable against nucleases and has improved pharmacokinetics compared to the unconjugated siRNA. Moreover, a robust and durable liver silencing of the targeted gene was observed after single or multiple subcutaneous administrations (up to 9 months) (Nair et al., 2014).

A new Alnylam proprietary chemical modification of the siRNA called ESC improves the efficacy and duration of the siRNA knockdown when conjugated with the same GalNac, and also allows subcutaneous administration of the conjugate (Manoharan, 2014).

Another conjugate strategy is developed by Arrowhead Research under the name of Dynamic Poly-Conjugates (DPCs) which aims at improving endosomal release by using multifunctional macromolecules that incorporate both targeting ligands and endosomal escape moieties (masked endosome-disrupting peptides or polymers) (Rozema et al., 2007). Briefly, a polymer grafted with PEG moieties and a ligand like GalNac is covalently linked to the 3′ end of the passenger strand via a reversible disulfide bond. In the acidic endosome, the polymer will become positively charged, allowing cytosolic escape of the siRNA. The resulting effect in the liver was 87% inhibition of the ApoB mRNA with a 2.5 mg/kg dose. The effect was observed for about 10 days, and resulted in lower serum cholesterol and ApoB protein levels.

Then, it appears that using chemical modifications in the design of a potent siRNA is critical, as it will mainly increase the siRNA stability in biological fluids, may improve the silencing efficiency and lower the different off-target effects by organ-specific targeting and avoiding immune response. However, there is no "magic recipe" for the design of a potent siRNA since different sets of modifications should be tried. Furthermore, the next great challenge for chemical modifications of siRNA will probably be to design conjugates able to target other tissues than the liver.

4 siRNA DELIVERY SYSTEMS: FOCUS ON LIPOSOMES AND LIPOPLEXES

Due to their instability, the efficient delivery of siRNA may be improved by vectorization that can be either viral or nonviral. Viral carriers present the advantage of being much more efficient in comparison to nonviral carriers, but the main drawback is their potential immunogenicity that prevents them from getting translated into clinical trial. To overcome this limitation, nonviral carriers have been developed during the last decades, often based on cationic polymers or liposomes, called polyplexes or lipoplexes, respectively. Typical siRNA-delivery systems are resumed in Fig. 26.3.

4.1 POLYMER-BASED siRNA CARRIERS

Negatively charged siRNA can bind via electrostatic interaction with cationic polymers and form complexes called polyplexes. The polymers used as formulating agents can be subdivided into two main categories: synthetic polymers like polyethyleneimine (PEI), dendrimers, and poly-L-lysine (PLL) and natural polymers, more easily degraded by the organism like chitosan and cyclodextrins (Mehrabadi et al., 2012; Pack et al., 2005).

4.1.1 Polyethylenimine

PEI was one of the first transfecting agent discovered and one of the cationic polymers that has been more explored as a polymeric agent for siRNA administration (Boussif et al., 1995; Taranejoo et al., 2015).

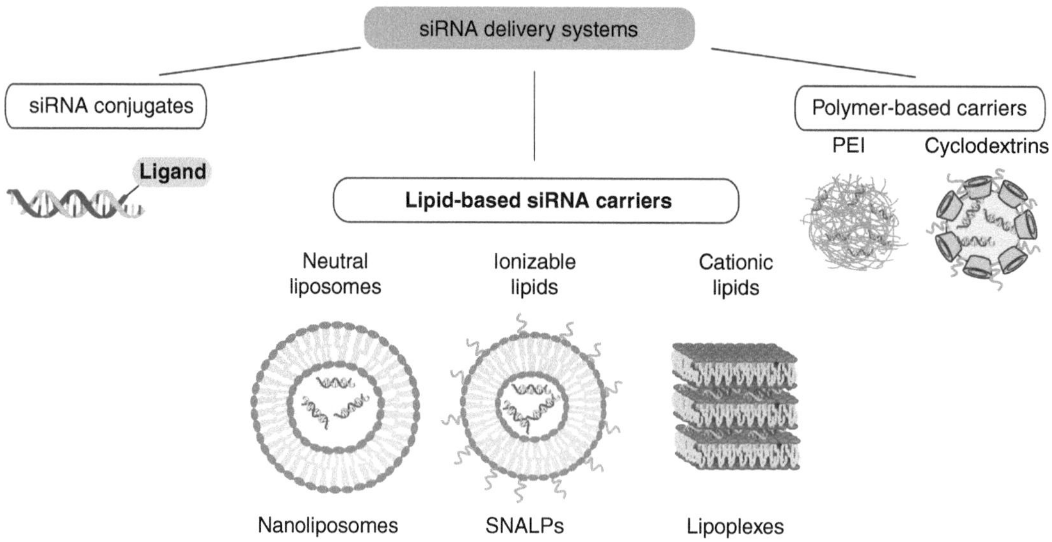

FIGURE 26.3 Schematic of siRNA Delivery Systems

Several siRNA vectorization systems have been developed for siRNA in vivo administration and can be resumed in these 3 categories: siRNA conjugates, polymer based-carriers, and lipid-based carriers.

PEI is an alkyl chain with primary, secondary, and tertiary amines. Only a portion of the amines of PEI is protonated at physiological pH, so the PEI has a high buffering capacity and allows the release of the nucleic acid in the acidic environment of endosome by the so-called mechanism of "proton sponge" (Akinc et al., 2005; Boussif et al., 1995). When polyplexes are absorbed by endocytosis into the cell the amine protonation increases from 20% to 45% when the endosomal pH drops from 7 to 5. Buffer power of PEI causes an influx of water to protons and chlorine ions in the endosome that makes the burst, releasing the contents of the PEI in the cytoplasm.

PEI exists in linear or ramified forms and in different molecular weights. PEI is an efficient transfecting agent, but suffers from high cytotoxicity, depending on its molecular weight. Several studies demonstrated that the association of PEI with other polymers, like hyaluronic acid or polyethylene glycol (PEG) can improve its safety while keeping its efficacy (Dahlman et al., 2014; Ito et al., 2010; Jiang et al., 2008). Moreover, the use of PEI to administer siRNA has been recently approved to translate into clinical trial (SNS01-T—NCT01435720) (Francis et al., 2014).

4.1.2 Poly-L-lysine

Poly-L-lysine is a polypeptide able to form stable complexes with nucleic acids allowing the delivery of nucleic acid into cancer cells (Emi et al., 1997; Wu and McMillan, 2009). But PLL is highly toxic and presents poor endosomal escape ability. To improve the efficacy and decrease the toxicity of such carriers, they are often used in combination with other polymers like PEG (Lopez-Fraga et al., 2009). This kind of combination has been demonstrated to allow the accumulation of vector in tumor and metastasis after IV injection (Kano et al., 2011).

4.1.3 Dendrimers

Cationic dendrimers like polyamidoamine (PAMAM, the more used) are branched multilayer polymers. Several concentric layers build up from the central core of the polymer. The size and number of layers can be easily controlled and influence the density of charge of the polymer, as well as its capacity to complex nucleic acids. Dendrimers can be modified either on the core composition or at the periphery of the structure, to improve the biocompatibility, the interaction with cell membrane, and the siRNA delivery to the cell (Mehrabadi et al., 2012). Several studies have demonstrated that in addition to protection of the formulated siRNA from degradation by nucleases, significant transfection of the cells was observed in vitro (Tang et al., 2012).

4.1.4 Cyclodextrins

Cyclodextrins are polymers composed of cyclic oligosaccharides. Cyclodextrins are biocompatible, have a low toxicity and a low immunogenicity. Cationic polymer of cyclodextrin can be associated with conjugates like adamantine-PEG to allow targeting moiety complexation and to cross biological barriers (Davis, 2009). Imidazole groups attached to the ends of the cyclodextrin polymer can facilitate the release of the endosome by their buffering effect. With a slightly positive charge ratio of 1 to 1.5, such cyclodextrin conjugates can be complexed with siRNA, leading to the total protection of siRNA against nucleases reaction. The complexes formed are in the range of 70 nm and have a surface charge from −10 to +10 mV. The size and charge of the complex are therefore compatible with the characteristics required to prevent renal excretion, to promote extravasation and distribution in the organs. RONDEL, a cyclodextrin-based delivery system was the first targeting carrier approved for clinical trial in humans in 2008 (CALAA-01, Calando Pharmaceuticals) (Davis et al., 2010).

4.1.5 Chitosan

Chitosan has been widely described in the formulation of nucleic acids, such as siRNA or DNA because of its cationic nature, low toxicity, and biocompatibility (Mao et al., 2010). Chitosan is a linear polysaccharide of randomly distributed *N*-acetylglucosamine and glucosamine units. SiRNA can form complexes with chitosan either by simple complexation or by ionic gelation (thanks to a cross-linker agent) resulting in chitosan-based nanoparticles loaded with siRNA. Chitosan is biodegradable in the human body. This property is highly recommended for further clinical development (Baldrick, 2010; Ragelle et al., 2013). It has been demonstrated that chitosan alone is not efficient enough to promote endosomal rupture and siRNA release in the cytoplasm (Kim et al., 2005). To overcome chitosan limitations and to improve or impart new properties to the polymer, further developments of chitosan-based delivery systems were developed by grafting or mixing additional materials to the formulation, such as peptides, polymers, or ligands (Ragelle et al., 2013). Recent studies demonstrate the promising potential of chitosan-based nanoparticles for siRNA for silencing specific gene in tumor-bearing mice (Han et al., 2010; Huh et al., 2010; Lee et al., 2012).

4.2 LIPIDIC siRNA DELIVERY SYSTEMS—LIPOPLEXES

Encapsulation of siRNA in lipid vectors presents many advantages: (1) protection of siRNA from enzymes present in the biological fluids, (2) increasing of siRNA stability and half-life in biological medium, (3) improving the properties of siRNA to cross the biological barriers, and (4) decreasing the quantity of siRNA necessary for effective gene-expression extinction (Wu and McMillan, 2009).

Lipids comprise a hydrophilic head, a spacer, and a hydrophobic tail. This amphiphilic nature permits them to adopt the complex three-dimensional structures of liposomes. Depending on the nature of the lipids used, the liposomes adopt different morphologies and sizes, the most common being unilamellar vesicle, analogous to the membrane of eukaryotes. This vesicle is characterized by a lipid bilayer, which contains an aqueous core. Several methods exist for forming liposomes, for example, hydration of a lipidic film or ethanol injection (Li and Szoka, 2007).

Two types of liposomes may be used for the transfection of nucleic acids: neutral liposomes and cationic liposomes. These two types of liposomes interact differently with nucleic acids either by encapsulation or by complexation. Several lipids have been developed for transfection of nucleic acids and are summarized in Fig. 26.4.

4.2.1 Neutral lipids

The neutral liposomes are formed of neutral lipids, such as DOPC (1,2-dioleoyl-*sn*-glycero-3-phosphatidylcholine). Mechanism of interaction between neutral liposomes and siRNA is encapsulation, with siRNA encapsulated into the aqueous core of liposome. SiRNA encapsulation may be realized either by lipid hydration or ethanolic injection and thin film evaporation, with an aqueous solution containing siRNA. This kind of carrier showed a relatively low siRNA encapsulation efficiency compared to cationic liposomes ($<10\%$) (Bruno, 2011).

However, neutral liposomes composed of DOPC have been demonstrated to form "nanoliposomes" with an average size of 65 nm, and allow the efficient administration of siRNA to tumor cells. A single IV injection of 150 μg/kg inhibits expression of the target protein (EphA2) for 48 h in the ovarian tumor (Landen et al., 2005, 2006; Mangala et al., 2009). This formulation is currently in Phase I clinical trial (NCT01591356).

4.2.2 Ionizable lipids

To overcome troubles about low encapsulation of siRNA in liposome, a key advance was the use of ionizable lipids as DODAP or DlinDMA (dilinoleoyl-DMA), associated with DSPC or DOPE (Heyes et al., 2005; Lin et al., 2011). At pH 4, the lipid is positively charged and interacts with the negatively charged oligonucleotides. At pH 7, the liposome becomes neutral and the associated siRNA dissociates. This technique allows encapsulation of up to 70% of oligonucleotides. This formulation significantly increases the circulation time in the blood of the siRNA relative to a formulation with a cationic liposome. This type of formulation is currently used in several clinical trials with the vectors SNALP (Patisiran ALN-TTR02—NCT01960348) (Ramanathan et al., 2014).

4.2.3 Cationic lipids

4.2.3.1 Complexation of cationic lipids with siRNA—mechanisms

Complexation of siRNA with cationic liposomes is a result of electrostatic interactions between the negatively charged phosphate backbone of the siRNA and the positively charged groups of the carrier. The interaction between lipid and nucleic acid leads to a self-assembled structure that can present either a multilamellar or a hexagonal structure, depending on the nature of lipid and the charges ratio (Dan and Danino, 2014). SiRNAs associated with cationic liposomes are called lipoplexes. The global positive charge of the lipoplexes facilitates their interaction with the components of the negatively charged cell membrane and allows the delivery of their contents inside the cell efficiently.

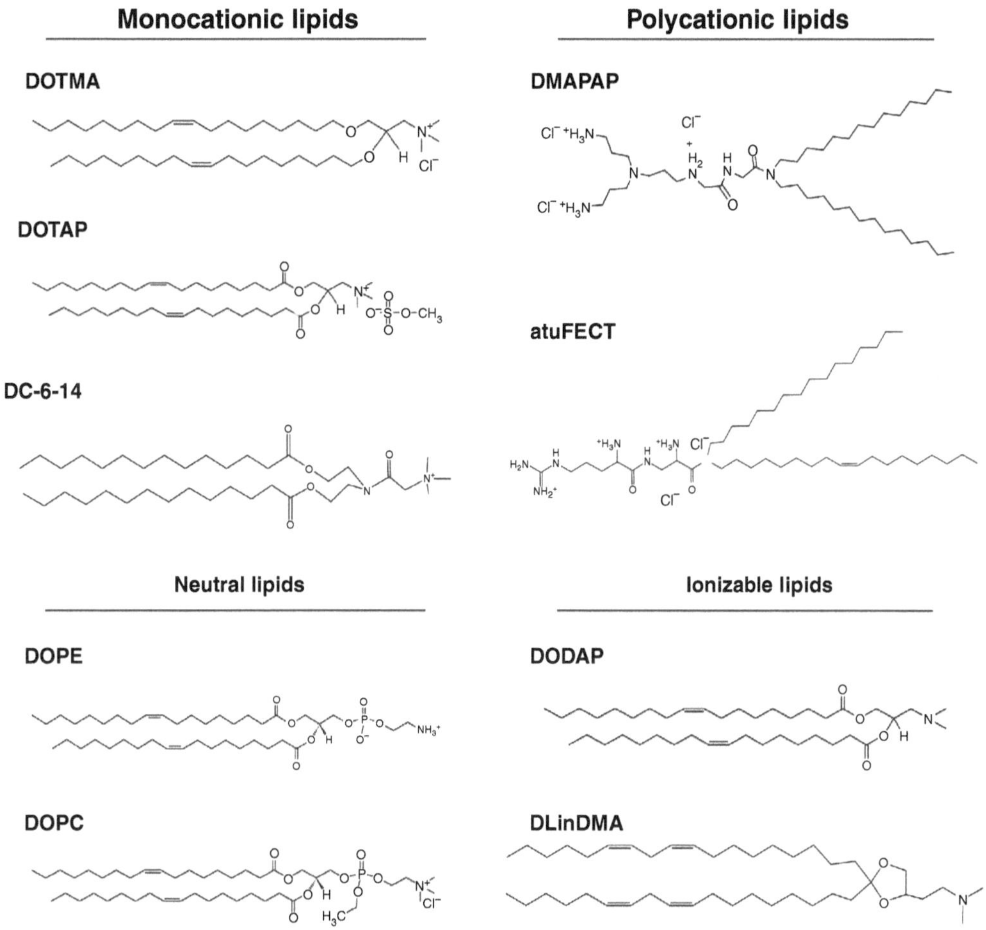

FIGURE 26.4 Examples of Lipids Used for siRNA Lipid Delivery System Design

Monocationic lipids: *DOTMA*, 1,2-di-O-octadecenyl-3-trimethylammonium propane; *DOTAP*, 1,2-dioleoyl-3-trimethylammonium-propane; *DC-6-14*, O,O-ditetradecanoyl-*N*-(a-trimethylammonioacetyl) diethanolamine chloride. Polycationic lipids: *DMAPAP*, Di-Miristyl AminoPropylAminoPropyl); *atuFECT*, β-L-Arginyl-2,3-L-diaminopropionic acid-*N*-palmityl-*N*-oleyl-amide trihydrochloride). Neutral lipids: *DOPC* (1,2-dioleoyl-*sn*-glycero-3-ethylphosphocholine; *DOPE*, 1,2-dioleoyl-*sn*-glycero-3-phosphoethanolamine. Ionizable lipids: *DODAP*, 1,2-dioleoyl-3-dimethylammonium-propane; *DlinDMA*, 2,2-dilinoleyl-4-dimethylaminoethyl-[1,3]-dioxolane.

A significant amount of work has been conducted to understand the formation of cationic lipids/siRNA complexes and their mechanism of action (Wasungu and Hoekstra, 2006).

The effect of lipid composition and headgroup charge on the biophysical and biological properties of cationic lipid-associated siRNA vectors was determined by X-ray diffraction. This study revealed that complexes exhibited lamellar and inverted hexagonal phases, qualitatively similar to DNA complexes, but that other nonlamellar structures are formed (Bouxsein et al., 2007). Consistent with these results, another study demonstrated that all lipoplexes prepared with different cationic liposomes showed multilayered structures with interlamellar spacing of 7 nm (sum of thickness of lipid bilayer and siRNA). But depending on the nature of lipids, irregular structures of size around 60 nm or concentric lamellar structures (200–500 nm) were observed. The authors determined that lipids that allowed forming large liposomes helped in forming lamellar structures over a long range (onion ring like) while smaller liposomes helped in forming short range ordered structures (Desigaux et al., 2007). Finally, cationic-lipid associated siRNA complex structures appeared more difficult to characterize and more delicate than DNA complexes.

The use of cationic liposomes was first envisaged with DOTMA lipid for the transfection of plasmids in 1987 (Felgner et al., 1987). Since the discovery of DOTMA, other cationic lipids were synthesized for siRNA association, such as DOTAP, DLinDMA, AtuFECT, DMAPAP, or DC-6-14 most of the case along with a neutral "helper" lipid, such as DOPE or DOPG (Kapoor and Burgess, 2012) (Fig. 26.4). Each cationic lipid has a different structure depending on the nature of the polar head and the hydrophobic tail. The influence of several parameters on gene-expression silencing by cationic lipid-based carrier have been investigated, including the lipid:nucleic acid ratio of mixing, the extent of cationic liposome/lipid–nucleic acid complex (lipoplex) formation plus medium used, the lipoplex particle size, the mode of delivery, and dose-response effects (Spagnou et al., 2004). Contrary to the other types of molecules, the efficacy of lipoplexes is not influenced by the size of the particle but rather by the properties of the overall self-assembled system and notably, concentrations during siRNA lipoplex formation are a crucial parameter for maximum knockdown (Ewert et al., 2005).

4.2.3.2 Internalization of lipoplexes

Internalization of lipoplexes occurs mainly by endocytosis, and the endosomal escape is a delicate and essential step of lipofection mechanism to avoid nucleic acid degradation into lysosomes. Addition of neutral "helper" lipids, such as dioleoylphosphatidylethanolamine (DOPE) in cationic liposomes promotes the endosomal membrane rupture, the release of siRNA in the cytosol, and improves the transfection efficacy (Farhood et al., 1995; Hafez et al., 2001).

In addition, it is likely that when the lipoplexes are transferred to the endosomal compartment, the lipid part of the lipoplex fuses with the endosomal membrane due to the formation of ion pairs between the positively and negatively charged lipids. This would cause the transition of lipoplexes to a hexagonal phase and would free the nucleic acid into the cytoplasm (Ma et al., 2007; Tseng et al., 2009; Zelphati et al., 1998).

Various commercial liposomal preparations were used or adapted for the siRNA transfection as Lipofectin (DOTMA/DOPE), DMRIE-C, or lipofectamine (DOSPA/DOPE). However, the use of such transfection agents is recommended in the absence of serum and no systemic administration has been reported so far (Bruno, 2011). Thus, the optimization of in vitro transfection conditions for siRNA lipoplexes was extensively studied (Dass, 2002; Spagnou et al., 2004). Several parameters influence transfection efficiency, such as the charge ratio +/−, the size of the complex, their structure, and the presence of serum (Escriou et al., 1998; Lin et al., 2003; Ma et al., 2007; Ross and Hui, 1999; Zelphati et al., 1998).

Due to their general positive charge, lipoplexes are rapidly cleared from the circulation when administered intravenously. This is due to opsonization leading to recognition by the reticuloendothelial system (liver, spleen) and to the distribution in the lungs (Bennett et al., 1996).

4.2.3.3 Lipoplexes for in vivo delivery

Various improvements have been made to liposomes to strengthen their potential in vivo, in particular, addition of polyethylene glycol at the surface of lipoplexes. This can be performed either by simple adsorption or bonding covalently with a neutral lipid (DSPE-PEG).

The presence of PEG reduces the clearance of lipoplexes and their aggregation; therefore pegylated lipoplexes show a better bioavailability. Due to decreased immune response and increase in circulation time associated to pegylated liposomes, these particles are called "stealth liposomes." However, the presence of PEG reduces the transfection efficiency by preventing the interaction of the lipid with the cell membrane upon internalization within endosomes (Shi et al., 2002). PEG prevents good dissociation by stabilizing the complex in lamellar phase, and it also prevents the necessary interaction between the lipoplexes and the endosome membrane and thus melting (Holland et al., 1996; Remaut et al., 2007; Song et al., 2002).

To develop stealthy but effective lipoplexes, several solutions have been proposed. The first one is the use of lipids PEG-ceramide with a little aliphatic tail so they can easily get out of the lipoplex and turn into an ineffective stealth lipoplex (Mok et al., 1999; Peeters et al., 2007; Shi et al., 2002). An alternative is to use a lipid in which the PEG is linked by a hydrolyzable bond at acidic pH. In the endosome, which has a pH of around 5–6, the link between PEG and lipoplex is cleaved, allowing lipids of the lipoplex to interact and fuse with the membrane, and to release the contents of the lipoplex in the cytosol (Guo and Szoka, 2001; Simoes et al., 2004).

However, these alternatives lack stability in vivo and a compromise must be found between stealth and efficiency. A final option is to use a cleavable spacer between PEG and lipid that would be cleaved in the targeting region, such as the tumoral environment (Rozema et al., 2015). The tumor stroma is acidic and enzymes, such as metalloproteinases are highly expressed, the substrate can be used as a hydrolyzable bond. The use of such liposomes shows better activity in the tumor after IV injection (Hatakeyama et al., 2007).

4.2.3.4 Specific targeting carriers

Active targeting can be achieved by functionalizing the siRNA lipoplexes with a ligand that specifically recognizes a receptor at the cell surface allowing internalization. Such approach has been reported for many different siRNA formulations in vivo, using ligands, such as small molecules, peptides, antibodies, or proteins.

Antibody directed against transferring receptor conjugated to lipoplexes showed effective targeting to primary tumor and metastasis in a breast cancer model. Moreover, these targeting lipoplexes accumulate less in the liver than nontargeting lipoplexes (Pirollo et al., 2006, 2007). Conjugation of lipoplexes with cyclo Cyclo(Arg-Gly-Asp-D-Phe-Lys) (cRGD), a well-known ligand to $\alpha_V\beta_3$ integrin, led to a selective extinction of target gene expression in tumor endothelial cells (TEC). A substantial delay in tumor growth was observed after three sequential RGD-conjugated lipoplexes injections. These lipoplexes represent a promising approach for the characterization of TECs and for antiangiogenic therapy (Sakurai et al., 2014).

Another interesting way to target solid tumor has been recently demonstrated by the use of hyaluronic acid as a ligand of CD44 receptor, which is overexpressed in a wide variety of solid tumors and is involved in cancer progression and metastasis. This study put in evidence that lipoplexes conjugated to hyaluronan can be delivered through blood circulation and extravasation, and target effectively CD44-positive tumor cells in mice (Qhattal et al., 2014).

Small molecules like vitamins, conjugated to siRNA lipid-carriers may also be used to target specific cell receptors with an interesting efficiency in vivo. Codelivery of Doxorubicin and Bmi1 siRNA to tumor cells by folate-conjugated liposomes was demonstrated to exhibit a higher efficacy of Bmi1 expression silencing both in vitro and in vivo on subcutaneous model of tumor in mice (Yang et al., 2014). Recently, another study using this strategy concerned the specific delivery of siRNA with liposomes conjugated to vitamin A as a ligand to target hepatic stellate cells (HSC). HSCs are known to play a central role in storing hepatic retinoid (vitamin A) inside lipid droplets and to be the major cell responsible for the excessive deposition of extracellular matrix in liver tissue during hepatic fibrosis. The authors demonstrated that these modified liposomes subsequently reduce the fibrosis process in three different fibrosis animal models (Li et al., 2015; Sato et al., 2008). This technology is really promising and led to the first clinical trial about siRNA-based therapeutic for hepatic fibrosis or cirrhosis therapy (METAVIR F3-4; NCT02227459).

In order to strengthen the specificity of siRNA delivery, some strategies of dual modification of siRNA delivery systems have been developed based on enzyme-sensitive or light-sensitive systems.

One really original strategy to target tumor cells was recently described, using dual-modified liposomes presenting a peptide motif Asparagine-Glycine-Arginine (NGR) to target cell receptor and a photosensitive cell-penetrating peptide. This dual-modified liposome is designed to allow a specific targeting via NGR ligand, and a photocontrolled activation of cell penetration. The authors demonstrated a higher efficacy of siRNA delivery under near infra-red illumination both in vitro and in vivo on a tumor model. This dual-modality approach of siRNA delivery looks very promising as it is intended to improve the selective delivery to cancer cells and to reduce intrinsic toxicity to healthy cells. However, the disadvantage is that the externally applied trigger is restricted to superficial tissues, although deep-seated tissue may be reached with the aid of laparoscopy in the clinic (Lin et al., 2015; Yang et al., 2015).

5 CLINICAL TRIALS

After a large number of preclinical assays, some siRNA-based drugs have entered clinical trials. Since the first clinical trial in 2004, 70 trials concerning 30 siRNA-based drugs have been filed by more than 15 companies or institutions. These clinical trials were carried out in various therapeutic areas ranging from eye diseases to cancer through genetic, metabolic, or infectious diseases. SiRNA-based drugs were delivered locally or systematically, either as an unformulated molecule (naked siRNA) or associated with a delivery system. Historically, siRNA-based therapies were first delivered locally, mainly through intravitreal injection to treat ocular lesions, such as age-related macular degeneration (AMD), diabetic macular edema (DME), or CNV (choroidal neovascularization). Then other routes of administration and/or the use of delivery systems were explored for other therapeutic applications than eye diseases.

5.1 CLINICAL TRIALS INVOLVING LOCALLY DELIVERED siRNA-BASED DRUGS (TABLE 26.1)

Five siRNA-based drugs (bevasiranib/Cand5, AGN211745/siRNA-027, PF-655, QPI-1007, bamosiran/SYL040012, and SYL1001) have been tested in clinical trials in the ocular disease area. Bevasiranib/

Table 26.1 siRNA-Based Drugs Tested in Clinical Trials—Local Administration

Route	Drug	Formulation	Indication	Sponsor	NCT	Status	Phase
IVT	Bevasiranib/ cand5	Naked	AMD/DME	Opko Health Inc	259753	Completed	II
			AMD/CNV	Allergan/siRNA Therapeutics Inc	306904	Completed	II
	AGN211745/ siRNA-027				499590	Terminated	III
					557791	Withdrawn	III
					722384	Completed	I
					363714	Completed	I/II
					395057	Terminated	II
	PF-04523655/ PF-655		AMD/DME/CNV/Diabetic retinopathy	Quark Pharma	701181	Terminated	II
					713518	Completed	II
					725686	Completed	I
					1445899	Completed	II
	QPI-1007		NAION/glaucoma/optic neuropathy		1064505	Completed	I
					1965106	Ongoing NR	II
					2341560	Not yet open	II/III
Ocular (eye drop)	SYL040012/ bamosiran		Ocular hypertension/open angle glaucoma	Sylentis SA	990743	Completed	I
					1227291	Completed	I/II
					1739244	Completed	II
					2250612	Recruiting	II
	SYL1001		Ocular pain/dry eye syndrome		1438281	Completed	I
					1776658	Completed	I/II
					2455999	Recruiting	II
Intra lesion	TD101		Pachyonychia congenita	PCP	716014	Completed	I
Intra tumor	siG12D Loder	LODER polymer	Pancreatic adenocarcinoma	Silenseed Ltd	1188785	Completed	I
					1676259	Not yet open	II
Nasal (nebulization)	ALN-RSV01	Naked	Respiratory syncytial virus	Alnylam Pharma	496821	Completed	II
					658086	Completed	II
					1065935	Completed	II
Subcutaneous	Revusiran/ALN-TTRsc	Conjugate	TTR-mediated amyloidosis cardiomyopathy		1814839	Ongoing NR	I
					1981837	Completed	II
					2292186	Ongoing NR	II
					2319005	Recruiting	III
	ALN-AT3sc		Hemophilia A and B		2035605	Recruiting	I
	ALN-PCSsc		Hypercholesteromia		2314442	Ongoing NR	I
	ALN-CC5		Paroxysmal nocturnal hemoglobinurea		2352493	Recruiting	I/II
	ALN-AS1		Acute intermittent prophyria		2452372	Recruiting	I
	ALN-AAT		α1-Antitrypsin deficiency		2503683	Recruiting	I/II

AMD, Age-related macular degeneration; CNV, choroidal neovascularization; DME, diabetic macular edema; IVT, intravitreal; LODER, local drug eluter; NAION, nonarteritic anterior ischemic optic neuropathy; NCT, National clinical trial; PCP, Pachyonychia Congenital Project; TTR, transthyretin.

Cand5, developed by Opko Health Inc., is an unmodified siRNA targeting vascular endothelial growth factor (VEGF) to inhibit retinal vascularization in patients with AMD and DME (Garba and Mousa, 2010). It was delivered by intravitreal injection and safety and efficiency were observed in phase I and II trials, from 2004 to 2007. However, two phase III trials were terminated early (COBALT trial) and withdrawn prior to enrollment (CARBON trial), respectively, due to poor therapeutic efficacy (COBALT trial) and the risk of not being able to meet its primary endpoint (CARBON trial). AGN211745/siRNA-027 from Allergan/siRNA Therapeutics Inc., a chemically modified siRNA, containing phosphorothioate and abasic nucleotides and targeting vascular endothelial growth factor receptor 1 (VEGFR1), was also administered by intravitreal injection in patients with AMD and CNV (Kaiser et al., 2010). After a completed phase I trial demonstrated its safety between 2004 and 2007 (Nguyen et al., 2012), the following phase II trial was terminated early because the drug failed to meet a key efficacy endpoint. PF-655 (or PF-04523655) from Quark Pharmaceuticals is a methylated siRNA that inhibits RTP801, a stress-induced adaptor protein inhibiting mTOR function upstream to TSC1/TSC2 complex in response to a variety of stresses. Intravitreal injection of PF-655 in preclinical animal models of laser-induced choroidal neovascularization (CNV) led to induction of expression of antiangiogenic and neurotrophic factors, and subsequent reduction of CNV volume, vessel leakage, and infiltration of inflammatory cells into the choroid. Quark Pharma also tested QPI-1007 (Solano et al., 2014), a chemically modified siRNA targeting caspase-2, initiator and effector of apoptosis, in patients with glaucoma, optic neuropathy, and nonarteritic anterior ischemic optic neuropathy (NAION). In phase I, intravitreal injection of QPI-1007 was found safe, a phase II trial is ongoing, and a phase II/III is not yet open.

In Spain, Sylentis S.A. has launched two drugs, SYL040012/bamosiran and SYL1001, administered as eye drops to patients with ocular hypertension or open angle glaucoma and ocular pain associated with dry eye syndrome, respectively. SYL040012 is a naked siRNA targeting the β2-adrenergic receptor and has been tested in three trials of phase I, I/II, and II, and is now investigated in a phase II recruiting trial (Moreno-Montañés et al., 2014). SYL1001, also a naked siRNA, targets the transient receptor potential cation channel subfamily V member 1 (TRPV1), has been tested in two completed phase I and phase I/II trials and is now investigated in a phase II recruiting trial.

In 2008, TD101, a siRNA-based drug that targets the keratin 6a (K6a) N171K mutant mRNA was administered locally (in plantar skin lesions) to a patient suffering from congenital Pachyonychia, a rare, disabling, autosomal dominant inherited disorder that affects the nails, skin, oral mucosae, hair, and teeth (Hickerson et al., 2011). The safety and efficacy of TD101 was tested in a single-patient phase I trial. However, intense pain was experienced and led to the search of less painful delivery system.

Two clinical trials launched by Silenseed Ltd., a completed phase I and an ongoing phase II, used an unmodified siRNA targeting the driving oncogene mutant KRAS encapsulated in a biodegradable poly(lactic-*co*-glycolic) acid-based matrix named LODER (for Local Drug EluteR). This polymer delivery system containing siKRAS(G12D), siG12D LODER, was inserted into the tumor in patients with locally advanced pancreatic cancer to release siRNA drug along 4 months (Golan et al., 2015).

A naked siRNA targeting nucleocapsid N gene of the respiratory syncytial virus (RSV), ALN-RSV, was developed by Alnylam Pharmaceuticals and administered in RSV-infected patients via nasal spray and nebulizer (DeVincenzo et al., 2010). Three phase II trials were completed between 2007 and 2012.

Over the past few years, Alnylam Pharmaceuticals has also launched a series of trials, which involve the subcutaneous injection of siRNA– *N*-acetylgalactosamine (GalNAc) conjugates to suppress gene expression of the targeted mRNA in liver (Nair et al., 2014). GalNAc, a ligand for the hepatic asialoglycoprotein receptor (ASGPR) is conjugated to chemically stabilized siRNA targeting six liver-specific genes, (1) proprotein convertase subtilisin/kexin type 9 (PCSK9) for patients with elevated LDL cholesterol (ALN-PCSsc), (2) antithrombin for the treatment of hemophilia and rare bleeding disorders (ALN-AT3), (3) complement component C5 (CC5) for the treatment of complement-mediated disease, such as paroxysmal nocturnal hemoglobinuria (ALN-CC5), (4) aminolevulinic acid synthase 1 (ALAS1), a liver enzyme involved in the biosynthesis of heme, for the treatment of hepatic porphyrias, including acute intermittent porphyria (AIP) (ALN-AS1), (5) α1-antitrypsin (AAT), a liver-produced serine proteinase inhibitor (ALN-AAT). Mutations in AAT cause the protein to misfold, leading to liver injury, fibrosis, cirrhosis, and hepatocellular carcinoma, (6) transthyretin (TTR), produced primarily in the liver, is a carrier for retinol-binding protein (ALN-TTRsc/revusiran). Mutations in the TTR gene cause misfolding of the protein and the formation of amyloid fibrils that deposit in the peripheral nerves and heart, resulting in intractable peripheral sensory neuropathy, autonomic neuropathy, and/or cardiomyopathy-mediated amyloidosis. ALN-AT3 and ALN-AS1 are investigated in recruiting phase I trials, whereas ALN-CC5 and ALN-AAT are tested in recruiting phase I/II trials. The clinical trial of ALN-PCSsc is ongoing. ALN-TTRsc/revusiran trials are more advanced since a phase II has been completed in 2015, a phase I on healthy volunteers and a phase II are ongoing and a phase III is recruiting patients with TTR-mediated familial amyloidotic cardiomyopathy.

5.2 CLINICAL TRIALS INVOLVING INTRAVENOUSLY DELIVERED siRNA-BASED DRUGS (TABLE 26.2)

Seventeen siRNA-based drugs have been tested upon systemic administration, one of them is a naked form of siRNA, two are conjugate forms of siRNA, and the others involve lipidic or polymeric delivery systems.

Quark Pharma initiated a phase I trial in 2007 in patients suffering from kidney injury and acute renal failure. QPI-1002/I5NP, a nuclease-resistant siRNA targeting proapoptotic p53, was able to accumulate in the kidneys (Thompson et al., 2012). A phase I/II trial was completed in 2014.

Arrowhead Research Corporation has developed siRNA-based conjugates called Dynamic Polyconjugates (DPC) composed of an amphipathic polymer to which shielding agents, such as polyethylene glycol, as well as targeting ligands are reversibly attached (Rozema et al., 2007). Two drugs ARC-250 and ARC-AAT are currently under trials, for the treatment of Hepatitis B (Wooddell et al., 2013) and α1-antitrypsin deficiency, respectively. Five phase I and II trials have been initiated for ARC-250 and a phase I trial is recruiting for ARC-AAT.

The first trial using a nanocarrier to systemically deliver siRNA was conducted by Calando Pharmaceuticals in 2008. CALAA-01 consisted of a siRNA targeting M2 subunit ribonucleotide reductase associated to a polymeric delivery system composed of cyclodextrin and tumor cell targeting ligand transferrin coupled to polyethyleneglycol (PEG) (Davis et al., 2010). Sevion Therapeutics (formerly Senesco Technologies) is developing a siRNA-based drug, designated SNS 01, as a treatment for multiple myeloma and B-cell malignancies. SNS 01 comprises a siRNA against Factor 5A and a DNA plasmid that modulates eukaryotic initiation factor 5A (eIF5A), formulated with cationic polymer PEI (polyethylenimine) that delivers the nucleic acid molecules (Francis et al., 2014). SNS 01 upregulates apoptotic pathways within cancer cells and also downregulates proinflammatory cytokines (NF-κB and ICAM). The phase Ib/IIa trial initiated in 2011 was put on hold by corporate decision.

Table 26.2 siRNA-Based Drugs Tested in Clinical Trials—Intravenous Administration

Route	Drug	Formulation	Indication	Sponsor	NCT	Status	Phase
Intravenous	ALN-TTR01	SNALP	TTR-mediated amyloidosis	Alnylam Pharma	1148953	Completed	I
	Patisiran/ALN-TTR02	LNP	cardiomyopathy		1559077	Completed	I
					1617967	Completed	II
					1960348	Recruiting	III
					1961921	Ongoing NR	II
					2053454	Completed	I
					2510261	Recruiting	III
	PRO-040201/TKM ApoB	SNALP	Hypercholesterolemia	Tekmira	927459	Terminated	I
	ALN-PCS02			Alnylam Pharma	1437059	Completed	I
	I5NP/QPI-1002	Naked	Acute renal failure/injury of kidney	Quark Pharma	554359	Completed	I
					683553	Terminated	I
					802347	Completed	I/II
	ALN-VSP02	SNALP	Solid tumors with liver involvement	Alnylam	882180	Completed	I
					1158079	Completed	I
	Atu027	Atuplex LNP	Pancreatic carcinoma	Silence Therapeutics	938574	Completed	I
					1808638	Ongoing NR	I/II
	CALAA-01	RONDEL	Solid tumor	Calando Pharm	689065	Completed	I
	DCR-MYC	EnCore LNP	Solid tumor/HCC	Dicerna	2110563	Recruiting	I
					2314052	Recruiting	I/II
	EphA2	Liposome	Advanced cancer	MD Anderson center	1591356	Recruiting	I
	SNS01-T	PEI	B-cell malignancies	Senesco Technologies Inc	1435720	Ongoing NR	I/II
	TKM-080301/TKM-PLk1	SNALP	Carcinoma	Tekmira	1262235	Completed	I/II
					1437007	Completed	I
					2191878	Recruiting	I/II
	ARC-AAT	DPC	α1-Antitrypsin deficiency	Arrowhead Res Corp	2363946	Recruiting	I
	ARC-250		Hepatitis B virus		1872065	Completed	I
					2065336	Recruiting	II
					2349126	Withdrawn	I
					2452528	Recruiting	II
					2535416	Not yet open	I
	TKM-100201/TKM-100802	SNALP	Ebola virus	Tekmira/Arbutus	1518881	Terminated	I
					2041715	Terminated	I
	ND-L02-s0201	vitA-LNP	Hepatic fibrosis	Nitto Denko Corp	1858935	Completed	I
					2227459	Recruiting	I

DPC, Dynamic polyconjugate; LNP, lipid nanoparticle; NCT, National clinical trial; PEI, polyethyleneimine; RONDEL, RNAi/oligonucleotide nanoparticle delivery; SNALP, stable nucleic acid-lipid particle; TTR, transthyretin.

From 2009, with advances made in the development of cationic lipid-based systemic nanocarrier technology, numerous clinical trials using systemic RNAi drugs have entered the clinic. Alnylam Pharmaceuticals and Tekmira Pharmaceuticals have developed a partnership for the use of SNALP (stable nucleic acid-lipid particle) as a carrier to several siRNA-based drugs. Alnylam has tested four drugs (ALN-TTR01, ALN-VSP, ALN-TTR02, and ALN-PCS). ALN-TTR01 and ALN-VSP utilize the same first-generation SNALP to encapsulate siRNA targeting TTR, for ALN-TTR01 and two siRNAs, for ALN-VSP, one targeting kinesin spindle protein (KSP) involved in cellular proliferation and one targeting VEGF (Tabernero et al., 2012). ALN-TTR02/patisiran and ALN-PCS are second-generation formulations of SNALP containing siRNA targeting TTR and PCSK9, respectively. First- and second-generation SNALP formulations have similar physicochemical properties but different ionizable lipid components, which have been shown to be key determinants of potency (Jayaraman et al., 2012). ALN-TTR01 and ALN-TTR02 have been intravenously administered to patients suffering from TTR-mediated amyloidosis (Coelho et al., 2013). ALN-TTR02/patisiran induced better efficiency and lower adverse reactions than ALN-TTR01, only tested in a phase I trial. On the basis of these results, six trials, from phase I to phase III, have been launched with patisiran. ALN-PCS02 and ALN-VSP02 have been tested in phase I trials in patients with elevated LDL-cholesterol (Fitzgerald et al., 2014) and in patients suffering from solid tumors with liver involvement, respectively. Meanwhile, Tekmira Pharmaceuticals conducted trials with three drugs (TKM ApoB/PRO-040201, TKM Plk1/TKM-080301, and TKM-100201/TKM-100802) on patients suffering from hypercholesterolemia, hepatocellular carcinoma (HCC) or on healthy volunteers against Ebola infection, respectively. TKM-ApoB was tested in a phase I trial, early terminated due to immune stimulation observed for high dose of the drugs. TKM-Plk1 has reached two phase I/II trials, including one currently recruiting. The two phase I trials initiated with TKM-100201 and TKM-100802, on healthy volunteers, have been early terminated by corporate decision. Silence Therapeutics is developing Atu027 (Schultheis et al., 2014), a cationic lipid (AtuPLEX) encapsulated siRNA targeting PKN3 (protein kinase N3) in patients with advanced solid cancer. After completion of phase I trial, a phase I/II trial is currently being conducted on patients with advanced or metastatic pancreatic adenocarcinoma in combination with gemcitabine chemotherapy. In 2014, Dicerna Pharmaceuticals launched two phase I trials for HCC, solid tumors, multiple myeloma, or non-Hodgkin lymphoma, concerning DCR-MYC, a MYC-targeted siRNA (Dudek et al., 2014) encapsulated in proprietary *EnCore* lipid nanoparticles. Nitto Denko Corporation tested ND-L02-s0201, a siRNA against heat shock protein 47 (HSP47) encapsulated in nanoparticles containing a cholesterol-based cationic lipid and vitamin-A coupled lipid-targeting stellar cells (Ishiwatari et al., 2013). After a completed phase I trial on healthy volunteers, ND-L02-s0201 is currently tested in patients with hepatic fibrosis on a currently recruiting phase I trial.

The last delivery system tested in a clinical trial is neutral lipid-based and developed by the MD Anderson Cancer center (Landen et al., 2006). A recruiting phase I trial is in progress in patients with ovarian cancer with a siRNA targeting EphA2, ephrin type-A receptor 2, a transmembrane protein from the ephrin family of receptor protein-tyrosine kinase, encapsulated in neutral DOPC liposomes.

6 CONCLUSIONS

SiRNA-based therapeutics represent a new class of drug that brings new hopes for the treatment of many diseases, such as cancer or infectious diseases. The ability of siRNAs to silence any gene gives them a huge therapeutic potential and access to all signaling pathways and/or all proteins. Several clinical trials, some of which have reached Phase III, have demonstrated the safety of these products and it is likely that in the near future, thanks to advances in chemical modification and delivery systems, a siRNA-based drug will be placed on the market. However, development of a siRNA drug might be a long process, since the right chemical modifications have to be determined, as well as their right position in the sequence. The next challenge now will be to target other organs than the liver.

REFERENCES

Akinc, A., Thomas, M., Klibanov, A.M., Langer, R., 2005. Exploring polyethylenimine-mediated DNA transfection and the proton sponge hypothesis. J. Gene Med. 7 (5), 657–663.

Baldrick, P., 2010. The safety of chitosan as a pharmaceutical excipient. Regul. Toxicol. Pharmacol. 56 (3), 290–299.

Bennett, C.F., Zuckerman, J.E., Kornbrust, D., Sasmor, H., Leeds, J.M., Crooke, S.T., 1996. Pharmacokinetics in mice of a H-3-labeled phosphorothioate oligonucleotide formulated in the presence and absence of a cationic lipid. J. Control. Release 41 (1–2), 121–130.

Borna, H., Imani, S., Iman, M., Azimzadeh Jamalkandi, S., 2015. Therapeutic face of RNAi: in vivo challenges. Expert Opin. Biol. Ther. 15, 269–285.

Boussif, O., Lezoualch, F., Zanta, M.A., Mergny, M.D., Scherman, D., Demeneix, B., Behr, J.P., 1995. A versatile vector for gene and oligonucleotide transfer into cells in culture and in vivo—polyethylenimine. Proc. Natl Acad. Sci. USA 92 (16), 7297–7301.

Bouxsein, N.F., McAllister, C.S., Ewert, K.K., Samuel, C.E., Safinya, C.R., 2007. Structure and gene silencing activities of monovalent and pentavalent cationic lipid vectors complexed with siRNA. Biochemistry 46 (16), 4785–4792.

Bramsen, J.B., Laursen, M.B., Nielsen, A.F., Hansen, T.B., Bus, C., Langkjær, N., Babu, B.R., Højland, T., Abramov, M., Van Aerschot, A., Odadzic, D., Smicius, R., Haas, J., Andree, C., Barman, J., Wenska, M., Srivastava, P., Zhou, C., Honcharenko, D., Hess, S., Müller, E., Bobkov, G.V., Mikhailov, S.N., Fava, E., Meyer, T.F., Chattopadhyaya, J., Zerial, M., Engels, J.W., Herdewijn, P., Wengel, J., Kjems, J., 2009. A large-scale chemical modification screen identifies design rules to generate siRNAs with high activity, high stability and low toxicity. Nucleic Acids Res. 37 (9), 2867–2881.

Bruno, K., 2011. Using drug-excipient interactions for siRNA delivery. Adv. Drug Deliv. Rev. 63 (13), 1210–1226.

Coelho, T., Adams, D., Silva, A., Lozeron, P., Hawkins, P.N., Mant, T., Perez, J., Chiesa, J., Warrington, S., Tranter, E., Munisamy, M., Falzone, R., Harrop, J., Cehelsky, J., Bettencourt, B.R., Geissler, M., Butler, J.S., Sehgal, A., Meyers, R.E., Chen, Q., Borland, T., Hutabarat, R.M., Clausen, V.A., Alvarez, R., Fitzgerald, K., Gamba-Vitalo, C., Nochur, S.V., Vaishnaw, A.K., Sah, D.W., Gollob, J.A., Suhr, O.B., 2013. Safety and efficacy of RNAi therapy for transthyretin amyloidosis. N. Engl. J. Med. 369, 819–829.

Dahlman, J.E., Barnes, C., Khan, O.F., Thiriot, A., Jhunjunwala, S., Shaw, T.E., Xing, Y., Sager, H.B., Sahay, G., Speciner, L., Bader, A., Bogorad, R.L., Yin, H., Racie, T., Dong, Y., Jiang, S., Seedorf, D., Dave, A., Sandhu, K.S., Webber, M.J., Novobrantseva, T., Ruda, V.M., Lytton-Jean, A.K.R., Levins, C.G., Kalish, B., Mudge, D.K., Perez, M., Abezgauz, L., Dutta, P., Smith, L., Charisse, K., Kieran, M.W., Fitzgerald, K., Nahrendorf, M., Danino, D., Tuder, R.M., von Andrian, U.H., Akinc, A., Panigrahy, D., Schroeder, A., Koteliansky, V., Langer, R., Anderson, D.G., 2014. In vivo endothelial siRNA delivery using polymeric nanoparticles with low molecular weight. Nat. Nanotechnol. 9 (8), 648–655.

Dan, N., Danino, D., 2014. Structure and kinetics of lipid-nucleic acid complexes. Adv. Colloid Interface Sci. 205, 230–239.

Dass, C.R., 2002. Biochemical and biophysical characteristics of lipoplexes pertinent to solid tumour gene therapy. Int. J. Pharm. 241 (1), 1–25.

Davis, M.E., 2009. The first targeted delivery of siRNA in humans via a self-assembling, cyclodextrin polymer-based nanoparticle: from concept to clinic. Mol. Pharm. 6 (3), 659–668.

Davis, M.E., Zuckerman, J.E., Choi, C.H.J., Seligson, D., Tolcher, A., Alabi, C.A., Yen, Y., Heidel, J.D., Ribas, A., 2010. Evidence of RNAi in humans from systemically administered siRNA via targeted nanoparticles. Nature 464 (7291), 1067–1700.

Deleavey, G.F., Damha, M.J., 2012. Designing chemically modified oligonucleotides for targeted gene silencing. Chem. Biol. 19 (8), 937–954.

Desigaux, L., Sainlos, M., Lambert, O., Chevre, R., Letrou-Bonneval, E., Vigneron, J.-P., Lehn, P., Lehn, J.-M., Pitard, B., 2007. Self-assembled lamellar complexes of siRNA with lipidic aminoglycoside derivatives promote efficient siRNA delivery and interference. Proc. Natl. Acad. Sci. USA 104 (42), 16534–16539.

DeVincenzo, J., Lambkin-Williams, R., Wilkinson, T., Cehelsky, J., Nochur, S., Walsh, E., 2010. A randomized, double-blind, placebo-controlled study of an RNAi-based therapy directed against respiratory syncytial virus. Proc. Natl. Acad. Sci. 107, 8800–8805.

Dirin, M., Winkler, J., 2013. Influence of diverse chemical modifications on the ADME characteristics and toxicology of antisense oligonucleotides. Expert Opin. Biol. Ther. 13, 875–888.

Doherty, G.J., McMahon, H.T., 2009. Mechanisms of endocytosis. Annu. Rev. Biochem. 78, 857–902.

Dudek, H., Wong, D.H., Arvan, R., Shah, A., Wortham, K., Ying, B., Diwanji, R., Zhou, W., Holmes, B., Yang, H., Cyr, W.A., Zhou, Y., Shah, A., Farkiwala, R., Lee, M., Li, Y., Rettig, G.R., Collingwood, M.A., Basu, S.K., Behlke, M.A., Brown, B.D., 2014. Knockdown of β-catenin with dicer-substrate siRNAs reduces liver tumor burden in vivo. Mol Ther. 22 (1), 92–101.

Emi, N., Kidoaki, S., Yoshikawa, K., Saito, H., 1997. Gene transfer mediated by polyarginine requires a formation of big carrier-complex of DNA aggregate. Biochem. Biophys. Res. Commun. 231 (2), 421–424.

Engels, J.W., 2013. Gene silencing by chemically modified siRNAs. Nat. Biotechnol. 30 (3), 302–307.

Escriou, V., Ciolina, C., Lacroix, F., Byk, G., Scherman, D., Wils, P., 1998. Cationic lipid-mediated gene transfer: effect of serum on cellular uptake and intracellular fate of lipopolyamine/DNA complexes. Biochim. Biophys. Acta 1368 (2), 276–288.

Ewert, K., Evans, H.M., Ahmad, A., Slack, N.L., Lin, A.J., Martin-Herranz, A., Safinya, C.R., 2005. Lipoplex structures and their distinct cellular pathways. Non Viral Vectors Gene Ther. 53, 119–155, second ed.: Part 1.

Farhood, H., Serbina, N., Huang, L., 1995. The role of dioleoyl phosphatidylethanolamine in cationic liposome-mediated gene transfer. Biochim. Biophys. Acta 1235 (2), 289–295.

Felgner, P.L., Gadek, T.R., Holm, M., Roman, R., Chan, H.W., Wenz, M., Northrop, J.P., Ringold, G.M., Danielsen, M., 1987. Lipofection—a highly efficient, lipid-mediated DNA-transfection procedure. Proc. Natl. Acad. Sci. USA 84 (21), 7413–7417.

Fitzgerald, K., Frank-Kamenetsky, M., Shulga-Morskaya, S., Liebow, A., Bettencourt, B.R., Sutherland, J.E., Hutabarat, R.M., Clausen, V.A., Karsten, V., Cehelsky, J., Nochur, S.V., Kotelianski, V., Horton, J., Mant, T., Chiesa, J., Ritter, J., Munisamy, M., Vaishnaw, A.K., Gollob, J.A., Simon, A., 2014. Effect of an RNA interference drug on the synthesis of proprotein convertase subtilisin/kexin type 9 (PCSK9) and the concentration of serum LDL cholesterol in healthy volunteers: a randomised, single-blind, placebo-controlled, phase 1 trial. Lancet 383, 60–68.

Francis, S.M., Taylor, C.A., Tang, T., Liu, Z., Zheng, Q., Dondero, R., Thompson, J.E., 2014. SNS01-T modulation of eIF5A inhibits B-cell cancer progression and synergizes with bortezomib and lenalidomide. Mol. Ther. 22 (9), 1643–1652.

Gallas, A., Alexander, C., Davies, M.C., Puri, S., Allen, S., 2013. Chemistry and formulations for siRNA therapeutics. Chem. Soc. Rev. 42, 7983–7997.

Gao, S., Dagnaes-Hansen, F., Nielsen, E.J., Wengel, J., Besenbacher, F., Howard, K.A., Kjems, J., 2009. The effect of chemical modification and nanoparticle formulation on stability and biodistribution of siRNA in mice. Mol. Ther. 17, 1225–1233.

Garba, A.O., Mousa, S.A., 2010. Bevasiranib for the treatment of wet, age-related macular degeneration. Ophthalmol. Eye Dis. 2, 75–83.

Golan, T., Khvalevsky, E.Z., Hubert, A., Gabai, R.M., Hen, N., Segal, A., Domb, A., Harari, G., David, E.B., Raskin, S., Goldes, Y., Goldin, E., Eliakim, R., Lahav, M., Kopleman, Y., Dancour, A., Shemi, A., Galun, E., 2015. RNAi therapy targeting KRAS in combination with chemotherapy for locally advanced pancreatic cancer patients. Oncotarget 6, 24560–245700.

Guo, X., Szoka, F.C., 2001. Steric stabilization of fusogenic liposomes by a low-pH sensitive PEG-diortho ester-lipid conjugate. Bioconjug. Chem. 12 (2), 291–300.

Hafez, I.M., Maurer, N., Cullis, P.R., 2001. On the mechanism whereby cationic lipids promote intracellular delivery of polynucleic acids. Gene Ther. 8 (15), 1188–1196.

Han, H.D., Mangala, L.S., Lee, J.W., Shahzad, M.M.K., Kim, H.S., Shen, D., Nam, E.J., Mora, E.M., Stone, R.L., Lu, C., Lee, S.J., Roh, J.W., Nick, A.M., Lopez-Berestein, G., Sood, A.K., 2010. Targeted gene silencing using RGD-labeled chitosan nanoparticles. Clin. Cancer Res. 16 (15), 3910–3922.

Hatakeyama, H., Akita, H., Kogure, K., Oishi, M., Nagasaki, Y., Kihira, Y., Ueno, M., Kobayashi, H., Kikuchi, H., Harashima, H., 2007. Development of a novel systemic gene delivery system for cancer therapy with a tumor-specific cleavable PEG-lipid. Gene Ther. 14 (1), 68–77.

Heil, F., Hemmi, H., Hochrein, H., Ampenberger, F., Kirschning, C., Akira, S., Lipford, G., Wagner, H., Bauer, S., 2004. Species-specific recognition of single-stranded RNA via toll-like receptor 7 and 8. Science 303, 1526–1529.

Heyes, J., Palmer, L., Bremner, K., MacLachlan, I., 2005. Cationic lipid saturation influences intracellular delivery of encapsulated nucleic acids. J. Control. Release 107 (2), 276–287.

Hickerson, R.P., Flores, M.A., Leake, D., Lara, M.F., Contag, C.H., Leachman, S.A., Kaspar, R.L., 2011. Use of self-delivery siRNAs to inhibit gene expression in an organotypic pachyonychia congenita model. J. Inv. Dermatol. 131, 1037–1044.

Holland, J.W., Cullis, P.R., Madden, T.D., 1996. Poly(ethylene glycol)-lipid conjugates promote bilayer formation in mixtures of non-bilayer-forming lipids. Biochemistry 35 (8), 2610–2617.

Huang, Y., Hong, J., Zheng, S., Ding, Y., Guo, S., Zhang, H., Zhang, X., Du, Q., Liang, Z., 2011. Elimination pathways of systemically delivered siRNA. Mol. Ther. 19, 381–385.

Huh, M.S., Lee, S.-Y., Park, S., Lee, S., Chung, H., Lee, S., Choi, Y., Oh, Y.-K., Park, J.H., Jeong, S.Y., Choi, K., Kim, K., Kwon, I.C., 2010. Tumor-homing glycol chitosan/polyethylenimine nanoparticles for the systemic delivery of siRNA in tumor-bearing mice. J. Control. Release 144 (2), 134–143.

Huotari, J., Helenius, A., 2011. Endosome maturation. EMBO J. 30, 3481–3500.

Ishiwatari, H., Sato, Y., Murase, K., Yoneda, A., Fujita, R., Nishita, H., Birukawa, N.K., Hayashi, T., Sato, T., Miyanishi, K., Takimoto, R., Kobune, M., Ota, S., Kimura, Y., Hirata, K., Kato, J., Niitsu, Y., 2013. Treatment of pancreatic fibrosis with siRNA against a collagen-specific chaperone in vitamin A-coupled liposomes. Gut 62, 1328–1339.

Ito, T., Yoshihara, C., Hamada, K., Koyama, Y., 2010. DNA/polyethyleneimine/hyaluronic acid small complex particles and tumor suppression in mice. Biomaterials 31 (10), 2912–2918.

Jackson, A.L., Burchard, J., Leake, D., Reynolds, A., Schelter, J., Guo, J., Johnson, J.M., Lim, L., Karpilow, J., Nichols, K., Marshall, W., Khvorova, A., Linsley, P.S., 2006. Position-specific chemical modification of siRNAs reduces "off-target" transcript silencing. RNA 12 (7), 1197–1205.

Jayaraman, M., Ansell, S.M., Mui, B.L., Tam, Y.K., Chen, J., Du, X., Butler, D., Eltepu, L., Matsuda, S., Narayanannair, J.K., Rajeev, K.G., Hafez, I.M., Akinc, A., Maier, M.A., Tracy, M.A., Cullis, P.R., Madden, T.D., Manoharan, M., Hope, M.J., 2012. Maximizing the potency of siRNA lipid nanoparticles for hepatic gene silencing in vivo. Angew. Chem. Int. Ed. Engl. 51, 8529–8533.

Jiang, G., Park, K., Kim, J., Kim, K.S., Oh, E.J., Kang, H., Han, S.-E., Oh, Y.-K., Park, T.G., Hahn, S.K., 2008. Hyaluronic acid-polyethyleneimine conjugate for target specific intracellular delivery of siRNA. Biopolymers 89 (7), 635–642.

Judge, A.D., Sood, V., Shaw, J.R., Fang, D., McClintock, K., MacLachlan, I., 2005. Sequence-dependent stimulation of the mammalian innate immune response by synthetic siRNA. Nat. Biotechnol. 23 (4), 457–462.

Juliano, R., Bauman, J., Kang, H., Ming, X., 2009. Biological barriers to therapy with antisense and siRNA oligonucleotides. Mol. Pharm. 6, 686–695.

Juliano, R.L., Ming, X., Nakagawa, O., 2012. Cellular uptake and intracellular trafficking of antisense and siRNA oligonucleotides. Bioconjug. Chem. 23, 147–157.

Kaiser, P.K., Symons, R.C.A., Shah, S.M., Quinlan, E.J., Tabandeh, H., Do, D.V., Reisen, G., Lockridge, J.A., Short, B., Guerciolini, R., Nguyen, Q.D., 2010. RNAi-based treatment for neovascular age-related macular degeneration by Sirna-027. Am. J. Ophthalmol. 150, 33–39.

Kanasty, R.L., Whitehead, K.A., Vegas, A.J., Anderson, D.G., 2012. Action and reaction: the biological response to siRNA and its delivery vehicles. Mol. Ther. 20 (3), 513–524.

Kano, A., Moriyama, K., Yamano, T., Nakamura, I., Shimada, N., Maruyama, A., 2011. Grafting of poly(ethylene glycol) to poly-lysine augments its lifetime in blood circulation and accumulation in tumors without loss of the ability to associate with siRNA. J. Control. Release 149 (1), 2–7.

Kapoor, M., Burgess, D.J., 2012. Physicochemical characterization of anionic lipid-based ternary siRNA complexes. Biochim. Biophys. Acta. 1818, 1603–1612.

Kato, H., Sato, S., Yoneyama, M., Yamamoto, M., Uematsu, S., Matsui, K., Tsujimura, T., Takeda, K., Fujita, T., Takeuchi, O., Akira, S., 2005. Cell type-specific involvement of RIG-I in antiviral response. Immunity 23, 19–28.

Kim, T.H., Kim, S.I., Akaike, T., Cho, C.S., 2005. Synergistic effect of poly(ethylenimine) on the transfection efficiency of galactosylated chitosan/DNA complexes. J. Control. Release 105 (3), 354–366.

Landen, C.N., Chavez-Reyes, A., Bucana, C., Schmandt, R., Deavers, M.T., Lopez-Berestein, G., Sood, A.K., 2005. Therapeutic EphA2 gene targeting in vivo using neutral liposomal small interfering RNA delivery. Cancer Res. 65 (15), 6910–6918.

Landen, C.N., Merritt, W.M., Mangala, L.S., Sanguino, A.M., Bucana, C., Lu, C., Lin, Y.G., Han, L.Y., Kamat, A.A., Schmandt, R., Coleman, R.L., Gershenson, D.M., Lopez-Berestein, G., Sood, A.K., 2006. Intraperitoneal delivery of liposomal siRNA for therapy of advanced ovarian cancer. Cancer Biol. Ther. 5 (12), 1708–1713.

Lee, J., Yun, K.-S., Choi, C.S., Shin, S.-H., Ban, H.-S., Rhim, T., Lee, S.K., Lee, K.Y., 2012. T cell-specific siRNA delivery using antibody-conjugated chitosan nanoparticles. Bioconjug. Chem. 23 (6), 1174–1180.

Li, W., Szoka, Jr., F.C., 2007. Lipid-based nanoparticles for nucleic acid delivery. Pharm. Res. 24 (3), 438–449.

Li, Y., Liu, F., Ding, F., Chen, P., Tang, M., 2015. Inhibition of liver fibrosis using vitamin A-coupled liposomes to deliver matrix metalloproteinase-2 siRNA in vitro. Mol. Med. Rep. 12 (3), 3453–3461.

Lin, A.J., Slack, N.L., Ahmad, A., George, C.X., Samuel, C.E., Safinya, C.R., 2003. Three-dimensional imaging of lipid gene-carriers: membrane charge density controls universal transfection behavior in lamellar cationic liposome-DNA complexes. Biophys. J. 84 (5), 3307–3316.

Lin, W., Xie, X., Yang, Y., Liu, H., Fu, X., Chen, Y., Liu, H., Yang, Y., 2015. Enhanced small interfering RNA delivery into cells by exploiting the additive effect between photo-sensitive peptides and targeting ligands. J. Pharm. Pharmacol. 67 (9), 1215–1231.

Lin, X., Li, L., Wang, R., Wilcox, D., Zhao, X., Song, J., Huang, X., Hansen, T.M., Dande, P., Wada, C., Hubbard, R.D., Kohlbrenner, W.M., Fesik, S.W., Shen, Y., 2011. A robust in vivo positive-readout system for monitoring siRNA delivery to xenograft tumors. RNA 17 (4), 603–612.

Lopez-Fraga, M., Martinez, T., Jimenez, A., 2009. RNA interference technologies and therapeutics from basic research to products. Biodrugs 23 (5), 305–332.

Lorenzer, C., Dirin, M., Winkler, A.M., Baumann, V., Winkler, J., 2015. Going beyond the liver: progress and challenges of targeted delivery of siRNA therapeutics. J. Control. Release 203, 1–15.

Lu, J.J., Langer, R., Chen, J., 2009. A novel mechanism is involved in cationic lipid mediated functional siRNA delivery. Mol. Pharm. 6, 763–771.

Ma, B.C., Zhang, S.B., Jiang, H.M., Zhao, B.D., Lv, H.T., 2007. Lipoplex morphologies and their influences on transfection efficiency in gene delivery. J. Control. Release 123 (3), 184–194.

Mangala, L.S., Han, H.D., Lopez-Berestein, G., Sood, A.K., 2009. Liposomal siRNA for ovarian cancer. Methods Mol. Biol. (Clifton, N.J) 555, 29–42.

Manoharan, M., 2014. GalNAc-siRNA with enhanced stabilization chemistry: ESC-GalNAc-siRNA. Alnylam [online]. http://www.alnylam.com/web/assets/ALNYESC-GalNAc-siRNA-TIDES-May2014-Capella.pdf

Mao, S., Sun, W., Kissel, T., 2010. Chitosan-based formulations for delivery of DNA and siRNA. Adv. Drug Deliv. Rev. 62 (1), 12–27.

Marques, J.T., Devosse, T., Wang, D., Zamanian-Daryoush, M., Serbinowski, P., Hartmann, R., Fujita, T., Behlke, M.A., Williams, B.R., 2006. A structural basis for discriminating between self and nonself doublestranded RNAs in mammalian cells. Nat. Biotechnol. 24, 559–565.

Mehrabadi, F.S., Fischer, W., Haag, R., 2012. Dendritic and lipid-based carriers for gene/siRNA delivery (a review). Cur. Opin. Solid State Mater. Sci. 16 (6), 310–322.

Mok, K.W.C., Lam, A.M.I., Cullis, P.R., 1999. Stabilized plasmid-lipid particles: factors influencing plasmid entrapment and transfection properties. Biochim. Biophys. Acta 1419 (2), 137–150.

Moreno-Montañés, J., Sádaba, B., Ruz, V., Gómez-Guiu, A., Zarranz, J., González, M.V., Pañeda, C., Jimenez, A.I., 2014. Phase I clinical trial of SYL040012, a small interfering RNA targeting β-adrenergic receptor 2, for lowering intraocular pressure. Mol. Ther. 22, 226–232.

Morrissey, D.V., Lockridge, J.A., Shaw, L., Blanchard, K., Jensen, K., Breen, W., Hartsough, K., Machemer, L., Radka, S., Jadhav, V., Vaish, N., Zinnen, S., Vargeese, S., Bowman, K., Shaffer, C.S., Jeffs, L.B., Judge, A., MacLachlan, I., Polisky, B., 2005. Potent and persistent in vivo anti-HBV activity of chemically modified siRNAs. Nat. Biotechnol. 23, 1002–1007.

Nair, J.K., Willoughby, J.L.S., Chan, A., Charisse, K., Alam, R., Wang, Q., Hoekstra, M., Kandasamy, P., Kel'in, A.V., Milstein, S., Taneja, N., O'Shea, J., Shaikh, S., Zhang, L., van der Sluis, R.J., Jung, M.E., Akinc, A., Hutabarat, R., Kuchimanchi, S., Fitzgerald, K., Zimmermann, T., van Berkel, T.J., Maier, M.A., Rajeev, K.G., Manoharan, M., 2014. Multivalent *N*-acetylgalactosamine-conjugated siRNA localizes in hepatocytes and elicits robust RNAi-mediated gene silencing. J. Am. Chem. Soc. 136, 16958–16961.

Ng, E.W.M., Shima, D.T., Calias, P., Cunningham, Jr., E.T., Guyer, D.R., Adamis, A.P., 2006. Pegaptanib, a targeted anti-VEGF aptamer for ocular vascular disease. Nat. Rev. Drug Discov. 5, 123–132.

Nguyen, Q.D., Schachar, R.A., Nduaka, C.I., Sperling, M., Basile, A.S., Klamerus, K.J., Chi-Burris, K., Yan, E., Paggiarino, D.A., Rosenblatt, I., Khan, A., Aitchison, R., Erlich, S.S., 2012. Phase 1 dose-escalation study of a siRNA targeting the RTP801 gene in age-related macular degeneration patients. Eye (Lond.) 26 (8), 1099–1105.

Nykänen, A., Haley, B., Zamore, P.D., 2001. ATP requirements and small interfering RNA structure in the RNA interference pathway. Cell 107 (3), 309–321.

O'Neill, M.J., Bourre, L., Melgar, S., O'Driscoll, C.M., 2011. Intestinal delivery of non-viral gene therapeutics: physiological barriers and preclinical models. Drug Discov. Today 16, 203–218.

Pack, D.W., Hoffman, A.S., Pun, S., Stayton, P.S., 2005. Design and development of polymers for gene delivery. Nat. Rev. Drug Discov. 4 (7), 581–593.

Pardridge, W.M., 2007. shRNA and siRNA delivery to the brain. Adv. Drug Deliv. Rev. 59, 141–152.

Peeters, L., Sanders, N.N., Jones, A., Demeester, J., De Smedt, S.C., 2007. Post-pegylated lipoplexes are promising vehicles for gene delivery in RPE cells. J. Control. Release 121 (3), 208–217.

Pirollo, K.F., Rait, A., Zhou, Q., Hwang, S.H., Dagata, J.A., Zon, G., Hogrefe, R.I., Palchik, G., Chang, E.H., 2007. Materializing the potential of small interfering RNA via a tumor-targeting nanodelivery system. Cancer Res. 67 (7), 2938–2943.

Pirollo, K.F., Zon, G., Rait, A., Zhou, Q., Yu, W., Hogrefe, R., Chang, E.H., 2006. Tumor-targeting nanoimmunoliposome complex for short interfering RNA delivery. Hum. Gene Ther. 17 (1), 117–124.

Qhattal, H.S.S., Hye, T., Alali, A., Liu, X., 2014. Hyaluronan polymer length, grafting density, and surface poly(ethylene glycol) coating influence in vivo circulation and tumor targeting of hyaluronan-grafted liposomes. ACS Nano 8 (6), 5423–5440.

Ragelle, H., Vandermeulen, G., Preat, V., 2013. Chitosan-based siRNA delivery systems. J. Control. Release 172 (1), 207–218.

Ramanathan, R.K., Hamburg, S.I., Halfdanarson, T.R., Borad, M.J., Seetharam, M., Curtis, K.K., Lee, P., Fredlund, P., Gilbert, M., Semple, S.C., Judge, A.D., Crowell, B., Vocila, L., MacLachlan, I., Northfelt, D.W., 2014. A phase I/II dose escalation study of TKM-080301, a RNAi therapeutic directed against PLK1, in patients with advanced solid tumors, with an expansion cohort of patients with NET or ACC. Pancreas 43 (3), 502.

Remaut, K., Lucas, B., Braeckmans, K., Demeester, J., De Smedt, S.C., 2007. Pegylation of liposomes favours the endosomal degradation of the delivered phosphodiester oligonucleotides. J. Control. Release 117 (2), 256–266.

Rose, S.D., Kim, D.H., Amarzguioui, M., Heidel, J.D., Collingwood, M.A., Davis, M.E., Rossi, J.J., Behlke, M.A., 2005. Functional polarity is introduced by Dicer processing of short substrate RNAs. Nucleic Acids Res. 33 (13), 4140–4156.

Ross, P.C., Hui, S.W., 1999. Lipoplex size is a major determinant of in vitro lipofection efficiency. Gene Ther. 6 (4), 651–659.

Rozema, D.B., Blokhin, A.V., Wakefield, D.H., Benson, J.D., Carlson, J.C., Klein, J.J., Almeida, L.J., Nicholas, A.L., Hamilton, H.L., Chu, Q., Hegge, J.O., Wong, S.C., Trubetskoy, V.S., Hagen, C.M., Kitas, E., Wolff, J.A., Lewis, D.L., 2015. Protease-triggered siRNA delivery vehicles. J. Control. Release 209, 57–66.

Rozema, D.B., Lewis, D.L., Wakefield, D.H., Wong, S.C., Klein, J.J., Roesch, P.L., Bertin, S.L., Reppen, T.W., Chu, Q., Blokhin, A.V., Hagstrom, J.E., Wolff, J.A., 2007. Dynamic PolyConjugates for targeted in vivo delivery of siRNA to hepatocytes. Proc. Natl. Acad. Sci. USA 104, 12982–12987.

Sakurai, Y., Hatakeyama, H., Sato, Y., Hyodo, M., Akita, H., Ohga, N., Hida, K., Harashima, H., 2014. RNAi-mediated gene knockdown and anti-angiogenic therapy of RCCs using a cyclic RGD-modified liposomal-siRNA system. J. Control. Release 173, 110–118.

Sandvig, K., Pust, S., Skotland, T., van Deurs, B., 2011. Clathrin-independent endocytosis: mechanisms and function. Curr. Opin. Cell Biol. 23, 413–420.

Sato, Y., Murase, K., Kato, J., Kobune, M., Sato, T., Kawano, Y., Takimoto, R., Takada, K., Miyanishi, K., Matsunaga, T., Takayama, T., Niitsu, Y., 2008. Resolution of liver cirrhosis using vitamin A-coupled liposomes to deliver siRNA against a collagen-specific chaperone. Nat. Biotechnol. 26 (4), 431–442.

Schultheis, B., Strumberg, D., Santel, A., Vank, C., Gebhardt, F., Keil, O., Lange, C., Giese, K., Kaufmann, J., Khan, M., Drevs, J., 2014. First-in-human phase I study of the liposomal RNA interference therapeutic Atu027 in patients with advanced solid tumors. J Clin. Oncol. 32 (36), 4141–4148.

Scomparin, A., Polyak, D., Krivitsky, A., Satchi-Fainaro, R., 2015. Achieving successful delivery of oligonucleotides—from physico-chemical characterization to in vivo evaluation. Biotechnol. Adv. 33, 1294–1309.

Shi, F.X., Wasungu, L., Nomden, A., Stuart, M.C.A., Polushkin, E., Engberts, J., Hoekstra, D., 2002. Interference of poly(ethylene glycol)-lipid analogues with cationic-lipid-mediated delivery of oligonucleotides; role of lipid exchangeability and non-lamellar transitions. Biochem. J. 366, 333–341.

Simoes, S., Moreira, J.N., Fonseca, C., Duzgunes, N., de Lima, M.C.P., 2004. On the formulation of pH-sensitive long circulation times. Adv. Drug Deliv. Rev. 56 (7), 947–965.

Solano, E.C., Kornbrust, D.J., Beaudry, A., Foy, J.W., Schneider, D.J., Thompson, J.D., 2014. Toxicological and pharmacokinetic properties of QPI-1007, a chemically modified synthetic siRNA targeting caspase 2 mRNA, following intravitreal injection. Nucleic Acid Ther. 24 (4), 258–266.

Song, L.Y., Ahkong, Q.F., Rong, Q., Wang, Z., Ansell, S., Hope, M.J., Mui, B., 2002. Characterization of the inhibitory effect of PEG-lipid conjugates on the intracellular delivery of plasmid and antisense DNA mediated by cationic lipid liposomes. Biochim. Biophys. Acta 1558 (1), 1–13.

Soutschek, J., Akinc, A., Bramlage, B., Charisse, K., Constien, R., Donoghue, M., Elbashir, S., Geick, A., Hadwiger, P., Harborth, J., John, M., Kesavan, V., Lavine, G., Pandey, R.K., Racie, T., Rajeev, K.G., Röhl, I., Toudjarska, I., Wang, G., Wuschko, S., Bumcrot, D., Koteliansky, V., Limmer, S., Manoharan, M., Vornlocher, H.P., 2004. Therapeutic silencing of an endogenous gene by systemic administration of modified siRNAs. Nature 432, 173–178.

Spagnou, S., Miller, A.D., Keller, M., 2004. Lipidic carriers of siRNA: differences in the formulation, cellular uptake, and delivery with plasmid DNA. Biochemistry 43 (42), 13348–13356.

Tabernero, J., Shapiro, G.I., LoRusso, P.M., Cervantes, A., Schwartz, G.K., Weiss, G.J., Paz-Ares, L., Cho, D.C., Infante, J.R., Alsina, M., Gounder, M.M., Falzone, R., Harrop, J., White, A.C., Toudjarska, I., Bumcrot, D., Meyers, R.E., Hinkle, G., Svrzikapa, N., Hutabarat, R.M., Clausen, V.A., Cehelsky, J., Nochur, S.V., Gamba-Vitalo, C., Vaishnaw, A.K., Sah, D.W., Gollob, J.A., Burris, H.A., 2012. First-in-humans trial of an RNA interference therapeutic targeting VEGF and KSP in cancer patients with liver involvement. Cancer Discov. 3, 406–417.

Takeuchi, O., Akira, S., 2010. Pattern recognition receptors and inflammation. Cell 140, 805–820.

Tang, Y., Li, Y.-B., Wang, B., Lin, R.-Y., van Dongen, M., Zurcher, D.M., Gu, X.-Y., Holl, M.M.B., Liu, G., Qi, R., 2012. Efficient in vitro siRNA delivery and intramuscular gene silencing using PEG-modified PAMAM dendrimers. Mol. Pharm. 9 (6), 1812–1821.

Taranejoo, S., Liu, J., Verma, P., Hourigan, K., 2015. A review of the developments of characteristics of PEI derivatives for gene delivery applications. J. Appl. Polym. Sci. 132 (25), 10.1002/app.42096.

Thompson, J.D., Kornbrust, D.J., Foy, J.W., Solano, E.C., Schneider, D.J., Feinstein, E., Molitoris, B.A., Erlich, S., 2012. Toxicological and pharmacokinetic properties of chemically modified siRNAs targeting p53 RNA following intravenous administration. Nucleic Acid Ther. 22 (4), 255–264.

Tseng, Y.-C., Mozumdar, S., Huang, L., 2009. Lipid-based systemic delivery of siRNA. Adv. Drug Deliv. Rev. 61 (9), 721–731.

Vaish, N., Chen, F., Seth, S., Fosnaugh, K., Liu, Y., Adami, R., Brown, T., Chen, Y., Harvie, P., Johns, R., Severson, G., Granger, B., Charmley, P., Houston, M., Templin, M.V., Polisky, B., 2011. Improved specificity of gene silencing by siRNAs containing unlocked nucleobase analogs. Nucleic Acids Res. 39, 1823–1832.

Wasungu, L., Hoekstra, D., 2006. Cationic lipids, lipoplexes and intracellular delivery of genes. J. Control. Release 116 (2), 255–264.

Weitzer, S., Martinez, J., 2007. The human RNA kinase hClp1 is active on 3′ transfer RNA exons and short interfering RNAs. Nature 447, 222–226.

Whitehead, K.A., Langer, R., Anderson, D.G., 2009. Knocking down barriers: advances in siRNA delivery. Nat. Rev. Drug Discov. 8, 129–138.

Wittrup, A., Lieberman, J., 2015. Knocking down disease: a progress report on siRNA therapeutics. Nat. Rev. Genet. 16 (9), 543–552.

Wooddell, C.I., Rozema, D.B., Hossbach, M., John, M., Hamilton, H.L., Chu, Q., Hegge, J.O., Klein, J.J., Wakefield, D.H., Oropeza, C.E., Deckert, J., Roehl, I., Jahn-Hofmann, K., Hadwiger, P., Vornlocher, H.P., McLachlan, A., Lewis, D.L., 2013. Hepatocyte-targeted RNAi therapeutics for the treatment of chronic hepatitis B virus infection. Mol. Ther. 21, 973–985.

Wu, S.Y., McMillan, N.A.J., 2009. Lipidic systems for in vivo siRNA delivery. AAPS J. 11 (4), 639–652.

Yang, T., Li, B., Qi, S., Liu, Y., Gai, Y., Ye, P., Yang, G., Zhang, W., Zhang, P., He, X., Li, W., Zhang, Z., Xiang, G., Xu, C., 2014. Co-delivery of doxorubicin and Bmi 1 siRNA by Folate receptor targeted liposomes exhibits enhanced anti-tumor effects in vitro and in vivo. Theranostics 4 (11), 1096–1111.

Yang, Y., Yang, Y., Xie, X., Wang, Z., Gong, W., Zhang, H., Li, Y., Yu, F., Li, Z., Mei, X., 2015. Dual-modified liposomes with a two-photon-sensitive cell penetrating peptide and NGR ligand for siRNA targeting delivery. Biomaterials 48, 84–96.

Zelphati, O., Uyechi, L.S., Barron, L.G., Szoka, F.C., 1998. Effect of serum components on the physico-chemical properties of cationic lipid/oligonucleotide complexes and on their interactions with cells. Biochim. Biophys. Acta 1390 (2), 119–133.

CHAPTER

siRNA-BASED NUCLEOCEUTICALS FOR TISSUE REGENERATION

27

Fatemeh Atyabi, Fatemeh Mottaghitalab, Ali Rastegari
Tehran University of Medical Sciences, Tehran, Iran

CHAPTER OUTLINE

1 INTRODUCTION

Tissue engineering applies different cells and strategies for restoring and maintaining the normal function and behavior of wide variety of tissues, such as heart, cartilage, bone, skin, nerve, ocular, etc. (Barry and Murphy, 2004; Leor et al., 2003). When first conceived tissue engineering approaches, many engineers considered de novo engineering of new tissues for implantations. For this, different biomaterials and methods drawn from chemical engineering, biomaterials science, and mechanical engineering principles have been involved (Fisher and Mauck, 2013). The main challenge of using these biomimetic tissue constructs is their potential in triggering and promoting cell proliferation, migration, differentiation, and other cellular processes, such as signaling pathways (Farokhi et al., 2013, 2014; Mottaghitalab et al., 2011, 2013). Despite of using different structures, methods and biomaterials for tissue engineering applications (Kim et al., 2011; Rosso et al., 2004), new methodologies based on nucleic acid constructs for enhancing tissue regeneration are introduced, recently. Among different types of nucleic acids that have many applications in these areas, short-interfering RNA (siRNA) is a relatively new gene-silencing mechanism that involves the use of double-stranded RNA (dsRNA) for regulation of multiple endogenous genes and can be utilized to target a specific gene that is responsible for multiple pathways/factors. Therefore, it could simultaneously manipulate diverse and interlinked cellular processes that lead to tissue regeneration (Bertrand et al., 2002). Due to their enormous potential, RNA-based therapeutics may be considered a fifth category of regeneration therapy even though this approach remains in a conceptual stage of development. They are based on the principle of siRNA, a novel mechanism of action by which RNAs silence gene expression. Novel therapeutic approaches may be envisioned whereby the expression of certain genes detrimental to the tissue regeneration process is silenced by RNAs (Amarzguioui et al., 2003; Chiu and Rana, 2003; Czauderna et al., 2003). Here, we highlight some of the high impact advances within the siRNA-based therapeutics in different tissues over the past several years. Additionally, we attempt to identify recent potential of siRNA as a new tool in modern regenerative medicine in detail.

2 HEART REGENERATION

Cardiovascular diseases are the leading cause of mortality and morbidity, worldwide. The most important cardiovascular diseases are acute myocardial infarction and heart failure. In these diseases, the damaged cardiomyocytes lost their contractile performance and regeneration capacity (Addis and Epstein, 2013; Aguirre et al., 2013; Laflamme and Murry, 2011; Lloyd-Jones et al., 2010; Mirotsou et al., 2011; Roger et al., 2012). Adult mammalian cardiomyocytes unlike the lower vertebrates, such as zebrafish do not have a primitive regenerative response upon injury (Bergmann et al., 2009; Beltrami et al., 2001; Kajstura et al., 2010; Reiss et al., 1993). Recently, siRNA technology opened a new avenue for effective manipulation of regenerative pathways including cell cycle and inflammatory reactions. One of the major issues for siRNA delivery to heart is the type of applied carrier. To date, viral-, lipid-, or polymeric-based vehicles are used for siRNA delivery to heart (Li et al., 2006; Poller et al., 2010). In this section, different delivery systems and regulatory pathways used for heart regeneration are discussed.

2.1 siRNA DELIVERY TO HEART

The most important issue that should be considered in heart regeneration is heart targeting that requires efficient delivery systems and targeting RNAi to heart tissue (Poller et al., 2010). Adeno-associated virus (AAV) is one of the major vehicles used for heart targeting (Asokan et al., 2010; Müller et al., 2003). Viral vectors can protect RNAi efficiently in vivo and some serotypes of AAV are highly cardiotropic, such as serotype 9 (Gray and Samulski, 2008; Pacak et al., 2006; Poller et al., 2010). Administration of viral vectors could stimulate the immune system. The use of engineered AAVs with acceptable safety profile is a good strategy for heart regeneration without considering their immunogenicity (Li et al., 2009a,b). Lipid-based (i.e., lipofectamine) and polymeric-based vehicles are other types of delivery systems in heart regeneration. Lipid vehicles, such as liposomes are effective for siRNA protection and transfection; however, the use of these types of vehicles is limited in the clinic as inherent toxicity (Jiang et al., 2009; Li et al., 2006; Lv et al., 2006). On other hand, polymeric-based carriers have been less used in heart regeneration. Somasuntharam et al. (2013) used an acid-degradable polyketals for siRNA targeting to the post-MI environment. In their research polyketals vehicles showed minimal inflammatory responses in the target unlike the lipid-based vehicles (Somasuntharam et al., 2013; Sy et al., 2008).

2.2 ROLE OF siRNA IN HEART REGENERATION

Heart regeneration is regulated by several regulatory pathways and proteins (Table 27.1). The use of siRNA technology against these regulatory factors could effect cardiac cell cycle and heart regeneration. The proliferation of cardiomyocytes is stopped after birth in mammals unlike the lower vertebrates (Ahuja et al., 2007; Bicknell et al., 2007; van Amerongen and Engel, 2008). The cardiomyocytes of adult mammalian have been shown an early regenerative response upon the heart injury; however, this regenerative response is not sufficient for complete recovery of heart function (Addis and Epstein, 2013; Aguirre et al., 2013; Bergmann et al., 2009; Kajstura et al., 2010; Laflamme and Murry, 2011; Reiss et al., 1993). siRNA technology can be used for reentering the cardiomyocytes to the cell cycle and regeneration pathways. Eulalio and coworkers in their research have shown that siRNA administration could significantly increase the cardiac function and eliminate the infracted zone in the injured heart without any induction of fibrosis. AAV serotype 9 (AAV9) has been used in their research to make an extended efficacy (Eulalio et al., 2012). Meis1 and phospholamban (PLB) are two important regulatory factors in heart regeneration. Meis1 protein is a key regulatory factor for cardiac differentiation during embryonic development (Paige et al., 2012; Wamstad et al., 2012). Using siRNA against Mesi1 could increase the cardiomyocytes proliferation about threefold (Mahmoud et al., 2013). Additionally, researchers showed that the siRNA against phospholamban (PLB), as a key regulator of cardiac Ca2+ homeostasis, can improve the cardiac function in heart failure by using AAV9 vector (Fechner et al., 2007; Lowes et al., 2002; MacLennan and Kranias, 2003; Watanabe et al., 2004). Another strategy for heart regeneration is targeting inflammatory reactions in infracted area of the myocardium. Reactive oxygen species (ROS) are the key factors in developing heart injury (i.e., cardiomyocytes apoptosis, fibroblast proliferation, and myocyte hypertrophy) (Bolli, 1998; Cave, 2009; Ferdinandy and Schulz, 2003; Looi et al., 2008). NADPH oxidase is a major source for superoxide production in infarcted myocardium. This enzyme has a catalytic subunit known as Nox2 (Cave, 2009;

Table 27.1 Regulatory Pathways and Proteins Presented in Heart Regeneration

Regulatory role	Target	Effect	References
Cardiac remodeling	MMPs[a]	Myocardial matrix degradation during cardiac remodeling	Vanhoutte and Heymans (2010)
TLR/IL-1R signaling and innate immune system	IRAK-4[b]	Cardiac inflammatory process	Maekawa et al. (2009)
TLR/IL-1R signaling and innate immune system	MyD88[c]	Cardiac inflammatory process	Fuse et al. (2005)
Chemotactic factors for mononuclear cells	MCP1[d]	Immune cell migration	Göser et al. (2005)
Chemotactic factors for mononuclear cells	MIP-1α[e]	Immune cell migration	Göser et al. (2005)
Regulation of cardiac Ca^{2+} homeostasis	PLB[f]	Controlling SERCA2a[g]	Poller et al. (2010)
Central regulators of cardiac differentiation during embryonic development	Meis1	Cardiac differentiation	Mahmoud et al. (2013)
Inflammatory reaction	Nox2-NADPH	Superoxide production	Somasuntharam et al. (2013)

[a] *Matrix metalloproteinases.*
[b] *Interleukin-1 receptor associated kinase.*
[c] *Myeloid differentiation factor.*
[d] *Monocyte chemoattractant protein.*
[e] *Macrophage inflammatory protein.*
[f] *Phospholamban.*
[g] *Sarcoplasmic reticulum Ca^{2+} ATPase pump.*

Cave et al., 2006; Heymes et al., 2003; Krijnen et al., 2003; Looi et al., 2008). Cardiomyocyte apoptosis and remodeling can be protected following MI by knockdown of Nox2-NADPH. For example, local delivery of polymeric nanoparticle containing siRNA against Nox2-NADPH could improve and recover the cardiac function (Somasuntharam et al., 2013) (Table 27.2).

3 KIDNEY REGENERATION

Kidneys have important regulatory roles in the body. In medical regeneration, the kidney can be considered in three conditions including renal transplantation, ischemic acute kidney injury, and renal fibrosis in chronic renal diseases.

In renal transplantation, the kidney is sensitive to ischemia at the time of organ collection and the cellular injuries following ischemia can be exacerbated after reperfusion. Therefore, suitable approaches are needed to preserve the function of kidney grafts. siRNA technology is one of the effective

Table 27.2 List of siRNA Products Under Clinical Trials for Tissue Regeneration

Tissue	Target	Product	Role	Clinical phase	References
Kidney	ICAM-1[a]	ISIS 2302	Ischemia reperfusion process	Phase II	Kahan et al. (2004)
Kidney	P53	15NP (QPI-1002)	Delayed graft function	Phase II	Ritter et al. (2009)
Eye	Caspase 2	QPI-1007	Retinal ganglion cells death	Phase II	Ahmed et al. (2011)
Eye	VEGF[b]	Bevasiranib	Corneal angiogenesis	Phase III	Garba and Mousa (2010)

[a]*The intercellular adhesion molecule.*
[b]*Vascular endothelial growth factor.*

approaches for limiting the ischemia–reperfusion injury in transplantation and kidney graft preservation (Bon et al., 2012). Acute kidney injury (AKI) and renal fibrosis are other issues for kidney regeneration. AKI is defined to loss of kidney function within seven days. The main reasons for this condition are ischemia or toxic agent (Hoste et al, 2008; Mehta et al., 2004). Apoptosis and inflammatory factors have a crucial role for developing ischemic or toxic acute kidney injury (AKI) (Du et al., 2004; Matthijsen et al., 2007). siRNA administration has been shown to be effective for kidney regeneration in AKI (Cantaluppi et al., 2012; Rácz and Hamar, 2008; Van de Water et al., 2006). In addition, the renal fibrosis is a common pathway in chronic renal diseases, such as glomerulosclerosis, diabetic nephropathy, end-stage renal disease, and chronic allograft dysfunction (Chapman et al., 2005; Eddy, 2000; Stahl and Felsen, 2001; Zeisberg et al., 2001). Studies have shown that siRNA treatment could be useful for restraining the renal fibrosis (Kushibiki et al., 2005). Generally, the administration of siRNA to the kidney can be categorized in systematic (intravenous) and local delivery. Different vehicles have been used for kidney regeneration including cationic lipid-base vehicles, polymeric nanocarriers and viral vectors. In this section, we discussed about these issues with more details.

3.1 siRNA DELIVERY TO KIDNEY

siRNA delivery to the kidney usually can be categorized into systematic (intravenous) and local delivery. Intravenous administration is one of the most used methods to deliver antisense oligonucleotides into the renal tissue (Kobayashi et al., 2001). Renal clearance of oligonucleotides creates a good condition for nonspecific kidney targeting (Hamar et al., 2004). In a variety of studies demonstrated that intravenous injection of naked siRNA leads to high accumulation of oligonucleotide in the kidney (Hamar et al., 2004; Van de Water et al., 2006). In systemic approach, the hydrodynamic delivery is used for targeting the tubular epithelium of kidney. In this manner, a relatively large volume of vehicle solution (about 10% of the body weight of experimental animal) injected into the animal vein for a short time (about 15–30 s). This sudden hypervolemia makes high accumulation of oligonucleotide in the parenchymal tissue cells which cause liver damage in most of the animals (Budker et al., 2006; Hamar et al., 2004; Höcherl et al., 2010; Lewis and Wolff, 2007; Ma et al., 2009). However, this method has shown better efficiency for uptake and retention of siRNA in the renal tissue when compared with other systematic administration methods including standard intravenous and intraperitoneal delivery for uptake and retention of siRNA in the renal tissue (Larson et al., 2007). Local delivery to the kidney is another approach that can be achieved by different routes including renal artery targeting glomeruli, renal vein targeting the tubulointerstitium and intrauretral administration into the renal pelvis (Hamar et al., 2004; Takabatake et al., 2007; Xia et al., 2008a). RNAi delivery to the kidney is performed by polymeric, lipid- and viral-based vehicles. Cationic lipid-based vehicles (e.g., liposomes) protect siRNA from nuclease degradation and increase uptake of siRNA with tubular epithelium. Different studies showed that 1,2-dioleoyl-sn-glycero-3-phosphatidylcholine (DOPC) can increase the tubular uptake of siRNA (Landen et al., 2005; Sioud and Sørensen, 2003). Polymeric nanocarriers can be more useful for glomerular targeting and protein knockdown compared with naked siRNA and viral vectors. PEG-poly-L-lysine copolymer has been used as a polyion complex (PIC) nanocarrier for glomerular targeting. These nanocarriers showed effective glomerular targeting compared with Hemaglutinating virus of Japan (HVJ) envelope vector (pseudovirion vehicle) (Shimizu et al., 2010). Adenoviral and lentiviral vectors are other effective vehicles for gene delivery to the kidney. Adenoviral vectors can be useful for siRNA targeting to the tubules and glomeruli located in the renal cortex. Lentiviral vectors are also useful for siRNA targeting to the cortical and corticomedullary area of kidney (Gusella et al., 2002; Kim et al., 2009; Nakamura et al., 2004; Shou et al., 2009). However, these viral vectors have shown severe side effects that limit their potential for siRNA delivery to the kidney (Manjunath et al., 2009; Nakamura et al., 2004; Ortiz et al., 2003; Seregin and Amalfitano, 2009). Unlike these vectors, AAV vectors have fewer side effects and they are suitable for local gene delivery to the kidney (Bessis et al., 2004; Ito et al., 2008; Lipkowitz et al., 1999;).

3.2 ROLE OF siRNA IN KIDNEY REGENERATION

3.2.1 Kidney transplantation

Kidney is sensitive to ischemia at the time of organ collection. The cellular injuries can be exacerbated after reperfusion in renal transplantation. Ischemia–reperfusion is a complex pathophysiological process that is associated with proinflammatory immune responses (Perico et al., 2004). Administration of antisense oligonucleotides during ischemia or in ex vivo condition could improve the kidney function after ischemia reperfusion (Chatauret et al., 2011). The intercellular adhesion molecule (ICAM)-1 plays a key role in ischemia–reperfusion process. The efficacy of antisense oligonucleotide against the expression of ICAM-1 (ISIS 2302) was evaluated in phase I and II clinical trials with combination of cyclosporine–prednisone regimen. Generally, ISIS 2302 showed a little improvement of renal function. However, the rates of acute rejection episodes were 38.1% in the ISIS 2302 group versus 20.0% in the placebo group after 6 months therapy and the graft survivals were similar in both groups after 3 years (Kahan et al., 2004). Studies have shown that the expression of caspase-3 was increased in ischemia reperfusion injury. Caspases are intracellular cysteine proteases that play a central role in mediating apoptosis and inflammation. In a study, the researchers limited the ischemia reperfusion injury with direct administration of caspase-3 siRNA into the renal artery. In this study, the expression of caspase-3 precursor and its active subunit was downregulated by 30 and 56%, respectively. This issue made significant reduction of apoptotic cells in the renal cortex by 47% (Yang et al., 2011). One of the most common complications in renal transplantation is delayed graft function (DGF). DGF often occur in ischemia–reperfusion condition and studies have shown that p53 protein plays an important role in cell apoptosis during DGF. The use of siRNA against p53 can reduce DGF after renal transplantation. This specific siRNA has been entered into the phase II clinical trial as 15NP (QPI-1002) by Quark pharmaceuticals (Molitoris et al., 2009; Ritter et al., 2009). Furthermore, the complement system is activated in renal ischemia reperfusion injury. Third complement component (C3) has been demonstrated to have a major role in ischemia–reperfusion injury. Since the kidney is the main source of extra hepatic production of C3, the use of siRNA against C3 could be effective for protection of kidney transplant (de Vries et al., 2004; Pratt et al., 2002). In one study, the researchers could successfully silence the C3 component with a hydrodynamic injection method. In their research, the blood urea nitrogen (BUN) and serum creatinine values were significantly improved after 24 hour C3-siRNA vector injection and also the survival of ischemic mice increased (Zheng et al., 2006).

3.2.2 Ischemia–reperfusion and acute kidney injury

Acute kidney injury (AKI) is referred to the loss of kidney function within seven days. The main reasons for this condition are ischemia or toxic agent. In this condition, cell stress pathways are activated in the tubular epithelium, which increase the expression of inflammatory factors or apoptosis agents (Du et al., 2004; Fire et al., 1998; Matthijsen et al., 2007; Simpson et al., 2008). siRNA administration could be effective for kidney regeneration in AKI with modulating the proliferation, angiogenesis, and apoptosis of renal cells (Cantaluppi et al., 2012; Rácz and Hamar, 2008; Van de Water et al., 2006). Activation of the complement pathway during ischemia–reperfusion injury cause renal cell injury, apoptosis and necrosis with C5a binding to C5aR receptor on the tubular epithelium cells (de Vries et al., 2003). Zheng et al. (2008) preserved the renal function with hydrodynamic injection of C5aR siRNA into the tail vein of mice. siRNA administration significantly prevented lethal ischemia reperfusion injury and decreased the level of BUN and serum creatinine from 97.33 to 41.6 mmol/L and 213 to 112 mol/L. Treatment with siRNA also improved the survival of the mice to 70% during 8 days after injury (Zheng et al., 2008). In the acute kidney injury, the p53 protein has a crucial role in the related apoptotic pathways (Cunningham et al., 2002; Daemen et al., 1999; Kelly et al., 2001). Proximal tubule cells (PTCs) are the first regions of kidney injury in ischemia or nephrotoxicity condition. The expression of p53 protein is very low under normal physiological conditions in PTCs (Kelly et al., 2003). In a rat model study, intravenous administration of p53-siRNA could significantly reduce serum creatinine from 3.7 ± 0.8 to 1.9 ± 0.2 mg/dL and protect renal cells from injury. In this study, the researchers efficiently delivered naked siRNA to the PTCs without using local delivery, hydrodynamic administration or complicated gene delivery systems (Molitoris et al., 2009). Furthermore, Fas protein has been shown to influence the tubular cell death in renal ischemia–reperfusion injury. Studies showed that the Fas expression was increased during ischemia specifically in the distal tubular region (Castaneda et al., 2003; Lee et al., 2003). Hydrodynamic (systemically) or low volume renal vein injection (local delivery) of siRNA against Fas could reduce Fas mRNA and protein expression in the kidney fourfold. The survival rate in both types of administration was same and was about 20%. It seems that increasing the half-life of siRNA could be effective for elevating the survival rate with using chemical modification or suitable gene delivery systems (Eddy, 2000; Hamar et al., 2004).

3.2.3 Chronic renal diseases

Fibrosis of the tubulointerstitial compartment is a main complication of the chronic renal diseases, such as glomerulosclerosis and diabetic nephropathy (Eddy, 2000). High accumulation and deposition of extracellular matrix (ECM) and myofibroblasts (MFs) observed in the chronic renal disease (Eikmans et al., 2003). Connective tissue growth factor (CTGF) is a matricellular protein that has a key role in fibrosis process (Leask, 2003). In a rat model study, caudal vein administration of CTGF-siRNA could decrease the renal fibrosis and the serum creatinine level. Moreover, the Banff score and the expression of CTGF were significantly decreased compared with control groups at 8 weeks (Luo et al., 2008). The transforming growth factor-β1 (TGF-β1) is a key factor in initiating and progressing the tubulointerstitial fibrosis and a strong correlation has been observed between the amount of TGF-β1 and degree of fibrosis (Eddy, 2000; Yoshioka et al., 1993). Administration of siRNA against TGF-β1 could significantly decrease protein expression of TGF-β1 and thereby the ECM accumulation in experimental glomerulonephritis models (Kushibiki et al., 2005; Miyagishi and Taira, 2002; Takabatake et al., 2005; Tsujie et al., 2000). Due to more accumulation of RNase in the kidney and brain in comparison with other tissues, the researchers in a study, increased the effectiveness time of siRNA in the kidney by using chemically modified siRNA. They observed a significant reduction in both glomerular matrix deposition and proteinuria within 4 and 6 weeks in a progressive glomerulosclerosis rat model (Takabatake et al., 2007). Furthermore, lipid-based vehicles have shown more accumulation in the liver and therefore designing an appropriate vehicle is necessary for kidney targeting (Soutschek et al., 2004; Zimmermann et al., 2006). Gene silencing of mitogen-activated protein kinase 1 (MAPK1) could reduce the glomerular injury and expression of genes associated with fibrotic disease including TGF-β1 (Shimizu et al., 2010). Shimizu et al. (2010) designed poly(ethylene glycol) (PEG)-poly(L-lysine) (PLL) copolymer-based delivery system, also termed polyion complex (PIC) nanocarriers for MAPK1-siRNA delivery to the murine glomeruli. PIC nanocarriers formed through the electrostatic interaction between charged poly(L-lysine) and negatively charged siRNA. The formed nanoparticles had a size about 13.7 nm and the whole body imaging of the mice showed high accumulation of siRNA into the kidneys, particularly in the cortical area where located the glomeruli. In their research, the BUN and urinary protein levels were significantly reduced after PIC administration (Shimizu et al., 2010).

4 LIVER REGENERATION

The liver has a considerable potential for regeneration in response to liver injuries and liver transplantation (Locker et al., 2003; Mitchell and Willenbring, 2008). Hepatic fibrosis (HF) is a very important medical condition with high rate of morbidity and mortality ,worldwide. Generally, chronic inflammation responses are the main reason for leading the liver tissue into the fibrotic condition (Friedman, 2000; Pinzani, 2000). HF is characterized by the accumulation of excess ECM proteins, such as collagen type I and III in liver parenchyma that form fibrotic scars (Gabrielli et al., 1997; Moreira, 2007; Ryhänen et al., 1996). The process of HF is related to the activation of hepatic stellate cells (HSCs) (the major collagen producer cells in liver) and the balance between synthesis and degradation of ECM. It has been observed in many studies that HF occurs when the ECM synthesis is increased compared with its degradation rate (Hu et al., 2009; Okazaki et al., 2000, 2001). The expression of collagen type I and III genes are upregulated in the HF by activating HSCs and their transition to MFs. The receptor for advanced glycation end products (RAGE) plays a key role in this process and liver fibrogenesis (Fehrenbach et al., 2001; Moreira, 2007; Reeves and Friedman, 2002). In a study, administration of specific siRNA against the RAGE effectively prevented liver fibrosis and promoted liver regeneration (Xia et al., 2008b). Furthermore, plasminogen activator inhibitor- 1 (PAI-1) was known as a powerful fibrosis-promoting molecule. The use of siRNA against PAI-1 could also prevent the progression of hepatic fibrogenesis and enhance the hepatocellular proliferation (Hu et al., 2009). A family of enzymes called the matrix metalloproteinases (MMPs) have a crucial role in degradation process of ECM (Benyon and Arthur, 2001). The serum level of tissue inhibitors of metalloproteinases (TIMPs) is increased in the liver fibrosis (Gomez et al., 1997). In another study, it was observed that downregulation of TIMP-2 expression could be effective for treatment of liver fibrosis (Hu et al., 2007). In this section, different aspects of siRNA technology in liver regeneration are discussed.

4.1 ROLE OF siRNA ROLE IN LIVER REGENERATION

4.1.1 Antiproliferative role in termination step of liver regeneration

The liver has a capacity to regenerate itself after injuries (Locker et al., 2003). Liver regeneration after partial hepatectomy can be divided into three phases including initiation, proliferation and termination (Fausto et al., 2006; Taub, 2004). NF-kB signaling pathway has an important regulatory role in termination stage (Marquez et al., 2010). In this signaling pathway, the miR-34a has a crucial role for the progression and termination steps of the liver regeneration. Studies have shown that Met and INHBB genes are involved in this regulatory pathway (Apte et al., 2009; Marquez et al., 2010). The effect of INHBB siRNA has been studied on hepatocyte proliferation in vitro. The researchers transfected the hepatocytes with siRNA and showed strong knockdown efficiency by real-time PCR ($p < 0.05$) (Chen et al., 2011). Therefore, this specific siRNA could be a good option for terminating the hepatocyte proliferation in the end stage of liver regeneration particularly in tissue engineering.

4.1.2 Hepatic fibrosis

HF occurs when the production and deposition of ECM proteins are increased. Collagen proteins (type I and III) are the major component of ECM (Friedman, 2003). Activation and transition of HSCs to MFs is an important step for HF. The receptor for advanced glycation end products (RAGE) plays a crucial role in liver fibrogenesis and spreading of activated HSCs/MFs (Fehrenbach et al., 2001). Overexpression of RAGE has been observed in the process of activation of HSCs and transition of HSCs to MFs (De Vriese et al., 2006; Fehrenbach et al., 2001; Oldfield et al., 2001). One study has demonstrated that the blockage of RAGE with several agents, such as Fab 2-fragments of IgG against RAGE and amphoterin made a major improvement in liver function and regeneration (Zeng et al., 2004). The NF-κB/IκB pathway has shown an effective ability in liver fibrosis and inhibiting this pathway that increase the recovery of fibrotic tissues (Oakley et al., 2005; Wright et al., 2001). The use of RAGE-siRNA could effectively inhibit the expression of collagen type I and NF-κB protein in rat model of HF. Suppressing the RAGE expression could significantly decrease the criteria factors of the liver function. For example, the serum levels of alanine transaminase (ALT) and total bilirubin (TBIL) in the treatment group were decreased by 12.67% ($P < 0.01$) and 39.6% ($P < 0.01$), respectively; while the level of albumin was about 1.5-fold higher than of the control group ($P < 0.01$) after administration of siRNA (Xia et al., 2008b). In HF, accumulation of ECM was increased and the degradation process was not balanced with matrix synthesis (Arthur, 2000; Moreira, 2007; Zhang et al., 1999). High accumulation of ECM lead to accelerated production of collagen by HSC (Hinz, 2006). As mentioned earlier, MMPs have a crucial role in degradation process of a wide variety of collagenous and noncollagenous matrix proteins (Hinz, 2006). The plasminogen activator (PA)/plasmin system could activate the MMPs and inhibit the deposition of ECM. Plasminogen activator inhibitor-1 (PAI-1) is recognized as a physiological inhibitor of PA which is upregulated during liver fibrosis (Hu et al., 2008, 2009). The use of

PAI-1–siRNA is able to significantly downregulate the PAI-1 in HSC-T6 cell line and decrease the content of collagen type I and III to 64 and 67% after 48 h posttransfection (Hu et al., 2008). Furthermore, studies have found that the serum level of TIMPs is increased in liver fibrosis (Gomez et al., 1997). Two types of TIMPs (TIMP-1 and TIMP-2) have been found in the liver tissue that is produced by activated HSCs (Arthur et al., 1992; Gomez et al., 1997; Théret et al., 1997, 1998). In a study, administration of chemically modified siRNA against TIPM-2 has shown a significant decrease in the amount of TIMP-2 mRNA and the related protein expression. Moreover, the content of ECM and activated HSCs were decreased while an increase was observed in the level of hepatocyte growth factor (HGF) (crucial mitogen for liver regenerations) in rat model. The liver showed better performance in the treatment group and the levels of ALT and AST were decreased by 47.7 and 43.8%, respectively (Hu et al., 2007).

5 OCULAR REGENERATION

RNA interference technology particularly siRNA therapy could be useful for medical regeneration in some of the ocular diseases including optic neuropathy, AMD and glaucoma. siRNA delivery are usually performed by viral vectors or chemically modified siRNA for ocular regeneration. The retinal ganglion cells (RGCs) could not regenerate injured axons after optic nerve (ON) injury. One of the main reasons of this failed regeneration is related to three myelin-associated proteins including Nogo-A, myelin-associated glycoprotein and oligodendrocyte myelin glycoprotein (McKerracher and Winton, 2002). Nogo receptor (NgR) is able to bind to inhibitor proteins and could mediate their inhibitory roles (Fournier et al., 2001; Liu et al., 2002; Wang et al., 2002). siRNA technology could be used against NgR and suppress this inhibitory system for inducing axon regeneration in eye (Chen et al., 2009). In addition, preservation of RGC is very important in glaucoma and ischemic optic neuropathy (ION) since the permanent loss of RGC and vision loss occur in these conditions (Kisiswa et al., 2010; Schmidt et al., 2008). Studies have shown that the apoptosis pathways have a key role in RGC death. The family of caspase enzymes has a crucial role for activating the apoptotic pathways. Caspase-2 is one of the most highly conserved caspases that have been observed highly expressed in the RGC of ischemic retinas (Kurokawa et al., 1999; Lamkanfi et al., 2002; Singh et al., 2001; Upton et al., 2008;). The specific siRNA against RGC has been used in the clinical studies for treating ON injuries (Ahmed et al., 2011). Choroidal neovascularization (CNV) could be other suitable choice for ocular regeneration. CNV occurred in the wet form of AMD and defined as the growth of new blood vessels below the retina. Several proangiogenic proteins are involved in different types of neovascularization including retinal and CNV (Kwak et al., 2000; Okamoto et al., 1997; Ozaki et al., 2000; Seo et al., 1999). Studies demonstrated that vascular endothelial growth factor (VEGF) has a critical role in pathophysiology of the disease. (Campochiaro, 2000; Lu and Adamis, 2002; Reich et al., 2003). Administration of siRNA against VEGF can be useful for treatment of these diseases. In this regards, naked form of VEGF siRNA is under phase III clinical trials as bevasiranib (Garba and Mousa, 2010; Reich et al., 2003). Furthermore, studies have shown that the deposition of ECM is increased in the primary open angle glaucoma (POAG). The ECM deposition has been regulated by TGF-β2 (transforming growth factor β)/Smad (signaling mothers against decapentaplegic) pathway. Studies have shown that siRNA therapy can be useful for inhibiting the expression of ECM proteins (Su et al., 2012). In the following, we discussed about the siRNA therapy for the mentioned diseases.

5.1 siRNA VEHICLES FOR OCULAR DELIVERY

siRNA delivery could be achieved by viral intraocular injection or direct administration of chemically modified siRNA into the target for ocular regeneration (Ahmed et al., 2011; Chen et al., 2009; Reich et al., 2003). Posterior segment of the eye is usually a target for ocular regeneration (e.g., retina and ON) (Ahmed et al., 2011; Chen et al., 2009; Reich et al., 2003; Shen et al., 2006). The common type of virus that could be used in ocular regeneration is AAV (Bennett et al., 1994; Chen et al., 2009; Gruss et al., 2003; Reich et al., 2003). The AAV is suitable for stable and long-term expression and also can target a variety of different ocular cell types depending on the capsid serotype. The most common serotypes are 1, 2, 4, and 5 that could be used for retinal or ON regeneration (Bennett et al., 1994; Gruss et al., 2003; Lebherz et al., 2008; Reich et al., 2003). Chemical modifications of siRNA can increase the resistance against nucleases degradation in serum and vitreous humor. siRNA modification can also suppress the induction of innate immune responses during treatment (Ahmed et al., 2011; Shen et al., 2006). For example, the researchers have prepared the chemically modified siRNA for retina delivery by specific features including resistant to nuclease degradation for up to 28 days after intravitreal administration, effective uptake by the cells in the rat retina and RGCs without the ability to stimulate any interferon or cytokine responses (Ahmed et al., 2011; Su et al., 2012).

5.2 ROLE OF siRNA IN OCULAR REGENERATION

5.2.1 Optic nerve regeneration

The retinal ganglion cells (RGCs) could not regenerate injured axons after ON injury. One of the main reasons of this failure is related to three myelin-associated proteins (McKerracher and Winton, 2002). Nogo receptor (NgR) binds to these inhibitor proteins and mediate their inhibitory roles (Fournier et al., 2001; Liu et al., 2002; Wang et al., 2002). RGC axon regeneration has been observed after several intraocular manipulations (Berry et al., 1996; Fischer et al., 2000; Leon et al., 2000; Yin et al., 2003). The application of siRNA against NgR activity could promote the axonal regrowth and provide better regeneration with trophic support after ON injury. Intraocular injections of siRNA against NgR with AAV vector showed efficient knockdown after 8 weeks. In this in vivo study, two treatment methods were compared. The first method was laid on the use of a trophic support by zymosan as a potent inflammation activator of macrophages, retinal astrocytes and Muller cells and the second method was based on the application of siRNA-NgR plus zymosan. In this investigation, the groups with siRNA-NgR plus zymosan showed much more axon growth (Chen et al., 2009).

5.2.2 Preservation of retinal ganglion cell

Ocular neuroprotection, particularly the preservation of RGC is very important issue in ocular diseases, such as glaucoma and ION (Kisiswa et al., 2010; Schmidt et al., 2008). As mentioned earlier, apoptosis pathways have a key role for losing the majority of RGC. The family of caspase enzymes has a crucial role for activation of apoptotic pathways. Caspase-2 is one of the most highly conserved caspases that highly expressed in RGC of ischemic retinas (Kurokawa et al., 1999; Lamkanfi et al., 2002; Singh et al., 2001; Upton et al., 2008). Intravitreally administration of chemically modified siRNA-caspase 2 could protect the RGC from death after ON injury. Researchers have confirmed the decrease in caspase-2 mRNA level about 50 %. They have also showed that single intravitreal administration of siRNA could significantly protect RGC from apoptosis for atleast 7 days. This type of siRNA has currently being evaluated in the clinic as a neuroprotective agent for treatment of sudden vision loss associated with nonarteritic anterior ION (NAION) (Ahmed et al., 2011).

5.2.3 Macular degeneration

CNV has been observed in age-related macular degeneration (AMD) (Garba and Mousa, 2010). CNV occurred in the wet form of AMD and is defined as growth of new blood vessels below the retina. Several proangiogenic proteins involved in different types of neovascularization including retinal and CNV (Kwak et al., 2000; Okamoto et al., 1997; Seo et al., 1999). Studies demonstrated that VEGF has a critical role in pathophysiology of the disease and is a key player in intraocular neovascularization (Aiello et al., 1995; Reich et al., 2003). Administration of siRNA against VEGF could significantly reduce the protein expression of VEGF under hypoxic conditions. Moreover, the injection of siRNA-VEGF with adenovirus vector could significantly reduce the VEGF protein level and the extent of neovascularization in the murine laser photocoagulation model of CNV (Reich et al., 2003). In addition, a naked form of siRNA-VEGF (bevasiranib) is under phase III clinical trials for wet AMD (Garba and Mousa, 2010). Studies have shown that VEGF receptor class 1 (VEGFR-1) has a key role for stimulating corneal angiogenesis (Gille et al., 2001). In a study, intravitreous injection of chemically modified siRNA against VEGFR-1 could decrease the level of VEGFR-1 in the retinas by 30% and the treated eyes with 1.5 or 0.5 μg dose of siRNA showed significantly smaller areas of CNV (about 56 and 66%, respectively) (Shen et al., 2006).

5.2.4 Glaucoma

Glaucoma is a type of eye disease that could damage the ON. The main reason of glaucoma is eye pressure elevation. This disease can be divided into two main groups including open-angle and closed-angle glaucoma (Casson et al., 2012). POAG is one of the most important reasons of blindness in the world (Lütjen-Drecoll, 2005; Quigley and Broman, 2006). In the POAG, high deposition of ECM has been observed (Hynes, 2009). The ECM deposition have been regulated by TGF-β2 (transforming growth factor β)/Smad (signaling mothers against decapentaplegic) pathway (Takai et al., 2012). The ECM components including collagen, fibronectin, and elastin are increased in the aqueous humor and trabecular meshwork of the glaucoma patients. TGF-β2 has also a regulatory role for ECM metabolism (Chung et al., 2012; Hynes, 2009). In a study, the use of Smad7-siRNA could effectively antagonize the TGF-β/Smad signaling and reduce the expression of ECM proteins including fibronectin and laminin in vitro. The researchers suggested that combination of this signaling pathway and siRNA could open a new avenue for targeting POAG (Su et al., 2012).

6 PERITONEAL FIBROSIS REGENERATION

Peritoneal fibrosis and high accumulation of collagen in peritoneum have been observed after long-term peritoneal dialysis in the patients with end-stage renal disease (Devuyst et al., 2002; Gandhi et al., 1980; Williams et al., 2002). At present, there are not an efficient therapy for peritoneal fibrosis and tissue regeneration based on siRNA therapy. Epithelial-mesenchymal transition (EMT) of mesothelial cells plays an important role in the development of peritoneal fibrosis. Some important pathways have a crucial role in this process. Some studies have shown that regulation of these pathways with siRNA therapy could be useful for peritoneal fibrosis regeneration (Lei et al., 2012; Xiao et al., 2010). In addition, accumulation of collagen is increased in peritoneal fibrosis. Applying siRNA could be useful for regulating the collagen synthesis in peritoneal fibrosis (Obata et al., 2012). In this section we discussed about these issues.

6.1 ROLE OF siRNA IN PERITONEAL FIBROSIS

Epithelial-mesenchymal transition (EMT) of mesothelial cells plays an important role in the development of peritoneal fibrosis. There are some important pathways for EMT process and expression of ECM, such as protein kinase C, polyalcohol, and amidohexose pathways. These pathways are activated by increasing the reactive oxygen species (ROS) in high-glucose condition. Researchers have claimed that blocking these pathways could inhibit the development of peritoneal fibrosis (Brownlee, 2001; Cohausz et al., 2008; Kiss and Szabó, 2005; Lei et al., 2012; Okada et al., 2008). Applying siRNA against PARP-1 (polyADP-ribose polymerase-1) as a key enzyme for ROS activity in high-level glucose condition could prevent the EMT process and accumulation of ECM in peritoneum (Lei et al., 2012). Furthermore, connective tissue growth factor (CTGF), a cysteine rich peptide, has a key role in fibrosis process. The mechanism of CTGF is regulated by elevating the level of ECM in various tissues especially in peritoneum (Chujo et al., 2004; Kobayashi et al., 2005; Leask and Abraham, 2004; Sakamoto et al., 2005; Szeto et al., 2005; Zarrinkalam et al., 2003). The use of retroviral siRNA vector against CTGF could inhibit the CTGF expression and decrease the major components of ECM (e.g., fibronectin, collagen 1, laminin) (Xiao et al., 2010). On the other hand, the collagen accumulation has a key role in development of peritoneal fibrosis. Researchers have shown that Heat shock protein 47 (HSP47), a collagen-specific molecular chaperone, is necessary for biosynthesis and secretion of collagen molecules (Nagata, 1996; Nagata et al., 1986, 1988). Suppression of this protein is useful for preventing peritoneal fibrosis (Nishino et al., 2003). Obata et al. (2012) employed this strategy for inhibiting the progression of peritoneal fibrosis in vivo. They have used cationized gelatin microspheres (CGMs) as a nonviral vector for sustained release of siRNA for a period of 21 days. These microspheres could efficiently protect siRNA in mouse model (Kushibiki et al., 2003; Obata et al., 2012).

7 CARTILAGE REGENERATION

Cartilage exhibits low response to injury which limits the rate of its regeneration and repair. However, some approaches are developed for improving cartilage regeneration including microfractures, abrasions, or chondrocyte implantation. While these approaches provide symptomatic relief, healing is typically incomplete leaving the surrounding tissue predisposed to damage and susceptible to degradative processes (Pi et al., 2015; Song et al., 2007; Sun, 2014). microRNA technology using miR-27, miR-140, etc. is other possible approach for cartilage repair (Song et al., 2007). Despite of progress in using microRNA, applying siRNA nucleoceuticals for promoting the signaling pathways involved in cartilage repair has been less investigated. Cartilage repair based on siRNA nucleoceuticals is usually achieved by intraarticular delivery of nanoparticles, nanotubes, transfection of siRNA duplexes, lentiviruses, and adenoviruses (Pi et al., 2015; Song et al., 2007; Sun, 2014). One of the important therapeutic targets for osteoarthritis (OA), as a common cartilage trauma, is hypoxia-inducible factor-2α (Hif-2α). However, using this target for delivering therapeutic agents face many challenges. In this regard, Pi et al. (2015) have introduced chondrocyte-targeting vector for siRNA delivery in order to silence the expression of Hif-2α. Silencing Hif-2α with siRNA technology could interfere the signaling pathway of IL-1β that break down the cascade of cartilage matrix and inhibit the ossification process (Pi et al., 2015). They have also showed that this system is capable to downregulate other specific genes, such as matrix metalloproteinases (MMP13-9), disintegrin, and metalloproteinase with thrombospondin motifs (ADAMTS-4, -5), VEGF, type X collagen, and nuclear factor (NF)-κB in interleukin-1-beta (IL-1β). It was claimed in this study that downregulation of the mentioned genes plays an important role in cartilage regeneration (Pi et al., 2015). Another useful delivery system for siRNA is Rosette nanotubes (RNTs). RNTs are novel self-assembled nanomaterials that are able to incorporate siRNA under physiological condition and subsequently deliver it to human chondrocytes. RNTs are introduced as bioactive and biocompatible biomaterials with the ability to improve chondrocyte adhesion and effective siRNA delivery for enhancing cartilage repair (Onyekwelu et al., 2009). Transfecting chondrocytes with lentivirus-mediated ADAMTS-5 (aggrecanases) siRNA is another strategy for promoting cartilage regeneration. It was reported by Chu et al. that single injection of lentivirus-mediated ADAMTS-5 siRNA could prevent the degradation of articular cartilage (Iliopoulos et al., 2008). ADAMTS are proteolytic enzymes that could cleave ECM for supporting the normal biomechanical and homeostatic characteristics of joint cartilage (Song et al., 2007). The other possible siRNA delivery vehicle used for cartilage regeneration is adenoviruses. Adenovirus mediated NF-kBp65-specific siRNA was developed as a powerful approach for restraining the induction of IL-1b and TNF-a, improving the inflammation of synovium and decreasing cartilage degradation in the early phase of OA model. Totally, it is appeared that using siRNA silencing could be useful in promoting the cartilage repair.

8 SKIN REGENERATION

Skin is multilayered organ with the ability to protect the organism against external aggressions. High turnover capacity of skin makes it potential for repairing different injuries. Similar to other tissues, the stem cells existed in skin tissue play an important role in regulating the balance between cell loss and cell division. Moreover, it was reported in 1980 that various growth factors specially TGF-β could stimulate the migration of keratinocyte and thus could mediate the regeneration of damaged skin (Bannasch et al., 2003). Additionally,an in vivo study on rats revealed the potential of TGF-b1 in promoting the healing process, increasing collagen deposition and improving tensile strength of skin wounds. Studies of transgenic and gene knockout mice provided an in-depth information about the role of TGF-b1 in skin regeneration. It is confirmed that gene silencing of Smad3 as a key signaling pathway for regulating the activity of TGF-b1 with siRNA could induce the skin repairing. However, another study showed that inhibiting the signaling pathway of TGF-b1 might not always result in beneficial outcomes. For example, delayed healing was observed in immune-deficient SCID1 (severe combined immunodeficiency 1)-knockout mice after silencing TGF-b1 gene (Lee et al., 2013; Mardaryev et al., 2011).

8.1 siRNA DELIVERY SYSTEMS FOR SKIN REGENERATION

Two common delivery vehicles established their reputation for siRNA delivery to skin including topical delivery and microneedles. In this regard, Hegde et al. (2014) reported that topical delivery of Accel siRNA is responsible for silencing luciferase (luc2p) gene in murine skin. They claimed that this method is noninvasive, simple and ready for clinical use in order to regenerate skin injuries (Hegde et al., 2014). In another study, the potential of SPACE-peptide in combination with a DOTAP-based ethosomal carrier system to enhance skin delivery of siRNA was evaluated. They showed that topical delivery of DOTAP-SES on mice skin lead to 63.2% ± 7.7% of GAPDH knockdown, which introduced novel method for cutaneous siRNA delivery (Chen et al., 2014). Steel microneedles and motorized microneedle array are also used for siRNA loading and consequent deposition on skin and silencing genes. These novel systems have unique potential as nucleic-based therapeutics for skin regeneration (Chong et al., 2013; Hickerson et al., 2013).

8.2 ROLE OF siRNA IN SKIN REGENERATION

siRNA therapeutics could be widely used for skin regeneration due to their accessibility in the body (Geusens et al., 2009). Skin repair based on siRNA therapy have yielded promising results for various skin injuries such as psoriasis (Jakobsen et al., 2009), allergic skin disease (Inoue et al., 2007), epidermolysis bullosa simplex (Atkinson et al., 2011), epidermolytic palmoplantar keratoderma (Addis and Epstein, 2013), and pachyonychia congenita (Hickerson et al., 2008). As mentioned earlier, TGF-bi is an important gene in skin regeneration. Using Thy-1 siRNA could strongly downregulate Thy-1 in wound area that leads to increase in expression rate of TGF-b1 and accumulation of fibroblast at the site of injury. Thy-1 is able to regulate the migration and proliferation of fibroblast cells by regulating the amount of TGF-b1 (Lee et al., 2013). In conclusion, most of the siRNA therapeutics used for skin regeneration applies the aforementioned mechanism for accelerating the repair process.

9 BONE REGENERATION

Bone has unique regeneration capacity which triggers skeletal homeostasis, fracture repair, and bone graft incorporation. However, different patterns of repair may appear in various degrees of injury. Application of allogeneic bone and even autogenous bone grafts has not fully provided solutions for all bone related difficulties due to their biological and biomechanical requirements. The most investigated biological molecules incorporated in bone repair are listed in Table 27.3. Developing recombinant proteins and gene therapy techniques introduce new opportunities for bone repair. In this regard, siRNA therapeutics has attracted many researches to develop new tools for accelerating bone regeneration (Cheema et al., 2007; Petite et al., 2000).

9.1 siRNA DELIVERY VEHICLES FOR BONE REGENERATION

Nanotechnology provides promising vehicles for siRNA delivery in order to improve various types of bone injuries like osteoporosis, bone tumors, and nonunion and critical size defects. Lipid nanoparticles were first developed for siRNA delivery in order to stimulate bone repair. However, due to some limitations including short half-life and provoking immune response and inducing opsinzation, some researchers examine new strategies, such as PEGylation in order to overcome these challenges. The improvements in designing lipid-based systems for siRNA delivery encouraged the development of liposomal-based delivery systems. Lentiviral vectors are other possible delivery system that could be used for siRNA delivery to bone. The improvements in the level of lentiviruses in terms of versatility, efficiency, and biosafety introduced them as proficient vehicles for transforming genetic information in targeted cells (Bassit et al., 2015; Zhang et al., 2012). It was reported in another study that injection of

Table 27.3 Most Investigated Biological Molecules Incorporated in Bone Repair

Biological molecules	Produced by	Phase	Activity	References
Growth factors				
TGF-β[a]	Platelets, monocytes, macrophages, osteoblasts, osteoclasts, and chondrocytes	Intermediate injury response	Promoting angiogenesis, stimulating bone formation, regulating cartilage matrix calcification, stimulating osteoblast activity, intraosseus wound regeneration	Chen et al. (2004)
FGF[b]	Inflammatory cells, osteoblasts, chondrocytes	Inflammatory	Blood vessel formation, autocrine, intracellular functions, stimulating type 4 collagenase	Nies et al. (2000)
PDGF[c]	Platelets, monocytes, activated tissue macrophages, endothelial cells	Inflammatory	Tyrosine kinase activity, stimulating mesenchymal cell proliferation, initiating fracture repair, helping to form cartilage and intramembranous bone, initiating callus formation	Caplan and Correa (2011)
IGF[d]	Bone cells, Chondrocytes	Inflammatory	Bone formation, modulating osteoclast function, bone remodeling	Chen et al. (2010)
BMP[e]	Bone cells	Inflammatory	Promoting osteoblasts and chondrocytes function	Kempen et al. (2009)
Cytokines				
IL-1[f]	Macrophages	Early inflammatory	Initiating a cascade of signaling pathways for promoting bone repair, stimulating the activity of natural proteases	Dimitriou et al. (2005)
IL-6[g]	Osteoblasts	Inflammatory	High constitutive activity early in the healing process	Dimitriou et al. (2005)
GMCSF[h]	T lymphocytes, osteoblasts	Inflammatory	Stimulate osteoclast formation, increasing of T-lymphocytes proliferation, stimulating cytokine secretion	Dimitriou et al. (2005)

[a]*Transforming growth factor* β.
[b]*Fibroblast growth factor.*
[c]*Platelet-derived growth factors.*
[d]*Insulin-like growth factors.*
[e]*Bone morphogenetic proteins.*
[f]*Interleukin-1.*
[g]*Interleukin-6.*
[h]*Granulocyte-macrophage colony-stimulating factor.*

GL3 luciferase siRNA that was complexed with atelocollagen could effectively reduce the luciferase expression of bone-metastatic prostate tumor cells. Moreover, systemic administration of this system has potential activity in enhancing the expression of zeste homolog 2 and phosphoinositide 3′-hydroxykinase p110- α subunit that are possible candidates for inhibiting bone metastasis (Takeshita et al., 2005).

9.2 ROLE OF siRNA IN BONE REGENERATION

Transcription factor HIF-1 has two common subunits (HIF-1a and HIF-1b) controls angiogenesis by silencing prolyl hydroxylase domain-containing protein 2 (PHD2). PHD2 is responsible for hydroxylating the key proline residues of HIF-1 that mediates ubiquitin-proteasome pathway destruction. Under hypoxic conditions, HIF-1a turns on the transcription of key genes for vasculogenesis including vascular endothelial growth factor (VEGF), erythropoietin (EPO), stromal derived factor 1 (SDF1), leptin, and inducible nitric oxide synthase (iNOS). The siRNA against PHD2 could block the binding of PHD2 with HIF-1a and enhance the expression of angiogenic proteins via the hypoxia pathway. Another important transcription factor for differentiating bone precursor cells and bone formation is Cbfa1. Cbfa1 activity is regulated by a chain of the heterotrimeric G protein, Gsa, which is transcribed by the guanine nucleotide-binding protein (G protein) alphastimulating activity polypeptide 1 (GNAS1) gene. Expression of Gsa could inhibit osteogenic differentiation by proteolytic degradation of Cbfa1. Downregulation of GNAS1 gene via siRNA results in increased expression of Cbfa. This process provokes the production of bone-differentiating proteins, such as osteopontin, collagen I, and osteocalcin (Niyibizi et al., 2004; Wang and Grainger, 2013).

10 NERVE REGENERATION

Nervous system has less regeneration capacity and undergoes a complex repair process after injury. The regeneration process of defected peripheral nerve can be divided in the following steps: proliferation of Schwann cells, migration of macrophages to the site of injury, wallerian degeneration, clearance of myelin debris, and formation of bands of Büngner structure. Generally, two surface molecules including Complement receptor type 3 (CR3) and galactose-specific lectin MAC-2 are involved in wallerian degeneration process and consequently the nerve regeneration. Moreover, injured adult central nervous system (CNS) has not the ability to regenerate correct axonal and dendritic connections. The consequences of injury are not just a break in communication between healthy neurons, but a cascade of events that can lead to neuronal degeneration and cell death. However, some growth factors (IGF, FGF, PDGF, VEGF), and neurotropic factors (NGF, BDNF, GDNF), and ECM components (Laminin, collagen, fibronectin, etc.) are involved in CNS regeneration (Stoll and Müller, 1999; Sunderland and Walshe, 1968).

10.1 siRNA DELIVERY SYSTEMS FOR NERVE REGENERATION

Progressive loss of structure and function of neurons is usually occurs after neurodegeneration. Several therapeutic methods and drugs are available to relieve the symptoms of these diseases. Using siRNA technology is one of the main strategies that have been applied for treating neurodegenerative diseases, nowadays. However, choosing an appropriate delivery system for siRNA delivery remains challenging but some vehicles are introduced for providing optimal conditions for delivering siRNA. For example, peptide (TAT and MGF) tagged PEGylated chitosan nanoparticles have been used for siRNA delivery with the ability to pass the blood–brain barrier (Malhotra et al., 2013). Nonviral (naked siRNAs, P2X3 siRNA, RNA-polymer complexes, lipofectamine, Cholesterol conjugation), viral [adenoassociated virus (AAV) and lentivirus (LV)], and plasmids (pSUPER, PEGylated immunoliposomes containing plasmid encoding shRNA targeting epidermal growth factor receptor) are other possible vehicle for siRNA delivery in neurodegenerative diseases (Boudreau et al., 2011; Chakraborty, 2007).

10.2 ROLE OF siRNA IN NEURODEGENERATIVE DISORDERS

The prospect of using siRNA as a therapeutic method provides an opportunity for treating many diseases for which effective options are currently unavailable or limited. siRNA-based therapies are being investigated for diseases affecting the CNS and peripheral nervous system (PNS). The application of siRNA for treating different diseases related to nervous systems, whether central or peripheral, is widely growing. Therefore, we just focus on siRNA-based therapy in two important neurodegenerative diseases including Alzheimer's and Parkinson (Farah, 2007). Alzheimer's disease (AD) is the most common neurodegenerative disorder that is characterized by the presence of amyloid plaques and neurofibrillary tangle (NFTs) in the brain. Amyloid plaques are composed of the misfolded Ab peptide, which is produced by proteolytic processing of the amyloid precursor protein (APP). By using siRNA strategies it is possible to reduce Ab peptides in vivo by targeting the expression of enzymes like BACE1 that is required for the proteolytic processing of APP, or by directly targeting the expression of APP (Miller et al., 2004). Parkinson's disease (PD) is another neorudegenarative disease that used siRNA therapeutics in order to target a-syn expression as a plausible therapeutic approach for treating PD (Sapru et al., 2006).

11 CONCLUSIONS

To date, many siRNA-based nucleoceuticals have been used for regenerating different tissues. In this chapter, the application of various delivery systems for siRNA delivery to different tissues and their mechanisms of action are reviewed. These systems have established their reputation in the field of nanomedicine with significant potential in tissue regeneration. However, further investigations are required to confirm the ability of siRNA-based necleuceuticals for tissue repair.

REFERENCES

Addis, R.C., Epstein, J.A., 2013. Induced regeneration [mdash] the progress and promise of direct reprogramming for heart repair. Nat. Med. 19 (7), 829–836.

Aguirre, A., Sancho-Martinez, I., Belmonte, J.C.I., 2013. Reprogramming toward heart regeneration: stem cells and beyond. Cell Stem Cell 12 (3), 275–284.

Ahmed, Z., et al., 2011. Ocular neuroprotection by siRNA targeting caspase-2. Cell Death Dis. 2 (6), e173.

Ahuja, P., Sdek, P., MacLellan, W.R., 2007. Cardiac myocyte cell cycle control in development, disease, and regeneration. Physiol. Rev. 87 (2), 521–544.

Aiello, L.P., et al., 1995. Suppression of retinal neovascularization in vivo by inhibition of vascular endothelial growth factor (VEGF) using soluble VEGF-receptor chimeric proteins. Proc. Natl. Acad. Sci. 92 (23), 10457–10461.

Amarzguioui, M., et al., 2003. Tolerance for mutations and chemical modifications in a siRNA. Nucleic Acids Res. 31 (2), 589–595.

Apte, U., et al., 2009. Enhanced liver regeneration following changes induced by hepatocyte-specific genetic ablation of integrin-linked kinase. Hepatology 50 (3), 844–851.

Arthur, M.J., 2000. Fibrogenesis II. Metalloproteinases and their inhibitors in liver fibrosis. Am. J. Physiol. Gastrointestin. Liver Physiol. 279 (2), G245–G249.

Arthur, M., et al., 1992. Secretion of 72 kDa type IV collagenase/gelatinase by cultured human lipocytes. Analysis of gene expression, protein synthesis and proteinase activity. Biochem. J. 287, 701–707.

Asokan, A., et al., 2010. Reengineering a receptor footprint of adeno-associated virus enables selective and systemic gene transfer to muscle. Nat. Biotechnol. 28 (1), 79–82.

Atkinson, S.D., et al., 2011. Development of allele-specific therapeutic siRNA for keratin 5 mutations in epidermolysis bullosa simplex. J. Invest. Dermatol. 131 (10), 2079–2086.

Bannasch, H., et al., 2003. Skin tissue engineering. Clin. Plastic Surg. 30 (4), 573–579.

Barry, F.P., Murphy, J.M., 2004. Mesenchymal stem cells: clinical applications and biological characterization. Int. J. Biochem. Cell Biol. 36 (4), 568–584.

Bassit, A.C.F., et al., 2015. The Potential Use of Nanoparticles for Noggin siRNA Delivery to Accelerate Bone Formation in Distraction Osteogenesis. J. Nanomed. Nanotechnol. 6 (1), 1.

Beltrami, A.P., et al., 2001. Evidence that human cardiac myocytes divide after myocardial infarction. N. Engl. J. Med. 344 (23), 1750–1757.

Bennett, J., et al., 1994. Adenovirus vector-mediated in vivo gene transfer into adult murine retina. Invest. Ophthalmol. Visual Sci. 35 (5), 2535–2542.

Benyon, R.C., Arthur, M., 2001. Extracellular matrix degradation and the role of hepatic stellate cells. Semin. Liver Dis. 21 (3), 373–384.

Bergmann, O., et al., 2009. Evidence for cardiomyocyte renewal in humans. Science 324 (5923), 98–102.

Berry, M., Carlile, J., Hunter, A., 1996. Peripheral nerve explants grafted into the vitreous body of the eye promote the regeneration of retinal ganglion cell axons severed in the optic nerve. J. Neurocytol. 25 (1), 147–170.

Bertrand, J.-R., et al., 2002. Comparison of antisense oligonucleotides and siRNAs in cell culture and in vivo. Biochem. Biophys. Res. Commun. 296 (4), 1000–1004.

Bessis, N., GarciaCozar, F., Boissier, M., 2004. Immune responses to Gene Ther. vectors: influence on vector function and effector mechanisms. Gene Ther. 11, S10–S17.

Bicknell, K.A., Coxon, C.H., Brooks, G., 2007. Can the cardiomyocyte cell cycle be reprogrammed? J. Mol. Cell. Cardiol. 42 (4), 706–721.

Bolli, R., 1998. Causative role of oxyradicals in myocardial stunning: a proven hypothesis. Basic Res. Cardiol. 93 (3), 156–162.

Bon, D., et al., 2012. New strategies to optimize kidney recovery and preservation in transplantation. Nat. Rev. Nephrol. 8 (6), 339–347.

Boudreau, R.L., Rodríguez-Lebrón, E., Davidson, B.L., 2011. RNAi medicine for the brain: progresses and challenges. Hum. Mol. Genet. 20 (R1), R21–R27.

Brownlee, M., 2001. Biochemistry and molecular cell biology of diabetic complications. Nature 414 (6865), 813–820.

Budker, V.G., et al., 2006. Mechanism of plasmid delivery by hydrodynamic tail vein injection. II. Morphological studies. J. Gene Med. 8 (7), 874–888.

Campochiaro, P.A., 2000. Retinal and choroidal neovascularization. J. Cell. Physiol. 184 (3), 301–310.

Cantaluppi, V., et al., 2012. Microvesicles derived from endothelial progenitor cells protect the kidney from ischemia–reperfusion injury by microRNA-dependent reprogramming of resident renal cells. Kidney Int. 82 (4), 412–427.

Caplan, A.I., Correa, D., 2011. PDGF in bone formation and regeneration: new insights into a novel mechanism involving MSCs. J. Orthop. Res. 29 (12), 1795–1803.

Casson, R.J., et al., 2012. Definition of glaucoma: clinical and experimental concepts. Clin. Exp. Ophthalmol. 40 (4), 341–349.

Castaneda, M.P., et al., 2003. Activation of mitochondrial apoptotic pathways in human renal allografts after ischemia–reperfusion injury. Transplantation 76 (1), 50–54.
Cave, A., 2009. Selective targeting of NADPH oxidase for cardiovascular protection. Curr. Opin. Pharmacol. 9 (2), 208–213.
Cave, A.C., et al., 2006. NADPH oxidases in cardiovascular health and disease. Antioxid. Redox Signal. 8 (5–6), 691–728.
Chakraborty, C., 2007. Potentiality of small interfering RNAs (siRNA) as recent therapeutic targets for gene-silencing. Curr. Drug Targets 8 (3), 469–482.
Chapman, J.R., O'Connell, P.J., Nankivell, B.J., 2005. Chronic renal allograft dysfunction. J. Am. Soc. Nephrol. 16 (10), 3015–3026.
Chatauret, N., Thuillier, R., Hauet, T., 2011. Preservation strategies to reduce ischemic injury in kidney transplantation: pharmacological and genetic approaches. Curr. Opin. Organ Transplant. 16 (2), 180–187.
Cheema, S.K., et al., 2007. Regulation and guidance of cell behavior for tissue regeneration via the siRNA mechanism. Wound Repair Regen. 15 (3), 286–295.
Chen, Y.-J., et al., 2004. Recruitment of mesenchymal stem cells and expression of TGF-β1 and VEGF in the early stage of shock wave-promoted bone regeneration of segmental defect in rats. J. Orthop. Res. 22 (3), 526–534.
Chen, C., et al., 2009. NgR RNA interference, combined with zymosan intravitreal injection, enhances optic nerve regeneration. J. Neurochem. 110 (5), 1628–1634.
Chen, L., et al., 2010. Insulin-like growth factor 2 (IGF-2) potentiates BMP-9-induced osteogenic differentiation and bone formation. J. Bone Miner. Res. 25 (11), 2447–2459.
Chen, H., et al., 2011. Mir-34a is upregulated during liver regeneration in rats and is associated with the suppression of hepatocyte proliferation. PloS one 6 (5), e20238.
Chen, M., et al., 2014. Topical delivery of siRNA into skin using SPACE-peptide carriers. J. Control. Rel. 179, 33–41.
Chiu, Y.-L., Rana, T.M., 2003. siRNA function in RNAi: a chemical modification analysis. RNA 9 (9), 1034–1048.
Chong, R.H.E., et al., 2013. Gene silencing following siRNA delivery to skin via coated steel microneedles: in vitro and in vivo proof-of-concept. J. Control. Rel. 166 (3), 211–219.
Chujo, S., et al., 2004. Connective tissue growth factor causes persistent proa2 (I) collagen gene expression induced by transforming growth factor-b in a mouse fibrosis model. Arthritis and Rheumatism. Wiley-Liss Div John Wiley & Sons Inc, NJ, USA.
Chung, E.J., et al., 2012. Transduction of PTEN proteins using the tat domain modulates TGF-β1–mediated signaling pathways and transdifferentiation in subconjunctival fibroblasts. Invest. Ophthalmol. Visual Sci. 53 (1), 379–386.
Cohausz, O., et al., 2008. The roles of poly(ADP-ribose)-metabolizing enzymes in alkylation-induced cell death. Cell. Mol. Life Sci. 65 (4), 644–655.
Cunningham, P.N., et al., 2002. Acute renal failure in endotoxemia is caused by TNF acting directly on TNF receptor-1 in kidney. J. Immunol. 168 (11), 5817–5823.
Czauderna, F., et al., 2003. Structural variations and stabilising modifications of synthetic siRNAs in mammalian cells. Nucleic Acids Res. 31 (11), 2705–2716.
Daemen, M.A., et al., 1999. Inhibition of apoptosis induced by ischemia–reperfusion prevents inflammation. J. Clin. Invest. 104 (5), 541.
de Vries, B., et al., 2003. Inhibition of complement factor C5 protects against renal ischemia–reperfusion injury: inhibition of late apoptosis and inflammation1. Transplantation 75 (3), 375–382.
de Vries, B., et al., 2004. The mannose-binding lectin-pathway is involved in complement activation in the course of renal ischemia–reperfusion injury. Am. J. Pathol. 165 (5), 1677–1688.
De Vriese, A.S., et al., 2006. Myofibroblast transdifferentiation of mesothelial cells is mediated by RAGE and contributes to peritoneal fibrosis in uraemia. Nephrol. Dialys. Transplant. 21 (9), 2549–2555.

Devuyst, O., Topley, N., Williams, J.D., 2002. Morphological and functional changes in the dialysed peritoneal cavity: impact of more biocompatible solutions. Nephrol. Dialys. Transplant. 17 (Suppl. 3), 12–15.

Dimitriou, R., Tsiridis, E., Giannoudis, P.V., 2005. Current concepts of molecular aspects of bone healing. Injury 36 (12), 1392–1404.

Du, C., et al., 2004. Renal tubular epithelial cell self-injury through Fas/Fas ligand interaction promotes renal allograft injury. Am. J. Transpl. 4 (10), 1583–1594.

Eddy, A.A., 2000. Molecular basis of renal fibrosis. Pediatr. Nephrol. 15 (3–4), 290–301.

Eikmans, M., et al., 2003. ECM homeostasis in renal diseases: a genomic approach. J. Pathol. 200 (4), 526–536.

Eulalio, A., et al., 2012. Functional screening identifies miRNAs inducing cardiac regeneration. Nature 492 (7429), 376–381.

Farah, M.H., 2007. RNAi silencing in mouse models of neurodegenerative diseases. Curr. Drug Deliv. 4 (2), 161–167.

Farokhi, M., et al., 2013. Sustained release of platelet-derived growth factor and vascular endothelial growth factor from silk/calcium phosphate/PLGA based nanocomposite scaffold. Int. J. Pharm. 454 (1), 216–225.

Farokhi, M., et al., 2014. Bio-hybrid silk fibroin/calcium phosphate/PLGA nanocomposite scaffold to control the delivery of vascular endothelial growth factor. Mater. Sci. Eng. 35, 401–410.

Fausto, N., Campbell, J.S., Riehle, K.J., 2006. Liver regeneration. Hepatology 43 (S1), S45–S53.

Fechner, H., et al., 2007. Highly efficient and specific modulation of cardiac calcium homeostasis by adenovector-derived short hairpin RNA targeting phospholamban. Gene Ther. 14 (3), 211–218.

Fehrenbach, H., et al., 2001. Up-regulated expression of the receptor for advanced glycation end products in cultured rat hepatic stellate cells during transdifferentiation to myofibroblasts. Hepatology 34 (5), 943–952.

Ferdinandy, P., Schulz, R., 2003. Nitric oxide, superoxide, and peroxynitrite in myocardial ischaemia-reperfusion injury and preconditioning. Br. J. Pharmacol. 138 (4), 532–543.

Fire, A., et al., 1998. Potent and specific genetic interference by double-stranded RNA in *Caenorhabditis elegans*. Nature 391 (6669), 806–811.

Fischer, D., Pavlidis, M., Thanos, S., 2000. Cataractogenic lens injury prevents traumatic ganglion cell death and promotes axonal regeneration both in vivo and in culture. Invest. Ophthalmol. Visual Sci. 41 (12), 3943–3954.

Fisher, M.B., Mauck, R.L., 2013. Tissue engineering and regenerative medicine: recent innovations and the transition to translation. Tissue Eng. Part B 19 (1), 1–13.

Fournier, A.E., GrandPre, T., Strittmatter, S.M., 2001. Identification of a receptor mediating Nogo-66 inhibition of axonal regeneration. Nature 409 (6818), 341–346.

Friedman, S.L., 2000. Molecular regulation of hepatic fibrosis, an integrated cellular response to tissue injury. J. Biol. Chem. 275 (4), 2247–2250.

Friedman, S.L., 2003. Liver fibrosis–from bench to bedside. J. Hepatol. 38, 38–53.

Fuse, K., et al., 2005. Myeloid differentiation factor-88 plays a crucial role in the pathogenesis of coxsackievirus B3–induced myocarditis and influences type I interferon production. Circulation 112 (15), 2276–2285.

Gabrielli, G.B., et al., 1997. Serum laminin and type III procollagen in chronic hepatitis C. Diagnostic value in the assessment of disease activity and fibrosis. Clin. Chim. Acta 265 (1), 21–31.

Gandhi, V.C., et al., 1980. Sclerotic thickening of the peritoneal membrane in maintenance peritoneal dialysis patients. Arch. Intern. Med. 140 (9), 1201–1203.

Garba, A.O., Mousa, S.A., 2010. Bevasiranib for the treatment of wet, age-related macular degeneration. Ophthalmol. Eye Dis. 2, 75.

Geusens, B., et al., 2009. Cutaneous short-interfering RNA therapy. Expert Opin. Drug Deliv. 6 (12), 1333–1349.

Gille, H., et al., 2001. Analysis of biological effects and signaling properties of Flt-1 (VEGFR-1) and KDR (VEGFR-2) A reassessment using novel receptor-specific vascular endothelial growth factor mutants. J. Biol. Chem. 276 (5), 3222–3230.

Gomez, D., et al., 1997. Tissue inhibitors of metalloproteinases: structure, regulation and biological functions. Eur. J. Cell Biol. 74 (2), 111–122.

Göser, S., et al., 2005. Critical role for monocyte chemoattractant protein-1 and macrophage inflammatory protein-1α in induction of experimental autoimmune myocarditis and effective anti–monocyte chemoattractant protein-1 gene therapy. Circulation 112 (22), 3400–3407.
Gray, S.J., Samulski, R.J., 2008. Optimizing gene delivery vectors for the treatment of heart disease. Expert Opin. Biol. Ther. 8 (7), 911–922.
Gruss, C.J., et al., 2003. Stroma formation and angiogenesis by overexpression of growth factors, cytokines, and proteolytic enzymes in human skin grafted to SCID mice. J. Invest. Dermatol. 120 (4), 683–692.
Gusella, G.L., et al., 2002. In vivo gene transfer to kidney by lentiviral vector. Kidney Int. 61, S32–S36.
Hamar, P., et al., 2004. Small interfering RNA targeting Fas protects mice against renal ischemia–reperfusion injury. Proc. Natl. Acad. Sci. USA 101 (41), 14883–14888.
Hegde, V., et al., 2014. In vivo gene silencing following non-invasive siRNA delivery into the skin using a novel topical formulation. J. Control. Rel. 196, 355–362.
Heymes, C., et al., 2003. Increased myocardial NADPH oxidase activity in human heart failure. J. Am. Coll. Cardiol. 41 (12), 2164–2171.
Hickerson, R.P., et al., 2008. Single-nucleotide-specific siRNA targeting in a dominant-negative skin model. J. Invest. Dermatol. 128 (3), 594–605.
Hickerson, R.P., et al., 2013. Gene silencing in skin after deposition of self-delivery siRNA with a motorized microneedle array device. Molecular Ther.— Nucleic Acids 2 (10), e129.
Hinz, B., 2006. Masters and servants of the force: the role of matrix adhesions in myofibroblast force perception and transmission. Eur. J. Cell Biol. 85 (3), 175–181.
Höcherl, K., et al., 2010. Inhibition of NF-κB ameliorates sepsis-induced downregulation of aquaporin-2/V2 receptor expression and acute renal failure in vivo. Am. J. Physiol. Renal Physiol. 298 (1), F196–F204.
Hoste, E.A., et al., 2008. Epidemiology of acute kidney injury. CJASN 3 (3), 881–886.
Hu, Y.B., Li, D.G., Lu, H.M., 2007. Modified synthetic siRNA targeting tissue inhibitor of metalloproteinase-2 inhibits hepatic fibrogenesis in rats. J. Gene Med. 9 (3), 217–229.
Hu, P.F., et al., 2008. Inhibition of plasminogen activator inhibitor-1 expression by siRNA in rat hepatic stellate cells. J. Gastroenterol. Hepatol. 23 (12), 1917–1925.
Hu, P.-F., et al., 2009. Adenovirus-mediated transfer of siRNA against PAI-1 mRNA ameliorates hepatic fibrosis in rats. J. Hepatol. 51 (1), 102–113.
Hynes, R.O., 2009. The extracellular matrix: not just pretty fibrils. Science 326 (5957), 1216–1219.
Iliopoulos, D., et al., 2008. Integrative microRNA and proteomic approaches identify novel osteoarthritis genes and their collaborative metabolic and inflammatory networks. PLoS One 3 (11), e3740.
Inoue, T., et al., 2007. Modulation of scratching behavior by silencing an endogenous cyclooxygenase-1 gene in the skin through the administration of siRNA. J. Gene Med. 9 (11), 994–1001.
Ito, K., et al., 2008. Adeno-associated viral vector transduction of green fluorescent protein in kidney: effect of unilateral ureteric obstruction. BJU Int. 101 (3), 376–381.
Jakobsen, M., et al., 2009. Amelioration of psoriasis by anti-TNF-α RNAi in the xenograft transplantation model. Molecular Ther. 17 (10), 1743–1753.
Jiang, G., et al., 2009. Target specific intracellular delivery of siRNA/PEI– HA complex by receptor mediated endocytosis. Molecular Pharm. 6 (3), 727–737.
Kahan, B.D., et al., 2004. Phase I and phase II safety and efficacy trial of intercellular adhesion molecule-1 antisense oligodeoxynucleotide (ISIS 2302) for the prevention of acute allograft rejection. Transplantation 78 (6), 858–863.
Kajstura, J., et al., 2010. Cardiomyogenesis in the adult human heart. Circ. Res. 107 (2), 305–315.
Kelly, K., Plotkin, Z., Dagher, P.C., 2001. Guanosine supplementation reduces apoptosis and protects renal function in the setting of ischemic injury. J. Clin. Invest. 108 (9), 1291.
Kelly, K., et al., 2003. P53 mediates the apoptotic response to GTP depletion after renal ischemia–reperfusion: protective role of a p53 inhibitor. J. Am. Soc. Nephrol. 14 (1), 128–138.

Kempen, D.H., et al., 2009. Effect of local sequential VEGF and BMP-2 delivery on ectopic and orthotopic bone regeneration. Biomaterials 30 (14), 2816–2825.

Kim, M., et al., 2009. Kidney-specific reconstitution of the A1 adenosine receptor in A1 adenosine receptor knockout mice reduces renal ischemia–reperfusion injury. Kidney Int. 75 (8), 809–823.

Kim, B.-S., et al., 2011. Design of artificial extracellular matrices for tissue engineering. Prog. Polym. Sci. 36 (2), 238–268.

Kisiswa, L., et al., 2010. Retinal ganglion cell death postponed: giving apoptosis a break? Ophthal. Res. 43 (2), 61–78.

Kiss, L., Szabó, C., 2005. The pathogenesis of diabetic complications: the role of DNA injury and poly(ADP-ribose) polymerase activation in peroxynitrite-mediated cytotoxicity. Memórias do Instituto Oswaldo Cruz 100, 29–37.

Kobayashi, N., et al., 2001. Hepatic uptake and gene expression mechanisms following intravenous administration of plasmid DNA by conventional and hydrodynamics-based procedures. J. Pharmacol. Exp. Therap. 297 (3), 853–860.

Kobayashi, T., et al., 2005. Connective tissue growth factor mediates the profibrotic effects of transforming growth factor-β produced by tubular epithelial cells in response to high glucose. Musculoskel. Disord.Exp. Nephrol. Clin. Exp. Nephrol. 9 (2), 114–121.

Krijnen, P., et al., 2003. Increased Nox2 expression in human cardiomyocytes after acute myocardial infarction. J. Clin. Pathol. 56 (3), 194–199.

Kurokawa, T., et al., 1999. BDNF diminishes caspase-2 but not c-Jun immunoreactivity of neurons in retinal ganglion cell layer after transient ischemia. Invest. Ophthalmol. Visual Sci. 40, 3006–3011.

Kushibiki, T., et al., 2003. In vivo release and gene expression of plasmid DNA by hydrogels of gelatin with different cationization extents. J. Control. Rel. 90 (2), 207–216.

Kushibiki, T., et al., 2005. Delivery of plasmid DNA expressing small interference RNA for TGF-β type II receptor by cationized gelatin to prevent interstitial renal fibrosis. J. Control. Rel. 105 (3), 318–331.

Kwak, N., et al., 2000. VEGF is major stimulator in model of choroidal neovascularization. Invest. Ophthalmol. Visual Sci. 41 (10), 3158–3164.

Laflamme, M.A., Murry, C.E., 2011. Heart regeneration. Nature 473 (7347), 326–335.

Lamkanfi, M., et al., 2002. Alice in caspase land. A phylogenetic analysis of caspases from worm to man. Cell Death Differ. 9 (4), 358–361.

Landen, C.N., et al., 2005. Therapeutic EphA2 gene targeting in vivo using neutral liposomal small interfering RNA delivery. Cancer Res. 65 (15), 6910–6918.

Larson, S.D., et al., 2007. Effectiveness of siRNA uptake in target tissues by various delivery methods. Surgery 142 (2), 262–269.

Leask, A., 2003. Transcriptional profiling of the scleroderma fibroblast reveals a potential role for connective tissue growth factor (CTGF) in pathological fibrosis. In: Presented at the 1333rd Meeting of the Keio Medical Society in Tokyo.

Leask, A., Abraham, D.J., 2004. TGF-β signaling and the fibrotic response. FASEB J. 18 (7), 816–827.

Lebherz, C., et al., 2008. Novel AAV serotypes for improved ocular gene transfer. J. Gene Med. 10 (4), 375–382.

Lee, P., et al., 2003. Fas pathway is a critical mediator of cardiac myocyte death and MI during ischemia–reperfusion in vivo. Am. J. Physiol. Heart Circ. Physiol. 284 (2), H456–H463.

Lee, M.-J., Shin, J.-O., Jung, H.-S., 2013. Thy-1 knockdown retards wound repair in mouse skin. J. Dermatol. Sci. 69 (2), 95–104.

Lei, P., et al., 2012. Poly(ADP-ribose) polymerase-1 in high glucose-induced epithelial-mesenchymal transition during peritoneal fibrosis. Int. J. Molecular Med. 29 (3), 472–478.

Leon, S., et al., 2000. Lens injury stimulates axon regeneration in the mature rat optic nerve. J. Neurosci. 20 (12), 4615–4626.

Leor, J., et al., 2003. Reprogramming cells for transplantation. Heart Fail. Rev. 8 (3), 285–292.

Lewis, D.L., Wolff, J.A., 2007. Systemic siRNA delivery via hydrodynamic intravascular injection. Adv. Drug Deliv. Rev. 59 (2), 115–123.
Li, C.X., et al., 2006. Delivery of RNA interference. Cell Cycle 5 (18), 2103–2109.
Li, C., et al., 2009a. Cytotoxic-T-lymphocyte-mediated elimination of target cells transduced with engineered adeno-associated virus type 2 vector in vivo. J. Virol. 83 (13), 6817–6824.
Li, C., et al., 2009b. Cellular immune response to cryptic epitopes during therapeutic gene transfer. Proc. Natl. Acad. Sci. 106 (26), 10770–10774.
Lipkowitz, M.S., et al., 1999. Transduction of renal cells in vitro and in vivo by adeno-associated virus Gene Ther. vectors. J. Am. Soc. Nephrol. 10 (9), 1908–1915.
Liu, B.P., et al., 2002. Myelin-associated glycoprotein as a functional ligand for the Nogo-66 receptor. Science 297 (5584), 1190–1193.
Lloyd-Jones, D., et al., 2010. Heart disease and stroke statistics—2010 update. A report from the American Heart Association. Circulation 121 (7), e46–e215.
Locker, J., et al., 2003. A common set of immediate-early response genes in liver regeneration and hyperplasia. Hepatology 38 (2), 314–325.
Looi, Y.H., et al., 2008. Involvement of Nox2 NADPH oxidase in adverse cardiac remodeling after myocardial infarction. Hypertension 51 (2), 319–325.
Lowes, B.D., et al., 2002. Myocardial gene expression in dilated cardiomyopathy treated with beta-blocking agents. N. Engl. J. Med. 346 (18), 1357–1365.
Lu, M., Adamis, A.P., 2002. Vascular endothelial growth factor gene regulation and action in diabetic retinopathy. Ophthalmol. Clin. North Am. 15 (1), 69–79.
Luo, G., et al., 2008. Inhibition of connective tissue growth factor by small interfering RNA prevents renal fibrosis in rats undergoing chronic allograft nephropathy. Transpl. Proc. 40 (7), 2365–2369.
Lütjen-Drecoll, E., 2005. Morphological changes in glaucomatous eyes and the role of TGFβ 2 for the pathogenesis of the disease. Exp. Eye Res. 81 (1), 1–4.
Lv, H., et al., 2006. Toxicity of cationic lipids and cationic polymers in gene delivery. J. Control. Rel. 114 (1), 100–109.
Ma, D., et al., 2009. Xenon preconditioning protects against renal ischemic-reperfusion injury via HIF-1α activation. J. Am. Soc. Nephrol. 20 (4), 713–720.
MacLennan, D.H., Kranias, E.G., 2003. Phospholamban: a crucial regulator of cardiac contractility. Nat. Rev. Molecular Cell Biol. 4 (7), 566–577.
Maekawa, Y., et al., 2009. Survival and cardiac remodeling after myocardial infarction are critically dependent on the host innate immune interleukin-1 receptor-associated kinase-4 signaling a regulator of bone marrow-derived dendritic cells. Circulation 120 (14), 1401–1414.
Mahmoud, A.I., et al., 2013. Meis1 regulates postnatal cardiomyocyte cell cycle arrest. Nature 497 (7448), 249–253.
Malhotra, M., et al., 2013. Intranasal, siRNA delivery to the brain by TAT/MGF tagged PEGylated chitosan nanoparticles. Journal of Pharm. 2013.
Manjunath, N., et al., 2009. Lentiviral delivery of short hairpin RNAs. Adv. Drug Deliv. Rev. 61 (9), 732–745.
Mardaryev, A.N., et al., 2011. Lhx2 differentially regulates Sox9 Tcf4 and Lgr5 in hair follicle stem cells to promote epidermal regeneration after injury. Development 138 (22), 4843–4852.
Marquez, R.T., et al., 2010. MicroRNA-21 is upregulated during the proliferative phase of liver regeneration, targets Pellino-1, and inhibits NF-κB signaling. Am. J. Physiol. Gastrointestin. Liver Physiol. 298 (4), G535–G541.
Matthijsen, R.A., et al., 2007. Myeloperoxidase is critically involved in the induction of organ damage after renal ischemia reperfusion. Am. J. Pathol. 171 (6), 1743–1752.
McKerracher, L., Winton, M.J., 2002. Nogo on the go. Neuron 36 (3), 345–348.
Mehta, R.L., et al., 2004. Spectrum of acute renal failure in the intensive care unit: the PICARD experience. Kidney Int. 66 (4), 1613–1621.

Miller, V.M., et al., 2004. Targeting Alzheimer's disease genes with RNA interference: an efficient strategy for silencing mutant alleles. Nucleic Acids Res. 32 (2), 661–668.

Mirotsou, M., et al., 2011. Paracrine mechanisms of stem cell reparative and regenerative actions in the heart. J. Mol. Cell. Cardiol. 50 (2), 280–289.

Mitchell, C., Willenbring, H., 2008. A reproducible and well-tolerated method for 2/3 partial hepatectomy in mice. Nat. Protoc. 3 (7), 1167–1170.

Miyagishi, M., Taira, K., 2002. U6 promoter–driven siRNAs with four uridine 3′ overhangs efficiently suppress targeted gene expression in mammalian cells. Nat. Biotechnol. 20 (5), 497–500.

Molitoris, B.A., et al., 2009. siRNA targeted to p53 attenuates ischemic and cisplatin-induced acute kidney injury. J. Am. Soc. Nephrol. 20 (8), 1754–1764.

Moreira, R.K., 2007. Hepatic stellate cells and liver fibrosis. Arch. Pathol. Lab. Med. 131 (11), 1728.

Mottaghitalab, F., et al., 2013. A biosynthetic nerve guide conduit based on silk/SWNT/fibronectin nanocomposite for peripheral nerve regeneration. PLoS One 8 (9), e74417.

Mottaghitalab, F., et al., 2011. Enhancement of neural cell lines proliferation using nano-structured chitosan/poly(vinyl alcohol) scaffolds conjugated with nerve growth factor. Carbohydr. Polym. 86 (2), 526–535.

Müller, O.J., et al., 2003. Random peptide libraries displayed on adeno-associated virus to select for targeted Gene Ther. vectors. Nat. Biotechnol. 21 (9), 1040–1046.

Nagata, K., 1996. Hsp47: a collagen-specific molecular chaperone. Trends Biochem. Sci. 21 (1), 23–26.

Nagata, K., Saga, S., Yamada, K.M., 1986. A major collagen-binding protein of chick embryo fibroblasts is a novel heat shock protein. J. Cell Biol. 103 (1), 223–229.

Nagata, K., et al., 1988. Biosynthesis of a novel transformation-sensitive heat-shock protein that binds to collagen. Regulation by mRNA levels and in vitro synthesis of a functional precursor. J. Biol. Chem. 263 (17), 8344–8349.

Nakamura, A., et al., 2004. β2-Adrenoceptor activation attenuates endotoxin-induced acute renal failure. J. Am. Soc. Nephrol. 15 (2), 316–325.

Nies, B., Dingeldein, E., Wahlig, H., 2000. Bone replacement material with FGF. Google Patents.

Nishino, T., et al., 2003. Antisense oligonucleotides against collagen-binding stress protein HSP47 suppress peritoneal fibrosis in rats. Kidney Int. 64 (3), 887–896.

Niyibizi, C., et al., 2004. Gene Ther. approaches for osteogenesis imperfecta. Gene Ther. 11 (4), 408–416.

Oakley, F., et al., 2005. Inhibition of inhibitor of κB kinases stimulates hepatic stellate cell apoptosis and accelerated recovery from rat liver fibrosis. Gastroenterology 128 (1), 108–120.

Obata, Y., et al., 2012. HSP47 siRNA conjugated with cationized gelatin microspheres suppresses peritoneal fibrosis in mice. Acta Biomater. 8 (7), 2688–2696.

Okada, H., et al., 2008. Poly(ADP-ribose) polymerase-1 enhances transcription of the profibrotic CCN2 gene. J. Am. Soc. Nephrol. 19 (5), 933–942.

Okamoto, N., et al., 1997. Transgenic mice with increased expression of vascular endothelial growth factor in the retina: a new model of intraretinal and subretinal neovascularization. Am. J. Pathol. 151 (1), 281.

Okazaki, I., et al., 2000. Molecular mechanism of the reversibility of hepatic fibrosis: with special reference to the role of matrix metalloproteinases. J. Gastroenterol. Hepatol. 15 (s1), 26–32.

Okazaki, I., et al., 2001. Reversibility of hepatic fibrosis: from the first report of collagenase in the liver to the possibility of Gene Ther. for recovery. Keio J. Med. 50 (2), 58–65.

Oldfield, M.D., et al., 2001. Advanced glycation end products cause epithelial-myofibroblast transdifferentiation via the receptor for advanced glycation end products (RAGE). J. Clin. Invest. 108 (12), 1853.

Onyekwelu, I., Goldring, M.B., Hidaka, C., 2009. Chondrogenesis, joint formation, and articular cartilage regeneration. J. Cell. Biochem. 107 (3), 383–392.

Ortiz, P.A., et al., 2003. An in vivo method for adenovirus-mediated transduction of thick ascending limbs. Kidney Int. 63 (3), 1141–1149.

Ozaki, H., et al., 2000. Blockade of vascular endothelial cell growth factor receptor signaling is sufficient to completely prevent retinal neovascularization. Am. J. Pathol. 156 (2), 697–707.

Pacak, C.A., et al., 2006. Recombinant adeno-associated virus serotype 9 leads to preferential cardiac transduction in vivo. Circ. Res. 99 (4), e3–e9.
Paige, S.L., et al., 2012. A temporal chromatin signature in human embryonic stem cells identifies regulators of cardiac development. Cell 151 (1), 221–232.
Perico, N., et al., 2004. Delayed graft function in kidney transplantation. Lancet 364 (9447), 1814–1827.
Petite, H., et al., 2000. Tissue-engineered bone regeneration. Nat. Biotechnol. 18 (9), 959–963.
Pi, Y., et al., 2015. Intra-articular delivery of anti-Hif-2α siRNA by chondrocyte-homing nanoparticles to prevent cartilage degeneration in arthritic mice. Gene Ther.
Pinzani, M., 2000. Liver fibrosis. Semin. Immunopathol. 21 (4), 475–490.
Poller, W., et al., 2010. Cardiac-targeted delivery of regulatory RNA molecules and genes for the treatment of heart failure. Card. Res., cvq056.
Pratt, J.R., Basheer, S.A., Sacks, S.H., 2002. Local synthesis of complement component C3 regulates acute renal transplant rejection. Nat. Med. 8 (6), 582–587.
Quigley, H.A., Broman, A.T., 2006. The number of people with glaucoma worldwide in 2010 and 2020. Br. J. Ophthalmol. 90 (3), 262–267.
Rácz, Z., Hamar, P., 2008. RNA interference in research and therapy of renal diseases. Contrib. Nephrol. 159, 78–95.
Reeves, H.L., Friedman, S.L., 2002. Activation of hepatic stellate cells-a key issue in liver fibrosis. Front. Biosci. 7 (4), 808–826.
Reich, S.J., et al., 2003. Small interfering RNA (siRNA) targeting VEGF effectively inhibits ocular neovascularization in a mouse model. Mol. Vis. 9 (5), 210–216.
Reiss, K., et al., 1993. Impairment of myocyte contractility following coronary artery narrowing is associated with activation of the myocyte IGF 1 autocrine system, enhanced expression of late growth related genes, DNA synthesis, and myocyte nuclear mitotic division in rats. Exp. cell Res. 207 (2), 348–360.
Ritter, T., Nosov, M., Griffin, M.D., 2009. Gene Ther. in transplantation: Toward clinical trials. Curr. Opin. Mol. Ther. 11 (5), 504–512.
Roger, V.L., et al., 2012. Heart disease and stroke statistics—2012 update a report from the American heart association. Circulation 125 (1), e2–e220.
Rosso, F., et al., 2004. From cell–ECM interactions to tissue engineering. J. Cell. Physiol. 199 (2), 174–180.
Ryhänen, L., et al., 1996. The effect of malotilate on type III and type IV collagen, laminin and fibronectin metabolism in dimethylnitrosamine-induced liver fibrosis in the rat. J. Hepatol. 24 (2), 238–245.
Sakamoto, N., et al., 2005. Influence of glucose and inflammatory cytokines on TGF-beta1 and CTGF mRNA expressions in human peritoneal mesothelial cells. Int. J. Mol. Med. 15 (6), 907–911.
Sapru, M.K., et al., 2006. Silencing of human α-synuclein in vitro and in rat brain using lentiviral-mediated RNAi. Exp. Neurol. 198 (2), 382–390.
Schmidt, K.-G., Bergert, H., Funk, R., 2008. Neurodegenerative diseases of the retina and potential for protection and recovery. Curr. Neuropharmacol. 6 (2), 164.
Seo, M.S., et al., 1999. Dramatic inhibition of retinal and choroidal neovascularization by oral administration of a kinase inhibitor. Am. J. Pathol. 154 (6), 1743–1753.
Seregin, S.S., Amalfitano, A., 2009. Overcoming pre-existing adenovirus immunity by genetic engineering of adenovirus-based vectors. Expert Opin. Biol. Ther. 9 (12), 1521–1531.
Shen, J., et al., 2006. Suppression of ocular neovascularization with siRNA targeting VEGF receptor 1. Gene Ther. 13 (3), 225–234.
Shimizu, H., et al., 2010. siRNA-based therapy ameliorates glomerulonephritis. J. Am. Soc. Nephrol. 21 (4), 622–633.
Shou, Z., et al., 2009. SHARP-2 gene silencing by lentiviral-based short hairpin RNA interference prolonged rat kidney transplant recipients' survival time. J. Int. Med.Res. 37 (3), 766–778.
Simpson, K.J., et al., 2008. Identification of genes that regulate epithelial cell migration using an siRNA screening approach. Nat. Cell Biol. 10 (9), 1027–1038.

Singh, M., et al., 2001. Cell-specific caspase expression by different neuronal phenotypes in transient retinal ischemia. J. Neurochem. 77 (2), 466–475.

Sioud, M., Sørensen, D.R., 2003. Cationic liposome-mediated delivery of siRNAs in adult mice. Biochem. Biophys. Res. Commun. 312 (4), 1220–1225.

Somasuntharam, I., et al., 2013. Delivery of Nox2-NADPH oxidase siRNA with polyketal nanoparticles for improving cardiac function following myocardial infarction. Biomaterials 34 (31), 7790–7798.

Song, R.H., et al., 2007. Aggrecan degradation in human articular cartilage explants is mediated by both ADAMTS-4 and ADAMTS-5. Arthritis Rheum. 56 (2), 575–585.

Soutschek, J., et al., 2004. Therapeutic silencing of an endogenous gene by systemic administration of modified siRNAs. Nature 432 (7014), 173–178.

Stahl, P.J., Felsen, D., 2001. Transforming growth factor-β, basement membrane, and epithelial-mesenchymal transdifferentiation: implications for fibrosis in kidney disease. Am. J. Pathol. 159 (4), 1187–1192.

Stoll, G., Müller, H.W., 1999. Nerve injury, axonal degeneration and neural regeneration: basic insights. Brain Pathol. 9 (2), 313–325.

Su, Y., et al., 2012. Smad7 siRNA inhibit expression of extracellular matrix in trabecular meshwork cells treated with TGF-β2. Mol. Vis. 18, 1881.

Sun, L., 2014. Self-assembled rosette nanotubes for bone tissue engineering and drug delivery applications.

Sunderland, S., Walshe, F., 1968. Nerves and nerve injuries.

Sy, J.C., et al., 2008. Sustained release of a p38 inhibitor from non-inflammatory microspheres inhibits cardiac dysfunction. Nat. Mater. 7 (11), 863–868.

Szeto, C.-C., et al., 2005. Differential effects of transforming growth factor-beta on the synthesis of connective tissue growth factor and vascular endothelial growth factor by peritoneal mesothelial cell. Nephron Musculoskel. Disord.Exp. Nephrol. 99 (4), e95–e104.

Takabatake, Y., et al., 2005. Exploring RNA interference as a therapeutic strategy for renal disease. Gene Ther. 12 (12), 965–973.

Takabatake, Y., et al., 2007. Chemically modified siRNA prolonged RNA interference in renal disease. Biochem. Biophys. Res. Commun. 363 (2), 432–437.

Takai, Y., Tanito, M., Ohira, A., 2012. Multiplex cytokine analysis of aqueous humor in eyes with primary open-angle glaucoma, exfoliation glaucoma, and cataract. Invest. Ophthalmol. Vis. Sci. 53 (1), 241–247.

Takeshita, F., et al., 2005. Efficient delivery of small interfering RNA to bone-metastatic tumors by using atelocollagen in vivo. Proc. Natl. Acad. Sci. USA 102 (34), 12177–12182.

Taub, R., 2004. Liver regeneration: from myth to mechanism. Nat. Rev. Mol. Cell Biol. 5 (10), 836–847.

Théret, N., et al., 1997. Activation of matrix metalloproteinase-2 from hepatic stellate cells requires interactions with hepatocytes. Am. J. Pathol. 150 (1), 51.

Théret, N., et al., 1998. Differential expression and origin of membrane-type 1 and 2 matrix metalloproteinases (MT-MMPs) in association with MMP2 activation in injured human livers. Am. J. Pathol. 153 (3), 945–954.

Tsujie, M., et al., 2000. Gene transfer targeting interstitial fibroblasts by the artificial viral envelope-type hemagglutinating virus of Japan liposome method1. Kidney Int. 57 (5), 1973–1980.

Upton, J.-P., et al., 2008. Caspase-2 cleavage of BID is a critical apoptotic signal downstream of endoplasmic reticulum stress. Mol. Cell. Biol. 28 (12), 3943–3951.

van Amerongen, M.J., Engel, F.B., 2008. Features of cardiomyocyte proliferation and its potential for cardiac regeneration. J. Cell. Mol. Med. 12 (6a), 2233–2244.

Van de Water, F.M., et al., 2006. Intravenously administered short interfering RNA accumulates in the kidney and selectively suppresses gene function in renal proximal tubules. Drug Metab. Dispos. 34 (8), 1393–1397.

Vanhoutte, D., Heymans, S., 2010. TIMPs and cardiac remodeling: 'Embracing the MMP-independent-side of the family'. J. Mol. Cell. Cardiol. 48 (3), 445–453.

Wamstad, J.A., et al., 2012. Dynamic and coordinated epigenetic regulation of developmental transitions in the cardiac lineage. Cell 151 (1), 206–220.

Wang, Y., Grainger, D.W., 2013. Developing siRNA therapies to address osteoporosis. Ther. Deliv. 4 (10), 1239–1246.
Wang, K.C., et al., 2002. Oligodendrocyte-myelin glycoprotein is a Nogo receptor ligand that inhibits neurite outgrowth. Nature 417 (6892), 941–944.
Watanabe, A., et al., 2004. Phospholamban ablation by RNA interference increases Ca 2+ uptake into rat cardiac myocyte sarcoplasmic reticulum. J. Mol. Cell. Cardiol. 37 (3), 691–698.
Williams, J.D., et al., 2002. Morphologic changes in the peritoneal membrane of patients with renal disease. J. Am. Soc. Nephrol. 13 (2), 470–479.
Wright, M.C., et al., 2001. Gliotoxin stimulates the apoptosis of human and rat hepatic stellate cells and enhances the resolution of liver fibrosis in rats. Gastroenterology 121 (3), 685–698.
Xia, Z., et al., 2008a. Suppression of renal tubulointerstitial fibrosis by small interfering RNA targeting heat shock protein 47. Am. J. Nephrol. 28 (1), 34–46.
Xia, J.-R., Liu, N.-F., Zhu, N.-X., 2008b. Specific siRNA targeting the receptor for advanced glycation end products inhibits experimental hepatic fibrosis in rats. Int. J. Mol. Sci. 9 (4), 638–661.
Xiao, L., et al., 2010. Connective tissue growth factor knockdown attenuated matrix protein production and vascular endothelial growth factor expression induced by transforming growth factor-β1 in cultured human peritoneal mesothelial cells. Ther. Apher. Dial. 14 (1), 27–34.
Yang, B., Hosgood, S.A., Nicholson, M.L., 2011. Naked small interfering RNA of caspase-3 in preservation solution and autologous blood perfusate protects isolated ischemic porcine kidneys. Transplantation 91 (5), 501–507.
Yin, Y., et al., 2003. Macrophage-derived factors stimulate optic nerve regeneration. J. Neurosci. 23 (6), 2284–2293.
Yoshioka, K., et al., 1993. Transforming growth factor-beta protein and mRNA in glomeruli in normal and diseased human kidneys. Lab. Invest. 68 (2), 154–163.
Zarrinkalam, K.H., et al., 2003. Connective tissue growth factor and its regulation in the peritoneal cavity of peritoneal dialysis patients. Kidney Int. 64 (1), 331–338.
Zeisberg, M., et al., 2001. Renal fibrosis: collagen composition and assembly regulates epithelial-mesenchymal transdifferentiation. Am. J. Pathol. 159 (4), 1313–1321.
Zeng, S., et al., 2004. Blockade of receptor for advanced glycation end product (RAGE) attenuates ischemia and reperfusion injury to the liver in mice. Hepatology 39 (2), 422–432.
Zhang, L.P., et al., 1999. Increased expression of plasminogen activator and plasminogen activator inhibitor during liver fibrogenesis of rats: role of stellate cells. J. Hepatol. 31 (4), 703–711.
Zhang, G., et al., 2012. A delivery system targeting bone formation surfaces to facilitate RNAi-based anabolic therapy. Nat. Med. 18 (2), 307–314.
Zheng, X., et al., 2006. Preventing renal ischemia–reperfusion injury using small interfering RNA by targeting complement 3 gene. Am. J. Transpl. 6 (9), 2099–2108.
Zheng, X., et al., 2008. Gene silencing of complement C5a receptor using siRNA for preventing ischemia/reperfusion injury. Am. J. Pathol. 173 (4), 973–980.
Zimmermann, T.S., et al., 2006. RNAi-mediated gene silencing in non-human primates. Nature 441 (7089), 111–114.

CHAPTER

BONE TISSUE REGENERATIVE MEDICINE VIA BIOACTIVE NANOMATERIALS

28

Paula V. Messina, Noelia L. D'Elía, Luciano A. Benedini
Southern National University, Bahía Blanca, Argentina

CHAPTER OUTLINE

1 INTRODUCTION

To explain the contribution of nanoscience to the tissue engineering, we will initiate by describing the nanocomposite nature of bone and the extracellular matrix (ECM). Then we will discuss about how the recreation of native tissues nanostructure could enhance functional tissue organization and its effect on the matrix properties after the implantation of engineered tissues.

1.1 BONE TISSUE AS A NANOSTRUCTURE MATERIAL

At the lower hierarchical levels, bone is a composite of collagen and mineral nanoparticles that closely resembles the synthetic hydroxyapatite ($Ca_{10}(PO_4)_6(OH)_2$, HA) crystals. In addition, over 200 different varieties of noncollagenous matrix proteins (i.e., glycoproteins, proteoglycans, and sialoprotein) promote to the abundance of signals in the immediate extracellular surroundings (Stevens, 2008). Its structure has been considered in many relevant reviews (Fratzl and Weinkamer, 2007; Muller, 2009; Rho et al., 1998), as well as its mechanical properties (Fratzl et al., 2004; Meyers et al., 2008; Rho et al., 1998). However, the essence of the bone is its nanostructured design; both in terms of mechanical and biological assets. Accordingly, to correctly understand bone's properties, it is essential to examine its most basic level of organization: the nanostructured array of HA crystals embedded within the intrafibrillar collagen matrix. The mineralized collagen fibrils are the basic building block of bone tissue. Collagen proteins ≅200 nm length and 2–3 nm in diameter self-assemble into fibrillar assemblies, which are then further aggregated into fibrils frameworks of about 500 nm in diameter. It was revealed by TEM (Landis et al., 1996) and neutron scattering (White et al., 1977) methods that the regular stacking pattern of collagen molecules in a fiber, leaves small gap zones with 35 nm length and overlap zones with 32 nm length. This striped structure is occupied by plate-shaped HA nanocrystals of 10–20 nm in length and 2–3 nm wide, that are mostly arranged parallel to each other and to the long axis of collagen fibrils. Crystals arise at regular intervals along the fibrils, with an estimated repetitive distance of 67 nm. In addition to crystals embedded in fibrils, there is also an extra-fibrillar mineral coating the whole structure (Fantner et al., 2005). Neutron scattering experiments (Lees and Mook, 1986) exhibited that the equatorial spacing between collagen molecules, *d*, also have nanometer dimensions, about 1.6 nm in nonmineralized wet fibrils, whereas in dried conditions the spacing of the molecules is reduced to 1.1 nm (Fratzl et al., 1993). At a macroscopic level, determined by the type of bone, these mineralized fibers are arranged into larger tissue-level structures (Fratzl et al., 2004; Fratzl and Weinkamer, 2007; Kane and Ma, 2013). From a materials viewpoint the complex nature of the organic–inorganic phases, difficulties the bone tissue classification in one specific type of composite. Generally, it was defined as a polymer composite of collagen reinforced with calcium phosphate nanocrystals, which is organized in a highly ordered structure over several size scales (Kane and Ma, 2013). However, a deep inspection of deproteinated bone exposes individual HA crystals of 25–50 nm, signifying that bone possibly will be better described as a nanoparticle-reinforced polymer–matrix composite (Olszta et al., 2007). Certainly, from numerous interpretations, the bone can be designated as a composite material, that is, it is, simultaneously, a porous substance, a polymer–ceramic blend, a lamellar, and a fiber-matrix material. Its properties will be therefore governed by each of these compositions' features but especially of its organization.

1.2 STRUCTURE AND FUNCTION OF BONE: THE ENGAGEMENT BETWEEN STRENGTH AND TOUGHNESS

Bone is frequently stereotyped as basically a protective and supportive framework for the body. Although it performs these roles, the bone tissue is essentially a highly dynamic organ that responds to physical, chemical, and biological signals, growing in response to obligatory rules, perceiving minor damaged areas and substituting them through a process called remodeling (Kane and Ma, 2013). Moreover, bone stores and supplies crucial nutrients, minerals, and lipids; it produces blood cells that sustain the body, and perform a vital role in its protection against infections and diseases. All these functions make the skeletal system of the human body an organ that is indispensable to our daily life. Skeletal system is designed to carry major biomechanical loads; its natural nanocomposite structure, made of flexible collagen fibers reinforced with HA crystals, is essential for the requisite of compressive strength and of high fracture toughness. The hierarchical architecture of the bone, which has characteristic structural features on multiple length-scales from molecular to near-macroscopic dimensions, provides it a superior stiffness than classic tissues; while at the same time offers a remarkably high fracture toughness and resistance to damage. The foundation of the intrinsic (plastic deformation) mechanisms tends to be on the smaller, submicrometer length-scales, analogous to the nanometer scale of dislocation Burgers' vectors in metals, while the processes of extrinsic toughening and fracture occur on much coarser length-scales classically well into the micrometer range (Launey et al., 2010; Ritchie, 2011). In human cortical bone, for example, intrinsic toughening, that is, plasticity, derives from a fibrillar sliding mechanism on the scale of tens to hundreds of nanometers, the length-scales linked with the mineralized collagen fibrils (Launey et al., 2010). With aging, irradiation, and certain diseases, however, the bone can embrittle due to the increase of the collagen cross-linking, which obstructs this structure (Bailey, 2001); the intrinsic toughening then switches to higher length-scales and is blocked by the lack of elasticity related to microcracking. Nevertheless, the major source of toughness in bone tissue is extrinsic and arises from crack bridging and crack deflection as a growing crack encounters the more-mineralized interfaces of the osteonal configurations (Koester et al., 2008; Nalla et al., 2003); as the size and spacing of the osteons are in the range of tens to hundreds of micrometers, the typical length-scales for these events approach to millimeter dimensions.

1.3 REGENERATION OF BONE AND SKELETAL RECONSTRUCTION: THE CLINICAL NEED OF SUPERIOR ORTHOPEDIC IMPLANTS

Bone tissue has partial regenerative abilities thus, some damages are poorly restore due to anatomical location, patient age, or simply because they are too large for the body to repair the defect by itself. Any break exceeding 5 mm will not naturally heal and will require a clinical intervention to ensure the proper reparation (Kane and Ma, 2013). While regular fractures rarely result in a gap of critical size, some traumatic injuries, cancers, and infections of the bone, or simply age-related degeneration, may result in areas where the bone would not regenerate naturally. The number of these conditions causes the bone to be the second transplanted tissue after blood, more than 100 million only in United States, citizens have musculoskeletal conditions with limitations of function that are chronic and permanent; therefore, these conditions exceeds a cost of about $800 billion every year (http://www.orthoinfo.aaos.org/). In order to address the requirements of these patients, specialists have acquired a diversity of methods and have used a selection of materials to repair the bone tissue, or at a minimum to restore the mechanical integrity of the wound site. Improvements in surgical practices implant design, and postoperative care, have considerably enhanced treatment outcomes of intricate fractures and defects caused by high trauma, disease, progressive deformity, revision surgery, and tumor remotion. However, a broad soft tissue damage, inadequate surgical techniques, infections, and biomechanical instability can bring the foundation of permanent large defects associated to restricted intrinsic regenerative potential (Woodruff et al., 2012). There are clinically established therapeutic methods centered on the implantation of autografts, allografts, metallic implements, and ceramic-based materials to support the restitution and/or the substitution of bone defects; all with inherent disadvantages. In this context, the restrictions of the present applied techniques have established the necessities for novel therapeutic concepts to design and engineering best quality structural and functioning bone implants to substitute the current treatments. The incentive of scientific research and the emphasis on turning bone engineering concepts from bench to bedside is embedded in the current restrictions that medicine has for resolve the increasing and rather challenging problems of the orthopedic, dental, and reconstructive surgery of the society. It is within this context that the discipline of bone engineering has emerged, through the combination of engineering, life sciences, molecular and cell biology, stem cell biology, and surgery (Woodruff et al., 2012). Commercial aspects are strongly associated to investigation, it is important to mention how vast the bone graft substitute market is. Synthetic materials represent only about 15% of the global bone graft substitute market, but the growth rate is increasing significantly (Bohner, 2010). According to the new report from the research and consulting firm "Global Data" (http://www.healthcare.globaldata.com), the worldwide bone grafts and substitutes (BGS) marketplace value will augment progressively over the next years, expanding from almost $2.1 billion in 2013 to about $2.7 billion by 2020, at a compounded annual growth rate (CAGR) of 3.8%. The latest report of Global Data indicated that out of the 10 major markets (USA, France, Germany, Italy, Spain, the United Kingdom, Japan, Brazil, China, and India), the United States will continue to hold the largest BGS market share by 2020, with a 65.6% of increase. The global BGS market include the rising volumes of spinal fusion and joint reconstruction procedures, along with an improved availability to the latest-generation BGS products in emerging economies and enlarged public responsiveness of bone regeneration achieved through biologics. Nevertheless, a number of obstacles will restrain any additional growth in the BGS marketplace, including deleterious publicity, increased examination of stem cell–derived products by the American Food and Drug Administration (FDA), reimbursement and pricing pressures, and variations in the product regulations. Thereon, the allografts, the bone morphogenetic proteins, the demineralized bone matrices (DBMs), the synthetic bone substitutes and the cell-based matrices are all legally marketed as alternatives to autographs, each of these products with a separate regulatory classification. This calls for different regulatory pathways that have to be taken into account to gain the approval prior to marketing. Moreover, the legal protocols for the commercialization of human tissue-derived bone graft products, such as DBMs, may vary between European and American countries, adding an extra obstacle for companies hoping to achieve effective market permeation.

2 SYNTHETIC APPROACHES FOR BONE TISSUE REPLACEMENT

The presently offered materials for bone graft, implant, or replacement display a basic similarity to resemble the body's own materials (Kane and Ma, 2013); they reproduce selected constituent of natural bone tissue in order to induce an analogous cellular response. Although these constituents are effective for ensuring bioactivity, limit the material selection to biological sources situates restrictions on the final properties of the material. Nonnatural materials offer significant advantages in cost, consistency, and ease of regulatory approval. Synthetic devices with biomimetic structures showing bioactive properties are obtained through a careful design, so that they are able to direct the body's own cells to initiate healing and repair from within, thus, they may be used to enhance the biological signaling and the fast healing at wound place better than any bone-based material would allow. Selections of the current techniques used to formulate engineering biomaterials are discussed below.

2.1 LAYER BY LAYER CHEMISTRY

Layer-by-layer (LbL) chemistry is a dip-coat approach based on the simple alternating adsorption of complementary multivalent species on a substrate via electrostatic, hydrogen bonding, or additional secondary interactions (Decher, 1997; Easton et al., 2014; Hammond, 2012). LbL assembly has a precise potential in the area of biomaterials (Boudou et al., 2010) and especially in bone regeneration because it facilitates the nanometer level control of the thin film composition, and simultaneously the creation of highly complex, tailor-made coating arrangements (Hammond, 2011). Conversely, of other molecular level deposition techniques, these thin films are fabricated in water at room temperature, preserving the activity of sensitive proteins, nucleic acids, and other functional biomacromolecules (Hammond, 2012). Through LbL chemistry, it is possible to incorporate different types of therapeutic molecules with diverse targets at high loadings without the stability topics that are a problem to overcome in traditional polymer blends. The prospective for corelease of two or more drugs with dissimilar release profiles, or even time-dependent staged delivery becomes possible through the use of top–down compartmentalized regions of LbL thin films (Hammond, 2012). LbL assembly offers a simplified methodology toward the construction of nanoscale biomaterials. Bioactivity may be obtained through incorporation of biomolecular species within the film architecture, and film mechanics may be controlled through postformation chemical cross-linking steps (Pauthe and Van Tassel, 2014). Incorporating a number of highly sensitive growth factors that regulate the differentiation, proliferation, and signaling of stem cells that determine the formation of new tissues and key processes, such as wound healing, multilayer assembly presents true potential in regenerative medicine and tissue engineering. By the use of this technique, it is achieved a further extended release of smaller, controlled quantities of growth factor (i.e., bone morphogenetic growth factor 2, 2BMP; vascular endothelial growth factor, VEGF; arginine–glycine–aspartic, RGD, adhesive motif) that would generally replicate the natural process of tissue generation (Easton et al., 2014; Hammond, 2012; Macdonald et al., 2011; Yang et al., 2009). Usually, these growth factors are administered directly to tissues of interest with bolus injections, or are encapsulated in depots that lead to bolus release in the body. In those conditions the mass released of growth factor proteins generally leads to a large amount of clearance of the active protein, with only a small amount of the expensive biologic therapeutic remaining to exact its influence on tissue development. Also, most polymeric carriers can only contain limited amounts of protein in a stable blend formulation due to thermodynamic and other practical limitations. The cell's response depends on the stimuli of, adhesive ligands (such as fibronectin, collagen, or short peptide motifs), surface roughness and topography, surface hydrophilicity, and charge (Bacakova et al., 2011). However, substrate rigidity appears to play a strong role even when other factors are present (Engler et al., 2006; Pauthe and Van Tassel, 2014; Zemel et al., 2010). Most tissue cells need attachment, that is, require adhesion to some type of solid entity (i.e., ECM, synthetic substrate, and other cells). Once adhered, cells detect and respond to local rigidity. Today it is starting to recognize that the cell response, which can range from a force, exerted on the substrate to an alteration in cell shape, or to an augmented rate of proliferation or differentiation, can be governed by the substrate rigidity (Pauthe and Van Tassel, 2014). In the case of LbL films, formation of chemical cross-links, between functional groups on the constituent macromolecular species, is the classical resource of controlling mechanical rigidity (Chaubaroux et al., 2012; Phelps et al., 2011). LbL cross-linking significantly increases the film rigidity; typical Young's moduli for native and cross-linked films are 3–10 kPa and 100–400 kPa, respectively. Literature findings (Chen et al., 2009; Olenych et al., 2005; Seo et al., 2009; Wittmer et al., 2007) reveal that the initial biological response, that is, cell attachment and spreading is observed to increase with film rigidity and density of surface adsorbed species. The surface functionalizations with species that promote adhesion have a greatest efficacy in the case of softer substrates, therefore, a significant cell adhesion may be promoted on even nonadherent substrates. LbL films of individually and well-regulated bioactivity and mechanical rigidity could have an influence in numerous applications comprising cell–material interactions, for example, as biomedical device coatings that promote a confluent endothelial cell layer, cell-based bioreactor supports that induce production of a therapeutic product, and tissue engineering scaffolds signaling fetal or even stem cells to form mature cellular structures. A continuing dispute is to better understand the biological response to film-embedded bioactive agents, that is, cell sensitivity to intrafilm concentration, surface cross-linked polymer layers, and strength of bioactive agent–film interaction. The required film loading likely differs significantly depending on the biomedical application, a broader understanding of what systems can benefit most from high and controllable local bioactive agent exposures would help guide the development of these promising materials. LbL films formed using diverse surface cross-linking (Phelps et al., 2011) or nanoparticles-templating strategies (Wu et al., 2013) could present contacting cells with optimized mechanical properties, and a mixture of locally delivered growth factors, in order to promote the formation of viable, functional bone tissue.

2.2 3D ARCHITECTURES FOR BONE REPLICATION

In general, tissues' functionality arises from their constituents (cells and extracellular molecules), as well as their interrelation. The association of the components of a tissue is known as histoarchitecture. Until now, numerous methods have been published that attempt to recreate in vitro the three-dimensional (3D) histoarchitecture found in vivo. In these approaches that offer complementary advantages and limitations, researchers use scaffolding molecules of natural or synthetic origin to support cell growth (Perez-Castillejos, 2010). Porous bone scaffolds can be prepared by a diversity of methods including chemical/gas foaming (Zigang et al., 2008), solvent casting, particle/salt leaching (Cao and Kuboyama, 2010), freeze drying (Sultana and Wang, 2008), thermally induced phase separation (Wei and Ma, 2004), and foam-gel (Yoshikawa and Myoui, 2005). However, scaffolds with designed porosity for specific defects are difficult to assembly with most of aforementioned tactics and must to be tailored and fabricated using additive manufacturing (AM) methods. Different AM methodologies, for example, 3D printings (3DP), solid freeform fabrication, rapid prototyping, are approaches that allow complex shapes for scaffolds' fabrication directly from a computer aided design file (Bose et al., 2013). In this chapter, we will put emphasis on the applications of 3DP scaffolds for bone tissue modeling. 3DP method arise through the control of fine features including interconnected porosity, no contamination issues related to any second material for support structures and the direct printing ability on a large variety of ceramic, metallic, polymeric, and composite materials (Bose et al., 2013; Gross et al., 2014; Murphy and Atala, 2014). By mean of a 3DP it can be created a simulated bone matrix with the precise shape of the bone imperfection (Bose et al., 2013). These engineered structures enable the regeneration of bone tissue by providing a framework for osteocyte cells to attach themselves to, and an environment for health tissue to grow. Starch-based binders are one of the candidates for bone replacement applications. These binders are biocompatible and produce structures that have a mechanical strength close to trabecular bone (Oliveira et al., 2009). 3D-printed polyethylene (PE) scaffolds with 22.3%–49.7% porosity have shown a tensile strength up to 4 MPa, and no toxicity to human osteoblasts (Suwanprateeb et al., 2012). Capillaries and vessel formation, just with a homogeneous osteoconduction from central channels, have been observed in 3D-printed HA blocks (Becker et al., 2012). The effect of pore size on human fetal osteoblasts (hFOB) was studied with 3D-printed tricalcium phosphate (TCP) scaffolds (Tarafder et al., 2013a). Histological evaluation and histomorphometric analysis reveal that the treatment group (doped TCP scaffolds) facilitated higher osteoid like bone at an early stage, and completely mineralized bone later, which could be essential for fast bone healing and mineralization in vivo (Tarafder et al., 2013a, 2013b). 3DP of TCP with diluted phosphoric acid as binder allowed the construction of custom made brushite/TCP implants with well-defined architecture and the capacity of being resorbed in vivo (Gbureck et al., 2007). Human adipose-derived stem cells (hASCs) were successfully seeded and cultured on 3DP anatomically shaped polycaprolactone (PCL) scaffolds with variable internal pore structures (Temple et al., 2014). The findings of these studies exemplify the abilities and potential of 3DP to modulate scaffolds to be used for engineering autologous, anatomically shaped, vascularized bone grafts.

2.3 IMITATING THE BONE'S NANOSTRUCTURE: CHEMICAL, MORPHOLOGICAL, AND MECHANICAL BIOMIMESIS

The success of an engineered bone implant involves the close reproduction of the chemical, morphological, and biomechanical features to guarantee the osteoconduction, osteoinduction, and osseintegration properties of native tissue (McMahon et al., 2013; Sprio et al., 2011). Osteoconduction means that host bone tissue grows on the implant's surface; an osteoconductive graft is one that permits to promote the attachment, survival, migration, and distribution of osteogenic cells on its surface or down into its networks (Albrektsson and Johansson, 2001). Osseointegration is histologically defined in *Dorland's Illustrated Medical Dictionary* (Sciences, 2011) as the direct anchorage of an implant by the formation of bony tissue around the graft without the manifestation of periimplant fibrosis; osseointegrative implants permits a direct contact between living bone and the implant. Osteoinduction is the process by which osteogenesis is induced; osteoinductive scaffolds present spatial, physical, and biochemical stimuli to initiate stem cells or progenitor cells toward osteoblastic lineage. To join these three requirements, bone tissue engineering exploits a selection of biomimetic strategies which may comprise bioscaffolding (Yarlagadda et al., 2005), biochemical and physical stimuli (Barrère et al., 2008; Brown, 2000), stem cells (Bianco and Robey, 2001; Sundelacruz and Kaplan, 2009), vascularization or adequate blood supply (Kaigler et al., 2006; Rouwkema et al., 2008), running through the hierarchical organization of natural ECM to create functional bone tissues. These biomimetic efforts include the selection of the constituents that are present in native bone [i.e., HA and collagen (Pek et al., 2008)], assembling of multiple scale architectures especially with nanoscale components (Akkouch et al., 2011), and incorporating growth factors to the scaffold structure [i.e., bone morphogenetic proteins, BMPs (Geiger et al., 2003; Liu et al., 2013; Nakashima and Reddi, 2003)] to provide a biomimetic function for stimulating bone repair and regeneration. The mechanical design of a framework is a critical point to the performance of the scaffold. This scaffold not only must provide a structural support for cell integration, but also influence the regulation of cell proliferation, differentiation, and migration to form functional tissues (Hollister, 2005). The materials selection and processing techniques can affect the chemical, physical, mechanical, and cellular recognition properties of biomaterials (Bose et al., 2012). Standard tissue engineering techniques have used a top down approach to restore the macro- and microscale properties of native tissues, but these conventional scaffolds ignore the nanoscale structures. Cells are habituated to interact with nanotopographies at cell–matrix interface and their manifestations are decisive to control cell functions, such as proliferation, migration, and ECM assembly (Bacakova et al., 2011). That is why conventional tissue engineering materials usually fail as a result of insufficient tissue regeneration; due to of many complications, including adverse local tissue responses, such as severe inflammatory response due to neutrophils, macrophage extravasation and adhesion, foreign body giant cell, and fibrous capsule formation. One strategy to condense the hierarchical organization of natural ECM is to include nanoscale topographies in the 3D scaffolds design. As it was previously discussed in Section 1.1, many extracellular proteins have a fibrous structure with diameters on the nanometer or submicrometer scales. Limited tools have been established, with varying degrees of success, to construct materials that really mimic the nanofibrous ECM proteins architecture: (1) electrospining, (2) self-assembly, and (3) phase separation (Ma, 2008). A selection of natural macromolecules [i.e., collagen, silk fibroin, and fibrinogen (Fb)] and synthetic polymers (i.e., PGA, PLLA, PLGA, and PCL) have been processed into fine filaments for tissue engineering research. Various cells have been reported to attach, proliferate, and differentiate into or maintain their functional phenotypes on the electrospun nanofibrous structure formed by them (Blakeney et al., 2011; Di Martino et al., 2011; Tillman et al., 2009). Nevertheless, a significant challenge still exists in using this technique to fabricate complex 3D scaffold shapes or to generate designed specific internal pore size and interconnectivity (Guimarães et al., 2010; Khorshidi et al., 2015; Phipps et al., 2012). Furthermore, in general the electrospun fibers often give rise to diameters in the micrometer instead of nanometer range.

Self-assembly patterns or structures are naturally molded and preserved by noncovalent bonds, such as hydrogen bonds, electrostatic, and hydrophobic interactions (Whitesides and Boncheva, 2002; Whitesides and Grzybowski, 2002). Even though noncovalent bonds are rather weak in isolation, when they act together they govern the self-assembly of biological macromolecules and possibly, this method is the best to successfully replicates the process by which bone tissue is formed (Matson et al., 2011; Stupp, 2010). By this method, the obtained tissue engineering scaffolds should ideally have sufficiently large pores to allow for cell incorporation, migration, and proliferation (Lutolf and Hubbell, 2005). The major limitations of the current self-assembled scaffolds are their inability to regulate the macro sized pores structure and to form a mechanically stable 3D geometry. Their degradation properties also need to be addressed. Nevertheless, current self-assembled materials used in tissue regeneration are hydrogels, which are convenient for injection.

The phase separation of polymer solutions has been explored to generate porous structures as tissue engineering scaffolds (Holzwarth and Ma, 2011; Wei and Ma, 2004, 2008). Phase separation arises when a homogeneous multicomponent system becomes thermodynamically unstable under certain conditions and tends to separate into a multiphase system in order to lower the system's free energy. The matrices obtained by this methodology have similar restrictions to those that are self-assembled or electrospun materials, lacking of an ordered and interconnected 3D macroporosity (Ma, 2008).

Scientific community still not succeeded to get the accurate material to reproduce in full the properties of bone tissue. The level of biological complexity that needs to be confined within a synthetic 3D environment is still undefined; however, the current techniques applied to the fabrication of bioactive bone scaffolds have the potential to improve the performance of orthopedic implants and that will lead obtain a genuine bone mimetic material.

2.4 BIOENHANCING TOPOGRAPHY: EFFECT ON CELLS AND PROTEIN INTERACTIONS

The modification of biomaterials' surface properties can income on different levels of complexity, from simple variations in the hydrophilicity to functionalization with ECM derived ligands, receptors, drugs, or full proteins (Bacakova et al., 2011; Stevens, 2008). In this chapter we will focus specifically on the physicochemical properties of the materials surface, to be precise its chemical composition, surface energy, polarity, wettability, roughness, topography, and mechanical properties. The importance of the ECM nanostructure in guiding the seeded cells and promoting morphogenesis stop being a possibility and is now recognized as a scientific fact (Dvir et al., 2011; Ott et al., 2008; Stevens and George, 2005). Nanostructures effect on cell growth is comparable to other highly advanced strategies, such as the incorporation of synthetic ligands of cell adhesion receptors [e.g., (Arg-Gly-Asp) RGD or sequences (Lys-Arg-Ser strategies -Arg) KRSR] on the material surface, the construction of 3D materials with controllable degradation speed, or the controlled release of growth factors (Vagaska et al., 2010). Material nanotopography and surface properties may be engineered to dictate stem cell fate assessments (Kingham et al., 2011; Murphy et al., 2014). Furthermore, the manipulation of biomaterials' topography allows its insertion and bonding with the host tissue, which affects the favorable implant integration. The effect of surface properties of several HA and Ce–TiO_2 nanostructured materials on their osteo-integration properties was studied in our laboratory (D'Elía et al., 2013; Gravina et al., 2014, 2015). Our studies point out that the regulation of surface microrough properties is an efficient platform to influence the material bioactivity. The bioactivity of the materials was investigated based on the method established by Kokubo et al. (1990). Four crystalline HA nanorods structured materials templated by different CTAB-micelle-block copolymer structured networks were evaluated. We establish that the topography of each surface follows a direct association with the micromorphology, chemical composition, and kinetic growth of the mineral Ca–P coating after soaking in simulated body fluid (SBF) (D'Elía et al., 2013). From our investigation, a biocompatible nanostructured HA material displaying a Ca–P globular coating on its surface exhibited a Ca/P ratio close to that for the trabecular bone (Vanzillotta et al., 2006) and a morphology essential for bone bonding (Uchida et al., 2003) was obtained. In a second work, the authors of this chapter used a bottom-up microemulsion-mediated strategy to obtain highly bioactive and biocompatible, striped Ce–TiO_2 nanocrystalline superstructures. Regulating the aligned fiber microarchitectures to mimic the ECM for bringing important biological functions is a crucial point to the successful tissue regeneration. The presence of Ce atoms in the TiO_2 crystalline lattice stabilized the anatase polymorph, and that fact promoted the rapid disposition of a HA coating comprised of sphere-like globules of 3–5 mm in diameter essential for bone bonding (Uchida et al., 2003). It is important to note that epitaxial growth of HA on rutile is expected to occur based on the two-dimensional (2D) similarity of the two crystal structures (Lindberg et al., 2008). In contrast, HA is not biomimetically formed on single crystal TiO_2 anatase (Li et al., 1994). One of the reasons given by the authors is that the density of –OH groups is not plentifully available at the appropriate crystal surface. They also noted that while the individual crystals of dense anatase lacks in bioactivity, porous anatase induces apatite deposition. Our obtained materials showed bioactivity (Gravina et al., 2014), regardless of their crystalline phases and the fact that they have not presented high surface areas or an ordered array of mesopores. Subsequently, our results show that differences in crystal polymorphisms, as well as porosity are not enough to induce bioactivity. Furthermore, this effect is related to the surface texture that influences surface energy, which in turn, affects the bioreactivity. The influence of the surface topography on the protein adsorption process is another point of great importance for designing biomaterial surfaces and the biocompatibility for particular biomedical applications. An adsorbed protein layer can be guided through careful design of surface structure, permitting the construction of materials/surface coatings with tailored bioactivity (Roach et al., 2006). Improved information of the interaction of proteins with the surfaces of implantable materials, particularly regarding Fb, is indispensable for the interpretation of cellular events and of the overall host tissue response. Considering the potential of Si-nanofibers prepared in our laboratory (Hassan et al., 2011) as 3D frameworks for the construction of implantable synthetic tools, we investigated the relation among the material surface features and the amount, structure and distribution of adsorbed Fb molecules (Hassan et al., 2012). It was proven that Fb adsorption process occurs in a stepwise routine with a first rapid adsorption, followed by a molecular transitional rearrangement and to end a second slower adsorption regime over a longer period of time. After the first adsorption process, there is a partial desorption of the protein, which shows that this stage is reversible until the first 2×10^4 sec. However, the complete process is irreversible, with a great distortion of the original material morphology. The Fb concentration adsorbed on the Si-nanofibers once reached equilibrium was of 270 ± 20 mg/dm^2; more than three times the adsorption capacity of nonfibrillar-2D or -3D, scaffolds (Scopelliti et al., 2010; Snopok and Kostyukevich, 2006; Wertz and Santore, 1999). From examination of the material morphology before and after Fb adsorption by scanning electron microscopy (SEM), it was appreciated that after protein interaction the original fibrous structure was damaged, giving rise to fibers of higher dimensions or to bicontinuous structures. The key to the enhanced adsorption process and the final acquired topography of the material is allocated to the fibrous structure and the similarity in size between the substrate (d = 30–50 nm) and the Fb molecules (47–50 nm). After adsorption, the αC-domains are presumably available for interaction with the domains of the neighboring adsorbed Fb molecules, thereby promoting lateral and equilateral associations and the formation of an extensive network similar to those that exists in nature between fibrin units. These studies have the potential to give out the basis for new material designs in which hemocompatibility is the key factor (Fig. 28.1).

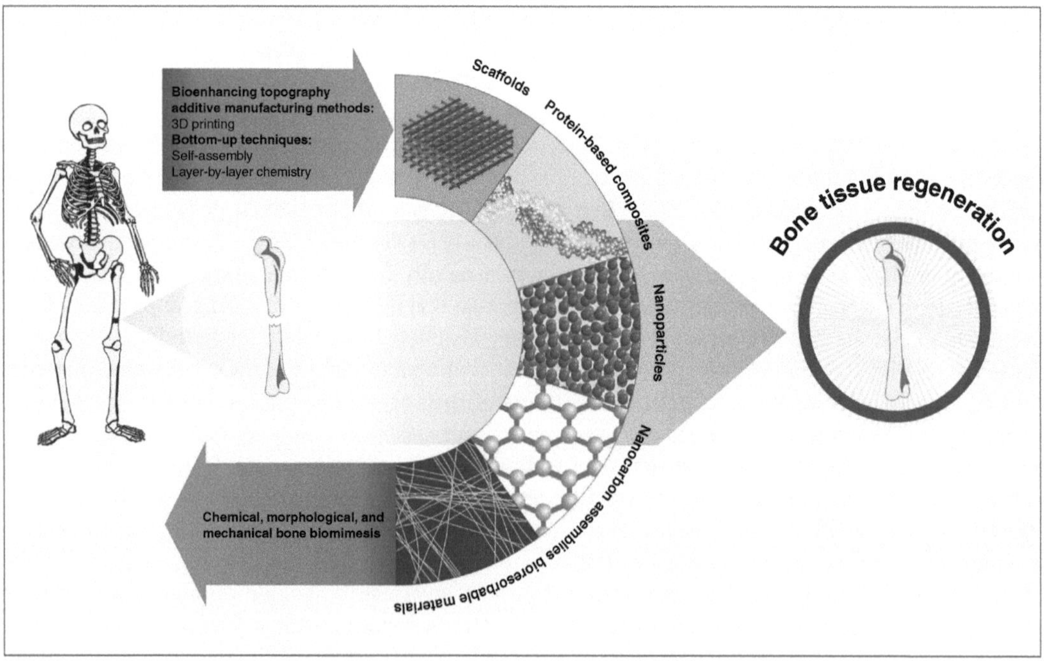

FIGURE 28.1 Nanotechnology and Nanomaterials: Promises for Enhanced Tissue Regeneration

3 NANOMATERIALS AS KEY ENABLERS OF BONE REGENERATION

3.1 BIOMIMETIC BEHAVIOR AND RELEVANT METHODOLOGIES FOR DEVELOPING MULTIFUNCTIONAL NANOMATERIALS

Investigations on bone tissue engineering over the past decade have stimulated a revolution in new materials, processing techniques, performance evaluation, and applications however; experimental therapies denoting the use of nanotechnology in bone regeneration are still in the initial stages. Generally, all attempts are being made to build in vitro materials that can mimic what happens in nature. The properties of the scaffold, the 3D porous matrix on which cells are grown (Marsh, 2001), are critical for the success of tissue regeneration.

3.1.1 Scaffolds

Porous 3D scaffolds are usually proposed as short-term templates for bone growth and as a specific environment and structural design for tissue improvement (Badylak et al., 2009; O'Brien, 2011) to impart the suitable environment for their restoration. These scaffolds fundamentally act as a support for the formation of novel tissue and are classically seeded with cells; occasionally act as delivery systems loaded with growth factors, or exposed to biophysical stimuli in the form of a bioreactor (Bose et al., 2012). These cell-seeded scaffolds are either cultured in vitro (Dupont et al., 2010; Pan et al., 2015) to create tissues which can then be introduced into the injured site, or are inserted directly into the damaged place, using the body's own systems, where restoration of tissues or organs is induced in vivo (Ko et al., 2013). Several scaffolds created from a selection of biomaterials and synthesized using a variety of fabrication procedures (Section 2) have been used in the field in attempts to regenerate different tissues (Badylak et al., 2009; Bose et al., 2012). Regardless of the tissue nature, a scaffold for use in tissue engineering must to be: (1) Biocompatible, cells must adhere, spread, and proliferate onto the surface and eventually through the scaffold; also, after implantation, the scaffold must provoke a negligible immune reaction. (2) Biodegradable, the scaffold must be biodegradable at a comparable rate to the healthy tissue reparation, so as to let cells to create their own ECM; products of this degradation should also be nontoxic at all sense. (3) Mechanically appropriate, the scaffold should have mechanical properties corresponding with the anatomical site into which it is to be grafted and, from a practical viewpoint, it must be robust enough to allow surgical handling during implantation. (4) Architecturally suitable, scaffolds should have an interconnected pore structure and high porosity to guarantee cellular penetration and the extracellular matrix formed by these cells. Additionally, a porous interconnected structure is required to allow diffusion of nutrients to cells within the construct and of waste products or the products of scaffold out of the scaffold, without interference with other organs and surrounding tissues (Zadpoor, 2015). (5) Commercially viable, it should be cost effective and potentially scalable to marketable batch production. Finally, the last principle for scaffolds design, and one of which all of the principles listed previously are reliant on, is the selection of the material from which the scaffold should be made. At that point, the uses of some of the representative materials in the construction of scaffolds are discussed.

3.2 RESORBABLE MATERIALS

The expression "resorbable" is frequently used as substitute for "bioresorbable," "biodegradable," "bioerodible," or "bioabsorbable." In this chapter, it is applied to define those materials that will disappear from the implantation site over time, regardless of the mechanism leading to its remotion. The construction of scaffolds for tissue engineering takes hold of many biological sources, such as collagen, several proteoglycans, alginate-based substrates, and chitosan. These natural polymers are biologically functional and typically stimulate cell adhesion, growth, and proliferation (O'Brien, 2011). Furthermore, they are also biodegradable and as a consequence allow that host cells will produce, over time, their own ECM replacing the degraded scaffold. However, there is a challenge in the fabrication of scaffolds from biological resources with identical and reproducible structures. The attained scaffolds frequently have limited mechanical properties, which reduce their use in, for example, load bearing orthopedic applications. Likewise, it is problematic the absolute control of degradation rate. An accurate resorption process of the graft occurs when the rate of degradation of the material is controlled and consistent with the growth rate of the healthy tissue. To overcome such difficulties synthetic polymer-based scaffolds are prepared and investigated. Many synthetic polymers have been employed in the effort to create bioengineered scaffolds, including polystyrene, poly-L-lactic acid (PLLA), polyglycolic acid (PGA), and poly-DL-lactic-*co*-glycolic acid (PLGA) (Place et al., 2009). To regulate the resorption rate of bone graft substitutes, generally, two main approaches are applied. In the first one, the structure of the material is optimized (Hing, 2005; Petite et al., 2000). An important point is the existence of interconnected pores or channels within the material. The channels must be large enough to allow the incursion of blood vessels and cells, hence permitting material biodegradation and bone ingrowth within the bone implant (Collins and Birkinshaw, 2013; Guan et al., 2005). In a second approach, the chemistry is reformed. A classic method, in the polymer field, is to include more degradable sequences via copolymerization (Mann et al., 2001) or cleavage sequences for targeted enzymes (West and Hubbell, 1999) and to modify the polymer crystallinity (Sabir et al., 2009). While synthetic polymer based materials have been revealed as a success because they can be designed and fabricated with a tailor-made architecture, and their degradation features can be modulated by changing the polymer itself or the arrangement of the individual polymer, they have disadvantages including the risk of rejection due to decreased bioactivity. Furthermore, fears exist about the degradation process of PLLA and PGA since they disintegrate by hydrolysis, producing carbon dioxide and therefore depressing the local pH which can result in cell and tissue necrosis (O'Brien, 2011).

3.3 PROTEIN-BASED COMPOSITES

Numerous original approaches of designing tissue repair and restoration resources using bioinspired strategies are currently being articulated to obtain in vivo-like cellular responses. For such attempts it has been used by ECM proteins as templates to recreate in vivo microenvironment. Nanoscale structured materials can be fabricated via the spontaneous organization of self-assembling proteins. The benefit of such a method is that polypeptides can be specifically designed as building blocks with molecular recognition characteristics and spatially distributed bioactive ligands that would provide a physiological environment for cells in vitro and in vivo (George and Ravindran, 2010). Several natural proteins have been investigated, displaying distinctive mechanical, chemical, electrical, electromagnetic, and optical properties. Elastins, collagens, silks, keratins, and resilins are some of the more common structural proteins considered for protein-based biomaterials (Hu et al., 2012). A common occurrence to all these families of structural proteins is that they are characterized by long range ordered molecular secondary structures (e.g., beta-pleated sheets, coiled coils, or triple helices) that arise due to the highly repetitive primary amino acid sequences within the proteins. These structures stimulate self-assembly, the formation of structural hierarchy and thus materials-related functional roles in nature. A large number of protein-based composite biomaterials can be formed into a wide range of biomaterials with tunable properties, including control of cell responses. Among these we can cite, protein–protein (Hu et al., 2012; Mandal et al., 2012), protein–inorganic ceramic (Liao et al., 2005), protein–synthetic polymer (Klok, 2009; Liao et al., 2005), or protein–polysaccharide (Hunt and Grover, 2010; Seliktar, 2012) composites that have played important roles in the recent development of biomaterials for biomedical research. The capacity to adapt the mechanical, surface, and regeneration features offers a great advance for biomaterials in general, to address the growing range of medical material and device needs. Particularly protein composites have demonstrated the potential for providing the next generation of functionalized biomaterials due to their intrinsic biocompatibility, biodegradability, and natural abundance, as well as their unique natural properties (e.g. mechanical, optical, electrical, chemical, thermal) (Hu et al., 2012). As the scientific community progresses in the understanding of the molecular-scale interactions between proteins, we will be able to create the next generation of multifunctional protein composite biomaterials that exhibit improved properties and functions to specifically match with the regenerative medicine's needs.

3.4 NANOHYDROXYAPATITE

Synthetic HA has a chemical resemblance to the inorganic component of bone matrix and displays strong similarity to host hard tissues, making it an ideal candidate for orthopedic and dental implants or components of implants (Zhou and Lee, 2011). HA has been extensively used in the restoration of hard tissues, and typical procedures include bone repair, bone augmentation, as well as coating of metallic implants or fillers in bone or teeth replacements to repair segmental defects caused by the removal of infected tissue or bone tumors (Dorozhkin, 2009; LeGeros, 2002). Nevertheless, the poor mechanical properties exhibited by regular HA crystals generally limits its use to low load-bearing applications (Zhou and Lee, 2011). Recent improvements in nanoscience and nanotechnology have renovating the interest in HA, specifically in the investigation of nanoscale HA formation and their properties. Nanocrystalline HA powders exhibit sinterability and enhanced densification due to their superior surface area, which can improve fracture toughness, as well as other mechanical properties (Dorozhkin, 2009). It has been proposed that nano-HA can be an ideal biomaterial due to its good biocompatibility, biodegradability, osteoconductivity, and bone integration ability (D'Elía et al., 2013). Current uses of nano-HA include surface modifications to modulate their colloid stability (Wilson, 2009; Wilson and Hull, 2008) and bioactivity (D'Elía et al., 2013), as an intermediate layer to allow strong bond formation between HA-polymer matrices avoiding dissolution and inflammation. Modification of calcium phosphate cement (CPC) is an alternative application of nano-HA. CPCs are self-setting synthetic bone graft materials that after mixed with an aqueous solution to form a paste; it can self-harden to form HA in situ in the bone defect. The incorporation of nano-HA crystals considerably augments the setting time and the transfiguration to apatite during the self-hardening of CPC, assists the nucleation of smaller crystals and promotes the nucleation of new nano-HA crystals in its surface. Another significant application of nanostructured HA is coating on metallic or other implants made of bioinert materials to improve their bioactivity and osteoconductivity. Nanoscale alteration of implant surfaces has been recommended to further intensify its biomimicry and bioactivity (Beltrán-Partida et al., 2015; Fernandez-Yague et al., 2015; Lee and Laurencin, 2008). Also, HA nanoparticles have functioned as nonviral carriers for drug delivery and gene therapy because of their recognized biocompatibility, ease of handling and well-known adsorption affinity (Bauer et al., 2008; Dasgupta et al., 2009; Ong et al., 2008; Palazzo et al., 2007; Victor and Kumar, 2008). Additionally, HA nanoparticles can be stably loaded with radioisotopes (Ong et al., 2008). After loading with genes or drugs by adsorption, HA nanoparticles offer a defensive atmosphere that protects them from degradation while giving a suitable trail for cell membrane penetration and the orderly release of the gene/drug (Palazzo et al., 2007). Summarizing, HA nanoparticles mimic the chemical composition and the dimensions of the mineral phase existing in the native bone tissue; offering superior surface-to-volume ratios, high wettability and enhanced mechanical properties, which heighten protein adsorption and other desirable properties, compared to their conventional counterparts. Nano-HA has been suggested to exhibit a favorable environment for cell attachment, spreading, proliferation, and to induce osteogenic differentiation optimizing the biological functionality, for which particle size is believed to play a key role.

3.5 MAGNESIUM PROCESSED NANOMATERIALS

Magnesium (Mg) is the 4th abundant element in the human body after Ca, K, and Na, and the 50% of the total Mg exists in osseous tissue while only 1.0% is present in extracellular fluid. Mg^{2+} has a special inference in numerous biological processes comprising cellular courses of proliferation and differentiation, the cell–matrix interaction, the normal function of parathyroid glands and metabolism of vitamin D (Cabrejos-Azama et al., 2014; Craciunescu et al., 2011). Mg^{2+} deficit may directly result in a decrease in osteoblastic bone formation and delay the onset of mineralization of newly formed cartilage matrix and bones (Enderle et al., 2005); its deficiency in bone appears to be a risk factor for osteoporosis in humans (Sader et al., 2009). Recently, it was informed that the amount of Mg^{2+} into calcified tissues is associated with the apatites phase decrease, indirectly has impacts on the mineral metabolism and calcification process. Strong calcification leads to changes of the bone matrix that determines the bone fragility (Raynaud et al., 2002; Seelig, 2012). Mineralized apatite phase in normal bone differ extensively in its chemical composition containing various trace elements, such as Na^+ (c 0.7 wt.%), K^+ (c 0.03 wt.%), Mg^{2+} (c 0.55 wt.%), Cl^- (c 0.10 wt.%), and F^- (c 0.02 wt.%); each of them takes significant roles in the physiological systems (Kannan et al., 2006; Zhang, 2013). An implant composed of or covered with pure HA crystal does not have this composition; therefore, its bioactivity is inferior to the bone mineral phase (Kalita et al., 2007), which could lead to unfavorably effects on the material osseointegration properties. Accordingly to that, a great attention is given in the developing of bone substitutes that incorporate magnesium (Mg^{2+}) ions into HA scaffolds in order to increase its morphological and chemical affinity toward natural bone tissue and therefore to improve its physico-chemical and structural properties (Craciunescu et al., 2011). Mg^{2+} is currently used as calcium switch; the incorporation of Mg^{2+} within CPC has attracted attention because of the potential beneficial effects of Mg^{2+} on the physicochemical properties of the minerals (Mestres and Ginebra, 2011; Wu et al., 2008) and also because the improvement of the bone metabolism (Jia et al., 2010; Waselau et al., 2007). At this time, most researchers are focusing on the preparation of cation-incorporated HA bulk, and the mechanical properties of bulk materials depend on the structures processed. For example, when Mg^{2+} ions were added into the apatite reactant solution to be introduced into a cement system, it resulted in a reduction in compressive strength of the cement (Mestres and Ginebra, 2011; Wu et al., 2008). The adhesion strength between Mg-incorporated apatite coating and Ti_6Al_4V substrate increased linearly as the concentration of Mg^{2+} ions incorporated in the apatite coatings increases. The chemical bonds Ca—P—H—O—Ti—Mg at the boundary of the coating and substrate promote to the improvement of adhesion strength. Additional Mg—Ti—O chemical bonds are formed as more Mg^{2+} are assimilated in the coating, giving rise to a superior adhesion strength in Mg-containing apatite coatings (Zhang, 2013). Mg^{2+}-containing apatite coatings prepared by the sol–gel method stimulated the formation of a biogenic apatite layer after soaking in a SBF solution and showed high-quality bioresponses with MG63 cells just like the HA coatings (Zhang, 2013).

3.6 NANOCARBON ASSEMBLIES: GRAPHENE

Structural design of nanocarbons, such as graphene (G), graphene oxide (GO), and reduced graphene oxide (rGO), can be arranged into assemblies of dissimilar dimensions and properties for biological applications including bone tissue regeneration. Graphene is mechanically robust, lightweight, and almost translucent, as well as an exceptional conductor of heat and electricity; properties which derive from the fact that it consists in a 2D honeycomb-network of a sp^2 carbon (Nakanishi et al., 2014). GO also has a 2D structure consisting in sp^2 and sp^3 carbon system including oxygen (and hydrogen), and defects in honeycomb structure. rGO has a 2D structure consisting mostly of a honeycomb-network of sp^2 carbon containing a malleable number of defects. The surfaces of G and its analogous have been used as platforms for cell culture. Although G and rGO are highly hydrophobic, several reports have shown that this environment may be suitable for cell culture (Lu et al., 2013b). Human osteoblasts and mesenchymal stem cells (MSCs) evolution on graphene was promoted compared to SiO_2 substrates (Kalbacova et al., 2010). Here we will discuss some selected examples. Sheets based on self-sustaining graphene-based hydrogels, created by chemical reduction of GO using hydrazine and NH_3, were exploited as a platform for bone regeneration (Lu et al., 2013a). It was established, from both in vitro and in vivo conditions, that the film stimulated osteogenic differentiation of rat bone marrow stromal stem cells without other additives. In addition, from the in vitro studies, it was detected that osteogenic differentiation and calcium deposits of stem cells occurred. In addition, folds of graphene can strongly adsorb proteins and enhance cell growth and differentiation (Lee et al., 2011; Mu et al., 2012); no inflammatory reactions were attained after material implantation to subcutaneous dorsum sites of rats. Nayak et al. (2011) prepared a 3D scaffold of graphene by the deposition of graphene sheet onto a substrate that has a 3D structure. A vast graphene piece of 1×1 cm dimensions was prepared by the chemical vapor deposition (CVD) method on copper foils and then transmitted to four different substrates, Si/SiO_2, glass, polyethylene terephthalate (PET), and polydimethylsiloxane (PDMS) films. Atomic force microscopic observations revealed that the graphene surface of each material had different undulations and folds at the nanometer scale. Graphene on glass or Si/SiO_2 showed good compatibility with human MSCs while graphene on PET or PDMS exhibited lower cell viability. Cell toxicity was exposed to have its origin on the 3D structure of the intrinsic substrates nature. Cell differentiation was tested after cultured in the presence of three of the previously mentioned substrate arrangements (graphene on Si/SiO_2, glass, or PET film) using typical markers: CD-44 for hMSCs and osteocalcin for osteoblasts. It was found that graphene's presence accelerated cell differentiation even in the absence of commonly used additional growth factors, such as 2BMP. The authors proposed that graphene was the driving force for bone cell formation, regardless of the underlying substrate (Nayak et al., 2011).

4 RISKS AND OBSTACLES OF NANOMATERIALS APPLICATIONS

Materials based on nanoparticles or exhibiting architectures displaying nanometric dimensions have definite exceptional characteristics which can be and have been exploited in numerous biomedical applications. However, these unique features are postulated to be the grounds for NP-induced biotoxicity, which arises from the complex interplay between particle characteristics (e.g., size, shape, surface chemistry, and charge), administered dose and host immunological integrity (Yildirimer et al., 2011). Although some studies have begun to inspect the effect of nanoparticulate systems on cell viability, the health properties of NPs incoming into human body through are still largely indefinite. A more comprehensive checking of the possible effects of novel designed and made-up nanomaterials under in vivo situations is an essential requirement. For example, several reports have showed possible negative effects of nanoparticles on cells (Braydich-Stolle et al., 2005; Verma and Stellacci, 2010) and their potential to incite the oxidative stress, inflammation, genetic damage, and long-term pathological problems (Medina et al., 2007; Nel et al., 2006). The biocompatibility and biodegradation properties of different nanomaterials also need to be methodically explored and analyzed both in vitro and in vivo conditions before they can be safe applied to bone regeneration in clinical trials (for a comprehensive review on toxicology of nanomaterials see Yildirimer et al., 2011). Wide investigation is being directed on the use of scaffolds seeded with stem cells to produce bone and cartilage. Nevertheless, the triumph of this technique is limited by the availability of stem cells and their efficiency in regeneration (Chaudhury et al., 2014). In addition, the interactions of cells with complex 3D porous nanoscale topographies have proved to be significant to bone regeneration; a significant challenge that must be addressed is the transference of these nanopatterning technologies complexity to scaled-up versions. Another point is the mechanical properties, current research focuses on elucidating the mechanistic underlying NP toxicity which is postulated to range from inflammatory cell infiltration and cellular necrosis to ROS-induced apoptosis; moreover it is fundamental to determine the key factor stimulating the assemblies of different tissues and to create specific microenvironment for such assemblies. Mechanical properties play a significant function in tissue regeneration, not only because of their temporary support role but also because of their effect at the cell level owing to the mechanotransduction phenomenon, by which mechanical stimuli are converted into biochemical signals and translated into changes in cell behavior. Actually, matching the mechanical properties of bone with a synthetic scaffold is still elusive; such point is crucial to avoid the stress-shielding effect and to promote bone regeneration.

5 PERSPECTIVES AND OUTLOOKS

Given that the cell–cell and cell–matrix communication in biological organism income at the nanoscale level, nanoscale materials, and composites of various chemical configurations has transfigured the designing of bone grafts and scaffolds. The design of various types of nanostructured scaffolds and the incorporation of nanoparticles into the matrices displaying optimal feature size and chemistry that enhance cell and protein interactions compared with their conventional counterparts are reported. Novel strategies claiming ease fabrication, the ability to incorporate bioactive factors, control over surface or bulk properties, excluding the use of toxic solvents, regulation of a uniform architecture, and optimal mechanical features are proposed to produce nanomaterials. In addition, new trends exploring the preparation of 3D rapidly prototyped scaffolds with 3D printing that are combined with living cells represent a promising advance to bone tissue regeneration. Nevertheless, the development of perfect materials capable of sending signals to the diseased or damaged cells and tissues to activate the regeneration process still remains a challenge. The complexity of biological systems has forced the interplay and interdisciplinary support from different fields, such as engineering, biology, materials science, chemistry, physics, and medicine to modulate and cautiously incorporate both micro/nanolevel topographies and chemical signals to induce the desired cellular response after implantation and to promote the effective tissue regeneration and restoration. The upcoming circumstances are optimistic for the scientific community, but great expectations have to be met.

ACKNOWLEDGMENTS

The authors acknowledge Universidad Nacional del Sur (PGI 24/Q064), Concejo Nacional de Investigaciones Científicas y Técnicas de la República Argentina (CONICET, PIP-11220130100100CO). NLD have doctoral fellowships of CONICET. PVM and LAB are researchers of CONICET.

REFERENCES

Akkouch, A., Zhang, Z., Rouabhia, M., 2011. A novel collagen/hydroxyapatite/poly(lactide-*co*-ε-caprolactone) biodegradable and bioactive 3D porous scaffold for bone regeneration. J. Biomed. Mater. Res. Part A 96A, 693–704.

Albrektsson, T., Johansson, C., 2001. Osteoinduction, osteoconduction and osseointegration. Eur. Spine J. 10, S96–S101.

Bacakova, L., Filova, E., Parizek, M., Ruml, T., Svorcik, V., 2011. Modulation of cell adhesion, proliferation and differentiation on materials designed for body implants. Biotechnol. Adv. 29, 739–767.

Badylak, S.F., Freytes, D.O., Gilbert, T.W., 2009. Extracellular matrix as a biological scaffold material: structure and function. Acta Biomater. 5, 1–13.

Bailey, A.J., 2001. Molecular mechanisms of ageing in connective tissues. Mech. Ageing Dev. 122, 735–755.

Barrère, F., Mahmood, T.A., De Groot, K., Van Blitterswijk, C.A., 2008. Advanced biomaterials for skeletal tissue regeneration: instructive and smart functions. Mater. Sci. Eng. R: Rep. 59, 38–71.

Bauer, I., Li, S.-P., Han, Y.-C., Yuan, L., Yin, M.-Z., 2008. Internalization of hydroxyapatite nanoparticles in liver cancer cells. J. Mater. Sci. 19, 1091–1095.

Becker, S.T., Bolte, H., Schünemann, K., Seitz, H., Bara, J.J., Beck-Broichsitter, B.E., Russo, P.A.J., Wiltfang, J., Warnke, P.H., 2012. Endocultivation: the influence of delayed vs. simultaneous application of BMP-2 onto individually formed hydroxyapatite matrices for heterotopic bone induction. Int. J. Oral Maxillofac. Surg. 41, 1153–1160.

Beltrán-Partida, E., Moreno-Ulloa, A., Valdez-Salas, B., Velasquillo, C., Carrillo, M., Escamilla, A., Valdez, E., Villarreal, F., 2015. Improved osteoblast and chondrocyte adhesion and viability by surface-modified Ti_6Al_4V alloy with anodized TiO_2 nanotubes using a super-oxidative solution. Materials 8, 867–883.

Bianco, P., Robey, P.G., 2001. Stem cells in tissue engineering. Nature 414, 118–121.

Blakeney, B.A., Tambralli, A., Anderson, J.M., Andukuri, A., Lim, D.-J., Dean, D.R., Jun, H.-W., 2011. Cell infiltration and growth in a low density, uncompressed three-dimensional electrospun nanofibrous scaffold. Biomaterials 32, 1583–1590.

Bohner, M., 2010. Resorbable biomaterials as bone graft substitutes. Mater. Today 13, 24–30.

Bose, S., Roy, M., Bandyopadhyay, A., 2012. Recent advances in bone tissue engineering scaffolds. Trends Biotechnol. 30, 546–554.

Bose, S., Vahabzadeh, S., Bandyopadhyay, A., 2013. Bone tissue engineering using 3D printing. Mater. Today 16, 496–504.

Boudou, T., Crouzier, T., Ren, K., Blin, G., Picart, C., 2010. Multiple functionalities of polyelectrolyte multilayer films: new biomedical applications. Adv. Mater. 22, 441–467.

Braydich-Stolle, L., Hussain, S., Schlager, J.J., Hofmann, M.-C., 2005. In vitro cytotoxicity of nanoparticles in mammalian germline stem cells. Toxicol. Sci. 88, 412–419.

Brown, T.D., 2000. Techniques for mechanical stimulation of cells in vitro: a review. J. Biomech. 33, 3–14.
Cabrejos-Azama, J., Alkhraisat, M.H., Rueda, C., Torres, J., Blanco, L., López-Cabarcos, E., 2014. Magnesium substitution in brushite cements for enhanced bone tissue regeneration. Mater. Sci. Eng. C 43, 403–410.
Cao, H., Kuboyama, N., 2010. A biodegradable porous composite scaffold of PGA/β-TCP for bone tissue engineering. Bone 46, 386–395.
Collins, M.N., Birkinshaw, C., 2013. Hyaluronic acid based scaffolds for tissue engineering—a review. Carbohyd. Polym. 92, 1262–1279.
Craciunescu, O., Tardei, C., Moldovan, L., Zarnescu, O., 2011. Magnesium substitution effect on porous scaffolds for bone repair. Cent. Eur. J. Biol. 6, 301–311.
Chaubaroux, C., Vrana, E., Debry, C., Schaaf, P., Senger, B., Voegel, J.-C., Haikel, Y., Ringwald, C., Hemmerlé, J., Lavalle, P., Boulmedais, F., 2012. Collagen-based fibrillar multilayer films cross-linked by a natural agent. Biomacromolecules 13, 2128–2135.
Chaudhury, K., Kumar, V., Kandasamy, J., Roychoudhury, S., 2014. Regenerative nanomedicine: current perspectives and future directions. Int. J. Nanomed. 9, 4153.
Chen, A.A., Khetani, S.R., Lee, S., Bhatia, S.N., Van Vliet, K.J., 2009. Modulation of hepatocyte phenotype in vitro via chemomechanical tuning of polyelectrolyte multilayers. Biomaterials 30, 1113–1120.
D'Elía, N.L., Gravina, A.N., Ruso, J.M., Laiuppa, J.A., Santillán, G.E., Messina, P.V., 2013. Manipulating the bioactivity of hydroxyapatite nano-rods structured networks: effects on mineral coating morphology and growth kinetic. Biochim. Biophys. Acta (BBA)V 1830, 5014–5026.
Dasgupta, S., Bandyopadhyay, A., Bose, S., 2009. Reverse micelle-mediated synthesis of calcium phosphate nanocarriers for controlled release of bovine serum albumin. Acta Biomater. 5, 3112–3121.
Decher, G., 1997. Fuzzy nanoassemblies: toward layered polymeric multicomposites. Science 277, 1232–1237.
Di Martino, A., Liverani, L., Rainer, A., Salvatore, G., Trombetta, M., Denaro, V., 2011. Electrospun scaffolds for bone tissue engineering. Musculoskelet. Surg. 95, 69–80.
Dorozhkin, S., 2009. Nanodimensional and nanocrystalline apatites and other calcium orthophosphates in biomedical engineering, biology and medicine. Materials 2 (4), 1975–2045.
Dupont, K.M., Sharma, K., Stevens, H.Y., Boerckel, J.D., García, A.J., Guldberg, R.E., 2010. Human stem cell delivery for treatment of large segmental bone defects. Proc. Natl. Acad. Sci. 107, 3305–3310.
Dvir, T., Timko, B.P., Kohane, D.S., Langer, R., 2011. Nanotechnological strategies for engineering complex tissues. Nat. Nano 6, 13–22.
Easton, C.D., Bullock, A.J., Gigliobianco, G., Mcarthur, S.L., Macneil, S., 2014. Application of layer-by-layer coatings to tissue scaffolds—development of an angiogenic biomaterial. J. Mater. Chem. B 2, 5558–5568.
Enderle, R., Götz-Neunhoeffer, F., Göbbels, M., Müller, F.A., Greil, P., 2005. Influence of magnesium doping on the phase transformation temperature of β-TCP ceramics examined by Rietveld refinement. Biomaterials 26, 3379–3384.
Engler, A.J., Sen, S., Sweeney, H.L., Discher, D.E., 2006. Matrix elasticity directs stem cell lineage specification. Cell 126, 677–689.
Fantner, G.E., Hassenkam, T., Kindt, J.H., Weaver, J.C., Birkedal, H., Pechenik, L., Cutroni, J.A., Cidade, G.A.G., Stucky, G.D., Morse, D.E., Hansma, P.K., 2005. Sacrificial bonds and hidden length dissipate energy as mineralized fibrils separate during bone fracture. Nat. Mater 4, 612–616.
Fernandez-Yague, M.A., Abbah, S.A., Mcnamara, L., Zeugolis, D.I., Pandit, A., Biggs, M.J., 2015. Biomimetic approaches in bone tissue engineering: integrating biological and physicomechanical strategies. Adv. Drug Deliv. Rev. 84, 1–29.
Fratzl, P., Fratzl-Zelman, N., Klaushofer, K., 1993. Collagen packing and mineralization. an x-ray scattering investigation of turkey leg tendon. Biophys. J. 64, 260–266.
Fratzl, P., Gupta, H.S., Paschalis, E.P., Roschger, P., 2004. Structure and mechanical quality of the collagen-mineral nano-composite in bone. J. Mater. Chem. 14, 2115–2123.
Fratzl, P., Weinkamer, R., 2007. Nature's hierarchical materials. Prog. Mater. Sci. 52, 1263–1334.

Gbureck, U., Hölzel, T., Klammert, U., Würzler, K., Müller, F.A., Barralet, J.E., 2007. Resorbable dicalcium phosphate bone substitutes prepared by 3D powder printing. Adv. Funct. Mater. 17, 3940–3945.

Geiger, M., Li, R.H., Friess, W., 2003. Collagen sponges for bone regeneration with rhBMP-2. Adv. Drug Deliv. Rev. 55, 1613–1629.

George, A., Ravindran, S., 2010. Protein templates in hard tissue engineering. Nano Today 5, 254–266.

Gravina, A.N., Ruso, J.M., Laiuppa, J.A., Santillan, G.E., Marco-Brown, J.L., D'elia, N.L., Messina, P.V., 2014. Striped, bioactive Ce–TiO_2 materials with peroxynitrite-scavenging activity. J. Mater. Chem. B 2, 834–845.

Gravina, N., Ruso, J.M., Mbeh, D.A., Yahia, L.H., Merhi, Y., Sartuqui, J., Messina, P.V., 2015. Effect of ceria on the organization and bio-ability of anatase fullerene-like crystals. RSC Adv. 5, 8077–8087.

Gross, B.C., Erkal, J.L., Lockwood, S.Y., Chen, C., Spence, D.M., 2014. Evaluation of 3D printing and its potential impact on biotechnology and the chemical sciences. Anal. Chem. 86, 3240–3253.

Guan, J., Fujimoto, K.L., Sacks, M.S., Wagner, W.R., 2005. Preparation and characterization of highly porous, biodegradable polyurethane scaffolds for soft tissue applications. Biomaterials 26, 3961–3971.

Guimarães, A., Martins, A., Pinho, E.D., Faria, S., Reis, R.L., Neves, N.M., 2010. Solving cell infiltration limitations of electrospun nanofiber meshes for tissue engineering applications. Nanomedicine 5, 539–554.

Hammond, P.T., 2011. Engineering materials layer-by-layer: challenges and opportunities in multilayer assembly. AIChE J. 57, 2928–2940.

Hammond, P.T., 2012. Building biomedical materials layer-by-layer. Mater. Today 15, 196–206.

Hassan, N., Verdinelli, V., Ruso, J.M., Messina, P.V., 2011. Mimicking natural fibrous structures of opals by means of a microemulsion-mediated hydrothermal method. Langmuir 27, 8905–9812.

Hassan, N., Verdinelli, V., Ruso, J.M., Messina, P.V., 2012. Assessing structure and dynamics of fibrinogen films on silicon nanofibers: towards hemocompatibility devices. Soft Matter 8, 6582–6592.

Hing, K.A., 2005. Bioceramic bone graft substitutes: influence of porosity and chemistry. Int. J. Appl. Ceramic Technol. 2, 184–199.

Holzwarth, J.M., Ma, P.X., 2011. Biomimetic nanofibrous scaffolds for bone tissue engineering. Biomaterials 32, 9622–9629.

Hollister, S.J., 2005. Porous scaffold design for tissue engineering. Nat. Mater. 4, 518–524.

Hu, X., Cebe, P., Weiss, A.S., Omenetto, F., Kaplan, D.L., 2012. Protein-based composite materials. Mater. Today 15, 208–215.

Hunt, N., Grover, L., 2010. Cell encapsulation using biopolymer gels for regenerative medicine. Biotechnol. Lett. 32, 733–742.

Jia, J., Zhou, H., Wei, J., Jiang, X., Hua, H., Chen, F., Wei, S., Shin, J.-W., Liu, C., 2010. Development of magnesium calcium phosphate biocement for bone regeneration. J. Royal Soc. Interface 7 (49), 1171–1180.

Kaigler, D., Wang, Z., Horger, K., Mooney, D.J., Krebsbach, P.H., 2006. VEGF scaffolds enhance angiogenesis and bone regeneration in irradiated osseous defects. J. Bone Miner. Res. 21, 735–744.

Kalbacova, M., Broz, A., Kong, J., Kalbac, M., 2010. Graphene substrates promote adherence of human osteoblasts and mesenchymal stromal cells. Carbon 48, 4323–4329.

Kalita, S.J., Bhardwaj, A., Bhatt, H.A., 2007. Nanocrystalline calcium phosphate ceramics in biomedical engineering. Mater. Sci. Eng. C 27, 441–449.

Kane, R., Ma, P.X., 2013. Mimicking the nanostructure of bone matrix to regenerate bone. Mater. Today 16, 418–423.

Kannan, S., Rocha, J., Ferreira, J., 2006. Synthesis and thermal stability of sodium, magnesium co-substituted hydroxyapatites. J. Mater. Chem. 16, 286–291.

Khorshidi, S., Solouk, A., Mirzadeh, H., Mazinani, S., Lagaron, J.M., Sharifi, S., Ramakrishna, S., 2015. A review of key challenges of electrospun scaffolds for tissue-engineering applications. J. Tissue Eng. Regen. Med.

Kingham, E.J., Tsimbouri, M., Gadegaard, N., Dalby, M.J., Oreffo, R.O.C., 2011. Nanotopography induced osteogenic differentiation of human stem cells. Bone 48 (Suppl. 2), S108–S109.

Klok, H.-A., 2009. Peptide/protein–synthetic polymer conjugates: quo vadis. Macromolecules 42, 7990–8000.

Ko, I.K., Lee, S.J., Atala, A., Yoo, J.J., 2013. In situ tissue regeneration through host stem cell recruitment. Exp. Mol. Med. 45, e57.

Koester, K.J., Ager, J.W., Ritchie, R.O., 2008. The true toughness of human cortical bone measured with realistically short cracks. Nat. Mater. 7, 672–677.

Kokubo, T., Kushitani, H., Sakka, S., Kisugi, T., Yamamuro, T., 1990. Solutions able to reproduce in vivo surface-structure changes in bioactive glass-ceramic A-W3. J. Biomed. Mat. Res. 24, 721–734.

Landis, W.J., Hodgens, K.J., Arena, J., Song, M.J., Mcewen, B.F., 1996. Structural relations between collagen and mineral in bone as determined by high voltage electron microscopic tomography. Microsc. Res. Tech. 33, 192–202.

Launey, M.E., Buehler, M.J., Ritchie, R.O., 2010. On the mechanistic origins of toughness in bone. Ann. Rev. Mater. Res. 40, 25–53.

Lee, D.A., Laurencin, C., 2008. Surface Modification of Biomaterials at the Nanoscale: Biomimetic Scaffolds for Tissue Engineering Applications. CRC Press, Boca Raton, FL.

Lee, W.C., Lim, C.H.Y., Shi, H., Tang, L.A., Wang, Y., Lim, C.T., Loh, K.P., 2011. Origin of enhanced stem cell growth and differentiation on graphene and graphene oxide. ACS Nano 5, 7334–7341.

Lees, S., Mook, H., 1986. Equatorial diffraction spacing as a function of water content in fully mineralized cow bone determined by neutron diffraction. Calcif. Tissue Int. 39, 291–292.

Legeros, R.Z., 2002. Properties of osteoconductive biomaterials: calcium phosphates. Clin. Orthop. Relat. Res. 395, 81–98.

Li, P., Ohtsuki, C., Kokubo, T., Nakanishi, K., Soga, N., De Groot, K., 1994. The role of hydrated silica, titania, and alumina in inducing apatite on implants. J. Biomed. Mater. Res. 28, 7–15.

Liao, S., Wang, W., Uo, M., Ohkawa, S., Akasaka, T., Tamura, K., Cui, F., Watari, F., 2005. A three-layered nano-carbonated hydroxyapatite/collagen/PLGA composite membrane for guided tissue regeneration. Biomaterials 26, 7564–7571.

Lindberg, F., Heinrichs, J., Ericson, F., Thomsen, P., Engqvist, H., 2008. Hydroxylapatite growth on single-crystal rutile substrates. Biomaterials 29, 3317–3323.

Liu, H., Peng, H., Wu, Y., Zhang, C., Cai, Y., Xu, G., Li, Q., Chen, X., Ji, J., Zhang, Y., Ouyang, H.W., 2013. The promotion of bone regeneration by nanofibrous hydroxyapatite/chitosan scaffolds by effects on integrin-BMP/Smad signaling pathway in BMSCs. Biomaterials 34, 4404–4417.

Lu, J., He, Y.-S., Cheng, C., Wang, Y., Qiu, L., Li, D., Zou, D., 2013a. Self-supporting graphene hydrogel film as an experimental platform to evaluate the potential of graphene for bone regeneration. Adv. Funct. Mater. 23, 3494–3502.

Lu, J., He, Y.S., Cheng, C., Wang, Y., Qiu, L., Li, D., Zou, D., 2013b. Self-supporting graphene hydrogel film as an experimental platform to evaluate the potential of graphene for bone regeneration. Adv. Funct. Mater. 23, 3494–3502.

Lutolf, M.P., Hubbell, J.A., 2005. Synthetic biomaterials as instructive extracellular microenvironments for morphogenesis in tissue engineering. Nat. Biotechnol. 23, 47–55.

Ma, P.X., 2008. Biomimetic Materials for Tissue Engineering. Adv. Drug Deliv. Rev. 60, 184–198.

Macdonald, M.L., Samuel, R.E., Shah, N.J., Padera, R., Beben, Y.M., Hammond, P.T., 2011. Tissue integration of growth factor-eluting layer-by-layer polyelectrolyte multilayer coated implants. Biomaterials 32, 1446–1453.

Mandal, B.B., Grinberg, A., Seok Gil, E., Panilaitis, B., Kaplan, D.L., 2012. High-strength silk protein scaffolds for bone repair. Proc. Natl. Acad. Sci. 109, 7699–7704.

Mann, B.K., Gobin, A.S., Tsai, A.T., Schmedlen, R.H., West, J.L., 2001. Smooth muscle cell growth in photopolymerized hydrogels with cell adhesive and proteolytically degradable domains: synthetic ECM analogs for tissue engineering. Biomaterials 22, 3045–3051.

Marsh, G., 2001. Tissue regeneration the material enablers. Mater. Today 4, 38–41.

Matson, J.B., Zha, R.H., Stupp, S.I., 2011. Peptide self-assembly for crafting functional biological materials. Curr. Opin. Solid State Mater. Sci. 15, 225–235.
McMahon, R.E., Wang, L., Skoracki, R., Mathur, A.B., 2013. Development of nanomaterials for bone repair and regeneration. J. Biomed. Mater. Res. Part B 101B, 387–397.
Medina, C., Santos-Martinez, M., Radomski, A., Corrigan, O., Radomski, M., 2007. Nanoparticles: pharmacological and toxicological significance. Br. J. Pharmacol. 150, 552–558.
Mestres, G., Ginebra, M.-P., 2011. Novel magnesium phosphate cements with high early strength and antibacterial properties. Acta Biomater. 7, 1853–1861.
Meyers, M.A., Chen, P.-Y., Lin, A.Y.-M., Seki, Y., 2008. Biological materials: structure and mechanical properties. Prog. Mater. Sci. 53, 1–206.
Mu, Q., Su, G., Li, L., Gilbertson, B.O., Yu, L.H., Zhang, Q., Sun, Y.-P., Yan, B., 2012. Size-dependent cell uptake of protein-coated graphene oxide nanosheets. ACS Appl. Mater. Interfaces 4, 2259–2266.
Muller, R., 2009. Hierarchical microimaging of bone structure and function. Nat. Rev. Rheumatol. 5, 373–381.
Murphy, S.V., Atala, A., 2014. 3D bioprinting of tissues and organs. Nat. Biotechnol. 32, 773–785.
Murphy, W.L., Mcdevitt, T.C., Engler, A.J., 2014. Materials as stem cell regulators. Nat. Mater. 13, 547–557.
Nakanishi, W., Minami, K., Shrestha, L.K., Ji, Q., Hill, J.P., Ariga, K., 2014. Bioactive nanocarbon assemblies: nanoarchitectonics and applications. Nano Today 9, 378–394.
Nakashima, M., Reddi, A.H., 2003. The application of bone morphogenetic proteins to dental tissue engineering. Nat. Biotechnol. 21, 1025–1032.
Nalla, R.K., Kinney, J.H., Ritchie, R.O., 2003. Mechanistic fracture criteria for the failure of human cortical bone. Nat. Mater. 2, 164–168.
Nayak, T.R., Andersen, H., Makam, V.S., Khaw, C., Bae, S., Xu, X., Ee, P.-L.R., Ahn, J.-H., Hong, B.H., Pastorin, G., Özyilmaz, B., 2011. Graphene for controlled and accelerated osteogenic differentiation of human mesenchymal stem cells. ACS Nano 5, 4670–4678.
Nel, A., Xia, T., Mädler, L., Li, N., 2006. Toxic potential of materials at the nanolevel. Science 311, 622–627.
O'brien, F.J., 2011. Biomaterials & scaffolds for tissue engineering. Mater. Today 14, 88–95.
Olenych, S.G., Moussallem, M.D., Salloum, D.S., Schlenoff, J.B., Keller, T.C.S., 2005. Fibronectin and cell attachment to cell and protein resistant polyelectrolyte surfaces. Biomacromolecules 6, 3252–3258.
Oliveira, A.L., Costa, S.A., Sousa, R.A., Reis, R.L., 2009. Nucleation and growth of biomimetic apatite layers on 3D plotted biodegradable polymeric scaffolds: effect of static and dynamic coating conditions. Acta Biomater. 5, 1626–1638.
Olszta, M.J., Cheng, X., Jee, S.S., Kumar, R., Kim, Y.-Y., Kaufman, M.J., Douglas, E.P., Gower, L.B., 2007. Bone structure and formation: a new perspective. Mater. Sci. Eng. 58, 77–116.
Ong, H., Loo, J.C., Boey, F.C., Russell, S., Ma, J., Peng, K.-W., 2008. Exploiting the high-affinity phosphonate "hydroxyapatite nanoparticle interaction for delivery of radiation and drugs. J. Nanopart. Res. 10, 141–150.
Ott, H.C., Matthiesen, T.S., Goh, S.-K., Black, L.D., Kren, S.M., Netoff, T.I., Taylor, D.A., 2008. Perfusion-decellularized matrix: using nature's platform to engineer a bioartificial heart. Nat. Med. 14, 213–221.
Palazzo, B., Iafisco, M., Laforgia, M., Margiotta, N., Natile, G., Bianchi, C.L., Walsh, D., Mann, S., Roveri, N., 2007. Biomimetic hydroxyapatite–drug nanocrystals as potential bone substitutes with antitumor drug delivery properties. Adv. Funct. Mater. 17, 2180–2188.
Pan, Z., Duan, P., Liu, X., Wang, H., Cao, L., He, Y., Dong, J., Ding, J., 2015. Effect of porosities of bilayered porous scaffolds on spontaneous osteochondral repair in cartilage tissue engineering. Regen. Biomater. 2, 9–19.
Pauthe, E., Van Tassel, P.R., 2014. Layer-by-layer films as biomaterials: bioactivity and mechanics. J. Biomater. Sci. Polym. Ed. 25, 1489–1501.
Pek, Y.S., Gao, S., Arshad, M.S.M., Leck, K.-J., Ying, J.Y., 2008. Porous collagen-apatite nanocomposite foams as bone regeneration scaffolds. Biomaterials 29, 4300–4305.
Perez-Castillejos, R., 2010. Replication of the 3D architecture of tissues. Mater. Today 13, 32–41.

Petite, H., Viateau, V., Bensaid, W., Meunier, A., De Pollak, C., Bourguignon, M., Oudina, K., Sedel, L., Guillemin, G., 2000. Tissue-engineered bone regeneration. Nat. Biotechnol. 18, 959–963.

Phelps, J.A., Morisse, S., Hindié, M., Degat, M.-C., Pauthe, E., Van Tassel, P.R., 2011. Nanofilm biomaterials: localized cross-linking to optimize mechanical rigidity and bioactivity. Langmuir 27, 1123–1130.

Phipps, M.C., Clem, W.C., Grunda, J.M., Clines, G.A., Bellis, S.L., 2012. Increasing the pore sizes of bone-mimetic electrospun scaffolds comprised of polycaprolactone, collagen I and hydroxyapatite to enhance cell infiltration. Biomaterials 33, 524–534.

Place, E.S., George, J.H., Williams, C.K., Stevens, M.M., 2009. Synthetic polymer scaffolds for tissue engineering. Chem. Soc. Rev. 38, 1139–1151.

Raynaud, S., Champion, E., Bernache-Assollant, D., Thomas, P., 2002. Calcium phosphate apatites with variable Ca/P atomic ratio I. Synthesis, characterisation and thermal stability of powders. Biomaterials 23, 1065–1072.

Rho, J.-Y., Kuhn-Spearing, L., Zioupos, P., 1998. Mechanical properties and the hierarchical structure of bone. Med. Eng. Phys. 20, 92–102.

Ritchie, R.O., 2011. The conflicts between strength and toughness. Nat. Mater. 10, 817–822.

Roach, P., Farrar, D., Perry, C.C., 2006. Surface tailoring for controlled protein adsorption: effect of topography at the nanometer scale and chemistry. J. Am. Chem. Soc. 128, 3939–3945.

Rouwkema, J., Rivron, N.C., Van Blitterswijk, C.A., 2008. Vascularization in tissue engineering. Trends Biotechnol. 26, 434–441.

Sabir, M., Xu, X., Li, L., 2009. A review on biodegradable polymeric materials for bone tissue engineering applications. J. Mater. Sci. 44, 5713–5724.

Sader, M., Legeros, R., Soares, G., 2009. Human osteoblasts adhesion and proliferation on magnesium-substituted tricalcium phosphate dense tablets. J. Mater. Sci. Mater. Med. 20, 521–527.

Sciences, E.H. (ed.), 2011. Dorland's Illustrated Medical Dictionary. Dorland's Medical Dictionary.

Scopelliti, P.E., Borgonovo, A., Indrieri, M., Giorgetti, L., Bongiorno, G., Carbone, R., Podesta, A., Milani, P., 2010. The effect of surface nanometre-scale morphology on protein adsorption. PloS One 5, e11862.

Seelig, M.S., 2012. Magnesium Deficiency in the Pathogenesis of Disease: Early Roots of Cardiovascular, Skeletal, and Renal Abnormalities, first ed. Springer Science & Business Media© 1980 Plenum Publishing Corporation, New York, NY.

Seliktar, D., 2012. Designing cell-compatible hydrogels for biomedical applications. Science 336, 1124–1128.

Seo, J., Lee, H., Jeon, J., Jang, Y., Kim, R., Char, K., Nam, J.-M., 2009. Tunable layer-by-layer polyelectrolyte platforms for comparative cell assays. Biomacromolecules 10, 2254–2260.

Snopok, B.A., Kostyukevich, E.V., 2006. Kinetic studies of protein–surface interactions: a two-stage model of surface-induced protein transitions in adsorbed biofilms. Anal. Biochem. 348, 222–231.

Sprio, S., Ruffini, A., Valentini, F., D'Alessandro, T., Sandri, M., Panseri, S., Tampieri, A., 2011. Biomimesis and biomorphic transformations: new concepts applied to bone regeneration. J. Biotechnol. 156, 347–355.

Stevens, M.M., 2008. Biomaterials for bone tissue engineering. Mater. Today 11, 18–25.

Stevens, M.M., George, J.H., 2005. Exploring and engineering the cell surface interface. Science 310, 1135–1138.

Stupp, S.I., 2010. Self-assembly and biomaterials. Nano Lett. 10, 4783–4786.

Sultana, N., Wang, M., 2008. Fabrication of HA/PHBV composite scaffolds through the emulsion freezing/freeze-drying process and characterisation of the scaffolds. J. Mater. Sci. 19, 2555–2561.

Sundelacruz, S., Kaplan, D.L., 2009. Stem cell- and scaffold-based tissue engineering approaches to osteochondral regenerative medicine. Semin. Cell Dev. Biol. 20, 646–655.

Suwanprateeb, J., Thammarakcharoen, F., Wongsuvan, V., Chokevivat, W., 2012. Development of porous powder printed high density polyethylene for personalized bone implants. J. Porous Mater. 19, 623–632.

Tarafder, S., Balla, V.K., Davies, N.M., Bandyopadhyay, A., Bose, S., 2013a. Microwave-sintered 3D printed tricalcium phosphate scaffolds for bone tissue engineering. J. Tissue Eng. Regen. Med. 7, 631–641.

Tarafder, S., Davies, N.M., Bandyopadhyay, A., Bose, S., 2013b. 3D printed tricalcium phosphate bone tissue engineering scaffolds: effect of SrO and MgO doping on in vivo osteogenesis in a rat distal femoral defect model. Biomater. Sci. 1, 1250–1259.

Temple, J.P., Hutton, D.L., Hung, B.P., Huri, P.Y., Cook, C.A., Kondragunta, R., Jia, X., Grayson, W.L., 2014. Engineering anatomically shaped vascularized bone grafts with hASCs and 3D-printed PCL scaffolds. J. Biomed. Mater. Res. Part A 102, 4317–4325.

Tillman, B.W., Yazdani, S.K., Lee, S.J., Geary, R.L., Atala, A., Yoo, J.J., 2009. The in vivo stability of electrospun polycaprolactone-collagen scaffolds in vascular reconstruction. Biomaterials 30, 583–588.

Uchida, M., Kim, H.-M., Kokubo, T., Fujibayashi, S., Nakamura, T., 2003. Structural dependence of apatite formation on titania gels in a simulated body fluid. J. Biomed. Mater. Res. Part A 64A, 164–170.

Vagaska, B., Bacakova, L., Filová, E., Balik, K., 2010. Osteogenic cells on bio-inspired materials for bone tissue engineering. Physiol. Res. 59, 309–322.

Vanzillotta, P.S., Sader, M.S., Bastos, I.N., Soares, G.D.A., 2006. Improvement of in vitro titanium bioactivity by three different surface treatments. Dental Mater. 22, 275–282.

Verma, A., Stellacci, F., 2010. Effect of surface properties on nanoparticle–cell interactions. Small 6, 12–21.

Victor, S.P., Kumar, T.S.S., 2008. Tailoring calcium-deficient hydroxyapatite nanocarriers for enhanced release of antibiotics. J. Biomed. Nanotechnol. 4, 203–209.

Waselau, M., Samii, V.F., Weisbrode, S.E., Litsky, A.S., Bertone, A.L., 2007. Effects of a magnesium adhesive cement on bone stability and healing following a metatarsal osteotomy in horses. Am. J. Vet. Res. 68, 370–378.

Wei, G., Ma, P.X., 2004. Structure and properties of nano-hydroxyapatite/polymer composite scaffolds for bone tissue engineering. Biomaterials 25, 4749–4757.

Wei, G., Ma, P.X., 2008. Nanostructured biomaterials for regeneration. Adv. Funct. Mater. 18, 3568–3582.

Wertz, C.F., Santore, M.M., 1999. Adsorption and relaxation kinetics of albumin and fibrinogen on hydrophobic surfaces: single-species and competitive behavior. Langmuir 15, 8884–8894.

West, J.L., Hubbell, J.A., 1999. Polymeric biomaterials with degradation sites for proteases involved in cell migration. Macromolecules 32, 241–244.

White, S.W., Hulmes, D., Miller, A., Timmins, P.A., 1977. Collagen-mineral axial relationship in calcified turkey leg tendon by X-ray and neutron diffraction. Nature 266, 421–425.

Whitesides, G.M., Boncheva, M., 2002. Beyond molecules: Self-assembly of mesoscopic and macroscopic components. Proc. Natl. Acad. Sci. USAm. 99, 4769–4774.

Whitesides, G.M., Grzybowski, B., 2002. Self-assembly at all scales. Science 295, 2418–2421.

Wilson, Jr., O.C., Hull, J.R., 2008. Surface modification of nanophase hydroxyapatite with chitosan. Mater. Sci. Eng. C 28, 434–437.

Wilson, O.C., 2009. Surface Modification of Hydroxyapatite: A Review. Advances in Biomedical and Biomimetic Materials. John Wiley & Sons, Inc., Hoboken, New Yersey.

Wittmer, C.R., Phelps, J.A., Saltzman, W.M., Van Tassel, P.R., 2007. Fibronectin terminated multilayer films: protein adsorption and cell attachment studies. Biomaterials 28, 851–860.

Woodruff, M.A., Lange, C., Reichert, J., Berner, A., Chen, F., Fratzl, P., Schantz, J.-T., Hutmacher, D.W., 2012. Bone tissue engineering: from bench to bedside. Mater. Today 15, 430–435.

Wu, C., Aslan, S., Gand, A., Wolenski, J.S., Pauthe, E., Van Tassel, P.R., 2013. Porous nanofilm biomaterials via templated layer-by-layer assembly. Adv. Funct. Mater. 23, 66–74.

Wu, F., Wei, J., Guo, H., Chen, F., Hong, H., Liu, C., 2008. Self-setting bioactive calcium–magnesium phosphate cement with high strength and degradability for bone regeneration. Acta Biomater. 4, 1873–1884.

Yang, G.-L., He, F.-M., Yang, X.-F., Wang, X.-X., Zhao, S.-F., 2009. In vivo evaluation of bone-bonding ability of RGD-coated porous implant using layer-by-layer electrostatic self-assembly. J. Biomed. Mater. Res. Part A 90A, 175–185.

Yarlagadda, P.K., Chandrasekharan, M., Shyan, J.Y.M., 2005. Recent advances and current developments in tissue scaffolding. Bio-Med. Mater. Eng. 15, 159–177.

Yildirimer, L., Thanh, N.T.K., Loizidou, M., Seifalian, A.M., 2011. Toxicology and clinical potential of nanoparticles. Nano Today 6, 585–607.

Yoshikawa, H., Myoui, A., 2005. Bone tissue engineering with porous hydroxyapatite ceramics. J. Artif. Organs 8, 131–136.

Zadpoor, A.A., 2015. Bone tissue regeneration: the role of scaffold geometry. Biomater. Sci. 3, 231–245.
Zemel, A., Rehfeldt, F., Brown, A.E.X., Discher, D.E., Safran, S.A., 2010. Optimal matrix rigidity for stress-fibre polarization in stem cells. Nat. Phys. 6, 468–473.
Zhang, S., 2013. Hydroxyapatite Coatings for Biomedical Applications. CRC Press. Taylor &Francis Group, Boca Raton, FL.
Zhou, H., Lee, J., 2011. Nanoscale hydroxyapatite particles for bone tissue engineering. Acta Biomater. 7, 2769–2781.
Zigang, G., Zhaoxia, J., Tong, C., 2008. Manufacture of degradable polymeric scaffolds for bone regeneration. Biomed. Mater. 3, 022001.

CHAPTER

29

TOXICITY OF NANOSTRUCTURES—A GENERAL APPROACH

Cristina Ş. Iosub*, Elena Olăreţ*, Alexandru Mihai Grumezescu*, Alina M. Holban, Ecaterina Andronescu***

**University Politehnica of Bucharest, Bucharest, Romania*

***Research Institute of the University of Bucharest (ICUB), Bucharest, Romania*

CHAPTER OUTLINE

1 INTRODUCTION

Humans are exposed to nanoparticles (NPs) contamination not only at the time they are being used, but also during the manufacturing period (Shi et al., 2013). The toxic effects of substances are usually described in terms of acute, subacute, subchronic, or chronic exposure conditions (Shi et al., 2013).

As a possible definition of nanotoxicology it could be said that nanotoxicology includes studies about the type and mechanism of toxic effects that nanoscale materials/nanoparticles (NPs) produce in biological systems. Due to the fact that NPs' dimension and the biological molecules' dimension are similar it could be simply estimated that NPs penetrate the cell and affect their functions, so there are some questions that must be answered before using NPs in human organisms: do they disturb the normal biochemical reactions in organism? Are these changes good or damaging for the organism? How could the harmful effect be avoided? This is why the toxic effects of nanostructures should be investigated (Zhao et al., 2012).

It is known that NPs induce reactive oxygen species known as ROS (Manke et al., 2013).

There are many possibilities in which cells can react when they are in contact with NPs and in Fig. 29.1 there is a suggestive image adapted from Schrand et al. (2009), which indicates the fact that NPs may be detected in different cell components and interact with biomolecules in such a way that the cell can dictate either a biocompatible or a toxic response.

Apart from realizing a direct connection with cells, the interface between NPs and cells is actively controlled by the biological environment, which induces NP to be superficially dressed with a corona of proteins that along with the decreased colloidal stability caused by salt, contribute to the recognition of NP surface (Monopoli et al., 2012).

Despite the wide array of scientific subjects related to the interdependence between the physico-chemical properties of NPs and their impact on cells, a complete description of NPs' responses is still required. To a certain extent, one reason would be the difficulty to empirically analyze the physico-chemical properties in such a manner that the majority of them are still confusing (one example would be the augmentation of the effective hydrodynamic diameter of the NPs which is caused by the decreased diffusion of NPs) (Rivera-Gil et al., 2012).

Generally, the way in which nanostructures may react in our biological system was described by Fischer and Chan (2007) in the following steps: (1) they get in contact with human organisms by intravenous, dermal, subcutaneous, inhalation, intraperitoneal, and oral administration; (2) they can be ingested mainly after their interaction with biological constituents (for instance with cells and proteins); (3) they can be diffused to different organs and there is a possibility for them to preserve their structure or they can be simply altered and metabolized; (4) they penetrate the cells of the organ and they can live inside the cell for an unspecific period of time and afterward they may prefer other organs or they can be released through excretion. In correspondence with those mentioned, they are introduced in different ways in the biological environment: direct routes, such as dermal or oral exposure or indirect routes, such as food (Bahadar et al., 2016). Studies showed that NPs are dispersed to different organs depending on the route of administration: after inhalation and oral exposure, NPs diffuse to liver, heart, spleen, and brain in addition to lungs and gastrointestinal tract. Not only does the biodistribution of NPs depend on the administration route, but so does the accumulation, metabolism, and excretion (Yildirimer et al., 2011).

Buzea et al. (2007) reiterated the impact of NPs on health disorders and mentioned that these are not generated by all types of NPs because toxic effects are influenced by the following properties: dimension, agglomeration, chemical structure, crystalin characteristics, surface chemistry and so on and so forth and, over and above, the toxic response of a NP is dictated by the genetic information of each person (Buzea et al., 2007).

In addition to that the influence of NPs in the biological organisms is controlled by physicochemical characteristics and due to this interdependence, before analyzing biological responses, researchers must completely evaluate several particularities of nanostructures (Fischer and Chan, 2007)

One of the problems encountered is that while toxicity assays are developed, the in vitro assay results do not always fit the in vivo ones because NPs could induce some biochemical interaction in vivo which may have passed unnoticed on cell cultures (Yildirimer et al., 2011).

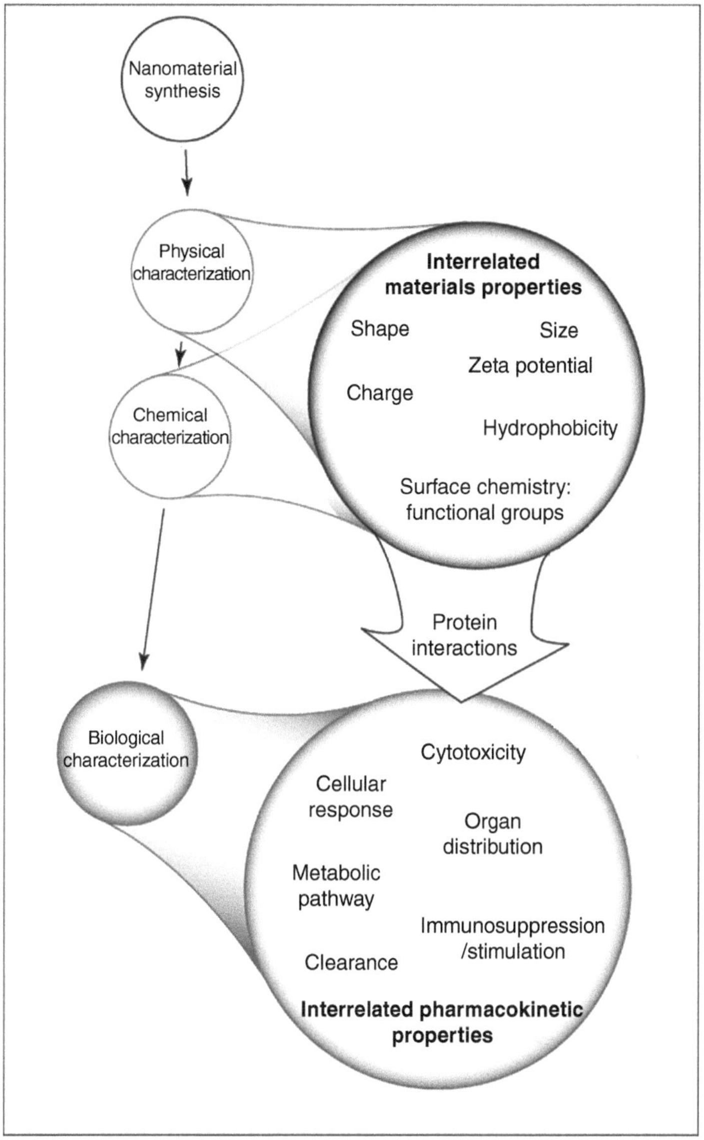

FIGURE 29.1 The Interdependence Between the Physicochemical Properties of Nanoparticles and Their In Vivo Reactions

From Fischer, H.C., Chan, W.C., 2007. Nanotoxicity: the growing need for in vivo study. Curr. Opin. Biotechnol. 18(6), 565–571, from Elsevier.

2 METALLIC NANOSTRUCTURES

2.1 SILVER NANOSTRUCTURES

Due to their antimicrobial characteristics, silver nanoparticles (AgNPs) have gained great attention especially in the fabrication of a wide range of medical and consumer devices (Chernousova and Epple, 2013) thus promoting them as one of the most popular nanomaterials (Sotiriou and Pratsinis, 2011). Moreover, their nanoscale dimensions increase their contact area thus promoting their reactivity but at the same time involve some toxicological effects when being used in biological applications (Wijnhoven et al., 2009).

Even if it is accepted that a significant reason for considering silver's toxicity represents the oxidative process followed by the release of Ag ions, the accurate source of AgNPs' toxicity needs to be rigorously clarified as well (Liu et al., 2010). This lack of evidence is generally caused by the fact that in their particular studies, scientists do not use the same type of NPs and do not work on the same cell type so this makes it difficult for them to compare and envision the conclusions (Caballero-Díaz et al., 2013).

The following study suggests that the surface chemistry of AgNPs, the way they aggregate and their assimilation by cells seem to influence their cytotoxicity (Caballero-Díaz et al., 2013). NIH/3T3 embryonic fibroblasts were used to investigate three sorts of AgNPs with similar Ag cores, but with different coatings realized either from mercaptoundecanoic acid (MUA) or from a polymer covering consisting of poly(isobutylene-alt-maleic anhydride) (PMA) which was further on adjusted with polyethyleneglycol (PEG) (Caballero-Díaz et al., 2013). It was concluded that all types of AgNPs caused less toxicity than single Ag(I) ions. Moreover, it has been demonstrated that the modification with PEG decreased the level of toxicity by restraining the NPs' assimilation by cells, whereas MUA-tailored NPs caused more toxic effects than PMA-tailored NPs (Caballero-Díaz et al., 2013).

AgNPs with particular dimensions between 5.7 and 20.4 nm were fabricated on a nanostructured silica support by flame aerosol technology without using any other methods of functionalization or covering in order to evaluate their interactions with murine macrophages, which are amid the first cells that defy nanosilver particles after consumption (Pratsinis et al., 2013). It seems that nanosilver particles with dimensions under 10 nm discharge Ag ions, the main factors that induce cytotoxicity, whereas for NPs of more than 10 nm the toxic effect caused especially by these ions is diminished but apparently is rather stimulated because of the direct interplay of AgNPs with murine macrophages. As a matter of fact, it is suggested that only the form of cytotoxicity (Ag ions and/or particles) is controlled by AgNPs dimensions (Pratsinis et al., 2013). Another study confirms the fact that for AgNPs with dimensions up to 10 nm, the antibacterial activity is particularly influenced by their interactions with bacterial cell (i.e., *Escherichia coli*) and the discharge of ions is less relevant in this process (Sotiriou and Pratsinis, 2010). The significance of AgNPs' sizes for their cytotoxic response was evaluated in vitro on various organisms, such as bacteria *E. coli* and *Pseudomonas fluorescens*, yeast *Saccharomyces cerevisiae*, algae *Pseudokirchneriella subcapitata*, crustacean *Daphnia magna*, and murine fibroblast cell line Balb/3T3 using NPs of 10, 20, 40, 60, and 80 nm (Ivask et al., 2014). They confirmed that NPs of less than 10 nm are more useful for antimicrobial applications, notably, because of their ability to interact with cells and their biodisponibility to diffuse in the proximity of cell membrane or inside the cell. However, further experiments on different organisms than bacteria are required in order to evaluate if Ag ion biodisponibility can be the cause of possible toxic reactions of 10 nm AgNPs (Ivask et al., 2014). AgNPs of distinct dimensions (10, 40, 100 nm) covered either with citrate (CT) or polyvinylpyrrolidone (PVP) were injected within a vein at a particular dosage of 10 mg/kg body weight in male CD-1(ICR) mice and analyzed after 24 h (Recordati et al., 2016). They discovered that unlike their coating (CV or PVP), the size of NPs dominates toxicity in such a way that the smallest NPs (10 nm) derived the highest concentrations of silver accumulation in tissues, such as spleen, lung, kidney, brain, and blood of mice and induced significant toxicity through hepatobiliary tract, so they assumed that the in vivo reactions of AgNPs are not only associated with the in vivo diffusion of Ag ions but exclusively assigned to the NP model (Recordati et al., 2016). Another study stated that in animal models, after being orally administrated, only 0.4%–18% Ag was injected but diffused in the entire analyzed organs, with increasing amounts especially in the intestine and the stomach, which explained that toxicity is influenced by silver ions delivered from the particle surface (Hadrup and Lam, 2014).

2.2 GOLD NANOSTRUCTURES

It is known that nanometer size changes the characteristics and properties of the bulk material. Even if the properties and the structures of gold nanoparticles (AuNPs) make them suitable for biological application, at high concentrations a certain toxicological effect seems to be identified. While some works suggest that AuNPs show no toxicity (Shukla et al., 2005) there are researches which suggest the opposite. On the one hand, toxicity of the AuNPs depends on their interactions with the cell membrane, a characteristic which is controlled by their strong electrostatic attraction to the negatively charged layers (Goodman et al., 2004) and on the other hand it depends on the NPs' size, concentrations, or aggregation state.

A group of researchers focused their attention on the observation that 20 nm AuNPs with different concentrations have distinct impact on MRC-5 human fetal lung fibroblast cell line (Li et al., 2008). Their results showed no difference in cell morphology between the treated and the control samples, but there could be observed clusters of AuNPs in fibroblasts whose size depends on the concentration: the lowest concentration expressed the smallest cluster (Li et al., 2008). They also performed viability and proliferation assays which pointed out that AuNPs do not influence cell viability, but at high concentrations they inhibit cell proliferation and cause oxidative DNA damage (Li et al., 2008). In order to prevent AuNPs toxicity, some researchers realized a biological synthesis of different concentrations of AuNPs using star anise (SA-AuNPs) and compared their cytotoxicity with AuNPs synthesized from chemical routes (Sathishkumar et al., 2015). Cell viability and toxicity assays were performed on A549 cell line. The results showed that the higher the concentration of AuNPs, the lower the cell viability response and that the SA-AuNPs promoted the highest cell viability compared with the control (Sathishkumar et al., 2015). They also compared NPs of the same size (20–50 nm) obtained from both chemical and biological synthesis in different concentrations and observed that the NPs obtained from chemical synthesis are relatively toxic than SA-AuNPs which also expressed a lower oxidative stress (Sathishkumar et al., 2015). An experimental study revealed that MgFe2O4-Au core–shell NPs showed no cytotoxic response to mouse areola fibroblast cell line (L-929) (Nonkumwong et al., 2016). On a human breast cancer cell line (MCF-7) a concentration of more than 200 μM of 20-nm sized AuNPs showed no important toxicity (Murugan et al., 2014), while on mouse fibroblast cell line (Balb/3T3) a concentration higher than 50 μM of 5-nm AuNPs showed a cytotoxic effect.

2.3 ALUMINUM OXIDE NANOPARTICLES

It was reported that aluminum oxide (Al_2O_3) at nanometric sizes has the possibility to enter through blood brain circulation (Verstraeten et al., 2008). Moreover, in agreement with Kumar and Gill (2009), aluminum agglomeration in the brain can cause numerous diseases and may promote the occurrence of dialysis encephalopathy, Alzheimer, and Parkinson disorders.

As reported by Dong et al. (2011), nanogamma aluminum oxide generates minor toxicity to neural stem cells (NSCs) C17.2 even if cells conveniently incorporate aluminum nanoparticles. They say that cell activity decreases according to the dose administrated and the type of sample meaning that, although for less than 100 g/mL of aluminum oxide no important toxic effects are detected, for concentrations of 200 g/mL cell viability diminishes. Besides, apparently ROS inside the cell may dictate the cytotoxicity of Al_2O_3NPs at increased concentrations and yet it seems that neural cells' death is not caused by them (Dong et al., 2011).

An interesting research study was realized by Huang et al. (2013) who studied the rhythmic activities in the antennal lobe of *Drosophila* sp. in the direction of evaluating the impact of Al_2O_3NPs on central nervous system. They concluded that after 15 min of treatment with aluminum oxide, side effects of the NPs could be observed on *Drosophila*'s nervous system concerning the frequencies of its movements which were considerably depreciated in comparison with controls (0.65 ± 0.13 Hz, 0.34 ± 0.07 Hz, $*P < 0.05$) (Huang et al., 2013).

In accordance with Lu et al. (2009), aluminum oxide particles of 7 nm diameter are responsible for lung inflammation when exposed to rat models and in agreement with this study, Wang et al. (2014a) claim that Al_2O_3NPs determine the lysis of red blood cells, thus producing inflammation.

2.4 TITANIUM OXIDE NANOPARTICLES

Due to their size and structure, titanium nanotubes could be loaded with biomolecules or growth factors in order to obtain drug-delivery systems, therefore biocompatibility and cytotoxicity of titanium nanotubes is an important aspect that needs to be investigated (Salou et al., 2015). Titanium nanotubes show no cytotoxic effect. Moreover, MC3T3-E1 cells cultured on pure titanium and titanium nanotubes show a better adhesion and a good spreading on the TiO_2 nanotubes (Oh et al., 2006). Currently it seems that the biocompatibility is not the main problem thus researches focused their attention on how to modify the surface of the implant and the orientation of TiO_2 nanotubes in order to obtain a superior cell adhesion and cell proliferation response; as a matter of fact, cells seem not to prefer the highly ordered nanoporous layer (Wang et al., 2014b).

Nowadays, titanium dioxide nanoparticles (TiO_2NPs) are widely used in commercial products, such as sunscreens, food, cosmetics, paint, paper, and toothpaste (Wang et al., 2014c). Until now, no toxic effects of TiO_2NPs have been described in human organisms (Shi et al., 2013). No cytotoxicity was observed during MTT assay using human alveolar cells (A549) for 6, 24, and 48 h, when low concentration were used, but for 150 and 200 μg/mL after 48 h it was observed that TiO_2NPs induce apoptosis (Kansara et al., 2015). The way that TiO_2NPs interact with cell membrane and gave a cytotoxic effect depends on cell size, aggregation state, or dispersion. A group of researchers studied the cytotoxic effect of TiO_2NPs on three different cell lines: human pulmonary microvascular endothelial cells (HPMEC-ST1.6R), alveolar macrophage cells (THP-1), and alveolar epithelial (A549), incubating them with different concentrations from 0 to 800 μg/mL of TiO_2NPs (Hanot-Roy et al., 2016). The only significant cytotoxic effect was observed on HPMEC-ST1.6R cells from 50 up to 800 μg/mL and the same cell line showed an apoptosis response in a dose-dependent manner (Hanot-Roy et al., 2016). In vitro studies on human astrocytes (D384) and human neuroblastoma (SH-SY5Y) revealed that NPs produced dose- and time-dependent alterations of the mitochondrial function and damaged the cell membrane starting at 125 mg/mL after 24 h exposure in both types of cells (Coccini et al., 2015). In order to determine the cytotoxicity there have been in vivo studies in rats or mouse models and most of the studies focused on respiratory system showing that TiO_2NPs produced both local and systemic symptoms (Shi et al., 2013). Moreover, researches demonstrated that TiO_2NPs absorbed in lungs can be distributed to various organs, such as liver, kidneys, or brain (Liu et al., 2009). To evaluate the systemic effects of TiO_2NPs, Sprague–Dawley rats were treated by intratracheal instillation with nano-TiO_2 at doses of 0.5, 4, and 32 mg/kg body weight (Sang et al., 2012). Results showed a negligible congestion in spleen and brown particulate accumulation. The immune response was characterized by increased proliferation of B- and T-lymphocyte (Sang et al., 2012). To conclude TiO_2NPs could produce moderate toxicity, causing pulmonary inflammatory response.

2.5 COPPER NANOSTRUCTURES

Even if copper oxide nanoparticles (CuONPs) have functional applications in semiconductors, antimicrobial products, heat transfer fluids, and intrauterine contraceptive devices they seem to produce severe damages in liver, kidney, and spleen in experimental animals (Bahadar et al., 2016). In one in vitro study, CuONPs with a size of 50 nm were exposed to human lung epithelial cells (A 459) and revealed a significantly toxic effect by generating damages in cells' membrane integrity and inducing oxidative stress (Ahamed et al., 2010). The toxicity of CuONPs evaluated in human bronchial epithelial cells (HBEC) showed no significant cytotoxic effect while in lung adenocarcinoma cells (A549 cells) showed a dose–response result (Jing et al., 2015; Kim et al., 2013). In order to determine the acute or subacute effects of CuONPs, in vivo studies have been made. Male rats were exposed for 5 consecutive days via nose-only short-term inhalation to a desired exposure concentration of 10 mg/m^3 (Gosens et al., 2016). Organs weight was evaluated one day after the last exposure and after the recovery period on day 28 when it was observed that the lungs have been completely cleaned. Toxicological results showed that CuONPs were acutely inflammogenic to the lungs at all applied dose-levels increasing the total number of macrophages but besides the nose and lung, no other organs under study were affected based on histological analysis and organ weights (Gosens et al., 2016).

2.6 ZINC OXIDE NANOPARTICLES

Zinc oxide nanoparticles (ZnONPs) have been valued in various applications, such as sunscreens, toothpastes, and cosmetics especially because of their ability to absorb ultraviolet radiation (Lopes et al., 2014).

Studies conducted by Liu et al. (2013) revealed that ZnONPs along with silver oxide and titanium oxide have been applied as products for disinfection, so this is the reason why they analyzed their respiratory immunotoxicity in vitro on alveolar macrophages (AMs) and in vivo on rat models after the intratracheal infusion with NPs. In vitro results showed that ZnO NPs displayed more toxic effects than the other NPs, especially according to WST-8 and LDH assays, whereas in vivo NPs generated oxidative damages to the lungs and thus induced immunotoxicity. (Liu et al., 2013)

In order to determine whether ZnONPs produce any cardiopulmonary harmful effects, Chuang et al. (2014) administrated ZnONPs to Sprague–Dawley rats through intratracheal inhalation and after 24 h of treatment they declared that NPs agglomerated especially in the lungs and insignificant levels of zinc were observed in the heart, liver, kidneys, and blood. Furthermore, after 7 days of treatment, fibrous tissue started to grow and cardiac inflammation was also detected whereas after 30 days, deterioration and necrosis of the myocardium were identified (Chuang et al., 2014). On the other side, Adamcakova-Dodd et al. (2014) pointed out that ZnONPs induce few sub-chronic toxic effects after being inhalated for 3 weeks by mice C57Bl and consequently caused them slightest pulmonary inflammation, cytotoxicity, or lung histopathologic changes.

2.7 IRON OXIDE NANOSTRUCTURES

Iron oxide NPs (IONPs) have been used in biomedical field, targeted drug delivery, and diagnostic field mostly due to their physicochemical properties. Despite being considered as generally safe, possessing a good biocompatibility, low or no toxicity as reported in many studies (Jeng and Swanson, 2006; Kunzmann et al., 2011; Paik et al., 2015), their potential toxicity effects should not be neglected. In vivo studies have proved that after penetrating the cells, IONPs remain in cell organelles and accumulate in the liver, spleen, lungs, and brain after inhalation, being capable of crossing the BBB (blood—brain barrier) generating a toxic effect (Bahadar et al., 2016). The capacity to cross BBB could make them suitable as drug delivery and imaging diagnostics on nervous system, but it could also cause serious damages to this system. The most in vitro studies evaluating IONPs toxicity are centered on cytotoxic effects that these NPs have on cell cultures which mainly show a reduction in viability, ROS production, and iron ion release. In order to evaluate the oxidative stress induced by IONPs and its bulk, a group of researchers used biomarkers, such as (GSH), malondialdehyde (MDA), antioxidant enzymes catalase (CAT), superoxide dismutase (SOD), glutathione peroxidase (GPx), glutathione reductase (GR), and glutathione S transferase (GST) and observed that the NPs accumulated in the liver followed by kidney and the brain, but just the highest dose, lead to pathological changes only in liver; NPs were eliminated in urine; and it clearly generates free radicals, leading to the initiation of oxidative stress condition in rats (Reddy et al., 2015). It was reported that needle-like IONPs (0.5–2 mg/kg) accumulate and remain in the lungs until day 90 after intratracheal instillation, and an important increase of apoptotic cells was observed in BAL (bronchial alveolar lavage) cells (Park et al., 2015). The cytotoxic effect of iron oxide nanorods was investigated. Results indicated that a greater concentration of iron oxide nanorods significantly decreased cell viability (Ghasempour et al., 2015). Modifying the surface of the IONPs makes them less toxic (Sabareeswaran et al., 2016; Shukla et al., 2015). Chitosan-coated IONPs decrease cellular damage and temperate ROS production resulting in a less harmful cytotoxic effect (Shukla et al., 2015).

3 NONMETALLIC NANOSTRUCTURES

3.1 CARBON NANOTUBES

Carbon nanotubes (CNTs) possess excellent physicochemical, electrical, and mechanical properties, so this is the reason why they have attracted the attention of many researchers from numerous fields including the biomedical one (due to their capacity to carry drugs, genes, and biomolecules, as well as their important applicability in bioimaging and diagnostic applications). It seems that the highest embedding efficacy belongs to CNT-based systems compared to liposomes, NPs, and dendrimers (Mody et al., 2014). One of the most frequently reported toxicity effects of CNT is the formation of ROS (Yildirimer et al., 2011). Another important aspect related to biocompatibility and stability in the biological environment is their functionalization which is also important when CNTs want to be conjugate with different biomolecules, contrast agents, or drugs (Kumar et al., 2014). The biocompatibility affair of CNT-based biomedical applications still is disputed. There are reports that confirm the assurance of functionalized CNTs (Mehra et al., 2015) and also reports which sustain their toxic effects on cells and tissues in both in vitro and in vivo systems (Bussy et al., 2015). It was noticed that in vivo toxicity is related to aggregation state and not to individual CNTs (Lee et al., 2015). In a recent study it was investigated that the toxic effects of different multiwalled CNTs (MWCNTs) coated with Pluronic F-68 (PF68) on rat bone marrow–derived MSCs proved that the polymeric coat influence in a favorable way biocompatibility to mesenchymal stem cells (Yao et al., 2016). As for most of the nanostructures, it seems that the toxicity of MWCNTs is size- and dose-dependent. MWCNTs, with diameter larger than 40 nm, have no negative effects on cell viability of macrophages while for WCNTs, with diameter between 15 and 40 nm, cytotoxicity was observed (Allegri et al., 2016). The influence of the different carbon-based nanomaterials (CNTs, nanographite, and carbon black) on cells have been studied. Functionalized CNT produces a high proinflammatory response and oxidative stress but a low cytotoxicity while functionalized carbon black performed the lowest biological response: no cytotoxicity, low inflammation, and equivalent oxidative stress (Figarol et al., 2015). Even if the situation about the biosafety is not yet elucidating, important observations can be used in future experiments for biomedical applications: toxicity is influenced by the type and method of functionalization, size, and dose of CNTs and biodistribution is influenced by functionalization agents (Sajid et al., 2016).

3.2 NANODIAMOND NANOPARTICLES

In agreement with Huynh et al. (2013), nanodiamonds are outstanding nanomaterials that can be used in drug delivery, notably, for the reason that they are biocompatible, inert, and accessible for being noncovalently or covalently grafted with biomolecules, such as DNA and proteins and especially with bioactive substances. Apart from that, according to Simpson et al. (2014) nanodiamonds' fluorescent properties provide applications for in vivo evaluation of intracellular activities.

Impressive results have been reported by Wehling et al. (2014) who state that nanodiamonds subjected to relative oxidation and charged with negative groups particularly with those containing acid anhydride present promising antibacterial properties against Gram-positive and Gram-negative bacteria. They analyzed six distinct nanodiamond materials which enclosed on their surface various oxygen groups and observed that they interacted with cellular elements whereas the anisotropic dispersion of charges on nanodiamonds' surface modified the bacterial surface. What is more, they noticed that the antibacterial properties of nanodiamonds were regulated by proteins, especially because these proteins dress the surface groups of nanodiamonds thus dictating their biocompatibility (Wehling et al., 2014).

In their research study, Paget et al. (2014), examined the potential toxicity of commercial carboxylated nanodiamonds with 20 and 100 nm diameters on lung, kidney, colon, and liver human cell lines and demonstrated that up to a concentration of 250 mg/mL nanodiamonds with identical surface chemistry but distinct dimensions do not generate serious cytotoxicity or genotoxicity even though they penetrate the cells and encourage the application of nanodiamonds in human medicine.

On the other side, after evaluating the toxic effects of nanodiamonds of less than 10 nm on human lymphocytes, Dworak et al. (2014) realized that nanodiamond structures up to a dose of 50 μg/mL proved their cytotoxic impact on cell growth and produced apoptosis. In addition, Dworak et al. (2014) admitted that nanodiamonds caused oxidative stress which affected DNA and chromatin resistence thus restricting their biocompatible properties and uncovering their genotoxic and mutagenic impact.

Finally, Schrand (2016) presented various in vivo tests realized on nanodiamond structures, which in their opinion do not seem to reveal considerable suspicion of nanodiamond cytotoxicity even if there appears to be some dissonance reckoned on the type of animal used for research and the estimated results. Schrand (2016) admitted that nanodiamonds are characterized by detoxification properties that along their substantial biological responses would represent promising particularities in biomedical and bioengineering domain.

3.3 SILICA NANOPARTICLES

Silica NPs have been widely used specially in drug-delivery system, being easily functionalized and being capable to accumulate in tumors. Due to their biocompatibility and its insignificant toxicity, mesoporous silica NPs (MSNs) were approved by Generally Recognized As Safe (GRAS) and by US Food and Drug Administration (FDA) (Beltrán-Osuna and Perilla, 2015). It was reported that MNPs lead to the formation of ROS and oxidative stress, and that MNPs are eliminated in urine and feces (Chen et al., 2013). It has been showed that MSNs are biodegradable, which is a convenience for their complete drug release and excretion from the body after therapy (Zhang et al., 2013). Acute toxicity of MSNs after intravenous injection in mice is mostly influenced by MSNs' porosity and surface characteristics. In one study, examination of long-term and short-term safety of the MNPs modified by phosphonate was realized using mouse model systems and it was administrated twice a week at 50 mg/kg on body weight (Lu et al., 2010). No significant adverse effects were observed with kidney, lung functions, and hematologic profiles (Lu et al., 2010). As it was shown, not only the concentration of silica but also particle size influences cellular toxicity as the smaller diameters cause a bigger harm than the bigger ones.

The sum up of some of the in vivo/in vitro results of the current research study can be observed in Table 29.1.

Table 29.1 In Vivo and In Vitro Citotoxicity Results Obtained for Various Clinically Relevant Nanomaterials

NPs, Size (nm), and Concentration	Cell Line/Cell Culture/Animal Model	Assay	Results	References
AuNPs, 20 nm, 0.1 nM; 0.5 nM; 1 nM	MRC-5 human fetal lung fibroblast	Trypan Blue	No significant difference in cell death; proliferation ↓ at 1 nM	Li et al. (2008)
AuNPs (biological synthesis and chemical synthesis), 20–50 nm, 10 nM; 20 nM; 40 nM; 80 nM; 100 nM; 200 nM	A549 human epithelial respiratory	Apotox-Glo triplex assay kit	Viability ↓ while concentration ↑ Viability ↑ for biological synthesis	Sathishkumar et al. (2015)
AuNPs, 20 nm, 10–200 μM	MCF-7 human breast cancer	MTT	Did not show any significant toxicity	Murugan et al. (2014)
		LDH	No ↑ LDH activity	
AuNPs, 5 and 15 nm, 10–300 μM	Balb/3T3	Trypan Blue	Did not affect cell viability	Murugan et al. (2014)
		CFE	Showed toxicity for 5 nm AuNPs with concentration higher than 50 μM	
Titanium nanotubes	MC3T3-E1	Trypan Blue	Good proliferation	Oh et al. (2006)
		MTT	Did not affect cell viability	
Titanium nanotubes	SaOS2	MTS	High biocompatibility	Wang et al. (2014b)
TiNPs, 100, 150, and 200 mg/mL	A549	MTT	No cytotoxicity	Kansara et al. (2015)
TiNPs, 0–800 μg/mL	A549	MTS Flow cytometry (Annexin V/PI staining)	No cytotoxicity No necrosis No apoptosis	Hanot-Roy et al. (2016)
	HPMEC-ST1.6R		Cytotoxic effect starts at 50 μg/mL Dose-dependent increase of apoptotic cells	
	THP-1		No cytotoxicity No necrosis No apoptosis	
TiNPs, 15 nm, 1.5–250 mg/mL	D384 human astrocytes	MTT assay Calcein AM/PI staining	After 48 h ↓ cell viability Cell membrane damage starting at 125 mg/mL Alterations of the mitochondrial function on 31 mg/mL	Coccini et al. (2015)

(Continued)

Table 29.1 In Vivo and In Vitro Citotoxicity Results Obtained for Various Clinically Relevant Nanomaterials (*cont.*)

NPs, Size (nm), and Concentration	Cell Line/Cell Culture/Animal Model	Assay	Results	References
TiNPs, 15 nm, 1.5–250 mg/mL	SH-SY5Y human neuroblastoma		Cell membrane damage starting at 125 mg/mL Alterations of the mitochondrial function on 15 mg/mL	
CuNPs, 50 nm, 10, 25, and 50 μg/mL	A549	MTT	Viability ↓ with ↑ concentrations	Ahamed et al. (2010)
CuNPs, 9.2–12 nm	HBEC	Alamar Blue assay	No significant differences	Jing et al. (2015)
CuNPs, 10 nm, 0.6, 2.4, 3.3, 6.3, and 13.2 mg/m^3	Eight-week-old male outbred rats (HsdCpb:WU)	Nose-only short-term inhalation	Dose-dependent increase in lung inflammation Cellular deterioration and nose epithelium damage After recovery period the lungs were cleaned Brain, liver, spleen, kidney have been not affected by exposure	Gosens et al. (2016)
Iron oxide needle-like nanoparticles lengths of 50–200 nm and width of 10 nm, 0.5–2 mg/kg	Five-week-old male ICR mice	LDH	↑ Apoptotic cells NPs accumulate and remained in the lungs	Park et al. (2015)
Iron oxide nanorods, 15 × 150 nm, 200 and 400 mg/mL	L929 mouse fibroblast cells	MTT	↓ Viability	Ghasempour et al. (2015)
Iron oxide nanorods, 15 × 150 nm, 200 mg/mL	Fourteen Wistar rats weighed 240–260 g	Injected via the tail	No changes on morphology of glomerulus and tubular structure No inflammation in kidney and liver cells	Ghasempour et al. (2015)
IONPs chitosan capped, 8 ± 2.7 nm	A549 cells (human lung carcinoma)	MTT	Toxicity time and dose dependent	Shukla et al. (2015)
	HeK293 (human embryonic kidney)		Toxicity time and dose dependent	
	HeLa cells (human cervix carcinoma)		No significant toxicity	

Table 29.1 In Vivo and In Vitro Citotoxicity Results Obtained for Various Clinically Relevant Nanomaterials (*cont.*)

NPs, Size (nm), and Concentration	Cell Line/Cell Culture/Animal Model	Assay	Results	References
Mesoporous silica nanoparticles, 40 mg/kg	Male ICR mice	Oral administration	No histological abnormality in liver and spleen In the jejunum and ileum aggregates were found	Li et al. (2015)
SiO_2NPs, 50–1000 μg/mL	Human placentas cells	LDH MTS	No increase in LDH Did not induce significant toxicity	Paris et al. (2016)
AgNPs, 5.7–20.4 nm	Murine macrophages	MTS		
Pluronic-MWCNT, MWCNT-COOH MWCNT-OH, 12.5 mg/mL	Three-week-old male rats	H&E staining (Hematoxylin/Eosin)	Inflammatory response in lung and accumulation of CNTs	Yao et al. (2016)
	Rat bone marrow–derived MSCs	MTT	Viability ↓ By comparing the viability ↑ for Pluronic-MWCNT	
MWCNT (diameter: 15 and 100 nm; length: 10 and 30 μm), 2.5–80 g/cm^2	Raw 264.7 macrophages	Resazurin assay	Dose-dependent decrease of cell viability	Allegri et al. (2016)
Short-cut SWCNTs (average size: 230 nm), 80 mg/mouse	C57BL/6 mice	Pharyngeal aspiration	↑ Lung weight	Shvedova et al. (2013)
P-HiPco-SWCNTs (length: 1–3 μm; diameter: 1–4 nm), 5 mg/m^3	Female C57BL/6 mice	Inhalation	Micronucleated cells in lungs; pneumonia, fibrosis	Shvedova et al. (2014)
SWCNTs (diameter: 1.4 nm, 89% purity), 10 μg	Male Wistar rats	Injection into the tooth socket	No effects on cardiac function	Joviano-Santos et al. (2014)

4 CONCLUSIONS

In conclusion, due to their enhanced properties, such as their nanometric dimensions, increased surface area, excellent surface chemistry, and their possibility to be functionalized, the nanostructured materials are used in a wide variety of different domains including the biomedical one. In order to develop biomedical devices, all nanostructures must be toxicologically tested. Even though results of these tests are controversial and contradictory, they are mandatory for the evaluation of nanomaterials' interactions with the biological systems. This study contributed to a general picture of the toxicological responses that nanomaterials can generate. Besides the size and the concentration of NPs, there are several factors that influence their cytotoxicity, such as aggregation state, shape, chemistry, biodistribution, etc. However, the toxic effects could be diminished by capping or coating the NPs.

Without any doubt, the nanotechnology area is a tremendous resource for research studies especially in view of the fact that it represents a reserve that provides nanomaterials with promising applications in biomedicine that still need to be exploited.

REFERENCES

Adamcakova-Dodd, A., et al., 2014. Toxicity assessment of zinc oxide nanoparticles using sub-acute and sub-chronic murine inhalation models. Part. Fibre Toxicol. 11 (1), 1–15.

Ahamed, M., et al., 2010. Genotoxic potential of copper oxide nanoparticles in human lung epithelial cells. Biochem. Biophys. Res. Commun. 396 (2), 578–583.

Allegri, M., et al., 2016. Toxicity determinants of multi-walled carbon nanotubes: the relationship between functionalization and agglomeration. Toxicol. Rep. 3, 230–243.

Bahadar, H., et al., 2016. Toxicity of nanoparticles and an overview of current experimental models. Iran. Biomed. J. 20 (1), 1.

Beltrán-Osuna, Á.A., Perilla, J.E., 2015. Colloidal and spherical mesoporous silica particles: synthesis and new technologies for delivery applications. J. Sol Gel Sci. Technol. 77, 1–17.

Bussy, C., et al., 2015. Microglia determine brain region-specific neurotoxic responses to chemically functionalized carbon nanotubes. ACS Nano 9 (8), 7815–7830.

Buzea, C., Pacheco, I.I., Robbie, K., 2007. Nanomaterials and nanoparticles: sources and toxicity. Biointerphases 2 (4), MR17–MR71.

Caballero-Díaz, E., et al., 2013. The toxicity of silver nanoparticles depends on their uptake by cells and thus on their surface chemistry. Part. Part. Syst. Charact. 30 (12), 1079–1085.

Chen, Y., Chen, H., Shi, J., 2013. In vivo bio-safety evaluations and diagnostic/therapeutic applications of chemically designed mesoporous silica nanoparticles. Adv. Mater. 25 (23), 3144–3176.

Chernousova, S., Epple, M., 2013. Silver as antibacterial agent: ion, nanoparticle, and metal. Angew. Chem. Int. Ed. 52 (6), 1636–1653.

Chuang, H.-C., et al., 2014. Cardiopulmonary toxicity of pulmonary exposure to occupationally relevant zinc oxide nanoparticles. Nanotoxicology 8 (6), 593–604.

Coccini, T., et al., 2015. Comparative cellular toxicity of titanium dioxide nanoparticles on human astrocyte and neuronal cells after acute and prolonged exposure. Neurotoxicology 48, 77–89.

Dong, E., et al., 2011. Toxicity of nano gamma alumina to neural stem cells. J. Nanosci. Nanotechnol. 11 (9), 7848–7856.

Dworak, N., et al., 2014. Genotoxic and mutagenic activity of diamond nanoparticles in human peripheral lymphocytes in vitro. Carbon 68, 763–776.

Figarol, A., et al., 2015. In vitro toxicity of carbon nanotubes, nano-graphite and carbon black, similar impacts of acid functionalization. Toxicol. Vitro 30 (1), 476–485.

Fischer, H.C., Chan, W.C., 2007. Nanotoxicity: the growing need for in vivo study. Curr. Opin. Biotechnol. 18 (6), 565–571.

Ghasempour, S., et al., 2015. Investigating the cytotoxicity of iron oxide nanoparticles in in vivo and in vitro studies. Exp. Toxicol. Pathol. 67 (10), 509–515.

Goodman, C.M., et al., 2004. Toxicity of gold nanoparticles functionalized with cationic and anionic side chains. Bioconjug. Chem. 15 (4), 897–900.

Gosens, I., et al., 2016. Organ burden and pulmonary toxicity of nano-sized copper (II) oxide particles after short-term inhalation exposure. Nanotoxicology, 1–32.

Hadrup, N., Lam, H.R., 2014. Oral toxicity of silver ions, silver nanoparticles and colloidal silver—a review. Regul. Toxicol. Pharmacol. 68 (1), 1–7.

Hanot-Roy, M., et al., 2016. Oxidative stress pathways involved in cytotoxicity and genotoxicity of titanium dioxide (TiO_2) nanoparticles on cells constitutive of alveolo-capillary barrier in vitro. Toxicol. Vitro 33, 125–135.

Huang, N., et al., 2013. Alumina nanoparticles alter rhythmic activities of local interneurons in the antennal lobe of *Drosophila*. Nanotoxicology 7 (2), 212–220.

Huynh, V.T., et al., 2013. Nanodiamonds with surface grafted polymer chains as vehicles for cell imaging and cisplatin delivery: enhancement of cell toxicity by POEGMEMA coating. ACS Macro Lett. 2 (3), 246–250.

Ivask, A., et al., 2014. Size-dependent toxicity of silver nanoparticles to bacteria, yeast, algae, crustaceans and mammalian cells in vitro. PLoS One 9 (7), pe102108.

Jeng, H.A., Swanson, J., 2006. Toxicity of metal oxide nanoparticles in mammalian cells. J. Environ. Sci. Health Part A 41 (12), 2699–2711.

Jing, X., et al., 2015. Toxicity of copper oxide nanoparticles in lung epithelial cells exposed at the air–liquid interface compared with in vivo assessment. Toxicol. In Vitro 29 (3), 502–511.

Joviano-Santos, J., et al., 2014. Evaluation of cardiovascular toxicity of carbon nanotubes functionalized with sodium hyaluronate in oral regenerative medicine. Braz. J. Med. Biol. Res. 47 (7), 560–566.

Kansara, K., et al., 2015. TiO2 nanoparticles induce DNA double strand breaks and cell cycle arrest in human alveolar cells. Environ. Mol. Mutagen 56 (2), 204–217.

Kim, J.S., et al., 2013. Validation of an in vitro exposure system for toxicity assessment of air-delivered nanomaterials. Toxicol. In Vitro 27 (1), 164–173.

Kumar, R., et al., 2014. Carbon nanotubes: a potential concept for drug delivery applications. Recent Pat. Drug Deliv. Formul. 8 (1), 12–26.

Kumar, V., Gill, K.D., 2009. Aluminium neurotoxicity: neurobehavioural and oxidative aspects. Arch. Toxicol. 83 (11), 965–978.

Kunzmann, A., et al., 2011. Efficient internalization of silica-coated iron oxide nanoparticles of different sizes by primary human macrophages and dendritic cells. Toxicol. Appl. Pharmacol. 253 (2), 81–93.

Lee, S., Khang, D., Kim, S.-H., 2015. High dispersity of carbon nanotubes diminishes immunotoxicity in spleen. Int. J. Nanomed. 10, 2697.

Li, J.J., et al., 2008. Gold nanoparticles induce oxidative damage in lung fibroblasts in vitro. Adv. Mater. 20 (1), 138–142.

Li, L., et al., 2015. Biodistribution, excretion, and toxicity of mesoporous silica nanoparticles after oral administration depend on their shape. Nanomed. Nanotechnol. Biol. Med. 11 (8), 1915–1924.

Liu, H., et al., 2009. Biochemical toxicity of nano-anatase TiO_2 particles in mice. Biol. Trace Elem. Res. 129 (1–3), 170–180.

Liu, H., et al., 2013. Comparative study of respiratory tract immune toxicity induced by three sterilisation nanoparticles: silver, zinc oxide and titanium dioxide. J. Hazard. Mater. 248–249, 478–486.

Liu, J., et al., 2010. Controlled release of biologically active silver from nanosilver surfaces. ACS Nano 4 (11), 6903–6913.

Lopes, S., et al., 2014. Zinc oxide nanoparticles toxicity to *Daphnia magna*: size-dependent effects and dissolution. Environ. Toxicol. Chem. 33 (1), 190–198.

Lu, J., et al., 2010. Biocompatibility, biodistribution, and drug-delivery efficiency of mesoporous silica nanoparticles for cancer therapy in animals. Small 6 (16), 1794–1805.

Lu, S., et al., 2009. Efficacy of simple short-term in vitro assays for predicting the potential of metal oxide nanoparticles to cause pulmonary inflammation. Environ. Health Perspect., 117.

Manke, A., Wang, L., Rojanasakul, Y., 2013. Mechanisms of nanoparticle-induced oxidative stress and toxicity. BioMed Res. Int.

Mehra, N.K., Jain, K., Jain, N.K., 2015. Pharmaceutical and biomedical applications of surface engineered carbon nanotubes. Drug Discov. Today 20 (6), 750–759.
Mody, N., et al., 2014. Dendrimer, liposomes, carbon nanotubes and PLGA nanoparticles: one platform assessment of drug delivery potential. Aaps PharmSciTech 15 (2), 388–399.
Monopoli, M.P., et al., 2012. Biomolecular coronas provide the biological identity of nanosized materials. Nat. Nano 7 (12), 779–786.
Murugan, M., et al., 2014. Biofabrication of gold nanoparticles and its biocompatibility in human breast adenocarcinoma cells (MCF-7). J. Ind. Eng. Chem. 20 (4), 1713–1719.
Nonkumwong, J., et al., 2016. Synthesis and cytotoxicity study of magnesium ferrite-gold core–shell nanoparticles. Mater. Sci. Eng. 61, 123–132.
Oh, S., et al., 2006. Significantly accelerated osteoblast cell growth on aligned TiO_2 nanotubes. J. Biomed. Mater. Res. Part A 78A (1), 97–103.
Paget, V., et al., 2014. Carboxylated nanodiamonds are neither cytotoxic nor genotoxic on liver, kidney, intestine and lung human cell lines. Nanotoxicology 8 (Suppl. 1), 46–56.
Paik, S.-Y.-R., et al., 2015. Characterization, quantification, and determination of the toxicity of iron oxide nanoparticles to the bone marrow cells. Int. J. Mol. Sci. 16 (9), 22243–22257.
Paris, J.L., et al., 2016. Decidua-derived mesenchymal stem cells as carriers of mesoporous silica nanoparticles. In vitro and in vivo evaluation on mammary tumors. Acta Biomater. 33, 275–282.
Park, E.-J., et al., 2015. Chronic pulmonary accumulation of iron oxide nanoparticles induced Th1-type immune response stimulating the function of antigen-presenting cells. Environ. Res. 143, 138–147.
Pratsinis, A., et al., 2013. Toxicity of silver nanoparticles in macrophages. Small 9 (15), 2576–2584.
Recordati, C., et al., 2016. Tissue distribution and acute toxicity of silver after single intravenous administration in mice: nano-specific and size-dependent effects. Part. Fibre Toxicol. 13 (1), 1–17.
Reddy, U.A., Prabhakar, P., Mahboob, M., 2015. Biomarkers of oxidative stress for in vivo assessment of toxicological effects of iron oxide nanoparticles. Saudi J. Biol. Sci.
Rivera-Gil, P., et al., 2012. The challenge to relate the physicochemical properties of colloidal nanoparticles to their cytotoxicity. Acc. Chem. Res. 46 (3), 743–749.
Sabareeswaran, A., et al., 2016. Effect of surface-modified superparamagnetic iron oxide nanoparticles (SPIONS) on mast cell infiltration: an acute in vivo study. Nanomed. Nanotechnol. Biol. Med. 12, 1523–1533.
Sajid, M.I., et al., 2016. Carbon nanotubes from synthesis to in vivo biomedical applications. Int. J. Pharm. 501 (1–2), 278–299.
Salou, L., et al., 2015. Bone apposition on nanoporous titanium implants. Handbook of Nanoceramic and Nanocomposite Coatings and Materials. Butterworth-Heinemann, pp. 427–444, (Chapter 20).
Sang, X., et al., 2012. The chronic spleen injury of mice following long-term exposure to titanium dioxide nanoparticles. J. Biomed. Mater. Res. Part A 100 (4), 894–902.
Sathishkumar, M., et al., 2015. Biosynthesis of gold nanoparticles and related cytotoxicity evaluation using A549 cells. Ecotoxicol. Environ. Saf. 114, 232–240.
Schrand, A.M., 2016. Perspectives on carbon nanomaterials in medicine based upon physicochemical properties: nanotubes, nanodiamonds, and carbon nanobombs. In: Zhang, M., Naik, R.R., Dai, L. (Eds.), Carbon Nanomaterials for Biomedical Applications. Springer International Publishing, Cham, Switzerland, pp. 3–29.
Schrand, A.M., Hens, S.A.C., Shenderova, O.A., 2009. Nanodiamond particles: properties and perspectives for bioapplications. Crit. Rev. Solid State Mater. Sci. 34 (1–2), 18–74.
Shi, H., et al., 2013. Titanium dioxide nanoparticles: a review of current toxicological data. Part. Fibre Toxicol. 10 (1), 1–33.
Shukla, R., et al., 2005. Biocompatibility of gold nanoparticles and their endocytotic fate inside the cellular compartment: a microscopic overview. Langmuir 21 (23), 10644–10654.
Shukla, S., et al., 2015. In vitro toxicity assessment of chitosan oligosaccharide coated iron oxide nanoparticles. Toxicol. Rep. 2, 27–39.

Shvedova, A.A., et al., 2013. Carbon nanotubes enhance metastatic growth of lung carcinoma via up-regulation of myeloid-derived suppressor cells. Small 9 (9–10), 1691–1695.

Shvedova, A.A., et al., 2014. Long-term effects of carbon containing engineered nanomaterials and asbestos in the lung: one year postexposure comparisons. Am. J. Physiol. Lung Cell. Mol. Physiol. 306 (2), L170–L182.

Simpson, D.A., et al., 2014. In vivo imaging and tracking of individual nanodiamonds in *Drosophila melanogaster* embryos. Biomed. Opt. Express 5 (4), 1250–1261.

Sotiriou, G.A., Pratsinis, S.E., 2010. Antibacterial activity of nanosilver ions and particles. Environ. Sci. Technol. 44 (14), 5649–5654.

Sotiriou, G.A., Pratsinis, S.E., 2011. Engineering nanosilver as an antibacterial, biosensor and bioimaging material. Curr. Opin. Chem. Eng. 1 (1), 3–10.

Verstraeten, S.V., Aimo, L., Oteiza, P.I., 2008. Aluminium and lead: molecular mechanisms of brain toxicity. Arch. Toxicol. 82 (11), 789–802.

Wang, X.Z., et al., 2014a. Principal component and causal analysis of structural and acute in vitro toxicity data for nanoparticles. Nanotoxicology 8 (5), 465–476.

Wang, Y., et al., 2014b. Biocompatibility of TiO_2 nanotubes with different topographies. J. Biomed. Mater. Res. A 102 (3), 743–751.

Wang, Y., et al., 2014c. Review of the progress in preparing nano TiO_2: an important environmental engineering material. J. Environ. Sci. 26 (11), 2139–2177.

Wehling, J., et al., 2014. Bactericidal activity of partially oxidized nanodiamonds. ACS Nano 8 (6), 6475–6483.

Wijnhoven, S.W.P., et al., 2009. Nano-silver—a review of available data and knowledge gaps in human and environmental risk assessment. Nanotoxicology 3 (2), 109–138.

Yao, M.-Z., et al., 2016. Toxicity analysis of various pluronic F-68-coated carbon nanotubes on mesenchymal stem cells. Chem. Biol. Interact. 250, 47–58.

Yildirimer, L., et al., 2011. Toxicology and clinical potential of nanoparticles. Nano Today 6 (6), 585–607.

Zhang, S., et al., 2013. Controllable drug release and simultaneously carrier decomposition of SiO_2-drug composite nanoparticles. J. Am. Chem. Soc. 135 (15), 5709–5716.

Zhao, Y., et al., 2012. Nanotoxicology: Toxicological and Biological Activities of Nanomaterials. Nanoscience and Nanotechnologies. Encyclopedia of Life Support Systems (EOLSS) Publishers, Oxford, UK, pp. 1–53.

CHAPTER

30

ROLE OF EXCIPIENTS IN FORMULATION DEVELOPMENT AND BIOCOMPATIBILITY OF LIPID NANOPARTICLES (SLNs/NLCs)

Slavomira Doktorovova*, Ranjita Shegokar, Eliana B. Souto***

**University of Coimbra, Coimbra, Portugal*

***Freie Universität Berlin, Berlin, Germany*

CHAPTER OUTLINE

1 INTRODUCTION

There is an increasing interest for lipid nanostructures in the development of new drug-delivery systems. Their versatility relies on the improved physicochemical properties of lipids and surfactants to develop systems for several administration routes with optimized drug loading and release properties, longer shelf life and lower cytotoxicity. The extent and duration of drug action depends on how much of the drug reaches the target site and how long it stays there. The concentration time relationship at the site of action is a function of the amount and rate of dose delivered to the absorption site and the drug's pharmacokinetics. To achieve and keep the therapeutic levels, lipid nanostructures have been proposed as these can be tailored to deliver the drug at controlled rates for extend periods of time, which is pharmaceutically relevant for short acting drugs, with shorter plasma half-lives and narrower therapeutic ranges.

Solid lipid nanoparticles (SLNs) and nanostructured lipid carriers (NLCs) are in the solid state (either amorphous or crystalline), and can have a diameter between 50 nm and 1 μm, whereby the size depends mainly on the production method and on the composition of lipid matrix (i.e., lipid and surfactant).

Generally, the mean particle size ranges between 150 and 300 nm (Keck et al., 2014), which influences the colloidal stability, the drug loading, the pharmacokinetics and pharmacodynamics. SLNs and NLCs are produced by variety of methods or combinations of them are as follows (Muller et al., 2011):

1. microemulsion and double emulsion technique method
2. high pressure homogenization (HPH)
3. solvent evaporation or diffusion method
4. solvent injection method
5. microfluidization
6. sonication or ultrasonication
7. membrane contactor technique
8. supercritical fluid technology

Depending upon the method, the use and amount of excipient varies, for example, in case of microemulsion method, the amount of lipid excipient used in composition is less compared to the HPH method. Another, for example is the use of solvent in method 3 and 4, which is not used in other methods. Fig. 30.1 lists main components and steps of the methods used. For the production of stable dispersions, it is mandatory to have the adequate qualitative and quantitative composition of excipient for a particular production method. Although the chosen excipient may in certain cases be optimized, the production under excessive energy during particle disruption may result in an instable product. Researchers can take advantage of a design of experiment (DoE) approach and mathematical calculations to determine the right excipient concentration, drug–excipient ratio.

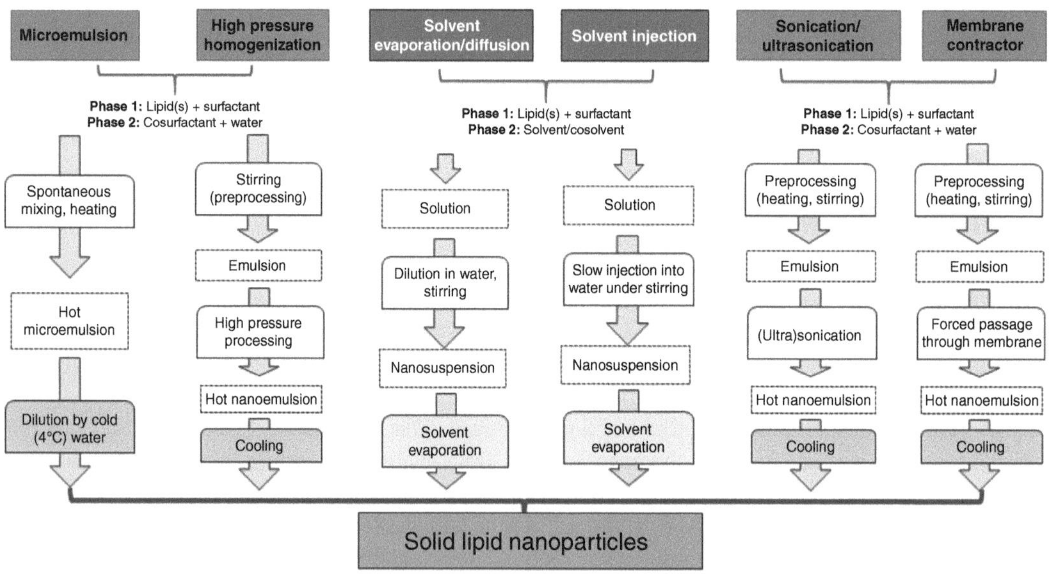

FIGURE 30.1 Schematic Representation of the Types of Commonly Used SLNs/NLCs Production Methods, Process Steps, and Excipient Type

Similar processes are applicable for NLC production; the only difference is Phase 1 contains mixture of liquid lipid and solid lipid along with surfactant.

For reproducible in vivo results and for stable dispersion, monodispersed populations are required, which can be obtained by selecting the appropriate excipient concentration especially lipid, surfactant, and their ratio followed by parameter settings of production method. Briefly, SLNs typically consist of three essential components, that is, solid lipid, surfactant, and water. These lipid particles are composed only of fats/lipids, which are solid at body and room temperature, while in NLCs a certain percentage of solid lipid is replaced by a liquid lipid (i.e., oil) but keeping the matrix also solid. SLNs and NLCs are able to absorb and/or encapsulate hydrophilic/lipophilic drugs, thus protecting them from chemical and enzymatic degradation. Furthermore, the encapsulated drug may be prevented from crystallization, thus, forming a solid solution. As a foreign material, the drug incorporation in the crystal lattice of the nanoparticle carrier material can happen to a limited extent, mainly due to its limited solubility in the stable crystal form. For that reason, the loading capacity of SLNs and NLCs is lower than that of an equally concentrated nanoemulsion (Bunjes et al., 1996; Westesen et al., 1997).

The main focus of this chapter is to discuss the role of major excipients, namely, lipids, oils, and surfactants in the formulation of stable lipid nanoparticles. The significance of other complimentary excipients is also addressed. Furthermore, an overview of commonly used and upcoming promising excipients is given. The biocompatibility and toxicity of lipid nanoparticles is reviewed at the end of the chapter.

2 ROLE OF EXCIPIENTS IN SLNs AND NLCs

The biological performance of lipid nanoparticle is greatly influenced by their overall composition and nature of excipients used. Lipid nanoparticles, namely, SLNs and NLCs, employ several types of excipients, for example, solid lipids, liquid lipids (oils), surfactants, and surface modifying agents. Compatibility of these excipients with each other and with the drug is of high importance. The main steps (from commonly used methods) for the selection of an excipient in the production of SLNs and NLCs are listed later:

1. solid lipid selection
2. liquid lipid selection
3. solid-liquid lipid compatibilities
4. solvent selection (as per production method used)
5. surfactants (anionic, cationic, and nonionic)
6. surface modifiers/ligand addition
7. auxiliary excipients—antioxidants, preservatives, tonicity adjuster
8. excipients for down scaling—production of dry mass—either by lyophilization (cryoprotectants) or spray drying (adsorbents)

Successful formulation of stable lipid nanoparticle depends on multiple factors. It starts from the selection of excipient, process optimization to selection of packaging. Selection of the most appropriate excipient further depends upon the several factors, starting with the nature of drug to be incorporated, its melting point, solubility determination of drug in oil/lipid, and drug recrystallization behavior are major factors. Based on the drug properties, excipients selection (mainly solid lipid) needs to be done. Route of drug administration, like, oral, dermal, and intravenous makes further differentiate excipient selection. For parenteral formulation, excipients used must be safe and have a FDA recognized GRAS (generally regarded as safe) status while dermal lipid nanoformulations can employ excipients from cosmetic market.

Lipid nanoparticles are more physiological and biocompatible compared to polymeric carriers. This makes lipid selection more critical as it comprises major part of excipients used. Many properties of lipid like polarity, mucoadhesiveness, melting point, degradation behavior by gut enzymes, recrystallization (precipitation) behavior, bulk properties, length of composite lipids, compatibility with other solid lipids (in case of combination used), degree lipophilicity, temperature sensitivity, contact angle with water/buffer are critical factors in production of stable dispersion. GRAS approved lipid from different categories is used in production of lipid nanoparticles. These include fatty acids, glycerides, and waxes. There is no generalized rule on use of specific lipid or oil with specific drug. As discussed previously, compatibility, polymorphism, and miscibility with drug are key parameters. Several cationic lipids are being employed for gene drug delivery due to enhanced transfection efficiencies. Branched cationic lipids are most common due to lower cytotoxicity potential than single branched (refer lipid section, 2.1.1). Chemical nature of lipid (modified chemical structure), miscibility of lipid with other used solid or liquid lipids offer further advantages. In general, polar oils due to the presence of hydroxyl groups further offer miscibility with other lipids and can improve miscibility thereby increasing thermodynamic stability. Miscibility can be an essential component for the formulation stability of SLN. Wendt et al. (2013) discussed the efforts by USP/NF in modernization of fixed oil excipient monograph. Understanding the complex nature of such excipient, USP/NF planned to make available reference material for such excipients. US pharmacopeia's excipient expert committee is evaluating oil excipient monographs which are registered as "oil" (vegetable oil, petrochemical oil, and essential oils). Currently, 31 monographs are under evaluation. Fixed oils (24 out of 31) included in USP/NF are mainly refined oils composed mainly of triglycerides and their derivatives. The main aim of modernizing these monographs is to control adulteration or contaminations by other low purity oils (Wendt et al., 2013). This type of adulteration can affect hydrolysis, oxidation, and degradation behavior of oil, as well as from lipid formulation. NLCs, which are based on oil part, are affected by this type of adulteration and can results in unstable product, causing expulsion of encapsulated drug thereby resulting in agglomeration or settling of formulation.

Surfactants play an important role in the prevention of agglomeration, and maintaining the dispersitivity. They can be selected based on compatibility and stability profile with particular lipid–oil matrix. In case of solvent emulsification methods, type and amount of solvent (type I, II, II) play an important role. Commonly used solvent is ethanol and other cosolvents like polyethylene glycol, glycerol, and propylene glycol are used in formulation of lipid nanoparticles. Additives, such as antioxidants are incorporated into SLN to reduce oxidation. The commonly used antioxidants are propyl gallate, α-tocopherol, and butylated hydroxy toluene. Based on the way of addition of antioxidants (during production or after production) can affect the stability of SLNs/NLCs. Besides that, if SLNs/NLCs dispersions are being incorporated into end dosage form like cream or gel a compatibility profile needs be checked. The role of excipients in final dosage form and its effect on SLNs/NLCs performance is out of the scope of this chapter. Toxicity potential based on the route of administration is equally determining factor in selection of lipid. In general, cationic surfactants have high toxicity than anionic and nonionic surfactants, while amphoteric surfactants have lowest toxicity.

2.1 CURRENTLY USED EXCIPIENTS IN SLNs AND NLCs

By definition, SLNs are composed from a solid lipid and stabilizing surfactant layer. NLC is composed of a lipid matrix formed by a lipid blend—most frequently mixture of oil and a solid lipid or eventually two-lipid matrix forming lipids. Solid lipid is meant to be a lipid that is in solid state at room, as well as at body temperature (i.e., has melting point > 37°C). There is evidence of melting point (T_m) decrease during SLNs/NLCs production, well described by Thompson–Gibbs equation—the decrease in individual particle diameter leads to depression of melting point. The higher decrease in particle size, greater T_m drop occurs. The practical consequence of this phenomenon is that many solid lipids traditionally used for SLN/NLC formulation do not form solid matrix at the end of SLNs/NLCs production.

One of the main challenge associated to SLNs and NLCs, and their physicochemical stability during shelf life, is the polymorphic transformation of the lipid matrix. It is the ability of the matrix to crystallize in more than one distinct crystalline species with different internal lattice structures. Lipid crystallization is an important point for the performance (both in vitro and in vivo) of the lipid nanoparticles. The internal structure of the lipid matrix changes in variety of ways, as lipids can crystallize in different three-dimensional structures. Crystallization of bulk triglycerides from the melt after rapid cooling usually occurs in the less stable α modification, which transforms via the β′, into the more stable β modification upon heating or during storage time. The α-form (hexagonal) is the least stable with a lower melting point and latent heat, whereas the β-form (triclinic) is the most stable with higher melting point and higher latent heat. The transformation of α to β′ (orthorhombic) and β is irreversible and occurs toward a more hydrodynamic stable system. In SLNs and NLCs, these transformations are faster than in the bulk material, which lead to a change in the relative fraction of the polymorphic forms. Depending on the chemical nature of the lipid excipient and on the production process parameters, different fractions of α and β′ modifications may occur. This phenomenon can lead to a reduction in the melting point, or more precisely, to changes in the form and shift of the melting peak. These created polymorphic forms are not long-term stable, leading to a gradual transformation to more stable modifications, which means increasing content of β′/βi and finally β. This is not desired because the change in lipid structure is responsible for drug expulsion during storage and changes in the release profile of incorporated drug, as well as changes in particle size parameters. There are many possibilities for loading drugs in lipid matrices. In general, when a crystal is formed, foreign molecules can be incorporated by replacing host molecules in the lattice by a guest molecule or by being incorporated in between host molecules.

Drugs may also be localized in between the lipid lamellae, resulting in increase of the lattice spacing "*d*" that can be analyzed by X-ray diffraction studies. The drug can also be present in form of amorphous clusters, mainly localized in the imperfections of the crystal. In this case, drug accommodation is improved when the lipid crystal has more imperfections. Thus, drug loading can be increased by using crude lipid mixtures or by controlled nanostructuring of the lipid matrix, by creating as many imperfections as possible.

Lipids employed for the production of nanoparticles include triglycerides and their mixtures, fatty acids, and waxes. The selection of the lipids for the production of the solid matrix is carried out by a solubility screening test based on the solubility of drug to give a visually clear solution in lipid melt under light against naked eye. The drug and varying quantities of selected lipid in glass vials are heated above the melting point of lipid in controlled temperature water bath. After melting the lipid in vials the solubility of drug is visually checked in the melt. The solubility of the drug in the lipid melt is determinant for higher loading capacity and higher encapsulation efficiency, that is, higher lipid solubility of the drug will result in higher encapsulation parameters.

Depending on the variety of factors, a fat may exist in one crystalline form or it may be a mixture of several different crystal polymorphs. Fats are polymorphic and transform systemically through a series of successive crystalline forms without change in chemical structure. The polymorphic transitions of lipid matrix can be influenced by the addition of one or more substances or other fats. The thermal characteristics, crystal habit, texture, and appearance of the lipid matrix are relevant when determining its suitability as drug-delivery system. Since the properties of suspended colloids or dispersed materials may differ significantly from those in the bulk, comprehensive structural investigations are mandatory. The incorporated drug participates in the microstructure of the system and may even influence it due to molecular interactions, especially if the drug possesses amphiphilic and/or mesogenic properties.

It is observed that SLNs composed of pure triglycerides (e.g., Dynasan (R) 114) could eventually maintain liquid or liquid crystalline state for considerably long periods of time, and only triglycerides with chains of C16 and longer could recrystallize and form solid lipid crystals (Bunjes et al., 2000; Bunjes and Westesen, 2001). In fact, most of the reported SLN/NLC formulations do not show 100% recrystallization index, which means that certain part of lipid excipients is always present in its liquid state. In case the formulation does not reach solid state in the end of production procedure, a supercooled melt is obtained. This will tend to crystallize within short period of time, which may disrupt the surfactant layer and lead to uncontrolled coalescence of lipid droplets and formation of large crystals. In some special cases, a structure with characteristics of liquid crystals can be obtained and if properly stabilized, may show excellent physicochemical stability. In some cases, intentionally or unintentionally, a nanoemulsion may be obtained. By intentional selection of lipids with low T_m, for example, for the purpose of formulation of thermosensitive drugs or peptides, a solid matrix is not obtained and major part of lipid excipient is present in its melted form. Unintentionally, the same may happen when selecting a lipid with low T_m and high T_m depression. Additionally, a lipid blend may create a eutectic blend, where an additional T_m drop occurs. These observations confirm, how critical is type and nature of lipid selection to produce stable SLNs/NLCs.

The other condition of selection of a lipid matrix-forming lipid is its safety status. The original SLN/NLC patented by Gasco and Müller relied only on excipients well known from other pharmaceutical oral solid forms, dermal semisolid dosage forms, or cosmetic products, and therefore have well known and favorable safety profile. Most of the excipient used for lipid nanoparticle formulation holds GRAS status, which gives SLN/NLC a great advantage over many others colloidal drug-delivery systems from the regulatory point of view. An issue to consider is the actual route of administration of the intended SLNs/NLCs, exactly because most of the excipients are accepted by dermal or oral route, but not other routes of administration, for example, intravenous route.

In this section, a comprehensive list of currently used excipients for SLN/NLC formulation and their role in SLN/NLC functionalities is discussed as follows:

1. Solid lipids—lipid matrix forming lipids, that is, lipids intended to form the (solid) lipid core.
2. Oils, that is, the lipids or lipophilic excipients that are intended to integrate the (solid) lipid core and decrease its crystallinity.
3. Surfactants—the material, often polymeric, intended to stabilize the lipid crystal suspension, maintain uniform particle size distribution and prevent particle growth.
4. Surface modifiers—the ingredients which add special function and impact in vivo effects.

2.1.1 Solid lipids

The most frequently used solid lipids for SLN/NLC formulation are from following category:

1. Triglycerides, that is, glycerol esters with three (monoacid) saturated, linear fatty acids.
2. Partial glycerides (mono- and diglycerides), that is, glycerol esters with 1–2 saturated, linear fatty acids, and 2 or 1 free hydroxyl groups, respectively.

3. Waxes, that is, esters of fatty acids and fatty alcohols.
4. Fatty acids, that is, saturated linear carboxylic acids with chain length of C12 and more.
5. Fatty alcohols, that is, saturated linear primary alcohols with chain length of C12 and more.

Often single solid lipids are employed in production of SLN/NLC. Based on the requirement of formulation or nature of API, combination or blend of lipids is used. Main limitations in using such blends are control over degradation behavior over the time, stability of chemical modification form, and concentration selection of each lipid in blend.

The role of the solid lipid is to form the lipid nanoparticle core and form a matrix that can accommodate the active ingredient (one or eventually more than one). All of these materials are intended to solidify (after processing at temperature > T_m) or precipitate (when processed by aid of an organic solvent that is removed in the final step of production process) and form crystalline platelets or spheres with reduced dimensions. As mentioned earlier, even though the lipids used for SLNs/NLCs production are solid at room temperature, the final SLN/NLC particles are rarely 100% recrystallized. The amount of recrystallized mass can be controlled to certain amount by selection of excipients with lower or higher T_m, as needed, and by using a blend of lipids, which usually leads to decreased crystallinity (Doktorovova and Souto, 2009).

Higher portion of solid lipid mass contribute to sustained release of the drug, provided that the drug is incorporated in particle core, and to drug protection from environmental factors that might deteriorate the drug. On the other hand, lipid nanoparticles with decreased crystallinity are believed to be more stable (i.e., maintain the average particle size and size distribution) and, more importantly, due to limited polymorphic phase transitions, are less prone to drug expulsion upon solid lipid crystalline structure changes.

It is generally accepted that SLN/NLC form spherical or rounded particles, as appears in various scanning electron microscope (SEM) images. As sample preparation for SEM requires drying the sample, freeze-dried SLN are often visualized. The freeze-dried SLN often require presence of a cryoprotectant, and often form redispersible aggregates. Evidence provided by transmission electron microscope (TEM) images shows that SLN prepared from pure glyceryl trimyristate form platelets composed of individual crystal layers, evidenced by a stair-like structure observed on platelet edge (Bunjes et al., 2007). In order to obtain a drug-delivery system that provides all the advantages that it is capable to provide, the selection of lipid matrix forming lipid is essential. The most important factors to consider for matrix forming lipids are:

1. Drug solubility in the lipid in question, to obtain high encapsulation efficiency and loading capacity.
2. T_m of the lipid, with respect to physicochemical characteristics of the active to be incorporated (temperature of decomposition, temperature at which degradation occurs).
3. T_m of the lipid, considering that this T_m will most probably decrease after processing, especially when using preparation methods based on processing lipid melt. This decrease may be advantageous (decreased crystallinity, decreased rates of drug expulsion) but can also compromise some of the functionality features of SLN/NLC, namely the capacity to control drug release and protect the drugs from light, humidity, or acidic/basic environment.

Typical pure lipids used for SLN/NLC formulations are indicated in Table 30.1, together with their most important physicochemical characteristics. Examples of SLNs/NLCs together with findings on physical state of solid lipid matrix and its influence on drug encapsulation capacity can be found in our previous review (Souto et al., 2011b). Briefly, SLN/NLCs composed of pure lipids are rarely the

Table 30.1 List of Commonly Used Lipids (in Pure Form) in the Formulation Development of Lipid Nanoparticles

Type	Lipids	Fatty Acid Chain Length	Melting Range (°C)	Availability
1	Glyceryl tribehenate	—	81–85	Dynasan (R) 122 (Io)
1	Glyceryl tristearate	C18	70–73	Dynasan (R) 118 (Io)
1	Glyceryl tripalmitate	C16	61–65	Dynasan (R) 116 (Io)
1	Glyceryl trimirystate	C14	55–58	Dynasan (R) 114 (Io)
1	Glyceryl trilaurate	C12	43–46	Dynasan (R) 112 (Io)
2	Glyceryl monostearate	C18	66–77	Imwitor (R) 491 (Io)
3	Cetyl palmitate, syn. Hexadecyl decanoate	C16(OOH)C16	43–57	Crodamol (R) CP (Cro), Kollicream (R) CP 15 (Ba)
4	Stearic acid	C18	56–72	Kolliwax (R) S (Ba)
5	Cetyl Alcohol	C16	46–52	Crodacol (R) C90 EP (Cro), Ginol (R) 16/95—(Io), Kolliwax (R) CA (Ba)
5	Stearyl Alcohol	C18	57–60	Crodacol (R) S95 EP (Cro) Ginol (R) 18/95—(Io), Kolliwax (R) SA (Ba)

The combination of lipids can be used to create imperfections in lipid lattice, to enhance drug loading. The types of lipid: (1) triglycerides, (2) partial glycerides (mono- and diglycerides), (3) waxes, (4) fatty acids, (5) fatty alcohols. Details on structure are explained in text. Ba, BASF GmbH, Germany; Io, Iol Oleo GmbH, former Cremer Oleo GmbH and Sasol GmbH; Cro, Croda Healthcare GmbH, Germany.

best lipid matrix to accommodate drugs, due to extensive polymorphic phase transitions that occur during storage and eventually lead to drug expulsion. Triglycerides tend to recrystallize in metastable α polymorph with loose fatty acid chains packing, which transitions into the most stable β polymorph, either directly ($\alpha \rightarrow \beta$) or through temporary formation of metastable β' polymorph ($\alpha \rightarrow \beta' \rightarrow \beta$). The most stable β polymorph is the one with most dense fatty acid chain packing. This structural reorganization leads often to premature drug expulsion from the solid lipid matrix. The rate of polymorphic phase transitions is mainly influenced by surfactants present in the system: both Span 65 (Aquilano and Sgualdino, 2001) and phosphatidylcholine (Bunjes and Koch, 2005) were shown to delay the $\alpha \rightarrow \beta$ transition for several months.

Fatty acids show different polymorphic behavior but also suffer from drug expulsion upon lipid crystal reorganization and formation of a more densely packed structure. Stearic acid, which is the most frequently used solid lipid for SLN/NLC production, shows four polymorphs (A triclinic, B monoclinic, C monoclinic, and E monoclinic) and, additionally, several protypes. C polymorph forms upon recrystallization from melt and is the most stable at temperatures >32°C. Below this temperature, the most stable polymorph is the most densely packed B polymorph. High drug loading capacity of stearic acid was reported in the earliest years of SLN development, the latest reports, however, usually indicate limited drug loading capacity, often resolved by addition of oleic acid into the lipid matrix. Oleic acid also shows three different polymorphs (α,β,γ), but the most stable β polymorph melts at 16°C.

The importance of lipid matrix physical state is related to drug encapsulation efficiency. Three models of drug incorporation were proposed in the early years of SLN development:

1. the drug-enriched core model, consisting of drug concentrated mostly in nanoparticle core and covered by a lipid shell,
2. the amorphous model, where the drug is molecularly dispersed within the lipid matrix, and
3. drug-enriched shell model, where the drug forms a layers condensed on a lipid core.

The first two models are desired to obtain a system with sustained release properties. Amorphous drug–lipid system formation is often reported, but the fact is that obtaining a system in which the drug is homogenously distributed within lipid matrix is not easy at all. The drug quantity used must be completely soluble in the lipid in question not only at production temperature but also at room temperature, and should not partition only into lipid phase during the production. Absence of drug peak in differential scanning calorimetry (DSC) thermograph or wide-angle X-ray diffractogram is often regarded as sufficient proof of homogenous drug distribution in SLN/NLC system, however, at the drug loading typically achieved in SLN/NLC (<10%) but frequently well below <2% related to solid lipid content), an eventual peak of solidified or even crystallized drug may be undetected. The drug-enriched shell model was proposed for special drug–lipid combinations were the lipid crystallized before the drug during cooling step of preparation. Recent evidence for a curcumin-loaded trimyristin SLN suggest that this model may actually occur much more often than it might seem. The (lipophilic) drug, incompletely dissolved in lipid matrix, or released prematurely during production or storage, accumulates at water–lipid interface, that is, at SLN surface (Noack et al., 2012). Interestingly, even lipid nanoparticle systems with drug "only" attached to the surface may provide controlled drug release. In short, lipid or lipid blends need to be optimized as per nature of drug, dose of drug, and production method used.

2.1.2 Commercially available mixtures of solid lipids and fats

The trend in SLN/NLC formulation is to prefer glycerides mixtures rather than pure triglycerides. The material is again inspired by established dermal and solid oral dosage forms and the fats used for their production. Glyceryl dibehenate (Ph. Eur) is, according to corresponding monograph, a mixture of tri-, di- and monoesters of glycerols of behenic acid (C22) and to minor extent stearic or palmitic acids. In Europe, Compritol 888 ATO (Gattefossé, S.A., France) is very well established in pharmaceutical industry, not only as release modifying agent but also as emulsifier (the mixture shows overall hydrophilic–lipophilic balance, HLB, of approx. 2) and viscosity enhancer (Aburahma and Badr-Eldin, 2014). It is one of the most frequently used lipids for lipid nanoparticle formulation. Another product that complies with the same pharmacopoeial monograph is Kolliwax GDB from BASF, Germany (former product name Speziol GDB PHARMA). Glyceryl distearate (Ph. Eur) is also a mixture of di- and triglycerides of stearic and palmitic acid (C16-C18, therefore also referred to as glyceryl palmito-stearate), required to contain 40%–60% diglycerides fraction and having overall melting point slightly lower than that of glyceryl dibehenate. Precirol ATO 5, is a product that most closely fulfils the pharmacopoeia definition for type I glyceryl distearate. Glyceryl monostearate 40–55 (Ph Eur) also requires a mixture that contains 40%–55% of monoglyceride and 30%–45% of diglycerides. Three types are defined according to the stearic acid content, being type I: the one with least stearic acid content (40%–60%, the other fatty acid summing 90% together with stearic acid is palmitic acid), type II requires 60%–80% glycerol substitution by stearic acid. The products that comply with definition of glyceryl monostearate type I are Geleol mono- and diglycerides NF from Gattefosse, France, and Imwitor 900P from IoI Oleo (former Cremer Oleo and Sasol, GmbH) Oleo (former Sasol, Germany). Kolliwax GMS II from BASF, Germany (known before as Cutina GMS V PH) and Imwitor 900K conform to pharmacopoeia type II definition. Apart from mixtures with 40%–55% content, IoI Oleo GmbH, Germany also provides Imwitor 491 containing 90% glyceryl monostearate.

Apart from well-defined lipid mixtures, chemical companies also provide a portfolio of fats with different characteristics, currently used in dermal and cosmetic formulations, that were adapted with success in SLNs/NLCs or simply lipid nanoparticle formulations. Gelucire series of lipids from Gattefossé) are PEG containing mixtures of acylglycerides: lauryl macrogol-32 glycerides (Gelucire 44/14) or stearoyl macrogol-32 glycerides (Gelucire 50/13). Gelucires are defined by their melting point/HLB value. As discussed previously, even though most Gelucires have $T_m > 37°C$, (the most frequently used for SLN/NLC formulation up to now is Gelucire 50/13), their T_m after processing to colloidal dimensions is expected to be decreased, leading to only partially crystallized lipid matrix. This can be advantageous for certain drugs or peptides and in preparation of paediatric/geriatric rectal and vaginal suppositories. There are already quite a few lipid nanoparticle formulations reported that are based on Gelucire 50/13 (Hazzah et al., 2016; Jeon et al., 2013), Gelucire 43/01 (Gonçalves et al., 2015; Nnamani et al., 2014) or Gelucire 44/14 with partially solid matrix. All of these lipid nanoparticle formulations resulted in low crystalline lipid matrix and were tested mostly for skin delivery of the encapsulated actives, some researchers successfully formulated then as a mucoadhesive lipid nanoparticle (Gonçalves et al., 2015; Hazzah et al., 2016).

Witepsol series from IoI Oleo GmbH present lipids with melting point ranging from 32 to 37°C (but there is also Witepsol S series offer solid lipids with melting range up to 70°C). Some of them were successfully adapted for lipid nanoparticle formulation with low crystallinity. Witepsol H15 (melting range 33–35°C) was used for formulation of a sensitive polyphenol compound and Witepsol E 85 (T_m of 42–44°C) for chemotherapeutic actives (Ferreira et al., 2015; Martins et al., 2012). Beeswax and carnauba waxes are also used in lipid nanoparticle formulation. A summary of lipid mixtures currently used in formulation of SLN/NLC, is presented in Table 30.2.

2.1.3 Oils

In order to solve the issues related to lipid nanoparticles composed only from solid lipids and solid lipid mixtures, NLCs were developed as a second generation lipid carrier with improved functionality. NLCs were proposed and proven to provide better physicochemical stability (i.e., maintenance of acceptable size distribution during longer period of time) and improved drug encapsulation. Theoretical models predicted oil droplet would integrate solid lipid matrix and thus decrease its crystallinity and provide better drug encapsulation efficiency. In reality, as proven by two independent research groups, the oil tends to accumulate on solid lipid surface, which at the same time contributes positively to NLC stabilization (Jores et al., 2004; Schwarz et al., 2012).

By far the most frequently employed oils are mixtures of medium chain (C6–C10) triglycerides and their mixtures. As pure substances, only oleic acid and squalene are being typically used. The use of oleic acid was adapted from conventional dermal formulations as it is a readily available excipient supplied by many companies. It is valued for its skin permeation enhancer properties. Under certain circumstances and preparation procedures, oleic acid can form vesicular structures (Aquilano and Sgualdino, 2001). Squalene and isopropyl myristate are two other liquid lipids frequently used in NLC formulation. Squalene is triterpene compound produced by skin cells (mainly as precursor for cholesterol). It is used as moisturizer in cosmetic and dermal products, but is also approved for other routes of administration. Owing to advantageous properties in dermal products, squalene is often selected

Table 30.2 List of Frequently Used and Commercially Available Lipid Excipients Employed in Formulation Development of Lipid Nanoparticle

Composition (Ph. Eur. Monograph)	Brand Name	Supplier	Melting Range (°C)	Regulatory Status
Glyceryl dibehenate	Compritol (R) 888 ATO	Ga	69–74	D, O
	Kolliwax (R) GDB	Ba	65–77	
Glyceryl distearate type I	Precirol (R) ATO 5	Ga	53–57	O
Glyceryl monostearate 40-55 type I	Geleol Mono and Diglycerides NF	Ga	55–58	D, O
	Imwitor (R) 900P	Io	54–64	—
Glyceryl monostearate 40-55 type II	Imwitor (R) 900K	Io	54–64	—
	Kolliwax (R) GMS II	Ba	54–64	—
Lauroyl macrogol-32 glycerides	Gelucire (R) 44/14	Ga	42–47	O
Stearoyl macrogol-32 glycerides	Gelucire (R) 50/13	Ga	46–51	O
Cetearyl Alcohol, syn. Cetostearyl alcohol (C16/C18)	Crodacol (R) CS90 EP	Cro		D
	Kolliwax (R) CSA 50	Ba	49–56	—
	Kolliwax (R) CSA 70	Ba	49–56	—
(Hard fat, adeps solidus)	Witepsol (R) Series, for example, Witepsol E85	Io	42–44 (E85)	
Hydrogenated Palm oil	Softisan (R) 154	Io	53–58	

Based on the route of drug administration selection can be done. Ba, BASF GmbH, Germany; Io, IoI Oleo GmbH, former Cremer Oleo GmbH, former Sasol GmbH, Germany; Cro, Croda Healthcare GmbH, Germany; Ga, Gattefossé, SA, France; Ph. Eur., European Pharmacopoeia.

as liquid lipid in NLC formulations for dermal delivery. Isopropyl myristate is available from BASF (Kollicream IPM, former isopropyl myristate PH) and is recommended for use in dermal formulations.

Medium chain triglycerides (triglycerides, medium chain as per Ph. Eur.) are defined as triglycerides of capric (C10) and caprylic (C8) acid, in total at least 95% of the fatty acids present in the mixture. The most frequently used commercial lipid of this type is Miglyol 812 (IoI Oleo GmbH), another brand products are Kollisolv MCT 70 from BASF (former Myritol 318 PH) or Labrafac Lipophile WL 1349 from Gattefossé, France. Gattefossé provides a great variety of medium chain glycerides (MCT) mixtures and ethoxylated medium chain glycerides mixtures, the most frequently used in SLN/NLC formulations being Labrasol, syn. caprylocaproyl macrogol-8 glycerides, which is in fact an oil–water (o/w) stabilizer (HLB 14), and other products from Labrafac line and Labrafil product lines. Several other oils like coconut oil, jojoba oil, and other super refined oils are being used in formation of SLN/NLC and are under investigation.

2.1.4 Surfactants

The use of surfactants is crucial to maintain the lipid nanoparticle stability. The success of SLN in drug delivery is partly due to improved stabilization of lipid nanoparticles composed of solid matrix for a period of time longer than few days. The concept of SLN in the very beginning of their development was the replacement of liquid lipids forming nano- or microemulsion by lipids that are solid at room temperature. Therefore, stabilization of SLN and NLC adapts many concepts for stabilization of o/w emulsions, including the selection of the surfactant. Being in fact a dispersion of solid matter in outer liquid phase, principles of suspension stabilizations are also valid for lipid nanoparticles. The behavior of lipid nanoparticles has partial resemblance to the behavior of both emulsions and suspensions. Lipid nanoparticles are of typical size, which regard them as a colloidal system.

Surfactants play an important role in drug formulation, determining various properties of the colloidal system, such as its viscosity, and capacity to solubilize water-insoluble material. Surfactants have amphiphilic structure that are used to reduce the surface tension and facilitate the particle partition, that is, their hydrophilic groups oriented toward the aqueous phase and the hydrophobic groups oriented to the lipid (Rosen, 2004). In aqueous media, the surfactant molecules are oriented with their polar heads predominantly toward the aqueous phase while keeping their hydrophobic groups away from it. In vesicles, an aqueous phase will also occur in the interior of the structure. The selection of the surfactant mainly depends on the chosen lipid, since they need to be physicochemically compatible (Severino et al., 2012). The HLB is directly related to the solubility, that is, it is the balance of the size and strength of the lipophilic and hydrophilic groups of the surfactant (Severino et al., 2011). The required HLB (rHLB) of a final dispersion is predominantly dependent on the HLB of the lipid and on the HLB of the surfactant (and cosurfactant, if applied), and is determined applying the following equation (Ferreira et al., 2012):

$$\mathrm{rHLB} = \left\lfloor \%_{\mathrm{Lipid}}\, x\mathrm{HLB}_{\mathrm{Lipid}} \right\rfloor = \left\lfloor \%_{\mathrm{Surfactant}}\, x\mathrm{HLB}_{\mathrm{Surfactant}} \right\rfloor + \left\lfloor \%_{\mathrm{Cosurfactant}}\, x\mathrm{HLB}_{\mathrm{Cosurfactant}} \right\rfloor$$

The rHLB is of high importance, as the selection of the optimized surfactant or combination of them contributes to the enhanced stability of the SLN and NLC dispersions. The blend of surfactants should adjust to the tails of the lipid chains in the interface to allow the coexistence of the oil droplets in a continuous aqueous phase. The interactions between lipids and surfactants are mainly electrostatic. These are physically adsorbed onto the surface or loaded within the lipid matrix. The hydrophilic groups of the surfactant increase the repulsive forces, depending on the volume size and chemical nature of the hydrophilic moieties of surfactants (Rosen, 2004). The surfactants typically used in the production of SLN and NLC are selected depending on their HLB) and also on the nature of the hydrophilic group. In general, nonionic emulsifiers do not show ionic charge (e.g., monoglycerides of long-chain fatty acids), such as Tweens, Spans, Mirjs tyloxapol, poloxamers, sugar esters and esters of stearic, palmitic, oleic, and lauric acids. Anionic or cationic surfactants may be important to improve the surface electrical charge to avoid particle aggregation and/or sedimentation. The cationic surfactants [e.g., stearylamine (Kuo and Chen, 2009; Pedersen et al., 2006; Vighi et al., 2010)], quaternary ammonium salts (Doktorovova et al., 2011; Doktorovova et al., 2012b; Tabatt et al., 2004), *N,N*-di-(β-stearoylethyl)-*N,N*-dimethyl-ammonium chloride (Vighi et al., 2007, 2010) have a molecule moiety with positive charge. Anionic surfactants (e.g., sodium cholate and sodium taurocholate) have a negative moiety in the molecule, which improve the absorption of particles in the gastrointestinal tract. Phospholipids derived from soy or egg phosphatidyl choline have a variable fatty chain composition, and may also be used in the production of SLN and NLC. Soybean phosphatidyl choline contains more saturated fatty acyl chains than egg phosphatidyl choline (Souto et al., 2011a), and it has been reported to improve the emulsions stability, decrease particle size due to the amphiphilic properties (Schubert and Müller-Goymann, 2005; Schubert et al., 2006), and increase the skin permeation of topically administered formulations (Cui et al., 2006; Dreher et al., 1997). When cosurfactants are used, stabilized emulsions are usually obtained when both surfactants are of the same hydrocarbon chain length (Schmidts et al., 2009, 2010). Different alkyl polyglucosides (APGs) were used as surfactants in the production of cetyl palmitate SLN (Keck et al., 2014).

Stabilization with APGs led to smaller SLN and contact angle analysis of the SLN dispersions thereby demonstrating good wettability, high surface activity, a high tendency to remain at the o/w interface, and the improved wetting of cetyl palmitate. The diffusion velocity and the ability to cover newly created surfaces in a shorter time, depends on the molecular weight of the surfactant. Lower molecular weight surfactants move faster therefore are expected to be more efficient than those of higher molecular. The authors also found that the crystallinity of SLN increased with an increase in the alkyl chain length of the surfactants.

Changes in temperature, concentration of surfactant, additives in the liquid phase, and structural groups in the surfactant may all cause changes in the mean size, shape, and aggregation number of nanoparticles. Aqueous SLN and NLC dispersions of low viscosity have potentially the risk of transformation into a viscous gel, resulting in loss of their colloidal size and increase of particle size, and risk of aggregation (Mehnert and Mäder, 2001). Usually, these changes are unpredictable and occur very fast. To avoid gel formation, the physicochemical stability must be improved by the addition of cosurfactants with high mobility (Westesen and Siekmann, 1997), but also storage at temperatures below room temperature, under dark conditions and under nitrogen atmosphere may prevent particle growth (Freitas and Müller, 1998).

The surfactants for lipid nanoparticle stabilization are mostly o/w emulsifiers/stabilizers, that is, hydrophilic surfactants, mostly polymeric, with HLB > 8. Only one surfactant may not be sufficient to provide sufficient stabilization, therefore combinations of surfactants are frequently used. The combinations often include an o/w together with a w/o surfactant. The selection of surfactant is purely empirical, although some systematic approaches were presented: (1) similarly to emulsion stabilization using the HLB principle. Severino and coworkers showed that a solid matrix with known required HLB value can predict its stabilization in an emulsion system, and may also require similar HLB when formulated as SLN. This method was used for SLN composed of stearic acid, material for which the required HLB value is well known and easily available from current literature (Severino et al., 2011). Keck et al. (2014) used the (2) contact angle approach: aqueous solution of intended surfactant was required to form droplets with low contact angle when deposited on solid lipid surface. Great majority of reported SLN/NLC formulation relies on three surfactants: Polysorbate 80, Poloxamer 188, and phosphatidylcholine (syn. lecithins) of different origin (soy, egg). Less frequently, other types of Poloxamer than the 188 are used, usually Poloxamer 407.

Polyoxyethylene sorbitan monooleate, (syn. polysorbate 80) is an o/w surfactant with HLB of approximately 15, accepted by regulatory agencies around the world for pharmaceutical products, including intravenous products. It is a partial ester of sorbitan, containing polyethylene glycol 20 chains and an oleic acid chain. Other types of polysorbates (sorbitans substituted with different fatty acids) are far less frequently used for SLN/NLC stabilization. It is marketed as stabilizer and holds GRAS status. Commercially, it is made available by various companies, for example, Kolliphor PS 80 from BASF, or Tween 80 from Croda. Tween trademark is property of Uniquema, but is used frequently as synonym for polysorbate 80 irrespective of its origin. Polysorbate 80 critical micellar concentration is as low as 0.02%, despite this number is not the most correct for polysorbate in presence of other (lipid) material, it is expected that apart from accumulating on lipid surface, it might form micelles on its own or mixed micelles with other surfactants present in the system. These colloidal systems might also contribute to solubilization of the encapsulated drug. Solubilization of the active ingredient by polysorbate present in lipid nanoparticle formulation is not a defect, but it may have consequences on drug release, showing biphasic behavior. There are several reports in the literature of drug being actually more soluble in polysorbate 80 than in the solid lipids considered for SLN/NLC formulation. Again, contribution of polysorbate might not be detrimental but it is worth verifying if the use of a lipid matrix actually brings advantage over a solubilization only by polysorbate or surfactant mixture. Special attention must be given to the concentration range used of polysorbate 80 in the formation of SLN/NLC.

Poloxamer 188 is a nonionic triblock copolymer consisting of polyoxyethylene (hydrophilic block) and polyoxypropylene (lipophilic) units, showing HLB of > 24 (BASF—Nutrition & Health). In many cases, Poloxamer 188 is sufficient on its own to stabilize SLN, especially those composed of Compritol 888 ATO. It should be notes that at higher concentrations (>10%) Poloxamers form hydrogels; the effect on the overall lipid nanoparticle formulation should therefore be carefully evaluated. Furthermore, Poloxamer 188 shows melting point of around 52°C, which can be confusing when evaluating SLN/NLC thermal behavior. If present in sufficient quantity, a small melting peak of Poloxamer may be observed in differential scanning thermographs, exactly in the temperature range where some triglycerol polymorphs melt. In aqueous environment, Poloxamer 188 also forms micelles. In addition, Poloxamer 188 shows thermoreversible gelling properties. Up to now scarcely used, the good stabilization and solubilization properties together with gelling properties provide an opportunity for formulation of thermosensitive systems by using a Poloxamer type that shows gel transition at body temperature. Apart from Poloxamer 188, Poloxamer of different molecular weights is less frequently used, for example, Poloxamer 407. BASF offers wide range of grades with acceptable pharmaceutical quality (Kolliphor P series, former brand name Lutrol F (BASF, 2013)). Another supplier of Poloxamers is Croda (Synperonic series).

Phosphatidylcholine, syn. lecithin, is also readily available in most research laboratories that deal with formulation and therefore is also frequently used for SLN stabilization. Phosphatodylcholine forms bilayer structures under wide range of hydration and temperature. It is generally accepted that phosphatidylcholine accumulates at solid lipid surface and acts as a stabilizing agent, but many authors also regard it as lipid matrix forming lipid. In case of lipid emulsion containing both triglycerides and phospholipids (e.g., Lipofundin MCT), it is known that the various colloidal structures coexist in the dispersion, including nanoemulsion droplets and multilamellar vesicles (liposomes). The only proof of phospholipids being actually associated with SLN surface is from Heiati et al., 1996. Considering later known facts about solid lipids recrystallization behavior in nanoparticles, these reports, however, has some drawbacks: SLN used for the experiments were based on trilaurine, which is known to form supercooled melts rather than solid matrix unless they are frozen after production and return to room temperature. As there is no cooling step mentioned in the report of Heiati, it is possible that the experiments were performed with supercooled trilaurine matrices, that is, with liquid droplets of trilaurine, which obviously have different behavior than suspensions of solid (crystalline) triglycerides. As adsorption on solid surfaces is not the same as adsorption on liquid–liquid interface, this theory should not be automatically considered true for suspensions.

Tocopheryl polyethylene glycol 1000 succinate (TPGS) is less frequently used in SLN/NLC formulation. Commercially it is available as Kolliphor TPGS from BASF (former Speziol TPGS Pharma) is solubilizing and stabilizing agent, and at the same time it is hydrophilic form of tocopherol, that is, vitamin E, providing thus some antioxidant properties, too. TPGS is sometimes regarded as solid lipid due to its Tm of 38°C, but the compound shows HLB of 13 making it suitable for stabilization of o/w of solid lipid-in-water systems.

Other less frequently used surfactants in SLN/NLC formulation include:

- Polyoxyethylene (20) stearyl ether, syn. steareth-20 (Brij S20 from Croda, former names Brij78 or Volpo S20), is a nonionic, o/w emulsifier (HLB 15) and liquid crystal stabilizer from group of ethoxylated fatty alcohols. Its use in SLN/NLC is limited mostly to nanoparticles produced by microemulsion template method, frequently in combination with TPGS and/or emulsifying wax (Dong et al., 2009; Howard et al., 2011; Ma et al., 2009). It was also used successfully for encapsulation improvement of doxorubicin (Siddiqui et al., 2012).
- Macrogol 15 hydroxy stearate, marketed as Kolliphor HS 15 (formerly known as Solutol HS 15) is a water soluble, nonionic solubilizer with HLB 14–16, making it suitable for o/w emulsion formulation and therefore obviously also interesting for stabilization of lipid/water colloidal suspensions. Its use is by far more prominent in lipid nanocapsules where it is one of crucial excipients (Huynh et al., 2009). Another important characteristic of this surfactant is the ability to inhibit P-glycoprotein (P-gp) function. Its use in SLN/NLC is limited, but has long tradition form the very first years of SLN/NLC development up to most recent examples (Baek and Cho, 2015; Soni et al., 2014).
- Polyglyceryl-3 dioleate, Plurol Oleique CC 497 from Gattefosse, a water insoluble w/o surfactant with HLB of approx. 3. It is marketed as a surfactant and bioavailability enhancer. In SLN/NLC, it was used both as a stabilizing agent (Shegokar and Singh, 2012a) and as an oil in NLC (Zhang et al., 2014).

Apart from the most frequently used surfactants, some research groups pointed their investigation in characterization of suitability of "new" surfactants like Solutol HS 15, Plurol oleic, and Plantacare series, that is, those that are not typically used in SLN/NLC formulations. These are listed in Table 30.3.

2.2 SURFACE MODIFICATION AGENT AND LIGANDS

Methods used for surface modification of SLN are in general deeply inspired by the methods used for liposome functionlization. This is mainly true for postinsertion method, adapted by many groups for SLN use, and inclusion of special functionalized lipids in the formulation. Another approach is linking the coating agent to an excipient believed to be an integral part of SLN/NLC by a covalent bond using *N*-hydroxy succinimide/1-ethyl-3-(3-dimethylaminopropyl) carbodiimide (NHS/EDC) chemistry, an approach well validated in polymeric particles (Yu et al., 2012). A summary of surface modifications and methodology used in provided in Table 30.4.

Polyethylene glycol (PEG) coating is used to obtain lipid nanoparticles with hydrophilic surface, which would enable longer circulation time by avoiding clearance by reticuloendothelial systems (RES) (Howard et al., 2008). Longer presence in systemic circulation improves bioavailability and is especially interesting for delivery of chemotherapeutic drugs by systems that rely on enhanced permeation and retention (EPR) effect (Maeda, 2001). In a similar manner than in liposomes, SLN/NLC are modified by inclusion of PEG conjugated either to distearoyl phosphatidyl ethanolamine (DPSE-PEG) or stearic acid residue (Polyoxyethylene-stearate, SA-PEG) in the formulation. DPSE is regarded by many authors as surfactant; indeed it was shown that inclusion of DSPE-PEG could improve mean particle size and stability. SA-PEG is used for its structural similarity to simple stearic acid, a common

Table 30.3 Less Frequently Used Surfactants and Their Types in Formulation Development of Lipid Nanoparticles

Surfactant Type	Brand Names	Composition/Important Characteristics	HLB	References
Alkyl polyglucosides.	Plantacare (R) 810 UP (Co)	C8-10 fatty alcohol glucoside	15–16	Keck et al. (2014)
	Glucopon (R) 425 N/HH (Co)	Alkyl polyglucosides of natural C8-14 fatty alcohols	12–13	
	Glucopon (R) 600 CS UP (Co)	Alkyl polyglucosides of natural C10-16 fatty alcohols	11–12	
	Plantacare (R) 1200 UP (Co)	C12-16 fatty alcohol glucoside	16–17	
Polyhydroxy surfactants	Plurol (R) Stearique WL1009 (Ga)	Polyglycerol 6-distearate	9–10	Kovacevic et al. (2011)
	Plantacare (R) 810 PL (Co)	Caprylyl/capryl glucoside	15–16	
Alkyl esters of PABA	NA	decyl, dodecyl, and tetradecyl esthers of p-amino benzoic acid	—	Chandra et al. (2011)
Sucrose esters	Surfhope SE Pharma D1216 (Mi)	Sucrose laurate HLB 16	16	Surajit et al. (2014a,b)
	Surfhope SE Cosme C1416 (Mi),	Surcrose mirystate HLB 16,		
	C1616 (Mi)	Sucrose palmitate HLB 16		
	C1816 (Mi)	Sucrose stearate HLB 16		
Quillaja saponins	Quillaja saponaria wood extract, composed mainly of predominantly of glycosides of quillaic acid. (An).	Quillaja saponaria wood extract	13.5	Salminen et al. (2014, 2016)
Chlorogenic acid[a]	N.A. (Genex Life Science Pvt Ltd, New Delhi, India)	Chlorogenic acid	—	Khan et al. (2016)
Cationic arginine ester	CAE (Ai)	PCA Ethyl Cocoyl Arginate syn. DL-Pyrrolidonecarboxylic acid salt of L-cocyl arginine ethyl ester	—	Patil-Gadhe and Pokharkar (2014)

Manufacturers: (Ai) Ajinomoto, Tokyo, Japan (An) Andean QDP Ultra Organic, represented by Desert King International, San Diego, CA, USA, (Co) Cognis, Germany, currently part of BASF, Ludwigshafen, Germany, (Ga) Gattefosse, Saint-Priest, France, (Mi) Mitsubishi-Kagaku Foods Corporation, Tokyo, Japan

[a]Used in combination with Poloxamer 188, PABA = p-amino benzoic acid.

Table 30.4 List of Excipients Used as Surface Modifiers in Lipid Nanoparticle Formulation to Offer Unique Functionality or Targeting Potential

Surface Material	Attachment Method	Outcomes	References
PEG 2000	SA-PEG2000 as surfactant	Higher conc. Of SA-PEG2000, more DOX transported into brain Significantly decreased DOX concentration in heart and liver	Zara et al. (2002)
PEG 2000	SA-PEG2000 as surfactant	Enhanced permeability through mucus, enhanced permeation through cell monolayer (Caco-2/HT29) Increased relative bioavailability versus simple SLN	Yuan, Fan et al. (2014)
PEG2000	SA-PEG2000 as surfactant	Enhanced biochanin A bioavailability after oral administration in Sprague-Dawley rats versus drug suspension	Wang et al. (2015)
PEG2000	DSPE-PEG as surfactant	Accelerated release of salbutamol sulphate from PEG-SLN compared to simple SLN	Hong et al. (2006)
PEG 1000, polysaccharide and serum albumin	Solutol as surfactant	Enhanced bioavailability of SLN when surface modified with polysaccharide > serum albumin > PEG. Superior safety profile over uncoated NP	Shegokar and Singh (2012)
PEG2000	DSPE-PEG as surfactant	Sustained release Increased $t_{1/2}$ Good protective effect in rat ischemia-reperfusion injury model.	Gao et al. (2008)
PEG2000	DSPE-N-PEG2000 as surfactant		Lee et al. (2007)
PEG 2000	PEG2000-peptides	Preferential biodistribution to tumor tissue, decreased distribution to lung, liver, and kidney (MMP cleavable PEG-peptide functionalized SLN)	Zheng et al. (2014)
Folate	Folate-PEG-Chol as surfactant	Improved survival of Balb/C mice bearing murine lung carcinoma (M109) tumor (Folate-targeted PTXL SLN)	Jain et al. (2008)
Folate	Folate-SA as part of formulation	Increased brain targeting versus uncoated formulation (DTX and ketoconazole). Higher brain permeation than marketed formulation of DTX (Taxotere)	Venishetty et al. (2013a)
Folate	FA-PEG-DSPE as part of formulation	Improved cisplatin anticancer effect in vitro (HeLa) and in vivo (BALB/c mice)	Zhang et al. (2016)
Biotin	SA + (SLN, amino group) + EDC. HCl (crosslinker) + NHS + Biotin (carboxyl group)	Enhanced bioavailability after oral administration to Sprague-Dawley rats	Zhou et al. (2015)

(Continued)

Table 30.4 List of Excipients Used as Surface Modifiers in Lipid Nanoparticle Formulation to Offer Unique Functionality or Targeting Potential (*cont.*)

Surface Material	Attachment Method	Outcomes	References
Ferritin	DSPE+ EDC. HCl + Ferritin (carboxyl group)	Enhanced cellular uptake, increased anticancer efficacy of encapsulated chemotherapeutic drug (5-fluoro-uracyl)	Jain et al. (2008)
Transferrin	DSPE (SLN, amino group) + EDC. HCl (crosslinker) + Transferrin (carboxyl group)	Enhanced bran uptake of quinine hydrochloride from Tf-SLN (i.v., albino rats) compared to solution or simple SLN	Gupta et al. (2007)
Transferrin	Transferrin-PEG-PE adsorption (on cationic SLN)	Enhanced in vitro and in vivo transfection in hepatocellular carcinoma (HepG2) cell line and Kupfer Cells in Balb/c mice	Jing et al. (2012)
Transferrin	Transferrin-PEG-PE adsorption (on cationic SLN)	Enhanced in vivo transfection and DOX delivery in model lung (A549) tumor in C57BL/6 mice	Han et al. (2014)
Ovalbumin	Adsorption from 5% mannitol solution to cationic SLN	Enhanced IL-12 release from murine BMDDC ex vivo enhanced immune response to ovalbumin in vivo (mice)	Patel et al. (2007)
β-D-Galactosides	β-D-galactosides with long fatty chain ($nC_{18}H_{37}$) and 2 or 10 Polyoxyethylene units as aglycon moiety (G2, G10) as parts of formulation	Increased accumulation in mice liver; increased $t_{1/2}$ in blood plasma	Lian et al. (2008)
Mannose	Covalent bond; Schiff base formation between mannose aldehyde group and secondary amine on SLN surface (containing SA^+)	Increased uptake of FITC-labeled mannosylated formulation in A549 and MCF-7 cells, compared to simple SLN and DOX solution Significantly increased bioavailability, $t_{1/2}$ and MRT in Balb/c mice versus DOX solution, decreased c_{max}	Jain et al. (2010)
Mannose	Covalent bond; Schiff base formation between mannose aldehyde group and secondary amine on SLN surface (containing SA^+ and DSPE)	Increased uptake of rhodamine-labeled mannosylated formulation in A549 Decreased hemolytic activity versus simple SLN and PTX solution	Sahu et al. (2015)
Mannose	Man-C6-Chol as part of LNP formulation	Targeting to alveolar macrophages Decreased Rifampicin encapsulation efficiency Similar pharmacokinetic parametrs versus simple LNP	Song et al. (2015)
Mannose	Monomannosyl-dioleoyl glycerol as part of SLN formulation		Mishra et al. (2010)

Mannan	Mannan-PEG-PE adsorption (on cationic SLN)	Enhanced in vitro and in vivo transfection in hepatocellular carcinoma (HepG2) cell line and Kupfer Cells in Balb/c mice	Jing et al. (2012)
Galactose	*N*-hexadecyl lactobionamide as part of SLN formulation	Improved liver targeting in Wistar rats (i.v.)	Wang et al. (2010)
Galactose	Covalent bond; Schiff base formation between galactose aldehyde group and secondary amine on SLN surface (containing SA^+)	More sustained DOX release in vitro Decreased hemolytic activity of galactosylated DOX-loaded formulation versus nongalactosylated Increased cytotoxicity toward A549 cells	Jain et al. (2015)
Concacavalin A	SA^+(SLN, amino group) + EDC (crosslinker) + Concacavalin A (carboxyl group)	Increased mitoxanthrone accumulation in lung tumor (A549 in Balb/C mice)	Mahor et al. (2010)
Phenylalanine	DSPE(SLN, amino group) + EDC (crosslinker) + Phenylala-nine + NHS (carboxyl group)	Enhanced bioavailability in rat brain	Vyas et al. (2015)
L-Arginine	Adsorption	Increased uptake by human brain-microvascular endothelial cells (HBMECs)	Kuo and Lin (2009)
Histidine tagged polyarginine (cell penetrating peptides)	Chelation to nickel complex (SLN component DOGS-NTA-Ni)	Enhanced drug (celecoxib, ketoprofen) permeation into skin	Desai et al. (2012), Shah et al. (2012)
Pentapeptide (Thr-Lys-Pro-Pro-Arg)	Pentapeptide-PEG_{2000}-SA as part of SLN formulation	Increased uptake by Wistar rat macrophages in vitro and in vivo	Zhao et al. (2013)
Pentapeptide	Pentapeptide-PEG_{2000}-SA as part of SLN formulation	–	Fan et al. (2014)
Oligo-chitosan (3–6 kDa)	Adsorption	Enhanced mucoadhesive properties compared to simple NLC	Luo et al. (2010)
Oligo-chitosan (8–42 kDa	Adsorption Secondary crosslinking by glutaraldehyde	Improved DOX loading capacity More sustained drug release (further improved by crosslinking)	Ying et al. (2011)
Chitosan (50 kDa)	Adsorption	Improved transcellular transport through cell monolayer (Caco-2) Enhanced absorption (of encapsulated insulin) from GIT versus simple SLN	Fonte et al. (2011)

(*Continued*)

Table 30.4 List of excipients used as surface modifiers in lipid nanoparticle formulation to offer unique functionality or targeting potential (*cont.*)

Surface Material	Attachment Method	Outcomes	References
Chitosan (50–190 kDa)	Inclusion in aqueous phase during SLN production	Improved SLN stability Enhanced mucoadhesive properties versus simple SLN Enhanced cellular uptake Evidence of thick chitosan layer formation around lipid core (FT-IR, TEM)	Luo et al. (2015)
Chitosan	Adsorption Secondary crosslinking by glutaraldehyd	—	Dharmala et al. (2008)
Alginate[a]	Inclusion in (outer) aqueous phase during SLN production	Improved stabilization (ZP) of SLN Enhanced mucoadhesive properties	Fangueiro et al. (2012)
Hyaluronic acid	Adsorption (on cationic SLN)	Increased cellular uptake of targeted SLN versus simple SLN (almost inexistent uptake). SK-OV-3 cells	Ghalaei et al. (2014)
Hyaluronic acid	Alendronate–hyaluronate graft polymer adsorption on (negatively charged) NLC surface	Enhanced irinotecan-NLC cytotoxicity toward multidrug resistant cell line (Colo-320) P-gp inhibition, tumor targeting in vivo (Balb/c mice) Decreased hemolytic activity versus uncoated NLC Decreased uptake by macrophages versus uncoated NLC	Negi et al. (2014a,b)
Hyaluronic acid	Hyaluronic acid-protamine-siRNA complex adsorption on cationic SLN surface (containing DOTAP)	Successful gene silencing in vitro (HepG2, ARPE-19) CD44-mediated endocytosis and cavelae-dependent cell internalization (with possible lysosome evasion)	Torrecilla et al. (2015), Apaolaza et al. (2014)
Wheat germ agglutinin	Covalent bond; PVA (surfactant) crosslinked with glutaraldehyd at acidic pH and linked to WGA amino group	Enhanced binding and uptake by cell monolayer (Caco-2) Enhanced adhesion to rat intestinal mucosa ex vivo Improved bioavailaility after oral administration (Sprague-Dawley rats)	Liu et al. (2010), Liu et al. (2011)
poly(Allylamine hydrochloride)$^+$ poly(diallyldimethy-ammonium chloride)$^+$ poly(acrylic acid)	Layer-by-layer deposition of polyelectrolytes—polyelectrolyte multilayers Adsorption (to SLN)	SLN (slightly negative ZP) bind best to multilayers with polycationic outmost layer PEGylated surfaces repelled SLN	Finke et al. (2013)
Beta-hydroxybutyric acid	Beta-hydroxybutyric acid-SA+ conjugate as part of formulation	Enhanced brain uptake of DTX versus unmodified DTX-SLN and Taxotere	Venishetty et al. (2013b)

BMDDC, Bone marrow derived dendritic cells; Conc., concentration, DOGS-NTA-Ni = 1,2-dioleoyl-sn-glycero-3-[(N-(5-amino-1-carboxypentyl) imidodiacetic acid] succinyl nickel salt, DOTAP, 1,2-dioleoyl-3-trimethylammonium-propane chloride
[a]*Used in combination with Poloxamer 188, PABA = p-amino benzoic acid.*

excipient for SLNs/NLCs. SA-PEG is an amphiphilic molecule; it might also contribute to stabilization of SLN/NLC itself. Due to the possibility of formation of separate colloidal object by DSPE-PEG itself (see earlier), SA-PEG might be more suitable excipient for PEGylated SLN/NLC formulation.

Despite encouraging data on colloidal carriers modified by PEG, recent reports show that PEGylation is not a straightforward solution for improvement of bioavailability. It was observed that repeated doses of PEG-coated liposomes (Ishida et al., 2005) or SLN (Zhao et al., 2012a) actually lead to increased clearance of the colloidal system. This is described as accelerated blood clearance (ABC) phenomenon. The first dose of PEGylated colloidal carrier induces IgM type antibody production, which reacts with PEGylated surfaces after second and subsequent doses, and marks such carrier for removal by RES (Zhao et al., 2012a,b). Another theory clarifies that crucial factor is PEG distribution on the surface of a colloidal carrier: while low density of PEG chains (typically $< 9\%$ PEG per mm^2) on the nanoparticle surface promote "mushroom-like" configuration of PEG chains, higher concentrations induce "brush-like" configuration. The mushroom-like configuration is recognized by phagocytic cells, leads to complement activation and to rapid clearance of colloidal system in question (Amin et al., 2015). Brush-like configuration is believed to repel opsonins and thus assure longer circulation time. This issue was not addressed in SLN/NLC. While there are many reports of PEGylated SLN/NLC showing improved $t_{1/2}$ and bioavailability, indeed most of the currently available reports are based on single dose administration. Therefore, PEGylation might not be a universal solution for RES avoidance. Furthermore, Shegokar and coworkers showed that PEG adsorption on a nanosuspension (stabilized by surfactants frequently used in SLN/NLC formulation) influences the quantitative distribution of adsorbed plasma proteins and apolipoproteins, but does not change their qualitative pattern. As a consequence, the protein adsorption pattern reveals good possibilities for macrophage recognition, confirmed by biodistribution study in rats which affirmed increased distribution to liver and spleen (Shegokar et al., 2011a).

As an alternative to PEG coating, different hydrophilic or amphiphilic polymers were tested. Polyvinyl pyrolidone (Ishihara et al., 2010) and polyglycerols (Abu Lila et al., 2013) we suggested as promising alternatives. Polyglycerol derivates can also serve as surfactant for SLN/NLC stabilization, as shown by Kovacevic et al. (2011). Interestingly, hydrophilic "coating" can also be achieved by using poloxamers, very frequently used in SLNs/NLCs. Many encouraging in vivo reports might owe their success exactly to poloxamer outer layer of the colloidal carrier. We have shown that Poloxamer 188 is effective in shielding the SLN (even with positive surface charge) from blood plasma protein adsorption. Only limited amount of bound plasma proteins were found by 2D-PAGE, none of them being known opsonin. Such protein corona promotes longer circulation times (Doktorovova et al., 2012a).

Surface modification by alginate coating aims at improved mucoadhesive properties, especially for formulations intended for oral administration. Alginate is a polysaccharide consisting of β-D-mannuronic acid and α-l-gluronic acid and typically bears negative surface charge (at neutral pH). Adsorption on SLN/NLC surface was assured by inclusion of alginate in outer aqueous phase of SLN prepared by double emulsion method (w/o/w). The authors observed that alginate could actually improve stability of SLN/NLC, probably by contribution to both steric and electrostatic repulsion between individual particles. (Fangueiro et al., 2012). Shegokar et al., studied in vivo effect of SLN after surface modification with PEGs, polysaccharide and serum albumin. The surface modification significantly enhanced the cellular uptake and bioavailability of stavudine encapsulated SLN. The type of surface modification showed different degree of bioavailability upon intravenous administration (Shegokar and Singh, 2012b).

In order to enhance mucoadhesive properties desirable for oral formulation or interaction with sialic acid on eye surface for ocular formulations, chitosan is frequently used in SLN/NLC formulations. Chitosan, syn. Poly-D-glucosamine or deacetylated chitin, has gained much attention in Pharm. Res. for its biocompatibility, biodegradability, mucoadhesive properties and capacity to form stable complexes with nucleic acids. Chitosan is a weak base (p*K*a ≈ 6) (Lavertu et al., 2006), therefore it is soluble at acidic pH and at that pH it is positively charged and binds readily to SLN surface, which is negative (unless modified by inclusion of cationic components). Furthermore, chitosan shows high affinity to lipid surfaces. Several chitosan types are commercially available, ranging from chitosan to various chitosan salts and including chitosan of different molecular weights. Oligochitosans and chitosan salts are soluble in water (at neutral pH). Simple incubation of SLN with chitosan solution was used by different research groups; the reported results indicate that adsorption of chitosan was successful. Luo et al. (2015) proved recently by Fourier-transform infrared spectroscopy (FT-IR) and transmission electron microscopy (TEM) that chitosan forms a thick layers around (solid) lipid core. Therefore, chitosan can also contribute to stabilization of SLNs/NLCs, by both steric and electrostatic mechanisms. Various authors also noted chitosan could improve drug-loading capacity.

In order to further reinforce chitosan surface layers, Dharmala et al. (2008) and Ying et al. (2011) used glutaraldehyde to promote crosslinking of chitosan hydroxyl groups. This approach (referred to as ionic gelation) is frequently used in formulation of chitosan nanoparticles, although with different crosslinking agents (e.g., tripolyphosphate) and usually leads to nanoparticles with better physicochemical properties and drug loading (Katas and Alpar, 2006).

Hyaluronic acid (HA), syn. D-glucuronic acid and D-*N*-acetyl glucosamine, typically with very high molecular weight, is an anionic polysaccharide frequently found in extracellular matrix of various tissues. Its biological activity includes participation in regulation of wound healing/tissue regeneration (including skin lesions), namely by promoting cell proliferation and migration. Rationale for its use may be improvement of skin disorders including dry skin conditions or skin repair (Sandri et al., 2013). Apart from its skin effects, HA might also regulate cancer progression and metastization as the major ligand to CD44 receptor. Indeed, many types of invasive carcinomas express CD44 receptor, controlling signaling pathways involved in tumor progression (Horta et al., 2015). Attachment of HA to SLN/NLC surface was mostly achieved by ionic bond between negatively charged HA and cationic SLN, especially formulated for the purpose of HA adsorption. Some of these formulations served for the purpose of gene delivery (namely siRNA delivery, Apaolaza et al., 2014; Torrecilla et al., 2015) and therefore contain multiple cationic components to assure sufficient positive charge density. In another work, Cetyl pyridinium Chloride (CPC) a water-soluble antimicrobial quaternary ammonium compound (QAC), having one long-chain alkyl group and pyridine moiety is used in lipid nanoparticle production. The feasibility of positively charged NLC production is performed using CPC as a stabilizer cum active for topical application (Shegokar et al., 2009).

Also for the purpose of targeted delivery to malignant cells, ligands of lectin receptors were explored. These are various carbohydrates, most frequently used as galactose or mannose residues coating on SLN/NLC surface, that should promote receptor mediated cellular internalization of the whole nanoparticle. Therefore, covalent link to a SLN component, most frequently a surfactant with available amino group (SA+ or DSPE) was used. Mannose was subject to ring opening reaction in acidic environment and subsequently reacted with the amine present on SLN/NLC surface, forming a Schiff base. Another approach used for mannose coating was prior synthesis of mannose derivates with high lipophilicity [e.g., 3(mannos-1-yl)cholesterol] that can be included in SLN/NLC formulation, where the hydrophilic mannose moiety should be oriented to SLN/NLC surface and the lipophilic part should prefer adsorption or mixing with the rest of excipient forming the lipid core.

For the purpose of enhanced skin penetration (Patlolla et al., 2010; Shah et al., 2012) or overcoming the blood–brain barrier (Venishetty et al., 2013a), cell-penetrating peptides are considered a promising strategy. Histidine-tagged polyarginine of different chain length (i.e., arginine residues) were adsorbed on SLN surface using the chelate complex formation, an approach used for isolation of histidine tagged peptides (Lauer and Nolan, 2002). In the case of SLNs, this approach resulted in increased transdermal delivery of two nonsteroidal antiinflammatory drugs. Another approach for attaching peptides to SLN surface is linking the peptide covalently to PEG chain of SA-PEG which then can be included in SLN formulation.

Another surface modification approach aims at a specific targeting to a selected cell population. Apart from functionlization with monoclonal antibodies and their fragments, transferrin and folate are frequently used for surface modifications. The rationale is in increased metabolic needs of malignant tissues, reflected in folate and transferrin (Tf) receptor over expression. Folate (folic acid, FA) attachment is typically facilitated by inclusion of FA-PEG or Tf-PEG functionalized lipids in SLN/NLC formulation. These functionalized lipids (SA-PEG-FA, DSPE-PEG-FA or Cholesterol-PEG-FA, Tf-PEG-SA) are commercially available.

2.2.1 Complementary excipients

Freeze-drying is often required for long-term storage of SLN/NLC, especially those with positive surface charge and special formulations that for any reason cannot guarantee suitable storage time in dispersion. Despite reports that it is possible to freeze-dry cationic SLN without use of any cryoprotectant (Vighi et al., 2007), most frequently the cryoprotectant is needed to assure good dispersion quality after rehydration of the product. Most frequently used cryoprotectants are carbohydrates; these include:

- monosaccharide: glucose or mannose
- disaccharide: lactose, sucrose, fructose, or the most efficient trehalose
- sugar alcohol: mannitol, sorbitol

Additionally, polymers like polyvinyl pyrollidone were successfully tested as cryoprotectants (Singh and Shegokar, 2011).

Selection of the most suitable cryoprotectant is most often based on empirical optimization, as there is no rule yet that would identify suitable excipient combinations. Apart from acceptable appearance of the product, freeze-dried SLN/NLC should also show ability to redisperse easily into dispersion with similar size distribution characteristics to those before freeze-drying. Furthermore, the product should assure that the redispersion capacity is maintained over time of storage, which can be an additional challenge for product development (Doktorovova et al., 2014a). Similar selection of carbohydrates, but mainly mannitol, lactose, and sucrose were tested for SLN spray drying. There is also a report on spray-dried SLN powder with excellent properties for pulmonary deposition prepared without any additional excipient. The same report also indicates mannitol as an especially suitable excipient (Dolatabadi et al., 2015). Further reports on SLN spray drying include the use of ethanol (Blasi et al., 2013) or methanol (Freitas and Mullera, 1998). The type and concentration of cryoprotectant can further affect plasma stability of redispersed system.

The selection of excipient is considered crucial in assuring overall biocompatibility and safety of SLN/NLC. The theory is that it is the surfactant that has major influence of SLN/NLC biocompatibility. Generally, most of the marketed surfactants are safe for topical or oral use and if used in moderate quantities, do not cause harm on cellular or tissue levels. As we have shown before, there is no traceable effect of the three most frequently used types of surfactants on cell viability (Doktorovova et al., 2014b). Caution is in place when using cationic surfactants or surface modifiers for their known induction of cell membrane damage, oxidative stress and interference with intracellular lipid trafficking. The potential toxicity, metal contamination, miscibility, solvent capacity, and disposability must be taken into consideration. Several reports confirm role of excipients in determination of bioavailability and toxicity potential, it should be very well noted that the route of administration plays critical role and affects the toxicity potential, for example, SLN injected intravenously will show different toxicity compared to the SLN applied dermally with same composition. Besides that scaling up nanoparticles and role of excipients is separate topic itself for discussions. The successful scale up of lipid nanoparticles using various capacity homogenizers using same excipient composition to lab scale production is reported. However, the production parameters at large scale significantly affects the final particle size distribution (Shegokar et al., 2011b).

2.3 NEW COMERS IN EXCIPIENT MARKET

In March 2015, Evonik has launched RSPO-certified preservative-free palm kernel oil–based surfactant Tego Betain P 50 C. Chemically, it is cocamidopropyl betaine and contains 38% active betaine. A patented manufacturing process assures high quality (with low levels of amidoamine and chloro acetic acids) and microbial purity of surfactant. Tego Betain P 50 C, an amphoteric surfactants can be used in preparation of lipid nanoparticles and can offer viscosity to skin and hair care formulations. For pharmaceutical application, the concentration optimization will be needed to exert optimal effect. Innospec Speciality Chemicals launched Pureact SLMI-85, a sulphate free isethionate ester based surfactant to expand their portfolio of taurates and isethionates based surfactants. It can be used as primary or secondary surfactant and has comparable flash foam properties to that sodium laureth sulphate and has excellent stability over wide range of pH. Another leading Surfactant manufacturer Stepan has launched in collaboration with Elevance surfactant Steposol Met-10U derived from natural oils mainly targeted to displace solvents. This product has main applications household, adhesive cleaning, and in paint removal.

Recently, Naturex launched Sapnov, a natural foaming agent and surfactant extracted from quillaia. Sapnov is a nonionic and water soluble surfactant. Other examples of surfactant introduction include Colonial chemical's 100% natural coconut based surfactant, Colafax PME, a monoalkyl phosphate, which can be used as alternative to sodium lauryl sulfate and ammonium laureth sulfate. It can be employed as primary or secondary surfactant and Cognis's, Plantacare, Plantaren, Plantapon LGC product line which can be employed in development of SLN-based products for skin care.

In 2014, Symrise launched a PEG-free solubilizer called SymSol PF-3 to be in competition of PEG-free products and this surfactant can solubilize lipophilic substances. It can be used as an emulsifier or coemulsifier in o/w formulations. It showed superior solubilizing performance in comparison to PEG-containing products (i.e., PEG-40 Hydrogenated Castor Oil) and other PEG-free solubilizer, The Company has applied for patent protection for SymSol PF-3. In 2007, Advanced Materials Inc. has launched innovative emulsifier named Pemulen, a polymeric emulsifiers based on high molecular weight polyacrylic acid polymers as o/w stabilizer. Various grades of Pemulen are available based on their oil emulsifying capability, for example, Pemulen TR-1, Pemulen TR-2. Another class of surfactants called Gemini surfactant is in boom. It combines monomers of two surfactant linked chemically. These surfactants offer improved surface active, wetting properties at typically exhibit much lower critical micelle concentration strong dependence on spacer structure, special aggregate morphology, and strong hydrophobic microdomain. These types of surfactants are being widely explored mainly in skincare, gene delivery compared to other drug delivery routes (Sekhon, 2013).

Pierre Potier prize winner sustainable green chemistry based Sepiclear G7 (heptyl glucoside) is launched by Seppic. It is a nonionic surfactant derived from sugar and castor seeds. It helps in solubilization of essential oils, perfumes and vitamin E, in aqueous media. Another example of PEG-free solubilizer includes TEGO Solve 61 a naturally derived solubilizer for lipophilic and natural oils by Evonik. It is polyglyceryl esters (polyglyceryl-6 caprylate; polyglyceryl-4 caprate; polyglyceryl-4 cocoate; polyglyceryl-6 ricinoleate) and is effective for the solubilization of very lipophilic ingredients like fatty or natural oils.

In 2014, Longvida launched new product from Douglas laboratories for nutritional supplement application based on solid lipid particle (SLPTM) technology. Formulation composed of curcumin with neurophenol. The composition showed 65 times higher bioavailability and exhibited longer half-life compared to regular curcumin.

Now, Capsugel's Lipidex Technology Platform allows integration of lipid, liquid, and semi-solid fill technologies. This platform allows customer to find tailored solutions for the many issues like low solubility drugs, poor therapeutic performance, and life cycle management. NanoLipid Restore CLR, a semi-finished product developed by Chemisches Laboratorium Dr. Kurt Richter, Germany and distributed by Pharmacos India. It is a NLC dispersion containing black current seed oil and is used in the cosmetic product line IOPE from Amore Pacific, South Korea. Other NLC incorporated products are Nanorepair Q10, Nanovital Q10, Cutanova (Dr. Rimpler, Germany) and Surmer (Isabelle Lancray, France). CEA-Leti, France developed two DDS platform based on lipids, namely Lipidots (lipid nanodroplet for diagnostic/drug delivery, 20-180 nm) and LipiVAC (vaccine DDS for antigens) has been developed. Most recent (September 2015) acquisition between the Lubrizol Corporation, a Berkshire Hathaway technology driven company, with Particle Sciences, a leading contract manufacturing and drug delivery solutions organization headquartered in Bethlehem, PA. This acquisition will now together offer variety of drug delivery solutions to the market across a variety of dosage forms.

2.4 LIQUID LIPIDS

40 years market player Fuchs, Germany offers highly refined white oils based on mineral hydrocarbons, consisting of saturated aliphatic hydrocarbons. The white oils are produced to meet the European Pharmacopoeia standards for light and liquid paraffin, and also meet the requirements for the USP/NF mineral and light mineral oil USP. In 2014, Exxon Mobil Corporation has launched two white oils called Marcol and Primol as an ideal solution for cosmetic and pharmaceutical applications. These oils offer excellent hydration and emolliency. Another market player Medisca offers variety of pharmaceutical, nutraceutical essential and fixed oils, for example, coconut oil, cod liver oil, eucalyptus oil (natural), lavender oil (natural).

Stability of drug is concern for pharmaceutical industry, some polar (monoglycerides, diglycerides), free fatty acids, plant sterols, coloring matter (chlorophyll, carotene), and oxidative impurities present in oil excipients can cause degradation of drug and ultimately endanger the stability of formulation.

Keeping this in mind Croda chemicals has launched Super Refined Oils, processed by flash chromatographic process. Super Refined Oils has reduced impurities and are suitable in pharmaceutical and health care formulations. Croda currently market, Super Refined Castor, cotton seed, olive, peanut, sunflower, sesame, soyabean oil, etc. Super Refining by chromatography helps to purity excipients and control the quality by controlling the moisture; residual catalyst (peroxide and aldehyde) without altering original chemical structure. Super Refined PEGs and are polysorbate, propylene glycol are also made available by Croda. Another supplier, Reitzer Pharmaceuticals offer Tea Tree Oil extracted from *Melaleuca alternifolia* has been used in many cosmetic and pharmaceutical applications. This oil has an antifungal, antimicrobial, and antiinflammatory healing oil and is suitable to use in eczema, psoriasis, and acne.

Thirty years of heritage makes Avanti polar lipids a leading supplier of lipids to pharma industry. Various grades of lipids are available starting from class antigenic lipids, biotinylated lipids, deuterium labeled lipids, diglycerols, diglyceride pyrophosphates (DGPP), fluorescent phospholipids, fluorescent sphingolipids, functionalized PEG-Lipids, glycerol based lipids, glycolipids, lanthanide chelating lipids, maleimido containing lipids, natural lipids, PEG-lipids, phosphatidylinositols, phosphatidylserine, phytosphingosine to several synthetic lipids.

Corden pharma has a wide phospholipid portfolio to offer for drug delivery, pharmaceutical, and biotechnological applications. Several lipids are available from monoglycerol, diglycerol, phosphatidyl acids, phosphoglycerols, phosphoserine, phosphoethanolamines, phosphocholines, methoxy PEGylated phospholipids, phosphoglycerols with heterogeneous fatty acid chains, cholesterol, 2-lyso phospholipids, ether lipids, cationic lipids, and fatty acid ate the main categories. As stated earlier, the choice of excipient is highly depend on the route of administration. Currently, most of the excipients are approved for topical application and some for oral use. Researchers are kindly advised to check their GRAS status before their use in intravenous applications. Overall leading excipient (mainly surfactant and lipid) companies are expanding their portfolio with innovative excipients to offer multiple and optimum choice of excipient for their drug delivery and final dosage form.

3 CONCLUSIONS

Lipid nanoparticles especially SLNs and NLCs are well trusted and long researched (almost 25 years) as a drug delivery systems in pharma for various biomedical applications. Literature has extensively discussed and recognized the role of excipient in formulation, stability, in vitro and in vivo effects of the lipid nanoparticles. Biocompatibility of lipid nanoparticles is highly dependent on nature and type of excipient used. The extensive use of wide ranged excipient in cosmetic and dermal application is reported, followed by average use in oral and pulmonary drug delivery. However, excipients administered via intravenous route are really limited and need to follow more strict guidelines on metal impurities, toxicity considerations beside their GRAS status. From the very first step of "drug–lipid" screening till the conversion of lipid nanoparticle dispersion to dry powder and scaling up, excipients play critical role and their performance varies on case to case basis based on nature of other excipients and production method. The excipient companies are well aware of customer needs and limitations during formation of "functional" lipid nanoparticles. They are trying to overcome these issues by offering innovative products like surfactant cum surface modifier, functionalized lipids, PEG free ingredients, or refined excipients. Industries are now also offering their knowledge on understanding of excipients and its functionality related critical attributes as a part of QbD (quality by design approach) along consistency information and elemental impurity profile. Overall, excipient is soul of formulation development and biocompatibility performance of lipid nanoparticles and must be selected intelligently to achieve optimum performance.

ACKNOWLEDGMENT

Doktorovova S. is recipient of postdoctoral scholarships from Portuguese Science and Technology Foundation (Fundação para a Ciência e Tecnologia) under ref. SFRH/BPD/101650/2014.

REFERENCES

Abu Lila, A.S., Nawata, K., Shimizu, T., Ishida, T., Kiwada, H., 2013. Use of polyglycerol (PG), instead of polyethylene glycol (PEG), prevents induction of the accelerated blood clearance phenomenon against long-circulating liposomes upon repeated administration. Int. J. Pharm. 456, 235–242.

Aburahma, M.H., Badr-Eldin, S.M., 2014. Compritol 888 ATO: a multifunctional lipid excipient in drug delivery systems and nanopharmaceuticals. Expert Opin. Drug Deliv. 11, 1865–1883.

Amin, M.L., Joo, J.Y., Yi, D.K., An, S.S., 2015. Surface modification and local orientations of surface molecules in nanotherapeutics. J. Control. Release 207, 131–142.

Apaolaza, P.S., Delgado, D., Pozo-Rodriguez, A.D., Gascon, A.R., Solinis, M.A., 2014. A novel gene therapy vector based on hyaluronic acid and solid lipid nanoparticles for ocular diseases. Int. J. Pharm. 456, 413–426.

Aquilano, D., Sgualdino, G., 2001. Fundamental aspects of equilibrium and crystallization kinetics. In: Garti, N., Sato, K. (Eds.), Crystalization Processes in Fats and Lipid Systems. Marcel Dekker AG, Basel, Switzerland.

Baek, J.-S., Cho, C.-W., 2015. Comparison of solid lipid nanoparticles for encapsulating paclitaxel or docetaxel. J. Pharm. Invest. 45 (7), 1–7.

BASF, 2013. BASF—Nutrition & Health Kolliphor P Grades. Technical Information leaflet. pp. 1–8.

Blasi, P., Schoubben, A., Romano, G.V., Giovagnoli, S., Di Michele, A., Ricci, M., 2013. Lipid nanoparticles for brain targeting II. Technological characterization. Colloids Surf. B Biointerface 110, 130–137.

Bunjes, H., Koch, M.H.J., 2005. Saturated phospholipids promote crystallization but slow down polymorphic transitions in triglyceride nanoparticles. J. Control. Release 107, 229–243.

Bunjes, H., Westesen, K., 2001. Influences of colloidal state on physical properties of solid fats. In: Garti, N., Sato, K. (Eds.), Processes in Fats and Lipid Systems. Marcel Dekker, Inc, New York, pp. 177–210.

Bunjes, H., Westesen, K., Koch, M.H.J., 1996. Crystallization tendency and polymorphic transitions in triglyceride nanoparticles. Int. J. Pharm. 129, 159–173.

Bunjes, H., Koch, M.H.J., Westesen, K., 2000. Effect of particle size on colloidal solid triglycerides. Langmuir 16, 5234–5241.

Bunjes, H., Steiniger, F., Richter, W., 2007. Visualizing the structure of triglyceride nanoparticles in different crystal modifications. Langmuir 23, 4005–4011.

Chandra Shekhar, B., Jukanti, R., Shashank, A.S., Devraj, R., 2011. Improved ex vivo transcutaneous permeation of diclofenac from solid lipid nanoparticles in the presence of novel long-chain alkyl esters of PABA. J. Dispers. Sci. Technol. 32, 1158–1164.

Cui, Z., Qiu, F., Sloat, B.R., 2006. Lecithin-based cationic nanoparticles as a potential DNA delivery system. Int. J. Pharm. 313, 206–213.

Desai, P., Shah, P., Patlolla, R., Singh, M., 2012. Dermal microdialysis technique to evaluate the trafficking of surface-modified lipid nanoparticles upon topical application. Pharm. Res. 29, 2587–2600.

Dharmala, K., Yoo, J.W., Lee, C.H., 2008. Development of Chitosan–SLN Microparticles for chemotherapy: In vitro approach through efflux-transporter modulation. J. Control. Release 131, 190–197.
Doktorovova, S., Souto, E.B., 2009. Nanostructured lipid carrier-based hydrogel formulations for drug delivery: a comprehensive review. Expert Opin. Drug Deliv. 6, 165–176.
Doktorovova, S., Shegokar, R., Rakovsky, E., Gonzalez-Mira, E., Lopes, C.M., Silva, A.M., Martins-Lopes, P., Muller, R.H., Souto, E.B., 2011. Cationic solid lipid nanoparticles (cSLN): structure, stability and DNA binding capacity correlation studies. Int. J. Pharm. 420, 341–349.
Doktorovova, S., Jansch, M., Shegokar, R., Macedo, A., Martins-Lopes, P., Silva, A.M., Lopes, C.M., Müller, R.H., Souto, E.B., 2012a. Qualitative analysis of protein adsorption pattern on cationic solid lipid nanoparticles. 8th World Meeting on Pharmaceutics and Biopharmaceutics, Istambul, Turkey.
Doktorovova, S., Shegokar, R., Martins-Lopes, P., Silva, A.M., Lopes, C.M., Müller, R.H., Souto, E.B., 2012b. Modified Rose Bengal assay for surface hydrophobicity evaluation of cationic solid lipid nanoparticles (cSLN). Eur. J. Pharm. Sci. 45 (5), 606–612.
Doktorovova, S., Shegokar, R., Fernandes, L., Martins-Lopes, P., Silva, A.M., Müller, R.H., Souto, E.B., 2014a. Trehalose is not a universal solution for solid lipid nanoparticles freeze-drying. Pharm. Dev. Technol. 19, 922–929.
Doktorovova, S., Souto, E.B., Silva, A.M., 2014b. Nanotoxicology applied to solid lipid nanoparticles and nanostructured lipid carriers—a systematic review of in vitro data. Eur. J. Pharm. Biopharm. 87, 1–18.
Dolatabadi, J.E.N., Hamishehkar, H., Valizadeh, H., 2015. Development of dry powder inhaler formulation loaded with alendronate solid lipid nanoparticles: solid-state characterization and aerosol dispersion performance. Drug Dev. Ind. Pharm. 41, 1431–1437.
Dong, X., Mattingly, C.A., Tseng, M., Cho, M., Adams, V.R., Mumper, R.J., 2009. Development of new lipid-based paclitaxel nanoparticles using sequential simplex optimization. Eur. J. Pharm. Biopharm. 72, 9–17.
Dreher, F., Walde, P., Walther, P., Wehrli, E., 1997. Interaction of a lecithin microemulsion gel with human stratum corneum and its effect on transdermal transport. J. Control. Release 45, 131–140.
Fan, T., Chen, C., Guo, H., Xu, J., Zhang, J., Zhu, X., Yang, Y., Zhou, Z., Li, L., Huang, Y., 2014. Design and evaluation of solid lipid nanoparticles modified with peptide ligand for oral delivery of protein drugs. Eur. J. Pharm. Biopharm. 88, 518–528.
Fangueiro, J.F., Andreani, T., Egea, M.A., Garcia, M.L., Souto, S.B., Souto, E.B., 2012. Experimental factorial design applied to mucoadhesive lipid nanoparticles via multiple emulsion process. Colloids Surf. B Biointerface 100, 84–89.
Ferreira, S.F., Andreani, T., Fangueiro, J.F., Doktorovová, S., Silva, A.M., Souto, E.B., 2012. "PARSOL® MCX LOADED LIPID-SILICA NANOPARTICLES: AN INNOVATIVE SUNSCREEN APPROACH", 9th Central European Symposium of Pharmaceutical Technology, In 9th Central European Symposium of Pharmaceutical Technology, Dubrovnik, Croatia.
Ferreira, M., Chaves, L.L., Lima, S.A.C., Reis, S., 2015. Optimization of nanostructured lipid carriers loaded with methotrexate: a tool for inflammatory and cancer therapy. Int. J. Pharm. 492, 65–72.
Finke, J.H., Schmolke, H., Klages, C.P., Muller-Goymann, C.C., 2013. Controlling solid lipid nanoparticle adhesion by polyelectrolyte multilayer surface modifications. Int. J. Pharm. 449, 59–71.
Fonte, P., Nogueira, T., Gehm, C., Ferreira, D., Sarmento, B., 2011. Chitosan-coated solid lipid nanoparticles enhance the oral absorption of insulin. Drug Deliv. Transl. Res. 1, 299–308.
Freitas, C., Müller, R.H., 1998. Effect of light and temperature on zeta potential and physical stability in solid lipid nanoparticles (SLN™) dispersions. Int. J. Pharm. 168, 221–229.
Freitas, C., Mullera, R.H., 1998. Spray-drying of solid lipid nanoparticles (SLN TM). Eur. J. Pharm. Biopharm. 46, 145–151.
Gao, Y., Gu, W.W., Chen, L.L., Xu, Z.H., Li, Y.P., 2008. The role of daidzein-loaded sterically stabilized solid lipid nanoparticles in therapy for cardio-cerebrovascular diseases. Biomaterials 29, 4129–4136.

Ghalaei, P.M., Varshosaz, J., Aliabadi, H.S., 2014. Evaluating cytotoxicity of hyaluronate targeted solid lipid nanoparticles of etoposide on SK-OV-3 Cells. J. Drug Deliv. 2014, 746325.

Gonçalves, V.S.S., Matias, A.A., Rodríguez-Rojo, S., Nogueira, I.D., Duarte, C.M.M., 2015. Supercritical fluid precipitation of ketoprofen in novel structured lipid carriers for enhanced mucosal delivery—a comparison with solid lipid particles. Int. J. Pharm. 495 (1), 302–311.

Gupta, Y., Jain, A., Jain, S.K., 2007. Transferrin-conjugated solid lipid nanoparticles for enhanced delivery of quinine dihydrochloride to the brain. J. Pharm. Pharmacol. 59, 935–940.

Han, Y., Zhang, Y., Li, D., Chen, Y., Sun, J., Kong, F., 2014. Transferrin-modified nanostructured lipid carriers as multifunctional nanomedicine for codelivery of DNA and doxorubicin. Int. J. Nanomed. 9, 4107–4116.

Hazzah, H.A., Farid, R.M., Nasra, M.M.A., Zakaria, M., Gawish, Y., El-Massik, M.A., Abdallah, O.Y., 2016. A new approach for treatment of precancerous lesions with curcumin solid–lipid nanoparticle-loaded gels: in vitro and clinical evaluation. Drug Deliv. 23 (4), 1409–1419.

Heiati, H., Phillips, N.C., Tawashi, R., 1996. Evidence for phospholipid bilayer formation in solid lipid nanoparticles formulated with phospholipid and triglyceride. Pharm. Res. 13, 1406–1410.

Hong, Y., Hu, F.Q., Yuan, H., 2006. Effect of PEG2000 on drug delivery characterization from solid lipid nanoparticles. Pharmazie 61, 312–315.

Horta, S., Agostinho, A.L., Mateus, R., Pereira, L., Pereira, C., Capinha, L., Doktorovova, S., Brito, A., Videira, M., 2015. Looking out for cancer stem cells' properties: the value-driving role of CD44 for personalized medicines. Curr. Cancer Drug Targets 14, 832–849.

Howard, M.D., Jay, M., Dziublal, T.D., Lu, X.L., 2008. PEGylation of nanocarrier drug delivery systems: state of the art. J. Biomed. Nanotechnol. 4, 133–148.

Howard, M.D., Lu, X., Rinehart, J.J., Jay, M., Dziubla, T.D., 2011. Physicochemical characterization of nanotemplate engineered solid lipid nanoparticles. Langmuir 27, 1964–1971.

Huynh, N.T., Passirani, C., Saulnier, P., Benoit, J.P., 2009. Lipid nanocapsules: a new platform for nanomedicine. Int. J. Pharm. 379, 201–209.

Ishida, T., Harada, M., Wang, X.Y., Ichihara, M., Irimura, K., Kiwada, H., 2005. Accelerated blood clearance of PEGylated liposomes following preceding liposome injection: effects of lipid dose and PEG surface-density and chain length of the first-dose liposomes. J. Control. Release 105, 305–317.

Ishihara, T., Maeda, T., Sakamoto, H., Takasaki, N., Shigyo, M., Ishida, T., Kiwada, H., Mizushima, Y., Mizushima, T., 2010. Evasion of the accelerated blood clearance phenomenon by coating of nanoparticles with various hydrophilic polymers. Biomacromolecules 11, 2700–2706.

Jain, A., Agarwal, A., Majumder, S., Lariya, N., Khaya, A., Agrawal, H., Majumdar, S., Agrawal, G.P., 2010. Mannosylated solid lipid nanoparticles as vectors for site-specific delivery of an anti-cancer drug. J. Control. Release 148, 359–367.

Jain, A., Kesharwani, P., Garg, N.K., Jain, A., Jain, S.A., Jain, A.K., Nirbhavane, P., Ghanghoria, R., Tyagi, R.K., Katare, O.P., 2015. Galactose engineered solid lipid nanoparticles for targeted delivery of doxorubicin. Colloids Surf. B Biointerface 134, 47–58.

Jain, S.K., Chaurasiya, A., Gupta, Y., Jain, A., Dagur, P., Joshi, B., Katoch, V.M., 2008. Development and characterization of 5-FU bearing ferritin appended solid lipid nanoparticles for tumour targeting. J. Microencapsul. 25, 289–297.

Jeon, H.S., Seo, J.E., Kim, M.S., Kang, M.H., Oh, D.H., Jeon, S.O., Jeong, S.H., Choi, Y.W., Lee, S., 2013. A retinyl palmitate-loaded solid lipid nanoparticle system: effect of surface modification with dicetyl phosphate on skin permeation in vitro and anti-wrinkle effect in vivo. Int. J. Pharm. 452, 311–320.

Jing, F., Li, J., Liu, D., Wang, C., Sui, Z., 2012. Dual ligands modified double targeted nano-system for liver targeted gene delivery. Pharm. Biol. 51, 643–649.

Jores, K., Mehnert, W., Drechsler, M., Bunjes, H., Johann, C., Mäder, K., 2004. Investigations on the structure of solid lipid nanoparticles (SLN) and oil-loaded solid lipid nanoparticles by photon correlation spectroscopy, field-flow fractionation and transmission electron microscopy. J. Control. Release 95, 217–227.

Katas, H., Alpar, H.O., 2006. Development and characterisation of chitosan nanoparticles for siRNA delivery. J. Control. Release 115, 216–225.

Keck, C.M., Kovačević, A., Müller, R.H., Savić, S., Vuleta, G., Milić, J., 2014. Formulation of solid lipid nanoparticles (SLN): the value of different alkyl polyglucoside surfactants. Int. J. Pharm. 474, 33–41.

Khan, S., Baboota, S., Ali, J., Narang, R.S., Narang, J.K., 2016. Chlorogenic acid stabilized nanostructured lipid carriers (NLC) of atorvastatin: formulation, design and in vivo evaluation. Drug Dev. Ind. Pharm. 42 (2), 209–220.

Kovacevic, A., Savic, S., Vuleta, G., Müller, R.H., Keck, C.M., 2011. Polyhydroxy surfactants for the formulation of lipid nanoparticles (SLN and NLC): effects on size, physical stability and particle matrix structure. Int. J. Pharm. 406, 163–172.

Kuo, Y.-C., Chen, H.-H., 2009. Entrapment and release of saquinavir using novel cationic solid lipid nanoparticles. Int. J. Pharm. 365, 206–213.

Kuo, Y.-C., Lin, C.-W., 2009. Impact of arginine-modified solid lipid nanoparticles on the membrane charge of human brain-microvascular endothelial cells. Colloids Surf. B Biointerface 72, 201–207.

Lauer, S.A., Nolan, J.P., 2002. Development and characterization of Ni-NTA-bearing microspheres. Cytometry 48, 136–145.

Lavertu, M., Methot, S., Tran-Khanh, N., Buschmann, M.D., 2006. High efficiency gene transfer using chitosan/DNA nanoparticles with specific combinations of molecular weight and degree of deacetylation. Biomaterials 27, 4815–4824.

Lee, M.-K., Lim, S.-J., Kim, C.-K., 2007. Preparation, characterization and in vitro cytotoxicity of paclitaxel-loaded sterically stabilized solid lipid nanoparticles. Biomaterials 28, 2137–2146.

Lian, J.F., Zhang, S.Q., Wang, J.W., Fang, K.Q., Zhang, Y.M., Hao, Y., 2008. Novel galactosylated SLN for hepatocyte-selective targeting of floxuridinyl diacetate. J. Drug Target. 16, 250–256.

Liu, Y., Wang, P., Sun, C., Feng, N., Zhou, W., Yang, Y., Tan, R., Chen, Z., Wu, S., Zhao, J., 2010. Wheat germ agglutinin-grafted lipid nanoparticles: preparation and in vitro evaluation of the association with Caco-2 monolayers. Int. J. Pharm. 397, 155–163.

Liu, Y., Wang, P., Sun, C., Zhao, J., Du, Y., Shi, F., Feng, N., 2011. Bioadhesion and enhanced bioavailability by wheat germ agglutinin-grafted lipid nanoparticles for oral delivery of poorly water-soluble drug bufalin. Int. J. Pharm. 419, 260–265.

Luo, Q., Zhao, J., Zhang, X., Pan, W., 2010. Nanostructured lipid carrier (NLC) coated with chitosan oligosaccharides and its potential use in ocular drug delivery system. Int. J. Pharm. 403, 185–191.

Luo, Y., Teng, Z., Li, Y., Wang, Q., 2015. Solid lipid nanoparticles for oral drug delivery: chitosan coating improves stability, controlled delivery, mucoadhesion and cellular uptake. Carbohydr. Polym. 122, 221–229.

Ma, P., Dong, X.W., Swadley, C.L., Gupte, A., Leggas, M., Ledebur, H.C., Mumper, R.J., 2009. Development of idarubicin and doxorubicin solid lipid nanoparticles to overcome Pgp-mediated multiple drug resistance in leukemia. J. Biomed. Nanotechnol. 5, 151–161.

Maeda, H., 2001. The enhanced permeability and retention (EPR) effect in tumor vasculature: the key role of tumor-selective macromolecular drug targeting. Adv. Enzyme Regul. 41, 189–207.

Mahor, A., Alok, S., Gupta, Y., Jain, S.K., 2010. Body distribution and stability studies on mitoxantrone loaded solid lipid nanoparticles conjugated with concanavalina. Int. J. Pharm. Pharm. Sci. 2, 39–42.

Martins, S., Tho, I., Reimold, I., Fricker, G., Souto, E., Ferreira, D., brandl, M., 2012. Brain delivery of camptothecin by means of solid lipid nanoparticles: formulation design, in vitro and in vivo studies. Int. J. Pharm. 439, 49–62.

Mehnert, W., Mäder, K., 2001. Solid lipid nanoparticles—production, characterization and applications. Adv. Drug Deliv. Rev. 47, 165–196.

Mishra, D., Mishra, H., Mishra, P., Nahar, M., Dubey, V., Jain, N., 2010. Evaluation of solid lipid nanoparticles as carriers for delivery of hepatitis B surface antigen for vaccination using subcutaneous route. J. Pharm. Pharm. Sci. 13, 495–509.

Muller, R.H., Shegokar, R., Keck, C.M., 2011. 20 years of lipid nanoparticles (SLN and NLC): present state of development and industrial applications. Curr. Drug Discov. Technol. 8, 207–227.

Negi, L.M., Talegaonkar, S., Jaggi, M., Verma, A.K., Verma, R., Dobhal, S., Kumar, V., 2014a. Surface engineered nanostructured lipid carriers for targeting MDR tumor: part I. Synthesis, characterization and in vitro investigation. Colloids Surf. B Biointerface 123, 600–609.

Negi, L.M., Talegaonkar, S., Jaggi, M., Verma, A.K., Verma, R., Dobhal, S., Kumar, V., 2014b. Surface engineered nanostructured lipid carriers for targeting MDR tumor: part II. In vivo biodistribution, pharmacodynamic and hematological toxicity studies. Colloids Surf. B Biointerface 123, 610–615.

Nnamani, P.O., Hansen, S., Windbergs, M., Lehr, C.-M., 2014. Development of artemether-loaded nanostructured lipid carrier (NLC) formulation for topical application. Int. J. Pharm. 477, 208–217.

Noack, A., Hause, G., Mader, K., 2012. Physicochemical characterization of curcuminoid-loaded solid lipid nanoparticles. Int. J. Pharm. 423, 440–451.

Patel, J.D., Gandhapudi, S., Jones, J., Ronan, O., Woodward, G.J., Mumper, R.J., 2007. Cationic nanoparticles for delivery of CpG oligodeoxynucleotide and ovalbumin: in vitro and in vivo assessment. J. Biomed. Nanotechnol. 3, 97–106.

Patil-Gadhe, A., Pokharkar, V., 2014. Montelukast-loaded nanostructured lipid carriers: part I Oral bioavailability improvement. Eur. J. Pharm. Biopharm. 88, 160–168.

Patlolla, R.R., Desai, P.R., Belay, K., Singh, M.S., 2010. Translocation of cell penetrating peptide engrafted nanoparticles across skin layers. Biomaterials 31, 5598–5607.

Pedersen, N., Hansen, S., Heydenreich, A.V., Kristensen, H.G., Poulsen, H.S., 2006. Solid lipid nanoparticles can effectively bind DNA, streptavidin and biotinylated ligands. Eur. J. Pharm. Biopharm. 62, 155–162.

Rosen, M.J., 2004. Surfactants and Interfacial Phenomena. John Wiley & Sons, Inc, NJ, USA.

Sahu, P.K., Mishra, D.K., Jain, N., Rajoriya, V., Jain, A.K., 2015. Mannosylated solid lipid nanoparticles for lung-targeted delivery of Paclitaxel. Drug Dev. Ind. Pharm. 41, 640–649.

Salminen, H., Aulbach, S., Leuenberger, B.H., Tedeschi, C., Weiss, J., 2014. Influence of surfactant composition on physical and oxidative stability of Quillaja saponin-stabilized lipid particles with encapsulated ω-3 fish oil. Colloids Surf. B Biointerface 122, 46–55.

Salminen, H., Gömmel, C., Leuenberger, B.H., Weiss, J., 2016. Influence of encapsulated functional lipids on crystal structure and chemical stability in solid lipid nanoparticles: Towards bioactive-based design of delivery systems. Food Chem. 190, 928–937.

Sandri, G., Bonferoni, M.C., D'Autilia, F., Rossi, S., Ferrari, F., Grisoli, P., Sorrenti, M., Catenacci, L., Fante, C.D., Perotti, C., Caramella, C., 2013. Wound dressings based on silver sulfadiazine SLN for tissue repairing. Eur. J. Pharm. Biopharm. 84, 84–90.

Schmidts, T., Dobler, D., Nissing, C., Runkel, F., 2009. Influence of hydrophilic surfactants on the properties of multiple W/O/W emulsions. J. Colloid Interface Sci. 338, 184–192.

Schmidts, T., Dobler, D., Schlupp, P., Nissing, C., Garn, H., Runkel, F., 2010. Development of multiple W/O/W emulsions as dermal carrier system for oligonucleotides: effect of additives on emulsion stability. Int. J. Pharm. 398, 107–113.

Schubert, M.A., Müller-Goymann, C.C., 2005. Characterisation of surface-modified solid lipid nanoparticles (SLN): influence of lecithin and nonionic emulsifier. Eur. J. Pharm. Biopharm. 61, 77–86.

Schubert, M.A., Harms, M., Mã¼Ller-Goymann, C.C., 2006. Structural investigations on lipid nanoparticles containing high amounts of lecithin. Eur. J. Pharm. Sci. 27, 226–236.

Schwarz, J.C., Weixelbaum, A., Pagitsch, E., Low, M., Resch, G.P., Valenta, C., 2012. Nanocarriers for dermal drug delivery: influence of preparation method, carrier type and rheological properties. Int. J. Pharm. 437, 83–88.

Sekhon, B.S., 2013. Surfactants: Pharmaceutical and medicinal aspects. J. Pharm. Technol. Res. Manage. 1, 11–36.

Severino, P., Pinho, S.C., Souto, E.B., Santana, M.H., 2011. Polymorphism, crystallinity and hydrophilic–lipophilic balance of stearic acid and stearic acid-capric/caprylic triglyceride matrices for production of stable nanoparticles. Colloids Surf. B Biointerfaces 86, 125–130.

Severino, P., Andreani, T., Macedo, A., Fangueiro, J.F., Santana, M.H., Silva, A.M., Souto, E.B., 2012. Current state-of-art and new trends on lipid nanoparticles (SLN and NLC) for oral drug delivery. J. Drug Deliv. 2012, 750891.
Shah, P., Desai, P., Channer, D., Singh, M., 2012. Enhanced skin permeation using polyarginine modified nanostructured lipid carriers. J. Control. Release 161, 735–745.
Shegokar, R., Singh, K.K., 2012. Preparation, characterization and cell based delivery of stavudine surface modified lipid nanoparticles. J. Nanomed. Biother. Discov. 2, 1–9.
Shegokar, R., Singh, K.K., 2012a. Preparation, characterization and cell based delivery of stavudine surface modified lipid nanoparticles. Nanomed. Biother. Discov. 2, 1000105.
Shegokar, R., Singh, K.K., 2012b. Preparation, Characterization and Cell Based Delivery of Stavudine Surface Modified Lipid Nanoparticles. J. Nanomed. Biother. Discov. 2, 1–9.
Shegokar R., Singh K.K., Müller, R.H., 2009. Cationic Nanostructured Lipid carriers (NLC) for Mucosal Drug Delivery. 5th German-Polish Symposium on New Challenges for Pharmaceutical Sciences, Poland.
Shegokar, R., Jansch, M., Singh, K.K., Muller, R.H., 2011a. In vitro protein adsorption studies on nevirapine nanosuspensions for HIV/AIDS chemotherapy. Nanomedicine 7, 333–340.
Shegokar, R., Singh, K.K., Muller, R.H., 2011b. Production & stability of stavudine solid lipid nanoparticles—from lab to industrial scale. Int. J. Pharm. 416, 461–470.
Siddiqui, A., Gupta, V., Liu, Y.-Y., Nazzal, S., 2012. Doxorubicin and MBO-asGCS oligonucleotide loaded lipid nanoparticles overcome multidrug resistance in adriamycin resistant ovarian cancer cells (NCI/ADR-RES). Int. J. Pharm. 431, 222–229.
Singh, K.K., Shegokar, R., 2011. Conversion of stavudine lipid nanoparticles into dry powder. Int. J. Pharm. Biosci. 2, P443–P457.
Song, X., Lin, Q., Guo, L., Fu, Y., Han, J., Ke, H., Sun, X., Gong, T., Zhang, Z., 2015. Rifampicin loaded mannosylated cationic nanostructured lipid carriers for alveolar macrophage-specific delivery. Pharm. Res. 32, 1741–1751.
Soni, M.P., Shelkar, N., Gaikwad, R.V., Vanage, G.R., Samad, A., Devarajan, P.V., 2014. Buparvaquone loaded solid lipid nanoparticles for targeted delivery in theleriosis. J. Pharm. Bioallied Sci. 6, 22–30.
Souto, E.B., Doktorovova, S., Boonme, P., 2011a. Lipid-based colloidal systems (nanoparticles, microemulsions) for drug delivery to the skin: materials and end-product formulations. J. Drug Del. Sci. Tech. 21, 43–54.
Souto, E.B., Doktorovová, S., Boonme, P., 2011b. Lipid-based colloidal systems (nanoparticles, microemulsions) for drug delivery to the skin: materials and end-product formulations. J. Drug Deliv. Sci. Technol. 21, 43–54.
Surajit, D., Wai Kiong, N., Reginald, B.H.T., 2014a. Sucrose ester stabilized solid lipid nanoparticles and nanostructured lipid carriers: I. Effect of formulation variables on the physicochemical properties, drug release and stability of clotrimazole-loaded nanoparticles. Nanotechnology 25, 105101.
Surajit, D., Wai Kiong, N., Reginald, B.H.T., 2014b. Sucrose ester stabilized solid lipid nanoparticles and nanostructured lipid carriers: II. Evaluation of the imidazole antifungal drug-loaded nanoparticle dispersions and their gel formulations. Nanotechnology 25, 105102.
Tabatt, K., Kneuer, C., Sameti, M., Olbrich, C., Müller, R.H., Lehr, C.-M., Bakowsky, U., 2004. Transfection with different colloidal systems: comparison of solid lipid nanoparticles and liposomes. J. Control. Release 97, 321–332.
Torrecilla, J., Del Pozo-Rodríguez, A., Apaolaza, P.S., Solinís, M.Á., Rodríguez-Gascón, A., 2015. Solid lipid nanoparticles as non-viral vector for the treatment of chronic hepatitis C by RNA interference. Int. J. Pharm. 478, 181–188.
Venishetty, V.K., Komuravelli, R., Kuncha, M., Sistla, R., Diwan, P.V., 2013a. Increased brain uptake of docetaxel and ketoconazole loaded folate grafted solid lipid nanoparticles. Nanomed. Nanotechnol. Biol. Med. 9, 111–121.
Venishetty, V.K., Samala, R., Komuravelli, R., Kuncha, M., Sistla, R., Diwan, P.V., 2013b. Beta-hydroxybutyric acid grafted solid lipid nanoparticles: a novel strategy to improve drug delivery to brain. Nanomed. Nanotechnol. Biol. Med. 9, 388–397.

Vighi, E., Ruozi, B., Montanari, M., Battini, R., Leo, E., 2007. Re-dispersible cationic solid lipid nanoparticles (SLNs) freeze-dried without cryoprotectors: characterization and ability to bind the pEGFP-plasmid. Eur. J. Pharm. Biopharm. 67, 320–328.

Vighi, E., Ruozi, B., Montanari, M., Battini, R., Leo, E., 2010. pDNA condensation capacity and in vitro gene delivery properties of cationic solid lipid nanoparticles. Int. J. Pharm. 389, 254–261.

Vyas, A., Jain, A., Hurkat, P., Jain, A., Jain, S.K., 2015. Targeting of AIDS related encephalopathy using phenylalanine anchored lipidic nanocarrier. Colloids Surf. B Biointerface 131, 155–161.

Wang, L., Luo, Q., Lin, T., Li, R., Zhu, T., Zhou, K., Ji, Z., Song, J., Jia, B., Zhang, C., Chen, W., Zhu, G., 2015. PEGylated nanostructured lipid carriers (PEG–NLC) as a novel drug delivery system for biochanin A. Drug Dev. Ind. Pharm. 41, 1204–1212.

Wang, W., Zhao, X., Hu, H., Chen, D., Gu, J., Deng, Y., Sun, J., 2010. Galactosylated solid lipid nanoparticles with cucurbitacin B improves the liver targetability. Drug Deliv. 17, 114–122.

Wendt, R.H., Apte, S.P., Munson, E.J., Moreton, R.C., Block, L.H., Wang, H., Sheehan, C., 2013. Fixed-oil excipient monographs: development of usp fixed-oil reference standards. Pharm. Technol. 37 (4), 37.

Westesen, K., Bunjes, H., Koch, M.H.J., 1997. Physicochemical characterization of lipid nanoparticles and evaluation of their drug loading capacity and sustained release potential. J. Control. Release 48, 223–236.

Westesen, K., Siekmann, B., 1997. Investigation of the gel formation of phospholipid-stabilized solid lipid nanoparticles. Int. J. Pharm. 151, 35–45.

Ying, X.-Y., Cui, D., Yu, L., Du, Y.-Z., 2011. Solid lipid nanoparticles modified with chitosan oligosaccharides for the controlled release of doxorubicin. Carbohydr. Polym. 84, 1357–1364.

Yu, M.K., Park, J., Jon, S., 2012. Targeting strategies for multifunctional nanoparticles in cancer imaging and therapy. Theranostics 2, 3–44.

Zara, G.P., Cavalli, R., Bargoni, A., Fundaro, A., Vighetto, D., Gasco, M.R., 2002. Intravenous administration to rabbits of non-stealth and stealth doxorubicin-loaded solid lipid nanoparticles at increasing concentrations of stealth agent: pharmacokinetics and distribution of doxorubicin in brain and other tissues. J. Drug Target. 10, 327–335.

Zhang, C., Peng, F., Liu, W., Wan, J., Wan, C., Xu, H., Lam, C.W., Yang, X., 2014. Nanostructured lipid carriers as a novel oral delivery system for triptolide: induced changes in pharmacokinetics profile associated with reduced toxicity in male rats. Int. J. Nanomed. 9, 1049–1063.

Zhang, G., Liu, F., Jia, E., Jia, L., Zhang, Y., 2016. Folate-modified, cisplatin-loaded lipid carriers for cervical cancer chemotherapy. Drug Deliv. 23 (4), 1393–1397.

Zhao, C., Fan, T., Yang, Y., Wu, M., Li, L., Zhou, Z., Jian, Y., Zhang, Q., Huang, Y., 2013. Preparation, macrophages targeting delivery and anti-inflammatory study of pentapeptide grafted nanostructured lipid carriers. Int. J. Pharm. 450, 11–20.

Zhao, Y., Wang, C., Wang, L., Yang, Q., Tang, W., She, Z., Deng, Y., 2012a. A frustrating problem: accelerated blood clearance of PEGylated solid lipid nanoparticles following subcutaneous injection in rats. Eur. J. Pharm. Biopharm. 81, 506–513.

Zhao, Y., Wang, L., Yan, M., Ma, Y., Zang, G., She, Z., Deng, Y., 2012b. Repeated injection of PEGylated solid lipid nanoparticles induces accelerated blood clearance in mice and beagles. Int. J. Nanomed. 7, 2891–2900.

Zheng, J., Wan, Y., Elhissi, A., Zhang, Z., Sun, X., 2014. Targeted paclitaxel delivery to tumors using cleavable PEG-conjugated solid lipid nanoparticles. Pharm. Res. 31, 2220–2233.

Zhou, X., Zhang, X., Ye, Y., Zhang, T., Wang, H., Ma, Z., Wu, B., 2015. Nanostructured lipid carriers used for oral delivery of oridonin: an effect of ligand modification on absorption. Int. J. Pharm. 479, 391–398.

Index

E

F

H

I

J

K

L

O

P

Q

R

S

T

U

V

Kolibri 360 GmbH
Ferdinand-Jühlke-Straße 7
99095 Erfurt, Deutschland
produktsicherheit@kolibri360.de